Advanced Structured Materials

Volume 260

Common engineering materials are reaching their limits in many applications, and new developments are required to meet the increasing demands on engineering materials. The performance of materials can be improved by combining different materials to achieve better properties than with a single constituent, or by shaping the material or constituents into a specific structure. The interaction between material and structure can occur at different length scales, such as the micro, meso, or macro scale, and offers potential applications in very different fields.

This book series addresses the fundamental relationships between materials and their structure on overall properties (e.g., mechanical, thermal, chemical, electrical, or magnetic properties, etc.). Experimental data and procedures are presented, as well as methods for modeling structures and materials using numerical and analytical approaches. In addition, the series shows how these materials engineering and design processes are implemented and how new technologies can be used to optimize materials and processes.

Advanced Structured Materials is indexed in Google Scholar and Scopus.

Holm Altenbach · Dmitriy S. Lisovenko ·
Sergey A. Lurie · Alexander O. Vatulyan
Editors

Stability Problems in Mechanics of Continua

Springer

Editors
Holm Altenbach
Institut für Werkstoffe
Technologien und Mechanik
Otto-von-Guericke-Universität Magdeburg
Magdeburg, Sachsen-Anhalt, Germany

Sergey A. Lurie
Institute of Applied Mechanics
Russian Academy of Sciences
Moskva, Russia

Dmitriy S. Lisovenko
Ishlinsky Institute for Problems
in Mechanics
Russian Academy of Sciences
Moskva, Russia

Alexander O. Vatulyan
Institute of Mathematics, Mechanics,
and Computer Science named after of I.I.
Vorovich
Southern Federal University
Rostov-na-Donu, Russia

ISSN 1869-8433 ISSN 1869-8441 (electronic)
Advanced Structured Materials
ISBN 978-3-032-31031-6 ISBN 978-3-032-31032-3 (eBook)
https://doi.org/10.1007/978-3-032-31032-3

This Springer imprint is published by the registered company Springer Nature Switzerland AG
The registered company address is: Gewerbestrasse 11, 6330 Cham, Switzerland

Preface

This volume "Stability Problems in Mechanics of Continua" is dedicated to the 60th birthday of Prof. DSc. Dmitri Vladimirovich Georgievskii, who has devoted much of his academic career to continuum mechanics, its applications and related fields. The honouree has numerous friends and colleagues in the Russian Federation and abroad, resulting in a total of 38 contributions being submitted, 37 of which have undergone successful a peer-review process. Russian authors predominate, but the volume also includes contributions from colleagues outside the Russian Federation with whom the honoree has collaborated at various times.

Dmitri Georgievskii on the roof of the Duomo Cathedral (Milano, Italy).

Dmitri Georgevskii was born on August 19th, 1966, in Moscow (Soviet Union). His parents Vladimir Pavlovich and Nina Kuzminichna are engineers who graduated from Bauman Moscow High Technical Institute. Dmitri Georgievskii and his wife Irina have three children: Egor, Kristina and Daniil.

He enrolled in the Faculty of Mechanics and Mathematics at Moscow State University in 1981 and graduated with honors in 1986. From 1986 to 1989, he studied in the graduate program of the Department of Mechanics at the Faculty of Mechanics and Mathematics at Moscow State University. In 1989, he defended his candidate's dissertation (PhD thesis) on "The Stability of Deformation Processes in Heavy Layered Inelastic Bodies" which was supervised by Alexei Antonovich Ilyushin. In 1996, he defended his doctoral dissertation on "The Stability of Viscoelastic Flows with Arbitrary Hardening". D.V. Georgievskii's academic development and formation was largely influenced by his advisor, A.A. Il'yushin, as well as by Prof. B.E. Pobedrya (both of Lomonosov University in Moscow).

Since 1989, he has been working at the Faculty of Mechanics and Mathematics of Moscow State University. In 1999, he was appointed a professor. From 1991 to 2015, he worked at the Department of Composite Mechanics. From 1994 to 2015, he was its scientific secretary. From 1989 to 1991 and from 2015 to the present, he has been at the Department of Elasticity Theory, Since 2015, he has been the head of the department. From 2012 to 2014, he headed the Laboratory of Strength and Creep at High Temperatures at the Research Institute of Mechanics of Moscow State University. Since 2020, he has headed the Laboratory of Elasticity and Plasticity at this institute. Since 2021, he has been acting director of the Research Institute of Mechanics at Moscow State University. From 1999 to the present, he has been working part-time at the A.Yu. Ishlinskii Institute of Mechanics Problems of the Russian Academy of Sciences (RAS) in the Laboratory of Strength and Fracture Mechanics of Materials and Structures, and then in the Laboratory of Mechanics of Technological Processes.

From 2005 to 2014 and from 2019 to the present, he has been the deputy dean of the Faculty of Mechanics and Mathematics at Moscow State University. From 2014 to 2015, he was the vice rector of Moscow State University. He is a member of the Academic Council of the Moscow State University and of the Academic Council of the Faculty of Mechanics and Mathematics of Moscow State University. He served as a chairman of the Academic Council of the Research Institute of Mechanics of Moscow State University and as a member of the Dissertation Council MGU.01.14 in the specialty 01.02.04 (Mechanics of Deformable Solids) at the Faculty of Mechanics and Mathematics of Moscow State University. At the same time he is a member of the Dissertation Council D.002.059.06 in the specialty 01.02.06 (Dynamics, Strength of Machines, Devices, and Equipment) at the A.A. Blagonravov Institute of Machine Science of the Russian Academy of Sciences and a member of the Dissertation Council D.212.262.07 in the specialties 05.02.04 (Friction and Wear in Machines) and 06.05.06 (Mining Machines) at Tver State Technical University.

Dmitri Georgevskii's teaching activities are related to the following basic and special courses: "Mechanics of Continuous Media", "Mechanics of Deformable Solids", "Theory of Elasticity (Selected Chapters)", "Theory of Plasticity", "Mechanics of

Composites", "Classical Mechanics", "Theory of Constitutive Relations", and "Stability of Deformation Processes". Since 1999, he is a co-organizer of the scientific seminar "Current Problems in Geometry and Mechanics" at the Faculty of Mechanics and Mathematics of Moscow State University. Since 2005, he has co-organized the scientific seminar "Problems of Continuum Mechanics" at the A.Yu. Ishlinskii Institute of Mechanics Problems of the Russian Academy of Sciences.

His work for various scientific journals includes: executive secretary of the editorial board of the journal "Vestnik Moskovskogo Universiteta. Seriya 1. Matematika. Mekhanika" (published as *Moscow University Mechanics Bulletin* and *Moscow University Mechanics Bulletin* by Pleiades Publishing/Springer), he serve as a member of the editorial board of the journals "Izvestiya RAN. Mechanics of Solids", "Differential Equations", "Vestnik MSTU im. N.E. Bauman. Series: Natural Sciences, Composites and Nanostructures", "Chebyshev Collection", and a number of other editorial boards.

In 2016, Dmitri Georgevskii was appointed as a professor of the Russian Academy of Sciences. In 2008, 2011, 2016, 2019, and 2022, he participated in the RAS elections for the Department of Energy, Mechanical Engineering, Mechanics, and Control Processes for the position of candidate for corresponding member of the RAS. In 2025, he was elected as a Corresponding Member of the Russian Academy of Sciences. He is a member of the Russian National Committee for Theoretical and Applied Mechanics and a member of the International Association of Applied Mathematics and Mechanics (GAMM, Germany), of the American Mathematical Society (AMS), and of the International Society for Analysis, Applications and Computation (ISAAC). He is a full member of the International Academy of Sciences of Higher Education of the Russian Federation, a member of the Expert Council of the Russian Foundation for Basic Research (RFBR), and a expert of the Department of Energy, Mechanical Engineering, Mechanics, and Control Processes of Russian Academy of Sciences. Dmitri Georgevskii has been awarded the following prizes for his scientific achievements

- I.I. Shuvalov Prize ("Stability of viscoplastic flows with arbitrary hardening") (1997).
- European Academy of Sciences Award for Young Scientists of the CIS (1999).
- International Society for Analysis, Applications and Computation (ISAAC) Award for Young Scientists (2005).
- Academician H.A. Rakhmatulin Medal (2016).

The main areas of scientific interest of Dmitri Georgevskii include the following topics:

- The theory of constitutive relations in continuum mechanics, tensor calculus, integral representations, theory of setting experiments for finding material functions;
- Phenomenological description of stress-strain states in multi-scale modeling;
- Asymptotic methods in the theory of thin bodies, deformation of bodies with widely varying characteristic dimensions;

- Stability of deformation processes in the mechanics of deformable solids and composites, the Lyapunov–Movchan method and its development;
- Hydrodynamic stability of flows with complex rheology, spectral stability problems, generalized Orr–Sommerfeld problems, methods of integral relations and energy (variational) estimates of stability;
- Non-Newtonian and viscoplastic flows, analytical and numerical solutions, automodality and the Stefan problem;
- Weakly inhomogeneous flows., description of transitions to mixing; and
- Heavy gravitationally unstable multilayer systems, dominant waves, salt diapirs.

These and other issues are addressed in the following monographs and textbooks:

1. Georgievskii DV (1998) Ustoychivost' protsessov deformirovaniya vyazkoplasticheskikh tel (in Russ., Stability of Deformation Processes in Viscoplastic Bodies). Moscow, URSS
2. Pobedrya BE, Georgievskii DV (1999) Lektsii po teorii uprugosti (in Russ., Lectures on the Theory of Elasticity. Moscow, Editorial URSS (2nd edit. Moscow: LENAND, 2018, 3rd edit. Moscow: LENAND, 2020)
3. Klimov DM, Petrov AG, Georgievskii DV (2005) Vyazkoplasticheskiye techeniya. Dinamicheskiy khaos, ustoychivost', peremeshivaniye (in Russ., Viscoelastic Flows. Dynamic Chaos, Stability, Mixing). Moscow, Nauka
4. Pobedrya BE, Georgievskii DV (2006) Osnovy mekhaniki sploshnoy sredy (in Russ., Fundamentals of Continuum Mechanics). Moscow, Fizmatlit
5. Klimov DM, Petrov AG, Georgievskii DV (2018) Mekhanika sploshnoy sredy: vyazkoplasticheskiye techeniya (in Russ., Mechanics of Continuous Media: Viscoelastic Flows). Moscow, Yurait, 2018
6. Georgievskii DV (2018) Izbrannyye zadachi mekhaniki sploshnoy sredy (in Russ., Selected Problems in Mechanics of Continuous Media). Moscow: LENAND (2nd edit. Moscow: URSS, 2020)
7. Georgievskii DV (2023) Modeli teorii vyazkouprugosti (in Russ., Models of Viscoelasticity Theory). Moscow, LENAND
8. Georgievskyii DV (2025) Tenzornyye funktsii v teorii opredelyayushchikh sootnosheniy (in Russ., Tensor Functions in the Theory of Constitutive Relations). Moscow, LENAND

In addition, he published approximately 300 scientific papers (190 without coauthors, 180 of them in academic journals *Proceedings of the Russian Academy of Sciences*, *Izvestiya RAN. Ser. Mechanics of Solids*, *Series Mechanics of Fluids and Gases*, *Applied Mathematics and Mechanics*, *Applied Mechanics and Technical Physics*, *Mechanics of Composite Materials*, *Russian Journal of Mathematical Physics*, *Journal of Applied Mathematics and Mechanics/ZAMM*, *Journal of Fluids and Structures*, *Computers and Structures*). The main scientific results are reflected in these papers:

- It were created new areas of research in the theory of stability of flows of ideally rigid-plastic, viscoplastic, and nonlinear viscous bodies with respect to given classes of perturbations, in particular, perturbations of material functions determining relations. Developed a method of integral relations for linearized stability analysis of both steady and unsteady deformation processes of bodies with tensor-nonlinear and scalar-nonlinear determining relations and analytical derivation of integral sufficient stability estimates.

- It was developed a mathematical apparatus of singular asymptotic methods applicable to a number of important technological problems concerning the pressing, spreading, and extrusion of plastic material from thin layers and confusers, generalizing the problems of Prandtl and Hamel.
- It was classified the manifestation of dynamic effects in thin-layer plastic flows and during neck formation in the process of rapid deformation.
- Based on the apparatus of tensor nonlinear functions of one or more tensor arguments and the theory of invariants, he described and theoretically substantiated new tensor nonlinear effects ("orthogonal effects") of the stress-strain state of initially isotropic bodies.
- He proposed a trajectory asymptotic method of Eulerian description of the motion of weakly inhomogeneous continuous media, which served as a stage in the study of the mixing process in viscous, ideally rigid-plastic, and viscoplastic media.
- In the kinematics of continuous media, he introduced the concept of a high-rank deformator and generalized Cauchy relations in multidimensional Euclidean space, which allowed for a unified approach to the equations of compatibility known in mechanics, which are involved, in particular, in the formulation of stress problems.

This book contains articles on various topics in mechanics that are currently of great importance and are the focus of intense debate. Not all of the papers deal with issues of stability in continuum mechanics. The reader is thus provided with an excellent overview of current research in the field of continuum mechanics. This is particularly important at present, as face-to-face contact has decreased due to COVID and other events.

After discussion among the editorial team, it was decided to arrange the contributions in alphabetical order. The idea of dividing the book into thematic parts was rejected because the parts would have varied greatly in length. It must be emphasized, however, that the papers presented here encompass all aspects of modern mechanics research:

- modeling,
- simulation,
- numerical analysis, and
- experiments.

The presentations in the individual works vary greatly in terms of structure and notation. The editors did not consider it appropriate to standardize them here. Consequently, readers can also gain insight into the different schools of thought in mechanics. Numerous works contain direct references to publications by D.M. Georgievskii. This demonstrates that his scientific findings also inspired other researchers.

The editors and authors wish the honouree continued good health and all the best. Provided this remains the case, we can certainly look forward to further significant scientific achievements from him and his research group. We hope that his administrative duties will allow him time for that. At the same time, we hope that his work responsibilities will leave him enough time for his family.

Magdeburg, Germany *Holm Altenbach*
Moscow, Russian Federation *Dmitry S. Lisovenko*
Moscow, Russian Federation *Sergey A. Lurie*
Rostov on Don, Russian Federation *Alexander O. Vatulyan*

May 2026

Contents

List of Contributors

Bilen Emek Abali
Uppsala University, Division of Applied Mechanics, Department of Materials Science and Engineering, Uppsala, Sweden
e-mail: b.emek.abali@angstrom.uu.se

Reza Afshar
Uppsala University, Division of Applied Mechanics, Department of Materials Science and Engineering, Uppsala, Sweden
e-mail: reza.afshar@angstrom.uu.se

Kamalutdin G. Akhmedov
Russian University of Medicine, Delegatskaya ulitsa 20/1, 127437 Moscow, Russian Federation,
e-mail: kama05doc@gmail.com

Alexander V. Aksenov
Faculty of Mechanics and Mathematics, Lomonosov Moscow State University, 1 Leninskie Gory, 119991 Moscow, Russian Federation,
e-mail: aksenov@mech.math.msu.su

Sergei Alexandrov
Ishlinsky Institute for Problems in Mechanics RAS, Prospekt Vernadskogo 101-1, 119526 Moscow, Russian Federation,
e-mail: sergei_alexandrov@yahoo.com

Sergey D. Arutyunov
Russian University of Medicine, Delegatskaya ulitsa 20/1, 127437 Moscow, Russian Federation,
e-mail: sd.arutyunov@mail.ru

Dmitry V. Balandin
Sirius University of Science and Technology, Sochi, Russian Federation,
e-mail: dbalandin@yandex.ru

Nikolay V. Banichuk
Ishlinsky Institute for Problems in Mechanics RAS, Prospekt Vernadskogo 101-1, 119526 Moscow, Russian Federation,
e-mail: banichuk@ipmnet.ru

Aleksandra S. Begun
Institute of Automation and Control Processes FEB RAS, 5 Radio St., 690041 Vladivostok, Russian Federation,
e-mail: ustinova@iacp.dvo.ru

Aleksandr Belov
Research Institute for Mechanics, National Research Lobachevsky State University of Nizhny Novgorod, 23, Gagarin av., bld. 6, 603950 Nizhny Novgorod, Russian Federation,
e-mail: belov_a2@mech.unn.ru

Petr Belov
Institute of Applied Mechanics of RAS, Leningradsky prospekt 7, 125040 Moscow, Russian Federation, e-mail: belovpa@yandex.ru

Alexander K. Belyaev
Institute for Problems in Mechanical Engineering of the Russian Academy of Sciences, St. Petersburg, Russian Federation,
e-mail: vice.ipme@gmail.com

Aleksandr A. Bobylev
Faculty of Mechanics and Mathematics, Moscow State University, Leninskie gory, 119992 Moscow, Russian Federation,
e-mail: abobylov@gmail.com

Svetlana A. Bochkareva
Institute of Strength Physics and Materials Science SB RAS, Akademicheskiy str., 2/4, 634021 Tomsk, Russian Federation,
e-mail: svetlanab7@yandex.ru

Yuli D. Chashechkin
Ishlinsky Institute for Problems in Mechanics RAS, Prospekt Vernadskogo 101-1, 119526 Moscow, Russian Federation,
e-mail: chakin@ipmnet.ru, yulidch@gmail.com

Vladimir P. Chuev
Belgorod National Research University, ulitsa Pobedy 85, 308015 Belgorod, Russian Federation,
e-mail: chuev@vladmiva.ru

Leonid V. Fomin
Institute of Mechanics, Lomonosov Moscow State University, Michurinskii prosp. 1, 119192 Moscow, Russsian Federation,
e-mail: fleonid1975@mail.ru

Linar Gareev
Institute of Mechanics, Lomonosov Moscow State University, Michurinskii prosp. 1, 119192 Moscow, Russsian Federation, e-mail: gareev@imec.msu.ru

Natalia Golovina
Tyumen Industrial University (TIU), Surgut Branch, Surgut, Tyumen Region, Russian Federation,
e-mail: ntgolovina@rambler.ru

Gleb Gorynin
Polytechnic Institute, Surgut State University, Lenin ave. 1, 628412 Surgut, Russian Federation,
e-mail: ggorynin@list.ru

Dmitry I. Grachev
Russian University of Medicine, Delegatskaya ulitsa 20/1, 127437 Moscow, Russian Federation,
e-mail: dr.grachev@mail.rus

Leonid Igumnov
Research Institute for Mechanics, National Research Lobachevsky State University of Nizhny Novgorod, 23, Gagarin av., bld. 6, 603950 Nizhny Novgorod, Russian Federation,
e-mail: igumnov@mech.unn.ru

Oleg Ivanov
Institute of Mechanics, Lomonosov Moscow State University, Michurinskii prosp. 1, 119192 Moscow, Russsian Federation, e-mail: ivanov@imec.msu.ru

Svetlana Y. Ivanova
Ishlinsky Institute for Problems in Mechanics RAS, Prospekt Vernadskogo 101-1, 119526 Moscow, Russian Federation,
e-mail: ivanova@ipmnet.ru

Vagid Kadymov
MIREA - Russian Technological University, Moscow, Russian Federation,
e-mail: vkadymov@yandex.ru

Alexander V. Kaptsov
Ishlinsky Institute for Problems in Mechanics RAS, Prospekt Vernadskogo 101-1, 119526 Moscow, Russian Federation,
e-mail: kaptsov@ipmnet.ru

Karen G. Karakov
Stavropol State Medical University, ul. Mira 310, 355017 Stavropol, Russian Federation,
e-mail: terstomsgma@yandex.ru

Nikolay I. Khokhlov
Moscow Institute of Physics and Technology, Institutskiy Pereulok 9, 141701 Dolgoprudny, Moscow Region, Scientific Research Institute for System Analysis of the National Research Centre "Kurchatov Institute" & Autonomous noncommercial organization of higher education "Innopolis University", Innopolis, Tatarstan, Russian Federation,
e-mail: khokhlov.ni@mipt.ru

Dmitrii Khristich
Tula State University, pr. Lenina 92, 300012 Tula, Russian Federation,
e-mail: dmitrykhristich@rambler.ru

Mikhail Kirsanov
Faculty of Mechanics and Mathematics, Moscow State University, Leninskie gory, 119992 Moscow, Russian Federation,
e-mail: c216@ya.ru

Aleksander Konstantinov
National Research Lobachevsky State University of Nizhny Novgorod, 603022 Nizhny Novgorod, Gagarina ave., 23, Russian Federation,
e-mail: konstantinov@mech.unn.ru

Yuliya Konstantinova
National Research Lobachevsky State University of Nizhny Novgorod, 603022 Nizhny Novgorod, Gagarina ave., 23, Russian Federation,
e-mail: ullichka@list.ru

Larisa V. Kovtanyuk
Institute of Automation and Control Processes FEB RAS, 5 Radio St., 690041 Vladivostok, Russian Federation,
e-mail: lk@iacp.dvo.ru

Georgii Lavrikov
Faculty of Mechanics and Mathematics, Moscow State University, Leninskie gory, 119991 Moscow, Russian Federation,
e-mail: george.lavrikov@yandex.ru

Vladimir A. Levin
Faculty of Mechanics and Mathematics, Lomonosov Moscow State University, Leninskie gory 1, 119992 Moscow, Russian Federation,
e-mail: v.a.levin@mail.ru

Dmitry S. Lisovenko
Ishlinsky Institute for Problems in Mechanics RAS, Prospekt Vernadskogo 101-1, 119526 Moscow, Russian Federation,
e-mail: lisovenk@ipmnet.ru

Svetlana Litvinchuk
Research Institute for Mechanics, National Research Lobachevsky State University of Nizhny Novgorod, 23, Gagarin av., bld. 6, 603950 Nizhny Novgorod, Russian Federation,
e-mail: litivnchuk@mech.unn.ru

Fedor F. Losev
Central Research Institute of Dental and Maxillofacial Surgery, Timur Frunze Str. 16, 119991 Moscow, Russian Federation,
e-mail: losev@df.ru

Sergey Lurie
Ishlinsky Institute for Problems in Mechanics RAS, Prospekt Vernadskogo 101-1, 119526 Moscow, Russian Federation,
e-mail: salurie@mail.ru

Elena Lyamina
Ishlinsky Institute for Problems in Mechanics RAS, Prospekt Vernadskogo 101-1, 119526 Moscow, Russian Federation,
e-mail: lyamina@inbox.ru

Sergey A. Lychev
Ishlinsky Institute for Problems in Mechanics RAS, Prospekt Vernadskogo 101-1, 119526 Moscow, Russian Federation,
e-mail: lychevsa@mail.ru

Alexander K. Lyubimov
Lobachevsky State University of Nizhni Novgorod, Nizhni Novgorod, Russian Federation,
e-mail: ljubimov@mm.unn.ru

Vladimir Metrikin
National Research Lobachevsky State University of Nizhny Novgorod, 603022 Nizhny Novgorod, Gagarina ave., 23, Russian Federation,
e-mail: vsmetrikin@mail.ru

Evegnii V. Murashkin
Ishlinsky Institute for Problems in Mechanics RAS, Prospekt Vernadskogo 101-1, 119526 Moscow, Russian Federation,
e-mail: murashkin@ipmnet.ru, evmurashkin@gmail.com

Duc Dinh Nguyen
VNU University of Engineering and Technology, 144 D. Xuan Thy, 100000 Hanoi, Vietnam,
e-mail: guyenpd@vnu.edu.vn

Pham Dinh Nguyen
VNU University of Engineering and Technology, 144 D. Xuan Thy, 100000 Hanoi, Vietnam,
e-mail: nguyenpd@vnu.edu.vn

Mikhail Nikabadze
Lomonosov Moscow State University, Leninskie Gory, Bldg. 1, Institute of Mechanics, LMSU, Michurinsky prospekt, Bldg. 1 & Bauman Moscow State Technical University, 2nd Baumanskaya Ulitsa, Bldg. 5, Str. 1, Moscow, Russian Federation,
e-mail: munikabadze@yandex.ru

Anatolij Nikonov
Faculty of Mechanical Engineering, University of Ljubljana, Aškerčeva 6, 1000 Ljubljana, Slovenia,
e-mail: anatolij.nikonov@gmail.com

Kirill Y. Osipenko
Ishlinsky Institute for Problems in Mechanics RAS, Prospekt Vernadskogo 101-1, 119526 Moscow, Russian Federation,
e-mail: osipenko@ipmnet.ru

Galina L. Panchenko
Institute of Automation and Control Processes FEB RAS, 5 Radio St., 690041 Vladivostok, Russian Federation,
e-mail: panchenko@iacp.dvo.ru

Sergey V. Panin
Institute of Strength Physics and Materials Science SB RAS, Akademicheskiy str., 2/4, 634021 Tomsk, Russian Federation,
e-mail: svp@ispms.ru

Iliya L. Panov
Panin Institute of Strength Physics and Materials Science SB RAS, Tomsk, Russian Federation,
e-mail: panov.iliya@mail.ru

Alexander G. Petrov
Ishlinsky Institute for Problems in Mechanics RAS, Prospekt Vernadskogo 101-1, 119526 Moscow, Russian Federation,
e-mail: petrovipmech@gmail.com

Andrey Petrov
Research Institute for Mechanics, National Research Lobachevsky State University of Nizhny Novgorod, 23, Gagarin av., bld. 6, 603950 Nizhny Novgorod, Russian Federation,
e-mail: andrey.petrov@mech.unn.ru

Igor B. Petrov
Moscow Institute of Physics and Technology, Institutskiy Pereulok 9, 141701 Dolgoprudny, Moscow Region, Russian Federation,
e-mail: petrov@mipt.ru

Anton A. Pivovarov
Russian University of Medicine, Delegatskaya ulitsa 20/1, 127437 Moscow, Russian Federation,
e-mail: anton_pivovarov@bk.ru

Elizaveta Pozharskaya
Don State Technical University, Gagarin Square 1, 344000, Rostov-on-Don, Russian Federation,
e-mail: pozharskaya.elizaveta@rambler.ru

Dmitry Pozharskii
Don State Technical University, Gagarin Square 1, 344000, Rostov-on-Don, Russian Federation,
e-mail: pozharda@rambler.ru

Sergey G. Pshenichnov
Institute of Mechanics, Lomonosov Moscow State University, Michurinsky prospect 1, 119192 Moscow, Russian Federation,
e-mail: serp56@yandex.ru

Anna Pul
Research Institute for Mechanics, National Research Lobachevsky State University of Nizhny Novgorod, 23, Gagarin av., bld. 6, 603950 Nizhny Novgorod, Russian Federation,
e-mail: anna.iudintseva@gmail.com

Yuri N. Radayev
Ishlinsky Institute for Problems in Mechanics RAS, Prospekt Vernadskogo 101-1, 119526 Moscow, Russian Federation,
e-mail: radayev@ipmnet.ru, y.radayev@gmail.com

Nadezhda Rautian
Faculty of Mechanics and Mathematics, Moscow State University, Leninskie gory, 119992 Moscow, Russian Federation,
e-mail: nrautian@mail.ru

Tatiana V. Ryabikova
Sirius University of Science and Technology, Sochi, Russian Federation,
e-mail: tanya.dovid@gmail.com

Vasily Saurin
Ishlinsky Institute for Problems in Mechanics RAS, Prospekt Vernadskogo 101-1, 119526 Moscow, Russian Federation,
e-mail: saurin@ipmnet.ru

Efim I. Shifrin
Ishlinsky Institute for Problems in Mechanics RAS, Prospekt Vernadskogo 101-1, 119526 Moscow, Russian Federation,
e-mail: shifrin@ipmnet.ru

Vera Snigireva
Polytechnic Institute, Surgut State University, Lenina ave., 1, 628412, Surgut, Russian Federation,
e-mail: snow-vera@mail.ru

Evgeny Sosenushkin
Moscow State Technological University "STANKIN", Vadkovskiy Pereulok 1, 127055 Moscow, Russian Federation,
e-mail: sen@stankin.ru

Polina P. Tkachenko
Sirius University of Science and Technology, Sochi, Russian Federation,
e-mail: PTkachen@gmail.com

Tatyana P. Tovstik
Institute for Problems in Mechanical Engineering of the Russian Academy of Sciences, St. Petersburg, Russian Federation,
e-mail: tovstik_t@mail.ru

Vladimir Trifinov
Institute of Mechanics, Lomonosov Moscow State University, Michurinskii prosp. 1, 119192 Moscow, Russsian Federation,
e-mail: trifonov@imec.msu.ru

Alexander Vatulyan
Southern Federal University, 344049 Rostov-on-Don, Russian Federation,
e-mail: aovatulyan@sfedu.ru

Vasily Vedeneev
Institute of Mechanics, Lomonosov Moscow State University, Michurinskii prosp. 1, 119192 Moscow, Russsian Federation,
e-mail: vasily@vedeneev.ru

Andrey Volkov
Research Institute for Mechanics, National Research Lobachevsky State University of Nizhny Novgorod, 23, Gagarin av., bld. 6, 603950 Nizhny Novgorod, Russian Federation,
e-mail: Vollovandr@gmail.com

Ivan Volkov
Research Institute for Mechanics, National Research Lobachevsky State University of Nizhny Novgorod, 23, Gagarin av., bld. 6, 603950 Nizhny Novgorod, Russian Federation,
e-mail: ivan.volkov@mech.unn.ru

Maksim Vtyurin
National Research Lobachevsky State University of Nizhny Novgorod, 603022 Nizhny Novgorod, Gagarina ave., 23, Russian Federation,
e-mail: vtyurin@unn.ru

Elena Yanovskaya
Moscow State Technological University “STANKIN”, Vadkovskiy Pereulok 1, 127055 Moscow, Russian Federation,
e-mail: elena_yanovskaya@bk.ru

Victor Yurov
Southern Mathematical Institute, Vladikavkaz Scientific Center of the Russian Academy of Sciences, 362027 Vladikavkaz, Russian Federation,
e-mail: vyurov@sfedu.ru

Dmitry Zakharov
Institute of Computer Aided Design, Russian Academy of Sciences, 2nd Brestskaya Str, Moscow, 123056, Russian Federation,
e-mail: dmitrii.zakharov@gmail.com

Eleonora Zavoychinskaya
Faculty of Mechanics and Mathematics, Moscow State University, Leninskie gory, 119991 Moscow, Russian Federation,
e-mail: elen@velesgroup.com

Andrey Zhelnov
Moscow State Technological University “STANKIN”, Vadkovskiy Pereulok 1, 127055 Moscow, Russian Federation,
e-mail: a.zhelnov@stankin.ru

Konstantin M. Zingerman
Department of Applied Mathematics and Cybernetics, Tver State University, Zhelyabov Street, 33, 170100 Tver, Russian Federation,
e-mail: zingerman@rambler.ru

Chapter 1
Accumulated Damage Related Fatigue Description by Means of the Phase Field Method

Bilen Emek Abali and Reza Afshar

Abstract Damage modeling by a phase-field approach allows to extend the fracture mechanics to complex crack formation and propagation as well as branching. Despite its modeling strength, the approach is limited in predicting fatigue related phenomena. Empirical relations are widely used in design for inolving fatigue but a continuum mechanics formulation is still missing for this well-known phenomenon. We describe a simple approach and discuss its benefits and drawbacks from a theoretical perspective.

Key words: Phase-field method, Fatigue modeling, Accumulated damage, Fracture mechanics, Crack propagation, Continuum mechanics formulation

1.1 Continuum Length Scale Damage Modeling

Damage is described as a reduction of the integrity between neighboring particles. Depending on the material properties, it is called a rupture (ductile) or fracture (brittle) once the integrity is lost completely. Hence, in the case of continuum setup, we model this process as microcracks modeled by a damage variable [2, 3, 33, 34] that increases monotonously. In other words, caused damage is not healing. Therefore, it is expected that a damage accumulation is a possible prediction of a fatigue response that occurs in a cyclic loading with relatively low amplitude that fails to cause a fracture or rupture.

Damage variable based models go back to [37] and we begin with an initial 1 increasing to infinity [25, 28]; or equivalently, we may start from 1 to 0 (inverse of the damage variable); or even one minus the latter variable makes starting from 0 until 1 that denotes fracture. Phase-field approach is a possible method to model damage

Bilen Emek Abali (corresponding author) · Reza Afshar
Uppsala University, Division of Applied Mechanics, Department of Materials Science and Engineering, Uppsala, Sweden, e-mail: b.emek.abali@angstrom.uu.se, reza.afshar@angstrom.uu.se

H. Altenbach et al. (eds.), *Stability Problems in Mechanics of Continua*,
Advanced Structured Materials 260, https://doi.org/10.1007/978-3-032-31032-3_1

variable as a diffusive fracture modeling by means of a staggered [26, 27, 31, 38] or monolithic approach [6, 8, 15, 29]. In all cases, the model uses a toughness parameter to measure [1] and therefore gives the same result in each cycle. An accumulation of cyclic loading is needed for a fatigue modeling but the method is not yet clear. Recent advancements [9, 21] have suggested a possible formulation that we develop further in the basic setup.

1.2 Brittle Damage Formulation

We use index notation in Cartesian coordinates; and understand EINSTEIN's summation convention over repeated indices. An evolution equation governs the damage variable and the phase-field is mathematically an order parameter based on a potential [32] developed in [17] as a minimization problem or by a Γ-convergence analysis [5]. The method is a solution of a bump function that is 1 within a finite domain and 0 otherwise. With sufficient regularity, the solution of this differential equation gives the order parameter. Hence, a sharp crack is regularized by a crack surface density,

$$\gamma = \gamma(d, d_{,i}),$$

where a comma notation denotes a (partial) spatial derivative leading to a damage parameter, d, as a smeared out function representing the sharp crack as illustrated in Fig. 1.1. We obtain in a continuum body, $\mathcal{B}$, an integral relation with an accumulated

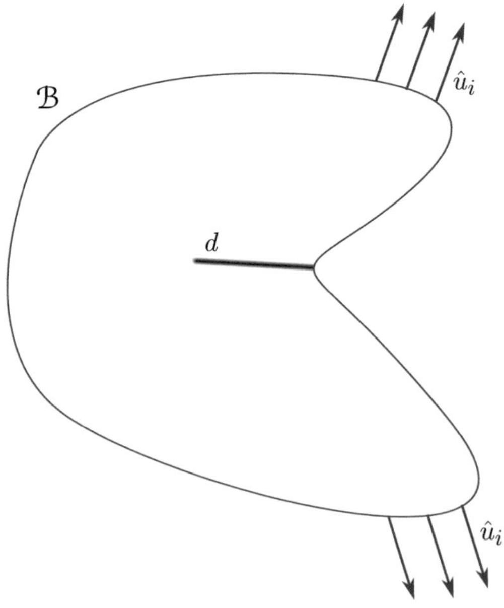

Fig. 1.1 Illustrative drawing of a continuum body, $\mathcal{B}$, under prescribed loading, say, given displacement, $\hat{u}_i$, on boundaries causing a stress localization and a crack initiation as well as propagation represented by d.

(dissipative) energy density, ${}^{\mathrm{d}}\psi$, and so-called critical energy release rate per surface, G_c, as follows:

$$\int_{\mathcal{B}} {}^{\mathrm{d}}\psi \,\mathrm{d}V = \int_{\mathcal{B}} G_\mathrm{c} F \gamma \,\mathrm{d}V \,. \tag{1.1}$$

We model G_c as a given constant, discuss the fatigue related degradation, F, in what follows, and assume an exponential decay [14] from the crack tip [40] that is the solution of the following differential equation:

$$\gamma = \gamma(d, d_{,i}) = \frac{1}{c_\alpha \ell}\left(\alpha + \ell^2 d_{,i} d_{,i}\right) . \tag{1.2}$$

With a normalized constant

$$c_\alpha = 4 \int_0^1 \alpha^{0.5} \,\mathrm{d}d,$$

we involve a crack geometry function,

$$\alpha = \alpha(d) \in [0, 1] .$$

Fatigue is modeled by an internal variable, ξ, due to the deformation by the available (free) energy,

$$\mathscr{f} = g\, {}^{\mathrm{e}}\psi \,, \tag{1.3}$$

which is given by a degradation function, $g = g(d)$, and stored energy density, ${}^{\mathrm{e}}\psi$. The internal variable is the accumulation of the available energy rate during the cyclic loading condition

$$\xi = \begin{cases} \int |\dot{\mathscr{f}}| \,\mathrm{d}t, & \text{if } \mathscr{f}\dot{\mathscr{f}} \geq 0 \\ 0, & \text{else} \end{cases} , \tag{1.4}$$

where the condition

$$\mathscr{f}\dot{\mathscr{f}} \geq 0$$

determines the condition that the loading direction stays the same. In other words, ξ is an increasing variable during continuous loading but remains constant during unloading. This internal variable increases and surpasses a threshold, ξ_th, where we follow [10] and introduce

$$F = \begin{cases} \left(\dfrac{2\xi_\mathrm{th}}{\xi + \xi_\mathrm{th}}\right)^2, & \text{if } \xi > \xi_\mathrm{th} \\ 1, & \text{else} \end{cases} . \tag{1.5}$$

By considering

$$\xi_{\text{th}} = \frac{G_{\text{c}}}{K_{\text{f}}\ell},$$

we may introduce a stress measure, K_{f}, steering the onset of fatigue degradation. Herein, we give the threshold variable directly. In order to couple this damage formulation in a variational setting, we begin with with a LAGRANGE function:

$$\mathcal{L} = \mathcal{L}(d, u_i, u_{i,j}) = -\not{f} + \rho f_i u_i \ . \tag{1.6}$$

A potential energy is caused by gravitation, where $|f_i| = 9.81\,\text{N/kg}$ is the specific body force. Free energy density, as given in Eq. (1.3), depends on the damage via $g = g(d)$ and strain,

$$\varepsilon_{ij} = \frac{1}{2}\Big(u_{i,j} + u_{j,i}\Big) , \tag{1.7}$$

via ${}^{\text{e}}\psi$. Let us consider the case of linear elasticity, in order to prevent crack growth under compression, the strain energy density ${}^{\text{e}}\psi$ is decomposed into a tensile/positive part and a compressive/negative part, where only the positive part affects damage, by following [4], we employ

$$\not{f} = g\,{}^{\text{e}}\psi^{+} + {}^{\text{e}}\psi^{-} \ . \tag{1.8}$$

The consistent elastic stiffness tensor C_{ijkl} is given by

$$C_{ijkl} = \frac{\partial \not{f}}{\partial \varepsilon_{ij} \partial \varepsilon_{kl}} = g \frac{\partial^2\, {}^{\text{e}}\psi^{+}}{\partial \varepsilon_{ij} \partial \varepsilon_{kl}} + \frac{\partial^2\, {}^{\text{e}}\psi^{-}}{\partial \varepsilon_{ij} \partial \varepsilon_{kl}} , \tag{1.9}$$

and the Cauchy stress tensor reads

$$\sigma_{ij} = C_{ijkl} \varepsilon_{kl} .$$

In the literature many decompositions have been suggested, cf. [7, 16, 23, 24, 30]. Here, we apply a hydrostatic-deviatoric decomposition [7], which models a crack growth as driven by shear and tensile bulk deformation. Assuming a linear elastic isotropic material and using the bulk modulus, K, and shear modulus, G, the strain energy density is written

$${}^{\text{e}}\psi = \frac{1}{2} K \varepsilon_{kk}^2 + G \varepsilon_{ij}^{\text{dev}} \varepsilon_{ji}^{\text{dev}} , \tag{1.10}$$

where the deviatoric strain tensor

$$\varepsilon_{ij}^{\text{dev}} = \varepsilon_{ij} - \frac{1}{\delta_{ll}} \varepsilon_{kk} \delta_{ij} .$$

We decompose ${}^{\text{e}}\psi$ into a negative and a crack-driving positive part, as follows:

$$
\begin{aligned}
{}^{\mathrm{e}}\psi^{-} =& \frac{1}{2} K \left\langle \varepsilon_{kk} \right\rangle_{-}^{2} \ , \\
{}^{\mathrm{e}}\psi^{+} =& \frac{1}{2} K \left\langle \varepsilon_{kk} \right\rangle_{+}^{2} + G \varepsilon_{ij}^{\mathrm{dev}} \varepsilon_{ji}^{\mathrm{dev}} \ .
\end{aligned} \tag{1.11}
$$

Here, $\langle\cdot\rangle_+$ and $\langle\cdot\rangle_-$ selects the positive and negative parts of their argument, i.e.

$$
\langle \varepsilon_{kk} \rangle_+ = \begin{cases} \varepsilon_{kk}, & \text{if } \varepsilon_{kk} \geq 0 \\ 0, & \text{else} \end{cases} \ , \ \langle \varepsilon_{kk} \rangle_- = \begin{cases} \varepsilon_{kk}, & \text{if } \varepsilon_{kk} \leq 0 \\ 0, & \text{else} \end{cases} \ . \tag{1.12}
$$

By asserting a right-hand side to the action in a quasistatic case (energy formulation instead of involving a time integral)

$$
\int_{\mathcal{B}} \mathcal{L} \, \mathrm{d}V = \int_{\mathcal{B}} {}^{\mathrm{d}}\psi \, \mathrm{d}V \ , \tag{1.13}
$$

we obtain

$$
\begin{aligned}
\delta \int_{\mathcal{B}} \mathcal{L} \, \mathrm{d}V =& \delta \int_{\mathcal{B}} {}^{\mathrm{d}}\psi \, \mathrm{d}V \ , \\
\int_{\mathcal{B}} \left(\frac{\partial \mathcal{L}}{\partial d} \delta d + \frac{\partial \mathcal{L}}{\partial u_i} \delta u_i + \frac{\partial \mathcal{L}}{\partial u_{i,j}} \delta u_{i,j} \right) \mathrm{d}V =& \int_{\mathcal{B}} \left(\frac{\partial {}^{\mathrm{d}}\psi}{\partial d} \delta d + \frac{\partial {}^{\mathrm{d}}\psi}{\partial d_{,i}} \delta d_{,i} \right) \mathrm{d}V \ ,
\end{aligned} \tag{1.14}
$$

and thus,

$$
\begin{aligned}
\int_{\mathcal{B}} \left(\frac{\partial \mathcal{L}}{\partial u_i} \delta u_i + \frac{\partial \mathcal{L}}{\partial u_{i,j}} \delta u_{i,j} \right) \mathrm{d}V =& \int_{\mathcal{B}} \left(\frac{\partial {}^{\mathrm{d}}\psi}{\partial d} \delta d + \frac{\partial {}^{\mathrm{d}}\psi}{\partial d_{,i}} \delta d_{,i} - \frac{\partial \mathcal{L}}{\partial d} \delta d \right) \mathrm{d}V \ , \\
\int_{\mathcal{B}} \left(\rho f_i \delta u_i - \frac{\partial \ell}{\partial u_{i,j}} \delta u_{i,j} \right) \mathrm{d}V =& \int_{\mathcal{B}} \left(\frac{\partial {}^{\mathrm{d}}\psi}{\partial d} \delta d + \frac{\partial {}^{\mathrm{d}}\psi}{\partial d_{,i}} \delta d_{,i} + \frac{\partial g}{\partial d} {}^{\mathrm{e}}\psi^{+} \delta d \right) \mathrm{d}V \ .
\end{aligned} \tag{1.15}
$$

Since kinematic variables, d and u_i, are independent, we may solve this weak form in two subsequent steps, as follows:

$$
\begin{aligned}
\int_{\mathcal{B}} \left(\rho f_i \delta u_i - \frac{\partial \ell}{\partial u_{i,j}} \delta u_{i,j} \right) \mathrm{d}V =& 0 \Rightarrow u_i \ , \\
\int_{\mathcal{B}} \left(\frac{\partial {}^{\mathrm{d}}\psi}{\partial d} \delta d + \frac{\partial {}^{\mathrm{d}}\psi}{\partial d_{,i}} \delta d_{,i} + \frac{\partial g}{\partial d} {}^{\mathrm{e}}\psi^{+} \delta d \right) \mathrm{d}V =& 0 \Rightarrow d \ .
\end{aligned} \tag{1.16}
$$

We call the first one the elasticity problem and the second one the damage problem. One usual choice for the degradation function is a quadratic one, $g = (1-d)^2$, and $\alpha = d$ as in [36] leading to $c_\alpha = 8/3$ such that we obtain

$$
{}^{\mathrm{d}}\psi = \frac{3G_cF}{8\ell}\Big(d + \ell^2 d_{,i} d_{,i}\Big) . \tag{1.17}
$$

After inserting this choice, called Ambrosio–Tortorelli 1 (AT1), we obtain a weak form of the damage problem

$$
\int_{\mathcal{B}} \Big(\frac{3G_cF}{8\ell} \delta d + \frac{3G_cF\ell}{4} d_{,i} \delta d_{,i} \Big) \mathrm{d}V = - \int_{\mathcal{B}} \frac{\partial g}{\partial d} {}^{\mathrm{e}}\psi^{+} \delta d \, \mathrm{d}V ,
$$

$$
\int_{\mathcal{B}} \Big(\frac{3G_cF}{8\ell} \delta d + \frac{3G_cF\ell}{4} d_{,i} \delta d_{,i} - 2(1-d) H_{\max} \delta d \Big) \mathrm{d}V = 0 , \tag{1.18}
$$

where $H_{\max} = \max({}^{\mathrm{e}}\psi^{+})$ is used algorithmically in order to ensure a monotonously increasing damage variable.

Another common choice is to use a quadratic degradation function, $g = (1-d)^2$, and a quadratic geometric relation, $\alpha = d^2$, resulting in $c_\alpha = 2$ with

$$
{}^{\mathrm{d}}\psi = \frac{G_cF}{2\ell}\Big(d^2 + \ell^2 d_{,i} d_{,i}\Big) . \tag{1.19}
$$

In this case, called Ambrosio–Tortorelli 2 (AT2), we acquire the weak form

$$
\int_{\mathcal{B}} \Big(\frac{G_cF}{\ell} d \delta d + G_cF\ell d_{,i} \delta d_{,i} - 2(1-d) H_{\max} \delta d \Big) \mathrm{d}V = 0 . \tag{1.20}
$$

The implementation of AT1 or AT2 is almost identical. We focus on the fatigue adition herein in order to demonstrate high-cycle fatigue results relevant for design.

1.3 Computational Implementation

We solve the abovementioned weak form numerically in space and time. For discretization in space, we utilize the finite element method, and in time, the finite difference method. For the sake of simlicity and a possible time step convergence study, we use equidistant time steps, Δt. Time herein is a load step in the sense that there are no time derivatives in the formulation. Cyclic nature of the loading makes it easier to introduce time and a loading rate. However, owing to the lack of viscoelasticiy or rate dependent plasticity, the loading rate makes no difference in the outcome. In each position, unknowns are displacement, u_i, and damage variable, d. We use a staggered scheme for implementing Eq. (1.16) in order to select their solvers more easily, although monolithic computation is possible as well, where use of a mobility is encouraged for the damage variable due to numerical reasons. For explaining the structure of the implementation, we describe it in Algorithm 1.

There are key issues to allow a convergence by means of the finite element method (FEM) [41] that the function spaces are bump functions around their corresponding

Algorithm 1 Implementing a staggered computational algorithm

Data: $t = 0$: $u_i = 0$, $d = 0$
while $t \leq t_{\text{end}}$ **do**
  err $= 1$
  while err > 0.01 **do**
    $$\int_{\mathcal{B}} \left(\rho f_i \delta u_i - \frac{\partial f}{\partial u_{i,j}} \delta u_{i,j} \right) \mathrm{d}V = 0 \Rightarrow u_i$$
    $$\int_{\mathcal{B}} \left(\frac{\partial^{\mathrm{d}}\psi}{\partial d} \delta d + \frac{\partial^{\mathrm{d}}\psi}{\partial d_{,i}} \delta d_{,i} + \frac{\partial g}{\partial d} {}^{\mathrm{e}}\psi^{+} \delta d \right) \mathrm{d}V = 0 \Rightarrow d$$
    if $d \geq 0$ **then**
      $d \leftarrow d$
    else
      $d \leftarrow 0$
    if $d \leq 1$ **then**
      $d \leftarrow d$
    else
      $d \leftarrow 1$
    $$\text{err} = \int \left(\frac{|d - d^{\text{last}}|}{d} \right) \mathrm{d}V$$
    $\boldsymbol{u}^{\text{last}} \leftarrow \boldsymbol{u}$, $d^{\text{last}} \leftarrow d$
  if $H_{\text{max}} \geq \psi^{+}$ **then**
    $H_{\text{max}} \leftarrow H_{\text{max}}$
  else
    $H_{\text{max}} \leftarrow \psi^{+}$
  if $f\dot{f} \geq 0$ **then**
    $\xi \leftarrow \xi + |f - f^{0}|$
  else
    $\xi \leftarrow \xi$
  Result: $\boldsymbol{u}$, d write out for postprocessing
  $\boldsymbol{u}^{0} \leftarrow \boldsymbol{u}$, $d^{0} \leftarrow d$
  $t \leftarrow t + \Delta t$

nodes with a compact support to the nearest neighbor. The space discretization is employed for trial and test functions by spanning over nodes after a triangulation of the computational domain, $\mathcal{B}$, with its closure, $\partial\mathcal{B}$. The discrete representation of unknowns are written with the same notation as their analytical counterpart, since they never appear together. The solution space is Hilbertian–Sobolev, $\mathcal{H}^1$, on a triangulation τ, as follows:

$$
\begin{aligned}
\mathcal{V}_u &= \left\{ \{\boldsymbol{u}\} \in \mathcal{H}^1(\Omega) : \{\boldsymbol{u}\}\Big|_{\tau} \mathcal{P}_1(\tau)\ \forall\tau \in \mathcal{T} \right\}, \\
\mathcal{V}_d &= \left\{ \{d\} \in \mathcal{H}^1(\Omega) : \{d\}\Big|_{\tau} \mathcal{P}_1(\tau)\ \forall\tau \in \mathcal{T} \right\}, \\
\mathcal{V}_\psi &= \left\{ \{\psi\} \in \mathcal{H}^1(\Omega) : \{\psi\}\Big|_{\tau} \mathcal{Q}_4(\tau)\ \forall\tau \in \mathcal{T} \right\}, \\
\mathcal{V}_\xi &= \left\{ \{\xi\} \in \mathcal{H}^1(\Omega) : \{\xi\}\Big|_{\tau} \mathcal{Q}_4(\tau)\ \forall\tau \in \mathcal{T} \right\}, \\
\mathcal{V}_H &= \left\{ \{H_{\max}\} \in \mathcal{H}^1(\Omega) : \{H_{\max}\}\Big|_{\tau} \mathcal{Q}_4(\tau)\ \forall\tau \in \mathcal{T} \right\}.
\end{aligned}
\tag{1.21}
$$

These standard LAGRANGE elements are utilized over the triangulation $\mathcal{T}$ consisting of non-overlapping triangles, τ. In simpler words, displacement and damage variables are modeled by polynomial, linear elements, $\mathcal{P}_1$. We select LAGRANGE elements with polynomials within the elements and C^0 in general. This compact support guarantees a monotonous convergence and a mesh may be chosen by an h-convergence with an error estimation. As known as the GALERKIN approach [35], the same space is applied for all test functions, $\delta\boldsymbol{u}$, δd. Standard FEM elements are continuous but not differentiable such that we choose to save all quantities related to their derivatives in a quadrature elements, $\mathcal{Q}_4$, evaluating on quadrature points. For numerical integration (quadrature), we choose 4 quadrature points in all directions. For interpolation of energy, ψ, $H_{\max}$, and thus, ξ, we use this quadrature elements that increase the accuracy by reducing the interpolation errors during assembly.

For the implementation, we utilize Firedrake open-source packages [22] with PetSc solvers [11–13]. Displacement is solved by a direct solver based on lower upper decomposition from MUMPS and damage is solved efficiently (owing to its symmetric matrix) by CHOLESKY algorithm. The framework is based on Python scripting that is then compiled in C++ and solved efficiently via mpi parallelization. We use a docker environment with ParMMG libraries. For creating the geometry and mesh, Python scripting is used in Gmsh. Firedrake can import msh from Gmsh directly.

1.4 Results and Discussion

A notched specimen is generated by Gmsh with a fine mesh around the notch. The mesh is visualized in Fig. 1.2. For the simulation, the damage lengthscale is chosen $\ell = 0.1$ mm and the mesh is using the half of this value around the notch. However, after 1.5 mm from the notch end, the coarsening is allowed. Therefore, as seen in Fig. 1.2(c) the mesh is not exactly symmetric along the horizontal direction. Damage

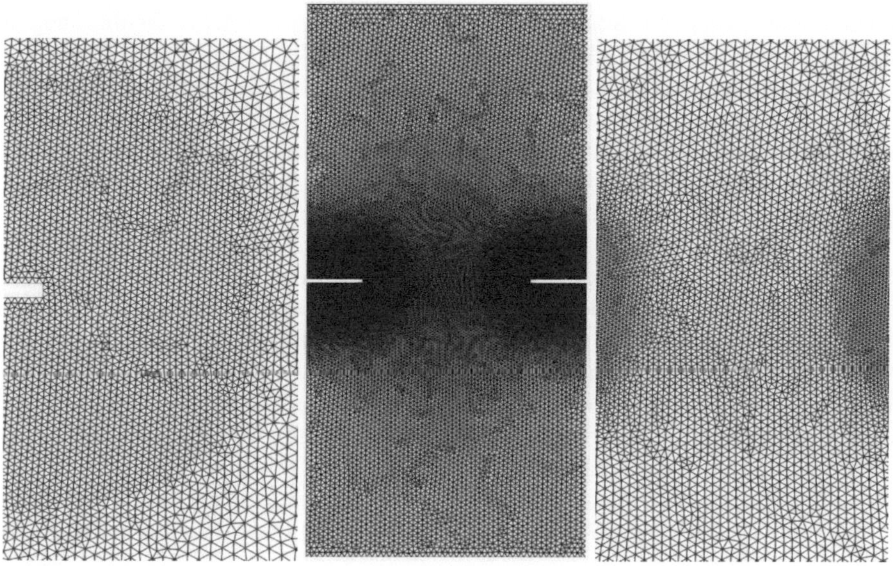

(a) Zoom: left notch (b) Specimen mesh (c) Zoom: middle

Fig. 1.2: Specimen of 10×20 mm with notches of 2 mm deep and 0.1 mm thick, around the notch global mesh length is chosen $h = 0.05$ mm while at far away around $h = 0.2$ mm, mesh is generated by Gmsh with top and bottom marking.

propagation is mesh dependent for the formulation without numerical stabilization (such as higher orders). We use fictitious but realistic parameters as compiled in Table 1.1. An important parameter is ξ_{th} that has the role in accumulating and altering the internal variable F to enable a crack propagation. We use 3 cases with the same material parameters in order to demonstrate the formulation's capacity in modeling the fatigue phenomenon. A displacement (equivalently, strain) controlled cyclic loading is applied. We stress that the formulation is rate-independent such that we use a triangular half loading that a given strain amplitude is reached in one time step and then zero is obtained in the next time step. Only stretching without

Table 1.1: Parameters and their units as used in the simulation for cyclic loading and damage elasticity in the notched specimen.

Parameter	Unit
Phase field length scale	$\ell = 0.1$ mm
Critical release rate	$G_c = 10$ J/m^2
Young's modulus	$E = 200$ GPa
Poisson's ratio	$\nu = 0.3$
Fatigue variable threshold	$\xi_{\text{th}} = 200$ kJ/m^3

compression is applied. Sometimes a half-cycle is a given name for this fatigue loading. Without using F reducing release rate, the crack may be initiated depending on the amplitude but the crack is not propagating since the deformation pattern is the same in each cycle of loading. In the case of choosing 0.002 mm displacement on top of the 20 mm long specimen, in other words, with a strain amplitude of $\Delta\varepsilon = 0.01\%$, we observe a crack propagation already within 10 cycles in Fig. 1.3. Exactly the same parameters but 3/4 of the strain reads around 20 times more lifetime. We demonstrate this case with $\Delta\varepsilon = 0.0075\%$ in Fig. 1.4. By halving the strain amplitued from the first case, we reach almost 100 times more lifetime as seen in Fig. 1.5. All cases are similar in the crack propagation that is as expected in mode I as shown in Fig. 1.6. We hold the bottom as clamped that causes at the right and left bottom corners a stress localization, more significant in the case 3. In all cases, we observe a slight deviation from mode I crack propagation due to the aforementioned mesh irregularities. In the case 3, as sizeable cycles are needed to build the fracture, we clearly see that the ξ parameter has an effect already for microcracks where the damage variable is less than one indicating that the crack is not physically initiated but numerically built up ξ leading to a decrease of G_c. Thus, we see $d \approx 0.5$ within a larger domain in the case 3.

Fatigue modeling is often done by the strain amplitude, $\Delta\varepsilon$, and an empirical relation. Until the crack initiation, Coffin–Manson type of power law is used to model the number of cycles until failure,

$$N_f = a\Delta\varepsilon^b .$$

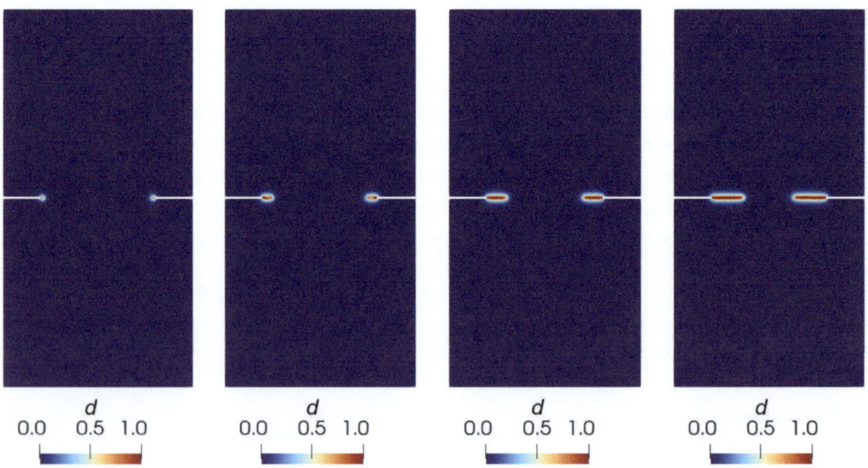

(a) $N = 1$ (b) $N = 3$ (c) $N = 5$ (d) $N = 7$

Fig. 1.3: Case 1: tensile loading with $\Delta\varepsilon = 0.01\%$ strain and damage after N cycles.

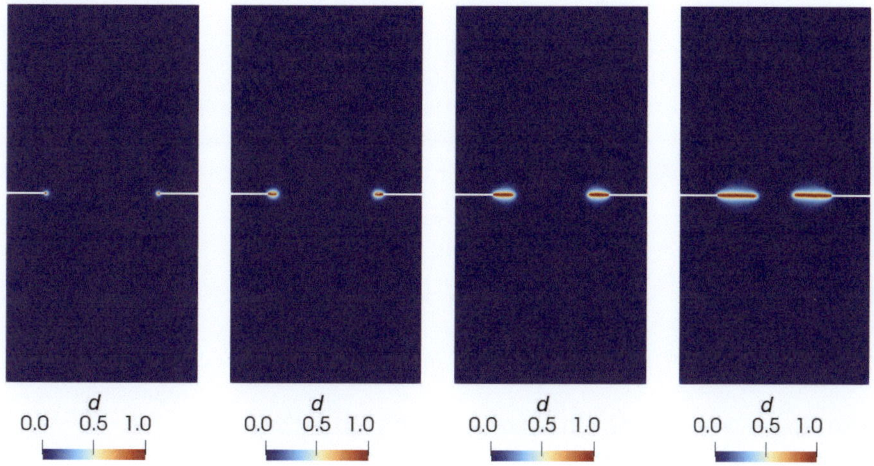

(a) $N = 50$ (b) $N = 100$ (c) $N = 150$ (d) $N = 190$

Fig. 1.4: Case 2: tensile loading with $\Delta\varepsilon = 0.0075\%$ strain and damage after N cycles.

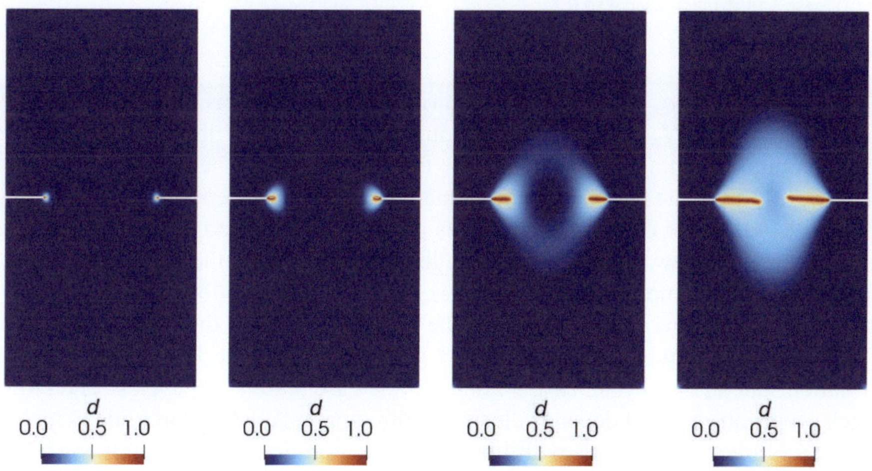

(a) $N = 200$ (b) $N = 400$ (c) $N = 600$ (d) $N = 800$

Fig. 1.5: Case 3: tensile loading with $\Delta\varepsilon = 0.005\%$ strain and damage after N cycles.

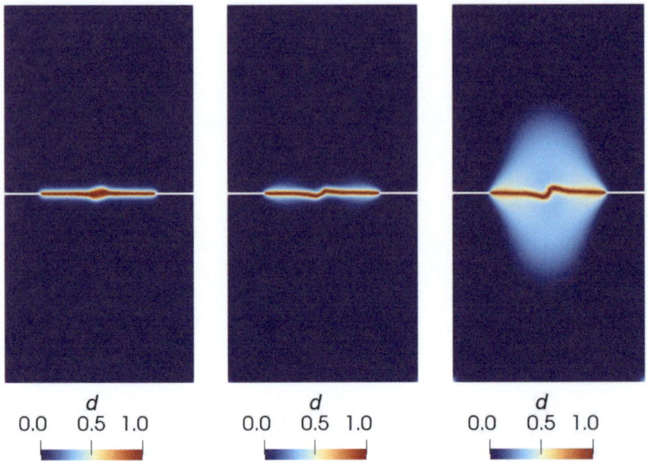

(a) $\Delta\varepsilon = 0.01\%$ $N_\mathrm{f} = 9$ (b) $\Delta\varepsilon = 0.0075\%$ $N_\mathrm{f} = 200$ (c) $\Delta\varepsilon = 0.005\%$ $N_\mathrm{f} = 900$

Fig. 1.6: Tensile loading with different strain amplitudes and resulting failure in N_f cycles, case 1 in (a), case 2 in (b), and case 3 in (c).

These parameters, a, b, match the empirical log-log datasets where for reduced amplitudes longer N_f is observed. For longer than $N_\mathrm{f} > 10^6$ the loading case has infinite fatigue length. We realize that the formulation shown here has the ability to qualitatively simulate the phenomenon that the full fracture depends on the loading amplitude. Once the crack is initiated, its propagation has also a speed sometimes given by a Paris law, relating the crack length rate to a similar power-law model depending on the amplitude. In simpler words, crack propagation speed is given by how many cycles are needed to obtain the full fracture, after the crack is initiated. Crack initiation may be seen as the $d = 1$ at the notches. Case 1 reaches that crack initiation state after 1 cycle, case 2 after 50, and case 3 after 200 cycles. Therefore, crack propagation speed depends clearly on the (applied) strain amplitude, which is again in full-correspondence with the well-known observation in metals for High-Cycle-Fatigue (HCF) where the elasticity causes fatigue.

1.5 Conclusions

By using an internal variable for reducing the toughness, we have developed and implemented a numerical elasticity and damage variable that leads to a crack propagation under cyclic loading. We have used a standard notched specimen for different amplitudes that is in fully accordance with empirical, well-established fatigue formulations. We leave possible extensions to the future studies such as involving plasticity, either for a perfect plastic case [20] or with unloading conditions [19]. The formulation has an important benefit that only one new material parameter is introduced, ξ_{th}, the simplicity has its potential, yet the methodology [18] how to determine this parameter remains open.

References

[1] Abali BE, Afshar R, Khaksar N, Segersten D, Vedin T (2024) Damage behavior in additive manufacturing based on infill pattern and density with carbon particle filled polylactic acid (cf-pla) polymer filaments. In: Altenbach H, Öchsner A (eds) State of the Art and Future Trends in Materials Modelling 2, Advanced Structured Materials, Vol 200. Cham: Springer, pp 1–16, DOI 10.1007/978-3-031-72900-3_1

[2] Altenbach H, Kolupaev VA (2015) Classical and non-classical failure criteria. In: Altenbach H, Sadowski T (eds) Failure and Damage Analysis of Advanced Materials, CISM International Centre for Mechanical Sciences, Vol 560, Vienna: Springer, pp 1–66, DOI 10.1007/978-3-7091-1835-1_1

[3] Altenbach H, Sadowski T (eds) (2015) Failure and Damage Analysis of Advanced Materials, CISM International Centre for Mechanical Sciences, vol 560. Vienna: Springer, DOI 10.1007/978-3-7091-1835-1

[4] Ambati M, Gerasimov T, De Lorenzis L (2015) A review on phase-field models of brittle fracture and a new fast hybrid formulation. Computational Mechanics 55(2):383–405, DOI 10.1007/s00466-014-1109-y

[5] Ambrosio L, Tortorelli VM (1990) Approximation of functional depending on jumps by elliptic functional via t-convergence. Communications on Pure and Applied Mathematics 43(8):999–1036, DOI 10.1002/cpa.3160430805

[6] Amirian B, Jafarzadeh H, Abali BE, Reali A, Hogan JD (2022) Phase-field approach to evolution and interaction of twins in single crystal magnesium. Computational Mechanics 70(4):803–818, DOI 10.1007/s00466-022-02209-3

[7] Amor H, Marigo JJ, Maurini C (2009) Regularized formulation of the variational brittle fracture with unilateral contact: Numerical experiments. Journal of the Mechanics and Physics of Solids 57(8):1209–1229, DOI 10.1016/j.jmps.2009.04.011

[8] Babaei H, Levitas VI (2018) Phase-field approach for stress-and temperature-induced phase transformations that satisfies lattice instability conditions. Part 2. Simulations of phase transformations Si I ↔ Si II. International Journal of Plasticity 107:223–245, DOI 10.1016/j.ijplas.2018.04.006

[9] Baktheer A, Martínez-Pañeda E, Aldakheel F (2024) Phase field cohesive zone modeling for fatigue crack propagation in quasi-brittle materials. Computer Methods in Applied Mechanics and Engineering 422:116834, DOI 10.1016/j.cma.2024.116834

[10] Baktheer A, Gopakumar M, Kenjo G, Aldakheel F (2026) Corrosion-fatigue degradation in reinforced concrete structures: A multiphysics phase-field mod-

eling approach. Computer Methods in Applied Mechanics and Engineering 452:118693, DOI 10.1016/j.cma.2025.118693
[11] Balay S, Gropp WD, McInnes LC, Smith BF (1997) Efficient management of parallelism in object oriented numerical software libraries. In: Arge E, Bruaset AM, Langtangen HP (eds) Modern Software Tools in Scientific Computing, Boston, MA: Birkhäuser Press, pp 163–202, DOI 10.1007/978-1-4612-1986-6_8
[12] Balay S, Abhyankar S, Adams MF, Brown J, Brune P, Buschelman K, Dalcin L, Dener A, Eijkhout V, Gropp WD, Karpeyev D, Kaushik D, Knepley MG, May DA, McInnes LC, Mills RT, Munson T, Rupp K, Sanan P, Smith BF, Zampini S, Zhang H, Zhang H (2019) PETSc, Web page: `https://www.mcs.anl.gov/petsc`
[13] Balay S, Abhyankar S, Adams MF, Brown J, Brune P, Buschelman K, Dalcin L, Dener A, Eijkhout V, Gropp WD, Karpeyev D, Kaushik D, Knepley MG, May DA, McInnes LC, Mills RT, Munson T, Rupp K, Sanan P, Smith BF, Zampini S, Zhang H, Zhang H (2020) PETSc users manual. Tech. Rep. ANL-95/11 - Revision 3.13, Argonne National Laboratory, Web page: `https://www.mcs.anl.gov/petsc`
[14] Chirkov AY, Nazarenko L, Altenbach H (2024) Plane crack problems within strain gradient elasticity and mixed finite element implementation. Computational Mechanics 74(3):703–721, DOI 10.1007/s00466-024-02451-x
[15] Clayton J, Knap J (2016) Phase field modeling and simulation of coupled fracture and twinning in single crystals and polycrystals. Computer Methods in Applied Mechanics and Engineering 312:447–467, DOI 10.1016/j.cma.2016.01.023
[16] van Dijk NP, Espadas-Escalante JJ, Isaksson P (2020) Strain energy density decompositions in phase-field fracture theories for orthotropy and anisotropy. International Journal of Solids and Structures 196:140–153, DOI 10.1016/j.ijsolstr.2020.04.022
[17] Francfort GA, Marigo JJ (1998) Revisiting brittle fracture as an energy minimization problem. Journal of the Mechanics and Physics of Solids 46(8):1319–1342, DOI 10.1016/S0022-5096(98)00034-9
[18] Georgievskii DV (2016) Establishing experiments in tensor nonlinear theories of continuum mechanics. Moscow University Mechanics Bulletin 71(2):49–50
[19] Georgievskii DV, Rautian NA (2024) Measure of unloading disproportion in the theory of small elastoplastic deformations. Moscow University Mechanics Bulletin 79(2):69–73, DOI 10.3103/S0027133024700092
[20] Georgievskii DV, Müller WH, Abali BE (2019) Thin-layer inertial effects in plasticity and dynamics in the Prandtl problem. ZAMM-Journal of Applied Mathematics and Mechanics/Zeitschrift für Angewandte Mathematik und Mechanik 99(12):e201900184, DOI 10.1002/zamm.201900184
[21] Golahmar A, Niordson CF, Martinez-Paneda E (2023) A phase field model for high-cycle fatigue: Total-life analysis. International Journal of Fatigue 170:107558, DOI 10.1016/j.ijfatigue.2023.107558

[22] Ham DA, Kelly PHJ, Mitchell L, Cotter CJ, Kirby RC, Sagiyama K, Bouziani N, Vorderwuelbecke S, Gregory TJ, Betteridge J, Shapero DR, Nixon-Hill RW, Ward CJ, Farrell PE, Brubeck PD, Marsden I, Gibson TH, Homolya M, Sun T, McRae ATT, Luporini F, Gregory A, Lange M, Funke SW, Rathgeber F, Bercea GT, Markall GR (2023) Firedrake User Manual. Imperial College London and University of Oxford and Baylor University and University of Washington, first edition edn, DOI 10.25561/104839
[23] Hesammokri P, Isaksson P (2024) Numerical analyses of brittle crack growth experiments in compression using a modified phase-field theory. International Journal of Solids and Structures 296:112815, DOI 10.1016/j.ijsolstr.2024.112815
[24] Hesammokri P, Yu H, Isaksson P (2023) An extended hydrostatic–deviatoric strain energy density decomposition for phase-field fracture theories. International Journal of Solids and Structures 262:112080, DOI 10.1016/j.ijsolstr.2022.112080
[25] Kachanov L (1986) Introduction to Continuum Damage Mechanics, Mechanics of Elastic Stability, Vol 10. ordrecht: Springer Science & Business Media, DOI 10.1007/978-94-017-1957-5
[26] Kästner M, Hennig P, Linse T, Ulbricht V (2016) Phase-field modelling of damage and fracture—convergence and local mesh refinement. In: Naumenko K, Aßmus M (eds) Advanced Methods of Continuum Mechanics for Materials and Structures, Advanced Structured Materials, Vol 60, Cham: Springer, pp 307–324, DOI 10.1007/978-981-10-0959-4_17
[27] Kiendl J, Ambati M, De Lorenzis L, Gomez H, Reali A (2016) Phase-field description of brittle fracture in plates and shells. Computer Methods in Applied Mechanics and Engineering 312:374–394, DOI 10.1016/j.cma.2016.09.011
[28] Lemaitre J, Desmorat R (2005) Engineering Damage Mechanics: Ductile, Creep, Fatigue and Brittle Failures. Berlin, Heidelberg: Springer Science & Business Media, DOI 10.1007/b138882
[29] Levitas VI (2018) Phase field approach for stress-and temperature-induced phase transformations that satisfies lattice instability conditions. Part I. General theory. International Journal of Plasticity 106:164–185, DOI 10.1016/j.ijplas.2018.03.007
[30] Miehe C, Hofacker M, Welschinger F (2010) A phase field model for rate-independent crack propagation: Robust algorithmic implementation based on operator splits. Computer Methods in Applied Mechanics and Engineering 199(45-48):2765–2778, DOI 10.1016/j.cma.2010.04.011
[31] Miehe C, Welschinger F, Hofacker M (2010) Thermodynamically consistent phase-field models of fracture: Variational principles and multi-field FE implementations. International Journal for Numerical Methods in Engineering 83(10):1273–1311, DOI 10.1002/nme.2861
[32] Mumford DB, Shah J (1989) Optimal approximations by piecewise smooth functions and associated variational problems. Communications on Pure and Applied Mathematics 42(5):577–685, DOI 10.1002/cpa.3160420503

[33] Murakami S (2012) Continuum Damage Mechanics: a Continuum Mechanics Approach to the Analysis of Damage and Fracture, Vol 185. Dordrecht: Springer Science & Business Media, DOI 10.1007/978-94-007-2666-6
[34] Öchsner A (2016) Continuum damage mechanics. In: Continuum Damage and Fracture Mechanics, Singapore: Springer, pp 65–84, DOI 10.1007/978-981-287-865-6_4
[35] Ottosen NS, Petersson H (1992) Introduction to the Finite Element Method. New York: Prentice Hall
[36] Pham K, Amor H, Marigo JJ, Maurini C (2011) Gradient damage models and their use to approximate brittle fracture. International Journal of Damage Mechanics 20(4):618–652, DOI 10.1177/1056789510386852
[37] Rabotnov YN (1959) Mekhanizm dlitelnogo razrusheniya (in Russ., The mechanism of gradual destruction). Sbornik "Voprosy prochnosti materialov i konstruktsii", pp 5–7
[38] Teichtmeister S, Kienle D, Aldakheel F, Keip MA (2017) Phase field modeling of fracture in anisotropic brittle solids. International Journal of Non-Linear Mechanics 97:1–21, DOI 10.1016/j.ijnonlinmec.2017.06.018
[39] Vălean C, Marşavina L, Mărghitaş M, Linul E, Razavi J, Berto F (2020) Effect of manufacturing parameters on tensile properties of fdm printed specimens. Procedia Structural Integrity 26:313–320, DOI 10.1016/j.prostr.2020.06.040
[40] Wu JY (2017) A unified phase-field theory for the mechanics of damage and quasi-brittle failure. Journal of the Mechanics and Physics of Solids 103:72–99, DOI 10.1016/j.jmps.2017.03.015
[41] Zohdi TI (2018) Finite Element Primer for Beginners. Cham: Springer, DOI 10.1007/978-3-319-70428-9

Chapter 2
Influence of Reinforcing Framework Shapes on Load-Bearing Capacities of Complete Denture Prostheses of the Maxilla at Different Adhesion Levels and Support Conditions

Kamalutdin G. Akhmedov, Fedor F. Losev, Svetlana A. Bochkareva, Sergey D. Arutyunov, Karen G. Karakov, Dmitry I. Grachev, Vladimir P. Chuev, Iliya L. Panov, Anton A. Pivovarov, and Sergey V. Panin

Abstract This study compares the effectiveness of different shapes of reinforcing perforated polyetheretherketone (PEEK) frameworks of the hybrid complete denture prostheses (CDPs) under conditions of weak adhesion between the polymethyl methacrylate (PMMA) base and the PEEK framework. To increase the adhesion levels, "dovetail" cuts were made along the framework edges. Based on the results of computer simulations, stress-strain states (SSSs) of CDPs with the frameworks of different shapes were analyzed and their load-bearing capacities were assessed for different (static) loading options. It was shown that the framework shapes should be selected considering multiple factors that influence the probability of failure of CDPs under static loads, including the stiffness, the adhesion levels between frameworks and bases, their properties, as well as supports on the prosthetic foundations. The key parameter is the elastic moduli of the frameworks, which determine the load-bearing capacities of CDPs for the suggested shapes, regardless of the adhesion levels.

Kamalutdin G. Akhmedov · Sergey D. Arutyunov · Dmitry I. Grachev · Anton A. Pivovarov
Russian University of Medicine, Delegatskaya ulitsa 20/1, 127437 Moscow, Russian Federation, e-mail: kama05doc@gmail.com, sd.arutyunov@mail.ru, dr.grachev@mail.rus, anton_pivovarov@bk.ru

Fedor F. Losev
Central Research Institute of Dental and Maxillofacial Surgery, Timur Frunze Str. 16, 119991 Moscow, Russian Federation, e-mail: losev@df.ru

Svetlana A. Bochkareva · Iliya L. Panov · Sergey V. Panin (corresponding author)
Institute of Strength Physics and Materials Science SB RAS, Akademicheskiy str., 2/4, 634021 Tomsk, Russian Federation, e-mail: svetlanab7@yandex.ru, panov.iliya@mail.ru, svp@ispms.ru

Karen G. Karakov
Stavropol State Medical University, ul. Mira 310, 355017 Stavropol, Russian Federation, e-mail: terstomsgma@yandex.ru

Vladimir P. Chuev
Belgorod National Research University, ulitsa Pobedy 85, 308015 Belgorod, Russian Federation, e-mail: chuev@vladmiva.ru

H. Altenbach et al. (eds.), *Stability Problems in Mechanics of Continua*,
Advanced Structured Materials 260, https://doi.org/10.1007/978-3-032-31032-3_2

Key words: Computer aided engineering (CAE), Structural integrity, Failure, Complete denture prosthesis (CDP), Polymethylmethacrylate (PMMA), Polyetheretherketone (PEEK), Additive manufacturing, Prosthesis framework, Finite element method (FEM), Failure criterion, Adhesion

2.1 Introduction

The increase of edentulous population and the need for conventional complete denture prostheses (CDPs) is exponentially increasing globally, particularly among geriatric individuals [34, 36]. Current aging trends are associated with increased budget expenditures for social obligations, such as pensions, health insurance, and care for infirm citizens. These activities increase the risks of insufficiently comprehensive and targeted consideration of the requirements of elderly and senile people [11, 15, 20].

The progression of this issue for European adults and the factors influencing the frequency of using CDPs have been analyzed in [36]. It has been reported that all their types are in operation, especially among the geriatric population. Estimated adults with CDPs are within 13–29%, while 3–13% wear them on both jaws (older age groups with lower both socioeconomic status and education levels, living in rural areas). Complete tooth loss disfigures a person's appearance, interferes with proper mastication and speech, as well as disrupts both social communication and quality of life [9].

The selection of polymers for CDPs depends on their biocompatibility with the oral cavity ecosystems, the vulnerability of the soft/hard tissues of the prosthetic foundations and fields [19, 33]. The operational capabilities of CDPs are limited by intolerance to plastic, mechanical trauma to the soft tissues contacting with bases and hard plaque components [29], allergic reactions and cytotoxic effects on the mucous membranes of the prosthetic foundations, free (unreacted) monomers, systemic diseases, as well as psychogenic factors [6]. Both physical and mechanical characteristics of potential materials should be considered in their selection for the bases of CDPs [3, 13]. Their elastic moduli and impact strengths should be correlated with the features and complexity of clinical cases (dental occlusions [19, 33], anatomical and topographic characteristics of the prosthetic foundations, severity of the alveolar bones/jaw parts, etc.) [5]. It is well known that acrylic plastics are not strong enough under both tensile and bending loads, so cracks can initiate and propagate upon biting off and mastication, causing fractures of the bases [4, 27]. Therefore, even minor errors in the production routs of CDPs reduce their mechanical properties [35]. Among numerous suitable materials, PMMA is far from ideal, but it combines some advantages that contribute to its use in dental practice [13, 14, 35].

Wearing CDPs by patients is often associated with failure of their bases [18], especially for geriatric individuals [18, 34]. The main causes are fatigue limits, depending on the maximum stresses at which such failures do not occur over large numbers of cycles, and fall, when dynamic loads exceed the impact toughness of the applied materials [21].

In 22–31% cases, failures of CDPs are observed during the first year of their wearing at repair rates of 18.7–20.3% [2, 18, 20]. Sometimes, their replacements are required, which entail financial burden for doctors and/or patients. Their unsuccessful repairing deteriorates the patient's quality of life [9, 10]. For properly manufactured CDPs, the main causes of failure are static and cyclic operational loads, some imbalances due to atrophy of the alveolar bones/jaw crests or unaligned occlusions of the dental arches [4, 19, 33], as well as combined fatigue and impact effects [12, 17]. Particular attention should be paid to their both pronounced bulges (>1.5 mm) and the minimum base thicknesses (>2.3 mm) after processing and polishing [27]. According to [4], the maximum critical stresses are observed near the vestibular notches of the frenulum of the upper lips, as well as in areas of sharp changes in the base thicknesses along the boundaries of the torus insulation zones. In these regions, main cracks initiate typically.

CDPs can be reinforced with metal plates, meshes, wires [10, 21], both polyethylene [31] and fiberglass [23, 28] frameworks, aramid fibers [16], as well as by filling with zirconium dioxide particles [5]. It should be noted that metal meshes may be useless due to the lack of their adhesion with polymer matrices [7], preserving the mechanical properties of CDPs at similar levels [10].

The flexural strengths of composites based on hot-cured PMMA reinforced with glass, aramid, or nylon fibers have been tested in three-point bending tests [16]. The greatest results have been achieved after reinforcing with glass fibers, followed by aramid and nylon. In [5], an advantage of reinforcing hot-cured PMMA with polyethylene fibers has been reported. Basically, failures of the bases of CDPs are associated with both physico-chemical characteristics and mechanical properties of their material (PMMA), which should meet the requirements for biocompatibility, strength, stability and durability [23], errors in their design [30] and production [19], as well as the patient's frailty and inability to follow the dentist's instructions for oral hygiene and prosthetic structures [2, 18, 20, 36].

The durability of the bases of CDPs has been predicted by computer simulation of operational loads caused by biting and mastication of food [12]. In this way, the finite element method (FEM) has been implemented and an algorithm for considering edentulous patients has been proposed. In [24], the effects of the frenulum notch sizes and the palate (vault) depths (complexity in Gothic) on stress raisers in CDPs of the maxilla have been assessed. In such scenarios, strategies for reinforcing with frameworks of various chemical and physical natures should be taken into account. For example, fractures of the medial bases of CDPs from polymer acrylic plates are common due to the initiation and propagation of cracks near stress raisers [12].

The influence of placements of high-performance polyethylene (HPPE) fibers on stress distributions in the bases of CDPs of the maxilla and their performance characteristics has been investigated by the FEM [7]. Without reinforcements, stress raisers have been observed on the incisal notch in addition to both anterior and posterior palatal surfaces. However, they have been reduced by reinforcing, especially when arranging fibers on the polished side. The experiment has enabled to identify "weak" areas in CDPs of the maxilla and the effectiveness of the reinforcement of HPPE with oriented fibers.

In this study, different shapes of the (reinforcing) frameworks with perforations were designed for the bases of CDPs of the maxilla (Fig. 2.1, a–g).

Also, different operational conditions were simulated (the mechanical properties, the adhesion levels between the frameworks and the bases, both characteristics and shapes of "virtual supports", as well as loading options) for assessing the possibility of failure, including due to the bone tissue resorption and changes in the relief of the alveolar bone, as noted above [19, 33]. By analyzing previously obtained results on the load-bearing capacities and stress-strain states (SSSs) of CDPs with (reinforcing) framework № 1 (original) from polyetheretherketone (PEEK) for different loading options [1], the authors suggested to make "dovetail" cuts along its perimeter (Fig. 2.1, a) for increasing the adhesion level and to modify its shape (new designs are shown in Fig. 2.1, b–d and f–h).

The updated frameworks were characterized by the following features:

- Comparing with framework № 1 (original) [1], framework № 2 (3-petal) was truncated to the dental arch (the alveolar part of the base) and possessed three attaching petals.
- Framework № 3 (truncated) was an analog of framework № 1 (original), reduced uniformly along the entire perimeter by 3 mm. As a"average" version between frameworks № 1 (original) and № 2 (3-petal). In addition, it was moved closer to the inner surface of the base (teeth).
- Framework № 4 (5-petal) was a truncated version of framework № 3 (truncated) with five attaching petals, but partially located under the dental arch (on the alveolar part of the base).

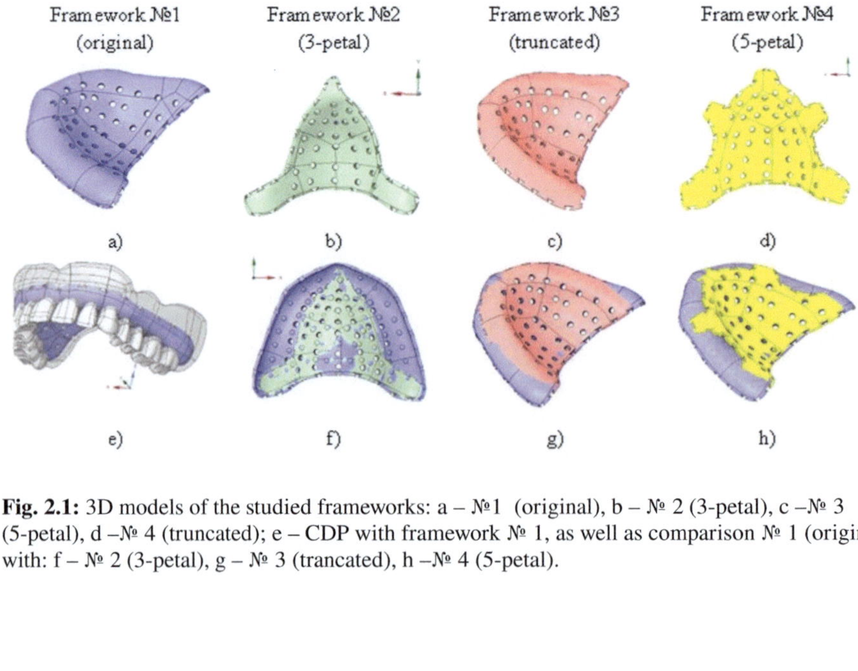

Fig. 2.1: 3D models of the studied frameworks: a – №1 (original), b – № 2 (3-petal), c –№ 3 (5-petal), d –№ 4 (truncated); e – CDP with framework № 1, as well as comparison № 1 (original) with: f – № 2 (3-petal), g – № 3 (trancated), h –№ 4 (5-petal).

Thus, framework № 1 (original) had the largest surface area among all the studied ones, while framework № 2 (3-petal) was the smallest version. Holes (perforations) 2 mm in diameter were considered across their entire surfaces, primarily to increase the adhesion levels (to the bases) by infusing PMMA (Fig. 2.1). Some stoppers (spikes 0.5 mm in height and 2.0 mm in diameter) were located on the outer framework surfaces *s* for their precise installations.

For computer simulations, the following properties of PEEK were applied: the *E* elastic modulus of 3.8 GPa and the tensile strength ultimate tensile strength of 92 MPa [26]. As noted above, both heights and volumes of the alveolar bones could decrease locally over time due to resorption. Accordingly, reliefs of the prosthetic foundation also changed (according to type 5 as an instance [33], when the alveolar bone was pronounced in the lateral regions and significantly atrophied in the frontal zones (Fig. 2.2, a).

To assess the impact of the alveolar bone degradation on the load-bearing capacities of CDPs of the maxilla, such changes were taken into account by the corresponding variations in the front parts of the initial "virtual supports", simulating the prosthetic foundation. An example of the deformed one is shown in Fig. 2.2, b. The 3D model of the deformed "virtual support" (Fig. 2.2, b) reflected the type 5 alveolar bone atrophy in the anterior region, that was, its partial damage, considering the impact of disease progression on the load-bearing capacities of CDPs.

As a result, it did not fit tightly to the bases in the deformed areas (Fig. 2.3, b), in contrast to the initial one (Fig. 2.3, a) that completely corresponded to the outer base surfaces, similar to framework №1 located at the dome center (Fig. 2.3).

In addition, the following assumptions were taken into account:

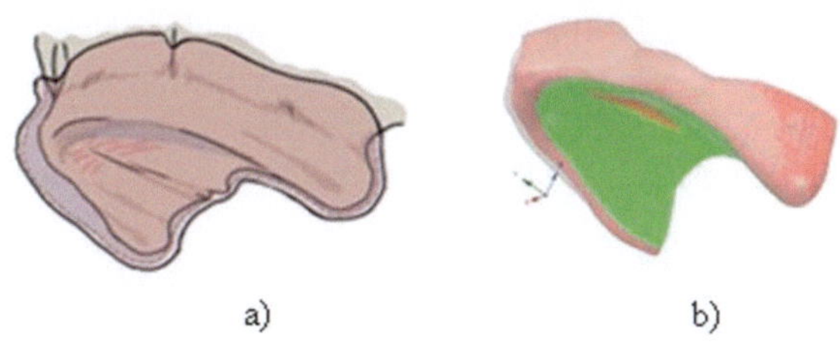

Fig. 2.2 The (type 5) alveolar bone atrophy in the frontal region: a – schematic representation, b – 3D model of the deformed "virtual support".

Fig. 2.3: Cross-sections of the simulated CDPs: a – with framework №1 (original) and the "virtual support", b – with framework № 2 (3-petal) and the deformed "virtual support".

- the minimum base thickness in the dome area was about 2–3 mm, while the framework thickness was 1.0–1.5 mm;
- framework displacements from the inner base surfaces (from the side of both palate and alveolar bone) to the center were 0.5 mm (Fig. 2.3, b).

To assess the effects of the frameworks (Fig. 2.1) on the load-bearing capacities of CDPs, their SSSs were calculated using the above-described FEM-based models (both with the frameworks and without them). Their shapes, the physical and mechanic properties, the adhesion levels, and the characteristics of the "virtual supports" were considered for various loading options.

2.2 Research Methodology

A CDP model consisted of a base with attached denture teeth, forming an individual dentition (Fig. 2.4). At the attachment points, bridges were considered between the teeth (Fig. 2.4, c) due to filling the space between them with PMMA. The base had a cutout for the frenulum in the anterior part (highlighted by a circle in Fig. 2.4, a) and individually shaped cutouts for anatomical structures (the lateral cords marked by an oval in Fig. 2.4a). The simulated denture teeth included a visible part (protruding above the base) and a hidden one, recessed in it (in the sockets shown in Fig. 2.4, c and d). The contact area between the teeth and the base could be observed in Fig. 2.4, d. Real joints of dental arches and bases enabled to simulate their fastening conditions as the "ideal adhesion". Both tensile and compressive elastic moduli of the teeth were 2.2 GPa, corresponding to those for additively manufactured PMMA/GF composites. For the base, the tensile elastic modulus of 0.9 GPa and the tensile strength of 22 MPa reflected those for the "NOLATEK 3D LCD/DLP PMMA" feedstock ("VladMiVa" LLC, Belgorod, Russian Federation) [8]. According to the authors, the tensile strength propertied of materials had to be used in the computer simulations because they were typically several times lower than those determined

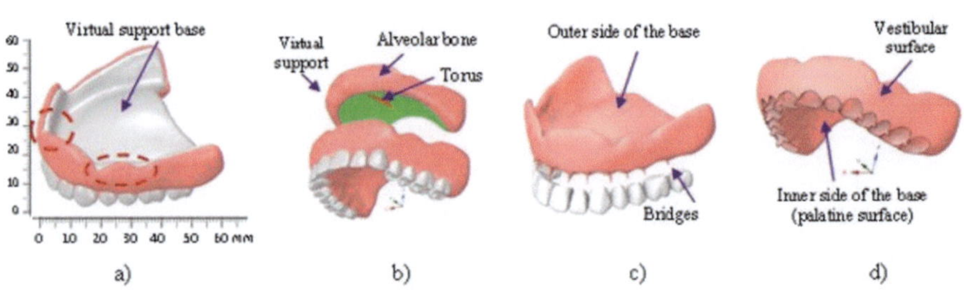

Fig. 2.4: The CDP model: a – with a dental row of set teeth, b – with the "virtual support" indicating the alveolar bone and torus areas, c and d – the base from the external (c) and internal (d) sides, respectively.

under compressive and shear loads [8]. So, this scheme allowed for a certain safety margin. All materials were assumed to be elastic and isotropic, with the exception of PMMA in the base, for which nonlinear behavior was taken into account.

Similar to the earlier research [1], the base attachment conditions were considered to be close to real ones. In particular, the prosthetic bed was simulated as a "virtual support" possessing "buffer zones" with uneven compliance of the oral mucosa [19, 33]. The alveolar bone area (along the support perimeter) and the torus (in the center) were characterized by negligible mucosal compliance (Fig. 2.4, b), i.e., lower damping than other regions [19, 33]. So, the elastic modulus and Poisson ratio of the alveolar bone of the upper jaw and torus were 30 MPa and 0.25, respectively, in contrast to 5 MPa and 0.45 for all other support zones, reflecting the different compliance of the palatal mucosa [22].

The load-bearing capacities of CDPs were assessed by analyzing their SSSs under operational loads. In this way, the "ABAQUS" FEM-based software package, v. 2019 (Dassault Systemes, France) was used according to the static formulation and geometric nonlinearity in the Abaqus/Standard module. The FEM-based model was developed using C3D8R tetrahedral volume elements with linear displacement approximation. The number of elements of 959486 with 206441 nodes (excluding the frameworks consisting of 1176187 elements and 421657 nodes) were determined based on mesh convergence. The loads were applied stepwise, evenly to several teeth (Fig. 2.5a). Taking into account the "virtual support", the immobility conditions were preset (no movements and/or rotations according to Fig. 2.5a).

The ideal adhesion was specified between the contacting surfaces of the base with the teeth and the "virtual support" (Fig. 2.5b, highlighted in green). At the non-contact region of the deformed "virtual support", no adhesion was preset (Fig. 2.5b). In the Abaqus software package, this fact corresponded to the "Normal behavior"/"Hard contact" option, preventing both surface separation and mutual penetration of one material into another, as well as "Frictionless"/"Tangential behavior" one, averting their slippage. To ensure the contact constraints, the direct method was used by default, guaranteeing rigid connections between the contacting surfaces due to the equality of forces and moments. As a result, they could not penetrate each other. At the contact between the base and the frameworks, both the ideal adhesion,

Fig. 2.5: The loading options (boundary conditions) for the CDP model and the "virtual support" at the attachment surfaces (a) in additon to the contacting surfaces between the base with the teeth (b), as well as both the initial and deformed "virtual supports" (c).

as well as its both minimum and maximum levels of 0.1 and 5.0 MPa, respectively, were specified (the "cohesive behavior" option). At the separation stress and contact layer stiffness values of 0.1 MPa, detachment of the frameworks from the base began during the first loading steps.

The initiation and propagation of cracks in the base was simulated using the extended FEM (XFEM) approach. Reaching the maximum principal stress of 22 MPa (the ultimate tensile strength of PMMA) was taken as a criterion for loss of the load-bearing capacities of CDPs, which corresponded to the initiation of cracks.

2.3 Computer Simulation Results

2.3.1 Effects of both Framework Shapes and Adhesion Levels on the Load-Bearing Capacities of CDPs Without "Virtual Supports"

2.3.1.1 Loading Options

To reduce the number of parameters influencing the load-bearing capacities of CDPs and more objectively evaluate the effects of the framework shapes, the first stage was carried out without considering the "virtual supports". Their rigid attachments complied with the boundary conditions, prohibiting movement along all axes and rotation in all planes on the outer surface of the base.

The base attachment areas (Fig. 2.6) varied from its partial adherence to the prosthetic foundation due to changes in its shapes (mucous membrane of the alveolar ridge of the maxilla) to complete fixation, that was, the ideal adhesion of the entire inner surface of the base.

Simultaneous loading on several teeth normal to their surfaces was considered, as shown in Fig. 2.6. A force of 0.2 N was applied to each tooth incrementally until 22 MPa was reached (the ultimate tensile strength of PMMA) and a crack initiated. The base attachment areas were selected considering the most dangerous combination with the loading option that determined low load-bearing capacities of CDPs. This problem formulation (without considering "virtual supports") did

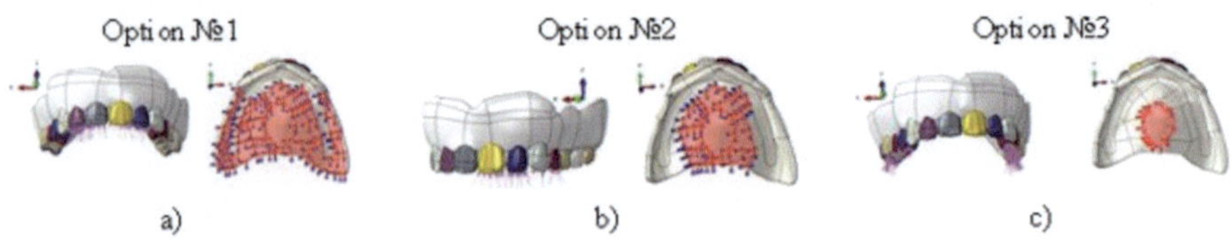

Fig. 2.6: The base attachment schemes and the loading options: № 1 - on four incisors (a); № 2 - on two incisors, one canine, and one molar (b); № 3 - on four molars and four premolars (c).

not correspond to real conditions, as it excluded the influence of parameters related to their both properties and shapes. However, it allowed for a comparison of the framework shapes on deformations of CDPs and their load-bearing capacities for different loading options. Since the research goal was to optimize both framework shapes and adhesion levels, the simplified formulation was acceptable for such trial calculations of SSSs.

2.3.1.2 Influence of the Framework Shapes on the Load-Bearing Capacities of CDPs Without "Virtual Supports" for Loading Option № 1

When attaching the entire outer surfaces of CDPs with framework № 1 (original), their load-bearing capacities increased slower with enhancing the elastic modulus (of the framework, hereafter) than those for attached locally (according to loading option № 1 shown in Fig. 2.6, a), both with the ideal adhesion and its minimum level of 0.1 MPa.

At the elastic modulus above 50 GPa, its role remained virtually unchanged when the ideal adhesion was assumed. An increase in the elastic modulus by two orders of magnitude (from 3.8 up to 400 GPa) enhanced the load-bearing capacities of CDPs with framework № 1 by only 5–10% with the ideal adhesion compared to those without frameworks. At the minimum adhesion, it was lower by 10% compared to those of CDPs with framework № 1 (original) and without it.

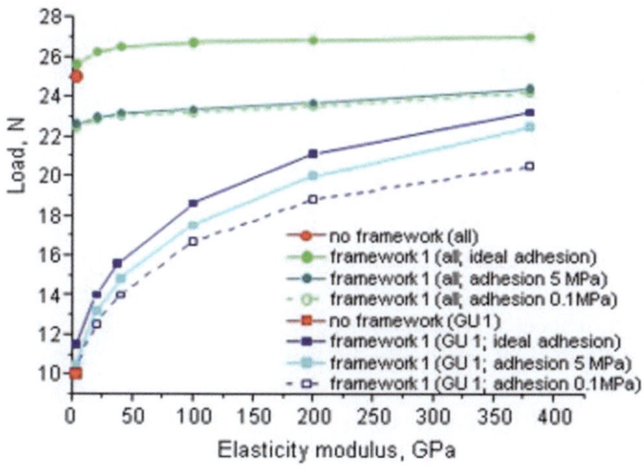

Fig. 2.7: The load-bearing capacities of CDPs with framework № 1 (original) and without it at different adhesion levels and attachment areas for loading option № 1 (Fig. 2.6, a).

For the locally attached CDPs (Fig. 2.6, a), the load-bearing capacities increased nonlinearly by up to two times with enhancing the elastic modulus by two orders of magnitude at different adhesion levels (Fig. 2.7). Thus, the larger were the attachment areas of CDPs, the lesser the influence of the elastic modulus on their load-bearing capacities, and the more significant the effect of the adhesion level.

For loading option № 1 (Fig. 2.6, a), all other framework shapes (Fig. 2.6) also increased the load-bearing capacities of the locally attached CDPs (Fig. 2.8) at different adhesion levels. The effect of increasing the elastic modulus from 0.38 GPa (for neat PEEK) up to 20 GPa (for composites reinforced with fiberglass fabrics) was evaluated as well.

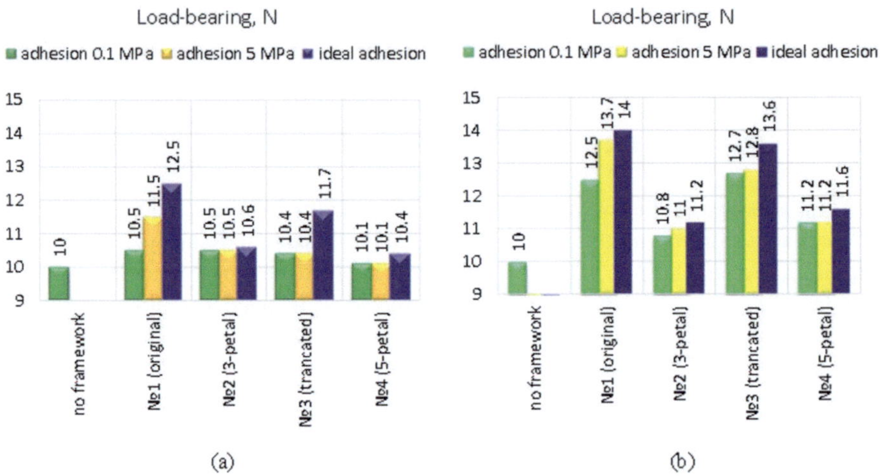

Fig. 2.8: The load-bearing capacities of CDPs with framework № 1 (original) and without it at different adhesion levels and attachment areas for loading option № 1 (Fig. 2.6, a).

At the elastic moduli of 0.38 and 20 GPa, the use of the studied frameworks increased the load-bearing capacities of CDPs by 1–25% and 10–40%, respectively (Fig. 2.8, a and b). With the ideal adhesion, the greatest effects of 25–40% were provided by framework № 1 (original), in contrast to 1–5% and 8–15% for ones № 2 (3-petal) and № 4 (5-petal), respectively. It should be noted that for the latter two, changing the adhesion levels did not affect greatly. With the minimum adhesion, framework № 4 (5-petal) enhanced the load-bearing capacities of CDPs by only 1%, which was the minimum value in all the studied cases. With the elastic modulus of 3.8 GPa (Fig. 2.8, a), framework № 3 (truncated) also slightly improved the load-bearing capacities of CDPs by 5% with the minimum adhesion and 15% with the ideal level, but the results of its use of 25–35% were comparable to those for framework № 1 (original) at 20 GPa (Fig. 2.8b).

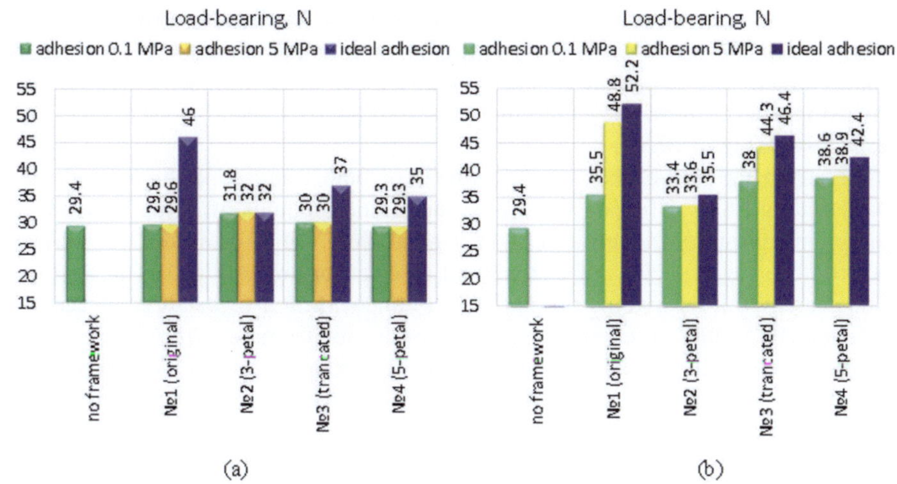

Fig. 2.9: The load-bearing capacities of CDPs, with the studied frameworks and without them, at different adhesion levels for loading option № 2 considering the elastic moduli of 0.38 GPa (a) and 20 GPa (b).

2.3.1.3 Influence of the Framework Shapes on the Load-Bearing Capacities of CDPs for Loading Option № 2

For loading option № 2, the load-bearing capacities of CDPs with framework № 1 (original) was lower than those for both № 2 (3-petal) and № 3 (truncated) with the minimum adhesion and the elastic modulus of 3.8 GPa (Fig. 2.9a), compared to the above discussed cases (Fig. 2.8a). Only framework № 2 (3-petal) enhanced it by 5%. With the ideal adhesion, the improvements were more significant: by 55% for framework № 1 (original) and by 25% for framework № 3 (truncated). At the elastic modulus of 20 GPa (Fig. 2.9b) the obtained results were similar to those shown in Fig. 2.8b: both frameworks № 1 (original) and № 3 (truncated) turned out to be the most effective (by 20–80% and 30–55%, respectively) regardless of the adhesion levels.

2.3.1.4 Influence of the Framework Shapes on the Load-Bearing Capacities of CDPs for Loading Option № 3

For loading option № 3 at the minimum adhesion, the load-bearing capacities of CDPs with framework № 1 (original) decreased by 35–40% compared to those without it at the elastic modulus of 0.38 GPa, and by 5–13% at 20 GPa (Fig. 2.10, a and b, respectively). With the ideal adhesion, it increased by 25% and 120% at the elastic moduli of 3.8 and 20 GPa, respectively, as in the previous cases.

At the elastic modulus of 3.8 GPa (Fig. 2.10, a), the greatest increase in the load-bearing capacities of CDPs by 15% (with the minimum adhesion) was provided by frameworks № 2 (3-petal) and № 4 (5-petal). At 20 GPa, the most suitable were ones № 3 (truncated) and № 4 (5-petal) that increased the parameter by 70% and 60%, respectively (Fig. 2.10, b).

Based on the above, the complex effects of both framework shapes and their elastic moduli had to be considered. At the elastic modulus of 3.8 GPa (with the minimum adhesion), frameworks № 1 (original) and № 4 (5-petal) did not increase the load-bearing capacities of CDPs in all cases according to Fig. 10 (the first one even decreased it for loading option № 3). Framework № 2 (3-petal) enhanced this parameter by 5–20% for all loading options, or did not decrease it (the changes in the adhesion levels had virtually no influence on the obtained results). Similar patterns were found for framework № 3 (truncated), but these variations were limited by 2–15%.

With the ideal adhesion, framework № 1 (original) increased the load-bearing capacities of CDPs by 20–120%, including with varying the elastic modulus, while framework № 3 (truncated) was less effective, improving it by 15–70%. At the elastic modulus of 20 GPa, it proved to be the greatest one, increasing the load-bearing capacities of CDPs by 15–70% in all cases, regardless of the adhesion levels.

In order to identify the factors determining the load-bearing capacities of CDPs with the studied frameworks at the minimum adhesion (Figs. 2.9 and 2.10), their SSSs were analyzed in detail for different loading options.

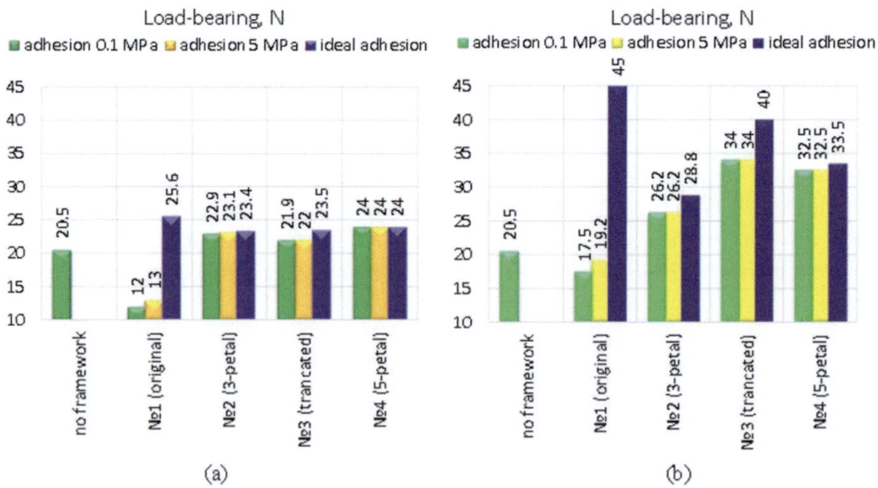

Fig. 2.10: The load-bearing capacities of CDPs, with the studied frameworks and without them, at different adhesion levels for loading option № 3 considering the elastic moduli of 0.38 GPa (a) and 20 GPa (b).

2.3.1.5 Comparative Analysis of SSSs of CDPs Depending on the Framework Shapes at the Minimum Adhesion for Different Loading Options

Figure 2.11 shows distributions of the maximum principal stresses in CDPs with framework № 1 (original) at the minimum adhesion for different loading options. In these images and below, strain visualization scales were magnified by several times using the corresponding function in the “ABAQUS” software package, as their calculated values did not exceed 0.5 mm. In Fig. 2.11, the CDP cross-sections are shown without scaling, since it was clear that the presented shapes did not change for different loading options. Nevertheless, the highlighted sections are given with increased strain visualization scales. In the latter cases, changes in the shapes of both deformed CDPs and each of the components (teeth, base, and frameworks), locations of the greatest strains (bending), the framework detachment areas with the minimum adhesion, could be clearly observed. Due to its curvilinear shape, the base was bent in different planes and experienced complex deformations for different loading options.

In all studied cases, the loaded teeth of CDPs exerted torque forces on framework № 1 (original) that caused bending of the alveolar region (Fig. 2.11). Consequently, it was detached from the base at the bending point, further increasing the deformation (rotation) of the base and leading to the achievement of ultimate stresses that resulted in failure. The initiated cracks are highlighted by red and blue arrows in Fig. 11. The red ones are oriented in the direction of the tensile force normal to the crack plane, while the blue arrows are parallel to the momentum acting in the crack plane. For loading options № 1 and № 2, the greatest concentrations of tensile stresses were observed near the bridges between the tooth sockets (Fig. 2.11, a and b). For loading option № 3 with the minimum adhesion, a crack initiated and propagated on the inner side of the base (Fig. 2.11, c) due to extreme tensile stresses near the framework hole in the alveolar zone, where it detached at both considered elastic moduli. With the maximum adhesion of 5 MPa, a crack initiated in the “bridge” between the tooth sockets under similar conditions due to a different stress distribution, as with the ideal adhesion. At the same time, locations of the most pronounced detachment areas remained virtually unchanged.

The conducted computer simulations visualized the transformation of round holes (perforations) on the inner side of the base into ovals under tension (Fig. 2.11). The greatest concentrations of compressive stresses were observed on its outer side along the attachment perimeter. In these areas, the base was bent, causing detachment at the minimum adhesion. If the holes were in these areas, they additionally contributed to the negative impact of stress raisers, accelerating failure of CDPs. At the elastic modulus of 20 GPa, the distributions of stresses in CDPs were similar with all the studied frameworks (therefore not duplicated). In these cases, the detachment patterns were similar but the frameworks allowed the base to withstand higher loads due to their greater rigidity. When using other frameworks, the stress levels and the pattern of their distributions in CDPs differed insignificantly at failure (Figs. 2.11, b and 2.12).

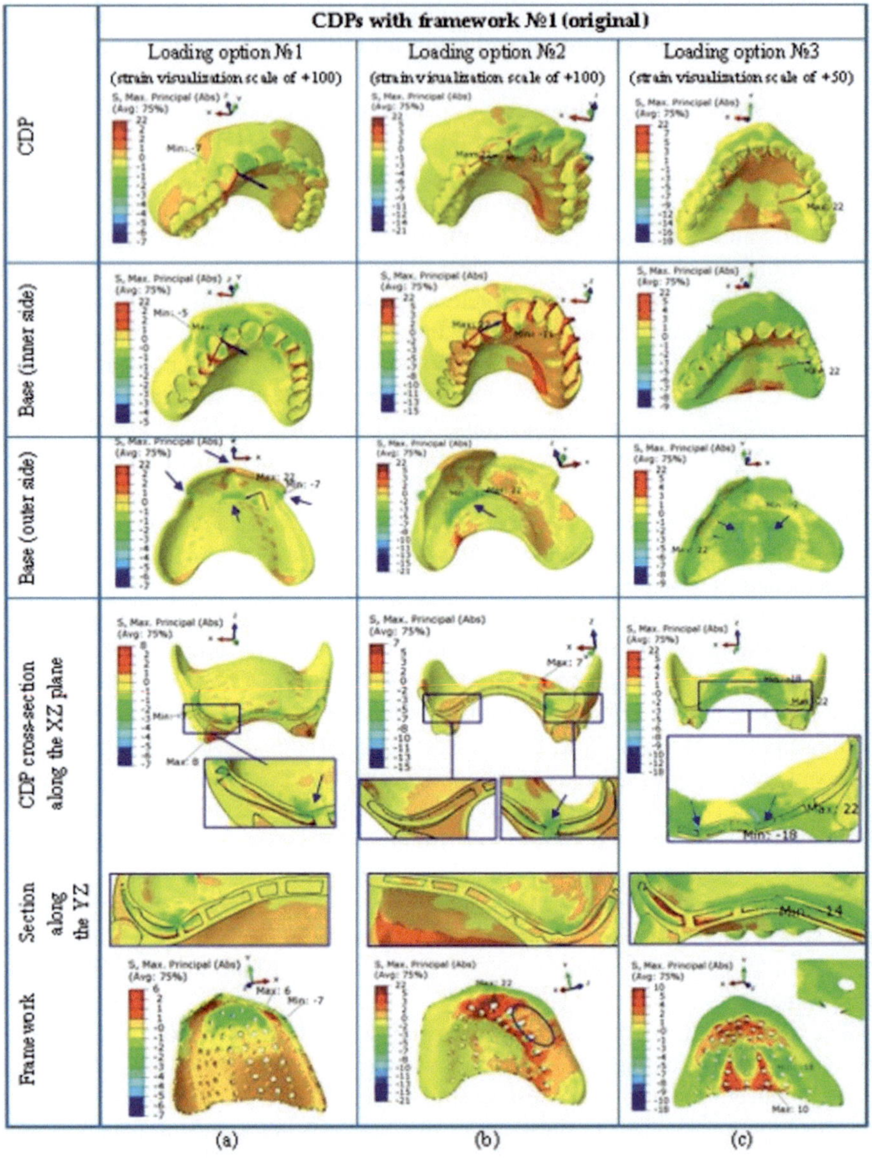

Fig. 2.11: The distributions of the maximum principal stresses in CDPs with framework №1 (original) but without "virtual supports" for different loading options (Fig. 2.6): (a) № 1; (b) № 2; (c) № 3.

For loading option № 3, both base and frameworks were bent in the alveolar zone and along the edge of the base attachment area. In the CDP cross-sections along the XZ-plane (Figs. 2.11, c and 2.12), an interface between the areas experiencing

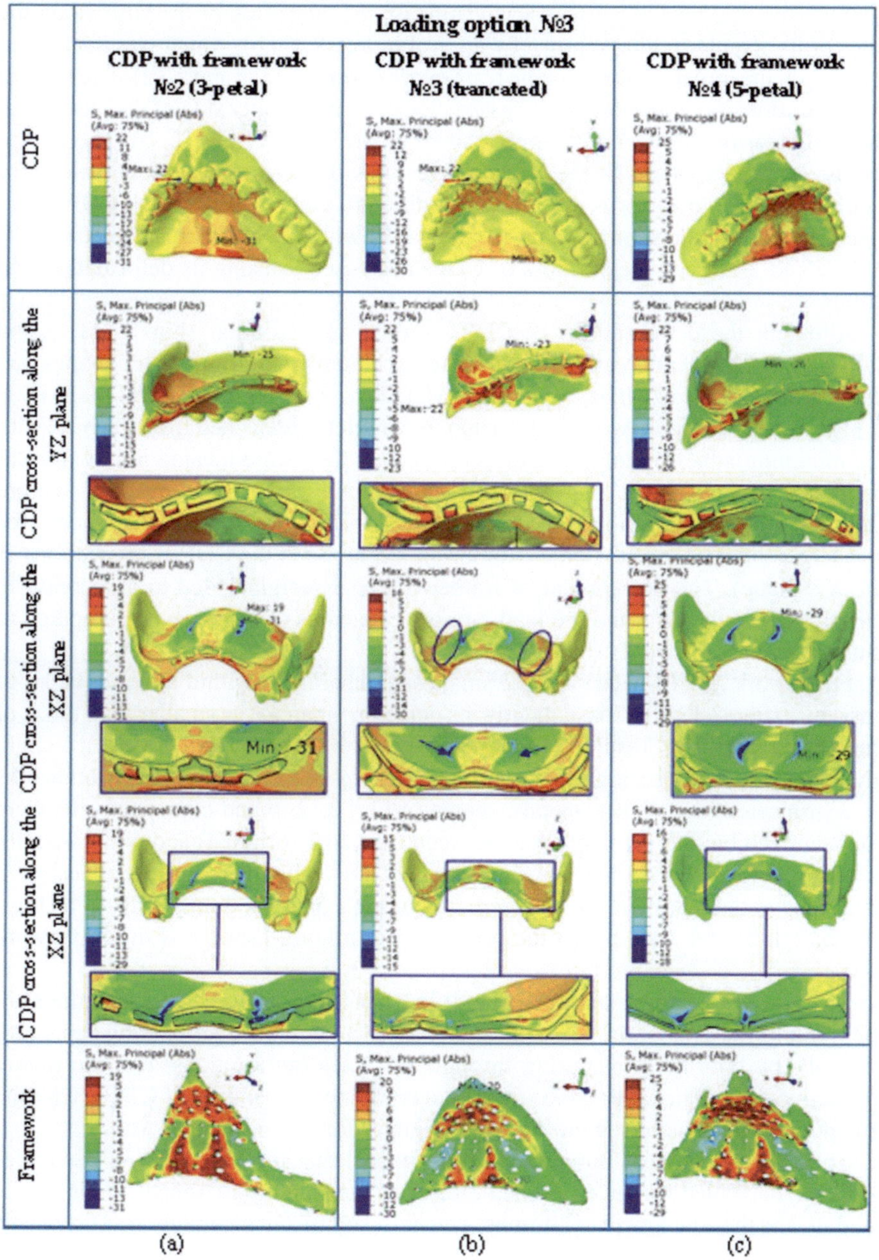

Fig. 2.12: The distributions of the maximum principal stresses in CDPs for loading option № 3 (strain visualization scale of +50).

compressive and tensile stresses was visible on the outer side of the base along the edge of its alveolar region (shown by an oval in Fig. 2.12, b), which was bent when loading on the molars deforming the frameworks (Figs. 2.11, c and 2.12). The highest concentration of compressive stresses was located on the outer side of the base along the perimeter of its attachment area (shown by arrows in Fig. 2.12, b). The reason for its occurrence could be the structural features of the prosthetic foundation (for example, a protruding torus under real conditions), as a result of which CDPs were also bent, causing detachment of the frameworks with the minimum adhesion (Figs. 2.11, c and 2.12). So, the stiffness of CDPs decreased, causing its deformation and failure.

Although the deformations of CDPs were visually similar (Figs. 2.11, c and 2.12), the detachment areas differed due to variations in the framework shapes and their load-bearing capacities. According to CDP cross-sections in the XZ-plane (Fig. 2.11), the edges of framework № 1 (original) completely detached from the base with the minimum adhesion for all loading options, as they were located in the alveolar bone area. It was the most strongly curved part due to the rotation of the denture teeth, facilitating the framework detachment. For this reason, the detachment area was larger than those of the other frameworks (Fig. 2.12, the CDP cross-sections in the XZ-plane). The detachment of framework № 1 (original) led to a decrease in the stiffness of the base and the load-bearing capacities of CDPs, even compared to those without frameworks (Fig. 2.10, a).

Framework № 2 (3-petal) was the smallest and barely fit within the alveolar ridge curvature (Fig. 2.12, a). Consequently, its detachment area was smaller than those of the other frameworks, resulting in a lesser reduction in the base stiffness but a slight increase in the load-bearing capacities of CDPs (Figs. 2.8–2.10). In turn, frameworks № 3 (truncated) and № 4 (5-petal), shown in Fig. 2.12, b and c, respectively, were characterized by intermediate areas between ones № 1 (original) and № 2 (3-petal), better providing the load-bearing capacities of CDPs with the minimum adhesion. It should be noted that their effects decreased at the elastic modulus of 20 GPa (Figs. 2.8, b; 2.9, b; and 2.10, b), so the primary determining factor was the framework shape.

Thereby, despite the reduction in the stiffness of the base due to the detachments of the frameworks at the minimum adhesion, they increased the load-bearing capacities of CDPs by both their anchoring through the holes (perforations) and the higher elastic modulus. The latter factor increased resistance to bending of CDPs, with the exception of ones with framework № 1 (original) for loading option № 3.

Thus, the computer simulation results showed that the load-bearing capacities of CDPs depended on the combined effects of the framework shapes, their elastic moduli, and the attachment conditions. The most significant parameter was the elastic modulus, as its increase improved the load-bearing capacities of CDPs (Fig. 2.7) with all the designed framework shapes, even with the minimum adhesion. At the elastic modulus of 3.8 GP, the most significant factors were both adhesion levels and framework shapes.

The comparison of the effects of the framework shapes revealed that framework № 2 (3-petal) was the most suitable at the elastic modulus of 3.8 GPa and the minimum adhesion, while one № 1 (original) was preferable with the ideal adhesion. For the elastic modulus of 20 GPa, framework № 3 (truncated) proved to be the most effective. In all the studied cases, it increased the load-bearing capacities of CDPs by 15–70%, regardless of the adhesion levels. Therefore, the framework shapes gave the greatest effect on the framework stiffness, followed by the adhesion levels. Based on the above, the preset attachment conditions enabled to compare the influence of different framework shapes on the load-bearing capacities of CDPs without "virtual supports".

2.3.2 Effects of the "Virtual Supports" on the Load-Bearing Capacities of CDPs with Frameworks № 1 (Original) and № 2 (3-Petal)

The diversity of individual CDP designs was determined, among other things, by the shapes and properties of the prosthetic foundations (simulated in this study by the "virtual supports"), affecting both contact conditions with bases and load-bearing capacities of CDPs. In order to evaluate the effects of the "virtual supports" (Fig. 2.4), both initial and deformed (according to the type 5 alveolar bone atrophy, Fig. 2.2, a), on the load-bearing capacities of CDPs, their SSSs were calculated considering frameworks № 1 (original) and № 2 (3-petal) with at the elastic modulus of 3.8 GPa. Options of ideal as well as minimum (0.1 MPa) adhesion of frames with a base are considered. In these cases, the loading options shown in Fig. 2.6 were taken into account.

2.3.2.1 Comparison of the Load-Bearing Capacities of CDPs with Frameworks № 1 (Original) and № 2 (3-Petal), both with and Without the "Virtual Supports"

When the deformed "virtual support" was used, the load-bearing capacities of CDPs, both with frameworks № 1 (original) and № 2 (3-petal), decreased by 5–30% compared to those for the initial one, regardless of the adhesion levels and the framework shapes (Fig. 2.13).

Frameworks № 1 (original) and № 2 (3-petal) increased the load-bearing capacities of CDPs by 5–30% and 0–15%, respectively (Fig. 2.13). The exception was loading option № 3, when framework № 1 (original) decreased it by 50% with the minimum adhesion (Fig. 2.13, c), compared to those without frameworks, regardless of the use of the "virtual supports" (Fig. 10, a). In general, these results were similar to those obtained without the use of the "virtual supports" (Figs. 2.8–2.10, a), but the values were several times higher since the areas of their contacts with the base were larger. However, the effects of the frameworks were smaller due to the large support areas at CDPs.

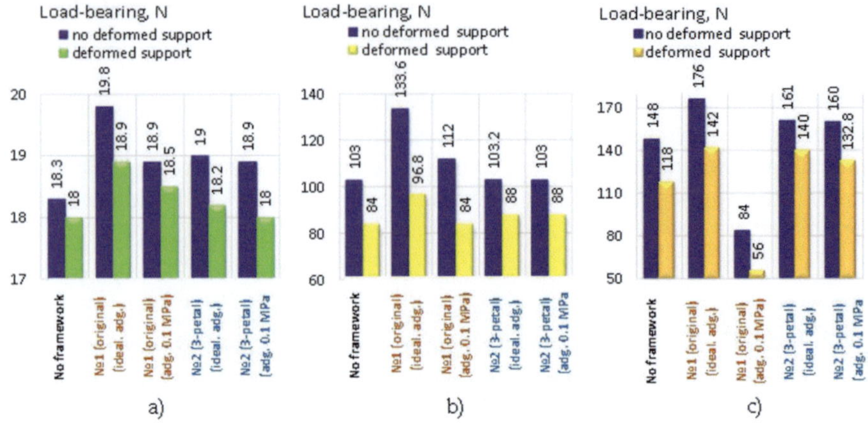

Fig. 2.13: The load-bearing capacities of CDPs with frameworks № 1 (original) and № 2 (3-petal), as well as without them, using both initial and deformed "virtual supports" for different loading options shown in Fig. 2.6: a – № 1, b – № 2, c – № 3.

2.3.2.2 Analysis of SSSs of CDPs with Framework № 1 (Original) and the "Virtual Supports"

As showed the computer simulation results (Fig. 2.14), the levels of compressive and tensile stresses in CDPs were maximal simultaneously near both canine and premolar, arising due to their rotation.

According to the CDP cross-section in the YZ-plane, compressive stresses were transferred to the vestibular area of the base, while tensile ones to the palatal region. As a result, both base and framework № 1 (original) were bent in the alveolar zone located beneath the loaded teeth, being an interface on both outer and inner surfaces of the base. In turn, the CDP cross-section in the XZ-plane revealed compressive stresses from the loaded incisor transmitted to framework № 1 (original) through the base. On its palatal side, located under the incisor, stresses were also compressive. However, they were tensile on its outer surface and the adjacent base, distorting both framework № 1 (original) and the base (Fig. 2.14).

On the inner side of the base, the maximum stresses were arisen in the bridges between the tooth sockets due to the rotation of the teeth and the base, as well as near both bands and frenulum on the outer side. They contributed to bending of the base in these areas and the initiation of cracks at the bridges. When using the initial "virtual support", the primary effect gave tensile stresses, but compressive ones when it was deformed.

Stress raisers were also observed in framework № 1 (original) around the holes (perforations) and in the "dovetail" cuts. Upon bending, the holes become oval. However, these stresses were not critical for both framework № 1 (original) and the base in the considered cases, as they were much lower than the strength properties of PEEK.

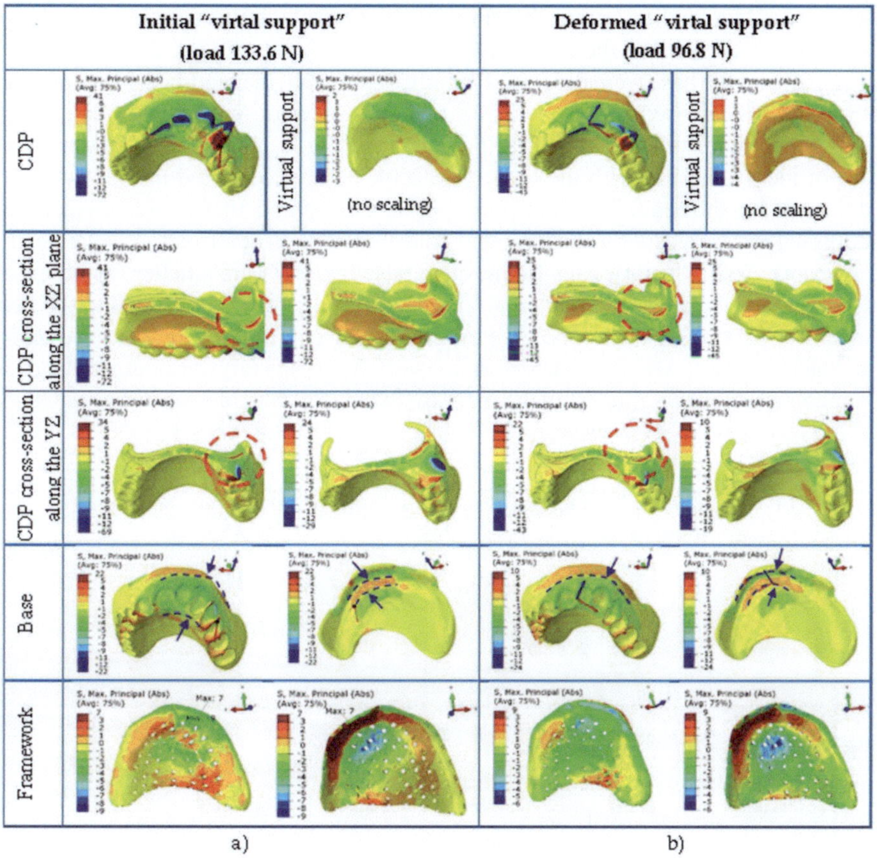

Fig. 2.14: The distributions of the maximum principal stresses in CDPs, the base, framework № 1 (original) (a) and (b) the "virtual supports" with the ideal adhesion (strain visualization scale of +100).

Due to the variations in the contact areas between the initial and deformed "virtual supports" with the base, stresses were distributed differently. According to the cross-sections shown in Fig. 2.14, b they were minimal in the non-contact region, increasing significantly in the contact areas and thus reducing the load-bearing capacities of CDPs.

2.3.2.3 Influence of the Absence of Adhesion on SSSs of CDPs with Frameworks № 1 (Original) and № 2 (3-Petal), as well as the Deformed "Virtual Support", for Loading Option № 2

Figure 2.15 shows the effect of the minimum adhesion on SSSs of CDPs with frameworks № 1 (original) and № 2 (3-petal), as well as the deformed "virtual support", was assessed for loading option № 2 (Fig. 2.6, b). Overall, the distributions of stresses were similar to those in the above-discussed case (Fig. 2.14), but framework № 1 (original) and the base were deformed greatly. As a result, the load-bearing capacities of the CDPs decreased, and a “dent” formed in framework № 1 (original).

The maximum stresses were three times higher (Fig. 2.15a), as compared with the case of ideal adhesion with the base (Fig. 2.14) and concentrated in the “dovetail” cuts located in the bending zone. Upon a possible detaching, they could have acted as additional stress raisers for the base and contributed to its deformation. When using framework № 2 (3-petal), gaps in the same base regions were smaller (or absent at all) due to its negligible surface area. Therefore, it only partially resisted to the bent “alveolar” part of the base. The greatest compressive stresses were observed under the loaded incisor. All other parts of framework № 2 (3-petal) were not involved, so their contribution was modest to the increase in the load-bearing capacities of CDPs.

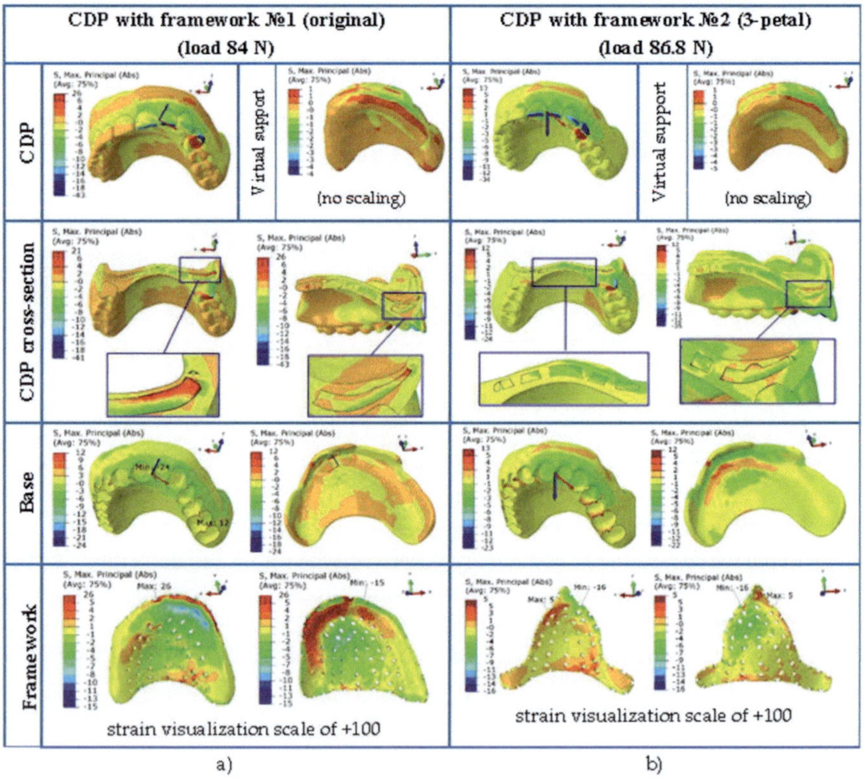

Fig. 2.15: The distributions of the maximum principal stresses in CDPs, the base, frameworks № 1 (original)- (a) and (b) № 2 (3-petal), as well as the deformed “virtual support” (strain visualization scale of +50).

2.3.2.4 Effect of the Absence of Adhesion on SSSs of CDPs with Framework № 1 (Original) and the Deformed "Virtual Support" for Loading Option № 3

For loading option № 3, framework № 1 (original) decreased the load-bearing capacities of CDPs more than twice at the minimum adhesion (Fig. 2.16) when using both initial and deformed "virtual supports" (Fig. 2.13, c), just as without them (Figs. 2.10, a and 2.11, c). A crack initiated in the base (Figs. 2.16 and 2.11, c) because of stress raisers near the holes (perforations) in framework № 1 (original), coinciding with the base curvature at the alveolar bone of the upper jaw. Without adhesion, shear stresses developed near the holes (perforations) in framework № 1 (original). Under the loaded molars, predominantly compressive stresses developed in both base and framework № 1 (original), pressing their alveolar areas and straightening due to the lack of "virtual supports" (as could be observed in the cross-section). Tensile stresses concentrated along the base and framework edges in the attachment area. The interface between these zones is a deflection line of the alveolar region (Fig. 2.16). Thus, the "virtual support" shapes significantly altered deformations of CDPs under the same loads, being a significant factor influencing their load-bearing capacities and the reinforcement efficiency.

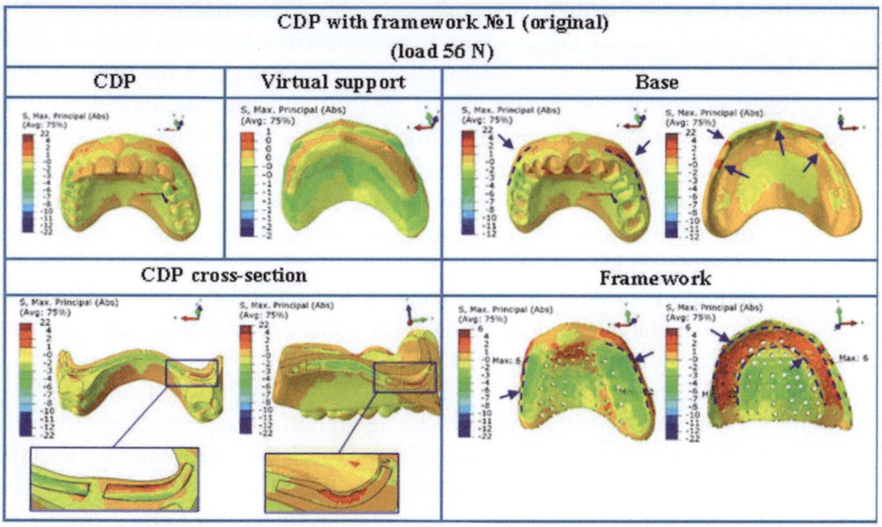

Fig. 2.16: The distributions of the maximum principal stresses in CDPs, the the base, framework № 1 (original) and the deformed "virtual support" with the minimum adhesion (strain visualization scale of +100).

2.4 Conclusions

The performed computer simulations of the load-bearing capacities of CDPs of the upper jaws reinforced with the designed frameworks of different shapes showed that all the considered options improved them by 5–120%. This parameter depended simultaneously on three most significant interrelated factors, namely both shapes and materials of the frameworks, as well as the adhesion levels. In addition, the prosthetic foundation configurations and their mechanical properties played important roles.

With local attachments of CDPs at small support areas, weak adhesion exerted the lesser influence on their load-bearing capacities as the elastic modulus increased, so the framework shapes became the determining factor. For the completely attached CDPs with the frameworks possessing the increased elastic modulus but with the minimum adhesion, their load-bearing capacities were lower than those without frameworks for some loading options. Therefore, the elastic modulus had the lesser impact on the load-bearing capacities of CDPs as the attachment areas were enlarged, increasing the significance of the adhesion levels. For the considered loading options, the deformed "virtual support" was less effective by 5–40% compared to the initial one, regardless of the presence of the frameworks and the adhesion levels.

The performed comparison of different framework shapes showed that they reduced the load-bearing capacities of CDPs due to their detachment from the base at the minimum adhesion of 0.1 MPa. This phenomenon reflected the main difference from the results obtained with the ideal adhesion, which (nevertheless) increased the load-bearing capacities of CDPs without frameworks in most of the studied cases. The reasons were the framework attachment to the base through the holes (perforations), the presence of the "dovetail" cuts, and the higher elastic modulus of the frameworks than that of the base, increasing resistance to bending of CDPs. The analysis of the SSSs of CDPs revealed that the considered loading options caused both rotation and bend of their base in the alveolar bone region. As a result, the base detached from the frameworks with the minimum adhesion, reducing both rigidities and load-bearing capacities of CDPs.

At the minimum adhesion, the placement of framework № 1 (original) in the alveolar bone area enlarged its detachment zone compared to smaller analogs. The cause was the base bending in this area, reducing both stiffness and load-bearing capacities of CDPs for some loading options compared to those of ones without frameworks. As a result, the frameworks with negligible overlapping of the alveolar bone area were more effective. The locations of the holes (perforations) and the "dovetail" cuts in the bending area of CDPs with framework № 1 (original) also reduced their load-bearing capacities of during detachment due to the great stress raisers in the base.

At the elastic modulus of 3.8 GPa and the minimum adhesion, framework № 2 (3-petal) should be used, but one № 1 (original) with the ideal adhesion. In turn, the most effective was the framework № 3 (truncated) at 20 GPa, which increased the load-bearing capacities of CDPs by 15–70% in all the studied cases. For the rigid frameworks, the most significant factor was their shapes, followed by the adhesion levels.

With the minimum adhesion adhesion, frameworks № 4 (5-petal) and № 1 (original) with the elastic modulus of 3.8 GPa did not increase the load-bearing capacities of CDPs in all the studied cases. For loading option № 3, the use of framework № 1 (original) even decreased it. In contrast, framework № 2 (3-petal) enhanced this parameter by 5–20% (or did not reduce it) for all loading options (changing the adhesion levels had virtually no effect). Similar results were obtained for framework № 3 (truncated), but the improvements by 2–15The use of framework № 1 (original) increased the load-bearing capacities of CDPs by 20–120% with the ideal adhesion at all the considered elastic moduli; framework № 3 (truncated) had the lesser effect (15–70%).

The obtained results were also valid in the presence of both initial and deformed "virtual supports". With the ideal adhesion, framework № 1 (original) was also preferred. It increased the load-bearing capacities of CDPs by 5–30%. With the minimum adhesion, framework № 2 (3-petal) enhanced it by 0–15%. Accordingly, the effectiveness of the used frameworks was determined by their shapes and the properties of the "virtual supports". Thus, the conducted computer simulations of the SSSs of CDPs made it possible to identify the main factors affecting their load-bearing capacities taking into account both the framework shapes and the loading options.

Acknowledgements Sergey Panin, Svetlana Bochkareva and Iliya Panov acknowledge support from the project FWRW-2026-0008 through the government research assignment for ISPMS SB RAS.

References

[1] Akhmedov K, Arutyunov D, Lomakin M, Bochkareva S, Panov I, Panin S, Mustafaev M (2025) Improving mechanical properties of complete removable dentures for the maxilla by reinforcing their bases with perforated PEEK frameworks. Fracture and Structural Integrity 19(72):280-294, DOI 10.3221/igf-esis.72.20

[2] Akhmetov EN, Fedorov VE, Bukhkamer NY, Akhmetov SE, Baranchuk iY, Tokarev KN, Shtol' AI, Stepanenko EV, Kogut NN, Semenyuk VM (2019) Prichiny obrashcheniy patsiyentov za restavratsiyami syemnykh zubnykh protezov v period garantiynogo sroka (in Russ., Reasons for patients' requests for removable denture restorations during the warranty period). Klinicheskaya stomatologiya 2:32-34,

[3] Alqutaibi AY, Baik A, Almuzaini SA (2023) Polymeric Denture Base Materials: A Review. Polymers 15(15):3258, DOI 10.3390/polym15153258

[4] Arutyunov SD, Grachev DI, Nikitin AD (2020) Mathematical modelling on the fracture of a laminar prosthesis basis under natural chewing loads. IOP Conference Series: Materials Science and Engineering 747:012065, DOI 10.1088/1757-899X/747/1/012065

[5] Barapatre D, Somkuwar S, Mishra S, Chowdhary R (2022) The effects of reinforcement with nanoparticles of polyetheretherketone, zirconium oxide and its mixture on flexural strength of PMAA resin. European Oral Research 56(2):61-66, DOI 10.26650/eor.2022904564

[6] Bural C, Aktaş E, Deniz G, Ünlüçerçi Y, Bayraktar G (201)1 Effect of leaching residual methyl methacrylate concentrations on in vitro cytotoxicity of heat polymerized denture base acrylic resin processed with different polymerization cycles. Journal of Applied Oral Science 19(4)306-312, DOI 10.1590/s1678-77572011005000002

[7] Cheng YY, Li JY, Fok SL, Cheung WL, Chow TW (2010) 3D FEA of high-performance polyethylene fiber reinforced maxillary dentures. Dental Materials 26(9):e211-e219, DOI 10.1016/j.dental.2010.05.002

[8] Chizhmakov EA, Arutyunov AS, Muslov SA, Bochkareva SA, Panov IL, Akhmedov GD, Buslovich DG, Panin SV, Arutyunov SD (2024) Application of polyethylene terephthalate as a denture base material for manufacturing temporary removable complete dentures. Mechanics of Composite Materials 60(2):227-242, DOI 10.1007/s11029-024-10186-2

[9] El Osta N, Haddad E, Fakhouri J, Saad, R, El Osta L (2021) Comparison of psychometric properties of GOHAI, OHIP-14, and OHIP-EDENT as measures of oral health in complete edentulous patients aged 60 years and more. Quality of Life Research: an International Journal of Quality of Life Aspects of Treatment, Care and Rehabilitation 30(4):1199-1213, DOI 10.1007/s11136-020-02709-w

[10] Geerts AVM, Overturf J-H, Oberholzer TG (2008) The effect of different reinforcements on the fracture toughness of materials for interim restorations. The Journal of Prosthetic Dentistry 99(6)461-467, DOI 10.1016/S0022-3913(08)60108-0

[11] Grachev DI, Martynenko AV, Perekhodov SN, Kostyrin EV, Mustafaev MS, Akhmedov KG, Deshev AV, Rozanov DG, Korotkova NL, Kerasov SN, Arutyunov SA (2024) New assessment model of financing treatment of patients with complete tooth loss. Emerging Science Journal 8(5):1898-1916, DOI 10.28991/ESJ-2024-08-05-014

[12] Grachev DI, Ruzuddinov NS, Arutyunov GD, Akhmedov GD, Dubova LV, Kharakh YN, Panin SV, Arutyunov SD (2022) Algorithm for designing a removable complete denture (RCD) based on the FEM analysis of its service life. Materials 17(15):20, DOI 10.3390/ma15207246

[13] Grachev DI, Tsarev VN, Majidova ER, Malginov NN, Zolotnisky IV, Tsimbalistov AV, Voytyatskaya IV, Chuev VP, Arutyunov SD (2021) Fiziko-mekhanicheskiye i mikrobiologicheskiye kharakteristiki pervogo otechestvennogo bazisnogo materiala gibridnoy polimerizatsii (in Russ., Physico-mechanical and microbiological characteristics of the first domestic basic material of hybrid polymerization). Klinicheskaya Stomatologiya 1(97):144-148, DOI 10.37988/1811-153X_2021_1_144

[14] Grachev DI, Zolotnitsky IV, Stepanov DY (2023) Ranking technologies of additive manufacturing of removable complete dentures (RCD) by the

results of their mechanical testing. Dentistry Journal 11(11):265, DOI 10.3390/dj11110265
[15] Jain N, Dutt U, Radenkov I, Jain S (2022) WHO's global oral health status report 2022: Actions, discussion and implementation. Oral Diseases 30(2):73-79, DOI 10.1111/odi.14516
[16] John J, Gangadhar SA, Shah I (2001) Flexural strength of heat-polymerized polymethyl methacrylate denture resin reinforced with glass, aramid, or nylon fibers. The Journal of Prosthetic Dentistry 86(4):424-427, DOI 10.1067/mpr.2001.118564
[17] Kim SH, Watts DC (2004) The effect of reinforcement with woven E-glass fibers on the impact strength of complete dentures fabricated with high-impact acrylic resin. The Journal of Prosthetic Dentistry 91(3):274-280, DOI 10.1016/j.prosdent.2003.12.023
[18] Kumari R, Bala S (2021) Assessment of Cases of Complete Denture Fracture. Journal of Pharmacy & Bioallied Sciences 13(Supplement 2):S1558-S1560, DOI 10.4103/jpbs.jpbs_284_21
[19] Lebedenko IY, Arutyunov SD, Ryakhovsky AN (2022) Prosthetic Dentistry. Vol 1: National Guide: in 2 vols, GEOTAR-Media, Moscow. `https://medbase.ru/book/ISBN9785970463666.html`, cited 18 Nov 2025
[20] Matsuyama Y, Aida J, Takeuchi K, Tsakos G, Watt RG, Kondo K, Osaka K (2014) Inequalities of dental prosthesis use under universal healthcare insurance. Community Dentistry and Oral Epidemiology 42(2):122-128, DOI 10.1111/cdoe.12074
[21] Murthy HB, Shaik S, Sachdeva H, Khare S, Haralur SB, Roopa KT (2015) Effect of reinforcement using stainless steel mesh, glass fibers, and polyethylene on the impact strength of heat cure denture base resin - An in vitro study. Journal of International Oral Health 7(6):71-79
[22] Muslov SA, Arutyunov SD, Chizhmakov EA, Em AV, Karakov KG (2023) K voprosu o biomekhanike slizistoy obolochki rta i tverdykh tkaney zuba: obzor i sobstvennyye issledovaniya (in Russ., Bioengineering aspects of oral mucosa: review and authors' research). Modern Issues of Biomedicine 7(4):1:14, DOI 10.24412/2588-0500-2023_07_04_38
[23] Nagai E, Otani K, Satoh Y, Suzuki S (2001) Repair of denture base resin using woven metal and glass fiber: effect of methylene chloride pretreatment. The Journal of Prosthetic Dentistry 85(5):496-500, DOI 10.1067/mpr.2001.115183
[24] Nejatidanesh F, Peimannia E, Savabi O. (2009) Effect of labial frenum notch size and palatal vault depth on stress concentration in a maxillary complete denture: a finite element study. J Contemp Dent Pract, DOI:10.5005/jcdp-10-3-59
[25] de Oliveira Limírio JPJ, Gomes JML, Alves Rezende MCR, Araújo Lemos CA, Del Rei Daltro Rosa CD, Pellizzer EP (2021) Mechanical properties of polymethyl methacrylate as a denture base: Conventional versus CAD-CAM resin - A systematic review and meta-analysis of in vitro studies. The Journal of Prosthetic Dentistry 128(6)1221-1229, DOI 10.1016/j.prosdent.2021.03.018

[26] PEEK. Matbase: the free and independent online materials properties resource, `http://www.matbase.com/material/polymers/engineering/peek/properties`, cited 10 Nov 2024

[27] Petrosyan AF, Lebedenko IY (2020) Profilaktika perelomov polnykh syemnykh plastinochnykh zubnykh protezov pri rezko vyrazhennom toruse (in Russ., Prevention of fractures of complete removable laminar dentures with pronounced torus). Klinicheskaya stomatologiya 1(93):86-89, DOI 10.37988/1811-153X_2020_1_86

[28] Rana MH, Shaik S, Hameed MS, Al-Saleh S, AlHamdan EM, Alshahrani A, Alqahtani A, Albaqawi AH, Vohra F, Abduljabbar T (2021) Influence of dental glass fibers and orthopedic Mesh on the failure loads of polymethyl methacrylate denture base resin. Polymers 13(16):2793, DOI 10.3390/polym13162793

[29] Rashid H, Sheikh Z, Vohra F (2015) Allergic effects of the residual monomer used in denture base acrylic resins. European Journal of Dentistry 9(4):614-619, DOI 10.4103/1305-7456.172621

[30] Somani MV, Khandelwal M, Punia V, Sharma V (2019) The effect of incorporating various reinforcement materials on flexural strength and impact strength of polymethylmethacrylate: A meta-analysis. Journal of Indian Prosthodontic Society 19(2):101-112, DOI 10.4103/jips.jips_313_18

[31] Tuominen R (2003) Clinical quality of removable dentures provided by dentists, denturists and laboratory technicians. Journal of Oral Rehabilitation 30(4):347-352, DOI 10.1046/j.1365-2842.2003.01055.x

[32] Venkat R, Gopichander N, Vasantakumar M (2013) Comprehensive analysis of repair/reinforcement materials for polymethyl methacrylate denture bases: mechanical and dimensional stability characteristics. The Journal of Indian Prosthodontic Society 13(4):439-449, DOI 10.1007/s13191-012-0249-z

[33] Voronov AP, Lebedenko IY, Voronov IA (2006) Ortopedicheskoe lechenie bol'nyh s polnym otsutstviem zubov (in Russ., Orthopedic treatment of patients with complete absence of teeth). MEDpress-inform, Moscow

[34] World Health Organization (ed) (2022) Global oral health status report: towards universal health coverage for oral health by 2030. `https://www.who.int/publications/i/item/9789240061484`

[35] Zafar MS (2020) Prosthodontic Applications of Polymethyl Methacrylate (PMMA): An Update. Polymers 12(10):2299, DOI 10.3390/polym12102299

[36] Zitzmann NU, Hagmann E, Weiger R (2007) What is the prevalence of various types of prosthetic dental restorations in Europe? Clinical Oral Implants Research 18(Supplement 3):20-33, DOI 10.1111/j.1600-0501.2007.01435.x

Chapter 3
Symmetries and Fundamental Solutions of the Transversely Isotropic Elastic Medium Equations

Alexander V. Aksenov

Abstract A fourth-order linear elliptic partial differential equation governing the displacements of a transversely isotropic linear elastic medium is considered. Its Lie symmetries, as well as the symmetries of the corresponding inhomogeneous equation with a delta function source term on the right-hand side, are determined. The latter symmetries are then employed to construct an invariant fundamental solution of the basic equation in terms of elementary functions.

Key words: Transversely isotropic elastic medium, Fundamental solution, Dirac delta function, Symmetry operator, Determining equations, Biharmonic equation

3.1 Introduction

In [4], the system of equilibrium equations for the displacement field in a transversely isotropic linear elastic medium is reduced to a system of three linear inhomogeneous equations for the three displacement components. The associated homogeneous equations correspond to canonical linear partial differential equations of the fourth order. These canonical equations represent a generalization of the biharmonic equation, which describes the displacements of an isotropic linear elastic medium. To determine the displacements of a transversely isotropic linear elastic medium subjected to a given body force, knowledge of the fundamental solutions to these canonical equations is required.

It is worth noting that in 3D elasticity, reducing the system of displacement equations to systems of higher-order equations using operators more amenable to numerical-analytical study than the Lamé operator is a classical approach, often

Alexander V. Aksenov
Faculty of Mechanics and Mathematics, Lomonosov Moscow State University, 1 Leninskie Gory, 119991 Moscow, Russia, e-mail: `aksenov@mech.math.msu.su`

H. Altenbach et al. (eds.), *Stability Problems in Mechanics of Continua*,
Advanced Structured Materials 260, https://doi.org/10.1007/978-3-032-31032-3_3

referred to as solution representations. Specifically, a reduction to tetraharmonic equations was discussed in [5].

The main result of this paper is the construction, in terms of elementary functions, of an invariant fundamental solution to the basic equation of a transversely isotropic linear elastic medium.

3.2 Symmetries and Fundamental Solutions

Fundamental solutions of linear partial differential equations are often invariant under transformations admitted by the original equation [6]. In the present paper, a fundamental solution is constructed using the algorithm proposed in [1] for finding fundamental solutions of linear partial differential equations. This algorithm utilizes the symmetries admitted by a linear partial differential equation with a delta function on its right-hand side. Let us briefly outline the main result of that work. Consider the pth-order linear partial differential equation

$$Lu \equiv \sum_{|\alpha|=1}^{p} A_\alpha(x) D^\alpha u = 0, \qquad x \in \mathbb{R}^m, \tag{3.1}$$

where the following standard notation is employed: $\alpha = (\alpha_1, \ldots, \alpha_m)$ is a multi-index with nonnegative integer entries, $|\alpha| = \alpha_1 + \cdots + \alpha_m$, and

$$D^\alpha \equiv \left(\frac{\partial}{\partial x^1}\right)^{\alpha_1} \cdots \left(\frac{\partial}{\partial x^m}\right)^{\alpha_m}.$$

A fundamental solution of Eq. (3.1) is a solution of the equation

$$Lu = \delta(x - x_0), \tag{3.2}$$

where δ denotes the Dirac delta function.

It was shown in [2] that for $p \geq 2$ and $m \geq 2$, Eq. (3.1) can only admit symmetry operators of the form

$$Z = \sum_{i=1}^{m} \xi^i(x) \frac{\partial}{\partial x^i} + \eta(x, u) \frac{\partial}{\partial u}, \qquad \frac{\partial^2 \eta}{\partial u^2} = 0.$$

As a vector space, the Lie algebra of symmetry operators of Eq. (3.1) decomposes into a direct sum of two subalgebras: one consisting of operators of the form

$$X = \sum_{i=1}^{m} \xi^i(x) \frac{\partial}{\partial x^i} + \zeta(x)\, u \frac{\partial}{\partial u}, \tag{3.3}$$

and the infinite-dimensional subalgebra generated by the operators

$$X_\infty = \varphi(x)\frac{\partial}{\partial u}, \tag{3.4}$$

where $\varphi(x)$ is an arbitrary solution of Eq. (3.1). Note that operators of the form (3.4) are symmetry operators of Eq. (3.2). In what follows, we consider only symmetry operators of the form (3.3). Let $\mathrm{pr}^{(p)} X$ denote the pth-order prolongation of the symmetry operator (3.3).

Proposition 3.1. *(A.V. Aksenov [1]) The infinitesimal operator given by* (3.3) *is a symmetry operator of Eq.* (3.1) *if and only if there exists a function* $\lambda = \lambda(x)$ *satisfying the identity*

$$\mathrm{pr}^{(p)} X\,(Lu) \equiv \lambda(x)\,Lu \tag{3.5}$$

for any function $u = u(x)$ *from the domain of Eq.* (3.1).

Theorem 3.1. *(A.V. Aksenov [1]) The Lie algebra of symmetry operators of Eq.* (3.2) *is a subalgebra of the Lie algebra of symmetry operators of Eq.* (3.1) *and is determined by the conditions*

$$\begin{gathered} \xi^i(x_0) = 0, \qquad i = 1, \dots, m, \\ \lambda(x_0) + \sum_{i=1}^{m} \frac{\partial \xi^i(x_0)}{\partial x^i} = 0. \end{gathered} \tag{3.6}$$

3.2.1 Algorithm for Constructing Fundamental Solutions Using Symmetries

We now describe an algorithm for constructing fundamental solutions by means of symmetry methods [1]:

1. Determine a general symmetry operator of Eq. (3.1) and the corresponding function $\lambda(x)$ satisfying identity (3.5).
2. Using this operator together with conditions (3.6), obtain a basis for the Lie algebra of symmetry operators of Eq. (3.2).
3. Construct invariant fundamental solutions with the aid of the symmetries of Eq. (3.2).
4. Generate new fundamental solutions from known ones by means of the symmetries of Eq. (3.2) (solution generation).

Remark 3.1. To construct generalized invariant fundamental solutions, it is necessary to seek invariants within the class of generalized functions.

3.2.2 Example of Applying the Algorithm

Consider the two-dimensional biharmonic equation

$$\Delta\Delta u \equiv \frac{\partial^4 u}{\partial x^4} + 2\frac{\partial^4 u}{\partial x^2 \partial y^2} + \frac{\partial^4 u}{\partial y^4} = 0. \tag{3.7}$$

Fundamental solutions of the two-dimensional biharmonic equation satisfy

$$\frac{\partial^4 u}{\partial x^4} + 2\frac{\partial^4 u}{\partial x^2 \partial y^2} + \frac{\partial^4 u}{\partial y^4} = \delta(x, y). \tag{3.8}$$

A finite basis of the Lie algebra of symmetry operators of equation (3.7) is given by [3]

$$X_1 = \frac{\partial}{\partial x}, \qquad X_2 = \frac{\partial}{\partial y}, \qquad X_3 = x\frac{\partial}{\partial x} + y\frac{\partial}{\partial y},$$
$$X_4 = y\frac{\partial}{\partial x} - x\frac{\partial}{\partial y}, \qquad X_5 = (x^2 - y^2)\frac{\partial}{\partial x} + 2xy\frac{\partial}{\partial y} + 2xu\frac{\partial}{\partial u},$$
$$X_6 = 2xy\frac{\partial}{\partial x} + (y^2 - x^2)\frac{\partial}{\partial y} + 2yu\frac{\partial}{\partial u}, \qquad X_7 = u\frac{\partial}{\partial u}.$$

To determine the symmetry operators admitted by equation (3.8), we write the general form of the symmetry operator admitted by equation (3.7) as $X = \sum_{i=1}^{7} a_i X_i$, or

$$\begin{aligned} X = {} & \left[a_1 + a_3 x + a_4 y + a_5(x^2 - y^2) + 2a_6 xy\right]\frac{\partial}{\partial x} + \\ & + \left[a_2 + a_3 y - a_4 x + 2a_5 xy + a_6(y^2 - x^2)\right]\frac{\partial}{\partial y} + \\ & + (2a_5 x + 2a_6 y + a_7)u\frac{\partial}{\partial u}, \end{aligned} \tag{3.9}$$

where a_i ($i = 1, \dots, 7$) are arbitrary constants. The symmetry operator (3.9) corresponds to the function $\lambda(x, y)$ given by

$$\lambda = a_7 - 4a_3 - 6a_5 x - 6a_6 y.$$

Applying Theorem 3.1, we obtain the conditions

$$a_1 = a_2 = 0, \qquad a_7 - 2a_3 = 0.$$

Proposition 3.2. *Equation* (3.8) *admits the following basis of Lie algebra symmetry operators:*

$$Y_1 = y\frac{\partial}{\partial x} - x\frac{\partial}{\partial y}, \qquad Y_2 = x\frac{\partial}{\partial x} + y\frac{\partial}{\partial y} + 2u\frac{\partial}{\partial u},$$
$$Y_3 = (x^2 - y^2)\frac{\partial}{\partial x} + 2xy\frac{\partial}{\partial y} + 2xu\frac{\partial}{\partial u},$$
$$Y_4 = 2xy\frac{\partial}{\partial x} + (y^2 - x^2)\frac{\partial}{\partial y} + 2yu\frac{\partial}{\partial u}.$$

The fundamental solution of the two-dimensional biharmonic equation is known to be [8]

$$u = \frac{x^2 + y^2}{16\pi} \ln(x^2 + y^2). \tag{3.10}$$

Solution (3.10) is invariant under the one-parameter group of rotations corresponding to the symmetry operator Y_1. The symmetry operator Y_2 generates a one-parameter group of inhomogeneous scaling transformations $x' = e^a x$, $y' = e^a y$, $u' = e^{2a} u$, where a is the group parameter. Under the action of this group, the fundamental solution (3.10) is transformed into the family

$$u = \frac{x^2 + y^2}{16\pi} \left[\ln(x^2 + y^2) + 2a\right].$$

Now consider the symmetry operator Y_3. It generates the one-parameter transformation group

$$\begin{aligned} x' &= \frac{x - a(x^2 + y^2)}{1 - 2ax + a^2(x^2 + y^2)}, \\ y' &= \frac{y}{1 - 2ax + a^2(x^2 + y^2)}, \\ u' &= \frac{u}{1 - 2ax + a^2(x^2 + y^2)}, \end{aligned} \tag{3.11}$$

where a is the group parameter. Under the action of this group, the fundamental solution (3.10) is mapped to the nontrivial fundamental solution

$$u = \frac{[x - a(x^2 + y^2)]^2 + y^2}{16\pi[1 - 2ax + a^2(x^2 + y^2)]} \ln\left(\frac{\left(x - a(x^2 + y^2)\right)^2 + y^2}{\left(1 - 2ax + a^2(x^2 + y^2)\right)^2}\right). \tag{3.12}$$

Similarly, the symmetry operator Y_4 corresponds to the one-parameter transformation group

$$\begin{aligned} x' &= \frac{x}{1 - 2ay + a^2(x^2 + y^2)}, \\ y' &= \frac{y - a(x^2 + y^2)}{1 - 2ay + a^2(x^2 + y^2)}, \\ u' &= \frac{u}{1 - 2ay + a^2(x^2 + y^2)}, \end{aligned} \tag{3.13}$$

where a is the group parameter. Under the action of this group, the fundamental solution (3.10) is transformed into

$$u = \frac{1}{16\pi} \frac{x^2 + \left(y - a(x^2 + y^2)\right)^2}{1 - 2ay + a^2(x^2 + y^2)} \ln\left(\frac{x^2 + \left(y - a(x^2 + y^2)\right)^2}{\left(1 - 2ay + a^2(x^2 + y^2)\right)^2}\right). \tag{3.14}$$

Remark 3.2. By considering the composition of transformations (3.11) and (3.13), one can obtain a nontrivial two-parameter family of fundamental solutions, rather than just the one-parameter families (3.12) and (3.14).

3.3 Governing Equations

Consider the following fourth-order linear differential equations, introduced in [4]:

$$\begin{aligned} L_1 u &\equiv u_{xxxx} + 2u_{xxyy} + u_{yyyy} + B_1(u_{xxzz} + u_{yyzz}) + B_2 u_{zzzz} = 0, \\ L_2 u &\equiv B_3(u_{xxxx} + 2u_{xxyy} + u_{yyyy}) + B_4(u_{xxzz} + u_{yyzz}) + u_{zzzz} = 0. \end{aligned} \tag{3.15}$$

Here B_1, B_2, B_3, and B_4 are positive constants characterizing the linear elastic medium. The fundamental solutions of Eqs. (3.15) satisfy

$$L_1 u = \delta(x)\delta(y)\delta(z), \qquad L_2 u = \delta(x)\delta(y)\delta(z). \tag{3.16}$$

We now show that Eqs. (3.15) and (3.16) can be reduced to identical equations by a change of variables. To this end, for the equations involving the differential operator L_1, we introduce the new variables

$$\bar{z} = \frac{z}{\sqrt[4]{B_2}}, \qquad \bar{u} = \sqrt[4]{B_2}\, u.$$

Omitting the bars over the new variables, the corresponding equations (3.15) and (3.16) become

$$L_3 u \equiv u_{xxxx} + 2u_{xxyy} + u_{yyyy} + b(u_{xxzz} + u_{yyzz}) + u_{zzzz} = 0, \tag{3.17}$$

$$L_3 u = \delta(x)\delta(y)\delta(z), \tag{3.18}$$

where $b = B_1/\sqrt{B_2}$. Similarly, by the change of variables

$$\bar{x} = \frac{x}{\sqrt[4]{B_3}}, \qquad \bar{y} = \frac{y}{\sqrt[4]{B_3}}, \qquad \bar{u} = \sqrt{B_3}\, u,$$

the equations for the differential operator L_2 are also reduced to Eqs. (3.17) and (3.18) with $b = B_4/\sqrt{B_3}$.

From the condition of positive definiteness of the quadratic strain energy density, it follows that $b \geq 2$ [4]. In what follows, it is assumed that $b \geq 2$.

Equation (3.17) is henceforth considered as the basic equation. Its axisymmetric solutions $u = \bar{u}(r, z)$ satisfy

$$L_4 \bar{u} \equiv \bar{u}_{rrrr} + b\bar{u}_{rrzz} + \bar{u}_{zzzz} + \frac{2}{r}\bar{u}_{rrr} + \frac{b}{r}\bar{u}_{rzz} - \frac{1}{r^2}\bar{u}_{rr} + \frac{1}{r^3}\bar{u}_r = 0, \tag{3.19}$$

while the axisymmetric fundamental solutions (i.e., solutions of Eq. (3.18)) satisfy

$$rL_4\bar{u} = \frac{1}{\pi}\,\delta(r)\delta(z). \tag{3.20}$$

Here $r = \sqrt{x^2 + y^2}$ and

$$\int_0^\infty \delta(r)\,dr = \frac{1}{2}.$$

Equation (3.20) can be rewritten in conservative form as

$$\left(r\bar{u}_{rrr} + br\bar{u}_{rzz} + \bar{u}_{rr} - \frac{1}{r}\bar{u}_r\right)_r + \left(r\bar{u}_{zzz}\right)_z = \frac{1}{\pi}\,\delta(r)\delta(z). \tag{3.21}$$

3.4 Symmetries of the Basic Equation

3.4.1 Symmetries of the Basic Eq. (3.17)

We are looking for the symmetry operators of the Eq. (3.17) in the form

$$X = \xi^1(x, y, z, u)\frac{\partial}{\partial x} + \xi^2(x, y, z, u)\frac{\partial}{\partial y} + \xi^3(x, y, z, u)\frac{\partial}{\partial t} + \eta(x, y, z, u)\frac{\partial}{\partial u}.$$

3.4.1.1 Case 1. $b \neq 2$

In this case applying the invariance criterion [6], we obtain the following overdetermined linear homogeneous system of determining equations

$$\begin{aligned}
&\xi^1_z = 0, \quad \xi^1_u = 0, \quad \xi^1_{xx} = 0, \quad \xi^1_{xy} = 0,\\
&\xi^1_{yy} = 0, \quad \xi^2_z = 0, \quad \xi^2_u = 0,\\
&\xi^1_y + \xi^2_x = 0, \quad \xi^1_x - \xi^2_y = 0, \quad \xi^3_x = 0,\\
&\xi^3_y = 0, \quad \xi^3_u = 0, \quad \xi^1_x - \xi^3_z = 0,\\
&\eta_{xu} = 0, \quad \eta_{yu} = 0, \quad \eta_{zu} = 0, \quad \eta_{uu} = 0,\\
&\eta_{xxxx} + 2\eta_{xxyy} + \eta_{yyyy} + b(\eta_{xxzz} + \eta_{yyzz}) + \eta_{zzzz} = 0.
\end{aligned} \tag{3.22}$$

It is not difficult to show that the joint solution of the overdetermined system of equations (3.22) is given by the formulas

$$\begin{aligned}
\xi^1 &= c_1 x + c_4 y + c_5 x,\\
\xi^2 &= c_2 - c_4 x + c_5 y,\\
\xi^3 &= c_3 + c_5 z,\\
\eta &= c_6 u + \varphi(x, y, z),
\end{aligned}$$

where c_j $(j = 1, \ldots, 6)$ are arbitrary constants and $u = \varphi(x, y, z)$ is an arbitrary solution of Eq. (3.17).

Proposition 3.3. *Equation (3.17) with an arbitrary parameter b admits the following the finite-dimensional part of the basis of the Lie algebra of symmetry operators:*

$$X_1 = \frac{\partial}{\partial x}, \qquad X_2 = \frac{\partial}{\partial y}, \qquad X_3 = \frac{\partial}{\partial z}, \qquad X_4 = u\,\frac{\partial}{\partial u},$$
$$X_5 = y\,\frac{\partial}{\partial x} - x\,\frac{\partial}{\partial y}, \qquad X_6 = x\,\frac{\partial}{\partial x} + y\,\frac{\partial}{\partial y} + z\,\frac{\partial}{\partial z}. \tag{3.23}$$

3.4.1.2 Case 2. $b = 2$

In this case applying the invariance criterion, we obtain the following overdetermined linear homogeneous system of determining equations

$$\begin{gathered}
\xi^1_u = 0, \quad \xi^2_u = 0, \quad \xi^3_u = 0, \quad \xi^1_x - \xi^3_z = 0,\\
\xi^1_y + \xi^2_x = 0, \quad \xi^1_z + \xi^3_x = 0, \quad \xi^2_y - \xi^3_z = 0,\\
\xi^2_z + \xi^3_y = 0, \quad \xi^3_{xy} = 0, \quad \xi^1_{yy} + \xi^3_{xz} = 0, \quad \eta_{uu} = 0,\\
\xi^3_{xx} + \xi^3_{zz} = 0, \quad \xi^3_{yy} + \xi^3_{zz} = 0, \quad \xi^3_{xz} - 2\eta_{xu} = 0,\\
\xi^3_{yz} - 2\eta_{yu} = 0, \quad \xi^3_{zz} - 2\eta_{zu} = 0, \quad \xi^3_{yzz} = 0, \quad \xi^3_{zzz} = 0,\\
\eta_{xxxx} + \eta_{yyyy} + \eta_{zzzz} + 2(\eta_{xxyy} + \eta_{xxzz} + \eta_{yyzz}) = 0.
\end{gathered} \tag{3.24}$$

It is not difficult to show that the joint solution of the overdetermined system of equations (3.24) is given by the formulas

$$\begin{aligned}
\xi^1 &= c_1 x + c_4 y + c_5 x + c_7 z + c_9(x^2 - y^2 - z^2) + 2c_{10}xy + 2c_{11}xz,\\
\xi^2 &= c_2 - c_4 x + c_5 y + c_8 z + 2c_9 xy + c_{10}(y^2 - x^2 - z^2) + 2c_{11}yz,\\
\xi^3 &= c_3 + c_5 z - c_7 x - c_8 y + 2c_9 xz + 2c_{10}yz + c_{11}(z^2 - x^2 - y^2),\\
\eta &= c_6 u + c_9 xu + c_{10} yu + c_{11} zu + \varphi(x, y, z),
\end{aligned}$$

where c_j $(j = 1, \ldots, 11)$ are arbitrary constants and $u = \varphi(x, y, z)$ is an arbitrary solution of Eq. (3.17).

Proposition 3.4. *For* $b = 2$, *the basis of the Lie algebra of symmetry operators* (3.23) *is supplemented by symmetry operators*

$$X_7 = z\,\frac{\partial}{\partial x} - x\,\frac{\partial}{\partial z}, \qquad X_8 = z\,\frac{\partial}{\partial y} - y\,\frac{\partial}{\partial z},$$
$$X_9 = (x^2 - y^2 - z^2)\,\frac{\partial}{\partial x} + 2xy\,\frac{\partial}{\partial y} + 2xz\,\frac{\partial}{\partial z} + xu\,\frac{\partial}{\partial u},$$
$$X_{10} = 2xy\,\frac{\partial}{\partial x} + (y^2 - x^2 - z^2)\,\frac{\partial}{\partial y} + 2yz\,\frac{\partial}{\partial z} + yu\,\frac{\partial}{\partial u},$$

$$X_{11} = 2xz\,\frac{\partial}{\partial x} + 2yz\,\frac{\partial}{\partial y} + (z^2 - x^2 - y^2)\,\frac{\partial}{\partial z} + zu\,\frac{\partial}{\partial u}\,.$$

3.4.2 Symmetries of Eq. (3.18)

To determine the symmetries of Eq. (3.18), we employ the results of [1]. Using the finite-dimensional part of the Lie algebra of symmetry operators of Eq. (3.17), we consider the general symmetry operator

$$X = \sum_{i=1}^{6} a_i X_i,$$

where a_i $(i = 1, \ldots, 6)$ are arbitrary constants.

Proposition 3.5. *The following relation holds:*

$$\mathrm{pr}^{(4)}\, X\,(L_3 u) = (a_6 - 4a_5) L_3 u.$$

Then, applying Theorem 3.1, we obtain the conditions

$$a_1 = a_2 = a_3 = 0, \qquad a_5 - a_6 = 0.$$

Proposition 3.6. *For an arbitrary parameter b, Eq.* (3.18) *admits the following basis of the Lie algebra of symmetry operators:*

$$Y_1 = y\frac{\partial}{\partial x} - x\frac{\partial}{\partial y}, \qquad Y_2 = x\frac{\partial}{\partial x} + y\frac{\partial}{\partial y} + z\frac{\partial}{\partial z} + u\frac{\partial}{\partial u}. \tag{3.25}$$

Remark 3.3. It can also be shown that Eq. (3.18) admits, in addition to the symmetry operators (3.25), the symmetry operator

$$Y_3 = z\frac{\partial}{\partial x} - x\frac{\partial}{\partial z}, \tag{3.26}$$

as well as the symmetry operators X_8, X_9, X_{10}, and X_{11}.

3.5 Fundamental Solution

We now construct a solution of Eq. (3.17) that is invariant under the symmetry operators (3.25). The invariants of the admitted transformation group are $J_1 = r^2/z^2 = \tau$ and $J_2 = u/z$. Accordingly, an invariant solution is sought in the form

$$u = zf(\tau). \tag{3.27}$$

Substituting (3.27) into Eq. (3.19) yields the fourth-order linear ordinary differential equation

$$\begin{aligned} 4\tau^2(\tau^2 + b\tau + 1)\frac{d^4 f}{d\tau^4} + 2\tau(14\tau^2 + 11b\tau + 8)\frac{d^3 f}{d\tau^3} + \\ + (39\tau^2 + 22b\tau + 8)\frac{d^2 f}{d\tau^2} + 2(3\tau + b)\frac{df}{d\tau} = 0. \end{aligned} \tag{3.28}$$

Proposition 3.7. *The linear PDE* (3.17) *admits the factorization*

$$\left(\frac{\partial^2}{\partial x^2} + \frac{\partial^2}{\partial y^2} + a\frac{\partial^2}{\partial z^2}\right)\left(\frac{\partial^2}{\partial x^2} + \frac{\partial^2}{\partial y^2} + \frac{1}{a}\frac{\partial^2}{\partial z^2}\right)u = 0,$$

where the parameter a satisfies $b = a + 1/a$.

Consider the two second-order linear PDEs

$$u_{1xx} + u_{1yy} + \frac{1}{a}\,u_{1zz} = 0, \tag{3.29}$$

$$u_{2xx} + u_{2yy} + a\,u_{2zz} = 0. \tag{3.30}$$

Solutions of Eqs. (3.29) and (3.30) are also solutions of the governing Eq. (3.17). Equations (3.29) and (3.30) admit solutions of the form (3.27):

$$u_1 = zg_1(\tau), \qquad u_2 = zg_2(\tau),$$

where

$$g_1(\tau) = \sqrt{\frac{\tau}{a} + 1} - \operatorname{arccoth}\sqrt{\frac{\tau}{a} + 1}, \tag{3.31}$$

$$g_2(\tau) = \sqrt{a\tau + 1} - \operatorname{arccoth}\sqrt{a\tau + 1}. \tag{3.32}$$

Thus, we have established that in addition to the trivial solution $f_0(\tau) = 1$, the linear ODE (3.28) admits two further particular solutions, namely (3.31) and (3.32).

The following proposition holds.

Proposition 3.8. *Equation* (3.28) *possesses an intermediate integral of the form*

$$4\tau^2(\tau^2 + b\tau + 1)\frac{d^3 f}{d\tau^3} + 2\tau(6\tau^2 + 5b\tau + 4)\frac{d^2 f}{d\tau^2} + \tau(3\tau + 2b)\frac{df}{d\tau} = c, \tag{3.33}$$

where c is an arbitrary constant.

The particular solutions (3.31) and (3.32) of the linear ODE (3.28) satisfy the homogeneous version of Eq. (3.33) (i.e., with $c = 0$). Using these solutions, one can construct a particular solution to the inhomogeneous Eq. (3.33).

Proposition 3.9. *For $c = (a^2 - 1)/(2a)$, the inhomogeneous Eq.* (3.33) *admits the particular solution*

$$
\begin{aligned}
f(\tau) = & \sqrt{\frac{\tau}{a}+1}\, \operatorname{arccoth}\sqrt{\frac{\tau}{a}+1} - \frac{1}{2}\operatorname{arccoth}^2\sqrt{\frac{\tau}{a}+1} - \\
& - \sqrt{a\tau+1}\, \operatorname{arccoth}\sqrt{a\tau+1} + \frac{1}{2}\operatorname{arccoth}^2\sqrt{a\tau+1}.
\end{aligned}
\tag{3.34}
$$

Remark 3.4. Solution (3.34) can also be derived by applying order-reduction techniques for ODEs [7] to Eq. (3.28), using the known solutions (3.31) and (3.32).

Proposition 3.10. *For $a > 0$ with a satisfying $b = a + 1/a$, the linear ODE* (3.28) *possesses the following fundamental system of solutions:*

$$
\begin{aligned}
f_1 &= 1, \\
f_2 &= \sqrt{\frac{\tau}{a}+1} - \operatorname{arccoth}\sqrt{\frac{\tau}{a}+1} + \sqrt{a\tau+1} - \operatorname{arccoth}\sqrt{a\tau+1}, \\
f_3 &= \frac{a}{a^2-1}\left(\sqrt{\frac{\tau}{a}+1} - \operatorname{arccoth}\sqrt{\frac{\tau}{a}+1} - \sqrt{a\tau+1} + \operatorname{arccoth}\sqrt{a\tau+1}\right), \\
f_4 &= \frac{a}{a^2-1}\left(\sqrt{\frac{\tau}{a}+1}\, \operatorname{arccoth}\sqrt{\frac{\tau}{a}+1} - \frac{1}{2}\operatorname{arccoth}^2\sqrt{\frac{\tau}{a}+1} - \right. \\
&\qquad \left. - \sqrt{a\tau+1}\, \operatorname{arccoth}\sqrt{a\tau+1} + \frac{1}{2}\operatorname{arccoth}^2\sqrt{a\tau+1}\right).
\end{aligned}
\tag{3.35}
$$

Remark 3.5. Solutions (3.35) are invariant under the discrete transformation $a \to 1/a$.

Remark 3.6. In the special case $a = 1$, the fundamental system of solutions (3.35) reduces to

$$
\begin{aligned}
f_1 &= 1, \\
f_2 &= 2\left(\sqrt{\tau+1} - \operatorname{arccoth}\sqrt{\tau+1}\right), \\
f_3 &= -\frac{1}{2}\sqrt{\tau+1}, \\
f_4 &= \frac{1}{2}\left(1 - \sqrt{\tau+1}\, \operatorname{arccoth}\sqrt{\tau+1}\right).
\end{aligned}
$$

Consider the general solution of Eq. (3.28):

$$
f = \sum_{i=1}^{4} c_i f_i, \tag{3.36}
$$

where c_i ($i = 1, \ldots, 4$) are arbitrary constants. Among the solutions of the form (3.36), we seek those that remain finite, together with their first two derivatives, at $\tau = 0$.

Proposition 3.11. *As $\tau \to 0$, the following asymptotic expansions hold:*

$$f = \left[c_2 + \frac{a\ln a}{2(a^2-1)}\,c_4\right]\ln\tau + O(1),$$

$$\frac{df}{d\tau} = \left[c_2 + \frac{a\ln a}{2(a^2-1)}\,c_4\right]\frac{1}{\tau} + \frac{c_4}{8}\ln\tau + O(1),$$

$$\frac{d^2 f}{d\tau^2} = -\left[c_2 + \frac{a\ln a}{2(a^2-1)}\,c_4\right]\frac{1}{\tau^2} + \frac{c_4}{8\tau} + O(1).$$

It follows that the finiteness conditions require $c_2 = 0$ and $c_4 = 0$. Setting also $c_1 = 0$ (which corresponds to discarding the trivial solution), we obtain the following one-parameter family of solutions to Eq. (3.28):

$$f = \frac{c_3 a}{a^2-1}\left(\sqrt{\frac{\tau}{a}+1} - \sqrt{a\tau+1} - \operatorname{arccoth}\sqrt{\frac{\tau}{a}+1} + \operatorname{arccoth}\sqrt{a\tau+1}\right).$$

Finally, using (3.27), we arrive at a one-parameter family of solutions to Eq. (3.19):

$$\bar{u}_0(r,z) = \frac{c_3 a}{a^2-1}\left[\sqrt{\frac{r^2}{a}+z^2} - \sqrt{ar^2+z^2} + \right.$$
$$\left. + \frac{z}{2}\ln\frac{\left(\sqrt{\frac{r^2}{a}+z^2}-z\right)\left(\sqrt{ar^2+z^2}+z\right)}{\left(\sqrt{\frac{r^2}{a}+z^2}+z\right)\left(\sqrt{ar^2+z^2}-z\right)}\right]. \tag{3.37}$$

We now demonstrate that the family of solutions (3.37) includes the fundamental solution. To this end, we integrate both sides of Eq. (3.21) with $u = \bar{u}_0(r,z)$ over the rectangular domain

$$\Pi = \{0 \leqslant r \leqslant r_0,\ -z_1 \leqslant z \leqslant z_2,\ r_0 > 0,\ z_1 > 0,\ z_2 > 0\}.$$

This yields

$$\iint\limits_{\Pi}\left[\left(r\bar{u}_{0rrr} + br\bar{u}_{0rzz} + \bar{u}_{0rr} - \frac{1}{r}\bar{u}_{0r}\right)_r + \left(r\bar{u}_{0zzz}\right)_z\right] dr\,dz = \frac{1}{2\pi}.$$

Applying Stokes' theorem, the integral on the left-hand side can be transformed into a contour integral over the boundary $\partial\Pi$:

$$\oint\limits_{\partial\Pi} \boldsymbol{A}\cdot d\boldsymbol{l} = \frac{1}{2\pi}, \tag{3.38}$$

where

$$\begin{aligned}
&\boldsymbol{A} = (A_1(r,z),\ A_2(r,z)), \qquad \boldsymbol{dl} = (dr,\ dz),\\
&A_1(r,z) = -r\bar{u}_{0zzz},\\
&A_2(r,z) = r\bar{u}_{0rrr} + br\bar{u}_{0rzz} + \bar{u}_{0rr} - \frac{1}{r}\bar{u}_{0r}.
\end{aligned}$$

The left-hand side of relation (3.38) decomposes as

$$\oint\limits_{\partial\Pi} \boldsymbol{A}\cdot \boldsymbol{dl} = I_1 + I_2 + I_3 + I_4,$$

where

$$I_1 = \int_0^{r_0} A_1(r,-z_1)\,dr = \frac{c_3}{a^2-1}\left(-\frac{a^2 z_1}{\sqrt{\frac{r_0^2}{a}+z_1^2}} + \frac{z_1}{\sqrt{ar_0^2+z_1^2}} + a^2 - 1\right),$$

$$I_3 = -\int_0^{r_0} A_1(r,z_2)\,dr = \frac{c_3}{a^2-1}\left(-\frac{a^2 z_2}{\sqrt{\frac{r_0^2}{a}+z_2^2}} + \frac{z_2}{\sqrt{ar_0^2+z_2^2}} + a^2 - 1\right),$$

$$I_2 = \int_{-z_1}^{z_2} A_2(r_0,z)\,dz = 2c_3 - I_1 - I_3,$$

$$I_4 = -\int_{-z_1}^{z_2} A_2(0,z)\,dz = 0.$$

Summing these contributions, we obtain

$$I_1 + I_2 + I_3 + I_4 = 2c_3.$$

Consequently, the constant c_3 is determined as $c_3 = 1/(4\pi)$.

We now state the main result of this work.

Theorem 3.2. *The axisymmetric fundamental solution of Eq.* (3.17) *is given by*

$$u_f = \frac{a}{4\pi(a^2-1)}\left[\sqrt{\frac{r^2}{a}+z^2} - \sqrt{ar^2+z^2} + \frac{z}{2}\ln\frac{\left(\sqrt{\frac{r^2}{a}+z^2}-z\right)\left(\sqrt{ar^2+z^2}+z\right)}{\left(\sqrt{\frac{r^2}{a}+z^2}+z\right)\left(\sqrt{ar^2+z^2}-z\right)}\right], \tag{3.39}$$

where the parameter a satisfies $b = a + 1/a$.

Remark 3.7. In the special case $a = 1$ (equivalently $b = 2$), the fundamental solution (3.39) reduces to

$$u_f = -\frac{1}{8\pi}\sqrt{r^2+z^2},$$

which coincides with the fundamental solution of the three-dimensional biharmonic equation [8].

Remark 3.8. For $b = 2$, the symmetry-based construction of a fundamental solution is particularly efficient. In this case, the solution of Eq. (3.17) that is invariant under the symmetry operators (3.25) and (3.26) is determined uniquely up to a multiplicative constant and takes the form

$$u = c\sqrt{r^2 + z^2}. \tag{3.40}$$

Proceeding as described above, one finds that (3.40) represents a fundamental solution with $c = -1/(8\pi)$.

3.6 Conclusion

The main result of this paper is the construction, in terms of elementary functions, of an invariant fundamental solution to the basic equation for a transversely isotropic linear elastic medium. In conclusion, we note that the symmetry-based approach can also be effectively employed to construct fundamental solutions of linear partial differential equations with variable coefficients, as well as for equations of higher order.

References

[1] Aksenov AV (1995) Symmetries of linear partial differential equations and fundamental solutions. Doklady Mathematics 51(3):329-331 (translation from Doklady Akademii Nauk, Rossiiskoi Akademii Nauk 342(2):151-153)
[2] Bluman G (1990) Simplifying the form of Lie groups admitted by a given differential equation. Journal of Mathematical Analysis and Applications 145(1):52-62, DOI 10.1016/0022-247X(90)90431-E
[3] Bluman G, Anco SC (2002) Symmetry and Integration Methods for Differential Equations. Applied Mathematical Sciences, Vol 154. Springer, New York
[4] Georgievskii DV (2015) An extended Galerkin representation for a transversely isotropic linearly elastic medium. Journal of Applied Mathematics and Mechanics 79(6):618-621, DOI 10.1016/j.jappmathmech.2016.04.010
[5] Georgievskii DV (2015) The Galerkin tensor operator, reduction to tetraharmonic equations, and their fundamental solutions. Doklady Physics 60(8):364-367, DOI 10.1134/S1028335815080030
[6] Ovsiannikov LV (1982) Group Analysis of Differential Equations. Academic Press, New York
[7] Polyanin AD, Zaitsev VF (2018) Handbook of Ordinary Differential Equations: Exact Solutions, Methods, and Problems. CRC Press, Boca Raton
[8] Vladimirov VS (1979) Generalized Functions in Mathematical Physics. Mir Publishers, Moscow

Chapter 4
A New Work Function for Plastically Compressible Materials

Sergei Alexandrov and Elena Lyamina

Abstract Using the work function as a constitutive equation in place of the yield criterion and its associated flow rule significantly simplifies the applications of the upper bound theorem in plasticity. However, no publication has attempted to apply this technique to plastically compressible materials. In particular, no work function has been proposed for such materials, except the functions that can be derived using available yield criteria. This paper presents such a generalized work function involving the linear and quadratic invariants of the strain rate tensor. The function is a first-order homogeneous function. The yield criterion associated with this work function is calculated numerically. At a specific value of a parameter within the work function, the yield criterion reduces to Green's yield criterion. The upper bound theorem based on this new work function is used to determine the force required to initiate plastic compression of a porous cylinder between two parallel rough platens.

Key words: Work function, Plastic compressibility, Upper bound theorem, Rigid plasticity, Compression of porous cylinder

4.1 Introduction

The phenomenological theory of plasticity is an efficient tool for describing the behavior of porous materials at large strains. In many cases, the material can be regarded as rigid plastic [5]. These materials are plastically compressible. The upper bound theorem supplies a convenient method for evaluating the load required to deform plastically incompressible materials [4]. The upper bound theorem has also

Sergei Alexandrov · Elena Lyamina
Ishlinsky Institute for Problems in Mechanics RAS, Prospekt Vernadskogo 101-1, 119526 Moscow, Russian Federation, e-mail: sergei_alexandrov@yahoo.com, lyamina@inbox.ru

H. Altenbach et al. (eds.), *Stability Problems in Mechanics of Continua*,
Advanced Structured Materials 260, https://doi.org/10.1007/978-3-032-31032-3_4

been derived for plastically compressible materials [5]. However, the corresponding method is not widely used. The upper bound theorem presented in [5] is an extension of the classical upper bound theorem [8]. Its use has the same inconvenience as using the classical theorem. In particular, it is impossible to calculate the functional involved in the theorem using a kinematically admissible velocity field without calculating the stress components by means of the associated flow rule. As a result, the overall calculation is lengthy and cumbersome (for example [3]). An exception is material models based on some particular yield criteria, such as Green's criterion [6]. An alternative approach to deriving the upper bound theorem has been proposed in [7]. This approach is based on the work function, which can be easily computed using kinematically admissible velocity fields. The material is rate-independent if the work function is a homogeneous function of the first order. A yield criterion corresponds to each work function. Typically, the yield criterion must be determined numerically, except in simple cases such as Green's yield criterion. However, the yield criterion is not required for applying the theorem.

The present paper proposes a new work function for porous materials, assuming it depends on the linear and quadratic strain rate invariants. The corresponding yield surface lies between the conical and cylindrical yield surfaces [5], assuming that the yield stresses in the pure shear and hydrostatic compression are given. Green's yield criterion is a particular case of the generalized yield criterion that can be derived using this work function.

4.2 Upper Bound Theorem, Work Function, and Yield Criterion

The concept of work functions has been introduced in [7]. In the case of isotropic materials, it is a convex function of strain rate invariants. Moreover, it is a homogeneous function of the first order for rate-independent materials. One of the simplest work functions corresponding to the von Mises yield criterion has been derived in [7]. Other work functions for pressure-independent materials have been proposed in [1, 9]. The new work function developed in this section depends on the linear and quadratic strain rate invariants. Therefore, it is suitable for plastically compressible materials.

The linear and quadratic strain rate invariants are denoted as ξ and χ, respectively. These invariants are defined as

$$\xi = \xi_{ij}\delta_{ij} \quad \text{and} \quad \chi = 2\chi_{ij}\chi_{ij}, \tag{4.1}$$

where ξ_{ij} are the strain rate components, δ_{ij} is Kronecker's symbol, and

$$\chi_{ij} = \xi_{ij} - \xi\frac{\delta_{ij}}{3}.$$

Alternatively, these invariants can be expressed in terms of the principal strain rates, ξ_1, ξ_2, and ξ_3, as

$$\xi = \xi_1 + \xi_2 + \xi_3 \quad \text{and} \quad \chi = 2\left(\chi_1^2 + \chi_2^2 + \chi_3^2\right), \tag{4.2}$$

where

$$\chi_1 = \frac{2\xi_1 - \xi_2 - \xi_3}{3}, \quad \chi_2 = \frac{2\xi_2 - \xi_3 - \xi_1}{3}, \quad \chi_3 = \frac{2\xi_3 - \xi_1 - \xi_2}{3}. \tag{4.3}$$

Let E be the work function. Then, the principal stresses are determined as [7]

$$\sigma_1 = \frac{\partial E}{\partial \xi_1}, \quad \sigma_2 = \frac{\partial E}{\partial \xi_2}, \quad \sigma_3 = \frac{\partial E}{\partial \xi_3}. \tag{4.4}$$

The theorem in [7] can be represented as

$$\int E^* dV - \int \mathbf{F} \cdot \mathbf{u}^* dS_F \geq \int E dV - \int \mathbf{F} \cdot \mathbf{u} dS_F. \tag{4.5}$$

Here, V is the volume of a body, and S_F is a part of its surface where the traction $\mathbf{F}$ is given. Moreover, $\mathbf{u}$ is the actual velocity field, $\mathbf{u}^*$ is a kinematically admissible velocity field, and E^* is calculated using the kinematically admissible velocity field. If E is a homogeneous function of the first order, equation (4.5) simplifies to

$$\int E^* dV - \int \mathbf{F} \cdot \mathbf{u}^* dS_F \geq \int \mathbf{F} \cdot \mathbf{u} dS_U, \tag{4.6}$$

where S_U is a part of the body's surface on which the velocity is given. Equation (4.6) allows for the traction over S_U to be evaluated. The proposed work function is

$$E = \sqrt[n]{A(m\xi)^n + B\chi^{n/2}}, \tag{4.7}$$

where A, B and $n > 1$ are constants. Moreover, $m = +1$ if $\xi \geq 0$ and $m = -1$ if $\xi \leq 0$. Substituting (4.7) into (4.4) and using (4.2) and (4.3) yields

$$\begin{aligned}
\sigma_1 &= \left[Am(m\xi)^{(n-1)} + 2B\chi_1\chi^{(n/2-1)}\right]\left[A(m\xi)^n + B\chi^{n/2}\right]^{(-1+1/n)}, \\
\sigma_2 &= \left[Am(m\xi)^{(n-1)} + 2B\chi_2\chi^{(n/2-1)}\right]\left[A(m\xi)^n + B\chi^{n/2}\right]^{(-1+1/n)}, \\
\sigma_3 &= \left[Am(m\xi)^{(n-1)} + 2B\chi_3\chi^{(n/2-1)}\right]\left[A(m\xi)^n + B\chi^{n/2}\right]^{(-1+1/n)}.
\end{aligned} \tag{4.8}$$

The proposed work function reduces to the work function corresponding to Green's yield criterion if $n = 2$ [2].

The linear and quadratic stress invariants are denoted as σ and τ, respectively, and are defined as

$$\begin{aligned} \sigma &= \frac{\sigma_1 + \sigma_2 + \sigma_3}{3}, \\ \tau &= \frac{\sqrt{(\sigma_1 - \sigma)^2 + (\sigma_2 - \sigma)^2 + (\sigma_3 - \sigma)^2}}{\sqrt{2}}, \end{aligned} \tag{4.9}$$

where σ_1, σ_2, and σ_3 are the principal stresses. These equations and (4.8) enable the calculation of the yield criterion corresponding to the proposed work function. Let p_0 be the yield stress in hydrostatic compression, and τ_0 be the yield stress in pure shear. Their values are supposed to be known functions of the relative density, ϑ. In particular,

$$p_0 \to \infty \quad \text{and} \quad \tau_0 \to k \tag{4.10}$$

as $\vartheta \to 1$. Here, k is the shear yield stress of the solid phase of the porous material. Substituting (4.8) into (4.9) supplies

$$\begin{aligned} \sigma &= Am(m\xi)^{(n-1)}\left[A(m\xi)^n + B\chi^{n/2}\right]^{(1/n-1)}, \\ \tau &= B\chi^{[(n-1)/2]}\left[A(m\xi)^n + B\chi^{n/2}\right]^{(1/n-1)}. \end{aligned} \tag{4.11}$$

Since $\sigma = -p_0$ at $\chi = 0$ and $\tau = \tau_0$ at $\xi = 0$, it follows from (4.11) that

$$A = p_0^n \quad \text{and} \quad B = \tau_0^n. \tag{4.12}$$

Eliminating A and B in (4.7) using these expressions yields the work function as

$$E = \tau_0 \sqrt[n]{\left(\frac{p_0}{\tau_0}\right)^n (m\xi)^n + \chi^{n/2}} = \tau_0 \sqrt[n]{(mq\xi)^n + \chi^{n/2}}, \tag{4.13}$$

where $q = p_0/\tau_0$. Also, the principal stresses follow from (4.8) as

$$\begin{aligned} \sigma_1 &= \tau_0\left[mq^n(m\xi)^{(n-1)} + 2\chi_1\chi^{(n/2-1)}\right]\left[(mq\xi)^n + \chi^{n/2}\right]^{(-1+1/n)}, \\ \sigma_2 &= \tau_0\left[mq^n(m\xi)^{(n-1)} + 2\chi_2\chi^{(n/2-1)}\right]\left[(mq\xi)^n + \chi^{n/2}\right]^{(-1+1/n)}, \\ \sigma_3 &= \tau_0\left[mq^n(m\xi)^{(n-1)} + 2\chi_3\chi^{(n/2-1)}\right]\left[(mq\xi)^n + \chi^{n/2}\right]^{(-1+1/n)}. \end{aligned} \tag{4.14}$$

One must eliminate A and B in (4.11) employing (4.12) to derive the yield criterion. In particular,

$$\begin{aligned} \frac{\sigma}{\tau_0} &= mq(m\xi)^{(n-1)}\left[(m\xi)^n + q^{-n}\chi^{n/2}\right]^{(1/n-1)}, \\ \frac{\tau}{\tau_0} &= \chi^{[(n-1)/2]}\left[q^n\xi^n + \chi^{n/2}\right]^{(1/n-1)}. \end{aligned} \tag{4.15}$$

Setting $\xi = R\sin\theta$ and $\chi = R^2\cos^2\theta$, one transforms (4.15) to

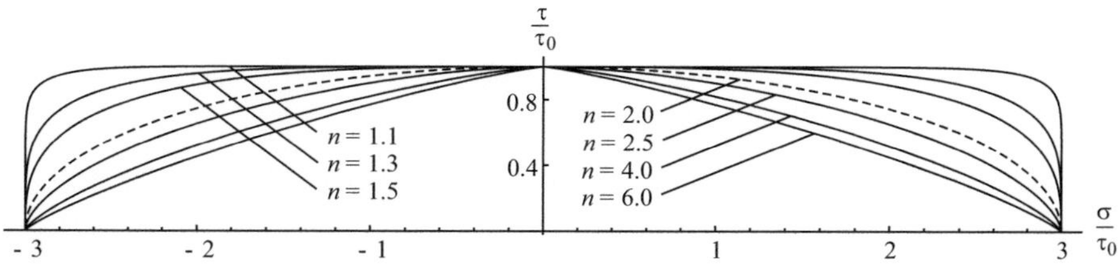

Fig. 4.1: Effect of the $n-$value on the shape of the yield surface at $q = 3$.

$$\frac{\sigma}{\tau_0} = q\sin^{n-1}\theta[\sin^n\theta + q^{-n}\cos^n\theta]^{(1/n-1)},$$
$$\frac{\tau}{\tau_0} = \cos^{n-1}\theta[q^n\sin^n\theta + \cos^n\theta]^{(1/n-1)} \tag{4.16}$$

in the range

$$0 \leq \theta \leq \pi/2.$$

Similar expressions can be derived in the range

$$-\pi/2 \leq \theta \leq 0.$$

The yield criterion is represented in parametric form when θ varies in the range

$$-\pi/2 \leq \theta \leq \pi/2.$$

The corresponding yield surface is depicted in Fig. 4.1 for several values of n at $q = 3$. It coincides with Green's yield surface at $n = 2$, approaches the cylindrical yield surface as $n \rightarrow 1$, and the conical yield surface as $n \rightarrow \infty$. The broken line represents Green's yield surface.

4.3 Compression of a Cylinder Between Parallel Platens

This section applies the inequality in (4.6) and the work function in (4.13) to evaluate the force required to compress a cylinder between two parallel, rough platens.

4.3.1 Statement of the Problem

A cylinder of radius R and half-height H is compressed between parallel, rough platens (Fig. 4.2). The velocity of each platen is U. The lateral surfaces are traction-free. It is required to evaluate the applied load F. It is natural to use a cylindrical coordinate system (r, θ, z) whose $z-$axis coincides with the axis of symmetry of the process and the plane $z = 0$ with the plane of symmetry. Then, it is sufficient to

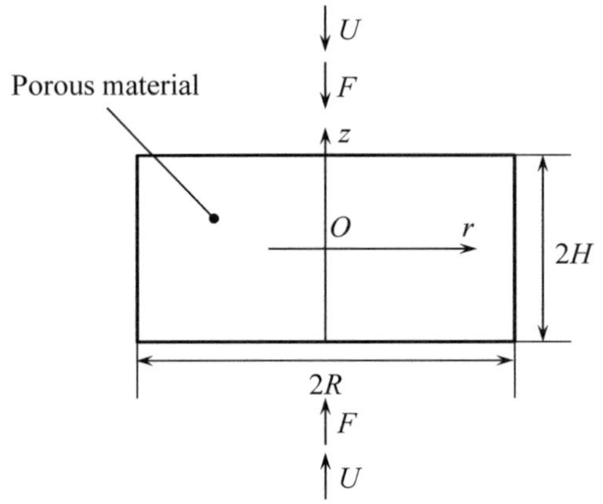

Fig. 4.2 Schematic diagram of the process.

consider the domain $z \geq 0$. The boundary value problem is axisymmetric. Therefore, the solution is independent of θ. It is assumed that the relative density is uniformly distributed. Therefore, p_0 and τ_0 are constant. At sliding, the friction law reads [5]

$$\tau_f = k\vartheta^2, \tag{4.17}$$

where τ_f is the friction stress. Let u_r and u_z be the radial and axial velocity, respectively. Then, the non-zero strain rate components in the cylindrical coordinate system are expressed as

$$\xi_{rr} = \frac{\partial u_r}{\partial r}, \quad \xi_{zz} = \frac{\partial u_z}{\partial z}, \quad \xi_{\theta\theta} = \frac{u_r}{r}, \quad \xi_{rz} = \frac{1}{2}\left(\frac{\partial u_z}{\partial r} + \frac{\partial u_r}{\partial z}\right). \tag{4.18}$$

It is then straightforward to calculate the linear and quadratic strain rate invariants using (4.1). The velocity boundary conditions are

$$u_z = -U \tag{4.19}$$

for $z = H$,

$$u_z = 0 \tag{4.20}$$

for $z = 0$, and

$$u_r = 0 \tag{4.21}$$

for $r = 0$.

In what follows, the following dimensionless quantities will be used:

$$h = \frac{H}{R}, \quad \rho = \frac{r}{R}, \quad \zeta = \frac{z}{H}. \tag{4.22}$$

4.3.2 Exact Solution With no Friction

This section supplies an exact analytical solution, assuming that the platens are frictionless. This solution is used in the next section for constructing a kinematically admissible velocity field.

The velocity is sought in the form:

$$u_z = -\frac{U}{H}z, \qquad u_r = \frac{U}{H}rw, \tag{4.23}$$

where w is constant. It is natural to assume that

$$w \le \frac{1}{2}. \tag{4.24}$$

This assumption must be verified *a posteriori*. It is evident that the velocity field in (4.23) satisfies the boundary conditions in (4.19) – (4.21). Substituting (4.23) into (4.18) and the resulting strain rate components in (4.1) yields

$$\xi = \frac{U}{H}(2w-1), \qquad \chi = \frac{4}{3}\frac{U^2}{H^2}(w+1)^2. \tag{4.25}$$

Also, it follows from (4.18) and (4.23) that the normal strain rates in the cylindrical coordinate system are the principal strain rates. One may set without loss of generality $\xi_r \equiv \xi_1$, $\xi_\theta \equiv \xi_2$, and $\xi_z \equiv \xi_3$. Then, the equations in (4.3) provide

$$\chi_1 = \frac{U}{H}\frac{(w+1)}{3}, \quad \chi_2 = \frac{U}{H}\frac{(w+1)}{3}, \quad \chi_3 = -\frac{2}{3}\frac{U}{H}(w+1). \tag{4.26}$$

The normal stresses in the cylindrical coordinate system are denoted as $\sigma_r \equiv \sigma_1$, $\sigma_\theta \equiv \sigma_2$, and $\sigma_z \equiv \sigma_3$. The equilibrium equations are satisfied if the principal stresses are constant and $\sigma_2 = \sigma_1$. Since the lateral surface of the cylinder is traction-free, $\sigma_1 = 0$. Then, substituting (4.25) and (4.26) into the first equation in (4.14) and taking into account (4.24) leads to

$$-q^n(1-2w)^{(n-1)} + \frac{2^{(n-1)}}{3^{n/2}}(w+1)^{(n-1)} = 0. \tag{4.27}$$

This equation determines w as

$$w = \frac{1-c}{2+c}, \tag{4.28}$$

where

$$c = 2\left(3^{n/2}q^n\right)^{1/(1-n)}. \tag{4.29}$$

Equation (4.28) shows that the inequality in (4.24) is satisfied if $c \geq 0$. Equation (4.29) confirms the latter inequality. The third equation in (4.14) using (4.24) supplies

$$\sigma_3 = \tau_0 \left[-q^n (-\xi)^{(n-1)} + 2\chi_3 \chi^{(n/2-1)} \right] \left[(-q\xi)^n + \chi^{n/2} \right]^{(-1+1/n)}. \tag{4.30}$$

Equations (4.25), (4.26), (4.28), and (4.29) allow for σ_3 to be calculated from (4.30). The final expression is cumbersome and is not provided here. Note, however, that the ratio U/H involved in (4.25) and (4.26) is cancelled. Equation (4.30) simplifies for certain values of n. For example,

$$\sigma_3 = -\frac{3q}{\sqrt{1+3q^2}} \tau_0 \tag{4.31}$$

for $n = 2$. Equation (4.31) can be derived using Green's yield criterion and its associated flow rule.

Let f be the dimensionless compression force defined as

$$f = \frac{F}{\pi R^2 k}. \tag{4.32}$$

Since the stress distribution is uniform, it follows from (4.30) and (4.32) that

$$f = -\frac{\tau_0}{k} \left[-q^n (-\xi)^{(n-1)} + 2\chi_3 \chi^{(n/2-1)} \right] \left[(-q\xi)^n + \chi^{n/2} \right]^{(-1+1/n)}. \tag{4.33}$$

4.3.3 Kinematically Admissible Velocity Field

Using (4.23) and (4.28), a reasonable kinematically admissible velocity field is

$$u_z = -\frac{U}{H} z, \qquad u_r = \frac{Ur}{H} \left[\left(\frac{1-c}{2+c} \right) + \alpha \left(\frac{z}{H} \right)^2 \right], \tag{4.34}$$

where α is constant. The expression for the radial velocity takes into account that it is an even function of z. The velocity field in (4.34) satisfies the boundary conditions in (4.19) – (4.21). Substituting (4.34) into (4.18) yields

$$\xi_{zz} = -\frac{U}{H}, \quad \xi_{rr} = \xi_{\theta\theta} = \frac{U}{H} \left[\left(\frac{1-c}{2+c} \right) + \alpha \left(\frac{z}{H} \right)^2 \right], \quad \xi_{rz} = \frac{Urz\alpha}{H^3}. \tag{4.35}$$

Introduce the following dimensionless quantities:

$$\varepsilon_{zz} = \frac{H}{U} \xi_{zz}, \quad \varepsilon_{rr} = \frac{H}{U} \xi_{rr}, \quad \varepsilon_{\theta\theta} = \frac{H}{U} \xi_{\theta\theta}, \quad \varepsilon_{rz} = \frac{H}{U} \xi_{rz}. \tag{4.36}$$

Equations (4.22), (4.35), and (4.36) combine to give

$$\varepsilon_{zz} = -1, \quad \varepsilon_{rr} = \varepsilon_{\theta\theta} = \left(\frac{1-c}{2+c}\right) + \alpha\zeta^2, \quad \varepsilon_{rz} = \frac{\rho\zeta\alpha}{h}. \tag{4.37}$$

Substituting (4.37) into (4.2) supplies

$$\varepsilon = \frac{H}{U}\xi = 2\left(\frac{1-c}{2+c}\right) + 2\alpha\zeta^2 - 1,$$

$$\eta = \frac{H^2}{U^2}\chi = \frac{4\alpha^2\rho^2\zeta^2}{h^2} + \frac{4}{3}\left[1 + \left(\frac{1-c}{2+c}\right) + \alpha\zeta^2\right]^2.$$

Equation (4.13) can be rewritten as

$$E^* = \tau_0 \frac{U}{H} \sqrt[n]{(-q\varepsilon)^n + \eta^{n/2}}. \tag{4.38}$$

Considering the boundary conditions, including the friction law (4.17), and equation (4.38), one can transform (4.6) to

$$\tau_0 \frac{U}{H} \int \sqrt[n]{(-q\varepsilon)^n + \eta^{n/2}} dV + k\vartheta^2 \int |u_r|\, dS_F \geq U \int F dS_U. \tag{4.39}$$

It is understood here that the radial velocity is calculated at $z = H$. Therefore, equations (4.22), (4.34), and (4.39) combine to give

$$\frac{\tau_0}{H} \int \sqrt[n]{(-q\varepsilon)^n + \eta^{n/2}} dV + \frac{k\vartheta^2}{h} \int \left|\rho \left[\left(\frac{1-c}{2+c}\right) + \alpha\right]\right| dS_F \geq F. \tag{4.40}$$

Integrating here and employing (4.32) yields

$$\frac{2\tau_0}{k} \int_0^1 \int_0^1 \sqrt[n]{(-q\varepsilon)^n + \eta^{n/2}} \rho d\rho d\zeta + \frac{2\vartheta^2}{h} \int_0^1 \left|\rho \left[\left(\frac{1-c}{2+c}\right) + \alpha\right]\right| \rho d\rho \geq f. \tag{4.41}$$

The first integral must be evaluated numerically. Then, the left-hand side of the inequality in (4.41) is a function of α for a given set of geometrical and physical parameters. This function must be minimized numerically to find the best upper bound on f.

4.4 Numerical Results

The yield stresses required for the work function (4.13) are taken as [5]

$$\begin{aligned} \tau_0 &= \vartheta^{3/2} k, \\ p_0 &= \frac{2}{\sqrt{3}} \frac{\vartheta^2 k}{\sqrt{1-\vartheta}}. \end{aligned} \tag{4.42}$$

It is evident that the conditions in (4.10) are satisfied. It follows from (4.42) that

$$q = \frac{2}{\sqrt{3}} \frac{\sqrt{\vartheta}}{\sqrt{1-\vartheta}}. \tag{4.43}$$

Using (4.42) and (4.43), one can rewrite (4.41) as

$$2\vartheta^{3/2} \int_0^1 \int_0^1 \sqrt[n]{(-2q\varepsilon)^n + \eta^{n/2}}\, \rho d\rho d\zeta + \frac{2\vartheta^2}{h} \int_0^1 \left| \rho \left[\left(\frac{1-c}{2+c} \right) + \alpha \right] \right| \rho d\rho \geq f. \tag{4.44}$$

The left-hand side of this inequality has been minimized at $h = 0.5$. The solid lines in Fig. 4.3 depict the variation of f with the relative density for several values of n. The broken lines represent the solution in (4.33). Therefore, the difference between the two lines for the same value of n shows the effect of friction on the compression force. As expected, the compression force increases with relative density independently of the n-value. The effect of friction also increases with relative density. The solution shows that the compression force increases with n-value. This

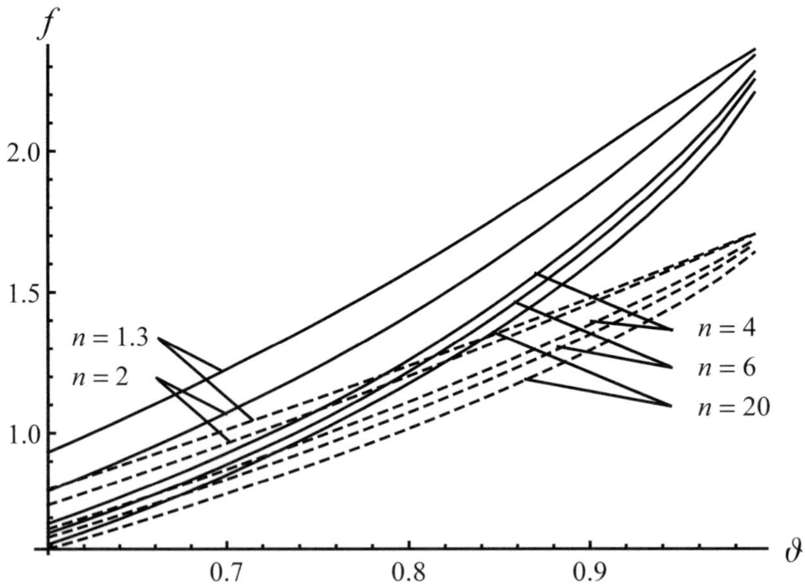

Fig. 4.3: Variation of the dimensionless compression force with the relative density at $h = 0.5$ for several n-values. The broken lines correspond to the frictionless plates.

Table 4.1: Dependence of ϑ_{cr} on n at h = 1/2.

n	20	6	4	2
ϑ_{cr}	0.7	0.776	0.795	0.776

effect is more pronounced for smaller n-values. The curves at $n = 2$ correspond to Green's yield criterion. Note that there exists a critical value of the relative density denoted as ϑ_{cr} such that the regime of sticking occurs if $\vartheta \leq \vartheta_{cr}$. Table 4.1 shows the dependence of ϑ_{cr} on n. The present solution predicts no sticking at $n = 1.3$.

4.5 Conclusions

A new work function for plastically compressible materials, such as porous and powder materials, has been proposed and used to analyze the compression of a porous cylinder between two parallel, rough platens. It is probably the first work function that accounts for plastic compressibility, except for those derived from available yield criteria and the associated flow rule. The yield criterion linked to this new yield function has been determined numerically. This yield criterion coincides with Green's yield criterion for a certain value of a parameter involved in the generalized work function. The numerical example shows the effect of the work function, which is also the yield criterion's effect, on the force required to initiate plastic deformation of a porous cylinder compressed between two parallel, rough platens.

Acknowledgements The authors would like to acknowledge the financial support from the Russian Ministry of Science and Higher Education under contract 124013000674-0.

References

[1] Alexandrov S, Lyamina E, Jeng Y-R (2023) Application of the upper bound theorem for metal forming processes considering an arbitrary isotropic pressure-independent yield criterion with no strength differential effect. The International Journal of Advanced Manufacturing Technology 126(7):3311-3321, DOI 10.1007/s00170-023-11312-5

[2] Alexandrov S, Lyamina E, Jeng Y-R (2025) Selected issues in the upper bound theorem of plasticity. ZAMM - Journal of Applied Mathematics and Mechanics / Zeitschrift für Angewandte Mathematik und Mechanik 105(2):e202401219, DOI 10.1002/zamm.202401219

[3] Alexandrov S, Strashnov S, Li Y (2023) An upper bound solution for axisymmetric extrusion and drawing considering a generalized yield criterion. Metals 13(3):602, DOI 10.3390/met13030602

[4] Avitzur B (1977) Metal forming: the application of limit analysis. Annual Review of Materials Research 7, 261–300, DOI 10.1146/annurev.ms.07.080177.001401

[5] Druyanov B (1993) Technological Mechanics of Porous Bodies. Clarendon Press, New-York

[6] Green RJ (1972) A plasticity theory for porous solids. International Journal of Mechanical Sciences 14(4):215-224, DOI 10.1016/0020-7403(72)90063-X

[7] Hill R (1956) New horizons in the mechanics of solids. Journal of the Mechanics and Physics of Solids 5(1):66-74, DOI 10.1016/0022-5096(56)90009-6

[8] Hill R (1950) The Mathematical Theory of Plasticity. Clarendon Press, Oxford

[9] Jeng Y-R, Alexandrov S, Lyamina E (2025) A new work function for pressure-independent materials exhibiting the strength-differential effect. Journal of Mechanics 41:547-558, DOI 10.1093/jom/ufaf041

Chapter 5
On Calculations of Large Strains in Materials with Elastic, Plastic and Viscous Properties

Aleksandra S. Begun, Larisa V. Kovtanyuk, and Galina L. Panchenko

Abstract The paper presents the results of calculations of a solution of the coupled boundary value problem on large strains during the pushing of a finite length plug through a round non-deformable pipe under the action of variable pressure difference. The plug has a cylindrical shape, and its material is incompressible and has elastic, viscous and plastic properties. The pressure difference is set at the boundary transverse surfaces of the plug and initially increases over time, then becomes constant. The case is considered when, before the onset of viscoplastic flow in the plug material, irreversible deformations accumulate due to creep under conditions of rigid adhesion of the plug material to the pipe walls, and the heating of the plug occurs only due to its irreversible deformation. With the onset of viscoplastic flow, the material begins to slip in the vicinity of the pipe wall, and its heating occurs both due to the accumulation of plastic deformations and due to friction against the boundary surface. Reversible and irreversible strains in the material are determined by differential equations of their change. The parameters of the adopted governing laws of creep and plastic flow are considered to be temperature-dependent. Strains, stresses and temperature are calculated at each moment of the deformation process. The conditions for the occurrence and development of viscoplastic flow and the pattern of movement of the elastic-plastic boundary under increasing and constant pressure are indicated.

Key words: Large strains, Creep, Viscoplastic flow, Deformation heat production, Thermal conductivity, Coupled thermal deformation

---

Aleksandra S. Begun · Larisa V. Kovtanyuk (corresponding author) · Galina L. Panchenko
Institute of Automation and Control Processes FEB RAS, 5 Radio St., 690041 Vladivostok, Russian Federation, e-mail: ustinova@iacp.dvo.ru, lk@iacp.dvo.ru, panchenko@iacp.dvo.ru

H. Altenbach et al. (eds.), *Stability Problems in Mechanics of Continua*,
Advanced Structured Materials 260, https://doi.org/10.1007/978-3-032-31032-3_5

5.1 Introduction

An actual problem in modern fundamental mechanics and modern technological practice of pressure treatment of materials is accounting for complex rheological properties of materials under coupled deformation and heat generation and heat transfer processes. It is the technological practice of manufacturing metal products with enhanced functional properties that has created an urgent challenge to fundamental mechanics in modeling the process of deformation of materials when they acquire large strains, possessing elastic, viscous, and plastic properties under thermomechanical influences. The elastic properties of the deformed material determine the conservative part of the deformation process, while the viscous and plastic properties determine its dissipative component. In this case, along with temperature, the reversible (thermoelastic) and irreversible components [9, 10] of the total strains should be taken as thermodynamic parameters of the state.

Reversible and irreversible deformations in any deformation process are interconnected and interdependent. They cannot be measured experimentally, but together they constitute experimentally measurable total deformations. The main requirement for dividing total deformations into components and, consequently, determining reversible and irreversible deformations is to ensure geometrically consistent constructions of the kinematics of the deformation process [4, 14, 15, 22]. Based on the consistent kinematics of large deformations, various mathematical models have been developed that are geometrically and thermodynamically correct [4, 8, 14, 15, 17, 18, 22].

Here we will use a mathematical model of large strains, which takes into account the elastic, viscous, and plastic properties of deformed materials under conditions of coupling of deformation and thermal conductivity processes [5, 6, 13].

For the thermodynamic parameters of the state in a nonequilibrium deformation process, according to the formalism of thermodynamics of irreversible processes [4, 17, 22], differential equations for their change (transfer) should be written. According to nonequilibrium thermodynamic approaches, irreversible deformations are not divided into creep deformations and plastic flow deformations. Taking into account the viscous properties of a material is possible both during plastic flow and in the stages of deformation preceding it [1–3]. That is, we will assume that before the onset of plastic flow, the material is irreversibly deformed due to creep. With such non-isothermal deformation, the creep parameters of the material, the yield stress and viscosity are assumed to be temperature-dependent.

In [13], a solution was obtained for a problem in which an incompressible elastoviscoplastic material is in conditions of rigid adhesion to the walls of the pipe and is under the action of a time-varying pressure difference specified on the ends of the plug. The case was considered in which heat is generated internally within the plug solely due to irreversible deformation. A problem with a similar setup, but incorporating wall slip according to the Coulomb–Mohr dry friction law, was studied in [12]. Here, we present a solution where material slippage occurs by near-wall viscoplastic flow.

5.2 Basic Relationships of the Model of Large Strains

Differential equations for the change (transfer) of tensors of reversible **m** and irreversible **p** strains in [4] were proposed in the form

$$\frac{D\mathbf{p}}{Dt} = \frac{d\mathbf{p}}{dt} - \mathbf{r}\cdot\mathbf{p} + \mathbf{p}\cdot\mathbf{r} = \boldsymbol{\gamma} - \mathbf{p}\cdot\boldsymbol{\gamma} - \boldsymbol{\gamma}\cdot\mathbf{p},$$
$$\frac{D\mathbf{m}}{Dt} = \boldsymbol{\varepsilon} - \boldsymbol{\gamma} - \frac{1}{2}\left((\boldsymbol{\varepsilon} - \boldsymbol{\gamma} + \mathbf{z})\cdot\mathbf{m} + \mathbf{m}\cdot(\boldsymbol{\varepsilon} - \boldsymbol{\gamma} - \mathbf{z})\right). \quad (5.1)$$

Relations (5.1) are written in the Euler spatial coordinate system, and designations were introduced into them

$$\mathbf{v} = \frac{\partial\mathbf{u}}{\partial t} + \mathbf{v}\cdot\nabla\mathbf{u}, \quad \nabla\mathbf{v} = \frac{1}{2}\left(\nabla\mathbf{v} + \nabla^{\top}\mathbf{v}\right) + \frac{1}{2}\left(\nabla\mathbf{v} - \nabla^{\top}\mathbf{v}\right) = \boldsymbol{\varepsilon} + \mathbf{w},$$
$$\mathbf{r} = -\mathbf{r}^{\top} = \mathbf{w} + \mathbf{z}, \quad \mathbf{m} = \mathbf{e} + \beta\theta\mathbf{I}, \quad \beta = \alpha T_0, \quad \theta = T_0^{-1}(T - T_0),$$
$$\mathbf{z} = -\mathbf{z}^{\top} = A^{-1}\Big(B^2(\boldsymbol{\varepsilon}\cdot\mathbf{m} - \mathbf{m}\cdot\boldsymbol{\varepsilon}) + B\Big(\boldsymbol{\varepsilon}\cdot\mathbf{m}^2 - \mathbf{m}^2\cdot\boldsymbol{\varepsilon}\Big) + \mathbf{m}\cdot\boldsymbol{\varepsilon}\cdot\mathbf{m}^2 - \mathbf{m}^2\cdot\boldsymbol{\varepsilon}\cdot\mathbf{m}\Big),$$
$$A = 8 - 8L_1 + 3L_1^2 - L_2 - \frac{1}{3}L_1^3 + \frac{1}{3}L_3, \quad B = 2 - L_1,$$
$$L_1 = \operatorname{tr}\mathbf{m}, \quad L_2 = \operatorname{tr}\mathbf{m}^2, \quad L_3 = \operatorname{tr}\mathbf{m}^3. \quad (5.2)$$

In Eqs. (5.1) and (5.2) **u**, **v** are the displacement and velocity vectors; t is the time; **I** is the identity tensor of the second rank. We assume that the reversible strains **m** are thermoelastic, so that T, T_0 are the current temperature and the free state temperature (room temperature), respectively; α is the coefficient of linear expansion, **e** is the linear part of the elastic strain tensor; **r** is the rotation tensor, $\mathbf{z}(\boldsymbol{\varepsilon}, \mathbf{m})$ is its nonlinear component. When $\mathbf{z}(\boldsymbol{\varepsilon}, \mathbf{m}) = \mathbf{0}$ the objective time derivative $\frac{D}{Dt}$ in equalities (5.1) coincides with the Jaumann derivative. The sources in the differential equations of transfer (5.1) are $\boldsymbol{\varepsilon} - \boldsymbol{\gamma}$ and $\boldsymbol{\gamma}$ respectively for reversible and irreversible strains. Consequently, the position on the additive decomposition of the total strain rate tensor $\boldsymbol{\varepsilon}$ into the sum of the tensors of the rates of the reversible $\boldsymbol{\varepsilon} - \boldsymbol{\gamma}$ and irreversible $\boldsymbol{\gamma}$ components is accepted. The objective time derivative of the tensors from (5.1) is not a matter of assumption or choice. Its specific form is a necessary consequence of the geometric correctness of kinematic constructions. According to (5.1), at $\boldsymbol{\gamma} = \mathbf{0}$ only the case of conservative deformation is possible, when the irreversible strain tensor is unchanged. In this case, the first dependence from (5.1) specifies patterns of change of the components of the tensor **p** while the tensor does not change. In any other case, a change in the tensor **e** entails changes in the tensor **p**, and vice versa. The reversible and irreversible strain tensors, defined by dependencies (5.1) and (5.2), are related to the Almansi total strain tensor by the relation

$$\mathbf{d} = \frac{1}{2}\left(\nabla\mathbf{u} + \nabla^{\top}\mathbf{u} - \nabla^{\top}\mathbf{u}\cdot\nabla\mathbf{u}\right) = \mathbf{m} + \mathbf{p} - \frac{1}{2}\mathbf{m}^2 - \mathbf{m}\cdot\mathbf{p} - \mathbf{p}\cdot\mathbf{m} + \mathbf{m}\cdot\mathbf{p}\cdot\mathbf{m}. \quad (5.3)$$

For the quasi-stationary deformation processes under consideration, we choose free energy as the thermodynamic potential

$$\psi = e - Ts,$$

where e and s are the mass distribution densities of internal energy and entropy. In general,

$$\psi = \psi(\mathbf{d}, T),$$

but when solving specific boundary value problems of the theory of large strains [1, 16, 18] it is most often accepted that $\psi(\mathbf{m}, T)$. This makes it possible to separate the conservative and dissipative parts of the deformation process. Using the local consequence of the law of conservation of energy

$$\rho\frac{de}{dt} - \boldsymbol{\sigma}\cdot\cdot\boldsymbol{\varepsilon} + \nabla\cdot\mathbf{q} = 0,$$

where ρ is the density of the deformable medium, $\mathbf{q}$ is the heat flow vector, $\boldsymbol{\sigma}$ is the Euler-Cauchy stress tensor, and assuming that a change in the allocated volume of a mechanically incompressible material is possible due to linear expansion

$$1 - 2J_1 + 2J_1^2 - 2J_2 - \frac{4}{3}J_1^3 + 4J_1J_2 - \frac{8}{3}J_3 = (1+3\beta\theta)^{-2}, \tag{5.4}$$
$$J_1 = \operatorname{tr}\mathbf{d}, \quad J_2 = \operatorname{tr}\mathbf{d}^2, \quad J_3 = \operatorname{tr}\mathbf{d}^3.$$

we obtain relations

$$\boldsymbol{\sigma} = -p\mathbf{I} + (1+3\beta\theta)^{-1}\frac{\partial W}{\partial\mathbf{m}}\cdot(\mathbf{I}-\mathbf{m}), \tag{5.5}$$
$$\frac{\partial(\rho S)}{\partial t} = -\nabla\cdot\mathbf{J} - T^{-2}\mathbf{q}\cdot\nabla T + T^{-1}\boldsymbol{\sigma}\cdot\cdot\boldsymbol{\gamma} \tag{5.6}$$

Here $\mathbf{J}$ is the vector of the total entropy flow. Dependence (5.5) is an analogue of the Murnaghan formula, known in the theory of elasticity [16], which allows, when specifying the thermodynamic potential $\psi(\mathbf{m}, T)$, to connect the stress with elastic strains and temperature by an unambiguous dependence. According to the entropy balance equation (5.6), the entropy production is associated with two nonequilibrium processes, the first of which is the nonequilibrium process of thermal conductivity (the penultimate term on the right-hand side of (5.6)), and the second is the nonequilibrium process of irreversible deformation (the last term on the right-hand side of (5.6)). Such an entropy source in (5.6), responsible for the contribution of irreversible deformation to entropy production, must be non-negative $\boldsymbol{\sigma}\cdot\cdot\boldsymbol{\gamma} \geq 0$.

Assuming that the medium is isotropic, the elastic potential $W(J_1, J_2, \theta)$ is taken as a decomposition of this function into a Taylor series with respect to the free state of the deformed material at room temperature T_0

$$W = -2\mu I_1 - \mu I_2 + \kappa I_1^2 + (\kappa - \mu) I_1 I_2 - \chi I_1^3 + \nu_1\theta I_1 +$$

$$+\nu_2\theta^2 - \nu_3 I_1\theta^2 - \nu_4 I_1^2\theta - \nu_5 I_2\theta - \nu_6\theta^3, \tag{5.7}$$
$$I_1 = \mathrm{tr}\,\mathbf{c}, \quad I_2 = \mathrm{tr}\,\mathbf{c}^2, \quad \mathbf{c} = \mathbf{m} - \tfrac{1}{2}\mathbf{m}^2.$$

In (5.7) μ is the shear modulus, κ, χ are higher-order elastic moduli, ν_k ($k = 1, 2, ..., 6$) are thermomechanical constants.

Taking the Fourier law of thermal conductivity for the heat flow vector

$$\mathbf{q} = -\lambda \nabla T,$$

where λ is the thermal conductivity coefficient, from the entropy balance equation (5.6), we obtain the heat equation in the form:

$$(1 + \beta_1\theta + \beta_2 \mathrm{tr}\,\mathbf{c})\,\frac{d\theta}{dt} + \beta_3\,(\boldsymbol{\varepsilon} - \boldsymbol{\gamma})\cdot\cdot\,\mathbf{c} = a\Delta\theta - \frac{1}{2\nu_2}\boldsymbol{\sigma}\cdot\cdot\boldsymbol{\gamma}, \tag{5.8}$$
$$\beta_1 = \frac{\nu_2\,(1 - 3\beta) - 3\nu_6}{\nu_2}, \quad \beta_2 = -\frac{\nu_3}{\nu_2}, \quad \beta_3 = -\frac{\nu_1 + \nu_5}{\nu_2}, \quad a = -\frac{\lambda}{2\nu_2}.$$

Here a is the thermal diffusivity coefficient. Equation (5.8) is valid both in areas where irreversible strains accumulate in the material, and in areas where they are present but do not change, which corresponds to the zero source of irreversible strains $\boldsymbol{\gamma}$.

When deformation begins, irreversible strains in the solid are produced due to the irreversible creep process. The source of irreversible strains $\boldsymbol{\gamma}$ in (5.1) and (5.6) should be identified with the creep strain rate tensor $\boldsymbol{\gamma} = \boldsymbol{\varepsilon}^{v}$. It is sufficient to choose any creep law as the defining law. This choice may be related to the deformation properties of the material being studied, the properties of the mathematical or algorithmic apparatus used, and, finally, personal preferences. The approach being developed does not impose any restrictions in this regard. As an example, we choose a constitutive law in the form of the classical Norton power law [19]

$$\boldsymbol{\gamma} = \boldsymbol{\varepsilon}^{v} = \frac{\partial V\,(\Sigma)}{\partial \boldsymbol{\sigma}}, \quad V\,(\Sigma) = B\Sigma^n\,(\sigma_1, \sigma_2, \sigma_3)\,, \quad \Sigma = \max\left|\sigma_i - \sigma_j\right|. \tag{5.9}$$

In relations (5.9) σ_i are the principal stresses, B and n are the creep parameters of the material. When the stress state reaches the loading surface, a region of plastic flow appears, at the boundary of which the governing creep law (5.9) must be replaced by a governing law based on the plastic flow law. The stresses and strains on this boundary surface are continuous, and we also require the continuity of the rates of production of irreversible strains.

Therefore, if $f\,(\boldsymbol{\sigma}, k) = 0$ is the loading surface equation in the stress space and k is the yield stress, then the plastic strain rates are related to stresses by the associated law of plastic flow

$$\mathbf{y} = \boldsymbol{\varepsilon}^{p} - \boldsymbol{\varepsilon}^{v_0} = \phi\frac{\partial f\,(\boldsymbol{\sigma}, k)}{\partial \boldsymbol{\sigma}}, \quad f\,(\boldsymbol{\sigma}, k) = 0, \quad \phi > 0. \tag{5.10}$$

Here $\boldsymbol{\varepsilon}^{v_0}$ is the creep strain rate tensor at the onset of plastic flow. Creep strains and their rates $\boldsymbol{\varepsilon}^{v_0}$ influence the production of irreversible deformations in the flow region only in the form of initial conditions for the development of the plastic flow region and the growth of irreversible deformations due to the flow process.

As a loading surface, we adopt the Tresca plasticity condition taking into account the viscous resistance to plastic flow [7]:

$$f(\sigma_i, y_k, k) = \max\left|\sigma_i - \sigma_j\right| - 2k - 2\eta \max\left|y_k\right|. \tag{5.11}$$

In (5.11) y_k are the principal values of the tensor $\mathbf{y}$, η is the viscosity.

We assume that the material parameters are temperature dependent. The creep parameters B and n are accepted in the form [11]

$$B = \frac{c_1}{\sigma_0^{n-1}} \exp\left(-\frac{Q}{R_u T_0 (1+\theta)}\right), \quad n = b_1 + \frac{b_2}{T_0 (1+\theta)}. \tag{5.12}$$

In (5.12) c_1, σ_0, b_1 and b_2 are constants of the material, Q is the activation energy, R_u is the gas constant. For the yield stress and the coefficient of plastic flow viscosity we will use the relations [20]

$$k = k_0\left(1 - \theta\theta_m^{-1}\right)^2, \quad \eta = \eta_0 Exp\left(-\upsilon T_0 \theta\right), \quad \theta_m = (T_m - T_0)T_0^{-1}. \tag{5.13}$$

In (5.13) T_m is the melting temperature of the deformed material, k_0, η_0 are the yield stress and the viscosity of the material at room temperature, υ is the exponential rate.

We emphasize that the selected dependencies (5.9), (5.11)-(5.13) are the simplest, and other laws, including more complex ones, can be used instead of them.

By adding the equilibrium equations

$$\nabla \cdot \boldsymbol{\sigma} = 0 \tag{5.14}$$

to the written relations, we obtain a closed system of equations of quasi-stationary non-isothermal deformation.

5.3 Statement of the Problem and Deformation until Viscoplastic Flow

We consider the formulation of a boundary value problem within the framework of the presented model. There is a non-deformable cylindrical tube of radius R containing a material layer of a finite length l (Fig. 5.1). This layer has the same radius as the pipe and forms a tight plug. The plug material is incompressible and has elastic, viscous, and plastic properties. In this case, the material exhibits viscous properties at all stages of deformation, first in the form of creep, then when the stress state reaches the loading surface in the form of viscous resistance to plastic flow.

For the antiplane motion under consideration in a cylindrical coordinate system r, φ, z, we assume

$$\theta = \theta\left(r,t\right), u = u_z\left(r,t\right), v = v_z\left(r,t\right), P = P\left(r,z,t\right),$$

where u and v are the non-zero components of the displacement and velocity vectors.

A variable pressure $p\left(t\right)$ is set on the boundary surface of the plug $z = u(r,t)$. It is assumed that there is no punching shear resistance at $r = 0$ (the coordinate of maximum displacement for the plug boundary points) on the free surface of the plug $z = l + u(r,t)$. Thus, the boundary conditions for $\sigma_{zz}\left(r,z,t\right)$ have the form

$$\sigma_{zz}\left(0,u\left(0,t\right),t\right) = -p\left(t\right), \quad \sigma_{zz}\left(0,l+u\left(0,t\right),t\right) = 0. \tag{5.15}$$

We consider that from the onset of deformation, irreversible strains accumulate in the material as a result of the slow creep process. As long as there is no viscoplastic flow, we assume the material remains adherent to the pipe wall:

$$u|_{r=R} = 0, \quad v|_{r=R} = 0. \tag{5.16}$$

Irreversible deformation of the material causes an increase in temperature within it. We consider a case in which heat is created internally only due to irreversible deformation, that is, there is no heat influx into the medium through its boundary. We specify this temperature change by the following conditions

$$\theta\left(r,0\right) = 0, \quad \left.\frac{\partial\theta\left(r,t\right)}{\partial r}\right|_{r=0} = 0, \quad \left.\frac{\partial\theta\left(r,t\right)}{\partial r}\right|_{r=R} = 0. \tag{5.17}$$

For the motion under consideration, the non-zero components of the Almansi strain tensor are d_{rr} and d_{rz}. From formula (5.3), we obtain the following relations for them

$$d_{rr} = -\frac{1}{2}u'^2, \quad d_{rz} = \frac{1}{2}u', \quad u' = \frac{\partial u}{\partial r}. \tag{5.18}$$

It follows from the incompressibility condition (5.4) and relations (5.18) that in the case of antiplane motion under consideration, there is no thermal expansion in the

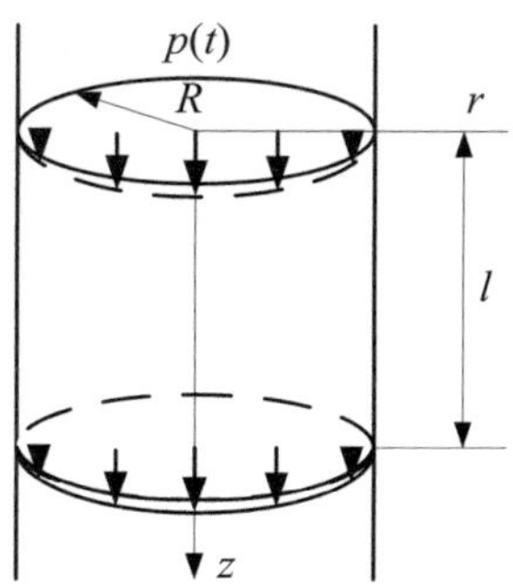

Fig. 5.1 Geometry of the problem.

material; thus, dependencies (5.4), (5.5) and (5.8) are valid when coefficient β is zero.

The stresses are determined by reversible deformations and, according to relations (5.5) and (5.7), have the form

$$\begin{aligned} \sigma_{\varphi\varphi} &= -\left(P_1 + 2\mu\right) + 2\left(\kappa - \nu_4\theta\right)\left(m_{rr} + m_{zz}\right) - 2\left(\mu - \nu_4\theta\right) m_{rz}^2 \\ &\quad + \nu_1\theta - \nu_3\theta^2 = -p_1, \\ \sigma_{rz} &= 2\left(\mu - l_1\theta\right) m_{rz}, \quad l_1 = \nu_1 + \nu_5, \\ \sigma_{rr} &= -p_1 + 2\left(\mu - l_1\theta\right) m_{rr} + \left(3\mu + l_1\theta\right) m_{rz}^2, \\ \sigma_{zz} &= -p_1 + 2\left(\mu - l_1\theta\right) m_{zz} + \left(3\mu + l_1\theta\right) m_{rz}^2. \end{aligned} \tag{5.19}$$

Relations (5.19) for the stress tensor components are given with accuracy up to the first-order terms in the reversible strain components $m_{rr}$and m_{zz} and the second-order terms in m_{rz} [6, 13].

The equilibrium equations (5.14) in the case under consideration take the form

$$\begin{aligned} \frac{\partial \sigma_{rr}}{\partial r} + \frac{\partial \sigma_{rz}}{\partial z} + \frac{\sigma_{rr} - \sigma_{\varphi\varphi}}{r} = 0, \\ \frac{\partial \sigma_{rz}}{\partial r} + \frac{\partial \sigma_{zz}}{\partial z} + \frac{\sigma_{rz}}{r} = 0. \end{aligned} \tag{5.20}$$

It follows from the solution of the second equation (5.20) that p_1 is a linear function of z:

$$p_1 = c\left(t\right) z + p_0\left(r, t\right),$$

and the stress component σ_{rz} has the form

$$\sigma_{rz} = \frac{c\left(t\right) r}{2} + \frac{c_1\left(t\right)}{r}. \tag{5.21}$$

The integration function $c_1\left(t\right)$ in (5.21) should be set equal to zero so that the stress σ_{rz} has a finite value at $r = 0$. To find the integration function $c\left(t\right)$, we use boundary conditions (5.15). Thus, for the component σ_{rz} we obtain the relation

$$\sigma_{rz} = -\frac{pr}{2l}, \quad p = p\left(t\right). \tag{5.22}$$

The creep potential (5.9) in this case takes the form

$$V\left(\sigma_{ij}\right) = B\left(\left(\sigma_{rr} - \sigma_{zz}\right)^2 + 4\sigma_{rz}\sigma_{zr}\right)^{\frac{n}{2}}. \tag{5.23}$$

In potential (5.23) we restrict ourselves to terms up to order n in stresses, then for the creep strain rates from (5.9), (5.19) and (5.23) we obtain

$$\varepsilon_{rz}^{v} = -Bn\left(\frac{pr}{l}\right)^{n-1}, \quad \varepsilon_{rr}^{v} = -\varepsilon_{zz}^{v} = \frac{\varepsilon_{rz}^{v}}{2}\frac{m_{rr} - m_{zz}}{m_{rz}}. \tag{5.24}$$

In view of dependencies (5.19) and (5.24), the heat equation (5.8) takes the form

$$\left(1+\beta_1\theta+\frac{\beta_3 l_1 p^2 r^2}{8l^2\left(\mu-l_1\theta\right)^3}\right)\frac{\partial\theta}{\partial t}+\frac{\beta_3 p r^2}{8l^2\left(\mu-l_1\theta\right)^2}\frac{\partial p}{\partial t}$$
$$=a\left(\frac{\partial^2\theta}{\partial r^2}+\frac{1}{r}\frac{\partial\theta}{\partial r}\right)-\frac{Bn}{2\nu_2}\left(\frac{pr}{l}\right)^n. \tag{5.25}$$

The heat equation (5.25) under conditions (5.17) was solved numerically.

It follows from relations (5.1) and (5.2) that the kinematics of the medium in this case is described by the dependencies

$$\begin{aligned}
\varepsilon_{rz} &= \frac{dd_{rz}}{dt}=\frac{\partial d_{rz}}{\partial t}=\frac{1}{2}\frac{\partial v}{\partial r}=\varepsilon_{rz}^e+\gamma_{rz}=\frac{\partial m_{rz}}{\partial t}+\frac{\partial p_{rz}}{\partial t},\\
r_{rz} &= -r_{zr}=2w_{rz},\\
\gamma_{rr} &= \frac{dp_{rr}}{dt}+2p_{rz}\left(r_{zr}+\gamma_{rz}\right),\\
\gamma_{zz} &= \frac{dp_{zz}}{dt}+2p_{rz}\left(r_{rz}+\gamma_{rz}\right),\\
\gamma_{rr} &= -\gamma_{zz}=\frac{\gamma_{rz}}{2}\frac{m_{rr}-m_{zz}}{m_{rz}}
\end{aligned} \tag{5.26}$$

in which the rates of irreversible strains are equal to the rates of creep strains: $\gamma_{rr}=\varepsilon_{rr}^v$, $\gamma_{rz}=\varepsilon_{rz}^v$ and $\gamma_{zz}=\varepsilon_{zz}^v$.

From (5.19), (5.22), (5.24), and (5.27), we obtain the expressions for the components of the reversible and irreversible strain tensors m_{rz} and p_{rz} and for the displacement gradient of the plug points

$$m_{rz}=-\frac{pr}{4l\left(\mu-l_1\theta\right)},\quad p_{rz}=-\int_0^t Bn\left(\frac{pr}{l}\right)^{n-1}dt,\quad u'=2\left(m_{rz}+p_{rz}\right). \tag{5.27}$$

By integrating the last equation (5.27) under the adhesion condition, we find the displacement of the points of the plug material. For the components of reversible strains m_{rr}, m_{zz} and irreversible strains p_{rr}, p_{zz}, from dependencies (5.27) we obtain a system of equations

$$\frac{\partial p_{zz}}{\partial t}=\gamma_{rz}\left(m_{rz}-\frac{p_{zz}}{m_{rz}}\right)-2p_{rz}r_{rz},\quad p_{rr}=-p_{zz}-2p_{rz}^2,$$
$$m_{zz}=-p_{zz}+\frac{m_{rz}^2}{2}+2m_{rz}p_{rz},\quad m_{rr}=-m_{zz}-m_{rz}^2. \tag{5.28}$$

Taking into account relations (5.19), boundary conditions (5.15) and integrating the first equilibrium equation (5.20), we obtain dependencies for calculating the stress tensor components based on the previously found distributions of temperature, displacements and reversible strain components

$$
\begin{aligned}
\sigma_{zz} &= \frac{p}{l}\left(z - l - u\left(0,t\right)\right) - \int\limits_0^r \frac{2\left(\mu - l_1\theta\right) m_{rr} + \left(3\mu + l_1\theta\right) m_{rz}^2}{r} dr \\
&+ 2\left(\mu - l_1\theta\right)\left(m_{zz} - m_{rr}\right) \\
&- 2\left(\mu - l_1\theta\left(0,t\right)\right)\left(m_{zz}\left(0,t\right) - m_{rr}\left(0,t\right)\right), \\
\sigma_{rr} &= \sigma_{zz} + 2\left(\mu - l_1\theta\right)\left(m_{rr} - m_{zz}\right), \\
\sigma_{\varphi\varphi} &= -p_1 = \sigma_{zz} - 2\left(\mu - l_1\theta\right) m_{zz} - \left(3\mu + l_1\theta\right) m_{rz}^2.
\end{aligned} \tag{5.29}
$$

The obtained solution of the non-isothermal boundary value problem under the increasing pressure drop remains valid until the time t_1, when the plasticity condition (5.11) is first satisfied on the lateral boundary of the plug $r = R$; in the case under consideration, this condition takes the form:

$$
\left|\sigma_{rz}\right|\Big|_{r=R} = k\left(\theta\left(R, t_1\right)\right). \tag{5.30}
$$

The yield stress in (5.30) depends on temperature (the first formula (5.13)). The time of the flow onset t_1 is determined from the equation $p\left(t_1\right) R = 2lk\left(\theta\left(R, t_1\right)\right)$.

5.4 Viscoplastic Flow

Starting from the time t_1, a viscoplastic flow region $m(t) \le r \le R$ develops within the material. The moving boundary $r = m(t)$ separates the flow region from the region $0 \le r \le m(t)$ in which the material continues to deform viscoelastically.

The viscoplastic flow developing from the pipe boundary leads to material slippage, which we define by the following boundary condition:

$$
\left(\left|\sigma_{rz}\right| - k\left(\theta\left(r, t_1\right)\right) - \frac{\eta v}{R - m\left(t\right)}\right)\Bigg|_{r=R} = 0. \tag{5.31}
$$

From time point t_1, the material begins to heat up also due to friction at the rigid pipe wall. In this case, we replace the first condition (5.17) with the temperature continuity condition at time t_1, and the third condition (5.17) with the following boundary condition:

$$
\theta\left(R, t\right) = \gamma_1 u\left(R, t\right), \tag{5.32}
$$

in which γ_1 is the specified constant of heat production due to friction. Additionally, we assume that the temperature θ does not reach the melting temperature during the deformation process.

Relations (5.19) for stresses apply within the region $0 \le r \le m(t)$. Using formulas (5.5) and (5.7), we establish that in the flow region $m(t) \le r \le R$ the components of the stress tensor also satisfy relations (5.19). From the second equilibrium equation (5.20) and boundary conditions (5.15), relation (5.22) for the stress tensor component σ_{rz} follows in both regions.

In the region $0 \le r \le m(t)$, the components of the creep strain rate tensor take the form (5.24), while the heat equation in this region is given by (5.25). It follows from the associated law of plastic flow (5.10) that the plasticity condition (5.11) for this case under consideration ($\sigma_{rz} < 0$, $\varepsilon^p_{rz} < 0$) takes the form:

$$\sigma_{rz} = -k + \eta \left(\varepsilon^p_{rz} - \varepsilon^{v_0}_{rz} \right). \tag{5.33}$$

Solving equation (5.33) for ε^p_{rz} using relation (5.22), we obtain

$$\varepsilon^p_{rz} = \frac{1}{\eta} \left(k - \frac{pr}{2l} \right) + \varepsilon^{v_0}_{rz}. \tag{5.34}$$

It should be noted again that $\varepsilon^{v_0}_{rz}$ in formulas (5.33) and (5.34) is the component of the creep strain rate tensor ε^v_{rz} at a point at the moment it reaches the elastoplastic boundary; at the current time t, it is calculated from the relation:

$$\varepsilon^{v_0}_{rz} = -Bn \left(\frac{pm}{l} \right)^{n-1}. \tag{5.35}$$

From the continuity condition for irreversible strain rates at the boundary $r = m(t)$ we obtain an equation for determining its position

$$\frac{pm}{2l} = k_0 \left(1 - \frac{\theta(m,t)}{\theta_m} \right)^2. \tag{5.36}$$

In the viscoplastic flow region $m(t) \le r \le R$, the heat equation (5.8) takes the following form by using relations (5.19) and (5.34):

$$\begin{aligned} &\left(1 + \beta_1 \theta + \frac{\beta_3 l_1 p^2 r^2}{8l^2 (\mu - l_1 \theta)^3} \right) \frac{\partial \theta}{\partial t} + \frac{\beta_3 p r^2}{8l^2 (\mu - l_1 \theta)^2} \frac{\partial p}{\partial t} \\ &= a \left(\frac{\partial^2 \theta}{\partial r^2} + \frac{1}{r} \frac{\partial \theta}{\partial r} \right) + \frac{pr}{2l\nu_2} \left(\frac{1}{\eta} \left(k - \frac{pr}{2l} \right) + \varepsilon^{v_0}_{rz} \right). \end{aligned} \tag{5.37}$$

Equations (5.25), (5.35)-(5.37) form a system of equations for unknown functions: θ in the regions $0 \le r \le m(t)$ and $m(t) \le r \le R$, $m(t)$ and $\varepsilon^{v_0}_{rz}$. We add to this system the second boundary condition (5.17), the boundary condition (5.32) and the initial conditions

$$m(t_1) = R, \quad \varepsilon^{v_0}_{rz}(R, t_1) = -B(\theta(R,t_1))\, n(\theta(R,t_1)) \left(p(t_1) R/l \right)^{n(\theta(R,t_1))-1}.$$

We also assume the continuity of the desired function θ and its derivative $\partial\theta/\partial r$ at the boundary $r = m(t)$ and the continuity of the temperature at the moment of the onset of viscoplastic flow t_1. To solve this system of equations, an algorithm based on the finite difference method was developed [21].

The kinematics of the medium in the viscoelastic region $0 \le r \le m(t)$ is described by relations (5.27) with

$$\gamma_{rr} = \varepsilon_{rr}^{v}, \qquad \gamma_{rz} = \varepsilon_{rz}^{v}, \qquad \gamma_{zz} = \varepsilon_{zz}^{v}.$$

The same relations also hold in the flow region $m(t) \leq r \leq R$ with

$$\gamma_{rr} = \varepsilon_{rr}^{p}, \qquad \gamma_{rz} = \varepsilon_{rz}^{p}, \qquad \gamma_{zz} = \varepsilon_{zz}^{p}.$$

Dependencies (5.27) are valid for the components m_{rz}, p_{rz} and the displacement gradient in the region $0 \leq r \leq m(t)$. The first and third formulas (5.27) are also hold in the region $m(t) \leq r \leq R$. The irreversible strain component in the flow region is found by integrating the differential equation following from (5.27) and (5.34):

$$\frac{\partial p_{rz}}{\partial t} = \frac{1}{\eta}\left(k - \frac{pr}{2l}\right) + \varepsilon_{rz}^{v_0}. \tag{5.38}$$

In solving equation (5.38), the continuity condition of the component of the irreversible strain tensor at time t_1 is used.

The components of reversible strains m_{rr}, m_{zz} and irreversible strains p_{rr}, p_{zz} in both regions are calculated from the system of equations (5.28). In the region $0 \leq r \leq m(t)$, $\gamma_{rz} = \varepsilon_{rz}^{v}$, while in the region $m(t) \leq r \leq R$, $\gamma_{rz} = \varepsilon_{rz}^{p}$.

Using the calculated distributions of temperature and deformations, the stresses in the region $0 \leq r \leq m(t)$ are found from dependencies (5.29). The stress continuity conditions at the boundary $r = m(t)$ lead to the same relations for determining stresses in the flow region $m(t) \leq r \leq R$. The displacements are found by integrating the last equation (5.27) using the boundary condition

$$u(R,t) = \int_{t_1}^{t} \frac{R - m(t)}{\eta(\theta(R,t))}\left(\frac{pR}{2l} - k(\theta(R,t_1))\right) dt,$$

which follows from condition (5.31).

Assume that at some time $t_2 > t_1$, the pressure becomes constant and equal to $p(t_2)$. This change in the loading force does not cause qualitative changes in the deformation process. Both the viscoelastic deformation region $0 \leq r \leq m(t)$ and the evolving viscoplastic flow region $m(t) \leq r \leq R$ persist within the material. All the relationships of this section remain valid in this case. Only in the heat equations (5.25) and (5.37) it is necessary to take into account that

$$\frac{\partial p}{\partial t} = 0.$$

The calculations were carried out in dimensionless variables

$$\frac{r}{R}$$

and

$$\tau = \frac{\alpha_1 t}{\mu}$$

with the function $p(t) = \alpha_1 t$ and the values of dimensionless constants:

$\beta_1 = 0.098$, $\beta_3 = -4.5$,
$k_0\mu^{-1} = 2.298 \cdot 10^{-3}$, $l_1\mu^{-1} = 0.452$, $\nu_2\mu^{-1} = 0.1$,
$\theta_m = 2.074$, $lR^{-1} = 5$, $a\mu\alpha_1^{-1}R^{-2} = 2.638 \cdot 10^4$,
$b_1 = 2.518$, $b_2T_0^{-1} = 1.482$,
$c_1\mu\alpha_1^{-1} = 4.887 \cdot 10^4$, $\sigma_0\mu^{-1} = 7.199 \cdot 10^{-6}$,
$QR_u^{-1}T_0^{-1} = 52.37$, $\upsilon T_0 = 1.212$,
$\alpha_1\eta_0\mu^{-2} = 3.257 \cdot 10^{-6}$, $\gamma_1 R = 20$.

Figure 5.2 (a) shows a graph of the change in the boundary of the viscoplastic flow region depending on the dimensionless time in the interval from τ_1 to $\tau_3 > \tau_2$. The temperature distribution in the plug material as a function of the radius at time τ_3 is shown in Fig. 5.2 (b). Figure 5.3 (a) and 5.3 (b) show graphs of the component p_{rz} of the irreversible strain tensor and the displacements of the points of the plug material depending on the radius at times τ_2 (solid line) and τ_3 (dotted line).

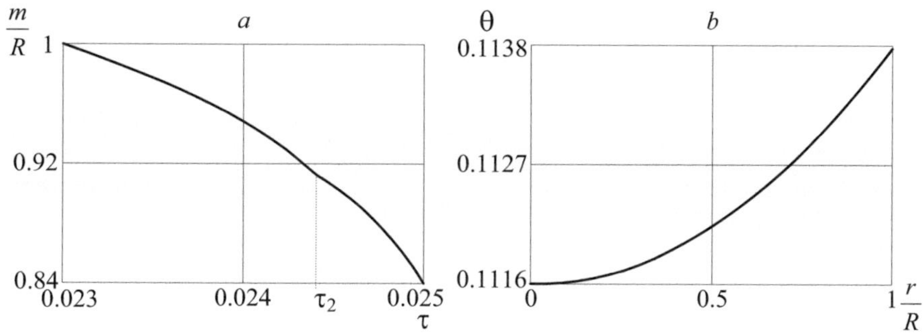

(a) (b)

Fig. 5.2: Graph of the boundary of the viscoplastic flow region as a function of time (a) and temperature distribution depending on the radius (b).

5.5 Conclusion

As previously noted, the solution of the problem under conditions of adhesion of the material to the pipe wall was obtained in [13], and under conditions of slippage using the Coulomb-Mohr dry friction law was investigated in [12]. In the first case, the heating of the material is insignificant (in the absence of external heat sources). The graph of the elastoplastic boundary is close to linear and with a constant pressure drop the position of the boundary of the viscoplastic flow region stops changing. In the second case, intense heating of the material is observed due to friction. The graph of the boundary is also close to linear, but the flow region continues to expand

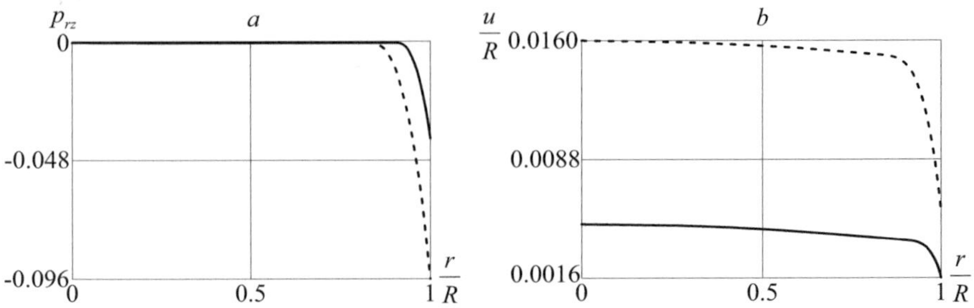

(a) (b)

Fig. 5.3: Graphs of the component p_{rz} of the irreversible strain tensor (a) and displacements (b) depending on the radius.

with a constant pressure drop. In the presented work, the case of material slippage due to near-wall viscoplastic flow is considered. In this case, the material undergoes significant heating compared to the results in [13]; however, the temperature rise in [12] remains more substantial. The graph of the elastoplastic boundary has a pronounced nonlinear appearance, and the flow region, similar to the second case, continues to grow with a constant pressure drop.

Acknowledgements The work was carried out within the framework of the state assignment of the Institute of Automation and Control Processes of the Far Eastern Branch of the Russian Academy of Sciences (topic No. FWFW-2026-0010).

References

[1] Begun AS, Burenin AA, Kovtanyuk LV (2016) Large irreversible deformations under conditions of changing mechanisms of their formation and the problem of definition of plastic potentials. Doklody Physics 61(9):463-466, DOI 10.1134/S102833581609007X

[2] Begun AS, Kovtanyuk LV, Lemza AO (2018) Change of accumulation mechanisms of irreversible deformations of materials in an example of viscometric deformation. Mechanics of Solids 53(1:85-92, DOI 10.3103/S0025654418010107

[3] Burenin AA, Galimzyanova KN, Kovtanyuk LV, Panchenko GL (2018) Matching growth mechanisms of irreversible deformation of a hollow sphere under uniform compression. Doklady Physics 63(10):407-410, DOI 10.1134/S1028335818100026

[4] Burenin AA, Kovtanyuk LV (2013) Bol'shiye neobratimyye deformatsii i uprugoye posledeystvie (in Russ., Large Irreversible Deformations and Elastic Aftereffect). Dal'nauka, Vladivostok

[5] Burenin AA, Kovtanyuk LV, Panchenko GL (2010) Modeling of large elastoviscoplastic deformations with thermophysical effects taken into account. Mechanics of Solids 45(4):583-594, DOI 10.3103/S0025654410040084
[6] Burenin AA, Kovtanyuk LV, Panchenko GL (2023) Thermomechanical loading of an elastoviscoplastic heavy layer held by an inclined plane. Continuum Mechanics and Thermodynamics 35(4):1325-1341, DOI 10.1007/s00161-022-01131-6
[7] Bykovtsev GI, Ivlev DD (1998) Teoriya plastichnosti (in Russ., Theory of Plasticity). Dal'nauka, Vladivostok
[8] Golovanov AI, Sultanov LU (2009) Matematicheskie modeli nelineynoy vychislitelnoy mekhaniki (in Russ., Mathematical Models of Nonlinear Computational Mechanics). KSU Publishing House, Kazan
[9] de Groot SR, Mazur P (1962) Non-Equilibrium Thermodynamics. North-Holland, Amsterdam
[10] Gyarmati I (1970) Non-equilibrium Thermodynamics. Springer-Verlag, Berlin, Heidelberg, DOI 10.1007/978-3-642-51067-0
[11] Iost A (1998) The correlation between the power-law coefficients in creep: the temperature dependence. Journal of Materials Science 33(12):3201-3206, DOI 10.1023/A:1004368511595
[12] Kovtanyuk LV, Panchenko GL, Popova EO (2024) Deformation and heating due to surface friction of the plug material in a round pipe as it moves under the action of a changing pressure difference. Scientific Notes of Komsomolsk-on-Amur State Technical University 7(79):68-76
[13] Kovtanyuk LV, Panchenko GL, Popova EO (2024) Heat production due to creep strains and wall viscoplastic flow in the plug material in a round pipe under the action of variable pressure difference. Mechanics of Solids 59(5):2694-2705, DOI 10.1134/S0025654424604270
[14] Lee EH (1969) Elastic-plastic deformation at finite strains. Transaction ASME. Journal of Applied Mechanics 36(1):1-6, DOI 10.1115/1.3564580
[15] Levitas VI (1987) Bol'shie uprugoplasticheskie deformatsii materialov pri vysokom davlenii (in Russ., Large Elastoplastic Deformations of Materials at High Pressure). Naukova Dumka, Kiev
[16] Lurie AI (1990) Non-Linear Theory of Elasticity, North-Holland Series in Applied Mathematics and Mechanics Vol 36. Elsevier, Amsterdam
[17] Myasnikov VP (1996) Uravneniya dvizheniya uprugoplasticheskikh materialov pri bolshikh deformatsiyakh (in Russ., Equations of motion of elastic-plastic materials under large deformations). Bulletin of the Far Eastern Branch of the Russian Academy of Sciences 4:8-13
[18] Naghdi PM (1984) Recent developments in finite deformation plasticity. In: Plasticity Today: Modeling, Methods and Applications (Eds. A. Sawczuk and G. Bianchi), pp. 75–83. Elsevier Applied Science Publishers Ltd., Essex, England
[19] Norton FH (1929) The Creep Steel of High Temperature. McGraw Hill, New York

[20] Pla F, Mancho AM, Herrero H (2009) Bifurcation phenomena in a convection problem with temperature dependent viscosity at low aspect ratio. Physica D: Nonlinear Phenomena 238(5):572-580, DOI 10.1016/j.physd.2008.12.015
[21] Samarsky AA (1989) Teoriya raznostnykh skhem (in Russ., Theory of Difference Schemes). Nauka, Moscow
[22] Shitikov AV, Bykovtsev GI (1990) Konechnyye deformatsii uprugoplasticheskikh sred (in Russ., Finite deformations in an elastoplastic mediua). Doklady Akademii Nauk SSSR 311(1):59–62

Chapter 6
On Variational Models of Constitutive Equations of Elastic-Plastic Materials

Petr Belov, Sergey Lurie, and Natalia Golovina

Abstract A modeling algorithm for constitutive relations of elastoplastic materials is proposed. It is postulated that the material model is generally defined by a non-integrable variational form in the one-dimensional strain space, involving derivatives of stress with respect to strain of various differential orders. The constant parameters of this variational form determine the physical properties of the material. The stress-strain curve is interpreted as a solution arising from the stationarity condition of the variational formulation. According to the proposed procedure, the stationarity condition yields the Euler equation, which carries the physical meaning of a generalized Hooke's law for elastoplastic materials. A system of fundamental solutions for the obtained Euler equation is determined. The general solution is constructed as an expansion in a basis of the functional space of fundamental solutions. The set of expansion coefficients of the stress in the found basis defines the general form of the constitutive relations. The characteristic equation corresponding to the Euler equation and its roots are used in a spectral-variational interpretation of the yield point phenomenon and the Portevin-Le Châtelier (PLC) effect as a loss of stability of segments of the deformation diagram.

Key words: Variational model, Constitutive equations, Gradient model, Euler equations, Smoothness of interpolation, Stability of constitutive relations, Yield tooth, Portevin-Le Châtelier effect

---

Petr Belov
Institute of Applied Mechanics of RAS, Leningradsky prospekt 7, 125040 Moscow, Russian Federation, e-mail: belovpa@yandex.ru

Sergey Lurie (corresponding author)
Ishlinsky Institute for Problems in Mechanics RAS, Prospekt Vernadskogo 101-1, 119526 Moscow, Russian Federation, e-mail: belovpa@yandex.ru, salurie@mail.ru

Natalia Golovina
Tyumen Industrial University (TIU), Surgut Branch, Surgut, Tyumen Region, Russian Federation, e-mail: ntgolovina@rambler.ru

H. Altenbach et al. (eds.), *Stability Problems in Mechanics of Continua*,
Advanced Structured Materials 260, https://doi.org/10.1007/978-3-032-31032-3_6

6.1 Introduction

The problem of constructing analytically convenient representations of experimental stress-strain diagrams $\sigma(\varepsilon)$ for applications is a significant issue in applied mechanics of materials [1, 9]. Test data are typically provided as a discrete set of points and contain noise, an irregular strain grid, and local features (yield plateau, yield point, etc.), which may be physically significant but are challenging to model. Furthermore, to date, the task of formulating theoretically substantiated constitutive relations remains essentially unsolved. Phenomenological author-specific dependencies, convenient for describing experimental results for a particular class of materials, are commonly used [7, 9].

In practical calculations, a compromise inevitably arises between the accuracy of reproducing the experimental data set, the computational simplicity of subsequent operations (differentiation, integration, functional evaluation, use in parameter identification algorithms), and the stability of the result with respect to noise [11, 12]. Two aspects should be noted in this context. The first pertains to a formal problem of applied mathematics and is related to constructing an interpolation function of an appropriate continuity class. Unlike the discrete set of experimental points, this function can be termed the experimental deformation curve [2]. The second aspect, which is usually hardly discussed, is determined by the problem of structural analysis in materials mechanics and involves constructing an approximating function of an appropriate smoothness class, based on some material model [4].

It is assumed that this function should correspond, with a chosen accuracy, to the interpolation curve defined by the instrumental measurement errors. The structure of the present work consists of the following sections. First, it is shown that specific constitutive relations, whose form is known, can be defined within the framework of implementing a certain variational principle. Here, the physical relations (generalized equations of Hooke's law) are solutions of the corresponding Euler equation. Subsequently, generalized variational formulations are presented, demonstrating the possibility of theoretical modeling of interpolation curves for constitutive relations for single-segment and, especially, for multi-segment representation in the strain space across different classes of continuity. A procedure of *adaptive segment division* is proposed as an algorithm that ensures minimizing the discrepancy between the experimental and model curve down to the level of instrumental error. Classes of approximating curves in the strain space with enhanced smoothness are proposed, enabling the modeling of materials with a continuous tangent modulus and, with further increased smoothness, continuous curvature, etc.

In conclusion, the problems of the monotonicity of the stress-strain diagram $\sigma(\varepsilon)$ and the mechanistic interpretation of the yield point effects, the Portevin-Le Châtelier (PLC) effect [6, 8, 16] are discussed. These effects were associated with delays in relaxation processes during deformation and were later investigated within the framework of dynamic strain aging theory and using dislocation models [5]. In this paper, the yield tooth and the PLC effect are considered as the implementation of the effect of local loss of stability of the deformation curve in the deformation space.

6.2 Variational Models of Constitutive Relations

A variational approach for constructing deformation curves is developed using a variational model in which the deformation curve is determined by solving the corresponding Euler equation, and the boundary-value variational problem ensures correct conjugation conditions for solutions constructed for different segments (when segmentation of the deformation curve is employed). The deformation curve is represented in the one-dimensional strain space. In the general case, stress and strain can be understood as certain invariant quantities. Consider two examples of constructing variational models for known particular types of constitutive relations, thereby demonstrating that variational models of constitutive equations can indeed be formulated.

6.2.1 Variational Model of Linear Processes

Consider a deformation curve that, in particular, describes the process of linear material behavior. It is shown that a variational model of such a linear process exists. For reversible processes, there exists a potential in the strain space, which can be written as

$$2w = \frac{\partial \sigma}{\partial \varepsilon}\frac{\partial \sigma}{\partial \varepsilon}. \tag{6.1}$$

Taking into account (6.1), the following variational principle can be formulated, whose extrema provide a description of the linear deformation process:

$$\begin{gathered}\delta J_1 = 0, \qquad J_1 = \int\limits_{\varepsilon} w\, d\varepsilon = \frac{1}{2}\int\limits_{\varepsilon} \frac{\partial \sigma}{\partial \varepsilon}\frac{\partial \sigma}{\partial \varepsilon}\, d\varepsilon, \\ \delta J_1 = \int\limits_{\varepsilon} \frac{\partial \sigma}{\partial \varepsilon}\,\delta\,\frac{\partial \sigma}{\partial \varepsilon}\, d\varepsilon = \frac{\partial \sigma}{\partial \varepsilon}\,\delta\sigma\Big|_{\varepsilon_0}^{\varepsilon} - \int\limits_{\varepsilon} \frac{\partial^2 \sigma}{\partial \varepsilon^2}\,\delta\sigma\, dI = 0.\end{gathered} \tag{6.2}$$

Thus, the Euler equation takes the form

$$\frac{d^2\sigma}{d\varepsilon^2} = 0 \tag{6.3}$$

and fully corresponds to the linear process

$$\sigma(\varepsilon) = c_0 + c_1\varepsilon, \tag{6.4}$$

The boundary conditions in the variational equation (6.2) are analogous to the static and kinematic conditions in continuum mechanics. It is noted that the stationarity condition of the functional (6.2) corresponds to the minimum of the tangent to the deformation curve. In this case, these boundary conditions can be used to formulate the interface boundary conditions for constructing a piecewise linear approximation of a multi-segment deformation curve, which ensures continuous conjugation.

6.2.2 On Variational Models of Nonlinear Constitutive Relations

We will demonstrate that there also exists a variational model of constitutive relations with power-law dependencies between stresses and strains. The functional of such a model and the corresponding solution parameters will be denoted by the subscripts "RO". Consider a quadratic functional in which the generalized arguments include not only the value $\sigma' = d\sigma/d\varepsilon$, but also the nonlinear expression $(\varepsilon\sigma''(\varepsilon))$:

$$J_{RO} = \int_{\varepsilon_{\min}}^{\varepsilon_{\max}} [\sigma'(\varepsilon), (\varepsilon\sigma''(\varepsilon)) = = \frac{1}{2}\int_{\varepsilon} [A_{11}\sigma'\sigma' + 2A_{12}\sigma'(\varepsilon\sigma'') + A_{22}(\varepsilon\sigma'')(\varepsilon\sigma'')]d\varepsilon. \tag{6.5}$$

The requirement of stationarity of the functional J_{RO}, (6.5) leads to the following variational equation:

$$\delta J_{RO} = 0. \tag{6.6}$$

From (6.6), the Euler–Cauchy equation follows:

$$\left(\frac{A_{11} - A_{12}}{A_{22}} - 2\right)\sigma'' - 4\varepsilon\sigma''' - \varepsilon^2\sigma'''' = 0 \tag{6.7}$$

along with the corresponding boundary/contact conditions.

The exact solution of equation (6.7) is constructed using a power-law substitution $\sigma = \varepsilon^n$. The characteristic equation corresponding to () has the form

$$n(n-1)\left[\left(\frac{A_{11} - A_{12}}{A_{22}} - 2\right) - (n-2)(n+1)\right] = 0. \tag{6.8}$$

Consequently, the general solution of equation (6.7), which determines the extremals of the functional (6.5), is represented as an expansion

$$\sigma = c_0 + c_1\varepsilon + C_1^{RO}\varepsilon^{n_1} + C_2^{RO}\varepsilon^{n_2}, \tag{6.9}$$

where c_0, c_1, C_1^{RO}, C_2^{RO} are constants that determine the specific form of the constitutive relations (the form of the approximating dependence constructed from the extremals of the functional (6.5) and n_1, n_2 are the roots of the equation (see Eq. (6.8))

$$[(A_{11} - A_{12})/A_{22} - 2] - (n-2)(n+1) = 0.$$

In expression (6.9), the first two terms with constants c_0, c_1 determine the linear part of the segment of the simulated loading diagram. Overall, together with the terms containing parameters C_1^{RO}, C_2^{RO} the theoretically derived constitutive relation (6.9) corresponds to the known generalized Ramberg–Osgood type law [3, 9].

Similarly, it is easy to establish that if a gradient-type variational model is considered for the constitutive relation using the functional

$$J_{OR} = \int_{\varepsilon_{\min}}^{\varepsilon_{\max}} [\sigma'(\varepsilon), \sigma''(\varepsilon)]\, d\varepsilon. \tag{6.10}$$

Then the Euler equation of this variational problem for functional (6.10) defines a solution that is a linear combination of a linear function and two exponential solutions. This fully corresponds to the known exponential Voce law [15].

Thus, it has been shown that the idea of using variational models to describe constitutive relations, analogous to variational models in mechanics, proves to be highly fruitful in modeling multi-segment deformation curves, since the variational formulation provides consistent boundary conditions for the conjugation of model curves on individual segments. Moreover, the proposed approach of using variational models allows for obtaining a theoretical deformation curve from the stationarity conditions of the corresponding functional, which always ensures "energetic" consistency of the solution, following a mechanistic analogy. This is important for the theoretical prediction of limiting fracture relations expressed in terms of stress invariants, which are determined by the specific deformation curve. The presented examples provide sufficient justification for the use, development, and analysis of variational models of deformation curves for nonlinear materials.

6.3 Variational Models for Interpolation Curves of Various Smoothness Classes

The problem of mathematical description of the deformation curve has not only applied but also fundamental significance: a successful model should not merely approximate experimental points but, if possible, reveal the internal structure of the deformation mechanism and indicate the reasons for regime changes. The traditional approach consists of constructing phenomenological or semi-empirical stress-strain relationships. Such relations are convenient for engineering calculations and often provide high approximation accuracy on individual sections of the diagram. However, their general methodological drawback is that they typically fix the already formed shape of the curve without answering the question of why this shape turns out to be exactly as it is. This applies particularly to sections where the curve loses its monotonic smoothness. A successful model should not only approximate experimental points but also reveal the internal structure of the deformation mechanism and indicate the reasons for regime changes. We proceed from the assumption that the deformation curve should be considered not only as an empirical dependence but also as a solution to a special variational-differential problem. From a methodological point of view, this implies a transition from a purely approximative description to a spectral-variational analysis and the construction of a reference solution describing the loading branch of a given smoothness..

We will consider interpolation curves within function classes of different smoothness C^0, C^1 and C^2 for sets of experimental data points $\sigma(\varepsilon)$ in order to subsequently examine algorithms for obtaining an analytical representation of the experimental curve suitable for further analysis and computations. We will refer to a material model of class C^l, $l = 0, 1, 2$.

Each class of interpolation functions is defined as a spline, each segment of which is an extremal of some variational principle and belongs to the corresponding smoothness class. The contact conditions between spline segments are determined by the requirements of the corresponding smoothness of the junction, also following from the variational principle defining the model. The algorithm for constructing splines is universal for all smoothness classes (for all material models). All experimental data samples are normalized by the strain and stress values at the ultimate tensile strength point. Thus, the initial and final points of the interpolation curves for any material coincide by construction. This allows considering any interpolation curve as a variation of a certain reference curve (extremal) of a variational principle, be it the principle of stationarity of the functional for nonlinearly elastic materials, or the principle of stationarity of a non-integrable variational form for plastic materials. The interpolation curve is constructed according to the following algorithm. First, it is assumed to pass through the initial and final points. The accuracy of the interpolation is assessed by its root mean square error (RMSE) from the data sample:

$$\mathrm{RMSE} = \sqrt{\frac{1}{N}\sum_{i=1}^{N}(\sigma_i^* - \sigma(\varepsilon_i^*))^2}.$$

Here $(\varepsilon_i^*, \sigma_i^*)$ are the coordinates of the experimental data points, and $(\varepsilon, \sigma(\varepsilon))$ are the coordinates of points on the interpolation curve, N denotes the number of experimental points.

If the deviation exceeds the instrumental error, each data point not lying on the interpolation curve is considered a candidate for a division point of the original spline segment into two. The RMSE of a trial two-segment spline is computed. Consequently, the trial point that yields the maximum reduction in RMSE is selected. This procedure continues until all data points lie on the interpolation curve or until the RMSE reaches the instrumental accuracy:

$$\mathrm{RMSE} \le \delta_{\mathrm{instr}},$$

where δ_{instr} denotes the estimate of the instrumental measurement error. Such adaptive node selection strategies are similar in spirit to polyline simplification algorithms like Ramer–Douglas–Peucker [10], widely used in computational geometry.

6.3.1 Variational Model of a Material of Class C^0

Consider a variational model of a material of class C^0, which requires the minimality of stress variation along the deformation diagram. A quadratic form in the first derivative of stress (6.1) is adopted as the smoothness functional C^0, (see functional J_1, Eq. (6.1)), and variational statement (6.2)–(6.4) the resulting Euler equation provides a theoretical justification for piecewise linear interpolation of the deformation diagram. The constants c_0 and c_1 are determined from the boundary conditions or the compatibility conditions of the interpolation curve segments. Thus, in the smoothness class C^0 the spline is a polyline (piecewise linear curve). The piece/wise linear curve consists of sections, on each of which (the Euler equation) holds. The transitions between segments localize "kinks" (jumps in the tangent modulus), i.e., the junction points of the spline segments. Therefore, the task of reducing the number of segments here is the problem of controlled removal of kinks.

For the class C^0 model, a technologically simple and reproducible algorithm is constructed for building a continuous piecewise linear approximation of the experimental $\sigma(\varepsilon)$ diagram. The algorithm is organized as a controlled multiplication of the model's spline segments. Thereby, a practical compromise between the accuracy and complexity of the model spline is achieved: the increase in the number of segments stops exactly at the moment when further detailization ceases to be physically meaningful due to the limited reliability of the measurements.

Note that the experimental deformation curve is given by a set of nodes:

$$\{(\varepsilon_1, \sigma_i)\}_{i=0}^{N}, \qquad \varepsilon_0 < \varepsilon_1 < \cdots < \varepsilon_N, \tag{6.11}$$

where ε is strain, σ is stress

The sample is normalized. All points having the same abscissa are replaced by a single point whose ordinate is set equal to the arithmetic mean of the ordinates of these points. For each initial engineering diagram, the node with the maximum stress value $\sigma_{\max}$ is determined. If there are several such nodes, the rightmost one (with the largest strain) is selected – its coordinates $(\varepsilon_{\max}, \sigma_{\max})$ are used for normalization: $\varepsilon^* = \varepsilon/\varepsilon_{\max}$ and $\sigma^* = \sigma/\sigma_{\max}$. All points lying to the right of the node $(\varepsilon_{\max}, \sigma_{\max})$, are discarded and not considered further. Thus, for each sample, a unified dimensionless domain $\varepsilon^* \in [0, 1]$, $\sigma^* \in [0, 1]$.

In what follows, we will write normalized quantities without an asterisk, and associate the number of sample points N with the number of points falling within the normalized segment. Sample (6.11) generates a piecewise linear interpolation with $K = N - 1$ segments (a polyline through the sample nodes).

6.3.2 Variational Model of a Material of Class C^1

Consider a variational model of a material of class C^1, where the principle of stationarity requires that the variation of the tangent modulus over the entire domain be minimal:

$$J_2[\sigma] = \frac{1}{2} \int_{\varepsilon_{\min}}^{\varepsilon_{\max}} (\sigma''(\varepsilon))^2 \, d\varepsilon.$$

The stationarity requirement leads to:

$$\delta J_2[\sigma] = \sigma'' \delta\sigma' \Big|_{\varepsilon_{\min}}^{\varepsilon_{\max}} - \sigma''' \delta\sigma \Big|_{\varepsilon_{\min}}^{\varepsilon_{\max}} + \int_{\varepsilon_{\min}}^{\varepsilon_{\max}} (\sigma'''' \delta\sigma) \, d\varepsilon = 0.$$

The Euler equation

$$\sigma''''(\varepsilon) = 0.$$

yields the solution

$$\sigma(\varepsilon) = c_0 + c_1\varepsilon + c_2\varepsilon^2 + c_3\varepsilon^3.$$

At the ends of each segment, boundary conditions must be satisfied: continuity conditions for $\sigma^+(\varepsilon_{\min}) = \sigma^-(\varepsilon_{\min})$, $\sigma^+(\varepsilon_{\max}) = \sigma^-(\varepsilon_{\max})$ and differentiability conditions for $\sigma^{+\prime}(\varepsilon_{\min}) = \sigma^{-\prime}(\varepsilon_{\min})$, $\sigma^{+\prime}(\varepsilon_{\max}) = \sigma^{-\prime}(\varepsilon_{\max})$. This includes conditions at the fixed ends $\sigma(0) = 0$, $\sigma(1) = 1$, and $\sigma'(0) = E_0$, $\sigma'(1) = 0$, E_0 is the elastic modulus of the linear part of the deformation curve.

6.3.3 Variational Model of a Material of Class C^2

Consider a variational model of a material of class C^2. In this case, the interpolation curve must possess continuity of stress, tangent modulus, and its derivative. In other words, the curvature of the deformation diagram must be continuous.

The curvature of the diagram is determined by the second derivative of stress:

$$\kappa(\varepsilon) = \sigma''(\varepsilon).$$

A natural requirement is the minimality of curvature variation along the entire strain range. This leads to the functional:

$$J_3[\sigma] = \frac{1}{2} \int_{\varepsilon_{\min}}^{\varepsilon_{\max}} (\sigma'''(\varepsilon))^2 \, d\varepsilon.$$

The stationarity requirement leads to

$$\delta J_3[\sigma] = \sigma^{(3)} \delta\sigma'' \Big|_{\varepsilon_{\min}}^{\varepsilon_{\max}} - \sigma^{(4)} \delta\sigma' \Big|_{\varepsilon_{\min}}^{\varepsilon_{\max}} + \sigma^{(5)} \delta\sigma \Big|_{\varepsilon_{\min}}^{\varepsilon_{\max}} - \int_{\varepsilon_{\min}}^{\varepsilon_{\max}} (\sigma^{(6)} \delta\sigma) \, d\varepsilon = 0.$$

The Euler equation

$$\sigma''''''(\varepsilon) = 0$$

has the solution

$$\sigma(\varepsilon) = c_0 + c_1\varepsilon + c_2\varepsilon^2 + c_3\varepsilon^3 + c_4\varepsilon^4 + c_5\varepsilon^5.$$

The considered C^0, C^1 and C^2 models are proposed not as independent material models, but as successive stages of a unified variational approach to modeling deformation diagrams. Increasing the smoothness class of the interpolation function corresponds to incorporating higher-order stress derivatives into the variational formulation.

Note that while for the class C_0 material, the segment of the interpolation curve was determined completely, without arbitrariness, for the smoothness class C_1 material the situation is somewhat different. The global segment, from which the curve reduction begins, is determined up to the values of the tangent modulus at the initial and final points $\sigma'(0) = E_0$, $\sigma'(1) = E_1$.

6.3.4 Generalized Higher-Order Variational Model

The models C^0, C^1 and C^2 considered above can be interpreted as special cases of a more general variational principle. In these models, quadratic functionals depending respectively on the first, second, and third derivatives of stress with respect to strain are minimized.

A natural generalization is a functional containing quadratic forms in all these derivatives simultaneously. Such a functional allows accounting for the contributions of various mechanisms of diagram shape change and thus describing a wider class of experimental $\sigma(\varepsilon)$ dependencies.

Let us combine the previous formulations into a functional:

$$J_4[\sigma] = \frac{1}{2}\int\limits_{\varepsilon_{\min}}^{\varepsilon_{\max}} \left(A_{11}(\sigma')^2 + A_{22}(\sigma'')^2 + A_{33}(\sigma''')^2\right) d\varepsilon$$

and then introduce a complete bilinear functional.

This functional accounts for independent quadratic contributions of derivatives of various orders. However, in the general case, the variational functional may also contain mixed terms describing the interaction of these contributions. This leads to the following bilinear functional:

$$J_5[\sigma] = \frac{1}{2}\int\limits_{\varepsilon_{\min}}^{\varepsilon_{\max}} \left(A_{11}(\sigma')^2 + A_{22}(\sigma'')^2 + A_{33}(\sigma''')^2 + \right.$$
$$\left. + 2A_{12}\sigma'\sigma'' + 2A_{13}\sigma'\sigma''' + 2A_{23}\sigma''\sigma'''\right) d\varepsilon. \qquad (6.12)$$

The coefficients A_{11}, A_{22}, A_{33} characterize compliances associated with changes in the slope, curvature, and higher derivatives of the deformation diagram, respectively. The mixed coefficients A_{12}, A_{13}, A_{23} describe the interaction of the corresponding mechanisms of diagram shape change.

Note that functionals of this type can be called functionals of gradient type. They are completely analogous to the gradient variational models widely used in continuum mechanics.

It should be noted that, following the mechanical analogy, conditions for existence and uniqueness can also be introduced, which reduces to the requirement of positive definiteness of the solution to the variational problem. Concerning the problem of constructing theoretical models of constitutive relations, it would be difficult to establish a direct analogy with deformable solid mechanics. Nevertheless, the condition of positive definiteness can be associated with the fact that such models describe deformation curves that, although nonlinear, are reversible from the standpoint of loading-unloading.

6.3.5 Variational Principle for Irreversible Processes

To construct a dissipative generalization, a generalization of the variational principle of Sedov [13, 13], previously proposed by the authors, is used. According to this, to obtain a variational equation, in addition to the variation of the original functional (the reversible part), corresponding non-integrable variational forms must be introduced, the number of which depends on the possible dissipative pairs of arguments. In the case under consideration (for the functional $J_5[\sigma]$) the stationarity condition for the non-integrable variational form is:

$$\delta J_6[\sigma] = \delta J_5[\sigma] + \frac{1}{2} \int\limits_{\varepsilon_{\min}}^{\varepsilon_{\max}} \big(B_{12}(\sigma'\delta\sigma'' - \sigma''\delta\sigma') + \\ + B_{13}(\sigma'\delta\sigma''' - \sigma'''\delta\sigma') + B_{23}(\sigma''\delta\sigma''' - \sigma'''\delta\sigma'')\big)\, d\varepsilon'. \quad (6.13)$$

Here, the coefficients B_{12}, B_{13}, B_{23} are additional parameters of the deformation curve model, reflecting the dissipative properties of the material.

To account for dissipation processes (6.13), the energy stationarity condition is generalized to the requirement of stationarity of a non-integrable variational form:

$$\delta J_6[\sigma] = \delta J_5[\sigma] + \frac{1}{2} \int\limits_{\varepsilon_{\min}}^{\varepsilon_{\max}} \big(B_{12}(\sigma'\delta\sigma'' - \sigma''\delta\sigma') + \\ + B_{13}(\sigma'\delta\sigma''' - \sigma'''\delta\sigma') + B_{23}(\sigma''\delta\sigma''' - \sigma'''\delta\sigma'')\big)\, d\varepsilon. \quad (6.14)$$

For the variational model (6.14), the corresponding Euler equation takes the form:

$$A_{33}\sigma'''''' - B_{23}\sigma''''' + (A_{13} - A_{22} - 2B_{13})\sigma'''' - 2B_{12}\sigma''' + A_{11}\sigma'' = 0. \quad (6.15)$$

The general solution for equation (6.15) is written as:

$$\sigma(\varepsilon) = f_1 g_1(\varepsilon) + f_2 g_2(\varepsilon) + f_3 g_3(\varepsilon) + f_4 g_4(\varepsilon) + f_5 g_5(\varepsilon) + f_6 g_6(\varepsilon). \quad (6.16)$$

The functions $g_i(\varepsilon)$ are fundamental solutions of the Euler differential equation for the corresponding material model, i.e., they are eigenfunctions of the linear differential operator of this equation. These functions are determined by the roots of the corresponding characteristic equation arising from the solution of the Euler equation. The coefficients f_i are constants of integration and are determined from the boundary conditions of the problem, as well as from the conjugation conditions of the interpolation curve segments, ensuring the required class of smoothness of the spline.

The interpolation curve is smooth, with continuous derivatives up to and including the third order. At the ends of each segment, boundary conditions must be satisfied, including: conditions for the continuity of stress, continuity of the first and second derivatives, as well as three conditions at each of the diagram boundaries.

Let us note another quality of higher-order models of the type (6.12) and (6.14)–(6.16). Models of this kind allow for the reduction of a multi-segment higher-order model, decreasing the number of segments while preserving the quadratic error. This is achieved by including, among the sought parameters during error minimization, the roots of the characteristic equations, i.e., the parameters of the deformation curve model.

6.4 On the Stability of Constitutive Relations and Modeling of Portevin-Le Châtelier Type Effects

Let us briefly consider one simple version of a variational generalized model of constitutive relations, which is defined by the stationarity condition of a variational form:

$$\begin{aligned}
&\delta J[\sigma] + \delta W_B = 0,\\
&J[\sigma] = \frac{1}{2}\int_{\varepsilon} W(\sigma', \sigma'', \sigma''', \sigma) =\\
&\quad = \frac{1}{2}\int_{\varepsilon} [A_{11}(\sigma)(\sigma')^2 + A_{22}(\sigma)(\sigma'')^2 + A_{12}(\sigma')(\sigma'')]\, d\varepsilon, \qquad (6.17)\\
&\delta W_B = \frac{1}{2} B_{12} \int_{\varepsilon} [\sigma'\delta\sigma'' - \sigma''\delta\sigma']\, d\varepsilon.
\end{aligned}$$

In the expression for the functional (6.17), the physical characteristics of the process being described, i.e., the coefficients in nonlinear form, are in general a function of stress (or the corresponding stress invariant in the general case), as well as a function of strain (the strain invariant). Double subscripts indicate the value of the differential order of the derivatives in the bilinear expression of the corresponding factor. Note that the first two terms are completely analogous to the functional describing the stability problem for rods with variable characteristics in the direction of the longitudinal x-coordinate. Indeed, the second term corresponds to the deformation energy, and the first to the work of the compressive force. It suffices to assume

$$\sigma' \sim w', \qquad \sigma'' \sim w'' \ (w' = dw/dx, w'' = d^2w/dx^2)$$

and

$$\varepsilon \sim x.$$

This analogy allows us to hope for the possibility of using the variational model of constitutive relations to formulate the problem of stability of the deformation process with respect to small deviations from the equilibrium position, for modeling such fluctuation effects as the yield point phenomenon and the Portevin–Le Châtelier effect observed in experiments.

The variational model and the corresponding boundary value problem are inherently nonlinear. We will use the following linearization procedure. Suppose there is a function of general form $L(\sigma)$ which could be scalar, vector, or tensor-valued. The variable is also a scalar (though it could be vector or tensor) quantity. Then the function $L(\sigma)$ can be approximated in the neighborhood of a point $\sigma = \sigma_0$ in the direction D, using a tensor-valued Taylor expansion:

$$L(\sigma_0 + D) = L(\sigma_0) + DL(\sigma_0, D) + o(D). \tag{6.18}$$

Here we have defined the directional derivative of the function $L(\sigma)$ at the point$\sigma = \sigma_0$ in the direction D as follows:

$$DL(\sigma_0, D) = \frac{d}{ds}[L(\sigma_0 + sD)]_{s=0}$$

provided that

$$\lim_{D \to o} \frac{s(D)}{|D|} = 0.$$

Expansion (6.18) allows us to define the linear part of the function $L(\sigma)$ or linearize the function $L(\sigma)$ in the neighborhood of the point σ_0 in the direction D in the form of the following relation

$$L(\sigma, D)|_{\sigma_0} = L(\sigma_0, D) + DL(\sigma_0, D).$$

Using the linearization of the density of the functional (6.1) for the considered neighborhoodîñòè σ_0 allows us to consider the functional (6.17) in which the coefficients A_{ij}, B_j are constant and defined at $\sigma = \sigma_0$: $A_{ij} = A_{ij}(\sigma_0)$, $B_{ij}(\sigma_0)$. Essentially, the usual assumption for such problems is used: in the neighborhood of $\sigma = \sigma_0$ the solution changes slowly, and the coefficients $A_{ij}(\sigma_0(\varepsilon))$, $B_{ij}(\sigma_0(\varepsilon))$ are constant. In this case, for the linearized variational problem (6.17), it is not difficult to write the Lagrange equation in explicit form as a linear equation

$$F(\sigma) = \frac{d}{d\varepsilon}\frac{\partial W}{\partial \sigma'} - \frac{d^2}{d\varepsilon^2}\frac{\partial W}{\partial \sigma''} - \frac{\partial W}{\partial \sigma}$$
$$= -A_{22}\sigma'''' + A_{11}\sigma'' - 2\frac{d}{d\varepsilon}(B_{12}\sigma'') = 0, \tag{6.19}$$

where

$$W = W(\sigma, \sigma', \sigma'')$$

and

$$\sigma' = \frac{d\sigma}{d\varepsilon},$$

etc.

We assume that the boundary conditions for this segment correspond to the compatibility conditions with adjacent parts of the deformation curve and have the form

$$\begin{aligned} \sigma(\varepsilon_1) = \sigma'(\varepsilon_1) = 0, \\ \sigma(\varepsilon_2) = \sigma'(\varepsilon_2) = 0 \end{aligned} \tag{6.20}$$

with $\varepsilon_1 < \varepsilon_0 < \varepsilon_2$, $\sigma_0 = \sigma(\varepsilon_0)$. The solution to the boundary value problem (6.19), (6.20) is constructed as a linear combination of fundamental solutions $\exp(\lambda_i \varepsilon)$

$$\sigma(\varepsilon) = \sum_{i=1}^{4} c_i \exp(\lambda_i \varepsilon), \tag{6.21}$$

where λ_i are the roots of the characteristic equation:

$$-A_{22}\lambda^4 + A_{11}\lambda^2 - 2B_{12}\lambda^3 = 0,$$

corresponding to the Lagrange equation (6.19). We rewrite the characteristic equation in the following form

$$-\bar{A}_{22}\lambda^4 + \lambda^2 - 2\bar{B}_{12}\lambda = 0, \tag{6.22}$$

where

$$\bar{A}_{22} = \frac{A_{22}}{A_{11}}, \quad \bar{B}_{12} = \frac{B_{12}}{A_{11}}, \quad A_{11} \neq 0, \bar{A}_{22}, \bar{B}_{12}$$

are parameters of the deformation curve that determine the possible shape of this curve on the selected segment.

Substituting the solution (6.21) into the homogeneous boundary conditions (6.20) leads to a system of six homogeneous equations for the four unknowns c_i,

$$\begin{cases} \sum_{i=1}^{4} c_i \exp(\lambda_i \varepsilon_1) = 0, \\ \sum_{i=1}^{4} c_i \lambda_i \exp(\lambda_i \varepsilon_1) = 0, \end{cases} \qquad \begin{cases} \sum_{i=1}^{4} c_i \exp(\lambda_i \varepsilon_2) = 0, \\ \sum_{i=1}^{4} c_i \lambda_i(p) \exp(\lambda_i \varepsilon_2) = 0. \end{cases} \tag{6.23}$$

The requirement for a non-trivial solution reduces to the condition that the determinant of the system of linear equations (6.23) with respect to the arbitrary constants c_i equals zero, which obviously leads to an equation, generally transcendental, with respect to the parameters of the deformation curve. Any solution to this equation corresponds to effects of local instability, i.e., the presence of non-uniqueness in the shape of the deformation curve on the considered segment in the vicinity of the considered point $\sigma_0 = \sigma(\varepsilon_0)$. If all roots λ_i of the characteristic equation (6.22) are real, then localization regions are characterized by an exponential change in the vicinity of the investigated point on the deformation curve. On the other hand, if there is at least one pair of complex conjugate roots of the characteristic equation (6.22) in the vicinity of the studied point $\sigma_0 = \sigma(\varepsilon_0)$ localization with oscillation effects occurs, which is sometimes observed in experiments. It can be shown that including a dissipative part in the definition of the variational form leads to the possibility of complex roots appearing.

6.5 Conclusion

A variational approach to the construction of constitutive relations for elastoplastic materials in the strain space is proposed. It is shown that the deformation curve can be considered not only as an empirical approximation of experimental data but also as a solution to the corresponding variational-differential problem. It is established that some known constitutive relations can be interpreted as extremals of variational formulations, and examples are provided. Consequently, the variational approach is proposed as a tool for constructing deformation models, including multi-segment representations with consistent boundary and contact conditions. This enables a transition from phenomenological formula selection to a theoretically organized construction of a family of models of various structures and complexities. A hierarchy of interpolation models is formulated, in which an increase in smoothness is achieved by sequentially incorporating higher-order stress derivatives into the variational formulation. An adaptive algorithm for segment division is proposed, linking the number of spline segments not to the arbitrary choice of the researcher, but to the required accuracy of reproducing the experimental data down to the level of instrumental error. A dissipative generalization of the variational formulation based on non-integrable variational forms is introduced, which allows describing both the nonlinearity and irreversibility of the deformation process without destroying the variational consistency of the equations and coupling conditions.

A variational-spectral interpretation of the yield drop and the Portevin–Le Châtelier (PLC) effect is proposed, treating them as various forms of local loss of stability of a smooth deformation curve in the strain space.

References

[1] Bell JF (1973) Mechanics of Solids. Volume I: The Experimental Foundations of Solid Mechanics. Berlin, Heidelberg: Springer
[2] de Boor CA (1978) Practical Guide to Splines. New York: Springer-Verlag
[3] Douglas DH, Peucker TK (1973) Algorithms for the reduction of the number of points required to represent a digitized line or its caricature. Cartographica (The Canadian Cartographer) 10(2):112-122, DOI 10.3138/FM57-6770-U75U-7
[4] Golovina NY (2021) Stress-strain curve as extremal of some functional. Journal of Materials Engineering and Performance 30(6):4641-4650, DOI 10.1007/s11665-021-05768-3
[5] Kovács Z, Lendvai J, Vörös G (2000) Localized deformation bands in Portevin-Le Châtelier plastic instabilities at a constant stress rate. Materials Science and Engineering A 279(1-2):179-184, DOI 10.1016/S0921-5093(99)00628-0
[6] Kubin LP, Estrin Y (1990) Evolution of dislocation densities and the critical conditions for the Portevin-Le Châtelier effect. Acta Metallurgica et Materialia 38(5):697-708, DOI 10.1016/0956-7151(90)90021-8
[7] Mendelson A (1968) Plasticity: Theory and Application. New York, London: Macmillan
[8] Portevin A, Le Châtelier F (1923) Sur un phénomène observé lors de l'essai de traction d'alliages en cours de transformation. Comptes Rendus de l'Académie des Sciences (Paris) 176:507-510
[9] Ramberg W, Osgood WR (1943) Description of Stress-Strain Curves by Three Parameters. Technical Note. 902. National Advisory Committee for Aeronautics, Washington, DC
[10] Ramer U (1972) An iterative procedure for the polygonal approximation of plane curves. Computer Graphics and Image Processing 1(3):244-256, DOI 10.1016/S0146-664X(72)80017-0
[11] Roux S, Hild F (2020) Optimal procedure for the identification of constitutive parameters from experimentally measured displacement fields. International Journal of Solids and Structures 184:14-23, DOI 10.1016/j.ijsolstr.2018.11.008
[12] Rossi M, Pierron F (2012) On the use of simulated experiments in designing tests for material characterization from full-field measurement. International Journal of Solids and Structures 49(3-4):420-435, DOI 10.1016/j.ijsolstr.2011.09.025
[13] Sedov LI, Eglit ME (1962) Postroyeniye negolonomnykh modeley sploshnykh sred s uchetom konechnosti deformatsiy i nekotorykh fiziko-khimicheskikh effektov (in Russ., Construction of nonholonomic models of continuous me-

dia with allowance for the finite nature of deformations and certain physico-chemical effects). Doklady Akademii Nauk 142(1),54-57
[14] Sedov LI (1965) Mathematical methods for constructing new models of continuous media. Russian Mathematical Surveys 20(5):123-182, DOI 10.1070/RM1965v020n05ABEH001191
[15] Voce E (1948) The relationship between stress and strain for homogeneous deformation. Journal of the Institute of Metals 74:537-562
[16] Yilmaz A (2011) The Portevin-Le Châtelier effect: a review of experimental findings. Science and Technology of Advanced Materials 12(6):063001, DOI 10.1088/1468-6996/12/6/063001

Chapter 7
On the Application of Poincaré-Steklov Operators in Three-Dimensional Problems of Unilateral Discrete Contact of Elastic Bodies

Aleksandr A. Bobylev

Abstract Poincaré-Steklov operators, which transform boundary conditions from one type to another on a part of the domain boundary, are applied in this paper to solve three-dimensional problems of unilateral discrete contact. Using the Poincaré-Steklov operator, a variational formulation of these problems is established in the form of a boundary variational inequality. A minimization problem equivalent to the variational inequality is presented. Approximating this formulation yields a quadratic programming problem with both equality and inequality constraints. To solve the resulting minimization problem, a computational algorithm based on the conjugate gradient method is employed. The indentation process of punches featuring a surface microrelief into a stratified elastic layers has been investigated.

7.1 Introduction

Poincaré–Steklov operators (PSOs) map boundary conditions from one type to another on a part of the domain boundary. These operators are used to solve various classes of boundary value problems in mathematical physics, in particular, in domain separation methods [31], compositions [26], topology optimization, and sensitivity analysis of elastic bodies [28].

The application of PSOs for solving problems of unilateral discrete contact of elastic bodies allows us to derive variational formulations of these problems in the form of boundary variational inequalities and boundary functional minimization problems. The numerical solution of these boundary variational problems requires discretization of only a part of the domain boundary, which is the zone of possible contact. This significantly reduces the dimension of the resulting discrete problems while lowering computational costs.

Aleksandr A. Bobylev
Faculty of Mechanics and Mathematics, Moscow State University, Leninskie gory, 119992 Moscow, Russian Federation, e-mail: `abobylov@gmail.com`

H. Altenbach et al. (eds.), *Stability Problems in Mechanics of Continua*,
Advanced Structured Materials 260, https://doi.org/10.1007/978-3-032-31032-3_7

A computational algorithm developed using the PSO to solve the two-dimensional problem of unilateral discrete contact between a homogeneous elastic half-plane and a rigid punch was presented in [5]. Subsequently, this algorithm was modified and applied to solve the three-dimensional problem of unilateral discrete contact between a homogeneous elastic half-space and a rigid punch, as described in [13].

One of the most effective approaches to constructing PSOs for homogeneous and piecewise-homogeneous finite bodies is the use of boundary integral equations in combination with the boundary element method [21, 22, 32]. The lack of fundamental solutions to equations for inhomogeneous media requires the development of other analytical or numerical methods.

In [8] is presented a novel methodology for constructing PSO applied to a stratified elastic strip bonded to a rigid foundation. This methodology was subsequently generalized to the case of a stratified elastic strip bonded to a homogeneous elastic half-plane [10]. It was also generalized to the case of the three-dimensional problem for a stratified elastic layer bonded to a rigid foundation [14]. It should be noted that an elastic strip or layer bonded to an elastic or non-deformable base are often chosen as a calculation scheme when studying local contact interaction of bodies with coatings. The computational algorithm developed in [8] was used to solve two-dimensional problems of unilateral discrete contact for multilayer [11] and functionally graded [12] elastic strips.

In this paper, we apply the algorithm developed using the PSO to solve three-dimensional problems of unilateral discrete contact for a stratified elastic layer.

7.2 Problem Statement

Let us formulate the three-dimensional unilateral discrete contact problem for a stratified elastic layer. Let an isotropic elastic layer of thickness h occupy the domain $\Omega = \{\vec{x} = (x_1, x_2, x_3) \in R^3 : |x_1| \leq \infty,\ |x_2| \leq \infty,\ 0 \leq x_3 \leq h\}$. Denote the layer boundary $x_3 = 0$ by Γ_0, and the boundary $x_3 = h$ by Γ_1. By $u_k(\vec{x})$ and $\sigma_{kl}(\vec{x})$, $k, l = 1, 2, 3$, we mean the components of the displacement vector and of the stress tensor, respectively, at the point $\vec{x} \in \Omega$. It is assumed that the deformations are small, and the mass forces and stresses in the undeformed state are absent. We consider the Lamé parameters to be functions of the coordinate x_3: $\lambda = \lambda(x_3)$ and $\mu = \mu(x_3)$. The requirements for the smoothness of these functions will be formulated below. It follows from physical considerations that there exist constants $\lambda_0 > 0$ and $\mu_0 > 0$ such that

$$\lambda(x_3) \geq \lambda_0, \quad \mu(x_3) \geq \mu_0, \quad 0 \leq x_3 \leq h. \tag{7.1}$$

The stress-strain state of the layer Ω is described by the system of equations

$$\sigma_{11,1} + \sigma_{12,2} + \sigma_{13,3} = 0, \quad \sigma_{21,1} + \sigma_{22,2} + \sigma_{23,3} = 0, \quad \sigma_{31,1} + \sigma_{32,2} + \sigma_{33,3} = 0,$$
$$\sigma_{11} = \lambda\theta + 2\mu u_{1,1}, \quad \sigma_{22} = \lambda\theta + 2\mu u_{2,2}, \quad \sigma_{33} = \lambda\theta + 2\mu u_{3,3},$$
$$\sigma_{12} = \sigma_{21} = \mu(u_{1,2} + u_{2,1}), \quad \sigma_{13} = \sigma_{31} = \mu(u_{1,3} + u_{3,1}),$$

$$\sigma_{23} = \sigma_{32} = \mu(u_{2,3} + u_{3,2}), \tag{7.2}$$

where $\theta = u_{1,1} + u_{2,2} + u_{3,3}$.

The layer Ω is pressed by a smooth rigid punch, whose base has a surface relief. The part of the boundary Γ_1, along which the contact of the layer with the punch is possible, is denoted by Γ_p. The position and limit dimensions Γ_p are set a priori based on geometric considerations. In problems of unilateral discrete contact, the maximum possible contact area Γ_p includes a set of individual actual contact spots, the positions and dimensions of which are unknown in advance and must be determined in the process of solving the problem.

The macro geometry of the punch base and its surface relief are described by the function $\Phi(x_1, x_2)$. The value of this function at point $\vec{x} \in \Gamma_p$ is equal to the distance from this point to the surface of the punch, measured along the direction of the outer normal to the boundary Γ_p. Distance is measured relative to the undeformed state of the layer Ω. In the case of a punch equipped with a surface relief pattern, the function $\Phi(x_1, x_2)$ is multimodal (multi-extremal). For definiteness, we will assume $\min\limits_{\Gamma_p} \Phi(x_1, x_2) = 0$. Hereafter, the function $\Phi(x_1, x_2)$ will be referred to as the initial gap.

The punch position is determined by the displacement vectors $\vec{\delta} = (\delta_1, \delta_2, \delta_3)$ and rotation angles $\vec{\varphi} = (\varphi_1, \varphi_2, \varphi_3)$. The resultant vector $\vec{F} = (F_1, F_2, F_3)$ and the moment $\vec{M} = (M_1, M_2, M_3)$ of the external forces applied to the punch are considered to be given. The point $\vec{x}^c = \left(x_1^c, x_2^c, x_3^c\right)$ is chosen as the center of reduction. Next, we consider the problem of normal contact between the layer and the punch. Therefore, we will assume

$$\delta_1 = \delta_2 = 0, \quad F_1 = F_2 = 0, \quad M_3 = 0. \tag{7.3}$$

The linearized conditions of unilateral smooth contact describe the interaction between the elastic layer and the rigid punch:

$$u_3 \le \Phi + \delta_3 + (x_2 - x_2^c)\varphi_1 - (x_1 - x_1^c)\varphi_2 \quad \text{on} \quad \Gamma_p, \tag{7.4}$$

$$\sigma_{33} \le 0 \quad \text{on} \quad \Gamma_p, \tag{7.5}$$

$$\sigma_{33}\left[u_3 - \Phi - \delta_3 - (x_2 - x_2^c)\varphi_1 + (x_1 - x_1^c)\varphi_2\right] = 0 \quad \text{on} \quad \Gamma_p, \tag{7.6}$$

$$\sigma_{31} = \sigma_{32} = 0 \quad \text{on} \quad \Gamma_p. \tag{7.7}$$

The rest of the boundary Γ_1 is free from external loads:

$$\sigma_{31} = \sigma_{32} = \sigma_{33} = 0 \quad \text{on} \quad \Gamma_1 \setminus \Gamma_p. \tag{7.8}$$

The equilibrium equations for the rigid punch have the form as follows:

$$\int\limits_{\Gamma_p} \sigma_{33} d\Gamma_p = F_3; \quad \int\limits_{\Gamma_p} \sigma_{33}(x_2 - x_2^c) d\Gamma_p = M_1; \quad \int\limits_{\Gamma_p} \sigma_{33}(x_1 - x_1^c) d\Gamma_p = -M_2. \tag{7.9}$$

It should be noted that relations (7.9) represent essentially nonlocal boundary conditions.

To ensure the existence of a solution for the considered contact problem, we further assume that the external forces and moments acting on the rigid punch satisfy the following conditions:

- the magnitude of the vector component F_3 is bounded and $F_3 < 0$;
- the values of the force F_3 and moments M_1 and M_2 are coordinated with each other in such a way that there exists a distribution of normal stresses $\sigma_{33} \le 0$ on Γ_p that satisfies the equilibrium equations of the punch (7.9).

The layer Ω is bonded to the rigid base along the boundary Γ_0. For ideal contact, the boundary conditions take the form

$$u_1 = u_2 = u_3 = 0 \quad \text{on} \quad \Gamma_0. \tag{7.10}$$

To close the statement of the contact problem, it is necessary to set conditions at infinity. Usually, as such, conditions are used that characterize a certain order of change in displacements and stresses at infinity. As a rule, these conditions are purely mathematical in nature. For a layer bonded to a non-deformable base, the condition of finiteness of the elastic strain energy of the layer is more natural

$$\begin{aligned} \int\limits_{\Omega_1} [\lambda\theta^2 + \mu(2u_{1,1}^2 + 2u_{2,2}^2 + 2u_{3,3}^2 + (u_{1,2} + u_{2,1})^2 + (u_{1,3} \\ + u_{3,1})^2 + (u_{2,3} + u_{3,2})^2)]d\Omega_1 < \infty. \end{aligned} \tag{7.11}$$

In the differential formulation, the problem involves determining the components of the displacement vector $u_k(\vec{x})$ and the stress tensor $\sigma_{kl}(\vec{x})$, which satisfy Eqs. (7.2), unilateral smooth contact conditions (7.4)-(7.7), punch equilibrium equations (7.9), boundary conditions (7.8) and (7.10), and condition (7.11). The displacement δ_3 and rotations angles φ_1, φ_2 of the rigid punch must also be determined.

We emphasize that in the discrete contact problems under consideration, only the maximum permissible contact area Γ_p is specified a priori. The position and shape of the actual contact spots are not known in advance and must be determined.

7.3 Auxiliary Boundary Value Problem

We consider an auxiliary boundary value problem. Its formulation differs from the formulation of the contact problem discussed in Sect. 7.2 only in the boundary conditions on Γ_p. In this case, the unilateral contact conditions (7.4)-(7.6) and the punch equilibrium equations (7.9) are replaced by the boundary condition

$$\sigma_{33} = q \quad \text{on} \quad \Gamma_p. \tag{7.12}$$

7.3.1 Integral Representation of the Solution

If the conditions (7.1) are satisfied, then for any element $q \in L_2(\Gamma_p)$, the auxiliary boundary value problem has a unique solution, which can be represented as follows:

$$\vec{u}(\vec{x}) = \int\limits_{\Gamma_p} \vec{g}(\vec{x},\vec{y})q(\vec{y})d\Gamma_p(\vec{y}), \quad \vec{x} \in \Omega, \tag{7.13}$$

where $\vec{g}(\vec{x},\vec{y})$ is the Green's function that represents the displacement field and is the solution to the boundary value problem of the action of a normal concentrated unit force applied at a point $\vec{y}$ on Γ_p.

7.3.2 System of Inequalities and Equations Related to Contact Stresses

The solution given by (7.13) to the auxiliary boundary value problem satisfies all the conditions of the unilateral discrete contact problem presented in Sect. 7.2, except for the unilateral contact conditions (7.4)-(7.6) and the punch equilibrium equations (7.9). Therefore, the use of the integral representation (7.13) enables the reduction of the solution to determining normal stresses $\sigma_{33} \in L_2(\Gamma_p)$ on Γ_p, which satisfy the following system of inequalities and equations:

$$\sigma_{33} \leq 0 \quad \text{on} \quad \Gamma_p, \tag{7.14}$$

$$u_3(\sigma_{33}) \leq \Phi + \delta_3 + (x_2 - x_2^c)\varphi_1 - (x_1 - x_1^c)\varphi_2 \quad \text{on} \quad \Gamma_p, \tag{7.15}$$

$$\sigma_{33}\left[u_3(\sigma_{33}) - \Phi - \delta_3 - (x_2 - x_2^c)\varphi_1 + (x_1 - x_1^c)\varphi_2\right] = 0 \quad \text{on} \quad \Gamma_p, \tag{7.16}$$

$$\int\limits_{\Gamma_p} \sigma_{33} d\Gamma_p = F_3; \int\limits_{\Gamma_p} \sigma_{33}(x_2 - x_2^c) d\Gamma_p = M_1; \int\limits_{\Gamma_p} \sigma_{33}(x_1 - x_1^c) d\Gamma_p = -M_2, \tag{7.17}$$

where $u_3(\sigma_{33})$ are the normal displacements on Γ_p corresponding to the normal stresses σ_{33}.

7.4 Poincaré-Steklov Operator

The system (7.14)-(7.17) should be solved by means of the constructed the PSO $S : q \mapsto w$. This operator uses solution (7.13) to map the normal stresses

$$q(x_1, x_2) \equiv \sigma_{33}(x_1, x_2, h)$$

on the section Γ_p of the boundary Γ_1 of the elastic layer Ω into the normal displacements $w(x_1, x_2) \equiv u_3(x_1, x_2, h)$ on Γ_p.

The expression corresponding to the solution (7.13) for the PSO $S : q \mapsto w$, where $q \in L_2(\Gamma_p)$, has the form

$$w(x_1, x_2) = \int\limits_{\Gamma_p} g(x_1 - y_1, x_2 - y_2) q(y_1, y_2) dy_1 dy_2. \tag{7.18}$$

The explicit expression for the kernel in the integral representation (7.18) for an arbitrary stratified elastic layer is still unknown. The right-hand side of (7.18) is a convolution-type integral operator. By applying the two-dimensional Fourier integral transform with respect to the spatial coordinates x_1 and x_2, we obtain the algebraic relationship between the Fourier transforms of the normal displacements $\tilde{w}(\alpha_1, \alpha_2)$ and normal stresses $\tilde{q}(\alpha_1, \alpha_2)$:

$$\tilde{w}(\alpha_1, \alpha_2) = \tilde{g}(\alpha_1, \alpha_2) \tilde{q}(\alpha_1, \alpha_2), \tag{7.19}$$

where α_1, α_2 are the parameters of the Fourier transform, and $\tilde{g}(\alpha_1, \alpha_2)$ is the transform of the kernel of the integral representation (7.18). This function $\tilde{g}(\alpha_1, \alpha_2)$ will henceforth be referred to as the transfer function (TF). Thus, the action of the PSO S is reduced to performing direct and inverse Fourier transforms and multiplying the transforms.

7.5 Contact of two Elastic Bodies

In the modelling of contact interaction between two elastic bodies, where the stress-strain state of each body can be determined using Boussinesq's solution for a concentrated normal force applied to the boundary of an elastic half-space, the calculation scheme assumes that one body is non-deformable (a rigid punch) and the other is an elastic half-space with a composite modulus [24]:

$$\frac{1}{E^*} = \frac{1 - \nu_1^2}{E_1} + \frac{1 - \nu_2^2}{E_2},$$

where E_1 and E_1 are Young's moduli, and ν_1 and ν_1 are Poisson's ratios.

The same approach can be applied to model the local contact interaction between two stratified elastic bodies, Ω^1 and Ω^2, for which the PSOs are known and characterised by the TFs $G^1(\alpha)$ and $G^2(\alpha)$, respectively. In this case, the calculation scheme assumes that one body is non-deformable (a rigid punch) and the other is a stratified elastic body. The TF $G(\alpha)$ of the PSO for the latter is computed as:

$$G(\alpha) = G^1(\alpha) + G^2(\alpha).$$

7.6 Transfer Function of the Poincaré-Steklov Operator

In the case of a homogeneous layer, an expression for the TF can be obtained analytically [1]. A numerical-analytical method can be used to construct this function for a functionally graded layer with a special distribution of elastic properties through its thickness, such as power-law or exponential distribution [3]. In the case of a piecewise homogeneous layer, another numerical-analytical method is employed, which is based on solving systems of functional equations [27]. In the case of an arbitrary law of change in elastic properties through the thickness, approximate approaches are used. These approaches are based both on replacing a continuous inhomogeneous layer with a multilayer one (featuring a piecewise constant dependence of elastic moduli on the coordinate [27]) and on direct numerical integration of boundary value problems for systems of differential equations over the transverse coordinate. The use of these approaches is associated with computational problems that are due to the presence of exponential components in the fundamental solutions of the corresponding systems of differential equations. This leads to ill-conditioning in systems of linear algebraic equations that emerge during the implementation of boundary conditions, as well as to the instability of numerical solutions to Cauchy problems and their discrete counterparts. The primary method for overcoming these difficulties involves explicitly extracting exponential components, thereby reducing the problem to identifying functions of bounded variation. This is the basis of the modulating function method [2].

In the paper [8], a novel methodology for calculating the TF of the PSO in the two-dimensional problem for a stratified elastic strip was proposed. This methodology is based on a variational formulation of the boundary value problem for displacement transforms. In the paper [14], the methodology was generalized to the three-dimensional problem for a stratified elastic layer bonded to a rigid foundation. Furthermore, in [15], it is demonstrated that the computational algorithm previously developed for constructing the TF of the PSO in a plane deformation problem for a functionally graded strip can be applied to calculate the TF of the PSO in a three-dimensional boundary value problem for a functionally graded layer.

For the numerical solution of unilateral discrete contact problems, sufficiently fine meshes are required. Thus, in [7], when solving the problem of pressing a rigid punch with a regular microrelief (consisting of 128×128 microprotrusions) into an elastic layer, a mesh of $2^{26} = 67108864$ boundary elements was used. Therefore, the problem of reducing the cost of calculating the TF values is relevant.

In [9], a combined algorithm was developed to calculate the TF of the PSO for a two-dimensional boundary value problem for an elastic strip using an asymptotic series for large values of the Fourier transform parameter. It is demonstrated that for a functionally graded strip, the use of a three-term asymptotic expansion or its Padé approximations results in a significant reduction of computational costs compared to utilizing only the leading term of the expansion. In [15], a three-term asymptotic expansion of TF of the PSO was derived for a three-dimensional boundary value problem for a functionally graded layer.

7.7 Boundary Variational Formulations of the Contact Problem

The system (7.14)-(7.17), which contains equations and inequalities, is solved using a variational approach. Let us introduce the functional spaces necessary for constructing the variational formulations of the problem: the Hilbert triplet $H^{1/2}(\Gamma_p) \subset L_2(\Gamma_p) \subset \tilde{H}^{-1/2}(\Gamma_p)$ [7]. Taking into account the embedding density $L_2(\Gamma_p) \subset \tilde{H}^{-1/2}(\Gamma_p)$, we can extend the PSO S, which is defined by formula (7.18), by continuity to $\tilde{H}^{-1/2}(\Gamma_p)$. Following the approach described in [7], we can consider this operator as acting from $\tilde{H}^{-1/2}(\Gamma_p)$ in $H^{1/2}(\Gamma_p)$.

Let $\langle \cdot, \cdot \rangle$ denote the duality relation between $\tilde{H}^{-1/2}(\Gamma_p)$ and $H^{1/2}(\Gamma_p)$. This relation is generated by the extension by continuity of the scalar product in $L_2(\Gamma_p)$. We introduce continuous bilinear and linear forms in $\tilde{H}^{-1/2}(\Gamma_p)$

$$a(p,q) = \langle p, Sq \rangle \,, \quad b(p) = \langle p, \Phi \rangle \,, \tag{7.20}$$

under the assumption that $\Phi \in H^{1/2}(\Gamma_p)$. Similarly to [7], it can be shown that, if conditions (7.1) and conditions

$$\lambda(x_3) \in L_\infty(0,h), \quad \mu(x_3) \in L_\infty(0,h) \tag{7.21}$$

are fulfilled, then the bilinear form $a(\cdot,\cdot)$ is symmetric and coercive on $\tilde{H}^{-1/2}(\Gamma_p)$.

Let us define the set of statically admissible normal stresses on Γ_p

$$\begin{gathered} \Sigma = \{\, p \in \tilde{H}^{-1/2}(\Gamma_p) : \; q \le 0, \\ \langle q, 1 \rangle = F_3, \quad \langle q, x_2 - x_2^c \rangle = M_1, \quad \langle q, x_1 - x_1^c \rangle = -M_2 \,\} \,. \end{gathered} \tag{7.22}$$

It is easy to show that Σ is a closed convex set in $\tilde{H}^{-1/2}(\Gamma_p)$. In addition, based on the assumptions in the problem statement, this set is non-empty.

Theorem 7.1. *The solution $q \in L_2(\Gamma_p)$ of the system (7.14)-(7.17) satisfies the boundary variational inequality: find an element $q \in \Sigma$ such that*

$$a(p-q,q) - b(p-q) \ge 0 \quad \forall p \in \Sigma. \tag{7.23}$$

Proof. From (7.14) and (7.17) it follows that $q \in \Sigma$. Then for an arbitrary $p \in \Sigma$ we obtain the following estimate:

$$\begin{aligned} a(p-q,q) - b(p-q) &= \langle p-q, Sq - \Phi \rangle \\ &= \langle p-q, Sq - \Phi - \delta_3 - (x_2 - x_2^c)\varphi_1 + (x_1 - x_1^c)\varphi_2 \rangle \\ &+ \langle p-q, \delta_3 + (x_2 - x_2^c)\varphi_1 - (x_1 - x_1^c)\varphi_2 \rangle \\ &= \langle p, Sq - \Phi - \delta_3 - (x_2 - x_2^c)\varphi_1 + (x_1 - x_1^c)\varphi_2 \rangle \\ &- \langle q, Sq - \Phi - \delta_3 - (x_2 - x_2^c)\varphi_1 + (x_1 - x_1^c)\varphi_2 \rangle \\ &+ \delta_3 \langle p, 1 \rangle + \varphi_1 \langle p, x_2 - x_2^c \rangle - \varphi_2 \langle p, x_1 - x_1^c \rangle \\ &- \delta_3 \langle q, 1 \rangle - \varphi_1 \langle q, x_2 - x_2^c \rangle + \varphi_2 \langle q, x_1 - x_1^c \rangle \\ &= \langle p, Sq - \Phi - \delta_3 - (x_2 - x_2^c)\varphi_1 + (x_1 - x_1^c)\varphi_2 \rangle \ge 0. \end{aligned} \tag{7.24}$$

□

Note that variational inequality (7.23) does not involve the unknown displacement δ_3 or rotation angles φ_1 and φ_2 of the rigid punch. As can be readily seen from transformations (7.24), this result follows from the fact that the elements of the set of statically admissible normal stresses Σ satisfy the punch equilibrium equations.

Application of the well-known methods [25] allows us to prove the converse statement.

Theorem 7.2. *The solution q to the variational inequality (7.23) satisfies the system of equations and inequalities given by (7.14)-(7.17) at least in the generalized sense. Specifically, conditions (7.14) and (7.16) are fulfilled in the sense of the space $\tilde{H}^{-1/2}(\Gamma_p)$, condition (7.15) is met in the sense of the space $H^{1/2}(\Gamma_p)$, and conditions (7.17) are satisfied after their continuous extension to $\tilde{H}^{-1/2}(\Gamma_p)$:*

$$\langle q, 1\rangle = F_3, \quad \langle q, x_2 - x_2^c\rangle = M_1, \quad \langle q, x_1 - x_1^c\rangle = -M_2.$$

Definition 7.1. The solution q of the variational inequality (7.23) is referred to as the generalized solution to the system of equations and inequalities (7.14)-(7.17).

The next two statements follow from the fact that bilinear form $a(\cdot,\cdot)$ is continuous and positive definite, linear form $b(\cdot)$ is continuous, and the set Σ is a nonempty, closed, convex set in $\tilde{H}^{-1/2}(\Gamma_p)$ [20, 25].

Theorem 7.3. *The variational inequality (7.23) is equivalent to the boundary functional minimization problem: find an element $q \in \Sigma$ such that*

$$J(q) = \inf_{p\in\Sigma} \{J(p) = a(p,p)/2 - b(p)\}. \tag{7.25}$$

Theorem 7.4. *There exists a unique solution to the variational inequality (7.23) and to the minimization problem (7.25).*

7.8 Approximation of the Problem

The selection of an approximation method for solving the minimization problem (7.25) is primarily determined by the possibility of constructing a computationally effective approximation of the PSO S, defined by (7.18). From (7.19) it follows that the operation of the PSO consists of performing a pair of Fourier transforms (direct and inverse) and multiplying transforms. In numerical analysis, the Fourier transform is most effective for periodic functions because of the discreteness of their spectrum and the resulting possibility to transition from continuous to discrete transform. Considering that the TF of the PSO is constructed numerically for a stratified elastic layer, the main idea of the approach used is to approximate the desired normal stresses by periodic grid functions and utilize the two-dimensional fast Fourier transform (2-D FFT) algorithm. To reduce the resulting periodicity error, an extended computational domain Γ_c is introduced, such that $\Gamma_p \subset \Gamma_c \subset \Gamma_1$ [34, 35].

To apply the 2-D FFT, we select a rectangular computational domain Γ_c with dimensions of $b_1 \times b_2$. The sides of this domain are aligned with the coordinate axes Ox_1 and Ox_2.

Furthermore, we introduce the space $H^{1/2}(\Gamma_c)$ and its dual space $\tilde{H}^{-1/2}(\Gamma_c)$. According to the results of [22], it can be demonstrated that the restriction of a function from $H^{1/2}(\Gamma_c)$ to Γ_p belongs to the space $H^{1/2}(\Gamma_p)$, and the trivial extension q_0 of a function $q \in \tilde{H}^{-1/2}(\Gamma_p)$ to Γ_c belongs to the space $\tilde{H}^{-1/2}(\Gamma_c)$. We denote by $\hat{H}^{-1/2}(\Gamma_c)$ the subspace of functions from $\tilde{H}^{-1/2}(\Gamma_c)$ that are supported on $\bar{\Gamma}_p$. Similarly to the approach in [33], we periodically extend functions from $\hat{H}^{-1/2}(\Gamma_c)$ to the entire boundary Γ_1 with period Γ_c, and form a space $\hat{H}^{-1/2}_{per}(\Gamma_1)$ of Γ_c-periodic functions. Hereafter, we will identify functions from $\hat{H}^{-1/2}(\Gamma_c)$ with their periodic extensions to Γ_1. Thus, we define an operator $\Pi : \tilde{H}^{-1/2}(\Gamma_p) \to \hat{H}^{-1/2}_{per}(\Gamma_1)$, which maps a function $q \in \tilde{H}^{-1/2}(\Gamma_p)$ to a periodic extension $q_{per} \in \hat{H}^{-1/2}_{per}(\Gamma_1)$ on Γ_1. This extension is obtained from the trivial extension $q_0 \in \tilde{H}^{-1/2}(\Gamma_c)$ of a function $q \in \tilde{H}^{-1/2}(\Gamma_p)$ to Γ_c.

We denote by $\hat{L}_2(\Gamma_c)$ the subspace of functions in $L_2(\Gamma_c)$, which are supported on $\bar{\Gamma}_p$. Based on the dense embedding $L_2(\Gamma_p) \subset \tilde{H}^{-1/2}(\Gamma_p)$, one can prove the existence of a dense embedding of the space $\hat{L}_2(\Gamma_c)$ into $\tilde{H}^{-1/2}(\Gamma_c)$. Functions in the space $\hat{L}_2(\Gamma_c)$ can be considered as Γ_c-periodic. Consequently, the system of functions

$$\exp\left[2\pi \mathrm{i}(nx_1/b_1 + mx_2/b_2)\right], \quad n, m = 0, \pm 1, \pm 2, \ldots,$$

forms a dense set in $\hat{L}_2(\Gamma_c)$ [33]. Therefore the function $q \in \hat{H}^{-1/2}(\Gamma_c)$ can be approximated to a given accuracy by a finite segment of the Fourier series

$$q_f = \sum_{|n| \le L_1} \sum_{|m| \le L_2} c_{nm} \exp\left[2\pi \mathrm{i}(nx_1/b_1 + mx_2/b_2)\right]. \tag{7.26}$$

To calculate the Fourier coefficients $c_{nm} \in Z$ via 2-D FFT, grid functions are introduced. Construct on Γ_c a regular (uniform) grid Υ_c, consisting of $N_c = N_{c1} \times N_{c2}$ one-node, zero-order rectangular boundary elements of size $h_{e1} \times h_{e2}$, where $h_{e1} = b_1/N_{c1}$ and $h_{e2} = b_2/N_{c2}$. The nodes of Υ_c are denoted by s_{nm}, where $n = \overline{1, N_{c1}}$ and $m = \overline{1, N_{c2}}$. The portion of Υ_c located on Γ_p and containing N elements is denoted as Υ_p. To simplify the calculations, we will assume that N_{c1} and N_{c2} are even.

Let us introduce the grid functions (matrices) of normal stresses $\mathbf{Q}_c$ and normal displacements $\mathbf{W}_c$, defined on Υ_c:

$$\mathbf{Q}_c = [q_{nm}], \quad \mathbf{W}_c = [w_{nm}], \quad n = \overline{1, N_{c1}}, \quad m = \overline{1, N_{c2}}$$

together with their corresponding vectors:

$$\mathbf{Q}_{cv} = \mathbf{vec}(\mathbf{Q}_c), \quad \mathbf{W}_{cv} = \mathbf{vec}(\mathbf{W}_c), \tag{7.27}$$

where $\mathbf{vec}(\cdot)$ is the matrix vectorization operator that transforms a matrix into a vector by arranging its columns one under the other. We also introduce reductions of grid functions on Υ_p:

$$\mathbf{Q}_{pv} = \mathbf{R}_{pc}\mathbf{Q}_{cv}, \quad \mathbf{W}_{pv} = \mathbf{R}_{pc}\mathbf{W}_{cv}, \tag{7.28}$$

where $\mathbf{R}_{pc}$ is a rectangular matrix of dimensions $N \times N_c$, each row of which has exactly one non-zero element equal to 1.

The inverse operations of trivial continuation (zero padding) of $\mathbf{Q}_{pv}$ and $\mathbf{W}_{pv}$ on Υ_c can be performed using the transposed matrix $\mathbf{R}_{pc}^T$:

$$\mathbf{Q}_{cv} = \mathbf{R}_{pc}^T\mathbf{Q}_{pv}, \quad \mathbf{W}_{cv} = \mathbf{R}_{pc}^T\mathbf{W}_{pv}. \tag{7.29}$$

It is known [17, 23] that the spectrum of a grid function is periodic and consists of an infinite series of shifted copies of the spectrum of the function being approximated. The distance in the frequency domain between adjacent copies of the spectrum equals the sampling frequency $\omega_0 = 2\pi/\tau$, where τ is the grid spacing. To eliminate the effect of frequency aliasing and to recover the periodic function from its grid approximation, we assume that the number of harmonics in (7.26) and the number of nodes of the grid Υ_c satisfy the conditions:

$$L_i < N_{ci}/2, \quad i = 1, 2, \tag{7.30}$$

which follow from the Kotelnikov-Shannon theorem [17, 23].

Let us introduce Fourier matrices $\mathbf{F}_1$ and $\mathbf{F}_2$ of orders N_{c1} and N_{c2}, respectively:

$$\mathbf{F}_1 = (N_{c1})^{1/2}[\omega_1^{(n-1)(m-1)}], \quad \mathbf{F}_2 = (N_{c2})^{1/2}[\omega_2^{(n-1)(m-1)}], \tag{7.31}$$

where

$$\omega_1 = \exp(-2\pi \mathrm{i}/N_{c1}), \quad \omega_2 = \exp(-2\pi \mathrm{i}/N_{c2}). \tag{7.32}$$

The Fourier matrices are invertible, and the inverse matrices are of the form:

$$\mathbf{F}_1^{-1} = \mathbf{F}_1^*, \quad \mathbf{F}_2^{-1} = \mathbf{F}_2^*,$$

where $\mathbf{F}_1^*$ and $\mathbf{F}_2^*$ are Hermitian conjugate matrices.

Using the matrices $\mathbf{F}_1$ and $\mathbf{F}_2$, we calculate the Fourier images $\tilde{\mathbf{Q}}_c$ and $\tilde{\mathbf{W}}_c$ of the grid functions of normal stresses and displacements:

$$\tilde{\mathbf{Q}}_c = \mathbf{F}_1\mathbf{Q}_c\mathbf{F}_2, \quad \tilde{\mathbf{W}}_c = \mathbf{F}_1\mathbf{W}_c\mathbf{F}_2. \tag{7.33}$$

For the vectors corresponding to the matrices $\mathbf{Q}_c$ and $\mathbf{W}_c$, formulas (7.33) take the form:

$$\tilde{\mathbf{Q}}_{cv} = (\mathbf{F}_2 \otimes \mathbf{F}_1)\mathbf{Q}_{cv}, \quad \tilde{\mathbf{W}}_{cv} = (\mathbf{F}_2 \otimes \mathbf{F}_1)\mathbf{W}_{cv}, \tag{7.34}$$

where the symbol $\otimes$ denotes the Kronecker product of the matrices $\mathbf{F}_2$ and $\mathbf{F}_1$.

The grid functions $\mathbf{Q}_c$ and $\mathbf{W}_c$ are real, so it follows from the properties of the 2-D FFT [17, 23] that the elements of the matrices $\tilde{\mathbf{Q}}_c$ and $\tilde{\mathbf{W}}_c$ satisfy the following conditions:

$$\tilde{q}_{nm} = \tilde{q}_{ij}^*, \quad \tilde{w}_{nm} = \tilde{w}_{ij}^*, \tag{7.35}$$

where $n = \overline{2, N_{c1}}$, $m = \overline{2, N_{c2}}$, $i = N_{1c} - n + 2$ and $j = N_{2c} - m + 2$. For grid functions, relation (7.19) takes the form:

$$\tilde{\mathbf{W}}_c = \tilde{\mathbf{G}} \circ \tilde{\mathbf{Q}}_c, \tag{7.36}$$

where $\tilde{\mathbf{G}}$ is the grid TF, defined on Υ_c, and the symbol " $\circ$ " denotes the Hadamard product of the matrices. The elements of the matrix $\tilde{\mathbf{G}} = [\tilde{g}_{nm}]$ for the indices from the ranges $1 \le n \le N_{c1}/2+1$ and $1 \le m \le N_{c2}/2+1$ are calculated by the formula:

$$\tilde{g}_{nm} = \tilde{g}((n-1)\omega_1, (m-1)\omega_2), \tag{7.37}$$

where ω_1 and ω_2 are defined by (7.32).

Taking into account (7.35), the remaining elements of the matrix $\tilde{\mathbf{G}}$ are determined using the equalities:

$$\tilde{g}_{nm} = \tilde{g}_{ij}, \tag{7.38}$$

where $n = \overline{N_{c1}/2+2, N_{c1}}$, $m = \overline{N_{c2}/2+2, N_{c2}}$, $i = N_{1c} - n + 2$ and $j = N_{2c} - m + 2$. From (7.27)-(7.29), (7.33), (7.34), and (7.36) it follows that the grid approximation $\mathbf{S}_p : \mathbf{Q}_{pv} \mapsto \mathbf{W}_{pv}$ of the PSO S on Υ_p has the form:

$$\mathbf{S}_p = \mathbf{R}_{pc}(\mathbf{F}_2^* \otimes \mathbf{F}_1^*)\tilde{\mathbf{G}}_v \circ (\mathbf{F}_2 \otimes \mathbf{F}_1)\mathbf{R}_{pc}^T, \tag{7.39}$$

where $\tilde{\mathbf{G}}_v = \mathbf{vec}(\tilde{\mathbf{G}})$.

Considering that all elements of the matrix $\tilde{\mathbf{G}}$ are positive [14], it can be shown that $\mathbf{S}_p$ is a symmetric positive-definite square matrix of order N. In the numerical implementation, it is not required to form the matrix $\mathbf{S}_p$ in an explicit form, since it is enough to programmatically calculate the products of each of the matrices on the right side of (7.39) by vectors. Given that the matrices $\mathbf{R}_{pc}$ and $\mathbf{R}_{pc}^T$ contain only N non-zero elements, which are equal to 1, the calculation of the products of these matrices and vectors is reduced to performing N assignment operations. Computing the Hadamard product of two vectors in (7.39) requires N_c multiplication operations. To calculate the products of matrices $\mathbf{F}_2 \otimes \mathbf{F}_1$ and $\mathbf{F}_2^* \otimes \mathbf{F}_1^*$ by vectors, it is advisable to use 2-D FFT algorithms. Therefore, hereafter we will assume $N_{c1} = 2^n$, $N_{c2} = 2^m$, where n and m are natural numbers. Under these conditions, the total number of operations to perform one transformation is of the order of $O(N_c \log_2 N_c)$ [17, 23].

Further, to approximate problem (7.25), we apply the boundary-element approach [21, 22, 32]. The computation of the bilinear form $a(\cdot,\cdot)$ and linear $b(\cdot)$ form, defined by formulas (7.20), is performed using the cubature formula, the nodes of which coincide with the nodes of the grid Υ_p. This yields the approximating bilinear form $a_h(\cdot,\cdot)$ and linear $b_h(\cdot)$ form within the space R^N:

$$a_h(\mathbf{P}_{pv}, \mathbf{Q}_{pv}) = \mathbf{P}_{pv}^T \mathbf{S}_p \mathbf{Q}_{pv} h_{e1} h_{e2}, \quad b_h(\mathbf{P}_{pv}) = \mathbf{\Phi}_v^T \mathbf{P}_{pv} h_{e1} h_{e2},$$

where $\mathbf{\Phi}_v = \mathbf{vec}(\mathbf{\Phi})$ and $\mathbf{\Phi}$ represents the grid approximation on Υ_p of the function $\Phi(x_1, x_2)$ that describes the shape of the punch base.

A combined approach is employed when approximating the set of statically admissible normal stresses Σ, which is defined by formula (7.22). To approximate constraints in the form of inequalities, the collocation method is employed, while the above cubature formula is applied to handle constraints in the form of equalities.

The result is a closed convex set of statically admissible nodal normal stresses

$$\Sigma_h = \{\, \mathbf{Q} \in R^N : \ q_i \le 0, \ i = \overline{1, N};$$
$$\sum_{i=1}^{n} c_{3i} q_i = F_3; \ \sum_{i=1}^{n} c_{1i} q_i = M_1; \ \sum_{i=1}^{n} c_{2i} q_i = -M_2 \,\} .$$

The coefficients c_{1i}, c_{2i}, and c_{3i} are calculated using the following formulas:

$$c_{1i} = h_{e1} h_{e2} (x_2^i - x_2^c), \quad c_{2i} = h_{e1} h_{e2} (x_1^i - x_1^c), \quad c_{3i} = h_{e1} h_{e2}, \quad i = \overline{1, N},$$

where x_1^i and x_2^i are the coordinates of the i-th node of grid Υ_p.

For the minimization problem (7.25), the resulting grid approximation takes the form of a quadratic programming problem, where the vector of normal stresses $\mathbf{Q}_{pv} \in R^N$ satisfies

$$J_h(\mathbf{Q}_{pv}) = \inf_{\mathbf{Q} \in \Sigma_h} \left\{ J_h(\mathbf{Q}) = \mathbf{Q}^T \mathbf{S}_p \mathbf{Q}/2 - \mathbf{\Phi}_v^T \mathbf{Q} \right\} . \tag{7.40}$$

For simplicity, the multiplier $h_{e1} h_{e2}$ is intentionally omitted in the expression for $J_h(\cdot)$. This omission does not alter the solution of the problem (7.40).

It should be noted that the dimension of the problem (7.40) equals N, i.e., the number of grid nodes Υ_p on Γ_p. Furthermore, this dimension is independent of the size of the computational domain Γ_c.

7.9 Solving the Quadratic Programming Problem

To solve the quadratic programming problem (7.40), algorithms based on the conjugate gradient method (CGM) are currently considered the most efficient. The first algorithm for solving quadratic programming problems, based on the CGM, was proposed in [30]. The core concept of the algorithm involves maintaining a working list of active inequality constraints (active set) and using the CGM to determine the minimum within the current working subspace (an affine subspace). It was proved that the algorithm converges theoretically in a finite number of steps. However, practical application of this algorithm has demonstrated that the rate of convergence significantly depends on the number and type of constraints. Therefore, various modifications of the basic algorithm were subsequently proposed, which were adapted for solving specific classes of applied problems. The description and analysis of the CGM-based algorithms for solving unilateral contact problems can be found, for example, in [18, 19].

A prominent algorithm described in the frequently cited paper [29] differs from the basic algorithm [30] by accounting for the specific features of the constraint set in the quadratic programming problem. When the boundary variational formulation of the contact problem is discretised, the number of constraints in the form of inequalities (conditions for non-positivity of normal stresses in unilateral contact) is equal to the number of unknowns. Therefore, in the algorithm [29], when inactive constraints are violated, the method of projecting a point onto an admissible set is employed. This allows several constraints to be included in the working list simultaneously, which significantly reduces the number of restarts compared to the algorithm [30]. To reduce computational costs, it is also allowed, unlike in the algorithm [30], to exclude constraints from the working list until a minimum is reached within the current working subspace.

A key limitation of algorithm [29] is that it incorporates only the force equilibrium equation for a punch, neglecting rotational effects and thus restricting the analysis to translational displacements. Consequently, it cannot be used for contact problems under eccentric loading, where punch rotations must also be considered. Furthermore, at each iterative step, the algorithm performs the matrix-vector multiplication (Hessian of the objective function) twice, whereas [30] performs it only once. When the working subspace dimension changes, both [29] and [30] restart from the antigradient direction, significantly slowing convergence.

In [5], the author proposed a different algorithm based on the CGM for solving unilateral contact problems for an elastic half-plane. The main idea of this algorithm is to use a linear transformation of variables, which reduces the system of equality constraints – governing punch equilibrium and comprising all problem variables – to a set of single-variable constraints. Such simplified constraints can be efficiently incorporated into the CGM algorithm. This technique allows us to consider all the equations governing the punch equilibrium and specify the moments of external forces. As a restart direction, the developed algorithm uses a modification of the previous search direction to conform to the new working subspace. After a restart, to maintain conjugacy, the search directions computed using the three-term formula proposed in [4] for the algorithm with restarts for unconstrained minimization of nonquadratic functions. In [7], the algorithm was generalized for three-dimensional contact problems.

In [13], a new CGM algorithm for solving three-dimensional unilateral contact problems is proposed. Computational experiments have shown that this algorithm is less computationally costly compared to those described in [5] and [29].

7.10 Numerical Results

The developed algorithm, implemented in Fortran with the NVIDIA HPC SDK, used the cuFFT library for 2-D FFT computations. Numerical calculations were performed on a PC equipped with an Intel Core i7-7820X processor and a GeForce RTX 2080 Ti GPU. All computations were executed with double-precision accuracy.

A series of numerical solutions has been obtained for three-dimensional problems of unilateral contact. The following numerical results are presented for a stratified layer comprising $L = 10$ identical functionally graded sublayers. The Young's modulus E and Poisson's ratio ν of these sublayers vary according to one of the following laws:

$$
\begin{aligned}
&1)\ E(\xi_3) = E_1 + (E_2 - E_1)\xi_3^2, && \nu(\xi_3) = \nu_1 + (\nu_2 - \nu_1)\xi_3;\\
&2)\ E(\xi_3) = E_1 + (E_2 - E_1)\xi_3^4, && \nu(\xi_3) = \nu_1 + (\nu_2 - \nu_1)\xi_3^2;\\
&3)\ E(\xi_3) = E_2 + (E_1 - E_2)\xi_3^2, && \nu(\xi_3) = \nu_2 + (\nu_1 - \nu_2)\xi_3;\\
&4)\ E(\xi_3) = E_2 + (E_1 - E_2)\xi_3^4, && \nu(\xi_3) = \nu_2 + (\nu_1 - \nu_2)\xi_3^2,
\end{aligned}
$$

where $E_1 = E^*$, $E_1 = 4E^*$, E^* is the scale factor (the dimensional variable, MPa), $\nu_1 = 0.4$, $\nu_2 = 0.1$, $\xi_3 = \{Lx_3/h\}$ is the 'fast' coordinate, $\{\cdot\}$ denotes the fractional part of the number. Graphs of the above functions $E(\xi_3)$ and $\nu(\xi_3)$ are presented in Figs. 7.1 and 7.2, respectively. Curves 1–4 correspond to the respective stratified layers as shown in the legend.

The calculations were performed for punches having a rectangular footprint. The maximum possible contact area Γ_p was specified as:

$$
\Gamma_p = \{0 \le x_1 \le d_1,\ \ 0 \le x_2 \le d_2,\ \ x_3 = 0\}. \tag{7.41}
$$

Therefore, the condition $\Gamma_p \subset \Gamma_c$ is satisfied if

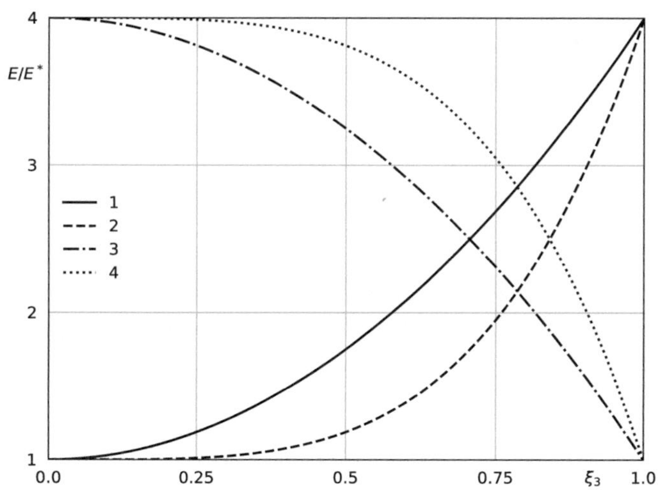

Fig. 7.1: Graphs of the function $E(\xi_3)$.

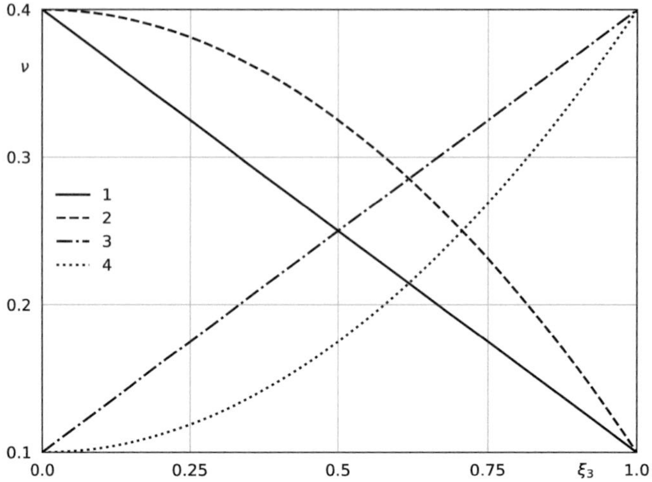

Fig. 7.2: Graphs of the function $\nu(\xi_3)$.

$$\frac{b_1}{d_1} = \chi_1 > 1, \quad \frac{b_2}{d_2} = \chi_2 > 1,$$

where χ_1 and χ_2 are the expansion coefficients of the computational domain.

To verify the algorithm and its implementation, we employed a test problem involving a punch whose base shape is described by a quadratic function:

$$\Phi^*(x_1, x_2) = \alpha_1 d_1 (2\frac{x_1}{d_1} - 1)^2 + \alpha_2 d_2 (2\frac{x_2}{d_2} - 1)^2.$$

The components of the resultant vector and the moment of external forces applied to the punch were specified as:

$$F_3 = -f d_1 d_2 E^*, \quad M_1 = e_2 F_3 d_2, \quad M_2 = -e_1 F_3 d_1, \tag{7.42}$$

where $f > 0$ is a dimensionless parameter and e_1, e_2 are dimensionless parameters determining the eccentricity of the resultant external load relative to the initial contact point $\vec{x}^c = (d_1/2, d_2/2, 0)$.

Assuming no eccentricity ($e_1 = e_2 = 0$), let the contact pressure distribution be described by the function $p^*(x_1, x_2)$. It can be shown analytically that if the eccentricity of the punch load is $(e_1 d_1, e_2 d_2)$ with respect to the point $\vec{x}^c$, then the distribution of the contact pressure has the form $p^*(x_1 - e_1 d_1, x_2 - e_2 d_2)$. The punch displacement δ_3 decreases by $4\alpha_1 d_1 e_1^2 + 4\alpha_2 d_2 e_2^2$, while the punch rotates through angles $\varphi_1 = -8\alpha_2 e_2$ and $\varphi_2 = 8\alpha_1 e_1$. Numerical results support these conclusions.

An analysis of discrete contact problem solutions is presented below. In these cases, the actual contact area is not continuous but consists of a collection of individual contact spots. Problems are considered for smooth rectangular punches with a regular surface microrelief consisting of $K_1 \times K_2$ microprotrusions. The possible contact area Γ_p has the shape given by (7.41), and the shape of the punch base is specified by the function

$$\Phi(x_1, x_2) = \Phi_0(x_1, x_2) + \Psi_1(\xi_1)/K_1 + \Psi_2(\xi_2)/K_2, \tag{7.43}$$

where $\Phi_0(x_1, x_2)$ is the convex function that determines the punch shape, $\Psi_1(\xi_1)$ and $\Psi_2(\xi_2)$ are convex functions describing the shape of the microprotrusions, $\xi_1 = \{K_1 x_1/d_1\}$ and $\xi_2 = \{K_2 x_2/d_2\}$ are fast coordinates. Here, $\{\cdot\}$ denotes the fractional part of a number.

During the computations, the required number of boundary elements was determined by comparing solutions obtained on nested grids subjected to double sequential refinement. Numerical solutions for punches featuring a regular surface microrelief were obtained using grids where each microprotrusion was discretized with 128×128 boundary elements. The expansion coefficients of the computational domain were equal to $\chi_1 = \chi_2 = 4$. The maximum number of grid elements for the punch with 32×32 microprotrusions was $2^{28} = 268435456$ boundary elements.

A numerical simulation was performed to investigate the indentation process of punches featuring a regular surface microrelief into a stratified elastic layer. The key characteristics describing the contact interaction between a punch and an elastic layer are defined as functions of the dimensionless load parameter f. These include the following:

- the relative actual contact area given by

$$s \equiv \frac{1}{\operatorname{meas}\Gamma_p} \int\limits_{\Gamma_p} [\sigma_{33} < 0]_a \, d\Gamma_p = B^0(f),$$

- the rotation angles $\varphi_1 = B^1(f)$ and $\varphi_2 = B^2(f)$,
- the punch displacement $\delta_3 = B^3(f)$.

Here $[\cdot]_a$ is the Iverson bracket (a function equal to 1 for a true argument and 0 otherwise).

We consider the process of indentation into an elastic layer of a punch for the external load parameter f changing in the interval $[0, f^*]$. To construct the functions $B^i(f), i = 0, 1, 2, 3$, the interval $[0, f^*]$ is partitioned into L segments such that their lengths form a geometric progression starting from the left end of the interval, with the ratio of the length of the largest segment to the length of the smallest one equal to l. As a result, we obtain the grid

$$T_f = \{f_0 = 0, f_1, \ldots, f_L = f^*\}.$$

Next, we define the grid functions

$$\mathbf{B}^i = R_f B^i, \quad i = 0, 1, 2, 3,$$

where R_f is the restriction operator on the grid T_f. An incremental loading scheme, aligned with the nodes of T_f, is used to compute $\mathbf{B}^i$. For the numerical implementation, we set the grid parameters as $L = 128$ and $l = 16$.

The dependencies $s = B^0(f)$, $\varphi_1 = B^1(f)$, $\varphi_2 = B^2(f)$, and $\delta_3 = B^3(f)$ are illustrated in Figs. 7.3–7.6 for a punch featuring 32×32 microprotrusions. The computations were performed for punches with a square base ($d_1 = d_2 = d$). The surface geometry is described by the following functions:

$$\Phi_0(x_1, x_2) = 10^{-5}d(2\frac{x_1}{d} - 1)^4 + 1.25 \times 10^{-5}d(2\frac{x_2}{d} - 1)^4;$$
$$\Psi_1(\xi_1) = 10^{-4}d(2\xi_1 - 1)^4, \quad \Psi_2(\xi_2) = 10^{-4}d(2\xi_2 - 1)^4.$$

The components of the vector and the moment of external forces applied to the punches are specified by formula (7.42). The punches are loaded with eccentricity, characterised by the parameters $e_1 = 0.05$ and $e_2 = -0.05$. The thickness of the elastic layer is equal to $h = d$.

7.11 Conclusions

The three-dimensional contact problems for a stratified elastic layer bonded to a non-deformable (rigid) foundation have been considered above. The presented results can be generalised to the case of a stratified elastic layer bonded to an elastic half-space

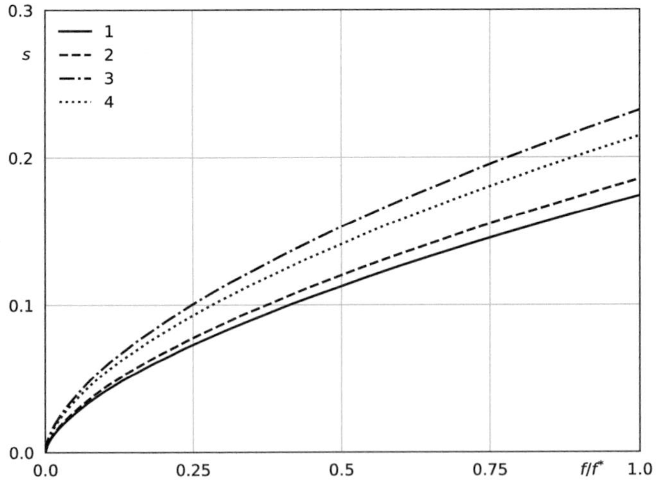

Fig. 7.3: Graphs of the dependence $s = B^0(f)$.

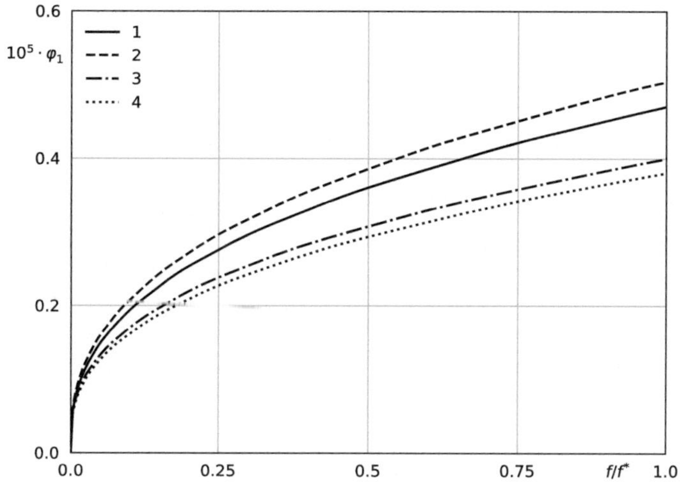

Fig. 7.4: Graphs of the dependence $\varphi_1 = B^1(f)$.

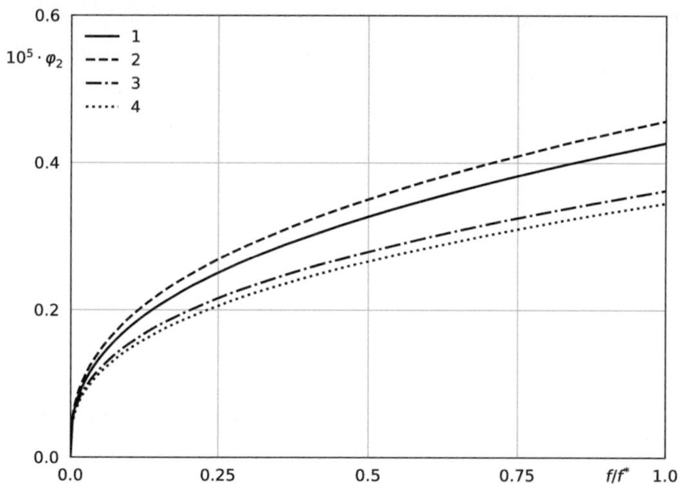

Fig. 7.5: Graphs of the dependence $\varphi_2 = B^2(f)$.

– that is, to the problem of local contact interaction between elastic bodies in the presence of stratified coatings.

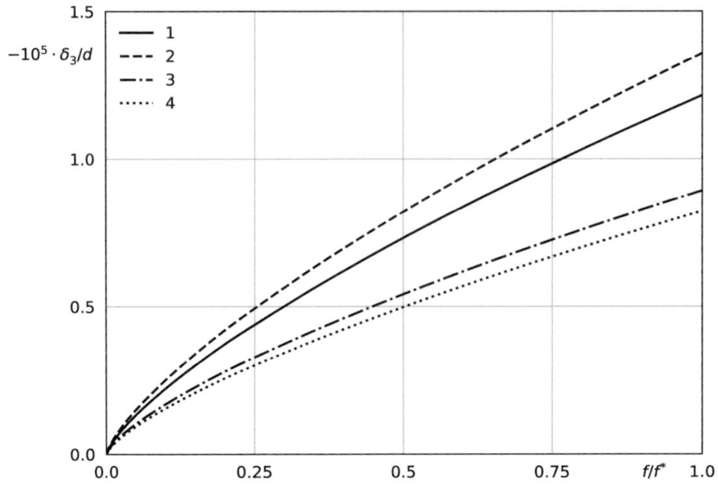

Fig. 7.6: Graphs of the dependence $\delta_3 = B^3(f)$.

Acknowledgements The study was conducted under the state assignment of Lomonosov Moscow State University.

References

[1] Alexandrov VM, Pozharskii DA (2001) Three-Dimensional Contact Problems. Dordrecht: Springer, DOI 10.1007/978-94-010-9893-9
[2] Babeshko VA, Glushkov EV, Glushkova NV (1987) Methods of constructing Green's matrix of a stratified elastic half-space. USSR Computational Mathematics and Mathematical Physics 27(1):60–65, DOI 10.1016/0041-5553(87)90119-4
[3] Barber JR (2018) Contact Mechanics. Cham: Springer, DOI 10.1007/978-3-319-70939-0
[4] Beale EML (1972) A derivative of conjugate gradients. In: Lootsma FA (ed) Numerical Methods for Nonlinear Optimization, pp 39–43. London: Academic Press
[5] Bobylev AA (2022) Application of the conjugate gradient method to solving discrete contact problems for an elastic half-plane. Mechanics of Solids 57:317–332, DOI 10.3103/S0025654422020029
[6] Bobylev AA (2022) Algorithm for solving discrete contact problems for an elastic strip. Mechanics of Solids 57:1766–1780, DOI 10.3103/S0025654422070068

[7] Bobylev AA (2023) Algorithm for solving discrete contact problems for an elastic layer. Mechanics of Solids 58:439–454, DOI 10.3103/S0025654422100296
[8] Bobylev AA (2023) Numerical construction of the transform of the kernel of the integral representation of the Poincaré-Steklov operator for an elastic strip. Differential Equations 59:119–134, DOI 10.1134/S0012266123010093
[9] Bobylev AA (2023) Computing a transfer function of the Poincaré-Steklov operator for a functionally graded elastic strip. Moscow University Mechanics Bulletin 78:134–142, DOI 10.3103/S0027133023050023
[10] Bobylev AA (2024) Construction of the transfer function of the Poincaré-Steklov operator for a coated elastic half-plane. Differential Equations 60:1232–1248, DOI 10.1134/S001226612409009X
[11] Bobylev AA (2024) Algorithm for solving unilateral discrete contact problems for a multilayer elastic strip. Journal of Applied Mechanics and Technical Physics 65:382–392, DOI 10.1134/S0021894424020202
[12] Bobylev AA (2024) The unilateral discrete contact problem for a functionally graded elastic strip. Moscow University Mechanics Bulletin 79:56–68, DOI 10.3103/S0027133024700080
[13] Bobylev AA (2024) Application of the conjugate gradient method for solving unilateral discrete contact problems for an elastic half-space. Computational Mathematics and Mathematical Physics 64:2680–2695, DOI 10.1134/S0965542524701525
[14] Bobylev AA (2026) Construction of the transfer function of the Poincaré-Steklov operator for an elastic layer. Differential Equations 62:115–130, DOI 10.7868/S3034503026010081
[15] Bobylev AA (2026) Calculation of a transfer function of the Poincaré-Steklov operator for a functionally graded elastic layer. Moscow University Mechanics Bulletin 81:77–81, DOI 10.55959/MSU0579-9368-1-67-1-14
[16] Bobylev AA (2026) Asymptotics of the transfer function of the Poincaré-Steklov operator for a functionally graded elastic layer. Russian Journal of Mathematical Physics 33:34–41, DOI 10.1134/S1061920825601193
[17] Brigham EO (1988) The Fast Fourier Transform and Its Applications. Englewood Cliffs: Prentice Hall
[18] Dostál Z (2009) Optimal Quadratic Programming Algorithms. With Applications to Variational Inequalities. New York: Springer, DOI 10.1007/b138610
[19] Dostál Z, Kozubek T, Sadowská M, Vondrák V (2016) Scalable Algorithms for Contact Problems. New York: Springer, DOI 10.1007/978-3-031-33580-8
[20] Eck C, Jarušek J, Krbec M (2005) Unilateral Contact Problems: Variational Methods and Existence Theorems. Boca Raton: CRC Press, DOI 10.1201/9781420027365
[21] Gwinner J, Stephan EP (2018) Advanced Boundary Element Methods. Treatment of Boundary Value, Transmission and Contact Problems. Cham: Springer, DOI 10.1007/978-3-319-92001-6
[22] Hsiao GC, Wendland WL (2008) Boundary Integral Equations. Berlin, Heidelberg: Springer, DOI 10.1007/978-3-540-68545-6

[23] Jain AK (1989) Fundamentals of Digital Image Processing. Englewood Cliffs: Prentice Hall
[24] Johnson KL (1985) Contact Mechanics. Cambridge: Cambridge University Press
[25] Kravchuk AS, Neittaanmäki PJ (2007) Variational and Quasi-Variational Inequalities in Mechanics. Dordrecht: Springer, DOI 10.1007/978-1-4020-6377-0
[26] Lebedev VI (1997) An Introduction to Functional Analysis and Computational Mathematics. Boston. Birkhäuser
[27] Nikishin VS (2001) Staticheskie kontaktnye zadachi dlya mnogosloinykh tel (in Russ., Static Contact Problems for Multilayer Bodies). In: Mekhanika kontaktnykh vzaimodeistvii (in Russ., Mechanics of Contact Interactions), pp 214–233. Moscow: Fizmatlit
[28] Novotny AA, Sokolowski J, Zochowski A (2019) Applications of the Topological Derivative Method. Cham:Springer, DOI 10.1007/978-3-030-05432-8
[29] Polonsky IA, Keer LM (1999) A numerical method for solving rough contact problems based on the multi-level multi-summation and conjugate gradient techniques. Wear 231:206–219, DOI 10.1016/S0043-1648(99)00113-1
[30] Polyak BT (1969) The conjugate gradient method in extremal problems. USSR Computational Mathematics and Mathematical Physics 9(4):94–112, DOI 10.1016/0041-5553(69)90035-4
[31] Quarteroni A, Valli A (1991) Theory and application of Steklov-Poincaré operators for boundary-value problems. In: Spigler R (ed) Applied and Industrial Mathematics, Mathematics and Its Applications, Vol 56, pp 179–203. Dordrecht: Springer, DOI 10.1007/978-94-009-1908-2_14
[32] Sauter SA, Schwab C (2011) Boundary Element Methods. Berlin, Heidelberg: Springer, DOI 10.1007/978-3-540-68093-2
[33] Serov V (2017) Fourier Series, Fourier Transform and Their Applications to Mathematical Physics. Cham: Springer, DOI 10.1007/978-3-319-65262-7
[34] Wang QJ, Sun L, Zhang X, Liu S, Zhu D (2020) FFT-Based Methods for Computational Contact Mechanics. Frontiers in Mechanical Engineering 6:92–113, DOI 10.3389/fmech.2020.00061
[35] Wang QJ, Zhu D (2019) Interfacial Mechanics: Theories and Methods for Contact and Lubrication. Boca Raton: CRC Press, DOI 10.1201/9780429131011

Chapter 8
Logical, Mathematical, and Observational Foundations of the Theory of Heterogeneous Media Flows

Yuli D. Chashechkin

Abstract The system of fundamental equations (SFE) for mechanics of heterogeneous fluids and gases is used to describe the dynamics and structure of flows based on the principles of advanced research logic. The density of fluids and gases depends on pressure, temperature, and the concentration of dissolved and suspended particles. SFE is a differential analogue of the laws of conservation of density, matter, momentum, and energy with the state equations for the Gibbs potential and the medium density. The impact of various mechanisms of momentum and energy transfer, including both macroscopic (with flows, waves, and vortices) and microscopic (dissipative and conversion processes) is taken into account. A reduced SFE, in which the stratification of the medium is preserved instead of the state equation, is employed to reduce the formulation in the experiment analysis. In a gravitational field, the stratification, which allows for the construction of complete solutions of SFE, enables the formation of non-stationary flows induced by diffusion on topography. The linearized equations of periodic flows are investigated by the methods of unified perturbation theory, which enables us to find the complete solutions to the problem with boundary conditions. The obtained regular components of the SFE solutions characterize the waves. The singular solutions, whose scales are determined by the dissipative and temporal parameters of flows, describe the evolving fine structure of the medium. The temporal and spatial scales of the SFE determine the requirements for the experimental methodology, which ensures the registration of all flow components. The results of calculations of periodic flows in stratified are in good agreement with the data of high-resolution shadow measurements.

Key words: Axiomatics, State equation, Thermodynamics, Fundamental system, Complete solutions, Waves, Ligaments, Experiment

Yuli D. Chashechkin
Ishlinsky Institute for Problems in Mechanics RAS, Prospekt Vernadskogo 101-1, 119526 Moscow, Russian Federation, e-mail: chakin@ipmnet.ru, yulidch@gmail.com

H. Altenbach et al. (eds.), *Stability Problems in Mechanics of Continua*,
Advanced Structured Materials 260, https://doi.org/10.1007/978-3-032-31032-3_8

8.1 Introduction

The study of fluid and gas flows has been of significant interest in recent decades as it traditionally takes an important place in mechanics due to its scientific content and practical value of results, which are applied in various fields of human activity. The development of computers and computational mathematics allowed solving more and more complex problems. The improvement of optical observation methods enables researchers to study flow patterns on a wide range of scales, from fractions of a millimeter in a laboratory to astronomical units in outer space.

Meanwhile, there are "alarming" assertions of the hydrodynamics state. The loss of interest among students in studying this discipline, the absence of new installations, the drop in funding, and the continued uncertainty in the theory of turbulence, which is treated "as an unsolved problem despite considerable intellectual and material resources spent on it, as well as on a number of other domains of theoretical hydrodynamics. . . " [23], despite the obvious achievements in aviation, shipbuilding, and the development of new technologies is noted. The hope for improvement is associated with "the ability of the community to educate some researchers who can easily cross scientific boundaries while trying to make sense of specific results in their own narrower domains" [23]. In the context of an increasing uncertainty in various domains of science, it is advisable to start exploring the direction of fluid mechanics advancement by considering the primary basis of scientific work, which is the logic of research formulation.

For a long period of time, the methodology of scientific research was based on Aristotle's principles of logic. They include the principles of "identities" (the concept should be used in the same meaning in the course of reasoning), "internal consistency" (it is impossible to exist and not to exist at the same time), and "the excluded middle" (either A or *not*-A is true, and there is no third option).

Over time, Ockham added the principle of "minimum sufficiency" (Ockham's razor — of two competing theories, the simpler explanation of an entity is to be preferred), and Leibniz added the principle of "sufficient foundation" (every an accomplished event has a sufficient reason). In the context of physical phenomena, Maxwell required "the mathematical classification of a system of physical quantities" that determines the phenomenon under study, while Mendeleev emphasized the importance of "measurability of the selected quantities" i.e. the possibility to determine the values with reliable error estimates. The active development of irreducible models of the same process, similar models of various types of fluid flows in fluid mechanics, in which the systems of equations with non-canonical physical quantities [15] were used, necessitated the introduction of an additional principle in the principles of logic: the principle of "*Meaningfulness*", i.e. definability of the essence and observability of the subject under study in the considered and independent categories [4]. Additionally, the principle of "*Completeness*" was formulated, i.e. the development of a unified description of all known properties of the phenomenon under study with a guaranteed error estimate and the openness to accept the newly discovered features [4].

In the context of the principles of *Meaningfulness* and *Sufficient Reason*, it should be noted that the basis of fluid mechanics is formed by axiomatically defined mathematical and physical concepts of number, set, continuous space and time, motion, matter, and flow.

The mathematical foundation of describing flows was consistently formed by the systems of the d'Alembert – Euler equations for an ideal fluid and the Navier – Stokes equations for a viscous medium [15]. When deriving the equations of fluid motion, Euler noted the necessity of taking into account the dependence of the fluid density on temperature, and the limitations of the mathematics that existed at the time [10]. Stokes further highly emphasized the need to employ the approximation of constant density: "As it is quite useless to consider cases of the utmost degree of generality, I shall suppose the fluid to be homogeneous. . . " [24].

Since the 19th century, many theoretical studies have been based on the Navier – Stokes equations, which describe the flows of a homogeneous incompressible fluid:

$$\mathrm{div}\vec{u} = 0; \qquad \rho\frac{\partial \vec{u}}{\partial t} + \rho\left(\vec{u}\nabla\right)\vec{u} = -\frac{\partial P}{\partial \vec{x}} + \mu\Delta\vec{u} + \rho\vec{g}. \tag{8.1}$$

The equations of motion (8.1), after division by the dimensional constant, turn into kinematic relations for an undefinable "liquid particle"

$$\mathrm{div}\vec{u} = 0; \qquad \frac{d\vec{u}}{dt} = -\vec{w}_P + \vec{w}_g + \vec{w}_\nu, \tag{8.2}$$

which do not contain any medium state parameters. The consideration is further carried out in the Cartesian coordinate system $(\vec{x})$, t is time. Here ρ is fluid density, $\vec{u}\,(\vec{x}, t)$ is its velocity, $\vec{w}$ is acceleration, P is pressure, $\mu = \rho\nu$ and ν are dynamic and kinematic viscosities, $\vec{g} = (0, 0, -g)$. The indices in (8.2) characterize the influence of pressure, gravity, dissipation.

In order to implement the principles of "Meaningfulness" and "Minimum Sufficiency" in research, it is necessary to define the medium under study and introduce equations of its motion in the problem formulation.

8.2 Physical Quantities of Flowing Media

The main feature that distinguishes fluids and gases from other physical states of matter is their fluidity, i.e. zero static friction and the ability to move in space under the influence of arbitrarily weak internal (gradients of physical quantities) or external (movement of boundaries) factors. In contrast, a liquid of a given mass is characterized by a finite volume, while gas or plasma fills the entire available space.

In classical mechanics, the main parameters of the medium state are density $\rho\,(\vec{x})$ i.e. the measure of inertia, gravitational interaction, amount of matter, and other properties, as well as pressure $P\,(\vec{x})$, moreover the dependence $\rho\,(\vec{x}) = \rho\,(P\,(\vec{x}))$ defines a barotropic fluid. The study is traditionally conducted in the "continuous medium" approximation, where the quantities under consideration are regarded to be continuous functions of coordinates and time [15].

Physical quantities such as density and pressure in fluid mechanics have a dual mechanical and thermodynamic nature, which is determined by the atomic and molecular structure of the substance under study. High-resolution investigation of flows in natural, industrial, and laboratory settings have revealed the heterogeneity of media, where the density value is determined by the distribution of matter, i.e. the concentration of the components of the medium S_i, pressure $P(\vec{x})$ and temperature $T(\vec{x})$. A large number of experimental and theoretical studies are dedicated to the investigation of the dependence of density on other thermodynamic quantities, i.e. the determination of the equation of state for liquids and gases $\rho(\vec{x}) = \rho(P, T, S_i)$ [11, 12].

Concerning the equation of state, the representations of the speed of sound are used as functions of density, temperature, pressure, and dissolved substance concentration along with the dependence of density on the physical parameters of the medium. Moreover, the dependencies of specific electrical conductivity and refractive index of the transparent medium have become widespread. Their measurements are employed for density calculation.

At the same time, The International Association for the Properties of Water and Steam (IAPWS) recommended using the Gibbs potential (Gibbs free energy, isobaric energy) as the basis for describing the thermodynamics of water. IAPWS is a non-profit association of national organizations dealing with the properties of water and steam, particularly thermophysical properties, recommendations on cycle chemistry, and other aspects of high-temperature steam, water, and aqueous mixtures relevant to thermal energy cycles, industrial and scientific applications [6]. The derivatives of the Gibbs potential with the differential $dG = -s_e dT + VdP$ determine the entropy s_e and specific volume $V = 1/\rho$, and indirectly — the fluid density. The Gibbs potential of solutions $G(P, T, S) = G_W(P, T) + G_S(P, T, S)$ is defined as the sum of potentials of the solvent (water) $G_W(P, T)$ and the dissolved substance $G_S(P, T, S)$.

For multicomponent fluid media, the Gibbs potential, as well as other thermodynamic potentials, depends on the value of the surface tension coefficient σ, the free surface area of the liquid S_d, and the chemical potential μ_i

$$dG = -s_e dT + VdP - S_d d\sigma + \mu_i dS_i \tag{8.3}$$

The inclusion of the Gibbs potential in the description of flows as an equation of state enables us to take into account the impact of the conversion processes of potential surface energy (PSE) in flow calculations. PSE conversion is the transition of PSE into other forms, i.e. thermal energy, pressure perturbations, redistribution of concentration, kinetic energy of flows when eliminating a free surface, and, conversely, the local absorption of thermal and kinetic energy when creating a new free surface [6]. The analysis of variations of thermodynamic potentials allows for the consideration of the influence of various mechanisms of energy transfer on the dynamics of flows and the structure of the medium. In practice, both the radiative transfer of energy by thermal radiation and the macroscopic transfer with flow, waves, and vortices, as well as the microscopic (dissipative and conversion transfer) are regarded [4, 15].

Modern scientific tools have made it possible to identify individual cells, fibers, and layers which are contoured by shells with their own physical properties in the distribution pattern of thermodynamic and other physical parameters of a liquid [1, 9, 18, 26]. These cells can be found both on a free surface and in the thickness of the liquid or gas. The formation of cells with minimal dimensions of an order of $\delta_\sigma \sim 10^{-6}$cm, which include associates of various physical and chemical nature, such as clusters, clathrates, bonds, voids, and others, is determined by the properties of the atomic and molecular structure of the substance and the physical conditions of the experiments. The distribution pattern of thermodynamic quantities is continuously changing. Some cells ("grains" or "granules") disappear, while others appear. The duration of existence and cell sizes, which depend on the composition of substances and the thermodynamic and mechanical conditions of the experiments, vary widely. Images of the solar photosphere serve as a visual representation of the physical model for the "granular" thermodynamic structure of a fluid or gas [25].

When the free surface of the merging liquids is annihilated, a part of the converted PSE, i.e. the available potential surface energy (APSE), is converted into the kinetic energy of the thinnest trickles. In turn, the formation of new contact surfaces is accompanied by the local absorption of a part of thermal and kinetic energy, which is converted into the PSE of new contact surfaces, and by the appearance of decelerated trickles and thin interfaces. Thus, an internally structured unperturbed medium is characterized by heterogeneous distributions of the thermodynamic potential $G_S(P, T, S)$ and its derivatives, i.e. as density, pressure, temperature, concentration of dissolved substances and suspended particles, and other physical parameters (such as sound velocity, refractive index, and others), which reflect the domain composition structure of liquids and gases. However, it is assumed here and further that all physical quantities are described by continuous functions of space and time, which are characterized by significantly different values of gradients at the boundaries and within the cells located in the thickness and free surface of the fluid. In other words, the analysis is conducted at scales that considerably exceed the discrete dimensions of the atomic and molecular components of the investigated media. It should be noted that the equations of state for liquids and gases are continuously being refined and supplemented [12].

8.3 System of Fundamental Equations for Mechanics of Heterogeneous Fluids

The properties of fluid or gas flow are characterized by the state equations for the Gibbs potential, density, and other derivatives. The definition of this concept of fluid or gas flow proposed in [4] is based on the requirement of simultaneous following the laws of conservation of momentum, energy, and matter: "A flow is the transfer of momentum, energy, and matter, accompanied by self-consistent changes in physical quantities which determine the state and other physical properties of the medium."

The measurable physical values that characterize liquid media are such scalar quantities as density ρ, total energy $E_t = E_M + E_p + E_i$, which includes specific kinetic energy

$$E_M = \frac{\rho \vec{u}^2}{2},$$

potential energy E_p, and internal energy E_i; momentum vector

$$\vec{p} = \left(p_x, p_y, p_z\right) = \rho\vec{u},$$

as well as other physical values of the medium under study. The fluid velocity $\vec{u}\left(\vec{x}, t\right) = \vec{p}/\rho$ is defined as the local ratio of two instantaneous local values of conserved quantities, i.e. momentum $\vec{p}$ and density ρ. When describing the kinematics of homogeneous fluid flows, it is assumed that the fluid velocity $\vec{u}$ is identical to the velocity of the transformation of Euclidean space $\mathbb{R}^3$ into itself.

The axiomatic system of fundamental equations of fluid mechanics, which takes into account the choice of defining physical quantities, has the following form [5, 15, 20, 28]

$$\begin{cases} \rho = \rho\left(P, T, S_n\right), \quad G = G\left(x, y, z\right), \\ \dfrac{\partial \rho}{\partial t} + \nabla \cdot \vec{p} = Q_m, \\ \dfrac{\partial S_i}{\partial t} + \nabla \cdot \left(S_i \vec{u} + \vec{I}_i\right) = Q\left(S_i\right), \\ \dfrac{\partial p^i}{\partial t} + \nabla_j \Pi^{ij} = \rho g^i + 2\rho\varepsilon^{ijk} u_j \Omega_k + Q^i\left(f\right), \\ \dfrac{\partial E}{\partial t} + \nabla_i\left(E\nu u^i\right) + \nabla_i\left(q^i + P\nu^i - \sigma^{ij}u_j + \dfrac{\partial w}{\partial S_n} I_n^i\right) = Q\left(e\right), \end{cases} \tag{8.4}$$

where S_i, $\vec{I}_n$ are concentration and density vector of the diffusion flux of the n-th impurity, respectively; $\Pi^{ij} = \rho u^i u^j + P\delta^i j - \sigma^{ij}$ is the momentum flux density tensor,

$$\sigma^{ij} = \mu\left(\frac{\partial u^i}{\partial x^j} + \frac{\partial u^j}{\partial x^i} - \frac{2}{3}\delta^{ij}\frac{\partial u^k}{\partial x^k}\right) + \zeta\delta^{ij}\frac{\partial u^k}{\partial x^k}$$

is the symmetric viscous stress tensor; δ^{ij} is the fundamental metric tensor; μ, ζ are the first and second dynamic viscosities; $\vec{\Omega}$ is the global rotation angular velocity vector of the fluid;

$$E = \rho\left(\frac{u^2}{2} + \varepsilon + U\right)$$

is the total energy density; ε is the specific internal energy,

$$w = \varepsilon + \frac{P}{\rho}$$

is the specific enthalpy; U is the gravitational potential, $\vec{g} = -\nabla U$ is the acceleration of gravity, $\vec{q}$ is the vector of the heat flux density. Mass sources of n-th impurity, force and energy respectively with density Q_m, $Q(S_n)$, $Q^i(f)$, $Q(e)$ describe the influence of processes of non-hydrodynamic nature and may be absent completely or partially. To date, the system (8.4) has not been analyzed taking into account the compatibility condition, which ensures the construction of complete solutions. The inclusion of sources in the main system (8.4) allows us to consider the influence of the transformations of thermal and kinetic flow energy into the latent form of internal energy and vice versa, the transition of potential surface energy into thermal and kinetic energy of fluid flows at small scales, on the dynamics and structure of the flow.

Assuming that the gradients of potentials, physical quantities, and the intensity of external sources are small, the system (8.4) is transformed into a system of equations for describing the matter transfer, the concentration of individual components, temperature, and momentum [15], supplemented by the equations of state for density and the Gibbs potential.

$$\begin{cases} G = G(P,S,T) = G(\vec{x},t), \quad \rho = \rho(P,S,T) = \rho(\vec{x},t), \\ \dfrac{\partial \rho}{\partial t} + \nabla_j p^j = Q_\rho, \\ \dfrac{\partial p^i}{\partial t} + \left(\nabla_j \dfrac{p^j}{\rho}\right) p^i = -\nabla^i P + \rho g^i + \nu \Delta p^i + 2\varepsilon^{ijk} p_j \Omega_k + Q^i, \\ \dfrac{\partial \rho T}{\partial t} + \nabla_j \cdot \left(p^j T\right) = \Delta\left(\varkappa_T \rho T\right) + Q_T, \\ \dfrac{\partial \rho S_i}{\partial t} + \nabla_j \cdot \left(p^j S_i\right) = \Delta\left(\varkappa_S \rho S_i\right) + Q_{Si}. \end{cases} \tag{8.5}$$

The system of equations (8.5) is widely employed in describing flows in the environment, such as the atmosphere and ocean, as well as in technical hydrodynamics. It includes physical quantities of mechanical $(\rho, \vec{p})$, thermodynamic (P, T, S_i), and kinematic $\left(\vec{u}, \vec{\Omega}\right)$ nature. The system considers the processes of momentum dissipation, the impact of internal or external sources of matter, temperature, and impurities on the structure and dynamics of the flow. When solving practical problems, the system is supplemented with physically justified initial and boundary conditions. They involve the condition of no-slip for velocity or momentum on solid boundaries; the conditions of no-flux for density and components of substance impurity concentration in the fluid; the values of temperature or its flux; the equality of momentum fluxes on the contact surfaces of two fluids, as well as the attenuation of all perturbations at infinity.

Equations (8.5) form a system of algebraic-differential equations, the solution of which is constructed considering the compatibility condition. The rank of the system (the higher-order derivative, if the system can be transformed into a single equation), as well as the order of its linearized version and the degree of the characteristic (dispersion) equation, determine the minimum number of eigenfunctions that constitute the complete solution. The complete system with a diffusion equation for a single impurity has a rank of ten [5]. Accordingly, the flow pattern for this set of solutions is formed at least by ten independent functions with their own spatial and temporal scales. The multiple components of the flow, which differ in scale and structure, manifest themselves in the continuous changes of the observed flow pattern. Over time, the number of structural components may increase due to nonlinear interaction processes of basic solutions [2].

The consistency of the infinitesimal symmetries of system (8.5) with the basic principles of physics [5] is evidence of the validity of its choice as the basis for studying fluid flows. Practical recommendations for selecting the parameters of numerical modeling and experimental techniques are based on the analysis of the intrinsic space-time scales of the system (8.5) with initial and boundary conditions. Since the symmetries of the fundamental system and the equations of other flow models (numerous variations of turbulence theories, wave theories, vortex theories, jet theories, wake theories, etc.) differ significantly, the same embedded symbols have different physical meanings. As the general properties of the system (8.5) and its linearized version have not been investigated yet, the analysis results of the simpler reduced versions of the system (3), which retain important distinguishing features such as medium heterogeneity and dissipativity are of particular interest.

Real heterogeneous liquids and gases naturally stratify in an external gravitational field with free-fall acceleration $\vec{g} = (0, 0, -g)$, i.e. light volumes float, and heavy volumes sink. When simplifying the description of the influence of stratification on the flow dynamics, the system of equations (8.5) is usually significantly reduced, i.e. it retains only the fact of density heterogeneity without considering the impact of changes in other physical quantities that determine its value such as distributions of pressure, temperature, and impurity concentrations [17]. To simplify the description, the influence of only one stratifying impurity is taken into account, for example, the concentration, the heterogeneity of which causes the diffusion of matter, or the temperature in a heterogeneously heated medium. Each of the examined influencing factors occurs separately in natural conditions and is employed to create stratified media in the laboratory.

The degree of heterogeneity in the distribution of the variable density of an incompressible medium with a vertical profile $\rho(z)$ is characterized by the scale of stratification $\Lambda = |d \ln \rho / dz|^{-1}$, the frequency $N = \sqrt{g/\Lambda}$, and the buoyancy period $T_b = 2\pi/N$. To simplify the formulation, we will study the properties of solutions to the reduced system of fundamental equations (8.5), which retains the heterogeneity in density due to the stable distribution of salinity $\rho_0 = \rho_0(S_i(z))$, without considering the influence of other physical factors. The densities of many fluids, in particular metal salt solutions and seawater, are related to the concentration of the dissolved substance by an almost linear relationship, which facilitates the calculations.

8.4 A Reduced System of Fundamental Equations Describing the Flows of a Single-Component Incompressible Stratified Fluid

Many calculations and experimental techniques are based on a linearized reduced system of fundamental equations for mechanics of an incompressible stratified fluid in the Boussinesq approximation:

$$\begin{aligned} &\rho = \rho_{00}\left(\exp\left(-z/\Lambda\right) + s\right), \qquad \operatorname{div}\vec{u} = 0, \\ &\frac{\partial \vec{u}}{\partial t} + \frac{\nabla P}{\rho_{00}} = \nu\Delta\vec{u} - s\vec{g}, \\ &\frac{\partial s}{\partial t} = \varkappa_S \Delta s + \frac{u_z}{\Lambda}. \end{aligned} \tag{8.6}$$

Here, s is the salinity perturbation, ν is the kinematic viscosity, and $\varkappa_S$ is the diffusion coefficient of the salt, which is a stratifying impurity with a variable concentration. In a liquid with an exponential density distribution, the stratification scale Λ, frequency N, and buoyancy period T_b do not depend on the depth. The system (8.6) is supplemented by boundary conditions on the solid surface, such as no-slip for velocity and no-flux for matter, as well as the condition of perturbations attenuation at infinity and their absence at $t = 0$.

The reduced system (8.6) includes a number of scales — both large, compared to the size of the source L_S or the length of the internal wave excited by the source moving at velocity U, $\lambda_w = UT_b$ — Λ is the stratification scale, and shorter scales δ_ν^l, $\delta_{\varkappa_S}^l$, which include small dissipative coefficients.

The latter include the Rayleigh numbers characterizing the fluid, such as the combined in viscosity $L_N^{g\nu} = \sqrt[3]{g\nu}/N$ or diffusion coefficient $L_N^{g\varkappa_S} = \sqrt[3]{g\varkappa_S}/N$, as well as the singular microscales $\delta_N^\nu = \sqrt{\nu/N}$ and $\delta_N^{\varkappa_S} = \sqrt{\varkappa_S/N}$. Periodic flows with frequency ω correspond to similar Stokes scales, which include the frequency ω instead of the buoyancy frequency N: $L_\omega^{g\nu} = \sqrt[3]{g\nu}/\omega$, $L_\omega^{g\varkappa_S} = \sqrt[3]{g\varkappa_S}/\omega$, $\delta_\omega^\nu = \sqrt{\nu/\omega}$ and $\delta_\omega^{\varkappa_S} = \sqrt{\varkappa_S/\omega}$. As calculations show, the parameters of the fibers and interfaces are determined by the sets of Rayleigh and Stokes scales. The intrinsic scales of the system (8.6) also contain other combinations of time and length scales that arise during the calculation process. They could be found in [8]. The intrinsic time and spatial scales of the problem determine the key requirements for the experimental technique and calculation programs in terms of the size of the observation or calculation domain, the sensitivity, and the temporal and spatial resolution of the instruments employed to register all flow components.

The System (8.6) describes the flows of a thermodynamically nonequilibrium medium in which the salinity gradient creates a diffusion transfer of a stratifying component in the vertical direction, and the variability of salinity ensures the creation of an available potential gravitational energy reserve. The combined action of diffusion and energy conversion processes provides the creation of flows in a fluid at rest when the horizontal uniformity of the salt flow is disrupted, even in the absence of mechanical destabilizing factors. A salt deficit (excess) is formed in the vicinity of an unfolding (overhanging) with the height inclined lateral boundary at which the no-flux condition is satisfied.

The first mathematical model of such spontaneously formed two-dimensional perturbations, called "diffusion-induced flows on topography" on an inclined impermeable plate, was developed by Prandtl [22]. However the stationary solution, which was refined by Wunsch and Phillips later, does not matched with the long-time asymptotics of the fluid velocity at the exact solution to the problem of establishing a two-dimensional flow on an infinite lateral surface, inclined at an angle α to the horizon [14]. In the local orthogonal coordinate system (ξ, ζ), oriented along and normally to sloping plane the velocity distribution $u(\zeta)$ along the normal to the surface ζ, which was found in [14],

$$u(\zeta) = \frac{N^2 \delta_N^{\varkappa_S} \tau^{3/2}}{\nu - \varkappa_S} \left[\mathrm{i}^3\mathrm{erfc}\left(\frac{\zeta}{2\delta_\nu \sqrt{\tau}}\right) - \mathrm{i}^3\mathrm{erfc}\left(\frac{\zeta}{2\delta_S \sqrt{\tau}}\right)\right] \sin 2\alpha, \tag{8.7}$$

does not have a stationary limit in the long-time approximation of the process

$$\tau = t/T_b \to \infty.$$

Here

$$\mathrm{i}^n\mathrm{erfc}(z) = \int_z^\infty \mathrm{i}^{n-1}\mathrm{erfc}(x)\,dx,\ \mathrm{i}^{-1}\mathrm{erfc}(z) = \frac{2}{\sqrt{\pi}}\mathrm{e}^{-z^2},\ \mathrm{i}^0\mathrm{erfc}(z) = \frac{2}{\sqrt{\pi}}\int_z^\infty \mathrm{e}^{-x^2}dx$$

are various representations of the probability integral. The velocity of the non-stationary diffusion-induced flow reaches a maximum at a sidewall inclined at an angle of 45° to the horizon.

The calculations of the diffusion-induced flow, which ensures the self-motion of a free wedge balanced at the neutral buoyancy horizon, are in satisfactory agreement with the schlieren patterns of the flow [16]. The results of the calculation of the diffusion-induced flow, taken as an initial condition, accelerate the construction of numerical models of problems for the wave generation by oscillating bodies and the flow around obstacles by a stratified liquid. The system (8.6) is widely employed in the investigation of gravitational surface and internal gravitational and capillary waves in a stratified liquid as well [8]. The presence of large and small scales, the ratios of which allow for the creation of small parameters, explains the widespread implementation of perturbation theory methods in the study of periodic flows in such media.

8.5 Calculations of the Dynamics and Structure of Periodic Flows in a Stratified Medium

When calculating infinitesimal periodic perturbations, the solutions of the system (8.6) with physically justified boundary conditions are found employing the perturbation theory. In spectral methods, physical quantities are usually represented in the form of plane-wave decompositions with an exponential core u_i, p,

$$\rho \sim A_i \exp i\Theta = A_i e^{i\left(\vec{k}\vec{x}-\omega t\right)}.$$

In this case, the frequency of the process ω, which is a measure of energy, is considered to be a real, positive-definite quantity, while the wave number $\vec{k} = \vec{k}_1 + i\vec{k}_2$ is supposed to be complex. The imaginary part of the wave number allows us to estimate the attenuation of physical quantities as the wave propagates. In the traditional approach, where the frequency ω is assumed to be complex and the wave number is real, the problem of the flow stability is determined by the sign of the imaginary part of frequency. If the sign is positive, the amplitude of random perturbations increases.

If the frequency definiteness ω retains to be positive, the selection of solution roots is based on physical considerations, i.e. only those solutions that correspond to the attenuation of solutions in the direction of propagation under study are analyzed. For each possible direction of wave propagation, individual assessments are performed, and a physically justified solution is selected.

In the process of solution construction, the first step is to determine the small parameters of the problem ε and the order of actions when there are several of them. In the theory of regular perturbations, all quantities are represented in the form of series of type $p = p_0 + \varepsilon p_1 + \varepsilon^2 p_2 + \ldots$. Moreover, many problems end with the calculation of the quantities of the first-order smallness and the evaluation of the quality of the discarded terms. Thus, it is found that $\mathrm{Re}\Theta \gg \mathrm{Im}\Theta$, $\mathrm{Im}\Theta \sim \nu^\alpha$, $\alpha > 0$. As calculations show, the application of regular expansion methods allows us to find only two solutions to this eighth-order system.

The remaining six solutions of the system (8.6) are found by methods singular perturbation theory [19]. The calculation algorithm is based on maintaining the strictness of the logic of actions during operations where the series of the type $p = \varepsilon^{-\gamma}\left(p_0 + \varepsilon p_1 + \varepsilon^2 p_2 + \ldots\right)$, $\gamma > 0$ are employed. The exponent γ can be found from a weaker condition than in the calculation of regular perturbations, i.e. the degree of the small parameter ε in the leading term must be not less than in all other terms of the equation [19]. After solving the redefined system, the maximum positive root γ is chosen. It is found that $\Theta \sim \sqrt{i}$, $\mathrm{Re}\Theta \sim \mathrm{Im}\Theta$, $\mathrm{Im}\Theta \sim \alpha_S$, $\alpha_S > 0$.

The difference in the functional properties of regular and singular expansions enables us to employ them simultaneously in solving high-order systems and to find a complete solution that satisfies both the equations and the boundary conditions without the need for additional hypotheses, functional relationships and constants. In such operations, the physical meaning of the symbols used remains unchanged.

However, the changes in the problem parameters and the values ε correspondingly can have different effects on the behavior of the separated groups of solutions.

Another important consequence is that the number of regular components in the solution of the wave equations is kept the same as the form of the state equation becomes more complex. The number of singular solutions ensuring the implementation of the condition for a complete problem solution of calculating periodic flows in full formulation (i.e. the condition of equality of the number of roots and the order of the system being analyzed) increases as the state equation becomes more complex. The increase in the number of physical quantities that determine the value of density and the number of transfer equations for additional quantities in the system of governing equations is accompanied by the rise in the number of pairs of additional singular roots. The implementation of the compatibility condition between the equations and boundary conditions can lead to an increase in the order of the analyzed system and growth in the number of roots.

The complicating impact of the effects of stratification and viscous dissipation is illustrated by the dispersion relations that arise when solutions in the form of two-dimensional plane surface waves are substituted into the reduced SFE, which retains the variable density but does not consider the physical nature of stratification [8]:

$$\begin{cases} -N^2 k_x^2 + \mathrm{e}^{z/\Lambda}\left(k_x^2 - k_{z,l}^2\right)\left(\mathrm{i}\nu\left(k_x^2 - k_{z,l}^2\right) + \omega\right)\omega = 0, \\ g k_x^2 - 3 i k_x^2 k_{z,l}\left(\mathrm{i}\nu k_{z,l}^2 - \omega\right) + k_x^4 \gamma = 0, \end{cases} \tag{8.8}$$

Here, the index l corresponds to the singular solution, ν is the kinematic viscosity and $\gamma = \sigma/\rho_{00}$ is the surface tension coefficient normalized by density. In general, the roots of the dispersion relation (8.8) determine two types of relations between the components of the wave number. Regular solutions characterize waves. Singular solutions (two in 2D formulation and four in 3D consideration) describe a family of ligaments in a viscous stratified medium.

The dispersion relation for harmonic waves in a viscous homogeneous fluid also defines the link between various components of the wave vector, but within a single polynomial expression, as opposed to (8.8):

$$\begin{aligned} k_x\Big[2k_x\left(k_l\omega^2 - g k_x^2 - \gamma k_x^4 + i\omega\nu k_l\left(3k_x^2 - k_l^2\right)\right) - \\ - \left(k_x^2 + k_l^2\right)\left(\omega^2 - g k_x - \gamma k_x^3 + 2\mathrm{i}\omega\nu k_x^2\right)\Big] = 0. \end{aligned} \tag{8.9}$$

The dispersion relation for gravitational waves on the surface of a deep ideal fluid is extremely simplified and describes only the regular component of the periodic flow [15]: $\omega^2 = gk + \gamma k^3$. Substitution of spectral expansions transforms the system of differential equations (8.6) into a system of algebraic equations, the solvability condition of which generates a dispersion relation.

The disk with a diameter D located on a horizontal plane and performing small harmonic oscillations with an amplitude A and a frequency ω in a liquid with a constant buoyancy frequency N is chosen as a source of waves. The condition

$\omega < N$ assumes the formation of a conical beam of internal waves propagating in the vertical plane at an angle $\sin\theta = \omega/N$ to the horizon. The geometry of the problem of an axially symmetric beam of waves provides a preferred choice of a cylindrical coordinate frame (r, φ, z), in which, due to the axial symmetry of equations and boundary conditions, the problem becomes two-dimensional (r, z), facilitating calculations and the procedure of their comparison with experimental data.

For internal waves in the thickness of a continuously stratified fluid with an exponential density distribution, the dispersion equation is:

$$\varkappa_S \nu K^3 - \mathrm{i}\omega\,(\nu + \varkappa_S)\,K^2 - \omega^2 K + N^2 k^2 = 0,\ \ K = k^2 + k_j^2. \tag{8.10}$$

The regular root (8.10), which coincides, up to the replacement of the kinematic viscosity coefficient by the sum of the dissipative coefficients, with the expressions obtained in a number of independently conducted studies, has the form:

$$k_1 \simeq -k\cot\theta + \frac{\mathrm{i}\,(\nu + \varkappa_S)\,k^3}{\omega\sin^2\theta\sin 2\theta}. \tag{8.11}$$

One of the singular solutions (8.10) describes a thin ligament with a thickness of the order of the Stokes viscous scale $\delta_\omega^\nu = \sqrt{\nu/\omega}$:

$$k_2 \simeq \frac{1+\mathrm{i}}{\sqrt{2}}\delta_\omega^\nu, \tag{8.12}$$

the second solution is an even thinner concentration ligament with the thickness $\delta_\omega^{\varkappa_S} = \sqrt{\nu/\varkappa_S}$

$$k_3 \simeq \frac{1+\mathrm{i}}{\sqrt{2}}\delta_\omega^{\varkappa_S}. \tag{8.13}$$

Substituting spectral representations into the boundary conditions completely solves the problem of wave generation by a moving part of the plane in a linear and weakly nonlinear approximation [2]. It should be emphasized that all components of the solutions exist in the entire space, i.e. both the waves and the associated ligaments propagate together. The properties of the ligaments are most clearly manifested in the vicinity of solid surfaces at scales significantly smaller than the wavelength, where they are more pronounced than the internal waves. In the full three-dimensional formulation, singular terms with the Rayleigh scale $\delta_N^\nu = \sqrt{\nu/N}$ and the combinational scale $\delta_N^{\nu\varkappa_S} = \sqrt[4]{\nu\varkappa_S/N^2}$ appear in the solution.

The constructed solution of the problem for the generation of internal waves created by an oscillating disk on a horizontal plane has the simplest form in the associated coordinate system (p, q), where the axis q is oriented in the direction of wave beam propagation and the axis p is oriented along the phase velocity: $p = r\sin\theta - z\cos\theta$, $q = r\cos\theta + z\sin\theta$. For large values of the Schmidt numbers $(\mathrm{Sc} = \nu/\varkappa_S \gg 1)$, which are typical for most aqueous solutions of metal salts and seawater, the amplitude of density and pressure perturbations in the wave beam is expressed in terms of Bessel functions:

$$\rho(r,z) = -\frac{\mathrm{i}\rho_0 AD}{2\Lambda} \int_0^\infty \frac{dk}{k_2 - k_1} J_0(kr) J_1(kD/2) \left(k_2 \mathrm{e}^{\mathrm{i}k_1 z} - k_1 \mathrm{e}^{\mathrm{i}k_2 z}\right). \quad (8.14)$$

To illustrate the pattern of the obtained solutions, Fig. 8.1 shows the results of numerical solution of the system (8.6) visualizing the central sections of periodic conical beams in the lateral projection (central vertical plane). To speed up the calculations, the spatial resolution of codes is limited to the condition of selecting ten nodes on the Rayleigh scale. The pattern of the velocity field also provides wave components and ligaments, which are thin shells covering the beams from the outside (Fig. 8.1a). Their thickness and center position are determined by the wave phase. The shells covering the beams from the outside and inside are even more expressed in the field of the second spatial derivative of the velocity (Fig. 8.1b).

The combined images shown in Fig. 8.1c demonstrate a good agreement between the calculated and observed density gradient fields near the plate, in the flow formation zone, and at large distances, in the zone where the asymptotic solutions are applicable.

The schlieren images in Fig. 8.2 illustrate the flow pattern consisting of the internal wave beams and accompanying ligaments in more detail. The experiments were performed at the Stands of the Unique Research Installation [27]. Stands include stratified pools with devices for creating and controlling stratification and high-resolution schlieren instruments designed by Maksusov, which allow for the visualization of both internal waves and finer ligaments.

A compact two-dimensional source, whose dimensions are smaller than the viscous Rayleigh or Stokes wave scale, excites four unimodal waves with a maximum displacement in the beam center. They create an equilateral "St. Andrew's cross" (Fig. 8.2a). An extended plate that oscillates alongside its length excites a bimodal

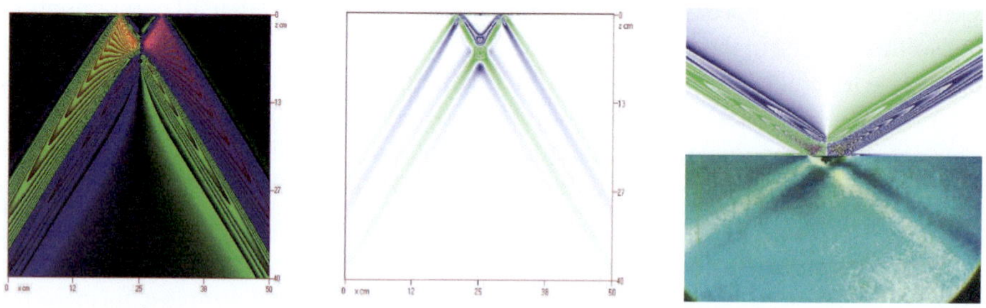

a) b) c)

Fig. 8.1: Calculated cross-section of a conical beam arising from oscillations of a disk with a diameter of $D = 8$ cm and an amplitude of $A = 0.25$ cm and a frequency of $\omega = 1.0\ \mathrm{s}^{-1}$ in a liquid with a buoyancy period of $T_b = 5.2$ s: a) and b) fields of the radial velocity component u_r and its second derivative $(\partial^2 u_r / \partial z^2)$, c) combined images of the calculated and observed unimodal beam of internal waves by the schlieren instrument $(T_b = 6.6\ \mathrm{s},\ D = 4\ \mathrm{cm},\ \omega = 0.56\ \mathrm{s}^{-1})$.

a) b) c)

Fig. 8.2: Schlieren images of periodic internal wave beams generated by an inclined band of width L_b oscillating along its surface with frequency $\omega = 1.0\ \mathrm{s}^{-1}$ in a stratified fluid with a buoyancy period of $T_b = 5.5$ s: a) $L_b = 1$ cm, $\omega = 1.01\ \mathrm{s}^{-1}$, $\varphi = 33°$, $\theta = 63°$; b) $L_b = 6$ cm, $\omega = 0.81\ \mathrm{s}^{-1}$, $\varphi = 45°$, critical angle $\varphi = \theta$; c) pattern of the conical internal wave beam reflection from sloping plane separated by an inclined ligament from a reflecting plane.

beam in two directions (left-up and right-down) with diversified maxima displacement positions and a reduced value in the center. In the first and third quadrants, a thin plate creates a unimodal beam (Fig. 8.2b).

The conical internal wave beam is created by the oscillating sphere in the lower part of the picture in Fig. 8.2b. The change of phases in the distribution of the horizontal component of the refractive coefficient gradient (the visualizating slit and Foucault knife are praced vertically) and the density gradient, respectively, is clearly expressed in the schlieren image of the conical internal wave beam. A thin ligament that forms a sharp angle with the plate plane separates the incident and outgoing beams from its surface.

The change in the relation between the contrast of the shadow images of the internal wave beams and thin ligaments is illustrated by a series of photographs of the flow pattern created by a vertically oscillating sphere in a continuously stratified medium. At low oscillation amplitudes, the sphere produces two conical beams of internal waves, the central section of which forms four wave beams in the shadow image.

Short ligaments (horizontal interfaces) are observed in the domains where the wave beams intersect (Fig. 8.3a). As the amplitude of the source oscillations increases, the schlieren instrument begins to visualize ligaments, which are thin strips that outline the outer and inner boundaries of the wave cone and its sections, respectively (Fig. 8.3b). In the flow field created by a long-lasting source, there are waves, ligaments, and thin, highly gradient shells that separate various flow components (Fig. 8.3c).

Since the vertical profile of the buoyancy frequency is usually heterogeneous in natural conditions, a beam of traveling harmonic internal waves that reaches the critical level $\omega = N$ is reflected and returned to the medium with the allowed propagation conditions, where $\omega < N$. A typical shadow pattern of such a flow is shown in Fig. 8.4.

The wave amplitude decline with the depth of penetration into the medium forbidden for propagation and the reflection coefficient from the critical level were

a) b) c)

Fig. 8.3: Schlieren images of conical beams of periodic internal waves created by a vertically oscillating sphere (D = 4.5 cm, "slit-thread" method): a)- c) $T_b = 11.2, 7.3, 11.2$ s; $A = 1, 2.8, 2.8$ cm, $\omega/N = 0.73, 0.8, 0.8$).

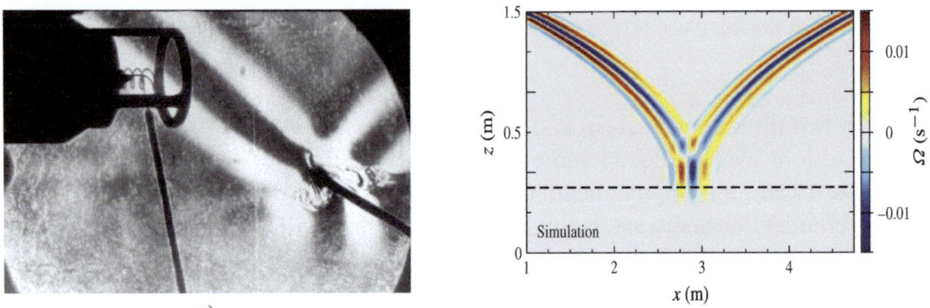

a) b)

Fig. 8.4: Reflection of a beam of two-dimensional periodic internal waves from a critical level $\omega = N_{cr}(z)$ in a fluid with a variable buoyancy period profile: a) shadow visualization, b) representation of the numerical solution [21].

calculated in [13] by the methods of the combined theory of regular and singular [19] perturbations, taking into account the properties of ligaments formed at the horizon of the beam center reflection. Subsequent detailed measurements of the wave velocity profiles in the incident, penetrated into the forbidden area, and reflected beams, as well as the accompanying numerical solutions of the governing equations, showed that the pre-calculated results [13] are consistent with the experimental and numerical modeling data [21] with a difference of a few percent.

8.6 Ligaments in Vortex Flows of Different-Density Fluids

Ligaments are even more clearly visualized in drop flows, where they demonstrate the properties of active dynamic components that generate other types of flows. Fig. 8.5 shows a sequence of video samples illustrating the generation of a singular vortex by a flow on the contracting cavity shell formed by the coalescence of a freely falling drop of colored ethanol with target water (time is measured from the initial contact of the falling drop with the target fluid).

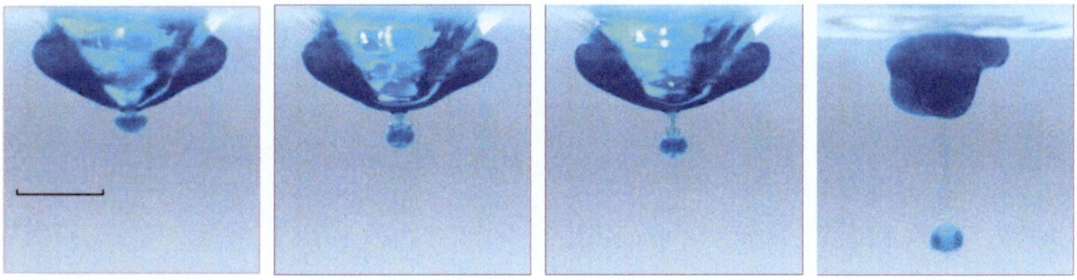

t = 15.25ms t = 16.5ms t = 17.5ms t = 40.5ms

Fig. 8.5: Immersion of a lighter singular vortex saving own dimensions.

A compact singular vortex comprising the fluid of the cavity shell, which contains both water and lighter ethanol, sinks, leaving a colored cylindrical wake. The vortex practically maintains its sizes until it stops [7]. After hovering, the vortex transforms into an intrusion of light fluid that floats in a higher-density fluid, and then into a regular annular vortex whose diameter increases with time. Fig. 8.6 illustrates samples from a video that illustrates the further evolution of the flow pattern.

An annular vortex with a convex top is forming at the head of the emerging volume during movement. Its cylindrical shell remains connected to the conical remnant of the secondary intrusion (Fig. 8.6, t = 67.5 ms). The central region of the flow is less densely tinted than the shell. The primary intrusion continues to spread actively in a thin layer beneath the free surface, where colored loops are present.

The pattern of the emerging flow is transformed. In the interval of $133 < t <$ 413ms (Fig. 8.6), the head of the intrusion takes on a mushroom-shaped form with a convex front and concave backsides, while a slowly emerging shell of the secondary intrusion takes on a bottle-shaped form. Both elements of the flow are connected by a cylindrical wake behind the floating vortex. The diameter of the arising regular vortex increases as it moves.

The diameter of the cylindrical wake in the area of contact with the head vortex decreases progressively. At the same time, the diameter of the intrusion shell residue which creates the foundations of the wake with a more symmetrical, increases.

The role of ligaments is clearly manifested in the formation and restructuring of the flow pattern around uniform horizontally towing a sphere in a continuously stratified fluid. In the creeping mode, the density wake of a rectangular cross-section is transformed from a vertical strip to a square as the speed increases, then contracts into a horizontal strip, which is filled with attached and detached finely structured vortices as the sphere velocity continues to increase. Ligaments limit the propagation domains of internal waves [3].

High-resolution observations show that flows of liquids and gases at all scales (from astronomical units in the evolution of stars and galaxies to fractions of a millimeter in the laboratory) have a fine structure, which is the optical image of ligaments, i.e. the most common component of flows of heterogeneous fluids. The mathematical image of the fine structures of stratified fluid flows is provided by

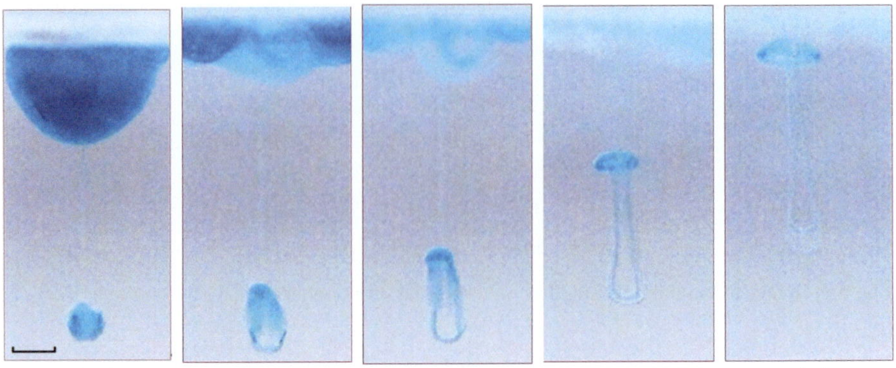

t = 67.5ms t = 133ms t = 176ms t = 281ms t = 413ms

Fig. 8.6: Transformation of a cylindrical intrusion into an up-floating regular vortex.

the families of singular components of the complete solutions to the system of fundamental equations of mechanics in stratified fluids or gases.

The physical basis for the formation of ligaments is the ongoing processes of reshaping the macromolecular fluid structure. In the course of these processes, the thermal and kinetic energy of flows is absorbed, which is converted into the potential energy of shells in associate generation, which in turn is returned to the finest components of flows during their decay. The change in the medium structure indicates the dependence of the flow of the transforming components of the total energy on the momentum and kinetic energy of the traveling wave, as well as the physical properties of the continuous medium.

8.7 Conclusion

High-resolution observations reveal the evolving fine structures of fluid and gas medium flows in the entire range of scales available for survey. All energy transfer processes play an important role in the formation of the flow pattern, i.e. radiation processes with thermal radiation, macroscopic processes with flows, waves, and vortices, and atomic and molecular processes with conversion and dissipation.

The system of fundamental equations of stratified fluid mechanics with state equations for the Gibbs potential and density allows for the systematization and unified characterization of observed phenomena in a wide range of scales and parameters of ongoing processes. The complete solutions of the system form the basis for a new classification of flow components. This classification includes traditional large-scale components such as jets, wakes, vortices, and waves, which are described by the regular components of the complete solutions, as well as the observed families of interfaces and fibers, which are corresponded to the families of singular solutions of the system, i.e. two-dimensional and three-dimensional ligaments.

Acknowledgements The research was carried out under the state assignment, registration number is 124012500442-3. The experiments were conducted at the stands of the Unique Research Facility, Hydrophysical Complex for Modeling Hydrodynamic Processes in the Environment and Their Impact on Underwater Technical Objects, as well as the Distribution of impurities in the Ocean and Atmosphere (HPC IPMech RAS).

References

[1] Bunkin NF, Suyazov NV, Shkirin AV, Ignat'ev PS, Indukaev KV (2008) Study of nanostructure of highly purified water by measuring scattering matrix elements of laser radiation. Physics of Wave Phenomena 16:243–260, DOI 10.3103/S1541308X08040018

[2] Chashechkin YD (2021) Conventional partial and new complete solutions of the fundamental equations of fluid mechanics in the problem of periodic internal waves with accompanying ligaments generation. Mathematics. 9(6):586, DOI 10.3390/math9060586

[3] Chashechkin YD (2022) Discrete and continuous symmetries of stratified flows past a sphere. Symmetry 14(6):1278, DOI 10.3390/sym14061278

[4] Chashechkin YD (2021) Foundations of engineering mathematics applied for fluid flows. Axioms 10(4):286, DOI 10.3390/axioms10040286

[5] Chashechkin YD (2018) Singularly perturbed components of flows – linear precursors of shock waves Mathematical Modelling of Natural Phenomena 13(2):1-29, DOI 10.1051/mmnp/2018020

[6] Chashechkin YD (2024) The laws of the matter distribution in a colored free-falling drop in a transparent target fluid (Review). Fluid Dynamics 59(6):1693-1734, DOI 10.1134/S0015462824604315

[7] Chashechkin YD, Il'inykh AY (2024) Submerging and floating-up vortices in the picture of smooth inflow of a free-falling ethanol drop into water. Fluid Dynamics 59:1789-1808, DOI 10.1134/S0015462824603772

[8] Chashechkin YD, Ochirov AA (2023) Periodic flows in a viscous stratified fluid in a homogeneous gravitational field. Mathematics 11(21):4443, DOI 10.3390/math11214443

[9] Eisenberg D, Kauzmann W (2005) The Structure and Properties of Water. Oxford University Press, Oxford

[10] Euler L (1757) Principes généraux du mouvement des fluids Mémoires de l'Academié royalle des sciences et belles letters. Berlin 11 (papers of 1755 year) 274-315 (Opera omnia. Series II. 12:54–91

[11] Feistel R (2018) Thermodynamic properties of seawater, ice and humid air: TEOS-10, before and beyond. Ocean Sciences 14:471–502, DOI 10.5194/os-14-471-2018

[12] Harvey AH, Hrubý J, Meier K (2023 Improved and always improving: reference formulations for thermophysical properties of water. Journal of Physical and Chemical Reference Data 52:011501, DOI 10.1063/5.0125524

[13] Kistovich YV, Chashechkin YD (1998) Linear theory of beams internal wave propagation an arbitrarily stratified liquid. Journal of Applied Mechanics and Technical Physics 39(5):302-309, DOI 10.1007/BF02468043
[14] Kistovich AV Chashechkin YD (1993) The structure of transient boundary flow along an inclined plane in a continuously stratified medium. Journal of Applied Mathematics and Mechanics 57(4):633–639, DOI 10.1016/0021-8928(93)90033-I
[15] Landau LD, Lifshitz EM (1987) Fluid Mechanics. Vol 6. Course of Theoretical Physics. Pergamon Press, Oxford, UK
[16] Levitsky VV, Dimitrieva NF, Chashechkin YD (2019) Visualization of the self-motion of a free wedge of neutral buoyancy in a tank filled with a continuously stratified fluid and calculation of perturbations of the fields of physical quantities putting the body in motion. Fluid Dynamics 54(7):948-957, DOI 10.1134/S0015462819070115
[17] Lighthill J (1978) Waves in fluids. CUP, Cambridge, UK
[18] Malenkov GG (2019) Structure and dynamics of surfaces of thin films and water microdroplets. Colloid Journal 72(5):649-659, DOI 10.1134/S1061933X1005011X
[19] Nayfeh AH (1993) Introduction to Perturbation Technique. John Wiley & Sons, New York
[20] Müller P (2006) The Equations of Oceanic Motions. CUP, Cambridge, UK
[21] Paoletti MS, Swinney HL (2012) Propagating and evanescent internal waves in a deep ocean model. Journal of Fluid Mechanics 706:571-583, DOI 10.1017/jfm.2012.284
[22] Prandtl L (1942) Führer durch die Strömungslehre. Vieweg und Sohn, Braunschweig
[23] Sreenivasan KR (2025) A brief perspective on fluid mechanics. Physics of Fluids. 37:081401, DOI 10.1063/5.0283288
[24] Stokes GG (1845) On the theories of the internal friction of fluids in motion and of the equilibrium and motion of elastic solids. Transactions of the Cambridge Philosophical Society 8:287-319, DOI 10.1017/CBO9780511702242.005
[25] Sun images: `https://sdo.gsfc.nasa.gov/data/Cited08Feb2025`
[26] Teschke O, de Souza EF 2005) Water molecule clusters measured at water/air interfaces using atomic force microscopy. Physical Chemistry Chemical Physics 7(22):3856-3865, DOI 10.1039/B511257E
[27] USI HPC IPMech RAS: Unique scientific installation "Hydrophysical complex for modeling hydrodynamic processes in the environment and their impact on underwater technical objects, as well as the distribution of impurities in the ocean and atmosphere, Ishlinsky Institute for Problems in Mechanics RAS" `http://ipmnet.ru/uniqequip/gfk`
[28] Vallis GK (2017) Atmospheric and Oceanic Fluid Dynamics. CUP, Cambridge, UK

Chapter 9
Kinetic Approach to Modeling Long-Term Failure of Materials and Structural Elements Under Creep Conditions and the Influence of an Active Medium

Leonid V. Fomin

Abstract This chapter of the book examines the application of a kinetic approach to modeling the long-term failure of structural elements, such as a composite rod and cylindrical shell, under creep and the influence of an active medium. The stress-strain states and times to failure of the elements under consideration are determined. Both a power-law model and a fractional-power-law model of creep and long-term failure are used. The kinetic approach is used to model the process of damage accumulation using a damage parameter that determines a long-term failure. Both scalar and vector forms of the parameter are under consideration. The influence of the active medium is taken into account in two variants: classical and non-classical diffusion processes. In the latter case, the active substance penetrating the material exists in two states – free and bound. A numerical calculation determines the characteristic parameters and times to failure.

Key words: Kinetic theory, Creep, Long-term failure, Active medium, Diffusion, Fractional-power-law model, Damage, Time to failure, Composite rod, Cylindrical shell

9.1 Introduction

The studies discussed in this chapter are based on the kinetic theory of creep and long-term strength of metallic materials and alloys proposed by Rabotnov [29] and Kachanov [12] and subsequently developed by their numerous students. This kinetic theory is based on the concept of the equation of mechanical state of a structurally stable material. Within this concept, the creep strain rate for a material at any given time t depends on the magnitude of the applied stress, temperature, and the structural

Leonid V. Fomin
Institute of Mechanics, Lomonosov Moscow State University, Michurinskii prosp. 1, 119192 Moscow, Russsian Federation, e-mail: fleonid1975@mail.ru

H. Altenbach et al. (eds.), *Stability Problems in Mechanics of Continua*,
Advanced Structured Materials 260, https://doi.org/10.1007/978-3-032-31032-3_9

state of the material at that time t. The structural state can be characterized by a set of values q_i called structural parameters. Kinetic equations are written for these structural parameters. Such parameters may include damage to the metal during creep, the work of stresses on creep strains (energy variant), the active medium concentration in the metal, etc.

In general, there are two approaches in the theory of long-term strength to determining failure (time to failure): kinetic, in which the accumulation of damage is considered as a process of gradual, long-term failure, and criterial, using the dependence of the time to failure on the invariant characteristics of the stress tensor or their combinations (equivalent stress). In this work, there will be used a kinetic approach, in which damage ω and the active medium concentration c in the material will be used as structural parameters

Ensuring the safe operation of materials and structural elements exposed to active media is undoubtedly an important and factual task for maintaining the structure's operability throughout its entire service life. The influence of an active medium can be caused by both diffusion penetration and more complex forms of interaction with the material, such as corrosion. The study of such processes under high-temperature, long-term loading of metals and alloys under creep conditions is of particular importance [1, 2, 9, 12, 16, 19–21, 25, 26, 28–33]. There is considered the application of the kinetic approach to modeling long-term failure due to creep in an active medium of such structural elements as a composite tensile rod and a cylindrical shell under an unsteady loading program with axial force and internal pressure.

9.2 Long-term Failure of a Composite Rod in an Active Medium

9.2.1 Problem Statement

The stress-strain state and time to failure under creep conditions of a long composite tensile rod of rectangular cross-section located in an active medium are determined [10]. The rod consists of three parts symmetrically located along the width (Fig. 9.1), connected by ideal adhesion without slippage. The length of the rod significantly exceeds its width and thickness. The influence of the active medium is considered in two variants: classical and non-classical diffusion processes [5]. In the case of a non-classical diffusion process, the active substance that has penetrated the material is in two states – free and bound. The process of such diffusion is described by a modified diffusion equation, which takes into account the two-phase state of the active substance in the material:

$$D\frac{\partial^2 c}{\partial z^2} = \frac{\partial c}{\partial t} + \frac{\partial S}{\partial t}, \qquad \frac{\partial S}{\partial t} = \beta c + \alpha S, \qquad c(z,0) = S(z,0) = 0$$

where c is concentration of the free (mobile) form of the active medium per unit volume (or per unit mass) of the sample; S is concentration of the bound form of the

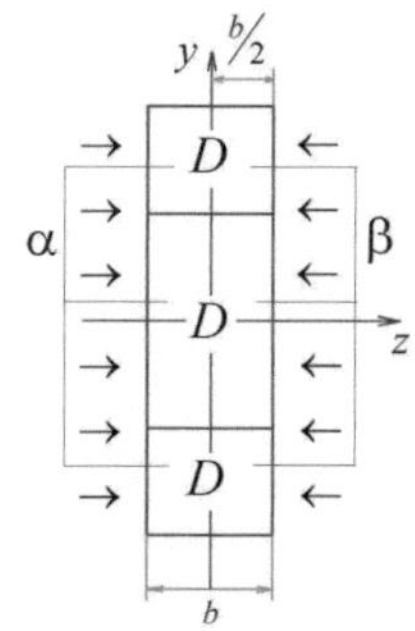

Fig. 9.1 Cross-section of a composite rod. The schematics of the active medium influence.

active medium per unit volume (or per unit mass) of the sample; D = const is the diffusion coefficient; α, β - are diffusion model parameters, z is a coordinate; t is time.

Since the width H of the cross-section significantly exceeds its thickness b, the influence of diffusion from the narrow sides of the rectangular cross-section can be neglected. The length of the rod L is many times greater than the characteristic dimensions of its cross-section, so the influence of the longitudinal coordinate of the rod x on the diffusion process can also be neglected. Thus, the diffusion process is a one-dimensional and occurs through the thickness b along the transverse coordinate z. The same boundary conditions for the diffusion process are adopted on the wide sides of the rectangular cross-section. Therefore, the diffusion process is symmetrical relative to the axis passing through the center of the rod cross-section. The same diffusion process characteristics are assumed for the central part and the two outer parts of the rod. Specifically, assume that the active medium penetrate the central and outer parts with a diffusion coefficient D = const. The probability α of the transition of the bound form of the active medium molecules to the free form and the probability β of the reverse process are assumed to be the same for all parts of the rod. The schematics of the effect of the active medium on the composite rod is shown in Fig. 9.1

Under the assumed condition of rigid connection of parts in a composite rod (without mutual slippage), the stress states near the junctions of the layers become non-uniform and non-homogeneous. Common joining methods include diffusion welding, soldering, gluing, and various gas-dynamic spraying techniques. Non-homogeneous stress states near joints can arise, among other things, due to inherent imperfections in the technological methods used to join parts together. Common joining methods include diffusion welding, soldering, gluing, and various gas-dynamic spraying techniques.

General trends in the development of methods for studying and determining the stress-strain state of typical structural elements (rods, plates, thin-walled shells) are leading to the reduction of three-dimensional problems to simpler two- and one-dimensional ones, thereby reducing the problem's dimensionality. Simplification methods based on some hypotheses and known principles are used. Implementation of methods using known and accepted hypotheses in mechanics typically does not include procedures for regular solution refining.

Returning to the problem statement, given the small transverse dimensions of the composite rod compared to its length and the additional condition of symmetry of the diffusion process relative to the central plane of symmetry, there is considered a one-dimensional form of the stress-strain state in the composite rod. In the solution process, there is used the hypothesis of plane sections and a generalization of Saint-Venant's principle for the problem under consideration, which allows assuming that the stress state is homogeneous far from the junctions of the composite rod parts.

9.2.2 Power-Law Model of Creep and Long-Term Failure

Power-law relationships are used as a model for creep and long-term failure – creep strain rates and damage accumulation are power-law functions of stress. Ratios (9.1) describe the creep strain rate $\dot{p}$ of the first (central) part and the two outer parts, respectively

$$\frac{dp_1}{dt} = B_1\left(\frac{\sigma_1}{1-\omega_1}\right)^n, \quad \frac{dp_2}{dt} = B_2\left(\frac{\sigma_2}{1-\omega_2}\right)^n \tag{9.1}$$

where σ_1 is the stress in the first (central) part of the rod, σ_2 is the stress in the two outer parts of the rod, ω_1 is the damage parameter for the central part of the rod, ω_2 is the damage parameter that is the same for the two outer parts of the rod, B_1, B_2, n are material constants, and t is time.

Most experimental data show that an active medium has a destructive effect on materials, reducing the time to failure compared to the operation of materials and structural elements under neutral conditions. Failure processes are initiated by the growth of damage in the material. The influence of the active medium is taken into consideration by introducing a function of the integral average concentration $\overline{c}_m$ into the kinetic equations for damage accumulation:

$$\begin{aligned} \frac{d\omega_1}{dt} &= A\left(\frac{\sigma_1}{1-\omega_1}\right)^m f(\overline{c}_m), \ \ \omega_1 = \omega_1(t), \ \ \omega_1(0) = 0, \ \ \omega_1^* \equiv \omega_1(t = t_1^*) = 1, \\ \frac{d\omega_2}{dt} &= A\left(\frac{\sigma_2}{1-\omega_2}\right)^m f(\overline{c}_m), \ \ \omega_2 = \omega_2(t), \ \ \omega_2(0) = 0, \ \ \omega_2^* \equiv \omega_2(t = t_2^*) = 1, \end{aligned} \tag{9.2}$$

where t_1^* is the time to failure of the central part of the rod, t_2^* is the time to failure of the outer parts of the rod, A, m are material constants determined on the basis of long-term strength experiments without an active medium under neutral conditions, while the function of the influence of the active medium is $f(\overline{c}_m) \equiv 1$.

When solving, the equilibrium equations, strain compatibility equations, and the ratio for the integral average concentration of the active medium in the material are additionally used. The resulting system of equations (9.3) models the creep of the composite rod under consideration, in which its parts are rigidly connected without slippage:

$$\begin{aligned}
&\frac{d\omega_1}{d\bar{t}} = C_1 \frac{\overline{\sigma}_1^m}{(1-\omega_1)^m} f\left(\overline{c}_m(K\bar{t})\right), \quad \omega_1(0) = 0, \\
&\frac{d\omega_2}{d\bar{t}} = C_1 \frac{(\Sigma_0 - \xi\overline{\sigma}_1)^m}{((1-\xi)(1-\omega_2))^m} f\left(\overline{c}_m(K\bar{t})\right), \quad \omega_2(0) = 0, \\
&\int\limits_0^{\bar{t}} \overline{\sigma}_1^n \frac{1}{(1-\omega_1)^n} d\bar{t} = k \int\limits_0^{\bar{t}} \left(\frac{(\Sigma_0 - \xi\overline{\sigma}_1)}{(1-\xi)(1-\omega_2)}\right)^n d\bar{t}, \\
&\overline{\sigma}_2 = \frac{(\Sigma_0 - \xi\overline{\sigma}_1)}{(1-\xi)}
\end{aligned} \tag{9.3}$$

where $\overline{\sigma}_1 = \sigma_1/\sigma_0$, $\overline{\sigma}_2 = \sigma_2/\sigma_0$ are dimensionless stresses in the central and two outer parts of the composite rod respectively; σ_0 is some characteristic stress, for example, half of the short-term tensile strength; ω_1, ω_2 are damages accumulated during creep in the materials of the central and outer parts,

$$C_1 = \frac{A}{B_1}\sigma_0^{m-n}, \quad k = \frac{B_2}{B_1}, \quad \xi = \frac{2h_1}{H}, \quad \Sigma_0 = \frac{P}{bH\sigma_0}$$

are dimensionless parameters; A, B_1, B_2, n, m are material parameters included into the kinetic and constitutive equations of the composite rod parts; P is an axial force; b, H are thickness and width of the composite rod, respectively; $2h_1$ is the width of the central part of the rod; $\bar{t} = B_1\sigma_0^n t$ is dimensionless time. The coefficient $K = D/(\sigma_0^n b^2 B_1)$ relates the characteristic dimensionless times of diffusion and creep.

A linear representation of function $f(\overline{c}_m(\bar{t})) = 1 + a_c\overline{c}_m(\bar{t})$ is used, where a_c is a constant determined from an experiment on the long-term strength of specimens in an active medium. In [6], the value of this constant $a_c = 9.5$ was determined for experiments on the long-term strength of carbon steel specimens in a high-temperature air medium [27]. In the resulting system of equations, the unknowns are damages $\omega_1 = \omega_1(\bar{t})$ and $\omega_2 = \omega_2(\bar{t})$, as well as stresses $\overline{\sigma}_1 = \overline{\sigma}_1(t)$ and $\overline{\sigma}_2 = \overline{\sigma}_2(t)$.

The failure criterion for each part of the rod is the condition for the damage parameter for the corresponding part to reach a value equal to unity

$$\omega_1^\star = \omega_1(\bar{t} = \bar{t}_1) = 1$$

and

$$\omega_2^\star = \omega_2(\bar{t} = \bar{t}_2) = 1.$$

The minimum of the resulting times to failure of the individual parts of the rod $\bar{t}^\star = \min[\bar{t}_1, \bar{t}_2]$ will be considered as the total time to failure of the entire composite rod $\bar{t}^\star$.

System (9.3) is solved numerically in time steps. Depending on the material parameters and the parameters of the diffusion process, the stress distributions and damage accumulation processes over time in various parts of the composite rod are analyzed.

Calculations were performed for two variants of the set of constants differing only in the values of the parameters $\overline{\alpha}$ and $\overline{\beta}$ – in the first variant $\overline{\alpha} = 1, \overline{\beta} = 1$, in the second variant $\overline{\alpha} = 1$, $\overline{\beta} = 0$. The second variant corresponds to the classical diffusion process, since the probability of the transition per unit time of the free form of the active medium molecules to the bound form equals zero $\overline{\beta} = 0$. Therefore, the concentration of the bound form of the active medium per unit volume (or per unit mass) of the sample is zero $S = 0$.

The calculations yielded dependencies for damage accumulation and stress distribution in the rod components over time. Figures 9.2 and 9.3 show the graphs of damage accumulation and stress distribution over time in the composite rod components for cases where classical and non-classical diffusion processes are taken into account. As a result, it was determined that the failure of the composite rod in the classical case occurs earlier than in the case of the non-classical diffusion process under consideration.

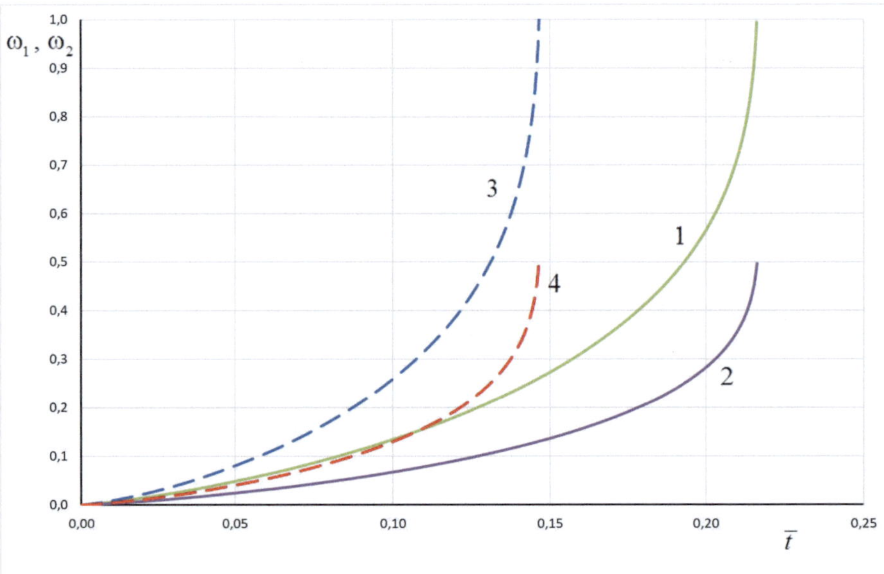

Fig. 9.2: Graphs of the dependences of damage accumulation over time in parts of a composite rod, where the dashed line corresponds to the case of a classical diffusion process and the solid line – to the case of a non-classical diffusion process.

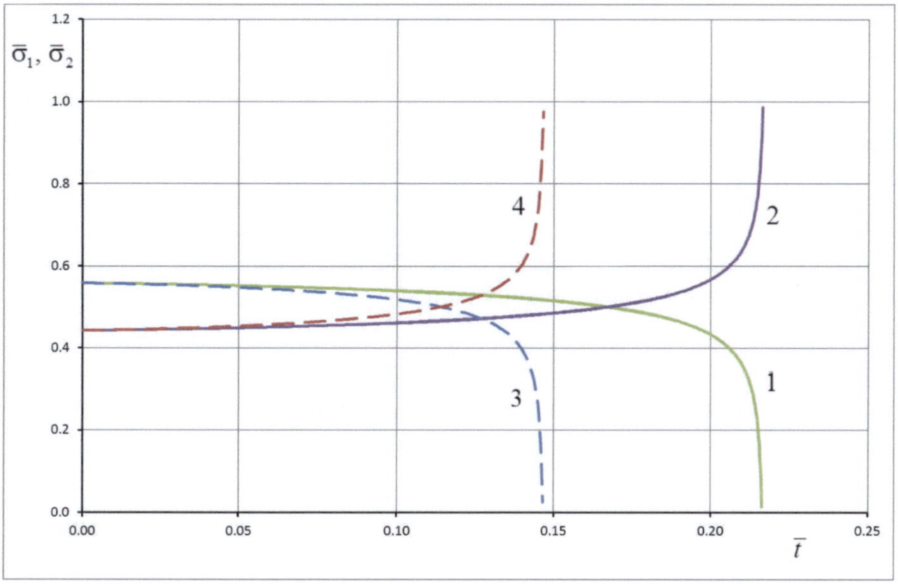

Fig. 9.3: Graphs of stress dependences over time in parts of a composite rod where the dashed line corresponds to the case of a classical diffusion process and the solid line – to the case of a non-classical diffusion process.

9.2.3 Fractional Power Model of Creep and Long-Term Failure

In [7], along with the use of a power-law model for creep and long-term failure, the use of a singular fractional-power-law model is considered [34]. It includes a natural mechanical characteristic – the short-term tensile strength of the material at the test temperature.

The constitutive and kinetic relations in dimensionless form for the central and two outer parts of the composite rod are as follows:

$$
\begin{aligned}
\frac{dp_1}{d\bar{t}} &= \left(\frac{\overline{\sigma}_1}{(1-\overline{\sigma}_1)(1-\omega_1)}\right)^n, \\
\frac{dp_2}{d\bar{t}} &= k\left(\frac{\overline{\sigma}_2}{(1-\overline{\sigma}_2)(1-\omega_2)}\right)^r, \\
\frac{d\omega_1}{d\bar{t}} &= C_1\left(\frac{\overline{\sigma}_1}{(1-\overline{\sigma}_1)(1-\omega_1)}\right)^m f(K\overline{c}_m), \\
\frac{d\omega_2}{d\bar{t}} &= C_2\left(\frac{\overline{\sigma}_2}{(1-\overline{\sigma}_2)(1-\omega_2)}\right)^v f(K\overline{c}_m),
\end{aligned}
$$

where

$$C_1 = \frac{A_1}{B_1}, C_2 = \frac{A_2}{B_1}, K = \frac{D}{b^2 B_1}, k = \frac{B_2}{B_1}, \xi = \frac{2h_1}{H}, \Sigma_0 = \frac{P}{bH\sigma_{b1}}, \zeta = \frac{\sigma_{b2}}{\sigma_{b1}}$$

are dimensionless parameters; A_1, B_1, A_2, B_2, n, r, m, v are material parameters included in the fractional-power kinetic and constitutive relationships of the parts of the composite rod; σ_{b1}, σ_{b2} are short-term strength limits at the test temperature of the material of the central part and the outer parts, respectively; $\bar{t} = B_1 t$ is dimensionless time. As a result of numerical calculations, the dependences of damage and stress on time in different parts of the rod were obtained for different ratios $\zeta = \sigma_{b2}/\sigma_{b1}$ of the short-term strength limits.

9.3 Long-Term Failure of a Cylindrical Shell in an Active Medium

9.3.1 Problem statement

A long, thin-walled, bottomless cylindrical shell with a mid-surface radius R and the wall thickness H is considered. The shell is exposed to a tensile axial force P and an internal pressure Q. The shell is placed in an active medium. The influence of the active medium is determined by the diffusion of its elements into the material. The curvature of the thin-walled shell is small, and the ratio of the shell wall thickness to its radius is much less than unity. This fact allows accurate consideration of the diffusion of the medium into the shell wall as one-dimensional diffusion into a flat plate. The shell is surrounded by the active medium on all sides (for example, air at high temperatures is an active medium), therefore a symmetric diffusion process relative to the mid-surface of the shell wall is considered. This study examines an unsteady loading program for the shell, which is implemented through the separate action of internal pressure, sequential in time, and an axial tensile force until failure.

9.3.2 Approximate Method of Solving the Diffusion Equation

Approximate methods for calculating stationary fields are described in a number of classical monographs. Apparently, the approximate method of solving parabolic problems with the introduction of a perturbation front was first considered by Lembke [15], and subsequently developed by Leibenzon [14], Charnyy [4], Barenblatt [3], Shesterikov and Yumasheva [35], and other scientists.

This approach is centered on the method of sequentially changing stationary states. The method under consideration is as follows. At each moment in time, the region is divided into perturbed and unperturbed parts, where in in the perturbed region the concentration distribution is assumed to follow a stationary law and the moving boundary of the perturbed and unperturbed regions is determined based on the material balance equations and boundary conditions

Khristianovich demonstrated the fundamental similarity between the problems of boundary layer theory and the theory of one-dimensional nonstationary filtration and diffusion, using the model of proven approximate methods, in particular, the generalized von Kármán-Polhausen method. The method is based on an approximate representation of the solution in the perturbed region by a polynomial in powers of the spatial variable with time-dependent coefficients. These coefficients are determined from the condition that the approximate solution satisfies several introduced "instant" (i.e. at the given moment in time) integral relations and conditions at the boundaries of the perturbed region are as follows [3]:

$$\frac{\partial c}{\partial t} = D\frac{\partial^2 c}{\partial x^2},$$

$$\frac{d}{dt}\int_{l_1}^{l_2} c(x,t)dx = D\left(\frac{\partial c}{\partial x}\right)\bigg|_{x=l_2} - D\left(\frac{\partial c}{\partial x}\right)\bigg|_{x=l_1} + c(l_2,t)\frac{dl_2}{dt} - c(l_1,t)\frac{dl_1}{dt},$$

where

$$c(x,t) = C_0(t) + C_1(t)\frac{x}{l(t)} + C_2(t)\frac{x^2}{l^2(t)} + \cdots + C_n(t)\frac{x^n}{l^n(t)} \quad \text{with} \quad 0 \le x \le l(t)$$

and

$$c(x,t) = 0 \quad \text{with} \quad x > l(t).$$

These relations are similar to the integral relations of boundary layer theory proposed by Loitsyansky [23].

If polynomials of increasingly higher degrees are used in the expressions for the concentration and a fixed number of conditions are satisfied at the boundaries of the perturbed region, as well as an increasing number of successive "instant" integral relations, then the resulting expressions for the concentration distribution will converge to an exact solution of the problem [3]. The convergence to an exact solution with increasing polynomial degree was demonstrated using a numerical experiment in the monograph [16].

In this chapter, an approximate method of solving the diffusion equation is proposed to analyze the diffusion of an active medium into the thin-walled cylindrical shell under consideration. This method is based on the introduction of a diffusion front propagating from the shell surface. Due to the small curvature of the shell wall, such a diffusion process can be considered as a one-dimensional process in a flat plate, which occurs symmetrically due to the geometric symmetry of the flat plate relative to the middle surface and the symmetry of the boundary conditions and diffusion characteristics.

Based on the symmetry condition, we consider half the shell cross-section along its thickness. We introduce dimensionless variables

$$\bar{z} = 2z/H \quad (0 \le \bar{z} \le 1), \qquad \bar{t} = 48Dt/H^2, \quad \bar{c} = c/c_0,$$

where z is the coordinate along the shell thickness ($z = 0$ at the boundary of the shell cross-section with the external medium, $z = 1$ at the midline of the shell cross-section), t is time, c_0 is the constant concentration at the boundary of the shell material and the external medium, D = const is the diffusion coefficient. The one-dimensional diffusion equation in these variables takes the following form:

$$\frac{\partial \bar{c}}{\partial \bar{t}} = \frac{1}{12}\left(\frac{\partial^2 \bar{c}}{\partial \bar{z}^2}\right) \qquad 0 < \bar{t} < \infty, \qquad 0 < \bar{z} < 1 \tag{9.4}$$

The initial and boundary conditions are as follows:

$$\bar{c}(\bar{z}, 0) = 0, \qquad \bar{c}(0, \bar{t}) = 1, \qquad \frac{\partial \bar{c}}{\partial \bar{t}}(1, \bar{t}) = 0$$

Here, a concentration burst is assumed on the shell surface when $\bar{t} = 0$. The latter condition (on the mid-surface at $\bar{z} = 1$) is adopted due to the symmetry of the diffusion process relative to the shell mid-surface.

The dependence of the concentration $\bar{c}$ on the coordinate $\bar{z}$ in this research is assumed to be a second-degree polynomial satisfying the boundary and initial conditions. Two stages of the diffusion process are considered – the front penetration stage and the saturation stage, separated by the time moment $\bar{t}_0$

$$\bar{c}(\bar{z}, \bar{t}) = \begin{cases} \left[1 - \left(\bar{z}/\bar{l}(\bar{t})\right)\right]^2 & \text{with} \quad 0 < \bar{t} \le \bar{t}_0, \quad 0 \le \bar{z} \le \bar{l}(\bar{t}) \\ 0 & \text{with} \quad 0 < \bar{t} \le \bar{t}_0, \quad \bar{l}(\bar{t}) < \bar{z} < 1 \\ B(\bar{t}) + \left[1 - B(\bar{t})\right] \cdot (1 - \bar{z})^2 & \text{with} \quad \bar{t} > \bar{t}_0 \end{cases} \tag{9.5}$$

where $\bar{l}(\bar{t})$ is the coordinate of the diffusion front, is the transition time between stages of the diffusion process, and $B(\bar{t})$ is the concentration at the center of the shell cross-section at $\bar{t} \ge \bar{t}_0$ (on the midline $\bar{z} = 1$).

The unknown dependencies $\bar{l}(\bar{t})$ and $B(\bar{t})$ are determined from the integral satisfaction of the parabolic function $\bar{c}(\bar{z}, \bar{t})$ from (9.5) with the diffusion equation (9.3). As a result, the coordinate of the diffusion front $\bar{l}(\bar{t})$ and the concentration $B(\bar{t})$ are obtained in the following form

$$\bar{l}(\bar{t}) = \sqrt{\bar{t}}, \qquad B = 1 - \exp\left[-(1/4)(\bar{t} - 1)\right]. \tag{9.6}$$

The diffusion front reaches the midline of the shell cross-section at time $\bar{t}_0$, with $\bar{l} = 1$. Based on this boundary condition, the value of time $\bar{t}_0 = 1$ is obtained. Using the concentration relations (9.5) and relation (9.6), there are obtained the expressions for in the following form

$$\overline{c}(\overline{z},\overline{t}) = \begin{cases} \left[1-\left(\overline{z}/(\sqrt{\overline{t}})\right)\right]^2 & \text{with } 0<\overline{t}\le 1,\quad 0\le\overline{z}\le\sqrt{\overline{t}} \\ 0 & \text{with } 0<\overline{t}\le 1,\quad \sqrt{\overline{t}}<\overline{z}<1 \\ 1-\left(1-(1-\overline{z})^2\right)\exp\left[-(1/4)(\overline{t}-1)\right] & \text{with } \quad \overline{t}>1 \end{cases}$$

Further, the integral average concentration $\overline{c}_m(\overline{t})$ will be used to analyze the influence of the active medium on the time to failure of the shell. It has the following form:

$$\overline{c}_m(\overline{t}) = \int_0^1 \overline{c}(\overline{z},\overline{t})dz = \begin{cases} (1/3)\sqrt{\overline{t}} & \text{with } \quad 0<\overline{t}\le 1 \\ 1-(2/3)\cdot\exp\left[-(1/4)(\overline{t}-1)\right] & \text{with } \quad \overline{t}>1 \end{cases} \tag{9.7}$$

[16] presents the results of a numerical experiment comparing the exact and proposed approximate solutions of the diffusion equation. The equality (9.8) below is used in [16] as a measure of the error of the approximate solution $\delta(\overline{t})$ compared to the exact one $\overline{c}_{\text{exact}}(\overline{z},t)$:

$$\delta(\overline{t}) = \sqrt{\int_0^1 \left(\overline{c}(\overline{z},\overline{t}) - \overline{c}_{\text{exact}}(\overline{z},t)\right)^2 d\overline{z}} \tag{9.8}$$

As a result, the error measure $\delta(\overline{t})$ in (9.8) is found to vary from 1% to 4% and decreases with increasing exponent of the approximation polynomial used $\overline{c}(\overline{z},\overline{t})$. Considering the well-known fact that experimental creep curves and times to failure typically exhibit significant statistical scatter for the same values of mechanical stress and temperature, the effect of a slight deviation in the active medium concentration on creep and long-term strength in accordance with the approximate solution compared to the exact solution is quite acceptable.

9.3.3 Determination of Stress-Strain State Components Using a Fractional-Linear Creep Model

The problem under consideration is statically determinable, and the stress state is determined only by external loads. The stress state in the shell is plane (with zero radial stress σ_{rr}). The components of the stress tensor are obtained from the equilibrium equations of the elementary volume of the solid [36]:

$$\sigma_{zz} = \frac{P}{2\pi RH}, \qquad \sigma_{\theta\theta} = \frac{RQ}{H} \tag{9.9}$$

The stress intensity and the average normal stress in the problem under consideration are, respectively, equal to

$$\sigma_u = \sqrt{(\sigma_{zz})^2 + (\sigma_{\theta\theta})^2 - \sigma_{zz}\sigma_{\theta\theta}},$$

$$\sigma = \frac{1}{3}\left(\sigma_{zz} + \sigma_{rr} + \sigma_{\theta\theta}\right), \quad \sigma_{rr} = 0 \quad \sigma = \frac{1}{3}\left(\sigma_{zz} + \sigma_{\theta\theta}\right)$$

A variant of the description of the creep process using the fractional-linear relation to be considered is proposed by Shesterikov and Yumasheva [34].

To describe the creep of metals and alloys in the one-dimensional case, the most widely used is the power-law dependence of the steady-state creep rate $\dot{p}$ on stress σ , which has the form:

$$\dot{p} = B\sigma^n, \quad B = \text{const}, \quad n = \text{const}$$

The dot above the corresponding symbol denotes the time derivative. It approximates experimental data fairly well within the selected narrow stress range for a wide class of metallic materials and alloys, but, as Rabotnov [28] noted, if a sufficiently wide stress range is taken, the exponent becomes dependent on stress σ.

One possible relationship for describing creep processes is a fractional-linear dependence [34], which in some cases allows for a fairly good representation of the creep rate as a function of stress over a wide stress range. Shesterikov and Yumasheva proposed using such a dependence in the following form [34]:

$$\dot{p} = k_0 \frac{\sigma - \sigma_0}{\sigma_b - \sigma}, \quad \sigma > 0, \quad k_0 = \text{const},$$

where σ_0 is the creep limit (at $\sigma < \sigma_0$, creep strain is virtually absent), in many cases $\sigma_0 = 0$, can be taken as $\sigma_0 = 0$, and σ_b is the short-term strength limit at the corresponding test temperature.

At stresses close to σ_b , the creep strain rate $\dot{p}$ increases significantly, and failure occurs at the considered times. Thus, in the fractional-linear relationship, stress acts as an ultimate stress. Unlike the standard power-law dependence, where stress can be chosen arbitrarily large, in this singular dependence, stress is limited by σ_b, which is better justified from a physical and mechanical perspective.

The influence of the active medium can be taken into account by introducing a function of the integrally average concentration of the active medium $f\left(\overline{c}(\overline{t})\right)$ into the shell material

$$\dot{p}_u = A\left(\sigma_u/(\sigma_b - \sigma_u)\right) f\left(\overline{c}(\overline{t})\right) \tag{9.10}$$

where $\dot{p}_u$ is creep strain rate intensity, σ_u is stress intensity, σ_b short-term tensile strength limit at a given test temperature (here, it is assumed that the values $\overline{c}(\overline{t})$ for tension and compression are the same), $\overline{c}(\overline{t})$ is an integral average concentration of medium elements in the shell, A is a material parameter. It should be noted that relation (9.10) uses stress intensity σ_u, which is always a positive value regardless of the stress sign.

In further calculations, a linear form of the function $f\left(\overline{c}(\overline{t})\right)$ satisfying the condition $f(0) = 1$ will be used:

$$f\left(\overline{c}(\overline{t})\right) = 1 + a \cdot \overline{c}_m(\overline{t}) \tag{9.11}$$

where $a =$ const determined from long-term strength experiments taking into account the influence of the surrounding active medium [6, 27].

In this study, the following assumptions are taken:

1. the strains are small;
2. the shell theory [36] is used to determine the stress-strain state components;
3. the shell material is incompressible;
4. the elastoplastic strains are neglected compared to creep strains.

The dependence of the creep strain tensor components on the stress tensor components, assuming steady-state creep of the shell material and taking into account the incompressibility of the material, is determined based on the hypothesis of proportionality of stress deviators and creep strain rates for an incompressible solid in the form [16, 28]:

$$\dot{p}_{ij} = \frac{3}{2}\frac{f(\sigma_u)}{\sigma_u} s_{ij}, \qquad s_{ij} = \sigma_{ij} - \sigma\delta_{ij} \qquad \sigma = \frac{1}{3}\sigma_{kk}, \qquad \dot{p}_u = f(\sigma_u) \tag{9.12}$$

where $\dot{p}_{ij} = 0$ is a condition of incompressibility of the material, s_{ij} is a stress deviator, δ_{ij} is the Kronecker unit tensor.

Given the relation (9.10) for $\dot{p}_u$

$$\dot{p}_{zz} = \frac{3}{2}\frac{A\left(\sigma_u/(\sigma_b - \sigma_u)\right) \cdot f\left(\overline{c}_m(\overline{t})\right)}{\sigma_u} s_{zz}, \; s_{zz} = \sigma_{zz} - \sigma\delta_{zz}, \; \sigma = \frac{1}{3}\left(\sigma_{zz} + \sigma_{\theta\theta}\right)$$

$$s_{zz} = \frac{2}{3}\left(\sigma_{zz} - \frac{1}{2}\sigma_{\theta\theta}\right)$$

the expression for $\dot{p}_{zz}$ was obtained:

$$\dot{p}_{zz} = \frac{Af\left(\overline{c}_m(\overline{t})\right))}{(\sigma_b - \sigma_u)}\left(\sigma_{zz} - \frac{1}{2}\sigma_{\theta\theta}\right) \tag{9.13}$$

Similarly, the expression for $p_{\theta\theta}$ can be obtained:

$$p_{\theta\theta} = \frac{3}{2}\frac{A\left(\sigma_u/(\sigma_b - \sigma_u)\right) \cdot f\left(\overline{c}_m(\overline{t})\right)}{\sigma_u}\sigma_{\theta\theta}, \; s_{\theta\theta} = \sigma_{\theta\theta} - \sigma\delta_{\theta\theta} = \frac{2}{3}\left(\sigma_{\theta\theta} - \frac{1}{2}\sigma_{zz}\right)$$

$$\dot{p}_{\theta\theta} = \frac{Af\left(\overline{c}_m(\overline{t})\right))}{(\sigma_b - \sigma_u)}\left(\sigma_{\theta\theta} - \frac{1}{2}\sigma_{zz}\right) \tag{9.14}$$

As a result, relations (9.9), (9.13) and (9.14), taking into account (9.11), determine the stress-strain state of a thin-walled cylindrical shell given the use of a fractional-linear creep model.

9.3.4 Long-Term Failure of the Shell Taking into Account the Influence of the Active Medium. Scalar and Vector Damage Parameters

Currently, there is insufficient systematic experimental data on creep strains and long-term strength under arbitrary unsteady complex stress states. The random nature of accumulated damage as the type of complex stress state changes should also be noted. The question arises as to how to account for damage accumulation during creep. According to the author, the optimal method for modeling fracture processes is to use a vector damage accumulation parameter.

Using a scalar parameter does not always fully account for the sensitivity of creep rate and damage accumulation to the anisotropic properties of the material. It should be noted that anisotropy of properties can be the result of preliminary mechanical processing of materials and products. Using a tensor damage parameter typically requires determining a significant number of material constants and material functions, which can be a labor-intensive task due to the need to conduct a wide range of setup experiments. In these respects, a vector approach is optimal.

It should also be noted that the vector approach allows modeling the long-term strength and determining the time to failure both with a constant position of the principal axes and with changes in the orientation of the principal axes of the stress-strain state, and even when such a change in the orientation of the principal axes is repeated multiple times. For example, in [18], the results of well-known tests of tubular specimens under constant axial stress and constant or alternating shear stress are considered. Experiments show that cyclic changes in the sign of shear stresses lead to a significant increase in the time to failure. To describe this effect, Kachanov's kinetic approach [11, 13] with a vector damage parameter is used. Application of this approach leads to a good agreement between the experimental and theoretical values of the times to failure.

Furthermore, in [17], using the vector approach, a well-known effect is modeled, namely, that the time to failure under equiaxial plane stress is several times shorter than the time to failure under uniaxial tension at the same stress value. To this end, a variant of Rabotnov's kinetic theory of creep and long-term strength [28, 29] is proposed, additionally accounting for material anisotropy. Also in this article, modeling of long-term strength under abrupt changes in stress intensity is performed in two ways - both by accounting for damage accumulation only during creep and by additionally accounting for damage accumulation under instantaneous loading.

All variants of the proposed kinetic equations lead to a good agreement between the experimental and theoretical values of the times to failure. An analytical review of known experimental and theoretical studies of creep and long-term strength of metals under unsteady complex stress states, published over the past 60 years, is provided in the article [22].

Rabotnov's kinetic theory of creep and long-term strength is used and a function of the integrally average concentration level $\overline{c}_m(\overline{t})$ is introduced into the kinetic equations. The kinetic equation, taking into account the scalar damage parameter, are considered in the following form:

$$\dot{\omega} = d\omega/d\bar{t} = C\left(\sigma_u/(\sigma_b - \sigma_u)\right) \cdot f\left(\bar{c}_m(\bar{t})\right), \qquad \omega(\bar{t} = 0) = 0. \tag{9.15}$$

The criterion of failure is the damage parameter $\omega^{\star}(\bar{t} = \bar{t}_{\omega}^{\star}) = 1$ reaching a value equal to unity at time moment $\bar{t} = \bar{t}_{\omega}^{\star}$.

As an option for developing the modeling of damage accumulation under conditions of long-term high-temperature loading, it is possible to take the vector $\overrightarrow{\Omega}(\Omega_1, \Omega_2, \Omega_3)$ [11, 24] as a damage parameter. Its components are associated with the space of principal stresses σ_i ($i = 1, 2, 3$) and are determined in general by the following dependencies:

$$\frac{d\Omega_i}{dt} = \begin{cases} f_i(\Omega_i, \sigma_i, t) & \text{with } \sigma_i > 0 \\ 0 & \text{with } \sigma_i \leq 0 \end{cases}$$

In this paper, taking into account the separate program of shell loading by internal pressure and axial tensile force considered, the axes of the introduced cylindrical coordinate system are the principal axes of the stress-strain state. The kinetic equations, considering the vector damage parameter [24] $\overrightarrow{\Omega}$, will be adopted in the following form:

$$\dot{\Omega}_i = \frac{d\Omega_i}{d\bar{t}} = \begin{cases} C \cdot \left(\sigma_i/(\sigma_b - \sigma_i)\right) \cdot f\left(\bar{c}_m(\bar{t})\right) & \text{with } \sigma_i > 0 \\ 0 & \text{with } \sigma_i \leq 0 \end{cases} \quad i = z,\ \theta \tag{9.16}$$

where Ω_i is a projection of the damage vector onto the i-axis of the coordinate system. The failure criterion is the length Ω of vector $\overrightarrow{\Omega}$ reaching a value equal to unity:

$$\Omega = \sqrt{(\Omega_z)^2 + (\Omega_\theta)^2} \qquad \Omega(\bar{t} = 0) = 0 \qquad \Omega^{\star} = \Omega(\bar{t} = \bar{t}_{\Omega}^{\star}) = 1$$

Here, some commentary is needed on the determination of the parameters of the creep and long-term failure model equations. These parameters (material constants) are determined by processing experimental data on creep and long-term failure in a neutral medium. It should be also noted that to obtain qualitative characteristics of creep properties and differences in damage accumulation when taking into account scalar and vector parameters, certain model values of the specified material constants can be adopted.

The kinetic approach and constitutive relations discussed above were applied to modeling the long-term failure of a cylindrical shell under a given unsteady loading program [8]. The times to failure were determined using scalar and vector damage parameters. The following unsteady loading program was adopted. First, the internal pressure $Q > 0$ is applied to the long cylindrical shell under consideration for a time

$$0 < \bar{t} < \bar{t}_1.$$

The time is taken as $\bar{t}_1 = 0.5\bar{t}_1^{\star}$ (where $\bar{t}_1^{\star}$ is the time to failure of the shell exposed to only this internal pressure Q), while the axial tensile force is equal to zero $P = 0$. Then, for time $\bar{t}_1 < \bar{t} < \bar{t}^{\star}$ (where $\bar{t}^{\star}$ is the time to failure of the shell at the end of the loading program), the axial tensile force $P > 0$ is applied to the shell under consideration, while the internal pressure is equal to zero (i.e., the conditions $Q = 0$, $P > 0$ are satisfied). Such a program will be called a separate loading program [8].

9.3.5 Calculation Results

As a result, the following values of dimensionless times to failure were obtained [8]:

- $\bar{t}^{\star} \equiv \bar{t}_{\omega}^{\star} = 0.614$ - when using a scalar damage parameter;
- $\bar{t}^{\star} \equiv \bar{t}_{\Omega}^{\star} = 0.817$ - when using a vector damage parameter.

Thus, the time to failure obtained using the vector damage parameter is greater than the time to failure obtained using the scalar damage parameter. The ratio of the times to failure is equal to $\xi = \bar{t}_{\Omega}^{\star}/\bar{t}_{\omega}^{\star} = 1.33$.

The undertaken study describes the difference in accounting for damage accumulation mechanisms when applying scalar and vector damage parameters using a fractional-linear model of creep and long-term failure under long-term high-temperature loading in an active medium. The obtained result confirms the similar results obtained in [19] in the study of long-term unsteady loading of a plate in mutually orthogonal planes. In [19], the ratios of times to failure $\bar{t}_{\Omega}^{\star}/\bar{t}_{\omega}^{\star}$, depending on the ratio of bending moments, are within the range $\bar{t}_{\Omega}^{\star}/\bar{t}_{\omega}^{\star} = 1.15 - 1.39$.

9.4 Conclusion

This paper examines the application of a kinetic approach to modeling the long-term failure of a composite rod and cylindrical shell under creep conditions and the influence of an active medium. The applied models utilized power-law and fractional-power-law constitutive and kinetic relations, including the influence of the medium function. When modeling the long-term failure of the composite rod, classical and non-classical diffusion processes were considered. In the latter case, the active substance that penetrated the material exists in the two states – free and bound. It was determined that the time to failure of the composite rod in the former case is shorter than in the latter. A comparison of the times to failure of the cylindrical shell under an unsteady loading program was conducted, taking into account the use of scalar and vector damage parameters. The time to failure using the vector damage parameter exceeds the time to failure obtained using the scalar damage parameter. This study is fairly fundamental in its development of constitutive and kinetic relations in the theory of creep and long-term strength of materials and structural elements, taking into account the impact of an active media. The study also has a practical aspect. The proposed approaches and results can be applied in power and petrochemical engineering.

Acknowledgements The study was conducted under the state assignment of Lomonosov Moscow State University

References

[1] Altenbach H, Knape K (2025) Creep of metals at elevated temperatures. Proceedings of I Vekua Institute of Applied Mathematics 75:11–28

[2] Altenbach H, Knape, K (2020) On the main directions in creep mechanics of metallic materials. Proceedings of the National Academy of Sciences of Armenia: Mechanics 73(3):24–43, DOI 10.33018/73.3.2

[3] Barenblatt GI (1954) O nekotorykh priblizhennykh metodakh v teorii odnomernoy neustanovivsheysya fil'tratsii zhidkosti pri uprugom rezhime (in Russ., On some approximate methods in the theory of one-dimensional unsteady filtration of liquid in elastic mode). Izvestiya AN SSSR. OTN (9):35-49

[4] Charnyy IA (1963) Podzemnaya gidrogazodinamika (in Russ., Underground Fluid Dynamics). GNTI Neftyannoi i gorno-toplivoi literatury, Moscow

[5] Crank J (1975) The Mathematics of Diffusion (2nd ed). Clarendon Press, Oxford

[6] Fomin LV (2013) Opisaniye dlitel'noy prochnosti rastyagivayemykh sterzhney pryamougol'nogo i kruglogo poperechnykh secheniy v vysokotemperaturnoy vozdushnoy srede (in Russs., Description of creep rupture strength of tensile rod with rectangular and circular cross-section at high temperature air media). Vestnik Samarskogo Gosudarstvennogo Tekhnicheskogo Universiteta. Seriya Fiziko-Matematicheskie Nauki 3(32):87-97, DOI 10.14498/vsgtu1228

[7] Fomin LV (2024) On the application of a fractional-power rheological model in the study of creep and long-term failure of a composite rod taking into account the non-classical diffusion effect of the active medium. Mechanics of Solids 59(7):3747-3755, DOI 10.1134/S0025654424606414

[8] Fomin LV (2025) On the application of a linear fractional model in the problem of long-term destruction of a cylindrical shell under creep conditions in an active medium. Mechanics of Solids 60(3):1650-1665, DOI 10.1134/S0025654424606967

[9] Fomin LV (2021) Problem statement options for the problem of the corrosion processes influence on the delayed fracture of the rod under creep conditions. In: Altenbach, H., Eremeyev, V.A., Igumnov, L.A. (eds) Multiscale Solid Mechanics. Advanced Structured Materials, Vol 141. Springer, Cham. DOI 10.1007/978-3-030-54928-2_11

[10] Fomin LV, Dalinkevich AA, Basalov YG (2024) On the influence of a non-classical diffusion process on the long-term fracture of a compos-

ite tensile rod during creep. Mechanics of Solids 59(5):2735-2745, DOI 10.1134/S0025654424603471

[11] Kachanov LM (1974) Osnovy mekhaniki razrusheniya (in Russ., Fundamentals of Fracture Mechanics). Nauka, Moscow

[12] Kachanov LM (1958) O vremeni razrusheniya v usloviyakh polzuchesti (in Russ., On the time of failure under creep conditions). Izverstiya AN SSSR. OTN (8):26-36

[13] Kachanov LM (1972) Razrusheniya v usloviyakh polzuchesti pri slozhnom nagruzhenii (in Russ., Fracture in creep under complex loading). Izvestiya Akademii Nauk SSSR. Mekhanika Tverdogo Tela (5):11-15

[14] Leibenzon LS (1947) (in Russ., Movement of Natural Liquids and Gases in Porous Media). Gostekhizdat, Moscow, Leningrad

[15] Lembke KE (1887) Dvizheniye gruntovykh vod i teoriya vodosbornykh sooruzheniy (in Russ., Movement of groundwater and the theory of drainage structures). Gosudarstvennaya publichnaya nauchno-tekhnicheskaya biblioteka Rossii Elektronnyi Arkhiv, St. Petersburg

[16] Lokoshchenko AM (2017) Creep and Long-Term Strength of Metals. CRC Press, Routledge, Boca Raton

[17] Lokoshchenko AM (2018) Modeling the long-term strength of metals in an unsteady complex stress state. Mechanics of Solids 53(Supplement 1):88-100, DOI 10.3103/S0025654418030081

[18] Lokoshchenko AM (2016) Use of a vector damage parameter in modeling of long-term strength of metals. Mechanics of Solids 51(3):315-320, DOI 10.3103/S0025654416030080

[19] Lokoshchenko AM, Fomin LV (2018) Delayed fracture of plates under creep condition in unsteady complex stress state in the presence of aggressive medium. Applied Mathematical Modelling 60:478-489, DOI 10.1016/j.apm.2018.03.031

[20] Lokoshchenko A, Fomin L (2027) Kinetic theory of creep and long-term strength of metals. In: Kyzas, GZ, Mitropoulus AC (eds) Kinetic Theory. InTechOpen, Rijeka, DOI 10.5772/intechopen.70768

[21] Lokoshchenko AM, Fomin LV (2018) Modelirovanie povedeniya materialov i elementovkonstrukcii, nakhodyashchiesya pod vozdeistviem agressivnykh sred (obzor) (in Russ., Modeling the behavior of materials and structural elements under the impact of aggressive mediums (review)). Problemy Prochnosti i Plastichnosti 80(2):145-179, DOI 10.32326/1814-9146-2018-80-2-145-179

[22] Lokoshchenko AM, Fomin LV, Teraud VV, Basalov YG, Agababyan VS (2020) Polzuchest' i dlitel'naya prochnost' metallov pri nestatsionarnykh slozhnykh napryazhennykh sostoyaniyakh (obzor) (in Russ., Creep and long-term strength of metals under non-stationary complex stress states (review)). Journal of Samara State Technical University, Series Physical and Mathematical Sciences 24(2):275-318, DOI 10.14498/vsgtu1765

[23] Loitsyansky LG (1949) Priblizhennyy metod integrirovaniya uravneniy laminarnogo pogranichnogo sloya v neszhimayemom gaze (in Russ., An approxi-

mate method for integrating the equations of a laminar boundary layer in an incompressible gas). Prikladnaya Matematica i Mekhanika XIII(5):513-524

[24] Namestnikova IV, Shesterikov SA (1985) Vektornoye predstavleniye parametra povrezhdennosti (in Russ., Vector representation for a damage parameter), In: Deformation and Fracture for Solids. Collection of Scientific Works of Institute of Mechanics Lomonosov Moscow State University (MGU), Moscow, pp. 43-52

[25] Naumenko K, Altenbach H (2016)Modeling High Temperature Materials Behavior for Structural Analysis Part I: Continuum Mechanics Foundations and Constitutive Models, Advanced Structured Materials, Vol 28. Cham, Springer, DOI 10.1007/978-3-319-31629-1

[26] Naumenko K, Altenbach H (2020)Modeling High Temperature Materials Behavior for Structural Analysis Part II: Solution Procedures and Structural Analysis Examples, Advanced Structured Materials, Vol 112. Cham, Springer, DOI 10.1007/978-3-030-20381-8

[27] Oding IA, Friedman ZG (1959) Rol' poverkhnostnykh sloyev pri dlitel'nom razrushenii metallov v usloviyakh polzuchesti (in Russ., The role of surface layers during long-term destruction of metals under creep conditions). Zavodskaya Laboratoriya 25(3):329-332

[28] Rabotnov YN (1969) Creep Problems in Structural Members. North-Holland, Amsterdam

[29] Rabotnov YN (1959) O mekhanizme dlitel'nogo razrusheniya (in Russ., On the mechanism of long-term failure). Voprosy prochnosti materialov i konstruktsiy. Publishing house of the USSR Academy of Sciences, Moscow, pp. 5-7

[30] Radchenko VP, Afanasyeva EA, Saushkin MN (2025) Matematicheskaya model' polzuchesti i dlitel'noi prochnosti navodorozhennogo titanovogo splava VT6 pri temperature 600°C (in Russ., Mathematical Model of Creep and Long-Term Strength of Hydrogenated Titanium Alloy VT6 at a Temperature of 600°C), Vestnik Samarskogo Gosudarstvennogo Tekhnicheskogo Universiteta. Seriya Fiziko-Matematicheskie Nauki 4(29):750–762, DOI 10.14498/vsgtu2256

[31] Radchenko VP, Afanasyeva EA, Saushkin MN (2025) Prognozirovanie polzuchestu i dlitel'noi prochnosti navodorozhennogo titanovogo splava VT6 po izdeliyu-lideru (in Russ., Prediction of Creep and Long-Term Strength of Hydrogenated Titanium Alloy VT6 Based on a Leading Product), Vestnik Samarskogo Gosudarstvennogo Tekhnicheskogo Universiteta. Seriya Fiziko-Matematicheskie Nauki 3(29):579–590, DOI 10.14498/vsgtu2202

[32] Radchenko VP, Zoteev VE, Afanasyeva EA (2024) Chislennuy metod structurnoi i parametricheskoi identifikazcii matematicheskoi modeli nepolnoi obratimosti deformazcii polzuchesti (Iin Russ., Numerical Method for Structural and Parametric Identification of a Mathematical Model of Incomplete Reversibility of Creep Deformation), Vestnik Samarskogo Gosudarstvennogo Tekhnicheskogo Universiteta. Seriya Fiziko-Matematicheskie Nauki, 1(28):73–95, DOI 10.14498/vsgtu2071

[33] Saitova R, Arutyunyan A, Altenbach H (2024) High temperature creep and embrittlement in metals and alloys under conditions of the long-term usage. Acta Mechanica 235(10):6033–6055, DOI 10.1007/s00707-024-04026-6
[34] Shesterikov SA, Yumasheva MA (1984) Konkretizatsiya uravneniya sostoyaniya v teorii polzuchesti (in Russ., Specification of the equation of state in the theory of creep). Izvestiya Akademiya Nauk SSSR. Mekhanika Tverdogo Tela (1):86-91.
[35] Shesterikov SA, Yumasheva MA (1983) K probleme termorazrusheniya pri bystrom nagreve (in Russ., On the problem of thermal destruction during rapid heating). Izvestiya Akademii Nauk SSSR. Mekhanika Tverdogo Tela (1):128-135
[36] Timoshenko SP, Woinowsky-Krieger S (1959) Theory of Plates and Shells (2nd ed). McGraw-Hill, New York

Chapter 10
Improvement of Concrete-Filled Steel Tubular Structures to Prevent Local Buckling of the Tube Wall

Gleb Gorynin and Vera Snigireva

Abstract This study investigates the mechanism of local buckling of the steel tube wall in conventional concrete-filled steel tube (CFST) columns. A modified CFST configuration in the form of a prestressed column is proposed. The suggested design eliminates the above-mentioned limitation inherent in conventional CFST columns and provides enhanced load-bearing capacity. The conclusions are based on three-dimensional analytical modeling that accounts for the nonlinear behavior of concrete, including the variation of Poisson's ratio during the deformation process.

Key words: Concrete-filled steel tube (CFST) column, Nonlinear properties of concrete, Poisson's ratio of concrete, Local buckling

10.1 Introduction

Concrete-filled steel tubular (CFST) structures have been extensively studied since the twentieth century and continue to attract sustained research interest worldwide. This is largely driven by the complex mechanics of steel–concrete interaction and by several unresolved issues that directly affect the reliability and design efficiency of CFST columns. In practice, conventional CFST structures are most commonly used as columns and piles. A typical configuration comprises a steel tube of outer radius R_{ext} filled with concrete forming a cylindrical core of radius R_{int} (Fig. 10.1). Under concentric compression, the axial load is commonly applied to the entire cross-section, i.e., to both the concrete core and the steel tube simultaneously (Fig. 10.2).

CFST columns may be manufactured with or without longitudinal reinforcement within the concrete core. A key constructional advantage of CFST technology is that the steel tube serves both as a structural component and as permanent formwork,

Gleb Gorynin (corresponding author) · Vera Snigireva
Polytechnic Institute, Surgut State University, Lenin ave. 1, 628412 Surgut, Russian Federation,
e-mail: ggorynin@list.ru, snow-vera@mail.ru

H. Altenbach et al. (eds.), *Stability Problems in Mechanics of Continua*,
Advanced Structured Materials 260, https://doi.org/10.1007/978-3-032-31032-3_10

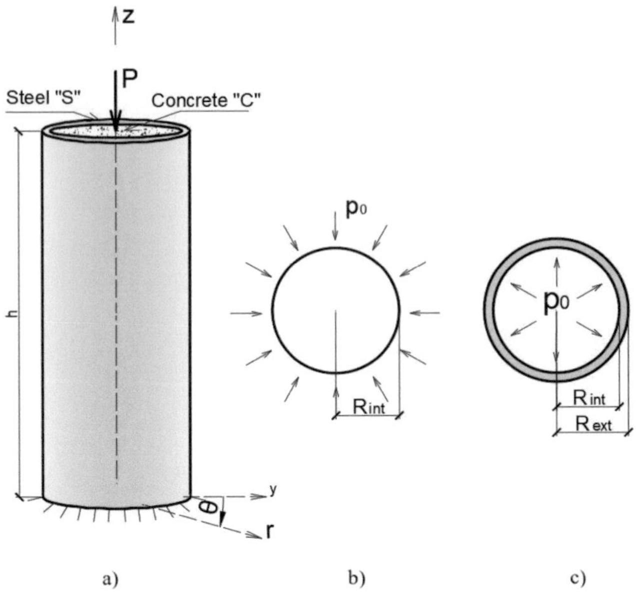

Fig. 10.1: Design scheme of a CFST column: a) general view of the CFST column; b) cross-section of the concrete core; c) cross-section of the steel tube.

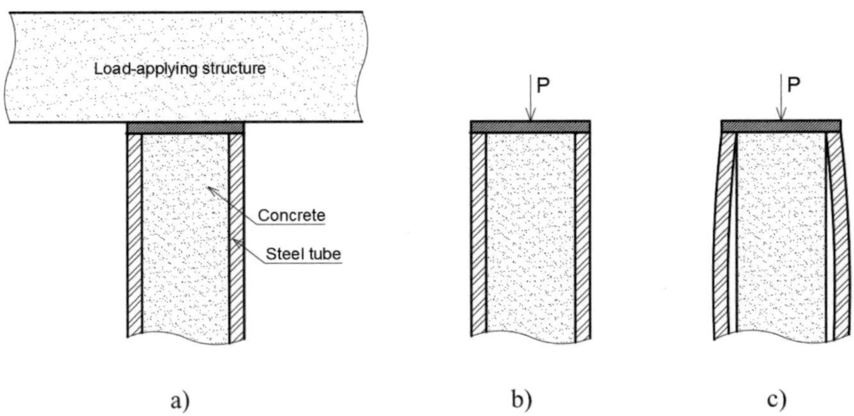

Fig. 10.2: Schematic loading of a conventional concrete-filled steel tube column: (a, b) – column up to the separation of the steel tube from the concrete core, (c) – column after the separation of the steel tube from the concrete core..

eliminating the need for removable formwork and thereby reducing construction time.

Early design concepts assumed that the steel tube provides effective confinement, restraining lateral expansion of concrete and thus enhancing concrete strength while simultaneously preventing local buckling of the steel wall. However, subsequent experimental evidence has demonstrated a fundamental limitation of this assumption: at an early loading stage, radial separation (debonding) between the steel tube and the concrete core may occur, which has also been confirmed in more recent studies [5, 13, 16, 24, 38]. Once separation initiates, the steel tube wall - already subjected to substantial axial compression - loses the stabilizing effect of concrete support and becomes susceptible to local buckling, while the concrete core does not receive the expected lateral confinement. Numerous experimental investigations indicate that, for massive CFST columns under concentric compression, the failure process is often triggered by local buckling of the steel tube wall [7, 12, 25, 28, 34].

To mitigate this problem, various measures have been proposed to restrain transverse deformations of the steel tube, including the use of internal anchors [2], internal bolts tying opposite tube walls [5], and internal stiffeners or ribs [22, 33]. Despite their effectiveness in certain cases, these solutions are typically construction-intensive and, importantly, they do not fully address the underlying mechanism that promotes steel–concrete separation.

In addition to these structural issues, rigorous analytical modeling of the three-dimensional stress–strain state in CFST structures remains challenging. Many existing design approaches evaluate the axial capacity by superposition of the ultimate contributions of the concrete core and the steel tube, combined with empirical coefficients or finite-element calibration [9, 10, 15, 17, 18, 21, 27, 32, 34, 37], including provisions in design standards [2–4]. Recently, data-driven models - particularly machine-learning approaches trained on large experimental datasets - have also been proposed for capacity prediction [1, 8, 14, 20, 31, 39]. Nevertheless, most of these methods implicitly assume sustained composite action in the transverse direction and therefore do not explicitly capture the initiation conditions for radial debonding, nor the associated onset of local buckling of the steel tube wall.

Concrete exhibits pronounced nonlinear behavior; in addition to the evolution of its tangent modulus with increasing load, Poisson's ratio of concrete is also known to vary during deformation [19]. Many practical calculation procedures either adopt linear elasticity for concrete or incorporate nonlinearity only under uniaxial stress conditions. As a result, the nonlinear response under triaxial stress states - together with the evolution of Poisson's ratio - remains insufficiently represented in most analytical formulations for CFST columns.

10.2 Linear Modelling of the Three-Dimensional Behaviour of a Conventional Concrete-Filled Steel Tube Column

Let us consider the behaviour of a conventional concrete-filled steel tube (CFST) column at the initial stage of deformation (when the concrete elastic modulus $E = \text{const}$ and the concrete Poisson's ratio $\nu = \text{const}$). This behaviour was partly studied in [29].

We perform a qualitative analysis of the reasons for local buckling of the steel tube wall based on modelling of the three-dimensional behaviour of a conventional CFST column. In order to simplify the problem and obtain analytical results in a simple form, we consider the case without longitudinal reinforcing bars in the concrete core.

In the modelling, we use the cylindrical coordinate system shown in Fig. 10.1. Due to axisymmetry of the problem, one of the equilibrium equations is satisfied identically, while the other two are simplified and take the form:

$$\frac{\partial \sigma_{rr}}{\partial r} + \frac{\sigma_{rr} - \sigma_{\theta\theta}}{r} = 0, \quad \frac{\partial \sigma_{zz}}{\partial z} = 0. \tag{10.1}$$

A load P is applied to the entire cross-section of a column of height h (Fig. 10.2). However, the load distribution between the steel shell and the concrete core is not known in advance. In addition, longitudinal displacements of points W are absent at the lower end of the column, because this end is rigidly fixed (Fig. 10.1). Therefore, the boundary conditions at the column ends can be written as:

$$W = 0 \quad (\text{when } z = 0), \qquad \int_F \sigma_{zz}\, dF = -P \quad (\text{when } z = h). \tag{10.2}$$

Further, physical quantities related to the steel tube will be denoted by the superscript "S" (steel), and quantities related to concrete by the superscript "C" (concrete).

We assume that there is no lateral pressure on the outer surface of the steel tube. Then, the boundary condition on the lateral surface (Fig. 10.1) is:

$$\sigma^{\mathrm{S}}_{rr} = 0 \quad (\text{when } r = R_{\text{ext}}). \tag{10.3}$$

To ensure the composite action of the steel tube and the concrete core, the steel-concrete interface conditions must be satisfied:

$$\sigma^{\mathrm{C}}_{rr} = \sigma^{\mathrm{S}}_{rr}, \quad U^{\mathrm{C}} = U^{\mathrm{S}}, \quad W^{\mathrm{C}} = W^{\mathrm{S}} \quad (\text{when } r = R_{\text{int}} \text{ (Fig. 10.4)}). \tag{10.4}$$

Let us introduce the following notation: p^{C}, p^{S} are the longitudinal (along the z-axis) compressive pressures acting on the concrete core and the steel tube, respectively; p_0 is the radial lateral pressure at the steel-concrete interface (confinement pressure). Then:

$$p_0 = -\sigma^{\mathrm{C}}_{rr} = -\sigma^{\mathrm{S}}_{rr} \quad (\text{when } r = R_{\text{int}}).$$

The stress-strain state of the tube material and concrete at the initial loading stage obeys the generalized Hooke's law:

$$\begin{cases} \sigma_{rr} = \dfrac{E}{(1-2\nu)(1+\nu)}\big((1-\nu)\varepsilon_{rr} + \nu\varepsilon_{zz} + \nu\varepsilon_{\theta\theta}\big); \\ \sigma_{zz} = \dfrac{E}{(1-2\nu)(1+\nu)}\big((1-\nu)\varepsilon_{zz} + \nu\varepsilon_{rr} + \nu\varepsilon_{\theta\theta}\big); \\ \sigma_{\theta\theta} = \dfrac{E}{(1-2\nu)(1+\nu)}\big((1-\nu)\varepsilon_{\theta\theta} + \nu\varepsilon_{rr} + \nu\varepsilon_{zz}\big), \end{cases} \tag{10.5}$$

where E and ν are Young's modulus and Poisson's ratio for concrete or steel, respectively. Strains are related to displacements by:

$$\begin{aligned} &\varepsilon_{rr}^{S} = \frac{dU^{S}}{dr}, \quad \varepsilon_{\theta\theta}^{S} = \frac{U^{S}}{r}, \quad \varepsilon_{zz}^{S} = \frac{dW^{S}}{dz}, \\ &\varepsilon_{rr}^{C} = \frac{dU^{C}}{dr}, \quad \varepsilon_{\theta\theta}^{C} = \frac{U^{C}}{r}, \quad \varepsilon_{zz}^{C} = \frac{dW^{C}}{dz}. \end{aligned} \tag{10.6}$$

The problem (10.1)–(10.6) is a three-dimensional elasticity problem in the Saint-Venant formulation. The solutions of the equilibrium equations (10.1) for the steel tube and the concrete core, considered separately and taking into account the stress interface conditions (10.4) and the boundary condition (10.3), can be written as:

$$\begin{cases} \sigma_{rr}^{S} = \dfrac{p_0 \cdot R_{\text{int}}^2 \cdot (r^2 - R_{\text{ext}}^2)}{(R_{\text{ext}}^2 - R_{\text{int}}^2) \cdot r^2}; \\ \sigma_{zz}^{S} = -p^{S}; \\ \sigma_{\theta\theta}^{S} = \dfrac{p_0 \cdot R_{\text{int}}^2 \cdot (r^2 + R_{\text{ext}}^2)}{(R_{\text{ext}}^2 - R_{\text{int}}^2) \cdot r^2}, \end{cases} \qquad \begin{cases} \sigma_{rr}^{C} = -p_0; \\ \sigma_{zz}^{C} = -p^{C}; \\ \sigma_{\theta\theta}^{C} = -p_0. \end{cases} \tag{10.7}$$

If expressions (10.7) are substituted into Hooke's law (10.5), and the strains are expressed in terms of displacements using (10.6), then two differential equations are obtained for concrete and steel. After integration, using the second and third equalities of the steel–concrete interface conditions (10.4) and the equalities (10.2), we obtain the formulas for the pressures:

$$
\begin{aligned}
p^{\mathrm{C}} &= p_0 \left(\frac{(1-2\nu^{\mathrm{C}})(1+\nu^{\mathrm{C}})}{(\nu^{\mathrm{C}}-\nu^{\mathrm{S}})} + 2\nu^{\mathrm{C}} + \right. \\
&\left. + \frac{E^{\mathrm{C}}}{E^{\mathrm{S}}} \cdot \frac{R_{\mathrm{int}}^2(1-2\nu^{\mathrm{S}})(1+\nu^{\mathrm{S}}) + R_{\mathrm{ext}}^2(1+\nu^{\mathrm{S}})}{(\nu^{\mathrm{C}}-\nu^{\mathrm{S}})(R_{\mathrm{ext}}^2 - R_{\mathrm{int}}^2)} \right), \\
p^{\mathrm{S}} &= p_0 \left(\frac{E^{\mathrm{S}}}{E^{\mathrm{C}}} \cdot \frac{(1-2\nu^{\mathrm{C}})(1+\nu^{\mathrm{C}})}{(\nu^{\mathrm{C}}-\nu^{\mathrm{S}})} + \right. \\
&\left. + \frac{R_{\mathrm{int}}^2(1-\nu^{\mathrm{S}}) + R_{\mathrm{ext}}^2(1+\nu^{\mathrm{S}}) - 2R_{\mathrm{int}}^2\nu^{\mathrm{S}}\nu^{\mathrm{C}}}{(\nu^{\mathrm{C}}-\nu^{\mathrm{S}})(R_{\mathrm{ext}}^2 - R_{\mathrm{int}}^2)} \right), \\
p_0 &= P \frac{E^{\mathrm{C}}E^{\mathrm{S}}(\nu^{\mathrm{C}}-\nu^{\mathrm{S}})(R_{\mathrm{ext}}^2 - R_{\mathrm{int}}^2)}{\alpha},
\end{aligned}
\tag{10.8}
$$

where

$$
\begin{aligned}
\alpha &= \pi R_{\mathrm{int}}^2 E^{\mathrm{C}} \Big(E^{\mathrm{S}}(1-\nu^{\mathrm{C}} - 2\nu^{\mathrm{C}}\nu^{\mathrm{S}})(R_{\mathrm{ext}}^2 - R_{\mathrm{int}}^2) + \\
&+ E^{\mathrm{C}}(1+\nu^{\mathrm{S}})(R_{\mathrm{int}}^2(1-2\nu^{\mathrm{S}}) + R_{\mathrm{ext}}^2) \Big) + \\
&+ \pi(R_{\mathrm{ext}}^2 - R_{\mathrm{int}}^2)E^{\mathrm{S}} \Big(E^{\mathrm{S}}(1-2\nu^{\mathrm{C}})(1+\nu^{\mathrm{C}})(R_{\mathrm{ext}}^2 - R_{\mathrm{int}}^2) + \\
&+ E^{\mathrm{C}}(R_{\mathrm{int}}^2(1-\nu^{\mathrm{S}} - 2\nu^{\mathrm{S}}\nu^{\mathrm{C}}) + R_{\mathrm{ext}}^2(1+\nu^{\mathrm{S}})) \Big).
\end{aligned}
$$

According to expression (10.8), the sign of the confinement pressure p_0 is completely determined by the difference in the Poisson's ratios of concrete and steel $(\nu^{\mathrm{C}} - \nu^{\mathrm{S}})$. The Poisson's ratio of concrete at the initial loading stage is $\nu^{\mathrm{C}} = 0.2$, and that of steel is $\nu^{\mathrm{S}} = 0.3$; consequently, the confinement pressure p_0 is negative, and the radial stresses in the concrete core are tensile (10.7). It is known that the bond strength between concrete and steel in tension is low. Therefore, under negative confining pressure of the concrete by the tube p_0 , it can be assumed that separation of the steel tube from the concrete core occurs. This circumstance is a prerequisite for local buckling of the tube wall. This structural disadvantage does not depend on the column height, cross-section dimensions, or tube wall thickness, because it is related only to the material properties. Therefore, for conventional CFST columns, it is inherent and cannot be eliminated.

10.3 Nonlinear Modelling of the Triaxial Stress–Strain Behaviour of Concrete

In the general case, modelling the stress–strain state of concrete is a difficult problem, because concrete exhibits inherently nonlinear deformation. In addition, in concrete-filled steel tube (CFST) structures the concrete is subjected to a complex three-dimensional stress–strain state due to its interaction with the steel tube.

It is known that the concrete stress–strain diagram is nonlinear (curvilinear). However, in order to simplify calculations, Russian design codes [23] recommend using a bilinear diagram instead of the curvilinear uniaxial stress–strain diagram. According to this diagram, the tangent modulus of deformation is a piecewise-constant function, as shown in Fig. 10.3a and 10.3b.

10.3.1 Variation of Poisson's Ratio of Concrete

By the tangent Poisson's ratio we mean the ratio of infinitesimal increments of the transverse strains of concrete to the longitudinal strain increments, taken with a negative sign, under axial loading.

$$\nu_{\tan}^{C} = -\frac{d\varepsilon_{rr}}{d\varepsilon_{zz}}.$$

It is known that the Poisson's ratio of concrete is a variable quantity. Karpenko proposed analytical relationships that make it possible to determine the tangent Poisson's ratio of concrete using empirical coefficients that account for nonlinear deformation [19]. Figure 10.4 shows a graphical dependence of the tangent Poisson's ratio of concrete strength class B35 (according to Russian design codes [23]) on axial stresses, with an initial value equal to 0.2. We approximate this curve by a two-step piecewise constant function with a discontinuity at the point $\sigma_{zz} = -0.6R_b$ (R_b is the design value of the cylindrical compressive strength of concrete), where the uniaxial compression diagram of concrete undergoes a kink according to [23] (Fig. 10.3a). Thus, we obtain $\nu = 0.255$ and $\nu_1 = 0.41$.

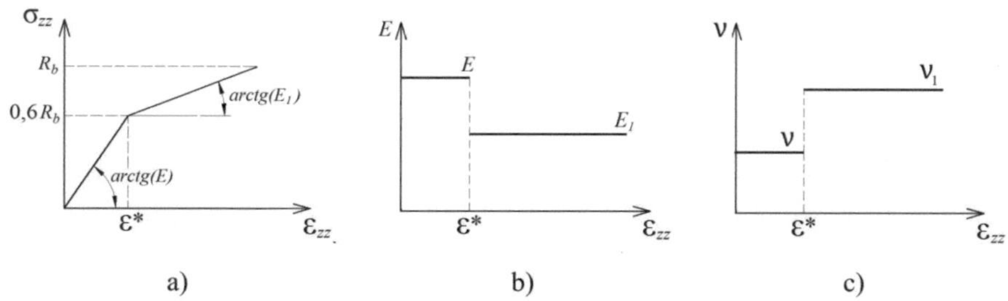

Fig. 10.3: a) Stress-strain diagram of concrete under uniaxial compression (stresses and strains are used without sign; R_b is the design value of the cylindrical compressive strength of concrete), (b) Concrete tangent modulus versus strain; (c) Concrete Poisson's ratio versus strain.

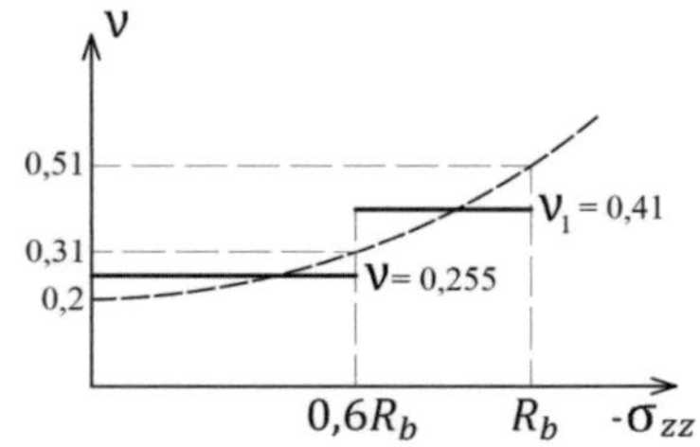

Fig. 10.4 Dependence of the concrete Poisson's ratio on axial stress (dash-dot curve – analytical relationship by Karpenko; solid straight lines – piecewise constant approximation).

10.3.2 Consideration of Nonlinear Deformation of Concrete

To describe the triaxial nonlinear behavior of concrete, a hypoelastic material model is adopted [35, 36]. A distinguishing feature of this formulation is that all constitutive parameters are defined in terms of tangent moduli. Subsequently, we assume that the load increases uniformly with time t. If time is measured in specially chosen units of measurement, then differentiation with respect to time is equivalent to differentiation with respect to load.

Within the hypoelastic framework, the stress-rate–strain-rate relations are written as:

$$\dot{\sigma}_{\alpha\beta} = \lambda(S_2^{\mathrm{D}}, \theta)\dot{\theta}\delta_{\alpha\beta} + 2\mu(S_2^{\mathrm{D}})\dot{\varepsilon}_{\alpha\beta}, \tag{10.9}$$

where λ, μ are the tangent Lamé moduli, $\delta_{\alpha\beta}$ is the Kronecker delta, $\varepsilon_{\alpha\beta}$ is the strain tensor, $\theta = \varepsilon_{\alpha\alpha}$ is the first invariant of the strain tensor (volumetric strain), $\dot{(\,)} = d/dt$, S_2^{D} is the second invariant of the stress deviator tensor, for which the following formula holds under an axisymmetric stress state:

$$S_2^{\mathrm{D}} = \frac{1}{3}\left[(\sigma_{rr} - \sigma_{\theta\theta})^2 + (\sigma_{rr} - \sigma_{zz})^2 + (\sigma_{zz} - \sigma_{\theta\theta})^2\right].$$

The following relations hold between the tangent moduli of elasticity (Fig. 10.3b, 10.3c) and the tangent bulk modulus k and the Lamé moduli μ and λ (Fig. 10.5):

$$\mu = \frac{E}{2(1+\nu)}, \quad k = \frac{E}{3(1-2\nu)}, \quad \lambda = k - \frac{2}{3}\mu.$$

For a uniaxial axisymmetric stress state, the following equalities hold:

$$\theta = (1-2\nu)\varepsilon_{zz}, \quad S_2^{\mathrm{D}} = \frac{2}{3}\sigma_{zz}^2.$$

And at the kink point of the uniaxial compression diagram (Fig. 10.3a), these invariants take the following values:

$$\theta^* = (1-2\nu)\varepsilon_{zz}^*, \quad S_2^{D*} = \frac{2}{3}(0.6R_b)^2.$$

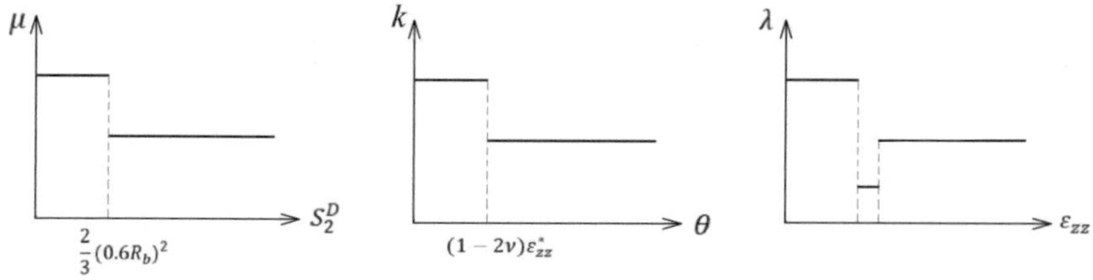

Fig. 10.5: Stress-strain diagrams of concrete under triaxial compression within a CFST column.

It is precisely at these points that the quantities $k(\theta)$ and $\mu(S_2^{\mathrm{D}})$ in Fig. 10.5 are discontinuous, as a result of which $\lambda(k;\mu)$ also exhibits two discontinuities. Moreover, the positions of these discontinuity points are determined by the locations of the kink points on the diagrams in Fig. 10.3 and by the column parameters.

Equation (9), assuming constant tangent Lamé moduli, can be integrated with respect to time, yielding the incremental stress–strain relations:

$$\Delta\sigma_{\alpha\beta} = \lambda\Delta\theta\delta_{\alpha\beta} + 2\mu\Delta\varepsilon_{\alpha\beta} \tag{10.10}$$

According to Fig. 10.5, the entire process of concrete deformation is divided into three phases ($i = 0, 1, 2$, where i is the sequential number of the deformation phase). Then, the values of stresses and strains of the concrete core can be expressed in terms of increments:

$$\begin{aligned} \sigma_{rr}^{\mathrm{C}} &= \sigma_{rr_{i*}}^{\mathrm{C}} + \Delta\sigma_{rr_i}^{\mathrm{C}}, \quad \sigma_{\theta\theta}^{\mathrm{C}} = \sigma_{\theta\theta_{i*}}^{\mathrm{C}} + \Delta\sigma_{\theta\theta_i}^{\mathrm{C}}, \quad \sigma_{zz}^{\mathrm{C}} = \sigma_{zz_{i*}}^{\mathrm{C}} + \Delta\sigma_{zz_i}^{\mathrm{C}}, \\ \varepsilon_{rr}^{\mathrm{C}} &= \varepsilon_{rr_{i*}}^{\mathrm{C}} + \Delta\varepsilon_{rr_i}^{\mathrm{C}}, \quad \varepsilon_{\theta\theta}^{\mathrm{C}} = \varepsilon_{\theta\theta_{i*}}^{\mathrm{C}} + \Delta\varepsilon_{\theta\theta_i}^{\mathrm{C}}, \quad \varepsilon_{zz}^{\mathrm{C}} = \varepsilon_{zz_{i*}}^{\mathrm{C}} + \Delta\varepsilon_{zz_i}^{\mathrm{C}}, \end{aligned} \tag{10.11}$$

where $\sigma_{\alpha\alpha_{i*}}^{\mathrm{C}}, \varepsilon_{\alpha\alpha_{i*}}^{\mathrm{C}}$ are the values of stresses and strains in concrete at the beginning of deformation phase number i (it is obvious that $\sigma_{\alpha\alpha_{0*}}^{\mathrm{C}} = \varepsilon_{\alpha\alpha_{0*}}^{\mathrm{C}} = 0$); $\Delta\sigma_{\alpha\alpha_i}^{\mathrm{C}}, \Delta\varepsilon_{\alpha\alpha_i}^{\mathrm{C}}$ are the increments of stresses and strains at deformation phase number i; ($\alpha \in \{r, \theta, z\}$ is the cylindrical coordinate system introduced in Fig. 10.1).

Taking into account the variation of Poisson's ratio and the modulus of deformation of concrete (Fig. 10.3b, 10.3c), the following relations hold for the various phases of concrete deformation:

$$\begin{aligned} &i = 0 \quad (\theta \le \theta^*, S_2^{\mathrm{D}} \le S_2^{D*}): \\ &\mu_0^{\mathrm{C}} = \frac{E^{\mathrm{C}}}{2(1+\nu^{\mathrm{C}})}, \quad k_0 = \frac{E^{\mathrm{C}}}{3(1-2\nu^{\mathrm{C}})}, \quad \lambda_0^{\mathrm{C}} = \frac{E^{\mathrm{C}}\nu^{\mathrm{C}}}{(1-2\nu^{\mathrm{C}})(1+\nu^{\mathrm{C}})}; \\ &i = 1 \quad (\theta > \theta^*, S_2^{\mathrm{D}} \le S_2^{D*}): \\ &\mu_1^{\mathrm{C}} = \frac{E^{\mathrm{C}}}{2(1+\nu^{\mathrm{C}})}, \quad k_1 = \frac{E_1^{\mathrm{C}}}{3(1-2\nu_1^{\mathrm{C}})}, \end{aligned} \tag{10.12}$$

$$\lambda_1^{C} = \frac{E_1^{C}(1+\nu^{C})(1+3\nu_1^{C}) - E^{C}(1+\nu_1^{C})}{3(1-2\nu_1^{C})(1+\nu^{C})(1+\nu_1^{C})};$$

$$i = 2 \quad (\theta > \theta^*, S_2^{D} > S_2^{D*}):$$

$$\mu_2^{C} = \frac{E_1^{C}}{2(1+\nu_1^{C})}, \quad k_2 = \frac{E_1^{C}}{3(1-2\nu_1^{C})}, \quad \lambda_2^{C} = \frac{E_1^{C}\nu_1^{C}}{(1-2\nu_1^{C})(1+\nu_1^{C})}.$$

Equation (10.10) for stresses at each of the deformation phases takes the form:

$$\begin{cases} \Delta\sigma_{rr_i}^{C} = (\lambda_i^{C} + 2\mu_i^{C})\Delta\varepsilon_{rr_i}^{C} + \lambda_i^{C}(\Delta\varepsilon_{\theta\theta_i}^{C} + \Delta\varepsilon_{zz_i}^{C}), \\ \Delta\sigma_{\theta\theta_i}^{C} = (\lambda_i^{C} + 2\mu_i^{C})\Delta\varepsilon_{\theta\theta_i}^{C} + \lambda_i^{C}(\Delta\varepsilon_{rr_i}^{C} + \Delta\varepsilon_{zz_i}^{C}), \\ \Delta\sigma_{zz_i}^{C} = (\lambda_i^{C} + 2\mu_i^{C})\Delta\varepsilon_{zz_i}^{C} + \lambda_i^{C}(\Delta\varepsilon_{rr_i}^{C} + \Delta\varepsilon_{\theta\theta_i}^{C}). \end{cases} \tag{10.13}$$

10.4 Modification of the CFST Column Design

To address the problem of separation of the steel tube from the concrete core and the associated local buckling of the tube wall, a new type of structure is proposed—the prestressed concrete-filled steel tube (CFST) column. The key idea is to divide the application of the axial load into two stages. Initially, the load is applied only to the concrete core, while a gap of height d is maintained between the loading structure and the end of the steel tube. Subsequently, the load is applied to the entire cross-section, including the steel tube (Fig. 10.6). This loading scheme differs fundamentally from that used for conventional CFST columns, where the load is applied to the full cross-section from the outset (Fig. 10.2). Such two-stage loading of the prestressed column is achieved through a design feature of the column head, namely the provision of an

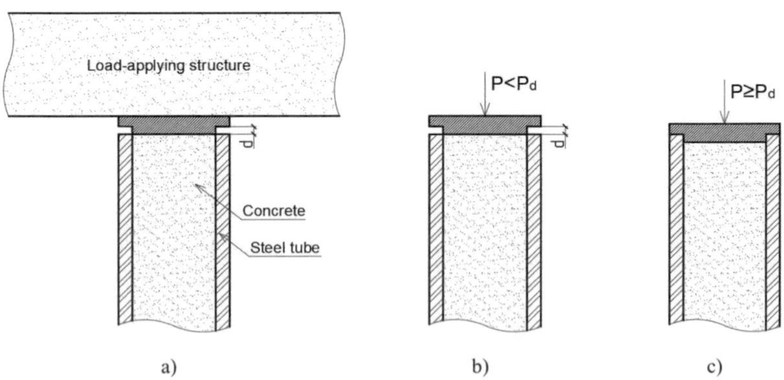

Fig. 10.6: Schematic loading of loading of a prestressed CFST column: (a, b) – at loading stage 1, (c) – at loading stage 2.

initial gap of height d between the loading structure and the end of the steel tube (Fig. 10.6).

Let us consider in more detail the features of the proposed structure's behavior at each of the loading stages:

- **Loading Stage 1 "Pre-compression of the concrete core":** at this stage, the column behaves as a concrete column in a steel cage. The external compressive load is applied only to the concrete core; there is no longitudinal action of the steel tube. However, in the transverse direction, the tube restrains the deformations of the concrete core, exerting a reactive compressive pressure p_0 on the concrete in the radial direction, i.e., it acts as a cage (Fig. 10.6b). In this case, local buckling of the tube wall is precluded.
- **Loading Stage 2 "Design working state":** this stage commences when the concrete core has shortened by the amount d. From this point onward, the load is applied to the full cross-section of the column, including the steel tube (Fig. 10.6c). The load at which Stage 2 commences will be denoted as P_d. At this stage, the column behaves as a conventional CFST column, but with a precompressed concrete core, which ensures protection against local buckling of the tube wall when it is engaged in resisting longitudinal load.

The strain increments that appear in equalities (12) are determined from the displacement increments using formulas (10.6):

$$\Delta\varepsilon^{\mathrm{C}}_{rr_i} = \frac{d\Delta U^{\mathrm{C}}_i}{dr}, \quad \Delta\varepsilon^{\mathrm{C}}_{\theta\theta_i} = \frac{\Delta U^{\mathrm{C}}_i}{r}, \quad \Delta\varepsilon^{\mathrm{C}}_{zz_i} = \frac{d\Delta W^{\mathrm{C}}_i}{dz}, \tag{10.14}$$

where U^{C}_i are the radial (along the r-axis) displacements in concrete at concrete deformation phase $i = 0, 1, 2$; W^{C}_i are the longitudinal (along the z-axis) displacements in concrete at concrete deformation phase $i = 0, 1, 2$ (the coordinate system is introduced in Fig. 10.1).

The deformation of steel is assumed to be linear until the stresses in it reach the design resistance, and therefore the following stress-strain relations for steel hold:

$$\begin{cases} \sigma^{\mathrm{S}}_{rr} = (\lambda^{\mathrm{S}} + 2\mu^{\mathrm{S}})\varepsilon^{\mathrm{S}}_{rr} + \lambda^{\mathrm{S}}(\varepsilon^{\mathrm{S}}_{\theta\theta} + \varepsilon^{\mathrm{S}}_{zz}), \\ \sigma^{\mathrm{S}}_{\theta\theta} = (\lambda^{\mathrm{S}} + 2\mu^{\mathrm{S}})\varepsilon^{\mathrm{S}}_{\theta\theta} + \lambda^{\mathrm{S}}(\varepsilon^{\mathrm{S}}_{rr} + \varepsilon^{\mathrm{S}}_{zz}), \\ \sigma^{\mathrm{S}}_{zz} = (\lambda^{\mathrm{S}} + 2\mu^{\mathrm{S}})\varepsilon^{\mathrm{S}}_{zz} + \lambda^{\mathrm{S}}(\varepsilon^{\mathrm{S}}_{rr} + \varepsilon^{\mathrm{S}}_{\theta\theta}), \end{cases} \tag{10.15}$$

$$\lambda^{\mathrm{S}} = \frac{E^{\mathrm{S}}\nu^{\mathrm{S}}}{(1 - 2\nu^{\mathrm{S}})(1 + \nu^{\mathrm{S}})}, \quad \mu^{\mathrm{S}} = \frac{E^{\mathrm{S}}}{2(1 + \nu^{\mathrm{S}})},$$

and also the strain-displacement relations hold:

$$\varepsilon^{\mathrm{S}}_{rr} = \frac{dU^{\mathrm{S}}}{dr}, \quad \varepsilon^{\mathrm{S}}_{\theta\theta} = \frac{U^{\mathrm{S}}}{r}, \quad \varepsilon^{\mathrm{S}}_{zz} = \frac{dW^{\mathrm{S}}}{dz}, \tag{10.16}$$

where U^{S} are the radial (along the r-axis) displacements in the tube; W^{S} are the longitudinal (along the z-axis) displacements in the tube.

For the prestressed column, the following boundary conditions hold:

$$W^{\mathrm{C}} = W^{\mathrm{S}} = 0 \quad (\text{when } z = 0); \quad \sigma^{\mathrm{S}}_{rr} = 0 \quad (\text{when } r = R_{\mathrm{ext}}); \tag{10.17}$$

and the steel-concrete interface conditions hold:

$$\sigma^{c}_{rr} = \sigma^{\mathrm{S}}_{rr} = -p_0; \quad W^{\mathrm{C}} = W^{\mathrm{S}} \quad (\text{only at loading stage 2}); \quad U^{\mathrm{C}} = U^{\mathrm{S}};$$

$$\sigma^{\mathrm{C}}_{rz} = \sigma^{\mathrm{C}}_{r\theta} = \sigma^{\mathrm{C}}_{\theta z} = \sigma^{\mathrm{S}}_{rz} = \sigma^{\mathrm{S}}_{r\theta} = \sigma^{\mathrm{S}}_{\theta z} = 0 \quad (\text{when } r = R_{\mathrm{int}}). \tag{10.18}$$

In addition to the conditions described above, the following conditions hold at the end face $z = h$:

- at loading stage 1:

$$\int\limits_{F^{\mathrm{C}}} \sigma^{\mathrm{C}}_{zz}\, dF^{\mathrm{C}} = -p^{\mathrm{C}} \cdot F^{\mathrm{C}} = -P; \tag{10.19}$$

- at loading stage 2:

$$\int\limits_{F} \sigma_{zz}\, dF = -p^{\mathrm{S}} \cdot F^{\mathrm{S}} - p^{\mathrm{C}} \cdot F^{\mathrm{C}} = -P, \tag{10.20}$$

where F^{C} and F^{S} are the cross-sectional areas of the concrete core and the steel tube, respectively. Note that expressions (10.7) for the stresses in the concrete core and the steel tube are a consequence of the equilibrium equations (10.1) and therefore also hold for the prestressed column at both loading stages.

For the prestressed column at loading stage 1, the concrete undergoes three deformation phases ($i = 0, 1, 2$). The solution of problem (10.1), (10.11)-(10.19) at this loading stage, omitting intermediate derivations, can be written as:

$$p^{\mathrm{S}} = 0, \quad P = P_{i*} + \Delta P, \quad p^{\mathrm{C}} = p^{\mathrm{C}}_{i*} + \Delta p^{\mathrm{C}}_i, \quad p_0 = p_{0_{i*}} + \Delta p_{0_i}, \tag{10.21}$$

$$\Delta p_{0_i} = \Delta p^{\mathrm{C}}_i \frac{J_i}{L_i + K}, \quad \Delta p^{\mathrm{C}}_i = \frac{\Delta P_i}{F^{\mathrm{C}}},$$

where

$$K = \frac{\mu^{\mathrm{S}}(3\lambda^{\mathrm{S}} + 2\mu^{\mathrm{S}}) + (\lambda^{\mathrm{S}})^2}{2\mu^{\mathrm{S}}(\lambda^{\mathrm{S}} + \mu^{\mathrm{S}})(3\lambda^{\mathrm{S}} + 2\mu^{\mathrm{S}})} \cdot \frac{R^2_{\mathrm{int}}}{R^2_{\mathrm{ext}} - R^2_{\mathrm{int}}} + \frac{R^2_{\mathrm{ext}}}{2\mu^{\mathrm{S}}(R^2_{\mathrm{ext}} - R^2_{\mathrm{int}})},$$

$$L_i = \frac{\mu^{\mathrm{C}}_i(3\lambda^{\mathrm{C}}_i + 2\mu^{\mathrm{C}}_i) + (\lambda^{\mathrm{C}}_i)^2}{2\mu^{\mathrm{C}}_i(3\lambda^{\mathrm{C}}_i + 2\mu^{\mathrm{C}}_i)(\lambda^{\mathrm{C}}_i + \mu^{\mathrm{C}}_i)}, \quad J_i = \frac{\lambda^{\mathrm{C}}_i}{2\mu^{\mathrm{C}}_i(3\lambda^{\mathrm{C}}_i + 2\mu^{\mathrm{C}}_i)}, \quad i = 0, 1, 2,$$

where p^{C}_{i*}, $p_{0_{i*}}$ are the axial and radial pressures on the concrete at the beginning of concrete deformation phase number i; P_{i*} are the values of the external compressive force at which concrete deformation phase number i begins; ΔP is the increment of the axial compressive force during concrete deformation phase number i; $P_{0*} = p^{\mathrm{C}}_{0*} = p_{0_{0*}} = 0$.

Relations (10.21) are valid only at loading stage 1, i.e., until the moment the steel tube becomes engaged in longitudinal action. For this column at loading stage 2, the state of concrete continues to correspond to the second deformation phase ($i = 2$). However, instead of problem (10.1), (10.11)-(10.19), another problem (10.1), (10.11)-(10.18), (10.20) is solved in terms of increments to the quantities corresponding to the completion of loading stage 1. The solution of this problem can be represented as:

$$P = P_d + \Delta P, \quad p^{\mathrm{C}} = p_d^{\mathrm{C}} + \Delta p^{\mathrm{C}}, \quad p_0 = p_{0d} + \Delta p_0, \quad p^{\mathrm{S}} = \Delta p^{\mathrm{S}},$$

where $P_d, p_d^{\mathrm{C}}, p_{0d}$ are the values of the compressive force on the column, the pressure on the concrete core, and the confinement pressure of the concrete core, respectively, at the moment of transition of the column from loading stage 1 to loading stage 2. The pressure increments are determined according to the formulas:

$$\Delta p_0 = \Delta P \frac{YJ_2 - H_2 D}{F^{\mathrm{C}}(Y(K + L_2) - D(2J_2 + \alpha)) + F^{\mathrm{S}}(H_2(K + L_2) - J_2(2J_2 + \alpha))},$$

$$\Delta p^{\mathrm{C}} = \Delta P \frac{Y(K + L_2) - D(2J_2 + \alpha)}{F^{\mathrm{C}}(Y(K + L_2) - D(2J_2 + \alpha)) + F^{\mathrm{S}}(H_2(K + L_2) - J_2(2J_2 + \alpha))},$$

$$\Delta p^{\mathrm{S}} = \Delta P \frac{H_2(K + L_2) - J_2(2J_2 + \alpha)}{F^{\mathrm{C}}(Y(K + L_2) - D(2J_2 + \alpha)) + F^{\mathrm{S}}(H_2(K + L_2) - J_2(2J_2 + \alpha))},$$

$$Y = \frac{\lambda^{\mathrm{S}} + \mu^{\mathrm{S}}}{\mu^{\mathrm{S}}(3\lambda^{\mathrm{S}} + 2\mu^{\mathrm{S}})}, \quad D = \frac{\lambda^{\mathrm{S}}}{2\mu^{\mathrm{S}}(3\lambda^{\mathrm{S}} + 2\mu^{\mathrm{S}})}, \quad \alpha = \frac{\lambda^{\mathrm{S}}}{\mu^{\mathrm{S}}(3\lambda^{\mathrm{S}} + 2\mu^{\mathrm{S}})} \cdot \frac{R_{\mathrm{int}}^2}{R_{\mathrm{ext}}^2 - R_{\mathrm{int}}^2},$$

$$H_i = \frac{(\lambda_i^{\mathrm{C}} + \mu_i^{\mathrm{C}})}{\mu_i^{\mathrm{C}}(3\lambda_i^{\mathrm{C}} + 2\mu_i^{\mathrm{C}})} \quad (i = 2 \text{ at loading stage 2}),$$

K, L_2, J_2 are determined by the formulas described in (10.21) for $i = 2$.

Note that in a conventional CFST column, separation of the tube from the concrete occurs at the initial loading stage, as shown previously. Consequently, when modeling the three-dimensional behavior of a conventional column, the concrete and steel do not interact in the transverse direction. Then, following reasoning analogous to that described above, we obtain the solution for the conventional column, accounting for the nonlinear deformation of concrete, in incremental form:

$$p^{\mathrm{C}} = \Delta p_i^{\mathrm{C}} + p_{i*}^{\mathrm{C}}, \quad p^{\mathrm{S}} = \Delta p_i^{\mathrm{S}} + p_{i*}^{\mathrm{S}},$$

$$\Delta p_i^{\mathrm{C}} = \Delta P \frac{Y}{F^{\mathrm{C}} Y + F^{\mathrm{S}} H_i}, \quad \Delta p_i^{\mathrm{S}} = \Delta P \frac{H_i}{F^{\mathrm{C}} Y + F^{\mathrm{S}} H_i},$$

where $p_{i*}^{\mathrm{C}}, p_{i*}^{\mathrm{S}}$ are the pressures at the beginning of concrete deformation phase number $i = 0, 1, 2$.

10.5 Strength Criterion for Concrete

When designing building structures, it is convenient to use the following concrete strength criterion for concrete subjected to compression only in the longitudinal direction z:

$$|\sigma^{\mathrm{C}}_{zz}| \leq R_b,$$

where R_b is the design value of the cylinder compressive strength of concrete (used without sign).

In [26], a strength criterion for triaxially compressed concrete was proposed. For the case when concrete is compressed in the longitudinal direction (along the z-axis) while being subjected to a constant lateral pressure p_0 in the transverse direction, this criterion can be written as follows:

$$|\sigma^{\mathrm{C}}_{zz}| \leq R_b + 4p_0.$$

Then we introduce notation for the magnitude of the ultimate stress (along the z-axis, without sign) in the concrete core of the CFST column, accounting for triaxial compression:

$$R_{b,3} = R_b + 4p_0, \quad |\sigma^{\mathrm{C}}_{zz}| \leq R_{b,3}, \tag{10.22}$$

10.6 Yield Criterion for the Steel Tube

The steel tube in CFST structures is subjected to a complex stress state. In this case, the yielding of steel is not determined by the value of a single stress component, as in a uniaxial stress state, but depends on the complete stress tensor. It is assumed that the von Mises yield criterion [6] is applicable to the steel tube, according to which the material transitions to the plastic state when the octahedral shear stress τ_0 reaches a limiting value:

$$\tau_0 = \frac{1}{3}\sqrt{(\sigma^s_{zz} - \sigma^s_{rr})^2 + (\sigma^s_{rr} - \sigma^s_{\theta\theta})^2 + (\sigma^s_{\theta\theta} - \sigma^s_{zz})^2} \leq \frac{\sqrt{2}}{3}\sigma_{sd}, \tag{10.23}$$

where σ_{sd} is the yield strength of steel under uniaxial loading.

10.7 Examples

A CFST column with the following parameters is considered: $R_{\mathrm{int}} = 0.255\,\mathrm{m}$, $R_{\mathrm{ext}} = 0.265\,\mathrm{m}$, $h = 6\,\mathrm{m}$, $d = 0.015\,\mathrm{m}$, $E^{\mathrm{C}} = 34500\,\mathrm{MPa}$, $E^{\mathrm{C}}_1 = 4698.8\,\mathrm{MPa}$, $\nu^{\mathrm{C}} = 0.225$, $\nu^{\mathrm{C}}_1 = 0.41$, $E^{\mathrm{S}} = 206000\,\mathrm{MPa}$, $\nu^{\mathrm{S}} = 0.3$, design value of the cylinder compressive strength of concrete (used without sign) $R_b = 19.5\,\mathrm{MPa}$, design value of the yield strength of structural steel (used without sign) $\sigma_{sd} = 240\,\mathrm{MPa}$.

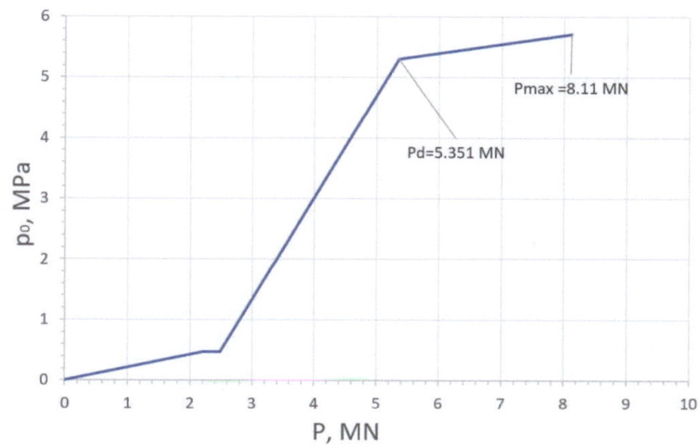

Fig. 10.7: Dependence of concrete confinement pressure p_0 on axial compressive force P (P_{max} is the ultimate load on the column).

The results of solving the problem of compression of a prestressed column are presented in Figs. 10.7 and 10.8. At loading stage 1 of the prestressed column ($P < P_d$), the lateral confinement pressure p_0 exerted on the concrete by the tube increases (Fig. 10.7), due to the fact that the tube acts only as a confining element and is not subjected to axial load. At loading stage 2 ($P \geq P_d$), after the application of a portion of the axial load to the steel tube, the confinement pressure p_0 also continues to increase (Fig. 10.7). This is due to the fact that the Poisson's ratio of

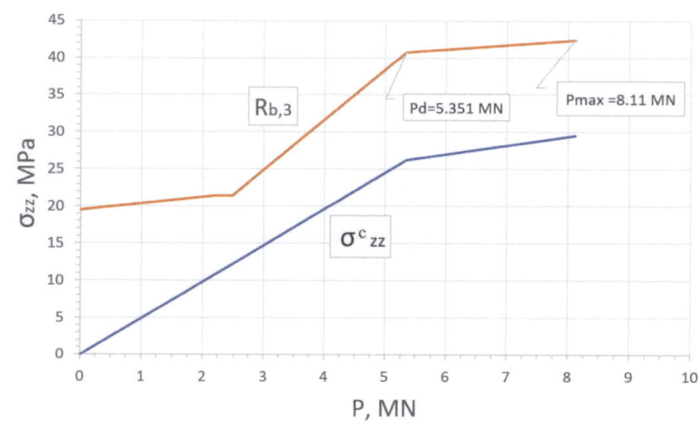

Fig. 10.8: Dependence of the magnitude of axial stresses σ^{C}_{zz} in the concrete core (blue line) and the compressive strength of confined concrete $R_{b,3}$ (under triaxial compression) on the axial compressive force P (red line).

concrete increases ($\nu_1^{\mathrm{C}} = 0.41$) and exceeds the Poisson's ratio of steel ($\nu^{\mathrm{S}} = 0.3$). Thus, at all loading stages of the prestressed column, the concrete exerts pressure on the inner surface of the tube, which practically precludes local buckling of the tube wall. This represents a significant advantage of this column type compared to the conventional CFST column.

According to Fig. 10.8, the magnitude of the ultimate stresses in concrete increases; this is caused by the increase in the lateral pressure p_0 (Fig. 10.7). Furthermore, it follows from Fig. 10.8 that as the load increases, the axial longitudinal stresses in concrete cannot reach the magnitude of the ultimate strength $R_{b,3}$, i.e., the concrete in such a structure cannot begin to fail before the failure of the steel. The stress-strain state in the steel tube eventually reaches yielding with increasing load, which triggers the onset of failure of the CFST column.

10.7.1 Comparison of Prestressed and Conventional CFST Columns

Consider an example of a comparative analysis of a prestressed concrete-filled steel tube column and a conventional column for different cross-sectional dimensions and different values of the initial concrete core protrusion d in the prestressed column. In this case, to analyze the influence of the tube wall thickness for a constant concrete core diameter, tubes with non-standard dimensions are considered in some instances. The column height is taken as $h = 6\,\mathrm{m}$, $E^{\mathrm{C}} = 34500\,\mathrm{MPa}$, $E_1^{\mathrm{C}} = 4698.8\,\mathrm{MPa}$, $\nu^{\mathrm{C}} = 0.225$, $\nu_1^{\mathrm{C}} = 0.41$, $E^{\mathrm{S}} = 206000\,\mathrm{MPa}$, $\nu^{\mathrm{S}} = 0.3$, $R_b = 19.5\,\mathrm{MPa}$, $\sigma_{sd} = 240\,\mathrm{MPa}$. The analysis is performed taking into account the nonlinear deformation of concrete within the CFST columns. The results are presented in Table 10.1, where $P_{\max 1}$ is the value of the ultimate load on the prestressed CFST column (according to condition (10.23), and $P_{\max 2}$ is the value of the ultimate load on the conventional CFST column (according to the condition $|\sigma_{zz}^{s}| \leq \sigma_{sd}$), considering the absence of interaction between the steel tube and the concrete core in the transverse direction due to separation.

According to Table 10.1, the load-bearing capacity of the prestressed CFST column exceeds that of a conventional column by at least 12-18%. It should be noted that the actual ultimate load on the conventional column may be less than the calculated value due to local buckling of the tube wall.

10.8 Conclusions

1. The developed analytical model demonstrates that radial separation between the steel tube and the concrete core in conventional CFST columns initiates already at the early stage of loading. This phenomenon is caused by the negative difference between the Poisson's ratios of concrete and steel. As a result of this separation,

Table 10.1: Calculation examples of a prestressed CFST column and a conventional CFST column taking into account the nonlinear deformation of concrete.

External diameter of the tube × wall thickness, mm	d, mm	$P_{\max 1}{}^{*}$, MN	$P_{\max 2}$, MN	$P_{\max 1}/P_{\max 2}$
	5	7.59		
530×10	10	7.94	7.10	
	15	8.11		1.14
	20	8.03		
	10	8.72		
534×12	15	8.87	7.90	1.12
	20	8.77		
	10	10.45		
630×10	15	10.71	9.22	1.16
	20	10.64		
	10	11.36		
634×12	15	11.59	10.18	1.14
	20	11.48		
	10	13.00		
720×10	15	13.39	11.34	1.18
	20	13.32		
	10	16.15		
732×16	15	16.42	14.63	1.12
	20	16.23		

* $P_{\max 1}$ is calculated when $\nu^{C} = 0.255$, $\nu_1^{C} = 0.41$

the steel tube wall becomes susceptible to local buckling. Numerous experimental studies confirm that the failure of conventional CFST columns under concentric compression is typically initiated by local buckling of the steel tube wall.

2. An improved CFST column design has been proposed, in which transverse confinement of the concrete core by the steel tube is maintained throughout all loading stages. This design prevents radial separation and effectively protects the tube wall against local buckling
3. The analysis is based on a three-dimensional model of the stress–strain state of CFST columns that accounts for transverse interaction between the steel tube and the concrete core, as well as for the nonlinear evolution of the concrete modulus of deformation and Poisson's ratio.
4. Under concentric compression, the load-bearing capacity of the proposed prestressed CFST column exceeds that of a conventional CFST column by approximately 12–18%.

References

[1] Abbasi A, Lork A, Rostami V (2025) An innovative machine learning model to formulate the bearing capacity of concrete-filled steel tube column. Arabian Journal for Science and Engineering 50:1921–1943, DOI 10.1007/s13369-024-09148-6

[2] ACI Committee 318-14 (2014) Building code requirements for structural concrete and commentary. American Concrete Institute

[3] AISC (1999) Load and Resistance Factor Design Specification for Structural Steel Buildings. American Institute of Steel Construction, Chicago

[4] AISC 360-16 (2016) Specification for structural steel buildings. American Institute of Steel Construction

[5] Alrebeh S, Ahmed A, Al-Asad A, et al. (2023) Experimental performance evaluation of concrete-filled steel tube columns confined by high-strength steel bolts. International Journal of Steel Structures 23:1135–1147, DOI 10.1007/s13296-023-00755-x

[6] Altenbach H, Kolupaev V (2014) Classical and non-classical failure criteria. In: Altenbach H, Sadowski T (eds) Failure and Damage Analysis of Advanced Materials, Springer, Heidelberg, pp 1–66

[7] Bajad M (2024) Influence of axial compression on a steel tubular stub column filled with reinforced concrete for sustainable infrastructure. Innovative Infrastructure Solutions 9(399), DOI 10.1007/s41062-024-01714-z

[8] de Carvalho A, Rossi A, Morkhade S, et al. (2023) Machine learning-based design approach for concrete-filled stainless steel tubular columns. Arabian Journal for Science and Engineering 48:14105–14118, DOI 10.1007/s13369-023-08090-3

[9] Chang Y, Chen W, Xiao Q, Rong E, Peng L (2021) Theoretical and experimental study on axial compression concrete-filled tubes with different confinements. Journal of Constructional Steel Research 185:106862

[10] Chen S, Zhang R, Jia L, Wang J, Gu P (2018) Structural behavior of uhpc filled steel tube columns under axial loading. Thin-Walled Structures 130:550–563

[11] EN 1994-1-1 (2004) Eurocode 4: Design of composite steel and concrete structures— Part 1-1: General rules and rules for buildings

[12] Garg V, Patel A, Purohit S (2022) Experimental investigations on axial flexural behaviour of slender concrete filled steel tubular columns. Journal of The Institution of Engineers (India) Series A 103(4):1089–1100

[13] Giakoumelis G, Lam D (2004) Axial capacity of circular concrete-filled tube columns. Journal of Constructional Steel Research 60:1049–1068

[14] Gupta M, Prakash S, Ghani S (2025) An interactive ensemble learning approach for predicting ultimate axial load capacity in concrete-filled steel tube column. Modeling Earth Systems and Environment 11:185, DOI 10.1007/s40808-025-02348-5

[15] Han L (2016) Concrete Filled Steel Tubular Structures-Theory and Practice. Science Press, Beijing

[16] Han L, Yao G, Zhao X (2005) Tests and calculations for hollow structural steel (hss) stub columns filled with self-consolidating concrete (scc). Journal of Constructional Steel Research 61:1241–1269
[17] Han L, Yao G, Tao Z (2007) Performance of concrete-filled thin-walled steel tubes under pure torsion. Thin-Walled Structures 45:24–36
[18] Hu H, Lin K, Shahrooz B, Guo Z (2021) Revisiting the composite action in axially loaded circular cfst columns through direct measurement of load components. Engineering Structures 235:112066
[19] Karpenko N (1996) Obshchie modeli mekhaniki zhelezobetona (General models of reinforced concrete mechanics). Stroyizdat, Moscow, 416 p. (in Russian)
[20] Lai Z, Jiang H, Cai Y (2022) A new design equation to estimate the axial compressive strength of circular concrete-filled steel tubular stub columns. Structures 46:1043–1054
[21] Liu J, Gao P, Lin X, Wang X, Zhou X, Chen Y (2023) Experimental assessment on the size effects of circular concrete-filled steel tubular columns under axial compression. Engineering Structures 275:115247
[22] Luo K, Yu L, Shu G, et al. (2024) Research on the axial compressive behavior of concrete-filled steel tubular column reinforced with annular stiffener. International Journal of Steel Structures 24:1265–1277, DOI 10.1007/s13296-024-00837
[23] Minstroy Rossii (2018) SP 63.13330.2018 Betonnye i zhelezobetonnye konstruktsii. Osnovnye polozheniya. Aktualizirovannaya redaktsiya SNiP 52-01-2003 (in Russ., Concrete and reinforced concrete structures. General provisions)
[24] O'Shea M, Bridge R (2000) Design of circular thin-walled concrete filled steel tubes. Journal of Structural Engineering 126:1295–1303
[25] Patel P, Patel A, Purohit S (2025) Confinement studies on concrete filled steel tube composite columns: Numerical and experimental investigations. In: Goel M, Gupta L, Jaiswal O (eds) Recent Developments in Structural Engineering, Volume 3. SEC 2023, Springer, Singapore, Lecture Notes in Civil Engineering, vol 548, DOI 10.1007/978-981-97-9885-8_24
[26] Richart F, Brown A, Brandraeg A (1928) A study of failure of concrete under combined compressive stresses. University of Illinois, Engineering Experiment Station, bulletin No. 185, 105 p.
[27] Sakino K, Nakahara H, Morino S, Nishiyama I (2004) Behavior of centrally loaded concrete-filled steel-tube short columns. Journal of Structural Engineering 130:180–188
[28] Sangeetha P, Ashwin M, Gunasekaran D, Dhivya M, Janani S, Madumathi S (2018) Behaviour of concrete filled steel tubes. Journal of Informatics and Mathematical Sciences 10:297–304, DOI 10.26713/jims.v10i1-2.1056
[29] Snigireva V, Gorynin G (2018) The nonlinear stress-strain state of the concrete-filled steel tube structures. Magazine of Civil Engineering 7(83):74–83
[30] Standards Australia (2017) AS/NZS 2327 Composite structures—composite steel-concrete construction in buildings

[31] Su B, Wu Z, Jiang D, Bi G, Fan J, Yan K (2025) Feature importance on axial compressive strength prediction of concrete-filled steel tubular columns using random forest algorithm. Communications in Computer and Information Science 2451:189–200
[32] Subedi N, Obara T, Kono S (2022) Noncompact and slender concrete-filled steel tubes under axial compression: Finite-element modeling and evaluation of stress-strain models for fiber-based analysis. Journal of Constructional Steel Research 196:107353
[33] Tao Z, Han L, Wang D (2007) Experimental behaviour of concrete-filled stiffened thin-walled steel tubular columns. Thin-Walled Structures 45(5):517–527, DOI 10.1016/j.tws.2007.04.003
[34] Tao Z, Wang Z, Yu Q (2013) Finite element modelling of concrete-filled steel stub columns under axial compression. Journal of Constructional Steel Research 89:121–131
[35] Truesdell C (1955) Hypo-elasticity. Journal of Rational Mechanics and Analysis 4:83–133, 1019-1020
[36] Truesdell C (1963) Remarks on hypo-elasticity. Journal of Research of the National Bureau of Standardss - B. Mathematics and Mathematical Physics 67B(3):141-143
[37] Wang Y, Chen P, Liu C, Zhang Y (2017) Size effect of circular concrete-filled steel tubular short columns subjected to axial compression. Thin-Walled Structures 120:397–407
[38] Yu Q, Tao Z, Liu W, Chen Z (2010) Analysis and calculations of steel tube confined concrete (stcc) stub columns. Journal of Constructional Steel Research 66:53–64
[39] Zhou Z (2023) Unified model for axial bearing capacity of concrete-filled steel tubular circular columns based on hoek–brown failure criterion. Buildings 13:2408, DOI 10.3390/buildings13102408

Chapter 11
On the Existence of a Set of Symmetric and Asymmetric Oscillations in a Symmetric Electromechanical System with Impact Interactions

Leonid Igumnov, Vladimir Metrikin, Aleksander Belov, and Aleksander Lyubimov

Abstract The dynamics of a vibroimpact system consisting of a mass M subjected to a harmonic action $F \sin \Omega t$, in the cavity of which a second charged mass is located, is investigated. The free movement of the second charged mass in the direction of the external force is limited by two obstacles of the first mass. Upon contact with these obstacles, an instantaneous collision with a coefficient of restitution R occurs , and the second mass is recharged. Using the Poincaré surface point mapping method, a numerical and analytical study of the three-dimensional phase space structure is performed, and the existence domains of stable symmetric, asymmetric, and chaotic motions of the second mass are found in the parameter space. Bifurcation diagrams for the main parameters of the system are presented, which allow us to qualitatively evaluate and specify specific values of the bifurcation parameters for the generation of arbitrarily complex periodic motion regimes and establish a scenario for the transition to chaotic motion.

Key words: Electromechanical vibro-impact system, Stability, Point transformation method, Symmetric and asymmetric motion modes, Bifurcation diagrams, Chaos

11.1 Introduction

The article investigates the dynamics of an electromechanical vibroimpact system consisting of a mass M with a cavity subjected to a harmonic action $F \sin \Omega t$, in the cavity of which a second charged mass is located. The free movement of the latter in the direction of the external force is limited by two charged obstacles

---

Leonid Igumnov (corresponding author) · Vladimir Metrikin · Aleksander Belov · Aleksander Lyubimov
National Research Lobachevsky State University of Nizhny Novgorod, 603022 Nizhny Novgorod, Gagarina ave., 23, Russian Federation, e-mail: igumnov@mech.unn.ru, vsmetrikin@mail.ru, belov_a2@mech.unn.ru, ljubimov@mm.unn.ru

H. Altenbach et al. (eds.), *Stability Problems in Mechanics of Continua*,
Advanced Structured Materials 260, https://doi.org/10.1007/978-3-032-31032-3_11

attached to the mass. Upon contact with these obstacles, an instantaneous collision with a velocity restoration coefficient R and a charge exchange of the second mass occurs. A detailed numerical and analytical study of the dynamics is carried out using the mathematical apparatus of the Poincaré surface point mapping method [7, 8, 10–12, 21]. Analytical equations for the Poincaré surface point mappings are given, along with the coordinates of fixed points in parametric form corresponding to periodic symmetric and asymmetric motions of the second mass. The domains of existence of stable periodic motions of various structures are presented in the parameter space of the system, and bifurcation diagrams were constructed using the developed software product, which made it possible to determine in more detail the values of the bifurcation parameters indicating what changes occur in the phase space structure [15, 23]. The presented bifurcation diagrams for the main parameters of the system also made it possible to qualitatively evaluate and indicate specific values of the bifurcation parameters for the generation of arbitrarily complex periodic motion regimes and to establish a scenario for the transition to chaotic motion [23]. It was proven that periodic stable symmetric and asymmetric motion regimes have a period greater than the period of the harmonic effect on the first mass.

Research into dynamic models of this type continues to generate considerable interest (see [1–6, 9, 13, 14, 16–20, 22, 24, 25] and the literature cited therein). Examples of practical applications of such research include shock dampers [1, 7, 8, 10–12], designed to absorb vibrations and impacts, making them indispensable in various industries. In particular, hydraulic shock dampers are used in vehicle suspensions to absorb shocks from road imperfections; pendulum-based bridge dampers play a key role in ensuring structural stability under strong wind loads and earthquakes.

11.2 Mathematical Model

Assuming that the contact of the second mass with the oppositely charged limiters of its oscillations occurs instantaneously, and at the same time it also has time to recharge and change its speed upon impact, using Newton's hypothesis [8, 11] with the coefficient of speed restoration upon impact $0 < R < 1$, and also assuming that the masses interact only upon impact, and excluding from consideration cases of "stuck" motion of the masses, the mathematical model, with the usual simplifications, can be represented in dimensionless form as

$$\begin{cases} \ddot{x} = PA(\tau) - e\sin(\tau), & |x| < 1 \\ \dot{x}^{+} = -R\dot{x}^{-}, \quad \text{recharge}, & |x| = 1 \end{cases} \tag{11.1}$$

where the relative displacement of the masses is chosen as the coordinate, and the position corresponding to the symmetrical mutual arrangement of the masses is taken as the origin = 0, and the introduced dimensionless time and parameters mean the following: $\tau = \omega t$ dimensionless time, $\mathrm{A}(\tau) = 2q\,(\tau/\omega)\,(E/hm\omega_0^2)$ is the dimensionless charge of the second mass; $P = \omega_0^2/\omega^2$ is the dimensionless frequency parameter; $e = 2\varepsilon/h$ – dimensionless amplitude of oscillations of the first mass. Dimensional parameters of the system under consideration have the following meaning; E – electric field strength, $q(t)$– charge of the second mass, ε – amplitude of the external force, ω – frequency of the external action, $\dot{x}^-, \dot{x}^+$ -impact and post-impact velocities of the second mass during the impact interaction of the second mass with the limiters, h - distance between the limiters of oscillations of the second mass. The resulting system (11.1), as is known [7, 8, 10–12, 21], is essentially nonlinear, with a variable structure of the phase space. From the extensive class of dynamic systems with collisions [10–12], the so-called symmetric systems are usually distinguished, the study of which is usually limited to the study of symmetric periodic motions. The full period of such motions obviously consists of completely identical intervals – half-periods. In the present work, the existence and stability of asymmetric two-impact modes, the period of which is greater than the period of the external harmonic action, are proved for a two-mass electromechanical system. Some of the regions of the parameter space corresponding to these modes do not have common boundaries with the region of the symmetric mode. Oscillograms of asymmetric modes and corresponding bifurcation diagrams are presented, allowing us to prove the existence of chaotic modes and determine the bifurcation values of the parameters and the scenario for the emergence of chaos. To study motions of this type, it is sufficient to consider the point mapping of one of the impact interaction surfaces onto itself, realized by the trajectories of the system of equations (11.1) [12, 21].

11.3 Structure of Phase Space

The phase space of the system is three-dimensional, in coordinates $F(-1 \leq x \leq 1, -\infty < \dot{x} < +\infty, \tau \geq 0)$. According to the mathematical model, it follows that the representing point always falls on either the plane $\Pi_+(x = 1)$ or the plane $\Pi_-(x = -1)$. In this regard, it can be argued that these planes $x = \pm 1$ are Poincaré surfaces [10, 21] and, therefore, the dynamics of the system can be studied using a point mapping of the Poincaré plane onto itself.

The movement of the representative point between planes $|X| < 1$ is described by the equation: $\ddot{X} = PA(\tau) - e\sin(\tau)$, and the movement on the planes $|X| = 1$ – equation $\dot{X}^+ = -R\dot{X}^-$ and recharging. The representative point, emerging from the plane Π_+, moves towards the plane Π_-, and the one emerging from Π_- -– towards Π_+. The dynamics of a system can be studied (11.1) by analyzing the behavior of the points of a point mapping of a plane Π_+ (Π_-) into itself.

11.4 General Point Transformation

Let the initial point of the point transformation $M_0(x_o = 1, \dot{x}_o; \tau_o)$ correspond to the moment of time immediately before the impact, $M_1(x_1 = -1, \dot{x}_1; \tau_1)$ be the moment of time immediately before the next impact on the surface $x = -1$, and $M_2(x_2 = 1, \dot{x}_2; \tau_2)$ let the point be the moment of time immediately before the next impact interaction on the initial surface $x = 1$. The relationship between the coordinates of the initial M_0 and final M_2 points of the point mapping is obtained by solving Eqs. (11.1) and is written in the following form

$$\begin{cases} \dot{X}_1^- = -R\dot{X}_0^- - PA(\tau_1 - \tau_0) - e\cos(\tau_0) + e\cos(\tau_1), \\ \dfrac{PA(\tau_1 - \tau_0)^2}{2} + e\sin(\tau_0) - e\sin(\tau_1) + \left(R\dot{X}_0^- + e\cos(\tau_0)\right)(\tau_1 - \tau_0) = 2, \\ \dot{X}_2^- = -R\dot{X}_1^- + PA(\tau_2 - \tau_1) + e\cos(\tau_2) - e\cos(\tau_1), \\ \dfrac{PA(\tau_2 - \tau_1)^2}{2} + e\sin(\tau_2) - e\sin(\tau_1) - \left(R\dot{X}_1^- + e\cos(\tau_1)\right)(\tau_2 - \tau_1) = 2. \end{cases} \tag{11.2}$$

If (11.2) is valid, we make formal transformations in the system: $\tau_0 = -\tau_0'$, $\tau_1 = \tau_1, \tau_2 = -\tau_2'$, then the display of the initial point M_0 the impact interaction surface to a point M_2 on the same surface is a doubly applied "half-period" point mapping [10, 12, 15, 21]. Therefore, to study the properties of symmetrical motions it is sufficient to use two point mapping Eqs. (11.2):

$$\begin{cases} \dot{X}_1^- = -R\dot{X}_0^- - PA(\tau_1 - \tau_0) - e\cos(\tau_0) + e\cos(\tau_1), \\ \dfrac{PA(\tau_1 - \tau_0)^2}{2} + e\sin(\tau_0) - e\sin(\tau_1) + \left(R\dot{X}_0^- + e\cos(\tau_0)\right)(\tau_1 - \tau_0) = 2 \end{cases} \tag{11.3}$$

11.5 Finding the Coordinates of Fixed Points Corresponding to Symmetric Periodic Motions

The coordinates of the fixed point are found from the equations of the half-period point transformation (11.3) [8, 12, 21]

$$\begin{cases} -\dot{X}_0^* = -R\dot{X}_0^* - PA\pi n + e\cos(\tau_0^*)\left((-1)^n - 1\right), \\ \dfrac{PA\pi^2 n^2}{2} + e\sin(\tau_0^*)\left(1 - (-1)^n\right) + \left(R\dot{X}_0^* + e\cos(\tau_0^*)\right)\pi n = 2 \end{cases} \tag{11.4}$$

with the addition of periodicity conditions

$$\tau_1 - \tau_0 = \pi n, \quad \dot{X}_1 = -\dot{X}_0 \tag{11.5}$$

where is n - the multiplicity of the periodic mode of motion of the second mass to the period of external influence.

For even numbers, ($n = 2,\ 4,\ 6$) from the system of equations (11.4) we obtain the coordinates of the fixed point in parametric form:

$$\begin{cases} \dot{X}_0^* = \dfrac{PA\pi n}{(1-R)}, \\ \cos\left(\tau_0^*\right) = \dfrac{4\,(1-R) - PA\pi^2 n^2\,(1+R)}{2\pi ne\,(1-R)} \end{cases} \tag{11.6}$$

From here one can find the initial phase τ_0^* of periodic symmetrical motion

$$-2\pi ne(1-R) \le 4(1-R) - PA\pi^2 n^2 (1+R) \le 2\pi ne(1-R)$$

For odd ones, ($n = 1,\ 3,\ 5$) from the system of equations (11.4) we obtain the coordinates of the fixed point in the following form

$$\begin{cases} \dot{X}_0^* = \dfrac{(PA\pi n + 2e\cos(\tau_0^*))}{(1-R)}, \\ \dfrac{PA\pi^2 n^2}{2} + 2e\sin(\tau_0^*) + R\dot{X}_0^*\pi n + e\cos(\tau_0^*)\pi n = 2 \end{cases} \tag{11.7}$$

11.6 Stability of Symmetric Periodic Motion

To study the stability of symmetric periodic motion, it is necessary to linearize the system of equations (11.3) in the vicinity of a fixed point (11.4) or (11.7) [11, 21], then, assuming $\delta\dot{X}_1 = -Z\delta\dot{X}_0, \delta\tau_1 = Z\delta\tau_0, \tau_1 = \tau_0 + \pi n$,we obtain the characteristic equation

$$A_1\, Z^2 + A_2\, Z + A_3 = 0, \tag{11.8}$$

the magnitude of the roots of which $Z_i, i = 1, 2$ determines the stability of the fixed points of the point symmetric mapping corresponding to the symmetric periodic mode of motion. In (11.8), the coefficients depend on the system parameters as follows

$$\begin{cases} A_1 = (-1)^n e\cos(\tau_0) - e\cos(\tau_0) - PA\pi n - R\dot{X}_0, \\ A_2 = -(-1)^n Re\cos(\tau_0) + Re\cos(\tau_0) + R^2\dot{X}_0 + PA\pi n + e\pi n\sin(\tau_0) + R\dot{X}_0, \\ -(-1)^n Re\pi n\sin(\tau_0), \\ A_3 = -R^2\dot{X}_0, \end{cases}$$

where instead of τ_0 and $\dot{X}_0$ in the coefficients A_1, A_2, A_3 and (11.10), (11.11) τ_0^* and $\dot{X}_0^*$ from (11.7) for even n, or from (11.7) for odd n are substituted.

For the stability of a fixed point, it is necessary and sufficient [8, 21] that the roots of the characteristic equation be located inside the unit circle $|Z_i| < 1$ for $i = 1,\ 2$. In particular, in the case of even, n the characteristic polynomial looks as follows:

$$P_2(Z) = \underbrace{(-PA)}_{A_1} Z^2 + \underbrace{(PA(R^2+1) + e\sin(\tau_0)(1-R)^2)}_{A_2} Z^1 + \underbrace{(-R^2PA)}_{A_3} Z^0 = 0 \tag{11.9}$$

and the boundaries of the stability region in the parameter space of a fixed point corresponding to symmetric periodic motion, as is known [8, 10–12, 21], are determined from the equations

$$N_- : PA + PA(R^2+1) + e\sin(\tau_0)(1-R)^2 + R^2PA = 0 \tag{11.10}$$

$$N_+ : PA - PA(R^2+1) - e\sin(\tau_0)(1-R)^2 + R^2PA = 0 \tag{11.11}$$

$$N_\varphi : PA - R^2PA = 0 \tag{11.12}$$

which, using Eqs. (11.5), can be rewritten

$$N_- : e^2 = \frac{\left(4(1-R) - PA\pi^2n^2(1+R)\right)^2}{4\pi^2n^2(1-R)^2} + \frac{4P^2A^2\left(1+R^2\right)^2}{(1-R)^4}, \tag{11.13}$$

$$N_+ : e^2 = \frac{\left(4(1-R) - PA\pi^2n^2(1+R)\right)^2}{4\pi^2n^2(1-R)^2}, \tag{11.14}$$

$$N_\varphi : R = 1. \tag{11.15}$$

11.7 Construction of the Boundaries of the Stability Region in the Case of Even *n*

In Figs. 11.1 - 11.4 in the plane of parameters, the stability (p, e) regions N - in the small (between curves) are shown for the set of the following parameter values

1. $A = 0.5, R = 0.6, n = 2$ and $A = 0.5, R = 0.9, n = 2$
2. $A = 5, R = 0.6, n = 2$ and $A = 5, R = 0.9, n = 2$

From Figs. 11.1 - 11.4 it follows:

- An increase in the parameter R leads Ato an increase in the size of the stability region and a shift of its boundaries into the region of lower frequencies and higher amplitudes of external influence,
- Calculations also showed that the increase n leads to a sharp decrease in the size of the stability region, which has been repeatedly noted previously [1–6, 8, 9, 11–14, 16–22, 24, 25], when studying various designs of vibration-impact systems.

11.8 Equations of Boundaries N_-, N_+, N_φ for odd *n*

Substituting in (11.4) $n = 2m + 1$, we obtain the form of the characteristic equation:

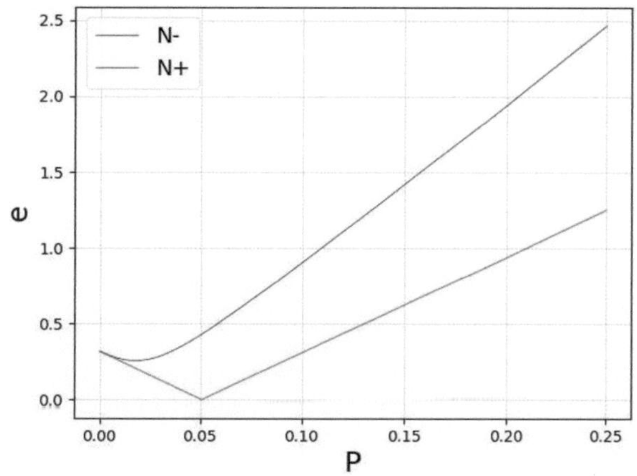

Fig. 11.1 Stability region: $A = 0.5, R = 0.6, n = 2.$

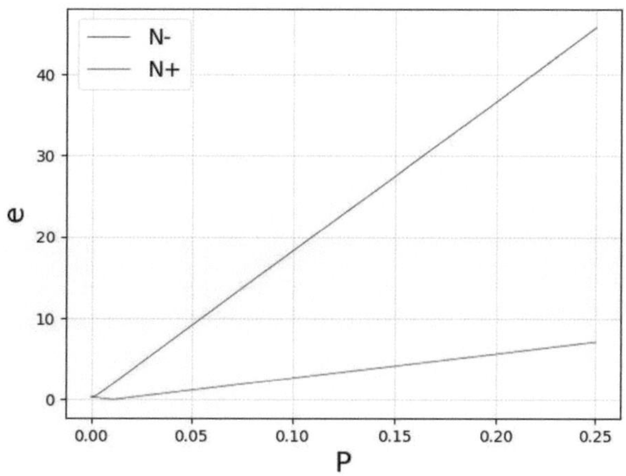

Fig. 11.2 Stability region: $A = 0.5, R = 0.9, n = 2.$

$$\begin{aligned}(-2e\cos(\tau_0) - PA\pi n - R\dot{X}_0)Z^2 + \big(2Re\cos(\tau_0) + R\dot{X}_0(R+1) + PA\pi n\big)\\ + e\pi n\sin(\tau_0)\,(1+R)\,Z - R^2\dot{X}_0 = 0\end{aligned} \tag{11.16}$$

or, using (11.7), we obtain the characteristic polynomial

$$\begin{aligned}P_2(Z) = \underbrace{(-\dot{X}_0)}_{A_1}\; Z^2 + \underbrace{\left(2R\dot{X}_0 + \pi n\,(PA\,(1-R) + (1+R)) - \frac{\dot{X}_0\pi^2 n^2(1+R)^2}{4}\right)}_{A_2} Z^1\\ + \underbrace{(-R^2\dot{X}_0)}_{A_3}\; Z^0 = 0.\end{aligned} \tag{11.17}$$

From where, after a series of formations, we will have

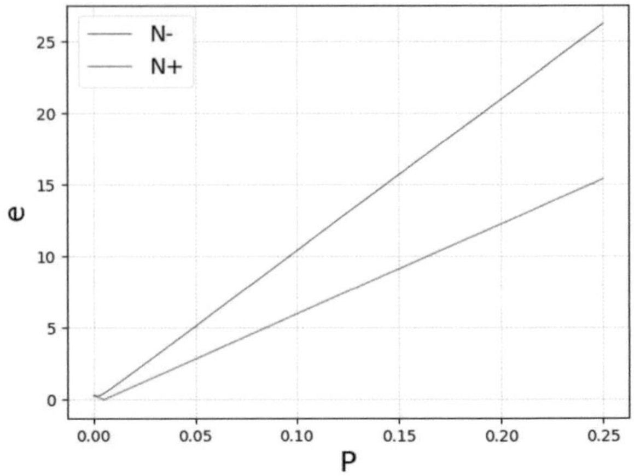

Fig. 11.3 Stability region: $A = 5, R = 0.6, n = 2$.

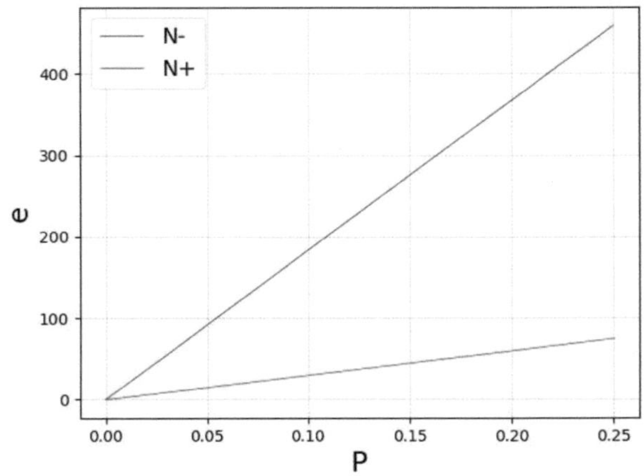

Fig. 11.4 Stability region: $A = 5, R = 0.9, n = 2$.

$$N_- : e^2 = \left(1 - \frac{\pi^2 n^2 (PA(1-R) + (1+R))}{(1+R)(\pi^2 n^2 - 4)}\right)^2 + \frac{1}{4}\left(\frac{4\pi n(PA(1-R)^2 + (1-R^2))}{(1+R)^2(\pi^2 n^2 - 4)} - PA\pi n\right)^2, \tag{11.18}$$

$$N_+ : e^2 = \left(1 - \frac{\pi^2 n^2 (PA\left(1-R^2\right) + (1+R)^2)}{4(1-R)^2 + \pi^2 n^2 (1+R)^2}\right)^2 + \frac{1}{4}\left(\frac{4\pi n(PA(1-R)^2 + (1-R^2))}{4(1-R)^2 + \pi^2 n^2 (1+R)^2} - PA\pi n\right)^2 \tag{11.19}$$

$$N_\varphi : R = 1 \tag{11.20}$$

11.9 Construction of the Baries of the Stability Region in the Case of odd n

In Figs. 11.5 - 11.8, the (p, e) stability regions in the small plane are shown in the parameter plane. N - for a set of the following parameter values

1. $A = 0.5, R = 0.6, n = 1$ and $A = 0.5, R = 0.9, n = 1,.$
2. $A = 5, R = 0.6, n = 1$ and $A = 5, R = 0.9, n = 1$

Figures 11.5 - 11.8 indicate that the dependencies of the location of the regions of existence and stability when changing the parameters are similar to the case of even n.

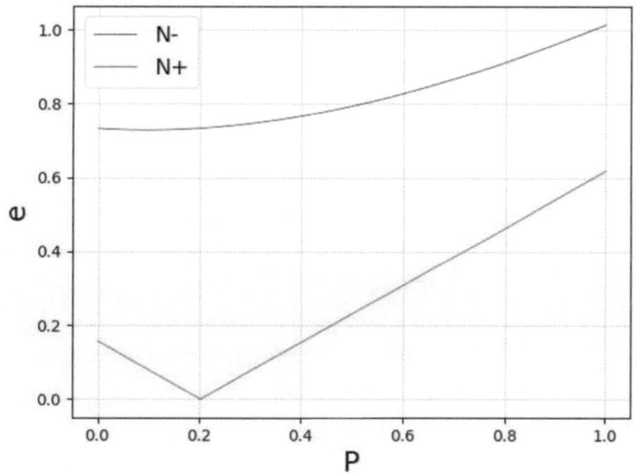

Fig. 11.5 Stability region: $A = 0.5, R = 0.6, n = 1.$

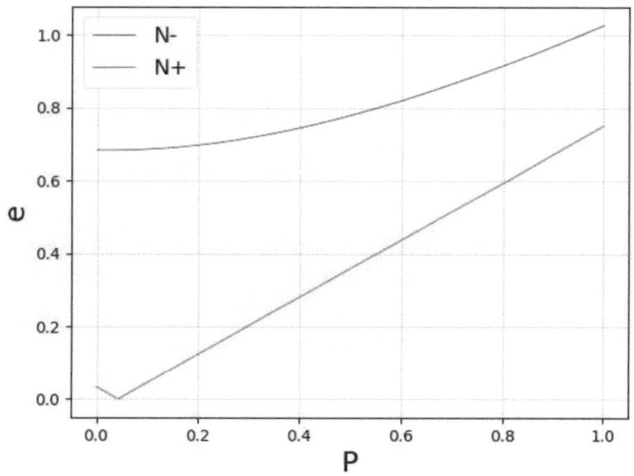

Fig. 11.6 Stability region: $A = 0.5, R = 0.9, n = 1.$

Fig. 11.7 Stability region: $A = 5, R = 0.6, n = 1$.

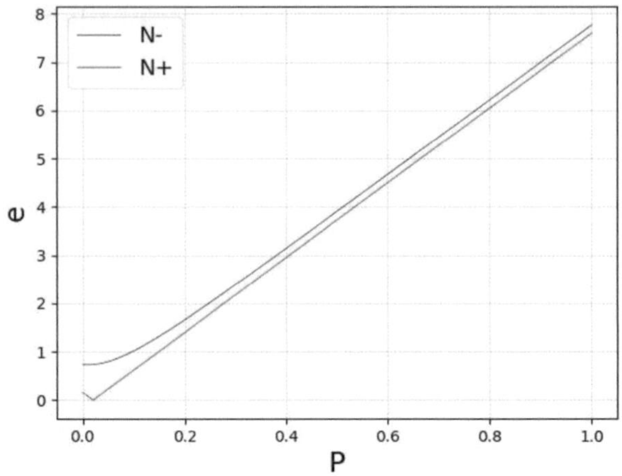

Fig. 11.8 Stability region: $A = 5, R = 0.9, n = 1$.

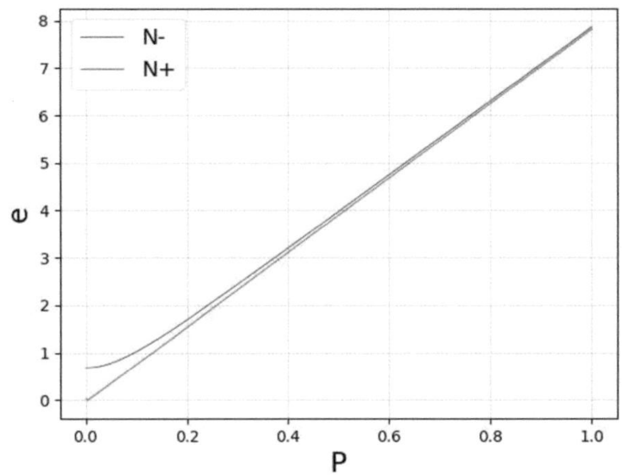

11.10 Bifurcation boundary of the existence of a point mapping. Numerical construction

It is known [8, 12, 21] that when studying the dynamics of highly nonlinear (vibro-impact) systems in regions of existence of stable periodic motions constructed using boundaries $N_{-}, N_{+}, N_{\varphi}$, it is necessary to construct a bifurcation boundary of the "existence" of a point transformation. As an example, Fig. 11.9, $R = 0.6$, $A = 1, n = 2$ shows the appearance of the bifurcation curve of the "existence" of a point mapping for the values of the parameters. The region of existence of stable symmetrical periodic motions is highlighted in green, and the bifurcation boundary is highlighted in orange. From Fig. 11.9 it follows that the bifurcation boundary of "existence" cuts off from the stability region a significant part.

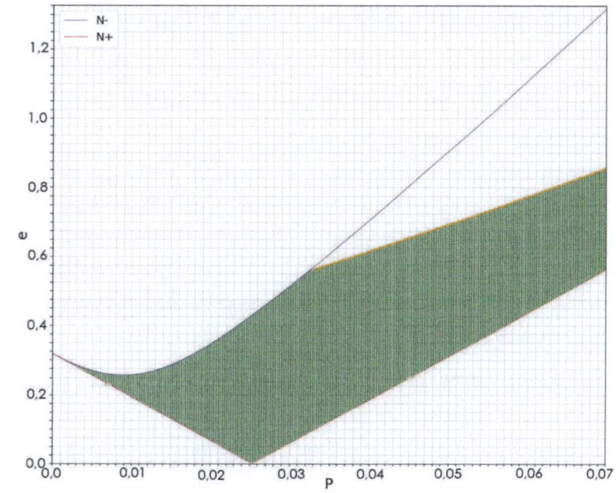

Fig. 11.9 Bifurcation boundary of existence of a point transformation: $A = 1, R = 0.6, n = 2$.

11.11 Equations of Existence of Asymmetric Two-Impact Periodic Motions

The equations for the existence of asymmetric two-impact periodic motions corresponding to fixed points of the twice applied "half-period" mapping after introducing into consideration the asymmetry parameter $\varepsilon = ((\tau_1 - \tau_0)/\pi n) - 1)$, using the general point mapping (11.3), take the form [10, 12, 21]

$$\begin{aligned}&2\pi n(1+R)e[(-1)^n C - R\cos\tau_0] - (1+R)^2 e[(-1)^n C - \cos\tau_0](1+\varepsilon)\pi n\\&+ 2(\pi n)^2 Ap(1+R)^2\varepsilon = 0\end{aligned} \tag{11.21}$$

$$\begin{aligned}&(1-R^2)e[(-1)^n S - \sin\tau_0] - (1+R)e[\cos\tau_0\\&- R(-1)^n C][(1+\varepsilon)\pi n] - (pA/2)[(1-R^2)(1+\varepsilon) - 4R](1+\varepsilon)(\pi\,\mathrm{n})^2 + 2 = 0\end{aligned} \tag{11.22}$$

where

$$S = \sin(\tau_0 + \varepsilon\pi n), C = \cos(\tau_0 + \varepsilon\pi n). \tag{11.23}$$

The stability of a fixed point corresponding to two-impact asymmetric movements is determined depending on the magnitude of the roots $Z_i, i = 1, 2$ of the following characteristic equation

$$\begin{aligned}&\dot{x}_0(Z - R^2)^2 + [(1+R)e(\sin\tau_1 - \sin\tau_0) - (1-R)pA][2\pi nR\dot{x}_0\\&+ (2\pi n - \delta\tau)\delta\tau(1-R)pA]Z - 2\pi nR\dot{x}_0(1-R)pAZ\\&+ (2\pi n - \delta\tau)\delta\tau(1+R)^2 e^2 \sin\tau_0 \sin\tau_1 Z = 0\end{aligned} \tag{11.24}$$

where $\delta\tau = \tau_1 - \tau_0$.

11.12 Results of Numerical Experiments

More complex motion regimes were studied using numerical experiments, utilizing the theoretical results presented. For this purpose, detailed bifurcation diagrams were constructed for all system parameters, leading to conclusions about the regions of existence of stable periodic motion regimes of the second mass with varying numbers of impacts, including chaotic behavior of mapping points on Poincaré planes.

Below in Figs. 11.10 and 11.11 are shown bifurcation diagrams on planes $X = 1$ and $X = -1$ by parameter A. From them it is clear that the following modes of the system are observed:

- Symmetrical periodic movements with one impact on each plane are observed; $0 < A < 0.55$.
- Since $A = 0.55$ chaotic movements have been observed [16, 20, 23].

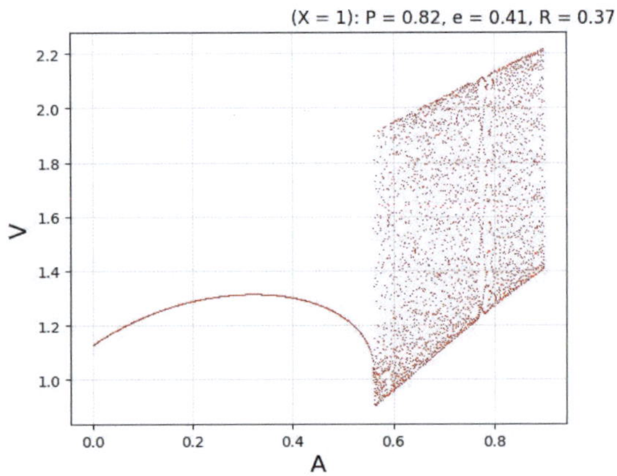

Fig. 11.10 Bifurcation diagram.

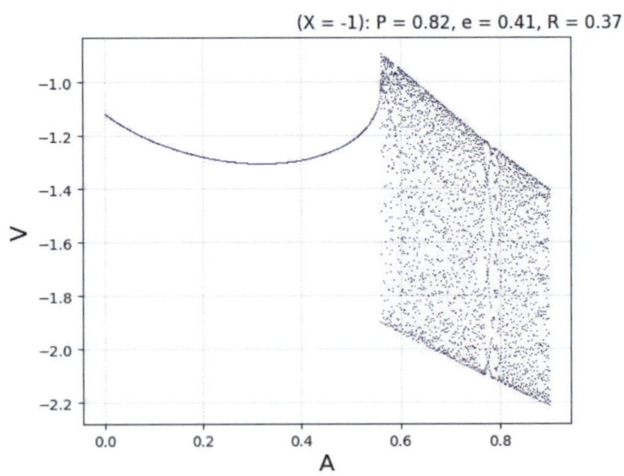

Fig. 11.11 Bifurcation diagram.

Calculations of bifurcation diagrams for the parameter also showed (see Figs. 11.12 and 11.13) that the following modes of the system are observed on the planes $X = 1$ and $X = -1$ on the plane $(e, V_i = \dot{x}_i)$:

- For $1.8 < e < 2.5$ complex non-periodic movements with a different number of impacts on each plane,
- With $2.5 < e < 3.1$ symmetrical periodic movements with two impacts on each plane,
- With $5 < P < 5.5$ symmetrical periodic movements with three impacts on each plane,
- With $7 < P < 7.5$ symmetrical periodic movements with four impacts on each plane.

Oscillograms of the motion of the imaging point for the described modes are presented below in Figs. 11.14 and 11.15.

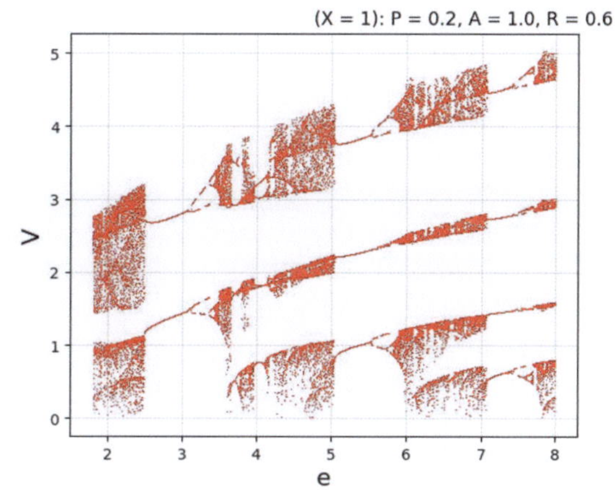

Fig. 11.12 Bifurcation diagram

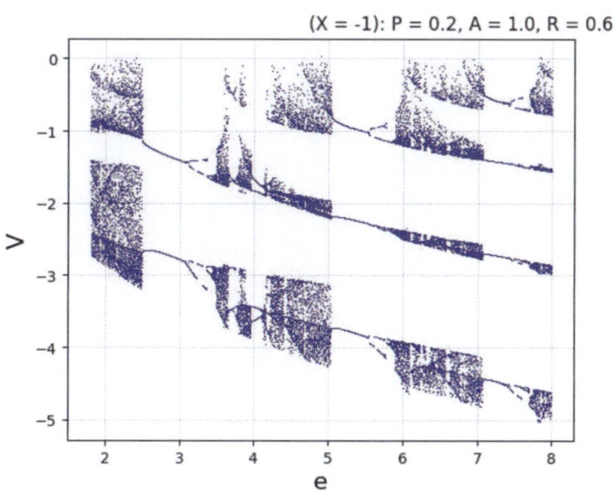

Fig. 11.13 Bifurcation diagram.

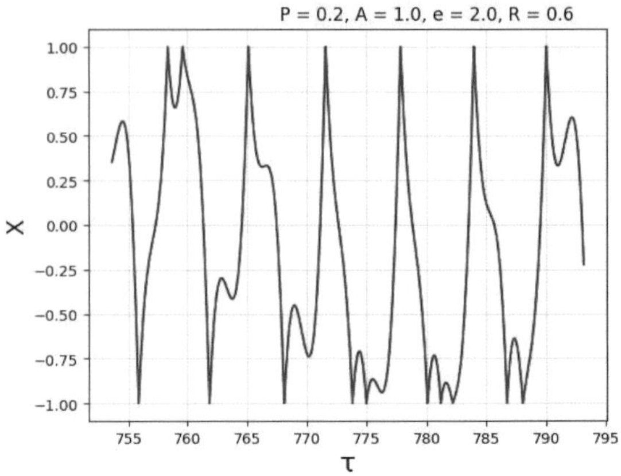

Fig. 11.14 Bifurcation diagram.

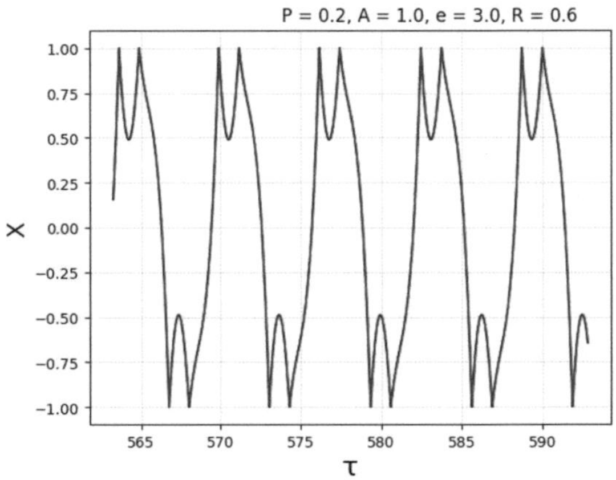

Fig. 11.15 Bifurcation diagram.

11.13 Conclusion

1. The paper studies the dynamics of an electromechanical system, which is a highly nonlinear dynamic system with a variable structure.
2. A mathematical model is constructed and the complex structure of the phase space is described.
3. Parametric relations are obtained for determining the coordinates of fixed points of point mappings corresponding to symmetric and asymmetric periodic motions, and their stability is investigated.
4. Bifurcation boundaries of "existence" in the parameter space of the constructed point mappings of Poincaré surfaces are constructed. It is proven that they cut off a significant portion of the stability regions.
5. Using bifurcation diagrams, a scenario for the emergence of chaotic motion was identified. It turned out to be a period-doubling process.

Acknowledgements The work is carried out with the financial support of the Ministry of Science and Higher Education of the Russian Federation (project No. FSWR-2026-0019).

References

[1] Astashev VK, Babitsky VI (2007) Ultrasonic Processes and Machines. Dynamics, Control and Applications. Springer, Berlin, DOI 10.1007/978-3-540-72061-4

[2] Awrejcewicz J (1990) Bifurcation portrait of the human vocal cord oscillations. Journal of Sound and Vibration 136(1):151-156, DOI 10.1016/0022-460X(90)90945-V

[3] Awrejcewicz J (1991) Numerical analysis of the oscillations of human vocal cords. Nonlinear Dynamics 2(1):35-52, DOI 10.1007/BF00045054

[4] Awrejcewicz J, Kudra G, Lamarque, C-H (2003) Dynamics investigation of three coupled rods with a horizontal barrier. Meccanica 38(6):687-698, DOI 10.1023/A:1025881409419

[5] Awrejcewicz J, Kudra G, Lamarque, C-H (2004) Investigation of triple pendulum with impacts using fundamental solution matrices. International Journal of Bifurcation and Chaos 14(12):4191-4213, DOI 10.1142/S0218127404011818

[6] Awrejcewicz J, Pyryev Y (2006) Chaos prediction in the dung-type system with friction using Melnikov's function. Nonlinear Analysis: Real World Applications 7(1):12-24, DOI 10.1016/j.nonrwa.2005.01.002

[7] Babitsky VI (1978) Teoriya vibroudarnykh sistem: priblizheniye metody (in Russ., Territory of vibro-impact systems: approximation methods). Moscow: Nauka

[8] Bespalova LV, Neimark YI, Feigin MI (1966) Dinamicheskiye sistemy s udarnymi vzaimodeystviyami i teoriya nelineynykh kolebaniy (in Russ., Dynamic systems with shock interactions and the theory of nonlinear oscillations). Inzhenernyy zhurnal, Mekhanika tverdogo tela (1):134-149.

[9] Demeio L, Lenci S (2006) Asymptotic analysis of chattering oscillations for an impacting inverted pendulum. Quarterly Journal of Mechanics and Applied Mathematics 59(3):419-434, DOI 10.1093/qjmam/hbl009

[10] Feigin MI (1967) O nesimmetrichnykh periodicheskikh rezhimakh v simmetrichnoy sisteme s udarnymi vzaimodeystviyami (in Russ., On asymmetric periodic regimes in a symmetric system with shock interactions). Izvestiya Vuzov, Radiofizika X(3):389-392

[11] Feigin MI (1960) O vynuzhdennykh kolebaniyakh dvukh mass, sochlenennykh s zazorom (in Russ., On forced oscillations of two masses articulated with a gap). Izvestiya AN SSSR, OTN. Mekhanika i Mashinostroenie (5):122-130

[12] Feigin MI (1994) Vynuzhdennyye kolebaniya sistem s razryvnymi nelineynostyami (in Russ., Forced oscillations of systems with discontinuous nonlinearities). Moscow: Nauka

[13] Guo B, Páez Chávez J, Liu Y, Liu C (2021) Discontinuity-induced bifurcations in a piecewise-smooth capsule system with bidirectional drifts. Communications in Nonlinear Science and Numerical Simulation 102:105909, DOI 10.1016/j.cnsns.2021.105909
[14] Gao Q, Cao X (2014) The stick-slip vibration and bifurcation of a vibro-impact system with dry friction. The Open Mechanical Engineering Journal 8:308-313, DOI 10.2174/1874155X01408010308
[15] Goryachenko VD (2001) Elementy teorii kolebaniy (in Russ., Elements of the Theory of Vibrations). Moscow: Vysshaya Shkola
[16] Igumnov L, Metrikin SV, Nikiforova VI, Fevral'skikh LN (2021) The dynamics of eccentric vibration mechanism (Part 2). In: dell'Isola F, Igumnov L (eds) Dynamics, Strength of Materials and Durability in Multiscale Mechanics. Advanced Structured Materials, Vol 137, Springer, Cham, pp 173-190, DOI 10.1007/978-3-030-53755-5_12
[17] Korendiy V, Gursky V, Kachur O, Gurey V, Havrylchenko O, Kotsiumbas O (2021) Mathematical modeling of forced oscillations of semidefinite vibro-impact system sliding along rough horizontal surface. Vibroengineering Procedia 39:164–169, DOI 10.21595/vp.2021.2229
[18] Korendiy V, Lanets O, Kachur O, Dmyterko P, Kachmar R (2021) Determination of inertia-stiffness parameters and motion modeling of three-mass vibratory system with crank excitation mechanism. Vibroengineering Procedia 36:7–12, DOI 10.21595/vp.2021.21924
[19] Lanets O, Kachur O, Korendiy V, Lozynskyy V (2021) Controllable crank mechanism for exciting oscillations of vibratory equipment. In: In: Ivanov V, Pavlenko I, Liaposhchenko O, Machado J, Edl M (eds) Advances in Design, Simulation and Manufacturing IV. DSMIE 2021. Lecture Notes in Mechanical Engineering. Cham, Springer, pp 43–52, DOI 10.1007/978-3-030-77823-1_5
[20] Metrikin SV, Igumnov L, Nikiforova VI (2017) The dynamics of eccentric vibration mechanism (Part 1). Journal of Vibroengineering 19(7):4854–4865, DOI 10.21595/jve.2017.18346
[21] Neimark YI (2010) Metod tochechnogo otobrazheniya v teorii nelineynykh kolebaniy (in Russ., The Point Mapping Method in the Theory of Nonlinear Oscillation). LIBROKOM, Moscow
[22] Nguyen K-T, La N-T, Ho K-T, Ngo Q-H, Chu N-H, Nguyen V-D (2021) The effect of friction on the vibro-impact locomotion system: modeling and dynamic response. Meccanica 56(8):2121–2137, DOI 10.1007/s11012-021-01348-w
[23] Shuster HG, Just W (2005) Deterministic Chaos: An Introduction (4th ed). Wiley, WeinheimDOI 10.1002/3527604804
[24] de Souza SLT, Caldas IL, Viana RL, Balthazar JM (2008) Control and chaos for vibro-impact and non-ideal oscillators. Journal of Theoretical and Applied Mechanics 46(3):641-664
[25] Yan Y, Liu Y, Liao M (2017) A comparative study of the vibro-impact capsule systems with one-sided and two-sided constraints. Nonlinear Dynamics 89(2):1063–1087, DOI 10.1007/s11071-017-3500-7

Chapter 12
A Generalized Model of Plastic Flow Under Multiaxial, Non-Proportional Monotonic and Cyclic Loading

Leonid Igumnov, Ivan Volkov, Andrey Volkov, and Anna Pul

Abstract The processes of complex plastic deformation of structural alloys along arbitrary complex trajectories of disproportionate deformation are considered. The review of the main directions of development of the mathematical theory of plasticity is carried out. The variant of Korotkov's elastic plasticity equations used in this article is based on the concept of a yield surface and the principle of gradiency of the velocity vector of plastic deformations to the yield surface at the loading point. This version of the equations of state reflects the main effects of elastoplastic deformation of a material for arbitrary complex deformation trajectories. The model is built on the principle of "nested" models: when the corresponding material parameters are zeroed out from the general version of the equations, as a special case, all its basic forms of equations of plastic flow with small deformations are obtained (a system of "nested" models). To assess the degree of reliability and determine the limits of applicability of the determining plasticity ratios, numerical studies of complex plastic deformation of 45 steel along flat trajectories of disproportionate deformation were carried out: flat two-link deformation trajectories of the "fan" type with different fracture angles; flat smooth deformation trajectories with a displaced center of constant curvature. The results of the study demonstrate that the Korotkov model provides both a qualitative and quantitative description of the behavior of metals for flat loading trajectories of any curvature, and its accuracy is sufficient to perform practical calculations.

Key words: Modeling, Plasticity, Complex loading, Two-link trajectories, Theory of plastic flow, Material parameters

Leonid Igumnov (corresponding author) · Ivan Volkov · Andrey Volkov · Anna Pul
Research Institute for Mechanics, National Research Lobachevsky State University of Nizhny Novgorod, 23, Gagarin av., bld. 6, 603950 Nizhny Novgorod, Russian Federation, e-mail: igumnov@mech.unn.ru, ivan.volkov@mech.unn.ru, Vollovandr@gmail.com, anna.iudintseva@gmail.com

H. Altenbach et al. (eds.), *Stability Problems in Mechanics of Continua*,
Advanced Structured Materials 260, https://doi.org/10.1007/978-3-032-31032-3_12

12.1 Introduction

The study of the laws of elastic-plastic deformation of structural materials (metals and their alloys) is of fundamental importance for the development of the fundamental principles of solid mechanics. In the mathematical theory of plasticity there is no single physical law similar to the generalized Hooke's law in the theory of elasticity, and to date several groups of mathematical models of elastic-plastic media have been developed [3, 4, 6–8, 10, 11, 13]:

a) A class of theories of plasticity based on general laws of generation and propagation of dislocations and static averaging of the value "smearing" all dislocations into a uniform distribution with a certain density function and average velocity, which are functionals (functions) of the process (R.I. Nigmatulin, N.N. Kholin, G.I. Taylor, U. Lindholm, S. Kuriyama, K. Kowata and others). Analyzing this class of models as a whole, it should be noted that at present these models can be applied reliably enough only under uniaxial stress states.
b) A class of local-macroscopic ("physical") theories of plasticity based on a phenomenological description of the elastic-plastic deformation of a polycrystalline aggregate based on microstructural mechanisms of plastic shear in the grain and static averaging over an ensemble of grains (V.P. Tamuzh, E.G. Lin, V. Kafka, S.B. Batdorf, B. Budyansky, A.K. Malmeister, V.D. Klyushnikov, and others). This class of plasticity theories also considers microstructural processes, but at a less microscopic level than dislocation theories (it considers not individual stress fields around dislocations, but their manifestations in an integral sense at the level of grain sizes). However, despite all the achievements in this area, the difficulties encountered in using these theories, of a mathematical, physical and experimental nature, even when using a number of simplifying hypotheses, make it difficult to widely apply these theories for practical calculations.
c) The class of "structural" models of plasticity, in which the unevenness of plastic deformation, caused by the uneven structure of the polycrystal and the uneven distribution of defects in the atomic lattices, is taken into account approximately by representing the plastic deformation tensor as a sum of elementary plastic deformations, each of which has its own yield surface and system of internal microelastic forces (S.A. Khristianovich, Yu.I. Kadashevich, J.F. Besseling , G. Masing , etc.). In this approach, the statistics of anisotropic crystals are replaced by the statistics of anisotropic particles with different properties. Analyzing this class of models, we note that structural models are very complex for solving applied problems of plasticity theory.
d) A class of theories of elastic-plastic processes developed by A.A. Ilyushin and his students (A.A. Ilyushin, R.A. Vasin, V.S. Lensky, V.I. Maly, V.G. Zubchaninov, and others). The law of relationship between stress and strain tensors in the viscoplastic region is based on the following principles: the uniqueness condition, the isotropy postulate, the principle of lag of scalar and vector material properties, the unloading hypothesis, and the plasticity postulate. These relationships are applicable to initially isotropic materials, for which the influence of the type of

stress state (the third invariant of the stress tensor) on the stress-strain equation is insignificant. Within the framework of this direction, general patterns have been established, confirmed by experiment, and a number of special and practically important classes of deformation processes have been identified, for which, based on the experimental-theoretical method of "SN-EVM", specific relationships between stresses and deformations with experimentally determined functional connections have been obtained.

d) The class of so-called "differential" theories of plasticity , establishing the defining relationships between infinitesimal increments of stresses and strains, the stresses themselves and some parameters of the plastic state ("flow" theory) (L.M. Kachanov, W. Prager, R. Hill, P.G. Hodge, A.Yu. Ishlinsky, Yu.I. Kadashevich, V.V. Novozhilov, V.S. Bondar, Yu.G. Korotkikh, P. Perzina, W. Nowacki, J.L. Chaboche and others). The basis of differential theories of plasticity is Drucker's postulate, the consequence of which is the convexity of the yield surface and the normality of the plastic deformation velocity vector to the yield surface at the loading point. Within the framework of the specified direction, many specific applied problems were solved for both static and dynamic effects; based on the obtained results, it can be concluded that even with the simplest assumptions about the nature of the change in the yield surface, this class of plasticity theories can be successfully applied to solve specific applied problems of plasticity theory for complex loading trajectories of arbitrary curvature.

When selecting a particular mathematical model of plastic deformation of a material for a wide range of thermomechanical loads with the widest range of applicability, it is inevitably necessary to seek a compromise between the model's correspondence to the actual behavior of the material under operating conditions and the feasibility of effectively using the model to solve applied problems on a modern computer, the possibility of experimentally determining material parameters, scalar functions of the constitutive relations, and the accuracy of determining these parameters. The inability to accurately determine the material parameters included in the model can lead to zero practical value for the model [5].

In this paper, Korotkikh's model of thermoplasticity with combined translational-isotropic hardening is used to describe complex plastic deformation of structural alloys under monotonic and cyclic non-proportional loading conditions. By comparing calculated values of elastic-plastic deformation characteristics with experimental data, the validity of the proposed constitutive relations for elastic-plasticity is demonstrated.

12.2 The Main Constitutive Relations of the Theory of Flow with Translational Isotropic Hardening

The formulated equations of thermoplasticity are aimed at describing the effects:

- monotonic and cyclic hardening under proportional and disproportionate loading;

- local anisotropy of plastic deformation at a bend in the deformation trajectory;
- non-isothermal elastic-plastic deformation of the material under various laws of temperature change and mechanical deformation.

The defining relationships of thermoplasticity are based on the following provisions:

- strain tensors e_{ij} and strain rates $\dot{e}_{ij}$ include elastic deformations $e^e_{ij}, \dot{e}^e_{ij}$ and plastic deformations, $e^p_{ij}, \dot{e}^p_{ij}$;
- the initial yield surface is described by the von Mises hypersphere;
- the evolution of the change in the yield surface is described by the change in its radius C_p and the displacement of its center ρ^p_{ij};
- the principle of the gradient of the plastic deformation rate vector to the yield surface is valid;
- change in the volume of the body is elastic $e^p_{ii} \equiv 0$;
- initially isotropic media are considered;
- only anisotropy caused by processes of irreversible deformation of the material is taken into account.

Experimental and theoretical studies show that:

- strengthening during monotonic deformation has both a kinematic and isotropic nature;
- the hardening modulus depends on the direction of the additional load vector at the loading point (local anisotropy);
- the patterns of cyclic hardening differ significantly from the patterns of monotonic hardening;
- under disproportionate cyclic loading, additional cyclic hardening occurs. Transient cyclic processes primarily depend on the isotropic component of hardening. Isotropic hardening consists of three types: monotonic, cyclic proportional, and cyclic disproportionate;
- the parameters of the stabilized hysteresis loop are mainly determined by the intensity of the deformation amplitude and the parameters of disproportionate cyclic loading;
- the patterns of cyclic deformation are mainly determined by the anisotropic part of hardening. The patterns of anisotropic hardening for monotonic and cyclic processes may be different;
- under rigid cyclic loading and initial anisotropy in stress amplitude, relaxation of the average cycle stresses to zero is observed.

These regularities of cyclic deformation within the framework of the considered version of the theory of plasticity are described using evolutionary equations for C_p and ρ^p_{ij}. The basic equations of the thermoplasticity model with translational isotropic hardening by Korotkov are as follows:

1. Yield Surface Equation and the Cyclic "Memory" Surface
   The effects of monotonic and cyclic deformation in stress space are taken into account using the von Mises yield surface:

$$F_p = S_{ij}S_{ij} - C_p^2 = 0, \quad S_{ij} = \sigma'_{ij} - \rho p_{ij}^p, \tag{12.1}$$

where σ'_{ij} are the components of the stress deviator. To describe complex cyclic deformation modes in stress space, a cyclic "memory" surface is introduced:

$$F_\rho = \rho_{ij}^p \rho_{ij}^p - \rho_{\max}^2 = 0, \tag{12.2}$$

where $\rho_{\max}$ is the maximum modulus of the variable ρ_{ij}^p over the loading history.

2. Evolution Equation for the Yield Surface Radius
   An evolutionary equation for the yield surface radius is postulated as follows [9, 12]:

$$\dot{C}_p = \left[q_\chi H\left(F_\rho\right) + a\left(Q_s - C_p\right)\Gamma\left(F_\rho\right)\right]\dot{\chi} + q_3\dot{T}, \tag{12.3}$$

$$C_p = C_p^0 + \int_0^t \dot{C}_p dt\,,\ \dot{\chi} = \left(\frac{2}{3}\dot{e}_{ij}^p\dot{e}_{ij}^p\right)^{1/2},\ \chi_m = \int_0^t \dot{\chi} H\left(F_\rho\right)dt, \chi = \int_0^t \dot{\chi}dt, \tag{12.4}$$

$$q_\chi = \frac{q_2 A\psi_1 + (1-A)q_1}{A\psi_1 + (1-A)}, Q_s = \frac{Q_2 A\psi_2 + (1-A)Q_1}{A\psi_2 + (1-A)}, 0 \le \psi_i \le 1 (i = 1, 2), \tag{12.5}$$

$$A = 1 - \cos^2\theta, \cos\theta = n^e n_{ij}^s, n_{ij}^e = \frac{\dot{e}'_{ij}}{(\dot{e}'_{ij}\dot{e}'_{ij})^{1/2}}, n_{ij}^s = \frac{S_{ij}}{(S_{ij}S_{ij})^{1/2}}, \tag{12.6}$$

$$H(F_\rho) = \begin{cases} 1, F_\rho = 0 \wedge \rho_{ij}^p \dot{\rho}_{ij}^p > 0 \\ \\ 0, F_\rho < 0 \vee \rho_{ij}^p \dot{\rho}_{ij}^p \le 0 \end{cases}, \Gamma(F_\rho) = 1 - H(F_\rho) \tag{12.7}$$

   Here q_1, q_2, q_3 are the moduli of isotropic hardening corresponding to monotonic radial loading paths (q_1), a 90° kink in the deformation trajectory (q_2), and a temperature change in the radius of the yield surface (q_3); a is a constant that determines the rate of the process of stabilization of the shape of the hysteresis loop of cyclic deformation of the material; Q_s is the stationary value of the radius of the yield surface for given $\rho_{\max}$ and T; χ and χ_m are the lengths of the trajectories of plastic deformation of the material under cyclic and monotonic loading; C_p^0 is the initial value of the radius of the yield surface.
   The first term of Eq. (12.1) describes isotropic hardening due to monotonic plastic deformation ($H(F_\rho) = 1$ and $\Gamma(F_\rho) = 1$), the second term (12.2) describes cyclic hardening of the material ($H(F_\rho) = 0$ and $\Gamma(F_\rho) = 0$), and the third (12.3) describes the change in the yield surface radius with temperature. Operators $H(F_\rho)$ and $\Gamma(F_\rho)$ allow automatic separation of monotonic and cyclic deformation processes.

3. Evolution Equation for ρ_{ij}^p
   The equation for the displacement of the yield surface is based on the hypothesis of Ilyushin, which is that hardening depends on the history of deformation only on some immediate part of the trajectory (delay of vector properties) [1, 2, 4, 9, 12]:

$$\dot{\rho}_{ij}^{p} = g_1^p \dot{e}_{ij}^p - g_2^p \rho_{ij}^p \dot{\chi} - g_3^p \rho_{ij}^p \dot{T}, \rho_{ij}^p = \int_0^t \dot{\rho}_{ij}^p dt, \tag{12.8}$$

where $g_1^p > 0$, $g_2^p > 0$, $g_3^p > 0$ are the anisotropic hardening moduli. Equation (12.8) describes the well-known spatial Bauschinger effect and the anisotropy of vector properties when the deformation direction changes (the deformation trajectory breaks). The introduction of a second term in this relation models the fading memory of the internal variable ρ_{ij}^p.
To characterize the behavior of the "memory" surface, it is necessary to formulate an evolutionary equation for $\rho_{\max}$:

$$\dot{\rho}_{\max} = \frac{(\rho_{ij}^p \dot{\rho}_{ij}^p) H(F_\rho)}{(\rho_{mn}^p \rho_{mn}^p)^{1/2}} - g_2 \rho_{\max} \dot{\chi} - g_3 \rho_{\max} \left\langle \dot{T} \right\rangle . \tag{12.9}$$

Equations (12.1)–(12.9) as a special case contain all the main known forms of equations of the theory of plastic flow under small deformations.

a) Isotropic Hardening
At $g_1^p \equiv g_2^p \equiv g_3^p \equiv 0$ in (12.8) and $\rho_{ij}^p \equiv 0$ we have a case of purely isotropic nonlinear hardening, described by equation (12.3) (in this case $H(F_\rho) = 1$, cyclic hardening is also not taken into account). If in (12.5) ψ_1 and $\psi_2 \equiv 0$, then the change in isotropic and kinematic hardening during disproportionate deformation is not taken into account: $q_\chi = q_1$, and $Q_s = Q_1$. If $q_1 \equiv const$, then linear isotropic hardening is described. If $q_1 \equiv 0$, then an ideal elastic-plastic Prandtl-Reuss material with a constant yield surface $C_p \equiv C_p^0$ is described.
b) Kinematic Hardening
At $q_1 = q_2 = q_3 = a \equiv 0$ in (12.3) we have a case of purely kinematic nonlinear hardening, described by equation (12.8). If $g_2^p \equiv 0$ and $g_1^p \equiv const$, we have a case of linear kinematic hardening.
c) Translational-Isotropic Hardening
With a certain connection between the material parameters of the models of Korotkikh and Chaboche, from (12.1)–(12.9) the constitutive relations of Chaboche are obtained [2], describing nonlinear kinematic and isotropic hardening (without cyclic hardening) [9, 12].

12.3 Numerical Results

The papers [14–16] present experimental data on proportional and disproportionate monotonic and cyclic elastic-plastic deformation of structural materials (metals and their alloys). The tests were conducted in a mechanical testing laboratory using a high-precision automated computational and experimental complex.

Experimental studies [14–16] were carried out using the "rigid" loading scheme (controlled deformation) at room temperature (20°C) on thin-walled tubular specimens made of grade 45 steel (with a yield plateau). The specimen geometry was characterized by the following parameters: length of the working part: $l = 110$ mm, wall thickness: $h = 1$ mm, radius of the middle surface: $R = 15.5$ mm.

The loading of the samples was carried out according to a given test program using a combination of axial force P, internal pressure q and torque M. The calculation of the components of the stress and strain tensors in the working area of the sample was performed using standard calculation formulas for the given geometry of the laboratory sample:

$$\begin{aligned} &\sigma_{11} = \frac{P}{2\pi Rh}, \quad \sigma_{22} = \frac{qR}{h}, \quad \sigma_{12} = \frac{M}{2\pi R^2 h}, \quad \sigma_{33} = \sigma_{13} = \sigma_{23} = 0, \\ &e_{11} = \frac{\Delta l}{l}, \quad e_{22} = \frac{\Delta R}{R}, \quad e_{12} = \frac{\varphi R}{2l}, \end{aligned} \tag{12.10}$$

where φ is the twist angle.

In the experiments, strains e_{11}, e_{22}, e_{12} and stresses were measured $\sigma_{11}, \sigma_{22}, \sigma_{12}$. The stress $\overline{\sigma}$ and strain vectors $\overline{e}$ in the three-dimensional combined vector subspace of deviators with a common fixed reference frame $\{\overline{e}_k\}$ have the form:

$$\overline{\sigma} = S_k \overline{e}_k, \quad \overline{e} = 3_k \overline{e}_k \quad (k = 1, 2, 3) \tag{12.11}$$

When processing experimental data for the components of stress and strain vectors, the following formulas were used:

$$\begin{aligned} &S_1 = \sqrt{\frac{3}{2}}\left(\sigma_{11} - \sigma_0\right) = \sqrt{\frac{2}{3}}\left[\sigma_{11} - \frac{1}{2}\left(\sigma_{22} + \sigma_{33}\right)\right], \quad S_2 = \frac{1}{\sqrt{2}}\left(\sigma_{22} + \sigma_{33}\right), \\ &S_3 = \sqrt{2}\sigma_{12}, S_4 = \sqrt{2}\sigma_{23} = 0, \quad S_5 = \sqrt{2}\sigma_{13} = 0, \\ &E_1 = \sqrt{\frac{3}{2}}\left(e_{11} - e_0\right), E_2 = \sqrt{2}\left(e_{22} + \frac{1}{2}e_{11} - e_0\right), \\ &E_3 = \sqrt{2}e_{12}, E_4 = \sqrt{2}e_{23} = 0, E_5 = \sqrt{2}e_{13}, \\ &\sigma = \sqrt{S_1^2 + S_2^2 + S_3^2}, \quad e = \sqrt{E_1^2 + E_2^2 + E_3^2}, \end{aligned} \tag{12.12}$$

where

$$e_0 = \frac{1}{3}(e_{11} + e_{22} + e_{33}), \quad \sigma_0 = \frac{1}{3}(\sigma_{11} + \sigma_{22} + \sigma_{33}).$$

When processing experimental data, the incompressibility condition was adopted $e_{11} + e_{22} + e_{33} = 0$ and it was believed that $\sigma_{33} = 0$ [14–16].

The first example examined a series of experiments involving complex loading of grade 45 steel (with a yield plateau) along two-link, fan-shaped, broken trajectories with varying fracture angles. The influence of the fracture angle of the deformation trajectory on the depth of the "dive" during partial complex unloading of the material was studied. The test program was implemented in the vector space of deformations in the plane $E_1 \sim E_3$, and the corresponding response was analyzed in the $S_1 \sim S_3$ vector space of the stress deviator.

The tests contain the results of four experiments with a break point in $E_1^* = 2\%$ two-link deformation trajectories with break angles of 90°, 135°, 150°, and 180°. After breaking the trajectory at an angle of 90°, the specimen was subjected to torsion; at angles of 135°and 150°, to elastic unloading after tension with torsion; at an angle of 180°, to uniaxial compression (curves 1, 2, 3, 4 in Figs. 12.1-12.4).

The main physical and mechanical characteristics, material parameters and scalar functions of the thermoplasticity model for steel 45 are given in Tables 12.1 and 12.2. Figure 12.2 shows the results of a comparison of the calculated and experimental diagrams $\sigma \sim e$ characterizing the scalar properties of the material. Here and in all subsequent illustrations, the experimental data are indicated by a dotted line, while the calculated results are shown by a solid line. Analysis reveals that at the moment of change in the deformation direction, the diagrams exhibit characteristic zones of abrupt stress changes ("dive"), the magnitude of which depends on the bending angle of the deformation trajectory. The descending portion of the "dive" corresponds to

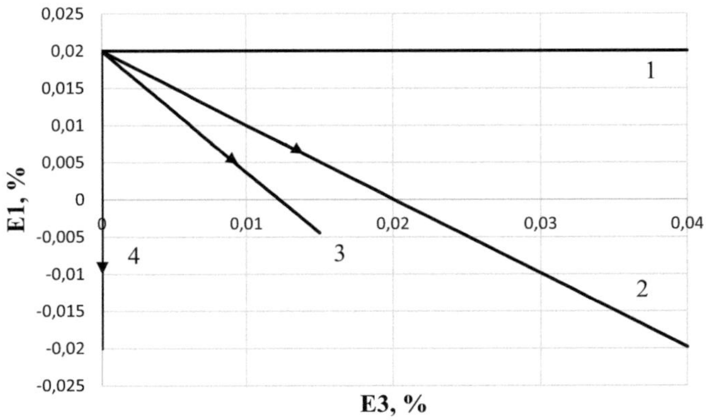

Fig. 12.1: Test program for the type trajectory "fans" in vector space $E_1 \sim E_3$..

Table 12.1: Physical mechanical properties of steel 45 at $T = 20$°C.

Property	Value
K, MPa	168500
G, MPa	78500
C_p^0, MPa	300
g_1^p, MPa	19,000
g_2^p	225
g_3^p	0
a	0

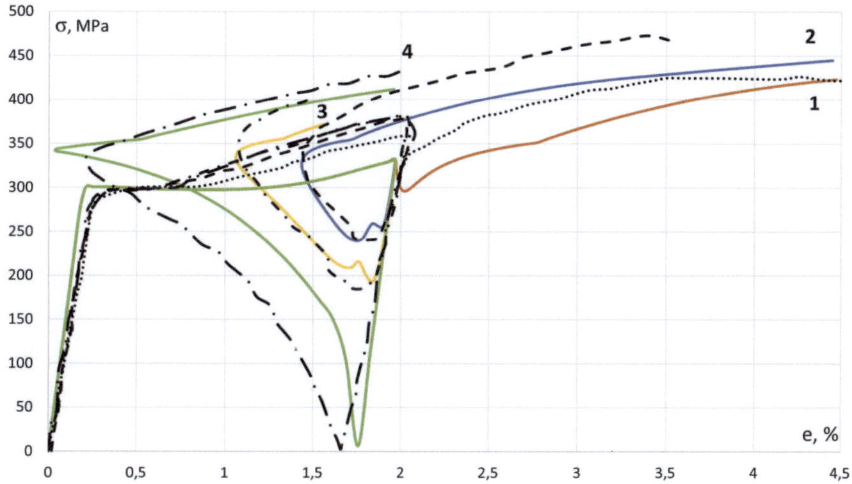

Fig. 12.2: Global deformation diagram $\sigma \sim e$.

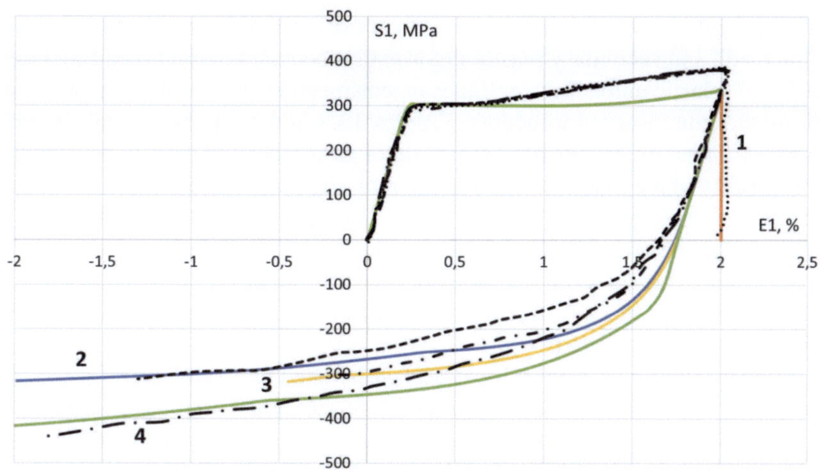

Fig. 12.3: Local deformation diagram $S_1 \sim E_1$.

Table 12.2: Dependence of the isotropic hardening modulus q_1 on the length of the plastic deformation trajectory χ ($q_2 = q_3 = 0$).

χ, MPa	0	0.004	0.008	0.012	0.02	0.03	0.04	0.05	0.06
q_1, MPa	-22070	-9500	-1970	5890	5000	2100	1850	1600	1350

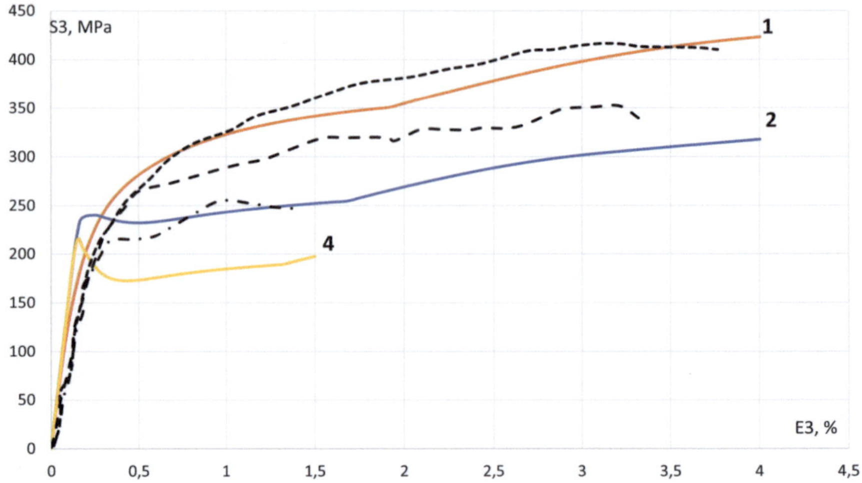

Fig. 12.4: Local deformation diagram $S_3 \sim E_3$.

partial complex unloading to the minimum stress value, at which the yield surface is "pierced" and secondary plastic deformation occurs. On the ascending branch of the "dive", elastic unloading continues in tension and compression, but active plastic deformation due to torsion occurs. Figures 12.3 and 12.4 show local deformation diagrams. Thus, the results of comparison of numerical and calculated data confirm the reliability of the mathematical model of plastic flow with combined translational-isotropic hardening.

The following example presents the results of a study of complex loading of grade 45 steel along flat curved trajectories of constant curvature in Ilyushin's deviatoric space. The effect of shifting the center of constant curvature of the deformation trajectory was investigated.

In experiments a 45 steel sample was pre-twisted to a value of $E_3^* = 0.62\%$, and then $E_1 \sim E_3$ a curved trajectory of constant curvature was realized on the plane ($R = 0.62\%$). Based on the detected delay trace, it was determined that this trajectory belongs to trajectories of large curvature.

Figure 12.5 shows a diagram $\sigma \sim e$ characterizing the scalar properties of the material. Figure 12.6 shows the $S_1 \sim S_3$ stress-plane response to the implemented complex deformation program. The solid line represents the numerical results, and the dashed line represents the experimental results. Figures 12.7 and 12.8 show the local deformation diagrams $S_1 \sim E_1$ and $S_3 \sim E_3$. The qualitative and quantitative agreement between the experimental and calculated data is evident. Thus, within the framework of these model concepts, it is possible to take into account the effect of "additional cyclic hardening" of materials during deformation along closed trajectories of disproportionate loading with an accuracy sufficient for engineering

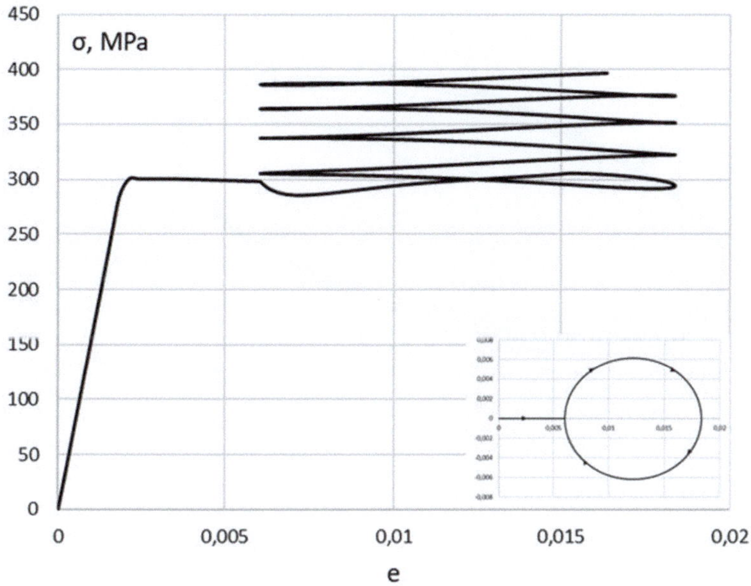

Fig. 12.5: Global stress-strain diagram $\sigma \sim e$ characterizing the scalar properties of the material.

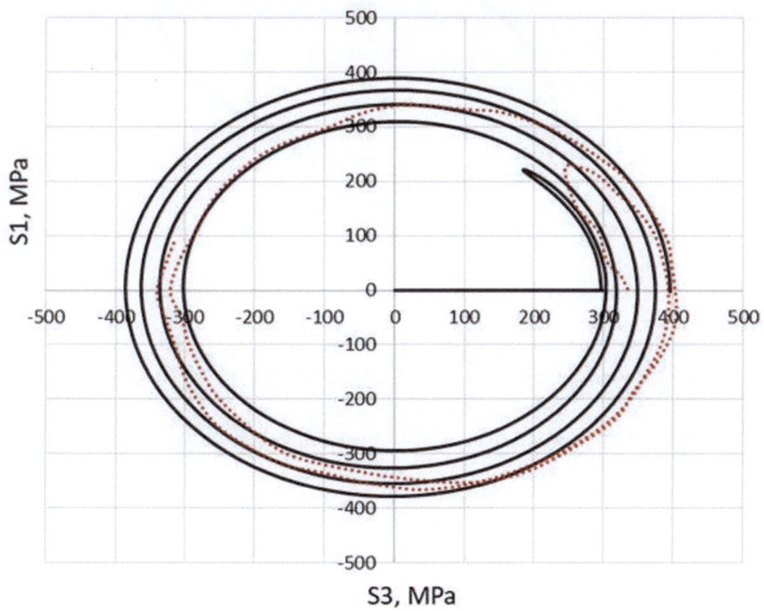

Fig. 12.6: Response to the implemented deformation program in space $S_1 \sim S_3$.

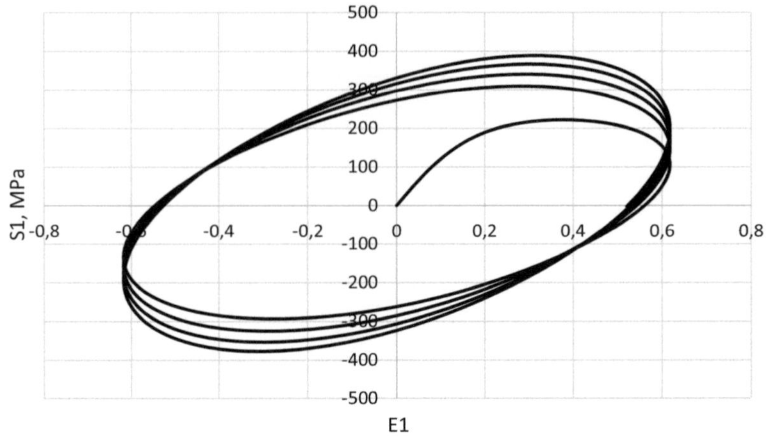

Fig. 12.7: Local deformation diagram $S_1 \sim E_1$.

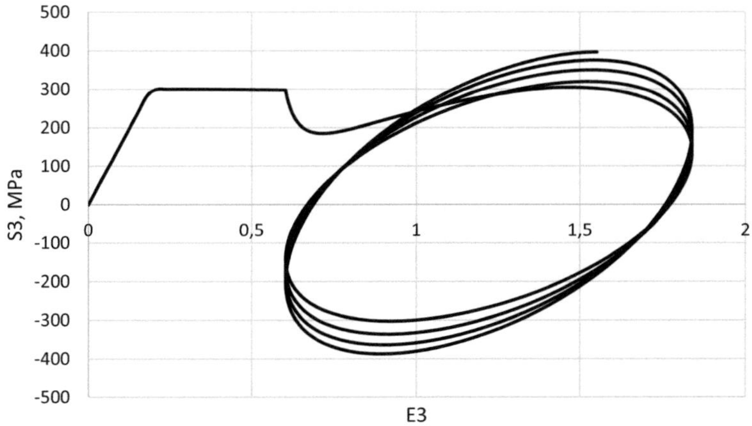

Fig. 12.8: Local deformation diagram $S_3 \sim E_3$.

calculations, in the first approximation, without introducing additional terms into the law of isotropic hardening (12.3).

12.4 Conclusion

The constitutive relations of plastic flow theory, which take into account isotropic and kinematic hardening, were validated by comparing numerical simulation results with experimental data on monotonic and cyclic deformation of grade 45 steel along planar trajectories with varying curvatures. The model's accuracy in describing both proportional and disproportionate deformation of structural alloys under loading along arbitrary complex deformation trajectories was confirmed.

An analysis of the applicability of the model allows us to recommend it for calculations of structural units of engineering objects operating under conditions of non-stationary thermomechanical loading.

Acknowledgements The work is carried out with the financial support of the Ministry of Science and Higher Education of the Russian Federation (project No. FSWR-2026-0019).

References

[1] Bondar VS (2004) Neuprugost'. Varianty teorii (in Russ., Inelasticity. Theory variants). Moscow: Fizmatlit

[2] Chaboche JL (1989) Constitutive equation for cyclic plasticity and cyclic viscoplasticity. International Journal of Plasticity 5(3):247-–302 DOI 10.1016/0749-6419(89)90015-6

[3] Hill R (1998) Mathematical Theory of Plasticity. Oxford: Oxford University Press

[4] Ilyushin AA (1963) Plasticity. Plastichnost'. Osnovy obshchey matematicheskoy teorii (in Russ., Fundamentals of General Mathematical Theory). Moscow: Publ. of the USSR Academy of Sciences

[5] Ilyushin AA, Lensky VS (1975) Model' i algoritm (in Russ., Model and algorithm). In. Applied Problems of Strength and Plasticity: All-Union Inter-University Collection. Gorky University, 1975

[6] Ishlinsky AY, Ivlev DD (2001) Matematicheskaya teoriya plastichnosti (in Russ., Mathematical Theory of Plasticity). Moscow: Fizmalit

[7] Klyushnikov VD (1979) Matematicheskaya teoriya plastichnosti (in Russ., Mathematical Theory of Plasticity). – Moscow: Moscow State University

[8] Knets IV (1971) Osnovnyye sovremennyye napravleniya v matematicheskoy teorii plastichnosti (in Russ., Main Modern Trends in the Mathematical Theory of Plasticity). Riga: Zinatne

[9] Mitenkov FM, Volkov IA, Igumnov LA, Kapliyenko AV, Korotkikh YG, Panov VA (2015) Prikladnaya teoriya plastichnosti (in Russ., Applied Theory of Plasticity). Moscow: Fizmatlit

[10] Olszak W, Mróz Z, Perzina P (1963) Recent Trends in the Development of the Theory of Plasticity. Oxford: Pergamon Press

[11] Prager W, Hodge PG (1951) Theory of Perfectly Plastic Solids, In: Sokolnikoff IS (ed) Applied Mathematics Series. New York: John Wiley & Sons

[12] Volkov IA, Igumnov LA, Korotkikh YG (2015) Prikladnaya teoriya vyazkoplastichnosti (in Russ., Applied Theory of Viscoplasticity). Nizhny Novgorod: Publishing house of Nizhny Novgorod State University

[13] Zubchaninov VG (2010) Mekhanika protsessov plasticheskikh sred (in Russ., Mechanics of Processes of Plastic Media). Moscow: Fizmatlit
[14] Zubchaninov VG (2008) Ustoychivost' i plastichnost' (in Russ., Stability and Plasticity), in 2 volumes: Vol 2 - Plastichnost' (in Russ., Plasticity). Moscow: Fizmalit
[15] Zubchaninov VG, Okhlopkov NL, Garannikov VV (2003) Eksperimental'naya plastichnost' (in Russ., Experimental plasticity). Book 1 - Protsessy slozhnogo deformirovaniya (in Russ., Complex deformation processes). Tver: TSTU
[16] Zubchaninov VG, Okhlopkov NL, Garannikov VV (2004) Eksperimental'naya plastichnost' (in Russ., Experimental plasticity). Book 2 - Protsessy slozhnogo nagruzheniya (in Russ., Complex loading processes). Tver: TSTU

Chapter 13
Control of Transition to Turbulence in a Laminar Jet by Radial Minijets

Oleg Ivanov, Linar Gareev, Vladimir Trifinov, and Vasily Vedeneev

Abstract This paper experimentally examines active control of a submerged jet. A set of six minijets positioned in a single plane perpendicular to the flow axis near the jet orifice is used for the introduction of perturbations into the main jet. We consider control laws with a constant minijets' flow rate, as well as with a periodic flow rate in the form of a standing or progressive azimuthal waves with different control frequencies that create steady and unsteady axisymmetric and helical perturbations. Introduced perturbations grow according to non-modal (steady) and modal (unsteady) growth mechanisms that interact with each other while the perturbation progresses downstream. It is shown that in all control cases, the entrainment rate and velocity fluctuations in the main jet are significantly increased. Among all the strategies considered, a low-frequency progressive control waves are the most advantageous for a rapid transition to turbulence.

Key words: Jet, Control, Instability, Laminar-turbulent transition

13.1 Introduction

Jet flows are often observed in nature and widely used in technological processes so that their control is an important scientific problem. Researchers are interested in such features as mixing [11], combustion [16], acoustic radiation [1, 15], spraying [23], etc. Optimizing these processes will lead to greater efficiency in the devices employing jet flows, but this requires a thorough understanding of the physics of the corresponding phenomena, particularly the laminar-turbulent transition. It is

Oleg Ivanov · Linar Gareev · Vladimir Trifinov · Vasily Vedeneev (corresponding author)
Institute of Mechanics, Lomonosov Moscow State University, Michurinskii prosp. 1, 119192 Moscow, Russsian Federation, e-mail: ivanov@imec.msu.ru, gareev@imec.msu.ru, trifonov@imec.msu.ru, vasily@vedeneev.ru

H. Altenbach et al. (eds.), *Stability Problems in Mechanics of Continua*,
Advanced Structured Materials 260, https://doi.org/10.1007/978-3-032-31032-3_13

essential to know how the jet can respond to various external excitations to obtain a specific scenario of transition to turbulence.

Flow controls are usually divided into passive and active. Active control requires energy from an external source, while passive control utilizes the flow energy or does not require any energy source at all. Passive control includes the installation of tabs [3] or chevrons [26] at the jet nozzle exit. Such devices alter the stability properties of mixing layer and lead to deformation and accelerated destruction of Kelvin-Helmholtz vortex rings, which results, for example, in reduction of the jet noise. Creating a swirling flow using guide vanes in the nozzle block creates conditions favorable for flame stabilization in combustion chambers and improving mixing of the jet with the surrounding air [7, 17]. Change of the nozzle geometry to create a non-round jet, i.e., a lobed or an elliptic jet, also can lead to faster transition to turbulence [12, 13].

Active control implies external excitation of the jet that results in modification of its properties. Acoustic excitation was studied by Vlasov and Ginevskii [25], who found that turbulent jet mixing can be intensified by weak acoustic excitation. Crow and Champagne [5] investigated the jet excitation by controlled acoustic perturbation and showed the existence of the optimal jet response called the "preferred" mode. Later, Zaman and Hussain [27] and Petersen and Samet [22], and other researchers discussed whether it is the only mode or two distinct preferred modes.

The idea of using minijets (a means of influencing the main jet by radially injecting fluid) to improve jet mixing was proposed by Davis [6]. Some of the studies were devoted to constant-flow control minijets. Seidel et al. [24] simulated the characteristics of non-circular jets by placing several stationary radial minijets around a circular main jet at locations corresponding to the corners or vertices of a non-circular nozzle. Their results were in good agreement with the similar results of non-circular jets [18], indicating that the jet can be controlled using similar means to achieve optimized performance under different operating conditions. The control minijets can be operated in a pulsed mode to improve the control efficiency [8, 19]. Pulsed microjets improve both jet range and spread under certain excitation conditions [21] and can be used not only to create large-scale flow structures but also to enhance jet instability. This was demonstrated by Choi et al. [4], who applied pulsed microjets to supersonic jet noise control and achieved noise reduction similar to that achieved with steady-state control, using only ~40% of the mass flow rate of steady microjets.

Machine learning (ML), a technique distinguished by its capabilities for multi-parameter and nonlinear optimization, has recently experienced rapid development and has achieved significant success in flow control. ML allows one to find a globally optimal flow control strategy even with a large number of control parameters (more than 10), outperforming traditional active control methods. The capabilities of machine learning are presented, for example, in [29], where a control system based on six independently operating pulsed radial minijets for a round jet control was developed. Each minijet has its own air supply frequency, mass flow rate coefficient, phase, and duty cycle. A genetic programming algorithm was used to search for optimal control laws. During the training process, classical control laws described in the literature, such as axisymmetric, helical, and flapping actions, were sequentially tested in order to increase control effectiveness, and ultimately a globally optimal control law, previously unknown, was generated. The resulting control structure was characterized by vortical motions and three-dimensional flapping motions with a much higher jet mixing rate than that of classical turbulent jets. Perumal et al. [20] developed a hybrid artificial intelligence control system to improve round jet mixing. Optimal control strategy enhanced turbulent mixing even for jet flows with varying Reynolds numbers Re.

The study of the fundamental mechanisms of laminar-turbulent transition in jets is associated with the development of linear stability theory and non-modal theory of perturbation growth in a submerged jet. Although the stability of a round submerged jet was calculated long ago, its experimental confirmation was performed recently. In [10], a jet with a long laminar region was used, and controlled harmonic disturbances were introduced by a thin oscillating ring, which resulted in excellent agreement between theoretical predictions and experimental results. Non-modal perturbation theory for jets has appeared quite recently [2], demonstrating the possibility of algebraic growth of disturbances, and was experimentally confirmed in [14] and further in [9]. In the latter study, small thin-walled structures were placed in the jet flow, generating disturbances that developed according to the lift-up growth mechanism.

In the present experimental study, we are interested in the development of perturbations that occupy a position between growing purely modally (Kelvin-Helmholtz waves) and non-modally (axial vortices evolving into axial streaks). Namely, we introduce combined perturbations that grow according to both modal and non-modal mechanisms. To this end, we use minijets that can create non-axisymmetric perturbations of desired frequency. As the flow rate in the minijet can vary from zero to a certain value, it can be divided into two components: constant flow rate (producing non-modally growing perturbation) and harmonic zero-mean blowing/suction (exciting jet eigenmode), whose simultaneous action creates the combined perturbation.

13.2 Experimental Setup

13.2.1 Laminar Jet and Method of its Excitation

We consider the evolution of perturbations created by small radially directed jets (called hereinafter minijets) in a main jet. The main air jet is created by a special jet-forming device (Fig. 13.1). The jet diameter is $D = 120$ mm, we study the flow regime with the axial and average velocities $V_{0,\,\mathrm{max}} = 1.25$ and $V_{0,\,\mathrm{avg}} = 0.56$ m/s, and the incoming turbulence intensity of 0.6%. The Reynolds number based on the radius and the maximum velocity is $\approx 5\,000$, and that based on diameter and average velocity is $\approx 4\,480$. The unperturbed jet stays laminar at least up to $x/D = 6$. The setup and the jet characteristics are described in detail in [28].

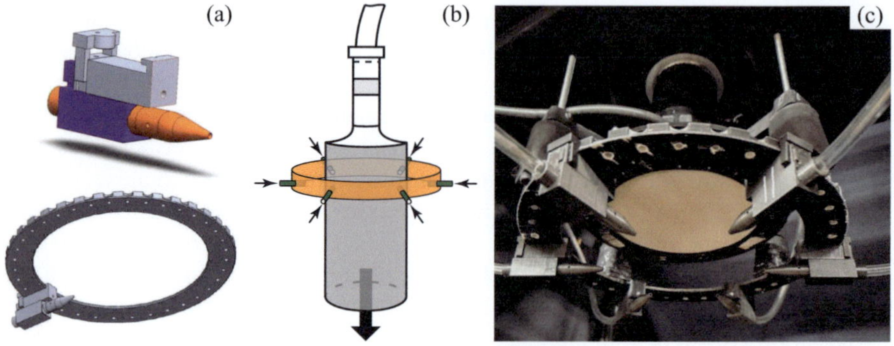

Fig. 13.1: Minijet's nozzle with its mount and the supporting ring (a); scheme of minijets' position relative to the main jet (b); a photography of the set-up with minijets installed (c).

Six minijets are located in a single plane at the distance of $x/D = 0.4$ from the main jet orifice (Fig. 13.1). All nozzles forming minijets have identical axisymmetrical geometry and were manufactured from PLA plastic by 3D-printer. The inner surface of the nozzle is a truncated cone with an input diameter of 5.3 mm (equal to the diameter of the air supply tube) and an output diameter of 2.3 mm. The scheme of the minijets pneumatic system is shown in Fig. 13.2. First, air flows from the compressor through the valve 1, which roughly controls the flow rate, and the flowmeter 2 to the chamber 4. The manometer 3 indicates the chamber pressure. Then, six pipes connect the chamber 4, needle valve 5, the electromagnetic valve 6, one more needle valve 7 (to finally adjust the flow rate and create a pressure drop for proper operations of the valve 6), and the minijet nozzle 8.

The distance from the main flow axis to every minijet nozzle is $0.54D$. All minijets can be manipulated independently so that various control patterns can be organized.

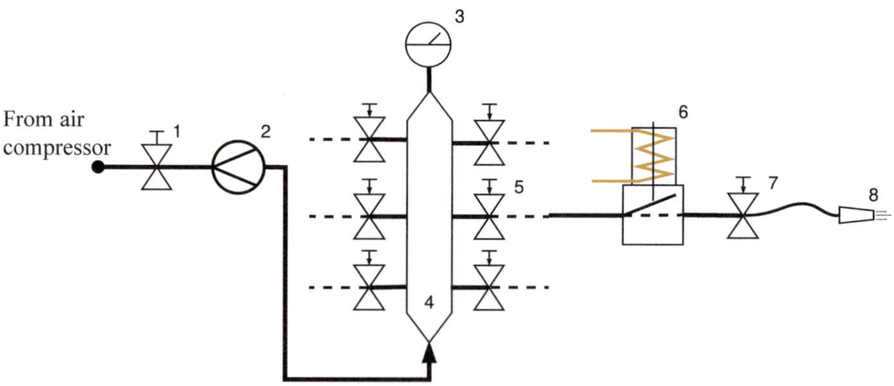

Fig. 13.2: Pneumatic system scheme of air delivery to minijets. 1 – valve, 2 – flowmeter, 3 – manometer, 4 – chamber, 5 – needle valves, 6 – electromagnetic valve, 7 – needle valve, 8 – minijet's nozzle.

Each minijet can work in continuous or periodic mode thanks to the electromagnetic valve, which can quickly open and close. To ensure that all minijets have similar flow properties (to have the same effect on the main flow), hot-wire measurements of their velocity profiles were conducted at different distances from their nozzles. Measurements were performed with a DISA 56C01 CTA thermoanemometer. A small 55P11 probe was mounted on a one-axis traverse system to measure the minijet velocity profiles. The needle valve 7 was used to set the same flow rate for each of the minijets.

The following control regimes were investigated in this work. The first and second regimes, denoted hereinafter CB6 and CB6-D, correspond to the constant blowing through all six minijets with the non-dimensional flow rates

$$C_Q = Q_{\text{blow}}/(V_{0,\,\text{avg}}\pi D^2/4) \approx 3.0 \times 10^{-3}$$

and 6.0×10^{-3}, respectively. Other tested regimes were unsteady with the flow rate $C_Q = 6.0 \times 10^{-3}$. In the third and fourth regimes, SW6-2.5 and SW6-5, minijets worked in a pulsed manner by switching on and off the electromagnetic valve with frequencies of $f_{\text{KH}}/2 = 2.5$ and $f_{\text{KH}} = 5$ Hz (Strouhal numbers $\text{St} = fD/V_{0,\,\text{max}} \approx 0.23$ and 0.46), respectively, where f_{KH} is a frequency of the fastest growing mode of the main jet [10]. In these regimes, close-to-axisymmetric velocity pertubation was introduced to the main flow. The minijets array generates a standing wave-like perturbation with azimuthal number $n = 0$: all minijets work simultaneously.

In the last two regimes, denoted PW6-2.5 and PW6-5, progressive wave control was used to switch on/off the minijets: at any given moment, only three adjacent minijets were in operation, and after 1/6 of the running wave rotation period, the outermost minijet was switched off, and the one opposite to it was switched on. The corresponding frequencies in these regimes were also 2.5 and 5 Hz. The PW6-2.5 and PW6-5 generate helical velocity perturbations in the main flow.

Note that in an unsteady regime, the minijet flow rate changes from 0 to the prescribed value, so the mean flow rate is non-zero in contrast to synthetic jets and equals $C_Q/2 \approx 3.0 \times 10^{-3}$. Consequently, the mean and maximum flow rates in unsteady regimes are similar to the steady CB6 and CB6-D regimes, respectively. Each unsteady regime can be represented as steady CB6 blowing superimposed by harmonic blowing/suction with zero mean flow rate. The steady blowing excites the non-modal lift-up growth mechanism [14], while the unsteady blowing/suction additionally excites instability modes [10]. Consequently, unsteady regimes result in "combined" perturbations that grow according to both mechanisms with less and more pronounced modal growth in -2.5 and -5 regimes, respectively. It is also worth mentioning that due to discrete nature of six minijets, they inevitably introduce perturbation with azimuthal wave number $n = 6$; however, we assume that this small-scale perturbation rapidly damps out compared to axisymmetric and helical perturbations with $n = 0$ and 1, respectively.

13.2.2 Flow Visualisation and Velocity Measurements

We use visualization and panoramic measurement methods to describe the flow structure in a perturbed jet. To show the characteristic properties of the introduced perturbations, the jet flow was seeded with an aerosol of glycerol and a continuous laser was used to create a laser sheet. Visualization was performed in the transverse and longitudinal planes.

To measure the instantaneous three-dimensional velocity fields, stereo particle image velocimetry (Stereo-PIV) was used (Fig. 13.3). Fine glycerol particles were added to the air before the jet-forming device whereas surrounding air in the experimental room was not seeded. The Beamtech Vlite-Hi-100 laser with a light-guide arm equipped with a cylindrical lens created the laser sheet, and two high-speed cameras Evercam 1000-16-M filmed the main jet cross-section. The light-guide arm and cameras were mounted on the three-axis traverse system, so we obtained velocity fields from $x/D = 2$ to $x/D = 5$ along the main jet axis with a step of $0.25D$ (x is the axial coordinate). Each measurement lasted 4 seconds with the sampling rate of 50 Hz, so 200 instantaneous velocity fields were filmed at each position along the axial coordinate. ActualFlow software was used to process the data and construct three-dimensional velocity fields in different planes.

13.3 Experimental Results

13.3.1 Visualisation of the Perturbed Jet

Flow visualizations were performed to see how the jet reacts to the introduced disturbance. Fig. 13.4 shows the jet cross-section at $x/D = 0.56$ under different

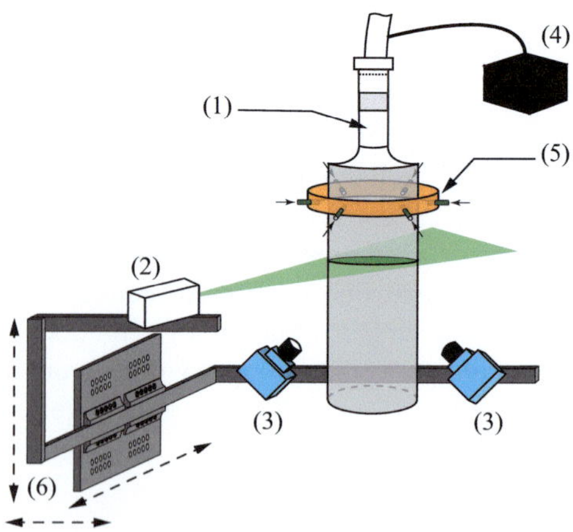

Fig. 13.3 Scheme of PIV measurements. 1 – jet-forming device, 2 – end of laser-guide arm with cylindrical lens, 3 – high-speed cameras, 4 – aerosol generator, 5 – six minijets, 6 – traverse system.

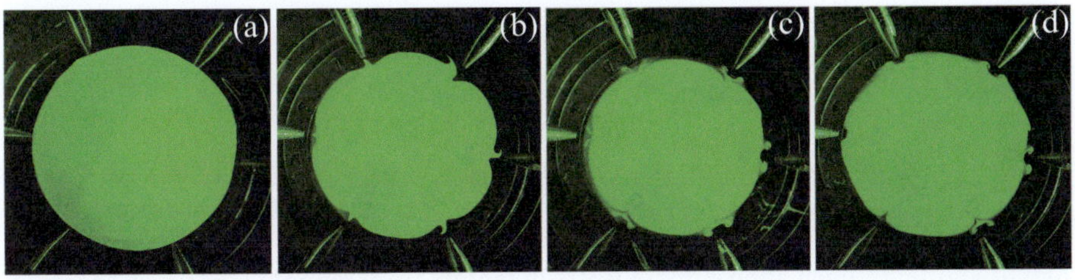

Fig. 13.4: Visualization of the jet cross-section at $x/D = 0.56$ with different control laws: no control (unperturbed jet) (a); CB6-D (b); SW6-5 (c); PW6-2.5 (d).

control laws. In Fig. 13.4b, corresponding to CB6-D excitation, six pairs of vortices generated by minijets are clearly visible. The jet boundary is no longer round and is now curved according to the location of the minijets. For the SW6-5 case, the cross-section changes its shape with time, but remains symmetrical at any given time. It is possible to identify the stage of the jet's introduction, similar to Fig. 13.4b. When the minijet stops working, a pair of vortices is carried downstream, and an additional disturbance appears on the main jet boundary, resembling a splash pattern of body-in-water impact. In the PW6-2.5 case, three pairs of vortices are clearly visible at the switched-on minijets (right-hand side), but at the same time, the jet disturbance near the switched-off minijets (left-hand side) has not yet gone completely downstream.

Figure 13.5 shows the longitudinal cut of the jet, in which some features of the introduced perturbations can also be seen. For example, in Fig. 13.5b, a quite long steady laminar region of the jet is still visible without any unsteady waves at its boundary, and in Fig. 13.5c,d, these waves are already noticeable.

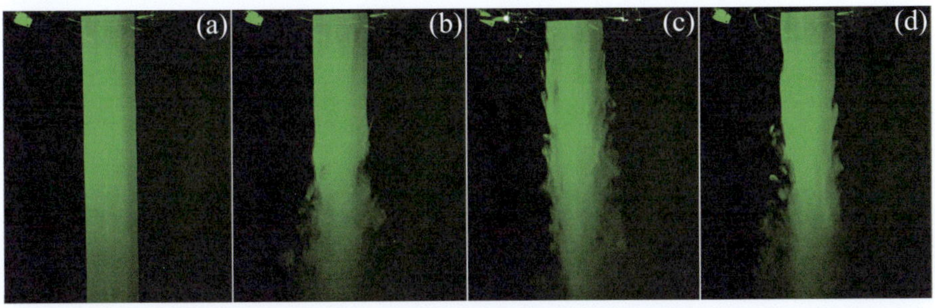

Fig. 13.5: Visualization of the jet flow in a longitudinal plane with different control laws: no control (unperturbed jet) (a); CB6-D (b); SW6-5 (c); PW6-2.5 (d).

13.3.2 Extraction of Laminar-Turbulent Transition Point from Visualizations

To extract the axial position corresponding to laminar-turbulent transition from longitudinal jet cut movies, we used the following automated procedure. First, a movie with minijets turned off was recorded. Next, for each type of perturbation, movie segments, each approximately 30 s in length, were recorded. For each segment, video stream averaging was performed to obtain an averaged frame (Fig. 13.6). Then, for the averaged frame, at each axial position x the brightness distribution $b(x, y)$ was extracted from the pixel intensity (which varies from 0 to 255). Next, brightness was rescaled to [0; 1] range to get rid of light intensity variations within the laser sheet:

$$\tilde{b}(x, y) = \frac{b(x, y) - b_{\min}(x)}{b_{\max}(x) - b_{\min}(x)},$$

Finally, the "momentum thickness" $B(x)$ was calculated for each brightness "profile" $b(x, y)$:

$$B(x) = \int_{y_1}^{y_2} \tilde{b}(x, y)(1 - \tilde{b}(x, y))dy,$$

which provides the degree of the jet spread for each x.

The length of the integration interval was $y_2 - y_1 = 1.25D$. Although the brightness profile does not represent an actual velocity profile, it reflects the degree of jet blur, indicating the level of fluctuations at its boundary. The transition point was determined when a certain threshold of $B(x)$ was reached (chosen to be $B(L_{\mathrm{lam}}) = 42$ based on the laminar region of a base flow, which is usually $\sim 6D$). The resulting laminar region lengths are shown in Table 13.1. All excitation regimes lead to sig-

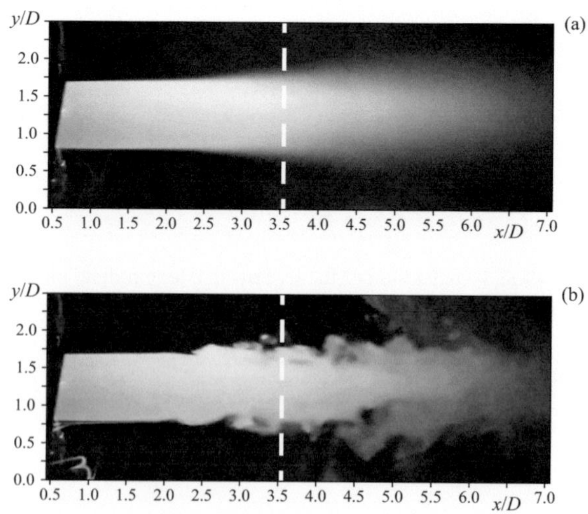

Fig. 13.6 Example of averaged (a) and instantaneous (b) jet frames and estimated position of laminar-turbulent transition (the flow is from the left to the right).

Table 13.1: Laminar region lengths L_{lam}/D for various excitations.

Regime	Unperturbed	CB6	CB6-D	SW6-2.5	PW6-2.5	SW6-5	PW6-5
L_{lam}/D	6.21	3.57	3.27	3.24	3.09	2.58	2.87

nificant reduction of the laminar region. Larger frequency yields the shorter laminar region, since the larger frequency provides the faster growth of Kelvin-Helmholtz instability mode. The SW6-5 regime shows the greatest impact on the main jet.

13.3.3 Evolution of the Jet Velocity Fields

Here we analyze the axial velocity fields, because they provide the jet evolution patterns and are directly related to the jet flow rate. The left column in Fig. 13.7 shows the time-averaged axial velocity fields for the undisturbed jet. Slight asymmetry is evident due to the small sample size, but the main flow features are clearly visible,

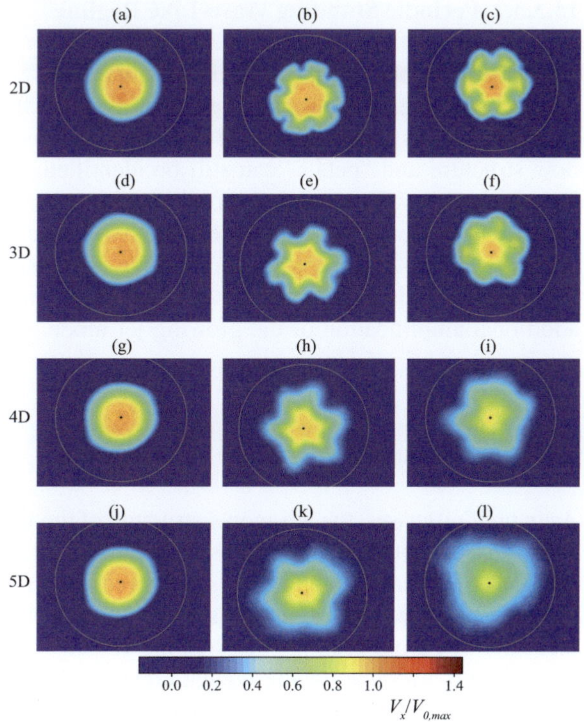

Fig. 13.7 Axial velocity fields of the main jet without and with steady blowing through six minijets. Each row corresponds to a fixed distance from the main jet nozzle's orifice (shown at the left). The left column corresponds to the unperturbed jet, the middle column corresponds to a smaller flow rate through minijets (CB6), and the right column corresponds to the one twice as large (CB6-D).

and it is also clear that the flow remains laminar and virtually unchanged up to $x/D = 5$. The effect of minijet excitations is discussed below.

13.3.3.1 Stationary Control Regimes

First, we consider constant flow rates $C_Q = 3.0\times10^{-3}$ and 6.0×10^{-3} that correspond to CB6 and CB6-D regimes (Fig. 13.7). The visualization shows that the cross-sectional shape of the jet spot clearly displays "petals", formed by the entraining slowly moving air into the main flow. Downstream, the spot gradually widens and blurs, losing any trace of the original jet shape. It can be seen that for cases CB6 and CB6-D, the flow patterns are generally similar in the initial cross-sections, but there are also clear differences. At the higher flow rate, the jet core occupies a smaller area, and the "petals" have less clear boundaries. For the case CB6, the flow core is still visible at a distance of 5D, while for the case CB6-D, it has become very small, and the maximum velocity at the jet axis has noticeably decreased. It is also worth noting that at the lower flow rate, a hexagonal flow structure is visible, while at the higher flow rate, the jet has become quite close to axisymmetric, indicating an earlier transition to turbulent flow.

13.3.3.2 Periodic Standing Wave-Like Regime

Let us consider the effect of six minijets that form a standing wave control pattern, i.e., they were periodically switched on and off at a frequency of 2.5 (SW6-2.5) or 5 Hz (SW6-5). As shown in Fig. 13.8, in the sections $x/D = 2$ and 3, the hexagonal flow structure and "petals" can still be identified, but unlike the CB6 case, these features become significantly blurred in subsequent cross-sections. As a result, the velocity field at $x/D = 5D$ is close to axisymmetric, and some flow core is retained only in the SW6-2.5 case, while for the SW6-5 regime, the axial velocity has dropped significantly. This may indicate that in the latter case an essential role is played by additional amplification of the growing jet mode, whose maximum growth rate is reached at $\approx 4.5 - 5$ Hz.

13.3.3.3 Azimuthal Progressive Wave Regime

As in previous cases, the hexagonal structure of the velocity field is visible in the first cross-sections, but it becomes less noticeable after $x/D = 3$ (Fig. 13.9). At further stations, significant jet blurring is observed across the cross-section. The control effect in the PW6-2.5 regime is stronger than in the PW6-5. Already in the first measurement cross-section $x/D = 2$, the velocity at the flow axis is 7% lower than for the PW6-5 regime. In subsequent cross-sections, it is clear that this control law leads to the most rapid blurring of the jet core and, consequently, faster jet spreading.

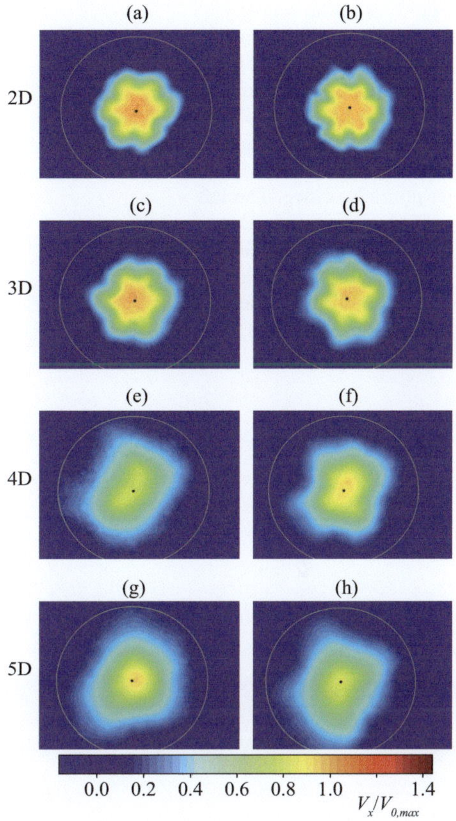

Fig. 13.8 Axial velocity fields of the main jet in cases of standing wave-like control regimes ($n = 0$), the flow rate is fixed. Each row corresponds to a fixed distance from the main jet nozzle's orifice (shown at the left). The left column corresponds to a switching frequency of 2.5 Hz (SW6-2.5), the right column corresponds to 5 Hz (SW6-5).

13.3.4 The Jet Flow Entrainment

To characterize the jet entrainment rate under different control regimes, we calculate the flow rate of the main jet $Q(x)$, whose growth is related to the surrounding fluid entrained by the jet:

$$Q(x) = \int_{S_{\text{int}}} V_x ds,$$

where S_{int} is the integration area. We chose it to be a circle with a diameter of $1.5D$ to contain the jet flow in all cases shown in the Figs. 13.7–13.9 above by a yellow circle. All measurements start at $x/D = 2$, so in order to compare them with each other, we non-dimensionalize all flow rates by the flow rate Q_0 of the unperturbed jet at $x/D = 2$.

The experimental results are shown in Fig. 13.10. In general, all jet control methods have a similar trend to increase the jet flow rate due to increased ambient air entrainment. However, it can be noted that the greatest difference is achieved in the part of the transition region $x/D = 3-4.25$, where the curves related to PW6-2.5 and

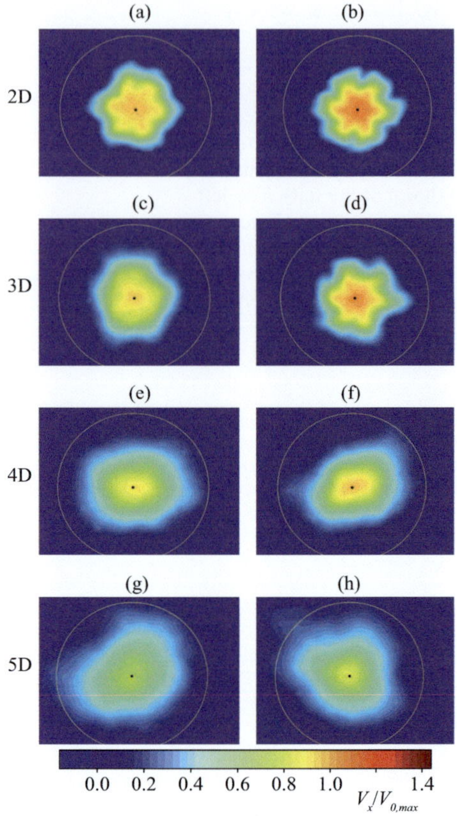

Fig. 13.9 Axial velocity fields of the main jet in cases of progressive wave regimes with an azimuthal number $n = 1$, the flow rate is fixed. Each row corresponds to a fixed distance from the main jet nozzle's orifice. The left column corresponds to a frequency of 2.5 Hz (PW6-2.5), and the right column corresponds to a frequency of 5 Hz (PW6-5).

SW6-5 regimes are located above all. In contrast, the CB6 and CB6-D continuous blowing regimes are located below all other lines, indicating their lower efficiency compared to unsteady control cases. When the $x/D = 4.5$ coordinate is reached, all control methods have a similar effect.

13.3.5 The Turbulence Intensity

Another measure of the control efficiency can be given by velocity fluctuations. Let us introduce axial velocity fluctuation $v_x(x, y, z)$ and the integral turbulence intensity $Tu(x)$ in a given x-cross-section:

$$v_x(x, y, z) = \left(\frac{1}{N} \sum_{i=1}^{N} \left(V_x(x, y, z, t_i) - \bar{V}_x(x, y, z) \right)^2 \right)^{1/2},$$

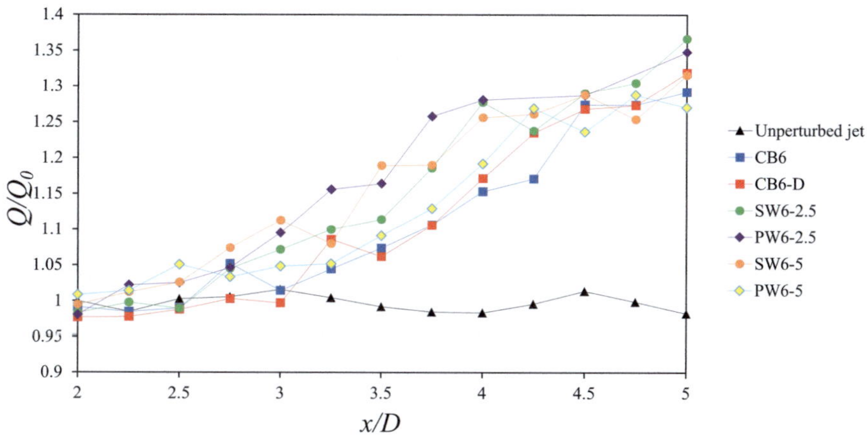

Fig. 13.10: The flow rate of the main jet in different control regimes.

$$Tu(x) = \frac{1}{V_{0,\text{mean}}} \left(\frac{1}{S_{\text{int}}} \int\limits_{S_{\text{int}}} v_x^2(x, y, z) ds \right)^{1/2} .$$

Here, $V_x(x, y, z, t_i)$ and $\bar{V}_x(x, y, z)$ are instantaneous and time-averaged axial velocity fields, N is the number of the time samples, and $V_{0,\text{mean}} = Q_0/S_{int}$ is the mean velocity in the unperturbed jet at $x/D = 2$. Measurement results are shown in Fig. 13.11. Despite the small number of time samples ($N = 200$) used to find velocity fluctuations in every spatial point, the results in terms of Tu are generally monotonous curves, which indicate the reliability of measurements.

Figure 13.11 shows that constant blowing CB6 produces a minimal velocity fluctuation level, but the CB6-D regime produces the same Tu level as other excitations. The PW6-2.5 regime is the most notable, which produces significantly higher velocity fluctuations than other control methods.

13.4 Conclusions

The evolution of a jet excited by six radially blown minijets was considered. The unperturbed main jet had a long laminar region so that any changes in the jet flow are attributed to the control. Six different control regimes were investigated:

a) two regimes (CB6 and CB6-D) of steady blowing through all minijets (with two different flow rate coefficients);

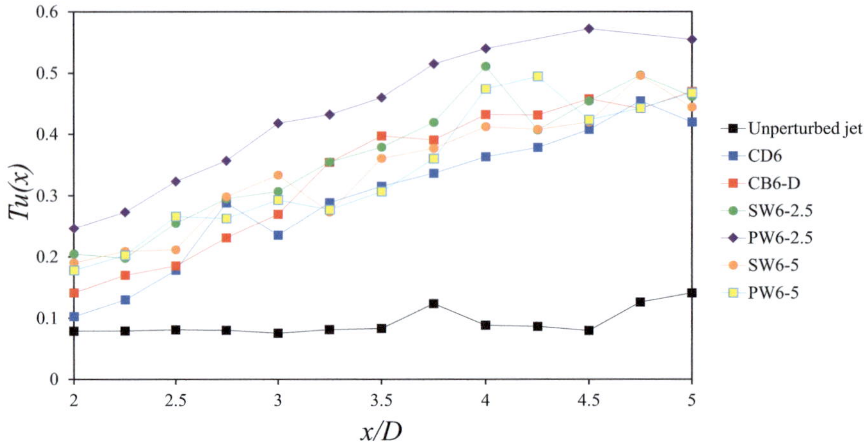

Fig. 13.11: The integral velocity fluctuation level Tu of the main jet in different control regimes.

b) two regimes (SW6-2.5 and SW6-5) with standing wave-like control with azimuthal number $n = 0$ and different switch on/off Strouhal numbers St $= 0.23$ and 0.46;
c) two regimes (PW6-2.5 and PW6-5) with a pattern of minijets switch on/off as an azimuthal progressive wave with wave number $n = 1$ and the same Strouhal numbers.

All the control laws had a serious impact on the flow in the main jet, its laminar region was significantly reduced. The fastest transition to turbulence is observed for SW6-5 control, which combines non-modal growth of steady perturbation and the fastest growing Kelvin-Helmholtz wave.

After the transition, when controlling with constant blowing, the regime with the lowest flow rate (CB6) showed the weakest effect relative to the others. When the minijet excitation amplitude was doubled (CB6-D), the flow rate of the main jet remained almost unchanged compared to the CB6 case, while the integral turbulence intensity Tu noticeably increased. However, similar results at a lower flow rate can be obtained for unsteady conditions (for example, the PW6-5 regime). The highest rate of jet destruction after transition was achieved by the PW6-2.5 unsteady control law. In terms of the flow rate increase, it is almost identical to the SW6-5 regime, but in terms of Tu, it far exceeds the others, which is possibly explained by the helical nature of induced perturbation, which at every x-section creates a local, almost solid-state shift of the jet (similar to the shift-up mechanism in [14]). From the visualizations of the axial velocity field, it is clear that the core of the main jet is the most blurred also for the PW6-2.5 regime compared to other control cases.

Acknowledgements This work is supported by the Russian Science Foundation under grant 25-19-00278.

References

[1] Belyaev IV, Bychkov OP, Zaitsev MY, Kopiev VA, Kopiev VF, Ostrikov NN, Faranosov GA, Chernyshev SA (2018) Development of the strategy of active control of instability waves in unexcited turbulent jets. Fluid Dynamics 53:347–360, DOI 10.1134/s0015462818030047

[2] Boronin SA, Healey JJ, Sazhin SS (2013) Non-modal stability of round viscous jets. Journal of Fluid Mechanics 716:96-119, DOI 10.1017/jfm.2012.521

[3] Bradbury LJS, Khadem AH (1975) The distortion of a jet by tabs. Journal of Fluid Mechanics 70:801–813, DOI 10.1017/s0022112075002352

[4] Choi JJ, Annaswamy AM, Lou H, Alvi FS (2006) Active control of supersonic impingement tones using steady and pulsed microjets. Experiments in Fluids 41:841-855, DOI 10.1007/s00348-006-0189-7

[5] Crow SC, Champagne FH (1971) Orderly structure in jet turbulence. Journal of Fluid Mechanics 48:547–591, DOI 10.1017/s0022112071001745

[6] Davis MR (1982) Variable control of jet decay. AIAA Journal 20:606–609, DOI 10.2514/3.7934

[7] Dulin VM, Lobasov AS, Chikishev LM, Markovich DM, Hanjalic K (2019) On impact of helical structures on stabilization of swirling flames with vortex breakdown. Flow Turbulence and Combustion 103:887–911, DOI 10.1007/s10494-019-00063-7

[8] Freund JB, Moin P (2000) Jet mixing enhancement by High-Amplitude fluidic Actuation. AIAA Journal 38:1863–1870, DOI 10.2514/2.839

[9] Gareev LR, Ivanov OO, Vedeneev VV, Ashurov DA (2024) Effect of the amplitude of a stationary perturbation on its non-modal growth in a laminar submerged jet. Journal of Applied Mechanics and Technical Physics 65(1):61–65 DOI 10.1134/S0021894424010073

[10] Gareev LR, Zayko JS, Chicherina AD, Trifonov VV, Reshmin AI, Vedeneev VV (2022) Experimental validation of inviscid linear stability theory applied to an axisymmetric jet. Journal of Fluid Mechanics 934:A3, DOI 10.1017/jfm.2021.1022

[11] Hilgers A, Boersma BJ (2001) Optimization of turbulent jet mixing. Fluid Dynamics Research 29:345–368, DOI 10.1016/s0169-5983(01)00035-1

[12] Ho C-M, Gutmark E (1987) Vortex induction and mass entrainment in a small-aspect-ratio elliptic jet. Journal of Fluid Mechanics 179:383–405, DOI 10.1017/s0022112087001587

[13] Hu H, Saga T, Kobayashi T, Taniguchi N (2002) Mixing process in a lobed jet flow. AIAA Journal 40:1339–1345, DOI 10.2514/2.1793

[14] Ivanov OO, Ashurov DA, Gareev LR, Vedeneev VV (2023) Non-modal perturbation growth in a laminar jet: an experimental study. Journal of Fluid Mechanics 963:A8, DOI 10.1017/jfm.2023.286

[15] Jordan P, Colonius T (2012) Wave packets and turbulent jet noise. Annual Review of Fluid Mechanics 45:173–195, DOI 10.1146/annurev-fluid-011212-140756

[16] Kozlov VV, Grek GR, Litvinenko YA (2015) Visualization of Conventional and Combusting Subsonic Jet Instabilities, SpringerBriefs in Applied Sciences and Technology. Cham: Springer, DOI 10.1007/978-3-319-26958-0
[17] Liang H, Maxworthy T (2005) An experimental investigation of swirling jets. Journal of Fluid Mechanics 525:115–159, DOI 10.1017/s0022112004002629
[18] Mi J, Nathan GJ, Luxton RE (2000) Centreline mixing characteristics of jets from nine differently shaped nozzles. Experiments in Fluids 28:93–94, DOI 10.1007/s003480050012
[19] Parekh D, Kibens V, Glezer A, et al (1996) Innovative jet flow control - Mixing enhancement experiments. AIAA Paper 96-0308, DOI 10.2514/6.1996-308
[20] Perumal AK, Wu Z, Fan DW, Zhou Y (2022) A hybrid artificial intelligence control of a turbulent jet: Reynolds number effect and scaling. Journal of Fluid Mechanics 942:A47, DOI 10.1017/jfm.2022.341
[21] Perumal AK, Zhou Y (2018) Parametric study and scaling of jet manipulation using an unsteady minijet. Journal of Fluid Mechanics 848:592–630, DOI 10.1017/jfm.2018.376
[22] Petersen RA, Samet MM (1988) On the preferred mode of jet instability. Journal of Fluid Mechanics 194:153–173, DOI 10.1017/s0022112088002939
[23] Sazhin S (2014) Droplets and Sprays. London: Springer, DOI 10.1007/978-1-4471-6386-2
[24] Seidel J -F, Pappert C, New TH, Tsai HM (2005) Effects of multiple radial blowing around a circular jet. AIAA Paper 2005-866, DOI 10.2514/6.2005-866
[25] Vlasov EV, Ginevskii AS (1967) Acoustic modification of the aerodynamic characteristics of a turbulent jet. Fluid Dynamics 2:93–96, DOI 10.1007/bf01019549
[26] Zaman KBMQ, Bridges JE, Huff DL (2011) Evolution from 'Tabs' to 'Chevron Technology' - A Review. International Journal of Aeroacoustics 10:685–709, DOI 10.1260/1475-472x.10.5-6.685
[27] Zaman KBMQ, Hussain AKMF (1980) Vortex pairing in a circular jet under controlled excitation. Part 1. General jet response. Journal of Fluid Mechanics 101:449–491, DOI 10.1017/s0022112080001760
[28] Zayko J, Teplovodskii S, Chicherina A, Vedeneev V, Reshmin A (2018) Formation of free round jets with long laminar regions at large Reynolds numbers. Physics of Fluids 30:043603, DOI 10.1063/1.5021017
[29] Zhou Y, Fan D, Zhang B, Li R, Noack BR (2020) Artificial intelligence control of a turbulent jet. Journal of Fluid Mechanics 897:A27, DOI 10.1017/jfm.2020.392

Chapter 14
Experimental Study of the Influence of Mechanical Characteristics of Structured Metamaterials made of Flexible Plastic TPU 95A on the Process of Penetration with a Rigid Striker

Svetlana Y. Ivanova, Kirill Y. Osipenko, Nikolay V. Banichuk, and Dmitry S. Lisovenko

Abstract The influence of the mechanical properties of metamaterials with a cellular internal structure upon normal penetration by a rigid spherical impactor was experimentally studied. Metamaterial samples were 3D-printed from TPU 95A plastic (thermoplastic polyurethane). They had auxetic and non-auxetic cell structures, respectively, in the form of concave and convex hexagons. Additionally, samples with a chiral structure were prepared. The results of experiments on penetration of metamaterial samples are presented to determine the relative loss of kinetic energy of the impactor and to test the influence of the metamaterial's internal structure on the deviation of the impactor's direction of motion from the approach direction after penetration. The experiments were conducted for various speed and temperature conditions. The influence of the rotation angle of the internal structure, its chirality, and the presence of a viscous filler in the samples were studied.

Key words: Metamaterials, Auxetics, Chiral structures, Experimental studies, Penetration, Rigid strikers, TPU 95A

14.1 Introduction and Description of Experimental Conditions

Among the structured metamaterials being developed, metamaterials with auxetic properties stand out. Auxetic materials (materials with a negative Poisson's ratio) have been actively studied in recent decades [2, 12–14, 16, 17, 20, 22, 25]. Auxetics exhibit unique mechanical properties. For example, they have higher indentation resistance than conventional materials and high specific mechanical energy

Svetlana Y. Ivanova · Kirill Y. Osipenko · Nikolay V. Banichuk · Dmitry S. Lisovenko (corresponding author)
Ishlinsky Institute for Problems in Mechanics RAS, Prospekt Vernadskogo 101-1, 119526 Moscow, Russian Federation, e-mail: ivanova@ipmnet.ru, osipenko@ipmnet.ru, banichuk@ipmnet.ru, lisovenk@ipmnet.ru

H. Altenbach et al. (eds.), *Stability Problems in Mechanics of Continua*,
Advanced Structured Materials 260, https://doi.org/10.1007/978-3-032-31032-3_14

absorption. These properties make auxetics promising for the creation of shock- and energy-absorbing structures [1, 3–6, 6–11, 15, 18, 19, 21, 23, 24, 26–28].

This paper reflects experimental studies conducted by the authors' team over a number of years, which are a continuation of the study of the mechanical properties of the created metamaterials with a cellular structure made of metal [10] or e-PLA plastic [4–6, 9] during penetration by a rigid spherical striker. Using a 3D printer, auxetic and non-auxetic metamaterial samples (including those with a chiral structure) were produced from TPU 95A plastic (thermoplastic polyurethane), which provides high impact strength, consisting of cells in the shape of concave or convex hexagons, respectively. In appearance and geometric parameters, these samples completely corresponded to the samples shown in the photographs and diagrams in [4, 6], but, unlike rigid metamaterials made of e-PLA plastic with a cellular structure, this paper investigated a flexible metamaterial that allows significant elastic deformations during penetration by a striker. For individual series of experiments, samples were prepared with an internal structure rotated relative to the vertical axis at a certain angle, as well as samples with a viscous filler (gelatin).

The aim of the study was to investigate the properties of flexible metamaterials in terms of their ability to resist penetration by assessing the relative loss of impactor kinetic energy under various experimental conditions. The feasibility of deflecting the impactor's direction of motion after penetration from the initial approach direction was also tested for the chiral structures and structures rotated about the vertical axis.

Auxetic (Fig. 14.1a, b) and non-auxetic (Fig. 14.2a, b) samples with the same number of layers in the cellular structure (S = 6 mm, L = 3 mm, h = 0.4 mm) had approximately equal masses (as far as the 3D printing process and sample structure allowed), allowing for a comparison of their properties. The samples were maintained in a thermostat for at least 5 hours at the specified temperature. To study the effect of viscous filler, the samples were filled with gelatin, prepared at a ratio of 50 g gelatin per 500 g water (Fig. 14.3).

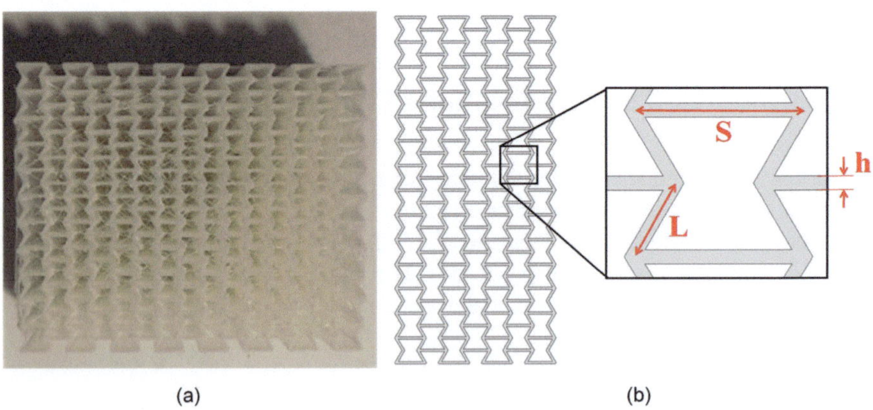

Fig. 14.1: Auxetic re-entrant honeycomb sample (a); 3D model (b).

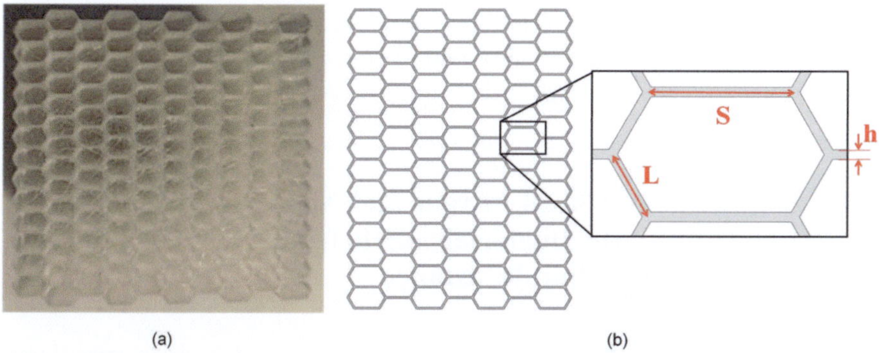

Fig. 14.2: Non-auxetic honeycomb sample (a); 3D model (b).

Fig. 14.3: Filling samples with gelatin..

The samples were penetrated using a 10 mm diameter steel spherical striker weighing 4.06 g at velocities of 150 m/s and 190 m/s, respectively. The side of the samples subjected to impact measured 70 × 73 mm in all cases. The striker's entry velocity was measured using an IBKh-ASS-0021 "Strelets" optical frame chronograph, as well as an oscilloscope and a FASTCAM mini AX200 high-speed video camera, which recorded the sample penetration process. The camera was positioned such that the video recording showed the striker penetrating the sample moving from right to left. It was also used to determine the striker's exit velocity from the barrier.

14.2 Comparison of Chiral Auxetic and Non-Auxetic (Honeycomb) Metamaterials made from TPU 95A Plastic in Terms of Penetration Resistance

Previous studies [4, 6] of metamaterials made from rigid e-PLA plastic have shown that the presence of chirality in the hexagonal structure enhances the penetration resistance and provides a significant advantage to auxetics over chiral non-auxetic cellular metamaterials. The first comparative study of auxetic and non-auxetic metamaterials based on flexible TPU 95A plastic was also conducted for chiral structures (see Figs. 14.4 and 14.5, where $S = 6$ mm, $L = 3$ mm, $h = 0.4$ mm, $r = 0.8$ mm was assumed).

Figures 14.6a, b demonstrate the advantage in damping the kinetic energy of the striker when penetrating chiral auxetic samples compared to non-auxetics. In the figures, markers of different types show the experimentally established dependencies of the value

$$\delta = \frac{v_1^2 - v_2^2}{v_1^2} (\%)$$

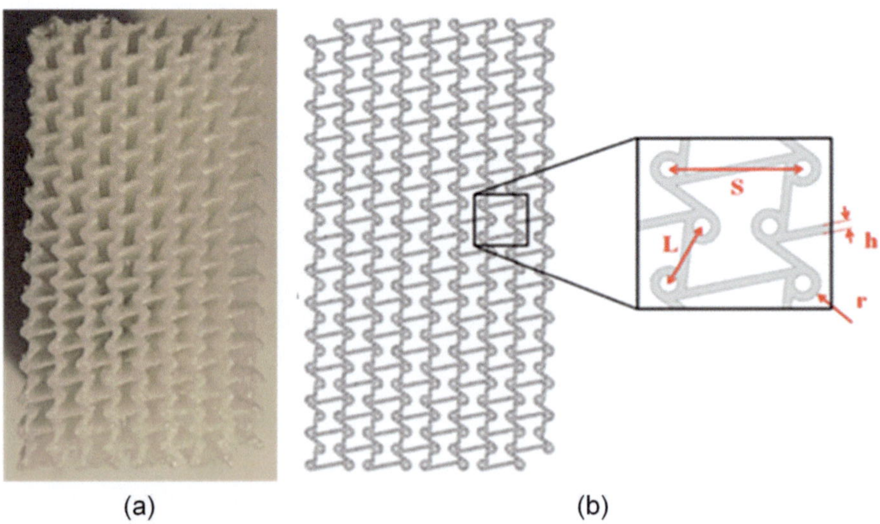

Fig. 14.4: Auxetic with chirality of the type "/" (a); 3D model (b).

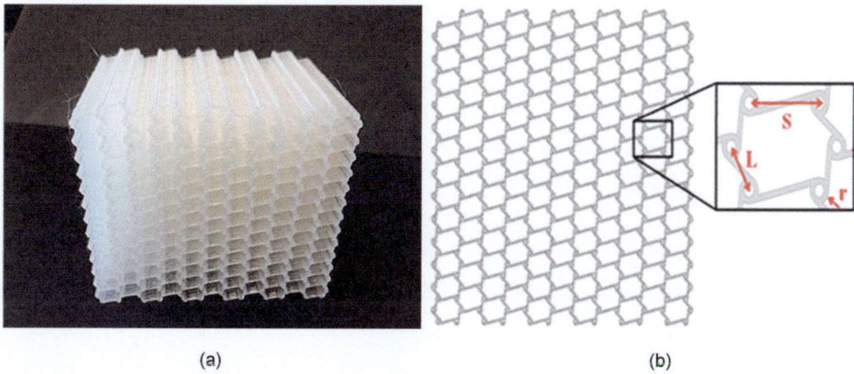

Fig. 14.5: Chiral non-auxetic honeycomb sample of the "/" type (a); 3D model (b).

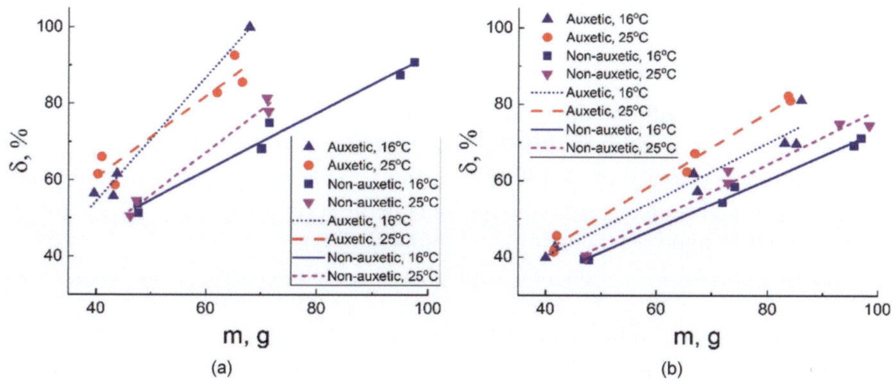

Fig. 14.6: Dependence of the relative loss of kinetic energy of the striker δ (%) on the mass m [g] of the penetrated chiral auxetic and non-auxetic samples: at approach velocities of ~ 150 m/s (a) and ~ 190 m/s (b).

of the relative loss of kinetic energy of the striker (υ_1 is the value of the striker velocity at the entrance and υ_2 at the exit from the sample) on the mass m [g] of the samples. For clarity, the corresponding linear trend lines are added to the figures. Data for the velocity mode ~ 150 m/s are shown in Fig. 14.6a, for the approach velocity ~ 190 m/s – in Fig. 14.6b. It is clear that the advantage of auxetics is greater in the first case. The experiments were carried out for sample temperatures of 16 and 25^0C.

In Fig. 14.7a, the dependencies of the relative loss of the kinetic energy of the impactor δ on the sample mass m corresponding to the velocity of ~ 150 m/s are compared for chiral auxetic samples, where the points are designated by triangular markers with the apex upward for the sample temperature of 16^0C and by round markers for 25^0C. Below in the figure, in the same coordinate axes, graphs are given for the approach velocity of ~ 190 m/s: square markers for the temperature for 16^0C,

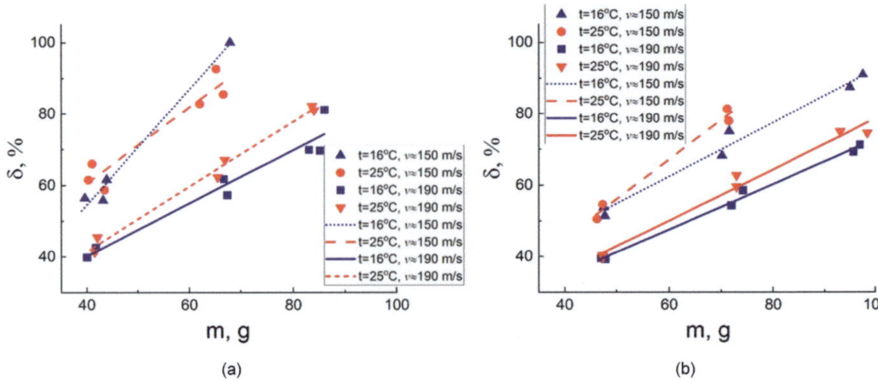

Fig. 14.7: Dependence of the relative loss of kinetic energy of the striker δ (%) on the mass m [g] of the penetrated chiral auxetic (a) and non-auxetic (b) samples: at approach velocities of $\sim$ 150 m/s and $\sim$ 190 m/s.

triangular markers with the apex downward for 25^0C. For clarity, four linear trend lines are added. It is evident that the relative energy loss is significantly greater at the approach velocity of $\sim$ 150 m/s in the considered temperature range. For non-auxetic chiral samples, similar dependencies are presented in Fig. 14.7b (for the same regimes of approach velocities and temperatures).

Figures 14.6 and 14.7 show that with increasing temperature in the range under consideration, the ability of metamaterials based on TPU 95A plastic to resist impact increases. This can be explained by the fact that with increasing temperature, deformations in the sample also increase, requiring additional energy. Section 14.5 is devoted to a more detailed study of the temperature dependence of metamaterials based on TPU 95A plastic.

Figure 14.8 demonstrates the better ability of metamaterials made from TPU 95A plastic to resist the penetration of impactors compared to the samples made from PLA plastic studied earlier in [4, 6] at the same values of approach velocity and temperature.

Previously [4, 6, 9], in experiments with chiral structures made of rigid e-PLA plastic, a deviation in the motion of the striker upon exiting the sample was observed. The direction of motion (upward or downward) could not always be explained by the type of chirality of the structure. The deviation could often be caused by brittle fracture of the sample due to defects, as well as by the ejection of sample fragments along with the striker. In this study, in individual experiments, the striker did not change its initial approach direction after exiting the perforated chiral sample. However, in most cases, an upward or downward deviation in the direction of motion was observed. Moreover, the deviation corresponded to the type of chirality of samples with both auxetic and non-auxetic structures. Figs. 14.9a and b show four positions of the striker during the perforation of samples, recorded by a video camera. For clarity, lines were added: a dashed-dotted straight line connects two images of the striker before entering the sample (to the right of it), and a solid line connects two

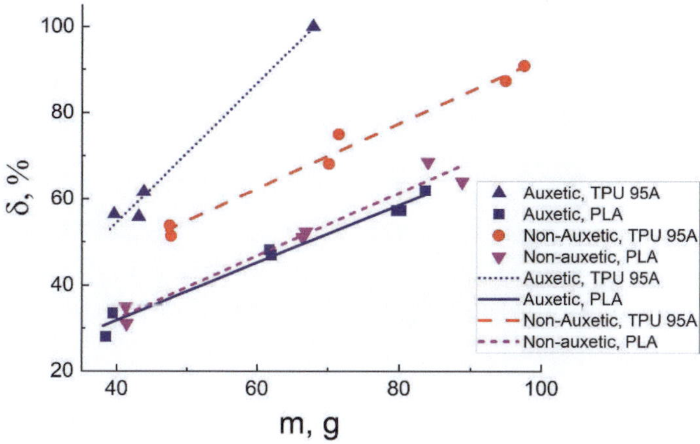

Fig. 14.8: Dependence of the relative loss of kinetic energy of the striker δ (%) on the mass m [g] of the penetrated chiral auxetic and non-auxetic samples made of TPU 95A and PLA plastics at approach speeds of ~150 m/s and a temperature of 16^0C.

images of it to the left, after exiting the sample. Figure 14.9a demonstrates the process of penetration of an auxetic with a chirality of the "/" type and a downward deflection of the striker, while Fig. 14.4b corresponds to the penetration of a non-auxetic with a chiral structure of the "\" type and an upward deflection of the striker. Characteristic asymmetric deformations of the samples in the observed plane immediately after the striker exits them were recorded in all experiments and reflect the peculiarity of chiral structures with a chirality of the "/" type (the convexity of the rear surface of the sample is directed downward, Fig. 14.4a) and with a chiral structure of the "\" type (the convexity of the rear surface of the sample is directed upward, Fig. 14.4b).

14.3 On the Influence of Viscous Filler on the Impact Resistance of Flexible Metamaterials with Auxetic Properties

In experiments with auxetic metamaterials made of rigid e-PLA plastic [5, 6] (it was found that when filling the metamaterial structure with gelatin, the resistance to penetration of auxetics was significantly increased compared to non-auxetics with the same increase in sample mass. For flexible metamaterials with auxetic properties with a chiral structure based on a concave hexagon, made of TPU 95A plastic, similar experiments were carried out on punching samples for two speed modes of ~ 150 m/s and ~ 190 m/s normal to the side surface, which in all cases had dimensions of 70×72 mm. The samples were filled with food-grade gelatin, prepared at the rate of 50 g gelatin per 500 g water. The temperature of the samples was $\sim 16 \div 17^0$C.

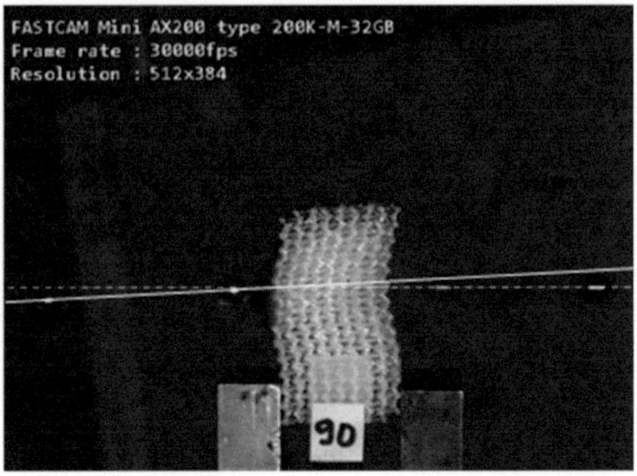

(a)

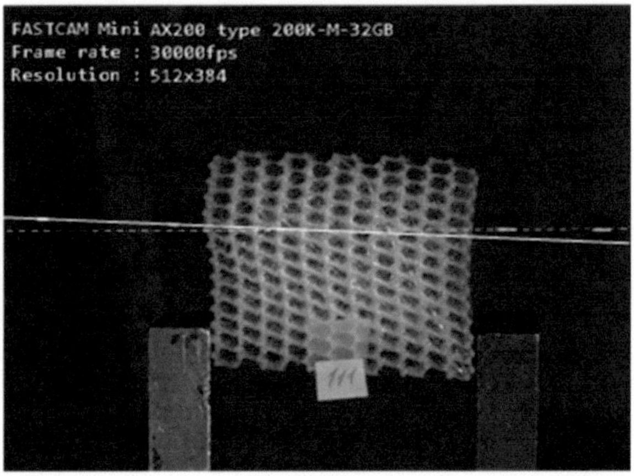

(b)

Fig. 14.9 Deviation of the direction of the impactor movement at the exit from the samples from the approach: chiral ("/") auxetic sample, downward deviation ~ 3° (a); chiral ("\") non-auxetic sample, upward deviation ~ 3° (b).

In Fig. 14.10a, markers of various types represent the experimental dependencies of the relative loss of kinetic energy δ of the impactor on the mass m of samples with gelatin or air for an entry velocity of ~ 150 m/s. The data for chiral auxetic metamaterial samples based on TPU 95A plastic are compared with the data for PLA-based samples with the same geometric parameters. For clarity, the corresponding trend lines 1–4 are added. It can be seen that the addition of a viscous filler in the form of gelatin (trend line 1) in the case of the PLA-based metamaterial resulted in an increase of the value δ compared to the unfilled samples (trend line 3). At the same time, the mass of the samples increased by approximately 3.5 times, and the energy loss δ increased by 1.6 times. Adding filler to a rigid PLA structure reduces its brittleness, increases its density, and leads to an increase in resistance to penetration by a striker, but also significantly increases the weight of the sample. For

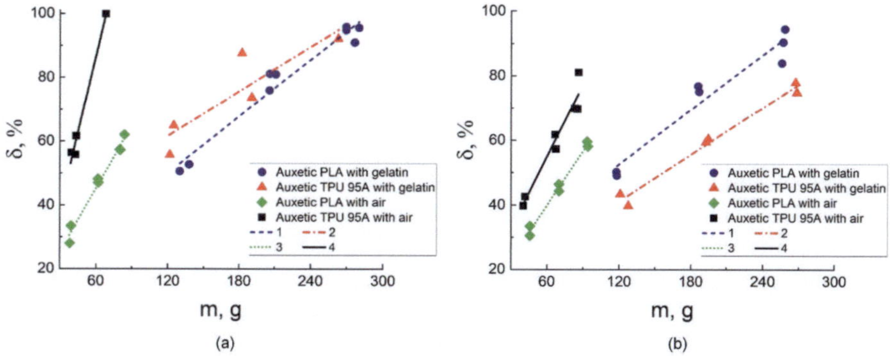

Fig. 14.10: Dependence of the relative loss of kinetic energy δ of the striker on the mass m of samples with gelatin and air for an entry velocity of ~ 150 m/s (a) and ~ 190 m/s (b).

the auxetic TPU 95A sample with air weighing 67.9 g and a thickness of 38.5 mm, the ballistic limit was reached at a striker impact velocity of 154.5 m/s (δ = 100%, trend line 4). Filling the flexible samples with gelatin, which is virtually incompressible, significantly reduced sample deformation during penetration and reduced the relative energy loss by a factor of 1.25, while the samples were approximately three times heavier (trend line 2). This allows us to conclude that flexible and lightweight air-filled TPU 95A plastic (thermoplastic polyurethane) samples proved to be the most effective in terms of resistance to striker penetration.

Fig. 14.10b corresponds to a higher entry velocity of ~ 190 m/s. In this case, when filled with gelatin, the most effective auxetic is PLA (trend line 1), not TPU 95A (trend line 2), as at 150 m/s. However, when filled with air, the resistance of the thermoplastic polyurethane samples is significantly higher than that of PLA, as is the case at 150 m/s (lines 3, 4). Given the low mass of the air-filled samples, we also conclude that the metamaterial based on TPU 95A plastic with air-filled cells is the most effective in terms of penetration resistance.

We also note that in the overwhelming majority of cases, for flexible samples filled with gelatin, no deviation in the direction of movement of the striker during take-off was observed (in contrast to the experiments with unfilled samples from Sect. 14.2 [8]).

14.4 On the Influence of the Chirality of the Structure of Auxetic Metamaterials on the Resistance to Penetration by a Striker

In [5, 6], it was concluded that the presence of chirality in the cells of auxetic samples manufactured on a 3D printer from e-PLA plastic led to an increase in the ability of these samples to resist penetration by a spherical impactor, compared to non-chiral samples. In Sects. 14.2 and 14.3, metamaterials with a chiral internal structure were studied in more detail. In this case, a comparison was made between metamaterials based on rigid e-PLA plastic and flexible metamaterials made of thermoplastic polyurethane (TPU 95A plastic), which allow large elastic deformations. In this section, chiral auxetic samples made of TPU 95A and PLA plastics are compared based on the 3D model previously used in [4, 4, 5, 7, 8] (see Fig. 14.4), and auxetic samples of similar mass based on TPU 95A and PLA plastics without chirality (Fig. 14.1). Separate experiments were conducted with chiral and non-chiral auxetic structures rotated in the xy plane (see Fig. 14.1b, 14.4b) by 90°. Examples of such metamaterials are shown in Fig. 14.11a, b.

In Fig. 14.12a, various marker types are used to represent the experimental dependences of the relative loss of impactor kinetic energy δ on the mass m of chiral and non-chiral metamaterial samples made of PLA and TPU plastics at an impactor entry velocity of $\sim$ 190 m/s. Linear trend lines are also included for clarity. It is evident that for rigid PLA-based samples, the presence of structural chirality has an insignificant effect at the chosen velocity, but for more massive samples, it nevertheless leads to an increase in penetration resistance. For flexible chiral thermoplastic polyurethane samples, penetration resistance increased by 5-7% compared to samples without chirality.

With a 90° rotation of the internal structure, non-chiral metamaterial samples (both PLA and TPU) exhibit greater penetration resistance. This is clearly seen in Fig. 14.12b. Trend lines 2 and 4, corresponding to samples without chirality, pass above the corresponding lines 1 and 3. For TPU samples, the improvement in resistance is more than 10%. According to the results of the experiments, the most effective in terms of penetration resistance were samples made of TPU 95A plastic (thermoplastic polyurethane) without chirality with a 90° rotation of the

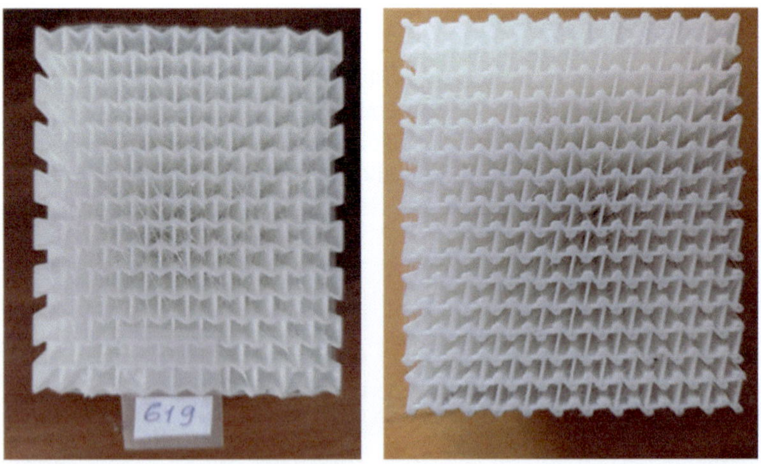

(a) (b)

Fig. 14.11: Samples of auxetic metamaterials made of TPU 95A plastic with structures rotated by 90°: without chirality (a) and chiral (b)..

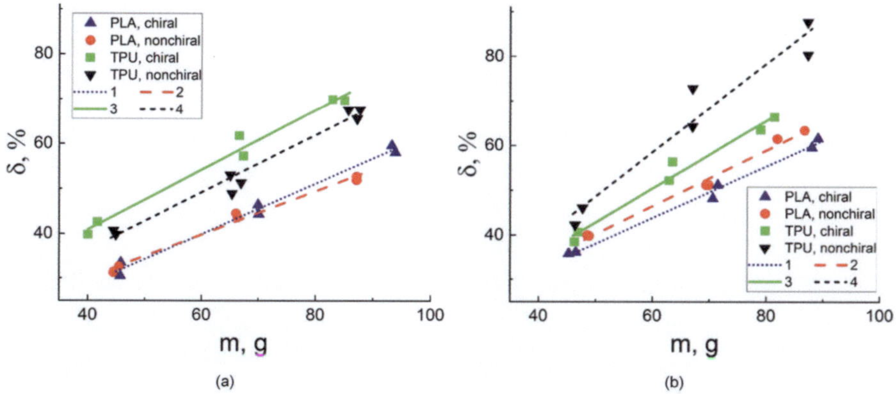

Fig. 14.12: Dependence of the relative loss of kinetic energy δ of the impactor on the mass m of chiral and non-chiral samples of e-PLA and TPU 95A for an entry velocity of ~ 190 m/s without structure rotation (a) and with structure rotation (b).

internal structure. For these samples, the highest increase in penetration resistance with increasing mass was also noted. The next section describes in more detail the dependence of the metamaterials' penetration resistance on the rotation angle of their internal auxetic structure.

14.5 On the Influence of Temperature of Flexible Auxetic Metamaterials on the Energy Loss of a Striker During Penetration

Presented here are the results of experiments on the penetration of flexible non-chiral auxetic metamaterial samples made of TPU 95A plastic, whose high efficiency in damping the impactor energy has been previously established. For the experiments, auxetic non-chiral metamaterial samples made of TPU 95A plastic were fabricated using a 3D printer with the internal structure rotated in the xy plane by an angle of α = 0, 30, 45, 60, and 90° from the vertical y-axis (see Fig. 14.13a, b). The angle α was assumed to take positive values for clockwise rotation of the structure and negative values for counterclockwise rotation. The samples were shot along the x-axis with a steel spherical impactor at a speed of ~ 190 m/s normal to the side surface with dimensions of ~70 × 75 mm. Before the experiment, the samples were kept in a thermostat for at least 4 hours at a temperature of 5, 16, 25 or 35^{0}C.

In Fig. 14.14a, markers of different types represent the experimental dependences of the relative loss of the kinetic energy δ of the striker on the mass m of non-chiral metamaterial samples made of TPU 95A plastic at the internal structure rotation angle α = 0° for temperatures of 5, 16, 25, and 35^{0}C. For clarity, linear trend lines are also added to the figure. The graph shows that the highest penetration resistance is

(a) (b)

Fig. 14.13: Examples of auxetic samples with a clockwise rotation of the structure at an angle of $\alpha = 60°$ (a) and a counterclockwise rotation at an angle of $\alpha = -45°$ (b).

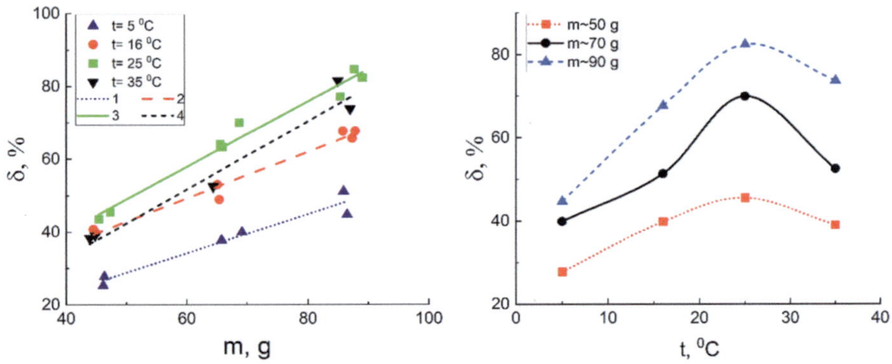

(a) (b)

Fig. 14.14: Dependence of the relative loss δ (%) of the kinetic energy of the impactor during penetration of metamaterial samples made of TPU 95A plastic at $\alpha = 0°$ on the mass m (g) of the samples at different temperature values (a) and on the temperature t (^{0}C) at different sample mass values (b).

exhibited by the samples at 25^{0}C, and the lowest – at 5^{0}C. In Fig. 14.14b, for clarity, the experimental points are connected by smooth curves illustrating the dependence of energy loss on temperature for samples weighing ~50, 70, and 90 g. It is evident that the maximum loss corresponds to a temperature of 25^{0}C.

Similar dependencies were obtained for the rotation angles of 30, 45, 60, 90°. The shape of these dependencies for the angles $\alpha = 30°$ and $\alpha = 90°$ differed little qualitatively from those shown in Fig. 14.14. Figs. 14.15a and 14.16a differ significantly from Fig. 14.14a. In them, the experimental dependencies δ on m for the temperature values of 5 and 16^0C are supplemented with polynomial trend lines (third-degree polynomial). Such a nonlinear nature of the dependencies may be due to the fact that with an increase in the mass of the samples in the cases under consideration, there is no increase in the energy loss of the impactors. It can be assumed that an increase in the energy costs for punching a thicker (and, consequently, more massive) sample is accompanied by a simultaneous decrease in the deformation energy losses (a thicker sample can perform smaller deformations). This phenomenon was observed for samples with $\alpha = 45, 60°$ at lower temperature conditions (16 and 5^0C) in the range under consideration.

A series of experiments with non-chiral TPU 95A plastic samples subjected to internal structure rotations of $\alpha = 0, 30, 45, 60$ and 90°, respectively, demonstrated the highest penetration resistance at 25^0C for all metamaterials examined. The best values were recorded for samples with $\alpha = 90°$, while those at rotation angles of 0 and 45° were minimal.

In Sect. 14.2, it was noted that during the penetration of chiral metamaterials, a deviation of the striker's departure direction from the approach direction is possible, depending on the internal structure of the samples. Such striker deviations were recorded not only for chiral samples made of TPU 95A plastic, but also for auxetic samples without chirality made of TPU with a rotated internal structure. In all such cases, at clockwise structure rotation angles $\alpha = 30, 45, 60°$, an upward striker deviation was observed upon exiting the sample (see Fig. 14.17a), while with counterclockwise rotation, a downward deviation was observed (Fig. 14.17b). Fig-

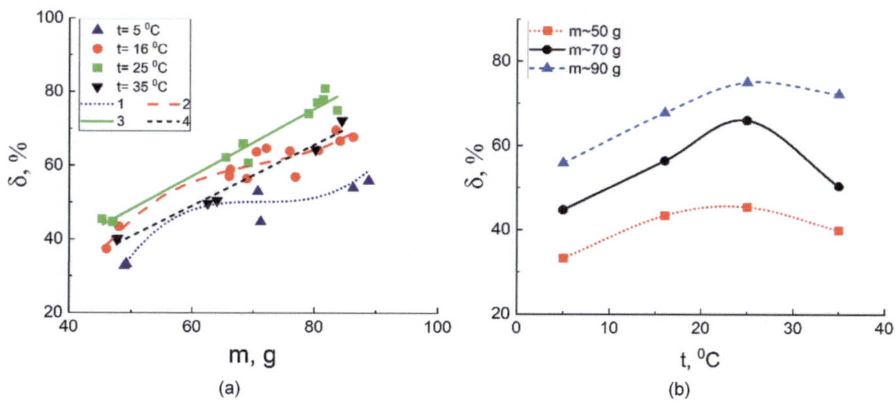

Fig. 14.15: Dependence of the relative loss δ (%) of the kinetic energy of the striker during penetration of samples of metamaterial made of TPU 95A plastic at $\alpha = 45°$ on the mass m (g) of the samples at different values of temperature (a) and on the temperature t (^{0}C) at different values of sample mass (b).

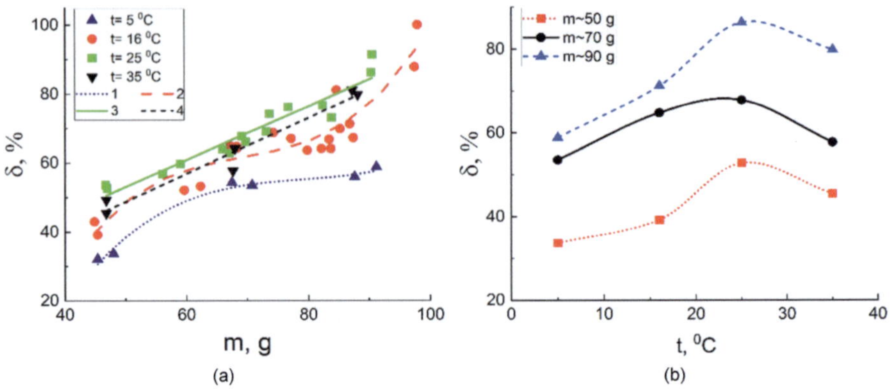

Fig. 14.16: Dependence of the relative loss δ (%) of the kinetic energy of the striker during penetration of samples of metamaterial made of TPU 95A plastic at $\alpha = 60°$ on the mass m (g) of the samples at different values of temperature (a) and on the temperature t (^{0}C) at different values of sample mass (b).

urs 14.17a, b, by analogy with Fig. 14.9, are a combination of four frames obtained using a high-speed camera, with the addition for clarity of lines connecting the positions of the striker in pairs before and after penetration of the sample.

The capabilities of a high-speed video camera make it possible to observe the deformation of the side surface of the penetrating sample. Fig.14.18 shows a photographic image of the sample with the internal auxetic structure rotated by 60° as the striker moves within it. The striker's entry velocity was 192.5 m/s. The filming was performed at 50000 frames per second. The light source was positioned behind the sample, so the striker and the trace of destroyed cells it left behind are visible as a dark area in the central portion of the sample.

Figure 14.19a shows the absolute velocity field for a selected set of lateral surface points, plotted using the GOM Correlate image processing program. Fig. 14.19b shows the acceleration field for these points. The regions of maximum velocity and acceleration values are highlighted in red in these figures. The light spot in the center of Figs. 14.19a and 14.19b corresponds to the dark spot in Fig. 14.18, where the image was not processed by the program. The presented distributions are asymmetrical with respect to the direction of impactor motion due to the peculiarities of the specimen's internal structure. They illustrate the action of forces arising during motion, which deflect the impactor from its original direction of motion.

14.6 Comments and Conclusions

Laboratory-developed penetration studies of auxetic metamaterials based on flexible TPU 95A plastic demonstrated the effectiveness of these metamaterials in damping

Fig. 14.17 Impactor deflection on exit from auxetic non-chiral samples based on TPU 95A at internal structure rotation angles: (a) upward at $\alpha = 45°$, (b) downward at $\alpha = -60°$.

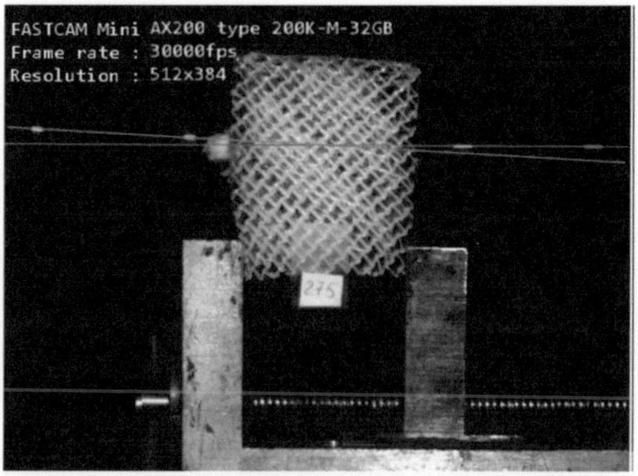

(a)

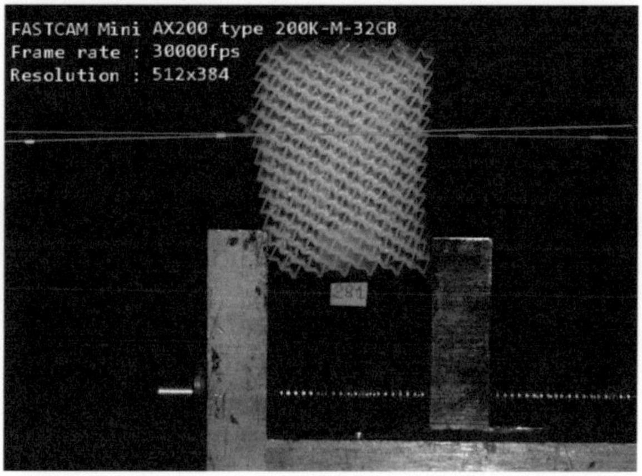

(b)

the kinetic energy of impactors in the studied range of velocities and temperatures. Energy damping was largely due to the fact that the studied metamaterials allowed for significant elastic deformations during impactor penetration. Auxetic non-chiral metamaterials with a 90° internal structure rotation exhibited the best damping properties. The studied metamaterials were sensitive to temperature changes. The greatest efficiency in damping impactor energy was achieved at temperatures close to 25^0C. When penetrating chiral samples and samples with a rotated internal structure, a deviation in impactor motion was observed during penetration, which depended on the type of chirality or structural rotation. A more detailed study of this phenomenon is of interest in the future.

Acknowledgements This work was completed within the scope of the State Assignment (state registration numbers 124013000674-0, 124012500437-9). The authors would like to thank A.I. Demin

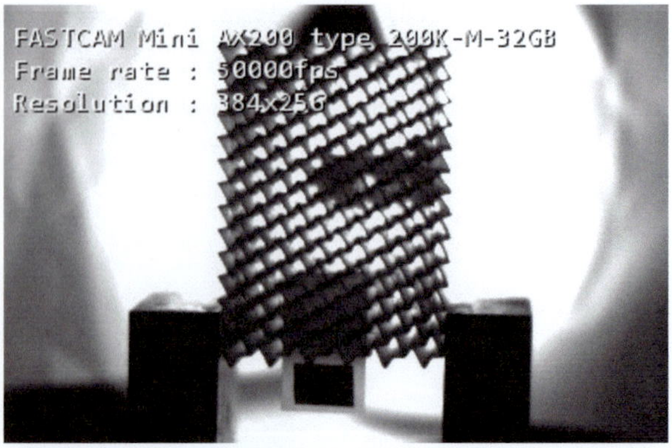

Fig. 14.18 Impactor penetration of an auxetic non-chiral sample based on TPU 95A with a 60° rotation of the internal structure.

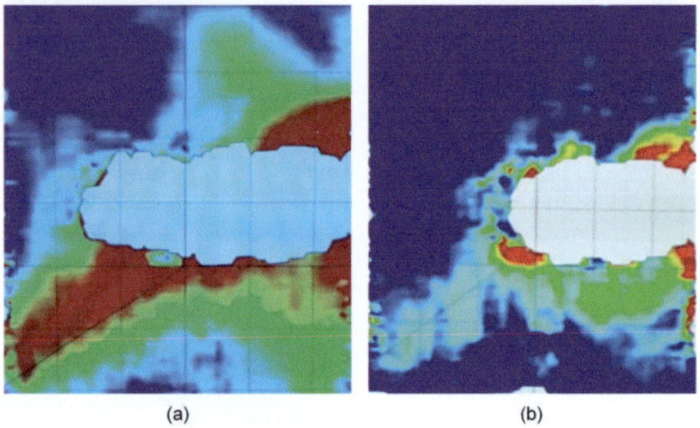

Fig. 14.19 The velocity field of points on the lateral surface of a sample with a 60° rotation of the internal structure (a) and the corresponding acceleration field (b) at the moment of penetration by a rigid spherical impactor

for assistance in preparing 3D models of the samples and A.V. Chentsov for preparation of video materials.

References

[1] Gao Y, Huang H (2022) Energy Absorption And Gradient Of Hybrid Honeycomb Structure With Negative Poisson's Ratio. Mech. Solids. 57(5):1118-1133, DOI 10.3103/S0025654422050053

[2] Gorodtsov VA, Lisovenko DS (2020) Auxetics among materials with cubic anisotropy. Mechanics of Solids 55(4):461-474, DOI 10.3103/S0025654420040044

[3] Imbalzano G, Tran P, Lee PVS, Gunasegaram D, Ngo TD (2016) Influences of material and geometry in the performance of auxetic composite structure under blast loading. Applied Mechanics and Materials 846:476-481, DOI 10.4028/www.scientific.net/amm.846.476

[4] Ivanova SY, Osipenko KY, Banichuk NV, Lisovenko DS (2024) Experimental study of the properties of metamaterials based on PLA plastic when

perforated by a rigid striker. Mechanics of Solids. 59(4):207-215, DOI 10.1134/S0025654424604695

[5] Ivanova SY, Osipenko KY, Banichuk NV, Lisovenko DS (2024) Issledovaniye vliyaniya vyazkogo zapolnitelya na mekhanicheskiye svoystva metamaterialov s otritsatel'nym i polozhitel'nym koeffitsiyentom Puassona pri probivanii zhestkim udarnikom (in Russ., Investigation of the effect of a viscous filler on the mechanical properties of metamaterials with negative and positive Poisson's ratio when punching with a rigid impactor). Vestnik Chuvashskogo gosudarstvennogo pedagogicheskogo universiteta im. I.YA. Yakovleva. Seriya: Mekhanika predel'nogo sostoyaniya 4(62):62-75, DOI 10.37972/chgpu.2024.62.4.005

[6] Ivanova SY, Osipenko KY, Banichuk NV, Lisovenko DS (2024) Investigation of the effect of a viscous filler on the punching process of auxetic and non-auxetic metamaterials. Mechanics of Solids 59(7):3727-3734, DOI 10.1134/S0025654424606633

[7] Ivanova SY, Osipenko KY, Banichuk NV, Lisovenko DS (2025) On the effect of the viscous filler on the impact resistance of flexible metamaterials with auxetic properties. Mechanics of Solids 60(2):958-964, DOI 10.1134/S0025654424607225

[8] Ivanova, SY, Osipenko KY, Banichuk NV, Lisovenko DS (2025) Temperature influence of metamaterials based on flexible TPU 95A plastic on resistance to penetration by a rigid striker. Mechanics of Solids 60(1):128-135, DOI 10.1134/S0025654424606797

[9] Ivanova SY, Osipenko KY, Demin AI, Banichuk NV, Lisovenko DS (2023) Studying the properties of metamaterials with a negative Poisson's ratio when punched by a rigid impactor. Mechanics of Solids 58(5):1536-1544, DOI 10.3103/S0025654423600897

[10] Ivanova SY, Osipenko KY, Kuznetsov VA, Solovyov NG, Banichuk NV, Lisovenko DS (2023) Experimental investigation of the properties of auxetic and non-auxetic metamaterials made of metal during penetration of rigid strikers. Mechanics of Solids 58(2):524-528, DOI 10.3103/S0025654422601616

[11] Jiang Q, Hao B, Chen G, et. al. (2024) Analysis of the penetration ability of exponential bullets on TPMS structures with variable density. Mechanics of Solids. 59(5):3198-3211, DOI 10.1134/S0025654424605640

[12] Joseph A, Manesh V, Harursampath D (2021) On the application of additive manufacturing methods for auxetic structures: a review. Advances in Manufacturing 9(3):342-368, DOI 10.1007/s40436-021-00357-y

[13] Kelkar PU, Kim HS, Cho K-H, Kwak JY, Kang C-Y, Song H-C (2020) Cellular auxetic structures for mechanical metamaterials: A Review. Sensors 20(11):3132, DOI 10.3390/s20113132

[14] Kolken HMA, Zadpoor AA (2017) Auxetic mechanical metamaterials. RSC Advances 7(9):5111-5129, DOI 10.1039/C6RA27333E

[15] Li C, Shen HS, Wang H (2020) Nonlinear Dynamic Response Of Sandwich Plates With Functionally Graded Auxetic 3D Lattice Core. Nonlinear Dynymics 100:3235-3252, DOI 10.1007/s11071-020-05686-4

[16] Lim T-C (2015) Auxetic Materials and Structures. Springer, Singapore, DOI 10.1007/978-981-287-275-3

[17] Novak N, Vesenjak M, Ren Z (2016) Auxetic cellular materials - A review. Strojniski vestnik - Journal of Mechanical Engineering 62(9):485-493, DOI 10.5545/sv-jme.2016.3656

[18] Novak N, Starcevic L, Vesenjak M, et. al. (2019) Blast response study of the sandwich composite panels with 3D chiral auxetic core. Composite Structures 210:167-178, DOI 10.1016/j.compstruct.2018.11.050

[19] Qiao JX, Chen CQ (2015) Impact resistance of uniform and functionally graded auxetic double arrowhead honeycombs. International Journal of Impact Engineering 83:47-58, DOI 10.1016/j.ijimpeng.2015.04.005

[20] Ren X, Das R, Tran P, Ngo TD, Xie YM (2018) Auxetic metamaterials and structures: a review. Smart Materials and Structures 27(2):023001, DOI 10.1088/1361-665X/aaa61c

[21] Shen ZY, Wen YK, Shen LY, et. al. (2024) Dynamic response and energy absorption characteristics of auxetic concave honeycomb pad for ballistic helmet under shock wave and bullet impact. Mechanics of Solids 59(5). 59(5):3050-3067, DOI 0.1134/S0025654424605159

[22] Shitikova MV (2022) Fractional operator viscoelastic models in dynamic problems of mechanics of solids: A review. Mechanics of Solids 57(1):1-33, DOI 10.3103/S0025654422010022

[23] Skripnyak VV, Chirkov MO, Skripnyak VA (2021) Modelirovanie mekhanicheskoi reakcii aukseticheskikh netamaterialov na dinamicheskie vozdeistviya (in Russ., Modeling the mechanical response of auxetic metamaterials to dynamic effects). PNRPU Mechanics Bulletin (2):144-152, DOI 10.15593/perm.mech/2021.2.13

[24] Usta F, Türkmen HS, Scarpa F (2022) High-velocity impact resistance of doubly curved sandwich panels with re-entrant honeycomb and foam core. International Journal of Impact Engineering 165:104230, DOI 10.1016/j.ijimpeng.2022.104230

[25] Wu W, Hu W, Qian G, Liao H, Xu X, Berto F (2019) Mechanical design and multifunctional applications of chiral mechanical metamaterials: A review. Materials & Design 180:107950, DOI 10.1016/j.matdes.2019.107950

[26] Xing Y, Deng B, Cao M, Qiu T, Wang L, Yang S (2024) Influence of structural and morphological variations in orthogonal trapezoidal aluminum honeycomb on quasi-static mechanical properties. Mechanics of Solids 59(1):445-458, DOI 10.1134/S0025654423602550

[27] Yu Y, Fu T, Wang S, Yang C (2025) Dynamic response of novel sandwich structures with 3D sinusoid-parallel-hybrid honeycomb auxetic cores: The cores based on negative Poisson's ratio of elastic jump. European Journal of Mechanics - A/Solids 109:105449, DOI 10.1016/j.euromechsol.2024.105449

[28] Zhao X, Gao Q, Wang L, Yu Q, Ma ZD (2018) Dynamic crushing of double-arrowed auxetic structure under impact loading. Materials & Design 160:527-537, DOI 10.1016/j.matdes.2018.09.041

Chapter 15
Some Problems of Contact Interaction of a Compressible Plastic Layer with External Bodies

Vagid Kadymov and Elena Yanovskaya

Abstract This work further developed Ilyushin's theory of flow in a thin plastic layer confined between two approaching surfaces of non-deformable external bodies. Later, in the works of Kiiko, this theory was generalized to include normal elastic displacements of the instrument's impacting bodies, which in the processes described are commensurate with the thickness of the flowing plastic layer.

This paper presents a generalized method of characteristics applied to flow problems in a plastic layer, presented in a simplified formulation described by first-order nonlinear partial differential equations. Solutions to a number of practical problems are presented in analytical form. An analogy is drawn between plastic flow problems in a thin layer and the problem of determining the surface shape of a homogeneous sand embankment.

The second part of the paper develops a general formulation of the contact boundary value problem for the flow of a thin plastic layer, described by second-order nonlinear partial differential equations. The paper concludes with various generalizations of Prandtl's classical problem of compression of a plastic strip and their solutions.

Key words: Plastic flow, Contact problem of spreading of a thin plastic layer, Generalizations of Prandtl's classical problem of compression of a plastic strip

---

Vagid Kadymov
MIREA - Russian Technological University, Moscow, Russian Federation, e-mail: vkadymov@yandex.ru

Elena Yanovskaya (corresponding author)
Moscow State Technological University "STANKIN", Moscow, Russian Federation, e-mail: elena_yanovskaya@bk.ru

H. Altenbach et al. (eds.), *Stability Problems in Mechanics of Continua*,
Advanced Structured Materials 260, https://doi.org/10.1007/978-3-032-31032-3_15

15.1 A General Case of an Analogue of Prandtl's Problem for Constrained Layer Shape Change in Domains of Non-Canonical Shape

Let us consider a perfectly plastic layer compressed by two parallel rigid plates under plane strain conditions. We will assume its length is significantly greater than its thickness. Yield strength of a material $k = k(z)$ we consider the coordinate function to be known z [21]. The plates are supposed to be rough:

$$[\sigma_{xz}]_{z=h} = mk(h), \; [\sigma_{xz}]_{z=-h} = -nk(h), \; 0 \le m \le 1, \; 0 \le n \le 1 \tag{15.1}$$

The stresses $\sigma_{xx}, \sigma_{zz}, \sigma_{xz}$ satisfy the partial differential equations of equilibrium:

$$\frac{\partial \sigma_{xx}}{\partial x} + \frac{\partial \sigma_{xz}}{\partial z} = 0, \qquad \frac{\partial \sigma_{xz}}{\partial x} + \frac{\partial \sigma_{zz}}{\partial z} = 0$$

and the condition of plasticity [6]

$$(\sigma_{xx} - \sigma_{zz})^2 + 4\sigma_{xz}{}^2 = 4k^2(z).$$

Let us assume that the shear stress σ_{xz} depends only on the coordinate z, then from the previous equations we find the values of the stresses

$$\begin{cases} \sigma_{xz} = az + b, \\ \sigma_{zz} = -ax - c, \\ \sigma_{xx} = -ax - c + 2\sqrt{k^2(z) - (az+b)^2}, \end{cases} \tag{15.2}$$

where a, b, c constant coefficients. From conditions (15.1) we find a and b:

$$a = \frac{1}{2h}\left[mk(h) + nk(-h)\right], \quad b = \frac{1}{2h}\left[mk(h) - nk(-h)\right].$$

The constant c is determined from the equilibrium condition of an element of width x. Since the left edge of this element is free from external stresses, then:

$$\int_0^x mk(h)dx + \int_0^x nk(-h)dx + \int_{-h}^{h} \sigma_{xx} dz = 0.$$

From here follows

$$c = \frac{1}{h}\int_{-h}^{h} \sqrt{k^2(z) - (az+b)^2} dz.$$

From (15.1) and (15.2) we find the equations of the slip lines:

$$x = \int \frac{\sqrt{k^2(z) - (az+b)^2}}{az+b \pm k} dz + \text{const.} \tag{15.3}$$

To determine the components of velocity v_x, v_z we have the incompressibility condition

$$\frac{\partial v_x}{\partial x} + \frac{\partial v_z}{\partial z} = 0 \tag{15.4}$$

and the condition of coaxiality of stress tensors and strain rates

$$\frac{\partial v_x}{\partial x} + \frac{\partial v_z}{\partial z} - \frac{2\sigma_{xz}}{\sigma_{xx} - \sigma_{zz}} \left(\frac{\partial v_x}{\partial x} - \frac{\partial v_z}{\partial z} \right) = 0, \tag{15.5}$$

as well as the kinematic condition at the contact boundary

$$[v_z]_{z=\pm h} = \pm V. \tag{15.6}$$

From equations (15.4), (15.5) and boundary conditions (15.6), we find the velocity distribution

$$\begin{aligned} v_z &= -\frac{V_z}{h}, \\ v_x &= -\frac{V_x}{h} + \frac{2V}{h} \int\limits_0^z \frac{az+b}{\sqrt{k^2(z) - (az+b)^2}} dz + C. \end{aligned} \tag{15.7}$$

We determine the constant C from the condition of conservation of mass of incompressible material: the flow of material through the section $x = \text{const}$ must be equal to the flow of material squeezed out by the plates to the right of the specified section:

$$-\int\limits_{-h}^{h} v_z dz = 2V(l - x). \tag{15.8}$$

The constant of integration follows

$$C = V \left(\frac{1}{h^2} \int\limits_{-h}^{h} \int\limits_0^z \frac{az+b}{\sqrt{k^2(z) - (az+b)^2}} dx dz - \frac{1}{h} \right).$$

This velocity distribution generalizes the well-known distribution obtained by Nadai for the case of a homogeneous material.

From (15.2) it follows that the ultimate pressure under the plates is distributed according to a linear law. Note that the stresses (15.2) do not satisfy the conditions at the end at $x = 0$ and the condition $\sigma_{xz} = 0$ at $x = l$. Velocities (15.7) in the center of the strip do not correspond to the actual velocity distribution, since the material does not transition to a plastic state near the plate surfaces in the center of the strip. Therefore, solution (15.2), (15.7), like the Prandtl–Nadai solution, is close to the exact solution only at a sufficient distance from the edge and center of the strip.

Let us consider a special case: dependence $k(z)$ – linear, and the plates are absolutely rough $(m = n = 1)$. Then

$$c = \frac{\pi}{2}\sqrt{k(h)k(-h)},$$

sliding lines are curves

$$\sqrt{1 - \frac{z^2}{h^2}} \pm \arcsin\frac{z}{h} = \frac{\left[\frac{k(h)}{k(-h)}\right]^{\pm\frac{1}{2}}}{h} + \text{const}. \tag{15.9}$$

The equations of slip lines (15.9) are the equations of cycloids. Solution (15.2), (15.6) makes sense if the following condition is satisfied:

$$|\sigma_{xz}| = |az + b| \leq k(z). \tag{15.10}$$

Otherwise, the stress σ_{xz} and the velocity component v_x become complex. Let us assume for simplicity that the distribution $k(z)$ is symmetrical about the midline $z = 0$ and that the friction on the plates is the same, $(b = 0)$. Shear stress σ_{xz} is determined only by the magnitude of friction on the plates and does not depend on the distribution of $k(z)$ by thickness. If $k(z)$ at $0 \leq z \leq h$ increases rapidly enough, then condition (15.10) means that the entire layer does not transition to a plastic state. In this case, rigid zones adjoin the slabs. We assume that the rigid zones are bounded by straight lines $z = \pm h_0$, where $h_0 \leq h$. In this case, the stresses and velocities in the plastic layer $|z| \leq h_0$ are determined by formulas (15.2) and (15.7), in which h should be replaced by a fixed value h_0. At the border of the plastic region $z = \pm h_0$ the maximum shear stress is applied, numerically equal to $k(h_0)$, so that the lines $z = \pm h_0$ are the envelopes of the slip lines. Otherwise, the solution (15.2) and (15.7) could be extended beyond the lines $z = \pm h_0$. Size h_0 is determined by the condition of touching of the graphs σ_{xz} and $k(z)$:

$$h_0 k'(h_0) = k(h_0). \tag{15.11}$$

Let us represent $k(z)$ in the form:

$$k(z) = k(0) + \Delta k\left(\frac{z}{h}\right)^{2p}, \text{ where } \Delta k = k(h) - k(z) > 0,\ 2p > 1.$$

Then:

$$\frac{h_0}{h} = \left(\frac{k(0)}{p\Delta k}\right)^{\frac{1}{2p}}.$$

Because

$$\frac{h_0}{h} < 1,$$

then the following condition must be met:

$$\frac{\Delta k}{k(0)} > \frac{1}{p}, \left(p > \frac{1}{2}\right).$$

From here

$$\frac{k(z)}{k(0)} > 3.$$

We have not obtained a limit on the amount of friction on the plates, however, due to the small thickness of the rigid layer, we can approximately consider the stress associated with overcoming friction to be equal to $k(h_0)$.

15.2 Method of Characteristics

The authors [15] proposed a simplified boundary value problem for flow in a thin plastic layer enclosed between two tool surfaces approaching each other according to a given law. The following assumptions were used:

- the plastic layer material is perfectly plastic;
- the layer material is volumetrically incompressible;
- elastic deformations of the tool bodies are neglected;
- Prandtl's classical flow law is adopted at the contact surfaces.

The problem is described by first-order nonlinear partial differential equations with respect to the contact pressure and the components of the flow velocity vector along the flow plane.

The boundary value problem equations are [7, 27]:

$$\operatorname{grad} p = -\frac{2\tau_s}{h}\frac{u}{|\vec{v}|}, \tag{15.12}$$

$$\frac{\partial u}{\partial x} + \frac{\partial v}{\partial y} + \frac{1}{h}\frac{dh}{dt} = 0. \tag{15.13}$$

The only boundary condition on the contour of a spreading plastic layer

$$p\,|_{\partial s} = \sigma_s. \tag{15.14}$$

In the general case of difficult spreading, where $p_\Gamma(\mu)$ – a given function of the parameter μ at the boundary of the domain formed by the grooves in one of the tool bodies, where the material of the layer flows freely

$$p\,|_{\partial s} = p_\Gamma(\mu). \tag{15.15}$$

The boundary value problem of flow in a thin plastic layer is considered in the formulation (15.12)–(15.15) with the condition at the boundary of the layer

$$p(\alpha,\ \beta, t)\,|_{\partial s_t} = k\sigma_s\ \ (k = 1, 2). \tag{15.16}$$

Moreover, it is assumed that elastic deformations of the bodies acting on the layer can be neglected $(h(\alpha, \beta, t) \equiv h_1)$.

Given the law of approach of external bodies, that is, given the known function $h(\alpha, \beta, t) \equiv h_1$ and the condition on the boundary of the layer in the form (15.16) in the problem is required for all $t > t_0$ determine the laws of distribution of contact pressure $p(\alpha,\beta,t)$, velocities $u(\alpha,\beta,t)$, $v(\alpha,\beta,t)$, the measure of accumulated deformation

$$\lambda(t) = \ln\left(\frac{h_0}{h(t)}\right),$$

total force in a given direction

$$P_{\text{gen}} = \iint\limits_S (p\vec{n}\,\vec{v})\,ds,$$

and also, the shape of the spreading area $s_t = s(\alpha, \beta, t)$, occupied by plastic material, according to a known initial area $s_0 = s(\alpha, \beta, t_0)$ on a geometrically unchanged base surface.

In general, friction factors μ_i $(0 \leq \mu_i \leq 1)$ represent complex functions that depend on the shape of the deformation zone, the magnitude of the contact pressure at the point of the contact surface under consideration, the presence of lubricant, its composition and intensity of application, the magnitude of the roughness of the contacting surfaces, etc. In the presence of dry friction (without lubrication) in the processes of flow of thin plastic layers, one can assume $\mu_1 = \mu_2 = 1$, which is experimentally confirmed in [22, 24–26]. Below, a method is proposed for solving problems of plastic layer flow in the formulation (15.12) and (15.13) with boundary conditions (15.14)–(15.15).

We write the equations of the boundary value problem (15.12)–(15.15) in the form [4], assuming that the slip lines α and β are taken as the axes of the curvilinear coordinate system:

$$\frac{\partial p}{A\partial\alpha} = -\frac{2\tau_s}{h} \cdot \frac{u}{\sqrt{u^2+v^2}}, \tag{15.17}$$

$$\frac{\partial p}{B\partial\beta} = -\frac{2\tau_s}{h} \cdot \frac{v}{\sqrt{u^2+v^2}}. \tag{15.18}$$

From vector equations (15.17) and (15.18) it follows that the level lines $p(\alpha,\beta) =$ const are orthogonal to the flow lines

$$\frac{Ad\alpha}{ds} = -\frac{u}{V}, \quad \frac{Bd\beta}{ds} = -\frac{v}{V};$$

i.e., the flow velocity vector is collinear to the vector grad $\mathbf{p}$. This condition allows us to move from system (15.17) and (15.18) to an equivalent system of two first-order partial differential equations with respect to $p(\alpha,\beta,t)$ [1] and the modulus of the velocity vector V:

$$\text{grad}^2 p = \frac{4\tau_s^2}{h^2(\alpha,\ \beta, t)}, \tag{15.19}$$

$$\frac{\partial h}{\partial t} - \frac{1}{AB}\frac{\partial(hBV\cos\gamma)}{\partial\alpha} - \frac{1}{AB}\frac{\partial(hAV\sin\gamma)}{\partial\beta} = 0, \tag{15.20}$$

where $\vec{V} = -V\left(\cos\gamma\vec{i} + \sin\gamma\vec{j}\right)$; $\gamma = \gamma(\alpha,\ \beta, t)$ – the angle between the tangent to the streamline at the point in the flow region under consideration and the α axis. In Eq. (15.19), for ease of notation, we set $\mu_1 = \mu_2 = 1$, although the method proposed below remains valid for any pair of valid values μ_1, μ_2.

From the first equation:

$$\begin{gathered} F(\alpha,\ \beta, P, p, q) \equiv \left(\frac{p}{A}\right)^2 + \left(\frac{q}{B}\right)^2 - \Omega^2(\alpha,\ \beta) = 0, \\ \text{where } \Omega(\alpha,\ \beta) = \frac{2\tau_s}{h} > 0,\ \ p = \frac{\partial P}{\partial\alpha},\ \ q = \frac{\partial P}{\partial\beta}, \end{gathered} \tag{15.21}$$

and t, as an independent parameter, is omitted for now. For a nonlinear first-order partial differential equation (15.21), each set of functions

$$\begin{gathered} alpha = \alpha(S'),\ \beta = \beta(S'), \\ P = P(S'),\ p = p(S'),\ q = q(S'), \end{gathered} \tag{15.22}$$

satisfying a system of ordinary differential equations

$$\begin{gathered} \frac{d\alpha}{dS'} = F_p,\ \frac{d\beta}{dS'} = F_q,\ \frac{dP}{dS'} = pF_p + qF_q, \\ \frac{dp}{dS'} = -\left(pF_\rho + F_\alpha\right),\ \frac{dq}{dS'} = -\left(qF_\rho + F_\beta\right) \end{gathered} \tag{15.23}$$

describes, together with Eq. (15.21), the characteristic band.

Instead of the parameter S' along the characteristic band, the parameter S is taken, which is the arc length along the support of the characteristics $\alpha = \alpha(S), \beta = \beta(S)$, i.e., according to (15.23)

$$dS = \sqrt{A^2 d\alpha^2 + B^2 d\beta^2} = \sqrt{A^2 {F_p}^2 + B^2 {F_q}^2} dS'. \tag{15.24}$$

In this case, the characteristic system (15.23) takes the form

$$\begin{cases} \dfrac{d\alpha}{ds} = \dfrac{1}{A^2}\dfrac{p}{\Omega},\ \dfrac{d\beta}{ds} = \dfrac{1}{B^2}\dfrac{q}{\Omega} \\ \dfrac{dp}{ds} = \dfrac{p^2}{\Omega A^3}A_\alpha + \dfrac{q^2}{\Omega B^3}B_\alpha + \Omega_\alpha \\ \dfrac{dq}{ds} = \dfrac{p^2}{\Omega A^3}A_\beta + \dfrac{q^2}{\Omega B^3}B_\beta + \Omega_\beta \\ \dfrac{dP}{ds} = \Omega \end{cases} \tag{15.25}$$

Problem (15.25) is a Cauchy-type problem with corresponding boundary conditions on the contour of the domain S_t. Solutions of system (15.25) satisfying (15.21) are called the characteristic strip, and the curve $(\alpha(s), \beta(s), P(s))$ carrying this strip is called the characteristic curve. We can establish a relationship between solutions (15.21) and (15.25) and demonstrate the equivalence of their integration under certain conditions, for which we will use well-known theorems from the theory of partial differential equations.

Theorem 15.1. *If the characteristic strip has a common element (i.e.,(α, β, P, p, q) with the integral surface $P = P(\alpha, \beta)$, then this strip belongs entirely to the integral surface.*

Theorem 15.2. *Let a spatial curve C be given:*

$$\alpha = \alpha(\mu),\ \beta = \beta(\mu),\ P = P(\mu),$$

which can be supplemented with functions $p = p(\mu), q = q(\mu)$, up to the initial strip

$$C_1 :\ \alpha = \alpha(\mu),\ \beta = \beta(\mu),\ P = P(\mu), p = p(\mu),\ q = q(\mu),$$

where C_1 satisfies the strip ratio

$$P'_\mu = p\alpha'_\mu + q\beta'_\mu \neq 0 \tag{15.26}$$

and the equation

$$F(\alpha,\ \beta, P, p, q) = 0, \tag{15.27}$$

or along the strip C_1 :

$$\Delta \equiv F_p\beta'_\mu - F_q\alpha'_\mu \neq 0, \tag{15.28}$$

then in the neighborhood C_1 there is one and only one integral surface passing through this strip.

As a spatial curve $C\ (\alpha(\mu), \beta(\mu),\ P(\mu))$ the regular part of the contour is selected ∂S_t flow areas at the given moment in time $\alpha(\mu), \beta(\mu) \in C^2$,

$$\left(\alpha'_\mu\right)^2 + \left(\beta'_\mu\right)^2 \neq 0$$

and put it on it $P(\mu) = k\sigma_s$.

We further define $p(\mu), q(\mu)$ in such a way that conditions (15.27) – (15.29)) are satisfied. Then, in the neighborhood of smooth points of the flow region contour, all the conditions of Theorem 15.2 are satisfied, i.e., there exists a unique integral surface $P = P(\alpha, \beta)$ passing through the selected part of the contour. On the other hand, the Cauchy problem for system (15.25), under Lipschitz-type conditions [1] for the right-hand sides α, β, P, p, q, has a unique characteristic strip, which in turn has a common element (at the contour points) with the integral surface. Consequently, by Theorem 15.1, solutions (15.25) and (15.21) coincide. It was established that the carriers of the characteristics $\alpha(s), \beta(s)$ of system (15.25) coincide with the current lines, and system (15.25) turns out to be equivalent to a system of four ordinary differential equations with respect to

$$\begin{aligned}
\frac{d\alpha}{ds} &= \frac{\cos\gamma}{A},\\
\frac{d\beta}{ds} &= \frac{\cos\gamma}{B},\\
\frac{dP}{ds} &= \Omega,\\
\frac{d\gamma}{ds} &= \frac{1}{\Omega}\left(-\frac{\Omega_\alpha}{A}\sin\gamma + \frac{\Omega_\alpha}{B}\cos\gamma\right) + \frac{1}{AB}\left(A_\beta\cos\gamma - B_\alpha\sin\gamma\right).
\end{aligned} \tag{15.29}$$

Considering that along the characteristics of the system (15.29)

$$\frac{dP}{dS} = \Omega > 0,$$

the latter system is presented in equivalent form:

$$\begin{aligned}
&\frac{d\alpha}{dP} = \frac{\cos\gamma}{A\Omega},\ \frac{d\beta}{dP} = \frac{\sin\gamma}{B\Omega},\ \frac{dS}{dP} = \frac{1}{\Omega},\\
&\frac{d\gamma}{dP} = \frac{1}{\Omega^2}\left(-\frac{\Omega_\alpha}{A}\sin\gamma + \frac{\Omega_\alpha}{B}\cos\gamma\right) + \frac{1}{AB\Omega}\left(A_\beta\cos\gamma - B_\alpha\sin\gamma\right).
\end{aligned} \tag{15.30}$$

Solutions of system (15.29) or (15.30) with boundary conditions at the points of the region contour define a Riemann surface $(\alpha,\ \beta,\ P)$ from which, according to the principle of uniqueness of pressure (more precisely, the principle of minimum power of external forces [2, 17, 20, 23], which, for a given law of approach of external bodies $h = h(\alpha,\ \beta,\ t)$, coincides with the minimum of external forces), a covering of the region is composed of parts of the characteristic curves enclosed between the boundary of the region contour and an edge (i.e., a line composed of the intersection points of at least two different characteristic curves), and corresponding to the minimum values of the contact pressure. Other parts of this multi-sheet surface (extensions of the characteristic curves) are not physically realized. For solving such problems, both exact and approximate methods are known in the literature (for example, the Runge-Kutta method [20]).

Equation (15.20) was integrated to analyze the flow kinematics. The equation is rewritten in equivalent form:

$$\frac{\partial W}{A\partial\alpha}\cos\gamma + \frac{\partial W}{B\partial\beta}\sin\gamma + W\Psi + \bar{\omega} = 0$$

or

$$\frac{\partial W}{\partial s} = -W\Psi - \bar{\omega},\ \text{where}\ W(s) = h\left(\alpha(s), \beta(s)\right)V(s),\ \bar{\omega}(s) = -\frac{\partial h}{\partial t}, \tag{15.31}$$

$$\Psi = \frac{1}{AB}\left[\frac{\partial(B\cos\gamma)}{\partial\alpha} + \frac{\partial(A\sin\gamma)}{\partial\beta}\right].$$

If we calculate in some way the values of the function Ψ along the supports of the characteristics $\alpha(S), \beta(S)$ (i.e., we know the dependence $\Psi = \Psi(S)$, then equation (15.31) is a non-homogeneous linear differential equation. Consequently, the following theorem is formulated.

Theorem 15.3.

$$\Psi = -\frac{1}{R(S)} + \frac{1}{R_\beta}\cos\gamma + \frac{1}{R_\alpha}\sin\gamma, \tag{15.32}$$

where

$$R_\alpha = \left(\frac{1}{AB}\frac{\partial A}{\partial\beta}\right)^{-1}, \quad R_\beta = \left(\frac{1}{AB}\frac{\partial B}{\partial\alpha}\right)^{-1}$$

is the tangent radii of curvature of lines α, β on the main surface,

$$\frac{1}{R(s)} = \chi(s) = \frac{\partial\gamma}{A\partial\alpha}\sin\gamma + \frac{\partial\gamma}{B\partial\beta}\cos\gamma = \frac{\partial\gamma}{\partial\eta}$$

is the the tangent (geodetic) curvature of the lines of the equations $P(\alpha, \beta) = const$ at the point in the flow region under consideration.

In some particular problems of the flow of thin plastic layers, the radius of curvature $R = R(s)$ of the level line is found exactly (flow of flat plastic layers of constant thickness $h = h(t)$), when the streamlines are straight lines, orthogonal to the contour of the region, for Ψ we obtain an expression that coincides with that obtained in [5]:

$$\Psi = -\frac{1}{R(s) - s + s_0},$$

where $R(s)$ is the radius of curvature of the initial level (the contour of the region). Now assuming that $\Psi = \Psi(S)$ is a known function, we integrate equation (15.31)

$$W(s) = h\left(\alpha(s), \beta(s)\right) V(s) = \bar{A}\mathfrak{J}_1(s) - \mathfrak{J}_2(s), \tag{15.33}$$

where are the integrals

$$\mathfrak{J}_1(S) = \mathrm{e}^{-\int\limits_{s_0}^{s}\Psi(s')ds'}, \quad \mathfrak{J}_2(S) = \int\limits_{s_0}^{s}\bar{\omega}(s'')\mathrm{e}^{-\int\limits_{s''}^{s}\Psi(s')ds'}ds'',$$

and to determine the constant of integration $\bar{A}$we have the flow branching condition $W\left(s = s_{rib}\right) = 0$ at unknown points of the pressure surface edge trace, but determined during the solution of system (15.30). The original problem (15.29) and (15.30) is solved as follows. System (15.25) is integrated with the initial conditions at the points of the region contour (it is known [3] that the corner points of the contour belong to the pressure edge trace), representing the initial level line, with the simultaneous calculation of two quadratures $\mathfrak{J}_1\left(s(P)\right)$, $\mathfrak{J}_2\left(s(P)\right)$. By setting the step ΔP, we will construct the next level line point by point and determine the values on it $\mathfrak{J}_1(s(P))$, $\mathfrak{J}_2(s(P))$. This process continued until the entire area was covered by an orthogonal grid of level lines and flow lines, thereby determining the trace of the pressure surface edge (as a line of junctions of individual characteristics) and the values of the quadratures $\mathfrak{J}_1(P)$, $\mathfrak{J}_2(P)$, i.e., having established the kinematics of the flow. Thus, at each fixed moment of time, the position of the contour is known ∂s_t flow region and the component of the flow velocity vector normal to it.

$$V(s_0) = \frac{1}{h(s_0)}\left[\bar{A}\mathfrak{J}_1(s_0) - \mathfrak{J}_2(s_0)\right], \text{ where } \bar{A} = \frac{\mathfrak{J}_2(s_{\text{ribs}})}{\mathfrak{J}_1(s_{\text{ribs}})}.$$

This means that it is possible to determine the shape of the contour of a freely spreading domain at the moment $t + dt$. If the boundary of the domain is a groove into which metal flows freely, then the metal flow rate Vh during flow into the rib and, therefore, the rib height can be determined.

A number of non-stationary problems of plastic flow of thin layers have been solved using the method of characteristics. We restrict ourselves to solving the problem of the flow of a plastic layer of material between inclined plates in a fixed circular domain, when its boundary is formed by grooves in one of the tool bodies into which plastic material flows freely (Fig. 15.1):

$$s_t \equiv s_0 : x_0^2 + y_0^2 \le R^2, \; h(x,t) = h_0(t) + \alpha_1(t)x.$$

This problem is partially solved (more precisely, only with respect to the force characteristics of the flow) in [19].

What is new in the proposed solution is the construction of the flow kinematics [14]. As is known, all streamlines converge at a single point $(x_\rho(t), 0)$ and represent an elliptical pencil of circles; the level lines form a hyperbolic pencil of circles. The edge degenerates into a single point with coordinate:

$$x_\rho(t) = \frac{\sqrt{h_0^2(t) - \alpha_1^2(t)R^2} - h_0}{\alpha_1(t)}.$$

Integrals $\mathfrak{J}_1(s(x))$, $\mathfrak{J}_2(s(x))$ to determine the velocities along known streamlines, they are taken exactly. For a streamline emanating from a point on the contour $M_0(x_0, y_0)$, at the same time $x_0 > 0$, $y_0 > 0$, they look like this:

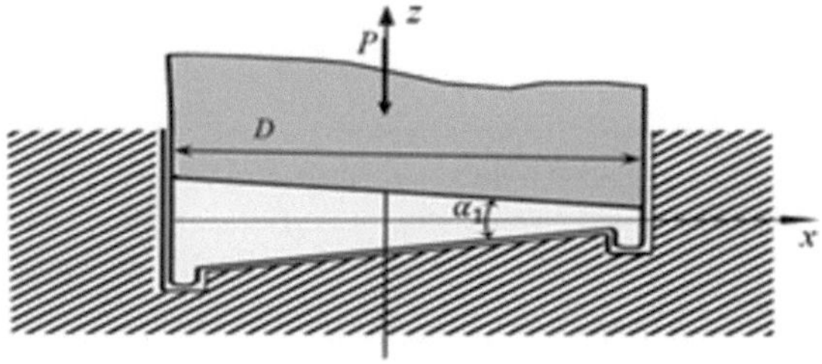

Fig. 15.1 Flow of a layer of variable thickness, with a circular boundary and flow into a contour rib.

$$\mathfrak{J}_1\left(s(x)\right) = \frac{y_0}{y(x)},\ \mathfrak{J}_2\left(s(x)\right) = \frac{1}{y(x)} \int\limits_{x_0}^{x} \left(\dot{h_0} + \dot{\alpha} x\right) y(x) \sqrt{1 + (y'(x))^2} dx,$$

Then the speed can be calculated

$$V(x) = \frac{1}{h(x)} \left(\bar{A}\mathfrak{J}_1(x) - \mathfrak{J}_2(x)\right),$$

where is the constant of integration

$$\bar{A} = \frac{1}{y_0} \int\limits_{x_0}^{x_p(t)} \left(\frac{dh_0(t)}{dt} + \frac{d\alpha_1(t)}{dt}\right) y(x) \sqrt{1 + (y'(x))^2} dx.$$

Flows of this kind in a layer are typical when considering the technology of manufacturing forging on forging and stamping presses with a hydraulic drive, where the speed of movement of the moving crosshead is low, and the metal, having reached a plastic state, is deformed between the approaching elements of the die, which have a given roughness. As was shown earlier, for a "correct" formulation of the problem in the two-dimensional case, it is sufficient to use a differential operator with respect to the contact pressure [27] from the condition

$$[V(s), \text{grad}\, p] = 0.$$

The formulation of the equivalent problem is reduced to [10] two partial differential equations of the first order with respect to the contact pressure $p(\alpha, \beta, t)$ and the modulus of the velocity vector V :

$$\text{grad}\, p = -\frac{2\tau_s}{h(\alpha,\ \beta, t)} \frac{\vec{V}}{\left|\vec{V}\right|}, \tag{15.34}$$

in this case, the components of speed are dependent on:

$$\frac{\partial h}{\partial t} - \frac{1}{AB} \frac{\partial (hBu)}{\partial \alpha} - \frac{1}{AB} \frac{\partial (hAv)}{\partial \beta} = 0. \tag{15.35}$$

Given the law of approach of external bodies (i.e., given the known function $h = h(\alpha, \beta, t)$ and the condition at the boundary of the layer

$$P\left(\alpha(\mu),\ \beta(\mu), t\right) = P_\Gamma(\mu) \tag{15.36}$$

required for everyone $t > t_0$ determine the laws of distribution of contact pressure $P(\alpha, \beta, t)$, velocities $u(\alpha, \beta, t)$, $v(\alpha, \beta, t)$and the shape of the region $s_t = s'(\alpha,\ \beta, t)$, occupied by plastic material, according to a known initial area$s_0 = s\left(\alpha,\ \beta, t_0\right)$ on the main surface.

The described method remains valid for volumetric deformation operations. A rationale for new approaches to solving specific problems is provided, one of which involves separately solving the problems of finding the contact pressure and analyzing the flow kinematics.

First, the contact pressures were found by solving the equation:

$$F(\alpha,\ \beta, P, p, q) \equiv \left(\frac{p}{A}\right)^2 + \left(\frac{q}{B}\right)^2 - \Omega^2(\alpha,\ \beta) = 0 \qquad (15.37)$$

taking into account the boundary condition in the general formulation (15.37):

$$p = \frac{dP}{d\alpha},\ q = \frac{dP}{d\beta},\ \Omega = \frac{2\tau_s}{h} > 0.$$

For singular points of the flow region boundary, the problem posed is reduced to solving the following equivalent system of ordinary differential equations

$$\begin{gathered}\frac{d\alpha}{dP} = \frac{\cos\gamma}{A\Omega},\ \frac{d\beta}{dP} = \frac{\sin\gamma}{B\Omega},\ \frac{dS}{dP} = \frac{1}{\Omega},\\ \frac{d\gamma}{dP} = \frac{1}{\Omega^2}\left(-\frac{\Omega_\alpha}{A}\sin\gamma + \frac{\Omega_\alpha}{B}\cos\gamma\right) + \frac{1}{AB\Omega}\left(A_\beta\cos\gamma - B_\alpha\sin\gamma\right),\end{gathered} \qquad (15.38)$$

at special points of the region's contour S_t let's write out the boundary conditions in more detail

$$\alpha = \alpha(\mu),\ \beta = \beta(\mu),\ P = P(\mu), p = p(\mu), q = q(\mu).$$

The uniqueness and uniqueness of $p(\mu), q(\mu)$ is determined by a set of conditions at regular points of the contour

$$\begin{gathered}\frac{dP_\gamma}{d\mu} = p\frac{d\alpha}{d\mu} + q\frac{d\beta}{d\mu},\\ F\left(\alpha(\mu),\ \beta(\mu), P(\mu), p, q\right) = 0,\\ \Delta = F_p\beta'_\mu - F_q\alpha'_\mu \neq 0.\end{gathered} \qquad (15.39)$$

From (15.40) follows the condition for determining $\gamma(\mu)$:

$$\frac{dP_\gamma}{d\mu} = -\Omega\cos\omega(\mu), \qquad \left(\frac{dP}{d\mu}\right)^2 < \Omega^2, \qquad (15.40)$$

where $\omega(\mu) = \measuredangle\left(\vec{V}^0,\ \vec{\tau}^0\right)$ – angle between the streamline $\vec{V}^0$ and the positive direction of the tangent $\vec{\tau}^0$ at the point of the contour under consideration

$$\gamma(\mu) = \pi + \left[\measuredangle\left(\alpha,\ \vec{\tau}^0\right) - \omega(\mu)\right]. \qquad (15.41)$$

For a fixed moment of time, the modulus of the flow velocity vector in the layer is determined

$$V(s) = \frac{W(s)}{h(s)}.$$

The resulting ratio is presented in the form

$$W(s) = V(s)h(s) = \bar{A}\mathfrak{J}_1(x) - \mathfrak{J}_2(x),$$

where are the integrals $\mathfrak{J}_1(x), \mathfrak{J}_2(x)$ and an arbitrary constant $\bar{A}$, are determined as proposed in the previous case. To confirm the adequacy of the proposed method, test examples are considered.

Example 15.1. It is assumed that in the plastic domain defined by the following relations

$$S = \left\{(x, y) \ : \ |x| < \infty, \ |y| < a + be^{-x^2}\right\},$$

the law of change in layer thickness is known $h = h(t)$ from time and pressure $P_\Gamma(x) = c - \Omega\, be^{-x^2}$ on the contact line. The flow lines are straight lines. The angle $\gamma(x)$ between the lines is found from relation (15.41)

$$\frac{dP_\Gamma}{dx} = p + qy'_x,$$
$$p = \Omega \cos\gamma(x), \quad q = \Omega \sin\gamma,$$

then:

$$2bxe^{-x^2} = \cos\gamma(x) - 2bxe^{-x^2} \sin\gamma(x).$$

At the same time for all $b \geq 0$, $\gamma(x) = \frac{\pi}{2}$, s are straight line segments that are parallel to the Ox axis, as shown in Fig. 15.2 [7] and, therefore, $\Psi(s) = 0$.

When integrating the system of ordinary differential equations (38) along the characteristics, the function was found $P(s, x) = P_\Gamma(x) + \Omega s$. Since the plastic region is symmetrical, the Ox axis coincides with the edge of the pressure surface and, taking into account the condition on the edge, from (2.32) we determine the kinematic parameters of the flow

$$-hV(s, x) = \frac{dh}{dt}\,(y(x) - s)\,.$$

Corollary 15.1. *Let the boundary of the plastic domain be known and the contours of the sides constructed on it. The height of the sides is determined from the relation*

$$z(x) = \frac{fP_\Gamma(x)}{\Omega}.$$

When filling a mold with quartz sand having a friction coefficient f, a fill is formed whose slope angle is constant, and a surface is formed with a vertex at the point $x = x_0$:

$$\frac{dP}{dy} = -\Omega < 0.$$

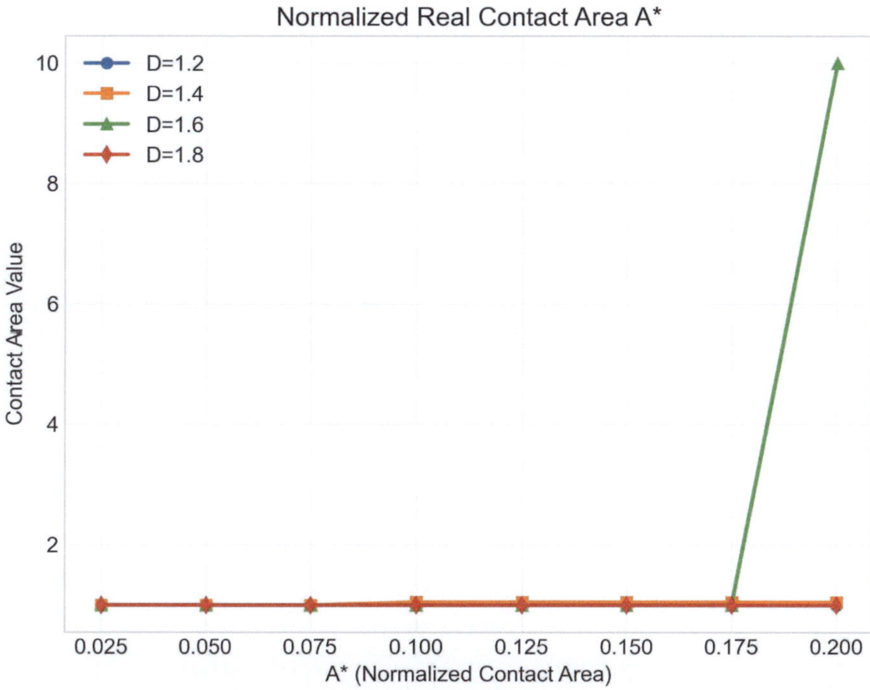

Fig. 15.2: Branching line of the current, level lines and current lines.

Example 15.2. Let a metal layer of thickness h = const be bounded by a circle:

$$s = \left\{(x, y) : x^2 + y^2 < R^2\right\},$$

Let, further, the pressure be set at the contact of the boundary of the circle with the tool

$$P_\Gamma(\varphi) = C - R\Omega\sqrt{\left(\frac{r}{R}\right)^2 + 1 - 2\frac{r}{R}\cos\varphi}. \tag{15.42}$$

In this case, a cylindrical coordinate system is used $x = R\cos\varphi$, $y = R\sin\varphi$*, * $z = z$, at the same time r $(0 \le r \le R)$ variable radius. Condition for determination $\omega = \omega(\varphi)$ obtained by substituting the parametric equation of the contour (15.42) into (15.40):

$$-\Omega\cos\omega = \frac{dP_\Gamma}{Rd\varphi} = -\frac{\frac{r}{R}\sin\varphi}{\sqrt{\left(\frac{r}{R}\right)^2 + 1 - 2\frac{r}{R}\cos\varphi}}\Omega, \; \gamma(\varphi) = \frac{3\pi}{2} + \varphi - \omega(\varphi).$$

From system (15.38), the streamlines are determined to be straight. The branching line of the currents is the only point with coordinates $(r,\ 0)$ at which the streamlines intersect (Fig. 15.3) [7].

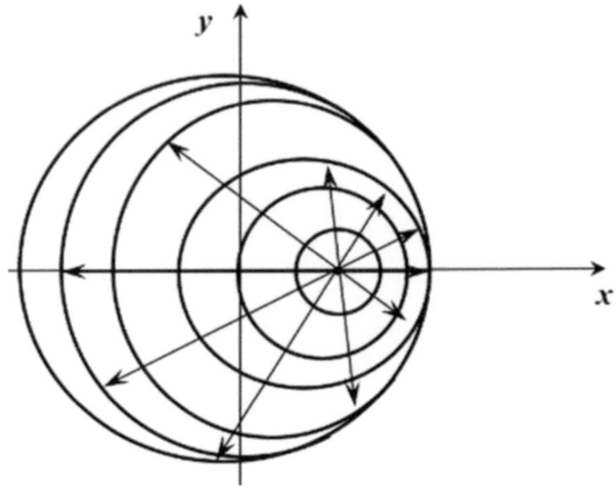

Fig. 15.3 Degeneration into a point of the edge of the pressure surface.

The equation of characteristics is analyzed in a more general form:

$$\begin{aligned} y(x,\varphi) &= y_\Gamma(\varphi) - \frac{x - x_\Gamma(\varphi)}{\sin(\varphi - \omega(\varphi))}\cos(\varphi - \omega(\varphi)) \\ &= R\sin\varphi - \frac{x - R\cos\varphi}{\sin(\varphi - \omega(\varphi))}\cos(\varphi - \omega(\varphi)), \end{aligned} \tag{15.43}$$

where, based on standard transformations of trigonometric functions, the following is presented:

$$\sin\gamma(\varphi) = -\cos(\varphi - \omega), \quad \cos\gamma(\varphi) = \pm\sin(\varphi - \omega).$$

For coordinate values $x = r, y = 0$, Eq. (15.43) becomes an identity, i.e., for the principal value φ ($\varphi \in [0,\ 2\pi]$), point with coordinates $(r,\ 0)$ will belong to the line (15.43). The level lines in this case are concentric circles centered at the edge point. The solution of the characteristic system (15.38) is represented as:

$$\begin{aligned} &\gamma(s) = \gamma_\Gamma(\varphi) = \text{const}, \\ &x(s) = x_\Gamma(\varphi) + s\cos\gamma(\varphi) = R\cos\varphi + s\cdot\cos\gamma(s), \\ &y(s) = y_\Gamma(\varphi) + s\sin\gamma(\varphi) = R\sin\varphi + s\cdot\sin\gamma(s), \\ &P(s) = P_\Gamma(s) + \Omega s. \end{aligned}$$

The function $\Psi(s)$ is equal to:

$$\Psi(s) = -\frac{1}{R(s)} = -\frac{1}{\sqrt{(x(s) - r)^2 + y^2(s)}}.$$

The condition of incompressibility along the found streamlines is transformed to the form:

$$\frac{dV}{ds} = -V\Psi(s) + \frac{1}{h}\frac{dh}{dt}$$

and leads to an exact solution

$$V(s) = \frac{1}{h}\left[\mathfrak{J}_1(s)A_0 - \mathfrak{J}_2(s)\right],$$

in which the corresponding integrals are determined by the relations:

$$\mathfrak{J}_1(s) = \mathrm{e}^{-\int\limits_0^s \Psi(s')ds'}, \quad \mathfrak{J}_2(s) = \int\limits_0^s \left(-\frac{dh}{dt}\right)\mathrm{e}^{-\int\limits_{s''}^s \Psi(s')ds'} ds''.$$

Edge condition

$$V\left(s = s_{rib}\right) = 0,$$

allow us to find the constant of integration A_0, where

$$s_{\text{ribs}}(\varphi) = \frac{x_p - R\cos\varphi}{\cos\gamma(\varphi)} = \frac{u - R\cos\varphi}{\cos\gamma(\varphi)}.$$

Corollary 15.2. *When constructing curbs on the boundary of a circular region with a height according to equation* (15.42) *and filling it with homogeneous sand with a friction coefficient f, the generatrix of the limiting surface will take the shape of a cone with its apex at the edge.*

The inverse problem of determining the height of the curbs of a circular region is considered, for the formation of a sand embankment in the form of a cone with its apex at an arbitrary point $(r,\ 0)$. Integration of condition (15.40) to determine $P_\Gamma(\varphi)$

$$\frac{dP_\Gamma}{Rd\varphi} = -\Omega\cos\omega(\varphi),$$

leads to a solution whose form is similar to (15.42), where

$$\cos\omega(\varphi) = \frac{\frac{r}{R}\sin\varphi}{\sqrt{\left(\frac{r}{R}\right)^2 + 1 - 2\frac{r}{R}\cos\varphi}}.$$

This formulation and solution of inverse problems for determining pressures at the boundaries of plastic regions excludes from consideration regions with contour inflection points. The existence of such points leads to the appearance of an edge on the pressure surface that does not degenerate into a point.

Example 15.3. Flow within a plastic layer of constant thickness, bounded by an ellipse.

$$S = \left\{(x, y) : \frac{x^2}{a^2} + \frac{y^2}{b^2} < 1\right\}.$$

On the border

$$\partial S = \{(x, y) : \ x = a \cos t, \ y = b \sin t\}$$

pressure:

$$P_\Gamma(\varphi) = -b\Omega\sqrt{\frac{a^2}{b^2} - \frac{\frac{a^2}{b^2} - 1}{1 + \frac{a^2}{b^2}\operatorname{tg}^2\varphi}} + c,$$

where

$$\operatorname{tg}\varphi = \frac{b}{a}\operatorname{tg} t.$$

From condition (15.41) to determine $\omega(\varphi)$: $\frac{dP_\Gamma}{d\mu} = -\Omega\cos\omega(\varphi)$ given that

$$d\mu = \sqrt{dx^2 + dy^2} = \sqrt{a^2 \sin^2 t + b^2 \cos^2 t}\,dt,$$

We get:

$$\cos\omega(\varphi) = \left(\frac{a^2}{b^2} - 1\right) b \sin t \cdot \left(1 + \frac{a^2}{b^2}\operatorname{tg}^2\varphi\right)^{-\frac{1}{2}} \cdot \left(a^2 \sin^2 t + b^2 \cos^2 t\right)^{-\frac{1}{2}}$$

the streamlines are a bundle of straight lines passing through the point $O(0, 0)$, the orthogonal trajectories for the bundle of straight lines are concentric circles with the center at the point $O(0, 0)$, (Fig. 15.4). Systems (15.25) and (15.30) lead to the exact solution of the problem.

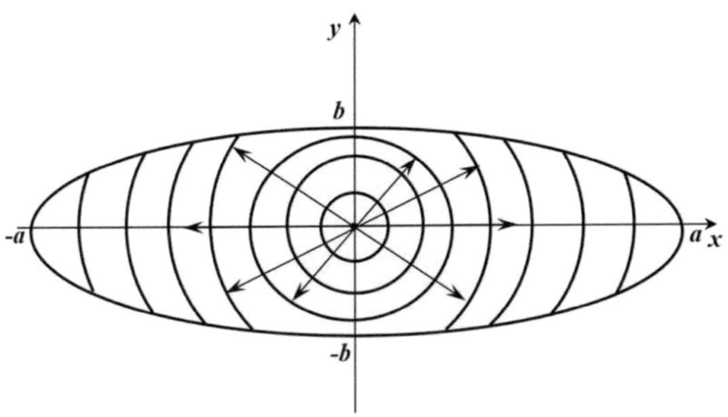

Fig. 15.4: For elliptical boundaries of the plastic region, the flow lines and the corresponding level lines are orthogonal.

15.3 A Mathematical Model of a "Viscous Fluid" in the Form of a Cauchy Problem with Boundary Conditions

Not all flow processes in a thin plastic layer can be described within the framework of the simplified "ideal fluid" model, which is confirmed by experimental studies [11, 27]. Therefore, for the correct description of problems that take into account the presence of shear stresses, anisotropy in flow processes [16], elastic deformation of the tool [8, 18] and other factors, it is necessary to use a more general "viscous fluid" model. Within the framework of the mathematical model of "viscous fluid", the boundary value problem of the flow of a plastic layer in a region with a moving boundary is formulated, described by nonlinear second-order partial differential equations with respect to three unknown functions - contact pressure and two components of the flow velocity. The basic equations of the boundary value problem of the flow of a plastic layer on a plane are of the form [4]:

$$\frac{\partial p}{\partial x} = \frac{\sigma_s}{3}\frac{dt}{d\lambda}\Delta u - \frac{2\tau_s}{h}\frac{u}{\sqrt{u^2+v^2}}, \tag{15.44}$$

$$\frac{\partial p}{\partial y} = \frac{\sigma_s}{3}\frac{dt}{d\lambda}\Delta v - \frac{2\tau_s}{h}\frac{v}{\sqrt{u^2+v^2}}, \tag{15.45}$$

$$\frac{\partial u}{\partial x} + \frac{\partial v}{\partial y} - \frac{d\lambda}{dt} = 0, \tag{15.46}$$

where p, u, v are the contact pressure and flow velocity components; $\sigma_s = \sqrt{3}\tau_s$ is the flow stress of plastic material in the area under consideration; $h = h(t)$ is a given law of change in layer thickness; value

$$\lambda(t) = \ln\frac{h_0}{h(t)}$$

is the degree of deformation; t_0 is the initial moment of time. On a previously unknown free border $\Gamma_\sigma : F(x, y, t) = 0$ the following boundary conditions are specified: scalar dynamic (15.47)

$$\vec{\sigma}^{(n)} = \tilde{\sigma}\cdot\vec{n} = \left(\sigma_{xx}n_x + \sigma_{xy}n_y;\ \sigma_{yy}n_y + \sigma_{xy}n_x\right) = \vec{0} \tag{15.47}$$

and one kinematic (15.48)

$$v_n \equiv \vec{v}\cdot\vec{n} = -\frac{\partial F}{\partial t}\cdot\frac{1}{|\operatorname{grad} F|}, \tag{15.48}$$

from which the border itself is determined Γ_σ. On the contact surface with the tool $\Gamma_c : G(x, y, t) = 0$, as a rule, known, the following two conditions are given: one dynamic with respect to the shear stress (15.49):

$$\left|\vec{\sigma}^{(n)}\cdot\vec{\tau}\right| = \left|\sigma_{xx}n_x n_y + \sigma_{xy}\left(n_y^2 - n_x^2\right) - \sigma_{yy}n_x n_y\right| = \mu\tau_s,\ (0 \le \mu \le 1) \tag{15.49}$$

and the kinematic condition in the form of the impermeability condition (15.50):

$$v_n \equiv \vec{v} \cdot \vec{n} = -\frac{\partial G}{\partial t} \frac{1}{|\operatorname{grad} G|}, \tag{15.50}$$

where $\vec{\tau}$ and $\vec{n}$ – unit vectors of the tangent and normal to the boundary.

In particular, at a fixed boundary the normal velocity is zero. Let

$$\sigma_s,\ p_0,\ L_0,\ h_0,\ \frac{dh(t_{0)}}{dt}$$

– characteristic values of the yield strength of the material, contact pressure, linear size and thickness of the layer, as well as the speed of approach of the deforming tools. Relative thickness of the plastic layer $\varepsilon = \frac{h_0}{L_0} \ll 1$ we take as a small parameter. When estimating the order of the terms in the incompressibility condition (15.46), we obtain that

$$t_0 = \frac{L_0}{v_0};\ v_0 = -\frac{dh\,(t_0)}{dt}\frac{1}{\varepsilon},$$

and from the quasi-static equilibrium equations (15.44) and (15.45), that

$$p_0 = \frac{\sigma_s}{\varepsilon}.$$

Therefore, we come to the obvious conclusion that the characteristic value of the flow velocity along the layer is much greater than the speed of approach of the tools and that the characteristic value of the contact pressure is much greater than the yield strength of the layer material. Assuming that the layer thickness h, unlike other linear dimensions (coordinates), when divided by $h(t_0)$ becomes dimensionless, the system of equations is rewritten in dimensionless form [22]:

$$\frac{\partial \bar{p}}{\partial \bar{x}} = \varepsilon \frac{d\bar{t}}{d\lambda} \bar{\Delta} u - \frac{2}{\sqrt{3}\bar{h}} \frac{\bar{u}}{\sqrt{\bar{u}^2 + \bar{v}^2}}, \tag{15.51}$$

$$\frac{\partial \bar{p}}{\partial \bar{y}} = \varepsilon \frac{d\bar{t}}{d\lambda} \bar{\Delta} v - \frac{2}{\sqrt{3}\bar{h}} \frac{\bar{v}}{\sqrt{\bar{u}^2 + \bar{v}^2}}, \tag{15.52}$$

$$\frac{\partial \bar{u}}{\partial \bar{x}} + \frac{\partial \bar{v}}{\partial \bar{y}} - \frac{d\lambda}{d\bar{t}} = 0, \tag{15.53}$$

where dimensionless quantities are marked with diacritics (note that in the equilibrium equations, a small parameter ε appears in the terms with the highest derivatives-Laplacians ε).

Dimensionless characteristics at the free boundary $\Gamma_\sigma : F(x, y, t) = 0$

$$\left[-\bar{p} + \left(\frac{2}{3} + \frac{2}{3}\frac{d\bar{t}}{d\lambda}\frac{\partial \bar{u}}{\partial \bar{x}}\right)\varepsilon\right] n_x + \left[\frac{1}{3}\frac{d\bar{t}}{d\lambda}\left(\frac{\partial \bar{u}}{\partial \bar{y}} + \frac{\partial \bar{v}}{\partial \bar{x}}\right)\varepsilon\right] n_y = 0, \tag{15.54}$$

$$\left[\frac{1}{3}\frac{d\bar{t}}{d\lambda}\left(\frac{\partial \bar{u}}{\partial \bar{y}} + \frac{\partial \bar{v}}{\partial \bar{x}}\right)\varepsilon\right] n_x + \left[-\bar{p} + \left(\frac{2}{3} + \frac{2}{3}\frac{d\bar{t}}{d\lambda}\frac{\partial \bar{u}}{\partial \bar{y}}\right)\varepsilon\right] n_y = 0, \tag{15.55}$$

$$\bar{v}_n = -\frac{\partial \overline{F}}{\partial \bar{t}} \frac{1}{\left|\text{grad}\,\overline{F}\right|}, \tag{15.56}$$

at the contact boundary with the tool Γ_c

$$\begin{gathered}\left[-\bar{p} + \left(\frac{2}{3} + \frac{2}{3}\frac{d\bar{t}}{d\lambda}\frac{\partial \bar{u}}{\partial \bar{x}}\right)\varepsilon\right] n_x n_y + \left[\frac{1}{3}\frac{d\bar{t}}{d\lambda}\left(\frac{\partial \bar{u}}{\partial \bar{y}} + \frac{\partial \bar{v}}{\partial \bar{x}}\right)\varepsilon\right]\left(n_y^2 - n_x^2\right) \\ - \left[-\bar{p} + \left(\frac{2}{3} + \frac{2}{3}\frac{d\bar{t}}{d\lambda}\frac{\partial \bar{u}}{\partial \bar{y}}\right)\varepsilon\right] n_x n_y = \pm\frac{\mu}{\sqrt{3}}\varepsilon,\end{gathered} \tag{15.57}$$

$$\bar{v}_n = -\frac{\partial \bar{G}}{\partial t}\left|\text{grad}\,\bar{G}\right|. \tag{15.58}$$

The resulting system of partial differential equations (15.51) – (15.53) together with the boundary conditions (15.54) – (15.58) constitute a gen boundary value problem of layer spreading in the "viscous fluid" model.

In the main approximation ($\varepsilon = 0$) the original equations are simplified with decreasing order of the system (15.58):

$$\frac{\partial \bar{p}}{\partial \bar{x}} = -\frac{2}{\sqrt{3}\bar{h}}\frac{\bar{u}}{\sqrt{\bar{u}^2 + \bar{v}^2}}; \quad \frac{\partial \bar{p}}{\partial \bar{y}} = -\frac{2}{\sqrt{3}\bar{h}}\frac{\bar{v}}{\sqrt{\bar{u}^2 + \bar{v}^2}}; \quad \frac{\partial \bar{u}}{\partial \bar{x}} + \frac{\partial \bar{v}}{\partial \bar{y}} - \frac{d\lambda}{d\bar{t}} = 0 \tag{15.59}$$

It is clear that the number of boundary conditions for a simplified problem within the framework of an "ideal fluid" is reduced.

15.4 Boundary Value Problem of Plastic Layer Flow in a new Formulation with Constraints in one or Several Directions

Let the current function be $\psi = \psi(x, y)$, the incompressibility condition (15.46) will take the form:

$$u = \frac{d\lambda}{dt}\left(\psi'_y + \frac{x}{2}\right); \quad v = -\frac{d\lambda}{dt}\left(\psi'_x - \frac{y}{2}\right). \tag{15.60}$$

We substitute the velocity components (15.60) into the quasi-static equilibrium equations (15.51) – (15.53) and as a result we obtain a system of two differential equations with respect to p and ψ:

$$\frac{p'_x - \left(\frac{\sigma_s}{3}\right)\Delta\psi'_y}{p'_y + \left(\frac{\sigma_s}{3}\right)\Delta\psi'_x} = -\frac{2\tau_s}{h}\frac{\psi'_y + \frac{x}{2}}{\psi'_x - \frac{y}{2}}, \tag{15.61}$$

$$\left(p'_x - \left(\frac{\sigma_s}{3}\right)\Delta\psi'_y\right)^2 + \left(p'_y - \left(\frac{\sigma_s}{3}\right)\Delta\psi'_x\right)^2 = \frac{4{\sigma_s}^2}{3h^2}. \tag{15.62}$$

As an example, the problem of flow of a plastic layer, which at the initial moment of deformation has the shape of a rectangle in plan (Fig. 15.5a), is considered in natural dimensions. In the Oxy-plane, this is a "two-dimensional flow region with movable (dashed lines) and impermeable ($y = \pm b_0$) boundaries" [6] (Fig. 15.5b).

Using the symmetry of the plastic region, we restrict our consideration to the layer element located in the first quadrant of the chosen coordinate system. At the stationary wall of the tool, ($G(x, y) \equiv y - b_0 = 0$):

$$v = 0 \Longrightarrow \psi'_x - \frac{b_0}{2} = 0, \tag{15.63}$$

$$\frac{\partial u}{\partial x} + \frac{\partial v}{\partial y} = -\sqrt{3}\lambda'\mu \Longrightarrow \psi''_{yy} - \psi''_{xx} = -\sqrt{3}\lambda'\mu. \tag{15.64}$$

Boundary conditions:

$$y = 0 : v = 0 \Longrightarrow \psi'_x = 0; \quad \frac{\partial u}{\partial y} = 0 \Longrightarrow \psi''_{yy} = 0, \tag{15.65}$$

$$x = 0 : u = 0 \Longrightarrow \psi'_y = 0; \quad \frac{\partial v}{\partial x} = 0 \Longrightarrow \psi''_{xx} = 0. \tag{15.66}$$

On an unknown free border $F(x, y, t) = 0$, at the initial moment having parameters $x = l_0$ $\left(n_x = 1, n_y = 0\right)$, boundary conditions are determined:

$$\sigma_{xx} = 0 \Longrightarrow -p + \sigma_s + \frac{2}{3}\sigma_s \psi''_{xy} = 0, \ \sigma_{xy} = 0 \Longrightarrow \psi''_{yy} - \psi''_{xx} = 0. \tag{15.67}$$

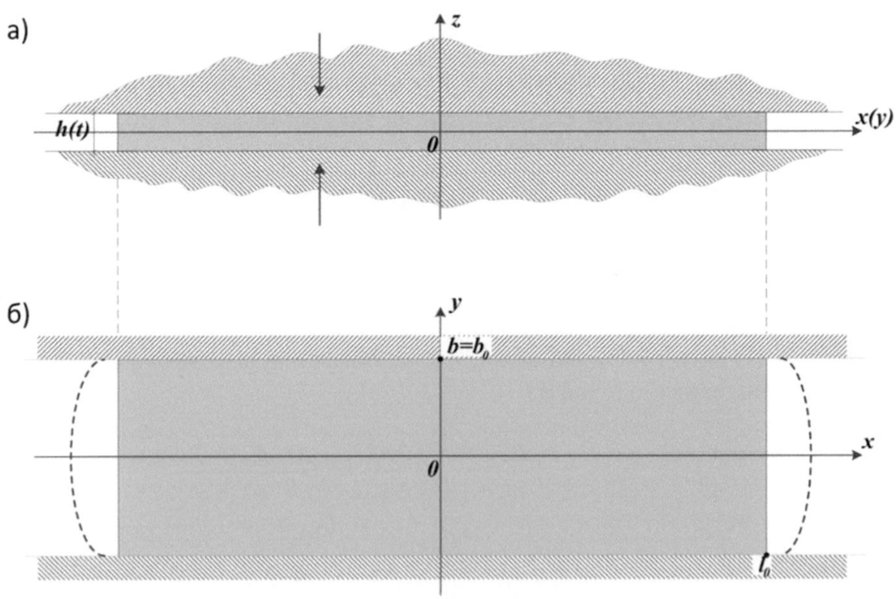

Fig. 15.5: Flow diagram of the deposited plastic layer with limitations on the lateral surfaces: a) - in the cross section (plane Oxz); b) – in section (plane Oxy).

To eliminate the influence of boundary effects, the analytical solution of the problem is located away from the $x = 0$ axis and the free boundary$x = 0,5l_0$*,* since the extent of the flow region in the x direction allows this to be done. For the section of the flow region under consideration $\upsilon\ (x,\ y)\ = 0$. After integrating the incompressibility condition (15.47):

$$u(x, y) = \lambda' \left[x + g(y)\right],$$

where the function $g(y)$ can be determined from other equations of the system: (15.56) gives $p = p(x)$ and then from (15.45):

$$p'(x) = \frac{\sigma_S}{3} - \frac{2\sigma_s}{\sqrt{3}\bar{h}}, \tag{15.68}$$

where the left-hand side is a function of the argument x, and the right-hand side is a function of y and the parameter $h(t)$. In other words, we have a differential equation with separable variables. As a result, we obtain a system with an arbitrary constant (a function of time) k:

$$g''(y) = k = \text{const};\ \ p'(x) = \frac{\sigma_S k}{3} - \frac{2\sigma_s}{\sqrt{3}\bar{h}}. \tag{15.69}$$

From the solution of the first equation (15.61) and substitution of boundary conditions (15.65), (15.66) we obtain

$$g(y) = c - \frac{\sqrt{3}\mu}{2b_0}y^2,\ k = -\frac{\sqrt{3}\mu}{b_0}, \tag{15.70}$$

$$u(x, y) = \lambda' \left[x - \frac{\sqrt{3}\mu}{2b_0}y^2 + c\right], \tag{15.71}$$

where c is an arbitrary constant of integration. The first term on the right-hand side of equation (15.71) corresponds to the flow velocity in the “ideal fluid” model, in which the shear stress at the fixed boundary, as well as at other boundaries, is absent. We assume that the cross-section$x = x_f$ represents a conditional free boundary ($x_f h = x_0 h_0$). Then from the integral condition of conservation of volume of incompressible material

$$-x_f(t)\frac{dh}{dt}b_0 = \int\limits_0^{b_0} hu\left(x_f, y\right)dy$$

an arbitrary constant included in the solution is defined

$$c = \frac{\sqrt{3}}{6}\mu b_0,$$

which makes it possible to obtain the final formula for the flow velocity:

$$u(x,y) = \lambda' \left[x + \frac{\sqrt{3}}{6b_0}(\mu b_0{}^2 - 3y^2) \right]. \tag{15.72}$$

By comparing the found velocity (15.72) and the boundary condition (15.66) at the value $x = 0$, we are convinced that it is fulfilled in integral form:

$$\int_0^{b_0} u(0,y)dy = \frac{\lambda' \sqrt{3}}{6b_0} \int_0^{b_0} (\mu b_0{}^2 - 3y^2)dy = 0.$$

From the equation of system (15.69) after integration we obtain (15.73):

$$p(x) = p_0 - \frac{\sigma_s x}{\sqrt{3}} \left(\frac{2}{h} + \frac{\mu}{b_0} \right), \tag{15.73}$$

with a constant of integration p_0, independent of coordinate y. Assuming that at the free boundary the normal stress is zero:

$$\sigma_{xx} = -p + \frac{4\sigma_s}{3} = 0,$$
$$p_0(t) = \frac{2\sigma_s x_f}{\sqrt{3}h} + \left(\frac{x_f}{\sqrt{3}b_0} + \frac{4}{3} \right) \sigma_s.$$

If large enough x_f, the third term on the right-hand side can be neglected, then

$$p_0(t) = \frac{\sigma_s x_f}{\sqrt{3}} \left(\frac{2}{h} + \frac{\mu}{b_0} \right), \quad p \approx \frac{\sigma_s}{\sqrt{3}} \left(\frac{2}{h} + \frac{\mu}{b_0} \right) (x_f - x). \tag{15.74}$$

We find the current function from the differential equation:

$$\psi'_x = \frac{y}{2};\ \psi'_y = \frac{x}{2} + \frac{\sqrt{3}}{6b_0} \left(\mu b_0{}^2 - 3y^2 \right) \Longrightarrow \psi(x,y) = \frac{\sqrt{3}}{6b_0} \left(\mu b_0{}^2 y - 3y^3 \right) + c_1.$$

Thus, within the framework of the general "viscous fluid" model, an approximate analytical solution is obtained. At the boundary $x = 0$, as well as at the free boundary $x = x_f(t)$ It satisfies the boundary conditions in integral form. Consequently, according to Saint-Venant's principle, the solution is valid in the central part of the flow domain. The nature of the flow in the layer is not one-dimensional, so the straight lines (x = const) become curved over time, and a noticeable lag in the longitudinal movement of particles near the stationary boundary is observed. This fact is confirmed by the results of experiments [28]. The obtained regularities cannot be described using the "ideal fluid" model, which does not take into account shear stresses.

15.5 Setting and Solving Practical Problems

A thin strip occupying the area

$$S(x, y) = \{(x, y)|l_1(t) < x < l_2(t); 0 \le y \le h(t)\},$$

freely spreads under the action of a deforming tool, and the differential equations of quasi-static equilibrium have the form [4]:

$$\frac{\partial p}{\partial x} = -\frac{2\tau_s}{h}\frac{u}{|u|}; \ \frac{\partial u}{\partial x} - \frac{d\lambda}{d\bar{t}} = 0, \tag{15.75}$$

where $h = h(t)$ – the well-known law of change in the thickness of the upset plastic strip,

$$\lambda(t) = \ln\frac{h_0}{h(t)} \ge 0$$

is the degree of deformation according to Ilyushin (analogous to the measure of deformation accumulated in a particle); $u = u\ (x,\ t)$; $p\ = p\ (x,\ t)$ – respectively, the flow velocity along the layer and the contact pressure in the strip. Boundary conditions at free boundaries:

$$x = l_1;\ p = \sigma_s\ ;\ x = l_2;\ \ p = \sigma_s\ . \tag{15.76}$$

There is an unknown branching line of the current $x = x_0 \epsilon (l_1, l_2)$ with $u\,(x_0, t) = 0$. After integrating equation (15.75) and substituting the boundary conditions (15.76), the value of the contact pressure in the strip is obtained (Fig. 15.6) [13]:

$$p(x,t) = \begin{cases} \sigma_s + \dfrac{2\tau_s}{h}(l_2 - x),\ x\epsilon\,(x_0, l_2) \\ \sigma_s + \dfrac{2\tau_s}{h}(x - l_1),\, x\epsilon\,(l_1,\ x_0) \end{cases}. \tag{15.77}$$

When integrating the incompressibility condition from (15.75) taking into account the condition on the branch line, it is found:

$$u(x,t) = -\frac{1}{h}\frac{dh}{dt}(x - x_0) = -\frac{1}{h}\frac{dh}{dt}\left(x - \frac{l_1 + l_2}{2}\right), \tag{15.78}$$

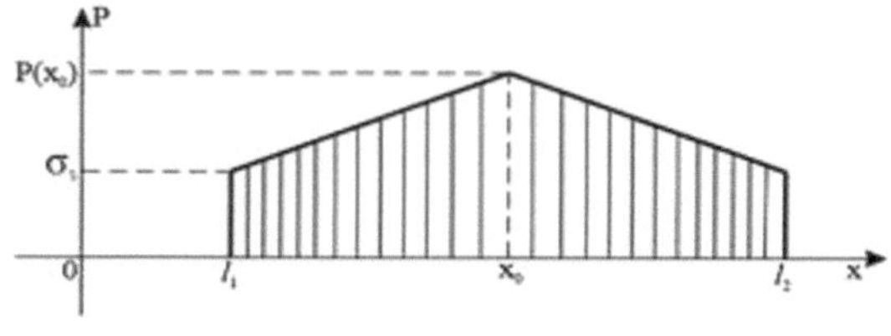

Fig. 15.6 Distribution diagram of contact pressure in the strip

in this case the branching line of the current $x = x_0(t)$ is determined from the condition of continuity of contact pressure. The specific force for achieving plastic upsetting of the strip is equal to the area of the diagram (see Fig. 15.5):

$$P(t) = \int_{l_1}^{l_2} p(x,t)dx = \int_{l_1}^{x_0} \left[\sigma_s + \frac{2\tau_s}{h}(x - l_1)\right] dx + \int_{x_0}^{l_2} \left[\sigma_s + \frac{2\tau_s}{h}(l_2 - x)\right] dx \tag{15.79}$$

$$= \left[\sigma_s x + \frac{2\tau_s}{h}\left(\frac{x^2}{2} - l_1 x\right)\right]\Bigg|_{l_1}^{x_0} + \left[\sigma_s x + \frac{2\tau_s}{h}\left(l_2 x - \frac{x^2}{2}\right)\right]\Bigg|_{x_0}^{l_2}$$

$$= \sigma_s (l_2 - l_1) + \frac{2\tau_s}{h}\frac{(l_2 - l_1)^2}{4}.$$

Various generalizations of the problem of free strip spreading are considered separately.

A) The strip material flows freely in both directions so that the contact area of the tool with the plastic strip, expanding in both directions, forms a segment $[l_1(t), l_2(t)]$ with movable ends (Fig. 15.7) [13].
In this case, the positions of the freely moving ends of the strip are found from the solution of the following Cauchy problem, described by a system of linear homogeneous differential equations:

$$\begin{aligned} \frac{dl_2}{dt} &= u(l_2, t) = -\frac{1}{h}\frac{dh}{dt}\left(\frac{l_2 - l_1}{2}\right), \\ \frac{dl_1}{dt} &= u(l_1, t) = -\frac{1}{h}\frac{dh}{dt}\left(\frac{l_1 - l_2}{2}\right) = \frac{1}{h}\frac{dh}{dt}\left(\frac{l_2 - l_1}{2}\right), \\ t &= t_0 : l_1(t) = l_{10}\,, l_2(t) = l_{20}\,. \end{aligned} \tag{15.80}$$

Then the solution to problem (15.80) has the form:

$$\begin{aligned} l_1(t) &= \frac{1}{2}\left[(l_{20} + l_{10}) - \frac{h_0}{h(t)}(l_{20} - l_{10})\right], \\ l_2(t) &= \frac{1}{2}\left[(l_{20} + l_{10}) + \frac{h_0}{h(t)}(l_{20} - l_{10})\right]. \end{aligned} \tag{15.81}$$

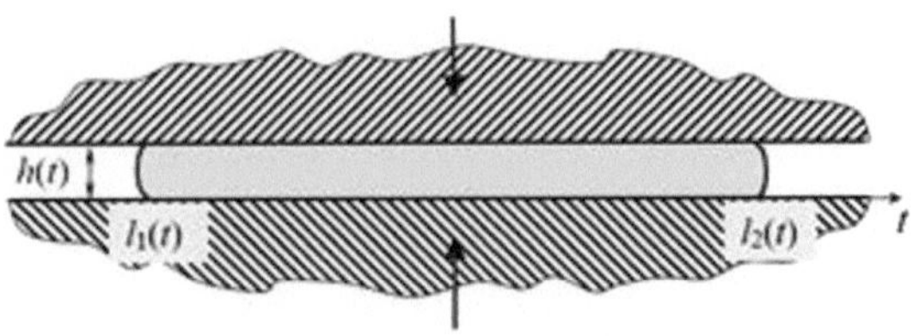

Fig. 15.7 Free flow of strip material with movable ends.

B) The plastic strip being deposited occupies the area of contact with the fixed left end so that it can move freely in the direction opposite to the fixed end (right) and is represented by a segment $[l_{10}, l_2(t)]$ with one movable end (Fig. 15.8).
In this case, it is assumed that the plastic material flows over the entire contact area [13]. The proposed solution remains valid. To determine $l_2(t)$ the following differential equation was used

$$l_1(t) = l_{10}\ ;\ \frac{dl_2}{dt} = u\,(l_2, t) = -\frac{1}{h}\frac{dh}{dt}\left(\frac{l_2 - l_{10}}{2}\right), \quad t = t_0;\ l_2(t) = l_{20}\ , \tag{15.82}$$

the solution of which has the form

$$l_2(t) = l_{10}\ + (l_{20} - l_{10})\sqrt{\frac{h_0}{h(t)}}. \tag{15.83}$$

The left fixed end of the contact domain is clamped so that plastic flow in the domain $[l_{10}, l_2(t)]$ happens in one direction, and $x_0 \equiv l_{10}$. In this case, the solution to the boundary value problem

$$\frac{\partial p}{\partial x} = -\frac{2\tau_s}{h};\ \frac{\partial \bar{u}}{\partial \bar{x}} - \frac{d\lambda}{d\bar{t}} = 0$$

under boundary conditions $x = l_{10};\ u\,(l_{10}, t) = 0;\ x = l_2(t);\ p = \sigma$ has the form:

$$p(x, t) = \sigma_s + \frac{2\tau_s}{h}\,(l_2 - x)\ ;\quad u(x, t) = \frac{1}{h}\frac{dh}{dt}\,(x - l_{10})\ ; \tag{15.84}$$

The magnitude of the specific force for the implementation of plastic upsetting in this case, of course, depends on the condition at the left end (Fig. 15.9) [9, 13], and takes the form:

$$p(t) \equiv p_3(t) = \int\limits_{l_{10}}^{l_2}\left[\sigma_s + \frac{2\tau_s}{h}\,(l_2 - x)\right]dx = \sigma_s\,(l_2 - l_{10}) + \frac{2\tau_s}{h}\frac{(l_2 - l_{10})^2}{2}. \tag{15.85}$$

A tensile stress is applied to the left fixed end of the contact area

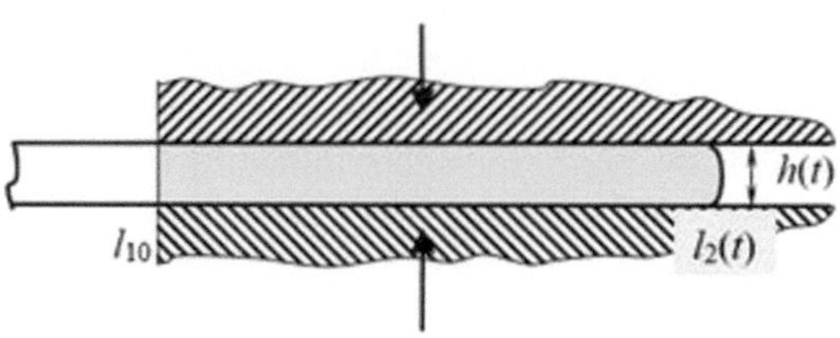

Fig. 15.8 Plastic upsetting of a strip with a movable right end.

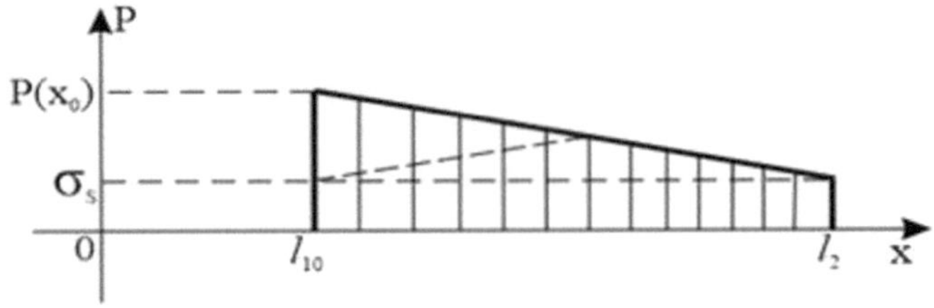

Fig. 15.9: Distribution of contact pressure in the problem of spreading of a plastic strip with a clamped left end.

$$x = l_{10} : \sigma_{xx} = -q_0 > 0,$$

such that

$$-\sigma_s < q_0 < 0.$$

According to the condition of full plasticity [9], for the contact pressure at the left end

$$p = q_0 + \sigma_s$$

the following condition is true:

$$0 < p < \sigma_s.$$

The limiting case is highlighted separately

$$q_0 = -\sigma_s,$$

which models the process of plastic stretching of a thin strip with simultaneous plastic subsidence of its clamped end parts [9]. The solution to the boundary value problem in this case is determined by the system:

$$p(x,t) = \begin{cases} \sigma_s + \dfrac{2\tau_s}{h}(l_2 - x), \ x \in (x_0, l_2), \\ \sigma_s + q_0 + \dfrac{2\tau_s}{h}(x - l_{10}), x \in (l_{10}, x_0), \end{cases} \tag{15.86}$$

$$u(x,t) = -\frac{1}{h}\frac{dh}{dt}(x - x_0), \text{ where } x_0 = \frac{l_{10} + l_2}{2} - \frac{q_0}{4\tau_s}h \geq \frac{l_{10} + l_2}{2}.$$

The magnitude of the specific force for the implementation of plastic settlement in this case is equal to the area of the shaded diagram shown in Fig. 15.10 [12, 13]. In this case, the pressure is determined by the ratio:

$$p(t) = \int_{l_{10}}^{l_2} p(x,t)dt = \left[(\sigma_s + q_0)\, x + \frac{2\tau_s}{h}\left(\frac{x^2}{2} - l_{10}x\right)\right] \Bigg|_{l_{10}}^{x_0}$$

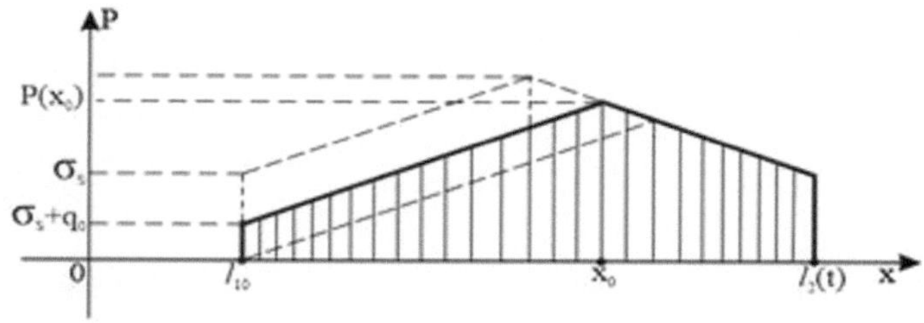

Fig. 15.10: Diagram of contact pressure in a plastic strip, at the left end of which a tensile stress is applied $\sigma_{xx} = -q_0 > 0$.

$$+ \left[\sigma_s x + \frac{2\tau_s}{h}\left(l_2 x - \frac{x^2}{2}\right)\right]\Bigg|_{x_0}^{l_2}$$
$$= \sigma_s\left(l_2 - l_{10}\right) + \frac{2\tau_s}{h}\frac{\left(l_2 - l_{10}\right)^2}{4} + q_0\left(\frac{l_2 + l_{10}}{2} - \frac{q_0}{2\tau_s}h\right) \leq P_1(t), \quad (15.87)$$

where

$$p_1(t) = \sigma_s\left(l_2 - l_{10}\right) + \frac{2\tau_s}{h}\frac{\left(l_2 - l_{10}\right)^2}{4}$$

is the specific force for achieving plastic upsetting of a strip with free ends.

An approximate analytical solution is obtained within the framework of a general mathematical model of a "viscous fluid". These various generalizations of Prandtl's classical problem are of interest not only for analyzing thin-layer compression, but also for studying problems involving the straightening of flat workpieces under uniaxial tension beyond the elastic limit, for which straightening by other methods yields unsatisfactory results. These solutions are also applicable to solving plane bending problems involving the tension of large aircraft cladding components and in shipbuilding. These solutions are universal, confirmed by experimental results, and can be extended to spatial contact problems involving flow in a thin plastic layer between approaching tool surfaces.

References

[1] Arnold VI (2012) Obyknovennyye differentsial'nyye uravneniya (in Russ., Ordinary Differential Equations). Moscow: MCNO

[2] Gun GY (1983) Matematicheskoye modelirovaniye protsessov formoobrazovaniya metallov: uchebnik dlya universitetov (in Russ., Mathematical Modeling of Metal Forming Processes: A Textbook for Universities). Moscow: Metallurgy

[3] Ilyushin AA (1955) Polnaya plastichnost' v protsessakh techeniya mezhdu zhestkimi poverkhnostyami, analogiya s peschanoy zasypkoy i nekotoryye

prilozheniya (in Russ., Complete plasticity in flow processes between rigid surfaces, analogy with a sand fill, and some applications). Applied Mathematics and Mechanics 19(6):693 – 713

[4] Ilyushin AA (2004) Trudy (1946–1966). Tom. 2. Plastichnost' (in Russ., Works (1946-1966). Vol 2. Plasticity ed by EA Ilyushina, MR Korotkina). Moscow: FIZMATLIT

[5] Ilyushin AA Voprosy teorii plasticheskogo techeniya veshchestv po poverkhnostyam (in Russ., Questions of the theory of plastic flow of substances over surfaces). Applied Mathematics and Mechanics 18(3):265 – 288

[6] Kadymov VA (2015) Kontaktnyye problemy plasticheskogo techeniya v tonkom sloye (in Russ., Contact Problems of Plastic Flow in a Thin Layer). Moscow: MGGEU

[7] Kadymov VA (1994) Nestatsionarnyye zadachi techeniya v tonkom plasticheskom sloye (in Russ., Non-stationary problems of flows in a thin plastic layer). Dissertation for the degree of Doctor of Physics and Mathematics. Specialty 01.02.04 Mechanics of Deformable Solids. Moscow: Moscow State University

[8] Kadymov VA, Bystrikov SK (2006) Nekotoryye novyye resheniya nestatsionarnykh zadach rastekaniya plasticheskogo sloya po uprugodeformiruyemym poverkhnostyam (in Russ., Some new solutions of non-stationary problems of spreading of a plastic layer over elastically deformable surfaces). Bulletin of Tula State University. Seriya Mechanics of Deformable Solids and Metal Forming 12(2):54–60

[9] Kadymov VA, Lisunets NL, Yanovskaya EA (2024) Plasticheskoye rastyazheniye polosy s nesimmetrichnymi usloviyami v oblasti zakhvatov (in Russ., Plastic stretching of a strip with asymmetric conditions in the grip area). Chernye metally (5):75-80, DOI 10.17580/chm.2024.05.10

[10] Kadymov VA, Maksyutov MS O novoy matematicheskoy formulirovke kontaktnoy krayevoy zadachi plasticheskogo techeniya v tonkom sloye (in Russ., On a new mathematical formulation of the contact boundary value problem of plastic flow in a thin layer). Nauka i Mir 7–1(35):20–23

[11] Kadymov VA, Sosenushkin EN, Belov NA (2016) Kontaktnaya zadacha nesvobodnogo rastekaniya plasticheskogo sloya po ploskosti: eksperiment i teoriya (in Russ., Contact problem of non-free spreading of a plastic layer on a plane: experiment and theory). Collection of scientific papers "Elasticity and Inelasticity". Moscow: MSU, 180 – 185

[12] Kadymov VA, Sosenushkin EN, Yanovskaya EA (2022) Contact problems of plastic flow in a thin layer: theory, analysis of solutions, and their applications. Journal of Machinery Manufacture and Reliability 51(3):206-215, DOI 10. 3103/S1052618822030062

[13] Kadymov VA, Sosenushkin EN, Yanovskaya EA (2024) Kontaktnyye zadachi plasticheskogo techeniya v tonkom sloye: teoriya, analiz resheniy i ikh primeneniye (in Russ., On the problem of plastic stretching of a strip taking into account conditions at the ends), Interuniversity conference "Mathematics in the constellation of sciences". Moscow: Moscow State University Press

[14] Kadymov VA, Sosenushkin EN, Yanovskaya EA (2019) Plasticheskiye techeniya v tonkom sloye: novaya postanovka krayevoy zadachi i yeye resheniye (in Russ., Plastic flows in a thin layer: a new statement of the boundary value problem and its solution). Collection of abstracts of the 3rd International Conference "Modeling of Nonlinear Processes and Systems". Moscow: Janus-K, pp 99-103

[15] Kadymov VA, Sosenushkin EN, Yanovskaya EA, Rogulin SA (2024) K probleme kombinirovannoy nagruzki polosy (in Russ., On the problem of combined loading of a strip). News of the Samara Scientific Center of the Russian Academy of Sciences 26(3):140 – 146, DOI 10.37313/1990-5378-2024-26-3-140-146

[16] Kiiko IA (2006) Anisotropy in the flow processes of a thin plastic layer. Applied Mathematics and Mechanics 70(2):344-351, DOI 10.1016/j.jappmathmech.2006.06.012

[17] Kiiko IA (ed) (2003) Plasticheskoye techeniye materialov. Fiziko-matematicheskiye osnovy tekhnologii obrabotki pod davleniyem (in Russ., Plastic Flow of Materials. Physical and Mathematical Foundations of Pressure Processing Technology). Moscow: Publishing House of the Center for Applied Research, Moscow State University

[18] Kiiko IA (1965) Techeniye tonkogo sloya plasticheskogo materiala nad uprugodeformiruyemymi ploskostyami (in Russ., Flow of a thin layer of plastic material over elastically deformable planes). Engineering journal 5(2)372 – 375

[19] Kiiko IA (1971) Teoriya plasticheskogo techeniya v tonkom sloye metalla (in Russ., Theory of plastic flow in a thin layer of metal). Scientific works of the Institute of Mechanics of Moscow State University, No 5

[20] Korn GA, Korn TM (2013) Mathematical Handbook for Scientists and Engineers: Definitions, Theorems, and Formulas for Reference and Review. New York: Dover Publications

[21] Prandtl L (1923) Anwendungsbeispiele zu einem Henckyschen Satz über das plastiche Gleichgewicht. ZAMM - Journal of Applied Mathematics and Mechanics / Zeitschrift für Angewandte Mathematik und Mechanik 3(6):401-406, DOI 10.1002/zamm.19230030601

[22] Privalov II (1984) Introduction to the Theory of Functions of a Complex Variable (in Russ., Introduction to the Theory of Functions of a Complex Variable). Moscow: Nauka

[23] Stepansky LG (1979) Raschety protsessov formoobrazovaniya metallov (in Russ., Calculations of Metal Forming Processes). Moscow: Mashinostroenie

[24] Subich VN (1986) Plasticheskoye techeniye materiala v tonkom sloye pri kombinirovannoy nagruzke (in Russ., plastic flow of material in a thin layer under combined loading). Forging and Stamping Production (10):5 – 7

[25] Tarnovsky IY, Levanov AN, Posevatkin MI (1966) Kontaktnyye napryazheniya pri plasticheskoy deformatsii (in Russ., Contact Stresses During Plastic Deformation). Moscow: Metallurgy

[26] Unksov EP (1955) Inzhenernyye metody rascheta sil pri obrabotke metallov (in Russ., Engineering Methods for Calculating Forces in Metal Forming). Moscow: Mashgiz

[27] Yanovskaya EA (2023) Ispol'zovaniye kharakteristik v nestatsionarnykh zadachakh plasticheskogo techeniya tonkogo sloya vdol' ploskostey (in Russ., Use of characteristics in non-stationary problems of plastic flow of a thin layer along planes). News of the Samara Scientific Center of the Russian Academy of Sciences 25(4):139 – 144, DOI 10.37313/1990-5378-2023-25-4-139-144

[28] Yanovskaya EA (2023) Modelirovaniye i sravneniye rezul'tatov tochnogo analiticheskogo resheniya krayevoy zadachi o techenii v tonkom sloye s eksperimental'nymi dannymi (in Russ., Modeling and comparison of the results of an exact analytical solution of a boundary-value problem of flow in a thin layer with experimental data). News of the Samara Scientific Center of the Russian Academy of Sciences 25(3)(113):130 – 139, DOI 10.37313/1990-5378-2023-25-3-130-139

Chapter 16
Numerical Modeling of Wave Processes and Fracture in Solids Using the Grid-Characteristic Method

Nikolay I. Khokhlov and Igor B. Petrov

Abstract This work presents an approach to numerical modeling of dynamic deformation and fracture of solids based on the grid-characteristic method and discrete fracture models. The fundamental equations of linear elasticity theory are formulated in matrix form for two-dimensional and three-dimensional cases. The implementation of boundary conditions for free surface, prescribed velocity, and surface traction using characteristic invariants is described. Discrete fracture models are considered: a shear failure model based on the von Mises criterion and the Mainchen-Sack crack model, which enable the description of crack formation and the transition of material into a fractured state. The results of verification calculations are presented, including the spallation problem, sphere-plate and plate-plate impact, as well as practical examples of seismic resistance assessment for dome structures and residential buildings. The obtained results demonstrate the correctness of the proposed approach and its applicability to a wide range of problems in solid mechanics.

Key words: Grid-characteristic method, Dynamic wave propagation, Dynamic collisions, Brittle fractures, Spallation phenomena

Nikolay I. Khokhlov (corresponding author)
Moscow Institute of Physics and Technology, Institutskiy Pereulok 9, 141701 Dolgoprudny, Moscow Region, Scientific Research Institute for System Analysis of the National Research Centre "Kurchatov Institute" & Autonomous noncommercial organization of higher education "Innopolis University", Innopolis, Tatarstan, Russian Federation, e-mail: `khokhlov.ni@mipt.ru`

Igor B. Petrov
Moscow Institute of Physics and Technology, Institutskiy Pereulok 9, 141701 Dolgoprudny, Moscow Region, Russian Federation, e-mail: `petrov@mipt.ru`

H. Altenbach et al. (eds.), *Stability Problems in Mechanics of Continua*,
Advanced Structured Materials 260, https://doi.org/10.1007/978-3-032-31032-3_16

16.1 Introduction

Mathematical modeling of wave processes in continuous media is one of the most important tools in modern solid mechanics. Problems of elastic wave propagation arise in seismology, geophysics, acoustics, non-destructive testing, as well as in the design and strength assessment of engineering structures [1]. Of particular interest are problems related to dynamic processes in heterogeneous media, including multilayer geological structures, porous and fractured rocks, as well as problems of material fracture under impact and seismic loading [10, 26]. The development of computing technology and numerical methods has enabled a transition from simplified one-dimensional and two-dimensional formulations to realistic three-dimensional modeling of complex physical processes [8]. However, despite significant progress, the development of efficient and accurate numerical methods for solving hyperbolic systems of equations describing wave propagation remains a relevant task.

A wide range of numerical methods is currently used for modeling elastic wave propagation. The finite difference method (FDM) is one of the most common approaches due to its relative simplicity of implementation and high efficiency on regular grids [5, 30, 31]. However, standard finite difference schemes have limited accuracy at discontinuities and require special treatment of material interfaces, and complex geometries are poorly approximated on rectangular grids. The finite element method (FEM) is widely used for the analysis of building structures and geological massifs, allowing for complex geometry and material heterogeneity [3, 25]. However, for solving dynamic problems, classical FEM requires very fine meshes to accurately describe high-frequency wave propagation, which leads to a significant increase in computational cost [20]. The spectral element method combines the accuracy of spectral methods with the geometric flexibility of the finite element method, successfully applied for global seismology modeling [17, 18, 23], but requires special meshes and is complex to implement. The discontinuous Galerkin method naturally describes solution discontinuities at element boundaries, making it attractive for problems with faults and cracks [19, 24], but is characterized by a large number of degrees of freedom and high computational cost.

The grid-characteristic method (GCM) was developed for solving hyperbolic systems of equations describing wave processes [15, 16]. Unlike traditional finite difference methods, GCM is based on the characteristic properties of the system, which allows for natural formulation of boundary and contact conditions and provides high accuracy at discontinuities [14, 28]. The main principle of the method consists of reducing the system of equations to diagonal form through spectral decomposition of the coefficient matrices, transitioning to characteristic variables (Riemann invariants), solving independent one-dimensional advection equations along characteristics, and subsequently returning to the original physical variables [8]. The method has been successfully applied to problems in gas dynamics, solid dynamics, seismology, and geophysics [7, 9], demonstrating high accuracy in calculating wave fields, efficiency on structured grids, and the ability to use high-order schemes [13, 14].

Modeling fracture in solids is a complex problem because the fracture process is discrete in nature (crack initiation and growth), while traditional continuum mechanics approaches operate with continuous fields. Currently, there are several main approaches to fracture modeling. Continuum models describe fracture as a process of damage accumulation at each point of the medium, introducing damage parameters that vary from 0 (undamaged material) to 1 (completely fractured material), with the elastic properties of the material being modified depending on the damage level [11]. The advantages of continuum models include their compatibility with traditional numerical methods and the ability to describe diffuse damage accumulation; however, they do not allow explicit description of individual cracks and their propagation paths.

Methods with explicit crack tracking, on the contrary, track the geometry of the discontinuity surface. These include the extended finite element method (XFEM) [2], particle methods (DEM, SPH) [4], and methods based on tracking the state at computational grid nodes [10]. Discrete fracture models, in which a parameter indicating the state (fractured or not) is stored for each grid node, naturally describe the formation and development of cracks without requiring mesh reconstruction [26]. The Mainchen-Sack model is designed to describe crack formation in brittle materials [21]. In this model, for each node, in addition to the fracture flag, a crack normal is stored indicating the orientation of the discontinuity surface, and stresses are corrected such that there are no stresses on the crack surface. The model allows describing the formation of multiple cracks in a node (up to $n - 1$ mutually perpendicular cracks, where n is the spatial dimension) and the transition of the material to a completely fractured state upon the formation of the n-th crack [10].

To describe fracture under shear loading, a model based on the von Mises criterion is used [22]. When the equivalent stress reaches the shear strength limit, the material is considered fractured. In the fractured material, tensile stresses are absent (positive principal stresses are set to zero) and the shear modulus is reduced [26]. This model allows describing the behavior of brittle materials under shear loads, which is important for seismic resistance problems where shear deformations predominate [27].

This work uses a combination of the grid-characteristic method and discrete fracture models, which allows efficient solving of problems of dynamic deformation and fracture of solids under impact and seismic loading [7, 29]. The proposed approach is adapted for use on structured grids and allows calculations for complex geometric configurations using the overset grid method [6, 12].

16.2 Mathematical Model

This section provides a mathematical description of wave propagation in a linear elastic isotropic medium. The model is formulated for three-dimensional and two-dimensional cases in a unified manner, which allows tracing the relationship between the formulations and simplifies the implementation of numerical methods.

16.2.1 Constitutive Relations

The primary unknown quantities describing the state of the medium at a fixed point in space and time are the components of the velocity vector and the stress tensor.

- **Three-dimensional case** (x, y, z): the state is determined by the velocity vector $\mathbf{v} = (v_x, v_y, v_z)$ and the symmetric Cauchy stress tensor $\boldsymbol{\sigma}$ with components $\sigma_{xx}, \sigma_{yy}, \sigma_{zz}, \sigma_{xy}, \sigma_{xz}, \sigma_{yz}$. Thus, the full unknown vector has dimension nine:

$$\mathbf{u} = (v_x, v_y, v_z, \sigma_{xx}, \sigma_{xy}, \sigma_{xz}, \sigma_{yy}, \sigma_{yz}, \sigma_{zz})^{\mathrm{T}}. \tag{16.1}$$

- **Two-dimensional case** (plane strain in the xOy plane): we assume derivatives with respect to the z-axis are zero. In this case, the three-dimensional system decouples, and five components are sufficient to describe P-SV waves (in-plane motion):

$$\mathbf{q} = (v_x, v_y, \sigma_{xx}, \sigma_{yy}, \sigma_{xy})^{\mathrm{T}}. \tag{16.2}$$

The medium parameters characterizing its elastic properties are density ρ and Lamé parameters λ and μ (shear modulus). The elastic wave propagation velocities are expressed through these parameters:

$$c_p = \sqrt{\frac{\lambda + 2\mu}{\rho}}, \quad c_s = \sqrt{\frac{\mu}{\rho}}, \tag{16.3}$$

where c_p is the longitudinal wave velocity (P-wave), and c_s is the shear wave velocity (S-wave).

16.2.2 Matrix Form of Equations

The system of linear elasticity equations is of hyperbolic type and can be written in matrix form as a system of first-order partial differential equations. This form is most convenient for applying the grid-characteristic method.

16.2.2.1 Three-Dimensional Case

In the three-dimensional formulation, the system has the form:

$$\frac{\partial \mathbf{u}}{\partial t} + \mathbf{A_x}\frac{\partial \mathbf{u}}{\partial x} + \mathbf{A_y}\frac{\partial \mathbf{u}}{\partial y} + \mathbf{A_z}\frac{\partial \mathbf{u}}{\partial z} = \mathbf{0}, \tag{16.4}$$

where $\mathbf{A_x}$, $\mathbf{A_y}$, $\mathbf{A_z}$ are square matrices of size 9×9, whose elements are combinations of the medium parameters ρ, λ, and μ. The explicit form of these matrices reflects the structure of conservation laws and constitutive relations. Matrix $\mathbf{A_x}$ is written as follows:

$$\mathbf{A_x} = \begin{pmatrix} 0 & 0 & 0 & -\frac{1}{\rho} & 0 & 0 & 0 & 0 & 0 \\ 0 & 0 & 0 & 0 & -\frac{1}{\rho} & 0 & 0 & 0 & 0 \\ 0 & 0 & 0 & 0 & 0 & -\frac{1}{\rho} & 0 & 0 & 0 \\ -(\lambda+2\mu) & 0 & 0 & 0 & 0 & 0 & 0 & 0 & 0 \\ 0 & -\mu & 0 & 0 & 0 & 0 & 0 & 0 & 0 \\ 0 & 0 & -\mu & 0 & 0 & 0 & 0 & 0 & 0 \\ -\lambda & 0 & 0 & 0 & 0 & 0 & 0 & 0 & 0 \\ 0 & 0 & 0 & 0 & 0 & 0 & 0 & 0 & 0 \\ -\lambda & 0 & 0 & 0 & 0 & 0 & 0 & 0 & 0 \end{pmatrix}. \quad (16.5)$$

Matrix $\mathbf{A_y}$ has the form:

$$\mathbf{A_y} = \begin{pmatrix} 0 & 0 & -\frac{1}{\rho} & 0 & 0 & 0 & 0 & 0 & 0 \\ 0 & 0 & 0 & -\frac{1}{\rho} & 0 & 0 & 0 & 0 & 0 \\ 0 & 0 & 0 & 0 & -\frac{1}{\rho} & 0 & 0 & 0 & 0 \\ 0 & -\lambda & 0 & 0 & 0 & 0 & 0 & 0 & 0 \\ -\mu & 0 & 0 & 0 & 0 & 0 & 0 & 0 & 0 \\ 0 & 0 & 0 & 0 & 0 & 0 & 0 & 0 & 0 \\ 0 & -(\lambda+2\mu) & 0 & 0 & 0 & 0 & 0 & 0 & 0 \\ 0 & 0 & -\mu & 0 & 0 & 0 & 0 & 0 & 0 \\ 0 & -\lambda & 0 & 0 & 0 & 0 & 0 & 0 & 0 \end{pmatrix}. \quad (16.6)$$

Matrix $\mathbf{A_z}$ is written as:

$$\mathbf{A_z} = \begin{pmatrix} 0 & -\frac{1}{\rho} & 0 & 0 & 0 & 0 & 0 & 0 & 0 \\ 0 & 0 & -\frac{1}{\rho} & 0 & 0 & 0 & 0 & 0 & 0 \\ 0 & 0 & 0 & -\frac{1}{\rho} & 0 & 0 & 0 & 0 & 0 \\ 0 & 0 & -\lambda & 0 & 0 & 0 & 0 & 0 & 0 \\ 0 & 0 & 0 & 0 & 0 & 0 & 0 & 0 & 0 \\ -\mu & 0 & 0 & 0 & 0 & 0 & 0 & 0 & 0 \\ 0 & 0 & -\lambda & 0 & 0 & 0 & 0 & 0 & 0 \\ 0 & -\mu & 0 & 0 & 0 & 0 & 0 & 0 & 0 \\ 0 & 0 & -(\lambda+2\mu) & 0 & 0 & 0 & 0 & 0 & 0 \end{pmatrix}. \quad (16.7)$$

It is easy to see that the first three rows of each matrix correspond to the three components of the equation of motion (momentum conservation law), while the remaining six rows correspond to the time derivatives of the stress tensor components, following from the generalized Hooke's law.

16.2.2.2 Two-Dimensional Case

For the two-dimensional case, the system of equations describing P-SV waves takes a similar form:

$$\frac{\partial \mathbf{q}}{\partial t} + \mathbf{A}_1 \frac{\partial \mathbf{q}}{\partial x} + \mathbf{A}_2 \frac{\partial \mathbf{q}}{\partial y} = \mathbf{0}, \tag{16.8}$$

where $\mathbf{A}_1$ and $\mathbf{A}_2$ are matrices of size 5×5. Matrix $\mathbf{A}_1$ (corresponding to the derivative with respect to x) has the form:

$$\mathbf{A}_1 = \begin{pmatrix} 0 & 0 & -\frac{1}{\rho} & 0 & 0 \\ 0 & 0 & 0 & 0 & -\frac{1}{\rho} \\ -(\lambda+2\mu) & 0 & 0 & 0 & 0 \\ -\lambda & 0 & 0 & 0 & 0 \\ 0 & -\mu & 0 & 0 & 0 \end{pmatrix}. \tag{16.9}$$

Matrix $\mathbf{A}_2$ (corresponding to the derivative with respect to y) is written as:

$$\mathbf{A}_2 = \begin{pmatrix} 0 & 0 & 0 & 0 & -\frac{1}{\rho} \\ 0 & 0 & 0 & -\frac{1}{\rho} & 0 \\ 0 & -\lambda & 0 & 0 & 0 \\ 0 & -(\lambda+2\mu) & 0 & 0 & 0 \\ -\mu & 0 & 0 & 0 & 0 \end{pmatrix}. \tag{16.10}$$

As in the three-dimensional case, the first two rows correspond to the components of the equation of motion, and the last three correspond to Hooke's law.

16.2.3 Operator Form

For theoretical analysis and construction of certain numerical schemes, it is convenient to use a more compact operator form. Let us introduce a unit vector $\mathbf{n}_k$ directed along the coordinate axis k (for 3D $k \in \{x, y, z\}$, for 2D $k \in \{x, y\}$). Then the action of the matrix $\mathbf{A}_k$ on the state vector can be represented as follows:

$$\mathbf{A}_k \begin{bmatrix} \mathbf{v} \\ \boldsymbol{\sigma} \end{bmatrix} = - \begin{bmatrix} \rho^{-1}(\boldsymbol{\sigma} \cdot \mathbf{n}_k) \\ \lambda(\mathbf{v} \cdot \mathbf{n}_k)\mathbf{I} + \mu(\mathbf{n}_k \otimes \mathbf{v} + \mathbf{v} \otimes \mathbf{n}_k) \end{bmatrix}. \tag{16.11}$$

This form of notation is universal and equally valid for both two-dimensional and three-dimensional cases. It explicitly separates the kinematic part (upper block, responsible for acceleration) and the dynamic part (lower block, responsible for the rate of stress change). Here $\mathbf{I}$ denotes the identity tensor of the corresponding dimension, and $\otimes$ denotes the tensor product operator.

16.2.4 Characteristic Analysis

The most important property of the system (16.4)–(16.8) is its hyperbolicity. The matrices $\mathbf{A}_k$ can be diagonalized:

$$\mathbf{A}_k = \Omega_k^{-1}\Lambda_k\Omega_k, \quad (16.12)$$

where Λ_k is a diagonal matrix containing real eigenvalues, and $\mathbf{\Omega}_k$ is a matrix composed of the corresponding eigenvectors.

The eigenvalues (characteristic propagation speeds) for all directions k coincide up to sign and multiplicity. For the three-dimensional case, the set of eigenvalues is:

$$\Lambda_k = \mathrm{diag}(c_p, -c_p, c_s, -c_s, c_s, -c_s, 0, 0, 0), \quad (16.13)$$

and for the two-dimensional case:

$$\Lambda_k = \mathrm{diag}(c_p, -c_p, c_s, -c_s, 0). \quad (16.14)$$

The presence of three (in 3D) or one (in 2D) zero characteristics reflects the fact that not all Riemann invariants are transported along the coordinate axes with finite speed. Characteristic analysis underlies the grid-characteristic method, allowing the reduction of the original system solution to a sequence of independent one-dimensional advection equations.

16.3 Boundary and Contact Conditions

Correct formulation of the dynamic deformation problem requires specifying conditions on the boundaries of the computational domain and on the interfaces between different media. Within the framework of the grid-characteristic method, boundary conditions are implemented based on the analysis of characteristic invariants, which naturally accounts for the wave nature of the equations being solved.

16.3.1 General Approach to Implementing Boundary Conditions

When solving a system of hyperbolic equations, characteristic variables propagate along characteristics with finite speeds. At the boundary of the computational domain, some characteristics are directed into the domain (incoming), and some are directed outward (outgoing). The values of incoming characteristics are determined from the solution inside the domain, while outgoing ones must be specified or calculated from boundary conditions.

Let the solution at a boundary node be represented as a sum of contributions from incoming and outgoing characteristics:

$$\vec{u}^{n+1} = \vec{u}^{n+1\,(\mathrm{in})} + \Omega^{(\mathrm{out})}\omega^{n+1\,(\mathrm{out})}, \tag{16.15}$$

where $\vec{u}^{n+1\,(\mathrm{in})}$ is the contribution from characteristics entering the computational domain, calculated from values at interior nodes; $\Omega^{(\mathrm{out})}$ is a matrix composed of columns of Ω^{-1} corresponding to outgoing characteristics; $\omega^{n+1\,(\mathrm{out})}$ are the unknown values of the outgoing characteristic invariants. Boundary conditions are generally formulated as a linear relation:

$$\mathbf{B}\vec{u} = \vec{b}, \tag{16.16}$$

where $\mathbf{B}$ is a rectangular matrix with the number of rows equal to the number of outgoing characteristics (spatial dimension), and $\vec{b}$ is a known right-hand side vector.

Substituting (16.15) into (16.16) yields a system for determining the unknown characteristic invariants:

$$\omega^{n+1\,(\mathrm{out})} = (\mathbf{B}\Omega^{(\mathrm{out})})^{-1}(\vec{b} - \mathbf{B}\vec{u}^{n+1\,(\mathrm{in})}). \tag{16.17}$$

The final formula for calculating boundary nodes takes the form:

$$\vec{u}^{n+1} = \vec{u}^{n+1\,(\mathrm{in})} + \Omega^{(\mathrm{out})}(\mathbf{B}\Omega^{(\mathrm{out})})^{-1}(\vec{b} - \mathbf{B}\vec{u}^{n+1\,(\mathrm{in})}). \tag{16.18}$$

16.3.2 Boundary Conditions for an Elastic Medium

16.3.2.1 Free Surface

The free surface condition assumes the absence of external forces:

$$\boldsymbol{\sigma} \cdot \vec{m} = \vec{0}, \tag{16.19}$$

where $\vec{m}$ is the unit vector of the outward normal to the boundary. This condition corresponds to the reflection of waves from the boundary without energy leakage outward.

In terms of characteristic variables, the free surface condition is implemented through setting combinations of stresses corresponding to outgoing characteristics to zero.

16.3.2.2 Prescribed Boundary Velocity

For a prescribed boundary velocity $\vec{V}$, the condition is:

$$\vec{v} = \vec{V}. \tag{16.20}$$

To implement this condition, the vector $\vec{g}^{n+1} = \vec{v}^{n+1\,(\mathrm{in})} - \vec{V}$ is first calculated, and then the stress tensor is corrected:

$$\vec{v}^{n+1} = \vec{V}, \tag{16.21}$$

$$\boldsymbol{\sigma}^{n+1} = \boldsymbol{\sigma}^{n+1(\text{in})} - \rho\left[(\vec{m}\cdot\vec{g}^{n+1})((c_p - 2c_s - c_3)\vec{m}\otimes\vec{m} + c_3\boldsymbol{I}) + c_s(\vec{g}^{n+1}\otimes\vec{m} + \vec{m}\otimes\vec{g}^{n+1})\right], \tag{16.22}$$

where $c_p = \sqrt{(\lambda + 2\mu)/\rho}$, $c_s = \sqrt{\mu/\rho}$, $c_3 = \sqrt{\lambda^2/(\rho(\lambda + 2\mu))}$.

16.3.2.3 Prescribed Surface Traction

For a prescribed surface traction density $\vec{f}$, the condition is:

$$\boldsymbol{\sigma} \cdot \vec{m} = \vec{f}. \tag{16.23}$$

Introduce the auxiliary vector $\vec{z}^{n+1} = \boldsymbol{\sigma}^{n+1(\text{in})} \cdot \vec{m} - \vec{f}$. Then the solution at the boundary is calculated using the formulas:

$$\vec{v}^{n+1} = \vec{v}^{n+1(\text{in})} - \frac{c_p\vec{z}^{n+1} - (c_p - c_s)(\vec{z}^{n+1} \cdot \vec{m})\vec{m}}{\rho c_p c_s}, \tag{16.24}$$

$$\boldsymbol{\sigma}^{n+1} = \boldsymbol{\sigma}^{n+1(\text{in})} - \vec{z}^{n+1} \otimes \vec{m} - \vec{m} \otimes \vec{z}^{n+1} - \frac{(\vec{z}^{n+1} \cdot \vec{m})}{\lambda + 2\mu}\big(\lambda\boldsymbol{I} - 2(\lambda + \mu)\vec{m} \otimes \vec{m}\big). \tag{16.25}$$

16.3.2.4 Mixed Boundary Conditions

In some formulations, only the normal component of velocity and the tangential components of surface tractions are prescribed:

$$\vec{v} \cdot \vec{m} = V_n, \quad (\boldsymbol{\sigma} \cdot \vec{m})_\tau = \vec{f}_\tau. \tag{16.26}$$

In this case, the total surface traction density is first calculated:

$$\vec{f} = \vec{f}_\tau + \vec{m}\big(\rho c_p V_n + \vec{m} \cdot (\boldsymbol{\sigma}^{n+1(\text{in})} \cdot \vec{m} - \rho c_p \vec{v}^{n+1(\text{in})})\big), \tag{16.27}$$

after which the formulas for prescribed traction (16.25) are used.

16.3.3 Contact Conditions at Material Interfaces

When modeling layered media or inclusions, contact conditions must be specified at the interface between two materials to ensure continuity of the relevant physical quantities.

16.3.3.1 Perfect Bonding (Rigid Contact)

Perfect bonding conditions imply equality of velocities and normal stresses at the contact interface:

$$\vec{v}^L = \vec{v}^R, \quad \boldsymbol{\sigma}^L \cdot \vec{m} = \boldsymbol{\sigma}^R \cdot \vec{m}, \tag{16.28}$$

where indices L and R refer to the media on the left and right of the interface, and $\vec{m}$ is the outward normal to the left medium (so for the right medium the normal is $-\vec{m}$). The velocity at the contact is calculated using the formula:

$$\begin{aligned}\vec{v}^{n+1} = \frac{1}{\rho_L c_{sL} + \rho_R c_{sR}} \Big\{ & \rho_L \left[(\vec{m} \cdot \vec{v}_L^{n+1(\text{in})})(c_{pL} - c_{sL})\vec{m} + c_{sL}\vec{v}_L^{n+1(\text{in})} \right] \\ & + \rho_R \left[(\vec{m} \cdot \vec{v}_R^{n+1(\text{in})})(c_{pR} - c_{sR})\vec{m} + c_{sR}\vec{v}_R^{n+1(\text{in})} \right] \\ & - (\boldsymbol{\sigma}_L^{n+1(\text{in})} - \boldsymbol{\sigma}_R^{n+1(\text{in})}) \cdot \vec{m} \\ & - \frac{\rho_L(c_{pL} - c_{sL}) + \rho_R(c_{pR} - c_{sR})}{\rho_L c_{pL} + \rho_R c_{pR}} \vec{m} \\ & \Big(\vec{m} \cdot \big[\rho_L c_{pL} \vec{v}_L^{n+1(\text{in})} + \rho_R c_{pR} \vec{v}_R^{n+1(\text{in})} \\ & - (\boldsymbol{\sigma}_L^{n+1(\text{in})} - \boldsymbol{\sigma}_R^{n+1(\text{in})}) \cdot \vec{m} \big] \Big) \Big\}. \end{aligned} \tag{16.29}$$

After finding the velocity, the stresses in each medium are corrected using the formulas for the prescribed velocity boundary condition.

16.3.3.2 Free Sliding

Free sliding conditions (no friction) imply equality of normal velocities and normal stresses, as well as the absence of tangential stresses:

$$\vec{v}^L \cdot \vec{m} = \vec{v}^R \cdot \vec{m} = V^{n+1}, \quad \boldsymbol{\sigma}^L \cdot \vec{m} = \boldsymbol{\sigma}^R \cdot \vec{m}, \quad [\vec{m}, \boldsymbol{\sigma}^L \cdot \vec{m}] = [\vec{m}, \boldsymbol{\sigma}^R \cdot \vec{m}] = \vec{0}, \tag{16.30}$$

where $[\vec{a}, \vec{b}]$ is the cross product.

The normal component of velocity at the contact is calculated as:

$$V^{n+1} = \frac{1}{\rho_L c_{sL} + \rho_R c_{sR}} \left[\rho_L c_{pL} \vec{v}_L^{n+1(\text{in})} + \rho_R c_{pR} \vec{v}_R^{n+1(\text{in})} - (\boldsymbol{\sigma}_L^{n+1(\text{in})} - \boldsymbol{\sigma}_R^{n+1(\text{in})}) \cdot \vec{m} \right] \cdot \vec{m}. \tag{16.31}$$

Afterwards, mixed boundary conditions (16.26) with $\vec{f}_\tau = \vec{0}$ are used.

16.4 Fracture Models

To model fracture processes in deformable solids, this work uses discrete models in which the fracture state is tracked at each node of the computational grid. This approach naturally allows describing the initiation and development of cracks, as well as the transition of the material to a fragmented state.

16.4.1 General Approach to Discrete Fracture Modeling

In discrete fracture models, a parameter indicating the state (fractured or not) is stored for each grid node. At each time step, for all unfractured nodes, a fracture criterion depending on the stress tensor components is checked. When the criterion is satisfied, the node is marked as fractured. In all fractured nodes, stresses are corrected according to the chosen model for fractured material behavior.

This approach allows modeling both brittle fracture and gradual damage accumulation under shear loading.

16.4.2 Shear Failure Model

The shear failure model describes material behavior when the shear strength limit is reached. The fracture criterion is based on stress intensity (von Mises criterion):

$$\sigma_{eq} = \frac{1}{\sqrt{2}}\sqrt{(\sigma_{xx} - p)^2 + (\sigma_{yy} - p)^2 + (\sigma_{zz} - p)^2 + 2\sigma_{xy}^2 + 2\sigma_{xz}^2 + 2\sigma_{yz}^2} > Y_s, \tag{16.32}$$

where $p = (\sigma_{xx} + \sigma_{yy} + \sigma_{zz})/3$ is the mean stress, and Y_s is the shear strength limit (material constant).

When the fracture criterion is satisfied, the material is considered fractured. This representation leads to two main properties:

1. **Absence of tensile stresses:** In a fractured material, tensile stresses cannot exist, so all positive principal stresses are set to zero.
2. **Reduction of shear modulus:** At the moment of fracture, the shear modulus decreases while the bulk modulus is preserved. The shear modulus at the new time level is calculated as:

$$\mu^{n+1} = \alpha \mu^n, \tag{16.33}$$

where α is the residual strength coefficient (in calculations, $\alpha = 0.1$ is used). The wave speeds are correspondingly adjusted:

$$c_p^{n+1} = \sqrt{(c_p^n)^2 - \frac{4}{3}(c_s^n)^2(1 - \alpha)}, \quad c_s^{n+1} = \sqrt{\alpha}\, c_s^n. \tag{16.34}$$

The stress correction algorithm in a fractured node is performed in several steps:

1. Transition to the principal stress basis:

$$\boldsymbol{D}^n = \mathbf{W}\boldsymbol{\sigma}^n\mathbf{W}^{\mathrm{T}}, \tag{16.35}$$

where $\mathbf{W}$ is the matrix composed of the eigenvectors of the stress tensor.

2. Setting positive principal stresses to zero:

$$D_i^{n+1} = \min(0, D_i^n), \quad i = x, y, z. \tag{16.36}$$

3. Inverse transformation to the original basis:

$$\boldsymbol{\sigma}^{n+1} = \mathbf{W}^{\mathrm{T}}\boldsymbol{D}^{n+1}\mathbf{W}. \tag{16.37}$$

This model is well suited for describing fracture of brittle materials under shear loading and is widely used in dynamic deformation problems.

16.4.3 Crack Model

The Mainchen-Sack model is designed to describe the process of crack initiation and development in brittle materials. Unlike the shear failure model, it additionally stores information about crack orientation.

16.4.3.1 Basic Principles

The model considers single-sided cracks. For each node, in addition to the fracture flag, a crack normal $\vec{d}$ is stored, indicating the orientation of the discontinuity surface. In nodes containing cracks, stresses must satisfy the condition:

$$\boldsymbol{\sigma} \cdot \vec{d} = \vec{0}, \tag{16.38}$$

meaning that there are no stresses on the crack surface. The crack initiation criterion is based on reaching the ultimate tensile stress:

$$\max(D_i) > \sigma_0, \tag{16.39}$$

where D_i are the principal stresses, and σ_0 is the tensile strength. When the criterion is satisfied, a crack is formed, the normal of which is aligned with the vector of maximum tensile stress (corresponding to the largest principal stress).

16.4.3.2 Accounting for Multiple Cracks

A single node can contain at most $n - 1$ mutually perpendicular cracks, where n is the spatial dimension. In the two-dimensional case ($n = 2$), the maximum number of cracks in a node is 1; in the three-dimensional case ($n = 3$), no more than 2 mutually perpendicular cracks. If the n-th crack forms in a node (third in 3D or second in 2D), the material is considered completely fractured (crushed), and further correction is performed using the shear failure model described above.

16.4.4 Application of Fracture Models

The described fracture models are successfully applied to solve a wide class of problems:

- **Impact interactions:** Modeling the spall effect in plate impact, fracture of glass barriers upon impact by steel projectiles.
- **Seismic resistance of structures:** Assessment of fracture zones in concrete structures (domes, residential buildings) under seismic loading.
- **Dynamic loading of brittle materials:** Investigation of crack formation in glass under local loading.

The combined application of both models enables the description of complex fracture scenarios, from the formation of individual cracks to complete material fracture under shear loading.

16.5 Examples of Calculations

This section presents the results of numerical modeling demonstrating the capabilities of the developed approaches for solving problems of dynamic deformation of solids with fracture. Problems of impact, seismic resistance of structures, as well as verification calculations are considered.

16.5.1 Verification of the Method on the Spallation Problem

To test the fracture calculation method, consider the classical problem of a steel plate impacting a glass plate. All quantities are given in dimensionless units: unit of time - 0.9 μs, unit of length - 5 mm, unit of stress - 70 MPa, unit of density - 1000 kg/m^3. The striker thickness is 0.1, the plate thickness is 1. The striker velocity is 0.003, and the lower boundary of the glass is free. Glass parameters: $\rho = 1200$, $c_1 = 1$, $c_2 = 0.5$. Steel parameters: $\rho = 3600$, $c_1 = 1$, $c_2 = 0.5$.

From experiments, the spallation effect is known: when a plate is impacted on one side, material spalls off from the opposite side. Physically, spallation occurs when the front of the wave reflected from the rear boundary meets the trailing front of the incident wave. The thickness of the spall plate in this impact configuration should exactly match the striker width.

Figure 16.1 shows the numerical modeling results at various times. It can be seen that the spall plate thickness corresponds to the striker width, which agrees well with experimental data. Black lines indicate formed cracks, white lines are velocity isolines, and color saturation corresponds to velocity magnitudes.

16.5.2 Impact with Various Velocities and Angles

To investigate the influence of kinematic parameters on the fracture pattern, plate impact calculations were performed at various velocities and impact angles. Figure 16.2 shows results for different combinations of normal and tangential components of striker velocity.

The results demonstrate a significant influence of the impact angle and the magnitude of the tangential velocity component on the fracture pattern. With increasing

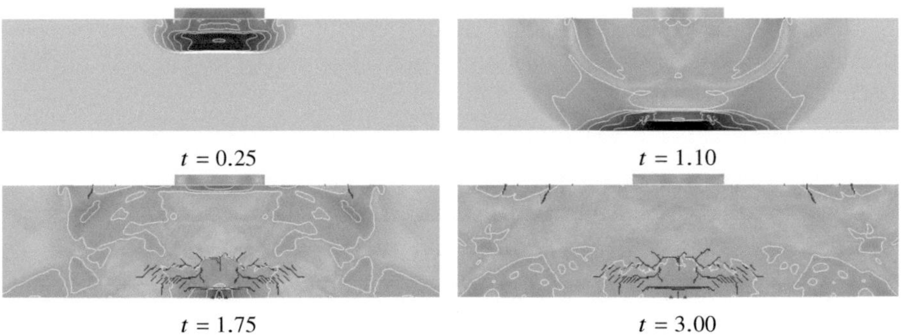

Fig. 16.1: Normal impact of a plate on glass at different times. Black lines indicate formed cracks, white lines are velocity isolines. Color saturation corresponds to velocity magnitudes.

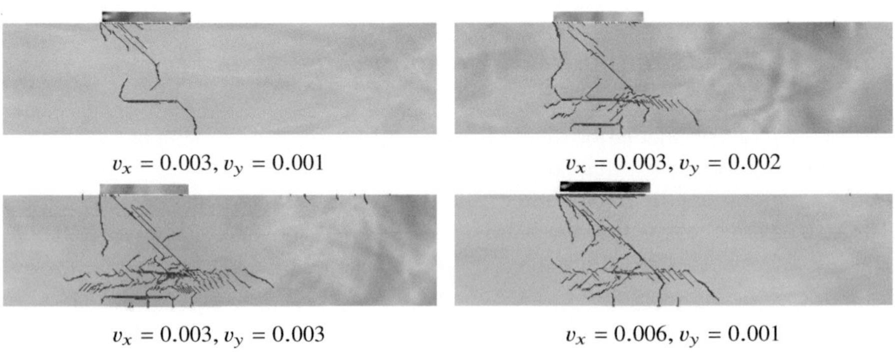

Fig. 16.2: Impact with various velocities and angles. Striker velocity components are indicated for each figure. Dynamic contact — bonding condition.

tangential component, more intense crack development is observed in the direction of striker motion.

16.5.3 Sphere Impact on a Glass Plate

The next problem is the impact of a steel sphere on a glass plate. The striker diameter is 0.5, velocity is 0.006, plate thickness is 1. Boundary conditions are similar to the previous calculation. Figure 16.3 shows fracture patterns at different times. It can be observed that the glass breaks approximately in half, which correlates well with experimental data. It is known that when glass is impacted, ring cracks form around the impact point. In the presented two-dimensional numerical experiment, similar effects are observed.

Figure 16.4 compares direct and sliding (frictionless) impacts. When a tangential velocity component is present, an asymmetric fracture pattern is observed with preferential crack development in the direction of striker motion.

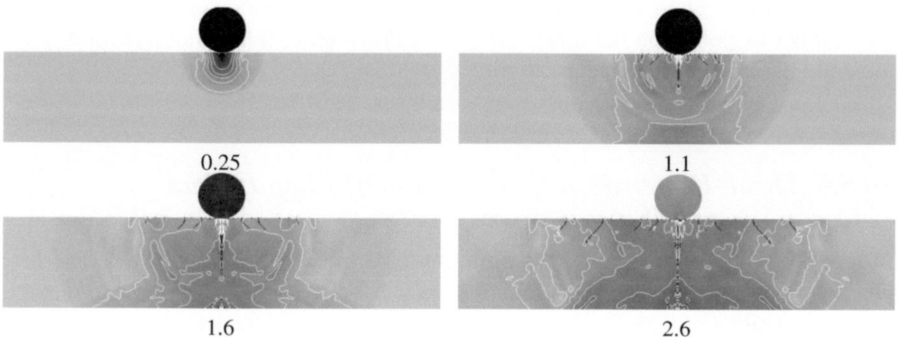

Fig. 16.3: Normal impact of a sphere on glass. Black lines indicate formed cracks, white lines are velocity isolines. Color saturation corresponds to velocity magnitudes.

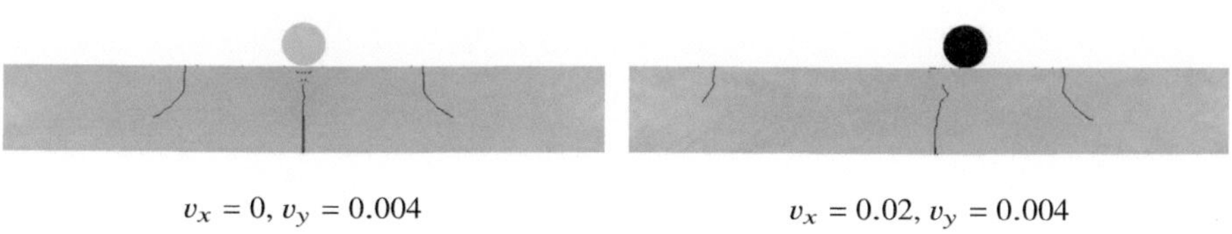

Fig. 16.4: Comparison of direct and frictionless sliding impact.

16.5.4 Two Spheres Impacting a Glass Plate

Consider the problem of two spheres impacting a glass plate. The diameter of each striker is 0.5, velocity is 0.004, plate thickness is 1. The second striker falls 0.0012 time units after the first. Figure 16.5 presents the modeling results. Note that the first cracks appear immediately after the fall of the second striker. An asymmetric fracture pattern is observed: there are more cracks on the right, which is associated with the interference of wave fields from two sequential impacts.

16.5.5 Modeling Crack Formation under Central Loading

The problem of glass fracture under central loading was investigated. A circular glass sample with a central cutout was considered, with stress prescribed on the inner surface. At each point, the stress was random, ranging from 150% to 225% of the material strength. This simulated the heterogeneity of the original material.

As can be seen from Fig. 16.6, the crack pattern resembles experimentally observed cracks in glass after impact. As the mesh resolution increases (increasing the number of points on the inner circle), the crack pattern tends to a certain limit, indicating convergence of the method.

16.5.6 Three-Dimensional Calculation of Plate Impact

In the three-dimensional case, two different experiments were conducted. The first was a striker impacting a plate. Striker geometry: $2.6 \times 2.6 \times 0.6$, elastic parameters: $C_p = 3000$ m/s, $C_s = 2000$ m/s, density $\rho = 5000$ kg/m^3. Plate geometry: $10 \times 10 \times 2$,

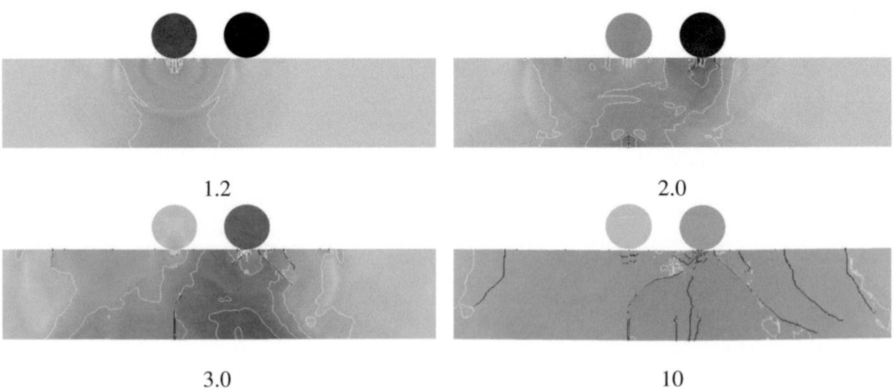

Fig. 16.5: Normal impact of two spheres with an interval of 0.0012.

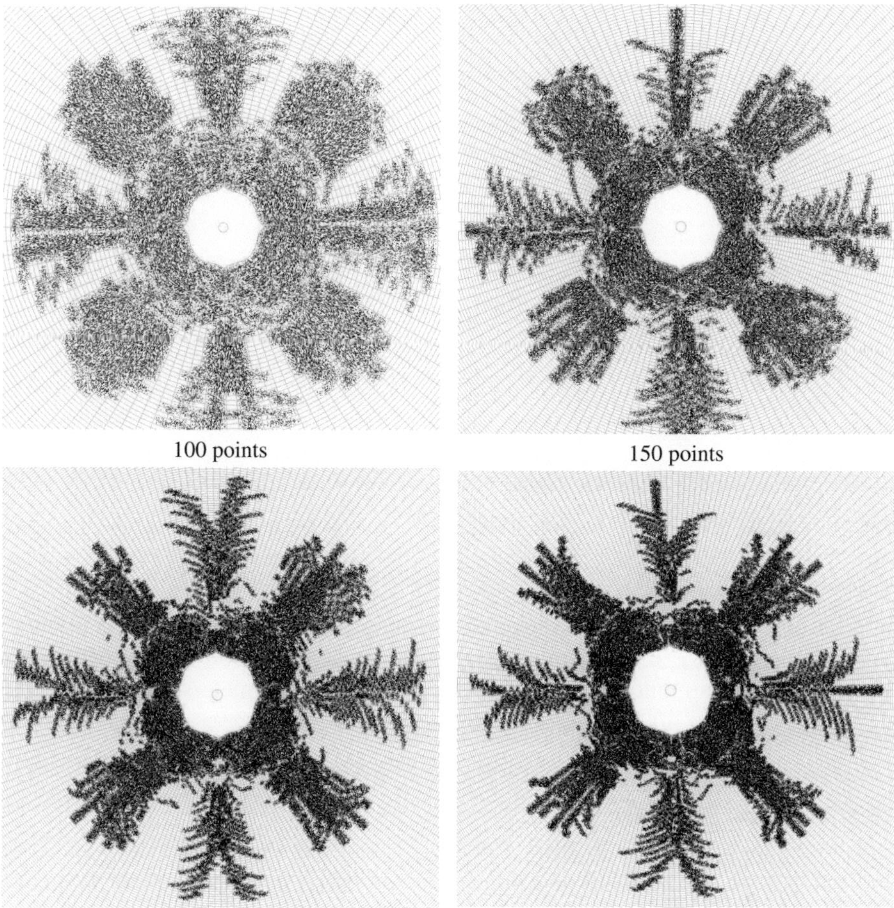

Fig. 16.6: Cracks in glass for different mesh resolutions. The number of points on the inner circle is indicated.

elastic parameters: $C_p = 3000$ m/s, $C_s = 2000$ m/s, density $\rho = 5000$ kg/m^3. Fracture parameter $y_s = 6 \times 10^7$, initial striker velocity $v_0 = 10$ m/s. The second experiment differed in that the metal striker had spherical symmetry. The fractured plates for both cases are shown in Fig. 16.7. Clearly distinct crack patterns are visible.

16.5.7 Assessment of Seismic Resistance of Surface Structures

Consider the problem of assessing the seismic resistance of surface structures. An earthquake is described as a single train of plane waves incident from the Earth's

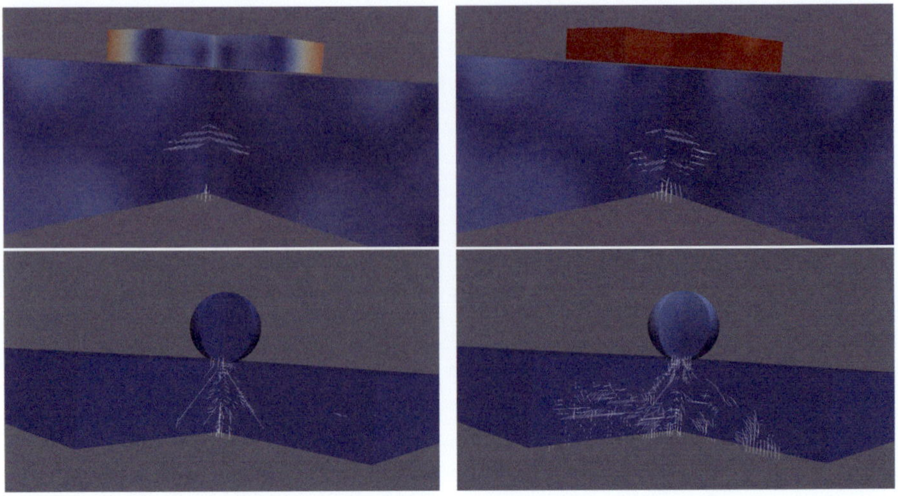

Fig. 16.7: Cracks in the glass plate. Full 3D problems. Cases of sphere and hammer were modeled.

interior perpendicular to the surface. The wavelength is 3 m, magnitude is 0.5 m/s. Soil parameters: $c_1 = 3000$ m/s, $c_2 = 1500$ m/s, $\rho = 2500$ kg/m^3. Structures are represented as monolithic concrete structures with parameters: $c_1 = 5000$ m/s, $c_2 = 3000$ m/s, $\rho = 4000$ kg/m^3.

16.5.7.1 Dome Structure

In the first case, the structure is a dome structure (e.g., a nuclear power plant containment) made of concrete. The reactor is located in the center and is mounted on beams through which seismic waves practically do not reach the reactor. The station roof has a dome shape for greater seismic stability. Station height is 30 m, wall thickness is 4 m. Figure 16.8 presents calculation results for two values of concrete

Fig. 16.8 Seismic resistance of a dome structure for different concrete shear moduli. Black color shows fracture zones

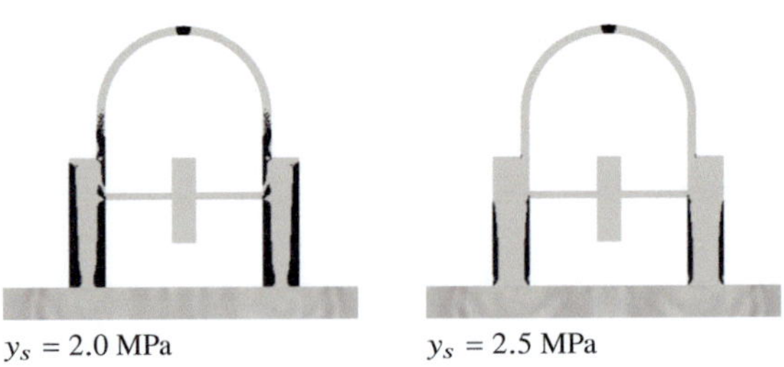

shear strength limit: $Y_s = 2.0$ MPa and $Y_s = 2.5$ MPa. Black color shows fracture zones. From the calculations, it can be seen that fracture does not reach the reactor, confirming the correctness of the chosen design.

A similar calculation was performed in a three-dimensional formulation. The calculations were performed on hexahedral curvilinear computational grids. To create this rather complex structure, only 22 computational grids in contact with each other were generated. The structure diameter is 20 m, wall thickness is 2 m, foundation height is 2 m. To model the earthquake, a force impulse of 0.05 s duration is applied to the Earth's surface, the wavelength of which is several times larger than the structure dimensions. Figure 16.9 shows examples of formed fractures according to the crack model.

Fig. 16.9 Seismic resistance of a dome structure, three-dimensional formulation. Black color shows fracture zones.

16.5.7.2 Two-Story Residential Building

The following is an example of calculating fracture in a two-story residential building with a strip foundation (three-dimensional calculation). The problem was formulated with the participation of Indian specialists from CDAC studying the seismic resistance of low-rise buildings in the Himalayan foothills. Wall thickness is 40 cm, earthquake magnitude is 0.1 m/s (5 times smaller than in the previous calculation).

Figure 16.10 shows a comparison of fracture patterns without accounting for corrections caused by fractured areas (left) and with accounting (right). It can be seen that accounting for changes in material properties in fractured areas significantly affects the propagation of fracture zones.

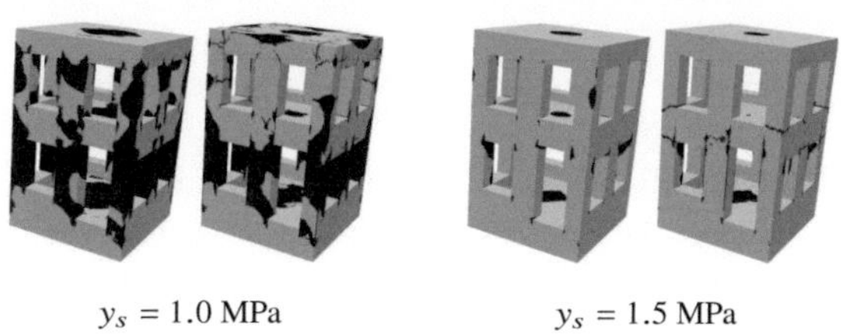

Fig. 16.10 Seismic resistance of a two-story residential building

16.6 Conclusion

This work presents an approach to modeling dynamic deformation and fracture of solids based on the combination of the grid-characteristic method and discrete fracture models. The use of characteristic properties of hyperbolic systems allows natural implementation of boundary and contact conditions, providing high accuracy in calculating wave fields on structured grids.

The proposed fracture models enable description of both damage accumulation and explicit crack formation with tracking of their orientation. The combined application of these models makes it possible to model the complete fracture cycle: from the initiation of individual cracks to complete material fragmentation.

The presented results of numerical experiments, including the spallation problem, sphere-plate and plate-plate impact, as well as the assessment of seismic resistance of structures, demonstrate the correctness and efficiency of the developed approach. Convergence of results with mesh refinement and adequate reproduction of known physical effects are shown.

The developed methods can be applied to solve a wide range of practical problems: from modeling impact interactions and fracture of brittle materials to assessing the seismic resistance of engineering structures. Further development of the work may be associated with the transition to fully three-dimensional calculations, the use of adaptive grids, and the consideration of more complex rheological properties of materials.

Acknowledgements The work was carried out within the framework of the state task of the NRC "Kurchatov Institute" - SRISA on the topic No. FNEF-2024-0002 "Mathematical modeling of multi-scale dynamic processes and virtual environment systems" (1023032900401-5-1.2.1).

References

[1] Aki K, Richards PG (2002) Quantitative Seismology, 2nd edn. University Science Books, Sausalito, California

[2] Belytschko T, Gracie R, Ventura G (2009) A review of extended/generalized finite element methods for material modeling. Modelling and Simulation in Materials Science and Engineering 17(4):043001, DOI 10.1088/0965-0393/17/4/043001

[3] Bittnar Z, Sejnoha J (1996) Numerical Methods in Structural Mechanics. ASCE Press, New York

[4] Cundall P, Strack O (1979) A discrete numerical model for granular assemblies. Geotechnique 29(1):47–65, DOI 10.1680/geot.1979.29.1.47

[5] Cyrus N, Fulton R (1966) Finite difference accuracy in structural analysis. Journal of the Structural Division 92(6):459–471, DOI 10.1061/JSDEAG.0001515

[6] Favorskaya A, Khokhlov N (2022) Using chimera grids to describe boundaries of complex shape. In: Smart Innovation, Systems and Technologies, vol 309, Springer, pp 249–258, DOI 10.1007/978-981-19-1292-8_26

[7] Favorskaya A, Khokhlov N, Petrov I (2020) Grid-characteristic method on joint structured regular and curved grids for modeling coupled elastic and acoustic wave phenomena in objects of complex shape. Lobachevskii Journal of Mathematics 41(4):512–525, DOI 10.1134/S1995080220040083

[8] Favorskaya AV, Zhdanov MS, Khokhlov NI, Petrov IB (2018) Modelling the wave phenomena in acoustic and elastic media with sharp variations of physical properties using the grid-characteristic method. Geophysical Prospecting 66(8):1485–1502, DOI 10.1111/1365-2478.12639

[9] Golubev V, Khokhlov N, Petrov I, Zhdanov M, Velikhov V (2015) Novel Approach to Modelling Elastic Wavefields in Fractured Media. DOI 10.3997/2214-4609.201413107

[10] Grigorievykh DP, Khokhlov NI, Petrov IB (2017)Raschet dinamicheskogo razrusheniya v tverdykh deformiruyemykh telakh (in Russ., Calculation of dynamic destruction in deformable bodies). Matematicheskoe modelirovanie 29(4):45–58

[11] Kachanov M (1993) Elastic solids with many cracks and related problems. Advances in Applied Mechanics 30:259–445, DOI 10.1016/S0065-2156(08)70176-5

[12] Khokhlov N, Favorskaya A, Stetsyuk V, Mitskovets I (2021) Grid-characteristic method using chimera meshes for simulation of elastic waves scattering on geological fractured zones. Journal of Computational Physics 446:110637, DOI 10.1016/j.jcp.2021.110637

[13] Khokhlov NI, Golubev VI (2019) On the Class of Compact Grid-Characteristic Schemes. pp 64–77, DOI 10.1007/978-3-030-06228-6_7

[14] Khokhlov NI, Petrov IB (2023) High-order grid-characteristic method for systems of hyperbolic equations with piecewise constant coefficients. Differential Equations 59(7):985–997, DOI 10.1134/s001226612307011x

[15] Kholodov AS, Kholodov YA (2006) Monotonicity criteria for difference schemes designed for hyperbolic equations. Computational Mathematics and Mathematical Physics 46(9):1560–1588, DOI 10.1134/s0965542506090089

[16] Kholodov YA, Kholodov AS, Tsybulin IV (2018) Construction of monotone difference schemes for systems of hyperbolic equations. Computational Mathematics and Mathematical Physics 58(8):1226–1246, DOI 10.1134/s0965542518080110

[17] Komatitsch D, Tromp J (1999) Introduction to the spectral element method for three-dimensional seismic wave propagation. Geophysical Journal International 139(3):806–822, DOI 10.1046/j.1365-246x.1999.00967.x

[18] Lee U (2009) Spectral Element Method in Structural Dynamics. John Wiley & Sons, Singapore

[19] Li X, Wiberg NE (1996) Structural dynamic analysis by a time-discontinuous galerkin finite element method. International Journal for Numerical Methods

in Engineering 39(12):2131–2152, DOI 10.1002/(SICI)1097-0207(19960630)39:12<2131::AID-NME943>3.0.CO;2-2

[20] Lian W, Yao W (2010) Fatigue life prediction of composite laminates by fea simulation method. International Journal of Fatigue 32(1):123–133, DOI 10.1016/j.ijfatigue.2009.02.022

[21] Maenchen G, Sack S (1964) The TENSOR code. In: Alder B., Fernbach S, Rotenberg M (eds) Methods in Computational Physics - Advances in Research and Applications, Vol 3 Fundamental Methods in Hydrodynamics, pp 185–211. New York and London: Academic Press

[22] von Mises R (1913) Mechanik der festen körper im plastisch-deformablen zustand. Nachrichten von der Gesellschaft der Wissenschaften zu Göttingen, Mathematisch-Physikalische Klasse 1913:582–592

[23] Nefovska-Danilovic M (2013) Dynamic analysis of soil-structure system using spectral element method. PhD thesis, University of Belgrade, Faculty of Civil Engineering, Belgrade

[24] Park SH, Antin N (2004) A discontinuous galerkin method for seismic soil–structure interaction analysis in the time domain. Earthquake Engineering & Structural Dynamics 33(2):285–293, DOI 10.1002/eqe.348

[25] Perelmuter A, Slivker V (2013) Numerical Structural Analysis: Methods, Models and Pitfalls. Springer, Berlin, Heidelberg, DOI 10.1007/978-3-642-36197-5

[26] Petrov IB, Chelnokov FB (2003) Chislennoye issledovaniye volnovykh protsessov i protsessov razrusheniya v mnogosloynykh pregradakh (in Russ, Numerical analysis of wave processes and fracture in layered targets. Zhurnal Vychislitel'noi Matematiki i Matematicheskoi Fiziki 43(10):1562–1579

[27] Petrov IB, Khokhlov NI (2014) Modeling 3d seismic problems using high-performance computing systems. Mathematical Models and Computer Simulations 6(4):342–350, DOI 10.1134/s2070048214040061

[28] Petrov IB, Kholodov AS (1984) Chislennoye issledovaniye nekotorykh dinamicheskikh zadach mekhaniki deformiruyemogo tvordogo tela setochno-kharakteristicheskim metodom (in Russ., Numerical investigation of some dynamical problems of mechanics of a deformable rigid body by the net-characteristic method). Zhurnal Vychislitel'noi Matematiki i Matematicheskoi Fiziki 24(5):722–739

[29] Stognii P, Khokhlov N, Petrov I (2019) Modeling vertical fractures using the schoenberg model on structured grids by grid-characteristic method. Radioelectronics Nanosystems Information Technologies 11(3):351–356, DOI 10.17725/rensit.2019.11.351

[30] Teixeira F, Sarris C, Zhang Y, Na DY, Berenger JP, Su Y, Okoniewski M, Chew W, Backman V, Simpson J (2023) Finite-difference time-domain methods. Nature Reviews Methods Primers 3(1):75, DOI 10.1038/s43586-023-00255-6

[31] Virdi K (2006) Finite difference method for nonlinear analysis of structures. Journal of Constructional Steel Research 62(11):1210–1218, DOI 10.1016/j.jcsr.2006.01.018

Chapter 17
Construction of Constitutive Relations of the Second and Third Orders for an Orthotropic Material

Dmitrii Khristich

Abstract The article presents variants of constitutive relations for a nonlinearly elastic anisotropic material in the form of dependencies between the components of strains and stresses, which meet the requirements of generalization of the particular postulate of Il'yushin on anisotropic materials. Constitutive relations for a nonlinear elastic orthotropic material are obtained on the basis of specifying the Gibbs potential for reversible processes of isothermal deformation of such materials. Models of elasticity of the second and third orders for an orthotropic material are considered. The models make it possible to describe nonlinear dependencies of strains on stresses in the processes of tension, compression, and shear known from experiments, as well as the difference in resistance of anisotropic materials. Experiments are indicated from which the parameters of the models can be determined. The values of the parameters of the second- and third-order models are found on the basis of experimental data.

Key words: Nonlinear constitutive relations, Anisotropic materials, Orthotropic materials, Identification, Composites

17.1 Introduction

Many materials, such as polymers and polymer composites, obviously exhibit nonlinear mechanical behavior. The physical nonlinearity of materials can be due to their inhomogeneity and the presence of initial defects caused by imperfections in manufacturing technologies. It requires a description of the relationship between stresses and strains on the basis of experimental data at the macroscopic level, that is, without taking into account the structural features. The processing of such exper-

---

Dmitrii Khristich
Tula State University, pr. Lenina, 92, 300012 Tula, Russian Federation, e-mail: dmitrykhristich@rambler.ru

H. Altenbach et al. (eds.), *Stability Problems in Mechanics of Continua*,
Advanced Structured Materials 260, https://doi.org/10.1007/978-3-032-31032-3_17

imental data is one of the most important stages in the construction of mathematical models of deforming of modern anisotropic materials.

Modern materials demonstrate nonlinear mechanical behavior in the elastic region, even at small deformations [2, 4, 10], and, at the same time, have a pronounced anisotropy of properties. Therefore, the development of substantiated models of nonlinear elasticity for anisotropic materials, as well as their identification and verification in experiments, is relevant.

In [1, 3, 5, 6, 9, 12–14] nonlinear models of orthotropic materials are presented. For example, in [13], a nonlinear model of orthotropic material describing biaxial tension of polyvinyl chloride (PVC) coated fabrics is developed. On the basis of the energy function, nonlinear dependencies of a general form between stresses and deformations are obtained, and nonlinear constitutive equations of the second and third order for a plane stressed state are proposed. The coefficients of nonlinear models are determined for three types of PVC-coated fabrics based on the results of experiments on uniaxial and biaxial tension. A more accurate description of experimental data using nonlinear models compared to linear models is noted.

To describe the mechanical behavior of a material, constitutive relations constructed in [11] are used in the present chapter. The constitutive relations are written in the form of deformations versus stresses. The purpose of the present work is to determine the material parameters included in the model based on experimental data known from the literature [2, 10].

17.2 Construction of Nonlinear Material Models

To describe the strained state, we use the Cauchy–Green strain tensor ε. The stressed state is described by the Cauchy true stress tensor S. The considered orthotropic materials have three linear invariants of the strain tensor [8]

$$
\begin{aligned}
e_0 &= \frac{1}{\sqrt{3}}(\varepsilon_{11} + \varepsilon_{22} + \varepsilon_{33}) \,, \\
e_1 &= \frac{1}{\sqrt{6}}(2\varepsilon_{33} - \varepsilon_{11} - \varepsilon_{22}) \,, \\
e_2 &= \frac{1}{\sqrt{2}}(\varepsilon_{11} - \varepsilon_{22})
\end{aligned}
\tag{17.1}
$$

and three quadratic invariants

$$
s_{(1)}^2 = \varepsilon_{12}^2 \,, \quad s_{(2)}^2 = \varepsilon_{23}^2 \,, \quad s_{(3)}^2 = \varepsilon_{31}^2 \,. \tag{17.2}
$$

The stress tensor invariants are defined similarly:

$$
\sigma_0 = \frac{1}{\sqrt{3}}(S_{11}+S_{22}+S_{33}) \,, \; \sigma_1 = \frac{1}{\sqrt{6}}(2S_{33}-S_{11}-S_{22}) \,, \; \sigma_2 = \frac{1}{\sqrt{2}}(S_{11}-S_{22}) \,, \tag{17.3}
$$

$$t_{(1)}^2 = S_{12}^2, \quad t_{(2)}^2 = S_{23}^2, \quad t_{(3)}^2 = S_{31}^2. \tag{17.4}$$

In [11], a variant of constitutive relations for a nonlinearly elastic anisotropic material was obtained on the basis of specifying the Gibbs potential for reversible processes of isothermal deformation of such materials. These relations for an orthotropic material can be represented in the form of a relationship between the invariants of stresses and strains (17.1)–(17.4) in the form:

$$\begin{aligned}
e_0 &= \left(c_0^{00} + 3c_1^{00}\sigma_0 + 6c_2^{00}\sigma_0^2\right)\sigma_0 \\
&+ \left(c_0^{01} + 2c_1^{01}\sigma_0 + c_1^{01}\sigma_1 + 3c_2^{01}\sigma_0^2 + 2c_2^{01}\sigma_0\sigma_1 + c_2^{01}\sigma_1^2\right)\sigma_1 \\
&+ \left(c_0^{02} + 2c_1^{02}\sigma_0 + c_1^{02}\sigma_2 3c_2^{02}\sigma_0^2 + 2c_2^{02}\sigma_0\sigma_2 + c_2^{02}\sigma_2^2\right)\sigma_2, \\
e_1 &= \left(c_0^{01} + 2c_1^{01}\sigma_1 + c_1^{01}\sigma_0 + 3c_2^{01}\sigma_1^2 + 2c_2^{01}\sigma_0\sigma_1 + c_2^{01}\sigma_0^2\right)\sigma_0 \\
&+ \left(c_0^{11} + 3c_1^{11}\sigma_1 + 6c_2^{11}\sigma_1^2\right)\sigma_1 \\
&+ \left(c_0^{12} + 2c_1^{12}\sigma_1 + c_1^{12}\sigma_2 + 3c_2^{12}\sigma_1^2 + 2c_2^{12}\sigma_1\sigma_2 + c_2^{12}\sigma_2^2\right)\sigma_2, \\
e_2 &= \left(c_0^{02} + 2c_1^{01}\sigma_2 + c_1^{02}\sigma_0 + 3c_2^{02}\sigma_2^2 + 2c_2^{02}\sigma_0\sigma_2 + c_2^{02}\sigma_0^2\right)\sigma_0 \\
&+ \left(c_0^{12} + 2c_1^{12}\sigma_2 + c_1^{12}\sigma_1 + 3c_2^{12}\sigma_2^2 + 2c_2^{12}\sigma_1\sigma_2 + c_2^{12}\sigma_1^2\right)\sigma_1 \\
&+ \left(c_0^{22} + 3c_1^{22}\sigma_2 + 6c_2^{22}\sigma_2^2\right)\sigma_2, \\
s_{(1)} &= 2\left(D_0^{(1)} + \frac{3}{2}D_1^{(1)}t_{(1)} + 2D_2^{(1)}t_{(1)}^2\right)t_{(1)}, \\
s_{(2)} &= 2\left(D_0^{(2)} + \frac{3}{2}D_1^{(2)}t_{(2)} + 2D_2^{(2)}t_{(2)}^2\right)t_{(2)}, \\
s_{(3)} &= 2\left(D_0^{(3)} + \frac{3}{2}D_1^{(3)}t_{(3)} + 2D_2^{(3)}t_{(3)}^2\right)t_{(3)},
\end{aligned} \tag{17.5}$$

where $c_0^{\alpha\beta}$, $c_1^{\alpha\beta}$, $c_2^{\alpha\beta}$, $D_0^{(\gamma)}$, $D_1^{(\gamma)}$, $D_2^{(\gamma)}$ are the elastic parameters of the material, constant in the selected basis of three-dimensional space. Constitutive relations (17.5) are a third-order elasticity model for the orthotropic material and contain 9 linear constants, 9 quadratic constants and 9 cubic constants. At small strains, these relations represent the generalized Hooke's law

$$e_0 = c_0^{00}\sigma_0 + c_0^{01}\sigma_1 + c_0^{02}\sigma_2,$$
$$e_1 = c_0^{01}\sigma_0 + c_0^{11}\sigma_1 + c_0^{12}\sigma_2, \quad e_2 = c_0^{02}\sigma_0 + c_0^{12}\sigma_1 + c_0^{22}\sigma_2,$$
$$s_{(1)} = 2D_0^{(1)}t_{(1)}, \quad s_{(2)} = 2D_0^{(2)}t_{(2)}, \quad s_{(3)} = 2D_0^{(3)}t_{(3)},$$

i.e. the constants $c_0^{\alpha\beta}$, $D_0^{(\gamma)}$ form a tensor of elastic compliance, the reciprocal of the elastic tensor of the material.

To write the constitutive relations in terms of the components of the strain and stress tensors in three-dimensional space, we express these components in terms of linear and quadratic invariants from relations (17.1)–(17.2) and (17.3)–(17.4), respectively:

$$\begin{gathered}\varepsilon_{11}=\frac{\sqrt{3}}{3}e_0-\frac{\sqrt{6}}{6}e_1+\frac{\sqrt{2}}{2}e_2\,,\\ \varepsilon_{22}=\frac{\sqrt{3}}{3}e_0-\frac{\sqrt{6}}{6}e_1-\frac{\sqrt{2}}{2}e_2\,,\quad \varepsilon_{33}=\frac{\sqrt{3}}{3}e_0+\frac{\sqrt{6}}{3}e_1\,,\\ |\varepsilon_{12}|=s_{(1)}\,,\quad |\varepsilon_{23}|=s_{(2)}\,,\quad |\varepsilon_{31}|=s_{(3)}\,;\end{gathered}\tag{17.6}$$

$$\begin{gathered}S_{11}=\frac{\sqrt{3}}{3}\sigma_0-\frac{\sqrt{6}}{6}\sigma_1+\frac{\sqrt{2}}{2}\sigma_2\,,\\ S_{22}=\frac{\sqrt{3}}{3}\sigma_0-\frac{\sqrt{6}}{6}\sigma_1-\frac{\sqrt{2}}{2}\sigma_2\,,\quad S_{33}=\frac{\sqrt{3}}{3}\sigma_0+\frac{\sqrt{6}}{3}\sigma_1\,,\\ |S_{12}|=t_{(1)}\,,\quad |S_{23}|=t_{(2)}\,,\quad |S_{31}|=t_{(3)}\,.\end{gathered}\tag{17.7}$$

From expressions (17.5), taking into account (17.6) and (17.7), we obtain the relationships between the components of the strain and stress tensors:

$$\begin{aligned}\varepsilon_{11}&=A_{1111}S_{11}+A_{1122}S_{22}+A_{1133}S_{33}\\ &+B_{1111}S_{11}^2+B_{1122}S_{22}^2+B_{1133}S_{33}^2+B_{1112}S_{11}S_{22}+B_{1113}S_{11}S_{33}+B_{1123}S_{22}S_{33}\\ &+C_{1111}S_{11}^3+C_{1112}S_{11}^2S_{22}+C_{1113}S_{11}^2S_{33}+C_{1122}S_{11}S_{22}^2+C_{1133}S_{11}S_{33}^2\\ &+C_{1123}S_{11}S_{22}S_{33}+C_{1222}S_{22}^3+C_{1223}S_{22}^2S_{33}+C_{1233}S_{22}S_{33}^2+C_{1333}S_{33}^3\,,\\ \varepsilon_{22}&=A_{2211}S_{11}+A_{2222}S_{22}+A_{2233}S_{33}\\ &+B_{2211}S_{11}^2+B_{2222}S_{22}^2+B_{2233}S_{33}^2+B_{2212}S_{11}S_{22}+B_{2213}S_{11}S_{33}+B_{2223}S_{22}S_{33}\\ &+C_{2111}S_{11}^3+C_{2112}S_{11}^2S_{22}+C_{2113}S_{11}^2S_{33}+C_{2122}S_{11}S_{22}^2+C_{2133}S_{11}S_{33}^2\\ &+C_{2123}S_{11}S_{22}S_{33}+C_{2222}S_{22}^3+C_{2223}S_{22}^2S_{33}+C_{2233}S_{22}S_{33}^2+C_{2333}S_{33}^3\,,\\ \varepsilon_{33}&=A_{3311}S_{11}+A_{3322}S_{22}+A_{3333}S_{33}\\ &+B_{3311}S_{11}^2+B_{3322}S_{22}^2+B_{3333}S_{33}^2+B_{3312}S_{11}S_{22}+B_{3313}S_{11}S_{33}+B_{3323}S_{22}S_{33}\\ &+C_{3111}S_{11}^3+C_{3112}S_{11}^2S_{22}+C_{3113}S_{11}^2S_{33}+C_{3122}S_{11}S_{22}^2+C_{3133}S_{11}S_{33}^2\\ &+C_{3123}S_{11}S_{22}S_{33}+C_{3222}S_{22}^3+C_{3223}S_{22}^2S_{33}+C_{3233}S_{22}S_{33}^2+C_{3333}S_{33}^3\,,\\ \varepsilon_{12}&=2\left(D_0^{(1)}+\frac{3}{2}D_1^{(1)}\,|S_{12}|+2D_2^{(1)}S_{12}^2\right)S_{12}\,,\\ \varepsilon_{23}&=2\left(D_0^{(2)}+\frac{3}{2}D_1^{(2)}\,|S_{23}|+2D_2^{(2)}S_{23}^2\right)S_{23}\,,\\ \varepsilon_{31}&=2\left(D_0^{(3)}+\frac{3}{2}D_1^{(3)}\,|S_{31}|+2D_2^{(3)}S_{31}^2\right)S_{31}\,.\end{aligned}\tag{17.8}$$

Constitutive relations (17.8) are a third-order elasticity model for the orthotropic material and contain linear constants A_{ijkl} and $D_0^{(i)}$, quadratic constants B_{ijkl}, $D_1^{(i)}$, and cubic constants C_{ijkl} and $D_2^{(i)}$. If the quadratic and cubic parameters B_{ijkl}, C_{ijkl}, $D_1^{(i)}$, $D_2^{(i)}$ are considered equal to zero, then relations (17.8) will be expressions of the generalized Hooke's law.

According to the data obtained in mechanical experiments with macrosamples, it is more convenient to determine the parameters A_{ijkl}, B_{ijkl} and C_{ijkl} included in relations (17.8). However, to identify the coefficients of model (17.5), it is required to find the values $c_0^{\alpha\beta}$, $c_1^{\alpha\beta}$, $c_2^{\alpha\beta}$. Coefficients A_{ijkl}, B_{ijkl}, C_{ijkl} are linearly related to constants $c_0^{\alpha\beta}$, $c_1^{\alpha\beta}$, $c_2^{\alpha\beta}$ respectively. To find these connections, we satisfy the conditions for the existence of the Gibbs potential in the form

$$\frac{\partial \varepsilon_{11}}{\partial S_{22}} = \frac{\partial \varepsilon_{22}}{\partial S_{11}}, \quad \frac{\partial \varepsilon_{11}}{\partial S_{33}} = \frac{\partial \varepsilon_{33}}{\partial S_{11}}, \quad \frac{\partial \varepsilon_{22}}{\partial S_{33}} = \frac{\partial \varepsilon_{33}}{\partial S_{22}}. \tag{17.9}$$

Let us find the partial derivatives of the strain tensor components with respect to the stress tensor components:

$$\begin{aligned}
\frac{\partial \varepsilon_{11}}{\partial S_{22}} &= A_{1122} + 2B_{1122}S_{22} + B_{1112}S_{11} + B_{1123}S_{33} \\
&+ C_{1112}S_{11}^2 + 2C_{1122}S_{11}S_{22} + C_{1123}S_{11}S_{33} + 3C_{1222}S_{22}^2 + 2C_{1223}S_{22}S_{33} + C_{1233}S_{33}^2, \\
\frac{\partial \varepsilon_{22}}{\partial S_{11}} &= A_{2211} + 2B_{2211}S_{11} + B_{2212}S_{22} + B_{2213}S_{33} \\
&+ 3C_{2111}S_{11}^2 + 2C_{2112}S_{11}S_{22} + 2C_{2113}S_{11}S_{33} + C_{2122}S_{22}^2 + C_{2133}S_{33}^2 + C_{2123}S_{22}S_{33}, \\
\frac{\partial \varepsilon_{11}}{\partial S_{33}} &= A_{1133} + 2B_{1133}S_{33} + B_{1113}S_{11} + B_{1123}S_{22} \\
&+ C_{1113}S_{11}^2 + 2C_{1133}S_{11}S_{33} + C_{1123}S_{11}S_{22} + C_{1223}S_{22}^2 + 2C_{1233}S_{22}S_{33} + 3C_{1333}S_{33}^2, \\
\frac{\partial \varepsilon_{33}}{\partial S_{11}} &= A_{3311} + 2B_{3311}S_{11} + B_{3312}S_{22} + B_{3313}S_{33} \\
&+ 3C_{3111}S_{11}^2 + 2C_{3112}S_{11}S_{22} + 2C_{3113}S_{11}S_{33} + C_{3122}S_{22}^2 + C_{3133}S_{33}^2 + C_{3123}S_{22}S_{33}, \\
\frac{\partial \varepsilon_{22}}{\partial S_{33}} &= A_{2233} + 2B_{2233}S_{33} + B_{2213}S_{11} + B_{2223}S_{22} \\
&+ C_{2113}S_{11}^2 + 2C_{2133}S_{11}S_{33} + C_{2123}S_{11}S_{22} + C_{2223}S_{22}^2 + 2C_{2233}S_{22}S_{33} + 3C_{2333}S_{33}^2, \\
\frac{\partial \varepsilon_{33}}{\partial S_{22}} &= A_{3322} + 2B_{3322}S_{22} + B_{3312}S_{11} + B_{3323}S_{33} \\
&+ C_{3112}S_{11}^2 + 2C_{3122}S_{11}S_{22} + C_{3123}S_{11}S_{33} + 3C_{3222}S_{22}^2 + 2C_{3223}S_{22}S_{33} + C_{3233}S_{33}^2.
\end{aligned}$$

Substituting the obtained expressions into conditions (17.9), we receive the relations between the model parameters:

$$\begin{aligned}
&A_{1122} = A_{2211},\ A_{1133} = A_{3311},\ A_{2233} = A_{3322},\ B_{2212} = 2B_{1122},\ B_{1112} = 2B_{2211}, \\
&B_{1123} = B_{2213},\ B_{3313} = 2B_{1133},\ B_{1113} = 2B_{3311},\ B_{1123} = B_{3312},\ B_{3323} = 2B_{2233}, \\
&B_{2223} = 2B_{3322},\ B_{2213} = B_{3312},\ C_{1112} = 3C_{2111},\ C_{1122} = C_{2112},\ C_{2122} = 3C_{1222}, \\
&C_{1113} = 3C_{3111},\ C_{1133} = C_{3113},\ C_{3133} = 3C_{1333},\ C_{2223} = 3C_{3222},\ C_{2233} = C_{3223}, \\
&C_{3233} = 3C_{2333}, \\
&C_{1123} = 2C_{2113} = 2C_{3112},\ C_{2123} = 2C_{1223} = 2C_{3122},\ C_{3123} = 2C_{1233} = 2C_{2133}.
\end{aligned} \tag{17.10}$$

Taking into account the equalities (17.10), the constitutive relations (17.8) take the form

$$
\begin{aligned}
\varepsilon_{11} &= A_{1111}S_{11} + A_{1122}S_{22} + A_{1133}S_{33} \\
&+ B_{1111}S_{11}^2 + B_{1122}S_{22}^2 + B_{1133}S_{33}^2 + 2B_{2211}S_{11}S_{22} + 2B_{3311}S_{11}S_{33} + B_{1123}S_{22}S_{33} \\
&+ C_{1111}S_{11}^3 + 3C_{2111}S_{11}^2S_{22} + 3C_{3111}S_{11}^2S_{33} + C_{1122}S_{11}S_{22}^2 + C_{1133}S_{11}S_{33}^2 \\
&+ 2C_{2113}S_{11}S_{22}S_{33} + C_{1222}S_{22}^3 + C_{1223}S_{22}^2S_{33} + C_{1233}S_{22}S_{33}^2 + C_{1333}S_{33}^3 , \\
\varepsilon_{22} &= A_{1122}S_{11} + A_{2222}S_{22} + A_{2233}S_{33} \\
&+ B_{2211}S_{11}^2 + B_{2222}S_{22}^2 + B_{2233}S_{33}^2 + 2B_{1122}S_{11}S_{22} + B_{1123}S_{11}S_{33} + 2B_{3322}S_{22}S_{33} \\
&+ C_{2111}S_{11}^3 + C_{1122}S_{11}^2S_{22} + C_{2113}S_{11}^2S_{33} + 3C_{1222}S_{11}S_{22}^2 + C_{1233}S_{11}S_{33}^2 \\
&+ 2C_{1223}S_{11}S_{22}S_{33} + C_{2222}S_{22}^3 + 3C_{3222}S_{22}^2S_{33} + C_{2233}S_{22}S_{33}^2 + C_{2333}S_{33}^3 , \qquad (17.11) \\
\varepsilon_{33} &= A_{1133}S_{11} + A_{2233}S_{22} + A_{3333}S_{33} \\
&+ B_{3311}S_{11}^2 + B_{3322}S_{22}^2 + B_{3333}S_{33}^2 + B_{1123}S_{11}S_{22} + 2B_{1133}S_{11}S_{33} + 2B_{2233}S_{22}S_{33} \\
&+ C_{3111}S_{11}^3 + C_{2113}S_{11}^2S_{22} + C_{1133}S_{11}^2S_{33} + C_{1223}S_{11}S_{22}^2 + 3C_{1333}S_{11}S_{33}^2 \\
&+ 2C_{1233}S_{11}S_{22}S_{33} + C_{3222}S_{22}^3 + C_{2233}S_{22}^2S_{33} + 3C_{2333}S_{22}S_{33}^2 + C_{3333}S_{33}^3 , \\
\varepsilon_{12} &= 2\left(D_0^{(1)} + \frac{3}{2}D_1^{(1)}\,|S_{12}| + 2D_2^{(1)}S_{12}^2\right)S_{12} , \\
\varepsilon_{23} &= 2\left(D_0^{(2)} + \frac{3}{2}D_1^{(2)}\,|S_{23}| + 2D_2^{(2)}S_{23}^2\right)S_{23} , \\
\varepsilon_{31} &= 2\left(D_0^{(3)} + \frac{3}{2}D_1^{(3)}\,|S_{31}| + 2D_2^{(3)}S_{31}^2\right)S_{31} .
\end{aligned}
$$

Constitutive relations (17.11) include 9 linear, 9 quadratic, and 9 cubic constants. All 6 parameters A_{ijkl}, all 9 parameters $D_0^{(1)}$, $D_0^{(2)}$, $D_0^{(3)}$, $D_1^{(1)}$, $D_1^{(2)}$, $D_1^{(3)}$, $D_2^{(1)}$, $D_2^{(2)}$, $D_2^{(3)}$, 6 of 10 parameters B_{ijkl}, and 6 of 15 parameters C_{ijkl} are linearly independent. If all parameters C_{ijkl}, $D_2^{(1)}$, $D_2^{(2)}$, $D_2^{(3)}$ are equal to zero, then relations (17.11) will be a quadratic model of the orthotropic material.

From expressions (17.10) and (17.5), taking into account the expressions for invariants (17.6), (17.7), we obtain

$$
\begin{aligned}
A_{1111} &= \frac{c_0^{00}}{3} - \frac{\sqrt{2}c_0^{01}}{3} + \frac{\sqrt{6}c_0^{02}}{3} + \frac{c_0^{11}}{6} - \frac{\sqrt{3}c_0^{12}}{3} + \frac{c_0^{22}}{2} , \\
A_{1122} &= \frac{c_0^{00}}{3} - \frac{\sqrt{2}c_0^{01}}{3} + \frac{c_0^{11}}{6} - \frac{c_0^{22}}{2} , A_{1133} = \frac{c_0^{00}}{3} + \frac{\sqrt{2}c_0^{01}}{6} + \frac{\sqrt{6}c_0^{02}}{6} - \frac{c_0^{11}}{3} + \frac{\sqrt{3}c_0^{12}}{3} , \\
A_{2222} &= \frac{c_0^{00}}{3} - \frac{\sqrt{2}c_0^{01}}{3} - \frac{\sqrt{6}c_0^{02}}{3} + \frac{c_0^{11}}{6} + \frac{\sqrt{3}c_0^{12}}{3} + \frac{c_0^{22}}{2} , \qquad (17.12) \\
A_{2233} &= \frac{c_0^{00}}{3} + \frac{\sqrt{2}c_0^{01}}{6} - \frac{\sqrt{6}c_0^{02}}{6} - \frac{c_0^{11}}{3} - \frac{\sqrt{3}c_0^{12}}{3} , \quad A_{3333} = \frac{c_0^{00}}{3} + \frac{2\sqrt{2}c_0^{01}}{3} + \frac{2c_0^{11}}{3} ,
\end{aligned}
$$

$$
\begin{aligned}
B_{1111} &= \frac{\sqrt{3}}{3}c_1^{00} + \frac{\sqrt{3}-\sqrt{6}}{6}c_1^{01} + \frac{\sqrt{2}+\sqrt{3}}{2}c_1^{02} - \frac{\sqrt{6}}{12}c_1^{11} + \frac{\sqrt{2}-\sqrt{6}}{4}c_1^{12} + \frac{3\sqrt{2}}{4}c_1^{22},\\
B_{1122} &= \frac{\sqrt{3}}{3}c_1^{00} + \frac{\sqrt{3}-\sqrt{6}}{6}c_1^{01} - \frac{\sqrt{2}+\sqrt{3}}{2}c_1^{02} - \frac{\sqrt{6}}{12}c_1^{11} + \frac{\sqrt{6}-\sqrt{2}}{12}c_1^{12} + \frac{3\sqrt{2}}{4}c_1^{22},\\
B_{1133} &= \frac{\sqrt{3}}{3}c_1^{00} + \frac{\sqrt{6}}{6}c_1^{01} + \frac{\sqrt{2}}{6}c_1^{02} - \frac{\sqrt{6}}{3}c_1^{11} + \frac{\sqrt{2}}{3}c_1^{12},\\
B_{1123} &= \frac{2\sqrt{3}}{3}c_1^{00} - \frac{\sqrt{3}}{3}c_1^{01} - \frac{\sqrt{3}}{3}c_1^{02} + \frac{\sqrt{6}}{3}c_1^{11} - \frac{\sqrt{6}}{3}c_1^{12},\\
B_{2211} &= \frac{\sqrt{3}}{3}c_1^{00} + \frac{\sqrt{3}-\sqrt{6}}{6}c_1^{01} + \frac{\sqrt{2}-\sqrt{3}}{6}c_1^{02} - \frac{\sqrt{6}}{12}c_1^{11} + \frac{\sqrt{2}+\sqrt{6}}{12}c_1^{12} - \frac{3\sqrt{2}}{4}c_1^{22},\\
B_{2222} &= \frac{\sqrt{3}}{3}c_1^{00} + \frac{\sqrt{3}-\sqrt{6}}{6}c_1^{01} + \frac{\sqrt{3}-\sqrt{2}}{2}c_1^{02} - \frac{\sqrt{6}}{12}c_1^{11} - \frac{\sqrt{2}+\sqrt{6}}{4}c_1^{12} - \frac{3\sqrt{2}}{4}c_1^{22},\\
B_{2233} &= \frac{\sqrt{3}}{3}c_1^{00} + \frac{\sqrt{6}}{6}c_1^{01} - \frac{\sqrt{2}}{6}c_1^{02} - \frac{\sqrt{6}}{3}c_1^{11} - \frac{\sqrt{2}}{3}c_1^{12},\\
B_{3311} &= \frac{\sqrt{3}}{3}c_1^{00} - \frac{\sqrt{3}}{6}c_1^{01} + \frac{2\sqrt{2}+\sqrt{3}}{6}c_1^{02} + \frac{\sqrt{6}}{6}c_1^{11} + \frac{\sqrt{6}-2\sqrt{2}}{6}c_1^{12},\\
B_{3322} &= \frac{\sqrt{3}}{3}c_1^{00} - \frac{\sqrt{3}}{6}c_1^{01} + \frac{\sqrt{3}-2\sqrt{2}}{6}c_1^{02} + \frac{\sqrt{6}}{6}c_1^{11} + \frac{2\sqrt{2}+\sqrt{6}}{6}c_1^{12},\\
B_{3333} &= \frac{\sqrt{3}}{3}c_1^{00} + \frac{\sqrt{6}+2\sqrt{3}}{3}c_1^{01} + \frac{2\sqrt{6}}{3}c_1^{11}.
\end{aligned}
\tag{17.13}
$$

In the quadratic part of the constitutive relations, 10 coefficients B_{iijk} are linearly expressed in terms of 6 parameters $c_1^{\alpha\beta}$, therefore, 6 constants B_{iijk} are linearly independent. We choose the following as independent parameters: B_{1111}, B_{1122}, B_{2211}, B_{2222}, B_{3311}, B_{3333}. The remaining 4 of 10 parameters B_{1123}, B_{1133}, B_{2233}, B_{3322} are linearly expressed through them.

In the cubic part of the constitutive relations (17.11), the parameters C_{1111}, C_{1222}, C_{2111}, C_{2222}, C_{3111}, C_{3333} are chosen as independent. The remaining 9 of 15 parameters are linearly expressed in terms of them. The parameters C_{1111}, C_{1222}, C_{2111}, C_{2222}, C_{3111} C_{3333} are expressed in terms of constants $c_2^{\alpha\beta}$ according to the following formulae:

$$
\begin{aligned}
C_{1111} &= \frac{2}{3}c_2^{00} + \frac{2-3\sqrt{2}}{9}c_2^{01} + \frac{6+5\sqrt{6}}{9}c_2^{02} + \frac{1}{6}c_2^{11} + \frac{3-4\sqrt{3}}{9}c_2^{12} + \frac{3}{2}c_2^{22},\\
C_{1222} &= \frac{2}{3}c_2^{00} + \frac{2-3\sqrt{2}}{9}c_2^{01} + \frac{\sqrt{6}}{18}c_2^{02} + \frac{1}{6}c_2^{11} - \frac{\sqrt{3}}{9}c_2^{12} - \frac{3}{2}c_2^{22},\\
C_{2111} &= \frac{2}{3}c_2^{00} + \frac{2-3\sqrt{2}}{9}c_2^{01} - \frac{\sqrt{6}}{18}c_2^{02} + \frac{1}{6}c_2^{11} + \frac{\sqrt{3}}{9}c_2^{12} - \frac{3}{2}c_2^{22},\\
C_{2222} &= \frac{2}{3}c_2^{00} + \frac{2-3\sqrt{2}}{9}c_2^{01} + \frac{6-5\sqrt{6}}{9}c_2^{02} + \frac{1}{6}c_2^{11} + \frac{3+4\sqrt{3}}{9}c_2^{12} + \frac{3}{2}c_2^{22},\\
C_{3111} &= \frac{2}{3}c_2^{00} + \frac{3\sqrt{2}-4}{36}c_2^{01} + \frac{4+3\sqrt{6}}{12}c_2^{02} - \frac{1}{3}c_2^{11} + \frac{\sqrt{3}-1}{3}c_2^{12},
\end{aligned}
\tag{17.14}
$$

$$C_{3333} = \frac{2}{3}c_2^{00} + \frac{8+12\sqrt{2}}{9}c_2^{01} + \frac{8}{3}c_2^{11} \ .$$

If the parameters A_{ijkl}, B_{ijkl}, C_{ijkl} are determined from experiments, then the elastic constants $c_0^{\alpha\beta}$, $c_1^{\alpha\beta}$, $c_2^{\alpha\beta}$ of the material can be found by inverting relations (17.12)–(17.14):

$$c_0^{00} = \frac{1}{3}(A_{1111} + 2A_{1122} + 2A_{1133} + A_{2222} + 2A_{2233} + A_{3333}) \ ,$$

$$c_0^{01} = -\frac{\sqrt{2}}{6}(A_{1111} + 2A_{1122} - A_{1133} + A_{2222} - A_{2233} - A_{3333}) \ ,$$

$$c_0^{02} = \frac{\sqrt{6}}{6}(A_{1111} + A_{1133} - A_{2222} - A_{2233}) \ ,$$

$$c_0^{11} = \frac{1}{6}(A_{1111} + 2A_{1122} - 4A_{1133} + A_{2222} - 4A_{2233} + 4A_{3333}) \ ,$$

$$c_0^{12} = -\frac{\sqrt{3}}{6}(A_{1111} - 2A_{1133} - A_{2222} + 2A_{2233}) \ , \quad c_0^{22} = \frac{1}{2}(A_{1111} - 2A_{1122} + A_{2222}) \ ;$$

$$\begin{aligned}
c_1^{00} = {} & \frac{29\sqrt{6} - 9\sqrt{2} - 28\sqrt{3}}{108}B_{1111} + \frac{3\sqrt{2} + 20\sqrt{3} - 7\sqrt{6}}{36}B_{1122} \\
& + \frac{11\sqrt{6} - 17\sqrt{2} - 12\sqrt{3} + 24}{12}B_{2211} + \frac{153\sqrt{2} + 140\sqrt{3} - 91\sqrt{6} - 216}{108}B_{2222} \\
& + \frac{4\sqrt{3} - 2\sqrt{6}}{9}B_{3311} + \frac{2\sqrt{6} - \sqrt{3}}{27}B_{3333} \ ,
\end{aligned}$$

$$\begin{aligned}
c_1^{01} = {} & \frac{9\sqrt{2} + 11\sqrt{3} - 9\sqrt{6} + 9}{36}B_{1111} + \frac{3\sqrt{6} - 3\sqrt{2} - \sqrt{3} - 3}{12}B_{1122} + \frac{2\sqrt{3}}{9}B_{3333} \\
& + \frac{5\sqrt{2} + 5\sqrt{3} - 3\sqrt{6} - 7}{4}B_{2211} + \frac{\sqrt{6} - 45\sqrt{2} - 37\sqrt{3} + 63}{36}B_{2222} - \frac{2\sqrt{3}}{3}B_{3311} \ ,
\end{aligned}$$

$$\begin{aligned}
c_1^{02} = {} & \frac{3\sqrt{2} - \sqrt{3} + \sqrt{6} - 3}{4}B_{1111} + \frac{\sqrt{3} - 3\sqrt{2} - \sqrt{6} + 3}{4}B_{1122} \\
& + \frac{3\sqrt{2} + \sqrt{3} - \sqrt{6} - 3}{4}B_{2211} + \frac{\sqrt{6} - 3\sqrt{2} - \sqrt{3} + 3}{4}B_{2222} \ ,
\end{aligned}$$

$$\begin{aligned}
c_1^{11} = {} & \frac{4\sqrt{6} - 27\sqrt{2} - 4\sqrt{3} - 36}{108}B_{1111} + \frac{9\sqrt{2} - 4\sqrt{3} - 8\sqrt{6} + 12}{36}B_{1122} \\
& + \frac{4 - 3\sqrt{2} - 4\sqrt{3}}{12}B_{2211} + \frac{27\sqrt{2} + 20\sqrt{3} - 20\sqrt{6} - 36}{108}B_{2222} \\
& + \frac{4\sqrt{3} + 2\sqrt{6}}{9}B_{3311} + \frac{4\sqrt{6} - 4\sqrt{3}}{27}B_{3333} \ ,
\end{aligned}$$

$$\begin{aligned}
c_1^{12} = {} & \frac{2\sqrt{3} - 3\sqrt{2} - 2\sqrt{6} + 6}{4}B_{1111} + \frac{3\sqrt{2} - 2\sqrt{3} + 2\sqrt{6} - 6}{4}B_{1122} \\
& + \frac{2\sqrt{6} - 3\sqrt{2} - 2\sqrt{3} + 6}{4}B_{2211} + \frac{3\sqrt{2} + 2\sqrt{3} - 2\sqrt{6} - 6}{4}B_{2222} \ ,
\end{aligned}$$

$$
\begin{aligned}
c_1^{22} &= \frac{\sqrt{2}}{12}(B_{1111} + 3B_{1122} - 3B_{2211} - B_{2222})\ ,\\
c_2^{00} &= \frac{111\sqrt{2} + 93\sqrt{3} - 248\sqrt{6} - 795}{1953}C_{1111} + \frac{568 - 40\sqrt{2}}{651}C_{3111}\\
&+ \frac{516\sqrt{2} + 954\sqrt{3} - 1088\sqrt{6} - 6546}{1953}C_{2111}\\
&+ \frac{248\sqrt{6} - 39\sqrt{2} - 93\sqrt{3} + 1335}{1953}C_{2222}\\
&+ \frac{1088\sqrt{6} - 444\sqrt{2} - 954\sqrt{3} + 7086}{1953}C_{1222} + \frac{97 - 16\sqrt{2}}{1302}C_{3333}\ ,\\
c_2^{01} &= \frac{124\sqrt{3} - 1665\sqrt{2} + 248\sqrt{6} - 444}{2604}C_{1111} + \frac{150\sqrt{2} + 40}{217}C_{3111}\\
&+ \frac{220\sqrt{3} - 1935\sqrt{2} + 608\sqrt{6} - 516}{2604}C_{2111}\\
&+ \frac{585\sqrt{2} - 124\sqrt{3} - 248\sqrt{6} + 156}{2604}C_{2222}\\
&+ \frac{1665\sqrt{2} - 220\sqrt{3} - 608\sqrt{6} + 156}{651}C_{1222} + \frac{30\sqrt{2} + 8}{217}C_{3333}\ ,\\
c_2^{02} &= \frac{\sqrt{6}}{4}(C_{1111} - 4C_{1222} + 4C_{2111} - C_{2222})\ ,\\
c_2^{11} &= \frac{4565\sqrt{3} - 4440\sqrt{2} + 1332\sqrt{6} - 14421}{3906}C_{1222} + \frac{134 - 40\sqrt{2}}{651}C_{3333}\\
&+ \frac{2220\sqrt{2} - 961\sqrt{3} - 186\sqrt{6} + 6234}{7812}C_{1111}\\
&+ \frac{961\sqrt{3} - 780\sqrt{2} + 186\sqrt{6} - 3246}{7812}C_{2222}\\
&+ \frac{5160\sqrt{2} - 4565\sqrt{3} - 1332\sqrt{6} + 15915}{3906}C_{2111} - \frac{200\sqrt{2} + 632}{651}C_{3111}\ ,\\
c_2^{12} &= \frac{\sqrt{3}}{4}(C_{1111} - 10C_{1222} + 10C_{2111} - C_{2222})\ ,\\
c_2^{22} &= \frac{6 - \sqrt{3} - 2\sqrt{6}}{36}C_{1111} + \frac{5\sqrt{3} + 4\sqrt{6} - 3}{18}C_{1222}\\
&- \frac{5\sqrt{3} + \sqrt{6} + 3}{18}C_{2111} + \frac{\sqrt{3} + 2\sqrt{6} + 6}{36}C_{2222}\ .
\end{aligned}
$$

Other sets of constants B_{ijkl}, C_{ijkl} can also act as independent ones. However, the choice made is explained by the fact that the parameters B_{1111}, B_{1122}, B_{2211}, B_{2222}, B_{3311}, B_{3333} and C_{1111}, C_{1222}, C_{2111}, C_{2222}, C_{3111}, C_{3333} can be determined from the results of the same mechanical experiments with macrosamples.

17.3 Identification of the Model from the Data of Experiments with Plates

Experiments on plates are widespread, since deformations in two directions in a plane can be measured in them. To determine the parameters A_{ijkl}, B_{ijkl}, C_{ijkl} samples for experiments should be cut in mutually perpendicular directions formed by the principal axes of anisotropy of the orthotropic material.

The parameters $A_{1111}, A_{1122}, B_{1111}, B_{2211}, C_{1111}, C_{2111}$ can be determined from the tension-compression experiment along the axis Ox_1 in the plane Ox_1x_2. The parameters $A_{2222}, B_{2222}, B_{1122}, C_{2222}, C_{1222}$ can be determined from the tension-compression experiment along the axis Ox_2 in the plane Ox_1x_2. The parameters A_{1133}, A_{3333}, B_{1133}, B_{3333}, C_{1333}, C_{3333} can be determined from the tension-compression experiment along the axis Ox_3 in the plane Ox_1x_3. The parameter A_{2233} can be determined from the tension-compression experiment along the axis Ox_3 in the plane Ox_2x_3. To determine some parameters, one can use the results of experiments on uniaxial tension-compression at the angle 45° to the principal axes of anisotropy. The coefficients $D_i^{(j)}$ of the model can be determined from the results of three shear experiments in the planes Ox_1x_2 ($D_0^{(1)}$, $D_1^{(1)}$, $D_2^{(1)}$), Ox_2x_3 ($D_0^{(2)}$, $D_1^{(2)}$, $D_2^{(2)}$), Ox_1x_3 ($D_0^{(3)}$, $D_1^{(3)}$, $D_2^{(3)}$) formed by the principal directions of anisotropy of the orthotropic material. In this case, shear experiments can be performed as biaxial tension-compression experiments at the angle 45° to the principal axes of anisotropy.

In [2], experimental diagrams of stress versus strains are given. These diagrams are built as results of tests for uniaxial tension, uniaxial compression along the principal axes of anisotropy $\vec{a}_1$ and $\vec{a}_3$, as well as shear in the planes $\vec{a}_1$, $\vec{a}_2$ and $\vec{a}_1$, $\vec{a}_3$ of carbon-carbon composite material AVCO Mod3a. It is noted in [2] that the mechanical behavior in the $\vec{a}_2$ direction is identical to the behavior in the $\vec{a}_1$ direction due to the manufacturing method of the material. Experimental data show that the composite material used has a diversity resistance in tension and compression.

Consider an orthotropic plate, in which one axis of anisotropy $\vec{a}_2$ is perpendicular to the plane, and two others $\vec{a}_1$, $\vec{a}_3$ are directed along the sides of the plate. Let us direct the axes of the Cartesian coordinate system along the sides of the plate. Assuming a stress state with $S_{12} = S_{22} = S_{23} = 0$, the constitutive relations (17.11) are written in the following form:

$$
\begin{aligned}
\varepsilon_{11} &= A_{1111}S_{11} + A_{1133}S_{33} + B_{1111}S_{11}^2 + B_{1133}S_{33}^2 + 2B_{3311}S_{11}S_{33} \\
&\quad + C_{1111}S_{11}^3 + 3C_{3111}S_{11}^2S_{33} + C_{1133}S_{11}S_{33}^2 \,, \\
\varepsilon_{22} &= A_{1122}S_{11} + A_{2233}S_{33} + B_{2211}S_{11}^2 + B_{2233}S_{33}^2 + B_{1123}S_{11}S_{33} \\
&\quad + C_{2111}S_{11}^3 + C_{2113}S_{11}^2S_{33} + C_{1233}S_{11}S_{33}^2 + C_{2333}S_{33}^3 \,, \\
\varepsilon_{33} &= A_{1133}S_{11} + A_{3333}S_{33} + B_{3311}S_{11}^2 + B_{3333}S_{33}^2 + 2B_{1133}S_{11}S_{33} \\
&\quad + C_{3111}S_{11}^3 + C_{1133}S_{11}^2S_{33} + 3C_{1333}S_{11}S_{33}^2 + C_{3333}S_{33}^3 \,, \\
\varepsilon_{31} &= 2\left(D_0^{(3)} + \frac{3}{2}D_1^{(3)}\,|S_{31}| + 2D_2^{(3)}S_{31}^2\right)S_{31} \,.
\end{aligned}
$$

Experimental data [2] made it possible, using the least squares method, to determine the material parameters included in the constitutive relations (17.11). When using the quadratic model

$$
\begin{aligned}
&A_{1111} = A_{2222} = 1.70 \times 10^{-10}\,\mathrm{Pa}^{-1}\,, && A_{3333} = 7.58 \times 10^{-11}\,\mathrm{Pa}^{-1}\,,\\
&B_{1111} = B_{2222} = 2.69 \times 10^{-13}\,\mathrm{Pa}^{-2}\,, && B_{3333} = -4.32 \times 10^{-13}\,\mathrm{Pa}^{-2}\,,\\
&D_0^{(1)} = 4.39 \times 10^{-11}\,\mathrm{Pa}^{-1}\,, && D_1^{(1)} = 6.44 \times 10^{-12}\,\mathrm{Pa}^{-2}\,,\\
&D_0^{(2)} = D_0^{(3)} = 8.97 \times 10^{-10}\,\mathrm{Pa}^{-1}\,, && D_1^{(2)} = D_1^{(3)} = -3.01 \times 10^{-11}\,\mathrm{Pa}^{-2}\,.
\end{aligned}
\tag{17.15}
$$

For the model of the third order

$$
\begin{aligned}
&A_{1111} = A_{2222} = 1.12 \times 10^{-10}\,\mathrm{Pa}^{-1}\,, && A_{3333} = 6.48 \times 10^{-11}\,\mathrm{Pa}^{-1}\,,\\
&B_{1111} = B_{2222} = 9.15 \times 10^{-13}\,\mathrm{Pa}^{-2}\,, && B_{3333} = -7.82 \times 10^{-13}\,\mathrm{Pa}^{-2}\,,\\
&C_{1111} = C_{2222} = 2.86 \times 10^{-14}\,\mathrm{Pa}^{-3}\,, && C_{3333} = 5.94 \times 10^{-15}\,\mathrm{Pa}^{-3}\,,\\
&D_0^{(1)} = 1.93 \times 10^{-10}\,\mathrm{Pa}^{-1}\,, && D_0^{(2)} = D_0^{(3)} = 8.78 \times 10^{-10}\,\mathrm{Pa}^{-1}\,,\\
&D_1^{(1)} = -1.08 \times 10^{-11}\,\mathrm{Pa}^{-2}\,, && D_1^{(2)} = D_1^{(3)} = -2.47 \times 10^{-11}\,\mathrm{Pa}^{-2}\,,\\
&D_2^{(1)} = 5.19 \times 10^{-13}\,\mathrm{Pa}^{-3}\,, && D_2^{(2)} = D_2^{(3)} = -3.69 \times 10^{-13}\,\mathrm{Pa}^{-3}\,.
\end{aligned}
\tag{17.16}
$$

Figure 17.1 shows experimental data for AVCO Mod3a material and curves corresponding to linear, quadratic, and cubic models (17.11). Analysis of the graphs in Fig. 17.1 allows us to conclude that the nonlinear constitutive relations (17.11) describe the results of experiments with anisotropic material more adequately than the linear relations of the generalized Hooke's law. They allow one to describe the experimentally observed tensile and compressive diversity resistance of an anisotropic material. In this case, the quadratic model with constants (17.15) incorrectly describes the data of the tensile experiment along the $\vec{a}_3$ axis, therefore, to take into account the nonlinearity, one should use the cubic model (17.11) with parameters (17.16). Note that in Fig. 17.1 (d) the curves constructed by the quadratic and cubic models practically coincide.

The experimental data given in [10] were also used in [4] to identify and to verify a model that takes into account the nonlinear dependence of shear deformations on tangential stresses, but postulates a linear relationship between strains and stresses under tension and compression along the main axes of anisotropy. The model (17.11) considered in the present work takes into account the nonlinearity in the elastic behavior of an anisotropic material not only at shear, but also at uniaxial loading, and its parameters are identified from the data of such experiments.

17.4 Conclusion

Elastic deformations of the orthotropic material at a constant temperature are considered. The elastic potential which coincides with the Gibbs potential is used to describe the isothermal process. On the basis of the specification of the Gibbs potential for reversible processes in orthotropic materials, nonlinear variants of constitutive relations are obtained. These relations satisfy the generalized postulate of Il'yushin for anisotropic media [7]. Models of elasticity of the second and third orders for orthotropic material are constructed.

Linear relations between the constants of the model, which follow from the condition for the existence of the Gibbs potential are established. Linear relationships between the coefficients of the quadratic and cubic parts of the relations are revealed. These linear dependencies are a consequence of an increase in the number of coefficients in the transition from the relations between the invariants of strains and stresses to the relations between the components of the strain tensor and the stress tensor.

Within the framework of the hypotheses used, the constitutive relations of the second and third orders contain 18 and 27 material constants respectively. The models allow one to describe the nonlinear dependencies of strains on stresses known from experiments in the processes of tension, compression and shear, as well as the difference in resistance of anisotropic materials.

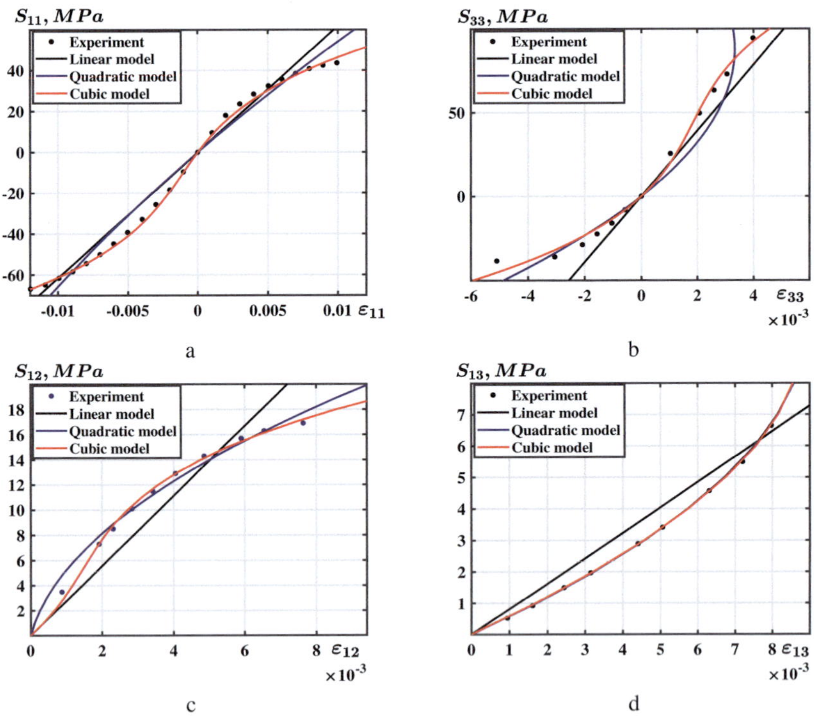

Fig. 17.1: Dependencies of stresses on strains: tension-compression experiment in the direction $\vec{a}_1$ (a), tension-compression experiment in the direction $\vec{a}_3$ (b), shear experiment in the plane $\vec{a}_1$, $\vec{a}_2$ (c), shear experiment in the plane $\vec{a}_1$, $\vec{a}_3$ (d) .

A system of experiments, according to which it is possible to determine the parameters of linear and nonlinear models is proposed. The system includes five uniaxial tension and compression experiments and three shear experiments. Tension and compression are performed along the main directions of anisotropy. Longitudinal and transverse deformations should be measured in each experiment. The shear experiments are performed in the planes formed by the main anisotropy vectors of the orthotropic material.

It is shown that the proposed model allows one to describe the experimentally observed nonlinearity of the relationship between stresses and strains not only in shear processes, but also in tension and compression processes along the main axes of anisotropy, as well as the difference in resistance of anisotropic materials. The identification of the model parameters is carried out according to the experimental data [2, 10]. Numerical values of the parameters of the constructed nonlinear models for two types of orthotropic materials are determined based on the data of experiments with composite materials known from the literature.

The models under consideration take into account the nonlinearity at small elastic deformations of an anisotropic material not only under shear, but also under uniaxial loading. At uniaxial tension and compression, the third-order model corresponds to experimental data better than the second-order one. The description of the nonlinearity in tension and compression distinguishes the proposed models from the known from the literature model of Lomakin [4]. Nonlinear shear deformation is described equally well by all of these models.

References

[1] Fedulov B, Fedorenko A, Safonov A, Lomakin E (2017) Nonlinear shear behavior and failure of composite materials under plane stress conditions. Acta Mechanica 228(6):2033–2040, DOI 10.1007/s00707-017-1817-4

[2] Jones RM (1980) Modeling nonlinear deformation of carbon-carbon composite materials. AIAA Journal 18(8):995–1001, DOI 10.1007/10.2514/3.50844

[3] Lellep J, Majak J (2000) Nonlinear constitutive behavior of orthotropic materials. Mechanics of Composite Materials 36(4):261–266, DOI 10.1007/BF02262803

[4] Lomakin EV, Fedulov BN (2015) Nonlinear anisotropic elasticity for laminated composites. Meccanica 50(6):1527–1535, DOI 10.1007/S11012-015-0104-5

[5] Lomakin E, Fedulov B (2018) Nonlinear deformation and failure analysis of laminated composites. Procedia Structural Integrity 13:664–669, DOI 10.1016/j.prostr.2018.12.110

[6] Lüders C (2020) Nonlinear-elastic orthotropic material modeling of an epoxy-based polymer for predicting the material behavior of transversely loaded fiber-reinforced composites. Procedia Structural Integrity 4:1–23, DOI 10.3390/jcs4020046

[7] Markin AA, Sokolova MYu, Khristich DV (2011) A.A. Il'yushin's postulate for anisotropic materials and a version of constitutive relations. Mechanics of Solids 46(1):30–35, DOI 10.3103/S0025654411010055
[8] Markin AA, Sokolova MYu (2015) Thermomechanics of Elastoplastic Deformation. Cambridge International Science Publishing, Cambridge
[9] Saliklis EP, Urbanik TJ, Tokyay B (2003) Bilinear modelling of cellulosic orthotropic nonlinear materials. Journal of Pulp and Paper Science 29(12):407–411
[10] Smith EW, Pascoe KJ (1977) The role of shear deformation in the fatigue failure of a glass fiber-reinforced composite. Composites 8(4):237–243, DOI 10.1016/0010-4361(77)90109-4
[11] Sokolova M, Rudakov V, Khristich D (2019) Strains of plates of nonlinear anisotropic materials. IOP Conference Series: Journal of Physics: Conference Series 1203(1):1–11, DOI 10.1088/1742-6596/1203/1/012024
[12] Tabiei A, Wu J (2020) Three-dimensional nonlinear orthotropic finite element material model for wood. Composite Structures 50(2):143–149, DOI 10.1016/S0263-8223(00)00089-1
[13] Yang B, Yu Z, Zhang Q, Shang Y, Yan Y (2020) The nonlinear orthotropic material model describing biaxial tensile behavior of PVC coated fabrics. Composite Structures 236(4):111850, DOI 10.1016/j.compstruct.2019.111850
[14] Yao Wj, Zhang C-h, Jiang X-f (2006) Nonlinear mechanical behavior of combined members with different moduli. International Journal of Nonlinear Sciences and Numerical Simulation 7(2):233–238, DOI 10.1515/IJNSNS.2006.7.2.233

Chapter 18
Diagrams of Statically Determinable Trusses Allowing Kinematic Variability

Mikhail Kirsanov

Abstract This review article presents diagrams of plane, statically determinate, regular trusses that allow kinematic variability for a given number of panels. Corresponding node velocity distributions are shown. Simple beam trusses with two supports, as well as more complex lattice trusses with additional supports and arched trusses, are considered. The geometric characteristics of variable trusses are presented. Two illustrative, sufficient criteria for determining the kinematic variability of a rod structure are provided, requiring no calculations. All diagrams were obtained by the author or his students by deriving analytical relationships between deflection and natural frequency and the number of panels.

Key words: Plane truss, Analytical solution, Deflection, Induction, Kinematic variability, External static indeterminacy, Multi-span truss

18.1 Introduction

Trusses are the most commonly used structural elements in both construction and mechanical engineering. To calculate deformations, forces, and natural frequencies, both numerical methods using the finite element method and analytical approaches are typically used. A separate challenge is the calculation of regular trusses with a periodic structure. For such problems, analytical calculation methods based on the induction method are acceptable. This method allows for obtaining relatively simple relationships between the desired quantity and the structural order. Here, order refers to the number of periodic elements, such as the number of panels or groups of panels. The importance of such calculations lies not only in the fact that the solution is obtained in the form of a short, easy-to-use and easy-to-analyze

Mikhail Kirsanov
Faculty of Mechanics and Mathematics, Moscow State University, Leninskie gory, 119992 Moscow, Russian Federation, e-mail: c216@ya.ru

H. Altenbach et al. (eds.), *Stability Problems in Mechanics of Continua*,
Advanced Structured Materials 260, https://doi.org/10.1007/978-3-032-31032-3_18

formula, but also in the fact that analytical solutions overcome the so-called "curse of dimensionality", which is associated with the inevitable accumulation of rounding errors. This accumulation of errors occurs in regular structures with a large or very large number of elements. These can include complex airport pavements, space antennas, and other truss structures with a large number of elements. However, the "curse of dimensionality" is not the main problem with regular systems. Practical calculations show that in some cases, a calculation that yields a result for one structural order fails for a different order of the same system—the determinant of the resolving system of equations vanishes. Schematics of such planar regular trusses are presented in this paper.

18.2 Schemes

The kinematic degeneracy effect occurs in regular beam structures. Among the first to address this issue were Hutchinson and Fleck [3, 4]. The author's reference books [5, 6] contain various schemes and calculation formulas for the deflections of statically determinate regular trusses, some of which allow for kinematic degeneracy at a certain number of panels. The main problem solved for regular trusses is finding analytical solutions by induction.

In [17], the dependence of the deflection of a spatial cantilever under an external vertical nodal load on the number of panels is determined. The simple formulas obtained in this work make it possible to estimate the deformations of a structure even with a very large number of panels, without the risk of loss of accuracy associated with the accumulation of rounding errors. In [13] the resonance safety zones of a truss structure with an arbitrary number of panels are studied. Similar problems related to the natural vibrations of regular trusses are solved in [14]. The deformations of a regular truss with an arbitrary number of panels are analyzed analytically in [2]. The vibrations of a two-span regular truss are considered in [15]. The case of kinematic variability for an even number of panels is considered. Formulas for a two-sided estimate of the natural frequency of truss vibrations are obtained in [18] using the Dunkerley and Rayleigh methods. Similar estimates are obtained in [8]. In [19], the deflections of a truss with a cross-shaped lattice are analyzed analytically. In [1], a diagram of a regular lattice truss with hexagonal panels is considered, allowing kinematic variability at a certain ratio of sizes and an arbitrary number of panels. The author's reference books [5, 6] contain analytical solutions for the dependence of truss deflection on the number of panels of various designs of planar statically determinate trusses, some of which allow for kinematic variability.

This article presents only fifteen regular truss designs that allow kinematic degeneration. In reality, there are many more. Over the past ten years, the author has collected over forty such designs. There are also spatial designs, but they are very few. It should be clearly understood that designs that exhibit kinematic degeneration are dangerous. Without understanding this design feature, an engineer can transfer a well-tested solution for a regular truss with one number of panels to a different number and unexpectedly obtain a kinematically variable system. Of course, this variability is instantaneous. After some slight displacement under load, the truss will "lock", and catastrophic failure will not occur. However, with knowledge of the cases described below, this can easily be avoided by changing the number of panels.

18.2.1 Trapezoidal Lattice Truss with Additional Lateral Supports

A truss with additional lateral supports is an externally statically indeterminate structure. The number of $2n$ panels is calculated based on the lower chord, Fig. 18.1. The truss consists of $8n + 4$ rods, not including spacer rods. It is noted that for an even number of n panels in half the span the truss becomes kinematically variable, Fig. 18.2. Rods 8-4 and 2-10 perform instantaneous translational motions, while rods 5-12, 4-12, 11-12, 1-6, 2-6, and 7-6 rotate around fixed hinges. The remaining movable rods exhibit planar motion.

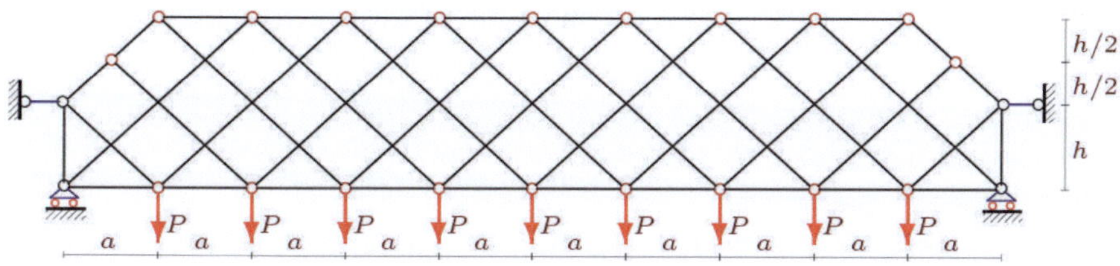

Fig. 18.1: Number of panels in half span $n = 5$. The design is kinematically unchangeable, statically determinate.

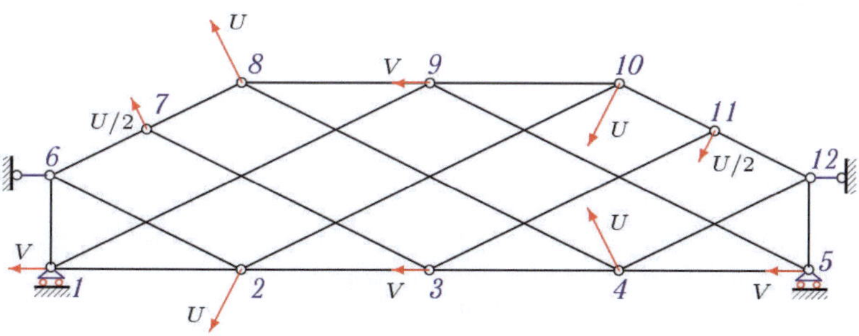

Fig. 18.2: Number of panels in half span $n = 2$. The design is kinematically variable, the velocities ratio of the nodes is $V/h = U/\sqrt{a^2 + h^2}$.

18.2.2 Multi-Porous Truss

The lattice truss has n panels of length a, n movable intermediate supports, and one fixed support [9]. The truss consists of $6n+6$ rods, not including spacer rods. For an odd number of panels, the truss is fixed (Fig. 18.3), while for an even number, it is variable, Fig. 18.4. The velocities ratios of the nodes of the variable truss are:

$$\vec{v}_1 = [-4v, 0], \vec{v}_2 = \vec{v}_6 = [-3v, 0], \vec{v}_3 = \vec{v}_7 = \vec{v}_{11} = [-2v, 0],$$
$$\vec{v}_4 = \vec{v}_{12} = [-v, 0], \vec{v}_8 = \vec{v}_{10} = \vec{v}' = [-2v, -av/h], \vec{v}_9 = \vec{u} = [-2v, -2av/h].$$

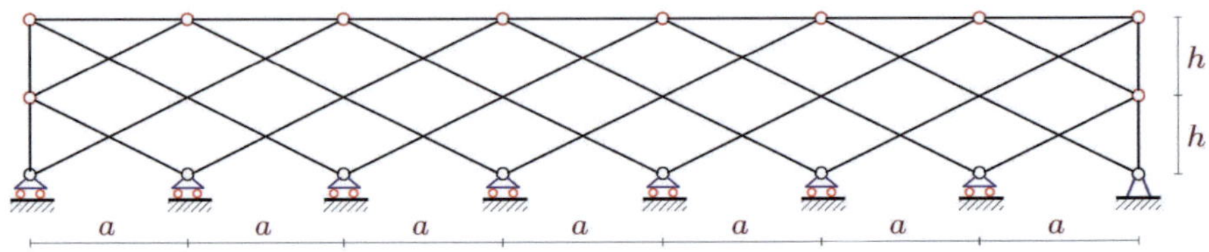

Fig. 18.3: Number of panels in the span $n = 7$. The design is kinematically unchangeable, statically determinate.

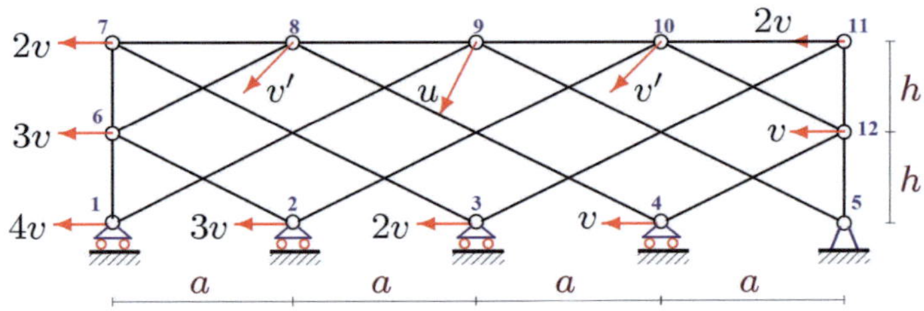

Fig. 18.4: The design is kinematically variable, $n = 4$.

18.2.3 Planar Arched Truss on Fixed Hinges

The arch rests on two fixed hinges (Fig. 18.5). The truss consists of $8n + 20$ rods, not including spacer rods [7]. It is externally statically indeterminate. The arch is kinematically invariant for an even number of panels. For an odd number of panels, the truss nodes acquire certain velocities, Fig. 18.6. The velocity ratio is equal to

$$\frac{U}{a} = 2\frac{V}{c}, \qquad c = \sqrt{a^2 + h^2}.$$

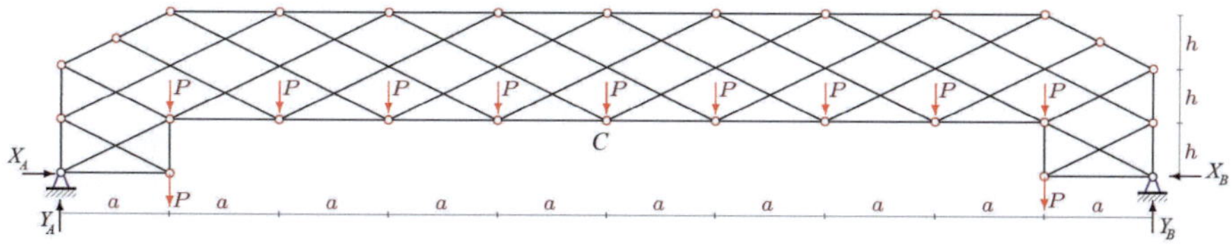

Fig. 18.5: Truss, number of panels in half span $n = 4$.

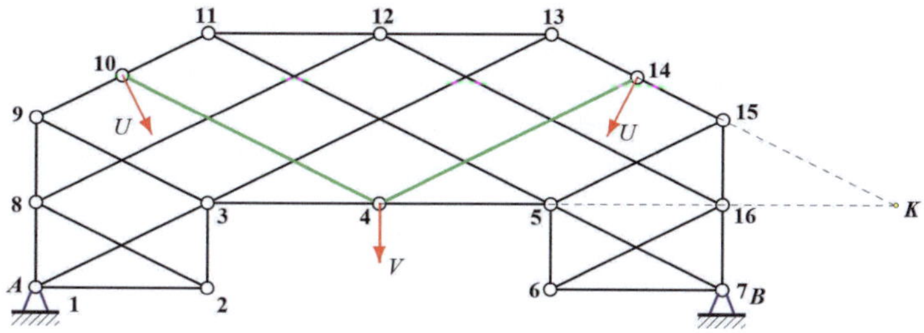

Fig. 18.6: Possible velocities of a variable truss, number of panels in half span $n = 1$, K is instantaneous center of velocities of link 4-14.

In [10], a sufficient criterion for kinematic variability is formulated, as applied to this case:

> *If a truss has an open chain of rods, each internal node of which is connected to four rods, two of which are not included in the chain and lie on the same line, and the end nodes connect three rods each, two of which are not included in the chain and lie on the same line, then such a truss is geometrically variable.*

For this case, it is a chain 10-4-14 of two rods.

18.2.4 Truss on Fixed Hinges

The truss has two fixed hinged supports and contains $2n$ panels. The truss consists of $8n + 12$ rods, not including spacer rods. It is externally statically indeterminate Fig. 18.7. The truss is invariant for an even number n of panels. For an odd number of panels, the truss nodes acquire specific velocities, Fig. 18.8. The velocity ratio is equal to

$$\frac{v'}{a} = \frac{v}{\sqrt{a^2 + h^2}} = \frac{u}{h}.$$

A similar picture of the velocity distribution can be constructed for any other odd value of n.

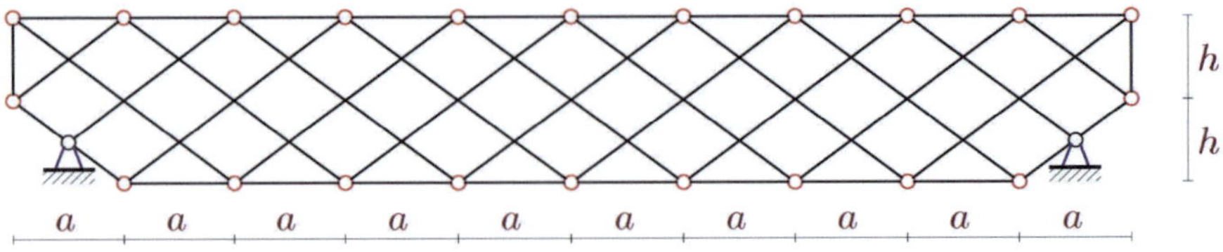

Fig. 18.7: Immutable statically determinate truss, n = 4.

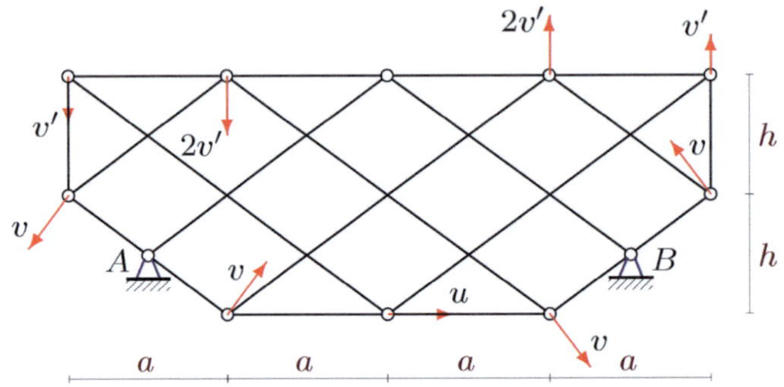

Fig. 18.8: Possible velocities of a variable truss, number of panels in half span n = 1.

18.2.5 Multi-Span Arched Truss 1

An arched multi-span truss made of n T-shaped hinged parts and rests on two fixed lateral hinged supports and $n-2$ intermediate movable supports, Fig. 18.9. The truss consists of 33n rods, not including spacer rods. In the case of an odd number of panels n, the truss allows for kinematic variability, Fig. 18.10.

All joints acquire velocity. Not all vectors are labeled. Only the outer supports are stationary. The velocities ratio of the nodes is

$$\frac{V}{\sqrt{a^2+4h^2}}=\frac{U}{4h}.$$

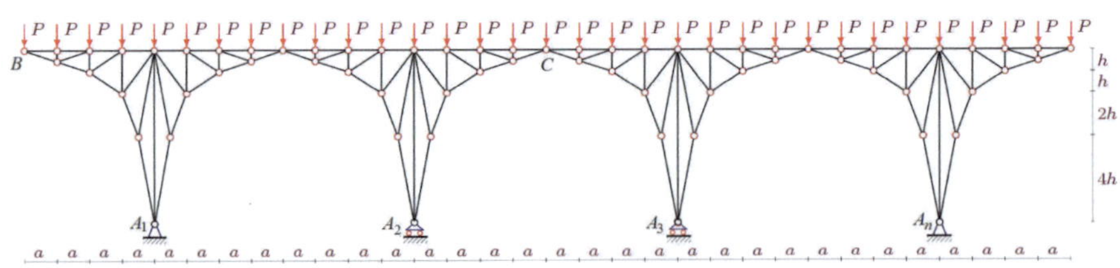

Fig. 18.9: Truss with three spans, n = 4.

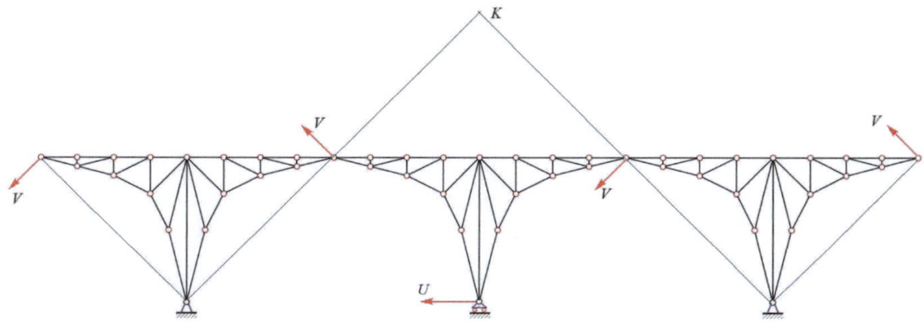

Fig. 18.10: Variable truss. Diagram of possible velocities of articulating joints and instantaneous center of velocities, $n = 3$.

The instantaneous center of velocity of the middle link of the truss is at point K. Both outer links rotate around the supports.

18.2.6 Arched Truss on Four Supports

The truss has four hinged supports, one of which is fixed in place, Fig. 18.11. The truss contains $2n$ panels of length $2a$ in its middle part [11]. The truss consists of $8n + 17$ rods, not including spacer rods. At $n = 1, n = 4, n = 7 \ldots$ or $n = (6k - (-1)^k + 1)/4$, $k = 1, 2, \ldots$ the truss is kinematically variable. At $n = 1$, the following hinge velocity relationships exist: $\vec{u} = [-u, 0]$, $\vec{v} = [-u/2, 0]$, $\vec{v}' = [-u/4, ua/(4h)]$, Fig. 18.12.

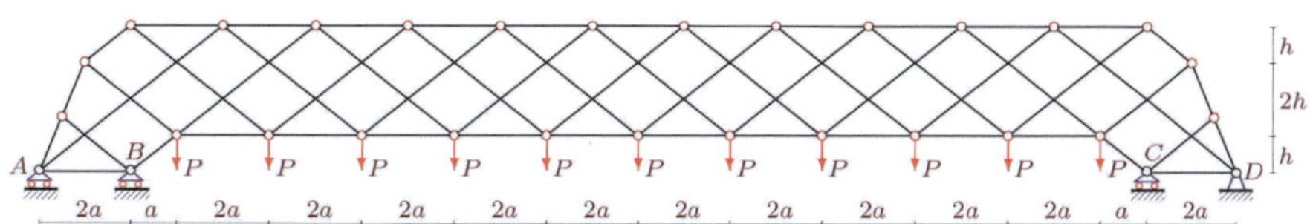

Fig. 18.11: Kinematically invariable truss, $n = 5$.

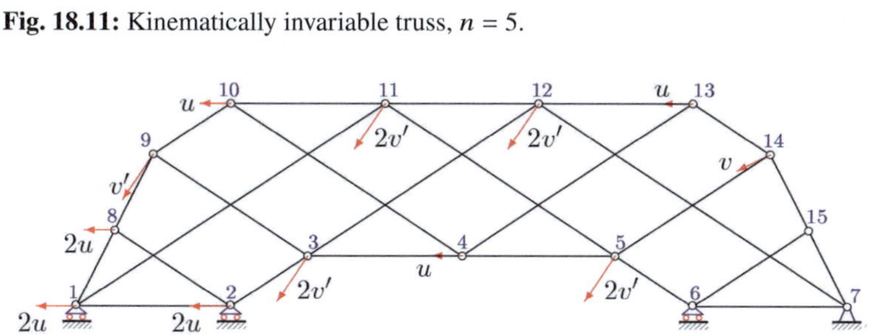

Fig. 18.12: Kinematically variable truss, $n = 1$.

18.2.7 Lattice Truss

The lattice truss has two supports and consists of $2n$ panels of length $2a$ each and two side panels of length a Fig. 18.13. The truss consists of $8n + 19$ rods, not including spacer rods. The lower chord of the panel is $2a$ long, the upper one consists of two rods of length a. The truss allows for kinematic variability at $n = 1, 2, 3, 5, 7, 8, 9, \ldots$ When $n = (1 - (-1)^k + 6k)/2$, $k = 1, 2, \ldots$ it is unchangeable, Fig. 18.14. The ratio of possible speeds of nodes is obvious when $n = 1$:

$$\frac{v}{a} = \frac{u}{h}.$$

The supporting nodes 1, 4 and nodes 2, 3, 5, 10, 12, 14, 8 remain stationary.

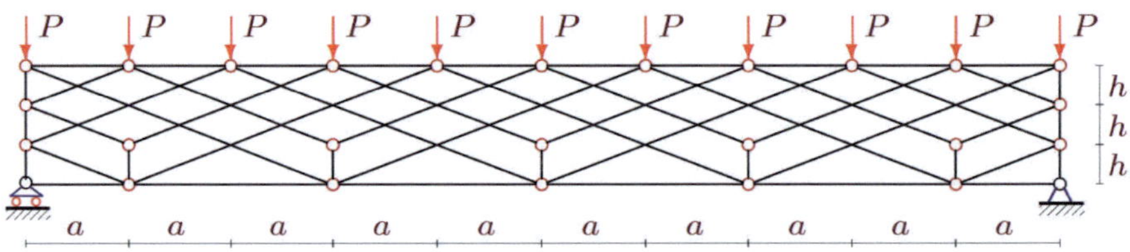

Fig. 18.13: Kinematically invariable truss, $n = 4$.

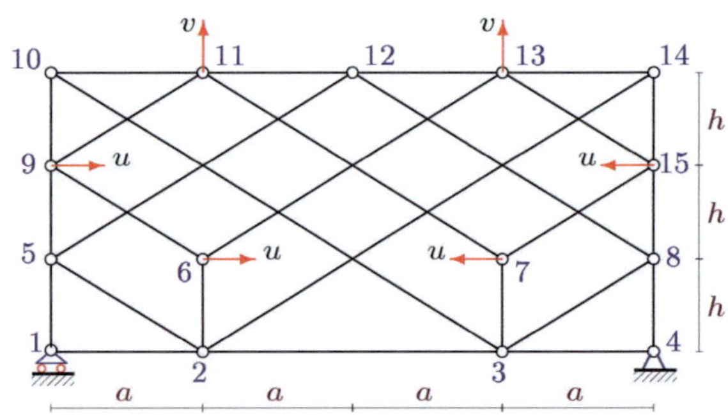

Fig. 18.14 Kinematically variable truss, $n = 1$.

18.2.8 Multi-Span Arched Truss 2

The truss consists of n arches, each with a span of of $6a$, Fig. 18.15. The two outermost hinged supports are fixed, the others $n - 1$ are movable. With an even number n of arches, the truss is kinematically variable. The truss consists of $21n + 1$

rods, not including spacer rods. The ratio of the velocities of the nodes for the variable truss is

$$\frac{u}{4h} = \frac{v}{\sqrt{a^2 + 4h^2}},$$

Fig. 18.16.

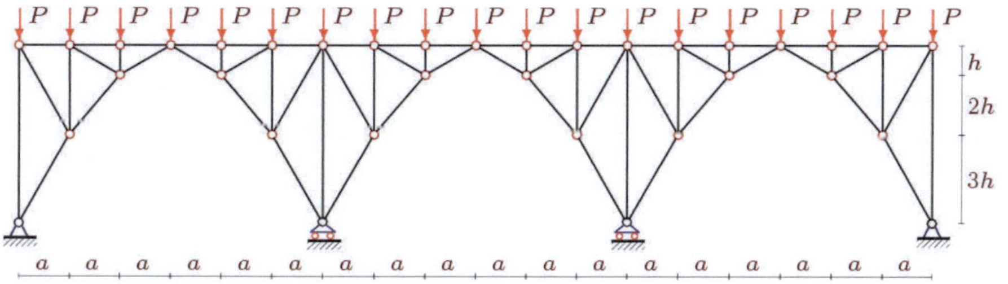

Fig. 18.15: Kinematically invariable truss, $n = 3$.

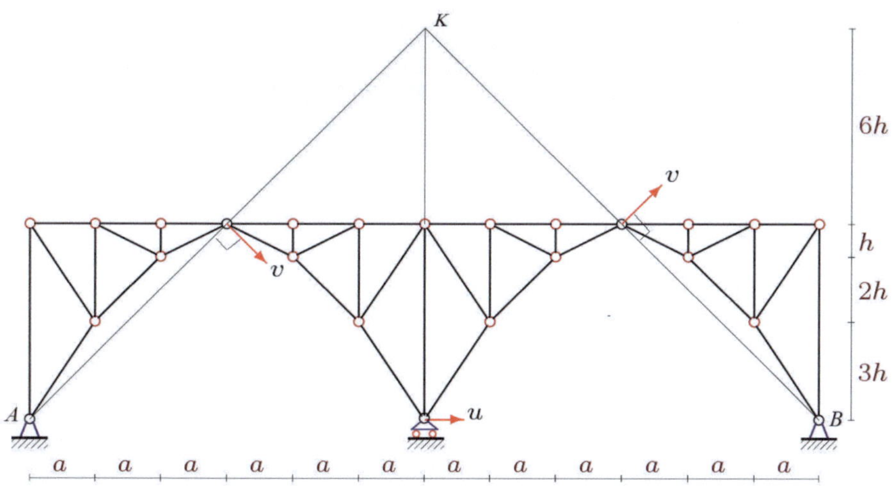

Fig. 18.16: Kinematically variable truss, $n = 2$.

18.2.9 Lattice Polygonal Truss with Four Supports

The two middle supports of a lattice truss with $2n$ panels of length $2a$ are movable hinges, and the side ones are fixed hinges, Fig. 18.17. The truss is kinematically

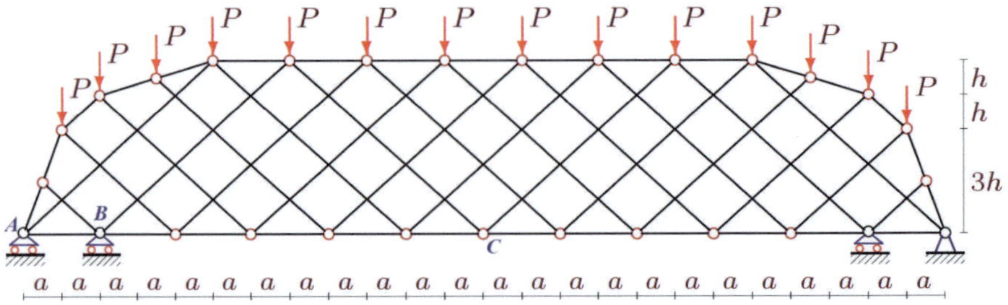

Fig. 18.17: Kinematically invariable truss, $n = 6$.

unchangeable at $n = 6, 7, 11, 12, 16, \ldots$ or at $n = (10k+11-3(-1)^k$, $k = 1, 2, 3, \ldots$, Fig. 18.18. The ratio of the velocities of the nodes of the variable truss are:

$$u'' = 4u'\frac{a}{d}, \quad u'\frac{3a-h}{d} = u\frac{3h-a}{f},$$

where $d = \sqrt{9a^2 + h^2}$, $f = \sqrt{a^2 + 9h^2}$.

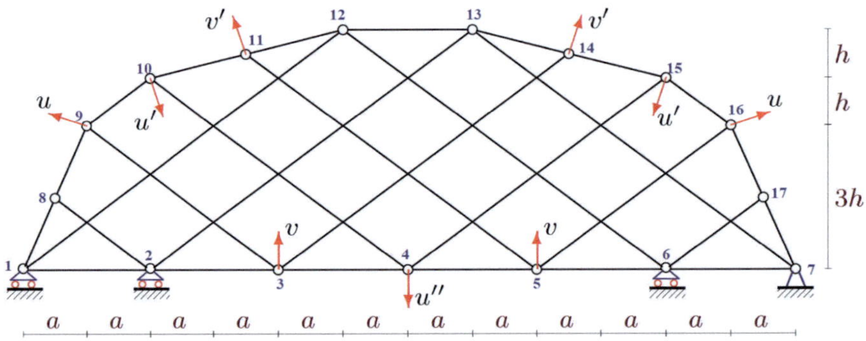

Fig. 18.18: Kinematically variable truss, $n = 3$.

18.2.10 Truss with a Diamond Lattice on two Fixed Supports

The truss has three rhombic panels of length $2a$, Fig. 18.19. The truss consists of $6n + 4$ rods, not including spacer rods. For an even number of panels, the truss is variable. The velocities ratio of the nodes is:

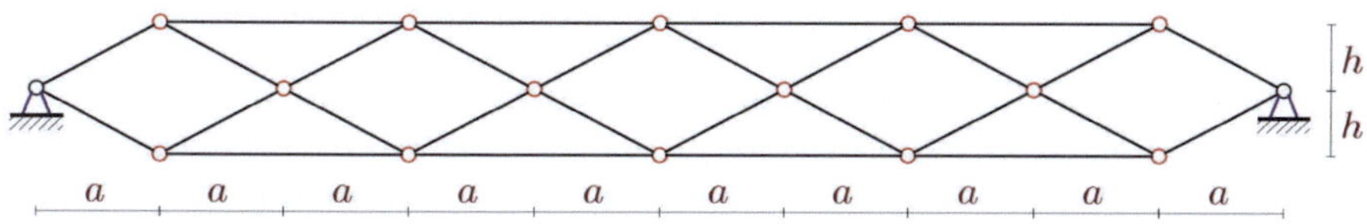

Fig. 18.19: Kinematically invariable truss, $n = 5$.

$$\frac{u}{h} = \frac{v}{\sqrt{a^2 + h^2}},$$

Fig. 18.20.

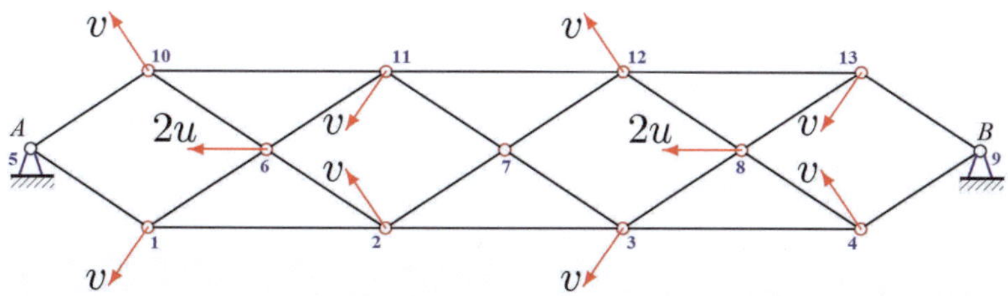

Fig. 18.20: Kinematically variable truss, $n = 4$.

18.2.11 *Multi-Span Truss with a Triangular Lattice on six Supports*

The truss lattice of the bridge with n panels is of the "fan" type, Fig. 18.21. Each conventional panel (a periodicity element of the structure) has 10 diagonals, 5 vertical elements in the upright, and 10 elements in the bottom chord. The truss consists of $22n + 3$ rods, not including spacer rods. Some rod elements belong to two panels simultaneously (the panels overlap). Calculations show that for trusses with an odd number of panels, the determinant of the system's equilibrium equation for nodes

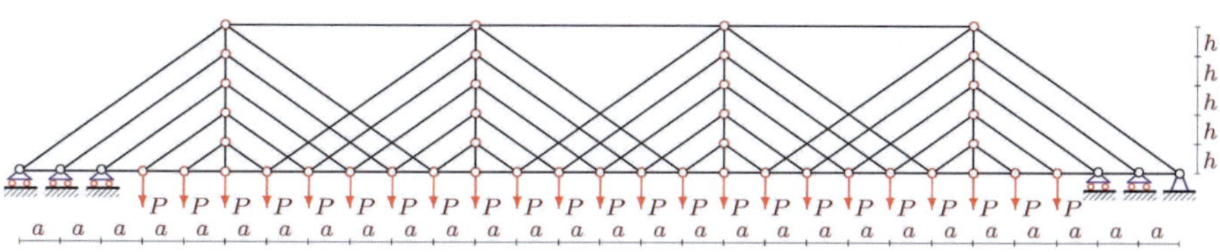

Fig. 18.21: Kinematically invariable truss, $n = 4$.

vanishes. This demonstrates the kinematic variability of a structure with this number of panels.

This is confirmed by the kinematically consistent distribution of possible node velocities. We introduce the designation ω for the angular velocity of the lateral link [23, 38]. The ratio of the velocities of the nodes of the variable truss at $n = 3$ is as follows (Fig. 18.22):

$$\vec{v}_5 = [0, 15a\omega], \vec{v}_6 = \vec{v}_{17} = [0, 5a\omega], \vec{v}_7 = \vec{v}_{18} = [0, -5a\omega],$$
$$\vec{v}_8 = \vec{v}_9 = \vec{v}_{10} = \vec{v}_{11} = [0, 10a\omega], \vec{v}_{13} = \vec{v}_{14} = \vec{v}_{15} = \vec{v}_{16} = [0, -10a\omega],$$
$$\vec{v}_{19} = [0, -15a\omega], \vec{v}_{24} = [10h, 5a]\omega, \vec{v}_{25} = \vec{v}_{31} = \vec{v}_{28} = \vec{v}_{34} = [10h, 0]\omega,$$
$$\vec{v}_{26} = [10h, -5a]\omega, \vec{v}_{27} = \vec{v}_{30} = \vec{v}_{33} = \vec{v}_{36} = 5[-h, a]\omega,$$
$$\vec{v}_{29} = \vec{v}_{32} = \vec{v}_{35} = \vec{v}_{38} = 5[-h, -a]\omega, \vec{v}_{37} = [-5h, 0]\omega.$$

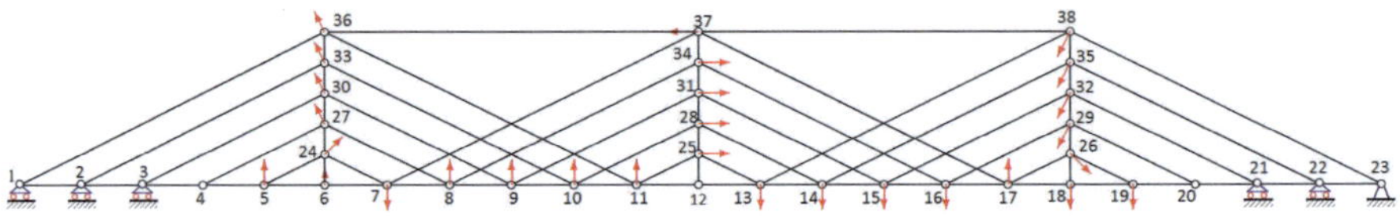

Fig. 18.22: Kinematically variable truss, $n = 3$.

To identify the kinematic variability of this truss, we apply the static criterion of an unbalanced chain of links. Consider a chain of rods 4-27-8-34-16-29-20 (Fig. 18.23). If node 4 is loaded with force P_1, then sequentially considering the equilibrium of all nodes in the chain, starting with node 4, we find that, regardless of the force P_2 applied to node 20, the forces in this chain are uniquely determined by the force P_1, and the equilibrium condition for node 20 projected onto the y-axis yields a contradiction. From the equilibrium condition for this node, it follows that the force P_2 depends on P_1, while in reality, the loads are independent of each other. For each node in the chain, it is sufficient to formulate only one equation: for nodes 4, 8, and 16, this equation is projected onto the y-axis, and for the remaining nodes, it is projected onto the x-axis. The presence of an unbalanced chain is a static criterion for the degeneration of the structure (Subsect. 18.2.3).

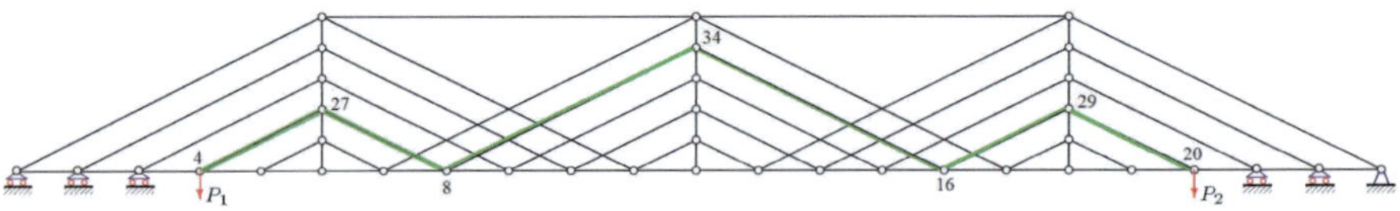

Fig. 18.23: Unbalanced truss chain, $n = 3$.

18.2.12 Trapezoidal Lattice Truss on two Supports

The truss consists of $2n$ panels Fig. 18.24. The outer panels have a size of a along the bottom chord, the middle ones have a size of $2a$. The truss consists of $8n+3$ rods, not including spacer rods. The truss becomes geometrically variable at $n = 2, 5, 8, ..$ or at $n = 3k - 1$, $k = 1, 2, \ldots$. The distribution pattern of possible velocities for these cases, confirming the effect found, is shown in the Fig. 18.25. Ratio of possible knot speeds velocities:

$$\frac{v}{a} = \frac{u}{h} = \frac{u'}{4\sqrt{a^2 + h^2}}.$$

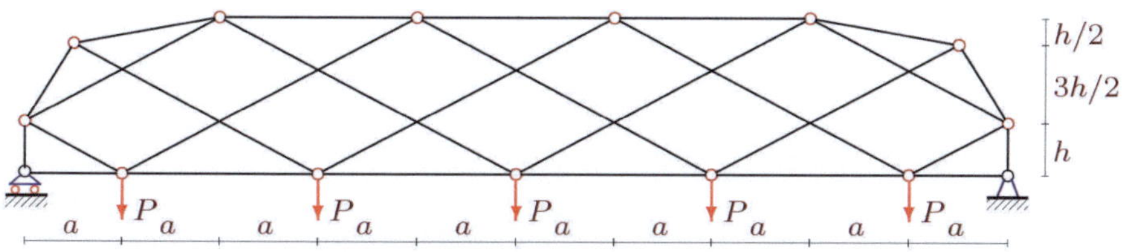

Fig. 18.24: Kinematically invariable truss, $n = 3$.

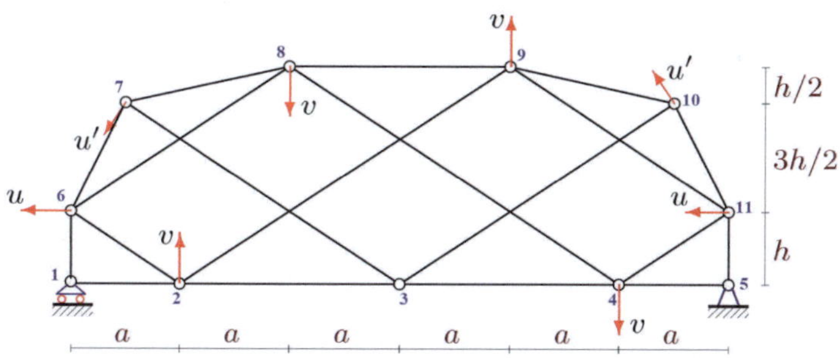

Fig. 18.25: Kinematically variable truss, $n = 2$.

18.2.13 Lattice Truss on two Supports

The truss consists of $2n$ panels Fig. 18.26. The truss design is statically determinate. The truss consists of $8n + 7$ rods, not including spacer rods. Kinematic variability is revealed when calculating the forces in the bars for $n = 2, 5, 6, 7, 10, 11, 12, \ldots$

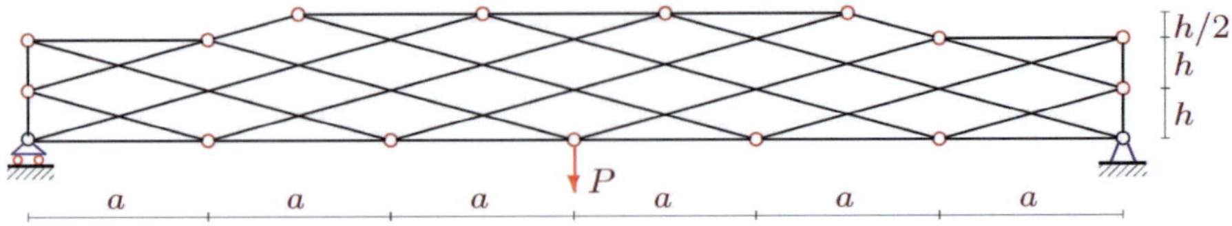

Fig. 18.26: Kinematically invariable truss, $n = 3$.

For these numbers of panels, the determinant of the system of linear equilibrium equations of nodes (including two support nodes) vanishes, which corresponds to kinematic degeneracy. This fact is confirmed by the diagram of possible node velocities in Fig. 18.27 for $n = 2$.

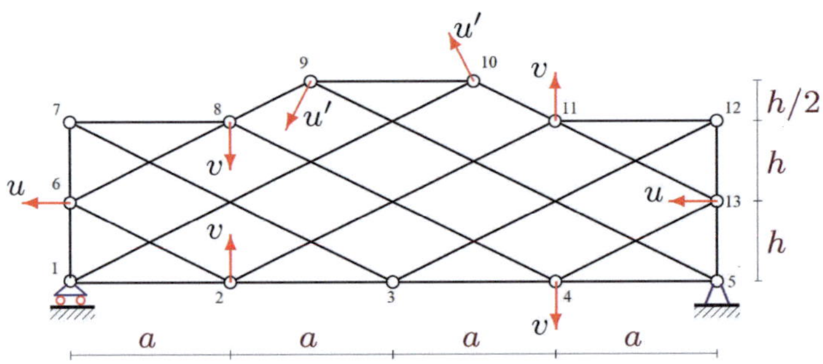

Fig. 18.27: Kinematically variable truss, $n = 2$.

Bars 1-2, 1-6, 6-8, 2-3, 7-8, 8-9, 9-5 and bars symmetrical to them perform rotational motions about fixed or instantly fixed points. Bars 8-4 and 2-11 have instantaneous translational motion. The relationship of velocities in the diagram follows from consideration of the instantaneous centers of velocities:

$$\frac{v}{a} = \frac{u}{h} = 2\frac{u'}{\sqrt{a^2 + h^2}}.$$

18.2.14 Asymmetrical Lattice Girder Truss

A planar statically determinate beam truss with a triple asymmetric lattice and $2n$ panels has two supports and rectilinear chords Fig. 18.28. The truss consists of $8n+5$ rods, not including spacer rods [6]. Regardless of the type of load, when the number of panels is a multiple of three ($n = 3k$, $k = 1, 2, 3, \ldots$), the determinant of the system of equilibrium equations for the truss nodes vanishes [8]. This means that the truss is kinematically variable. The corresponding diagram of possible velocities

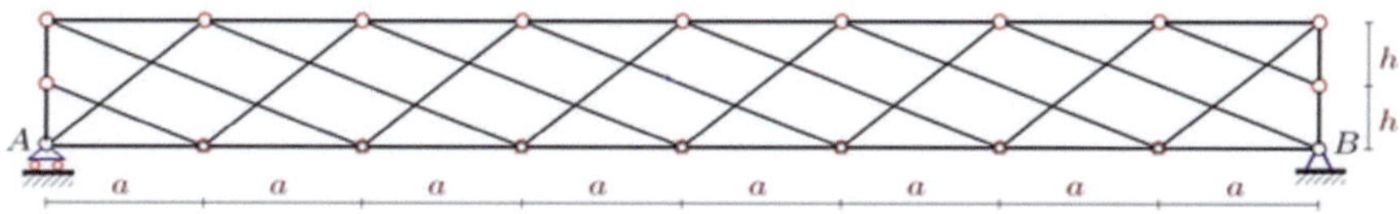

Fig. 18.28: Kinematically invariable truss, $n = 4$.

for $n = 3$ is shown in Fig. 18.29. Hinges 2, 5, 8, 11, 14, and 16 are displaced; the others are fixed. Similar diagrams confirming the variability of the truss for $n = 3k$, $k = 1, 2, 3, \ldots$, can be constructed for other values of n that are multiples of three. Ratio of knot velocities $v/h = u/a$.

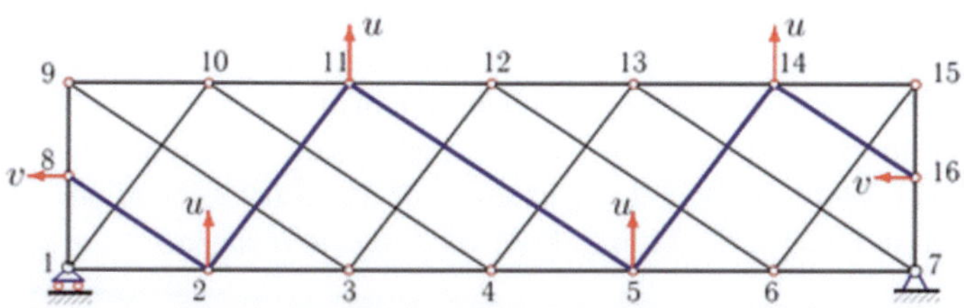

Fig. 18.29: Kinematically variable truss, $n = 3$.

There is also static confirmation of the degeneracy effect of the determinant in these cases. Indeed, if, for example, the truss is loaded with an arbitrary (non-vertical) force at node 16, and node 8 and all other nodes are unloaded, then, according to the lemmas on zero truss members, the forces in all members in the chain 8-2-11-5-14-16 are equal to zero, which contradicts the equilibrium condition for loaded node 16. The truss is kinematically invariable when $n = (6k - 3 - (-1)^k)/4$, $k = 1, 2, 3, \ldots$ For this truss, the sufficient criterion of variability formulated in Subsect. 18.2.3 is suitable.

18.2.15 Lattice Frame on two Supports

A planar statically determinate truss has a span of length $2(n + 3)a$ and height $(m + 2.5)h$, Fig. 18.30. Unlike the above-considered schemes, this truss has two independent integer parameters n and m that define its configuration. The truss consists of $8(n + m) + 31$ rods, not including spacer rods. When $m = 3j - 1$, $j = 1, 2, \ldots$, $n = (10k - 5 - 3(-1)k)/4$, $k = 1, 2, \ldots$ the determinant of the system of equilibrium equations is nonzero [12]. In other cases, the truss is kinematically variable.

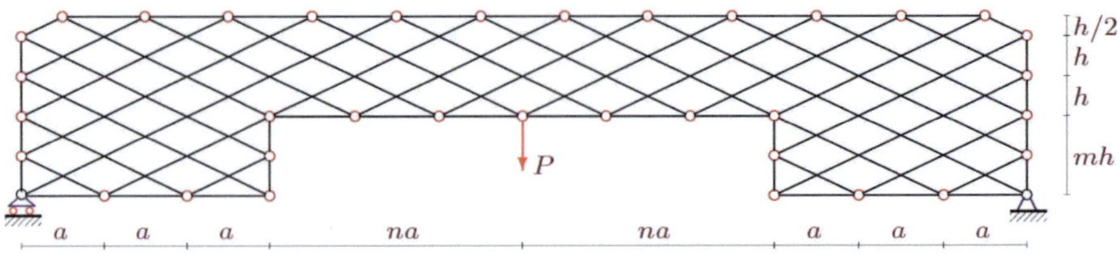

Fig. 18.30: Kinematically invariable truss, $n = 3$, $m = 2$..

In [10], a sufficient criterion for the kinematic variability of a truss is formulated, previously published in a different formulation by Rabinovich [16]. *If a truss has a closed chain of rods, each node of which connects four rods, two of which do not belong to the chain and lie on the same line, then such a truss is geometrically variable.*

Rectangular lattice truss that marked with a sufficient sign of the kinematic variability of the structure has the presence of a closed chain of rods that does not pass through the corner points of the structure. In the investigated truss this is confirmed by the chain 2-7-28-12-21-19-14-26-9-33-2. An example of the distribution of possible node speeds is shown in Fig. 18.31. Ratio of knot velocities $u/h = v/a$.

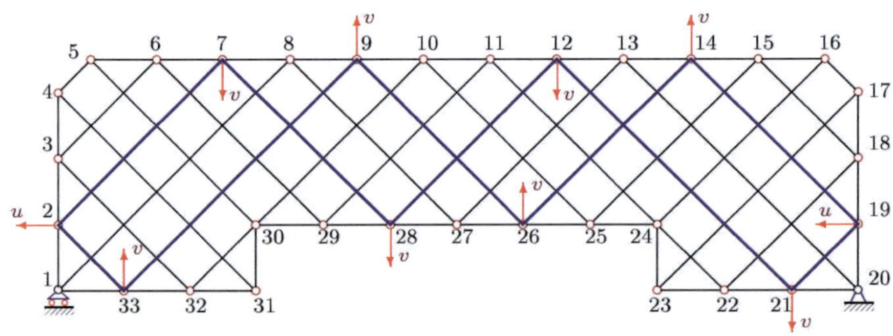

Fig. 18.31: Kinematically variable truss, $n = 3$, $m = 1$.

18.3 Conclusion

This article contains only a subset of the statically determinate trusses found by the author that allow for kinematic variability. All of these trusses are planar, regular structural models. Such spatial trusses are significantly fewer. In any case, kinematic variability, which manifests itself only at a certain number of panels, is a dangerous and insidious property of a truss, immediately invisible to the designer. With one number of panels, the structure performs its functions; with another, even if it differs by just one panel, it turns into a mechanism.

The trusses examined can be divided into distinct groups. For example, trusses Subsect. 18.2.7, 18.2.12 – 18.2.15, supported by two supports, one of which is a movable hinge and the other a fixed hinge, are statically determinate. The support reactions of such trusses are found from the equilibrium equations of the entire structure. These trusses are relatively easy to calculate. Another group of trusses are externally statically indeterminate. The forces in supports and members can only be found from a combined system of equilibrium equations for all structural components. Using computer mathematics systems significantly simplifies the solution. Analytical solutions are required to identify cases of kinematic variability, as numerical solutions can conceal the degeneracy of the determinant of the system of equilibrium equations due to rounding errors.

It should be noted that there is no universal and clear criterion for structural variability. It is difficult to detect this rather dangerous defect by external signs without a specific force calculation. Two criteria are presented here: Subsects. 18.2.3 and 18.2.15. These criteria are ineffective for the other examples given. It can be assumed that further analysis of cases of kinematic variability in trusses will help identify a universal criterion.

References

[1] Arutyunyan VB (2018) Raschet progiba dekorativnoy reshetki fermy s proizvol'nym chislom paneley pod deystviyem nagruzki v seredine proleta (in Russ., Calculation of the deflection of a decorative lattice of a truss with an arbitrary number of panels under the load in the middle of the span). Postulat 7(33):1-6

[2] Dai Q (2021) Analytical dependence of planar truss deformations on the number of panels. AlfaBuild 17:Article No 1701, DOI 10.34910/ALF.17.1

[3] Hutchinson RG, Fleck NA (2005) Microarchitectured cellular solidsthe hunt for statically determinate periodic trusses. ZAMM - Journal of Applied Mathematics and Mechanics / Zeitschrift für Angewandte Mathematik und Mechanik 85(9:607-617, DOI 10.1002/zamm.200410208

[4] Hutchinson RG, Fleck NA (2006) The structural performance of the periodic truss. Journal of the Mechanics and Physics of Solids 54(4):756-782, DOI 10.1016/j.jmps.2005.10.008

[5] Kirsanov, M. (2024) Planar Trusses: Schemes and Formulas. Cambridge Scholars Publishing, Cambridge.

[6] Kirsanov, M. (2020) Trussed Frames and Arches: Schemes and Formulas. Cambridge Scholars Publishing, Cambridge.

[7] Kirsanov MN (2025) Analytical calculation of an externally statically indeterminate truss oscillation frequency. Construction of Unique Buildings and Structures 116:Article No 11604, DOI 10.4123/CUBS.116.4

[8] Kirsanov, MN (2020) Estimation of the vibration frequency of a truss with an arbitrary number of panels. Structural Mechanics and Analysis of Constructions, 5, 47-52.
[9] Kirsanov MN (2024) Formuly dlya dvukhstoronney otsenki osnovnoy chastoty kolebaniy reshetchatoy fermy (in Russ., Formula for two-sided estimation of the fundamental frequency of oscillations of a lattice truss). Vestnik MGSU 19(3):377–386, DOI 10.22227/1997-0935.2024.3.377-386
[10] Kirsanov MN (2021) Priznaki geometricheskoy izmenyayemosti regulyarnykh ferm (in Russ., Signs of geometric variability of regular trusses). Stroitel'naya Mekhanika i Konstruktsii 1(28):22–31
[11] Kirsanov MN, Gribova OV (2025) Kinematicheskiy analiz i analiticheskiy raschet progibov pravil'noy fermy (in Russ., Kinematic analysis and analytical calculation of deflections of a regular truss). Mekhanika Konstruktsiy i Raschet Konstruktsiy (6):72–78, DOI 10.37538/0039-2383.2025.6.72.78
[12] Kirsanov M, Serdjuks D, Buka-Vaivade K (2019) Analytical dependence of deflection of the lattice truss on the number of panels. In: Proceedings of EECE 2019 (Borodinecs A, Vatin N, Sergeev V (eds)), Lecture Notes in Civil Engineering, vol 70, pp 25–35. Springer, Cham, DOI 10.1007/978-3-030-42351-33
[13] Luong CL (2024) Resonance safety zones of a truss structure with an arbitrary number of panels. Construction of Unique Buildings and Structures 113:Article No 11304, DOI 10.4123/CUBS.113.4
[14] Maslov A (2023) The first natural frequency of a planar regular truss. Analytical solution Construction of Unique Buildings and Structures 109:Article No 10912, DOI 10.4123/CUBS.109.12
[15] Petrenko VF (2021) The natural frequency of a two-span truss. AlfaBuild 20:Article No 2001, DOI 10.34910/ALF.20.1
[16] Rabinovich IM (1950) Kurs stroitel'noj mekhaniki sterzhnevyh sistem. Chast' 1. Staticheski opredelimye sistemy (in Russ., Course of Structural Mechanics of Rod Systems. Part 1. Statically Determined Systems). Stroyizdat, Moscow-Leningrad
[17] Sviridenko O, Komerzan E (2022) The dependence of the natural oscillation frequency of the console truss on the number of panels. Construction of Unique Buildings and Structures 101:Article No 10101, DOI 10.4123/CUBS.101.1
[18] Vorobev O (2020) Bilateral analytical estimation of first frequency of a plane truss. Construction of Unique Buildings and Structures 92:Article No 9204, DOI 10.18720/CUBS.92.4
[19] Voropai RA (2016) Analysis of the deflection of the regular truss with cross type lattice. Science Almanac 4-3(18):238-240, DOI 10.17117/na.2016.04.03.238

Chapter 19
Experimental Investigation and Modelling of Dynamic Deformation of High-Purity Copper

Aleksander Konstantinov, Yuliya Konstantinova, and Maksim Vtyurin

Abstract The paper presents the results of investigation of dynamic deformation of high-purity (99.9 %) copper C101 in the state of delivery (10 mm diameter rod) and after annealing (exposure to temperature). Two experimental schemes are used - the Split Hopkinson Pressure Bar method and the Taylor experiment. At first, the deformation diagrams in compression with strain rates of 1000 and 3000 s^{-1} are obtained using the Split Hopkinson Pressure Bar method. Based on the obtained diagrams, the parameters of the model describing the dependence of the yield surface radius on plastic strain were determined. The Taylor experiment was used to verify the mathematical model. Further, the procedure based on minimization of the deviation of the Taylor sample profile in the numerical and natural experiments is described. The results of the Taylor test were used to determine the approximation parameters of the deformation diagrams by the computational calibration method. The resulting diagrams are in good agreement with the diagrams determined using SHPB. A discussion of the pros and cons of using the direct and inverse approaches to the identification of model parameters to describe the plastic flow of materials under dynamic loading is given.

Key words: Experiment, Measuring bar, Strain rate, Dynamics, Numerical simulation, Taylor test, Identification

19.1 Introduction

Structural elements can be subjected to intensive dynamic loads of impact nature during their life cycle. When solving problems of strength assessment, one should

Aleksander Konstantinov (corresponding author) · Yuliya Konstantinova · Maksim Vtyurin
National Research Lobachevsky State University of Nizhny Novgorod, 603022 Nizhny Novgorod, Gagarina ave., 23, Russian Federation, e-mail: konstantinov@mech.unn.ru, ullichka@list.ru, vtyurin@unn.ru

H. Altenbach et al. (eds.), *Stability Problems in Mechanics of Continua*,
Advanced Structured Materials 260, https://doi.org/10.1007/978-3-032-31032-3_19

use reliable deformation and failure models and take into account the influence of strain rate on material characteristics. The large amount of experimental data should be obtained to construct and verify such models.

Constructing reliable deformation and failure models for materials of various physical nature in a wide range of loading conditions (strain rate, temperature, stress triaxiality factor) is a complex problem. The experimental-theoretical approach involving a combined use of experimental and numerical methods is the most effective and advanced methodology for analyzing deformation and failure processes of materials. Numerical modeling in this approach is used for determining material characteristics, based on the experimentally obtained data, when standard methods are not applicable (e.g., if the condition of the uniformity of stress and strain fields in the specimen is not satisfied, or a complex evolutionary process, e.g. damage accumulation, is analyzed).

The most commonly used method for studying the behavior of materials of different physical nature under dynamic loading is the Split Hopkinson Pressure Bar Method (SHPB) method [1, 9, 13, 14, 21]. This method makes it possible to create an almost constant strain rate in the process of loading the specimen in individual experiment, and thus to determine deformation diagrams at different strain rates in the set of experiments to study the strain rate dependences of material characteristics. The SHPB setup includes two measuring bars, a loading device (most often a pneumatic gun), an impactor, strain gauges with a power supply circuit and an oscilloscope.

To determine material characteristics at high strain rates, Taylor and Winnie [29] developed a method that allows estimating the flow stress based on the analysis of residual strains in a cylindrical specimen after hitting a rigid barrier. The minimum set of equipment for such an experiment is an accelerating device (gas gun), velocity meter and a rigid barrier. In a widespread approach, the characteristics of the strain hardening of a material are determined on the basis of the results of the Taylor experiment using numerical simulation.

In the present work, the dynamic behavior of high-purity copper (99.9 %) C101 in the delivery state (10 mm diameter rod) and after annealing is investigated. SPHB and Taylor methods are used. A comparison of the direct and inverse approach to the identification of the parameters of the plastic deformation model of the material using the above experimental schemes is given.

19.2 Experimental Methods

19.2.1 The Split Hopkinson Pressure Bar Method

One of the most widely used loading techniques at high strain rates is the Split Hopkinson Pressure Bar (SHPB) or the Kolsky method developed by Kolsky [21]. There are some review works [1, 13, 14] that describe a historical background of

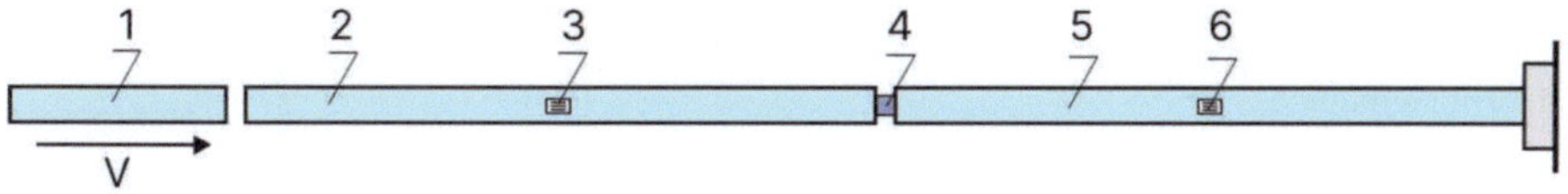

Fig. 19.1: SHPB installation configuration: 1 – striker, 2 – incident (or input) bar, 3, 6 – strain gauges, 4 – specimen, 5 – transmission (or output) bar, 7 – dumper.

SHPB method. This method can be easily implemented, it is well justified and realized theoretically.

Devis [9] and Kolsky [21] proposed to use SHPB for compression tests almost at the same time. SHPB consists of two stiff long bars (incident bar 2 and transmission bar 5) and a short soft specimen 4 sandwiched between them (Fig. 31.1). The striker 1 impacts the incident bar 2 and generates compressive pulse $\varepsilon^I(t)$, which propagates in the specimen direction with the bar sound velocity C. Upon reaching the interface between the incident bar and the specimen, a portion of the stress pulse $\varepsilon^T(t)$ travels through the specimen into transmission bar, while the remaining portion is reflected back into the incident bar as a tension pulse $\varepsilon^R(t)$. The ratio of transmitted and reflected pulses is determined by material properties of a specimen. The specimen exhibits elastic-plastic deformation while the measuring bars elastic one. The data from the strain gauges 3 and 4 allow us to calculate time histories of strain rate, strain and stress in the specimen by using the formulas proposed by Kolsky.

The main relations associating incident $\varepsilon^I(t)$, transmitted $\varepsilon^T(t)$ and reflected $\varepsilon^R(t)$ pulses in measuring bars with strain rate $\dot{\varepsilon}_s(t)$, strain $\varepsilon_s(t)$ and stress $\sigma_s(t)$ in specimen are given in [21]. Some high strain rate tensile testing schemes which make use of measuring bar technique can be found in [3, 6, 11, 15, 17, 22, 24]. The main difference of these schemes lies in the way they generate tension pulse.

19.2.2 The Taylor Method

The range of strain rates of the SHPB method is limited to values of the order of 10^4 s^{-1}. This is due to the fact that the measuring bars, which form the load on the specimen, must deform in the elastic region, which imposes conditions on the maximum load. To characterize materials at large values of strain rates, Taylor and Quinney [29] developed a method that allows the estimation of flow stress based on the analysis of residual strains in a cylindrical specimen after impact into a rigid barrier. The Taylor method for determining the dynamic yield stress, proposed in the late forties of the last century, is still widely used today because of its ease of implementation [12, 13, 30].

According to this technique, the flow stress of the specimen material can be estimated by the formula:

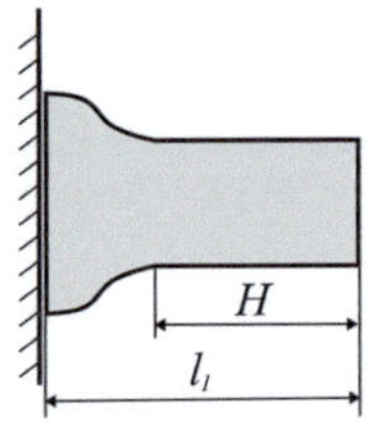

Fig. 19.2 The scheme for determining the yield strength in Taylor's method.

$$\sigma_y = \rho\, v_0^2 \,\frac{l_0 - H}{2(l_0 - l_1)}\,\frac{1}{\ln(l_0/H)} \tag{19.1}$$

with ρ - material density, v_0 – initial velocity of the sample, l_0 – initial sample length, l_1 – specimen length after testing, H – length of the undeformed part of the specimen (Fig. 19.2).

It should be noted that formula (19.1) is obtained under the assumption of rigid-plastic (without hardening) behavior of the material. Under the assumption of uniform deceleration of the undeformed part of the specimen, Taylor [28] obtained the following formula for estimating the average strain rate realized in the experiment:

$$\dot{\varepsilon}_{av} = \frac{v_0}{2\,(l_0 - H)} \tag{19.2}$$

Based on the analysis using two-dimensional finite-difference numerical scheme [31], it is noted that the Taylor method allows to determine the values of average dynamic yield strength at velocities of 10^4 s^{-1}. However, this method gives an overestimated value of yield strength when testing high-strength steels and alloys [25]. A symmetric Taylor impact method [10] allows avoiding the influence of friction at the specimen ends on the value of dynamic yield strength. Variants of the Taylor method that allow testing materials and alloys at elevated temperatures have been proposed in [2, 7, 27].

A modified scheme was proposed [4] in which specimen 1 impacts on a measuring bar 2 equipped with strain gauges 3 (Fig. 19.3). In this variant, in addition to measuring the residual shape of the specimen, the impulse in the measuring bar is recorded (Fig. 19.4), which makes it possible to determine the history of contact force at the boundary "impactor - measuring bar" during impact interaction. This force can be calculated by the formula:

$$F(t) = E\, S_b\, \varepsilon(t). \tag{19.3}$$

Fig. 19.3 The scheme of the experiment - a modified Taylor test.

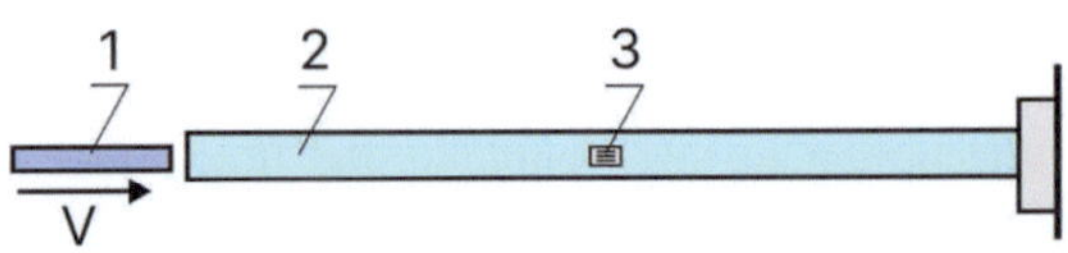

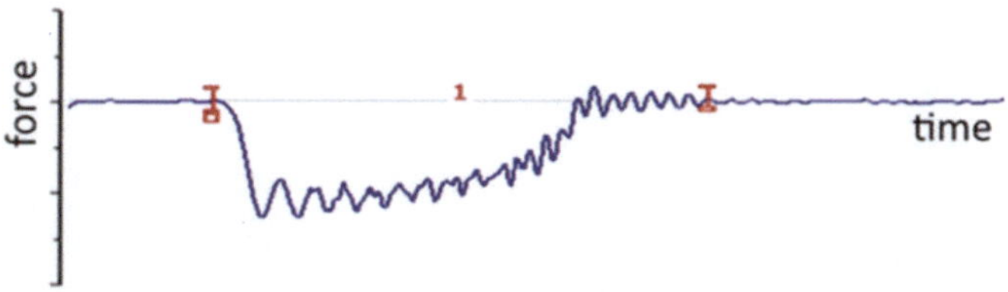

Fig. 19.4: Typical waveform obtained in the experiment on a modified Taylor test.

Here E and S_b - modulus of elasticity and cross-sectional area of the measuring bar, respectively. This information can be used for additional verification of model parameters when comparing numerical and natural experiments.

In the present study, visual registration of the process of deformation of the specimen during the interaction with the barrier is carried out using a *Phantom VEO 710 L* high-speed camera with high resolution. The characteristics of the camera are presented in Table 19.1.

To obtain high contrast, shadow imaging can be used to capture the contours of the specimen with high accuracy. Figure 19.5 shows the sequence of frames of interaction of an annealed copper impactor with a diameter of 9.9 mm and length of 70 mm with a steel measuring bar in the Taylor experiment. The interval between frames was 25 μs and the exposure time was 1 μs. The impact velocity was 125 m/s. To determine the dimensions of the sample after impact, the method of three-dimensional scanning of objects using the Range Vision Spectrum system (Fig. 19.6), which uses the principle of structured illumination, is used. When scanning objects with the size from 10 to 400 mm, this scanner has the following characteristics: 3D point accuracy 0.04 mm, 3D resolution 0.072 mm.

A 3D scanner with structured illumination illuminates the 3D scanning object with a projector. Simultaneously in synchronous mode, two cameras are shooting the 3D scanning object. Structural lines of projector illumination are displayed on the images from two cameras from different angles (stereo image). Images from two cameras are processed by software, forming a point cloud or polygonal model of the 3D object. A turntable is used to scan the object from different sides to obtain

Table 19.1: Characteristics of the high-speed camera.

Resolution	Maximum Frame Rate	Rec time @ max fps (72GB RAM)
1280 x 800	7400	6.5 seconds
1280 x 720	8200	6.5 seconds
640 x 480	21,000	7.5 seconds
256 x 256	75,000	9 seconds
64 x 8	80,000 (standard), 1,000,000 (optional)	62 minutes 90 seconds (optional)

Fig. 19.5: Taylor specimens after impact.

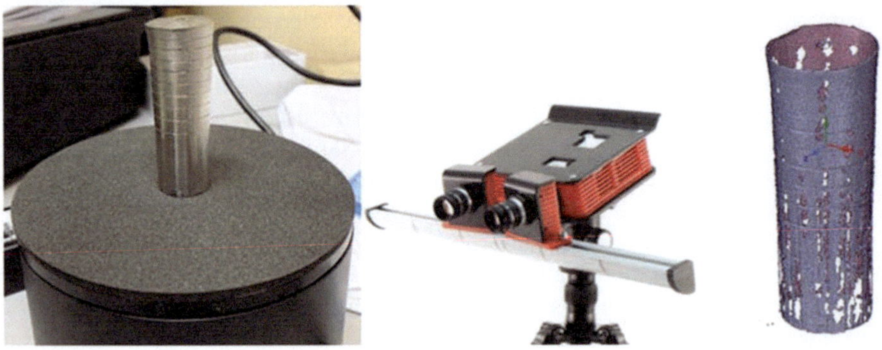

Fig. 19.6: Specimen scanning.

a complete three-dimensional model. A simpler and cheaper method, although less accurate, is to process a digital photograph of the specimen after the test, as shown in Fig. 19.7.

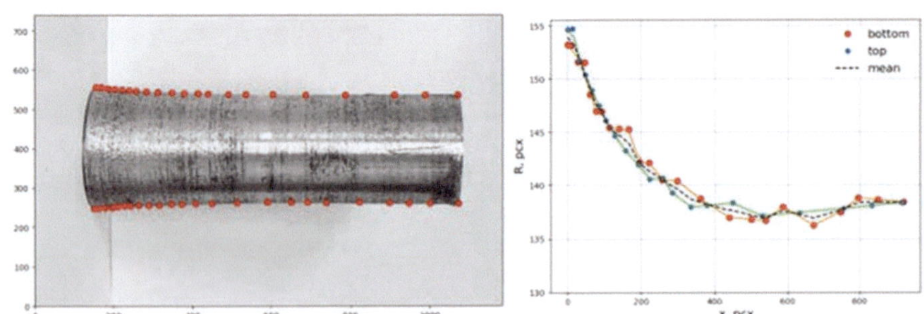

Fig. 19.7: Digital image processing.

19.3 Experimental Results

19.3.1 SHPB Experiments Results

The deformation diagrams obtained using the SHPB method are shown in Fig. 19.8. On the left are diagrams obtained on samples in the delivery state, on the right - on

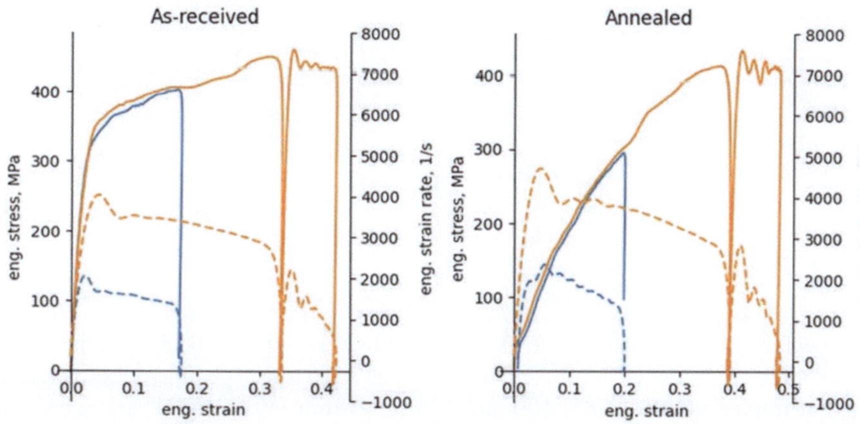

Fig. 19.8: Dynamic diagrams of C101.

samples after annealing. The experiments were performed at strain rates of 1000 and 3000 1/s. Solid lines show deformation diagrams, dashed lines show strain rates (right vertical axis). At a strain rate of 3000 1/s, the energy of the reflected pulse was sufficient for repeated plastic deformation of the specimen, so the deformation diagrams have two parts.

The material in the investigated range turned out to be insensitive to the strain rate. For numerical modeling, the deformation diagrams were approximated by the simplified Johnson-Cook model [18], which takes into account only strain hardening:

$$\sigma = A + B\,{\varepsilon_p}^n \tag{19.4}$$

Here σ - instantaneous radius of the von Mises yield surface, ε_p - effective plastic strain, A, B and n - material parameters. The approximation results compared to the experimental plots are shown in Fig. 19.9.

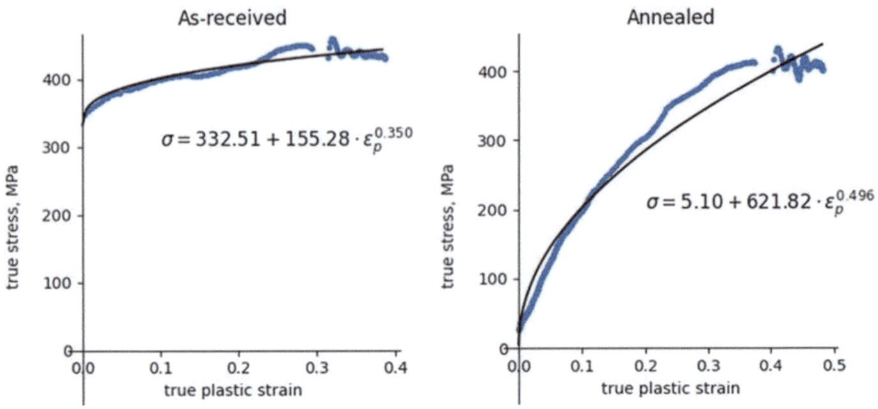

Fig. 19.9: Approximation of copper C101 diagrams.

19.3.2 Taylor Experiments Results

The specimens were tested in the as-received state and after annealing. The initial length of the specimen is 70 mm and the diameter is 10 mm. Specimen photographs, tested at impact velocities 85, 100 and 125 m/s (from left to right), are shown in the Figs, 19.10. The comparison of the specimen residual profiles after impact at different velocities are shown in Fig. 19.11. On the left copper specimens in the delivery state are presented, on the right - annealed copper specimens.

Fig. 19.10: Specimens after testing.

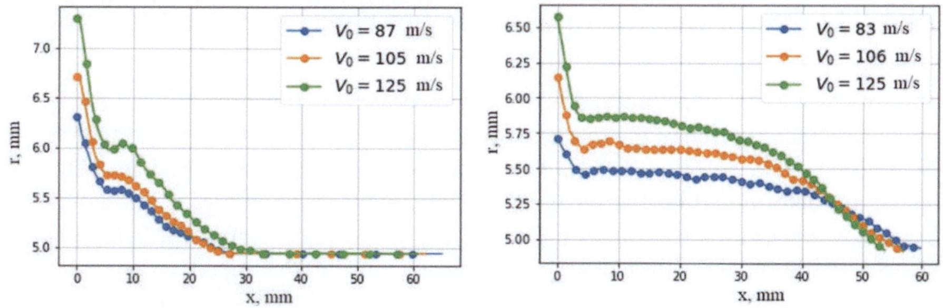

Fig. 19.11: Specimens after testing.

The results of the Taylor impact experiments (the residual length l_1 and the length of the non-deformed part H, as well as the material yield strength σ_y calculated by formula (19.1) and the average strain rate implemented in the experiments calculated by formula (19.2)) are summarized in Table 19.2 for copper specimens in delivered state and Table 19.3 for annealed copper specimens.

Table 19.2: The Taylor Test. Copper C101, as-received.

v_0, m/s	l_1, mm	H, mm	σ_y, MPa	$\dot{\varepsilon}_{av}$, s^{-1}
87	65.2	39.7	374	1435
105	62.3	37.6	334	1621
125	60.9	30.2	363	1570

Table 19.3: The Taylor Test. Copper C101, annealed.

v_0, m/s	l_1, mm	H, mm	σ_y, MPa	$\dot{\varepsilon}_{av}$, s^{-1}
83	59.7	3.3	65.5	622
106	56.9	2.29	75.8	783
125	53.8	1.35	75	910

19.4 The Use of the Taylor Method for Verifying Material Models

Traditionally, the Taylor method is widely used as a verification experiment in the process of construction of the mathematical models of viscoplastic material behavior [5, 8, 16, 18–20, 26, 30, 32, 33]. The verification of the model, the parameters of which are determined on the basis of data obtained, for example, using the SHPB method, is performed by comparing the shape of the sample after (or if high-speed video registration of the impact process was used, during) the impact with the barrier, obtained in real and virtual (numerical) experiments. In addition, the histories of contact forces between the specimen and the barrier can be compared, if a modified test scheme is used. Such verification, in addition to checking the reliability of the mathematical model in the investigated regime of loading conditions, allows one to justify its applicability in a wider range of strain rates and achieved plastic strains. Figure 19.12 shows a comparison of deformation processes of Taylor specimens during the impact interaction with a measuring bar in natural (dark profile, the result of high-speed shadowing) and virtual (colored profile) experiments. The experiments with the sample in the delivered state are shown on the left and of copper specimens after annealing - on the right.

Figures 19.13 and 19.14 show the comparison of the specimen profiles after impact in the computational and natural experiments (left part of the figure) and the comparison of the contact forces measured with the measuring bar. The relative error of the computational prediction of the specimen profile after impact is given

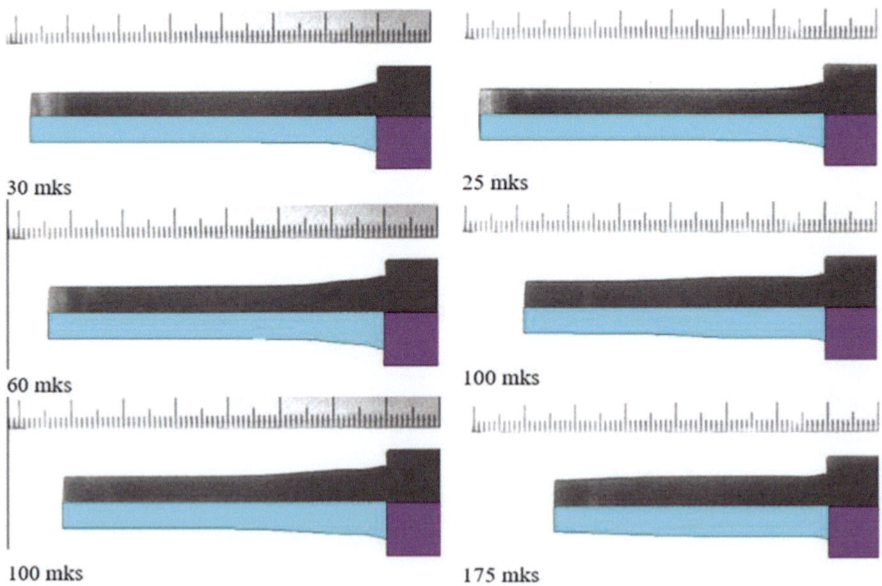

Fig. 19.12: The deformation process of samples in the Taylor test in natural and numerical experiments.

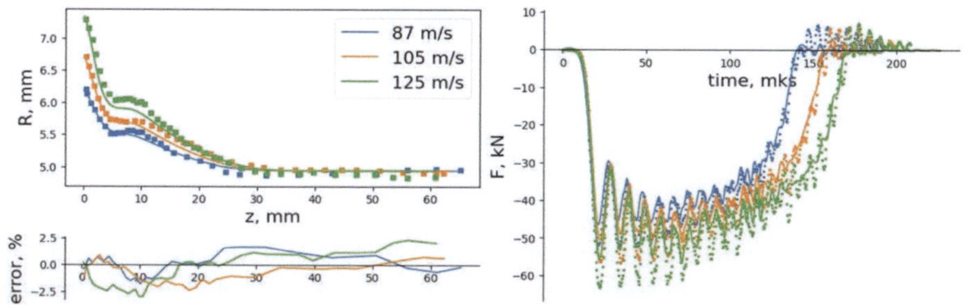

Fig. 19.13: Comparison of numerical simulation and experiment. As-received condition.

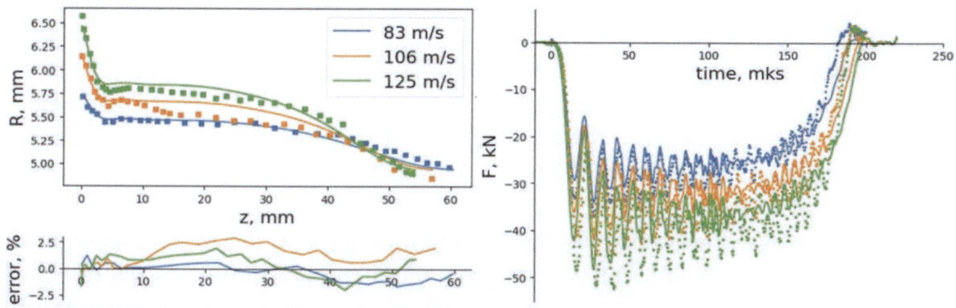

Fig. 19.14: Comparison of numerical simulation and experiment. Annealed samples.

by formula:

$$\text{err} = \frac{r^{\text{exp}} - r^{\text{calc}}}{r^{\text{exp}}} \cdot 100\ \% \tag{19.5}$$

A good agreement between the results of numerical modeling and experiment can be noted. The error in the sample size does not exceed 3 %. There is a good correspondence of amplitudes and durations of time dependences of contact forces.

19.5 The Use of the Taylor Method for Identification of the Material Models

In practice, materials are hardened (including nonlinear hardening) during plastic deformation, so estimates (19.1) and (19.2) give a large error. The approach in which the characteristics of strain hardening of a material are determined from the results of the Taylor experiment by means of numerical simulation is widespread. The set of parameters of the model describing the yield surface behavior is found by means of an iterative procedure minimizing the deviations of the experimental profile of the specimen from the profile obtained by numerical simulation. In the present work, a procedure for the inverse identification of parameters of the plasticity models based on the Taylor test is implemented. The open source program Calculix (`https://dhondt.de/`) is used as a solver. The objective function is defined as follows:

$$f = \sum_{j=1}^{M} w_j \sum_{i=1}^{N_j} \frac{\left| \left(r_{ji}^{\text{exp}}\right)^2 - \left(r_{ji}^{\text{calc}}\right)^2 \right|}{N_j} \tag{19.6}$$

where M is the number of experiments, w_j is weight factor of jth experiment, N_j is the number of point on specimen profile of jth experiment, r_{ji}^{exp} is experimental radius of the specimen in the point i, r_{ji}^{calc} is calculated radius of the specimen in the point i (Fig. 19.15). Thus, the objective function characterizes the sum of the differences in the cross-sectional areas of the specimen in the experiment and simulation.

Calculation control and search for the optimal solution were implemented in the python programming language (`https://www.python.org/`). The Nelder-Mead method [5] is used to search for the optimal set of parameters corresponding to the minimum of function (19.6). This is a simple and at the same time effective method that allows to optimize functions without using gradients. The method is robust and usually shows good results, although there is no theory of convergence. The algorithm consists in forming a simplex and then deforming it in the direction of the minimum, through three operations: reflection, expansion and contraction.

A simplex is a geometric figure, which is a multidimensional generalization of a triangle. An n-dimensional simplex has $n + 1$ vertices. For one-dimensional space it is a segment, for two-dimensional space it is a triangle. At the initial stage we choose a simplex and calculate the values of the objective function in its vertices: $f(V_1)$, $f(V_2)$, $f(V_3)$. The point where the target function has the minimum value is assigned the status of best (b), the point with the maximum value gets the status of worst (w). Other points are sorted by increasing values of the target function (Fig. 19.16). Then the average point p on the segment (plane, etc.), which is constructed without taking into account the worst point w, is determined. To replace the point w and construct a new simplex, the operations of reflection, compression and contraction are used.

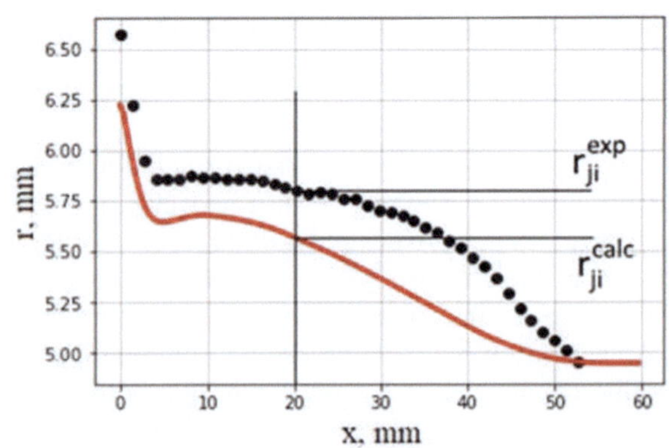

Fig. 19.15 Definition of the objective function.

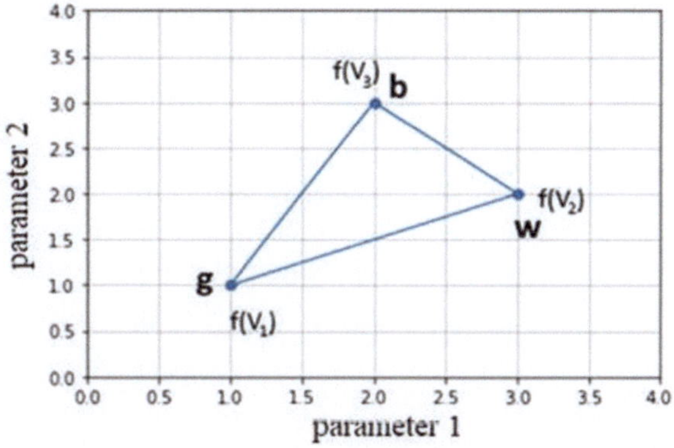

Fig. 19.16 Nelder-Mead algorithm illustration.

In the first case, the point w is reflected relative to the point p by different distances ($n1$, $n2$). If the values of the objective function in the new point $n2$ are smaller than in b, this point is used instead of the point w to build a new simplex (Figure 19.17a). If the reflection and stretching searches fail, the compression operation is applied (Figure 19.17b). If this operation has not yielded positive results, the simplex is reduced to the point b (Figure 19.17c). Then the search is repeated.

The searching procedure ends if:

- the simplex area has reached a certain value
- a certain number of iterations have been performed

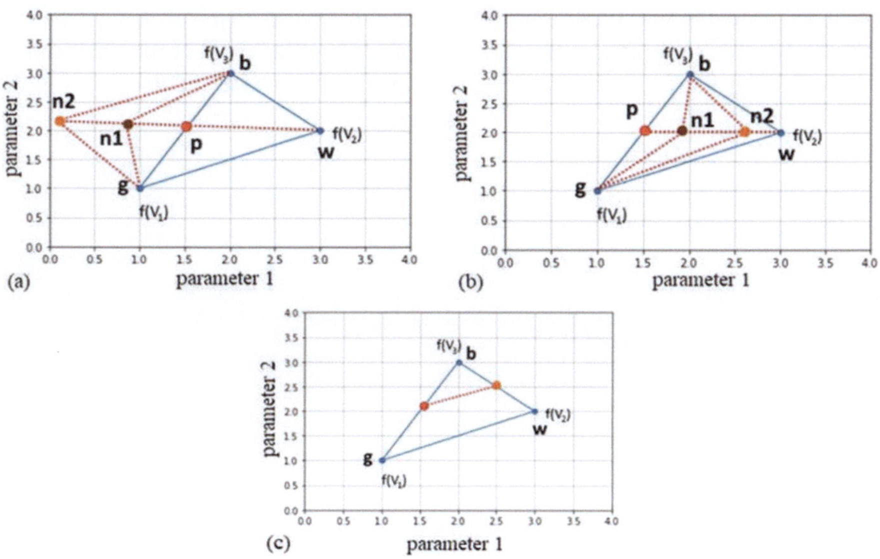

Fig. 19.17: (a) Reflection and tension operation, (b) Compression operation, (c) Contraction operation.

- the objective function value at point b does not change by more than a certain number per iteration (accuracy with respect to the objective function is achieved)
- parameter values do not change more than a certain amount per iteration (accuracy with respect to the parameter is achieved)

In the present study Nelder-Mead optimization algorithm is used to minimize the objective function (19.6). In order to determine the strain diagrams based on the inverse analysis, strain hardening of cupper C101 was described by the following model, which includes only two parameters k and n:

$$\sigma = k\, \varepsilon^{n} \tag{19.7}$$

where ε is total deformation, k and n are variable parameters of the model.

In Fig. 19.18 the experimental profiles (markers) are compared with the profiles obtained in calculation (solid lines) at the last iteration of the optimization procedure for copper specimens (as-received on the left and annealed on the right). The values of the obtained parameters of Eq. (19.7) are also presented. It can be seen that the implemented algorithm allows one to select the parameters of the strain hardening model in such a way that the profile obtained by numerical simulation is in good agreement with the experimental shape of the specimen after impact.

Figure 19.19 illustrates the point motion in the parameter space during profile optimization of the annealed copper specimen at a impact velocity of 125 m/s for three different initial values of model parameters. It can be seen that, regardless of the choice of the starting point, the solution converges to the same value.

A comparison of the diagrams determined by the inverse method using the Taylor test for copper with the curve obtained using the traditional version of the Kolsky method is shown in Fig. 19.20. The dotted line corresponds to the history of changes in the maximum strain rate (right vertical axis) during the impact process. It can be concluded that the diagrams obtained with direct and inverse method are in good agreement.

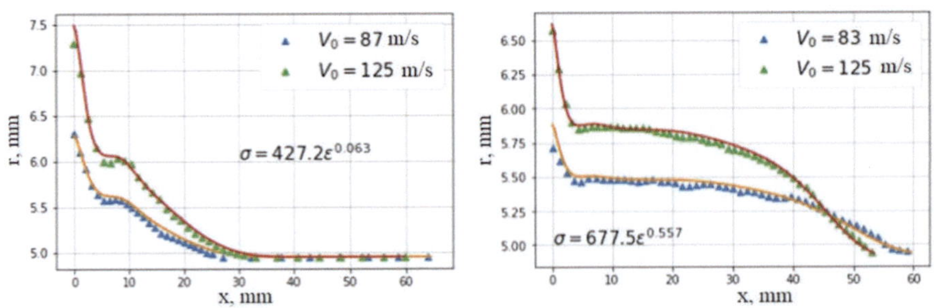

Fig. 19.18: The result of the procedure of inverse identification.

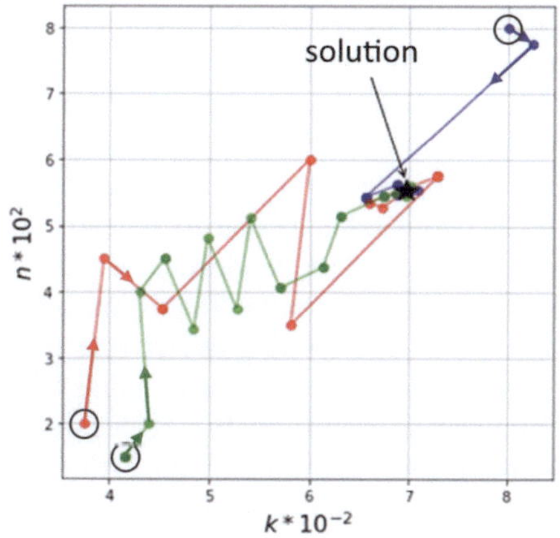

Fig. 19.19 Parameters trajectories during optimization.

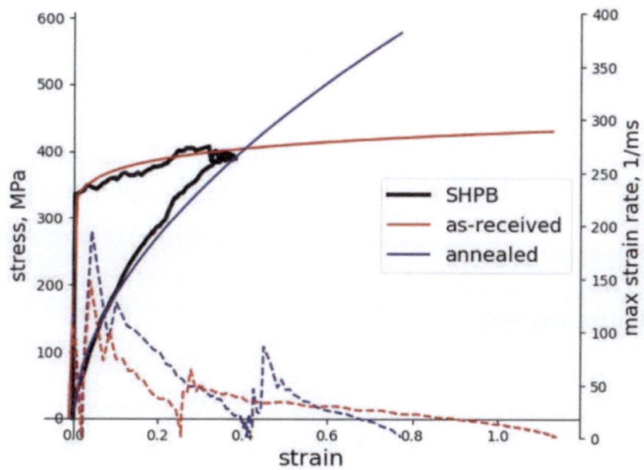

Fig. 19.20 Diagrams of copper, determined by the inverse method of the Taylor test in comparison with the curve obtained using the SHPB method.

19.6 Conclusions

The Taylor method is currently widely used as an experiment for verifying the behavior models of viscoplastic materials, identified on the basis of an independent system of basic experiments. Due to the simplicity of implementation, this method can be easily used to evaluate the dynamic yield strength of a material. However, the analytical determination of the deformation conditions (strain rate) and strain hardening is performed with a large error. The combination of experimental investigation methods and numerical modeling allows the inverse analysis to be used for evaluation of the deformation curves using the data of the Taylor test. The implementation of the identification procedure and the obtained data on copper in the delivered state and after annealing are presented in this paper. With account for the low influence of strain rate hardening of high-purity copper, the obtained curves agree quite well with the diagrams obtained using the SHPB method. However, the strain rates achieved in the Taylor tests are almost two orders of magnitude higher than those in experiments using the Kolsky method.

To perform tests using the SHPB method, the setup should include two measuring bars, a loading device (most often a pneumatic gun), an impactor, strain gauges with power supply circuitry, and an oscilloscope. The SHPB method allows experiments to be conducted at a controlled nearly constant strain rate and to measure with high accuracy the time dependencies of strain, strain rate, and stress in the specimen during loading. To perform the Taylor impact experiment, the minimum set of equipment includes an accelerating device (gas gun), a velocity meter, and a rigid barrier. The construction of the deformation diagram is possible using numerical simulation and optimization procedure. In a Taylor specimen, however, the strain and stress fields are non-uniform and the strain rate is time-varying. Nevertheless, for materials that have a weak dependence of the strain diagram on the strain rate, this approach can serve as a good and sufficiently accurate tool for determining the material's strain diagram. Anyway, the use of the Taylor method for both verification and identification of mathematical models describing the behavior of viscoplastic materials allows one to significantly expand the study range in terms of strain and strain rates.

Acknowledgements The work is carried out with the financial support of the Ministry of Science and Higher Education of the Russian Federation (project No. FSWR-2026-0019).

References

[1] Bacon C, Lataillade J-L (2001) Development of the Kolsky-Hopkinson techniques and applications for non-conventional testing. In: Nowacki WK, Klepaczko JR (eds) New Experimental Methods in Material Dynamics and Impact/Trends in Mechanics of Materials. Institute of Fundamental Technological Research, Warsaw, pp. 1-58

[2] Bardenhagen SG, Harstad EN, Maudlin PJ, Gray GT, Foster JC Jr (1998) Viscoelastic models for explosive binder materials. AIP Conference Proceedings 429(1):281-284, DOI 10.1063/1.55647

[3] Bragov A, Konstantinov A, Kruszka L, Lomunov A, Filippov A (2018) Dynamic properties of stainless steel under direct tension loading using a simple gas gun. EPJ Web of Conferences 183:02035, DOI 10.1051/epjconf/201818302035

[4] Bragov AM, Konstantinov AY, Lomunov AK (2018) Eksperimental'no-teoreticheskoye issledovaniye protsessov vysokoskorostnogo deformirovaniya i razrusheniya materialov razlichnoy fizicheskoy prirody s ispol'zovaniyem metoda Kol'skogo i yego modifikatsiy (in Russ., Experimental and theoretical investigation of dynamic deformation and fracture of materials of different physical nature using Kolsky method and its modifications). Publishing House of Nizhny Novgorod State University, Nizhny Novgorod

[5] Buchar J, Forejt M, Jopek M, Krivanek I (2000) Evaluation of constitutive relations for high strain rate behaviour using the Taylor test. Journal de Physique IV 10(PR9):75-80, DOI 10.1051/jp4:2000913.
[6] Caverzan A, Cadoni E, di Prisco M (2012) Tensile behaviour of high performance fibre-reinforced cementitious composites at high strain rates. International Journal of Impact Engineering 45:28-38, DOI 10.1016/j.ijimpeng.2012.01.006
[7] Chou PC, Clark W, Liang D-S (1995) Blunt cylinder impact tests for the determination of constitutive equation of explosives. Proceedings of the 15th International Symposium on Ballistics, Vol 1, Jerusalem, pp. 159-166
[8] Couque H (1998) On the use of the symmetric Taylor test to evaluate dynamic ductile compression fracture properties of metals. In: Brebbia CA (ed) Structures under Shock & Impact V. WIT Transactions on the Built Environment 32(11)579-589, DOI 10.2495/SU980521
[9] Davies RM (1948) A critical study of the Hopkinson pressure bar. Philosophical Transactions of the Royal Society of London, Series A: Mathematical and Physical Sciences 240(821):375-457, DOI 10.1098/rsta.1948.0001
[10] Erlich DC, Shockey DA (1984) Dynamic flow curve of 4340 steel as determined by the symmetric rod impact test. In: Asay JR, Graham RA, Straub GK (eds) Shock Waves in Condensed Matter 1983. North-Holland, Amsterdam, pp. 129-132, DOI 10.1016/B978-0-444-86904-3.50030-0
[11] Eskandari H, Nemes JA (2000) Dynamic testing of composite laminates with a tensile split Hopkinson bar. Journal of Composite Materials 34(4):260-273, DOI 10.1177/002199830003400402
[12] Field JE, Proud WG, Walley SM, Goldrein HT (2001) Review of experimental techniques for high rate deformation and shock studies. In: Nowacki WK, Klepaczko JR (eds) New Experimental Methods in Material Dynamics and Impact/Trends in Mechanics of Materials. Institute of Fundamental Technological Research, Warsaw, pp. 109-177
[13] Field JE, Walley SM, Bourne NK, Huntley JM (1994) Experimental methods at high strain rate. Journal de Physique IV 4(C8):3-22, DOI 10.1051/jp4:1994801
[14] Gama BA, Lopatnikov SL, Gillespie JW Jr (2004) Hopkinson bar experimental technique: A critical review. Applied Mechanics Revview 57(4):223-250, DOI 10.1115/1.1704626
[15] Hauser FE (1966) Techniques for measuring stress-strain relations at high strain rates. Experimental Mechanics 6(8):395-402, DOI 10.1007/BF02326284
[16] Holmquist TJ, Johnson GR (1991) Determination of constants and comparison of results for various constitutive models. Journal de Physique IV 1(C3):853-860, DOI 10.1051/jp4:19913119
[17] Jiang B, Zhang R (2006) Tensile properties in the through-thickness direction for a carbon fiber woven reinforced composite at impact loading rate. Journal de Physique IV 134:1071-1075, DOI 10.1051/jp4:2006134164
[18] Johnson GR, Cook WH (1983) A constitutive model and data for metals subjected to large strains, high strain rates and high temperatures. Proceedings of

the 7th International Symposium on Ballistics. The Hague, The Netherlands, pp 541–547

[19] Johnson GR, Holmquist TJ (1988) Evaluation of cylinder-impact test data for constitutive model constants. Journal of Applied Physics 64:3901-3910, DOI 10.1063/1.341344

[20] Jones SE, Drinkard JA, Rule WK, Wilson LL (1998) An elementary theory for the Taylor impact test. International Journal of Impact Engineering 21(1-2):1-13, DOI 10.1016/S0734-743X(97)00036-5

[21] Kolsky H (1949) An investigation of the mechanical properties of materials at very high rates of loading. Proceedings of the Physical Society London. Section B 62(11):676-700, DOI 10.1088/0370-1301/62/11/302

[22] Lindholm US, Yeakley LM (1968) High strain-rate testing: Tension and compression. Experimental Mechanics 8(1):1-9, DOI 10.1007/BF02326244

[23] Nelder JA, Mead R (1965) A simplex method for function minimization. The Computer Journal 7(4):308-313, DOI 10.1093/comjnl/7.4.308

[24] Nicholas T (1981) Tensile testing of materials at high rates of strain. Experimental Mechanics 21(5):177-185, DOI 10.1007/BF02326644

[25] Papirno R, Mescall J, Hansen A (1980) Proceedings of the Army Symposium on Solids Mechanics, 1980 - Designing for extremes: Environment, Loading, and Structural Behavior. Technical Report AMMRC MS 80-4, Watertown, MA, pp. 367-385

[26] Partom Y (1992) Modeling Taylor's anvil test for hardening and rate sensitive materials. In: Persson K (ed) Proceedings of the 13th International Symposium on Ballistics, Vol 3. Stockholm

[27] Quidot M., Racimor P., Chabin P. (1999) Constitutive models for PBX at high strain rate // Shock Compression of Condensed Matter. AIP Conference Proceedings 505(1):687-690, DOI 10.1063/1.1303566

[28] Taylor GI (1948) The use of flat-ended projectiles for determining yield stress. I: Theoretical considerations. Proceedings of the Royal Society of London. A. Mathematical and Physical Sciences 194(1038):289-299, DOI 10.1098/rspa.1948.0081

[29] Taylor GI, Quinney A (1934) The latent energy remaining in a metal after cold working. Proceedings of the Royal Society of London. Series A, Containing Papers of a Mathematical and Physical Character. 143(849):307-326 DOI 10.1098/rspa.1934.0004

[30] Walley SM, Church PD, Townsley R, Field JE (2000) Validation of a path-dependent constitutive model for FCC and BCC metals using "symmetric" Taylor impact. Journal de Physique IV 10(PR9):P. 69-74, DOI 10.1051/jp4:2000911

[31] Wilkins ML, Guinan MW (1973) Impact of cylinders on a rigid boundary. Journal of Applied Physics 44(3):1200-1206, DOI 10.1063/1.1662332

[32] Woodward RL, Lambert JP (1981) A discussion of the calculation of forces in the one-dimensional finite difference model of Hashmi and Thompson. International Journal of Mechanical Sciences 23(8):497-501, DOI 10.1016/0020-7403(81)90055-2

[33] Zerilli FJ, Armstrong RW (1987) Dislocation-mechanics-based constitutive relations for material dynamics calculations. Journal of Applied Physics 61(5):1816-1825, DOI 10.1063/1.338024

Chapter 20
Generalization of Poynting Effect for Superimposed Large Strains Based on the Exact Solution to the Problem of Torsion of a Composite Cylinder Made of Viscoelastic Materials

Vladimir A. Levin and Konstantin M. Zingerman

Abstract The generalization of Poynting effect is analyzed for the case of superimposed large strains. The pure torsion of a composite circular cylinder is considered. It is assumed that the cylinder is made of incompressible viscoelastic materials. The cylinder is formed sequentially. At the first stage the cylindrical bar undergoes an initial torsional deformation under the action of a torque and axial force. Then a hollow undeformed viscoelastic cylinder (layer) joins it. The cylinders are glued along their common boundary surface. Further the composite cylinder is subjected to additional deformation. It is also a deformation of pure torsion, which occurs as a result of the action of a torque together with an axial force. The problem is formulated for the quasi-static case on the foundation of the theory of superimposed large strains. The exact analytical solution of this problem is presented. The numerical results are given. Dependencies of axial force on time and the angle of twist per unit length are shown.

Key words: Viscoelasticity, Superposition of large strains, Torsion of cylinders, Poynting effect, Exact analytical solution, Discrete accretion

20.1 Introduction

A special attention in nonlinear solid mechanics is given to the investigation of effects that arise during torsion of cylindrical bodies. One of these effects is the Poynting effect [41]. This effect appears in two ways. One manifestation of this effect consists

---

Vladimir A. Levin (corresponding author)
Faculty of Mechanics and Mathematics, Lomonosov Moscow State University, Leninskie gory, 1, 119992 Moscow, Russian Federation, e-mail: `v.a.levin@mail.ru`

Konstantin M. Zingerman
Department of Applied Mathematics and Cybernetics, Tver State University, Zhelyabov Street, 33, 170100 Tver, Russian Federation, e-mail: `zingerman@rambler.ru`

H. Altenbach et al. (eds.), *Stability Problems in Mechanics of Continua*,
Advanced Structured Materials 260, https://doi.org/10.1007/978-3-032-31032-3_20

in changing the length of the cylinder during torsion in the absence of axial forces applied to the bases of the cylinder. Another manifestation is the occurrence of stresses directed along its axis during pure torsion of the cylinder, when the bases of the cylinder cannot move along this axis.

For theoretical explanation of the Poynting effect both exact and approximate analytical solutions are used. These solutions are obtained to the problems of torsion and tension-compression of cylinders. It is known that for isotropic linear-elastic materials under small strains this effect does not appear [46].

For incompressible isotropic nonlinear elastic materials one can use universal solutions in order to investigate the Poynting effect [14, 21, 23, 37, 43, 44, 47, 52]. Another exact analytical solutions are known for the case of pure torsion for some classes of compressible isotropic nonlinear-elastic materials [40].

The model developed in [48] employs the Gurtin–Murdoch theory to the analysis of torsion and tension of nonlinear-elastic rods. This model can be used to analyze the specific forms of the Poynting effect. Approximate analytical solutions that can be used to demonstrate the Poynting effect are presented in [8–10, 18, 19, 59]. It is interesting that for nano- and microtubes made of a specific class of anisotropic materials the Poynting effect takes place within the framework of linear elasticity under small strains [20].

In plasticity theory, the effect similar to the Poynting effect is known as the Swift effect [50]. An experimental study of this effect was carried out in [55] for glassy polymers and in [54] for metals. The theoretical investigation of this effect is performed using different models of elasto-plastitity and different deformation programs [4, 5, 7, 16, 25, 45]. The experimental analysis of large-strain torsion of viscoelastic polymers is performed in [26]. Phenomenological constitutive models, characterizing the complex and highly nonlinear thermo-mechanical behavior of these polymers, are developed in this investigation.

The promising trend in nonlinear solid mechanics is the large-strain analysis of surface growth (accretion) of deformable solids [38, 42, 49, 51, 53, 56, 57]. The practical use of results obtained within the framework of this lead is related with the description of growth of biological tissues [2, 17] and the strength analysis of additive manufacturing products [3, 34, 39].

For the case of discrete growth one can use the theory of multiple superposition of large deformations to model this phenomenon [27, 28, 30]. In [35], problems of tension-compression and torsion of composite cylinders made of incompressible isotropic nonlinear elastic materials were considered. These problems were solved for the case of discrete growth when a cylindrical layer is attached to the internal prestressed cylinder. The nontrivial effect related with the sign of Poynting effect is obtained. It is shown that for composite, preliminary strained cylinders made of Mooney-type materials with positive material constants C_1 and C_2 both positive (axial tension) and negative (axial compression) Poynting effect may take place. This is a non-trivial effect caused by the initial deformation of the inner part of a cylinder.

The generalization of solution obtained in [35] to the case of multiple cylinders is presented in [58]. In [32] this solution is generalized to the case of cylinders made of hypoelastic materials. The models of continuous accretion of viscoelastic bodies under finite and small strains are developed in [11]. In [29], the discrete accretion of a composite viscoelastic cylinder subjected to tension/compression and torsion is analyzed under finite strains. The cylinder is considered as a body subjected to the imposition of large deformations. The mechanical properties of the cylinder material are described by constitutive relations of integral type, which, in the absence of viscosity, are reduced to the constitutive relations for a neo-Hookean material [1, 15].

In the present paper, the detailed analysis of Poynting effect is performed for discretely growing composite viscoelastic cylinders. The computations are performed using the solution obtained in [29]. The case of pure torsion of the cylinder is considered.

20.2 Notation

In this work, the notation of the theory of multiple superposition of large deformations is used [28, 30, 32, 58]. The states (configurations) of the body are numbered from 0 (initial state) to n (final state). The following variables are introduced:

- $\sigma_{l,m}$ – the Cauchy stress tensor during transition from state l to m.
- $\mathbf{\Sigma}_{l,m}$ – generalized stress tensor during transition from state l to m.
- $\mathbf{C}_{l,m} = \mathbf{F}^T_{l,m} \cdot \mathbf{F}_{l,m}$ – the Green deformation tensor during transition from state l to m.
- $\mathbf{F}_{l,m}$ – deformation gradient during transition from state l to m.
- $\mathbf{E}$ – unit tensor.
- $\operatorname{tr}\mathbf{X} = \mathbf{X} \cdot \cdot \mathbf{E}$ — trace of a rank two tensor $\mathbf{X}$.
- p_m — pressure that is not expressed in terms of strain, in the m-th state .
- $x_s^{(m)}$ — Cartesian coordinates of particles in a compound cylinder in the m-th state.
- $\mathbf{i}_s^{(m)}$ — coordinate vectors of Cartesian coordinates in the m-th state.
- r_m, φ_m, z_m — cylindrical coordinates of particles in a compound cylinder in the m-th state.
- $R_m^{(l)}$ – the outer radius of the cylindrical layer l in the m-th state.

Relationships between cylindrical and Cartesian coordinates:

$$x_1^{(m)} = r_m \cos\varphi_m \ , \ x_2^{(m)} = r_m sin\varphi_m \ , \ x_3^{(m)} = z_m$$

$\mathbf{e}_r^{(m)}, \mathbf{e}_\varphi^{(m)}, \mathbf{e}_z^{(m)}$ — unit vectors of cylindrical coordinates in the state m:

$$\mathbf{e}_r^{(m)} = \mathbf{i}_1^{(m)} \cos\varphi_m + \mathbf{i}_2^{(m)} sin\varphi_m \ , \ \mathbf{e}_\varphi^{(m)} = -\mathbf{i}_1^{(m)} sin\varphi_m + \mathbf{i}_2^{(m)} \cos\varphi_m \ , \ \mathbf{e}_z^{(m)} = \mathbf{i}_3^{(m)}$$

20.3 Mechanical Problem Statement

The problem is formulated within the framework of the theory of repeated superposition of large strains [28, 30, 35]. Following [29], the combined torsion and tension (compression) of the composite cylinder is considered. There is a circular cylinder made of incompressible isotropic viscoelastic material. Initially it is undeformed. Then the cylinder undergoes an initial torsional deformation together with tension or compression in the axial direction under the action of a torque and axial force. This deformation can change over time according to a given rule. Due to the incompressibility of the material, tension or compression of the rod along its axis is accompanied, respectively, by compression or tension of the rod in the radial direction. It is assumed that at the end of the given time interval the cylinder is in the first intermediate state (configuration). In this state, a hollow undeformed viscoelastic cylinder (layer) joins it. The cylinders are joined along their common boundary surface, and it is assumed that the conditions of ideal contact are satisfied on this surface after the junction of cylinders.

Further (during the next time interval), the joined cylinders are subjected to additional deformation. It is also a deformation of torsion together with tension or compression along the axis, which occurs as a result of the action of a torque together with an axial force. At the end of the second time interval, the compound cylinder passes to the second intermediate state. Then the next hollow undeformed viscoelastic cylinder (layer) may be attached to it.

Sequential junction of the cylinders with their subsequent torsion and tension-compression can be repeated several times. The quasi-static processes are considered. It is assumed that each layer is added instantly, in sense of [6], and the dynamical effects are not taken into account.

20.4 Mathematical Problem Statement and Stress Analysis

We consider a multilayered circular cylinder deformed over several stages. The m–th stage ($m = 1, \ldots, n$) occurs over a given period of time $t_{m-1} \leq t \leq t_m$, $t_0 = 0$. At the beginning of the $m - th$ stage the next m–th undeformed layer with an outer radius $R_m^{(m-1)}$ is added to the multilayered cylinder. Further, during the stage, the multilayered cylinder is subjected to tension-compression and torsion, deforming as a single whole.

The coordinates of particles of the cylinder during the m–th stage of deformation are related to the coordinates of these particles in the previous state as follows [32, 58]:

$$r_m = \lambda_m (t)\, r_{m-1}, \quad \varphi_m = \varphi_{m-1} + \alpha_m (t)\,, \quad z_m = z_{m-1} \lambda_m^{-2} (t) \tag{20.1}$$

where $\lambda_m (t)$ is the specified elongation factor at stage m; $\alpha_m (t) = k_m (t)\, z_{m-1}$; $k_m (t)$ — specified characteristic of torsional deformation at stage m. Relations (20.1) ensure the fulfillment of the incompressibility condition. Note that when $t \le t_{m-1}$, relations $\lambda_m(t) = 1$, $k_m(t) = 0$ must be satisfied. When $t \ge t_m$, assumptions $\lambda_m(t) = \lambda_m(t_m)$, $k_m(t) = k_m(t_m)$ must be fulfilled.

Considering relations (20.1) for several successive stages, we obtain:

$$r_m = \Lambda_{l,m} r_{l-1}, \quad \varphi_m = \varphi_{l-1} + \alpha_{l,m}, \quad z_m = z_{l-1}\Lambda_{l,m}^{-2}, \tag{20.2}$$

where:

$$\Lambda_{l,m} = \lambda_l \lambda_{l+1} \ldots \lambda_{m-1}\lambda_m,$$

$$\alpha_{l,m} = k_{l,m} z_{l-1},$$

$$k_{l,m} = k_l + k_{l+1}\Lambda_{l,l}^{-2} + \cdots + k_m \Lambda_{l,m-1}^{-2}$$

To characterize the deformation of the l–th layer during the m–th stage, the deformation gradient is used:

$$\begin{gathered}\mathbf{F}_{l,m} = \Lambda_{l,m}\cos(\alpha_{l,m})\, \mathbf{e}_r^{(m)} \otimes \mathbf{e}_r^{(l)} - \Lambda_{l,m}\sin(\alpha_{l,m})\, \mathbf{e}_r^{(m)} \otimes \mathbf{e}_\varphi^{(l)} + \\ \Lambda_{l,m}\sin(\alpha_{l,m})\, \mathbf{e}_\varphi^{(m)} \otimes \mathbf{e}_r^{(l)} + \Lambda_{l,m}\cos(\alpha_{l,m})\, \mathbf{e}_\varphi^{(m)} \otimes \mathbf{e}_\varphi^{(l)} - \\ \Lambda_{l,m} k_{l,m} r_{l-1}\sin(\varphi_{l-1} + \alpha_{l,m})\, \mathbf{e}_r^{(m)} \otimes \mathbf{e}_z^{(l)} + \\ \Lambda_{l,m} k_{l,m} r_{l-1}\cos(\varphi_{l-1} + \alpha_{l,m})\, \mathbf{e}_\varphi^{(m)} \otimes \mathbf{e}_z^{(l)} + \Lambda_{l,m}^{-2}\mathbf{e}_z^{(m)} \otimes \mathbf{e}_z^{(l)}\end{gathered} \tag{20.3}$$

The Green deformation tensor during transition from state l to m is computed as follows:

$$\mathbf{C}_{l,m} = \mathbf{F}_{l,m}^T \cdot \mathbf{F}_{l,m}$$

It is assumed that the material of the multilayered cylinder is viscoelastic. The generalized stress tensor satisfies the constitutive relations:

$$\begin{gathered}\boldsymbol{\Sigma}_{l,m}\big|_{t=t_m} = -p_{l,m}\mathbf{C}_{l,m}^{-1} + \\ +\sum_{q=l}^{m} \int_{t^{(q-1)}}^{t^{(q)}} \left(\mu + \eta e^{-\beta(t_q-\tau)}\right)\left(\mathbf{E} - \frac{1}{3}\frac{\partial\left(\mathbf{C}_{l,q}^{-1} tr \mathbf{C}_{l,q}\right)}{\partial\tau}\right) d\tau,\end{gathered} \tag{20.4}$$

where μ, η, β are the mechanical parameters of the material, p_m is the Lagrange multiplier (pressure). If it is necessary to calculate $\boldsymbol{\Sigma}_{l,m}$ at an arbitrary moment of the $m-$ th stage of deformation, t_m should be replaced by t in the last integral (at $q = m$). The relations of this type are proposed by Adamov [1, 15] and used for solution of some stress concentration problems under finite strains [30, 31]. A more general constitutive relations of viscoelasticity are given in [24].

The true stress tensor is related to the generalized stress tensor as follows:

$$\boldsymbol{\sigma}_{l,m} = \mathbf{F}_{l,m} \cdot \boldsymbol{\Sigma}_{l,m} \cdot \mathbf{F}_{l,m}^T \tag{20.5}$$

When writing formula (20.5), the incompressibility of the material is taken into account. Using formulas (20.3)–(20.5), we can write down the stresses $\sigma_{l,m}$ at an arbitrary moment t from the interval $t_0 \leq t \leq t_m$ up to the factor p_m, which is determined from the equilibrium equations and boundary conditions of contact between the layers.

The equilibrium equation for the m-th state in radial direction is written as follows:

$$\frac{\partial\left(\Sigma_{l,m}\right)_{rr}}{\partial r_m}+\frac{\left(\Sigma_{l,m}\right)_{rr}-\left(\Sigma_{l,m}\right)_{\varphi\varphi}}{r_m}=0 \tag{20.6}$$

The equilibrium equations in other directions are satisfied identically due to symmetry. Differential equation (20.6) is supplemented with boundary conditions. For the outer layer $l = m$ this condition has the form:

$$\left(\Sigma_{m,m}\right)_{rr}\Big|_{r_m=R_m^{(m)}}=0 \tag{20.7}$$

For the internal layers the boundary conditions are written as follows:

$$\left(\Sigma_{l,m}\right)_{rr}\Big|_{r_m=R_m^{(l)}}=\left(\Sigma_{l+1,m}\right)_{rr}\Big|_{r_m=R_m^{(l)}} \tag{20.8}$$

By solving equation (20.6) with boundary condition (20.7) or (20.8) for each l–th layer, we determine p_m, and, as a consequence, $\Sigma_{l,m}$. Thus, we obtain a complete description of the deformation process at each stage. The axial force and torque in the m-th state are computed using the relations [35, 58]

$$N_0=2\pi\int\limits_0^{R_m^{(m)}}\sigma_{zz}r_m dr_m \tag{20.9}$$

$$M_0=2\pi\int\limits_0^{R_m^{(m)}}\sigma_{\varphi z}r_m^2 dr_m \tag{20.10}$$

20.5 Numerical Computations

Calculations were performed for the case of two layers. The following notation is used: R is the radius of the inner layer in the first state, T is the duration of the first stage of deformation ($R = R_1^{(1)}$, $T = t_1$); $r = r_1/R$. Material constants are taken as follows: $\eta = \mu$, $\beta = 1/T$. The outer radius of the second layer at the moment of attachment $R_2^{(1)} = 2R$. The formulas for the deformation parameters have the form

$$\begin{aligned} &\lambda_1(t)=1, && \lambda_2(t)=1,\\ &k_1(t)=0.5H(t), && k_2(t)=KH(t-t_1), \end{aligned}$$

where $H(t)$ is the Heaviside unit function; K is a twist parameter, the value of which is different for the different graphs below.

The distribution of true stresses in the cylinder is shown in Figs. 20.1 and 20.2. The stresses are referred to the modulus μ. The radial and axial stresses are given with the opposite sign.

The stress distribution for different time instants is presented in Fig. 20.1 for the case $KR = 0.5$. The stress distribution at the time $t = 2T$ is given in Fig. 20.2 for different values of twist parameter K.

The discontinuities in Figs. 20.1 and 20.2 are associated with the fact that the first (internal) layer is preliminarily strained, and the second (external) layer has no preliminary strains. So, there is a jump of total strains (and, respectively, a jump of total stresses) at the boundary between the layers. Note that unlike the general case

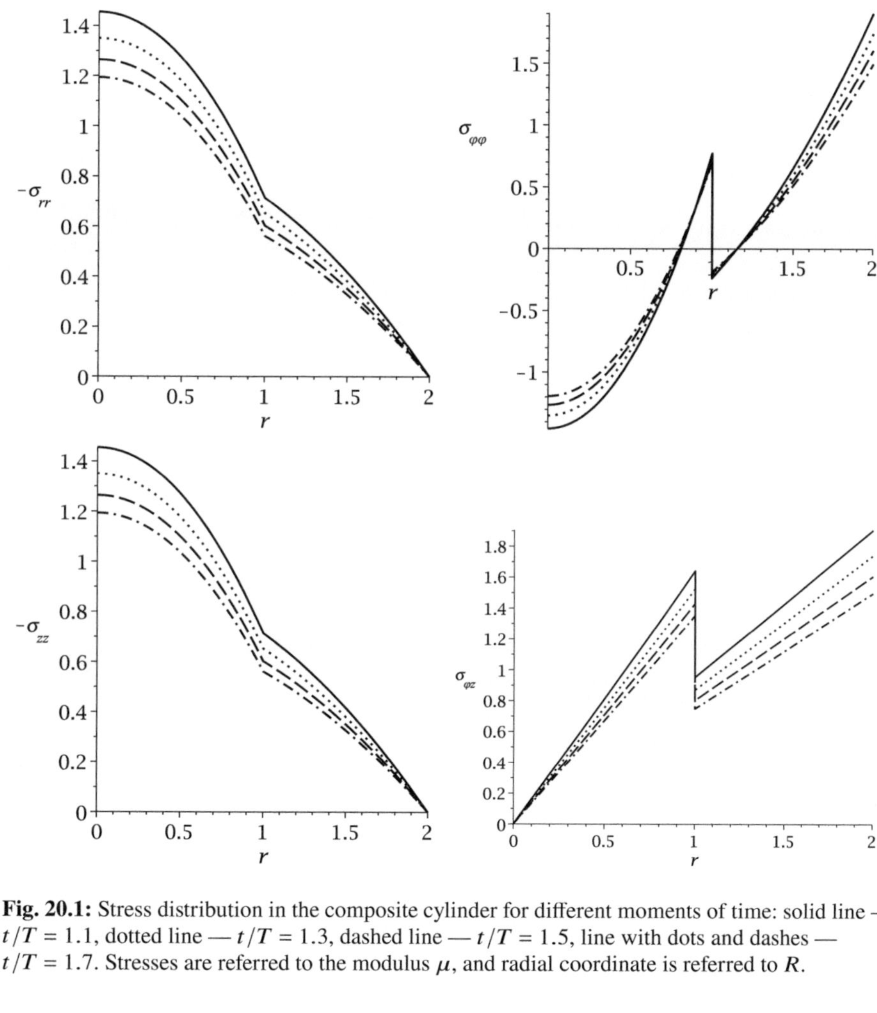

Fig. 20.1: Stress distribution in the composite cylinder for different moments of time: solid line — $t/T = 1.1$, dotted line — $t/T = 1.3$, dashed line — $t/T = 1.5$, line with dots and dashes — $t/T = 1.7$. Stresses are referred to the modulus μ, and radial coordinate is referred to R.

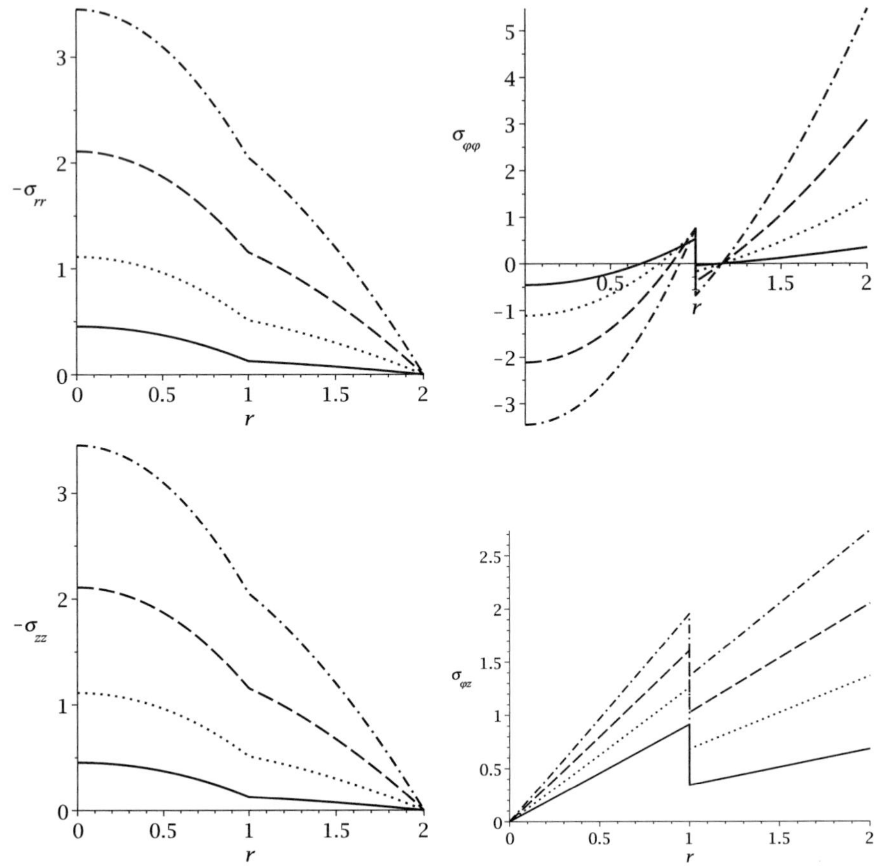

Fig. 20.2: Stress distribution in the composite cylinder for different values of twist parameter K: solid line — $KR = 0.25$, dotted line — $KR = 0.5$, dashed line — $KR = 0.75$, line with dots and dashes — $KR = 1$. Stresses are referred to the modulus μ, and radial coordinate is referred to R.

in which the torsion is combined with axial tension or compression (and with radial compression or tension, respectively) [29], in the case of pure torsion the axial stress is continuous. Naturally, the radial stress is continuous, too, as it follows from the boundary conditions.

One can see from Fig. 20.1 that the stresses are decreased in time in absolute value. This effect is evidently related with the relaxation. It may be seen from Fig. 20.2 that the growth of twist parameter leads to the growth of stresses in absolute value, and the dependencies of stresses on the twist parameter are nonlinear.

The dependencies of axial force on time and the twist parameter K are given in Figs. 20.3 and 20.4. Figure 20.3 shows the time dependence of axial force for different values of the twist parameter, and Fig. 20.4 presents the dependence of axial force on the twist parameter for different time instants. The axial force is given with an

Fig. 20.3 Time dependence of axial force for different values of twist parameter K: solid line — $KR = 0.25$, dotted line — $KR = 0.5$, dashed line — $KR = 0.75$, line with dots and dashes — $KR = 1$.

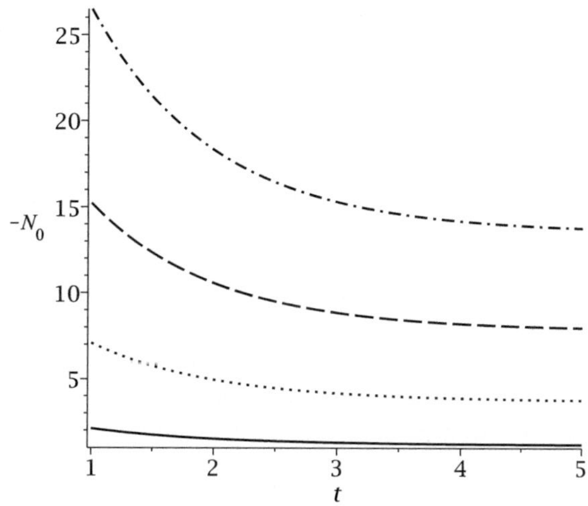

Fig. 20.4 Dependence of axial force on the twist parameter K for different values of time: solid line — $t/T = 1.1$, dotted line — $t/T = 1.3$, dashed line — $t/T = 1.5$, line with dots and dashes — $t/T = 1.7$

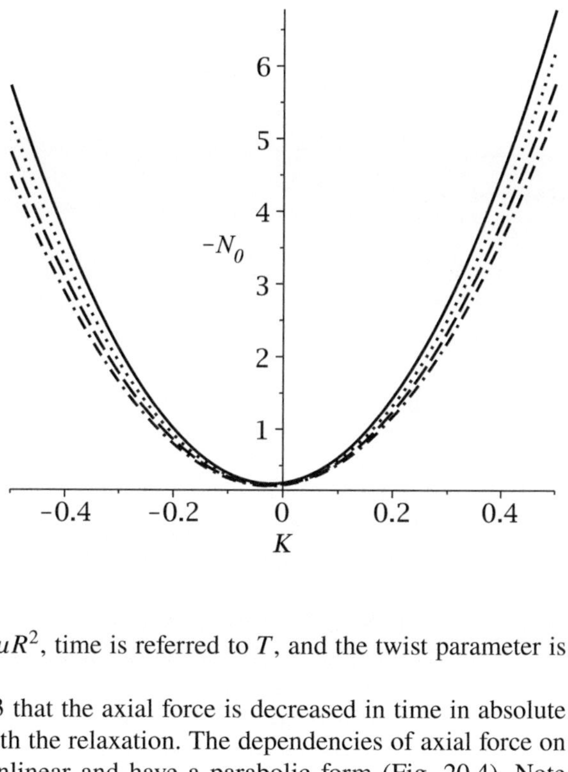

opposite sign and referred to μR^2, time is referred to T, and the twist parameter is referred to $1/R$.

One can see from Fig. 20.3 that the axial force is decreased in time in absolute value. This effect is related with the relaxation. The dependencies of axial force on the twist parameter K are nonlinear and have a parabolic form (Fig. 20.4). Note that the axial force is negative for the values of parameters for which the plot is constructed.

20.6 Conclusion

The generalized Poynting effect is investigated for growing cylinder made of incompressible isotropic viscoelastic materials. It is assumed that the cylinder is composed of some layers and these layers are added successively. In order to reveal this effect, an exact analytical solution is constructed for the problem of tension-compression and torsion of composite viscoelastic cylinder undergoing large strains. The problem is formulated using the theory of multiple superposition of large strains. The particular case of pure torsion of the cylinder is analyzed in detail. The results of numerical studies are obtained for this case. Significant nonlinear effects are found. The effects of viscosity are analyzed. It is determined that the axial force is negative for the values of parameters for which the analysis is performed. This force is decreased in time in absolute value due to relaxation.

The results of solution can be used for the verification and validation of software packages for engineering analysis. For example, the obtained results can be used to verify calculations of the stress-strain state of cylinders manufactured using additive technologies [33, 34]. It is of interest to take into account surface effects [12]. It may be useful to apply the proposed approach for the modeling of phase transitions in growing viscoelastic cylinders [13].

References

[1] Adamov AA, Matveenko VP, Trufanov NA, Shardakov IN (2003) Metody prikladnoi viazkouprugosti (in Russ., Methods of Applied Viscoelasticity). Russian Academy of Sciences, Ural Division, Ekaterinburg

[2] Ambrosi D, Ateshian GA, Arruda EM, Cowin SC, Dumais J, Goriely A, Holzapfel GA, Humphrey JD, Kemkemer R, Kuhl E, Olberding, JE, Taber LA, Garikipati K (2011) Perspectives on biological growth and remodeling. Journal of the Mechanics and Physics of Solids 59(4):863-883, DOI 10.1016/j.jmps.2010.12.011

[3] Bartlett JL, Li X (2019) An overview of residual stresses in metal powder bed fusion. Additive Manufacturing 27:131-149, DOI 10.1016/j.addma.2019.02.020

[4] Bruhns O, Xiao H, Meyers A (2001) Large-strain response of isotropic-hardening elastoplasticity with logarithmic rate: Swift effect in torsion. Archive of Applied Mechanics 71(6):389-404, DOI 10.1007/s00419000146

[5] Cazacu O, Revil-Baudard B, Barlat F (2014) New interpretation of cyclic Swift effects. European Journal of Mechanics A/Solids 44:82-90, DOI 10.1016/j.euromechsol.2013.10.005

[6] Christensen RM (1982) Theory of Viscoelasticity. An Introduction (2nd edn). Academic Press, New York

[7] Colak OU, Krempl E (2005) Modeling of the monotonic and cyclic Swift effects using an isotropic, finite viscoplasticity theory based on overstress (FVBO).

International Journal Plasticity 21(3):573-588, DOI 10.1016/j.ijplas.2004.04.010
[8] De Pascalis R (2010) The Semi-Inverse Method in Solid Mechanics: Theoretical Underpinnings and Novel Applications. PhD thesis, Paris, Salento
[9] Dell'Isola F, Ruta G, Batra R (1998) Generalized Poynting effects in predeformed prismatic bars. Journal of Elasticity 50:181-196, DOI 10.1023/A:1007480722570
[10] Dell'Isola F, Ruta G, Batra RA (2010) Second-order solution of Saint-Venant's problem for an elastic pretwisted bar using Signorini's perturbation method. Jornal of Elasticity 49:113-127, DOI 10.1023/A:1007498331650
[11] Drozdov AD (1998) Viscoelastic Structures: Mechanics of Growth and Aging. Academic Press, Cambridge
[12] Eremeyev VA (2024) Surface finite viscoelasticity and surface anti-plane waves. International Journal of Engineering Sciences 196:104029, DOI 10.1016/j.ijengsci.2024.104029
[13] Eremeyev VA, Pietraszkiewicz W (2009) Phase transitions in thermoelastic and thermoviscoelastic shells. Archives of Mechanics 61(1):41-67
[14] Ericksen JL (1954) Deformations possible in every isotropic, incompressible, perfectly elastic body. Journal of Applied Mathematics and Physics (ZAMP) 5:466-489, DOI 10.1007/BF01601214
[15] Fedorov AE, Adamov AA (2007) Modelirovaniye povedeniya kozhi cheloveka pri bol'shikh deformatsiyakh (in Russ., Modelling of behaviour of the human skin at finite deformations). Russian Journal of Biomechanics 11(1):74-81
[16] Feng B, Levitas VI (2017) Pressure self-focusing effect and novel methods for increasing the maximum pressure in traditional and rotational diamond anvil cells. Scientific Reports 7:45461, DOI 10.1038/srep45461
[17] Ganghoffer J-F (2010) Mechanical modeling of growth considering domain variation. Part II: Volumetric and surface growth involving Eshelby tensors. Journal of the Mechanics and Physics of Solids 58(9):1434-1459, DOI 10.1016/j.jmps.2010.05.003
[18] Georgievskii DV (2018) An order of smallness of the Poynting effect from the standpoint of the tensor nonlinear functions apparatus. Mechanics of Solids 53(4):381-384, doi10.3103/S0025654418040039
[19] Georgievskii DV (2019) Types of physical nonlinearity in the theory of constitutive relations and the generalized Poynting effect. In: Abali B, Altenbach H, dell'Isola F, Eremeyev V, Öchsner A (eds) New Achievements in Continuum Mechanics and Thermodynamics. Advanced Structured Materials, Vol 108. Springer, Cham DOI 10.1007/978-3-030-13307-8_10
[20] Goldstein RV, Gorodtsov VA, Lisovenko DS (2015) Linear Poynting's effect at torsion and extension of curvilinearly anisotropic tubes. Doklady Physics 60(9):396-399, DOI 10.1134/S1028335815090025
[21] Green AE, Adkins JE (1960) Large elastic deformations and non-linear continuum mechanics. Clarendon Press, Oxford

[22] Horgan CO (2024) Pure torsion for stretch-based constitutive models for incompressible isotropic hyperelastic soft materials. Journal of Elasticity 156:237-254, DOI 10.1007/s10659-024-10048-x

[23] Horgan CO, Murphy JG (2011) Extension and torsion of incompressible nonlinearly elastic solid circular cylinders. Mathematics and Mechanics of Solids 16(5):482-491, DOI 10.1177/1081286510387720

[24] Il'yushin AA, Pobedrya BE (1970) (in Russ., Foundations of mathematical theory of thermoelasticity). Nauka, Moscow

[25] Kamrani M, Levitas VI, Feng B (2017) FEM simulation of large deformation of copper in the quasi-constrain high-pressure-torsion setup. Materials Science and Engineering: A 705:219-230, DOI 10.1016/j.msea.2017.08.078

[26] Khan AS, Lopez-Pamies O, Kazmi R (2006) Thermo-mechanical large deformation response and constitutive modeling of viscoelastic polymers over a wide range of strain rates and temperatures. International Journal of Plasticity 22(4):581-601, DOI 10.1016/j.ijplas.2005.08.001

[27] Levin VA (1988) Kontsentratsiya napryazheniy okolo krugovogo v moment obrazovaniya otverstiya v tele iz vyazkouprugogo materiala (in Russ., Stress concentration near a hole, which is circular at the time of formation in a body made of viscoelastic material). Doklady Akademii Nauk SSSR 299(5):1079-1082

[28] Levin VA (1998) Theory of repeated superposition of large deformations. Elastic and viscoelastic bodies. International Journal of Solids and Structures 35(20):2585-2600, DOI 10.1016/S0020-7683(98)80032-2

[29] Levin VA, Biryukov DR, Zhang L, Zingerman KM (2025) Analytical solution of the problem of repeated imposition of large tensioncompression and torsion deformations in a multilayered viscoelastic cylinder. Mathematics and Mechanics of Solids DOI 10.1177/10812865251339741

[30] Levin VA, Zingerman KM (2008) A class of methods and algorithms for the analysis of successive origination of holes in a pre-stressed viscoelastic body. Finite strains. Communications in Numerical Methods in Engineering 24(12):2240-2251, DOI 10.1002/cnm.1080

[31] Levin VA, Zingerman KM, Vershinin AV, Freiman EI, Yangirova AV (2013) Numerical analysis of the stress concentration near holes originating in previously loaded viscoelastic bodies at finite strains. International Journal of Solids and Structures 50(20-21):3119-3135, DOI 10.1016/j.ijsolstr.2013.05.019

[32] Levin VA, Zingerman KM, Biryukov DR (2023) Detection of the Poynting effect in a composite hypoelastic cylinder under multiple imposition of large deformations. In: Altenbach H, Eremeyev V (eds) Advances in Linear and Nonlinear Continuum and Structural Mechanics. Advanced Structured Materials, Vol 198. Springer, Cham, DOI 10.1007/978-3-031-43210-1_17

[33] Levin VA, Zingerman KM, Vershinin AV, Konovalov DA (2022) Numerical simulation of superimposed finite strains using spectral element method. Continuum Mechechanics and Thermodynamics 34(5):1205-1217, DOI 10.1007/s00161-022-01115-6

[34] Levin VA, Zingerman KM, Vershinin AV, Konovalov DA (2022) Numerical modeling of residual stresses in additive manufacturing products using the theory of repeatedly superimposed finite strains. Mathematics and Mechanics of Solids 27(10):2099-2109, DOI 10.1177/10812865221096776

[35] Levin VA, Zubov LM, Zingerman KM (2014) The torsion of a composite, nonlinear-elastic cylinder with an inclusion having initial large strains. International Journal of Solids and Structures 51(6):1403-1409, DOI 10.1016/j.ijsolstr.2013.12.034

[36] Levin VA, Zubov LM, Zingerman KM (2013) Torsion of a composite nonlinear elastic cylinder with a prestressed inclusion. Doklady Physics 58(12):540-543, DOI 10.1134/S102833581310011X

[37] Lurie AI (1990) Nonlinear Theory of Elasticity. North-Holland, Amsterdam (1990)

[38] Lychev S, Koifman K, Djuzhev N (2021) Incompatible deformations in additively fabricated solids: discrete and continuous approaches. Symmetry 13(12):2331, DOI 10.3390/sym13122331

[39] Luo Z, Zhao Y (2018) A survey of finite element analysis of temperature and thermal stress fields in powder bed fusion Additive Manufacturing. Additive Manufacturing 21:318-332, DOI 10.1016/j.addma.2018.03.022

[40] Polignone DA, Horgan CO (1991) Pure torsion of compressible non-linearly elastic circular cylinders. Quarterly of Applied Mathematics 49(3):591-607, DOI 10.1090/qam/1121689

[41] Poynting JH (1909) On pressure perpendicular to the shear planes in finite pure shears, and on the lengthening of loaded wires when twisted. Proceedings of the Royal Society of London. Series A, Containing Papers of a Mathematical and Physical Character 82(557):546-559, DOI 10.1098/rspa.1909.0059

[42] Renzi D, Marfia S, Tomassetti G, Zurlo G (2024) A discrete model for layered growth. European Journal of Mechanics A/Solids 105:105232, DOI 10.1016/j.euromechsol.2024.105232

[43] Rivlin RS (1949) Large elastic deformations of isotropic materials. VI. Further results in the theory of torsion, shear, and flexure. Philosophical Transactions of the Royal Society of London, Series A: Mathematical and Physical Sciences 242(845):173-195, DOI 10.1098/rsta.1949.0009

[44] Saccomandi G (2001) Universal results in finite elasticity. In: Fu, Y.B., Ogden, R.W. (eds.) Non-linear elasticity: theory and applications. London Mathematical Society lecture notes, Vol 283, pp. 97–134. Cambridge Univ. Press, Cambridge

[45] Sevastyanov GM (2020) Analytical solution for high-pressure torsion in the framework of geometrically nonlinear non-associative plasticity. International Journal of Solids and Structures 206:383-395, DOI 10.1016/j.ijsolstr.2020.09.028

[46] Sedov LI (1970) (in Russ., Mechanics of Continuous Media). Vol 2. Moscow, Nauka

[47] Shield RT (1982) Extension and torsion of elastic bars with initial twist. Transactions ASME Journal of Applied Mechanics 49(4):779-786, DOI 10.1115/1.3162617

[48] Sigaeva T, Czekanski A (2016) Coupling of surface effect and hyperelasticity in combined tension and torsion deformations of a circular cylinder. International Journal of Solids and Structures 85-86:172-179, DOI 10.1016/j.ijsolstr.2016.02.019

[49] Sozio F, Yavari A (2017) Nonlinear mechanics of surface growth for cylindrical and spherical elastic bodies. Journal of the Mechanics and Physics of Solids 98:12-48, DOI 10.1016/j.jmps.2016.08.012

[50] Swift W (1947) Length changes in metals under torsional overstrain. Engineering 163:253

[51] Trincher VK (1984) (in Russ., Formulation of the problem of determining the stress-strain state of a growing body). Izvestiya AN SSSR. Mekhanika Tverdogo Tela 19(2):119-124

[52] Truesdell C (1972) A First Course in Rational Continuum Mechanics. The Johns Hopkins University, Baltimore

[53] Truskinovsky L, Zurlo G (2019) Nonlinear elasticity of incompatible surface growth. Physical Review E 99(5):053001 DOI 10.1103/PhysRevE.99.053001

[54] Valiev RZ, Estrin Y, Horita Z, Langdon TG, Zechetbauer MJ, Zhu YT (2006) Producing bulk ultrafine-grained materials by severe plastic deformation. JOM 58:33-39, DOI 10.1007/s11837-006-0213-7

[55] Wu PD, van der Giessen E (1993) On large-strain inelastic torsion of glassy polymers. International Journal of Mechanical Sciences 35(11):935-951, DOI 10.1016/0020-7403(93)90031-O

[56] Yavari A, Pradhan SP (2022) Accretion mechanics of nonlinear elastic circular cylindrical bars under finite torsion. Journal of Elasticity 152:29-60, DOI 10.1007/s10659-022-09957-6

[57] Yavari A, Safa Y, Soleiman Fallah A (2024) Finite extension of accreting nonlinear elastic solid circular cylinders. Continuum Mechechanics and Thermodynamics 36:1053-1069, DOI 10.1007/s00161-023-01208-w

[58] Zingerman KM, Zubov LM, Belkin AE, Biryukov DR (2023) Torsion of a multilayer elastic cylinder with sequential attachment of layers with multiple superposition of large deformations. Continuum Mechanics and Thermodynamics 35:1235-1244, DOI 10.1007/s00161-022-01110-x

[59] Zubov LM, Bogachkova LU (1995) The theory of torsion of elastic noncircular cylinders under large deformations. Transactions ASME Journal of Applied Mechanics 62(2):373-379, DOI 10.1115/1.2895941

Chapter 21
Geometric Theory for Relativistic Accretion of Elastic Solids

Sergey A. Lychev

Abstract This paper presents a systematic generalization of nonlinear continuum mechanics to the framework of general relativity, built upon the geometric theory of incompatible deformations. In the classical setting, deformations are viewed as embedding of a smooth material manifold into Euclidean physical space. The material manifold is endowed with a specific geometry (a metric and an affine connection) that encodes the distribution of sources of incompatibility, such as continuous fields of dislocations and disclinations. The deviation of this material geometry from the Euclidean one, quantified by the curvature, torsion and non-metricity of the connection, provides a measure of the density of these sources. The emergence of such sources in an elastic body can be attributed to surface growth processes, in which material surfaces are attached to the body in a state that is incompatible with the existing configuration. To extend this differential-geometric framework to general relativity, we propose to model deformations as embedding of the material manifold - now equipped with its own material geometry - into a four-dimensional spacetime manifold whose geometry is determined by the distribution of mass. This requires a refined understanding of the notions of deformation and of the deformed shapes of the body, which is systematically developed in the present work.

Key words: Nonlinear continuum mechanics, Incompatible deformations, Material connections, Curvature, Torsion, Spacetime, General relativity, Relativistic elasticity, Neutron star crust, Glitches

---

Sergey A. Lychev
Ishinsky Institute for Problems in Mechanics, Russian Academy of Sciences, Prospekt Vernadskogo, 101-1, Moscow 119526, Russian Federation, e-mail: lychevsa@mail.ru

H. Altenbach et al. (eds.), *Stability Problems in Mechanics of Continua*,
Advanced Structured Materials 260, https://doi.org/10.1007/978-3-032-31032-3_21

21.1 Introduction

Two fundamental branches of modern physics - nonlinear continuum mechanics and the theory of relativity - are almost contemporaries: their basic tenets were formulated at the turn of the 19th and 20th centuries. Although the development of mechanics took place on the stage of absolute space–time, in the spirit of the classical ideas of Euclid, Galileo and Newton, while relativity required a radical reconstruction of this stage in the revolutionary style of a four-dimensional smooth space–time manifold, methodologically these theories shared the common principles of stationary action and Noetherian (variational) symmetries, which is reflected in the unified style of their logical structure [16, 46]. In the middle of the 20th century, nonlinear continuum mechanics was substantially extended by the theory of distributed defects, in particular dislocations and disclinations, which revealed the unity of mechanics and relativity even more profoundly [4, 29, 30]. Indeed, the modeling of distributed defects necessitated the use of a space with a general affine connection [32] to describe the reference stress-free state of an elastic body, where the measures of deviation of this connection from the Euclidean one - curvature, torsion and non-metricity - quantitatively characterize the internal sources of stress: distributed fields of disclinations, dislocations and bulk defects [48, 57, 59]. Only one, albeit essential, difference remained: in the nonlinear theory of distributed defects, the non-Euclidean geometry was assigned to a three-dimensional, purely spatial manifold, whereas in relativity the non-Euclidean nature appears in the four-dimensional space-time. Soon, however, a more complete revision of continuum mechanics was required, which transformed it into a fully geometrized four-dimensional theory, thereby inviting it onto the common four-dimensional non-Euclidean stage together with general relativity as an equal player. This revision led to the construction of *Relasticity*, a continuum theory that combines the methods of nonlinear continuum mechanics and general relativity on the basis of common differential-geometric and variational principles [1, 3, 11, 24, 51, 54]. Specific objects of study of the new theory include, among others, the phenomena of gravitational waves and the radiative dynamics of neutron stars.

Let us dwell in more detail on the second problem, for it is precisely this problem that motivated the present study. It concerns supermassive compact objects, the existence of which was predicted long before their actual discovery [31]. Only after the development of radio and X-ray astronomy techniques, in 1967, were strictly periodic pulses recorded, emitted by the so-called pulsars, which turned out to be rotating neutron stars accreting matter from the surrounding gas and dust clouds. The deceleration of matter in the powerful electromagnetic field at the star's poles generates intense directed radiation in the X-ray range, the beam of which, due to the star's rotation, periodically points toward the Earth, thereby creating the effect of strictly periodic signals. Physicists and astronomers showed great interest in pulsars and other supermassive objects, which is not surprising: these objects represent gigantic laboratories created by Nature, revealing the secrets of space, field and matter. However, it soon became clear that the strict periodicity of the radiation of some pulsars is spontaneously disrupted, and at the moment of such a disruption the radiation intensity increases substantially. To explain this phenomenon, an astrophysical hypothesis of starquakes was put forward: the process of matter accretion causes the accumulation of stresses in the star's envelope (crust), leading to its local fracture, similar to the fracturing of lithospheric plates on a classical planet with a solid surface [8]. In this event, a part of the "exotic matter" (the so-called hadronic plasma, in which, due to enormous pressures, atomic structures are unstable and therefore do not exist in stationary forms [52]) is no longer shielded by the crust, which results in intense additional (aperiodic) radiation. Thus, distant and exotic astrophysics has led to a very "earthly" problem of the deformation and stability of a spherical shell, albeit complicated by the fact that the physical space in which the deformation occurs is no longer Euclidean, and time is no longer Galilean, because both space and time are curved by the giant mass of the star.

Within the framework of the present work, we shall not analyze the methodology of astronomical observations of pulsars, nor the physical models of their internal structure (a considerable number of publications are devoted to these questions, e.g., [12, 13, 49]), but for completeness of exposition, we present some observational data in the form of Table 21.1.

As can be seen from the table, glitches vary significantly in amplitude and recovery timescale. The Vela pulsar exhibits giant glitches with relative frequency jumps $\Delta\nu/\nu \sim 10^{-6}$, while the Crab pulsar shows smaller glitches with amplitudes typically two orders of magnitude lower [58]. The recovery times also differ substantially, indicating different internal mechanisms and possibly different locations of the glitch trigger within the neutron star structure.

The modeling of the mechanical behavior of an elastic star's crust was a pioneering problem of Relasticity and still remains an open issue. For the first time in a nonlinear formulation, this problem was considered in a series of works by Carter and Quintana [5, 6], for a special constitutive law characterizing the physical nonlinearity of the material under high pressures and hydrostatic compression. In their approach, to formalize the unstressed reference shape as a three-dimensional manifold of "ideal" particles, a projection of the four-dimensional spacetime onto a three-dimensional spacelike manifold was used. In the approximation of small velocities, this projection can be considered independent of the motion of a material point, since the splitting of its proper 4-frame into spatial and temporal parts can, up to small quantities, be identified with the splitting of the main coordinate frame (in other words, Lorentz–Fitzgerald contractions can be neglected). This, of course,

Table 21.1: Observed parameters of selected pulsars and their glitches.

Pulsar	Mass ($M_\odot$)	Radius (km)	Period (s)	Glitch amplitude ($\Delta\nu/\nu$)	Recovery time	Reference
Crab (PSR B0531+21)	~1.4	~10	0.033	3×10^{-8}	3–10 days	[58]
Vela (PSR B0833–45)	~1.5	~10	0.089	$\sim 10^{-6}$	6–60 days	[2, 15]
PSR J0537–6910	~1.4	~10	0.016	10^{-7}–10^{-6}	days–weeks	[47]

significantly simplifies the theory and excludes from its scope the classical problems of near-light-speed motion in relativity, such as the twin paradox, the problem of rigid motion (Born), the Ehrenfest paradox, *etc* [18]. This approximation can be characterized as quasistatic in the sense that relativistic effects manifest themselves not because of near-light speeds, but due to the curvature of spacetime caused by the large mass. Formally, this leads to a modification of classical nonlinear elasticity theory, in which the Euclidean metric of physical space is replaced by the spatial part of the general relativistic metric, for example, from the Schwarzschild or Kerr solutions [20]. This approach was developed in a series of works by Karlovini and Samuelsson [24–27], as well as in the monograph by Ferrarese and Bini [14]. A comprehensive numerical formulation of the relativistic elasticity equations as a conservation law system, suitable for implementation in numerical relativity codes, was later provided by Gundlach, Hawke, and Erickson [19].

A unified covariant framework that seamlessly interpolates between Newtonian and relativistic descriptions has been developed in a series of works by Carter and collaborators [7]. In particular, the 4-dimensional formulation of Newtonian spacetime via the Milne structure (degenerate metric and time covector fields) and the concept of material projection onto a three-dimensional base manifold provide an ideal language for transferring the growth model to a curved background.

These results are important milestones in the development of Relasticity; however, they do not answer the main question of pulsar dynamics: why does a glitch occur? In the present paper, we attempt to construct a theoretical explanation of this phenomenon. The key idea is as follows: being in a gas-dust cloud, the neutron star's envelope accretes extra material, which at the moment of accretion possesses significant energy (and velocity), thereby creating internal sources of stress in the crust, similar to growth stresses in classical bodies. When the internal stresses reach critical values, a loss of material stability occurs according to the Hadamard condition [17, 28], which entails the formation of a fracture - a starquake occurs [8]. An important factor here is the rapid rotation of the neutron star, causing significant centrifugal effects. This makes it necessary to take into account two spatial dimensions. Below, we attempt to construct a model that adequately accounts for these factors.

21.2 Newtonian Approach

The language of differential geometry allows us to formalize from a unified standpoint both compatible and incompatible deformations, and also naturally to take into account the curvature of the spacetime continuum when describing relativistic effects in a deformable solid body. Although to date this approach has been reflected in numerous publications and monographs [9, 10, 34, 44, 45, 50, 53], it seems appropriate first to demonstrate its implementation with a simple example of compatible deformations arising in a hollow hyperelastic ball when it is deformed in the field of central gravitational forces (still treated in the Newtonian approximation) and centrifugal forces caused by its rotation. The purpose of this demonstration is to show the duality in the interpretation of deformation: on the one hand, we can interpret it as a distortion of the shape of the body, i.e., of its image in physical space when replacing the reference configuration by the actual one; on the other hand, as a change of the metric and the associated geometric structure on the material manifold, which itself does not undergo any changes.

Given that the shapes of the bodies considered below are topologically equivalent to an open (possibly hollow) ball under any smooth configuration, it is convenient to take the topological space $\mathfrak{M} = \mathbb{R}^+ \times \mathbb{S}^2$ as the material manifold, equipping it with an atlas that includes the classical spherical coordinates (r, φ, θ) as a chart, where $r \in \mathbb{R}^+$ and $(\varphi, \theta) \in \mathbb{S}^2$. At this stage, the coordinate map appears very abstract:

$$\mathfrak{M} \ni \mathfrak{p} \mapsto (r, \varphi, \theta) \in \mathbb{R}^3,$$

and acquires concrete meaning only upon an explicit description of the embedding into the physical space, which will serve as the space of representations. This idea was allegorically expressed by Epstein [10, p. 274]:

> "The material body, whose points are called material particles, does not live permanently in the 3-dimensional space of our physical experience, but dwells in some Platonic world of similar entities, devoid of any particular ruler or clock. Nevertheless, it does manifest itself through visits to this world, a world equipped with rulers, protractors and clocks."

The next step is the formalization of the physical space as a smooth manifold, which for now carries a Euclidean geometric structure in the spirit of the Newtonian approximation of continuum mechanics.

Although we have been familiar with the concepts of Newtonian absolute space and Galilean absolute time since our school years, their complete mathematical formalization is not entirely trivial. When investigating classical problems of a deformable continuum, we can painlessly omit many mathematical subtleties of these definitions; however, when generalizing to the geometry of general relativity, such omissions lead to serious conceptual difficulties. In this regard, we must explain the Galilean–Newtonian spacetime in a full differential-geometric exposition, but we shall try to do this as concisely as possible, referring the reader to the monographs [10, 44, 53] for details.

The Galilean–Newtonian spacetime can be represented as an ordered pair of two affine Euclidean spaces $\mathcal{S}_{\text{G-N}} = (\mathcal{T}_{\text{G}} \times \mathcal{E})$, the first of which, $\mathcal{T}_{\text{G}} = (\mathcal{T}^p_{\text{G}}, \mathcal{T}^t_{\text{G}})$, defines the chronometry, and the second, $\mathcal{E} = (\mathcal{E}^p, \mathcal{E}^t)$, the spatial geometry. Their affine structure, determined by the linearity of the corresponding translation spaces $\mathcal{T}^t_{\text{G}}$, $\mathcal{E}^t$, formalizes the absoluteness of time and space: the coordinate representations of points from $\mathcal{T}_{\text{G}}$ and $\mathcal{E}$ are completely independent and are uniquely determined by the bases and origins chosen in corresponding translation and point spaces. It should be noted that the essence of these basis elements lies outside the formal structure of affine Euclidean spaces: they can be regarded as "external" elements having exclusively physical meaning and carrying physical dimension. In this sense, the basis elements constitute the empirical foundation of the theory, and it is precisely the revision of the structures associated with them that leads to more general models of spacetime, in particular, general relativity.

Thus, in order to prepare the stage for a physical action in the Newtonian spirit, one must specify, from some additional considerations, the origins and the primary basis elements (carriers of "dimensions of physical quantities"), in other words, the primary frames. The motivation for this can be diverse, ranging from purely empirical to theological (we refer to the remarkable physico-philosophical works of Max Jammer [21–23]). We shall assume that this choice has been made and that the affine Euclidean spaces $\mathcal{T}_G$, $\mathcal{E}$ are equipped with primary frames. For the spatial frame we shall use the standard notation $(\mathcal{O}, (\mathbf{i}, \mathbf{j}, \mathbf{k}))$.

This, however, is not sufficient. To reproduce the entire technique of synthetic Euclidean geometry, the concept of a spatial angle is required, and it is again introduced from considerations external to the formal theory: in the form of an expansion in the elements of a dyadic basis constructed from the elements of the frame dual to the primary one, a metric is assigned (an algebraic structure, not a field, as will be the case in general relativity). The question of what is primary, the frame or the metric, resembles the chicken-and-egg problem. Following the classical paradigm, we shall assume that these structures are dual and are given jointly, with the matrix of the metric expansion in the primary basis being the identity.

The assignment of the metric allows us to represent $\mathcal{E}$ as a trivial Riemannian manifold, whose atlas consists of a Cartesian chart and all possible locally equivalent chartings by means of curvilinear coordinates. Thus, $\mathcal{E}$, having acquired all the structures of a smooth manifold, becomes an equal player together with the material manifold, and even more so, since it carries, on top of the manifold structure, a geometry defined by the metric, which induces measures of lengths, areas and volumes, as well as a rule for parallel transport (a connection). The chronometric manifold can also be clothed in the garb of a smooth one-dimensional manifold, but as long as it is not metrically connected to the geometric space, this yields nothing useful, and is usually neglected in Newtonian physics (a more complete exposition of the "geometrization" of Newtonian spacetime can be found, e.g., in [16]).

Now, having two smooth manifolds at our disposal, namely, the material manifold $\mathfrak{M}$ and the manifold representing the physical space $\mathcal{E}^p$ (the Newtonian physical point space), we can define configurations as embeddings

$$\varkappa : \mathfrak{M} \hookrightarrow \mathcal{E}^p.$$

Among all possible configurations, a special one is distinguished, which we shall call the reference configuration and denote by the symbol $\varkappa_R$. The specific status of this configuration is due to the fact that its image — the reference shape — serves as the carrier of fields that determine the physical state of the body (energy, stresses, etc.), and the elementary volumes within the reference shape are used to quantify the influence of geometric shape distortions on this state. In classical nonlinear continuum mechanics, it is usually assumed that the reference shape is unstressed. However, in the general case, due to incompatible deformations and internal sources of stress, this assumption does not hold, and the choice of the reference configuration is dictated by the simplicity of its analytical description. As applied to the problem considered in the present paper, it is expedient to choose the reference configuration so that the reference shape is a hollow ball. Then the corresponding embedding takes the form of the canonical representation of $\mathbb{R}^2 \times \mathbb{R}^+$ into $\mathcal{E}^p$:

$$\varkappa_R : (r, \varphi, \theta) \mapsto \mathcal{O} + \mathbf{i}\, r c_\varphi s_\theta + \mathbf{j}\, r s_\varphi s_\theta + \mathbf{k}\, r c_\theta \in \mathcal{E}^p,$$

where $\mathbf{O}$ is the origin, $(\mathbf{i}, \mathbf{j}, \mathbf{k})$ is an orthonormal basis in $\mathcal{E}^t$, and (r, φ, θ) are the spherical coordinates inherited from the chart on $\mathfrak{M}$. From here on we use abbreviated notations $s_\theta = \sin\theta$, $c_\theta = \cos\theta$, $s_\varphi = \sin\varphi$, $c_\varphi = \cos\varphi$.

Embeddings whose images correspond to the deformed body will be called actual configurations (they can, of course, be parametrized by points of the chronometric manifold, but at this stage we need only one deformed shape, and we shall omit the notation associated with parametrization). Bearing in mind the axial symmetry, we write the expression for the actual configuration in explicit form up to two smooth functions $u(r, \theta)$ and $v(r, \theta)$ that specify the displacements (relative to the reference shape) along the radius and along the tangent to the meridian:

$$\begin{aligned}\varkappa(r, \varphi, \theta) = \mathbf{O} + \mathbf{i}\Big[(r+u)s_\theta c_\varphi + rvc_\theta c_\varphi\Big] + \mathbf{j}\Big[(r+u)s_\theta s_\varphi + rvc_\theta s_\varphi\Big]& \\ + \mathbf{k}\Big[(r+u)c_\theta - rvs_\theta\Big].&\end{aligned} \tag{21.1}$$

The deformation mapping the reference shape onto the actual shape is defined as the composition of configurations

$$\gamma = \varkappa \circ \varkappa_R^{-1} : \mathcal{E}^p \hookrightarrow \mathcal{E}^p,$$

which, under the assumption of axial symmetry, can be expressed explicitly (up to origin) in terms of the functions u and v as

$$\mathcal{E}^t \ni \mathbf{X} \to \mathbf{x} = \left(\mathbf{I} + (\mathbf{i}\otimes\mathbf{i} + \mathbf{j}\otimes\mathbf{j})\left(\frac{u}{r} + \alpha v\right) + \mathbf{k}\otimes\mathbf{k}\left(\frac{u}{r} - \frac{v}{\alpha}\right)\right)\cdot\mathbf{X} \in \mathcal{E}^t,$$

where $\mathbf{X}$ is the position vector of a point in the reference shape, $\mathbf{x}$ is its image in the actual shape, $r = |\mathbf{X}|$, and $\alpha = \mathbf{X}\cdot\mathbf{k}\, \left((\mathbf{X}\cdot\mathbf{i})^2 + (\mathbf{X}\cdot\mathbf{j})^2\right)^{-1/2} = \cot\theta$.

The coordinate representations of the configurations $\varkappa_R$ and $\varkappa$, defined as embeddings of the material manifold into physical space, incidentally perform the auxiliary problem of reparametrizing the physical space and generate curvilinear coordinates in it. In classical nonlinear mechanics they are called referential and actual coordinates, respectively (they should not be confused with Lagrangian and Eulerian coordinates; see, *e.g.*, Truesdell [56, p. 88]). This role of configurations presupposes their use only in composition with the charting map, and the fact that the arguments of the configurations are points of the material manifold remains completely hidden. In this regard, in the case of compatible deformations, when the need for an explicit description of the material manifold disappears, such a relationship between referential and actual coordinates is not mentioned at all (the ideology of two curvilinear coordinates is intensively used, for example, in the monograph [33], but the concept of a material manifold is not introduced in explicit form). For us, on the contrary, we must not "render" this relationship invisible even for compatible deformations, because it points the way to a non-Euclidean generalization of the theory.

Thus, the configurations introduced above generate curvilinear coordinates in $\mathcal{E}$. In the region occupied by the reference shape, the coordinates (r, θ, φ) give rise to the local basis

$$\begin{aligned}&\mathbf{E}_r = \varkappa_{R,r} = s_\theta c_\varphi\,\mathbf{i} + s_\theta s_\varphi\,\mathbf{j} + c_\theta\,\mathbf{k}, \quad \mathbf{E}_\varphi = \varkappa_{R,\varphi} = -r s_\theta s_\varphi\,\mathbf{i} + r s_\theta c_\varphi\,\mathbf{j},\\ &\mathbf{E}_\theta = \varkappa_{R,\theta} = r c_\theta c_\varphi\,\mathbf{i} + r c_\theta s_\varphi\,\mathbf{j} - r s_\theta\,\mathbf{k}.\end{aligned} \tag{21.2}$$

Similarly, for the actual configuration $\mathbf{x} = \varkappa(r, \theta, \varphi)$ we obtain

$$\begin{aligned}\mathbf{e}_r &= \varkappa_{,r} = (1 + u_{,r})\,\mathbf{E}_r + (v + r v_{,r}) r^{-1}\mathbf{E}_\theta,\\ \mathbf{e}_\theta &= \varkappa_{,\theta} = (u_{,\theta} - r v)\,\mathbf{E}_r + \left(r + u + r v_{,\theta}\right) r^{-1}\mathbf{E}_\theta,\\ \mathbf{e}_\varphi &= \varkappa_{\varphi} = \left((r + u) s_\theta + r v c_\theta\right)(r s_\theta)^{-1}\,\mathbf{E}_\varphi,\end{aligned} \tag{21.3}$$

where the subscripts after comma denote partial derivatives: $u_{,r} = \partial u/\partial r$, $u_{,\theta} = \partial u/\partial\theta$, and analogously for v.

The dual (contravariant) bases are defined by the relations $\mathbf{E}^\alpha \cdot \mathbf{E}_\beta = \delta^\alpha_\beta$ and $\mathbf{e}^\alpha \cdot \mathbf{e}_\beta = \delta^\alpha_\beta$. Their explicit form can be obtained by inverting the corresponding Gram matrices:

$$\begin{aligned}&\mathbf{E}^r = \mathbf{E}_r, \quad \mathbf{E}^\theta = r^{-2}\mathbf{E}_\theta, \quad \mathbf{E}^\varphi = (r\, s_\theta)^{-2}\,\mathbf{E}_\varphi.\\ &\mathbf{e}^r = \Delta^{-2}\left((Q^2 + S^2)\,\mathbf{e}_r - (AQ + PS)\,\mathbf{e}_\theta\right),\\ &\mathbf{e}^\theta = \Delta^{-2}\left(-(AQ + PS)\,\mathbf{e}_r + (A^2 + P^2)\,\mathbf{e}_\theta\right), \quad \mathbf{e}^\varphi = M^{-2}\,(r\, s_\theta)^2\,\mathbf{e}_\varphi,\end{aligned}$$

where

$$\begin{aligned}&A = 1 + u_{,r}, \quad P = v + r v_{,r}, \quad Q = u_{,\theta} - r v, \quad S = r + u + r v_{,\theta},\\ &R = r + u, \quad M = R\, s_\theta + r v\, c_\theta \quad \Delta = AS - PQ.\end{aligned}$$

Local bases allow us to write exact expressions for local deformations and configurations. It is only necessary to keep in mind that the local configurations $\mathfrak{K}_R$, $\mathfrak{K}$ map vectors from the tangent bundle of the material manifold into the translation space $\mathcal{E}^t$ of absolute space, while the local deformation $\mathbf{F}$ maps $\mathcal{E}^t$ into itself. So, for a point $\mathfrak{p} \in \mathfrak{M}$, local reference configuration is

$$\mathfrak{K}_R(\mathfrak{p}) = d\varkappa_R|_\mathfrak{p} = \mathbf{E}_r \otimes dr + \mathbf{E}_\theta \otimes d\theta + \mathbf{E}_\varphi \otimes d\varphi \;:\; T_\mathfrak{p}\mathfrak{M} \to \mathcal{E}^t, \tag{21.4}$$

where the basis vectors $\mathbf{E}_r, \mathbf{E}_\theta, \mathbf{E}_\varphi$ are given by (21.2) and the covectors $dr, d\theta, d\varphi$ constitute the dual basis to the coordinate frame $\partial_r, \partial_\theta, \partial_\varphi$ on $\mathfrak{M}$. Similarly, the local actual configuration is

$$\mathfrak{K}(\mathfrak{p}) = d\varkappa|_\mathfrak{p} = \mathbf{e}_r \otimes dr + \mathbf{e}_\theta \otimes d\theta + \mathbf{e}_\varphi \otimes d\varphi \;:\; T_\mathfrak{p}\mathfrak{M} \to \mathcal{E}^t, \tag{21.5}$$

with $\mathbf{e}_r, \mathbf{e}_\theta, \mathbf{e}_\varphi$ given by (21.3). Finally, the local deformation gradient $\mathbf{F}$ maps vectors from the reference shape to the actual shape:

$$\mathbf{F} = \mathfrak{K} \circ \mathfrak{K}_R^{-1} \, : \, \mathcal{E}^t \to \mathcal{E}^t.$$

In dyadic notation, using the spatial bases only, this can be written as

$$\mathbf{F} = \mathbf{e}_r \otimes \mathbf{E}^r + \mathbf{e}_\theta \otimes \mathbf{E}^\theta + \mathbf{e}_\varphi \otimes \mathbf{E}^\varphi.$$

Now everything is ready for transferring the metric to the material manifold. The metric thus transferred depends both on the "donor" — the physical space — and on the configuration that implements this transfer. It should be noted that the literature often uses the terms "referential metric" and "actual metric", which are associated only with different coordinates in the physical space, thereby creating the erroneous impression that different metrics exist in it. Actually, the metric of the physical space does not depend in any way on the coordinates used; only its representation in different bases generated by the coordinates changes. Therefore, it is incorrect to speak of a distinction between "referential" and "actual" metrics in the physical space. The situation is completely different for the metric induced on the material manifold: it indeed depends on the particular configuration, and the material manifold itself acts as a carrier of many different metrics. A more detailed explanation of this transfer can be given by the following reasoning.

Consider a smooth curve lying on the material manifold,

$$\gamma : (a, b) \ni t \mapsto \xi^{-1}\big(r(t), \theta(t), \varphi(t)\big) \in \mathfrak{M}.$$

Its image under the actual configuration is an Euclidean curve

$$\varkappa \circ \gamma : (a, b) \ni t \mapsto \varkappa_R\big(r(t), \theta(t), \varphi(t)\big) \in \mathcal{E}^p.$$

The length of this image within Euclidean metric is computed as

$$\begin{aligned}
L &= \int_a^b \big\|(\varkappa \circ \gamma)^{\cdot}\big\| \, dt = \int_a^b \Big\| \frac{\partial \mathbf{x}}{\partial r} \dot{r} + \frac{\partial \mathbf{x}}{\partial \theta} \dot{\theta} + \frac{\partial \mathbf{x}}{\partial \varphi} \dot{\varphi} \Big\| \, dt \\
&= \int_a^b \sqrt{\big(\dot{r}\mathbf{e}_r + \dot{\theta}\mathbf{e}_\theta + \dot{\varphi}\mathbf{e}_\varphi\big) \cdot \big(\dot{r}\mathbf{e}_r + \dot{\theta}\mathbf{e}_\theta + \dot{\varphi}\mathbf{e}_\varphi\big)} \, dt \\
&= \int_a^b \sqrt{g_{rr}\dot{r}^2 + g_{\theta\theta}\dot{\theta}^2 + g_{\varphi\varphi}\dot{\varphi}^2 + 2g_{r\theta}\dot{r}\dot{\theta} + 2g_{r\varphi}\dot{r}\dot{\varphi} + 2g_{\theta\varphi}\dot{\theta}\dot{\varphi}} \, dt,
\end{aligned}$$

where $g_{\alpha\beta} = \mathbf{e}_\alpha \cdot \mathbf{e}_\beta$, $\alpha, \beta \in \{r, \theta, \varphi\}$. With auxiliary functions introduced above we get

$$g_{rr} = A^2 + P^2, \qquad g_{\theta\theta} = Q^2 + S^2, \qquad g_{\varphi\varphi} = M^2,$$

$$g_{r\theta} = AQ + PS, \qquad g_{r\varphi} = 0, \qquad g_{\theta\varphi} = 0. \tag{21.6}$$

Setting $u \equiv 0$, $v \equiv 0$ (i.e., the reference configuration itself), we obtain the coefficients for the curve length in the reference shape:

$$\begin{aligned} &G_{rr} = 1, && G_{\theta\theta} = r^2, && G_{\varphi\varphi} = r^2 \sin^2\theta, \\ &G_{r\theta} = 0, && G_{r\varphi} = 0, && G_{\theta\varphi} = 0. \end{aligned} \tag{21.7}$$

These coefficients can be interpreted in three ways:

- as the elements of the matrix of the quadratic form arising when Cartesian coordinates are replaced by curvilinear coordinates;
- as the components of the spatial metric tensor decomposed in the basis dual to the local basis generated by the curvilinear coordinates:

$$\mathbf{g} = g_{rr}\, \mathbf{e}^r \otimes \mathbf{e}^r + g_{\theta\theta}\, \mathbf{e}^\theta \otimes \mathbf{e}^\theta + g_{\varphi\varphi}\, \mathbf{e}^\varphi \otimes \mathbf{e}^\varphi + g_{r\theta}\, (\mathbf{e}^r \otimes \mathbf{e}^\theta + \mathbf{e}^\theta \otimes \mathbf{e}^r);$$

- as the pull-back of the Euclidean metric to the material manifold by the embedding realised by the configuration (reference or actual):

$$\mathfrak{g}_{\mathfrak{p}}(\mathfrak{u}, \mathfrak{v}) = \mathbf{g}\big(d\varkappa|_{\mathfrak{p}}(\mathfrak{u}), d\varkappa|_{\mathfrak{p}}(\mathfrak{v})\big),$$

  or, in explicit form,

$$\mathfrak{g} = g_{rr}\, dr \otimes dr + g_{\theta\theta}\, d\theta \otimes d\theta + g_{\varphi\varphi}\, d\varphi \otimes d\varphi + g_{r\theta}\, (d^r \otimes d\theta + d\theta \otimes dr).$$

Thus, we have explicitly constructed two metric tensors on $\mathfrak{M}$: one induced by the reference configuration (with components (21.4)) and one induced by the actual one (with components (21.5)). The difference between these two metrics characterizes the deformation, and in the case of compatible deformations they are both flat (Euclidean) although their coordinate expressions differ. In the subsequent sections, when we introduce incompatible deformations, the metric induced by the actual configuration will no longer be flat, reflecting the presence of internal stresses.

Having at our disposal a metric (one of many possible, transferred by various configurations), we can proceed further towards the geometrization of the material manifold – namely, we can introduce a connection on it. Moreover, we can perform this step in different ways, which, generally speaking, lead to different results. We shall use two such methods: we shall define the Levi-Civita connection directly from the metric, and alternatively, we shall construct a connection from the field of local configurations that generates this metric, using the Krόner–Cartan formulae. Looking ahead, we note that in the case of a compatible deformation the result will be the same, while in the incompatible case they will differ, and this difference will have a deep physical meaning.

In Riemannian geometry, the Christoffel symbols are expressed through the derivatives of the basis vectors and the dual basis:

$$\Gamma^\alpha_{\beta\gamma} = \mathbf{e}^\alpha \cdot \frac{\partial \mathbf{e}_\beta}{\partial x^\gamma} \qquad \alpha, \beta, \gamma \in \{r, \theta, \varphi\}.$$

Direct calculations give

$$
\begin{aligned}
&\Gamma^r_{rr} = \Delta^{-1}\left(SA_{,r} - QP_{,r}\right), \quad \Gamma^r_{r\theta} = \Gamma^r_{\theta r} = \Delta^{-1}\left[S\,Q_{,r} + vrS_{,r} - u_{,\theta}(A + P_{,\theta})\right],\\
&\Gamma^r_{\theta\theta} = \Delta^{-1}\left(Q_{,\theta}S - QS_{,\theta} - S^2 - Q^2\right), \quad \Gamma^r_{\varphi\varphi} = \Delta^{-1}\left[M\left(Qc_\theta - Ss_\theta\right)\right],\\
&\Gamma^\theta_{rr} = \Delta^{-1}\left(-u_{,rr}P + A\,P_{,r}\right), \quad \Gamma^\theta_{r\theta} = \Gamma^\theta_{\theta r} = \Delta^{-1}\left(AP_{,\theta} + A^2 + P^2 - Pu_{,r\theta}\right),\\
&\Gamma^\theta_{\theta\theta} = \Delta^{-1}\left[A(S_{,\theta} + Q) + P(S - Q_{,\theta})\right] \quad \Gamma^\theta_{\varphi\varphi} = \Delta^{-1}M\left(Ps_\theta - Ac_\theta\right),\\
&\Gamma^\varphi_{r\varphi} = \Gamma^\varphi_{\varphi r} = M^{-1}\left(As_\theta + Pc_\theta\right), \quad \Gamma^\varphi_{\theta\varphi} = \Gamma^\varphi_{\varphi\theta} = M^{-1}\left(Qs_\theta + Sc_\theta\right),
\end{aligned}
$$

In the particular case $u \equiv 0$, $v \equiv 0$ (*i.e.*, for the reference configuration itself) we recover the standard Christoffel symbols for Euclidean space in spherical coordinates:

$$
\begin{aligned}
&\Gamma^r_{\theta\theta} = -r, \quad \Gamma^r_{\varphi\varphi} = -r\sin^2\theta, \quad \Gamma^\theta_{r\theta} = \Gamma^\theta_{\theta r} = \frac{1}{r}, \quad \Gamma^\theta_{\varphi\varphi} = -s_\theta c_\theta,\\
&\Gamma^\varphi_{r\varphi} = \Gamma^\varphi_{\varphi r} = \frac{1}{r}, \quad \Gamma^\varphi_{\theta\varphi} = \Gamma^\varphi_{\varphi\theta} = \cot\theta.
\end{aligned}
$$

The geodesics on the material manifold equipped with the metric induced by the actual configuration are determined by the equation

$$
\ddot{q}^\alpha + \Gamma^\alpha_{\beta\gamma}\dot{q}^\beta\dot{q}^\gamma = 0 \quad (q^1(s) = r(s),\, q^2(s) = \theta(s),\, q^3(s) = \varphi(s)), \tag{21.8}
$$

where the dot denotes differentiation with respect to an affine parameter s. Their images under the embedding $\varkappa$ are straight lines in the physical space if and only if the deformation is compatible (*i.e.*, the curvature of the material connection vanishes). Note that this property will be lost when we later consider incompatible deformations.

For the general expressions of the Christoffel symbols obtained in the previous subsection, finding an explicit solution of the geodesic equations (21.8) as a Cauchy problem is problematic. However, for the connection generated by the canonical embedding via the reference configuration, this problem is well studied and admits an effective solution. The geodesic equations become

$$
\begin{aligned}
\ddot{r} - r\dot{\theta}^2 - r\,s_\theta^2\,\dot{\varphi}^2 &= 0,\\
\ddot{\theta} + \frac{2}{r}\dot{r}\dot{\theta} - s_\theta c_\theta\,\dot{\varphi}^2 &= 0,\\
\ddot{\varphi} + \frac{2}{r}\dot{r}\dot{\varphi} + 2\cot\theta\,\dot{\theta}\dot{\varphi} &= 0,
\end{aligned}
$$

For the following initial values

$$
r|_{s=0} = r_0,\ \theta|_{s=0} = \theta_0,\ \varphi|_{s=0} = \varphi_0, \quad \dot{r}|_{s=0} = \dot{r}_0,\ \dot{\theta}|_{s=0} = \dot{\theta}_0,\ \dot{\varphi}|_{s=0} = \dot{\varphi}_0.
$$

these equations admit analytical solutions:

$$
\begin{aligned}
r(s) &= r_0\sqrt{1 + 2s\,\dot r_0/r_0 + s^2\left(\dot r_0^2/r_0^2 + \dot\theta_0^2 + \dot\varphi_0^2\sin^2\theta_0\right)},\\
\theta(s) &= \arccos\left[\left(\cos\theta_0 + s(\dot r_0/r_0\cos\theta_0 - \dot\theta_0\sin\theta_0)\right) r_0/r(s)\right],\\
\varphi(s) &= \arctan\left[\frac{\tan\theta_0\tan\varphi_0 + s(\dot\varphi_0 + \dot\theta_0 + \dot r_0/r_0\tan\varphi_0)\tan\theta_0}{\tan\theta_0 + s(\dot\theta_0 + (\dot r_0/r_0 - \dot\varphi_0\tan\varphi_0)\tan\theta_0)}\right].
\end{aligned}
$$

These geodesics correspond to straight lines in the Euclidean space when mapped back via the reference configuration $\varkappa_R$. In the actual (deformed) configuration, the geodesics of the pulled-back metric are no longer images of straight lines – a manifestation of the incompatibility that will be studied in the next sections.

The second way of constructing a connection is based on the idea of parallel transport. Specifically, the Cartan–Kröner connection on the material manifold $\mathfrak{M}$ is defined by the following condition for parallel transport of tangent vectors:

A vector $\mathfrak{y} \in T_{\mathfrak{p}}\mathfrak{M}$ is said to be parallelly transported along a curve $\gamma(t)$ (with $\gamma(0) = \mathfrak{p}$) if its image $w(t) = \mathfrak{K}(\gamma(t))(Y(t)) \in \mathcal{E}^t$ is constant in the sense of ordinary Euclidean parallel transport (i.e., its Cartesian components do not depend on t).

In other words, there exists a constant vector $\mathbf{w}_0 \in \mathcal{E}^t$ such that $\mathfrak{y}(t) = \mathfrak{K}(\gamma(t))^{-1}(\mathbf{w}_0)$ for all t. Differentiating the condition $\mathfrak{K}(\gamma(t))((t)) = \text{const}$ with respect to t yields

$$
\frac{d}{dt}(\mathfrak{K}(\mathfrak{y})) = \mathbf{0} \implies (\nabla_{\dot\gamma}\mathfrak{K})(\mathfrak{y}) + \mathfrak{K}(\nabla_{\dot\gamma}\mathfrak{y}) = \mathbf{0},
$$

where $\nabla_{\dot\gamma}\mathfrak{y}$ is the sought covariant derivative along the curve, obtained from the ordinary chain rule of differentiation.

In coordinates $q^a = (r, \varphi, \theta)$ we write $\mathfrak{y} = \mathfrak{y}^a\partial_a$, and $\mathfrak{K}$ in the Cartesian basis $\{\mathbf{i}, \mathbf{j}, \mathbf{k}\}$ can be represented as

$$
\mathfrak{K} = \mathfrak{K}^{i\cdot}_{\cdot\alpha}\mathbf{i}_i \otimes dq^\alpha, \quad \mathbf{i}_i \in (\mathbf{i}, \mathbf{j}, \mathbf{k}), \quad dq^\alpha \in (dr, d\theta, d\varphi).
$$

The constancy condition gives

$$
\frac{d}{dt}(\mathfrak{K}^{i\cdot}_{\cdot\alpha}\mathfrak{y}^a) = 0 \implies (\partial_\beta\mathfrak{K}^i_{\alpha})\,\mathfrak{y}^a\dot q^\beta + \mathfrak{K}^i_{\alpha}\dot{\mathfrak{y}}^a = 0.
$$

Multiplying by the inverse matrix $(\mathfrak{K}^{-1})^{\alpha\cdot}_{\cdot i}$ (satisfying $(\mathfrak{K}^{-1})^{\alpha\cdot}_{\cdot i}\mathfrak{K}^{i\cdot}_{\cdot\beta} = \delta^\alpha_\beta$), we obtain the equation of parallel transport:

$$
\dot{\mathfrak{y}}^\alpha + (\mathfrak{K}^{-1})^{\alpha\cdot}_{\cdot i}(\partial_\gamma\mathfrak{K}^{i\cdot}_{\cdot\beta})\,\mathfrak{y}^\beta\dot q^\gamma = 0.
$$

Hence the Christoffel symbols of the Cartan–Kröner connection are

$$
\tilde\Gamma^\alpha_{\beta\gamma} = (\mathfrak{K}^{-1})^{\alpha\cdot}_{\cdot i}\;\partial_\gamma\mathfrak{K}^{i\cdot}_{\cdot\beta}. \tag{21.9}
$$

In the case of axisymmetric deformation, the matrix $[\mathfrak{K}]$ takes the form

$$[\mathfrak{K}] = \begin{bmatrix} A s_\theta c_\varphi + r v_{,r} c_\theta c_\varphi & Q s_\theta c_\varphi + S c_\theta c_\varphi & -R s_\theta s_\varphi - r v c_\theta s_\varphi \\ A s_\theta s_\varphi + r v_{,r} c_\theta s_\varphi & Q s_\theta s_\varphi + S c_\theta s_\varphi & R s_\theta c_\varphi + r v c_\theta c_\varphi \\ A c_\theta - r v_{,r} s_\theta & Q c_\theta - S s_\theta & 0 \end{bmatrix}$$

Substituting (21.2) into formula (21.9), after lengthy but straightforward algebraic manipulations using trigonometric identities, one finds that these symbols coincide exactly with those obtained from the Levi-Civita connection of the metric $\mathfrak{g}$ induced by the actual configuration: $\tilde{\Gamma}^{\alpha}_{\beta\gamma} = \Gamma^{\alpha}_{\beta\gamma}$. This confirms that for a compatible deformation the two connections are identical, a consequence of the integrability of the field $\mathfrak{K} = d\varkappa$. In the case of incompatible deformations, which will be studied in the next section, the two connections differ. Their difference is measured by the contortion tensor, which encodes the relation between the density of dislocations and disclinations.

Having established the geometric framework for describing deformations, we now turn to the physical aspects. Following the classical variational approach, we formulate the action functional and derive the field equations and conservation laws from the principle of stationary action and Noether symmetries. To achieve a compact and unified description that naturally accommodates both spatial and temporal derivatives, we employ a (3+1)-dimensional formalism. This approach, which can be traced back to the fourfold integral formulas of Kirchhoff for the wave equation, provides an economical formulation in the sense of Mach's principle of economy of thought. Moreover, it prepares the ground for the relativistic generalization discussed in Section 21.5. The presentation follows the general scheme developed in [39].

The key idea is to introduce "body–time manifold"

$$\mathfrak{N} = \mathcal{T}^{p}_{\mathrm{G}} \times \mathfrak{M},$$

where $\mathfrak{M}$ is the three-dimensional material manifold introduced in Section 21.2 and $\mathcal{T}^{p}_{\mathrm{G}} \cong \mathbb{R}$ is the Galilean chronometry point space. Points of $\mathfrak{N}$ are denoted by $\mathfrak{q} = (\tau, \mathfrak{p})$ with $\tau \in \mathcal{T}^{p}_{\mathrm{G}}$ and $\mathfrak{p} \in \mathfrak{M}$. The manifold $\mathfrak{N}$ is equipped with a natural product structure; we denote by $\pi_\tau : \mathfrak{N} \to \mathcal{T}^{p}_{\mathrm{G}}$ and $\pi_{\mathfrak{M}} : \mathfrak{N} \to \mathfrak{M}$ the canonical projections. A fixed volume form ω on $\mathfrak{N}$ is chosen; in the Newtonian setting it is $(\omega = d\tau \wedge \varepsilon$ where ε is the volume form on $\mathfrak{M}$ induced by the reference metric.

The configuration of the body is now described by a submersion

$$\kappa : \mathfrak{N} \hookrightarrow \mathcal{E}^{p},$$

which assigns to each material point $\mathfrak{p}$ and each "time" τ its actual position in the Euclidean point space $\mathcal{E}^{p}$. In the Newtonian setting τ is the absolute time, but in the sequel it will be interpreted as a time coordinate (proper time of a distinguished observer, or any other convenient parameter). The restriction of κ to a fixed τ gives the instantaneous shape $\mathcal{S}_\tau = \kappa(\{\tau\} \times \mathfrak{M})$.

The action is taken as an integral over an open subset $V \subset \mathfrak{N}$ (which can be thought of as a "moving" domain that will later be identified with the world-tube). Following the standard first-order Lagrangian formalism, we postulate an action of the form

$$\mathfrak{A} = \int\limits_V \mathcal{L}\big(\mathfrak{q}, \kappa(\mathfrak{q}), d\kappa(\mathfrak{q})\big)\, \omega, \tag{21.10}$$

where the Lagrangian density $\mathcal{L}$ depends on the point $\mathfrak{q}$, on the value of the configuration, and on its exterior derivative $d\kappa$ (a 1-form on $\mathfrak{N}$ with values in $\mathcal{E}^t$). In local coordinates (τ, q^α) on $\mathfrak{N}$, we have

$$d\kappa = \dot{\varkappa}\, d\tau + \frac{\partial \varkappa}{\partial q^\alpha}\, dq^\alpha.$$

The dependence on $\dot{\varkappa}$ and on $\partial\varkappa/\partial q^\alpha$ is thus unified in the single object $d\kappa$.

To obtain the field equations and conservation laws we consider a simultaneous variation of the coordinates and the fields. Let $\hat{\mathfrak{q}}(\zeta)$ be a one-parameter family of points in $\mathfrak{N}$ (a flow) with $\hat{\mathfrak{q}}(0) = \mathfrak{q}$, and let $\hat{\kappa}(\zeta)$ be a one-parameter family of embeddings with $\hat{\kappa}(0) = \kappa$. The total variation of the coordinates is defined as the vector field $\delta\mathfrak{q} = \hat{\mathfrak{q}}'\big|_{\zeta=0} \delta\zeta$, and the total variation of the configuration is $\delta\kappa = (\hat{\kappa}(\hat{\mathfrak{q}}))'\big|_{\zeta=0} \delta\zeta$. It is convenient to split $\delta\kappa$ into a part that comes from the change of the field at a fixed base point (the "partial variation") and a part that comes from the motion of the base point:

$$\delta\kappa = \overline{\delta\kappa} + d\kappa \lrcorner \delta\mathfrak{q}, \qquad \overline{\delta\kappa} = \hat{\kappa}'(\mathfrak{q})\, \delta\zeta.$$

Here $d\kappa \lrcorner \delta\mathfrak{q}$ denotes the contraction of the 1-form $d\kappa$ with the vector $\delta\mathfrak{q}$. Substituting the varied fields into the action (21.10) we obtain a function $\mathfrak{A}(\zeta)$ whose derivative with respect to ζ at $\zeta = 0$ (*i.e.* variation $\delta\mathfrak{A}$) can be computed in two equivalent ways.

A straightforward computation (using the chain rule and the fact that the volume form is fixed) yields

$$\delta\mathfrak{A} = \int\limits_V \Big(\mathrm{div}_\omega \delta\mathfrak{J} + \mathcal{E}\cdot\overline{\delta\kappa}\Big)\omega, \tag{21.11}$$

where the Noether current $\delta\mathfrak{J}$ is a vector field (or, more intrinsically, a $(3+1)$-vector density) given by

$$\delta\mathfrak{J} = \mathcal{L}\, \delta\mathfrak{q} + \overline{\delta\varkappa}\cdot\frac{\partial \mathcal{L}}{\partial d\varkappa},$$

and the Euler–Lagrange operator $\mathcal{E}$ is a covector field on $T\mathcal{E}^p$:

$$\mathcal{E} = \frac{\partial \mathcal{L}}{\partial \kappa} - \mathrm{div}_\omega\Big(\frac{\partial \mathcal{L}}{\partial d\kappa}\Big). \tag{21.12}$$

The divergence operator div_ε is defined via Lie derivative L by the relation $\mathsf{L}_{\mathfrak{X}}\omega = (\mathrm{div}_\omega \mathfrak{X})\, \omega$ for any vector field $\mathfrak{X}$. This definition works on any oriented manifold with a fixed volume form, without requiring a metric. The dot $(\cdot)$, as above, denotes the scalar product in $\mathcal{E}^t$. For variations $\overline{\delta\kappa}$ that vanish on the boundary ∂V, the principle of stationary action $\delta\mathfrak{A} = 0$ implies the field equations $\mathcal{E} = \mathbf{0}$. When the field equations hold, the first term in (21.11) must also vanish for arbitrary $\overline{\delta\kappa}$, which leads to the weak conservation law $\mathrm{div}_\omega \delta\mathfrak{J} = 0$.

An alternative expression for the derivative of the action is obtained by expanding the Lagrangian with respect to the variations of the coordinates and fields directly, without separating the partial variation. Using the fact that $\delta(d\kappa) = \overline{\delta(d\kappa)} + (\nabla d\hat{\kappa})\big|_{\zeta=0} \lrcorner \delta\mathfrak{q}$, where $\nabla\delta\mathfrak{q}$ is the covariant derivative of the vector field $\delta\mathfrak{q}$ (on $\mathfrak{N} = \mathbb{R}\times\mathfrak{M}$ we use the flat product connection; in the relativistic generalization it will be replaced by the Levi-Civita connection of the spacetime metric), one obtains

$$\delta\mathfrak{A} = \int_V \left(\delta\mathfrak{q}\mathcal{L}\big|_{\mathrm{expl}} + \mathcal{T} \mathbin{\dot{\lrcorner}} \nabla\delta\mathfrak{q} + \frac{\partial\mathcal{L}}{\partial\kappa}\cdot\delta\kappa + \frac{\partial\mathcal{L}}{\partial d\kappa} \mathbin{\dot{\lrcorner}} \delta(d\kappa)\right)\omega. \tag{21.13}$$

Here $\delta\mathfrak{q}\mathcal{L}\big|_{\mathrm{expl}}$ denotes the derivative in the direction $\delta\mathfrak{q}$ while keeping the fields fixed, $\mathcal{T}$ is the energy–momentum tensor (a $(1, 1)$ tensor field on $\mathfrak{N}$) defined by

$$\mathcal{T} = \mathcal{L}\,\mathsf{I} - (d\kappa)^{\mathsf{T}}\cdot\frac{\partial\mathcal{L}}{\partial d\kappa}.$$

The symbol $\dot{\lrcorner}$ (double) denotes double contraction (over both indices of $\mathcal{T}$ and $\nabla\delta\mathfrak{q}$), and $\dot{\lrcorner}$ denotes contraction between the vector parts of $\partial\mathcal{L}/\partial d\kappa$, $\delta(d\kappa)$, and Euclidean scalar product between their values in $\mathcal{E}^t$.

For variations that are pure coordinate transformations, we set $\overline{\delta\kappa} = 0$. Then $\delta\kappa = d\kappa\cdot\delta\mathfrak{q}$ (the field is Lie-dragged along the flow of $\delta\mathfrak{q}$). Substituting this into (21.13) and using the field equations with (21.12), the terms involving $\delta\kappa$ and $\delta(d\kappa)$ combine into a divergence, yielding the identity

$$\mathrm{div}_\omega\big(\mathcal{T}\lrcorner\delta\mathfrak{q}\big) = \delta\mathfrak{q}\mathcal{L}\big|_{\mathrm{expl}} + \mathcal{T} \mathbin{\dot{\lrcorner}} \nabla\delta\mathfrak{q}. \tag{21.14}$$

When the Lagrangian does not depend explicitly on the event $\mathfrak{q}$ (i.e., $\delta\mathfrak{q}\mathcal{L}\big|_{\mathrm{expl}} = 0$ for a certain class of vector fields), this identity gives conservation laws for the corresponding flows. We consider two fundamental cases.

1. **Translations.** If $\delta\mathfrak{q}$ is a constant vector field (generating translations in $\mathfrak{N}$), then $\nabla\delta\mathfrak{q} = 0$ and (21.14) reduces to $\mathrm{div}_\omega\big(\mathcal{T}\lrcorner\delta\mathfrak{q}\big) = 0$. Since $\delta\mathfrak{q}$ is arbitrary, this implies the strong conservation law $\mathrm{div}_\omega\mathcal{T} = 0$. In the presence of explicit dependence on $\mathfrak{q}$, the right-hand side becomes $\nabla_{\mathrm{expl}}\mathcal{L}$, representing the configurational force $\mathfrak{f}$.
2. **Rotations.** Consider infinitesimal rotations in the tangent space of $\mathfrak{N}$ (*i.e.*, $\delta\mathfrak{q} = \Omega\lrcorner\mathfrak{q}$ with Ω an antisymmetric tensor). For such $\delta\mathfrak{q}$, the condition of invariance gives $\mathcal{T} \mathbin{\dot{\lrcorner}} \Omega = 0$, which is equivalent to the symmetry of $\mathcal{T}$ after lowering an index. This is the conservation of angular momentum.

It is important to emphasise that the variational formulation presented above is deliberately cast in a form that does not depend on the particular geometric interpretation of the fields. The action is written on a four-dimensional manifold $\mathfrak{N}$ (the body–time manifold) equipped with a fixed volume form ω. The only dynamical field is the configuration $\varkappa$, and the Lagrangian depends on it and its exterior derivative $d\varkappa$. The resulting Euler–Lagrange equations and Noether conservation laws are expressed in terms of the divergence div_ω, which is defined solely by the volume form and therefore does not rely on any additional geometric structure (metric or connection). This makes the formalism directly applicable to all cases considered in this paper:

- **Compatible deformations** (Section 21.2), where the reference metric and the actual metric are both flat and are obtained by pulling back the Euclidean metric via $\varkappa_R$ and $\varkappa$, respectively;
- **Incompatible deformations** (Section 21.3), where the reference metric $\mathfrak{g}$ is non-Euclidean and is synthesised from a family of local stress-free configurations, while the actual metric is again obtained from $\varkappa$;
- **Relativistic elasticity** (Section 21.5), where the ambient space is a curved Lorentzian manifold and the configuration is an embedding of $\mathfrak{N}$ into spacetime; the Lagrangian then depends on the induced metric and, possibly, on the four-velocity.

In all these situations the abstract variational machinery remains unchanged; only the interpretation of the geometric objects and the specific form of the Lagrangian vary. This unity is one of the main conceptual messages of the present work.

21.3 Incompatible Deformations

The geometric framework developed so far relies on the existence of a global reference shape $\mathcal{S}_R$ that is unstressed. This shape serves as the common "relaxed memory" of the material points: when the body is unloaded, it returns to $\mathcal{S}_R$. In the Platonic allegory of Epstein [10], the body dwells in a world of ideal unstressed configurations; what we observe are merely its "incarnations" in physical space. However, many materials — especially those that have undergone plastic deformation, phase transitions, or additive manufacturing — do not possess such a global unstressed shape. Their local relaxed states are incompatible: if we cut out an infinitesimal neighbourhood of a point and let it relax, it would attain a shape that cannot be reassembled into a continuous body without gaps or overlaps. In other words, the body suffers from a kind of "material paramnesia": it remembers an ideal relaxed state that never existed globally. The neutron star crust, continuously accreting new material and undergoing extreme pressures, is a paradigmatic example.

In continuum mechanics, two principal approaches have been developed to handle such situations. The first, purely phenomenological, postulates the impossibility of a global unstressed shape and replaces it with a continuous family of *locally* unstressed shapes, one for each infinitesimal neighbourhood. This is the essence of the **local unloading hypothesis** [4, 30, 34, 45, 55, 59]. The family of local stress-free shapes is then used to synthesize a non-Euclidean material manifold, whose curvature and torsion quantify the density of defects (disclinations and dislocations). This approach, while mathematically elegant and general, leaves the physical origin of the incompatibility unspecified.

The second, more constructive, approach derives the incompatibility from a specific evolutionary process [35, 36]. In the context of surface growth, which is central to the present work, the body is built up by accretion of new material layers. Each layer, at the moment of its deposition, possesses its own intrinsic (possibly stressed) state. After attachment, the layer interacts with the existing assembly, and the whole structure evolves. The final body, resulting from this history, generally cannot be relaxed to a single Euclidean shape. Instead, one can imagine *disassembling* it back into its constituent layers, each of which would then relax to its own stress-free shape. This **gedankenexperiment** (thought experiment, familiar from the theory of dislocations, see e.g. [30, 45]) directly provides the family of local stress-free shapes: the reference shapes of the individual layers. The local unloading hypothesis thus becomes a *consequence* of the growth process, not an independent postulate. This constructive viewpoint has been developed in detail for layered structures in [35, 36] and for continuous (laminated) bodies in [38, 40].

In the context of neutron stars, the accretion of matter from the surrounding medium continuously adds new layers to the crust. Each new layer, due to its different composition, density, and thermal history, would, if isolated, assume a different unstressed shape. The final, stressed configuration of the star is the result of assembling these geometrically incompatible layers.

In practical computations, it is convenient to keep a distinguished reference configuration $\varkappa_R : \mathcal{B} \to \mathcal{E}^p$ whose image $\mathcal{S}_R = \varkappa_R(\mathcal{B})$ is a region of the physical space with a simple geometry — for our axisymmetric problem, a hollow ball described by ordinary spherical coordinates (r, θ, φ). This shape is globally stressed; it serves as an intermediate assembly shape from which the actual configuration $\varkappa$ is reached by an elastic deformation $\gamma = \varkappa \circ \varkappa_R^{-1}$. However, $\varkappa_R$ itself is not stress-free: if we could cut it into infinitesimal pieces and let each piece relax, they would assume a variety of different shapes.

This intuitive picture is formalised by introducing a family of local configurations $\varkappa_R^{\mathfrak{p}}$ (one for each material point $\mathfrak{p} \in \mathcal{B}$) such that each $\varkappa_R^{\mathfrak{p}}$ embeds an open neighbourhood of $\mathfrak{p}$ into the physical space and its image is locally stress-free. The collection $\{\varkappa_R^{\mathfrak{p}}\}_{\mathfrak{p}\in\mathcal{B}}$ is the continuum analogue of the individual layers in a layered structure: each layer, if separated from the assembly, would relax to its own unstressed shape. The relation between the global reference configuration $\varkappa_R$ and the local ones can be written as a composition $\varkappa_R^{\mathfrak{p}} = \gamma^{\mathfrak{p}} \circ \varkappa_R$, where $\gamma^{\mathfrak{p}}$ is a local deformation that relaxes an infinitesimal neighbourhood of $\varkappa_R(\mathfrak{p})$. Differentiating this relation at the point $\mathfrak{p}$ we obtain the linear map

$$\mathfrak{K}_{\mathfrak{p}} := d\varkappa_R^{\mathfrak{p}}|_{\mathfrak{p}} \, : \, T_{\mathfrak{p}}\mathcal{B} \to \mathcal{E}^p,$$

which sends tangent vectors of the material manifold to vectors of the physical space. Its value at $\mathfrak{p}$ is precisely the tangent map of the local relaxing deformation. In the language of [45], $\mathfrak{K}_{\mathfrak{p}}$ sends the material neighbourhood to its crystal reference.

From $\mathfrak{K}_\mathfrak{p}$ and the derivative of the global reference configuration $d\varkappa_R|_\mathfrak{p} : T_\mathfrak{p}\mathcal{B} \to \mathcal{E}^p$ (which is a linear isomorphism onto its image), we can construct a linear map acting entirely in the physical space:

$$\mathbf{H}_\mathfrak{p} := \mathfrak{K}_\mathfrak{p} \circ (d\varkappa_R|_\mathfrak{p})^{-1} : \mathcal{E}^p \to \mathcal{E}^p.$$

This map is called the distortion field (or implant in the sense of [45]). It takes a vector attached to the reference shape $\mathcal{S}_R$ at the point $\varkappa_R(\mathfrak{p})$ and transforms it into the corresponding vector in the locally relaxed state. The two fields are related by $\mathfrak{K}_\mathfrak{p} = \mathbf{H}_\mathfrak{p} \circ d\varkappa_R|_\mathfrak{p}$.

Now we can define a geometric structure directly on the material manifold $\mathcal{B}$. For any point $\mathfrak{p}$ and any tangent vectors $\mathfrak{u}, \mathfrak{v} \in T_\mathfrak{p}\mathcal{B}$, we set

$$\mathfrak{g}_\mathfrak{p}(\mathfrak{u}, \mathfrak{v}) := \mathbf{g}(\mathfrak{K}_\mathfrak{p}(\mathfrak{u}), \mathfrak{K}_\mathfrak{p}(\mathfrak{v})) = \mathbf{g}(\mathbf{H}_\mathfrak{p}[d\varkappa_R|_\mathfrak{p}(\mathfrak{u})], \mathbf{H}_\mathfrak{p}[d\varkappa_R|_\mathfrak{p}(\mathfrak{v})]). \tag{21.15}$$

Here $\mathbf{g}$ is the spatial metric (in the Newtonian context, the Euclidean metric). The resulting symmetric positive-definite tensor field $\mathfrak{g}$ is called the material metric; it encodes the lengths and angles that infinitesimal line elements would possess in their locally relaxed state. Since the local configurations $\varkappa_R^\mathfrak{p}$ are generally incompatible (they cannot be reassembled into a global Euclidean shape), $\mathfrak{g}$ is typically non-Euclidean.

If one wishes to avoid the intermediate global reference $\varkappa_R$, the same metric can be written directly in terms of the family $\{\varkappa_R^\mathfrak{p}\}_{\mathfrak{p}\in\mathcal{B}}$:

$$\mathfrak{g}_\mathfrak{p}(\mathfrak{u}, \mathfrak{v}) = \mathbf{g}(d\varkappa_R^\mathfrak{p}|_\mathfrak{p}(\mathfrak{u}), d\varkappa_R^\mathfrak{p}|_\mathfrak{p}(\mathfrak{v})). \tag{21.16}$$

Both formulations are equivalent; the first emphasises the role of the convenient intermediate shape used in calculations, while the second highlights the abstract geometric idea of a non-Euclidean material manifold. In the relativistic generalization, $\mathbf{g}$ is replaced by the metric induced on a spacelike hypersurface of spacetime (e.g., the Schwarzschild or Kerr metric), while the material manifold and the distortion field remain unchanged. Thus the entire construction carries over to the curved setting almost verbatim.

Once the material metric $\mathfrak{g}$ is known, one can introduce the associated material connection $\nabla^\mathcal{B}$ (for instance, the Levi-Civita connection of $\mathfrak{g}$). This connection generally possesses non-zero torsion (related to the density of dislocations) and curvature (related to the density of disclinations).

To make the abstract construction of the previous subsection concrete, we now write explicit expressions for the family of local configurations $\varkappa_R^\mathfrak{p}$ in the axisymmetric setting. The coordinates (r, θ, φ) serve as global coordinates on the material manifold $\mathcal{B}$; they are fixed once and for all. For each material point $\mathfrak{p}$ with coordinates $(\mathfrak{r}, \mathfrak{t}, \mathfrak{p})$ (we use gothic letters to emphasize their role as parameters of the family), we consider a local configuration $\varkappa_R^\mathfrak{p}$ defined in a neighbourhood of that point. Its coordinate representation depends on two sets of arguments: the coordinates (r, θ, φ) of a point in the neighbourhood (varying) and the parameters $(\mathfrak{r}, \mathfrak{t}, \mathfrak{p})$ that label the particular local configuration. By axial symmetry, the dependence on φ and $\mathfrak{p}$ is trivial, so we focus on the (r, θ)-dependence.

The displacement functions now depend on both the point of the family (the parameters $(\mathfrak{r}, \mathfrak{t})$) and the point of evaluation:

$$\tilde{u} = \tilde{u}(r, \theta; \mathfrak{r}, \mathfrak{t}), \qquad \tilde{v} = \tilde{v}(r, \theta; \mathfrak{r}, \mathfrak{t}).$$

The semicolon separates the variables that are differentiated (the first pair) from the parameters that label the family (the second pair). For each fixed $(\mathfrak{r}, \mathfrak{t})$, the functions $\tilde{u}, \tilde{v}$ of (r, θ) define a smooth embedding of a neighbourhood of $\mathfrak{p}$ into $\mathcal{E}^p$. The embedding itself is given by the same formula as (21.1), but with u, v replaced by $\tilde{u}, \tilde{v}$. The local basis vectors for the family $\{\varkappa_R^{\mathfrak{x}}\}$ are obtained by differentiating with respect to the first argument (r, θ, φ) while keeping the parameters $(\mathfrak{r}, \mathfrak{t})$ fixed. Thus we obtain the following non-holonomic basis fields:

$$\begin{aligned}
\tilde{\mathbf{E}}_r &= (1 + \tilde{u}_{,r})\mathbf{E}_r + \left(\tilde{v} + r\,\tilde{v}_{,r}\right)r^{-1}\mathbf{E}_\theta\Big|_{\mathfrak{r}=r,\ \mathfrak{t}=\theta}, \\
\tilde{\mathbf{E}}_\theta &= (\tilde{u}_{,\theta} - r\,\tilde{v})\mathbf{E}_r + \left(r + \tilde{u} + r\,\tilde{v}_{,\theta}\right)r^{-1}\mathbf{E}_\theta\Big|_{\mathfrak{r}=r,\ \mathfrak{t}=\theta}, \\
\tilde{\mathbf{E}}_\varphi &= \left((r + \tilde{u})s_\theta + r\,\tilde{v}\,c_\theta\right)(r s_\theta)^{-1}\,\mathbf{E}_\varphi\Big|_{\mathfrak{r}=r,\ \mathfrak{t}=\theta}.
\end{aligned} \tag{21.17}$$

Here vertical bar indicates that after differentiation we substitute the parameters equal to the coordinates (i.e., we evaluate the resulting expressions at the point where the parameter point coincides with the point of the manifold). The vectors $\mathbf{E}_r, \mathbf{E}_\theta, \mathbf{E}_\varphi$ are the holonomic basis induced by the global reference configuration $\varkappa_R$; they are independent of the family parameter.

The field $\tilde{\mathbf{E}}_r, \tilde{\mathbf{E}}_\theta, \tilde{\mathbf{E}}_\varphi$ is generally non-holonomic; it becomes holonomic (i.e., integrable to a global embedding) only if the functions $\tilde{u}, \tilde{v}$ satisfy appropriate compatibility conditions (essentially, if they do not depend on the parameters $(\mathfrak{r}, \mathfrak{t})$ or depend on them in a very special way).

The material metric $\tilde{\mathfrak{g}}$ is obtained by pulling back the spatial metric $\mathbf{g}$:

$$\tilde{\mathfrak{g}}_{\mathfrak{p}}(\partial_\alpha, \partial_\beta) = \tilde{\mathbf{E}}_\alpha \cdot \tilde{\mathbf{E}}_\beta, \qquad \alpha, \beta \in \{r, \theta, \varphi\}. \tag{21.18}$$

The material metric $\tilde{\mathfrak{g}}$ defined in (21.18) encodes the lengths and angles of infinitesimal line elements in their locally relaxed state. With this metric alone we can already formulate constitutive relations, because the elastic response depends only on the metric (more precisely, on the difference between the metric induced by the actual deformation and the material metric). However, to transfer the entire physical description – including balance laws and conservation theorems like Stokes' theorem – from the Euclidean shapes to the material manifold, we need an affine connection that is compatible with the field of local frames ∂_α. The natural requirement is that the covariant derivative of the metric vanishes: $\tilde{\nabla}\tilde{\mathfrak{g}} = 0$. Together with the condition of zero torsion (which is natural when we want a unique connection determined solely by the metric), this yields the Levi-Civita connection $\tilde{\nabla}^L$. Its coefficients in a coordinate frame (∂_i) are given by the standard Christoffel formulae, but now with respect to the non-Euclidean metric $\tilde{\mathfrak{g}}$:

$$\tilde{\Gamma}^k_{ij} = \frac{\tilde{\mathfrak{g}}^{kl}}{2}\left(\partial_i \tilde{\mathfrak{g}}_{jl} + \partial_j \tilde{\mathfrak{g}}_{il} - \partial_l \tilde{\mathfrak{g}}_{ij}\right).$$

Although this expression looks identical to the classical one used in curvilinear coordinates, its meaning is different: here $\tilde{\mathfrak{g}}$ is not the pull-back of a Euclidean metric, and consequently the curvature of $\tilde{\nabla}^L$ does not vanish. The curvature is measured by the Riemann–Christoffel tensor

$$\tilde{\mathfrak{R}}^l_{\ ijk} = \partial_i \tilde{\Gamma}^l_{\ jk} - \partial_j \tilde{\Gamma}^l_{\ ik} + \tilde{\Gamma}^m_{jk}\tilde{\Gamma}^l_{\ im} - \tilde{\Gamma}^m_{ik}\tilde{\Gamma}^l_{\ jm}.$$

A non-zero $\tilde{\mathfrak{R}}$ signals the impossibility of flattening the material metric globally – a direct consequence of the non-holonomic nature of the local frames $\mathfrak{e}_\alpha$. It is important to note that the Levi-Civita connection, being derived solely from the metric, discards the rotational part of the local configurations $\mathfrak{K}_\mathfrak{p}$. In this respect it is analogous to the right Cauchy–Green tensor $\mathbf{C} = \mathbf{F}^\mathsf{T} \cdot \mathbf{F}$ in classical elasticity, which loses the orthogonal factor $\mathbf{R}$ of the polar decomposition $\mathbf{F} = \mathbf{R} \cdot \mathbf{C}^{1/2}$ [17]. The missing rotational information is precisely what distinguishes a medium with dislocations from one with only disclinations. To capture it we need a second, more refined connection.

In contrast to the Levi-Civita connection, the Cartan–Kröner connection uses the full information of the local configurations $\mathfrak{K}_\mathfrak{p}$. It is defined by the requirement that the vector fields $\mathfrak{K}^{-1}(\mathbf{i}_k)$ (the images of the Cartesian basis vectors) are parallel: $\tilde{\nabla}^W_u(\mathfrak{K}^{-1}(\mathbf{i}_k)) = 0$ for all $u \in T\mathcal{B}$, $k = 1, 2, 3$. In a coordinate frame (∂_i) this condition leads to the explicit expression

$$\tilde{\Gamma}^i_{\ jk} = (\mathfrak{K}^{-1})^i_{\ c}\, \partial_j \mathfrak{K}^c_{\ k},$$

where $\mathfrak{K}^c_k$ are the components of $\mathfrak{K}$ in the Cartesian basis $\{\mathbf{i}_c\}$. Unlike the Levi-Civita connection, $\tilde{\nabla}^W$ has zero curvature (by construction the frame $\mathfrak{K}^{-1}(\mathbf{i}_k)$ is globally parallel), but it generally possesses non-zero torsion $\mathfrak{T}$:

$$\mathfrak{T}^i_{\ jk} = \tilde{\Gamma}^i_{\ jk} - \tilde{\Gamma}^i_{\ kj}.$$

The Cartan–Kröner connection is metric-compatible with respect to $\tilde{\mathfrak{g}}$ (a direct computation using (21.9) and the definition of $\tilde{\mathfrak{g}}$ shows that $\tilde{\nabla}^W \tilde{\mathfrak{g}} = 0$, Theorem 3 of [41]). Hence it satisfies the same compatibility condition as the Levi-Civita connection, but it differs from it by a tensorial quantity called the contortion.

Since both connections are metric-compatible, their difference $\mathfrak{Q} := \tilde{\nabla}^W - \tilde{\nabla}^L$ is a tensor field of type $(1, 2)$. Expressing it in terms of the torsion yields the well-known formula

$$\mathfrak{Q}^i_{\ jk} = \frac{1}{2}\tilde{\mathfrak{g}}^{im}\left(\tilde{\mathfrak{T}}_{mjk} + \tilde{\mathfrak{T}}_{kmj} + \tilde{\mathfrak{T}}_{jmk}\right),$$

where $\tilde{\mathfrak{T}}_{ijk} = \tilde{\mathfrak{g}}_{im}\tilde{\mathfrak{T}}^m_{jk}$. Thus the contortion is completely determined by the torsion and the metric. Conversely, the torsion can be recovered from the contortion via $\mathfrak{T}^i_{jk} = \tilde{\mathfrak{Q}}^i_{jk} - \tilde{\mathfrak{Q}}^i_{kj}$.

Physically, the contortion encodes the additional information about the material inhomogeneity that is lost when one retains only the metric. In the language of defect theory, the Levi-Civita part describes the disclination content (curvature), while the contortion accounts for the dislocation content (torsion). Together they provide a complete geometric description of the incompatible deformations accumulated during the growth process.

It should be noted that the synthesis of the material metric $\tilde{\mathfrak{g}}$ or the Cartan–Kröner connection $\tilde{\nabla}^W$ requires the choice of an intermediate stressed reference shape $\mathcal{S}_R$. This choice is by no means unique; different intermediate shapes lead to different representations of the local configurations $\varkappa^{\mathfrak{X}}_R$. However, the geometric invariants — the curvature of the Levi-Civita connection and the torsion of the Cartan–Kröner connection — are independent of this choice. A rigorous proof of this fact can be found in Theorem 2 of [41]. Thus, the non-Euclidean geometry of the material manifold is an intrinsic characteristic of the growing body, unaffected by the particular intermediate configuration used in its construction.

21.4 Growing Body

So far we have considered deformations of a body with fixed material composition: its material structure, i.e., the set of material points and their adjacency, was assumed to be unchanging. This allowed us to formalise the body as a smooth manifold that uniquely determines its material content and topology, while only the geometric structures (metric, connection) placed on top of this manifold could vary. When we need to model a growth process, however, we must abandon the assumption of a fixed material description, simply because during such a process new material is added to the body. Moreover, the topological status of points changes – boundary points become interior points; figuratively speaking, "the boundary sinks into the depths of the growing body". To develop a suitable formalism we need, alongside the material manifold $\mathfrak{M}$, to introduce the notion of a body $\mathcal{B}$ as a submanifold of $\mathfrak{M}$ of the same dimension, and to consider families of such submanifolds ordered by inclusion. Such families provide a formal description of growth processes. When the index set of the family is at most countable, we speak of a *multilayered* growing body; when the index set is continuous, we speak of a *laminated* body [35, 36].

Following the approach developed in [35], we consider a family $\mathfrak{G} = \{\mathcal{B}_\alpha\}_{\alpha\in\mathbb{I}}$ of body manifolds, where the index set $\mathbb{I} \subset \mathbb{R}$ is compact and the bodies are ordered by inclusion:

$$\forall\alpha, \beta \in \mathbb{I}: \ (\alpha < \beta) \ \Rightarrow \ (\mathcal{B}_\alpha \subset \mathcal{B}_\beta).$$

We refer to each $\mathcal{B}_\alpha$ as an *assembly*. The minimal element $\mathcal{B}_*$ corresponds to the initial substrate, while the maximal element $\mathcal{B}^*$ contains all particles ever adhered.

The material points that are added between two successive assemblies form a *layer*. If $\mathbb{I} = \{0, 1, \dots, N\}$ is finite, the body is multilayered; if $\mathbb{I} = [0, 1]$ is an interval, the body is laminated.

To describe how layers are joined we use the *joining operation* $\vee$ introduced in [34]. For two bodies $\mathcal{B}_n$ and $\mathfrak{L}_{n+1}$ (the latter being a layer) that share a common boundary, the operation produces the next assembly:

$$\mathcal{B}_{n+1} = \mathcal{B}_n \vee \mathfrak{L}_{n+1} := \mathcal{B}_n \cup \mathfrak{L}_{n+1} \cup \partial_{\mathcal{B}_{n+1}}(\mathcal{B}_{n+1} \setminus \mathcal{B}_n).$$

In words, the points of the common boundary change their status from boundary to interior points, reflecting the intuitive idea that "the boundary sinks" as new material is attached.

For a continuous (laminated) growth, the family of assemblies can be organised as a smooth trivial fibre bundle. Specifically, let $\mathcal{B}^*$ be the final body and let $\mathbb{I} =]0, 1[$ be an open interval parametrising the layers. A smooth surjective map $\pi : \mathcal{B}^* \to \mathbb{I}$ (the projection) is such that each fibre $\mathfrak{L}_\tau := \pi^{-1}(\{\tau\})$ is a two-dimensional submanifold diffeomorphic to a fixed model lamina $\mathfrak{F}$. The bundle is trivial: there exists a diffeomorphism $\Psi : \mathcal{B}^* \to \mathbb{I} \times \mathfrak{F}$ satisfying $\mathrm{pr}_{\mathbb{I}} \circ \Psi = \pi$. The assembly at 'time' α is then $\mathcal{B}_\alpha = \Psi^{-1}(]0, \alpha[\times\mathfrak{F})$. In this picture, the evolution parameter α plays the role of a continuous label for the laminae, and the projection π assigns to each material point the instant of its accretion. This construction provides a precise geometric description of the intuitively appealing idea that the body grows by the continuous deposition of new material layers.

In the present paper we are primarily interested in the accretion process that builds up the crust of a neutron star. This process is naturally modelled as a multilayered or laminated growth, where each newly accreted layer comes with its own intrinsic state (density, composition, thermal history) and, upon attachment, contributes to the residual stress field. The formal framework outlined above provides the necessary language to describe such a process and will be used in the next sections to construct the distortion field $\mathbf{H}$ and the resulting material metric and connections. Further details, including the explicit treatment of layer shrinkage and the determination of reference radii, can be found in [35, 36].

We now turn to the central part of the modelling – the evolutionary problem. For each element of the family that formalises the stages of growth, we can determine (still in the Newtonian approximation) the stress-strain state caused by both external and internal sources of stress. The latter are characterised by a non-trivial family of reference shapes or, equivalently, by a non-Euclidean relaxed metric on the material manifold. The question of how to find this metric has not yet been answered. The answer is provided by the evolutionary problem.

The idea of the evolutionary problem consists in recursively determining the relaxed metric by simulating the entire growth process. It is easiest to explain for the case of discrete growth. At the initial instant, the assembly consists of a single element – the initial body – whose stress-strain state is either known or can be determined by standard calculations, since it possesses a global stress-free reference shape. At the next instant (the next element of the discrete process), the assembly already consists of two bodies: the initial one and the first layer, which was formed in such a way that its inner boundary coincides with the (deformed) outer boundary of the initial body. This is the **principle of coherent growth boundaries**. Thus the stress-free reference shape of the attached layer is determined by the boundary of the assembly on top of which it was formed, but its stress-strain state at this stage is determined only by the physical processes occurring within it (for example, shrinkage). At the same time, the initial body and the attached layer deform together, preserving the coherence of the boundary. From these conditions we determine the stress-free shape of the first layer. The calculations are then repeated for each subsequent growth stage, determining at each step the stress-free shape of the corresponding layer from the known shape of the boundary on which it was created (which is the deformed outer boundary of the growing assembly at the previous step) and the individual process occurring in the layer (*e.g.*, shrinkage). This recursive procedure yields the complete family of local stress-free shapes and, consequently, the material metric and connections that encode the residual stress state accumulated during accretion.

To give a precise mathematical formulation, we introduce the following notation (see also [35, 36]):

- $\mathfrak{M}$ is the material manifold, equipped with a global chart $\xi : \mathfrak{M} \to \mathbb{R}^3$ that assigns to each material point the coordinates (r, θ, φ). These coordinates serve as permanent labels throughout the growth process.
- $\mathfrak{L}_0 = \{\xi^{-1}(r, \theta, \varphi) \mid 0 < r < r_0,\ (\theta, \varphi) \in \mathbb{S}^2\}$ is the initial body.
- $\mathfrak{L}_k = \{\xi^{-1}(r, \theta, \varphi) \mid r_{k-1} < r < r_k,\ (\theta, \varphi) \in \mathbb{S}^2\}$, $k \geq 1$, where $r_k = r_0 + k\Delta$ and $\Delta > 0$ is a fixed thickness increment in the k-th layer.
- $\mathcal{B}_n = \bigcup_{k=0}^{n} \mathfrak{L}_k$ is the n-th assembly (the union of the initial body and the first n layers). Its coordinate range is $0 < r < r_n$.
- $\overset{n}{\varkappa} : \mathcal{B}_n \to \mathcal{E}^p$ is the actual configuration of the n-th assembly. In the axisymmetric case it is given by formulas analogous to (21.1), but with displacement functions $\overset{n}{u}(r, \theta)$ and $\overset{n}{v}(r, \theta)$ (we write $\overset{n}{u}, \overset{n}{v}$ instead of the previously used u, v to emphasise that they belong to the n-th stage).
- $\mathcal{S}_n = \overset{n}{\varkappa}(\mathcal{B}_n) \subset \mathcal{E}^p$ is the actual shape of the n-th assembly.
- $\mathcal{S}_0 = \varkappa_R(\mathfrak{L}_0)$ is the reference (stress-free) shape of the initial body; here $\varkappa_R$ is the fixed reference configuration introduced in Section 21.2.
- $\mathfrak{L}_n$ is the reference shape of the n-th layer. It is not prescribed a priori but constructed recursively during the evolutionary process.

The construction of $\mathfrak{L}_n$ consists of two steps. First, we take the actual outer boundary of the previous assembly, $\partial\mathcal{S}_{n-1}$, and extend it in the direction of the outward unit normal $\mathbf{n}$ by a distance $\delta > 0$ (the physical thickness of the layer before shrinkage). This gives an intermediate shape

$$\mathrm{L}_n^{(\mathrm{pre})} = \left\{ \mathbf{x} \in \mathcal{E}^p \;\middle|\; \mathbf{x} = \mathbf{y} + \mathbf{n}(\mathbf{y})\,(r - r_{n-1})\,\delta/\Delta,\ \mathbf{y} \in \partial\mathcal{S}_{n-1},\ r \in (r_{n-1}, r_n) \right\},$$

where the factor δ/Δ converts the material coordinate increment into a physical distance. Second, the layer undergoes a shrinkage (homothety) with coefficient $\mu > 0$ (*e.g.*, $\mu < 1$ for contraction, $\mu > 1$ for expansion). Thus $\mathrm{L}_n = \mu\,\mathrm{L}_n^{(\mathrm{pre})}$.

Once $\mathfrak{L}_n$ is defined, we have at our disposal the complete collection of reference shapes for all layers $\mathfrak{L}_0, \mathfrak{L}_1, \dots, \mathfrak{L}_n$ (note that $\mathfrak{L}_0 = \mathcal{S}_0$). The assembly $\mathcal{B}_n$ is now modelled as a hyperelastic body composed of these geometrically incompatible layers. The displacement fields $\overset{n}{u}(r,\theta), \overset{n}{v}(r,\theta)$ are obtained by solving the nonlinear boundary-value problem for the whole assembly under the given external loads (gravitational and centrifugal forces) and the condition of ideal contact at the interfaces. From the solution we extract the new outer boundary $\partial\mathcal{S}_n$ and proceed to the next step.

Repeating this procedure for $n = 1, 2, \dots, N$ yields a sequence of displacement fields $\{\overset{n}{u}, \overset{n}{v}\}_{n=0}^{N}$ and, consequently, the family of local configurations $\varkappa_R^{\mathfrak{p}}$ that characterise the stress-free states of infinitesimal neighbourhoods. As shown in Section 21.3, this family determines the material metric $\tilde{\mathfrak{g}}$ and, consequently, the material connections and the contortion tensor. Thus the evolutionary problem provides the missing link between the growth process and the non-Euclidean geometry of the final body.

In the continuous (laminated) case, the discrete steps are replaced by a continuous evolution. The recursive procedure then becomes a differential equation for the functions $\tilde{u}(r,\theta;t), \tilde{v}(r,\theta;t)$ with respect to the growth parameter t. This formulation is technically more involved but follows the same conceptual lines; details can be found in [38, 40].

21.5 Relasticity

In the Galilean–Newtonian setting we represented spacetime $\mathfrak{N} = \mathbb{R} \times \mathfrak{M}$ as an ordered pair of affine spaces, but time there served merely as an absolute parameter, independent of the geometry of the world, and therefore no explicit geometric union was required. Now our viewpoint has changed: not only do we combine the timelike coordinate with the spacelike ones into a single smooth manifold structure, but we are also ready to endow it with a metric geometry, exactly as we did earlier when describing deformation in terms of a varying metric. The only difference is that now the variation of the metric will correspond to a deformation of spacetime itself, whose deviation is measured relative to some reference metric. Staying within the classical paradigm, we take this reference metric to be the Minkowski metric. In coordinates (t, r, θ, ϕ) adapted to the product structure $\mathbb{R} \times \mathfrak{M}$ (with $\mathfrak{M} \simeq \mathbb{R}^+ \times \mathbb{S}^2$), it takes the form:

$$g = -dt \otimes dt + dr \otimes dr + d\Omega^2, \quad d\Omega^2 := r^2 d\theta \otimes d\theta + r^2 \sin^2\theta\, d\varphi \otimes d\varphi.$$

In more abstract terms, this is the unique metric of signature $(-, +, +, +)$ that is flat and respects the spherical symmetry of the problem.

The dynamics of material points is now described by world lines – smooth curves that are future-directed, timelike, and lie inside the light cone. For bodies of fixed material composition, their world lines form a congruence that fills a four-dimensional region called the world tube of the body. Growing bodies, on the other hand, are represented by what we may call world cones: as new material is accreted, additional world lines enter the region, so that the boundary of the world tube evolves. The part of the boundary that is crossed by newly entering world lines corresponds to the accretion front; it is the relativistic analogue of the phase boundary in a Stefan problem. In contrast, the lateral boundary of a world tube (for a body of constant composition) merely represents the evolution of the material surface.

This picture provides a natural geometric setting for the evolutionary problem: the history of a growing body is described by a one-parameter family of world tubes whose boundaries expand in a manner dictated by the accretion process. In the following sections we shall make this idea precise by constructing a suitable foliation of spacetime and defining the projection onto the material manifold, thereby extending the Carter–Quintana projection to the case of accretion.

A world line of a massive particle is a smooth curve $\gamma : \mathbb{I} \to \mathfrak{N}$ such that for every value of the parameter $s \in \mathbb{I}$ the tangent vector $\gamma_{,s}$ is timelike and future-directed. In the spherical chart (t, r, θ, φ) we write

$$\gamma(s) = \big(t(s), r(s), \theta(s), \varphi(s)\big), \qquad \gamma_{,s} = t_{,s}\partial_t + r_{,s}\partial_r + \theta_{,s}\partial_\theta + \varphi_{,s}\partial_\varphi.$$

The timelike and future-directed conditions are expressed invariantly as $g(\gamma_{,s}, \gamma_{,s}) < 0$, $g(\partial_t, \gamma_{,s}) < 0$, where ∂_t is the global timelike vector field. Among all possible parametrizations one can choose the proper time τ such that $g\left(\dot\gamma, \dot\gamma\right) = -1$ (the dot denotes differentiation with respect to τ). In the following we shall always assume that world lines are parametrized by proper time; the dot from now on will denote differentiation with respect to τ. The four-velocity field $\mathfrak{u} = \dot\gamma$ is then a unit timelike vector field tangent to the congruence of world lines that constitute the body.

Given the four-velocity, we introduce two complementary projection operators. The "temporal projector" onto the direction of $\mathfrak{u}$ is

$$\mathrm{T} = \mathfrak{u} \otimes \mathfrak{u}^\flat, \quad \forall \mathfrak{w}\ \mathfrak{u}^\flat(\mathfrak{w}) = g(\mathfrak{u}, \mathfrak{w}).$$

The spatial projector onto the subspace orthogonal to $\mathfrak{u}$ is simply the complement of the identity: $\mathrm{P} = \mathrm{I} - \mathrm{T}$, $\mathrm{I} = \partial_\alpha \otimes dx^\alpha$.

It is important to distinguish two different kinds of spacetime splitting. The first, which we call the coordinate splitting, is obtained by singling out the global time coordinate t. In our spherical chart the temporal direction is given by ∂_t (which is timelike for the metrics we shall consider), and the spatial platform is spanned by $\partial_r, \partial_\theta, \partial_\varphi$. The corresponding projectors are

$$\mathrm{T}_0 = \partial_t \otimes dt, \qquad \mathrm{P}_0 = \partial_r \otimes dr + \partial_\theta \otimes d\theta + \partial_\varphi \otimes d\varphi.$$

This splitting is natural when the metric is stationary and one works in adapted coordinates. The second, the observer-adapted splitting, is tied to the four-velocity $\mathfrak{u}$

of a particular material element. Here the time direction is $\mathfrak{u}$, and the spatial platform is the orthogonal complement of $\mathfrak{u}$, with projectors T and P defined above. For non-relativistic velocities the difference between these two splittings is negligible, and many works on relativistic elasticity simply replace the Euclidean spatial metric $\mathbf{g}$ (used in the Newtonian theory) by the spatial metric induced by the spacetime metric on a global hypersurface of constant coordinate time. However, for velocities close to the speed of light the distinction becomes essential, and one must work with the observer-adapted splitting.

To obtain an explicit orthonormal basis adapted to the observer, we start from the coordinate basis $(\partial_t, \partial_r, \partial_\theta, \partial_\varphi)$ and apply a Lorentz boost that sends ∂_t into $\mathfrak{u}$. Let $\gamma = -g(\partial_t, \mathfrak{u}) \geq 1$ be the Lorentz factor. A pure boost $\mathbf{\Lambda}$ satisfying $\mathbf{\Lambda}(\partial_t) = \mathfrak{u}$ and leaving the subspace orthogonal to both ∂_t and $\mathfrak{u}$ pointwise fixed can be written as

$$\mathbf{\Lambda} = I + (\gamma - 1)\mathbf{\Pi} + \sqrt{\gamma^2 - 1}\, \mathbf{\Xi},$$

where $\mathbf{\Pi}$ is a projection operator onto the plane spanned by ∂_t and $\mathfrak{u}$ while $\mathbf{\Xi}$ is a boost generator in this plane:

$$\mathbf{\Pi} = \frac{\mathfrak{u} \otimes \mathfrak{u}^\flat - \gamma\,(\mathfrak{u} \otimes \partial_t^\flat + \partial_t \otimes \mathfrak{u}^\flat) + \partial_t \otimes \partial_t^\flat}{\gamma^2 - 1}, \qquad \mathbf{\Xi} = \frac{\partial_t \otimes u^\flat - u \otimes \partial_t^\flat}{\sqrt{\gamma^2 - 1}}.$$

The operator Λ is an isometry of the Minkowski metric (i.e., a Lorentz transformation). Applying Λ to the coordinate basis yields a new frame

$$\mathfrak{e}_0 = u, \qquad \mathfrak{e}_\alpha = \mathbf{\Lambda}(\partial_i)/g(\mathbf{\Lambda}(\partial_i), \mathbf{\Lambda}(\partial_i)), \quad \alpha \in (r, \theta, \varphi).$$

These three spatial vectors form an orthonormal basis of the local rest space. The metric induced on this space is simply the Euclidean metric in this basis: $g(\mathfrak{e}_\alpha, \mathfrak{e}_\beta) = \delta_{\alpha\beta}$.

The Lorentz boost described above aligns the time axis of the local frame with the four-velocity $\mathfrak{u}$. However, for an accelerated or rotating observer, the spatial triad obtained by a simple boost is not non-rotating in the sense of Fermi–Walker. To construct a genuinely non-rotating (inertial) frame along a world line, one must use Fermi–Walker transport. A tetrad $(\mathfrak{e}_\alpha)$ is said to be Fermi–Walker transported if its spatial vectors satisfy $\nabla_u \mathfrak{e}_i = (\mathfrak{e}_i \cdot \mathfrak{a})\mathfrak{e} - (\mathfrak{e} \cdot \mathfrak{a})\mathfrak{e}_i$, where $\mathfrak{a} = \nabla_{\mathfrak{u}} \mathfrak{u}$ is the four-acceleration. This law guarantees that the spatial triad does not rotate with respect to a comoving inertial guidance system.

For a world line with non-zero acceleration and rotation, the Fermi–Walker tetrad differs from the tetrad obtained by a sequence of boosts. The deviation is measured by the rotation (or twist) of the congruence, which in the language of connections appears as non-zero torsion. Indeed, if one transports a vector around an infinitesimal closed loop formed by two neighbouring world lines and two connecting curves, the failure to return to its original orientation is precisely the torsion of the connection induced on the spatial platform.

In the context of our (3+1) formalism, adopting a Fermi–Walker transported frame has two important consequences:

- The metric induced on the spatial platform remains Euclidean, but the connection coefficients with mixed indices (involving time derivatives of spatial vectors) acquire additional terms proportional to the acceleration and rotation. These terms are responsible for the relativistic generalisations of the centrifugal and Coriolis forces.
- When the congruence is rotating (*i.e.*, $\omega_{\mu\nu} \neq 0$), the spatial platform is not integrable; there is no global hypersurface orthogonal to the world lines. The Fermi–Walker transport provides a way to compare spatial directions along the world line without needing such a global hypersurface.

Thus, the Fermi–Walker frame plays a role analogous to the non-holonomic basis $\tilde{\mathbf{E}}_\alpha$ introduced in Section 21.3 for incompatible deformations. In both cases, the connection (or the anholonomicity) encodes information that is lost when one works only with metrics. This parallel reinforces the deep unity between the geometric description of defects and the geometric description of relativistic motion.

The passage to general relativity implies a distortion of the flat spacetime metric such that gravitational effects become a consequence of this curvature. Since the logical structure of general relativity is similar to that of classical mechanics – the foundation is the stationarity of the action and variational symmetries, with the scalar curvature of the spacetime connection added to the list of Lagrangian arguments – the path to constructing a relativistic theory of a deformable continuum is clearly outlined. However, we do not have the opportunity to reproduce it here in full form, and we shall simplify the task by "decoupling" the effects of gravity and elastic deformation. This corresponds to a Lagrangian in which cross-terms between gravity and deformation are neglected – a sufficiently good approximation for the astrophysical problem under consideration: the growth, deformation or local failure of a neutron star crust produces a negligible change in the overall mass distribution (the bulk of the mass is concentrated in the hadron-plasma core). Consequently, the spacetime curvature can be determined in advance, without taking into account the processes in the crust, and treated as fixed during the modelling of deformation and growth. Of course, such a model should be regarded as a zeroth-order approximation, but its results may still reflect the qualitative physical properties of the process.

Under the decoupling assumption, the problem of finding the spacetime curvature can be considered for a non-deformable and non-growing sphere. For a non-rotating spherically symmetric star, the metric is obtained by continuously matching the interior and exterior Schwarzschild solutions. In Schwarzschild coordinates (t, r, θ, φ) it takes the form [20, 38]:

$$g = \begin{cases} -\dfrac{(b/r - 3a)^2}{4r}\, dt \otimes dt + \dfrac{r^3}{b^2}\, dr \otimes dr + d\Omega^2, & r \leqslant R, \\ -\left(1 - \dfrac{2M}{r}\right) dt \otimes dt + \dfrac{dr^2}{1 - \dfrac{2M}{r}} + d\Omega^2, & r > R, \end{cases}$$

where R is the stellar radius, M the total mass, and the auxiliary parameters are defined as $a = \sqrt{R - 2M}$, $b = \sqrt{R^3 - 2MR^2}$. For a rotating star, the situation is more delicate. The exterior field is described by the Kerr metric, which in Boyer–Lindquist coordinates (t, r, θ, φ) reads

$$
\begin{aligned}
g = &-\left(1 - \frac{2Mr}{\rho^2}\right) dt \otimes dt - \frac{4Mcr\sin^2\theta}{\rho^2}(dt \otimes d\varphi + d\varphi \otimes dt) \\
&+ \frac{\rho^2}{\Delta} dr \otimes dr + \rho^2 \, d\theta \otimes d\theta + \left(r^2 + c^2 + \frac{2Mc^2 r\sin^2\theta}{\rho^2}\right) \sin^2\theta \, d\varphi \otimes d\varphi,
\end{aligned}
$$

where $\rho^2 = r^2 + c^2\cos^2\theta$, $\Delta = r^2 - 2Mr + c^2$, $c = J/M$, and J is the stellar angular momentum.

In the relativistic description, a body of fixed material composition is represented by a congruence of world lines – a family of smooth, timelike, future-directed curves, each associated with a material point. The notion of an "instantaneous shape" is no longer absolute; it depends on a choice of a spacelike hypersurface that cuts the world tube. Such a hypersurface, by definition, is a three-dimensional submanifold of spacetime, everywhere spacelike, and intersecting each world line exactly once. It provides a one-to-one correspondence between material points and events, thereby playing the role of a "configuration" in the relativistic context. A one-parameter family of such hypersurfaces, smoothly varying with a parameter λ, gives a foliation of the world tube; each leaf is the shape of the body at the "instant" labelled by λ.

Several natural choices for the foliation exist, but they are not equivalent. One might be tempted to use the hypersurfaces of "constant proper time" along the world lines. However, as emphasised in our previous work [38], such surfaces are generally not orthogonal to the congruence and, more importantly, they do not provide a well-defined one-to-one mapping between world lines and points because the proper time is not a good global coordinate (different world lines can have the same proper time at different events). Another possibility is to use the "Poincaré–Einstein simultaneity surfaces" relative to a chosen reference world line; these are orthogonal to that particular world line but not to the whole congruence.

For the purpose of continuum mechanics, the most convenient choice is a family of "spacelike hypersurfaces that are transverse to the world lines" (not necessarily orthogonal). Such a family can always be constructed locally, for instance by taking the level sets of a time function t that is adapted to a distant observer (as in the Schwarzschild or Kerr metrics). In the flat Minkowski case, taking t = const yields the standard Newtonian foliation. More generally, any choice of a smooth time function that is everywhere timelike provides a foliation by spacelike surfaces. The leaves of this foliation inherit a Riemannian metric $h_{\alpha\beta} = g_{\alpha\beta} + n_\alpha n_\beta$, where n^α is the unit normal to the leaf (not to be confused with the four--elocity u^α of the matter). It is this induced metric that serves as the arena for measuring strains: the actual shape of the body at "time" t is exactly the leaf t = const, and its geometry is given by $h_{\alpha\beta}$.

In our construction of the body-time manifold $\mathfrak{N} = \mathbb{R} \times \mathfrak{M}$, the parameter s plays the role of this time function. The images $\mathbb{S}_s = \kappa(\{s\} \times \mathfrak{M})$ of the leaves under the history map $\kappa : \mathfrak{N} \to \mathcal{M}$ are precisely the shapes we need. They are not required to be surfaces of simultaneity in any absolute sense; they merely provide a convenient parametrisation of the evolution. In the Newtonian limit, s becomes the absolute Galilean time, and the induced metric $h_{\alpha\beta}$ reduces to the ordinary Euclidean metric on the shape. In the presence of strong gravitational fields, $h_{\alpha\beta}$ is the non-trivial three-dimensional metric obtained from the appropriate solution of Einstein's equations (e.g. the spatial part of the Schwarzschild or Kerr metric).

The discussion above leads to a refined understanding of what should be meant by the "shape" of a relativistic body. In the Newtonian limit, the shape is simply a region of absolute Euclidean space at a given instant of absolute time. In special relativity, one might be tempted to identify the shape with a hypersurface of constant proper time, but, as we have seen, such hypersurfaces are not orthogonal to the world lines and do not provide a natural platform for defining spatial distances. The correct local object is the "spatial platform" – the three-dimensional subspace orthogonal to the four-velocity $\mathbf{u}$ at each event. A global shape would then be a hypersurface that is everywhere tangent to these local platforms, i.e., an envelope of the spatial platforms.

However, as is well known from differential geometry, such an envelope exists globally only if the distribution of spatial platforms is integrable (Frobenius theorem). The condition for integrability is precisely the vanishing of the rotation (twist) tensor $\omega_{\mu\nu} = h^{\alpha}_{\mu} h^{\beta}_{\nu} \nabla_{[\alpha} u_{\beta]}$. When $\omega_{\mu\nu} \neq 0$ (*e.g.*, in a rotating star or, more generally, in any stationary axisymmetric spacetime), there is no global hypersurface orthogonal to the world lines. In such cases, the "shape" cannot be represented as a single Riemannian manifold; instead, one must work with a more general geometric structure.

This situation is strikingly analogous to the theory of distributed defects in solids. In a dislocated crystal, the material manifold cannot be equipped with a globally flat Riemannian metric; one needs a connection with torsion to describe the mismatch of lattice vectors. In the case of disclinations, the material manifold requires a connection with curvature. Here, the role of the "material" is played by the spatial platform, and the obstruction to integrability – the non-vanishing rotation – is precisely analogous to a continuous distribution of disclinations in the fabric of spacetime. The mathematical description therefore naturally leads to a Cartan geometry with torsion, where the envelope of spatial platforms is understood as a non-Riemannian three-dimensional manifold whose connection encodes the rotational structure of the congruence.

In [37, 42, 43] we avoided the problem of defining a global shape by working directly with the world tube and formulating the action as an integral over a four-dimensional region, without any foliation. This approach is perfectly legitimate and respects the principle of locality: all physical quantities are defined pointwise, and no global simultaneity is required. However, when we wish to speak of a "configuration" or a "deformation" – i.e., a mapping that compares the actual placement of material points with a reference placement – we inevitably need a three-dimensional object that serves as the domain of such a mapping. In the presence of rotation, that object cannot be a Riemannian manifold; it must be a more general space whose connection reflects the non-integrability of the local rest spaces.

Thus, the relativistic analogue of a shape is indeed the envelope of spatial platforms, but this envelope may be a non-Riemannian manifold, equipped with a connection that has torsion (and possibly curvature) determined by the rotation of the congruence. In the limit of slow rotation (or, equivalently, in the Newtonian limit where $\omega_{\mu\nu} \to 0$), this structure reduces to the familiar Riemannian shape of classical mechanics. The principle of localization, which lies at the heart of continuum mechanics, ensures that even in the absence of a global Riemannian shape, we can still define strain measures and stresses by working locally in the spatial platforms and using the connection to relate neighbouring points. This is exactly the philosophy we have already employed in Sects. 21.2 and 21.3: the material manifold $\mathfrak{M}$ (or its relativistic counterpart $\mathfrak{N}$) is endowed with a metric g and a connection Γ that may have torsion and curvature, reflecting the incompatibility of local deformations. In the relativistic context, the same manifold acquires an additional layer of structure coming from the rotation of the congruence, which can be absorbed into the connection.

21.6 Discussion and Conclusions

We have developed a unified differential-geometric framework that encompasses classical nonlinear elasticity, the theory of distributed defects, surface growth, and relativistic elasticity. The key idea is to view the deformation of a body as an embedding of a four-dimensional body–time manifold $\mathfrak{N}$ into the appropriate ambient space (Euclidean or Lorentzian). All physical information is encoded in geometric objects defined on $\mathfrak{N}$ or induced by the embedding (or submersion).

The action functional is written in a universal form (21.10) Variation of this action yields the Euler–Lagrange equations and Noether conservation laws expressed solely in terms of the divergence $\operatorname{div}_\omega$, which is defined by the volume form and does not require any metric or connection. This abstract formulation applies without change to all cases considered:

- **Classical compatible deformations** (Sect. 21.2) are recovered by choosing $\mathfrak{N} = \mathbb{R} \times \mathfrak{M}$ and taking the reference metric to be the pullback of the Euclidean metric via a fixed reference embedding $\varkappa_R$. The actual metric is the pullback via the time-dependent configuration $\varkappa_t$.
- **Incompatible deformations** (Sect. 21.3) arise when the reference metric $\mathfrak{g}$ is no longer flat. It is synthesised from a family of local stress-free configurations (the local unloading hypothesis). The non-Euclidean nature of $\mathfrak{g}$ is measured by its curvature and torsion, which correspond to distributions of disclinations and dislocations, respectively. The same variational principle, now with the Lagrangian depending on $\mathfrak{g}$ (or on the field of local frames $\tilde{\mathbf{E}}_\alpha$), produces the field equations for a body with residual stresses.
- **Surface growth** (Sect. 21.4) is modelled by a family of nested bodies $\mathcal{B}_\alpha$. The recursive construction of the reference shapes of newly added layers generates a

non-trivial material metric in the final assembly. This provides a concrete physical mechanism for the emergence of incompatibility.

- **Relativistic elasticity** (Sect. 21.5) is obtained by replacing the Euclidean ambient space with a curved Lorentzian manifold (e.g., Schwarzschild or Kerr) and interpreting the configuration as an embedding into spacetime. The body–time manifold $\mathfrak{N}$ now carries a natural product structure, but the ambient metric is no longer flat. The same variational principle, with the Lagrangian depending on the induced metric $h_{\alpha\beta} = g_{\alpha\beta} + u_\alpha u_\beta$ and the four-velocity u, yields the equations of relativistic elasticity. The fixed background approximation, justified by the negligible mass of the crust, decouples the gravitational field from the elastic deformation.

A profound analogy emerges between the description of material defects and the description of relativistic motion. In both cases, the relevant geometric structure is not a single metric but a pair consisting of a metric and a connection (or a field of frames). The connection may have torsion, curvature, or both, reflecting different physical sources of non-integrability:

- In a dislocated crystal, the material manifold cannot be globally embedded in Euclidean space with a flat metric; one needs a connection with torsion. The torsion tensor is directly related to the dislocation density.
- In a disclinated crystal, the material manifold requires a connection with curvature, corresponding to a continuous distribution of disclinations.
- For a rotating relativistic body, the spatial platforms orthogonal to the world lines are not integrable; the obstruction is measured by the rotation tensor $\omega_{\mu\nu}$. This non-integrability can be encoded in a connection with torsion, analogous to the dislocation case. The Fermi–Walker transport provides the appropriate frame, and the torsion appears when comparing vectors transported along different world lines.
- In a strongly gravitating object (e.g., a neutron star), the spatial metric itself is curved, reflecting the curvature of spacetime. This curvature is analogous to a disclination density.

Thus, the same mathematical language – Cartan geometry with torsion and curvature – describes both material defects and relativistic effects. The difference lies only in the physical origin of the non-integrability.

The ultimate goal of this work is to provide a theoretical foundation for understanding glitches in pulsars. The crust of a neutron star is continuously accreting matter from the surrounding medium, leading to the build-up of residual stresses. As the crust grows, the material metric $\mathfrak{g}$ evolves, and its curvature and torsion increase. When these geometric measures reach a critical value, the crust becomes unstable (Hadamard instability) and fractures locally. This releases a part of the stored elastic energy, causing a sudden change in the moment of inertia and hence a glitch in the pulsar's rotation period.

A detailed quantitative analysis of this process requires numerical simulations based on the evolutionary problem formulated in Sect. 21.4. The variational framework developed here provides the necessary weak formulation for such simulations.

In a first approximation, one can ignore the rotation of the star (since the maximum linear velocity of the crust is less than one per cent of the speed of light) and focus on the spherically symmetric accretion problem, using the Schwarzschild or Kerr metric as a fixed background. The weak form of the equations derived from the action principle is then ready for finite-element implementation. The effects of rotation can be included perturbatively via the Kerr metric and the Fermi–Walker transport, which would introduce additional terms in the connection and modify the balance of momentum.

The geometric framework presented here is far from exhausted. Several directions for future research suggest themselves:

- Inclusion of electromagnetic fields and magneto-elastic effects, which are crucial for understanding magnetar flares.
- A full dynamical treatment of the growth process, where the accretion rate is coupled to the stress state of the crust.
- Incorporation of more realistic equations of state for the neutron star matter, possibly derived from microscopic theories.
- Numerical implementation of the evolutionary problem and comparison with observational data on glitch amplitudes and recovery times.

We hope that the present work provides a solid foundation for these investigations and demonstrates the power of geometric methods in unifying seemingly disparate areas of physics.

Acknowledgements This research was supported by the Russian Science Foundation (grant No. 25-11-00333).

References

[1] Andersson L, Oliynyk TA, Schmidt B (2016) Dynamical compact elastic bodies in general relativity. Archive for Rational Mechanics and Analysis 220(2):849–887, DOI 10.1007/s00205-015-0944-0

[2] Ashton G, Lasky PD, Graber V, Palfreyman J (2019) Rotational evolution of the vela pulsar during the 2016 glitch. Nature Astronomy 3:1143–1148, DOI 10.1038/s41550-019-0844-6

[3] Bennoun JF (1965) Étude des milieux continus élastiques et thermodynamiques en relativité générale. Annales de l'Institut Henri Poincaré A. Physique Théorique 3(1):41-110

[4] Bilby BA, Gardner LRT, Stroh AN (1960) Continuous distributions of dislocations and the theory of plasticity. Proceedings of the 9th International Congress of Applied Mechanics 8:35–44

[5] Carter B (1973) Elastic perturbation theory in general relativity and a variation principle for a rotating solid star. Communications in Mathematical Physics 30(4):261–286, DOI 10.1007/BF01645505

[6] Carter B, Quintana H (1972) Foundations of general relativistic high-pressure elasticity theory. Proceedings of the Royal Society of London A 331:57–83

[7] Carter B, Chachoua E, Chamel N (2005) Covariant newtonian and relativistic dynamics of (magneto)-elastic solid model for neutron star crust. arXiv preprint astro-ph/0507006

[8] Chugunov AI, Horowitz CJ (2010) Breaking stress of neutron star crust. Monthly Notices of the Royal Astronomical Society: Letters 407(1):L54–L58, DOI 10.1111/j.1745-3933.2010.00903.x

[9] Clayton JD (2011) Differential Geometry and Kinematics of Continua. World Scientific, DOI 10.1142/9789814460017

[10] Epstein M (2010) The Geometrical Language of Continuum Mechanics. Cambridge University Press, DOI 10.1017/CBO9780511762679

[11] Epstein M, Burton D, Tucker R (2006) Relativistic anelasticity. Classical and Quantum Gravity 23(11):3545–3571, DOI 10.1088/0264-9381/23/11/001

[12] Espinoza CM, Lyne AG, Stappers BW, Kramer M (2011) A study of 315 glitches in the rotation of 102 pulsars. Monthly Notices of the Royal Astronomical Society 414(2):1679–1704, DOI 10.1111/j.1365-2966.2011.18503.x

[13] Facility ATN (2026) The ATNF pulsar catalogue. https://www.atnf.csiro.au/research/pulsar/psrcat/, version 2.7.0

[14] Ferrarese G, Bini D (2008) Introduction to Relativistic Continuum Mechanics, Lecture Notes in Physics, vol 727. Springer

[15] Flanagan CS (1990) Rapid recovery of the vela pulsar from a giant glitch. Nature 345:416–417, DOI 10.1038/345416a0

[16] Friedman M (1983) Foundations of Space-Time Theories: Relativistic Physics and Philosophy of Science. Princeton, NJ: Princeton University Press

[17] Georgievskii DV (2025) Tenzornyye funktsii v teorii opredelyayushchikh sootnosheniy (in Russ., Tensor Functions in the Theory of Constitutive Relations). Moscow: LENNAND

[18] Gourgoulhon É (2013) Special Relativity in General Frames: From Particles to Astrophysics. Springer, DOI 10.1007/978-3-642-37276-6

[19] Gundlach C, Hawke I, Erickson SJ (2012) A conservation law formulation of nonlinear elasticity in general relativity. Classical and Quantum Gravity 29:015005, DOI 10.1088/0264-9381/29/1/015005

[20] Heinicke C, Hehl FW (2015) Schwarzschild and Kerr solutions of Einstein's field equation: An introduction. International Journal of Modern Physics D 24(2):1530006, DOI 10.1142/S0218271815300062

[21] Jammer M (1954) Concepts of Space: The History of Theories of Space in Physics. Cambridge, MA: Harvard University Press

[22] Jammer M (1957) Concepts of Force: A Study in the Foundations of Dynamics. Cambridge, MA: Harvard University Press

[23] Jammer M (1974) The Philosophy of Quantum Mechanics: The Interpretations of Quantum Mechanics in Historical Perspective. New York: Wiley

[24] Karlovini M, Samuelsson L (2003) Elastic stars in general relativity: I. Foundations and equilibrium models. Classical and Quantum Gravity 20(16):3613–3648, DOI 10.1088/0264-9381/20/16/307

[25] Karlovini M, Samuelsson L (2004) Elastic stars in general relativity: II. Radial perturbations. Classical and Quantum Gravity 21(6):1559–1581, DOI 10.1088/0264-9381/21/6/018

[26] Karlovini M, Samuelsson L (2004) Elastic stars in general relativity: III. Stiff ultrarigid exact solutions. Classical and Quantum Gravity 21(19):4531–4548, DOI 10.1088/0264-9381/21/19/003

[27] Karlovini M, Samuelsson L (2007) Elastic stars in general relativity: IV. Axial perturbations. Classical and Quantum Gravity 24(13):3171–3189, DOI 10.1088/0264-9381/24/13/003

[28] Kondaurov VI (1991) On the rheological instability of an elastic damaging medium. Journal of Applied Mathematics and Mechanics 55(1):85–93, DOI 10.1016/0021-8928(91)90067-5

[29] Kondo K (1963) Non-Riemannian and Finslerian approaches to the theory of yielding. International Journal of Engineering Science 1(1):71–88, DOI 10.1016/0020-7225(63)90024-4

[30] Kröner E (1959) Allgemeine Kontinuumstheorie der Versetzungen und Eigenspannungen. Archive for Rational Mechanics and Analysis 4:273–334, DOI 10.1007/BF00281393

[31] Landau LD (1932) On the theory of stars. Physikalische Zeitschrift der Sowjetunion 1:285-288

[32] Lee JM (2012) Introduction to Smooth Manifolds (2nd ed), Graduate Texts in Mathematics, vol 218. New York, Springer, DOI 10.1007/978-1-4419-9982-5

[33] Lurie AI (1990) Nonlinear Theory of Elasticity, North-Holland Series in Applied Mathematics and Mechanics, vol 36. North-Holland, Amsterdam, translated from the Russian by KA Lurie

[34] Lychev S, Koifman K (2018) Geometry of Incompatible Deformations: Differential Geometry in Continuum Mechanics, De Gruyter Studies in Mathematical Physics, vol 46. De Gruyter, DOI 10.1515/9783110563214

[35] Lychev S, Koifman K (2019) Nonlinear evolutionary problem for a laminated inhomogeneous spherical shell. Acta Mechanica 230(11):3989–4020, DOI 10.1007/s00707-019-02480-7

[36] Lychev S, Koifman K, Djuzhev N (2021) Incompatible deformations in additively fabricated solids: Discrete and continuous approaches. Symmetry 13(12):2331, DOI 10.3390/sym13122331

[37] Lychev S, Koifman K, Bout D (2022) Finite incompatible deformations in elastic solids: Relativistic approach. Lobachevskii Journal of Mathematics 43(7):1670–1695, DOI 10.1134/S1995080222100250

[38] Lychev S, Koifman K, Pivovaroff N (2025) On the growing problem in conventional and relativistic elasticity. Lobachevskii Journal of Mathematics 46(6):2885–2904, DOI 10.1134/S1995080225608215

[39] Lychev SA (2007) Zakony sokhraneniya nedissipativnoy mikromorfnoy termouprugosti (in Russ., Conservation laws of nondissipative micromorphic thermoelasticity). Vestnik Samarskogo gosudarstvennogo universiteta Estestvennonauchnaya seriya 4(54):225–262

[40] Lychev SA, Koifman KG (2020) Material affine connections for growing solids. Lobachevskii Journal of Mathematics 41(8):2034–2052, DOI 10.1134/S1995080220080157
[41] Lychev SA, Koifman KG (2021) Contorsion of material connection in growing solids. Lobachevskii Journal of Mathematics 42(8):1852–1875, DOI 10.1134/S1995080221080187
[42] Lychev SA, Koifman KG, Pivovaroff NA (2023) Incompatible deformations in relativistic elasticity. Lobachevskii Journal of Mathematics 44(6):2352–2397, DOI 10.1134/S1995080223060343
[43] Lychev SA, Koifman KG, Pivovaroff NA (2024) Configurations and deformations in relativistic elasticity. Lobachevskii Journal of Mathematics 45(5):2342–2362, DOI 10.1134/S1995080224602613
[44] Marsden JE, Hughes TJR (1994) Mathematical Foundations of Elasticity. New York: Dover Publications, corrected reprint of the 1983 original
[45] Maugin GA (1993) Material Inhomogeneities in Elasticity, Applied Mathematics and Mathematical Computation, vol 3. London: Chapman & Hall
[46] Maugin GA (2013) Continuum Mechanics through the Twentieth Century: A Concise Historical Perspective, Solid Mechanics and Its Applications, vol 196. Dordrecht: Springer, DOI 10.1007/978-94-007-6353-1
[47] Middleditch J, Marshall FE, Wang QD, Gotthelf EV, Zhang W (2006) Predicting the Starquakes in PSR J0537–6910. The Astrophysical Journal 652(2):1531-1546, DOI 10.1086/508736
[48] Noll W (1967) Materially uniform simple bodies with inhomogeneities. Archive for Rational Mechanics and Analysis 27(1):1–32, DOI 10.1007/BF00276433
[49] Jodrell bank glitch catalogue. http://www.jb.man.ac.uk/pulsar/glitches.html.Accessed:2026
[50] Rakotomanana LR (2004) A Geometric Approach to Thermomechanics of Dissipating Continua. Modeling and Simulation in Science, Engineering and Technology, Birkhäuser, Boston
[51] Rayner CB (1963) Elasticity in general relativity. Proceedings of the Royal Society of London Series A 272(1348):44–53, DOI 10.1098/rspa.1963.0036
[52] Shapiro SL, Teukolsky SA (1983) Black Holes, White Dwarfs, and Neutron Stars: The Physics of Compact Objects. Wiley-VCH, Weinheim, DOI 10.1002/9783527617661, reprinted with corrections 2004, 2008
[53] Steinmann P (2015) Geometrical Foundations of Continuum Mechanics, Lecture Notes in Applied Mathematics and Mechanics, vol 2. Springer, DOI 10.1007/978-3-662-46460-3
[54] Synge JL (1959) A theory of elasticity in general relativity. Mathematische Zeitschrift 72:82–87, DOI 10.1007/BF01162934
[55] Teodosiu C (1970) A dynamic theory of dislocations and its applications to the theory of the elastic-plastic continuum. Fundamental Aspects of Dislocation Theory 317(II):837–876
[56] Truesdell C (1977) A First Course in Rational Continuum Mechanics, Pure and Applied Mathematics, vol 71. Academic Press, New York

[57] Wang CC (1967) On the geometric structures of simple bodies, a mathematical foundation for the theory of continuous distributions of dislocations. Archive for Rational Mechanics and Analysis 27(1):33–94, DOI 10.1007/BF00276434

[58] Wong T, Backer DC, Lyne AG (2000) Observations of a series of six recent glitches in the crab pulsar. The Astrophysical Journal 548(1):447, DOI 10.1086/318675

[59] Yavari A, Goriely A (2012) Riemann–Cartan geometry of nonlinear dislocation mechanics. Archive for Rational Mechanics and Analysis 205:59–118, DOI 10.1007/s00205-012-0500-0

Chapter 22
On Equilibrium Stability Criteria for Hemitropic Micropolar Elastic Media

Evgenii V. Murashkin and Yuri N. Radayev

Abstract In the present study the canonical representation of the energy form of hemitropic micropolar elastic medium is obtained in an attempt to establish mechanical moduli constitutive constraints. The strain energy is considered as depending on the asymmetric strain and wryness tensors treated as thermodynamic state variables. The first and the second variational principles proposed by J. W. Gibbs in 1875 at the aim of establishing equilibrium stability criteria are discussed. Both isentropic/isothermal processes are admittable for our study. However only isentropic ones are actually investigated. A-representation of the strain energy of hemitropic micropolar elastic medium characterized by 9 isentropic mechanical moduli is given. The strain energy quadratic form is diagonalized by the method of completing the squares. The canonical diagonal form enables us to explicitly formulate equilibrium stability criteria for a hemitropic micropolar elastic solid. Unilateral constraints imposed on the constitutive constants ensuring the positive definiteness of the quadratic form and nine constitutive inequalities providing the equilibrium stability of hemitropic micropolar elastic solid are obtained. For isotropic micropolar medium the five irreducible constitutive inequalities are found. In the both cases (hemitropic and isotropic media) none of constitutive constraints can be imposed on characteristic microlength L thus allowing us treat L as nano/micro/meso scale constitutive length.

Key words: Gibbs' variational principles, Strain energy, Internal energy, Entropy, Helmholtz free energy, Equilibrium stability, Hemitropic micropolar medium, Elasticity, Diagonalization, Canonical quadratic form, Positive definiteness, Constitutive inequalities, Auxetic

Evegnii V. Murashkin · Yuri N. Radayev
Ishlinsky Institute for Problems in Mechanics RAS, Prospekt Vernadskogo 101-1, 119526 Moscow, Russian Federation, e-mail: murashkin@ipmnet.ru, evmurashkin@gmail.com, radayev@ipmnet.ru, y.radayev@gmail.com

H. Altenbach et al. (eds.), *Stability Problems in Mechanics of Continua*,
Advanced Structured Materials 260, https://doi.org/10.1007/978-3-032-31032-3_22

22.1 Preliminaries and Introductory Remarks

Mathematical modelling mechanical behaviour of micropolar elastic solids begins with a derivation of the strain energy [1, 4–6, 9, 11, 17, 20–24]. Thermodynamical sense of the strain energy is similar to the internal energy since we are restricted to isentropic (adiabatic) processes. For isothermal processes the strain energy should be conventionally interpreted as the Helmholtz free energy. In the majority of available studies the internal energy is introduced as a thermodynamic state potential per invariant unit volume and is denoted by $\mathscr{U}$. In the course of our discussion only isentropic processes are considered. Thus, for the internal energy we have

$$\mathscr{U} = \overline{\mathscr{U}}(\overline{S}, \epsilon_{ij}, \kappa_{ij}), \tag{22.1}$$

wherein ϵ_{ij} is the asymmetric strain tensor, κ_{ij} is the wryness tensor, S is entropy per invariant unit volume, the overbar refers to the thermodynamic state potential contrary to the physical field quantity.

It is well known from thermomechanics that the Helmholtz free energy can be determined by the Legendre transformation of the internal energy according to the following equations

$$-\overline{\Psi}(\theta, \epsilon_{ij}, \kappa_{ij}) = \theta\overline{S} - \overline{\mathscr{U}}(\overline{S}, \epsilon_{ij}, \kappa_{ij}), \quad \theta = \frac{\partial\overline{\mathscr{U}}}{\partial\overline{S}},$$

where $\overline{\Psi}$ is the Helmholtz free energy state potential, θ is the absolute thermodynamic temperature. The fundamental Eq. (22.1) postulates the equality of the physical amount of the internal energy $\mathscr{U}$ per invariant unit volume (treating as an absolute scalar that invariant under the full group of isometric transformations) and the value of the thermodynamic state potential $\overline{\mathscr{U}}$.

In 1875, in [7] Gibbs proposed his theory of thermodynamic equilibrium based on his two variational principles. Actually, these principles establish criteria of equilibrium stability based on the concepts of thermodynamics.

The Gibbs' first principle states (see [7]): "for the local equilibrium it is necessary and sufficient that in all possible variations of the state of the system which do not alter its energy, the variation of its entropy shall either vanish or be negative: $\delta\overline{S} \leq 0$". The original formulation by Gibbs is unilateral with respect to entropy variations, however it is strengthened (see, for example, [2, 3]) as follows

$$\overline{S}(\overline{\mathscr{U}}, \epsilon_{ij}, \kappa_{ij}) \longrightarrow \max \qquad (\delta\overline{\mathscr{U}} = 0),$$

i.e. since $\delta\overline{\mathscr{U}} = 0$ then

$$\delta\overline{S} = 0, \qquad \delta^2\overline{S} < 0 \qquad (\delta^2\overline{S} = 0, \text{ iff } \delta\epsilon_{ij} = 0 \text{ and } \delta\kappa_{ij} = 0).$$

The Gibbs' second (dual) principle states (see [7]): "for the local equilibrium it is necessary and sufficient that in all possible variations of the stale of the system which do not alter its entropy, the variation of its energy shall either vanish or be positive: $\delta\overline{\mathcal{U}} \geq 0$". The original formulation of the dual principle can be similarly strengthened according to

$$\overline{\mathcal{U}}(\overline{S}, \epsilon_{ij}, \kappa_{ij}) \longrightarrow \min \qquad (\delta\overline{S} = 0),$$

i.e. since $\delta\overline{S} = 0$ then

$$\delta\overline{\mathcal{U}} = 0, \qquad \delta^2\overline{\mathcal{U}} > 0 \qquad (\delta^2\overline{\mathcal{U}} = 0, \text{ iff } \delta\epsilon_{ij} = 0 \text{ and } \delta\kappa_{ij} = 0). \tag{22.2}$$

In virtue of the bilateral constraint $\delta\overline{\mathcal{U}} = 0$, $\overline{\mathcal{U}}$ must not depend on linear terms with respect to state variables variations. This matter is demonstrated by estimating the internal energy increment:

$$\overline{\mathcal{U}} - \overline{\mathcal{U}}\Big|_{\text{equilibrium}} = \delta\overline{\mathcal{U}} + \frac{1}{2!}\delta^2\overline{\mathcal{U}} + \cdots,$$

where the first variation is obtained as

$$\delta\overline{\mathcal{U}}(\delta\epsilon_{ij}, \delta\kappa_{ij}) = \frac{\partial\overline{\mathcal{U}}}{\partial\epsilon_{ij}}\Bigg|_{\overline{S}=\text{fixed}} \delta\epsilon_{ij} + \frac{\partial\overline{\mathcal{U}}}{\partial\kappa_{ij}}\Bigg|_{\overline{S}=\text{fixed}} \delta\kappa_{ij},$$

and the second one is defined by

$$\begin{aligned}\delta^2\overline{\mathcal{U}}(\delta\epsilon_{ij}, \delta\kappa_{ij}) = \frac{\partial^2\overline{\mathcal{U}}}{\partial\epsilon_{ij}\partial\epsilon_{kl}}\Bigg|_{\overline{S}=\text{fixed}} \delta\epsilon_{ij}\delta\epsilon_{kl} + \frac{\partial^2\overline{\mathcal{U}}}{\partial\epsilon_{ij}\partial\kappa_{kl}}\Bigg|_{\overline{S}=\text{fixed}} \delta\epsilon_{ij}\delta\kappa_{kl} + \\ + \frac{\partial^2\overline{\mathcal{U}}}{\partial\kappa_{ij}\partial\kappa_{kl}}\Bigg|_{\overline{S}=\text{fixed}} \delta\kappa_{ij}\delta\kappa_{kl}.\end{aligned} \tag{22.3}$$

In view of the first formula of (22.2) we come to

$$\overline{\mathcal{U}} - \overline{\mathcal{U}}\Big|_{\text{equilibrium}} = \frac{1}{2}\delta^2\overline{\mathcal{U}} + \cdots. \tag{22.4}$$

Based on the dual principle the isentropic equilibrium stability criterion is formulated by the unilateral constraint with respect to the second variations of the internal energy $\delta^2\overline{\mathcal{U}}$:

$$\delta^2\overline{\mathcal{U}}(\delta\epsilon_{ij}, \delta\kappa_{ij}) > 0 \qquad (\delta^2\overline{\mathcal{U}} = 0, \text{ iff } \delta\epsilon_{ij} = 0 \text{ and } \delta\kappa_{ij} = 0). \tag{22.5}$$

The inequality (22.5) supplemented by the Eq. (22.3) can be treated as positive definiteness of the quadratic from with respect to variations $\delta\epsilon_{ij}$ and $\delta\kappa_{ij}$.

Gibbs considered his principles as the origin of equilibrium thermodynamics. These variational principles are usually presented in thermodynamics studies [2, 3] as some interesting extremal properties of equilibrium of thermodynamic states. Generalizations of the Gibbs' variational principles discussed in the monographs [2, 3] allow us to talk about criteria of the stable or unstable equilibrium.

The functional arguments of $\overline{\mathscr{U}}$ in the case of hemitropic micropolar solid are strain and wryness asymmetric second rank tensors, absolute temperature/entropy for isothermal/ isentropic processes. Instead of the asymmetric tensors ϵ_{kl} and κ_{kl} it is convenient to employ the new set of arguments comprising the symmetric parts and vectors associated to the antisymmetric parts that reads by

$$\epsilon_{(kl)}, \quad \kappa_{(kl)}, \quad \varphi_k, \quad \kappa_k. \tag{22.6}$$

For the linear theory of micropolar elasticity the strain energy is obtained as quadratic form of the state variables. In this case the inequality (22.5) enables us to discuss equilibrium stability criteria in a simple form of positiveness of the quadratic form (22.3).

The paper is arranged as follows. After the "Preliminaries and Introductory Remarks" in Sect. 22.2 an isometrically invariant A-representation of the quadratic strain energy form of hemitropic micropolar elastic medium determining by 9 isentropic (adiabatic) mechanical moduli is discussed. For isotropic micropolar medium the equilibrium stability criteria is obtained in form of unilateral constraints on micropolar moduli. In order to unify algebraic manipulations the formal state variables $Y_{\underline{1}},...,Y_{\underline{18}}$ are introduced.

Section 22.3 is devoted to algebraic operating with the A-representation of the strain energy. The diagonalization of quadratic form is carried out by the Lagrange method of completing the squares. The formal variables $Y_{\underline{1}},...,Y_{\underline{18}}$ are transformed to the canonical variables $X_{\underline{1}},...,X_{\underline{18}}$ in the course of applying the method.

In Sect. 22.4, constraints imposed on the constitutive mechanical moduli are obtained thus ensuring the positive definiteness of the quadratic form and the equilibrium stability of a micropolar solid. Then a simplification of the system of constitutive inequalities is discussed. After that the original 18 inequalities are replaced by ten equivalent irreducible ones. For isotropic micropolar medium the five irreducible constitutive inequalities are found. It is shown that none of constitutive constraints can be imposed on characteristic microlength L thus allowing us treat L as nano/micro/meso scale constitutive length.

The last section contains final discussions and conclusions to the paper. It is highlighted that a similar study can be carried out for isothermal processes by making replacement of the internal energy by the Helmholtz free energy.

This study extensively utilizes the terminology, notation, methods, and results developed in our previous papers [10, 12–19, 24–26].

22.2 Strain Energy of a Hemitropic Micropolar Solid

In this section considerations are restricted to three-dimensional space equipped by a Cartesian coordinate frame. In the rest of present discussion the overbars above symbols $\overline{\mathscr{U}}$ and $\overline{S}$ are omitted.

Micropolar solid is called as hemitropic if the constitutive tensors components are not changed under rotations of the coordinate frame, i.e. they are invariant under the group of semi-isometric transformations, namely, isometric transformations excepting mirror reflections and inversions of three-dimensional Euclidean space. In this case the correct script of the base equations is presented in pseudotensor framework, but the simple algorithm of pseudotensor transformation into corresponding absolute tensor helps us to put aside the pseudotensor formalism and then put into practice our study in terms of absolute tensors.

We introduce the strain energy $\mathscr{U}$ [12, 17, 19] calculated per unit invariant volume element at constant entropy (S = const) with the functional arguments (22.6) listed below

$$\mathscr{U}\Big|_{\text{adiabatic}} = \mathscr{U}\Big|_{S=\text{const}} = \mathscr{U}(\epsilon_{(ij)}, \kappa_{(ij)}, \varphi_i, \kappa_i), \tag{22.7}$$

wherein

$$\begin{aligned} &\epsilon_{ij} = \partial_i u_j - e_{ijk}\varphi_k, \quad 2\epsilon_{(ij)} = \partial_i u_j + \partial_j u_i, \quad \kappa_{ij} = \partial_i \phi_j, \\ &\kappa_{(ij)} = \partial_i \phi_j + \partial_j \phi_i, \quad \varphi_i = \phi_i - \frac{1}{2} e_{ikl}\partial_k u_l, \quad \kappa_i = \frac{1}{2} e_{ijs}\kappa_{[js]}. \end{aligned} \tag{22.8}$$

Hereafter instead of $\mathscr{U}\Big|_{\text{adiabatic}}$ the compact notation $\mathscr{U}$ is used.

In the Eqs. (22.7) and (22.8) we use the following notations: u_i are the translational displacements; ϕ_i are the spinor displacements; ∂_i denotes partial derivative with respect to the spatial coordinate x_i; e_{ijk} is the alternating tensor; φ_i is the relative microrotations; the brackets denote algebraic symmetrization; the square brackets denote algebraic alternation.

It should be noted, for the isothermal processes the analogues formula to (22.7) is furnished by

$$\mathscr{U}\Big|_{\text{isothermal}} = \Psi\Big|_{\theta=\text{const}} = \Psi(\epsilon_{(ij)}, \kappa_{(ij)}, \varphi_i, \kappa_i).$$

$\mathscr{U}$ is invariant under rotations and translations of space, as well as under spatial inversion transformations and mirror reflections, i.e. for the full isometric group. There are three different approaches to determining the strain energy, described in details in [17]. The first of them is the A-representation. A-representation [24] is based on the notions from the theory of rational algebraic invariants being a linear span of individual and joint integer rational algebraic invariants of the asymmetric strain tensor and the spatial gradient of the microrotation field. The foundations of the theory of integer rational algebraic invariants are thoroughly discussed in [8, 28].

Invariants belonging to the complete irreducible system of individual and joint integer rational algebraic quadratic invariants of symmetric components of the asymmetric strain tensors, the wryness tensor and the associated vectors for a hemitropic solid are calculated in Cartesian coordinates as follows [27]

$$\underset{1}{\mathrm{I}} = \delta_{ik}\delta_{lm}\epsilon_{(ik)}\epsilon_{(lm)},\ \underset{2}{\mathrm{I}} = \delta_{ik}\delta_{lm}\kappa_{(ik)}\kappa_{(lm)},\ \underset{3}{\mathrm{I}} = \delta_{ik}\delta_{lm}\epsilon_{(il)}\epsilon_{(km)},$$
$$\underset{4}{\mathrm{I}} = \delta_{ik}\delta_{lm}\kappa_{(il)}\kappa_{(km)},\ \underset{5}{\mathrm{I}} = \delta_{ik}\varphi_i\varphi_k,\ \underset{6}{\mathrm{I}} = \delta_{ik}\kappa_i\kappa_k, \tag{22.9}$$
$$\underset{7}{\mathrm{I}} = \delta_{ik}\delta_{lm}\epsilon_{(ik)}\kappa_{(lm)},\ \underset{8}{\mathrm{I}} = \epsilon_{(ik)}\kappa_{(ik)},\ \underset{9}{\mathrm{I}} = \kappa_k\varphi_k,$$

wherein δ_{lm} is the Kronecker delta.

The above reasonings lead us to A-representation of the strain energy of a hemitropic micropolar elastic solid as a linear span of the invariants (22.9):

$$\mathscr{U} = \sum_{\mathfrak{c}=1}^{9} \underset{\mathfrak{c}}{A}\,\underset{\mathfrak{c}}{\mathrm{I}}, \tag{22.10}$$

where $\underset{\mathfrak{c}}{A}$ $(\mathfrak{c} = 1, \ldots, 9)$ are arbitrary constants. Obviously these constants can be interpreted as the constitutive elastic moduli (absolute scalars) while specifying the model of a hemitropic micropolar elastic solid.

By virtue of the formulae (22.9) for the invariants $\underset{\mathfrak{c}}{\mathrm{I}}$ $(\mathfrak{c} = 1, \ldots, 9)$ the strain energy (22.7) is represented by (see [17, 19, 24]):

$$\begin{aligned}\mathscr{U} = \underset{1}{A}\delta_{is}\delta_{lm}\epsilon_{(is)}\epsilon_{(lm)} + \underset{2}{A}\delta_{is}\delta_{lm}\kappa_{(is)}\kappa_{(lm)} + \underset{3}{A}\delta_{is}\delta_{lm}\epsilon_{(il)}\epsilon_{(sm)} +\\ + \underset{4}{A}\delta_{is}\delta_{lm}\kappa_{(il)}\kappa_{(sm)} + \underset{5}{A}\delta_{is}\varphi_i\varphi_s + \underset{6}{A}\delta_{is}\kappa_i\kappa_s +\\ + \underset{7}{A}\delta_{is}\delta_{lm}\epsilon_{(is)}\kappa_{(lm)} + \underset{8}{A}\epsilon_{(is)}\kappa_{(is)} + \underset{9}{A}\kappa_i\varphi_i.\end{aligned} \tag{22.11}$$

For isotropic micropolar medium A-representation of the strain energy is compactly written as [10]

$$\mathscr{U} = \underset{1}{A}[\epsilon_{(ss)}]^2 + \underset{2}{A}[\kappa_{(ss)}]^2 + \underset{3}{A}[\epsilon_{(sl)}\epsilon_{(sl)}] + \underset{4}{A}[\kappa_{(sl)}\kappa_{(sl)}] + \underset{5}{A}[\varphi_s\varphi_s] + \underset{6}{A}[\kappa_s\kappa_s]. \tag{22.12}$$

Equations (22.11) and (22.12) are quite far from the canonical diagonal representations. However, the isotropic strain energy form (22.12) is the sum of squares. Therefore we are able to easily determine constraints imposed on the isotropic constitutive constants that are sufficient for ensuring the positive definiteness of the quadratic form (22.12) as follows

$$\underset{\mathfrak{a}}{A} > 0 \qquad (\mathfrak{a} = 1, \ldots, 6). \tag{22.13}$$

The mechanical moduli $\underset{\mathfrak{a}}{A}$ $(\mathfrak{a} = 1, \ldots, 9)$ in the Eq. (22.10) can be replaced by the alternative constitutive constants [24] somehow much more convenient for the present study:

$$\begin{aligned}&\underset{1}{A} = G\nu(1-2\nu)^{-1},\ \underset{2}{A} = GL^2c_3,\ \underset{3}{A} = G,\\ &\underset{4}{A} = GL^2,\ \underset{5}{A} = 2Gc_1,\ \underset{6}{A} = GL^2c_2,\\ &\underset{7}{A} = GLc_4,\ \underset{8}{A} = GLc_5,\ \underset{9}{A} = GLc_6,\end{aligned} \tag{22.14}$$

where G is the shear modulus of elasticity; ν is Poisson's ratio; L is the characteristic microlength of the micropolar theory; c_1, c_2, c_3, c_4, c_5, c_6 are dimensionless constants. In view of (22.14) the inequalities (22.13) are easily transformed into

$$G > 0, \qquad \frac{\nu}{1-2\nu} > 0, \qquad c_1 > 0, \qquad c_2 > 0, \qquad c_3 > 0, \tag{22.15}$$

that means that none of constitutive constraints is imposed on the characteristic microlength L.

For unifying in application of algebraic algorithms to diagonalization we introduce formal replacements of the independent state variables $\epsilon_{(ij)}, \kappa_{(ij)}, \varphi_i, \kappa_i$ of the strain energy $\mathscr{U}$ according to the following rules:

$$\begin{array}{llll}
\epsilon_{(11)} \longleftrightarrow Y_{\underline{1}}, & \varphi_1 \longleftrightarrow Y_{\underline{7}}, & \kappa_{(11)} \longleftrightarrow Y_{\underline{10}}, & \kappa_1 \longleftrightarrow Y_{\underline{16}}, \\
\epsilon_{(22)} \longleftrightarrow Y_{\underline{2}}, & \varphi_2 \longleftrightarrow Y_{\underline{8}}, & \kappa_{(22)} \longleftrightarrow Y_{\underline{11}}, & \kappa_2 \longleftrightarrow Y_{\underline{17}}, \\
\epsilon_{(33)} \longleftrightarrow Y_{\underline{3}}, & \varphi_3 \longleftrightarrow Y_{\underline{9}}; & \kappa_{(33)} \longleftrightarrow Y_{\underline{12}}, & \kappa_3 \longleftrightarrow Y_{\underline{18}}; \\
\epsilon_{(23)} \longleftrightarrow Y_{\underline{4}}, & & \kappa_{(23)} \longleftrightarrow Y_{\underline{13}}, & \\
\epsilon_{(13)} \longleftrightarrow Y_{\underline{5}}, & & \kappa_{(13)} \longleftrightarrow Y_{\underline{14}}, & \\
\epsilon_{(12)} \longleftrightarrow Y_{\underline{6}}; & & \kappa_{(12)} \longleftrightarrow Y_{\underline{15}}; &
\end{array} \tag{22.16}$$

where we underline a single 18-vector index in order to discriminate single two digit indices from two neighbouring one digit indices. Note that the formal variables of the theory of micropolar elasticity are not subject to physical interpretation at least at this stage. The vector of formal variables $[Y]$ belongs to a formal vector space of algebraic dimension 18.

Taking account of the replacements (22.16) the quadratic strain energy form (22.11) is written as:

$$\begin{aligned}
\mathscr{U} = {} & \underset{1}{A} Y_{\underline{1}}^2 + \underset{3}{A} Y_{\underline{1}}^2 + \underset{7}{A} Y_{\underline{1}} Y_{\underline{10}} + \underset{8}{A} Y_{\underline{1}} Y_{\underline{10}} + \underset{2}{A} Y_{\underline{10}}^2 + \underset{4}{A} Y_{\underline{10}}^2 + \underset{7}{A} Y_{\underline{1}} Y_{\underline{11}} + 2\underset{2}{A} Y_{\underline{10}} Y_{\underline{11}} + \\
& + \underset{2}{A} Y_{\underline{11}}^2 + \underset{4}{A} Y_{\underline{11}}^2 + \underset{7}{A} Y_{\underline{1}} Y_{\underline{12}} + 2\underset{2}{A} Y_{\underline{10}} Y_{\underline{12}} + 2\underset{2}{A} Y_{\underline{11}} Y_{\underline{12}} + \underset{2}{A} Y_{\underline{12}}^2 + \underset{4}{A} Y_{\underline{12}}^2 + \\
& + 2\underset{4}{A} Y_{\underline{13}}^2 + 2\underset{4}{A} Y_{\underline{14}}^2 + 2\underset{4}{A} Y_{\underline{15}}^2 + \underset{6}{A} Y_{\underline{16}}^2 + \underset{6}{A} Y_{\underline{17}}^2 + \underset{6}{A} Y_{\underline{18}}^2 + 2\underset{1}{A} Y_{\underline{1}} Y_{\underline{2}} + \underset{7}{A} Y_{\underline{10}} Y_{\underline{2}} + \\
& + \underset{7}{A} Y_{\underline{11}} Y_{\underline{2}} + \underset{8}{A} Y_{\underline{11}} Y_{\underline{2}} + \underset{7}{A} Y_{\underline{12}} Y_{\underline{2}} + \underset{1}{A} Y_{\underline{2}}^2 + \underset{3}{A} Y_{\underline{2}}^2 + 2\underset{1}{A} Y_{\underline{1}} Y_{\underline{3}} + \underset{7}{A} Y_{\underline{10}} Y_{\underline{3}} + \\
& + \underset{7}{A} Y_{\underline{11}} Y_{\underline{3}} + \underset{7}{A} Y_{\underline{12}} Y_{\underline{3}} + \underset{8}{A} Y_{\underline{12}} Y_{\underline{3}} + 2\underset{1}{A} Y_{\underline{2}} Y_{\underline{3}} + \underset{1}{A} Y_{\underline{3}}^2 + \underset{3}{A} Y_{\underline{3}}^2 + 2\underset{8}{A} Y_{\underline{13}} Y_{\underline{4}} + \\
& + 2\underset{3}{A} Y_{\underline{4}}^2 + 2\underset{8}{A} Y_{\underline{14}} Y_{\underline{5}} + 2\underset{3}{A} Y_{\underline{5}}^2 + 2\underset{8}{A} Y_{\underline{15}} Y_{\underline{6}} + 2\underset{3}{A} Y_{\underline{6}}^2 + \underset{9}{A} Y_{\underline{16}} Y_{\underline{7}} + \\
& + \underset{5}{A} Y_{\underline{7}}^2 + \underset{9}{A} Y_{\underline{17}} Y_{\underline{8}} + \underset{5}{A} Y_{\underline{8}}^2 + \underset{9}{A} Y_{\underline{18}} Y_{\underline{9}} + \underset{5}{A} Y_{\underline{9}}^2.
\end{aligned} \tag{22.17}$$

Note that the equation (22.17) can be represented in compact matrix notations as:

$$\mathscr{U} = [\mathrm{Y}]^{\mathrm{T}}[\mathrm{A}][\mathrm{Y}],$$

where

$$[\mathrm{Y}]^{\mathrm{T}} = [Y_{\underline{1}}, Y_{\underline{2}}, Y_{\underline{3}}, Y_{\underline{4}}, Y_{\underline{5}}, Y_{\underline{6}}, Y_{\underline{7}}, Y_{\underline{8}}, Y_{\underline{9}}, Y_{\underline{10}}, Y_{\underline{11}}, Y_{\underline{12}}, Y_{\underline{13}}, Y_{\underline{14}}, Y_{\underline{15}}, Y_{\underline{16}}, Y_{\underline{17}}, Y_{\underline{18}}];$$

$[\mathrm{Y}]^{\mathrm{T}}$ is the transposed row vector of algebraic dimension 18, composed from the formal variables of the theory of micropolar elasticity; $[\mathrm{A}]$ is a formal matrix of dimensions 18×18 of micropolar isentropic mechanical moduli. In the case of a hemitropic micropolar solid the formal matrix $[\mathrm{A}]$ has the following block structure (unfortunately not diagonal, but at least tridiagonal):

$$[\mathrm{A}] = \begin{bmatrix} [\mathrm{A}_{11}] & 0 & 0 & [\mathrm{A}_{14}] & 0 & 0 \\ 0 & [\mathrm{A}_{22}] & 0 & 0 & [\mathrm{A}_{25}] & 0 \\ 0 & 0 & [\mathrm{A}_{33}] & 0 & 0 & [\mathrm{A}_{36}] \\ [\mathrm{A}_{41}] & 0 & 0 & [\mathrm{A}_{44}] & 0 & 0 \\ 0 & [\mathrm{A}_{52}] & 0 & 0 & [\mathrm{A}_{55}] & 0 \\ 0 & 0 & [\mathrm{A}_{63}] & 0 & 0 & [\mathrm{A}_{66}] \end{bmatrix},$$

where $[\mathrm{A}_{hk}]$ $(h, k = 1, \ldots, 6)$ are symmetric block–matrices of algebraic dimensions 3×3, which are determined via the micropolar elastic mechanical moduli $\underset{\mathfrak{a}}{A}$ $(\mathfrak{a} = 1, \ldots, 9)$ according to the formulae:

$$[\mathrm{A}_{11}] = \begin{bmatrix} \underset{1}{A} + \underset{3}{A} & \underset{1}{A} & \underset{1}{A} \\ \underset{1}{A} & \underset{1}{A} + \underset{3}{A} & \underset{1}{A} \\ \underset{1}{A} & \underset{1}{A} & \underset{1}{A} + \underset{3}{A} \end{bmatrix}, \quad [\mathrm{A}_{14}] = [\mathrm{A}_{41}] = \frac{1}{2} \begin{bmatrix} \underset{7}{A} + \underset{8}{A} & \underset{7}{A} & \underset{7}{A} \\ \underset{7}{A} & \underset{7}{A} + \underset{8}{A} & \underset{7}{A} \\ \underset{7}{A} & \underset{7}{A} & \underset{7}{A} + \underset{8}{A} \end{bmatrix},$$

$$[\mathrm{A}_{22}] = 2 \begin{bmatrix} \underset{3}{A} & 0 & 0 \\ 0 & \underset{3}{A} & 0 \\ 0 & 0 & \underset{3}{A} \end{bmatrix}, \quad [\mathrm{A}_{25}] = [\mathrm{A}_{52}] = \begin{bmatrix} \underset{8}{A} & 0 & 0 \\ 0 & \underset{8}{A} & 0 \\ 0 & 0 & \underset{8}{A} \end{bmatrix}, \quad [\mathrm{A}_{33}] = 2 \begin{bmatrix} \underset{5}{A} & 0 & 0 \\ 0 & \underset{5}{A} & 0 \\ 0 & 0 & \underset{5}{A} \end{bmatrix},$$

$$[\mathrm{A}_{36}] = [\mathrm{A}_{63}] = \frac{1}{2} \begin{bmatrix} \underset{9}{A} & 0 & 0 \\ 0 & \underset{9}{A} & 0 \\ 0 & 0 & \underset{9}{A} \end{bmatrix}, \quad [\mathrm{A}_{44}] = \begin{bmatrix} \underset{2}{A} + \underset{4}{A} & \underset{2}{A} & \underset{2}{A} \\ \underset{2}{A} & \underset{2}{A} + \underset{4}{A} & \underset{2}{A} \\ \underset{2}{A} & \underset{2}{A} & \underset{2}{A} + \underset{4}{A} \end{bmatrix},$$

$$[\mathrm{A}_{55}] = 2 \begin{bmatrix} \underset{4}{A} & 0 & 0 \\ 0 & \underset{4}{A} & 0 \\ 0 & 0 & \underset{4}{A} \end{bmatrix}, \quad [\mathrm{A}_{66}] = \begin{bmatrix} \underset{6}{A} & 0 & 0 \\ 0 & \underset{6}{A} & 0 \\ 0 & 0 & \underset{6}{A} \end{bmatrix}.$$

22.3 The Diagonalization Method of Completing the Squares

In order to obtain the constitutive inequalities, we reduce the strain energy form (22.17) to a diagonal (or canonical) form by using the simplest and algorithmic method of linear algebra (the method of completing the squares proposed by Lagrange in 1759). In further discussion the underlines below the indices are omitted.

Note that all coefficients of the squares of the variables $Y_1,...,Y_{18}$ are assumed to be non-zero. Otherwise, this method is not applicable.

We transform the quadratic form (22.17) by separating a complete square for each formal variable, starting with the first. First, rearrange all terms including enter of the variable Y_1. Second separate the complete square containing this variable. Then take previously separated complete square as the new formal variable X_1^2. Next subtract the obtained complete square containing the variable Y_1 from the quadratic form (22.17). The new quadratic form $\mathcal{U}_{17}$ obtained in this way must not contain the variable Y_1. After that proceed similarly to transformations with respect to the formal variable Y_2. Then subtract the complete square containing Y_2 from the "shortened" (reduced) quadratic form obtained at the previous step. Next find a new quadratic form that does not contain the formal variables Y_1 and Y_2. Then continue the algorithm up to the formal variable Y_{18}. Eventually we come to the canonical quadratic form of diagonal type with respect to the new formal variables $X_1,...,X_{18}$ which are called as canonical variables. Let us proceed to the step by step implementation of completing the squares method.

Step 1. At the first step, as mentioned above, we separate the formal variable Y_1. It is easy to see that the variable X_1 can be obtained by differentiating the quadratic form (22.17) with respect to the variable Y_1 ($\mathcal{U} = \mathcal{U}_{18}$)[1], i.e.

$$X_1 = \frac{1}{2}\frac{\partial \mathcal{U}}{\partial Y_1} = \frac{1}{2}[2(\underset{1}{A} + \underset{3}{A})Y_1 + 2\underset{1}{A}Y_2 + 2\underset{1}{A}Y_3 + (\underset{7}{A} + \underset{8}{A})Y_{10} + \underset{7}{A}Y_{11} + \underset{7}{A}Y_{12}].$$

Then we introduce a new quadratic form $\mathcal{U}_{17}$ does not depend on the variable Y_1. It can be obtained in the following form:

$$\mathcal{U}_{17} = \mathcal{U}_{18} - \frac{1}{2}\left(\frac{\partial \mathcal{U}_{18}}{\partial Y_1}\right)^2\left(\frac{\partial^2 \mathcal{U}_{18}}{\partial Y_1^2}\right)^{-1}.$$

Hereafter we will not provide the exact representations of the quadratic forms $\mathcal{U}_1,...,\mathcal{U}_{17}$ due to their complexity and the irrelevance for aims of this study.

Step 2. At the second step we eliminate the formal variable Y_2 from the quadratic form $\mathcal{U}_{17}$:

$$X_2 = \frac{1}{2}\frac{\partial \mathcal{U}_{17}}{\partial Y_2} = \frac{1}{2(\underset{1}{A} + \underset{3}{A})}(\underset{3}{A}\underset{7}{A}Y_{10} - \underset{1}{A}\underset{8}{A}Y_{10} + \underset{3}{A}\underset{7}{A}Y_{11} + \underset{1}{A}\underset{8}{A}Y_{11} + \underset{3}{A}\underset{8}{A}Y_{11}$$
$$+ \underset{3}{A}\underset{7}{A}Y_{12} + 4\underset{1}{A}\underset{3}{A}Y_2 + 2\underset{3}{A}^2Y_2 + 2\underset{1}{A}\underset{3}{A}Y_3).$$

The quadratic form $\mathcal{U}_{16}$ does depend on the variables $Y_3,\ldots,Y_{18}$ and can be written as the equation:

$$\mathcal{U}_{16} = \mathcal{U}_{17} - \frac{1}{2}\left(\frac{\partial \mathcal{U}_{17}}{\partial Y_2}\right)^2\left(\frac{\partial^2 \mathcal{U}_{17}}{\partial Y_2^2}\right)^{-1}.$$

[1] The index of the energy form indicates the number of formal variables included in it.

We proceed to considerations of the next steps while omitting detailed descriptions. We are going to sequentially write down the formulae for the remaining formal variables and reduced quadratic forms.

Step 3.

$$X_3 = \frac{1}{2}\frac{\partial \mathscr{U}_{16}}{\partial Y_3} = \frac{1}{2(\underset{1}{A} + \underset{3}{A})}(\underset{3}{A}\underset{7}{A}Y_{10} - \underset{1}{A}\underset{8}{A}Y_{10} + \underset{3}{A}\underset{7}{A}Y_{11} - \underset{1}{A}\underset{8}{A}Y_{11} + \underset{3}{A}\underset{7}{A}Y_{12}$$
$$+ 2\underset{1}{A}\underset{8}{A}Y_{12} + \underset{3}{A}\underset{8}{A}Y_{12} + 6\underset{1}{A}\underset{3}{A}Y_3 + 2\underset{3}{A}^2 Y_3).$$

$$\mathscr{U}_{15} = \mathscr{U}_{16} - \frac{1}{2}\left(\frac{\partial \mathscr{U}_{16}}{\partial Y_3}\right)^2 \left(\frac{\partial^2 \mathscr{U}_{16}}{\partial Y_3^2}\right)^{-1}.$$

Step 4.

$$X_4 = \frac{1}{2}\frac{\partial \mathscr{U}_{15}}{\partial Y_4} = \underset{8}{A}Y_{13} + 2\underset{3}{A}Y_4.$$

$$\mathscr{U}_{14} = \mathscr{U}_{15} - \frac{1}{2}\left(\frac{\partial \mathscr{U}_{15}}{\partial Y_4}\right)^2 \left(\frac{\partial^2 \mathscr{U}_{15}}{\partial Y_4^2}\right)^{-1}.$$

Step 5.

$$X_5 = \frac{1}{2}\frac{\partial \mathscr{U}_{14}}{\partial Y_5} = \underset{8}{A}Y_{14} + 2\underset{3}{A}Y_5.$$

$$\mathscr{U}_{13} = \mathscr{U}_{14} - \frac{1}{2}\left(\frac{\partial \mathscr{U}_{14}}{\partial Y_5}\right)^2 \left(\frac{\partial^2 \mathscr{U}_{14}}{\partial Y_5^2}\right)^{-1}.$$

Step 6.

$$X_6 = \frac{1}{2}\frac{\partial \mathscr{U}_{13}}{\partial Y_6} = \underset{8}{A}Y_{15} + 2\underset{3}{A}Y_6.$$

$$\mathscr{U}_{12} = \mathscr{U}_{13} - \frac{1}{2}\left(\frac{\partial \mathscr{U}_{13}}{\partial Y_6}\right)^2 \left(\frac{\partial^2 \mathscr{U}_{13}}{\partial Y_6^2}\right)^{-1}.$$

Step 7.

$$X_7 = \frac{1}{2}\frac{\partial \mathscr{U}_{12}}{\partial Y_7} = \frac{1}{2}(\underset{9}{A}Y_{16} + 2\underset{5}{A}Y_7).$$

$$\mathscr{U}_{11} = \mathscr{U}_{12} - \frac{1}{2}\left(\frac{\partial \mathscr{U}_{12}}{\partial Y_7}\right)^2 \left(\frac{\partial^2 \mathscr{U}_{12}}{\partial Y_7^2}\right)^{-1}.$$

Step 8.

$$X_8 = \frac{1}{2}\frac{\partial \mathscr{U}_{11}}{\partial Y_8} = \frac{1}{2}(\underset{9}{A}Y_{17} + 2\underset{5}{A}Y_8).$$

$$\mathscr{U}_{10} = \mathscr{U}_{11} - \frac{1}{2}\left(\frac{\partial \mathscr{U}_{11}}{\partial Y_8}\right)^2 \left(\frac{\partial^2 \mathscr{U}_{11}}{\partial Y_8^2}\right)^{-1}.$$

Step 9.

$$X_9 = \frac{1}{2}\frac{\partial \mathcal{U}_{10}}{\partial Y_9} = \frac{1}{2}(\underset{9}{A}Y_{18} + 2\underset{5}{A}Y_9).$$

$$\mathcal{U}_9 = \mathcal{U}_{10} - \frac{1}{2}\Big(\frac{\partial \mathcal{U}_{10}}{\partial Y_9}\Big)^2\Big(\frac{\partial^2 \mathcal{U}_{10}}{\partial Y_9^2}\Big)^{-1}.$$

Step 10.

$$\begin{aligned} X_{10} = \frac{1}{2}\frac{\partial \mathcal{U}_9}{\partial Y_{10}} &= (24\underset{1}{A}\underset{2}{A}\underset{3}{A}\underset{5}{A} + 8\underset{2}{A}\underset{3}{A}^2\underset{5}{A} + 24\underset{1}{A}\underset{3}{A}\underset{4}{A}\underset{5}{A} + 8\underset{3}{A}^2\underset{4}{A}\underset{5}{A} - 6\underset{3}{A}\underset{5}{A}\underset{7}{A}^2 \\ &\quad - 4\underset{3}{A}\underset{5}{A}\underset{7}{A}\underset{8}{A} - 4\underset{1}{A}\underset{5}{A}\underset{8}{A}^2 - 2\underset{3}{A}\underset{5}{A}\underset{8}{A}^2)(8\underset{3}{A}(3\underset{1}{A} + \underset{3}{A})\underset{5}{A})^{-1}. \end{aligned}$$

$$\mathcal{U}_8 = \mathcal{U}_9 - \frac{1}{2}\Big(\frac{\partial \mathcal{U}_9}{\partial Y_{10}}\Big)^2\Big(\frac{\partial^2 \mathcal{U}_9}{\partial Y_{10}^2}\Big)^{-1}.$$

Step 11.

$$\begin{aligned} X_{11} = \frac{1}{2}\frac{\partial \mathcal{U}_8}{\partial Y_{11}} &= (4\underset{3}{A}\underset{4}{A} - \underset{8}{A}^2)(24\underset{1}{A}\underset{2}{A}\underset{3}{A}Y_{11} + 8\underset{2}{A}\underset{3}{A}^2Y_{11} + 12\underset{1}{A}\underset{3}{A}\underset{4}{A}Y_{11} + 4\underset{3}{A}^2\underset{4}{A}Y_{11} \\ &\quad - 6\underset{3}{A}\underset{7}{A}^2Y_{11} - 4\underset{3}{A}\underset{7}{A}\underset{8}{A}Y_{11} - \underset{1}{A}\underset{8}{A}^2Y_{11} - \underset{3}{A}\underset{8}{A}^2Y_{11} + 12\underset{1}{A}\underset{2}{A}\underset{3}{A}Y_{12} \\ &\quad + 4\underset{2}{A}\underset{3}{A}^2Y_{12} - 3\underset{3}{A}\underset{7}{A}^2Y_{12} - 2\underset{3}{A}\underset{7}{A}\underset{8}{A}Y_{12} + \underset{1}{A}\underset{8}{A}^2Y_{12})(4\underset{3}{A}(12\underset{1}{A}\underset{2}{A}\underset{3}{A} \\ &\quad + 4\underset{2}{A}\underset{3}{A}^2 + 12\underset{1}{A}\underset{3}{A}\underset{4}{A} + 4\underset{3}{A}^2\underset{4}{A} - 3\underset{3}{A}\underset{7}{A}^2 - 2\underset{3}{A}\underset{7}{A}\underset{8}{A} - 2\underset{1}{A}\underset{8}{A}^2 - \underset{3}{A}\underset{8}{A}^2))^{-1}. \end{aligned}$$

$$\mathcal{U}_7 = \mathcal{U}_8 - \frac{1}{2}\Big(\frac{\partial \mathcal{U}_8}{\partial Y_{11}}\Big)^2\Big(\frac{\partial^2 \mathcal{U}_8}{\partial Y_{11}^2}\Big)^{-1}.$$

Step 12.

$$\begin{aligned} X_{12} = \frac{1}{2}\frac{\partial \mathcal{U}_7}{\partial Y_{12}} &= (4\underset{3}{A}\underset{4}{A} - \underset{8}{A}^2)(36\underset{1}{A}\underset{2}{A} + 12\underset{2}{A}\underset{3}{A} + 12\underset{1}{A}\underset{4}{A} + 4\underset{3}{A}\underset{4}{A} - 9\underset{7}{A}^2 - 6\underset{7}{A}\underset{8}{A} - \underset{8}{A}^2) \\ &\times (4(24\underset{1}{A}\underset{2}{A}\underset{3}{A} + 8\underset{2}{A}\underset{3}{A}^2 + 12\underset{1}{A}\underset{3}{A}\underset{4}{A} + 4\underset{3}{A}^2\underset{4}{A} - 6\underset{3}{A}\underset{7}{A}^2 - 4\underset{3}{A}\underset{7}{A}\underset{8}{A} - \underset{1}{A}\underset{8}{A}^2 - \underset{3}{A}\underset{8}{A}^2))^{-1}Y_{12}. \end{aligned}$$

$$\mathcal{U}_6 = \mathcal{U}_7 - \frac{1}{2}\Big(\frac{\partial \mathcal{U}_7}{\partial Y_{12}}\Big)^2\Big(\frac{\partial^2 \mathcal{U}_7}{\partial Y_{12}^2}\Big)^{-1}.$$

Step 13.

$$X_{13} = \frac{1}{2}\frac{\partial \mathcal{U}_6}{\partial Y_{13}} = \frac{4\underset{3}{A}\underset{4}{A} - \underset{8}{A}^2}{2\underset{3}{A}}Y_{13}.$$

$$\mathcal{U}_5 = \mathcal{U}_6 - \frac{1}{2}\Big(\frac{\partial \mathcal{U}_6}{\partial Y_{13}}\Big)^2\Big(\frac{\partial^2 \mathcal{U}_6}{\partial Y_{13}^2}\Big)^{-1}.$$

Step 14.

$$X_{14} = \frac{1}{2}\frac{\partial \mathscr{U}_5}{\partial Y_{14}} = \frac{4\underset{3}{A}\underset{4}{A} - \underset{8}{A}^2}{2\underset{3}{A}} Y_{14}.$$

$$\mathscr{U}_4 = \mathscr{U}_5 - \frac{1}{2}\left(\frac{\partial \mathscr{U}_5}{\partial Y_{14}}\right)^2 \left(\frac{\partial^2 \mathscr{U}_5}{\partial Y_{14}^2}\right)^{-1}.$$

Step 15.

$$X_{15} = \frac{1}{2}\frac{\partial \mathscr{U}_4}{\partial Y_{15}} = \frac{4\underset{3}{A}\underset{4}{A} - \underset{8}{A}^2}{2\underset{3}{A}} Y_{15}.$$

$$\mathscr{U}_3 = \mathscr{U}_4 - \frac{1}{2}\left(\frac{\partial \mathscr{U}_4}{\partial Y_{15}}\right)^2 \left(\frac{\partial^2 \mathscr{U}_4}{\partial Y_{15}^2}\right)^{-1}.$$

Step 16.

$$X_{16} = \frac{1}{2}\frac{\partial \mathscr{U}_3}{\partial Y_{16}} = \frac{4\underset{5}{A}\underset{6}{A} - \underset{9}{A}^2}{4\underset{5}{A}} Y_{16}.$$

$$\mathscr{U}_2 = \mathscr{U}_3 - \frac{1}{2}\left(\frac{\partial \mathscr{U}_3}{\partial Y_{16}}\right)^2 \left(\frac{\partial^2 \mathscr{U}_3}{\partial Y_{16}^2}\right)^{-1}.$$

Step 17.

$$X_{17} = \frac{1}{2}\frac{\partial \mathscr{U}_2}{\partial Y_{17}} = \frac{4\underset{5}{A}\underset{6}{A} - \underset{9}{A}^2}{4\underset{5}{A}} Y_{17}.$$

$$\mathscr{U}_1 = \mathscr{U}_2 - \frac{1}{2}\left(\frac{\partial \mathscr{U}_2}{\partial Y_{17}}\right)^2 \left(\frac{\partial^2 \mathscr{U}_2}{\partial Y_{17}^2}\right)^{-1}.$$

Step 18.

$$X_{18} = \frac{1}{2}\frac{\partial \mathscr{U}_1}{\partial Y_{18}} = \frac{4\underset{5}{A}\underset{6}{A} - \underset{9}{A}^2}{4\underset{5}{A}} Y_{18}.$$

$$\mathscr{U}_0 = \mathscr{U}_1 - \frac{1}{2}\left(\frac{\partial \mathscr{U}_1}{\partial Y_{18}}\right)^2 \left(\frac{\partial^2 \mathscr{U}_1}{\partial Y_{18}^2}\right)^{-1} = 0.$$

As a result the original quadratic form transformed to the canonical variables $X_1,\ldots,X_{18}$ takes the canonical diagonal representation:

$$\mathscr{U}_{\text{diag}} = \sum_{s=1}^{18} \frac{X_s^2}{a_s}, \tag{22.18}$$

where the multipliers $1/a_s$ are defined according to

$$a_s = \frac{\partial X_s}{\partial Y_s} \quad \text{(no summation over the index } s\text{)}.$$

22.4 Constitutive Inequalities Ensuring the Equilibrium Stability for Hemitropic Micropolar Medium

The positive definiteness of (22.18) is ensured by the following inequalities

$$a_s > 0 \quad (s = 1, \ldots, 18). \tag{22.19}$$

The obtained system of inequalities (22.19) formulated in terms of the constitutive A-moduli reads by:

$$
\begin{gathered}
\underset{1}{A} + \underset{3}{A} > 0, \qquad \frac{\underset{3}{A}(2\underset{1}{A} + \underset{3}{A})}{\underset{1}{A} + \underset{3}{A}} > 0, \qquad \frac{\underset{3}{A}(3\underset{1}{A} + \underset{3}{A})}{2\underset{1}{A} + \underset{3}{A}} > 0, \\
\underset{3}{A} > 0, \qquad \underset{3}{A} > 0, \qquad \underset{3}{A} > 0, \\
\underset{5}{A} > 0, \qquad \underset{5}{A} > 0, \qquad \underset{5}{A} > 0, \\
\frac{4\underset{3}{A}(3\underset{1}{A} + \underset{3}{A})(\underset{2}{A} + \underset{4}{A}) - \underset{3}{A}\underset{7}{A}(3\underset{7}{A} + 2\underset{8}{A}) - (2\underset{1}{A} + \underset{3}{A})\underset{8}{A}^2}{4\underset{3}{A}(3\underset{1}{A} + \underset{3}{A})} > 0, \\
\frac{(4\underset{3}{A}\underset{4}{A} - \underset{8}{A}^2)(4\underset{3}{A}(3\underset{1}{A} + \underset{3}{A})(2\underset{2}{A} + \underset{4}{A}) - 2\underset{3}{A}\underset{7}{A}(3\underset{7}{A} + 2\underset{8}{A}) - (\underset{1}{A} + \underset{3}{A})\underset{8}{A}^2)}{4\underset{3}{A}(4\underset{3}{A}(3\underset{1}{A} + \underset{3}{A})(\underset{2}{A} + \underset{4}{A}) - \underset{3}{A}\underset{7}{A}(3\underset{7}{A} + 2\underset{8}{A}) - (2\underset{1}{A} + \underset{3}{A})\underset{8}{A}^2)} > 0, \\
\frac{(4\underset{3}{A}\underset{4}{A} - \underset{8}{A}^2)(4(3\underset{1}{A} + \underset{3}{A})(3\underset{2}{A} + \underset{4}{A}) - (3\underset{7}{A} + \underset{8}{A})^2)}{4(4\underset{3}{A}(3\underset{1}{A} + \underset{3}{A})(2\underset{2}{A} + \underset{4}{A}) - 2\underset{3}{A}\underset{7}{A}(3\underset{7}{A} + 2\underset{8}{A}) - (\underset{1}{A} + \underset{3}{A})\underset{8}{A}^2)} > 0, \\
\frac{4\underset{3}{A}\underset{4}{A} - \underset{8}{A}^2}{2\underset{3}{A}} > 0, \qquad \frac{4\underset{3}{A}\underset{4}{A} - \underset{8}{A}^2}{2\underset{3}{A}} > 0, \qquad \frac{4\underset{3}{A}\underset{4}{A} - \underset{8}{A}^2}{2\underset{3}{A}} > 0, \\
\frac{4\underset{5}{A}\underset{6}{A} - \underset{9}{A}^2}{4\underset{5}{A}} > 0, \qquad \frac{4\underset{5}{A}\underset{6}{A} - \underset{9}{A}^2}{4\underset{5}{A}} > 0, \qquad \frac{4\underset{5}{A}\underset{6}{A} - \underset{9}{A}^2}{4\underset{5}{A}} > 0.
\end{gathered}
\tag{22.20}
$$

The system (22.20) is a raw variant of the constitutive inequalities. After the elimination of repeated and obviously redundant inequalities from the system (22.20) the reduced system takes the following form:

$$
\begin{aligned}
&\underset{1}{A}+\underset{3}{A}>0, \qquad 2\underset{1}{A}+\underset{3}{A}>0, \qquad 3\underset{1}{A}+\underset{3}{A}>0,\\
&\underset{3}{A}>0, \qquad \underset{5}{A}>0,\\
&4\underset{3}{A}(3\underset{1}{A}+\underset{3}{A})(\underset{2}{A}+\underset{4}{A})-\underset{3}{A}\underset{7}{A}(3\underset{7}{A}+2\underset{8}{A})-(2\underset{1}{A}+\underset{3}{A})\underset{8}{A}^2>0,\\
&4\underset{3}{A}(3\underset{1}{A}+\underset{3}{A})(2\underset{2}{A}+\underset{4}{A})-2\underset{3}{A}\underset{7}{A}(3\underset{7}{A}+2\underset{8}{A})-(\underset{1}{A}+\underset{3}{A})\underset{8}{A}^2>0,\\
&4(3\underset{1}{A}+\underset{3}{A})(3\underset{2}{A}+\underset{4}{A})-(3\underset{7}{A}+\underset{8}{A})^2>0,\\
&4\underset{3}{A}\underset{4}{A}-\underset{8}{A}^2>0, \qquad 4\underset{5}{A}\underset{6}{A}-\underset{9}{A}^2>0.
\end{aligned}
$$

Then the latter system of inequalities is rearranged by employing the constitutive parameters (22.14):

$$
\begin{aligned}
&G\frac{\nu}{(1-2\nu)}+G>0, \qquad 3G\frac{\nu}{(1-2\nu)}+G>0,\\
&G>0, \qquad 2Gc_1>0,\\
&G^3L^2\Big(4(1+c_3)\frac{1+\nu}{1-2\nu}-c_4(3c_4+2c_5)-c_5^2\frac{1}{1-2\nu}\Big)>0,\\
&G^3L^2\Big(4(1+2c_3)\frac{1+\nu}{1-2\nu}-2c_4(3c_4+2c_5)-c_5^2\frac{1-\nu}{1-2\nu}\Big)>0,\\
&G^2L^2\Big(4(1+3c_3)\frac{1+\nu}{1-2\nu}-(3c_4+c_5)^2\Big)>0,\\
&4G^2L^2-G^2L^2c_5^2>0,\\
&8G^2L^2c_1c_2-G^2L^2c_6^2>0,
\end{aligned}
$$

which in virtue of $G>0$ and $L^2>0$ is simplified as follows:

$$
\begin{aligned}
&\frac{1-\nu}{1-2\nu}>0, \qquad \frac{1+\nu}{1-2\nu}>0,\\
&G>0, \qquad c_1>0,\\
&4(1+c_3)\frac{1+\nu}{1-2\nu}-c_4(3c_4+2c_5)-\frac{c_5^2}{1-2\nu}>0,\\
&4(1+2c_3)\frac{1+\nu}{1-2\nu}-2c_4(3c_4+2c_5)-c_5^2\frac{1-\nu}{1-2\nu}>0,\\
&4(1+3c_3)\frac{1+\nu}{1-2\nu}-(3c_4+c_5)^2>0,\\
&-2<c_5<2,\\
&8c_1c_2-c_6^2>0.
\end{aligned}
$$

Finally we come to the irreducible system of eight inequalities

$$
\begin{aligned}
&-1 < \nu < \frac{1}{2}, \qquad G > 0, \qquad c_1 > 0,\\
&4(1+c_3)\frac{1+\nu}{1-2\nu} - c_4(3c_4+2c_5) - \frac{c_5^2}{1-2\nu} > 0,\\
&4(1+2c_3)\frac{1+\nu}{1-2\nu} - 2c_4(3c_4+2c_5) - c_5^2\frac{1-\nu}{1-2\nu} > 0,\\
&4(1+3c_3)\frac{1+\nu}{1-2\nu} - (3c_4+c_5)^2 > 0,\\
&-2 < c_5 < 2, \qquad c_6^2 < 8c_1c_2.
\end{aligned}
\tag{22.21}
$$

It should be noted that none of the constitutive constraints can be imposed on characteristic microlength L. That allows us treat L as nano/micro/meso scale constitutive length. Obviously the constitutive microlength L is of great importance for micromechanics of materials.

The inequalities (22.15) corresponding to an isotropic micropolar solid may be weakened by reducing the inequalities (22.21) due to $c_4 \to 0$, $c_5 \to 0$, $c_6 \to 0$. At last for isotropic micropolar elastic medium we have

$$
-1 < \nu < \frac{1}{2}, \qquad G > 0, \qquad c_1 > 0, \qquad c_2 > 0, \qquad -\frac{1}{3} < c_3\,. \tag{22.22}
$$

It should be highlighted that the first and the last inequalities in (22.22) indicate the existence of auxetics among micropolar solids. The constitutive constant c_3 plays an important role similar to Poisson's ratio ν known from the Hookean elasticity.

22.5 Final Discussions and Conclusions

In the present study the representation of the strain energy for hemitropic micropolar elastic solid as a linear span of the irreducible quadratic algebraic invariants has been discussed. The first and the second variational principles known from Gibbs' studies and significant for the equilibrium thermodynamics development have been discussed and applied to formulation of equilibrium stability criterion for a hemitropic micropolar elastic solid. Summarizing the results of the study we arrive at the following conclusions:

1. An isometrically invariant A-representation of the quadratic strain energy form of a hemitropic micropolar elastic solid determined by 9 isentropic (adiabatic) mechanical moduli has been obtained.
2. Operating with the A-representation requires the formal variables $Y_1,\dots,Y_{18}$ are to be introduced and then to be transformed to the canonical variables $X_1,\dots,X_{18}$.
3. By the aid of the Lagrange completing the squares method the strain energy has been diagonalized.

4. The constraints on the constitutive mechanical moduli have been found out thus ensuring the positive definiteness of the quadratic strain energy form and the equilibrium stability of a micropolar elastic solid.
5. The rearrangements of system of constitutive inequalities have been discussed. Those make it possible to replace the system of 18 constitutive inequalities by nine equivalent irreducible ones.
6. A similar study can be carried out for isothermal processes by making replacement of the internal energy $\mathscr{U}$ by the Helmholtz free energy Ψ and in turn the strain energy $\mathscr{U}\Big|_{\text{adiabatic}}$ by $\mathscr{U}\Big|_{\text{isothermal}}$.
7. None of constitutive constraints can be imposed on characteristic microlength L thus allowing us treat L as nano/micro/meso scale constitutive length.
8. For isotropic micropolar medium the five irreducible constitutive inequalities ensuring equilibrium stability have been determined. In this case the constitutive constant c_3 plays an important role similar to Poisson's ratio ν known from the Hookean elasticity.

Acknowledgements The work was carried out on the topic of the state assignment (state registration number 124013000674-0).

References

[1] Altenbach H, Maugin GA, Erofeev V (eds.) (2011) Mechanics of Generalized Continua, Advanced Structured Materials, Vol 7. Springer, Berlin, DOI 10.1007/978-3-642-19219-7
[2] Arutyunyan NK, Drozdov AD, Naumov VE (1987) (in Russ., Mechanics of Growing Viscoelastoplastic Bodies). Nauka, Moscow
[3] Berdichevskii VL (2009) Variational Principles of Continuum Mechanics, Vol 1 & 2. Springer, Berlin, Heidelberg
[4] Besdo D (1974) Ein Beitrag zur nichtlinearen Theorie des Cosserat-Kontinuums. Acta Mechanica 20(1):105-131, DOI 10.1007/BF01374965
[5] Cosserat E, Cosserat F (1909) Théorie des corps déformables. A. Hermann et fils, Paris
[6] Dyszlewicz J (2004) Micropolar Theory of Elasticity, Lecture Notes in Applied and Computational Mechanics, Vol 15. Springer Science & Business Media, Berlin, DOI 10.1007/978-3-540-45286-7
[7] Gibbs JW (1875) On the equilibrium of heterogeneous substances. Transactions of the Connecticut Academy of Arts and Sciences III, Part 1:108-248, 343-524
[8] Gurevich GB (1964) Foundations of the Theory of Algebraic Invariants. Gröningen, Noordhoff
[9] Kessel S (1964) Lineare Elastizitätstheorie des anisotropen Cosserat-Kontinuums (1964) Abhandlungen der Braunschweigischen Wissenschaftlichen Gesellschaft 16:1-22. Friedrich Vieweg & Sohn, Braunschweig

[10] Krylova EY, Murashkin EV, Radaev Y N (2024) The Nye cells and figures for athermic hemitropic, isotropic, and ultraisotropic micropolar elastic solids. Mechanics of Solids 59(3):1311-1320, DOI 10.1134/s0025654424603719
[11] Maugin GA (2017) Non-classical Continuum Mechanics - A Dictionary. Springer Verlag, Singapore, DOI 10.1007/978-981-10-2434-4
[12] Murashkin EV (2023) O svyazi mikropolyarnykh opredelyayushchikh parametrov termodinamicheskikh potentsialov sostoyaniya (in Russ., On the relationship of micropolar constitutive parameters of thermodynamic state potentials. Vestnik Chuvashskogo gosudarstvennogo pedagogicheskogo universiteta im. IY Yakovleva. Seriya: Mekhanika predel'nogo sostoyaniya 1(55):110-121 DOI 10.37972/chgpu.2023.55.1.012
[13] Murashkin EV, Radayev YN (2023) A negative weight pseudotensor formulation of coupled hemitropic thermoelasticity, Lobachevskii Journal of Mathematics 44(6):2440–2449, DOI 10.1134/s1995080223060392
[14] Murashkin EV, Radayev YN (2023) Coupled thermoelasticity of hemitropic media. Pseudotensor formulation. Mechanics of Solids 58(3):802-813, DOI 10.3103/s0025654423700127
[15] Murashkin EV, Radayev YN (2023) Heat transfer in anisotropic micropolar solids. Mechanics of Solids 58(9):3111–3119, DOI 10.3103/S0025654423700255
[16] Murashkin EV, Radayev YN (2020) K teorii lineynykh gemitropnykh mikropolyarnykh sred (in Russ., On the theory of linear micropolar hemitropic media). Vestnik Chuvashskogo gosudarstvennogo pedagogicheskogo universiteta im. IY Yakovleva. Seriya: Mekhanika predel'nogo sostoyaniya 4(46):16-24, DOI 10.37972/chgpu.2020.89.81.031
[17] Murashkin EV, Radayev YN (2022) O dvukh osnovnykh yestestvennykh formakh potentsiala asimmetrichnykh tenzorov silovykh i momentnykh napryazhenii v mekhanike gemitropnykh tel (in Russ., On two base natural forms of asymmetric force and couple stress tensors of potential in mechanics of hemitropic solids). Vestnik Chuvashskogo gosudarstvennogo pedagogicheskogo universiteta im. IY Yakovleva. Seriya: Mekhanika predel'nogo sostoyaniya 3(53), 86-100, DOI 10.37972/chgpu.2022.53.3.010
[18] Murashkin EV, Radayev YN (2024) On algebraic triple weights formulation of micropolar thermoelasticity. Mechanics of Solids 59(1):555-580, DOI 10.1134/s0025654424700274
[19] Murashkin EV, Radayev YN (2022) Privedenie yestestvennykh form gemitropnykh energeticheskikh potentsialov k konventsional'nym (in Russ., Reducing natural forms of hemitropic energy potentials to conventional ones). Vestnik Chuvashskogo gosudarstvennogo pedagogicheskogo universiteta im. IY Yakovleva. Seriya: Mekhanika predel'nogo sostoyaniya 4(54):108-115, DOI 10.37972/chgpu.2022.54.4.009
[20] Neuber H (1968) On the effect of stress concentration in Cosserat continua. In: Kröner E (ed) Mechanics of Generalized Continua, pp. 109-113. Springer, Berlin, DOI 10.1007/978-3-662-30257-6_13

[21] Neuber H (1966) On the general solution of linear-elastic problems in isotropic and anisotropic Cosserat continua. In: Görtler H (eds) Applied Mechanics, pp. 153-158. Springer, Berlin, DOI 10.1007/978-3-662-29364-5_16
[22] Neuber H (1966) Über Probleme der Spannungskonzentration im Cosserat-Körper. Acta Mechanica 2(1):48-69, DOI 10.1007/BF01176729
[23] Nowacki W (1986) Theory of Asymmetric Elasticity. Pergamon Press, Oxford
[24] Radayev YN (2018) Pravilo mnozhiteley v kovariantnykh formulirovkakh mikropolyarnykh teoriy mekhaniki kontinuuma (in Russ., The Lagrange multipliers method in covariant formulations of micropolar continuum mechanics theories). Vestnik Samarskogo gosudarstvennogo tekhnicheskogo universiteta. Seriya "Fiziko-matematicheskiye nauki" 22(3):504-517, DOI 710.14498/vsgtu1635
[25] Radayev YN, Murashkin EV (2020) Psevdotenzornaya formulirovka mekhaniki gemitropnykh mikropolyarnykh sred (in Russ., Pseudotensor formulation of the mechanics of hemitropic micropolar media). Problems of Strength and Plasticity 82(4):399-412, DOI 10.32326/1814-9146-2020-82-4-399-412
[26] Murashkin EV, Radayev YN (2024) Theory of Poisson's ratio for a thermoelastic micropolar acentric isotropic solid. Lobachevskii Journal of Mathematics 45(5):2378-2390, DOI 10.1134/s1995080224602480
[27] Murashkin EV, Radayev YN (2025) Cubic approximation of stress potential for a hemitropic micropolar elastic solid. Lobachevskii Journal of Mathematics 46(5):2391-2400, DOI 10.1134/S1995080225606514
[28] Spencer AJM (1971) Theory of Invariants. In: Eringen AC (ed) Continuum Physics. Vol 1. New York, Academic Press, pp 239-353, DOI 10.1016/B978-0-12-240801-4.50008-X

Chapter 23
FSDT-PINN: A Physics-Informed Framework for Nonlinear Bending of Laminated Composite Plates on Elastic Foundations

Duc Dinh Nguyen and Pham Dinh Nguyen

Abstract This study presents FSDT-PINN, a specialized physics-informed neural network (PINN) framework for the nonlinear bending analysis of laminated composite plates resting on Winkler–Pasternak elastic foundations. By integrating first-order shear deformation theory (FSDT) and von Kármán's geometric nonlinearity, the proposed model embeds governing physical laws directly into the neural network architecture. Rather than relying on the strong form of partial differential equations, this framework adopts a deep energy method approach, where the total potential energy of the laminated plates is minimized within the loss function. The use of this variational principle reduces the required order of derivatives, thereby enhancing training stability and computational efficiency compared to traditional PINNs. FSDT-PINN enables high-fidelity prediction of deflection fields without requiring labeled datasets from expensive finite element simulations. The model explicitly accounts for transverse shear deformation, ensuring reliable results for both thin and moderately thick laminated composite plates. Numerical validations demonstrate that FSDT-PINN achieves excellent agreement with the finite element method while providing a mesh-free, highly interpretable approach for the nonlinear analysis of laminated composite plates.

23.1 Introduction

Laminated composite materials are widely used in modern structural engineering due to their superior mechanical properties, such as high strength-to-weight ratios, durability, and design flexibility. These materials consist of multiple layers with different properties and are commonly applied in aerospace, automotive, renewable energy, and civil engineering structures. However, their anisotropic nature, layered config-

Duc Dinh Nguyen · Pham Dinh Nguyen
VNU University of Engineering and Technology, 144 D. Xuan Thy, 100000 Hanoi, Vietnam, e-mail: ducnd@vnu.edu.vn, nguyenpd@vnu.edu.vn

H. Altenbach et al. (eds.), *Stability Problems in Mechanics of Continua*,
Advanced Structured Materials 260, https://doi.org/10.1007/978-3-032-31032-3_23

uration, and complex interactions between layers make their analysis challenging [10, 16].

In structural engineering, the FEM is widely regarded as the benchmark for analyzing laminated composites due to its capability to accurately capture layerwise interactions. Nevertheless, the high level of accuracy often comes at the expense of substantial computational cost, particularly in problems involving geometric nonlinearity, multiscale effects, or large-scale design optimization. From an analytical approach, the classical laminated plate theory (CLPT) provides computational efficiency but neglects transverse shear deformation, resulting in reduced accuracy for moderately thick plates [4, 17]. While first-order shear deformation theory (FSDT) addresses these limitations by incorporating shear effects, it traditionally necessitates empirical shear correction factor [7, 14, 20]. To further enhance kinematic accuracy, substantial efforts have been directed toward developing third-order shear deformation theories (TSDT) [15] and Higher-order Shear Deformation Theories (HSDTs) [18, 19]. While these advanced theories provide superior precision by accurately modeling the parabolic distribution of transverse shear stresses, they suffer from extreme mathematical complexity in the derivation of their governing equations and the intricate formulations required for numerical implementation

Physics-informed machine learning establishes a transformative paradigm by bridging data-driven architectures with fundamental physical principles [2, 3]. Domain knowledge is typically integrated into these frameworks through three strategic approaches: generating physics-guided synthetic data, tailoring physics-aware network architectures, or enforcing governing laws as constraints within the loss function. Pioneering the latter methodology, Raissi et al. [13] introduced Physics-Informed Neural Networks (PINNs). By embedding the residuals of partial differential equations (PDEs) directly into the loss function via automatic differentiation, PINNs incorporate physical laws as an inductive bias. This integration ensures that models remain physically consistent and significantly alleviates the reliance on large, computationally expensive labeled datasets. Physics-informed neural networks have demonstrated significant potential in computational mechanics, offering a mesh-free alternative for solving complex PDEs. In solid mechanics, these frameworks are generally implemented in two formats: the strong form, which minimizes the residuals of the governing PDEs, and the weak (energy) form, also known as the deep energy method (DEM), which is formulated on the principle of minimum total potential energy [6, 11, 12]. While the strong form is versatile, it frequently encounters numerical instabilities when addressing high-order derivatives (e.g., fourth-order equations) characteristic of plate bending. In contrast, the DEM approach utilizes a variational principle that reduces the required order of spatial derivatives, thereby enhancing training stability and computational efficiency [8, 22].

Recent studies have demonstrated growing interest in applying PINNs to laminated composite plates. Huang and Peng [9] used the Deep Energy Method (DEM) to geometrically nonlinear bending analysis of composite plates within the framework of classical plate theory by incorporating von Kármán's geometric nonlinearity and Green-Lagrange strain measures, achieving accurate predictions of large-deflection behavior. Similarly, Wang and Thai [21] proposed a PINN framework for linear bending analysis of laminated composite plates based on classical plate theory using an energy-based loss formulation. Ermis et al. [5] developed a physics-based machine learning approach for laminated composite plates using refined zigzag theory, demonstrating its capability in predicting structural responses.

Despite these advancements, the application of physics-informed frameworks to the nonlinear bending of laminated composite plates on Winkler-Pasternak elastic foundations remains relatively unexplored. This study introduces FSDT-PINN, a specialized physics-informed framework designed to fill this gap. By integrating FSDT with von Kármán's geometric nonlinearity, the proposed model embeds the total potential energy of the laminated composite plate resting on nonlinear elastic foundations into the learning process. The framework provides a mesh-free, highly interpretable, and computationally superior alternative that complements traditional FEM solvers, delivering high-fidelity predictions of deflection fields for both thin and moderately thick composite structures without requiring expensive labeled datasets.

23.2 Mathematical Formulation

23.2.1 Governing Equations of the Laminated Composite Plate

Consider a rectangular laminated composite plate resting on elastic foundations of length a, width b, and total thickness h. The plate comprises n layers with a stacking sequence of $[\theta_1/\theta_2/\ldots/\theta_n]$ and respective thicknesses of $[t_1/t_2/\ldots/t_n]$. As depicted in Fig. 23.1, a Cartesian coordinate system is defined with its xy-plane located at the mid-plane of the plate.

The displacement components of laminated composite plates according to the FSDT are shown as follows:

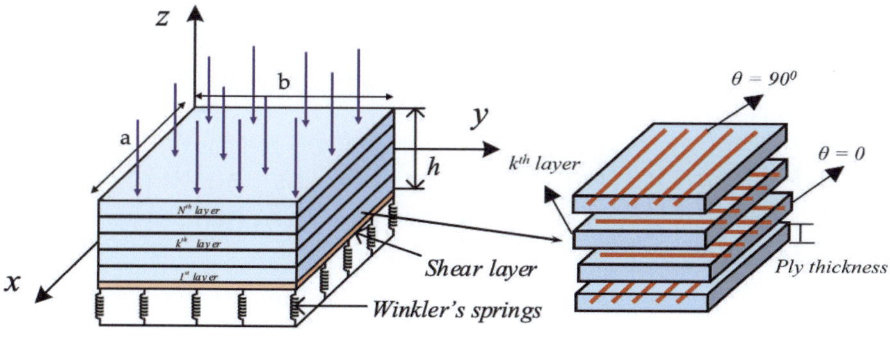

Fig. 23.1: Laminated composite plate resting on Winkler-Pasternak elastic foundation.

$$\begin{aligned} u(x,y,z,t) &= u_0(x,y,t) + z\phi_x \\ v(x,y,z,t) &= v_0(x,y,t) + z\phi_y \\ w(x,y,z,t) &= w_0(x,y,t) \end{aligned} \tag{23.1}$$

The von Kármán nonlinear strains with displacement fields for the FSDT are expressed as follows [16]:

$$\begin{pmatrix} \varepsilon_{xx} \\ \varepsilon_{yy} \\ \gamma_{xy} \end{pmatrix} = \begin{pmatrix} \dfrac{\partial u}{\partial x} + \dfrac{1}{2}\left(\dfrac{\partial w}{\partial x}\right)^2 \\ \dfrac{\partial v}{\partial y} + \dfrac{1}{2}\left(\dfrac{\partial w}{\partial y}\right)^2 \\ \dfrac{\partial u}{\partial y} + \dfrac{\partial v}{\partial x} + \dfrac{\partial w}{\partial x}\dfrac{\partial w}{\partial y} \end{pmatrix}, \begin{pmatrix} \gamma_{xz} \\ \gamma_{yz} \end{pmatrix} = \begin{pmatrix} \dfrac{\partial u}{\partial z} + \dfrac{\partial w}{\partial x} \\ \dfrac{\partial v}{\partial z} + \dfrac{\partial w}{\partial y} \end{pmatrix} \tag{23.2}$$

The nonlinear strain fields of the plates can be rewritten:

$$\varepsilon = \varepsilon^0 + zk^1, \gamma = \gamma^0 \tag{23.3}$$

where,

$$\begin{aligned} \varepsilon^0 &= \begin{pmatrix} \varepsilon_{xx}^0 \\ \varepsilon_{yy}^0 \\ \gamma_{xy}^0 \end{pmatrix} = \begin{pmatrix} \dfrac{\partial u_0}{\partial x} + \dfrac{1}{2}\left(\dfrac{\partial w_0}{\partial x}\right)^2 \\ \dfrac{\partial v_0}{\partial y} + \dfrac{1}{2}\left(\dfrac{\partial w_0}{\partial y}\right)^2 \\ \dfrac{\partial u_0}{\partial y} + \dfrac{\partial v_0}{\partial x} + \dfrac{\partial w_0}{\partial x}\dfrac{\partial w_0}{\partial y} \end{pmatrix}, \gamma^0 = \begin{pmatrix} \gamma_{xz}^0 \\ \gamma_{yz}^0 \end{pmatrix} = \begin{pmatrix} \phi_x + \dfrac{\partial w_0}{\partial x} \\ \phi_y + \dfrac{\partial w_0}{\partial y} \end{pmatrix} \\ k^1 &= \begin{pmatrix} k_x^1 \\ k_y^1 \\ k_{xy}^1 \end{pmatrix} = \begin{pmatrix} \dfrac{\partial \phi_x}{\partial x} \\ \dfrac{\partial \phi_y}{\partial y} \\ \dfrac{\partial \phi_x}{\partial y} + \dfrac{\partial \phi_y}{\partial x} \end{pmatrix} \end{aligned} \tag{23.4}$$

The stress-strain relationship for the k-th layer of the laminated composite plate is given by [16]:

$$\begin{Bmatrix} \sigma_{xx} \\ \sigma_{yy} \\ \sigma_{xy} \end{Bmatrix}^{(k)} = \begin{bmatrix} \bar{Q}_{11} & \bar{Q}_{12} & \bar{Q}_{16} \\ \bar{Q}_{12} & \bar{Q}_{22} & \bar{Q}_{26} \\ \bar{Q}_{16} & \bar{Q}_{26} & \bar{Q}_{66} \end{bmatrix}^{(k)} \begin{Bmatrix} \varepsilon_{xx} \\ \varepsilon_{yy} \\ \gamma_{xy} \end{Bmatrix}; \begin{Bmatrix} \sigma_{yz} \\ \sigma_{xz} \end{Bmatrix}^{(k)} = \begin{bmatrix} \bar{Q}_{44} & \bar{Q}_{45} \\ \bar{Q}_{45} & \bar{Q}_{55} \end{bmatrix}^{(k)} \begin{Bmatrix} \gamma_{yz} \\ \gamma_{xz} \end{Bmatrix} \tag{23.5}$$

The stiffness coefficients of the k-th layer $\bar{Q}_{ij}^{(k)}$ are defined by:

$$\begin{aligned}
\bar{Q}_{11} &= Q_{11}\cos^4\theta + 2(Q_{12} + 2Q_{66})\sin^2\theta\cos^2\theta + Q_{22}\sin^4\theta \\
\bar{Q}_{12} &= (Q_{11} + Q_{22} - 4Q_{66})\sin^2\theta\cos^2\theta + Q_{12}(\sin^4\theta + \cos^4\theta) \\
\bar{Q}_{22} &= Q_{11}\sin^4\theta + 2(Q_{12} + 2Q_{66})\sin^2\theta\cos^2\theta + Q_{22}\cos^4\theta \\
\bar{Q}_{16} &= (Q_{11} - Q_{12} - 2Q_{66})\sin\theta\cos^3\theta + (Q_{12} - Q_{22} + 2Q_{66})\sin^3\theta\cos\theta \\
\bar{Q}_{26} &= (Q_{11} - Q_{12} - 2Q_{66})\sin^3\theta\cos\theta + (Q_{12} - Q_{22} + 2Q_{66})\sin\theta\cos^3\theta \\
\bar{Q}_{66} &= (Q_{11} + Q_{22} - 2Q_{12} - 2Q_{66})\sin^2\theta\cos^2\theta + Q_{66}(\sin^4\theta + \cos^4\theta) \\
\bar{Q}_{44} &= Q_{44}\cos^2\theta + Q_{55}\sin^2\theta \\
\bar{Q}_{45} &= (Q_{55} - Q_{44})\cos\theta\sin\theta \\
\bar{Q}_{55} &= Q_{44}\sin^2\theta + Q_{55}\cos^2\theta
\end{aligned} \tag{23.6}$$

where,

$$Q_{11} = \frac{E_{11}}{1 - \upsilon_{12}\upsilon_{21}}, \quad Q_{12} = \frac{\upsilon_{12}E_{22}}{1 - \upsilon_{12}\upsilon_{21}},$$

$$Q_{22} = \frac{E_{22}}{1 - \upsilon_{12}\upsilon_{21}}, \quad Q_{44} = G_{23}, \quad Q_{55} = G_{13}, \quad Q_{66} = G_{12}$$

The governing equations of the first-order plate theory can be derived using the principle of virtual displacements [16]:

$$\delta U - \delta V = 0 \tag{23.7}$$

The virtual strain energy of the laminated plates:

$$\begin{aligned}
\delta U = \int_{\Omega}\Bigg\{\int_{-h/2}^{h/2}\Big[&\sigma_{xx}\left(\delta\varepsilon_{xx}^{(0)} + z\delta k_{xx}^{(1)}\right) + \sigma_{yy}\left(\delta\varepsilon_{yy}^{(0)} + z\delta k_{yy}^{(1)}\right) \\
&+\sigma_{xy}\left(\delta\gamma_{xy}^{(0)} + z\delta k_{xy}^{(1)}\right) + \sigma_{xz}\delta\gamma_{xz}^{(0)} + \sigma_{yz}\delta\gamma_{yz}^{(0)}\Big]\,dz\Bigg\}dxdy
\end{aligned} \tag{23.8}$$

The virtual work done by applied forces:

$$\delta V = \int_{\Omega}\left\{q + k_w w_0 + k_p\left(\frac{\partial^2 w_0}{\partial x^2} + \frac{\partial^2 w_0}{\partial y^2}\right)\right\}\delta w_0\,d\Omega \tag{23.9}$$

where q is the external distributed load, and k_w and k_p are the stiffness coefficients of the Winkler-Pasternak foundation model.

In this study, the laminated composite plates are subjected to clamped (C) and simply supported (S) boundary conditions. The four edges of the plate's mid-plane are defined (see Fig. 23.2) as: Edge 1 ($x = 0$), Edge 2 ($y = 0$), Edge 3 ($x = a$), and Edge 4 ($y = b$).

For a clamped boundary condition, representing the fully restrained case, all degrees of freedom (DOFs) along the specified edges vanish:

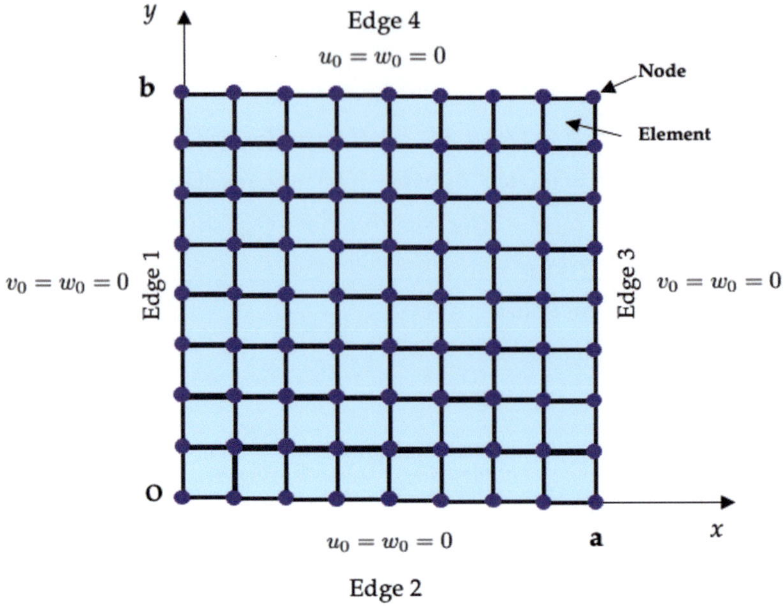

(**a**) SSSS.

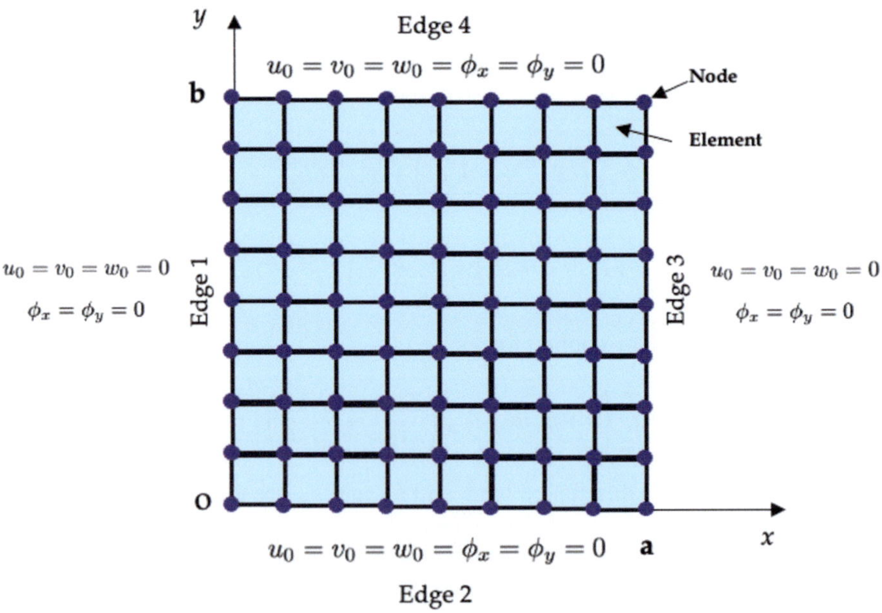

(**b**) CCCC.

Fig. 23.2: Boundary conditions applied to plates: **(a)** all boundaries simply supported (SSSS), **(b)** all boundaries clamped (CCCC).

$$u_0 = v_0 = w_0 = 0, \phi_x = \phi_y = 0 \quad \text{on all edges} \tag{23.10}$$

For the fully simply supported condition, the displacement constraints are given by:

$$\begin{aligned} &\text{Edges 1 and 3 (at } x = 0, a\text{): } v_0 = w_0 = 0, \\ &\text{Edges 2 and 4 (at } y = 0, b\text{): } u_0 = w_0 = 0 \end{aligned} \tag{23.11}$$

23.2.2 A Deep Energy Method

A multi-layer fully connected neural network (FNN) is utilized to approximate the displacement field of the laminated composite plate (Fig. 23.3). Pursuant to the universal approximation theorem, such architectures possess the theoretical capacity to represent any continuous function to an arbitrary degree of accuracy, provided they are structured with sufficient depth and width.

For an L-layer network, the output of the j-th neuron in the k-th layer is given by:

$$A_j^k = f\left(\sum_{i=1}^{Q_{k-1}} W_{ij}^k A_i^{k-1} - b_j^k\right) \tag{23.12}$$

where, W_{ij}^k and b_j^k denote the weight matrix and bias vector, respectively, and $f(\cdot)$ represents the nonlinear activation function. The neural network is subsequently defined in a recursive manner as follows:

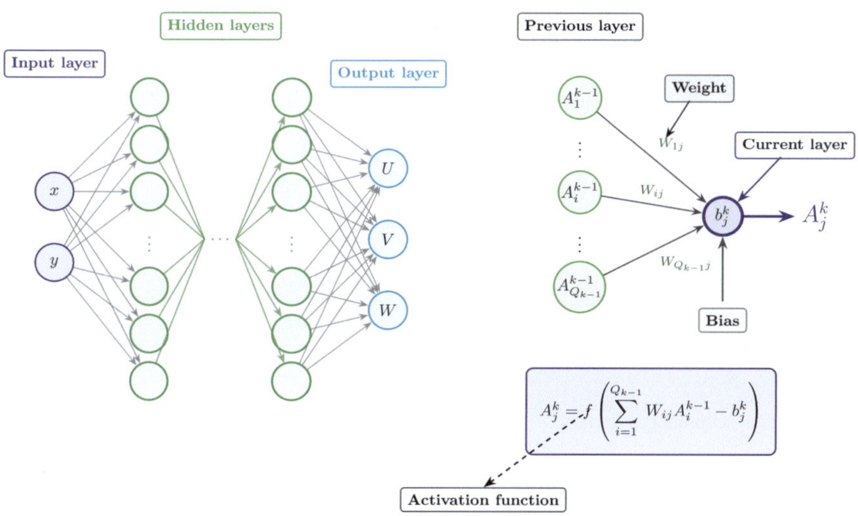

Fig. 23.3: Multilayer Neural Network Architecture.

- Input Layer: $A^0 = [x, y] \in \mathbb{R}^2$: accepts the input variables corresponding to the spatial coordinates (x, y).
- Hidden Layers: $A^k = f(W^{k^T} A^{k-1} + b^k) \in \mathbb{R}^{M_l}, 1 \leq k \leq L - 1$: comprise multiple fully connected neurons, which apply nonlinear transformations using the hyperbolic tangent (tanh) activation function.
- Output Layer: $A^L = [u, v] = W^{L^T} A^{L-1} + b^L \in \mathbb{R}^3$: produces the target value to be predicted (u, v).

Formulated on the principle of minimum potential energy, DEM utilizes the energy functional as the network's training bias. This approach optimizes the displacement field to achieve a minimum energy state, which is mathematically equivalent to fulfilling the system's equilibrium equations as shown in Fig 23.4.

The loss function of the laminated composite plates resting on elastic foundations is shown as

$$L_{\text{Energy-based}} = \Pi = U - V \tag{23.13}$$

The strain energy of the laminated plate is approximated over N_f sampling points as:

$$\begin{aligned} U &= \frac{1}{2} \int_\Omega \left(\varepsilon_x \sigma_x + \varepsilon_y \sigma_y + \gamma_{xy} \tau_{xy} + \gamma_{xz} \tau_{xz} + \gamma_{yz} \tau_{yz} \right) d\Omega \\ &\approx \frac{ab}{N_f} \sum_{i=1}^{N_f} \frac{1}{2} \left(\varepsilon_x^i \sigma_x^i + \varepsilon_y^i \sigma_y^i + \gamma_{xy}^i \tau_{xy}^i + \gamma_{xz}^i \tau_{xz}^i + \gamma_{yz}^i \tau_{yz}^i \right) \end{aligned} \tag{23.14}$$

The virtual work done by the applied forces is given by:

$$\delta V = \int_\Omega \left\{ q + k_w w_0 + k_p \left(\frac{\partial^2 w_0}{\partial x^2} + \frac{\partial^2 w_0}{\partial y^2} \right) \right\} \delta w_0 \, d\Omega \tag{23.15}$$

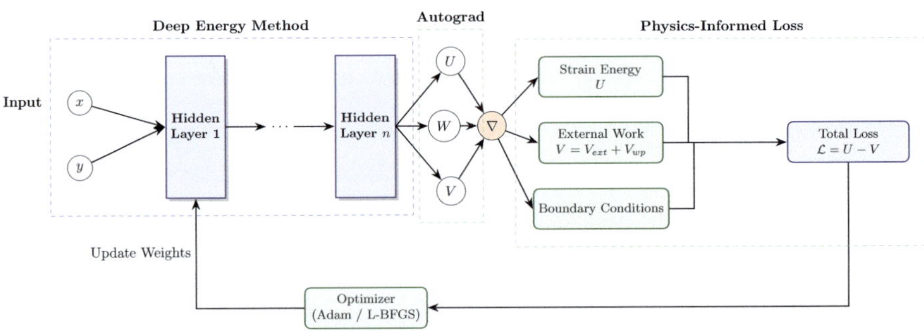

Fig. 23.4: The deep energy method.

23.3 Numerical Results and Discussions

To characterize the nonlinear response of the laminated composite plates resting on elastic foundations, the proposed framework utilizes a sequential incremental loading procedure where each step is resolved via a fully connected neural network (FNN). The computational architecture is configured as a deep solver consisting of five hidden layers, each with a uniform width of 5 neurons, to effectively approximate the complex kinematic fields of the composite system. For the training phase, the network parameters are optimized using the Adam algorithm, a first-order gradient-based method recognized for its efficiency in handling non-convex objective functions. A fixed learning rate of 1×10^{-3} is maintained throughout the optimization cycles to ensure stable parameter updates. Furthermore, to balance predictive accuracy with computational overhead, the maximum number of training iterations at each load increment is restricted to 10,000, ensuring the convergence of the energy-based loss function.

23.3.1 The Laminated Composite Plates Without Elastic Foundations

In this section, the material properties of the laminated composite plates are shown in Table 23.1: Figure 23.5 compares the deflection fields predicted by the proposed FSDT-PINN and the finite element analysis (FEA). The results show excellent agreement in both symmetry and magnitude, with accurate prediction of the maximum deflection at the plate center. The FSDT-PINN yields a maximum value of deflection $W = 0.09461$ (mm), closely matching the FEA result of $W = 0.09388$ (mm). These results demonstrate that FSDT-PINN is an effective mesh-free solver for nonlinear bending analysis of laminated composite plates.

Tables 23.2 and 23.3 present the maximum deflection W (mm) of laminated graphite-epoxy plates under uniformly distributed loads predicted by FSDT-PINN and compared with FEA results. In all cases, the CCCC plates exhibit significantly smaller deflections than SSSS plates. For example, at $a/h = 40$ with layup [0/0/0/0//0], the deflection of the SSSS plate (0.7179 mm) is approximately 4.3 times higher than that of the CCCC plate (0.1640 mm). This confirms that clamped boundary conditions provide higher bending stiffness and more effectively restrict deformation. Furthermore, the thickness ratio a/h has a pronounced influence on the

Table 23.1: Material properties of graphite/epoxy composite plates [1].

Module	E_1	E_2	ν_{12}	G_{12}	G_{13}	G_{23}
Value	130300	9720	0.308	5390	5390	3600
Unit	MPa	MPa		MPa	MPa	MPa

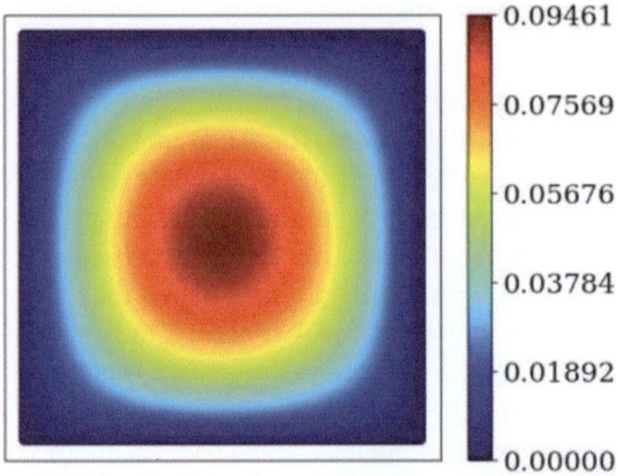

(a) FSDT-PINN (W_{max} = 0.09461 mm).

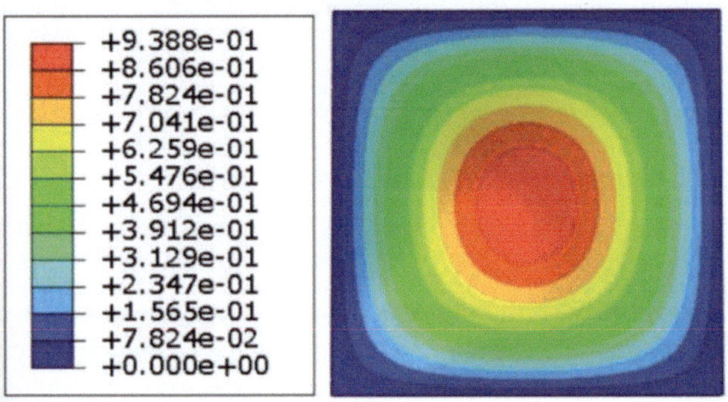

(b) FEA (W_{max} = 0.09388 mm).

Fig. 23.5: Comparison of the deflection fields of the laminated composite plate obtained by (a) FEA and (b) FSDT-PINN.

Table 23.2: Results of maximum deflection W_{max} (mm) of CCCC plates under uniformly distributed load using FSDT-PINN.

a/h	Ply orientation	FEA	FSDT-PINN
20	0/0/0/0//0	0.0249	0.0256
	0/90/0/90/0	0.0308	0.0293
	90/45/0/45/90	0.0317	0.0305
40	0/0/0/0//0	0.1640	0.1675
	0/90/0/90/0	0.1846	0.1838
	90/45/0/45/90	0.1857	0.1850

Table 23.3: Comparison of maximum deflection W_{max} (mm) of SSSS plates under uniformly distributed load using FSDT-PINN.

a/h	Ply orientation	FEA	FSDT-PINN
20	0/0/0/0//0	0.0939	0.0946
	0/90/0/90/0	0.1012	0.0981
	90/45/0/45/90	0.0907	0.0869
40	0/0/0/0//0	0.7179	0.7174
	0/90/0/90/0	0.7487	0.7284
	90/45/0/45/90	0.6656	0.6840

deflection behavior. As a/h increases from 20 to 40, corresponding to thinner plates, the deflection increases significantly. For instance, under CCCC conditions with layup [0/0/0/0//0], the deflection rises from 0.0249 mm to 0.1640 mm. This trend reflects the reduction in bending stiffness with increasing geometric slenderness, making thinner plates more susceptible to deformation.

Figure 23.6 depicts the convergence history of the objective loss function during the training phase. Analysis of the results indicates that the thinner plate geometry ($a/h = 40$) yields a higher convergence rate and enhanced stability relative to the $a/h = 20$ instance. Successful minimization of the energy functional in both cases rigorously validates the framework's ability to capture the inherent physics and equilibrium conditions of the composite system without the need for labeled data.

Table 23.4 presents the maximum deflection W (mm) of laminated plates under sinusoidal distributed loads predicted by FSDT-PINN and compared with FEA results. Overall, a good agreement between the two methods is observed across all cases. The results show that boundary conditions significantly influence the deflection response, with simply supported (SSSS) plates consistently exhibiting larger deflections than fully clamped (CCCC) plates.

23.3.2 The Laminated Plates Resting on Winkler-Pasternak Elastic Foundation

In this section, the material properties, non-dimensional deflection, applied external transverse load, stiffness coefficients of the elastic foundation are defined as follows:

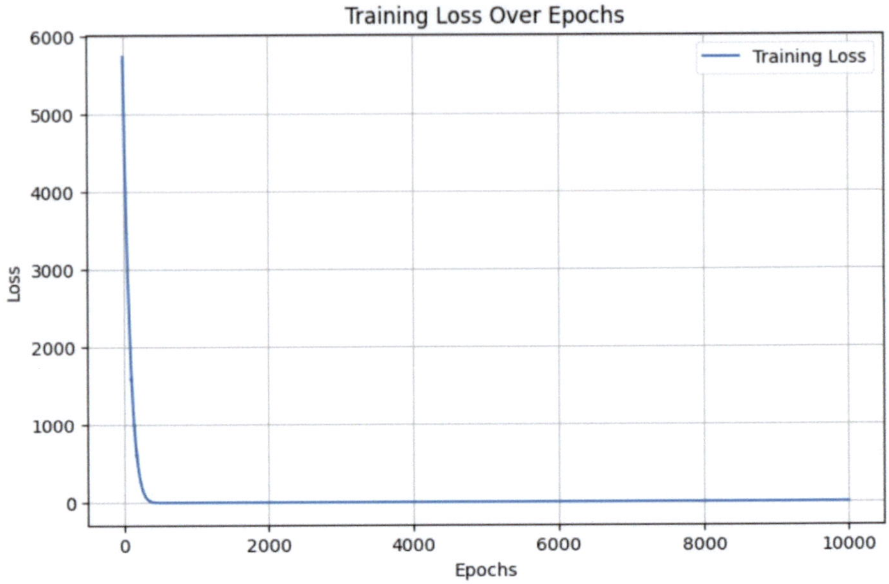

(a) $a/h = 20$.

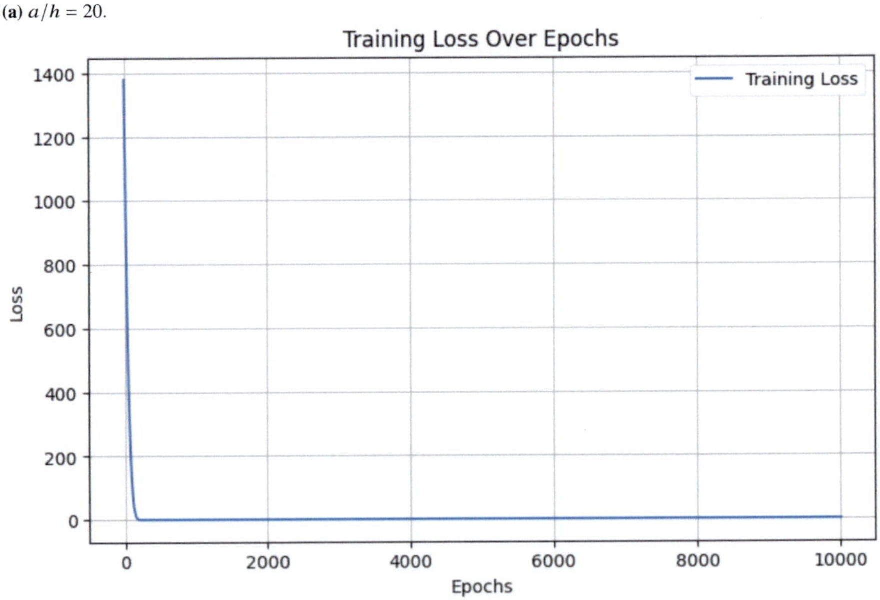

(b) $a/h = 40$.

Fig. 23.6: Training loss convergence history of the PINN model for simply supported laminated plates.

Table 23.4: Results of maximum deflection W_{max} (mm) of plates under sinusoidal distributed load using FSDT-PINN.

BCs	a/h	Ply orientation	FEA	FSDT-PINN
CCCC	20	0/0/0/0//0	0.0188	0.0204
		90/0/90/0/90	0.0221	0.0231
		90/45/0/45/90	0.0230	0.0240
	40	0/0/0/0//0	0.1245	0.1325
		90/0/90/0/90	0.1339	0.1387
		90/45/0/45/90	0.1363	0.1413
SSSS	20	0/0/0/0//0	0.0613	0.0663
		90/0/90/0/90	0.0643	0.0673
		90/45/0/45/90	0.0583	0.0627
	40	0/0/0/0//0	0.4667	0.4998
		90/0/90/0/90	0.4728	0.5008
		90/45/0/45/90	0.4259	0.4554

$$\begin{aligned}
\frac{E_1}{E_2} &= 25, \nu = 0.25, G_{12} = G_{13} = 0.5E_2, G_{23} = 0.2E_2 \\
\bar{w}(x, y, z) &= \frac{100E_2}{qhS^4} w\left(\frac{a}{2}, \frac{b}{2}, 0\right), S = a/h \\
q &= q_0 \sin(\frac{\pi x}{a}) \sin(\frac{\pi y}{b}) \\
K_W &= \frac{k_w b^4}{E_2 h^3}; \quad K_P = \frac{k_p b^2}{E_2 h^3}
\end{aligned} \tag{23.16}$$

Figure 23.7 illustrates the training loss convergence history of the FSDT-PINN framework for SSSS laminated plates resting on an elastic foundation. As observed, the training loss function experiences an abrupt and sharp decrease within the initial epochs, rapidly converging to near-zero values and remaining perfectly stable for the remainder of the 10,000 iterations.

Figure 23.8 depicts the evolution of the peak predicted deflection during the optimization process. The predicted value exhibits a sharp monotonic increase in the early stages before reaching a smooth plateau at a steady-state magnitude of approximately 0.200. This stable physical response rigorously validates the high fidelity of the FSDT-PINN framework in accurately capturing nonlinear structural behaviors and identifying a physically consistent equilibrium state. Tables 23.5, 23.6 and 23.7 provide a rigorous comparative evaluation of the non-dimensional central deflection for square laminated composite plates subjected to a sinusoidally distributed load across three distinct lamination schemes: two-layer (0°/90°), four-layer anti-symmetric (0°/90°/0°/90°), and four-layer symmetric (0°/90°/90°/0°) configurations.

Table 23.5: Effects of elastic foundation coefficients on the non-dimensional deflection of a square (0°/90°) laminated plates under SDL.

a/h	K_W	K_P	HSDT [18]	HSDT [19]	CPT [18]	Present study (FEA)	FSDT-PINN
5	0	0	1.6349	1.7583	1.0636	1.8666	1.2542
	100	0	0.6216	0.6374	0.3855	0.6662	0.5569
	100	10	0.2796	0.2822	0.1707	0.2886	0.2630
10	0	0	1.2089	1.2373	1.0636	1.2610	1.1320
	100	0	0.5487	0.5530	0.4754	0.5662	0.4827
	100	10	0.2640	0.2644	0.2273	0.2679	0.2559

* W stands for Winkler's elastic foundation ($K_W = 100$). P stands for Pasternak's elastic foundation ($K_W = 100$, $K_P = 10$)

Table 23.6: Effects of elastic foundation coefficients on the non-dimensional deflection of a square (0°/90°/0°/90°) laminated plates under SDL.

a/h	K_W	K_P	HSDT [18]	HSDT [19]	CPT [18]	Present study (FEA)	FSDT-PINN
5	0	0	1.2056	1.2013	0.5065	1.2712	1.2064
	100	0	0.5443	0.4920	0.2301	0.5694	0.4064
	100	10	0.2613	0.2561	0.1108	0.2681	0.2178
10	0	0	0.6847	0.6802	0.5065	0.7139	0.6892
	100	0	0.4069	0.4048	0.3015	0.4212	0.3439
	100	10	0.2260	0.2250	0.1676	0.2300	0.2074

Table 23.7: Effects of elastic foundation coefficients on the non-dimensional deflection of a square (0°/90°/90°/0°) laminated plates under SDL.

a/h	K_W	K_P	HSDT [19]	HSDT [18]	CPT [18]	Present study (FEA)	FSDT-PINN
5	0	0	1.1260	1.4208	0.4312	1.1560	1.0670
	100	0	0.5296	0.5853	0.2028	0.5446	0.4044
	100	10	0.2589	0.2709	0.0992	0.2623	0.2262
10	0	0	0.6049	0.7173	0.4312	0.6182	0.7120
	100	0	0.3769	0.4182	0.2687	0.3865	0.3211
	100	10	0.2161	0.2294	0.1541	0.2192	0.1987

Specifically, the inclusion of the Winkler foundation model ($K_W = 100$, $K_P = 0$) significantly reduces the transverse deflection of the plate compared to the unsupported case. This deflection is further suppressed when the Pasternak shear layer is incorporated ($K_W = 100$, $K_P = 10$), highlighting the enhanced stiffness provided by the two-parameter foundation model. The Pasternak component accounts for shear interactions between adjacent spring elements, leading to a more realistic representation of the foundation behavior and a more effective constraint on plate deformation.

Furthermore, the results highlight the necessity of shear deformation theories through the influence of the geometric slenderness ratio a/h. For moderately thick plates ($a/h = 5$), the classical plate theory significantly underestimates deflection compared to both HSDT and the FEA benchmark due to the neglect of transverse shear effects. This discrepancy diminishes as the plates become more slender ($a/h = 10$), in agreement with classical structural mechanics.

More importantly, by examining configurations from simple two-layer to more complex four-layer laminates, the proposed FSDT-PINN framework demonstrates high accuracy and strong generalization capability. The predictions show excellent agreement with both FEA and HSDT results. By employing a variational deep energy approach to minimize the total potential energy, the method ensures numerical stability and serves as an efficient mesh-free surrogate solver for nonlinear bending analysis of laminated composite plates.

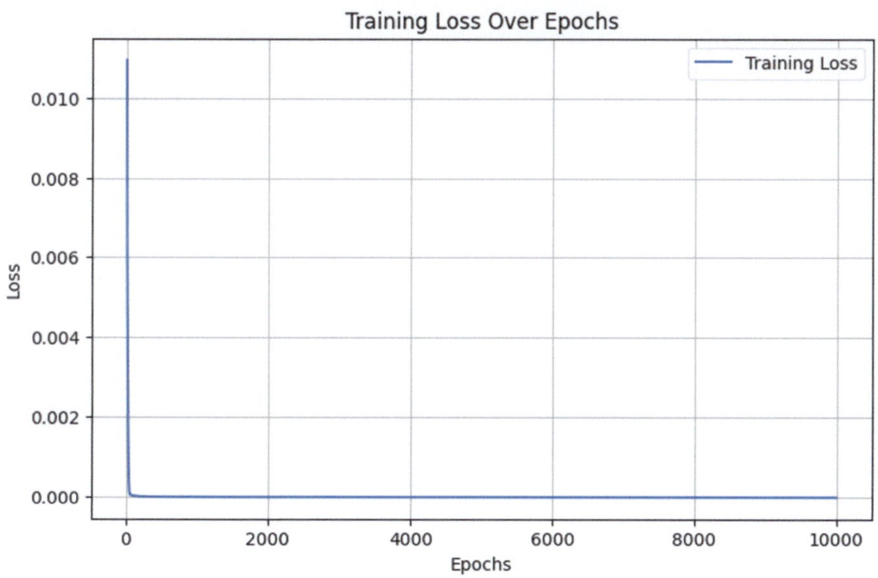

Fig. 23.7: Training loss convergence history of the FSDT-PINN model for SSSS laminated plates resting on elastic foundations.

Figure 23.9 illustrates the deflection field $\overline{w}$ on the mid-plane of a $(0°/90°/90°/0°)$ laminated composite plate resting on a Winkler–Pasternak elastic foundation. The results exhibit a smooth and symmetric distribution with respect to both axes, where the maximum deflection occurs at the plate center and gradually decreases toward the boundaries. This behavior is consistent with the applied loading and boundary conditions, as well as the stabilizing effect of the elastic foundation. The 3D surface plot further confirms the continuity and smoothness of the solution, indicating the capability of the proposed method to accurately capture the global deformation characteristics of laminated composite plates.

23.4 Conclusions

This study successfully established the FSDT-PINN framework for nonlinear bending analysis of laminated composite plates on Winkler–Pasternak elastic foundations. By integrating first-order shear deformation theory and von Kármán kinematics within a Deep Energy Method formulation, the model minimizes total potential energy, ensuring rigorous physical consistency while reducing the order of spatial derivatives and enhancing numerical stability. Numerical validations show that FSDT-PINN accurately predicts structural responses, closely matching finite element results, and explicitly captures transverse shear effects in thin and moderately thick plates. This

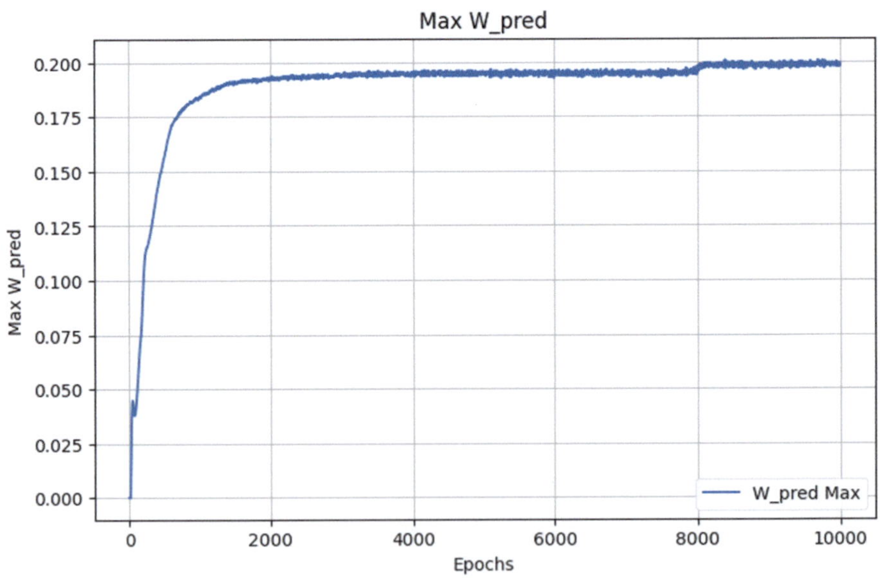

Fig. 23.8: Training loss convergence history of the maximum deflection of the plates.

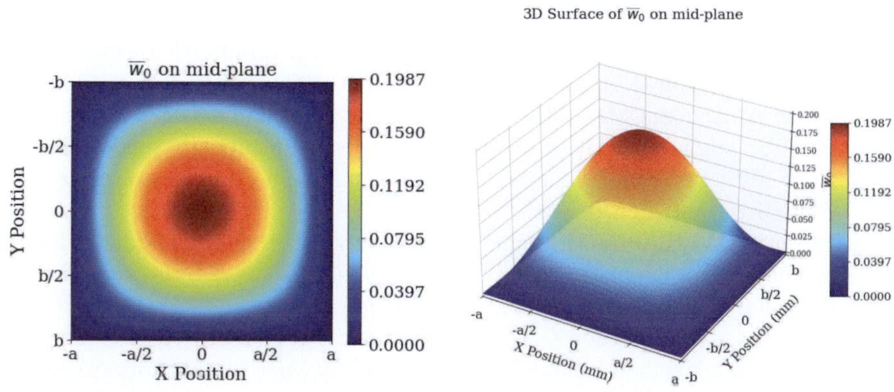

(a) 2D contour view.

(b) 3D surface view.

Fig. 23.9: The non-dimensional deflection fields ($\overline{w}$) of a $(0°/90°/90°/0°)$ laminated composite plate resting on a Winkler-Pasternak elastic foundation.

framework provides an efficient computational tool for bending analysis of laminated composite plates resting on elastic foundations.

Acknowledgements The authors would like to acknowledge the support by the Project "Research, apply and develop modern calculation methods in the design and manufacture of composite materials under special loads" of the National Foundation for Science and Technology Development of Vietnam (NAFOSTED) code ĐTĐLCN.08/25.

References

[1] Bowles DE, Post D, Herakovich CT, Tenny DR (1980) Thermal expansion of composites using moire interferometry. NASA Technical Report

[2] Chen Z, Andrejevic N, Smidt T, Ding Z, Xu Q, Chi Y-T, Nguyen QT, Alatas A, Kong J, Li M (2021) Direct prediction of phonon density of states with Euclidean neural networks. Advanced Science 8:2004214, DOI 10.1002/advs.202004214

[3] Darbon J, Langlois GP, Meng T (2020) Overcoming the curse of dimensionality for some Hamilton–Jacobi partial differential equations via neural network architectures. Research in the Mathematical Sciences 7:20, DOI 10.1007/s40687-020-00215-6

[4] Dawe DJ, Lam SSE, Azizian ZG (1992) Non-linear finite strip analysis of rectangular laminates under end shortening, using classical plate theory. International Journal for Numerical Methods in Engineering 35:1087-1110, DOI 10.1002/nme.1620350508

[5] Ermis M, Dorduncu M, Aydogan G (2025) Physics-based machine learning for modeling of laminated composite plates based on refined zigzag theory. Archive of Applied Mechanics 95:107, DOI 10.1007/s00419-025-02816-5

[6] Fallah A, Aghdam MM (2024) Physics-informed neural network for bending and free vibration analysis of three-dimensional functionally graded porous beam resting on elastic foundation. Engineering with Computers 40:437-454, DOI 10.1007/s00366-023-01799-7

[7] Ferreira AJM, Castro LM, Roque CMR, Reddy JN, Bertoluzza S (2011) Buckling analysis of laminated plates by wavelets. Computers & Structures 89:626-630, DOI 10.1016/j.compstruc.2011.01.007

[8] He J, Gong C, Anitescu C, Zhuang X, Rabczuk T (2023) A deep learning energy-based method for classical elastoplasticity. International Journal of Plasticity 162:103531, DOI 10.1016/j.ijplas.2023.103531

[9] Huang Z, Peng L (2024) Geometrically nonlinear bending analysis of laminated thin plates based on classical laminated plate theory and deep energy method. Computers & Structures 344:118314, DOI 10.1016/j.compstruct.2024.118314

[10] Kaw AK (2005) Mechanics of Composite Materials, 2nd edn. Boca Raton: CRC Press

[11] Mojahedin A, Salavati M, Rabczuk T (2021) A deep energy method for functionally graded porous beams. Journal of Zhejiang University-SCIENCE A 22(6):492-498, DOI 10.1631/jzus.A2000317

[12] Nguyen-Thanh VM, Zhuang X, Rabczuk T (2020) A deep energy method for finite deformation hyperelasticity. European Journal of Mechanics - A/Solids 80:103874, DOI 10.1016/j.euromechsol.2019.103874

[13] Raissi M, Perdikaris P, Karniadakis GE (2019) Physics-informed neural networks: a deep learning framework for solving forward and inverse problems involving nonlinear partial differential equations. Journal of Computational Physics 378:686-707, DOI 10.1016/j.jcp.2018.10.045

[14] Reddy JN, Chao WC (1981) A comparison of closed-form and finite-element solutions of thick, laminated, anisotropic rectangular plates. Nuclear Engineering and Design 64(2):153-167, DOI 10.1016/0029-5493(81)90001-7

[15] Reddy JN (1984) A simple higher-order theory for laminated composite plates. Journal of Applied Mechanics 51(4):745-752, DOI 10.1115/1.3167719

[16] Reddy JN (2003) Mechanics of Laminated Composite Plates and Shells: Theory and Analysis, 2nd edn. Boca Raton: CRC Press, DOI 10.1201/b12409

[17] Reddy JN (2007) Theory and analysis of elastic plates and shells. Boca Raton: CRC Press, DOI 10.1201/9780849384165

[18] Sayyad AS, Ghugal YM, Kant T (2024) Quasi-3D flexural analysis of laminated composite plates resting on elastic foundations using higher-order displacement model. Elsevier Progress in Engineering Science 1(2-3):100005, DOI 10.1016/j.proeng.2024.100005

[19] Setoodeh AR, Azizi A (2015) Bending and free vibration analyses of rectangular laminated composite plates resting on elastic foundation using a refined shear deformation theory. Iranian Journal of Materials Forming 2(2):1-13, DOI 10.22099/ijmf.2015.3182

[20] Wang J, Liew KM, Tan MJ, Rajendran S (2002) Analysis of rectangular laminated composite plates via FSDT meshless method. International Journal of Mechanical Sciences 44(7):1275-1293, DOI 10.1016/S0020-7403(02)00057-7

[21] Wang W, Thai H-T (2025) A physics-informed neural network framework for laminated composite plates under bending. Thin-Walled Structures 210:113014, DOI 10.1016/j.tws.2025.113014

[22] Zhuang X, Lou D, Hong JW, Rabczuk T (2021) Deep autoencoder based energy method for the bending, vibration, and buckling analysis of Kirchhoff plates with transfer learning. European Journal of Mechanics - A/Solids 87:104225, DOI 10.1016/j.euromechsol.2021.104225

Chapter 24
On some Issues of the First and Second Strain Tensor and Velocity Vector Gradient Theories of Elastic 3D Bodies and Thin Bodies

Mikhail Nikabadze

Abstract Expressions for specific energy, constitutive relations, equations of motion and equilibrium of the second strain tensor and velocity vector gradient theory for an arbitrary anisotropic bodies are given. From the latter, corresponding 3D relationships of the second strain tensor and velocity vector gradient 3D theory of elastic thin bodies are obtained, on the basis of which, using the method of orthogonal polynomials, the corresponding relationships in moments are derived. Constitutive relations for gradient bodies are written using tensor-block matrices. In addition, static boundary conditions and equations of motion and equilibrium in the case of gradient theories are represented by differential tensor-block matrix operators, for which, in the case of homogeneous bodies, differential tensor-block matrix operators of cofactors are constructed, which make it possible to split the initial-boundary value problems of some gradient theories (differential tensor-block matrix operator of static boundary conditions is called differential tensor-block matrix stress operator). Herewith static boundary conditions are not always split. The differential tensor-block matrix operator of cofactors for the differential tensor-block matrix operator of stress can be constructed only for a homogeneous body having a piecewise flat boundary, i.e. for each flat section of the boundary it is possible to construct a differential tensor-block matrix operator of cofactors and, therefore, for this flat section, the static boundary conditions can be split. Further, from 3D formulations of initial-boundary value problems for some gradient theories of elasticity, the formulations of initial-boundary value problems for gradient theories of elastic thin bodies are derived under the new parameterization of the regions of these bodies, from which, in turn, the initial-boundary value problems in moments of the displacement vector with respect to arbitrary systems of orthogonal polynomials and also Legendre polynomials are obtained. Several first approximations of formulations of boundary value problems for prismatic bodies are considered as special cases. The system of

---

Mikhail Nikabadze
Lomonosov Moscow State University, Leninskie Gory, Bldg. 1, Institute of Mechanics, LMSU, Michurinsky prospekt, Bldg. 1 & Bauman Moscow State Technical University, 2nd Baumanskaya Ulitsa, Bldg. 5, Str. 1, Moscow, Russian Federation, e-mail: munikabadze@yandex.ru

H. Altenbach et al. (eds.), *Stability Problems in Mechanics of Continua*,
Advanced Structured Materials 260, https://doi.org/10.1007/978-3-032-31032-3_24

equations in moments of the displacement vector of the 10th order approximation is considered in more detail. In particular, for each moment of the displacement vector included in these systems of equations, an equation of higher order elliptic type is obtained separately, for which, using the Vekua method, a way to obtain a general representation of the solutions is indicated. Furthermore, as a special case, some relations of the first strain tensor and velocity vector gradient (FST and VVG) elasticity theory are derived from the corresponding relations of the second strain tensor and velocity vector gradient (SST and VVG) elasticity theory. As a special case, using the procedure described above, the problems on a rectangular plate and a rectangle made of elastic and viscoelastic materials were considered. For brevity, these problems were not included in this paper. It should be noted that the methodologies outlined above can be easily extended to theories of other rheological media, including higher-order gradient theories, as well as to linear and nonlinear classical, micropolar and micromorphic theories of elastic and viscoelastic bodies.

Key words: Thin body, Deformation tensor, Stress tensor, gradient theory, second strain tensor and velocity vector gradient theory, Method of orthogonal polynomials, Initial-boundary value problems in moments

24.1 Introduction

Strain gradient elasticity theories and similar nonlocal theories have appeared in the literature (see [69]), as analytical tools capable of taking into account long-range interaction forces in the microstructure of materials and recording experimentally discovered phenomena (such as length scale effects, dispersion waves, localization of deformation, etc.), which classical theory is not able to take into account (see, for example, [28, 29, 42–44]). Strain gradient theories do not take into account the independent rotations of material particles, i.e. they are not polar in nature, and the moment stress tensor arises due to the gradient of the vortex vector of the continuous medium. In this context, of particular interest is the most important article by Mindlin [43], who constructed a gradient theory of second-order elasticity with respect to the strain tensor for statics, as well as for dynamics [42, 44]. Although it uses an excessive number of material constants, it is a benchmark because of its conceptual foundations and ability to capture length scale effects. Toupin and Gazis [72], exploring models of mono- and di-atomic lattices, pointed out the ability of the second strain tensor gradient (SSTG) theory to interpret the atomic structure of matter. Subsequently, Aifantis and his co-authors proposed a first strain tensor gradient (FSTG) theory of elasticity, which used only three material constants (two classical Lame constants and a length scale parameter) [2, 4, 71]. This theory was then extended to dynamics of higher order inertia [7, 9], which requires only one additional length scale parameter. This model of gradient elasticity, called the Aifantis model, has been used to solve a variety of structural problems, showing that strain singularities do not arise at crack tips and dislocation cores, and that dispersion effects in wave propagation can be captured [9]. However, as shown by Lazar and Maugin [33], the singularities of the double stress tensor cannot be eliminated using the mentioned Aifantis model.

An extension of above the FSTG theory to the SSTG one, characterized by four material constants: two Lame constants and two length scale parameters, was proposed in [66]. A similar theory, called bi-Helmholtz type strain gradient theory, was independently developed by Lazar and Maugin [34, 35] and applied to a number of dislocation problems in an infinite domain with the outstanding result that no singularities arise in the dislocation core and, of course, this material model has a high regularization ability. This result was largely confirmed by a number of applications to problems of defect interaction [78], dislocation analysis [31, 34, 35] and disclination analysis [15]. However, no attention has been paid to the boundary conditions and surface effects that typically characterize recent gradient theory models. The role of velocity gradients and their inherent higher order inertia terms in the equations of motion has been considered in a series of studies dealing with wave motion and related dispersion phenomena. This topic has been widely covered in the literature, but here we will limit ourselves to mentioning the following works: [3, 7, 8, 10, 11, 23, 41, 58], see also review article [9] and references therein. Recent literature suggests that higher order inertia models are capable of realistically describing wave dispersion phenomena. In the works [7, 8, 41] the authors put forward the concept of a "dynamically consistent" gradient model, namely a model endowed with gradient improvements in both stiffness and inertia characteristics, which makes it possible to eliminate deformation field singularities arising from e.g. near the sharp tip of a crack and realistically describe the dispersion characteristics of wave propagation in an inhomogeneous medium. However, the higher order inertial terms appearing in the governing equations were introduced heuristically in these studies, and their relationship to kinetic energy remained unclear.

In [1, 32] the authors provided useful information about the laws conservation and balance for gradient elastodynamics. A review of the historical developments of gradient elasticity and higher order inertia with relevant applications can be seen in [9]. It should be noted that the above brief overview of gradient theories is borrowed mainly from [69] (see also [49]). It should be especially noted that the works of S.A. Lurie and his co-authors [37–39, among others], as well as the following works [12–14, 20–22, 45, 59, 70, 76, 77, among others] of other authors, deserve great attention.

Note also that by extending the work [66] and, of course, the works mentioned above are the works [68–70], which set out some of the most important issues of the second strain tensor and velocity vector gradient elasticity theory (GET). Specifically, formulations of boundary-value and initial-boundary-value problems are given, which, as well as more general statements of initial-boundary-value problems formulated by the author of this work, will subsequently be used by him at obtaining of similar formulations of boundary-value and initial-boundary-value problems for the second strain tensor and velocity vector gradient theory of elastic and viscoelastic thin bodies.

24.2 The Specific Strain Energy and Kinetic Energy in the Second Strain Tensor and Velocity Vector Gradient Elasticity Theory

In this theory, the specific strain energy Φ and kinetic energy K are represented as (see [68, 69])

$$\Phi = \Phi(\underset{\sim}{\boldsymbol{\varepsilon}}, \nabla\underset{\sim}{\boldsymbol{\varepsilon}}, \nabla\nabla\underset{\sim}{\boldsymbol{\varepsilon}}), \quad K = K(\mathbf{v}, \nabla\mathbf{v}, \nabla\nabla\mathbf{v}), \tag{24.1}$$

where $\underset{\sim}{\boldsymbol{\varepsilon}}$ is strain tensor, $\mathbf{v}$ is velocity vector, ∇ is Hamilton nabla operator. Expanding functions $\Phi(\underset{\sim}{\boldsymbol{\varepsilon}}, \nabla\underset{\sim}{\boldsymbol{\varepsilon}}, \nabla\nabla\underset{\sim}{\boldsymbol{\varepsilon}})$ and $K(\mathbf{v}, \nabla\mathbf{v}, \nabla\nabla\mathbf{v})$ in the neighborhood of the natural state $(\underset{\sim}{\boldsymbol{\varepsilon}} = \underset{\sim}{\mathbf{0}}, \nabla\underset{\sim}{\boldsymbol{\varepsilon}} = \underset{\simeq}{\mathbf{0}}, \nabla\nabla\underset{\sim}{\boldsymbol{\varepsilon}} = \underset{\approx}{\mathbf{0}})$ and $(\mathbf{v} = \mathbf{0}, \nabla\mathbf{v} = \underset{\sim}{\mathbf{0}}, \nabla\nabla\mathbf{v} = \underset{\simeq}{\mathbf{0}})$ in the Maclaurin series, we will have

$$\begin{aligned}
\Phi(\underset{\sim}{\boldsymbol{\varepsilon}}, \nabla\underset{\sim}{\boldsymbol{\varepsilon}}, \nabla\nabla\underset{\sim}{\boldsymbol{\varepsilon}}) &= \sum_{k=0}^{\infty} \frac{1}{k!}\Big(\underset{\sim}{\boldsymbol{\varepsilon}} \overset{2}{\otimes} \frac{\partial}{\partial \underset{\sim}{\boldsymbol{\varepsilon}}} + \nabla\underset{\sim}{\boldsymbol{\varepsilon}} \overset{3}{\otimes} \frac{\partial}{\partial \nabla\underset{\sim}{\boldsymbol{\varepsilon}}} + \nabla\nabla\underset{\sim}{\boldsymbol{\varepsilon}} \overset{4}{\otimes} \frac{\partial}{\partial \nabla\nabla\underset{\sim}{\boldsymbol{\varepsilon}}}\Big)_0^{(k)}\Phi,\\
K(\mathbf{v}, \nabla\mathbf{v}, \nabla\nabla\mathbf{v}) &= \sum_{k=0}^{\infty} \frac{1}{k!}\Big(\mathbf{v}\cdot \frac{\partial}{\partial \mathbf{v}} + \nabla\mathbf{v} \overset{2}{\otimes} \frac{\partial}{\partial \nabla\mathbf{v}} + \nabla\nabla\mathbf{v} \overset{3}{\otimes} \frac{\partial}{\partial \nabla\nabla\mathbf{v}}\Big)_0^{(k)}K.
\end{aligned} \tag{24.2}$$

Here $\mathbf{0}$ is zero vector, $\underset{\sim}{\mathbf{0}}$ is zero tensor of the second rank, $\underset{\simeq}{\mathbf{0}}$ is zero tensor of the third rank and $\underset{\approx}{\mathbf{0}}$ is zero tensor of the fourth rank, $\overset{r}{\otimes}$ is the sign of the inner r-product, where r is an arbitrary non-negative integer. For $r = 0$, $\overset{0}{\otimes} \equiv \otimes$ is (direct) tensor multiplication, and for $r = 1$, $\overset{1}{\otimes} \equiv \cdot$ is single multiplication. For more details on its application, see [55, 56, 74]. The formulas below also show how the indices of the components of the factor tensors contract.

Neglecting terms higher than the second order of smallness in the (24.2), we obtain

$$\begin{aligned}
\Phi(\underset{\sim}{\boldsymbol{\varepsilon}}, \underset{\simeq}{\boldsymbol{\gamma}}, \underset{\approx}{\boldsymbol{\varkappa}}) = {} & \Phi(\underset{\sim}{\mathbf{0}}, \underset{\simeq}{\mathbf{0}}, \underset{\approx}{\mathbf{0}}) + \underset{\sim}{\boldsymbol{\varepsilon}} \overset{2}{\otimes} \Big(\frac{\partial\Phi}{\partial\underset{\sim}{\boldsymbol{\varepsilon}}}\Big)_{(0)} + \underset{\simeq}{\boldsymbol{\gamma}} \overset{3}{\otimes} \Big(\frac{\partial\Phi}{\partial\underset{\simeq}{\boldsymbol{\gamma}}}\Big)_{(0)} + \underset{\approx}{\boldsymbol{\varkappa}} \overset{4}{\otimes} \Big(\frac{\partial\Phi}{\partial\underset{\approx}{\boldsymbol{\varkappa}}}\Big)_{(0)}\\
& + \frac{1}{2}\Big[\underset{\sim}{\boldsymbol{\varepsilon}}\underset{\sim}{\boldsymbol{\varepsilon}} \overset{4}{\otimes} \Big(\frac{\partial^2\Phi}{\partial\underset{\sim}{\boldsymbol{\varepsilon}}\partial\underset{\sim}{\boldsymbol{\varepsilon}}}\Big)_{(0)} + \underset{\sim}{\boldsymbol{\varepsilon}}\underset{\simeq}{\boldsymbol{\gamma}} \overset{5}{\otimes} \Big(\frac{\partial^2\Phi}{\partial\underset{\sim}{\boldsymbol{\varepsilon}}\partial\underset{\simeq}{\boldsymbol{\gamma}}}\Big)_{(0)} + \underset{\sim}{\boldsymbol{\varepsilon}}\underset{\approx}{\boldsymbol{\varkappa}} \overset{6}{\otimes} \Big(\frac{\partial^2\Phi}{\partial\underset{\sim}{\boldsymbol{\varepsilon}}\partial\underset{\approx}{\boldsymbol{\varkappa}}}\Big)_{(0)}\\
& + \underset{\simeq}{\boldsymbol{\gamma}}\underset{\sim}{\boldsymbol{\varepsilon}} \overset{5}{\otimes} \Big(\frac{\partial^2\Phi}{\partial\underset{\simeq}{\boldsymbol{\gamma}}\partial\underset{\sim}{\boldsymbol{\varepsilon}}}\Big)_{(0)} + \underset{\simeq}{\boldsymbol{\gamma}}\underset{\simeq}{\boldsymbol{\gamma}} \overset{6}{\otimes} \Big(\frac{\partial^2\Phi}{\partial\underset{\simeq}{\boldsymbol{\gamma}}\partial\underset{\simeq}{\boldsymbol{\gamma}}}\Big)_{(0)} + \underset{\simeq}{\boldsymbol{\gamma}}\underset{\approx}{\boldsymbol{\varkappa}} \overset{7}{\otimes} \Big(\frac{\partial^2\Phi}{\partial\underset{\simeq}{\boldsymbol{\gamma}}\partial\underset{\approx}{\boldsymbol{\varkappa}}}\Big)_{(0)}\\
& + \underset{\approx}{\boldsymbol{\varkappa}}\underset{\sim}{\boldsymbol{\varepsilon}} \overset{6}{\otimes} \Big(\frac{\partial^2\Phi}{\partial\underset{\approx}{\boldsymbol{\varkappa}}\partial\underset{\sim}{\boldsymbol{\varepsilon}}}\Big)_{(0)} + \underset{\approx}{\boldsymbol{\varkappa}}\underset{\simeq}{\boldsymbol{\gamma}} \overset{7}{\otimes} \Big(\frac{\partial^2\Phi}{\partial\underset{\approx}{\boldsymbol{\varkappa}}\partial\underset{\simeq}{\boldsymbol{\gamma}}}\Big)_{(0)} + \underset{\approx}{\boldsymbol{\varkappa}}\underset{\approx}{\boldsymbol{\varkappa}} \overset{8}{\otimes} \Big(\frac{\partial^2\Phi}{\partial\underset{\approx}{\boldsymbol{\varkappa}}\partial\underset{\approx}{\boldsymbol{\varkappa}}}\Big)_{(0)}\Big];
\end{aligned} \tag{24.3}$$

$$
\begin{aligned}
K(\mathbf{v},\underset{\sim}{\mathbf{v}},\underset{\simeq}{\mathbf{v}}) = {} & K(\mathbf{0},\underset{\sim}{\mathbf{0}},\underset{\simeq}{\mathbf{0}}) + \mathbf{v}\cdot\left(\frac{\partial K}{\partial \mathbf{v}}\right)_{(0)} + \underset{\sim}{\mathbf{v}}\overset{2}{\otimes}\left(\frac{\partial K}{\partial \underset{\sim}{\mathbf{v}}}\right)_{(0)} + \underset{\simeq}{\mathbf{v}}\overset{3}{\otimes}\left(\frac{\partial K}{\partial \underset{\simeq}{\mathbf{v}}}\right)_{(0)} \\
& + \frac{1}{2}\left[\mathbf{v}\mathbf{v}\overset{2}{\otimes}\left(\frac{\partial^2 K}{\partial \mathbf{v}\partial \mathbf{v}}\right)_{(0)} + \mathbf{v}\underset{\sim}{\mathbf{v}}\overset{3}{\otimes}\left(\frac{\partial^2 K}{\partial \mathbf{v}\partial \underset{\sim}{\mathbf{v}}}\right)_{(0)} + \mathbf{v}\underset{\simeq}{\mathbf{v}}\overset{4}{\otimes}\left(\frac{\partial^2 K}{\partial \mathbf{v}\partial \underset{\simeq}{\mathbf{v}}}\right)_{(0)}\right. \\
& + \frac{1}{2}\left[\mathbf{v}\mathbf{v}\overset{2}{\otimes}\left(\frac{\partial^2 K}{\partial \mathbf{v}\partial \mathbf{v}}\right)_{(0)} + \mathbf{v}\underset{\sim}{\mathbf{v}}\overset{3}{\otimes}\left(\frac{\partial^2 K}{\partial \mathbf{v}\partial \underset{\sim}{\mathbf{v}}}\right)_{(0)} + \mathbf{v}\underset{\simeq}{\mathbf{v}}\overset{4}{\otimes}\left(\frac{\partial^2 K}{\partial \mathbf{v}\partial \underset{\simeq}{\mathbf{v}}}\right)_{(0)}\right. \\
& + \underset{\sim}{\mathbf{v}}\mathbf{v}\overset{3}{\otimes}\left(\frac{\partial^2 K}{\partial \underset{\sim}{\mathbf{v}}\partial \mathbf{v}}\right)_{(0)} + \underset{\sim}{\mathbf{v}}\underset{\sim}{\mathbf{v}}\overset{4}{\otimes}\left(\frac{\partial^2 K}{\partial \underset{\sim}{\mathbf{v}}\partial \underset{\sim}{\mathbf{v}}}\right)_{(0)} + \underset{\sim}{\mathbf{v}}\underset{\simeq}{\mathbf{v}}\overset{5}{\otimes}\left(\frac{\partial^2 K}{\partial \underset{\sim}{\mathbf{v}}\partial \underset{\simeq}{\mathbf{v}}}\right)_{(0)} \\
& \left. + \underset{\simeq}{\mathbf{v}}\mathbf{v}\overset{4}{\otimes}\left(\frac{\partial^2 K}{\partial \underset{\simeq}{\mathbf{v}}\partial \mathbf{v}}\right)_{(0)} + \underset{\simeq}{\mathbf{v}}\underset{\sim}{\mathbf{v}}\overset{5}{\otimes}\left(\frac{\partial^2 K}{\partial \underset{\simeq}{\mathbf{v}}\partial \underset{\sim}{\mathbf{v}}}\right)_{(0)} + \underset{\simeq}{\mathbf{v}}\underset{\simeq}{\mathbf{v}}\overset{6}{\otimes}\left(\frac{\partial^2 K}{\partial \underset{\simeq}{\mathbf{v}}\partial \underset{\simeq}{\mathbf{v}}}\right)_{(0)}\right],
\end{aligned}
\tag{24.4}
$$

where for brevity the following notations are introduced: $\underset{\simeq}{\boldsymbol{\gamma}} = \nabla\underset{\sim}{\boldsymbol{\varepsilon}}$, $\underset{\approx}{\boldsymbol{\varkappa}} = \nabla\nabla\underset{\sim}{\boldsymbol{\varepsilon}}$, $\underset{\sim}{\mathbf{v}} = \nabla\mathbf{v}$ and $\underset{\simeq}{\mathbf{v}} = \nabla\nabla\mathbf{v}$.

It should be noted that at $\underset{\sim}{\boldsymbol{\varepsilon}} = \underset{\sim}{\mathbf{0}}$, $\underset{\simeq}{\boldsymbol{\gamma}} = \underset{\simeq}{\mathbf{0}}$ and $\underset{\approx}{\boldsymbol{\varkappa}} = \underset{\approx}{\mathbf{0}}$ the specific strain energy, and at $\mathbf{v} = \mathbf{0}$, $\underset{\sim}{\mathbf{v}} = \underset{\sim}{\mathbf{0}}$ and $\underset{\simeq}{\mathbf{v}} = \underset{\simeq}{\mathbf{0}}$ the kinetic energy takes on a minimum value, which is equal to zero. In this regard, the values of the first derivatives of the specific deformation energy and kinetic energy from their arguments are also equal to zero. In virtue of the above in the (24.3) and (24.4) the first four terms are equal to zero, and therefore they are represented in the form[1]

$$
\begin{aligned}
\Phi(\underset{\sim}{\boldsymbol{\varepsilon}},\underset{\simeq}{\boldsymbol{\gamma}},\underset{\approx}{\boldsymbol{\varkappa}}) = {} & \frac{1}{2}\left({}^{4}\underset{\sim}{\mathbf{A}}^{(11)}\overset{4}{\otimes}\underset{\sim}{\boldsymbol{\varepsilon}}\underset{\sim}{\boldsymbol{\varepsilon}} + 2\,{}^{5}\underset{\sim}{\mathbf{A}}^{(12)}\overset{5}{\otimes}\underset{\sim}{\boldsymbol{\varepsilon}}\underset{\simeq}{\boldsymbol{\gamma}} + 2\,{}^{6}\underset{\sim}{\mathbf{A}}^{(13)}\overset{6}{\otimes}\underset{\sim}{\boldsymbol{\varepsilon}}\underset{\approx}{\boldsymbol{\varkappa}}\right. \\
& \left. + {}^{6}\underset{\sim}{\mathbf{A}}^{(22)}\overset{6}{\otimes}\underset{\simeq}{\boldsymbol{\gamma}}\underset{\simeq}{\boldsymbol{\gamma}} + 2\,{}^{7}\underset{\sim}{\mathbf{A}}^{(23)}\overset{7}{\otimes}\underset{\simeq}{\boldsymbol{\gamma}}\underset{\approx}{\boldsymbol{\varkappa}} + {}^{8}\underset{\sim}{\mathbf{A}}^{(33)}\overset{8}{\otimes}\underset{\approx}{\boldsymbol{\varkappa}}\underset{\approx}{\boldsymbol{\varkappa}}\right) \\
= {} & \frac{1}{2}\left(A^{(11)ijkl}\varepsilon_{ij}\varepsilon_{kl} + 2A^{(12)ijklm}\varepsilon_{ij}\gamma_{klm} + 2A^{(13)ijklmn}\varepsilon_{ij}\varkappa_{klmn}\right. \\
& \left. + A^{(22)ijklsn}\gamma_{ijk}\gamma_{lsn} + 2A^{(23)ijklsnp}\gamma_{ijk}\varkappa_{lsnp} + A^{(33)ijklsnpq}\varkappa_{ijkl}\varkappa_{snpq}\right);
\end{aligned}
\tag{24.5}
$$

$$
\begin{aligned}
K(\mathbf{v},\underset{\sim}{\mathbf{v}},\underset{\simeq}{\mathbf{v}}) = {} & \frac{1}{2}\left({}^{2}\underset{\sim}{\mathbf{B}}^{(11)}\overset{2}{\otimes}\mathbf{v}\mathbf{v} + 2\,{}^{3}\underset{\sim}{\mathbf{B}}^{(12)}\overset{3}{\otimes}\mathbf{v}\underset{\sim}{\mathbf{v}} + 2\,{}^{4}\underset{\sim}{\mathbf{B}}^{(13)}\overset{4}{\otimes}\mathbf{v}\underset{\simeq}{\mathbf{v}}\right. \\
& \left. + {}^{4}\underset{\sim}{\mathbf{B}}^{(22)}\overset{3}{\otimes}\underset{\sim}{\mathbf{v}}\underset{\sim}{\mathbf{v}} + 2\,{}^{5}\underset{\sim}{\mathbf{B}}^{(23)}\overset{5}{\otimes}\underset{\sim}{\mathbf{v}}\underset{\simeq}{\mathbf{v}} + {}^{6}\underset{\sim}{\mathbf{B}}^{(33)}\overset{6}{\otimes}\underset{\simeq}{\mathbf{v}}\underset{\simeq}{\mathbf{v}}\right),
\end{aligned}
\tag{24.6}
$$

where the following notations are introduced:

[1] The usual rules of tensor calculus are applying (see [36, 55, 56, 62, 74]). Uppercase and lowercase Latin indices run through the values 1, 2 and 1, 2, 3, respectively. Unless otherwise staded, summation occurs over repeating indices.

$$
\begin{aligned}
{}^4\underset{\sim}{\mathbf{A}}^{(11)} &= A^{(11)}_{ijkl}\mathbf{r}^i\mathbf{r}^j\mathbf{r}^k\mathbf{r}^l = A^{(11)}_{klij}\mathbf{r}^i\mathbf{r}^j\mathbf{r}^k\mathbf{r}^l = \left(\frac{\partial^2\Phi}{\partial\underset{\sim}{\boldsymbol{\varepsilon}}\partial\underset{\sim}{\boldsymbol{\varepsilon}}}\right)_{(0)},\\
{}^6\underset{\sim}{\mathbf{A}}^{(22)} &= A^{(22)}_{ijklmn}\mathbf{r}^i\mathbf{r}^j\mathbf{r}^k\mathbf{r}^l\mathbf{r}^m\mathbf{r}^n = A^{(22)}_{lmnijk}\mathbf{r}^i\mathbf{r}^j\mathbf{r}^k\mathbf{r}^l\mathbf{r}^m\mathbf{r}^n = \left(\frac{\partial^2\Phi}{\partial\underset{\simeq}{\boldsymbol{\gamma}}\partial\underset{\simeq}{\boldsymbol{\gamma}}}\right)_{(0)},\\
{}^8\underset{\sim}{\mathbf{A}}^{(33)} &= A^{(33)}_{ijklmnpq}\mathbf{r}^i\mathbf{r}^j\mathbf{r}^k\mathbf{r}^l\mathbf{r}^m\mathbf{r}^n\mathbf{r}^p\mathbf{r}^q = A^{(33)}_{mnpqijkl}\mathbf{r}^i\mathbf{r}^j\mathbf{r}^k\mathbf{r}^l\mathbf{r}^m\mathbf{r}^n\mathbf{r}^p\mathbf{r}^q = \left(\frac{\partial^2\Phi}{\partial\underset{\approx}{\boldsymbol{\varkappa}}\partial\underset{\approx}{\boldsymbol{\varkappa}}}\right)_{(0)},\\
{}^5\underset{\sim}{\mathbf{A}}^{(12)} &= A^{(12)}_{ijklm}\mathbf{r}^i\mathbf{r}^j\mathbf{r}^k\mathbf{r}^l\mathbf{r}^m = A^{(12)}_{klmij}\mathbf{r}^i\mathbf{r}^j\mathbf{r}^k\mathbf{r}^l\mathbf{r}^m = \frac{1}{2}\left[\left(\frac{\partial^2\Phi}{\partial\underset{\sim}{\boldsymbol{\varepsilon}}\partial\underset{\simeq}{\boldsymbol{\gamma}}}\right)_{(0)} + \left(\frac{\partial^2\Phi}{\partial\underset{\simeq}{\boldsymbol{\gamma}}\partial\underset{\sim}{\boldsymbol{\varepsilon}}}\right)^T_{(0)}\right],\\
{}^6\underset{\sim}{\mathbf{A}}^{(13)} &= A^{(13)}_{ijklmn}\mathbf{r}^i\mathbf{r}^j\mathbf{r}^k\mathbf{r}^l\mathbf{r}^m\mathbf{r}^n = A^{(13)}_{klmnij}\mathbf{r}^i\mathbf{r}^j\mathbf{r}^k\mathbf{r}^l\mathbf{r}^m\mathbf{r}^n = \frac{1}{2}\left[\left(\frac{\partial^2\Phi}{\partial\underset{\sim}{\boldsymbol{\varepsilon}}\partial\underset{\approx}{\boldsymbol{\varkappa}}}\right)_{(0)} + \left(\frac{\partial^2\Phi}{\partial\underset{\approx}{\boldsymbol{\varkappa}}\partial\underset{\simeq}{\boldsymbol{\gamma}}}\right)^T_{(0)}\right],\\
{}^7\underset{\sim}{\mathbf{A}}^{(23)} &= 6A^{(23)}_{ijklmnp}\mathbf{r}^i\mathbf{r}^j\mathbf{r}^k\mathbf{r}^l\mathbf{r}^m\mathbf{r}^n\mathbf{r}^p = A^{(23)}_{lmnpijk}\mathbf{r}^i\mathbf{r}^j\mathbf{r}^k\mathbf{r}^l\mathbf{r}^m\mathbf{r}^n\mathbf{r}^p\\
&= \frac{1}{2}\left[\left(\frac{\partial^2\Phi}{\partial\underset{\simeq}{\boldsymbol{\gamma}}\partial\underset{\approx}{\boldsymbol{\varkappa}}}\right)_{(0)} + \left(\frac{\partial^2\Phi}{\partial\underset{\approx}{\boldsymbol{\varkappa}}\partial\underset{\simeq}{\boldsymbol{\gamma}}}\right)^T_{(0)}\right];\ {}^2\underset{\sim}{\mathbf{B}}^{(11)} = B^{(11)}_{ij}\mathbf{r}^i\mathbf{r}^j = B^{(11)}_{ji}\mathbf{r}^i\mathbf{r}^j = \left(\frac{\partial^2K}{\partial\mathbf{v}\partial\mathbf{v}}\right)_{(0)},\\
{}^4\underset{\sim}{\mathbf{B}}^{(22)} &= B^{(22)}_{ijkl}\mathbf{r}^i\mathbf{r}^j\mathbf{r}^k\mathbf{r}^l = B^{(22)}_{klij}\mathbf{r}^i\mathbf{r}^j\mathbf{r}^k\mathbf{r}^l = \left(\frac{\partial^2K}{\partial\underset{\sim}{\mathbf{v}}\partial\underset{\sim}{\mathbf{v}}}\right)_{(0)},\\
{}^6\underset{\sim}{\mathbf{B}}^{(33)} &= B^{(33)}_{ijklmn}\mathbf{r}^i\mathbf{r}^j\mathbf{r}^k\mathbf{r}^l\mathbf{r}^m\mathbf{r}^n = B^{(33)}_{lmnijk}\mathbf{r}^i\mathbf{r}^j\mathbf{r}^k\mathbf{r}^l\mathbf{r}^m\mathbf{r}^n = \left(\frac{\partial^2K}{\partial\underset{\simeq}{\mathbf{v}}\partial\underset{\simeq}{\mathbf{v}}}\right)_{(0)},\\
{}^3\underset{\sim}{\mathbf{B}}^{(12)} &= B^{(12)}_{ijk}\mathbf{r}^i\mathbf{r}^j\mathbf{r}^k = B^{(12)}_{jki}\mathbf{r}^i\mathbf{r}^j\mathbf{r}^k = \frac{1}{2}\left[\left(\frac{\partial^2K}{\partial\mathbf{v}\partial\underset{\sim}{\mathbf{v}}}\right)_{(0)} + \left(\frac{\partial^2K}{\partial\underset{\sim}{\mathbf{v}}\partial\mathbf{v}}\right)^T_{(0)}\right],\\
{}^4\underset{\sim}{\mathbf{B}}^{(13)} &= B^{(13)}_{ijkl}\mathbf{r}^i\mathbf{r}^j\mathbf{r}^k\mathbf{r}^l = B^{(13)}_{jkli}\mathbf{r}^i\mathbf{r}^j\mathbf{r}^k\mathbf{r}^l = \frac{1}{2}\left[\left(\frac{\partial^2K}{\partial\mathbf{v}\partial\underset{\simeq}{\mathbf{v}}}\right)_{(0)} + \left(\frac{\partial^2K}{\partial\underset{\simeq}{\mathbf{v}}\partial\mathbf{v}}\right)^T_{(0)}\right],\\
{}^5\underset{\sim}{\mathbf{B}}^{(23)} &= B^{(23)}_{ijklm}\mathbf{r}^i\mathbf{r}^j\mathbf{r}^k\mathbf{r}^l\mathbf{r}^m = B^{(23)}_{klmij}\mathbf{r}^i\mathbf{r}^j\mathbf{r}^k\mathbf{r}^l\mathbf{r}^m = \frac{1}{2}\left[\left(\frac{\partial^2K}{\partial\underset{\sim}{\mathbf{v}}\partial\underset{\simeq}{\mathbf{v}}}\right)_{(0)} + \left(\frac{\partial^2K}{\partial\underset{\simeq}{\mathbf{v}}\partial\underset{\sim}{\mathbf{v}}}\right)^T_{(0)}\right],
\end{aligned}
\tag{24.7}
$$

where the index "T" in the upper right corner of the quantities means the transposition operation. It should be noted that the components of the tensors involved in the relations (24.5) and (24.6), in addition to those indicated in (24.7), have the properties:

$$
\begin{gathered}
A^{(12)}_{ijklm} = A^{(12)}_{jiklm} = A^{(12)}_{ijkml},\quad A^{(13)}_{ijklmn} = A^{(13)}_{jiklmn} = A^{(13)}_{ijlkmn} = A^{(13)}_{ijklnm},\\
A^{(23)}_{ijklmnp} = A^{(23)}_{ikjlmnp} = A^{(23)}_{ijkmlnp} = A^{(23)}_{ijklmpn},\quad A^{(11)}_{ijkl} = A^{(11)}_{jikl} = A^{(11)}_{ijlk} = A^{(11)}_{klij},\\
A^{(22)}_{ijklmn} = A^{(22)}_{ikjlmn} = A^{(22)}_{ijklnm} = A^{(22)}_{lmnijk};\quad B^{(13)}_{ijkl} = B^{(13)}_{ikjl},\quad B^{(23)}_{ijklm} = B^{(23)}_{ijlkm},\\
B^{(11)}_{ij} = B^{(11)}_{ji},\quad B^{(22)}_{ijkl} = B^{(22)}_{klij},\quad B^{(33)}_{ijklmn} = B^{(33)}_{ikjlmn} = B^{(33)}_{ijklnm} = B^{(33)}_{lmnijk}.
\end{gathered}
$$

24.2.1 Constitutive Relations of the Second Strain Tensor and Velocity Vector Gradient Elasticity

Due to (24.5) and (24.6) for the constitutive relations of the considered linear second strain tensor and velocity vector gradient elasticity theory will have

$$
\begin{aligned}
&\underset{\sim}{\mathbf{P}} = \partial\Phi/\partial\underset{\sim}{\boldsymbol{\varepsilon}} = {}^{4}\underset{\sim}{\mathbf{A}}^{(11)} \overset{2}{\otimes} \underset{\sim}{\boldsymbol{\varepsilon}} + {}^{5}\underset{\sim}{\mathbf{A}}^{(12)} \overset{3}{\otimes} \underset{\simeq}{\boldsymbol{\gamma}} + {}^{6}\underset{\sim}{\mathbf{A}}^{(13)} \overset{4}{\otimes} \underset{\approx}{\boldsymbol{\varkappa}},\\
&\underset{\simeq}{\mathbf{P}}^{(1)} = \partial\Phi/\partial\nabla\underset{\sim}{\boldsymbol{\varepsilon}} = ({}^{5}\underset{\sim}{\mathbf{A}}^{(12)})^{T} \overset{2}{\otimes} \underset{\sim}{\boldsymbol{\varepsilon}} + {}^{6}\underset{\sim}{\mathbf{A}}^{(22)} \overset{3}{\otimes} \underset{\simeq}{\boldsymbol{\gamma}} + {}^{7}\underset{\sim}{\mathbf{A}}^{(23)} \overset{4}{\otimes} \underset{\approx}{\boldsymbol{\varkappa}},\\
&\underset{\approx}{\mathbf{P}}^{(2)} = \partial\Phi/\partial\nabla\nabla\underset{\sim}{\boldsymbol{\varepsilon}} = ({}^{6}\underset{\sim}{\mathbf{A}}^{(13)})^{T} \overset{2}{\otimes} \underset{\sim}{\boldsymbol{\varepsilon}} + ({}^{7}\underset{\sim}{\mathbf{A}}^{(23)})^{T} \overset{3}{\otimes} \underset{\simeq}{\boldsymbol{\gamma}} + {}^{8}\underset{\sim}{\mathbf{A}}^{(33)} \overset{4}{\otimes} \underset{\approx}{\boldsymbol{\varkappa}};
\end{aligned}
\tag{24.8}
$$

$$
\begin{aligned}
&\boldsymbol{\pi} = \partial K/\partial\mathbf{v} = {}^{2}\underset{\sim}{\mathbf{B}}^{(11)} \overset{1}{\otimes} \mathbf{v} + {}^{3}\underset{\sim}{\mathbf{B}}^{(12)} \overset{2}{\otimes} \underset{\sim}{\mathbf{v}} + {}^{4}\underset{\sim}{\mathbf{B}}^{(13)} \overset{3}{\otimes} \underset{\simeq}{\mathbf{v}},\\
&\underset{\sim}{\boldsymbol{\pi}}^{(1)} = \partial K/\partial\nabla\mathbf{v} = ({}^{3}\underset{\sim}{\mathbf{B}}^{(12)})^{T} \overset{1}{\otimes} \mathbf{v} + {}^{4}\underset{\sim}{\mathbf{B}}^{(22)} \overset{2}{\otimes} \underset{\sim}{\mathbf{v}} + {}^{5}\underset{\sim}{\mathbf{B}}^{(23)} \overset{3}{\otimes} \underset{\simeq}{\mathbf{v}},\\
&\underset{\simeq}{\boldsymbol{\pi}}^{(2)} = \partial K/\partial\nabla\nabla\mathbf{v} = ({}^{4}\underset{\sim}{\mathbf{B}}^{(13)})^{T} \overset{1}{\otimes} \mathbf{v} + ({}^{5}\underset{\sim}{\mathbf{B}}^{(23)})^{T} \overset{2}{\otimes} \underset{\sim}{\mathbf{v}} + {}^{6}\underset{\sim}{\mathbf{B}}^{(33)} \overset{3}{\otimes} \underset{\simeq}{\mathbf{v}}.
\end{aligned}
\tag{24.9}
$$

Introducing tensor-block matrices and tensor columns

$$
\mathbb{M}=\begin{pmatrix} {}^{4}\underset{\sim}{\mathbf{A}}^{(11)} & {}^{5}\underset{\sim}{\mathbf{A}}^{(12)} & {}^{6}\underset{\sim}{\mathbf{A}}^{(13)}\\ ({}^{5}\underset{\sim}{\mathbf{A}}^{(12)})^{T} & {}^{6}\underset{\sim}{\mathbf{A}}^{(22)} & {}^{7}\underset{\sim}{\mathbf{A}}^{(23)}\\ ({}^{6}\underset{\sim}{\mathbf{A}}^{(13)})^{T} & ({}^{7}\underset{\sim}{\mathbf{A}}^{(23)})^{T} & {}^{8}\underset{\sim}{\mathbf{A}}^{(33)} \end{pmatrix},\quad
\mathbb{X}=\begin{pmatrix} \underset{\sim}{\boldsymbol{\varepsilon}}\\ \underset{\simeq}{\boldsymbol{\gamma}}\\ \underset{\approx}{\boldsymbol{\varkappa}} \end{pmatrix},\quad
\mathbb{Y}=\begin{pmatrix} \underset{\sim}{\mathbf{P}}\\ \underset{\simeq}{\mathbf{P}}^{(1)}\\ \underset{\approx}{\mathbf{P}}^{(2)} \end{pmatrix},
$$

$$
\mathbb{N}=\begin{pmatrix} {}^{2}\underset{\sim}{\mathbf{B}}^{(11)} & {}^{3}\underset{\sim}{\mathbf{B}}^{(12)} & {}^{4}\underset{\sim}{\mathbf{B}}^{(13)}\\ ({}^{3}\underset{\sim}{\mathbf{B}}^{(12)})^{T} & {}^{4}\underset{\sim}{\mathbf{B}}^{(22)} & {}^{5}\underset{\sim}{\mathbf{B}}^{(23)}\\ ({}^{4}\underset{\sim}{\mathbf{B}}^{(13)})^{T} & ({}^{5}\underset{\sim}{\mathbf{B}}^{(23)})^{T} & {}^{6}\underset{\sim}{\mathbf{B}}^{(33)} \end{pmatrix},\quad
\mathbb{U}=\begin{pmatrix} \mathbf{v}\\ \underset{\sim}{\mathbf{v}}\\ \underset{\simeq}{\mathbf{v}} \end{pmatrix},\quad
\mathbb{V}=\begin{pmatrix} \boldsymbol{\pi}\\ \underset{\sim}{\boldsymbol{\pi}}^{(1)}\\ \underset{\simeq}{\boldsymbol{\pi}}^{(2)} \end{pmatrix},
$$

the constitutive relations (24.8) and (24.9) can be written in short form

$$
\mathbb{Y} = \mathbb{M} \odot \mathbb{X}, \quad \mathbb{V} = \mathbb{N} \circledcirc \mathbb{U},
$$

where the following operations are introduced for the product of matrices:

$$
\odot=\begin{pmatrix} \overset{2}{\otimes} & 0 & 0\\ 0 & \overset{3}{\otimes} & 0\\ 0 & 0 & \overset{4}{\otimes} \end{pmatrix},\quad
\circledcirc=\begin{pmatrix} \overset{1}{\otimes} & 0 & 0\\ 0 & \overset{2}{\otimes} & 0\\ 0 & 0 & \overset{3}{\otimes} \end{pmatrix}.
$$

Naturally, the following problems arise:

$$
\mathbb{M} \odot \mathbb{X}=\lambda\mathbb{X}\,?, \quad \mathbb{N} \circledcirc \mathbb{U}=\lambda\mathbb{U}\,?,
\tag{24.10}
$$

i.e. eigenvalue problems for tensor block matrices. The eigenvalue problems for tensors and tensor-block matrices can be found in [55, 56].

Note that the terms involved in (24.5) and (24.6) can be written in the form

$$
\begin{aligned}
&{}^4\underset{\approx}{\mathbf{A}}^{(11)} \overset{4}{\otimes} \underset{\sim}{\boldsymbol{\varepsilon}}\underset{\sim}{\boldsymbol{\varepsilon}} = \underset{\approx}{\mathbf{A}} \overset{4}{\otimes} \underset{\sim}{\boldsymbol{\varepsilon}}\underset{\sim}{\boldsymbol{\varepsilon}}, \quad {}^6\underset{\approx}{\mathbf{A}}^{(22)} \overset{6}{\otimes} \underset{\simeq}{\boldsymbol{\gamma}}\underset{\simeq}{\boldsymbol{\gamma}} = \underset{\approx}{\mathbf{A}} \overset{4}{\otimes} (\overset{1}{\underset{\simeq}{\boldsymbol{\gamma}}}{}^{T} \cdot \underset{\sim}{\mathbf{a}}^{(1)} \cdot \underset{\simeq}{\boldsymbol{\gamma}}),\\
&{}^8\underset{\approx}{\mathbf{A}}^{(33)} \overset{8}{\otimes} \underset{\approx}{\boldsymbol{\varkappa}}\underset{\approx}{\boldsymbol{\varkappa}} = \underset{\approx}{\mathbf{A}} \overset{4}{\otimes} (\underset{\approx}{\boldsymbol{\varkappa}}^T \overset{2}{\otimes} \underset{\approx}{\mathbf{a}}^{(2)} \overset{2}{\otimes} \underset{\approx}{\boldsymbol{\varkappa}}), \quad {}^5\underset{\approx}{\mathbf{A}}^{(12)} \overset{5}{\otimes} \underset{\sim}{\boldsymbol{\varepsilon}}\underset{\simeq}{\boldsymbol{\gamma}} = \underset{\approx}{\bar{\mathbf{A}}} \overset{4}{\otimes} (\underset{\sim}{\boldsymbol{\varepsilon}}\mathbf{a} \cdot \underset{\simeq}{\boldsymbol{\gamma}}),\\
&{}^6\underset{\approx}{\mathbf{A}}^{(13)} \overset{6}{\otimes} \underset{\sim}{\boldsymbol{\varepsilon}}\underset{\approx}{\boldsymbol{\varkappa}} = \underset{\approx}{\bar{\mathbf{A}}} \overset{4}{\otimes} (\underset{\sim}{\boldsymbol{\varepsilon}}\underset{\sim}{\mathbf{a}}^{(3)} \overset{2}{\otimes} \underset{\approx}{\boldsymbol{\varkappa}}), \quad {}^7\underset{\approx}{\mathbf{A}}^{(23)} \overset{7}{\otimes} \underset{\simeq}{\boldsymbol{\gamma}}\underset{\approx}{\boldsymbol{\varkappa}} = \underset{\approx}{\bar{\mathbf{A}}} \overset{4}{\otimes} (\overset{1}{\underset{\simeq}{\boldsymbol{\gamma}}}{}^{T} \cdot \underset{\simeq}{\mathbf{a}}^{(4)} \overset{2}{\otimes} \underset{\approx}{\boldsymbol{\varkappa}});\\
&{}^2\underset{\sim}{\mathbf{B}}^{(11)} \overset{2}{\otimes} \mathbf{v}\mathbf{v} = \underset{\sim}{\mathbf{B}} \overset{2}{\otimes} \mathbf{v}\mathbf{v}, \quad {}^4\underset{\approx}{\mathbf{B}}^{(22)} \overset{3}{\otimes} \underset{\sim}{\mathbf{v}}\underset{\sim}{\mathbf{v}} = \underset{\sim}{\mathbf{B}} \overset{2}{\otimes} (\underset{\sim}{\mathbf{v}}^T \cdot \underset{\sim}{\mathbf{b}}^{(1)} \cdot \underset{\sim}{\mathbf{v}}),\\
&{}^6\underset{\sim}{\mathbf{B}}^{(33)} \overset{6}{\otimes} \underset{\simeq}{\mathbf{v}}\underset{\simeq}{\mathbf{v}} = \underset{\sim}{\mathbf{B}} \overset{2}{\otimes} (\overset{2}{\underset{\simeq}{\mathbf{v}}}{}^{T} \overset{2}{\otimes} \underset{\approx}{\mathbf{b}}^{(2)} \overset{2}{\otimes} \underset{\simeq}{\mathbf{v}}), \quad {}^3\underset{\sim}{\mathbf{B}}^{(12)} \overset{3}{\otimes} \mathbf{v}\underset{\sim}{\mathbf{v}} = \underset{\sim}{\bar{\mathbf{B}}} \overset{2}{\otimes} (\mathbf{v}\mathbf{b} \cdot \underset{\sim}{\mathbf{v}}),
\end{aligned}
\tag{24.11}
$$

$$
{}^4\underset{\sim}{\mathbf{B}}^{(13)} \overset{4}{\otimes} \mathbf{v}\underset{\simeq}{\mathbf{v}} = \underset{\sim}{\bar{\mathbf{B}}} \overset{2}{\otimes} (\mathbf{v}\underset{\sim}{\mathbf{b}}^{(3)} \overset{2}{\otimes} \underset{\simeq}{\mathbf{v}}), \quad {}^5\underset{\sim}{\mathbf{B}}^{(23)} \overset{5}{\otimes} \underset{\sim}{\mathbf{v}}\underset{\simeq}{\mathbf{v}} = \underset{\sim}{\bar{\mathbf{B}}} \overset{2}{\otimes} (\underset{\sim}{\mathbf{v}}^T \cdot \underset{\simeq}{\mathbf{b}}^{(4)} \overset{2}{\otimes} \underset{\simeq}{\mathbf{v}}),
$$

where $\underset{\approx}{\mathbf{A}} = A_{ijkl}\mathbf{e}_i\mathbf{e}_j\mathbf{e}_k\mathbf{e}_l$ is elastic modulus tensor, $A_{ijkl} = A_{klij} = A_{jikl}$, $\underset{\approx}{\bar{\mathbf{A}}} = \bar{A}_{ijkl}\mathbf{e}_i\mathbf{e}_j\mathbf{e}_k\mathbf{e}_l$, $\bar{A}_{ijkl} = \bar{A}_{jikl} = \bar{A}_{ijlk}$, $\bar{A}_{ijkl} \neq \bar{A}_{klij}$, $\mathbf{a} = a_i\mathbf{e}_i$, $\underset{\sim}{\mathbf{a}}^{(1)} = a^{(1)}_{ij}\mathbf{e}_i\mathbf{e}_j$, $\underset{\approx}{\mathbf{a}}^{(2)} = a^{(2)}_{ijkl}\mathbf{e}_i\mathbf{e}_j\mathbf{e}_k\mathbf{e}_l$, $\underset{\sim}{\mathbf{a}}^{(3)} = a^{(3)}_{ij}\mathbf{e}_i\mathbf{e}_j$ and $\underset{\simeq}{\mathbf{a}}^{(4)} = a^{(4)}_{ijk}\mathbf{e}_i\mathbf{e}_j\mathbf{e}_k$ are length scale tensors for statics, $a^{(1)}_{ij} = a^{(1)}_{ji}$, $a^{(2)}_{ijkl} = a^{(2)}_{klij} = a^{(2)}_{jikl}$, $a^{(3)}_{ij} = a^{(3)}_{ji}$, $a^{(4)}_{ijk} = a^{(4)}_{ikj}$; $\underset{\sim}{\mathbf{B}} = B_{ij}\mathbf{e}_i\mathbf{e}_j$, $\underset{\sim}{\bar{\mathbf{B}}} = \bar{B}_{ij}\mathbf{e}_i\mathbf{e}_j$, $\mathbf{b} = b_i\mathbf{e}_i$, $\underset{\sim}{\mathbf{b}}^{(1)} = b^{(1)}_{ij}\mathbf{e}_i\mathbf{e}_j$, $\underset{\approx}{\mathbf{b}}^{(2)} = b^{(2)}_{ijkl}\mathbf{e}_i\mathbf{e}_j\mathbf{e}_k\mathbf{e}_l$, $\underset{\sim}{\mathbf{b}}^{(3)} = b^{(3)}_{ij}\mathbf{e}_i\mathbf{e}_j$ and $\underset{\simeq}{\mathbf{b}}^{(4)} = b^{(4)}_{ijk}\mathbf{e}_i\mathbf{e}_j\mathbf{e}_k$ are tensors for inertia effects, $b^{(1)}_{ij} = b^{(1)}_{ji}$, $b^{(2)}_{ijkl} = b^{(2)}_{klij} = b^{(2)}_{jikl}$, $B_{ij} = B_{ji}$, $\bar{B}_{ij} \neq \bar{B}_{ji}$, $b^{(3)}_{ij} = b^{(3)}_{ji}$, $b^{(4)}_{ijk} = b^{(4)}_{ikj}$, $\overset{1}{\underset{\simeq}{\boldsymbol{\gamma}}}{}^{T} = (\gamma_{ijk}\mathbf{e}_i\mathbf{e}_j\mathbf{e}_k)^{\overset{1}{T}} = \gamma_{ijk}\mathbf{e}_j\mathbf{e}_k\mathbf{e}_i$, $\overset{2}{\underset{\simeq}{\mathbf{v}}}{}^{T} = (v_{ijk}\mathbf{e}_i\mathbf{e}_j\mathbf{e}_k)^{\overset{2}{T}} = v_{ijk}\mathbf{e}_k\mathbf{e}_i\mathbf{e}_j$, $\underset{\approx}{\boldsymbol{\varkappa}}^T = (\varkappa_{ijkl}\mathbf{e}_i\mathbf{e}_j\mathbf{e}_k\mathbf{e}_l)^T = \varkappa_{ijkl}\mathbf{e}_k\mathbf{e}_l\mathbf{e}_i\mathbf{e}_j$, $\mathbf{e}_i \cdot \mathbf{e}_j = \delta_{ij}$, δ_{ij} is Kronecker delta, $i, j, k, l = 1, 2, 3$. On repeating indices, summation occurs from one to three.

Taking into account (24.11), the specific strain energy (24.5), kinetic energy (24.6) and the constitutive relations (24.8) and (24.9) are represented in the form

$$
\begin{aligned}
\Phi(\underset{\sim}{\boldsymbol{\varepsilon}}, \underset{\simeq}{\boldsymbol{\gamma}}, \underset{\approx}{\boldsymbol{\varkappa}}) &= \frac{1}{2}\Big[\underset{\approx}{\mathbf{A}} \overset{4}{\otimes} \Big(\underset{\sim}{\boldsymbol{\varepsilon}}\underset{\sim}{\boldsymbol{\varepsilon}} + \overset{1}{\underset{\simeq}{\boldsymbol{\gamma}}}{}^{T} \cdot \underset{\sim}{\mathbf{a}}^{(1)} \cdot \underset{\simeq}{\boldsymbol{\gamma}} + \underset{\approx}{\boldsymbol{\varkappa}}^T \overset{2}{\otimes} \underset{\approx}{\mathbf{a}}^{(2)} \overset{2}{\otimes} \underset{\approx}{\boldsymbol{\varkappa}}\Big)\\
&\quad + 2\underset{\approx}{\bar{\mathbf{A}}} \overset{4}{\otimes} \Big(\underset{\sim}{\boldsymbol{\varepsilon}}\mathbf{a} \cdot \underset{\simeq}{\boldsymbol{\gamma}} + \underset{\sim}{\boldsymbol{\varepsilon}}\underset{\sim}{\mathbf{a}}^{(3)} \overset{2}{\otimes} \underset{\approx}{\boldsymbol{\varkappa}} + \overset{1}{\underset{\simeq}{\boldsymbol{\gamma}}}{}^{T} \cdot \underset{\simeq}{\mathbf{a}}^{(4)} \overset{2}{\otimes} \underset{\approx}{\boldsymbol{\varkappa}}\Big)\Big]\\
&= \frac{1}{2}\Big[A^{ijkl}\Big(\varepsilon_{ij}\varepsilon_{kl} + \gamma_{mij}a^{(1)mn}\gamma_{nkl} + \varkappa_{mnij}a^{(2)mnpq}\varkappa_{pqkl}\Big)\\
&\quad + 2\bar{A}^{ijkl}\Big(\varepsilon_{ij}a^{m}\gamma_{mkl} + \varepsilon_{ij}a^{(3)mn}\varkappa_{mnkl} + \gamma_{mij}a^{(4)mnp}\varkappa_{npkl}\Big)\Big];
\end{aligned}
\tag{24.12}
$$

$$
\begin{aligned}
K(\mathbf{v}, \underset{\sim}{\mathbf{v}}, \underset{\simeq}{\mathbf{v}}) &= \frac{1}{2}\Big[\underset{\sim}{\mathbf{B}} \overset{2}{\otimes} \Big(\mathbf{v}\mathbf{v} + \underset{\sim}{\mathbf{v}}^{T} \cdot \underset{\sim}{\mathbf{b}}^{(1)} \cdot \underset{\sim}{\mathbf{v}} + \underset{\simeq}{\mathbf{v}}^{\overset{2}{T}} \overset{2}{\otimes} \underset{\approx}{\mathbf{b}}^{(2)} \overset{2}{\otimes} \underset{\simeq}{\mathbf{v}}\Big) \\
&\quad + 2\underset{\sim}{\bar{\mathbf{B}}} \overset{2}{\otimes} \Big(\mathbf{v}\mathbf{b} \cdot \underset{\sim}{\mathbf{v}} + \mathbf{v}\underset{\sim}{\mathbf{b}}^{(3)} \overset{2}{\otimes} \underset{\simeq}{\mathbf{v}} + \underset{\sim}{\mathbf{v}}^{T} \cdot \underset{\simeq}{\mathbf{b}}^{(4)} \overset{2}{\otimes} \underset{\simeq}{\mathbf{v}}\Big)\Big] \\
&= \frac{1}{2}\Big[B^{ij}\Big(v_i v_j + v_{ki} b^{(1)kl} v_{lj} + v_{kli} b^{(2)klmn} v_{mnj}\Big) \\
&\quad + 2\bar{B}^{ij}\Big(v_i b^{k} v_{kj} + v_i b^{(3)kl} v_{klj} + v_{ki} b^{(4)klm} v_{lmj}\Big)\Big];
\end{aligned}
\tag{24.13}
$$

$$
\begin{aligned}
&\underset{\sim}{\mathbf{P}} = \partial\Phi/\partial\underset{\sim}{\boldsymbol{\varepsilon}} = \underset{\approx}{\mathbf{A}} \overset{2}{\otimes} \boldsymbol{\varepsilon} + \underset{\approx}{\bar{\mathbf{A}}} \overset{2}{\otimes} (\mathbf{a} \cdot \underset{\simeq}{\boldsymbol{\gamma}} + \underset{\sim}{\mathbf{a}}^{(3)} \overset{2}{\otimes} \underset{\approx}{\boldsymbol{\varkappa}}), \\
&\underset{\simeq}{\mathbf{P}}^{(1)} = \partial\Phi/\partial\nabla\underset{\sim}{\boldsymbol{\varepsilon}} = \mathbf{a}\underset{\approx}{\bar{\mathbf{A}}}^{T} \overset{2}{\otimes} \underset{\sim}{\boldsymbol{\varepsilon}} + \underset{\sim}{\mathbf{a}}^{(1)} \cdot \underset{\simeq}{\boldsymbol{\gamma}} \overset{2}{\otimes} \underset{\approx}{\mathbf{A}} + (\underset{\simeq}{\mathbf{a}}^{(4)} \overset{2}{\otimes} \underset{\approx}{\boldsymbol{\varkappa}}) \overset{2}{\otimes} \underset{\approx}{\bar{\mathbf{A}}}^{T}, \\
&\underset{\approx}{\mathbf{P}}^{(2)} = \partial\Phi/\partial\nabla\nabla\underset{\sim}{\boldsymbol{\varepsilon}} = \underset{\sim}{\mathbf{a}}^{(3)}\underset{\approx}{\bar{\mathbf{A}}}^{T} \overset{2}{\otimes} \underset{\sim}{\boldsymbol{\varepsilon}} + (\underset{\simeq}{\mathbf{a}}^{(4)})^{\overset{1}{T}} \cdot \underset{\simeq}{\boldsymbol{\gamma}} \overset{2}{\otimes} \underset{\approx}{\bar{\mathbf{A}}} + (\underset{\approx}{\mathbf{a}}^{(2)} \overset{2}{\otimes} \underset{\approx}{\boldsymbol{\varkappa}}) \overset{2}{\otimes} \underset{\approx}{\mathbf{A}}; \\
&\boldsymbol{\pi} = \partial K/\partial\mathbf{v} = \underset{\sim}{\mathbf{B}} \cdot \mathbf{v} + \underset{\sim}{\bar{\mathbf{B}}} \cdot (\mathbf{b} \cdot \underset{\sim}{\mathbf{v}} + \underset{\sim}{\mathbf{b}}^{(3)} \overset{2}{\otimes} \underset{\simeq}{\mathbf{v}}), \\
&\underset{\sim}{\boldsymbol{\pi}}^{(1)} = \partial K/\partial\nabla\mathbf{v} = \mathbf{b}\underset{\sim}{\bar{\mathbf{B}}}^{T} \cdot \mathbf{v} + \underset{\sim}{\mathbf{b}}^{(1)} \cdot \underset{\sim}{\mathbf{v}} \cdot \underset{\sim}{\mathbf{B}} + (\underset{\simeq}{\mathbf{b}}^{(4)} \overset{2}{\otimes} \underset{\simeq}{\mathbf{v}}) \cdot \underset{\sim}{\bar{\mathbf{B}}}^{T}, \\
&\underset{\simeq}{\boldsymbol{\pi}}^{(2)} = \partial K/\partial\nabla\nabla\mathbf{v} = \underset{\sim}{\mathbf{b}}^{(3)}\mathbf{v} \cdot \underset{\sim}{\bar{\mathbf{B}}} + \underset{\sim}{\bar{\mathbf{B}}}^{T} \cdot (\underset{\sim}{\mathbf{v}}^{T} \cdot \underset{\simeq}{\mathbf{b}}^{(4)}) + (\underset{\approx}{\mathbf{b}}^{(2)} \overset{2}{\otimes} \underset{\simeq}{\mathbf{v}}) \cdot \underset{\sim}{\mathbf{B}}.
\end{aligned}
\tag{24.14}
$$

If the material has a center of symmetry, then the material tensors of odd rank are equal to zero and the (24.12), (24.13) and (24.14) take the form

$$
\begin{aligned}
\Phi(\underset{\sim}{\boldsymbol{\varepsilon}}, \underset{\simeq}{\boldsymbol{\gamma}}, \underset{\approx}{\boldsymbol{\varkappa}}) &= \frac{1}{2}\underset{\approx}{\mathbf{A}} \overset{4}{\otimes} (\boldsymbol{\varepsilon}\boldsymbol{\varepsilon} + \underset{\simeq}{\boldsymbol{\gamma}}^{\overset{1}{T}} \cdot \underset{\sim}{\mathbf{a}}^{(1)} \cdot \underset{\simeq}{\boldsymbol{\gamma}} + \underset{\approx}{\boldsymbol{\varkappa}}^{T} \overset{2}{\otimes} \underset{\approx}{\mathbf{a}}^{(2)} \overset{2}{\otimes} \underset{\approx}{\boldsymbol{\varkappa}}) + \underset{\approx}{\bar{\mathbf{A}}} \overset{4}{\otimes} (\boldsymbol{\varepsilon}\underset{\sim}{\mathbf{a}}^{(3)} \overset{2}{\otimes} \underset{\approx}{\boldsymbol{\varkappa}}) \\
&= \frac{1}{2}A^{ijkl}(\varepsilon_{ij}\varepsilon_{kl} + \gamma_{mij} a^{(1)mn} \gamma_{nkl} + \varkappa_{mnij} a^{(2)mnpq} \varkappa_{pqkl}) + \bar{A}^{ijkl}\varepsilon_{ij} a^{(3)mn} \varkappa_{mnkl};
\end{aligned}
\tag{24.15}
$$

$$
\begin{aligned}
K(\mathbf{v}, \underset{\sim}{\mathbf{v}}, \underset{\simeq}{\mathbf{v}}) &= \frac{1}{2}\underset{\sim}{\mathbf{B}} \overset{2}{\otimes} (\mathbf{v}\mathbf{v} + \underset{\sim}{\mathbf{v}}^{T} \cdot \underset{\sim}{\mathbf{b}}^{(1)} \cdot \underset{\sim}{\mathbf{v}} + \underset{\simeq}{\mathbf{v}}^{\overset{2}{T}} \overset{2}{\otimes} \underset{\approx}{\mathbf{b}}^{(2)} \overset{2}{\otimes} \underset{\simeq}{\mathbf{v}}) + \underset{\sim}{\bar{\mathbf{B}}} \overset{2}{\otimes} \mathbf{v}\underset{\sim}{\mathbf{b}}^{(3)} \overset{2}{\otimes} \underset{\simeq}{\mathbf{v}} \\
&= \frac{1}{2}B^{ij}(v_i v_j + v_{ki} b^{(1)kl} v_{lj} + v_{kli} b^{(2)klmn} v_{mnj}) + \bar{B}^{ij} v_i b^{(3)kl} v_{klj};
\end{aligned}
\tag{24.16}
$$

$$
\begin{aligned}
&\underset{\sim}{\mathbf{P}} = \partial\Phi/\partial\underset{\sim}{\boldsymbol{\varepsilon}} = \underset{\approx}{\mathbf{A}} \overset{2}{\otimes} \underset{\sim}{\boldsymbol{\varepsilon}} + \underset{\approx}{\bar{\mathbf{A}}} \overset{2}{\otimes} (\underset{\sim}{\mathbf{a}}^{(3)} \overset{2}{\otimes} \underset{\approx}{\boldsymbol{\varkappa}}), \quad (\underset{\sim}{\mathbf{a}}^{(3)})^{T} = \underset{\sim}{\mathbf{a}}^{(3)}, \\
&\underset{\simeq}{\mathbf{P}}^{(1)} = \partial\Phi/\partial\nabla\underset{\sim}{\boldsymbol{\varepsilon}} = \underset{\sim}{\mathbf{a}}^{(1)} \cdot \underset{\simeq}{\boldsymbol{\gamma}} \overset{2}{\otimes} \underset{\approx}{\mathbf{A}}, \quad (\underset{\sim}{\mathbf{a}}^{(1)})^{T} = \underset{\sim}{\mathbf{a}}^{(1)}, \\
&\underset{\approx}{\mathbf{P}}^{(2)} = \partial\Phi/\partial\nabla\nabla\underset{\sim}{\boldsymbol{\varepsilon}} = \underset{\sim}{\mathbf{a}}^{(3)}\underset{\approx}{\bar{\mathbf{A}}}^{T} \overset{2}{\otimes} \underset{\sim}{\boldsymbol{\varepsilon}} + (\underset{\approx}{\mathbf{a}}^{(2)} \overset{2}{\otimes} \underset{\approx}{\boldsymbol{\varkappa}}) \overset{2}{\otimes} \underset{\approx}{\mathbf{A}}, \quad (\underset{\approx}{\mathbf{a}}^{(2)})^{T} = \underset{\approx}{\mathbf{a}}^{(2)}; \\
&\boldsymbol{\pi} = \partial K/\partial\mathbf{v} = \underset{\sim}{\mathbf{B}} \cdot \mathbf{v} + \underset{\sim}{\bar{\mathbf{B}}} \cdot (\underset{\sim}{\mathbf{b}}^{(3)} \overset{2}{\otimes} \underset{\simeq}{\mathbf{v}}), \quad (\underset{\sim}{\mathbf{b}}^{(3)})^{T} = \underset{\sim}{\mathbf{b}}^{(3)}, \\
&\underset{\sim}{\boldsymbol{\pi}}^{(1)} = \partial K/\partial\nabla\mathbf{v} = \underset{\sim}{\mathbf{b}}^{(1)} \cdot \underset{\sim}{\mathbf{v}} \cdot \underset{\sim}{\mathbf{B}}, \quad (\underset{\sim}{\mathbf{b}}^{(1)})^{T} = \underset{\sim}{\mathbf{b}}^{(1)}, \\
&\underset{\simeq}{\boldsymbol{\pi}}^{(2)} = \partial K/\partial\nabla\nabla\mathbf{v} = \underset{\sim}{\mathbf{b}}^{(3)}\mathbf{v} \cdot \underset{\sim}{\bar{\mathbf{B}}} + (\underset{\approx}{\mathbf{b}}^{(2)} \overset{2}{\otimes} \underset{\simeq}{\mathbf{v}}) \cdot \underset{\sim}{\mathbf{B}}, \quad (\underset{\approx}{\mathbf{b}}^{(2)})^{T} = \underset{\approx}{\mathbf{b}}^{(2)},
\end{aligned}
\tag{24.17}
$$

and if $\underset{\approx}{\bar{\mathbf{A}}} = 0$ and $\underset{\sim}{\bar{\mathbf{B}}} = 0$, then from (24.17) for homogeneous anisotropic media we get

$$\underset{\sim}{\mathbf{P}} = \underset{\approx}{\mathbf{A}} \overset{2}{\otimes} \underset{\sim}{\boldsymbol{\varepsilon}}, \quad \underset{\simeq}{\mathbf{P}}^{(1)} = \underset{\sim}{\mathbf{a}}^{(1)} \cdot \nabla \underset{\sim}{\mathbf{P}}, \quad \underset{\approx}{\mathbf{P}}^{(2)} = \underset{\approx}{\mathbf{a}}^{(2)} \overset{2}{\otimes} \nabla\nabla \underset{\sim}{\mathbf{P}};$$
$$\boldsymbol{\pi} = \underset{\sim}{\mathbf{B}} \cdot \mathbf{v}, \quad \underset{\sim}{\boldsymbol{\pi}}^{(1)} = \underset{\sim}{\mathbf{b}}^{(1)} \cdot \nabla \boldsymbol{\pi}, \quad \underset{\simeq}{\boldsymbol{\pi}}^{(2)} = \underset{\approx}{\mathbf{b}}^{(2)} \overset{2}{\otimes} \nabla\nabla \boldsymbol{\pi}. \tag{24.18}$$

24.2.2 Equations of Motion of the SST and VVG Elasticity Theory

Following [68, 69] (see also [18, 49, 66, 67, 70]), we introduce the symmetric total stress tensor $\underset{\sim}{\mathbf{T}}$ and the (basic) momentum vector (body momentum) $\mathbf{p}$ as a function of the velocity and velocity gradients, which due to (24.18) are represented as follows:

$$\underset{\sim}{\mathbf{T}} = \underset{\sim}{\mathbf{P}} - \nabla \cdot \underset{\simeq}{\mathbf{P}}^{(1)} + \nabla\nabla \overset{2}{\otimes} \underset{\approx}{\mathbf{P}}^{(2)} = (1 - \underset{\sim}{\mathbf{a}}^{(1)} \overset{2}{\otimes} \nabla\nabla + \underset{\approx}{\mathbf{a}}^{(2)} \overset{4}{\otimes} \nabla\nabla\nabla\nabla)\underset{\sim}{\mathbf{P}},$$
$$\mathbf{p} = \boldsymbol{\pi} - \nabla \cdot \underset{\sim}{\boldsymbol{\pi}}^{(1)} + \nabla\nabla \overset{2}{\otimes} \underset{\simeq}{\boldsymbol{\pi}}^{(2)} = (1 - \underset{\sim}{\mathbf{b}}^{(1)} \overset{2}{\otimes} \nabla\nabla + \underset{\approx}{\mathbf{b}}^{(2)} \overset{4}{\otimes} \nabla\nabla\nabla\nabla)\boldsymbol{\pi}, \tag{24.19}$$

where $\underset{\sim}{\mathbf{P}}$ is Cauchy stress tensor, $\boldsymbol{\pi} = \rho \mathbf{v}$, $\underset{\sim}{\mathbf{B}} = \rho \underset{\sim}{\mathbf{E}}$, ρ is density, $\underset{\sim}{\mathbf{a}}^{(1)}$ and $\underset{\approx}{\mathbf{a}}^{(2)}$ are length scale tensors for statics, $\underset{\sim}{\mathbf{b}}^{(1)}$ and $\underset{\approx}{\mathbf{b}}^{(2)}$ are length scale tensors for inertia effects (dynamics).

Having expressions for the total stress tensor and the momentum vector (see (24.19)), the motion equations for the considered body can be written as

$$\left(\nabla \cdot \underset{\sim}{\mathbf{T}} + \rho \mathbf{F} = \frac{d\mathbf{p}}{dt}\right) \Rightarrow \left(a\nabla \cdot \underset{\sim}{\mathbf{P}} + \rho \mathbf{F} = b\frac{d\boldsymbol{\pi}}{dt}\right),$$
$$a = 1 - \underset{\sim}{\mathbf{a}}^{(1)} \overset{2}{\otimes} \nabla\nabla + \underset{\approx}{\mathbf{a}}^{(2)} \overset{4}{\otimes} \nabla\nabla\nabla\nabla, \quad b = 1 - \underset{\sim}{\mathbf{b}}^{(1)} \overset{2}{\otimes} \nabla\nabla + \underset{\approx}{\mathbf{b}}^{(2)} \overset{4}{\otimes} \nabla\nabla\nabla\nabla. \tag{24.20}$$

Considering that in the case of an arbitrarily anisotropic linear elasticity theory $\underset{\sim}{\mathbf{P}} = \underset{\approx}{\mathbf{A}} \overset{2}{\otimes} \underset{\sim}{\boldsymbol{\varepsilon}} = \underset{\approx}{\mathbf{A}} \overset{2}{\otimes} \nabla \mathbf{u}$, where $\mathbf{u}$ is displacement vector, $\underset{\approx}{\mathbf{A}}$ is elastic modulus tensor and $(d\mathbf{v})/(dt) \approx (\partial^2 \mathbf{u})/(\partial t) = \partial_t^2 \mathbf{u}$, from (24.20) for $\rho = const$ we get

$$(1 - \underset{\sim}{\mathbf{a}}^{(1)} \overset{2}{\otimes} \nabla\nabla + \underset{\approx}{\mathbf{a}}^{(2)} \overset{4}{\otimes} \nabla\nabla\nabla\nabla)(\underset{\approx}{\mathbf{A}} \overset{3}{\otimes} \nabla\nabla\mathbf{u}) + \rho\mathbf{F}$$
$$= \rho(1 - \underset{\sim}{\mathbf{b}}^{(1)} \overset{2}{\otimes} \nabla\nabla + \underset{\approx}{\mathbf{b}}^{(2)} \overset{4}{\otimes} \nabla\nabla\nabla\nabla)\partial_t^2 \mathbf{u}. \tag{24.21}$$

From here in turn, taking into account

$$\underset{\approx}{\mathbf{A}} \overset{3}{\otimes} \nabla\nabla\mathbf{u} = \underset{\sim}{\mathbf{L}} \cdot \mathbf{u}, \quad \underset{\sim}{\mathbf{L}} = \mathbf{e}_i \mathbf{e}_l A_{ijkl} \partial_j \partial_k,$$

the equations (24.21) can be written in the short form

$$\underset{\sim}{\mathbf{M}}' \cdot \mathbf{u} + \rho \mathbf{F} = 0, \tag{24.22}$$

where the following notations are introduced:

$$\underset{\sim}{\mathbf{M}}' = \underset{\sim}{\mathbf{L}}' - \rho^* \underset{\sim}{\mathbf{E}} \partial_t^2, \quad \underset{\sim}{\mathbf{L}}' = a \underset{\sim}{\mathbf{L}}, \quad \rho^* = \rho b. \tag{24.23}$$

It should be noted that (24.22) is the equations of motion of the second strain tensor and velocity vector linear gradient theory for arbitrarily anisotropic homogeneous elastic media. $\underset{\sim}{\mathbf{M}}'$ is the differential tensor-operator of the second rank and 6th order of this vector equation, a is the scalar differential operator of the 4th order with length scale tensors $\underset{\sim}{\mathbf{a}}^{(1)}$ and $\underset{\approx}{\mathbf{a}}^{(2)}$ for statics and b is the scalar differential operator of the 4th order with length scale tensors $\underset{\sim}{\mathbf{b}}^{(1)}$ and $\underset{\approx}{\mathbf{b}}^{(2)}$ for dynamics.

24.2.3 Equations of Motion of the SST and VVG linear Theory for Isotropic Homogeneous Elastic Media

In the case of an isotropic material $\underset{\approx}{\mathbf{A}} = a_1\underset{\approx}{\mathbf{C}}_{(1)} + a_2(\underset{\approx}{\mathbf{C}}_{(2)} + \underset{\approx}{\mathbf{C}}_{(3)})$, $\underset{\approx}{\bar{\mathbf{A}}} = 0$, $\underset{\sim}{\mathbf{a}}^{(1)} = a^{(1)}\underset{\sim}{\mathbf{E}}$, $\underset{\approx}{\mathbf{a}}^{(2)} = a_1^{(2)}\underset{\approx}{\mathbf{C}}_{(1)} + a_2^{(2)}(\underset{\approx}{\mathbf{C}}_{(2)} + \underset{\approx}{\mathbf{C}}_{(3)})$; $\underset{\sim}{\mathbf{B}} = \rho\underset{\sim}{\mathbf{E}}$, $\underset{\sim}{\bar{\mathbf{B}}} = 0$, $\underset{\sim}{\mathbf{b}}^{(1)} = b^{(1)}\underset{\sim}{\mathbf{E}}$ and $\underset{\approx}{\mathbf{b}}^{(2)} = b_1^{(2)}\underset{\approx}{\mathbf{C}}_{(1)} + b_2^{(2)}(\underset{\approx}{\mathbf{C}}_{(2)} + \underset{\approx}{\mathbf{C}}_{(3)})$, due to which the constitutive relations (24.17) or (24.18) and the total stress tensor $\underset{\sim}{\mathbf{T}}$ and the momentum vector $\mathbf{p}$ (see (24.19)) are represented in the form

$$
\begin{aligned}
&\underset{\sim}{\mathbf{P}} = \underset{\approx}{\mathbf{A}} \overset{2}{\otimes} \underset{\sim}{\boldsymbol{\varepsilon}}, \quad \underset{\approx}{\mathbf{P}}^{(1)} = a^{(1)}\nabla\underset{\sim}{\mathbf{P}}, \quad \underset{\approx}{\mathbf{P}}^{(2)} = (a_1^{(2)}\underset{\sim}{\mathbf{E}}\Delta + 2a_2^{(2)}\nabla\nabla)\underset{\sim}{\mathbf{P}},\\
&\boldsymbol{\pi} = \underset{\sim}{\mathbf{B}}\cdot\mathbf{v}, \quad \underset{\sim}{\boldsymbol{\pi}}^{(1)} = b^{(1)}\nabla\boldsymbol{\pi}, \quad \underset{\simeq}{\boldsymbol{\pi}}^{(2)} = (b_1^{(2)}\underset{\sim}{\mathbf{E}}\Delta + 2b_2^{(2)}\nabla\nabla)\boldsymbol{\pi},\\
&\underset{\sim}{\mathbf{T}} = (1 - l_1^2\Delta + l_2^4\Delta^2)\underset{\sim}{\mathbf{P}}, \quad l_1^2 = a^{(1)}, \quad l_2^4 = a_1^{(2)} + 2a_2^{(2)},\\
&\mathbf{p} = (1 - d_1^2\Delta + d_2^4\Delta^2)\boldsymbol{\pi}, \quad d_1^2 = b^{(1)}, \quad d_2^4 = b_1^{(2)} + 2b_2^{(2)},
\end{aligned}
\tag{24.24}
$$

where l_1 and l_2 are length scale parameters for statics, and d_1 and d_2 analogous parameters for inertia effects. In practical application, one often considers the equalities $l_1 = l_2$ and $d_1 = d_2$ (see [35]). The relations (24.24) completely coincide with the similar formulas given in [68, 69].

It should be noted that a characteristic feature of the last constitutive equations (24.24) is that the ordinary stress tensor $\underset{\sim}{\mathbf{P}}$ obeys Hooke's law, and the ordinary impulse $\boldsymbol{\pi}$ is equal to the classical momentum vector $m\mathbf{v}$, while the double and triple stress and momentum tensors are expressed as the first and second gradient of the corresponding ordinary stress tensor and momentum vector, respectively. This feature is a distinctive feature of the so-called Aifantis model for elastostatics (first-order gradient theory with respect to the strain tensor (where $l_2 = 0$, $d_1 = 0$, $d_2 = 0$) (see [2, 4, 71]), and then for dynamics (the first strain tensor and velocity vector gradient elasticity theory (where $l_2 = 0$, $d_2 = 0$) (see [7, 9]). For elastostatics the second strain tensor gradient elasticity theory ($d_1 = 0$, $d_2 = 0$) was proposed by Polizzotto [66] and then independently developed by other authors (see [15, 31, 34, 35, 40]).

It is easy to see that in the case of an isotropic medium, due to (24.24) we will have

$$
\begin{aligned}
&\underset{\sim}{\mathbf{M}}'=\underset{\sim}{\mathbf{E}}Q_2'+(\lambda'+\mu')\nabla\nabla,\ \underset{\sim}{\mathbf{M}}_*'=(\underset{\sim}{\mathbf{E}}Q_1'-(\lambda'+\mu')\nabla\nabla)Q_2'=\underset{\sim}{\mathbf{N}}'Q_2',\\
&\underset{\sim}{\mathbf{N}}'=\underset{\sim}{\mathbf{E}}Q_1'-(\lambda'+\mu')\nabla\nabla,\ Q_1'=Q_2'+(\lambda'+\mu')\Delta=(\lambda'+2\mu')\Delta-\rho^*\partial_t^2,\\
&Q_2'=\mu'\Delta-\rho^*\partial_t^2,\quad a=1-l_1^2\Delta+l_2^4\Delta^2,\quad \rho^*=\rho(1-d_1^2\Delta+d_2^4\Delta^2),\\
&Q_1'=[(\lambda+2\mu)\Delta-\rho\partial_t^2]-\Delta[l_1^2(\lambda+2\mu)\Delta-d_1^2\rho\partial_t^2]+\Delta^2[l_2^4(\lambda+2\mu)\Delta\\
&-d_2^4\rho\partial_t^2],\ Q_2'=(\mu\Delta-\rho\partial_t^2)-\Delta(l_1^2\mu\Delta-d_1^2\rho\partial_t^2)+\Delta^2(l_2^4\mu\Delta-d_2^4\rho\partial_t^2),\\
&\mu'=\mu a,\ \lambda'=\lambda a,\ \underset{\sim}{\mathbf{N}}'\cdot\underset{\sim}{\mathbf{M}}'=\underset{\sim}{\mathbf{E}}Q_1'Q_2'.
\end{aligned}
\tag{24.25}
$$

Replacing $\underset{\sim}{\mathbf{M}}'$ in (24.22) with its expression due to the first relation from (24.25), we obtain the required vector equation in the form

$$
\underset{\sim}{\mathbf{M}}'\cdot\mathbf{u}+\rho\mathbf{F}=0,\quad \text{where}\quad \underset{\sim}{\mathbf{M}}'=\underset{\sim}{\mathbf{E}}Q_2'+(\lambda'+\mu')\nabla\nabla. \tag{24.26}
$$

Applying the tensor-operator $\underset{\sim}{\mathbf{N}}'$ to the equation (24.26) on the left with subsequent single multiplication, by virtue of the last formula (24.25) we obtain the following decomposed equation:

$$
Q_1'Q_2'\mathbf{u}+\underset{\sim}{\mathbf{N}}'\cdot(\rho\mathbf{F})=\mathbf{0}. \tag{24.27}
$$

Note that in more detail questions about the splitting of equations and boundary conditions can be found, for example, in [49, 52, 56], as well as in other works of the author on this topic.

24.2.4 *Equations of the Quasi-Static with Respect to Displacement Vector of the SSTG Linear Theory for Isotropic Homogeneous Elastic Media*

In the case of statics or quasi-statics from the corresponding formulas (24.25) we have

$$
\begin{aligned}
&Q_1'=(\lambda'+2\mu')\Delta=a(\lambda+2\mu)\Delta,\ \ Q_2'=\mu'\Delta=a\mu\Delta,\\
&\underset{\sim}{\mathbf{N}}'=a\underset{\sim}{\mathbf{N}},\ \ \underset{\sim}{\mathbf{N}}=(\lambda+2\mu)\Delta-(\lambda+\mu)\nabla\nabla,
\end{aligned}
$$

due to which from (24.27) we obtain the following required split vector equation:

$$
(1-l_1^2\Delta+l_2^4\Delta^2)\Delta^2\mathbf{u}+\mathbf{G}=0,\quad \text{or briefly}\quad a\Delta^2\mathbf{u}+\mathbf{G}=0, \tag{24.28}
$$

where $a=1-l_1^2\Delta+l_2^4\Delta^2$, $\mathbf{G}=1/[(\lambda+\mu)\mu]\underset{\sim}{\mathbf{N}}\cdot(\rho\mathbf{F})$. It is easy to see that the equation (24.28) can be written in the form

$$
l_2^4(\Delta-x_1)(\Delta-x_2)\Delta^2\mathbf{u}+\mathbf{G}=0,\ \ \text{where}\ x_{1,2}=\frac{l_1^2\pm\sqrt{l_1^4-4l_2^4}}{2l_2^2}. \tag{24.29}
$$

It should be noted that at equilibrium from (24.22) we obtain the following non-split vector equation

$$
\underset{\sim}{\mathbf{L}}'\cdot\mathbf{u}+\rho\mathbf{F}=0.
$$

Hence, taking into account that for an isotropic medium

$$\underset{\sim}{\mathbf{L}}' = a\underset{\sim}{\mathbf{L}} = (1 - l_1^2\Delta + l_2^4\Delta^2)\underset{\sim}{\mathbf{L}}, \quad \underset{\sim}{\mathbf{L}} = \underset{\sim}{\mathbf{E}}\mu\Delta + (\lambda + \mu)\nabla\nabla,$$

we obtain the non-split vector equation in the form

$$(1 - l_1^2\Delta + l_2^4\Delta^2)\underset{\sim}{\mathbf{L}} \cdot \mathbf{u} + \rho\mathbf{F} = 0 \quad \text{or} \quad a\underset{\sim}{\mathbf{L}} \cdot \mathbf{u} + \rho\mathbf{F} = \mathbf{0}. \tag{24.30}$$

In the case of a homogeneous medium, $a\underset{\sim}{\mathbf{L}} = \underset{\sim}{\mathbf{L}}a$, due to which the brief representation of the equation from (24.30) can be written in the following forms:

$$\begin{aligned} (a\underset{\sim}{\mathbf{L}}) \cdot \mathbf{u} + \rho\mathbf{F} = \mathbf{0}, \quad & (\underset{\sim}{\mathbf{L}}a) \cdot \mathbf{u} + \rho\mathbf{F} = \mathbf{0}, \\ \underset{\sim}{\mathbf{L}} \cdot (a\mathbf{u}) + \rho\mathbf{F} = \mathbf{0}, \quad & a(\underset{\sim}{\mathbf{L}} \cdot \mathbf{u}) + \rho\mathbf{F} = \mathbf{0}. \end{aligned} \tag{24.31}$$

By introducing the notation

$$\mathbf{U} = a\mathbf{u}, \quad \mathbf{V} = \underset{\sim}{\mathbf{L}} \cdot \mathbf{u}, \tag{24.32}$$

the last two equations from (24.31) can be written as

$$\underset{\sim}{\mathbf{L}} \cdot \mathbf{U} + \rho\mathbf{F} = \mathbf{0}, \quad a\mathbf{V} + \rho\mathbf{F} = \mathbf{0}. \tag{24.33}$$

From the first equation (24.33) it is clear that $\mathbf{U}$ satisfies the classical equation and if $\mathbf{U} = \mathbf{U}_0$ is a solution for the classical equation, by virtue of the first formula (24.32) we will have

$$l_2^4(\Delta - x_1)(\Delta - x_2)\mathbf{u} = \mathbf{U}_0, \quad \text{where } x_{1,2} = \frac{l_1^2 \pm \sqrt{l_1^4 - 4l_2^4}}{2l_2^2}. \tag{24.34}$$

Thus, the displacement vector $\mathbf{u}$ in the considering gradient theory at equilibrium satisfies the decomposed vector equation (24.34), provided that $\mathbf{U}_0$ is the solution of the classical equation. In other words, knowing the solution $\mathbf{U}_0$ of the classical vector equation, finding the solution $\mathbf{u}$ of the vector equation of the considering gradient theory at equilibrium is reduced to solving the decomposed equation (24.34), the order of which is less than the order of the original equation (see, say, third equation (24.31)). Note that the equation (24.34) can be solved analytically at a given $\mathbf{U}_0$.

Based on the second equation (24.33) we can say that $\mathbf{V}$ is the solution of decomposed equation with the scalar differential operator

$$a = l_2^4\Delta^2 - l_1^2\Delta + 1 = l_2^4(\Delta - x_1)(\Delta - x_2), \quad \text{where } x_{1,2} = \frac{l_1^2 \pm \sqrt{l_1^4 - 4l_2^4}}{2l_2^2}$$

at classical body force $\rho\mathbf{F}$ and if we assume that its solution is $\mathbf{V} = \mathbf{V}_0$, then by virtue of the second formula (24.32) we have

$$\underset{\sim}{\mathbf{L}} \cdot \mathbf{u} + (-\mathbf{V}_0) = \mathbf{0}, \tag{24.35}$$

i.e. the displacement vector $\mathbf{u}$ in the considering gradient theory at equilibrium satisfies the equation (24.35), where $\mathbf{V}_0$ is the solution of the second decomposed equation (24.33). In other words, the displacement vector $\mathbf{u}$ in the considering gradient theory at equilibrium is the solution of the classical equation, provided that in the classical equation the body force is replaced by $-\mathbf{V}_0$, where $\mathbf{V}_0$ is solution for the second decomposed equation (24.33).

24.2.5 Motion Equations with Respect to the Displacement Vector of the SST and VVG Linear Theory for Transversely Isotropic Homogeneous Elastic Bodies

The basis of the set of transversally isotropic tensors is the following linearly independent transversally isotropic tensors of the second rank [55, 62]:

$$\underset{\sim}{\mathbf{I}} = \mathbf{e}_I\mathbf{e}_I, \quad \underset{\sim}{\boldsymbol{\epsilon}} = \epsilon_{IJ}\mathbf{e}_I\mathbf{e}_J, \quad \underset{\sim}{\boldsymbol{\gamma}}^{(3)} = \mathbf{e}_3\mathbf{e}_3, \quad (I, J = 1, 2). \tag{24.36}$$

Therefore the general form of an arbitrary transversally isotropic tensor of the second rank $\underset{\sim}{\mathbf{a}}$ is a linear combination of tensors (24.36) (see [55, 62]), i.e.

$$\underset{\sim}{\mathbf{a}} = l_1\,\underset{\sim}{\mathbf{I}} + l\,\underset{\sim}{\boldsymbol{\epsilon}} + l_2\,\mathbf{e}_3\mathbf{e}_3. \tag{24.37}$$

If $\underset{\sim}{\mathbf{a}}$ is symmetric tensor ($\underset{\sim}{\mathbf{a}} = \underset{\sim}{\mathbf{a}}^T$), then $l = 0$, and from (24.37) we get the following representation:

$$\underset{\sim}{\mathbf{a}} = l_1\,\underset{\sim}{\mathbf{I}} + l_2\,\mathbf{e}_3\mathbf{e}_3. \tag{24.38}$$

Of course, the tensors $\underset{\sim}{\mathbf{a}}^{(1)}$ and $\underset{\sim}{\mathbf{b}}^{(1)}$ containing, for example, in (24.19) – (24.21), have representations similarly to (24.38)

$$\underset{\sim}{\mathbf{a}}^{(1)} = l_1^2\,\underset{\sim}{\mathbf{I}} + l_2^2\,\mathbf{e}_3\mathbf{e}_3, \quad \underset{\sim}{\mathbf{b}}^{(1)} = d_1^2\,\underset{\sim}{\mathbf{I}} + d_2^2\,\mathbf{e}_3\mathbf{e}_3. \tag{24.39}$$

Any transversally isotropic fourth-rank tensor $\underset{\approx}{\mathbf{A}}$ whose components have the symmetries $A_{ijkl} = A_{klij} = A_{ijlk}$, has five independent components and is represented in the form [55, 62]

$$\begin{aligned}\underset{\approx}{\mathbf{A}} &= A_1\underset{\approx}{\mathbf{C}}_{(1)} + 2(A_1 - A_2)(\underset{\approx}{\mathbf{C}}_{(2)} + \underset{\approx}{\mathbf{C}}_{(3)}) + A_3(\underset{\sim}{\mathbf{I}}\mathbf{e}_3\mathbf{e}_3 + \mathbf{e}_3\mathbf{e}_3\underset{\sim}{\mathbf{I}}) + \\ &+ A_4\mathbf{e}_3\mathbf{e}_3\mathbf{e}_3\mathbf{e}_3 + A_5(\mathbf{e}_I\mathbf{e}_3\mathbf{e}_I\mathbf{e}_3 + \mathbf{e}_I\mathbf{e}_3\mathbf{e}_3\mathbf{e}_I + \mathbf{e}_3\mathbf{e}_J\mathbf{e}_J\mathbf{e}_3 + \mathbf{e}_3\mathbf{e}_J\mathbf{e}_3\mathbf{e}_J).\end{aligned} \tag{24.40}$$

Since the components of the tensors $\underset{\approx}{\mathbf{a}}^{(2)}$ and $\underset{\approx}{\mathbf{b}}^{(2)}$ have the same properties as the components of the tensor $\underset{\approx}{\mathbf{A}}$, then they can be represented similarly to (24.40), i.e.

$$\begin{aligned}
&\underset{\approx}{\mathbf{a}}^{(2)} = l_1^{(2)}\underset{\approx}{\mathbf{C}}_{(1)} + 2(l_1^{(2)} - l_2^{(2)})(\underset{\approx}{\mathbf{C}}_{(2)} + \underset{\approx}{\mathbf{C}}_{(3)}) + l_3^{(2)}(\underset{\sim}{\mathbf{I}}\mathbf{e}_3\mathbf{e}_3 + \mathbf{e}_3\mathbf{e}_3\underset{\sim}{\mathbf{I}}) \\
&+l_4^{(2)}\mathbf{e}_3\mathbf{e}_3\mathbf{e}_3\mathbf{e}_3 + l_5^{(2)}(\mathbf{e}_I\mathbf{e}_3\mathbf{e}_I\mathbf{e}_3 + \mathbf{e}_I\mathbf{e}_3\mathbf{e}_3\mathbf{e}_I + \mathbf{e}_3\mathbf{e}_J\mathbf{e}_J\mathbf{e}_3 + \mathbf{e}_3\mathbf{e}_J\mathbf{e}_3\mathbf{e}_J), \\
&\underset{\approx}{\mathbf{b}}^{(2)} = d_1^{(2)}\underset{\approx}{\mathbf{C}}_{(1)} + 2(d_1^{(2)} - d_2^{(2)})(\underset{\approx}{\mathbf{C}}_{(2)} + \underset{\approx}{\mathbf{C}}_{(3)}) + d_3^{(2)}(\underset{\sim}{\mathbf{I}}\mathbf{e}_3\mathbf{e}_3 + \mathbf{e}_3\mathbf{e}_3\underset{\sim}{\mathbf{I}}) \\
&+d_4^{(2)}\mathbf{e}_3\mathbf{e}_3\mathbf{e}_3\mathbf{e}_3 + d_5^{(2)}(\mathbf{e}_I\mathbf{e}_3\mathbf{e}_I\mathbf{e}_3 + \mathbf{e}_I\mathbf{e}_3\mathbf{e}_3\mathbf{e}_I + \mathbf{e}_3\mathbf{e}_J\mathbf{e}_J\mathbf{e}_3 + \mathbf{e}_3\mathbf{e}_J\mathbf{e}_3\mathbf{e}_J).
\end{aligned} \tag{24.41}$$

It should be noted that in the relations (24.40) and (24.41) the tensors $\underset{\approx}{\mathbf{C}}_{(1)} = \underset{\sim}{\mathbf{I}}\underset{\sim}{\mathbf{I}} = \mathbf{e}_I\mathbf{e}_I\mathbf{e}_J\mathbf{e}_J$, $\underset{\approx}{\mathbf{C}}_{(2)} = \mathbf{e}_I\mathbf{e}_J\mathbf{e}_I\mathbf{e}_J$ and $\underset{\approx}{\mathbf{C}}_{(3)} = \mathbf{e}_I\mathbf{e}_J\mathbf{e}_J\mathbf{e}_I$ are isotropic tensors of the fourth rank.

Taking into account (24.39) and (24.41), from (24.18) and (24.19) we obtain

$$\begin{aligned}
&\underset{\sim}{\mathbf{P}} = \underset{\approx}{\mathbf{A}} \overset{2}{\otimes} \underset{\sim}{\boldsymbol{\varepsilon}}, \quad \underset{\simeq}{\mathbf{P}}^{(1)} = \underset{\simeq}{\mathbf{a}}^{(1)} \cdot \nabla\underset{\sim}{\mathbf{P}} = (l_1^2\bar{\nabla} + l_2^2\mathbf{e}_3\partial_3)\underset{\sim}{\mathbf{P}}, \\
&\underset{\approx}{\mathbf{P}}^{(2)} = \underset{\approx}{\mathbf{a}}^{(2)} \overset{2}{\otimes} \nabla\nabla\underset{\sim}{\mathbf{P}} = [(l_1^{(2)}\underset{\sim}{\mathbf{I}} + l_3^{(2)}\underset{\sim}{\mathbf{e}_3})\bar{\Delta} + 4(l_1^{(2)} - l_2^{(2)})\bar{\nabla}\bar{\nabla} \\
&\quad +2l_5^{(2)}(\mathbf{e}_3\bar{\nabla} + (\mathbf{e}_3\bar{\nabla})^T) + (l_3^{(2)}\underset{\sim}{\mathbf{I}} + l_4^{(2)}\underset{\sim}{\mathbf{e}_3})\partial_3^2]\underset{\sim}{\mathbf{P}}, \\
&\boldsymbol{\pi} = \underset{\sim}{\mathbf{B}} \cdot \mathbf{v}, \quad \underset{\sim}{\boldsymbol{\pi}}^{(1)} = \underset{\sim}{\mathbf{b}}^{(1)} \cdot \nabla\boldsymbol{\pi} = (d_1^2\bar{\nabla} + d_2^2\mathbf{e}_3\partial_3)\boldsymbol{\pi}, \\
&\underset{\simeq}{\boldsymbol{\pi}}^{(2)} = \underset{\approx}{\mathbf{b}}^{(2)} \overset{2}{\otimes} \nabla\nabla\boldsymbol{\pi} = [(d_1^{(2)}\underset{\sim}{\mathbf{I}} + d_3^{(2)}\underset{\sim}{\mathbf{e}_3})\bar{\Delta} + 4(d_1^{(2)} - d_2^{(2)})\bar{\nabla}\bar{\nabla} \\
&\quad +2d_5^{(2)}(\mathbf{e}_3\bar{\nabla} + (\mathbf{e}_3\bar{\nabla})^T) + (d_3^{(2)}\underset{\sim}{\mathbf{I}} + d_4^{(2)}\underset{\sim}{\mathbf{e}_3})\partial_3^2]\boldsymbol{\pi}, \quad \bar{\nabla} = \mathbf{e}_I\partial_I, \\
&\underset{\sim}{\mathbf{T}} = [1 + (l_3^4\bar{\Delta} - l_1^2)\bar{\Delta} + (l_4^4\bar{\Delta} - l_2^2)\partial_3^2 + l_5^4\partial_3^4]\underset{\sim}{\mathbf{P}} = a\underset{\sim}{\mathbf{P}}, \\
&\mathbf{p} = \rho[1 + (d_3^4\bar{\Delta} - d_1^2)\bar{\Delta} + (d_4^4\bar{\Delta} - d_2^2)\partial_3^2 + d_5^4\partial_3^4]\mathbf{v} = \rho^*\mathbf{v}, \\
&\bar{\Delta} = \partial_I\partial_I, \quad a = 1 + (l_3^4\bar{\Delta} - l_1^2)\bar{\Delta} + (l_4^4\bar{\Delta} - l_2^2)\partial_3^2 + l_5^4\partial_3^4, \\
&\rho^* = \rho[1 + (d_3^4\bar{\Delta} - d_1^2)\bar{\Delta} + (d_4^4\bar{\Delta} - d_2^2)\partial_3^2 + d_5^4\partial_3^4], \\
&l_3^4 = 5l_1^{(2)} - 4l_2^{(2)}, \quad l_4^4 = 2(l_3^{(2)} + 2l_5^{(2)}), \quad l_5^4 = l_4^{(2)}, \\
&d_3^4 = 5d_1^{(2)} - 4d_2^{(2)}, \quad d_4^4 = 2(d_3^{(2)} + 2d_5^{(2)}), \quad d_5^4 = d_4^{(2)}.
\end{aligned} \tag{24.42}$$

Knowing the expressions for the total stress tensor $\underset{\sim}{\mathbf{T}}$, the Cauchy stress tensor $\underset{\sim}{\mathbf{P}}$ and the total momentum vector $\mathbf{p}$ (see the corresponding formulas (24.42)), the vector equation of motion (24.21) in the case under consideration can be written as

$$(a\underset{\sim}{\mathbf{L}}) \cdot \mathbf{u} + \rho\mathbf{F} = \rho^*\partial_t^2\mathbf{u} \tag{24.43}$$

or similarly (24.22) can be written in short form

$$\underset{\sim}{\mathbf{M}}' \cdot \mathbf{u} + \rho\mathbf{F} = 0, \tag{24.44}$$

where the following notations are introduced:

$$\begin{aligned}
&\underset{\sim}{\mathbf{M}}' = \underset{\sim}{\mathbf{L}}' - \rho^*\underset{\sim}{\mathbf{E}}\partial_t^2, \quad \underset{\sim}{\mathbf{L}}' = a\underset{\sim}{\mathbf{L}}, \quad \underset{\sim}{\mathbf{L}} = \mathbf{e}_i\mathbf{e}_l A_{ijkl}\partial_j\partial_k, \\
&a = 1 + (l_3^4\bar{\Delta} - l_1^2)\bar{\Delta} + (l_4^4\bar{\Delta} - l_2^2)\partial_3^2 + l_5^4\partial_3^4, \\
&\rho^* = \rho b, \quad b = 1 + (d_3^4\bar{\Delta} - d_1^2)\bar{\Delta} + (d_4^4\bar{\Delta} - d_2^2)\partial_3^2 + d_5^4\partial_3^4]
\end{aligned} \tag{24.45}$$

and, in addition, due to (24.40) tensor-operator $\underset{\sim}{\mathbf{L}}$, tensor-operator of cofactors $\underset{\sim}{\mathbf{L}}_*$ for $\underset{\sim}{\mathbf{L}}$ and the determinant $|\underset{\sim}{\mathbf{L}}|$ in this case have the form

$$
\begin{aligned}
2\underset{\sim}{\mathbf{L}} &= \left[(A_1 - A_2)\underset{\sim}{\mathbf{I}} + 2A_5\underset{\sim}{\mathbf{e}}_3\right]\bar{\Delta} + (A_1 + A_2)\bar{\nabla}^0\bar{\nabla}^0 \\
&\quad +2(A_3 + A_5)[\mathbf{e}_3\bar{\nabla}^0 + (\mathbf{e}_3\bar{\nabla}^0)^T]\partial_3 + 2(A_5\underset{\sim}{\mathbf{I}} + A_4\underset{\sim}{\mathbf{e}}_3)\partial_3^2, \\
2\underset{\sim}{\mathbf{L}}_* &= 2\underset{\sim}{\mathbf{I}}\{A_1A_5\bar{\Delta}^2 + [A_1A_4 - A_3(A_3 + 2A_5)]\bar{\Delta}\partial_3^2 + A_4A_5\partial_3^4\} \\
&\quad -\{(A_1 + A_2)A_5\bar{\Delta} + [(A_1 + A_2)A_4 - 2(A_3 + A_5)^2]\partial_3^2\}\bar{\nabla}^0\bar{\nabla}^0 \\
&\quad -[(A_1 - A_2)(A_3 + A_5)\bar{\Delta} + 2(A_3 + A_5)A_5\partial_3^2][\mathbf{e}_3\bar{\nabla}^0 + (\mathbf{e}_3\bar{\nabla}^0)^T]\partial_3 \\
&\quad +\underset{\sim}{\mathbf{e}}_3[A_1(A_1 - A_2)\bar{\Delta}^2 + (3A_1 - A_2)A_5\bar{\Delta}\partial_3^2 + 2A_5^2\partial_3^4], \\
2|\underset{\sim}{\mathbf{L}}| &= (A_1-A_2)A_1A_5\bar{\Delta}^3+\{(A_1-A_2)[A_1A_4-A_3(A_3+2A_5)]+2A_1A_5^2\}\bar{\Delta}^2\partial_3^2 \\
&\quad +[(3A_1 - A_2)A_4A_5 - 2A_3A_5(A_3 + 2A_5)]\bar{\Delta}\partial_3^4 + 2A_4A_5^2\partial_3^6, \\
&\mathbf{e}_i \cdot \mathbf{e}_j = \delta_{ij},\ \ i,j = 1,2,3,\ \ \underset{\sim}{\mathbf{e}}_m \overset{2}{\otimes} \underset{\sim}{\mathbf{e}}_n = \delta_{mn},\ \ m,n = 1,2,\ldots,6, \\
&\underset{\sim}{\mathbf{e}}_1 = \mathbf{e}_1\mathbf{e}_1,\ \ \underset{\sim}{\mathbf{e}}_2 = \mathbf{e}_2\mathbf{e}_2,\ \ \underset{\sim}{\mathbf{e}}_3 = \mathbf{e}_3\mathbf{e}_3,\ \ \underset{\sim}{\mathbf{e}}_4 = (1/\sqrt{2})(\mathbf{e}_1\mathbf{e}_2 + \mathbf{e}_2\mathbf{e}_1), \\
&\underset{\sim}{\mathbf{e}}_5 = (1/\sqrt{2})(\mathbf{e}_2\mathbf{e}_3 + \mathbf{e}_3\mathbf{e}_2),\ \ \underset{\sim}{\mathbf{e}}_6 = (1/\sqrt{2})(\mathbf{e}_3\mathbf{e}_1 + \mathbf{e}_1\mathbf{e}_3).
\end{aligned}
$$

Here δ_{ij}, $i, j = 1, 2, 3$, is Kronecker delta; $\mathbf{e}_i$, $i = 1, 2, 3$, is orthonormal basis; $\underset{\sim}{\mathbf{e}}_m$, $m = \overline{1, 6}$, is orthonormal basis for symmetric tensors of the second rank.

The invariants $I_k(\underset{\sim}{\mathbf{L}})$, $k = 1, 2, 3$, of the tensor-operator $\underset{\sim}{\mathbf{L}}$ can be represented as

$$
\begin{aligned}
&2I_1(\underset{\sim}{\mathbf{L}}) = (3A_1 - A_2 + 2A_5)\bar{\Delta} + 2(A_4 + 2A_5)\partial_3^2,\ \ I_3(\underset{\sim}{\mathbf{L}}) = |\underset{\sim}{\mathbf{L}}|, \\
&2I_2(\underset{\sim}{\mathbf{L}}) = [(3A_1 - A_2)A_5 + A_1(A_1 - A_2)]\bar{\Delta}^2 + [(3A_1 - A_2)A_4 \\
&-2A_3(A_3+2A_5) + (2A_5+3A_1-A_2)A_5]\bar{\Delta}\partial_3^2 + 2(2A_4+A_5)A_5\partial_3^4.
\end{aligned}
$$

It is easy to see that the determinant of the tensor-operator $\underset{\sim}{\mathbf{L}}$ can also be written as a product of simple operators:

$$
|\underset{\sim}{\mathbf{L}}| = A\bar{\Delta}^3 + B\bar{\Delta}^2\partial_3^2 + C\bar{\Delta}\partial_3^4 + D\partial_3^6 = k(\bar{\Delta}+k_1\partial_3^2)(\bar{\Delta}+k_2\partial_3^2)(\bar{\Delta}+k_3\partial_3^2), \tag{24.46}
$$

where

$$
\begin{aligned}
&k = A,\ \ k_1 + k_2 + k_3 = B/A,\ \ k_1k_2 + k_1k_3 + k_2k_3 = C/A,\ \ k_1k_2k_3 = D/A, \\
&A = \frac{1}{2}(A_1 - A_2)A_1A_5,\ \ B = \frac{1}{2}\{(A_1 - A_2)[A_1A_4 - A_3(A_3+2A_5)] + 2A_1A_5^2\}, \\
&C = \frac{1}{2}[(3A_1 - A_2)A_4A_5 - 2A_3A_5(A_3+2A_5)],\ \ D = A_4A_5^2.
\end{aligned}
$$

The decomposed vector equation in the case of the classics will have the form

$$
|\underset{\sim}{\mathbf{L}}|\mathbf{u} + \underset{\sim}{\mathbf{L}}_* \cdot (\rho\mathbf{F}) = \mathbf{0},
$$

and if we will search $\mathbf{u}$ in the Galerkin form $\mathbf{u} = \underset{\sim}{\mathbf{L}}_* \cdot \mathbf{v}$, then we obtain the following decomposed equation

$$
|\underset{\sim}{\mathbf{L}}|\mathbf{v} + \rho\mathbf{F} = \mathbf{0}.
$$

24.2.6 Static or Quasi-Static Equation with Respect to the Displacement Vector of the SSTG Linear Theory for Transversely Isotropic Homogeneous Elastic Bodies

Neglecting the right side of Eq. (24.43), we obtain the vector equation of statics or quasi-statics of the considered gradient elasticity theory of transversely isotropic homogeneous solids, which due to $a\underset{\sim}{\mathbf{L}} = \underset{\sim}{\mathbf{L}}a$ similarly to (24.31) can be represented in various forms:

$$\begin{aligned}&\big((a\underset{\sim}{\mathbf{L}})\cdot\mathbf{u} + \rho\mathbf{F} = \mathbf{0}\big) \Leftrightarrow \big((\underset{\sim}{\mathbf{L}}a)\cdot\mathbf{u} + \rho\mathbf{F} = \mathbf{0}\big)\\ &\Leftrightarrow \big(\underset{\sim}{\mathbf{L}}\cdot(a\mathbf{u}) + \rho\mathbf{F} = \mathbf{0}\big) \Leftrightarrow \big(a(\underset{\sim}{\mathbf{L}}\cdot\mathbf{u}) + \rho\mathbf{F} = \mathbf{0}\big).\end{aligned} \tag{24.47}$$

By introducing the notation

$$\mathbf{U} = a\mathbf{u}, \quad \mathbf{V} = \underset{\sim}{\mathbf{L}}\cdot\mathbf{u}, \tag{24.48}$$

the last two equations (24.47) can be written as

$$\underset{\sim}{\mathbf{L}}\cdot\mathbf{U} + \rho\mathbf{F} = \mathbf{0}, \quad a\mathbf{V} + \rho\mathbf{F} = \mathbf{0}. \tag{24.49}$$

From the first equation (24.49) it is clear that $\mathbf{U}$ satisfies the classical equation and if $\mathbf{U} = \mathbf{U}_0$ is a solution for the classical equation, by virtue of the first formula (24.48) we will get

$$[1 + (l_3^4\bar{\Delta} - l_1^2)\bar{\Delta} + (l_4^4\bar{\Delta} - l_2^2)\partial_3^2 + l_5^4\partial_3^4]\mathbf{u} = \mathbf{U}_0. \tag{24.50}$$

Thus, the displacement vector $\mathbf{u}$ in the considered gradient theory of transversally isotropic homogeneous solids at equilibrium satisfies the decomposed vector equation (24.50), provided that $\mathbf{U}_0$ is a solution for the classical equation. In other words, knowing the solution $\mathbf{U}_0$ of the classical vector equation, finding the solution $\mathbf{u}$ of the vector equation of the above-mentioned gradient theory at equilibrium is reduced to solving the decomposed equation (24.50), the order of which is less than the order of the original equations (see, for example, the third equation (24.47)).

Based on the second equation (24.49) we can say that $\mathbf{V}$ is a solution for the decomposed equation with a scalar differential operator

$$a = 1 + (l_3^4\bar{\Delta} - l_1^2)\bar{\Delta} + (l_4^4\bar{\Delta} - l_2^2)\partial_3^2 + l_5^4\partial_3^4$$

at classical body force $\rho\mathbf{F}$ and if we assume that its solution is $\mathbf{V} = \mathbf{V}_0$, then by virtue of the second formula (24.48) we have

$$\underset{\sim}{\mathbf{L}}\cdot\mathbf{u} + (-\mathbf{V}_0) = \mathbf{0}, \tag{24.51}$$

i.e. the displacement vector $\mathbf{u}$ in the considering gradient theory of transversally isotropic homogeneous solids at equilibrium satisfies the equation (24.51), where $\mathbf{V}_0$ is the solution of the second decomposed equation (24.49). In other words, the displacement vector $\mathbf{u}$ in the considering gradient theory at equilibrium is the solution of the classical equation, provided that in the classical equation the body force is replaced by $-\mathbf{V}_0$, where $\mathbf{V}_0$ is solution for the second decomposed equation (24.49).

Applying to both sides of the first equation (24.47) the tensor-operator $\underset{\sim}{\mathbf{L}}_*$ on the left with single multiplication and taking into account that the formulas $\underset{\sim}{\mathbf{L}}_* \cdot \underset{\sim}{\mathbf{L}} = \underset{\sim}{\mathbf{L}} \cdot \underset{\sim}{\mathbf{L}}_* = \underset{\sim}{\mathbf{E}}|\underset{\sim}{\mathbf{L}}|$ and $\underset{\sim}{\mathbf{L}}_* a = a\underset{\sim}{\mathbf{L}}_*$ are valid for homogeneous bodies, we will obtain the following decomposed vector equation of statics or quasi-statics of the considered second gradient elasticity theory of transversally isotropic homogeneous solids

$$(a|\underset{\sim}{\mathbf{L}}|)\mathbf{u}+\underset{\sim}{\mathbf{L}}_*\cdot(\rho\mathbf{F})=\mathbf{0}, \quad \text{where } a=1+(l_3^4\bar{\Delta}-l_1^2)\bar{\Delta}+(l_4^4\bar{\Delta}-l_2^2)\partial_3^2+l_5^4\partial_3^4. \tag{24.52}$$

Taking into account (24.46), the equation (24.52) can be written in expanded form

$$\begin{aligned}&([1+(l_3^4\bar{\Delta}-l_1^2)\bar{\Delta}+(l_4^4\bar{\Delta}-l_2^2)\partial_3^2+l_5^4\partial_3^4](A\bar{\Delta}^3+B\bar{\Delta}^2\partial_3^2+C\bar{\Delta}\partial_3^4+D\partial_3^6)\mathbf{u}\\&+\underset{\sim}{\mathbf{L}}_*\cdot(\rho\mathbf{F}) = \mathbf{0}.\end{aligned} \tag{24.53}$$

It is easy to see that, if necessary, the differential operator of the equation (24.53), similarly to the operator (24.46), can be represented as a product of simple operators.

Let's write the equation (24.53) in expanded form. We'll have

$$\begin{aligned}&\{A(l_3^4\bar{\Delta}^2 - l_1^2\bar{\Delta} + 1)\bar{\Delta}^3 + [(Bl_3^4 + Al_4^4)\bar{\Delta}^2 - (Bl_1^2 + Al_2^2)\bar{\Delta} + B]\bar{\Delta}^2\partial_3^2\\&+[(Al_5^4 + Bl_4^4+Cl_3^4)\bar{\Delta}^2-(Bl_2^2+Cl_1^2)\bar{\Delta}+C]\bar{\Delta}\partial_3^4+[(Bl_5^4+Cl_4^4+Dl_3^4)\bar{\Delta}^2\\&-(Cl_2^2 + Dl_1^2)\bar{\Delta} + D]\partial_3^6 + Cl_5^4\bar{\Delta}\partial_3^8 + Dl_5^4\partial_3^{10}\} \cdot \mathbf{u} + \underset{\sim}{\mathbf{L}}_*\cdot(\rho\mathbf{F}) = 0,\end{aligned} \tag{24.54}$$

24.3 On some Issues the FST and VVG Elasticity Theory

Below, as a special case, some relations of the first strain tensor and velocity vector gradient elasticity theory are derived from the corresponding relations of the second strain tensor and velocity vector gradient elasticity theory.

24.3.1 Specific Strain Energy, Specific Kinetic Energy and Constitutive Relations in the FST and VVG Elasticity Theory

It is easy to see that in this case the specific strain energy (24.5), specific kinetic energy (24.6), and constitutive relations (24.8) are represented in the form

$$\begin{aligned}\Phi(\underset{\simeq}{\boldsymbol{\varepsilon}},\underset{\simeq}{\boldsymbol{\gamma}}) &= \frac{1}{2}\Big({}^4\underset{\sim}{\mathbf{A}}^{(11)} \overset{4}{\otimes} \underset{\simeq}{\boldsymbol{\varepsilon}}\underset{\simeq}{\boldsymbol{\varepsilon}} + 2\,{}^5\underset{\sim}{\mathbf{A}}^{(12)} \overset{5}{\otimes} \underset{\simeq}{\boldsymbol{\varepsilon}}\underset{\simeq}{\boldsymbol{\gamma}} + {}^6\underset{\sim}{\mathbf{A}}^{(22)} \overset{6}{\otimes} \underset{\simeq}{\boldsymbol{\gamma}}\underset{\simeq}{\boldsymbol{\gamma}}\Big)\\&= \frac{1}{2}\Big(A^{(11)ijkl}\varepsilon_{ij}\varepsilon_{kl} + 2A^{(12)ijklm}\varepsilon_{ij}\gamma_{klm} + A^{(22)ijklmn}\gamma_{ijk}\gamma_{lmn}\Big),\end{aligned} \tag{24.55}$$

$$
\begin{aligned}
K(\mathbf{v},\underset{\sim}{\mathbf{v}}) &= \frac{1}{2}\left({}^{2}\underset{\sim}{\mathbf{B}}^{(11)} \overset{2}{\otimes} \mathbf{v}\mathbf{v} + 2\,{}^{3}\underset{\sim}{\mathbf{B}}^{(12)} \overset{3}{\otimes} \mathbf{v}\underset{\sim}{\mathbf{v}} + {}^{4}\underset{\sim}{\mathbf{B}}^{(22)} \overset{3}{\otimes} \underset{\sim}{\mathbf{v}\mathbf{v}}\right)\\
&= \frac{1}{2}\left(B^{(11)ij}v_i v_j + 2B^{(12)ijk}v_i v_{jk} + B^{(22)ijkl}v_{ij}v_{kl}\right);
\end{aligned}
\tag{24.56}
$$

$$
\begin{aligned}
&\underset{\sim}{\mathbf{P}} = {}^{4}\underset{\sim}{\mathbf{A}}^{(11)} \overset{2}{\otimes} \underset{\sim}{\boldsymbol{\varepsilon}} + {}^{5}\underset{\sim}{\mathbf{A}}^{(12)} \overset{3}{\otimes} \underset{\simeq}{\boldsymbol{\gamma}} \;\; (P^{ij} = A^{(11)ijkl}\varepsilon_{ij}\varepsilon_{kl} + A^{(12)ijklm}\gamma_{klm}),\\
&\underset{\simeq}{\mathbf{P}}^{(1)} = ({}^{5}\underset{\sim}{\mathbf{A}}^{(12)})^{T} \overset{2}{\otimes} \underset{\sim}{\boldsymbol{\varepsilon}} + {}^{6}\underset{\sim}{\mathbf{A}}^{(22)} \overset{3}{\otimes} \underset{\simeq}{\boldsymbol{\gamma}}\\
&(P^{(1)ijk} = A^{(12)ijklm}\varepsilon_{lm} + A^{(22)ijklmn}\gamma_{ijk}\gamma_{lmn});\\
&\boldsymbol{\pi} = {}^{2}\underset{\sim}{\mathbf{B}}^{(11)} \overset{1}{\otimes} \mathbf{v} + {}^{3}\underset{\sim}{\mathbf{B}}^{(12)} \overset{2}{\otimes} \underset{\sim}{\mathbf{v}} \;\; (\pi^{i} = B^{(11)ij}v_j + B^{(12)ijk}v_i v_{jk}),\\
&\underset{\sim}{\boldsymbol{\pi}}^{(1)} = ({}^{3}\underset{\sim}{\mathbf{B}}^{(12)})^{T} \overset{1}{\otimes} \mathbf{v} + {}^{4}\underset{\sim}{\mathbf{B}}^{(22)} \overset{2}{\otimes} \underset{\sim}{\mathbf{v}} \;\; (\pi^{(1)ij} = B^{(12)ijk}v_k + B^{(22)ijkl}v_{kl}),
\end{aligned}
\tag{24.57}
$$

or, introducing notations

$$
\mathbb{M} = \begin{pmatrix} {}^{4}\underset{\sim}{\mathbf{A}}^{(11)} & {}^{5}\underset{\sim}{\mathbf{A}}^{(12)} \\ ({}^{5}\underset{\sim}{\mathbf{A}}^{(12)})^{T} & {}^{6}\underset{\sim}{\mathbf{A}}^{(22)} \end{pmatrix}, \quad \mathbb{X} = \begin{pmatrix} \underset{\sim}{\boldsymbol{\varepsilon}} \\ \underset{\simeq}{\boldsymbol{\gamma}} \end{pmatrix}, \quad \mathbb{Y} = \begin{pmatrix} \underset{\sim}{\mathbf{P}} \\ \underset{\simeq}{\mathbf{P}}^{(1)} \end{pmatrix},
$$

$$
\mathbb{N} = \begin{pmatrix} {}^{2}\underset{\sim}{\mathbf{B}}^{(11)} & {}^{3}\underset{\sim}{\mathbf{B}}^{(12)} \\ ({}^{3}\underset{\sim}{\mathbf{B}}^{(12)})^{T} & {}^{4}\underset{\sim}{\mathbf{B}}^{(22)} \end{pmatrix}, \quad \mathbb{U} = \begin{pmatrix} \mathbf{v} \\ \underset{\sim}{\mathbf{v}} \end{pmatrix}, \quad \mathbb{V} = \begin{pmatrix} \boldsymbol{\pi} \\ \underset{\sim}{\boldsymbol{\pi}}^{(1)} \end{pmatrix},
$$

the relations (24.55), (24.56), and (24.57) can be written in a short form

$$
\begin{aligned}
\Phi(\underset{\sim}{\boldsymbol{\varepsilon}},\underset{\simeq}{\boldsymbol{\gamma}}) &= \mathbb{X}^{T}\,{}^{2}\!\otimes_{3}\mathbb{M}\,{}^{2}\!\otimes_{3}\mathbb{X}, \qquad \mathbb{Y} = \mathbb{M}\,{}^{2}\!\otimes_{3}\mathbb{X},\\
K(\mathbf{v},\underset{\sim}{\mathbf{v}}) &= \mathbb{V}^{T}\,{}^{1}\!\otimes_{2}\mathbb{N}\,{}^{1}\!\otimes_{2}\mathbb{V}, \qquad \mathbb{V} = \mathbb{N}\,{}^{1}\!\otimes_{2}\mathbb{U},
\end{aligned}
$$

where the following operations are introduced for the matrix product:

$$
{}^{2}\!\otimes_{3} = \begin{pmatrix} \overset{2}{\otimes} & 0 \\ 0 & \overset{3}{\otimes} \end{pmatrix}, \qquad {}^{1}\!\otimes_{2} = \begin{pmatrix} \overset{1}{\otimes} & 0 \\ 0 & \overset{2}{\otimes} \end{pmatrix}.
$$

Naturally, similarly to (24.10), the following problems arise:

$$
\mathbb{M}\,{}^{2}\!\otimes_{3}\mathbb{X} = \lambda\mathbb{X}\,?, \qquad \mathbb{N}\,{}^{1}\!\otimes_{2}\mathbb{U} = \lambda\mathbb{U}\,?,
$$

i.e. eigenvalue problems for tensor-block matrices $\mathbb{M}$ and $\mathbb{N}$.

Further, from (24.18) it is clear that in the case under consideration the constitutive relations have the form

$$
\underset{\sim}{\mathbf{P}} = \underset{\approx}{\mathbf{A}} \overset{2}{\otimes} \underset{\sim}{\boldsymbol{\varepsilon}}, \quad \underset{\simeq}{\mathbf{P}}^{(1)} = \underset{\sim}{\mathbf{a}}^{(1)}\cdot\nabla\underset{\sim}{\mathbf{P}}; \quad \boldsymbol{\pi} = \underset{\sim}{\mathbf{B}}\cdot\mathbf{v}, \quad \underset{\sim}{\boldsymbol{\pi}}^{(1)} = \underset{\sim}{\mathbf{b}}^{(1)}\cdot\nabla\boldsymbol{\pi}. \tag{24.58}
$$

24.3.2 Motion Equations of the FST and VVG Elasticity Theory

In this case the symmetric total stress tensor $\underset{\sim}{\mathbf{T}}$ and the (basic) momentum vector (body momentum) $\mathbf{p}$ due to (24.58) are represented as follows:

$$\begin{aligned}\underset{\sim}{\mathbf{T}}&=\underset{\sim}{\mathbf{P}}-\nabla\cdot\underset{\approx}{\mathbf{P}}^{(1)}=(1-\underset{\sim}{\mathbf{a}}^{(1)}\overset{2}{\otimes}\nabla\nabla)\underset{\sim}{\mathbf{P}},\\ \mathbf{p}&=\boldsymbol{\pi}-\nabla\cdot\underset{\sim}{\boldsymbol{\pi}}^{(1)}=(1-\underset{\sim}{\mathbf{b}}^{(1)}\overset{2}{\otimes}\nabla\nabla)\boldsymbol{\pi},\end{aligned} \tag{24.59}$$

where $\underset{\sim}{\mathbf{P}}$ is Cauchy stress tensor, $\boldsymbol{\pi}=\rho\mathbf{v}$, $\underset{\sim}{\mathbf{B}}=\rho\underset{\sim}{\mathbf{E}}$, ρ is density, $\underset{\sim}{\mathbf{a}}^{(1)}$ is length scale tensors for statics, $\underset{\sim}{\mathbf{b}}^{(1)}$ is length scale tensors for inertia effects (dynamics).

Having expressions for the total stress tensor and the momentum vector (see (24.59)), the motion equations for the considered theory can be written as

$$\begin{aligned}&\left(\nabla\cdot\underset{\sim}{\mathbf{T}}+\rho\mathbf{F}=\frac{d\mathbf{p}}{dt}\right)\Rightarrow\left(a\nabla\cdot\underset{\sim}{\mathbf{P}}+\rho\mathbf{F}=b\frac{d\boldsymbol{\pi}}{dt}\right),\\ &a=1-\underset{\sim}{\mathbf{a}}^{(1)}\overset{2}{\otimes}\nabla\nabla,\quad b=1-\underset{\sim}{\mathbf{b}}^{(1)}\overset{2}{\otimes}\nabla\nabla.\end{aligned} \tag{24.60}$$

Taking into account that in the case of an arbitrarily anisotropic linear elasticity theory $\underset{\sim}{\mathbf{P}}=\underset{\approx}{\mathbf{A}}\overset{2}{\otimes}\underset{\sim}{\boldsymbol{\varepsilon}}=\underset{\approx}{\mathbf{A}}\overset{2}{\otimes}\nabla\mathbf{u}$, where $\mathbf{u}$ is displacement vector, $\underset{\approx}{\mathbf{A}}$ is elastic modulus tensor and $(d\mathbf{v})/(dt)\approx(\partial^2\mathbf{u})/(\partial t)=\partial_t^2\mathbf{u}$, from (24.60) under $\rho=const$ we get

$$(1-\underset{\sim}{\mathbf{a}}^{(1)}\overset{2}{\otimes}\nabla\nabla)(\underset{\approx}{\mathbf{A}}\overset{3}{\otimes}\nabla\nabla\mathbf{u})+\rho\mathbf{F}=\rho(1-\underset{\sim}{\mathbf{b}}^{(1)}\overset{2}{\otimes}\nabla\nabla)\partial_t^2\mathbf{u}. \tag{24.61}$$

From here, taking into account

$$\underset{\approx}{\mathbf{A}}\overset{3}{\otimes}\nabla\nabla\mathbf{u}=\underset{\sim}{\mathbf{L}}\cdot\mathbf{u},\quad \underset{\sim}{\mathbf{L}}=\mathbf{e}_i\mathbf{e}_l A_{ijkl}\partial_j\partial_k,$$

the equations (24.61) can be written in the short form

$$\underset{\sim}{\mathbf{M}}'\cdot\mathbf{u}+\rho\mathbf{F}=0, \tag{24.62}$$

where the following notations are introduced:

$$\underset{\sim}{\mathbf{M}}'=\underset{\sim}{\mathbf{L}}'-\rho^*\underset{\sim}{\mathbf{E}}\partial_t^2,\quad \underset{\sim}{\mathbf{L}}'=a\underset{\sim}{\mathbf{L}},\quad \rho^*=\rho b. \tag{24.63}$$

It should be noted that (24.62) is the equations of motion of the first strain tensor and velocity vector gradient linear theory for arbitrarily anisotropic homogeneous elastic media. $\underset{\sim}{\mathbf{M}}'$ is the differential tensor-operator of the second rank and 4th order of this vector equation, a is the scalar differential operator of the second order with length scale tensor $\underset{\sim}{\mathbf{a}}^{(1)}$ for statics and b is the scalar differential operator of the second order with length scale tensor $\underset{\sim}{\mathbf{b}}^{(1)}$ for dynamics.

24.3.3 Motion Equations of the FST and VVG Linear Theory for Isotropic Homogeneous Elastic Media

In the case of an isotropic material $\underset{\approx}{\mathbf{A}}=a_1\underset{\approx}{\mathbf{C}}_{(1)}+a_2(\underset{\approx}{\mathbf{C}}_{(2)}+\underset{\approx}{\mathbf{C}}_{(3)})$, $\underset{\sim}{\mathbf{a}}^{(1)}=a^{(1)}\underset{\sim}{\mathbf{E}}$, $\underset{\sim}{\mathbf{B}}=\rho\underset{\sim}{\mathbf{E}}$ and $\underset{\sim}{\mathbf{b}}^{(1)}=b^{(1)}\underset{\sim}{\mathbf{E}}$, due to which the constitutive relations (24.58) and the total stress tensor $\underset{\sim}{\mathbf{T}}$ and the momentum vector $\mathbf{p}$ (see (24.59)) are represented in the form

$$\begin{aligned} &\underset{\sim}{\mathbf{P}} = \underset{\approx}{\mathbf{A}} \overset{2}{\otimes} \boldsymbol{\varepsilon}, \quad \underset{\simeq}{\mathbf{P}}^{(1)} = a^{(1)}\nabla\underset{\sim}{\mathbf{P}}, \quad \boldsymbol{\pi} = \underset{\sim}{\mathbf{B}}\cdot\mathbf{v}, \quad \underset{\sim}{\boldsymbol{\pi}}^{(1)} = b^{(1)}\nabla\boldsymbol{\pi}, \\ &\underset{\sim}{\mathbf{T}} = (1 - l^2\Delta)\underset{\sim}{\mathbf{P}}, \quad l^2 = a^{(1)}, \quad \mathbf{p} = (1 - d^2\Delta)\boldsymbol{\pi}, \quad d^2 = b^{(1)}, \end{aligned} \tag{24.64}$$

where l is length scale parameter for statics, and d analogous parameter for inertia effects.

It is easy to see that in the case of an isotropic medium, due to (24.64) we will have

$$\begin{aligned} &\underset{\sim}{\mathbf{M}}' = \underset{\sim}{\mathbf{E}}Q_2' + (\lambda' + \mu')\nabla\nabla, \quad \underset{\sim}{\mathbf{M}}'_* = \underset{\sim}{\mathbf{N}}'Q_2', \quad Q_2' = \mu'\Delta - \rho^*\partial_t^2, \\ &\underset{\sim}{\mathbf{N}}' = \underset{\sim}{\mathbf{E}}Q_1' - (\lambda' + \mu')\nabla\nabla, \quad Q_1' = (\lambda' + 2\mu')\Delta - \rho^*\partial_t^2, \\ &a = 1 - l^2\Delta, \quad \rho^* = \rho(1 - d^2\Delta), \quad \mu' = \mu a \quad \lambda' = \lambda a; \\ &Q_1' = [(\lambda + 2\mu)\Delta - \rho\partial_t^2] - \Delta[l^2(\lambda + 2\mu)\Delta - d^2\rho\partial_t^2], \\ &Q_2' = (\mu\Delta - \rho\partial_t^2) - \Delta(l^2\mu\Delta - d^2\rho\partial_t^2), \quad \underset{\sim}{\mathbf{N}}'\cdot\underset{\sim}{\mathbf{M}}' = \underset{\sim}{\mathbf{E}}Q_1'Q_2'. \end{aligned} \tag{24.65}$$

Replacing $\underset{\sim}{\mathbf{M}}'$ in (24.62) with its expression due to the first relation from (24.65), we obtain the required vector equation in the form

$$\underset{\sim}{\mathbf{M}}'\cdot\mathbf{u} + \rho\mathbf{F} = 0, \quad \text{where} \quad \underset{\sim}{\mathbf{M}}' = \underset{\sim}{\mathbf{E}}Q_2' + (\lambda' + \mu')\nabla\nabla. \tag{24.66}$$

Applying the tensor-operator $\underset{\sim}{\mathbf{N}}'$ to the equation (24.66) on the left with subsequent single multiplication, by virtue of the last formula from (24.65) we obtain the following decomposed equation:

$$Q_1'Q_2'\mathbf{u} + \underset{\sim}{\mathbf{N}}'\cdot(\rho\mathbf{F}) = \mathbf{0}. \tag{24.67}$$

Note that in more detail questions about the splitting of equations and boundary conditions can be found, for example, in [50, 52, 56], as well as in other works of the author on this topic.

24.3.4 Equations of the Quasi-Static in Displacement Vector of the FSTG Theory for Isotropic Homogeneous Elastic Media

In the case of statics or quasi-statics from the corresponding formulas (24.65) we have

$$Q'_1 = (\lambda' + 2\mu')\Delta = a(\lambda + 2\mu)\Delta, \;\; Q'_2 = \mu'\Delta = a\mu\Delta,$$
$$\underset{\sim}{\mathbf{N}}' = a\underset{\sim}{\mathbf{N}}, \;\; \underset{\sim}{\mathbf{N}} = (\lambda + 2\mu)\Delta - (\lambda + \mu)\nabla\nabla,$$

due to which from (24.67) we obtain the following required split vector equation:

$$(1 - l^2\Delta)\Delta^2\mathbf{u} + \mathbf{G} = 0, \quad \text{or briefly} \quad a\Delta^2\mathbf{u} + \mathbf{G} = 0,$$

where

$$a = 1 - l^2\Delta, \quad \mathbf{G} = 1/[(\lambda + \mu)\mu]\underset{\sim}{\mathbf{N}} \cdot (\rho\mathbf{F}).$$

It should be noted that at equilibrium from (24.62) we obtain the following non-split vector equation

$$\underset{\sim}{\mathbf{L}}' \cdot \mathbf{u} + \rho\mathbf{F} = 0.$$

Hence, taking into account that for an isotropic medium

$$\underset{\sim}{\mathbf{L}}' = a\underset{\sim}{\mathbf{L}} = (1 - l^2\Delta)\underset{\sim}{\mathbf{L}}, \quad \underset{\sim}{\mathbf{L}} = \underset{\sim}{\mathbf{E}}\mu\Delta + (\lambda + \mu)\nabla\nabla,$$

we obtain the non-split vector equation in the form

$$(1 - l^2\Delta)\underset{\sim}{\mathbf{L}} \cdot \mathbf{u} + \rho\mathbf{F} = 0 \quad \text{or} \quad a\underset{\sim}{\mathbf{L}} \cdot \mathbf{u} + \rho\mathbf{F} = \mathbf{0}. \tag{24.68}$$

In the case of a homogeneous medium, $a\underset{\sim}{\mathbf{L}} = \underset{\sim}{\mathbf{L}}a$, due to which the brief representation of the equation from (24.68) can be written in the following forms:

$$\begin{aligned} &(a\underset{\sim}{\mathbf{L}}) \cdot \mathbf{u} + \rho\mathbf{F} = \mathbf{0}, \quad (\underset{\sim}{\mathbf{L}}a) \cdot \mathbf{u} + \rho\mathbf{F} = \mathbf{0}, \\ &\underset{\sim}{\mathbf{L}} \cdot (a\mathbf{u}) + \rho\mathbf{F} = \mathbf{0}, \quad a(\underset{\sim}{\mathbf{L}} \cdot \mathbf{u}) + \rho\mathbf{F} = \mathbf{0}. \end{aligned} \tag{24.69}$$

By introducing the notation

$$\mathbf{U} = a\mathbf{u}, \quad \mathbf{V} = \underset{\sim}{\mathbf{L}} \cdot \mathbf{u} \tag{24.70}$$

the last two equations from (24.69) can be written as

$$\underset{\sim}{\mathbf{L}} \cdot \mathbf{U} + \rho\mathbf{F} = \mathbf{0}, \quad a\mathbf{V} + \rho\mathbf{F} = \mathbf{0}. \tag{24.71}$$

From the first equation (24.71) it is clear that $\mathbf{U}$ satisfies the classical equation and if $\mathbf{U} = \mathbf{U}_0$ is a solution for the classical equation, by virtue of the first formula (24.70) we will have

$$(1 - l^2\Delta)\mathbf{u} = \mathbf{U}_0. \tag{24.72}$$

Thus, the displacement vector $\mathbf{u}$ in the considering gradient theory at equilibrium satisfies the decomposed vector equation (24.72), provided that $\mathbf{U}_0$ is the solution of the classical equation. In other words, knowing the solution $\mathbf{U}_0$ of the classical vector equation, finding the solution $\mathbf{u}$ of the vector equation of the considering gradient theory at equilibrium is reduced to solving the decomposed equation (24.72), the order of which is less than the order of the original equation (see, for example, third equation (24.69)). Note that the equation (24.72) can be solved analytically at a given $\mathbf{U}_0$.

Based on the second equation (24.71) we can say that $\mathbf{V}$ is the solution of decomposed equation with the scalar differential operator $a = 1 - l^2\Delta$, at classical body force $\rho\mathbf{F}$ and if we assume that its solution is $\mathbf{V} = \mathbf{V}_0$, then by virtue of the second formula (24.70) we have

$$\underset{\sim}{\mathbf{L}} \cdot \mathbf{u} + (-\mathbf{V}_0) = \mathbf{0}, \tag{24.73}$$

i.e. the displacement vector $\mathbf{u}$ in the considering gradient theory at equilibrium satisfies the equation (24.73), where $\mathbf{V}_0$ is the solution of the second decomposed equation (24.71). In other words, the displacement vector $\mathbf{u}$ in the considering gradient theory at equilibrium is the solution of the classical equation, provided that in the classical equation the body force is replaced by $-\mathbf{V}_0$, where $\mathbf{V}_0$ is solution for the second decomposed equation (24.71).

24.4 Equations at Quasi-Static of the SSTG Theory of Elastic Prismatic Bodies in Displacement Vector and its Moments

Let us consider a prismatic body of constant thickness $2h$. We take the middle plane as the base plane. Then the nabla-operator and the Laplacian are represented in the form (see [56])

$$\begin{aligned} &\nabla\mathbb{F} = (\mathbf{r}^P\partial_P + \mathbf{r}^3\partial_3)\mathbb{F} = (\mathbf{r}^P\partial_P + h^{-1}\mathbf{n}\partial_3)\mathbb{F}, \quad -1 \le x^3 \le 1,\\ &\Delta\mathbb{F} = (g^{PQ}\partial_P\partial_Q + g^{33}\partial_3^2)\mathbb{F} = (\bar{\Delta} + h^{-2}\partial_3^2)\mathbb{F}, \quad \bar{\Delta} = g^{PQ}\partial_P\partial_Q, \end{aligned} \tag{24.74}$$

where $\mathbf{n}$ is unit normal to the middle plane.

Taking into account the representation of the Laplacian (see (24.74)), Eq. (24.28) (see also (24.29)) for the prismatic homogeneous bodies can be represented as

$$\begin{aligned} &[(l_2^4\bar{\Delta}^2 - l_1^2\bar{\Delta} + 1)\bar{\Delta}^2 + h^{-2}(4l_2^4\bar{\Delta}^2 - 3l_1^2\bar{\Delta} + 2)\bar{\Delta}\partial_3^2\\ &+h^{-4}(6l_2^4\bar{\Delta}^2 - 3l_1^2\bar{\Delta} - 1)\partial_3^4 + h^{-6}(4l_2^4\bar{\Delta} - l_1^2)\partial_3^6 + 2h^{-8}l_2^4\partial_3^8]\mathbf{u} + \mathbf{G} = 0. \end{aligned} \tag{24.75}$$

Applying the kth moment operator of any system of orthogonal polynomials (Legendre, Chebyshev) (the theory of moments with respect to the systems of orthogonal polynomials can be seen in [56])) to the equation (24.75), we obtain the following equations in moments of displacement vector:

$$\begin{aligned} &(l_2^4\bar{\Delta}^2 - l_1^2\bar{\Delta} + 1)\bar{\Delta}^2\overset{(k)}{\mathbf{u}} + h^{-2}(4l_2^4\bar{\Delta}^2 - 3l_1^2\bar{\Delta} + 2)\bar{\Delta}\overset{(k)}{\mathbf{u}}{}''\\ &+h^{-4}(6l_2^4\bar{\Delta}^2 - 3l_1^2\bar{\Delta} - 1)\overset{(k)}{\mathbf{u}}{}^{IV} + h^{-6}(4l_2^4\bar{\Delta} - l_1^2)\overset{(k)}{\mathbf{u}}{}^{VI}\\ &+2h^{-8}l_2^4\overset{(k)}{\mathbf{u}}{}^{VIII} + \mathbf{G} = 0, \quad k \in \mathbb{N}_0 \ (\mathbb{N}_0 = \{0, 1, 2, \ldots\}). \end{aligned} \tag{24.76}$$

24.4.1 On Boundary Conditions in Gradient Theories

The boundary conditions for the first- and second-order gradient theories considered in this paper can, of course, be found, for example, in [12, 14, 18, 42, 44, 66, 68, 69, et al.]. The boundary conditions considered in [66, 68, 69] are most suitable for this work. By writing them relative to the displacement vector for the first- and second-order gradient theories considered in this work, and then adding them and the initial conditions to the above-considered equations of motion relative to the displacement vector for these theories, respectively, we obtain three-dimensional formulations of initial-boundary value problems for isothermal processes relative to the displacement vector for both theories separately. Further, by writing these formulations of initial boundary value problems for the considered (e.g., new) parameterization of the thin body domain, we obtain three-dimensional formulations of initial boundary value problems for isothermal processes relative to the displacement vector of first- and second-order gradient theories of elastic thin bodies. From the latter, using the method of orthogonal polynomials, we obtain the corresponding formulations of initial-boundary value problems at the moments of the displacement vector relative to the applied system of orthogonal polynomials (the procedure for obtaining the formulation of initial-boundary value and boundary value problems for the classical and micropolar theories of thin bodies is described in sufficient detail in [56], for the classical theory, see also [75]). I believe that the interested reader, using the procedure described above, can independently obtain the problem formulations, but I will not dwell on this in this paper.

Let us now consider more carefully the system of equations in moments (24.76). From this system changing $k = \overline{1, N}$ and neglecting the moments, order of which is more than N, we obtain a system of equation of Nth approximation. Giving N different values (starting from zero) we obtain the systems of equations of various approximation for prismatic bodies of constant thickness. It should be noted that with the method of deriving a system of equations in moments of displacement vector considered below with respect to a system of orthogonal polynomials, the resulting system of equations in moments, the order of approximation of which is greater than zero, is divided into two systems, one of which contains moments of even order of the unknown displacement vector, and the other contains moments of odd order. In addition, the resulting systems with even and odd moments of the displacement vector, in turn, are easily split separately and for each moment of the displacement vector an elliptic equation of higher order is obtained, for which, using the Vekua method, a general representation of the solutions can be written. Below, in the case of quasi-statics, equations of the tenth order of approximation are obtained in the moments of the displacement vector with respect to the system of Legendre polynomials and everything said above is demonstrated in this example.

24.4.2 Equations of the Quasi-Static SSTG Theory of Isotropic Prismatic Bodies in Moments of Displacement Vector

To obtain the desired systems of equations we need to find expression for the so-called "stroke-operators" $\overset{(k)}{\mathbf{u}}{}''$, $\overset{(k)}{\mathbf{u}}{}^{IV}$, $\overset{(k)}{\mathbf{u}}{}^{VI}$ and $\overset{(k)}{\mathbf{u}}{}^{VIII}$ when $-1 \leq x^3 \leq 1$. Note that in the application of the system of Legendre polynomials, the expressions for the stroke-operators $\overset{(k)}{\mathbf{u}}{}''$, $\overset{(k)}{\mathbf{u}}{}^{IV}$, $\overset{(k)}{\mathbf{u}}{}^{VI}$, $\overset{(k)}{\mathbf{u}}{}^{VIII}$, and so on are defined using the following relationship [56]:

$$\begin{aligned}
&\overset{(n)}{\mathbf{u}}{}^{(2m)}(x') = (2n+1)\sum_{k=1}^{\infty} C^{2m-1}_{k+2m-2}\prod_{s=1}^{2m-1}(2n+2k+2s-1)\overset{(n+2k+2m-2)}{\mathbf{u}}\\
&= \frac{2n+1}{2}\sum_{k=1}^{2m}(-1)^{k+1}[(\partial_3^{2m-k}\mathbf{u})^+ + (-1)^{n+k}(\partial_3^{2m-k}\mathbf{u})^-]P_n^{k-1}(1) + \overset{(k)}{\underline{\mathbf{u}}}{}^{(2m)},\\
&\overset{(k)}{\underline{\mathbf{u}}}{}^{(2m)} = (2n+1)\sum_{k=1}^{[n/2-m+1]} C^{2m-1}_{k+2m-2}\prod_{s=1}^{2m-1}(2n-2k-2s+3)\overset{(n-2k-2m+2)}{\mathbf{u}},\\
&n \in \mathbb{N}_0,\quad m \in \mathbb{N}.
\end{aligned} \tag{24.77}$$

Here $\mathbb{N}_0$ is the set of non-negative integers, but $\mathbb{N}$ is the set of natural numbers, C^{2m-1}_{k+2m-2} is binomial coefficient, $(\partial_3^s\mathbf{u})^- = (\partial_3^s\mathbf{u})\big|_{x^3=-1}$ and $(\partial_3^s\mathbf{u})^+ = (\partial_3^s\mathbf{u})\big|_{x^3=1}$, $s \in \mathbb{N}$.

It is seen from (24.77) that the so-called stroke-operators $\overset{(k)}{\mathbf{u}}{}''$, $\overset{(k)}{\mathbf{u}}{}^{IV}$, $\overset{(k)}{\mathbf{u}}{}^{VI}$, $\overset{(k)}{\mathbf{u}}{}^{VIII}$ and so on are represented as an infinite sum of moments of displacement vector, or as the final sum of moments of displacement vector, the sum of values of displacement vector and their partial derivatives in x^3 on the face surface, i.e. when $x^3 = -1$ and $x^3 = 1$. Therefore, taking into account (24.77), from (24.76) we get different representations of the systems of equations of the quasi-static second strain gradient theory of isotropic prismatic thin bodies of constant thickness in moments of displacement vector with respect to the system of Legendre polynomials.

Let's write down expressions for some stroke-operators $\overset{(k)}{\mathbf{u}}{}''$, $\overset{(k)}{\mathbf{u}}{}^{IV}$, $\overset{(k)}{\mathbf{u}}{}^{VI}$ and $\overset{(k)}{\mathbf{u}}{}^{VIII}$ when $-1 \leq x^3 \leq 1$, which are obtained based on the first equality (24.77). By virtue of simple calculations when considering, for example, the tenth order approximation, we obtain

$$\overset{(0)}{\mathbf{u}}{}''=3\overset{(2)}{\mathbf{u}}+10\overset{(4)}{\mathbf{u}}+21\overset{(6)}{\mathbf{u}}+36\overset{(8)}{\mathbf{u}}+55\overset{(10)}{\mathbf{u}},\quad \overset{(0)}{\mathbf{u}}{}^{IV}=105\overset{(4)}{\mathbf{u}}+1260\overset{(6)}{\mathbf{u}}+6930\overset{(8)}{\mathbf{u}}+25740\overset{(10)}{\mathbf{u}},$$

$$\overset{(0)}{\mathbf{u}}{}^{VI}=10395\overset{(6)}{\mathbf{u}}+270270\overset{(8)}{\mathbf{u}}+2837835\overset{(10)}{\mathbf{u}},\quad \overset{(0)}{\mathbf{u}}{}^{VIII}=2027025\overset{(8)}{\mathbf{u}}+91891800\overset{(10)}{\mathbf{u}};$$

$$\overset{(1)}{\mathbf{u}}{}''=15\overset{(3)}{\mathbf{u}}+42\overset{(5)}{\mathbf{u}}+81\overset{(7}{\mathbf{u}}+132\overset{(9)}{\mathbf{u}},\quad \overset{(1)}{\mathbf{u}}{}^{IV}=945\overset{(5)}{\mathbf{u}}+8316\overset{(7)}{\mathbf{u}}+38610\overset{(9)}{\mathbf{u}},$$

$$\overset{(1)}{\mathbf{u}}{}''=15\overset{(3)}{\mathbf{u}}+42\overset{(5)}{\mathbf{u}}+81\overset{(7}{\mathbf{u}}+132\overset{(9)}{\mathbf{u}},\quad \overset{(1)}{\mathbf{u}}{}^{IV}=945\overset{(5)}{\mathbf{u}}+8316\overset{(7)}{\mathbf{u}}+38610\overset{(9)}{\mathbf{u}},$$

$$\overset{(1)}{\mathbf{u}}{}^{VI}=135135\overset{(7)}{\mathbf{u}}+2432430\overset{(9)}{\mathbf{u}},\quad \overset{(1)}{\mathbf{u}}{}^{VIII}=34459425\overset{(9)}{\mathbf{u}};$$

$$\overset{(2)}{\mathbf{u}}{}''=35\overset{(4)}{\mathbf{u}}+90\overset{(6)}{\mathbf{u}}+165\overset{(8)}{\mathbf{u}}+260\overset{(10)}{\mathbf{u}},\quad \overset{(2)}{\mathbf{u}}{}^{IV}=3465\overset{(6)}{\mathbf{u}}+25740\overset{(8)}{\mathbf{u}}+107250\overset{(10)}{\mathbf{u}},$$

$$\overset{(2)}{\mathbf{u}}{}^{VI}=675675\overset{(8)}{\mathbf{u}}+9845550\overset{(10)}{\mathbf{u}},\quad \overset{(2)}{\mathbf{u}}{}^{VIII}=218243025\overset{(10)}{\mathbf{u}}. \tag{24.78}$$

In the case under consideration, the displacement vector is represented as

$$\mathbf{u}(x^1,x^2,x^3)=\overset{(0)}{\mathbf{u}}(x^1,x^2)+\overset{(1)}{\mathbf{u}}(x^1,x^2)P_1(x^3)+\ldots+\overset{(10)}{\mathbf{u}}(x^1,x^2)P_{10}(x^3),$$

where $P_k(x^3)$, $k=\overline{1,10}$ are Legendre polynomials.

24.4.3 Systems of Equilibrium Equations of the Tenth-Order Approximation of the SSTG Theory of Elastic Isotropic Prismatic Bodies in Odd- and Even-Order Moments

As mentioned above, by virtue of the method of orthogonal polynomials, a system of equations in moments is obtained, which is split into two systems of equations, one of which contains moments of even order of the unknown displacement vector, and the other contains moments of odd order. In this case, the resulting systems with even and odd moments of the displacement vector are separately split and for each moment of the displacement vector an elliptic equation of higher order is obtained, for which a general representation of solutions can be written by the Vekua method. In fact, it is easy to see that due to (24.77) (see also (24.78)) from (24.76) we obtain a system of equations with odd and even moments of the displacement vector relative to the system of Legendre polynomials tenth order approximation, from which the equations with odd moments of the displacement vector is represented in the form

$$
\begin{aligned}
&(l_2^4\bar{\Delta}^2-l_1^2\bar{\Delta}+1)\bar{\Delta}^2\overset{(1)}{\mathbf{u}}+15h^{-2}(4l_2^4\bar{\Delta}^2-3l_1^2\bar{\Delta}+2)\bar{\Delta}\overset{(3)}{\mathbf{u}}\\
&+[42h^{-2}(4l_2^4\bar{\Delta}^2-3l_1^2\bar{\Delta}+2)\bar{\Delta}+945h^{-4}(6l_2^4\bar{\Delta}^2-3l_1^2\bar{\Delta}-1)]\overset{(5)}{\mathbf{u}}\\
&+[81h^{-2}(4l_2^4\bar{\Delta}^2-3l_1^2\bar{\Delta}+2)\bar{\Delta}+8316h^{-4}(4l_2^4\bar{\Delta}^2-3l_1^2\bar{\Delta}+2)\\
&+135135h^{-6}(4l_2^4\bar{\Delta}-l_1^2)]\overset{(7)}{\mathbf{u}}+[132h^{-2}(4l_2^4\bar{\Delta}^2-3l_1^2\bar{\Delta}+2)\bar{\Delta}+38610h^{-4}(6l_2^4\bar{\Delta}^2\\
&-3l_1^2\bar{\Delta}-1)+2432430h^{-6}(4l_2^4\bar{\Delta}-l_1^2)+68918850h^{-8}l_2^4]\overset{(9)}{\mathbf{u}}+\overset{(1)}{\mathbf{G}}=0,\\
&(l_2^4\bar{\Delta}^2-l_1^2\bar{\Delta}+1)\bar{\Delta}^2\overset{(3)}{\mathbf{u}}+63h^{-2}(4l_2^4\bar{\Delta}^2-3l_1^2\bar{\Delta}+2)\bar{\Delta}\overset{(5)}{\mathbf{u}}+[154h^{-2}(4l_2^4\bar{\Delta}^2\\
&-l_1^2\bar{\Delta}+2)\bar{\Delta}+9009h^{-4}(6l_2^4\bar{\Delta}^2-3l_1^2\bar{\Delta}-1)]\overset{(7)}{\mathbf{u}}+[273h^{-2}(4l_2^4\bar{\Delta}^2-3l_1^2\bar{\Delta}+2)\bar{\Delta}\\
&+60060h^{-4}(6l_2^4\bar{\Delta}^2-3l_1^2\bar{\Delta}-1)+2297255h^{-6}(4l_2^4\bar{\Delta}-l_1^2)]\overset{(9)}{\mathbf{u}}+\overset{(3)}{\mathbf{G}}=0,\\
&(l_2^4\bar{\Delta}^2-l_1^2\bar{\Delta}+1)\bar{\Delta}^2\overset{(5)}{\mathbf{u}}+143h^{-2}(4l_2^4\bar{\Delta}^2-3l_1^2\bar{\Delta}+2)\bar{\Delta}\overset{(7)}{\mathbf{u}}\\
&+[330h^{-2}(4l_2^4\bar{\Delta}^2-3l_1^2\bar{\Delta}+2)\bar{\Delta}+36465h^{-4}(6l_2^4\bar{\Delta}^2-3l_1^2\bar{\Delta}-1)]\overset{(9)}{\mathbf{u}}+\overset{(5)}{\mathbf{G}}=0,\\
&(l_2^4\bar{\Delta}^2-l_1^2\bar{\Delta}+1)\bar{\Delta}^2\overset{(7)}{\mathbf{u}}+255h^{-2}(4l_2^4\bar{\Delta}^2-3l_1^2\bar{\Delta}+2)\bar{\Delta}\overset{(9)}{\mathbf{u}}+\overset{(7)}{\mathbf{G}}=0,\\
&(l_2^4\bar{\Delta}^2-l_1^2\bar{\Delta}+1)\bar{\Delta}^2\overset{(9)}{\mathbf{u}}+\overset{(9)}{\mathbf{G}}=0,
\end{aligned}
\tag{24.79}
$$

and the equations with even moments of the displacement vector has the form

$$
\begin{aligned}
&(l_2^4\bar{\Delta}^2-l_1^2\bar{\Delta}+1)\bar{\Delta}^2\overset{(0)}{\mathbf{u}}+3h^{-2}(4l_2^4\bar{\Delta}^2-3l_1^2\bar{\Delta}+2)\bar{\Delta}\overset{(2)}{\mathbf{u}}\\
&+[10h^{-2}(4l_2^4\bar{\Delta}^2-3l_1^2\bar{\Delta}+2)\bar{\Delta}+105h^{-4}(6l_2^4\bar{\Delta}^2-3l_1^2\bar{\Delta}-1)]\overset{(4)}{\mathbf{u}}\\
&+[21h^{-2}(4l_2^4\bar{\Delta}^2-3l_1^2\bar{\Delta}+2)\bar{\Delta}+1260h^{-4}(4l_2^4\bar{\Delta}^2-3l_1^2\bar{\Delta}+2)\\
&+10395h^{-6}(4l_2^4\bar{\Delta}-l_1^2)]\overset{(6)}{\mathbf{u}}+[36h^{-2}(4l_2^4\bar{\Delta}^2-3l_1^2\bar{\Delta}+2)\bar{\Delta}\\
&+6930h^{-4}(6l_2^4\bar{\Delta}^2-3l_1^2\bar{\Delta}-1)+270270h^{-6}(4l_2^4\bar{\Delta}-l_1^2)+4054050h^{-8}l_2^4]\overset{(8)}{\mathbf{u}}\\
&+[55h^{-2}(4l_2^4\bar{\Delta}^2-3l_1^2\bar{\Delta}+2)\bar{\Delta}+25740h^{-4}(6l_2^4\bar{\Delta}^2-3l_1^2\bar{\Delta}-1)\\
&+2837835h^{-6}(4l_2^4\bar{\Delta}-l_1^2)+183783600h^{-8}l_2^4]\overset{(10)}{\mathbf{u}}+\overset{(0)}{\mathbf{G}}=0,
\end{aligned}
$$

$$
\begin{aligned}
&(l_2^4\bar{\Delta}^2 - l_1^2\bar{\Delta} + 1)\bar{\Delta}^2 \overset{(2)}{\mathbf{u}} + 35h^{-2}(4l_2^4\bar{\Delta}^2 - 3l_1^2\bar{\Delta} + 2)\bar{\Delta} \overset{(4)}{\mathbf{u}} \\
&+[90h^{-2}(4l_2^4\bar{\Delta}^2 - 3l_1^2\bar{\Delta} + 2)\bar{\Delta} + 3465h^{-4}(6l_2^4\bar{\Delta}^2 - 3l_1^2\bar{\Delta} - 1)] \overset{(6)}{\mathbf{u}} \\
&+[165h^{-2}(4l_2^4\bar{\Delta}^2 - 3l_1^2\bar{\Delta} + 2)\bar{\Delta} + 25740h^{-4}(4l_2^4\bar{\Delta}^2 - 3l_1^2\bar{\Delta} + 2) \\
&+75675h^{-6}(4l_2^4\bar{\Delta} - l_1^2)] \overset{(8)}{\mathbf{u}} + [260h^{-2}(4l_2^4\bar{\Delta}^2 - 3l_1^2\bar{\Delta} + 2)\bar{\Delta} \\
&+107250h^{-4}(4l_2^4\bar{\Delta}^2 - 3l_1^2\bar{\Delta} + 2) + 9845550h^{-6}(4l_2^2\bar{\Delta} - l_1^2) \\
&+436486050h^{-8}l_2^4] \overset{(10)}{\mathbf{u}} + \overset{(2)}{\mathbf{G}} = 0, \\
&(l_2^4\bar{\Delta}^2 - l_1^2\bar{\Delta} + 1)\bar{\Delta}^2 \overset{(4)}{\mathbf{u}} + 99h^{-2}(4l_2^4\bar{\Delta}^2 - 3l_1^2\bar{\Delta} + 2)\bar{\Delta} \overset{(6)}{\mathbf{u}} \\
&+[234h^{-2}(4l_2^4\bar{\Delta}^2 - 3l_1^2\bar{\Delta} + 2)\bar{\Delta} + 19305h^{-4}(6l_2^4\bar{\Delta}^2 - 3l_1^2\bar{\Delta} - 1)] \overset{(8)}{\mathbf{u}} \\
&+[405h^{-2}(4l_2^4\bar{\Delta}^2 - 3l_1^2\bar{\Delta} + 2)\bar{\Delta} + 119340h^{-4}(6l_2^4\bar{\Delta}^2 - 3l_1^2\bar{\Delta} - 1) \\
&+6235715h^{-6}(4l_2^2\bar{\Delta} - l_1^2)] \overset{(10)}{\mathbf{u}} + \overset{(4)}{\mathbf{G}} = 0, \\
&(l_2^4\bar{\Delta}^2 - l_1^2\bar{\Delta} + 1)\bar{\Delta}^2 \overset{(6)}{\mathbf{u}} + 195h^{-2}(4l_2^4\bar{\Delta}^2 - 3l_1^2\bar{\Delta} + 2)\bar{\Delta} \overset{(8)}{\mathbf{u}} + [442h^{-2} \\
&(4l_2^4\bar{\Delta}^2 - 3l_1^2\bar{\Delta} + 2)\bar{\Delta} + 62985h^{-4}(6l_2^4\bar{\Delta}^2 - 3l_1^2\bar{\Delta} - 1)] \overset{(10)}{\mathbf{u}} + \overset{(6)}{\mathbf{G}} = 0, \\
&(l_2^4\bar{\Delta}^2 - l_1^2\bar{\Delta} + 1)\bar{\Delta}^2 \overset{(8)}{\mathbf{u}} + 323h^{-2}(4l_2^4\bar{\Delta}^2 - 3l_1^2 + 2)\bar{\Delta} \overset{(10)}{u} + \overset{(8)}{\mathbf{G}} = 0, \\
&(l_2^4\bar{\Delta}^2 - l_1^2\bar{\Delta} + 1)\bar{\Delta}^2 \overset{(10)}{\mathbf{u}} + \overset{(10)}{\mathbf{G}} = 0.
\end{aligned}
\tag{24.80}
$$

24.4.4 Decomposed Equations of the Tenth-Order Approximation of the SSTG Theory of Elastic Isotropic Prismatic Bodies in Odd-Order Moments

Introducing designations

$$
L^{(1)} = \begin{pmatrix} L_{11}^{(1)} & L_{12}^{(1)} & L_{13}^{(1)} & L_{14}^{(1)} & L_{15}^{(1)} \\ 0 & L_{22}^{(1)} & L_{23}^{(1)} & L_{24}^{(1)} & L_{25}^{(1)} \\ 0 & 0 & L_{33}^{(1)} & L_{34}^{(1)} & L_{35}^{(1)} \\ 0 & 0 & 0 & L_{44}^{(1)} & L_{45}^{(1)} \\ 0 & 0 & 0 & 0 & L_{55}^{(1)} \end{pmatrix}, \quad \mathbf{U}^{(1)} = \begin{pmatrix} \overset{(1)}{\mathbf{u}} \\ \overset{(3)}{\mathbf{u}} \\ \overset{(5)}{\mathbf{u}} \\ \overset{(7)}{\mathbf{u}} \\ \overset{(9)}{\mathbf{u}} \end{pmatrix}, \quad \mathbf{G}^{(1)} = \begin{pmatrix} \overset{(1)}{\mathbf{G}} \\ \overset{(3)}{\mathbf{G}} \\ \overset{(5)}{\mathbf{G}} \\ \overset{(7)}{\mathbf{G}} \\ \overset{(9)}{\mathbf{G}} \end{pmatrix},
$$

$$
\begin{aligned}
&L_{11}^{(1)} = L_{22}^{(1)} = L_{33}^{(1)} = L_{44}^{(1)} = L_{55}^{(1)} = (l_2^4\bar{\Delta}^2 - l_1^2\bar{\Delta} + 1)\bar{\Delta}^2; \\
&L_{12}^{(1)} = 15h^{-2}(4l_2^4\bar{\Delta}^2 - 3l_1^2\bar{\Delta} + 2)\bar{\Delta}; \ L_{13}^{(1)} = 42h^{-2}(4l_2^4\bar{\Delta}^2 - 3l_1^2\bar{\Delta} + 2)\bar{\Delta} \\
&+945h^{-4}(6l_2^4\bar{\Delta}^2 - 3l_1^2\bar{\Delta} - 1); \ L_{14}^{(1)} = 81h^{-2}(4l_2^4\bar{\Delta}^2 - 3l_1^2\bar{\Delta} + 2)\bar{\Delta} \\
&+8316h^{-4}(6l_2^4\bar{\Delta}^2 - 3l_1^2\bar{\Delta} - 1) + 10384h^{-6}(4l_2^4\bar{\Delta} - l_1^2); \\
&L_{15}^{(1)} = 132h^{-2}(4l_2^4\bar{\Delta}^2 - 3l_1^2\bar{\Delta} + 2)\bar{\Delta} + 386106h^{-4}(6l_2^4\bar{\Delta}^2 - 3l_1^2\bar{\Delta} - 1) \\
&+2432430h^{-6}(4l_2^4\bar{\Delta} - l_1^2) + 68918850h^{-8}l_2^4; \ L_{23}^{(1)} = 63h^{-2}(4l_2^4\bar{\Delta}^2 - 3l_1^2\bar{\Delta} + 2)\bar{\Delta};
\end{aligned}
$$

$$L^{(1)}_{24} = 154h^{-2}(4l_2^4\bar{\Delta}^2 - 3l_1^2\bar{\Delta} + 2)\bar{\Delta} + 9009h^{-4}(6l_2^4\bar{\Delta}^2 - 3l_1^2\bar{\Delta} - 1);$$
$$L^{(1)}_{25} = 154h^{-2}(4l_2^4\bar{\Delta}^2 - 3l_1^2\bar{\Delta} + 2)\bar{\Delta} + 9009h^{-4}(6l_2^4\bar{\Delta}^2 - 3l_1^2\bar{\Delta} - 1);$$
$$L^{(1)}_{34} = 143h^{-2}(4l_2^4\bar{\Delta}^2 - 3l_1^2\bar{\Delta} + 2)\bar{\Delta};\ L^{(1)}_{35} = 330h^{-2}(4l_2^4\bar{\Delta}^2 - 3l_1^2\bar{\Delta} + 2)\bar{\Delta}$$
$$+36465h^{-4}(6l_2^4\bar{\Delta}^2 - 3l_1^2\bar{\Delta} - 1);\ L^{(1)}_{45} = 255h^{-2}(4l_2^4\bar{\Delta}^2 - 3l_1^2\bar{\Delta} + 2)\bar{\Delta},$$

the system (24.79) can be written as a matrix equation

$$L^{(1)}\mathbf{U}^{(1)} + \mathbf{G}^{(1)} = 0. \tag{24.81}$$

It is not difficult to find the elements of the matrix of cofactors $L^{(1)}_*$ for the differential matrix $L^{(1)}$ of the equation (24.81). In fact, carrying out simple calculations, the matrix $L^{(1)}_*$ and its determinant are represented in the form

$$L^{(1)}_* = \begin{pmatrix} L^{(1)}_{*11} & 0 & 0 & 0 & 0 \\ L^{(1)}_{*21} & L^{(1)}_{*22} & 0 & 0 & 0 \\ L^{(1)}_{*31} & L^{(1)}_{*32} & L^{(1)}_{*33} & 0 & 0 \\ L^{(1)}_{*41} & L^{(1)}_{*42} & L^{(1)}_{*43} & L^{(1)}_{*44} & 0 \\ L^{(1)}_{*51} & L^{(1)}_{*52} & L^{(1)}_{*53} & L^{(1)}_{*54} & L^{(1)}_{*55} \end{pmatrix}, \quad |L^{(1)}_*| = (l_2^4\bar{\Delta}^2 - l_1^2\bar{\Delta} + 1)^{20}\bar{\Delta}^{40},$$

where, due to simple calculations, we have

$$L^{(1)}_{*\alpha\alpha} = (l_2^4\bar{\Delta}^2 - l_1^2\bar{\Delta} + 1)^4\bar{\Delta}^8,\ \langle\alpha = \overline{1,5}\rangle;\ L^{(1)}_{*21} = -15h^{-2}b_0^3 b_1\bar{\Delta}^7;$$
$$L^{(1)}_{*31} = (945h^{-4}b_0^2 b_1^2 - a_0 b_0^3)\bar{\Delta}^6;\ L^{(1)}_{*32} = -63h^{-2}b_0^3 b_1\bar{\Delta}^7,$$
$$L^{(1)}_{*41} = -\bar{\Delta}^5(135135h^{-6}b_0^2 b_1^3 + a_1\bar{\Delta}b_0^3 - 143h^{-2}a_0 b_0^2 b_1 - 15h^{-2}c_0 b_0^2 b_1),$$
$$L^{(1)}_{*42} = \bar{\Delta}^6 b_0^2(9009h^{-4}b_1^2 - b_0 c_0),\ L^{(1)}_{*43} = -143h^{-2}\bar{\Delta}^7 b_0^3 b_1,$$
$$L^{(1)}_{*51} = \bar{\Delta}^4(34459425h^{-8}b_1^4 - a_2\bar{\Delta}^4 b_0^3 - 36465h^{-4}a_0 b_0^2 b_1^2 + 255h{-}2a_1 b_0^2 b_1$$
$$-3825h^{-2}c_0 b_1^2 b_0 - a_0 b_0^2 d_0 + 15h{-}2c_1 b_0^2 b_1 - 945h^{-4}d_0 b_1^2 b_0),$$
$$L^{(1)}_{*52} = -\bar{\Delta}^5(32297295h^{-6}b_0 b_1^3 + \bar{\Delta}^4 b_0^3 c_1 - 255h^{-2}c_0 b_0^2 b_1 - 63h^{-2}d_0 b_0^2 b_1)$$
$$L^{(1)}_{*53} = \bar{\Delta}^6 b_0^2(36465h^{-4}b_1^2 - d_0 b_0),\ L^{(1)}_{*54} = -\bar{\Delta}^6 b_0^3(255h^{-2}b_1\bar{\Delta} - d_0).$$
$$b_0 = l_2^4\bar{\Delta}^2 - l_1^2\bar{\Delta} + 1,\ b_1 = 4l_2^4\bar{\Delta}^2 - 3l_1^2\bar{\Delta} + 2,\ b_2 = 6l_2^4\bar{\Delta}^2 - 3l_1^2\bar{\Delta} - 1,$$
$$b_3 = 4l_2^4\bar{\Delta} - l_1^2,\ a_0 = 42h^{-2}b_1\bar{\Delta} + 945h^{-4}b_2,\ a_1 = 81h^{-2}b_1\bar{\Delta} + 8316h^{-4}b_2$$
$$+10384h^{-6}b_3,\ a_2 = 132h^{-2}b_1\bar{\Delta} + 386106h^{-4}b_2 + 2432430h^{-6}b_3 + 68918850h^{-8}l_2^4,$$
$$c_0 = 273h^{-2}b_1\bar{\Delta} + 60060h^{-4}b_2 + 2297255h^{-6}b_3,\ c_1 = 154h^{-2}b_1\bar{\Delta} + 9009h^{-4}b_2,$$
$$d_0 = 330h^{-2}b_1\bar{\Delta} + 36465h^{-4}b_2,$$

where the notation $\langle\alpha = \overline{1,5}\rangle$ means that $\alpha = 1, \ldots, 5$ and there is no summation over α. It is easy to see that the matrix of cofactors $L^{(1)}_*$ can be represented as

$$L^{(1)}_* = \bar{\Delta}^4 N^{(1)},\quad N^{(1)} = matr(N^{(1)}_{ij}), i, j = \overline{1,5},$$
$$N^{(1)}_{1\alpha} = 0, \alpha = \overline{2,5}, N^{(1)}_{23} = N^{(1)}_{24} = N^{(1)}_{25} = 0, N^{(1)}_{34} = N^{(1)}_{35} = 0, N^{(1)}_{45} = 0,$$

$$
\begin{aligned}
&N^{(1)}_{\alpha\alpha} = (l_2^4\bar\Delta^2 - l_1^2\bar\Delta + 1)^4\bar\Delta^4,\ \langle\alpha = \overline{1,5}\rangle;\ N^{(1)}_{21} = -15h^{-2}b_0^3 b_1\bar\Delta^3;\\
&N^{(1)}_{31} = (945h^{-4}b_0^2 b_1^2 - a_0 b_0^3)\bar\Delta^2;\ N^{(1)}_{32} = -63h^{-2}b_0^3 b_1\bar\Delta^3,\\
&N^{(1)}_{41} = -\bar\Delta(135135h^{-6}b_0^2 b_1^3 + a_1\bar\Delta b_0^3 - 143h^{-2}a_0 b_0^2 b_1 - 15h^{-2}c_0 b_0^2 b_1),\\
&N^{(1)}_{42} = \bar\Delta^2 b_0^2(9009h^{-4}b_1^2 - b_0 c_0),\ N^{(1)}_{43} = -143h^{-2}\bar\Delta^3 b_0^3 b_1,\\
&N^{(1)}_{51} = (34459425h^{-8}b_1^4 - a_2\bar\Delta^4 b_0^3 - 36465h^{-4}a_0 b_0^2 b_1^2 + 255h{-}2a_1 b_0^2 b_1\\
&-3825h^{-2}c_0 b_1^2 b_0 - a_0 b_0^2 d_0 + 15h{-}2c_1 b_0^2 b_1 - 945h^{-4}d_0 b_1^2 b_0),\\
&N^{(1)}_{52} = -\bar\Delta(32297295h^{-6}b_0 b_1^3 + \bar\Delta^4 b_0^3 c_1 - 255h^{-2}c_0 b_0^2 b_1 - 63h^{-2}d_0 b_0^2 b_1)\\
&N^{(1)}_{53} = \bar\Delta^2 b_0^2(36465h^{-4}b_1^2 - d_0 b_0),\ N^{(1)}_{54} = -\bar\Delta^2 b_0^3(255h^{-2}b_1\bar\Delta - d_0).
\end{aligned}
$$

Applying the differential operator $N^{(1)T}$ to the equation (24.81) on the left, due to t $N^{(1)T}L^{(1)} = E(l_2^4\bar\Delta^2 - l_1^2\bar\Delta + 1)^5\bar\Delta^6$, where E is the identity matrix of the fifth order, we obtain the decomposed elliptic equation of higher order for each odd moment of the displacement vector in the form

$$
\begin{aligned}
&(l_2^4\bar\Delta^2 - l_1^2\bar\Delta + 1)^5\bar\Delta^6\mathbf{U}^{(1)} + N^{(1)T}\mathbf{G}^{(1)} = 0 \quad\text{or}\\
&(l_2^4\bar\Delta^2 - l_1^2\bar\Delta + 1)^5\bar\Delta^6\overset{(2k+1)}{\mathbf{u}} + \mathbf{\Phi}^{(2k+1)} = 0,\quad k = \overline{0,4}.
\end{aligned}
\tag{24.82}
$$

24.4.5 Decomposed Equations of the Tenth-Order Approximation of the SSTG Theory of Elastic Isotropic Prismatic Bodies in Even-Order Moments

Introducing designations (see (24.80))

$$
L^{(2)} = \begin{pmatrix}
L^{(2)}_{11} & L^{(2)}_{12} & L^{(2)}_{13} & L^{(2)}_{14} & L^{(2)}_{15} & L^{(2)}_{16}\\
0 & L^{(2)}_{22} & L^{(2)}_{23} & L^{(2)}_{24} & L^{(2)}_{25} & L^{(2)}_{26}\\
0 & 0 & L^{(2)}_{33} & L^{(2)}_{34} & L^{(2)}_{35} & L^{(2)}_{36}\\
0 & 0 & 0 & L^{(2)}_{44} & L^{(2)}_{45} & L^{(2)}_{46}\\
0 & 0 & 0 & 0 & L^{(2)}_{55} & L^{(2)}_{56}\\
0 & 0 & 0 & 0 & 0 & L^{(2)}_{66}
\end{pmatrix},\quad
\mathbf{U}^{(2)} = \begin{pmatrix}\overset{(0)}{\mathbf{u}}\\ \overset{(2)}{\mathbf{u}}\\ \overset{(4)}{\mathbf{u}}\\ \overset{(6)}{\mathbf{u}}\\ \overset{(8)}{\mathbf{u}}\\ \overset{(10)}{\mathbf{u}}\end{pmatrix},\quad
\mathbf{G}^{(2)} = \begin{pmatrix}\overset{(0)}{\mathbf{G}}\\ \overset{(2)}{\mathbf{G}}\\ \overset{(4)}{\mathbf{G}}\\ \overset{(6)}{\mathbf{G}}\\ \overset{(8)}{\mathbf{G}}\\ \overset{(10)}{\mathbf{G}}\end{pmatrix},
$$

$$L_{11}^{(2)} = L_{22}^{(2)} = L_{33}^{(2)} = L_{44}^{(2)} = L_{55}^{(2)} = L_{66}^{(2)} = (l_2^4\bar{\Delta}^2 - l_1^2\bar{\Delta} + 1)\bar{\Delta}^2;$$
$$L_{12}^{(2)} = 3h^{-2}(4l_2^4\bar{\Delta}^2 - 3l_1^2\bar{\Delta} + 2)\bar{\Delta};\ \ L_{13}^{(2)} = 10h^{-2}b_1\bar{\Delta} + 105h^{-4}b_2;$$
$$L_{14}^{(2)} = 21h^{-2}b_1\bar{\Delta} + 1260h^{-4}b_2 + 10395h^{-6}b_3;$$
$$L_{15}^{(2)} = 36h^{-2}b_1\bar{\Delta} + 69306h^{-4}b_2 + 270270h^{-6}b_3 + 4054050h^{-8}l_2^4;$$
$$L_{16}^{(2)} = 55h^{-2}b_1\bar{\Delta} + 75075h^{-4}b_2 + 2837835h^{-6}b_3 + 183783600h^{-8}l_2^4;$$
$$L_{23}^{(2)} = 35h^{-2}(4l_2^4\bar{\Delta}^2 - 3l_1^2\bar{\Delta} + 2)\bar{\Delta};\ \ L_{24}^{(2)} = 90h^{-2}b_1\bar{\Delta} + 3465h^{-4}b_2;$$
$$L_{25}^{(2)} = 165h^{-2}b_1\bar{\Delta} + 25740h^{-4}b_2 + 675675h^{-6}b_3;$$
$$L_{26}^{(2)} = 260h^{-2}b_1\bar{\Delta} + 107250h^{-4}b_2 + 9845550h^{-6}b_3;$$
$$L_{34}^{(2)} = 99h^{-2}(4l_2^4\bar{\Delta}^2 - 3l_1^2\bar{\Delta} + 2)\bar{\Delta};\ \ L_{35}^{(2)} = 234h^{-2}b_1\bar{\Delta} + 19305h^{-4}b_2;$$

$$L_{36}^{(2)} = 119340h^{-4}b_2 + 6235515h^{-6}b_3;\ \ L_{45}^{(2)} = 195h^{-2}(4l_2^4\bar{\Delta}^2 - 3l_1^2\bar{\Delta} + 2)\bar{\Delta};$$
$$L_{46}^{(2)} = 442h^{-2}b_1\bar{\Delta} + 62985h^{-4}b_2;\ \ L_{56}^{(2)} = 323h^{-2}(4l_2^4\bar{\Delta}^2 - 3l_1^2\bar{\Delta} + 2)\bar{\Delta};$$
$$b_0 = l_2^4\bar{\Delta}^2 - l_1^2\bar{\Delta} + 1,\ \ b_1 = 4l_2^4\bar{\Delta}^2 - 3l_1^2\bar{\Delta} + 2,\ \ b_2 = 6l_2^4\bar{\Delta}^2 - 3l_1^2\bar{\Delta} - 1,$$
$$b_3 = 4l_2^4\bar{\Delta} - l_1^2,\ \ a_0 = 10h^{-2}b_1\bar{\Delta} + 105h^{-4}b_2,\ \ a_1 = 21h^{-2}b_1\bar{\Delta} + 1260h^{-4}b_2$$
$$+10395h^{-6}b_3,\ \ a_2 = 36h^{-2}b_1\bar{\Delta} + 69306h^{-4}b_2 + 270270h^{-6}b_3 + 4054050h^{-8}l_2^4,$$
$$a_3 = 55h^{-2}b_1\bar{\Delta} + 75075h^{-4}b_2 + 2837835h^{-6}b_3 + 183783600h^{-8}l_2^4,$$
$$c_0 = 90h^{-2}b_1\bar{\Delta} + 3465h^{-4}b_2,\ \ c_1 = 165h^{-2}b_1\bar{\Delta} + 25740h^{-4}b_2 + 675675h^{-6}b_3,$$
$$c_2 = 260h^{-2}b_1\bar{\Delta} + 107250h^{-4}b_2 + 9845550h^{-6}b_3,\ \ d_0 = 234h^{-2}b_1\bar{\Delta} + 19305h^{-4}b_2,$$
$$d_1 = 119340h^{-4}b_2 + 6235515h^{-6}b_3,\ \ d_2 = 442h^{-2}b_1\bar{\Delta} + 62985h^{-4}b_2,$$

the system (24.80) can be written as the matrix equation

$$L^{(2)}\mathbf{U}^{(2)} + \mathbf{G}^{(2)} = 0. \tag{24.83}$$

It is not difficult to find the matrix of cofactors $L_*^{(2)}$ for the differential matrix $L^{(2)}$ of the equation (24.83). Indeed, by virtue of simple calculations we will have

$$L_*^{(2)} = \begin{pmatrix} L_{*11}^{(2)} & 0 & 0 & 0 & 0 & 0 \\ L_{*21} & L_{*22}^{(2)} & 0 & 0 & 0 & 0 \\ L_{1*31} & L_{1*32} & L_{*33}^{(2)} & 0 & 0 & 0 \\ L_{1*41} & L_{1*42} & L_{1*43} & L_{*44}^{(2)} & 0 & 0 \\ L_{1*51} & L_{1*52} & L_{1*53} & L_{1*54} & L_{*55}^{(2)} & 0 \\ L_{1*61} & L_{1*62} & L_{1*63} & L_{1*64} & L_{1*65} & L_{*66}^{(2)} \end{pmatrix},\ \ |L_*^{(2)}| = (l_2^4\bar{\Delta}^2 - l_1^2\bar{\Delta} + 1)^{30}\bar{\Delta}^{60},$$

$L^{(2)}_{*11}=L^{(2)}_{*22}=L^{(2)}_{*33}=L^{(2)}_{*44}=L^{(2)}_{*55}=L^{(2)}_{*66}=b_0^5\bar{\Delta}^{10},\ \ L_{*21}=-3h^{-2}b_0^4b_1\bar{\Delta}^9,$
$L_{*31}=b_0^2b_1h^{-2}(10395h^{-4}b_1^2-99a_0)\bar{\Delta}^7,\quad L_{*32}=-35h^{-2}b_0^4b_1\bar{\Delta}^9,$
$L_{*41}=-\bar{\Delta}^7b_0^2(10395h^{-6}b_1^3+a_1\bar{\Delta}b_0^2-99h^{-2}a_0b_0b_1-3h^{-2}c_0b_0b_1),$
$L_{*42}=\bar{\Delta}^8b_0^3(3465h^{-4}b_1^2-b_0c_0),\quad L_{*43}=-99h^{-2}\bar{\Delta}^9b_0^4b_1,$
$L_{*51}=\bar{\Delta}^6b_0(2027025h^{-8}b_1^4-a_2\bar{\Delta}b_0^3+195h^{-2}a_1b_0^2b_1^2+a_0b_0^2d_0$
$+3\bar{\Delta}h^{-2}c_1b_0^2b_1-19305h^{-4}a_0b_1^2b_0-585h^{-4}c_0^2b_1^2b_0-105h^{-4}d_0b_1^2b_0),$
$L_{*52}=-\bar{\Delta}^7b_0^2(675675h^{-6}b_1^3+\bar{\Delta}b_0^2c_1-195h^{-2}c_0b_0b_1-35h^{-2}d_0b_0b_1),$
$L_{*53}=\bar{\Delta}^8b_0^3(19305h^{-4}b_1^2-d_0b_0),$
$L_{*54}=-195h^{-2}\bar{\Delta}^9b_0^2b_1,\ \ L_{*54}=-195h^{-2}\bar{\Delta}^9b_0^4b_1,$
$L_{*61}=\bar{\Delta}^5(654729075h^{-10}b_1^5+a_3\bar{\Delta}^3b_0^4-6235515h^{-6}a_0b_0^2b_1^3-323h^{-2}\bar{\Delta}^2b_0^3b_1$
$-188955h^{-6}c_0b_1^3c_0b_0-3\bar{\Delta}^2c_2b_0^3b_1-\bar{\Delta}a_1b_0^3d_2-33915h^{-6}d_0b_1^3b_0$
$-10395h^{-6}d_2b_1^3b_0+62985h^{-4}a_1b_1^2b_0^2+969h^{-4}c_1b_1^2b_0^2\bar{\Delta}+105h^{-4}d_1b_1^2b_0^2\bar{\Delta}$
$+323h^{-2}a_0b_1b_0^2d_0+99h^{-2}a_0b_1b_0^2d_2+3h^{-2}c_0b_1b_0^2d_2),$
$L_{*62}=\bar{\Delta}^6b_0(218243025h^{-8}b_1^4-\bar{\Delta}^2b_0^4c_2$
$+323h^{-2}c_1b_0^3b_1\bar{\Delta}+35h^{-2}d_1b_0^2b_1\bar{\Delta}+b_0^2c_0d_2-62985h^{-4}c_0b_0^2b_1^2-11305h^{-4}d_0b_0^2b_1^2$
$-3465h^{-4}d_2b_0b_1^2),$
$L_{*63}=-\bar{\Delta}^7b_0^2(6235515h^{-6}b_1^3+\bar{\Delta}b_0^2d_1+33h^{-2}d_0b_0b_1$
$-99h^{-2}d_3b_0b_1),$
$L_{*64}=\bar{\Delta}^8b_0^3(62985h^{-4}b_1^2-d_2b_0),$
$L_{*65}=-323h^{-2}\bar{\Delta}^9b_0^4b_1.$

It is easy to see that the matrix of cofactors $L^{(2)}_*$ can be represented in the form

$$L^{(2)}_*=\bar{\Delta}^5N^{(2)},\quad N^{(2)}=\begin{pmatrix} N^{(2)}_{11} & 0 & 0 & 0 & 0 & 0\\ N^{(2)}_{21} & N^{(2)}_{22} & 0 & 0 & 0 & 0\\ N^{(2)}_{31} & N^{(2)}_{32} & N^{(2)}_{33} & 0 & 0 & 0\\ N^{(2)}_{41} & N^{(2)}_{42} & N^{(2)}_{43} & N^{(2)}_{44} & 0 & 0\\ N^{(2)}_{51} & N^{(2)}_{52} & N^{(2)}_{53} & N^{(2)}_{54} & N^{(2)}_{55} & 0\\ N^{(2)}_{61} & N^{(2)}_{62} & N^{(2)}_{63} & N^{(2)}_{64} & N^{(2)}_{65} & N^{(2)}_{66} \end{pmatrix},$$

$$N_{11}^{(2)} = N_{22}^{(2)} = N_{33}^{(2)} = N_{44}^{(2)} = N_{55}^{(2)} = N_{66}^{(2)} = b_0^5\bar{\Delta}^5;\ \ N_{21}^{(2)} = -3h^{-2}b_0^4 b_1\bar{\Delta}^4,$$
$$N_{31}^{(2)} = b_0^2 b_1 h^{-2}(10395h^{-4}b_1^2 - 99a_0)\bar{\Delta}^2,\ \ N_{32} = -35h^{-2}b_0^4 b_1\bar{\Delta}^4,$$
$$N_{41}^{(2)} = -\bar{\Delta}^2 b_0^2(10395h^{-6}b_1^3 + a_1\bar{\Delta}b_0^2 - 99h^{-2}a_0 b_0 b_1 - 3h^{-2}c_0 b_0 b_1),$$
$$N_{42}^{(2)} = \bar{\Delta}^3 b_0^3(3465h^{-4}b_1^2 - b_0 c_0),\quad N_{43}^{(2)} = -99h^{-2}\bar{\Delta}^9 b_0^4 b_1,$$
$$N_{51}^{(2)} = \bar{\Delta}b_0(2027025h^{-8}b_1^4 - a_2\bar{\Delta}b_0^3 + 195h^{-2}a_1 b_0^2 b_1^2 + a_0 b_0^2 d_0 + 3\bar{\Delta}h^{-2}c_1 b_0^2 b_1$$
$$-19305h^{-4}a_0 b_1^2 b_0 - 585h^{-4}c_0^2 b_1^2 b_0 - 105h^{-4}d_0 b_1^2 b_0),$$
$$N_{52}^{(2)} = -\bar{\Delta}^2 b_0^2(675675h^{-6}b_1^3 + \bar{\Delta}b_0^2 c_1 - 195h^{-2}c_0 b_0 b_1 - 35h^{-2}d_0 b_0 b_1),$$
$$N_{53}^{(2)} = \bar{\Delta}^3 b_0^3(19305h^{-4}b_1^2 - d_0 b_0),$$
$$N_{54}^{(2)} = -195h^{-2}\bar{\Delta}^9 b_0^2 b_1,$$
$$N_{61}^{(2)} = 654729075h^{-10}b_1^5 + a_3\bar{\Delta}^3 b_0^4 - 6235515h^{-6}a_0 b_0^2 b_1^3 - 323h^{-2}\bar{\Delta}^7 b_0^3 b_1$$
$$-188955h^{-6}c_0 b_1^3 c_0 b_0 - 3\bar{\Delta}^2 c_2 b_0^3 b_1 - \bar{\Delta}a_1 b_0^3 d_2 - 33915h^{-6}d_0 b_1^3 b_0$$
$$-10395h^{-6}d_2 b_1^3 b_0 + 62985h^{-4}a_1 b_1^2 b_0^2 + 969h^{-4}c_1 b_1^2 b_0^2\bar{\Delta} + 105h^{-4}d_1 b_1^2 b_0^2\bar{\Delta}$$
$$+323h^{-2}a_0 b_1 b_0^2 d_0 + 99h^{-2}a_0 b_1 b_0^2 d_2 + 3h^{-2}c_0 b_1 b_0^2 d_2),$$
$$N_{62}^{(2)} = \bar{\Delta}b_0(218243025h^{-8}b_1^4 - \bar{\Delta}^2 b_0^4 c_2 + 323h^{-2}c_1 b_0^3 b_1\bar{\Delta} + 35h^{-2}d_1 b_0^2 b_1\bar{\Delta}$$
$$+b_0^2 c_0 d_2 - 62985h^{-4}c_0 b_0^2 b_1^2 - 11305h^{-4}d_0 b_0^2 b_1^2 - 3465h^{-4}d_2 b_0 b_1^2),$$
$$N_{63}^{(2)} = -\bar{\Delta}^2 b_0^2(6235515h^{-6}b_1^3 + \bar{\Delta}b_0^2 d_1 + 33h^{-2}d_0 b_0 b_1 - 99h^{-2}d_3 b_0 b_1),$$
$$N_{64}^{(2)} = \bar{\Delta}^3 b_0^3(62985h^{-4}b_1^2 - d_2 b_0),\ \ N_{65}^{(2)} = -323h^{-2}\bar{\Delta}^4 b_0^4 b_1.$$

Applying the differential operator $N^{(2)T}$ to equation (24.83) on the left, due to the equality $N^{(2)T}L^{(2)} = Eb_0^6\bar{\Delta}^7$, where the E-unit matrix of the sixth order, we obtain the decomposed elliptic equation of higher order for each even moment of the displacement vector in the form

$$\begin{gathered}(l_2^4\bar{\Delta}^2 - l_1^2\bar{\Delta} + 1)^6\bar{\Delta}^7\mathbf{U}^{(2)} + N^{2T}\mathbf{G}^{(2)} = 0 \ \text{ or} \\ (l_2^4\bar{\Delta}^2 - l_1^2\bar{\Delta} + 1)^6\bar{\Delta}^7\overset{(2k)}{\mathbf{u}} + \mathbf{\Phi}^{(2k)} = 0,\ \ k = \overline{0,5}.\end{gathered} \tag{24.84}$$

It is not hard to see that in the relations (24.82) and (24.84) $\mathbf{\Phi}^{(2k+1)}$, $k = \overline{0,4}$ and $\mathbf{\Phi}^{(2k)}$, $k = \overline{0.5}$, are known functions that are determined using body force. In order to shorten the letter, their expressions were not written out.

24.4.6 Equations of the Quasi-Static SSTG Theory of Transversally Isotropic Prismatic Bodies in Displacement Vector and its Moments

As above, consider a prismatic body of constant thickness 2h. We take the middle plane as the base plane. Taking into account the representation of the Laplacian (see (24.74)), the equation (24.54) for the prismatic bodies can be represented as

$$
\begin{aligned}
&\{A(l_3^4\bar{\Delta}^2-l_1^2\bar{\Delta}+1)\bar{\Delta}^3+h^{-2}[(Bl_3^4+Al_4^4)\bar{\Delta}^2-(Bl_1^2+Al_2^2)\bar{\Delta}+B]\bar{\Delta}^2\partial_3^2\\
&+h^{-4}[(Al_5^4+Bl_4^4+Cl_3^4)\bar{\Delta}^2-(Bl_2^2+Cl_1^2)\bar{\Delta}+C]\bar{\Delta}\partial_3^4+h^{-6}[(Bl_5^4+Cl_4^4+Dl_3^4)\bar{\Delta}^2\\
&-(Cl_2^2+Dl_1^2)\bar{\Delta}+D]\partial_3^6+h^{-8}Cl_5^4\bar{\Delta}\partial_3^8+h^{-10}Dl_5^4\partial_3^{10}\}\cdot\mathbf{u}+\underset{\sim}{\mathbf{L}}_*\cdot(\rho\mathbf{F})=0.
\end{aligned}
\tag{24.85}
$$

Applying the kth moment operator of any system of orthogonal polynomials (Legendre, Chebyshev) (the theory of moments with respect to the systems of orthogonal polynomials can be seen in [56])) to the equation (24.85), we obtain the following equations in moments of displacement vector:

$$
\begin{aligned}
&A(l_3^4\bar{\Delta}^2-l_1^2\bar{\Delta}+1)\bar{\Delta}^3\overset{(k)}{\mathbf{u}}+h^{-2}[(Bl_3^4+Al_4^4)\bar{\Delta}^2-(Bl_1^2+Al_2^2)\bar{\Delta}+B]\bar{\Delta}^2\overset{(k)}{\mathbf{u}}{}''\\
&+h^{-4}[(Al_5^4+Bl_4^4+Cl_3^4)\bar{\Delta}^2-(Bl_2^2+Cl_1^2)\bar{\Delta}+C]\bar{\Delta}\overset{(k)}{\mathbf{u}}{}^{IV}+h^{-6}[(Bl_5^4+Cl_4^4\\
&+Dl_3^4)\bar{\Delta}^2-(Cl_2^2+Dl_1^2)\bar{\Delta}+D]\overset{(k)}{\mathbf{u}}{}^{VI}+h^{-8}Cl_5^4\bar{\Delta}\overset{(k)}{\mathbf{u}}{}^{VIII}+h^{-10}Dl_5^4\overset{(k)}{\mathbf{u}}{}^{X}\\
&+\underset{\sim}{\mathbf{L}}_*\cdot(\rho\mathbf{F})=0,\quad k\in\mathbb{N}_0.
\end{aligned}
\tag{24.86}
$$

Next, we note that to find a system of equations at the moments of a displacement vector of various orders of approximation, we need to find expressions for the stroke operators: $\overset{(k)}{\mathbf{u}}{}''$, $\overset{(k)}{\mathbf{u}}{}^{IV}$, $\overset{(k)}{\mathbf{u}}{}^{VI}$, $\overset{(k)}{\mathbf{u}}{}^{VIII}$ and $\overset{(k)}{\mathbf{u}}{}^{X}$ through the moments of the displacement vector at $-1\leq x^3\leq 1$. Note also that when using systems of Legendre polynomials, expressions for the above-listed stroke operators can be determined based on the relation (24.77), and then one can derive equations in the moments of the displacement vector relative to the system of Legendre polynomials of various orders of approximation, from which one can obtain just as above in the case of an isotropic medium, elliptic equations of higher order, for which, using the Vekua method, one can obtain the general representation of solutions. It should be especially noted that the boundary conditions are also represented in terms of the displacement vector moments. In this case, the orders of approximation of the equations and boundary conditions coincide. In order to shorten the letter, we will not dwell on these issues in this work. We'll leave them for future works.

24.5 On one Vekua Method for Solving Higher Order Elliptic Equations

Partial differential equation of the form

$$
\sum_{k=0}^{n}\mathbb{L}_k[\Delta^{n-k}u(x^1,x^2)]=f(x^1,x^2),\ \ \mathbb{L}_k=\sum_{p=0}^{k}\sum_{q=0}^{p}a_{kp-qq}(x^1,x^2)\partial_1^{p-q}\partial_2^{q},
$$

$$
\mathbb{L}_0=a_{000}=1,\quad \Delta^0=1,\quad \partial_1^0=1,\quad \partial_2^0=1,
$$

where $\Delta=\partial_1^2+\partial_2^2$ is Laplace operator, $\partial_1=\partial/\partial x_1$, $\partial_2=\partial/\partial x_2$, $a_{ijk}(x_1,x_2)$ are variable coefficients, which can be discontinuous functions, $u(x_1,x_2)$ is the required function and $f(x_1,x_2)$ is the known function, is called a differential equation of elliptic type of higher order (see [73]).

In particular, the equation

$$(\Delta^n + a_1\Delta^{n-1} + a_2\Delta^{n-2} + ... + a_{n-1})u(x^1, x^2) = f(x^1, x^2), \tag{24.87}$$

where the coefficients $a_1, a_2, ..., a_n$ are constant, is a differential equation of elliptic type of higher order.

Let us discuss one of the versions of the general representation of solutions of Eq. (24.87). Equation (24.87) in the most general case can be written as follows:

$$\Delta^k(\Delta + \varkappa_1)^{k_1}(\Delta + \varkappa_2)^{k_2} \cdot ... \cdot (\Delta + \varkappa_m)^{k_m} u = 0, \tag{24.88}$$

where $\varkappa_1, \varkappa_2, ... \varkappa_m$, are nonzero different roots of the equation

$$x^n - a_1 x^{n-1} + ... + (-1)^n a_n = 0,$$

and $k_1, k_2, ...\ k_m$ are their multiplicities respectively, k is the multiplicity of the root $\varkappa = 0$. It's clear that $k + k_1 + k_2 + ... + k_m = n$.

The following theorem is true:

Theorem 24.1. *All solutions to the equation* (24.88) *are obtained from the formula*

$$u = v + \sum_{j=1}^{m} \sum_{l=1}^{k_j-1} r^l \frac{\partial^l u_{jl}}{\partial r^l} \quad (r = |z|), \tag{24.89}$$

where v is any solution to the equation $\Delta^k u = 0$, and u_{jl} is any solution to the equation $(\Delta + \omega_j)u = 0$, $j = 1, 2, ..., m$, $l = 0, 1, ...k_j - 1$, and the functions v, u_{jl} are uniquely determined by u.

Formula (24.89) *remains valid for the space of any number of dimensions. In this case*

$$\Delta = \sum_{k=1}^{p} \frac{\partial^2}{\partial x_k^2}, \quad p \geq 2.$$

24.6 Conclusions and Future Steps

The issues of the second strain tensor and velocity vector gradient theory for an arbitrary anisotropic bodies are considered. For this case, expressions for specific and kinetic energies have been derived, and also equations of motion have been obtained with respect to the tensor of total stresses and the vector of total momentum within the framework of the considering second gradient theory. Besides, constitutive relations and equilibrium equations with respect to the tensor of total stresses in the considering second gradient theory are derived. The specific energy, constitutive relations, equilibrium equations, and the resulting relations are dependent on scaling tensors for statics. The kinetic energy depends on scaling tensors for dynamics. The equations of motion depend on both scaling tensors for statics and dynamics. The equations of motion and equilibrium are also derived with respect to the displacement vector in the case of the considered linear gradient theory. Isotropic and transversely isotropic media are considered as special cases. In particular, for these media, constitutive relations are derived, and split equations of motion and equilibrium equations with respect to the displacement vector are obtained using constructed cofactor tensors for the differential operators of these equations.

Further, in the case of isotropic and transversely isotropic media, from the split equilibrium equations with respect to the displacement vector, the corresponding equations for prismatic bodies are derived. From these relations, applying the method of orthogonal polynomials, equations in moments are derived with respect to an arbitrary system of orthogonal polynomials, including the Legendre polynomial system. In the case of an isotropic medium, as an example, a system of equations of the tenth-order approximation is obtained, which splits into two systems. One of these systems contains only moments of the displacement vector of odd order, and the other system contains moments of even order. Then, based on the constructed differential matrix cofactor operators (for the differential matrix operators of these systems of equations), each system of equations splits separately and for each moment of the displacement vector, a higher-order elliptic equation is obtained. For this equation, using the Vekua method, a path to obtaining the general representation of solutions is indicated. It should be noted that the described procedure for obtaining a general representation of solutions is valid for a system of equations of any arbitrary finite-order approximation. As a special case, some issues of first-order gradient theory are considered. In particular, expressions for the specific strain energy and kinetic energy, the constitutive relations and equations of motion and equilibrium for some anisotropic elastic bodies are obtained.

It should also be noted that all of the above can easily be extended to theories of other rheological bodies. In particular, it can be extended to linear and nonlinear classical and micropolar gradient theories of elastic and viscoelastic bodies, which present problems for further research.

In addition, the author proposes: to construct gradient theories of thin bodies with two small dimensions and to consider in more detail the issues concerning boundary conditions in gradient theories of thin bodies. In particular, the author proposes analogously to the equations to represent the static boundary conditions with respect to the displacement vector by means of differential tensor operators and to explore problems with their decomposition. It is especially worth noting that [5–9, 12–30, 45–70, 76, 77] will provide invaluable assistance in solving the problems listed above.

Acknowledgements The study was conducted under the state assignment of Lomonosov Moscow State University.

References

[1] Agiasofitou EK, Lazar M (2009) Conservation and balance laws in linear elasticity of grade three. Journal of Elasticity 94(1):69–85, DOI 10.1007/s10659-008-9185-x

[2] Aifantis EC (1992) On the role of gradients in the localization of deformation and fracture. International Journal of Engineering Science 30(10):1279–1299, DOI 10.1016/0020-7225(92)90141-3

[3] Altan BS, Aifantis EC (1997) On some aspects in the special theory of gradient elasticity. Journal of the Mechanical Behavior of Materials pp 231–282

[4] Altan SB, Aifantis EC (1992) On the structure of the mode III crack-tip in gradient elasticity. Scripta Metallurgica Et Materialia 26:319–324, DOI 10.1016/0956-716x(92)90194-j

[5] Altenbach H, Eremeyev VA, Morozov NF (2009) Linear theory of shells taking into account surface stresses. Doklady Physics 54(12):531–535, DOI 10.1134/S1028335809120039

[6] Altenbach H, Eremeyev VA, Lebedev LP (2011) Micropolar shells as two-dimensional generalized continua models. In: Altenbach H, Maugin GA, Erofeev V (eds) Mechanics of Generalized Continua, Advanced Structured Materials, vol 7, Springer Berlin Heidelberg, Berlin, Heidelberg, pp 23–55, DOI 10.1007/978-3-642-19219-7_2

[7] Askes H, Aifantis EC (2006) Gradient elasticity theories in statics and dynamics - a unification of approaches. International Journal of Fracture 139:297–304, DOI 10.1007/s10704-006-8375-4

[8] Askes H, Aifantis EC (2009) Gradient elasticity and flexural wave dispersion in carbon nanotubes. Physical Review B 80:195412, DOI 10.1103/PhysRevB.80.195412

[9] Askes H, Aifantis EC (2011) Gradient elasticity in statics and dynamics: an overview of formulations, length scale identification procedures, finite element implementations and new results. International Journal of Solids and Structures 48(13):1962–1990, DOI 10.1016/j.ijsolstr.2011.03.006

[10] Askes H, Suiker A, Sluys L (2002) A classification of higher-order strain-gradient models – linear analysis. Archive of Applied Mechanics pp 171–188

[11] Askes H, Bennett T, Aifantis EC (2007) A new formulation and c^0-implementation of dynamically consistent gradient elasticity. International Journal for Numerical Methods in Engineering 72(1):111–126

[12] dell'Isola F, Sciarra G, Vidoli S (2009) Generalized hooke's law for isotropic second gradient materials. Proceedings of the Royal Society A: Mathematical, Physical and Engineering Sciences 465(2107):2177–2196, DOI 10.1098/rspa.2008.0530

[13] dell'isola F, Andreaus U, Placidi L (2015) At the origins and in the vanguard of peri-dynamics, non-local and higher gradient continuum mechanics. an underestimated and still topical contribution of gabrio piola. Mathematics and Mechanics of Solids 20(8):887–928, DOI 10.1177/1081286513509811

[14] dell'Isola F, Seppecher P, Madeo A (2012) How contact interactions may depend on the shape of Cauchy cuts in Nth gradient continua: approach "à la D'Alembert". Zeitschrift für angewandte Mathematik und Physik 63(6):1119–1141, DOI 10.1007/s00033-012-0197-9

[15] Deng S, Liu J, Liang N (2007) Wedge and twist disclinations in second strain gradient elasticity. International Journal of Solids and Structures 44(11):3646–3665, DOI 10.1016/j.ijsolstr.2006.10.011
[16] Eremeyev V, Altenbach H (2014) Equilibrium of a second-gradient fluid and an elastic solid with surface stresses. Meccanica 49:2635–2643, DOI 10.1007/s11012-013-9851-3
[17] Eremeyev VA, Skrzat A, Vinakurava A (2016) Application of the micropolar theory to the strength analysis of bioceramic materials for bone reconstruction. Strength of Materials 48(10):573–582, DOI 10.1177/1081286522108l854
[18] Eremeyev VA, Skrzat A, Vinakurava A (2023) Strong ellipticity and infinitesimal stability within *n*th-order gradient elasticity. Mathematics 11(4), DOI 10.3390/math11041024
[19] Eringen AC (1999) Microcontinuum Field Theories. 1. Foundation and Solids. Springer, New York, DOI 10.1007/978-1-4612-0555-5
[20] Fedele R (2022) Approach à la Piola for the equilibrium problem of bodies with second gradient energies. Part II: Variational derivation of second gradient equations and their transport. Continuum Mechanics and Thermodynamics 34(5):1087–1111, DOI 10.1007/s00161-022-01100-z
[21] Fedele R (2022) Piola's approach to the equilibrium problem for bodies with second gradient energies. Part I: First gradient theory and differential geometry. Continuum Mechanics and Thermodynamics 34:445–474, DOI 10.1007/s00161-021-01064-6
[22] Forest S (2002) Homogenization methods and mechanics of generalized continua - part 2. Theoretical and Applied Mechanics pp 113–144, DOI 10.2298/TAM0229113F
[23] Georgiadis H, Vardoulakis I, Lykotrafitis G (2000) Torsional surface waves in a gradient-elastic half-space. Wave Motion 31(4):333–348, DOI 10.1016/S0165-2125(99)00035-9
[24] Georgievskii DV (2023) Models of Viscoelasticity Theory. Moscow: LENAND, (in Russian)
[25] Georgievskii DV (2023) Izbrannyye zadachi mekhaniki sploshnoy sredy (in Russ., Selected Problems in Continuum Mechanics). Moscow: URSS
[26] Georgievskii DV (2025) Tensor Functions in the Theory of Constitutive Relations. Moscow: LENAND, (in Russian)
[27] Ilyushin AA, Pobedria BE (1970) Osnovy matematicheskoy teorii termovyazko-uprugosti (in Russ., Fundamentals of the Mathematical Theory of Thermoviscoelasticity). Moscow: Nauka
[28] Kröner E (1967) Elasticity theory of materials with long range cohesive forces. International Journal of Solids and Structures 3(5):731–742, DOI https://doi.org/10.1016/0020-7683(67)90049-2
[29] Krumhansl JA (1968) Some considerations of the relation between solid state physics and generalized continuum mechanics. In: Kröner E (ed) Mechanics of Generalized Continua, Springer Berlin Heidelberg, Berlin, Heidelberg, pp 298–311

[30] Kupradze VD, Gegelia TG, Basheleishvili MO, Burchuladze TV (1976) Three-dimensional problems of the mathematical theory of elasticity and thermoelasticity. Moscow: Nauka, (in Russian)
[31] Lazar M (2013) The fundamentals of non-singular dislocations in the theory of gradient elasticity: Dislocation loops and straight dislocations. International Journal of Solids and Structures 50(2):352–362, DOI 10.1016/j.ijsolstr.2012.09.017
[32] Lazar M, Anastassiadis C (2007) Lie point symmetries, conservation and balance laws in linear gradient elastodynamics. Journal of Elasticity 88(1):5–25, DOI 10.1007/s10659-007-9105-5
[33] Lazar M, Maugin GA (2005) Nonsingular stress and strain fields of dislocations and disclinations in first strain gradient elasticity. International Journal of Engineering Science 43(13):1157–1184, DOI 10.1016/j.ijengsci.2005.01.006
[34] Lazar M, Maugin GA (2006) Dislocations in gradient elasticity revisited. Proceedings of the Royal Society A: Mathematical, Physical and Engineering Sciences 462(2075):3465–3480, DOI 10.1098/rspa.2006.1699
[35] Lazar M, Maugin GA, Aifantis EC (2006) Dislocations in second strain gradient elasticity. International Journal of Solids and Structures 43(6):1787–1817, DOI 10.1016/j.ijsolstr.2005.07.005
[36] Lurie AI (1990) Nonlinear Theory of Elasticity, North-Holland Series in Applied Mathematics and Mechanics, Vol 36. Amsterdam: North-Holland
[37] Lurie S, Belov P, Volkov-Bogorodsky D, Tuchkova N (2003) Nanomechanical modeling of the nanostructures and dispersed composites. Computational Materials Science 28(3):529–539, DOI 10.1016/j.commatsci.2003.08.010, twelfth International Workshop on Computational Mechanics of Materials
[38] Lurie S, Volkov-Bogorodskiy D, Solyaev Y, Rizahanov R, Agureev L (2016) Multiscale modelling of aluminium-based metal–matrix composites with oxide nanoinclusions. Computational Materials Science 116:62–73, DOI 10.1016/j.commatsci.2015.12.034, articles from EUROMECH Colloquium 577 "Micromechanics of Metal Ceramic Composites"
[39] Lurie SA, Kalamkarov AL, Solyaev YO, Ustenko AD, Volkov AV (2018) Continuum micro-dilatation modeling of auxetic metamaterials. International Journal of Solids and Structures 132-133:188–200, DOI 10.1016/j.ijsolstr.2017.09.022
[40] M Shodja H, Ahmadpoor F, Tehranchi A (2012) Calculation of the additional constants for fcc materials in second strain gradient elasticity: Behavior of a nano-size bernoulli-euler beam with surface effects. Journal of Applied Mechanics 79(2):0210081–0210088, DOI 10.1115/1.4005535
[41] Metrikine AV, Askes H (2002) One-dimensional dynamically consistent gradient elasticity models derived from a discrete microstructure. part 1: Generic formulation. European Journal of Mechanics - A/Solids 21(4):555–572, DOI 10.1016/S0997-7538(02)01218-4
[42] Mindlin RD (1964) Micro-structure in linear elasticity. Archive for Rational Mechanics and Analysis 16(1):51–78, DOI 10.1007/BF00248490

[43] Mindlin RD (1965) Second gradient of strain and surface-tension in linear elasticity. International Journal of Solids and Structures 1(4):417–438, DOI 10.1016/0020-7683(65)90006-5
[44] Mindlin RD, Eshel NN (1968) On first strain-gradient theories in linear elasticity. International Journal of Solids and Structures 4(1):109–124, DOI 10.1016/0020-7683(68)90036-X
[45] Misra AK, Poorsolhjouy P (2015) Granular micromechanics based micromorphic model predicts frequency band gaps. Continuum Mechanics and Thermodynamics 28:215–234, DOI 10.1007/s00161-015-0420-y
[46] Nazarenko L, Glüge R, Altenbach H (2021) Inverse hooke's law and complementary strain energy in coupled strain gradient elasticity. ZAMM - Journal of Applied Mathematics and Mechanics / Zeitschrift für Angewandte Mathematik und Mechanik 101(9):e202100005, DOI 10.1002/zamm.202100005
[47] Nazarenko L, Glüge R, Altenbach H (2021) Positive definiteness in coupled strain gradient elasticity. Continuum Mechanics and Thermodynamics 33(3):713–725, DOI 10
[48] Nazarenko L, Glüge R, Altenbach H (2022) On variational principles in coupled strain-gradient elasticity. Mathematics and Mechanics of Solids 27(10):2256–2274
[49] Nikabadze M (2024) On some issues of second strain tensor and velocity vector gradient theories of 3D bodies and thin bodies. Lobachevskii Journal of Mathematics 45(8):3683–3706, DOI 10.1134/S1995080224604387
[50] Nikabadze MU (2019) Splitting of initial boundary value problems in anisotropic linear elasticity theory. Moscow University Mechanics Bulletin 74(5):103–110, DOI 10.3103/S0027133019050017
[51] Nikabadze M, Ulukhanyan A (2020) Modeling of multilayer thin bodies. Continuum Mechanics and Thermodynamics 32(4):817–842, DOI 10.1007/s00161-019-00762-6
[52] Nikabadze M, Ulukhanyan A (2020) On the decomposition of equations of micropolar elasticity and thin body theory. Lobachevskii Journal of Mathematics 41(10):2060–2075, DOI 10.1134/S1995080220100145
[53] Nikabadze M, Ulukhanyan A (2022) Generalized reissner-type variational principles in the micropolar theories of multilayer thin bodies with one small size. Continuum Mechanics and Thermodynamics 34(2):1207–1221, DOI 10.1007/s00161-022-01091-x
[54] Nikabadze M, Ulukhanyan A (2022) On some variational principles in micropolar theories of single-layer thin bodies. Continuum Mechanics and Thermodynamics 34(2):1147–1164, DOI 10.1007/s00161-022-01089-5
[55] Nikabadze MU (2017) Topics on tensor calculus with applications to mechanics. Journal of Mathematical Sciences 225(1):1–194, DOI 10.1007/s10958-017-3467-4
[56] Nikabadze MU (2023) Razvitiye metoda ortogonal'nykh polinomov v mekhdsanike mikropolyarnykh i klassicheskikh uprugikh tonkikh tel (in Russ., Development of the method of orthogonal polynomials in the classical and micropolar mechanics of elastic thin bodies). Moscow University Publishing House

[57] Nikabadze MU, Ulukhanyan AR (2021) On the theory of multilayer thin bodies. Lobachevskii Journal of Mathematics 42(8):1900–1911, DOI 10.1134/S1995080221080217
[58] Papargyri-Beskou S, Polyzos D, Beskos D (2009) Wave dispersion in gradient elastic solids and structures: A unified treatment. International Journal of Solids and Structures 46(21):3751–3759, DOI 10.1016/j.ijsolstr.2009.05.002
[59] Placidi L, Rosi G, Giorgio I, Madeo A (2014) Reflection and transmission of plane waves at surfaces carrying material properties and embedded in second-gradient materials. Mathematics and Mechanics of Solids 19(5):555–578, DOI 10.1177/1081286512474016
[60] Pobedria BE (2006) Theory of thermomechanical processes. In: Kiyko IA, Vasin RA, Brovko GL (eds) Elasticity and Inelasticity, URSS, Moscou, pp 70–85, proceedings of the International Scientific Symposium on Problems of Mechanics of Deformable Solids, dedicated to the 95th anniversary of A. A. Ilyushin's birth, Moscow, January 19-20, 2006. 480 pp. (in Russian)
[61] Pobedria BE (1984) Mekhanika kompozitsionnykh materialov (in Russ., Mechanics of Composite Materials). Moscow University Publishing House
[62] Pobedria BE (1986) Lektsii po tenzornomu analizu (in Russ., Lectures on tensor analysis). Moscow University Publishing House
[63] Pobedria BE (1995) Chislennyye metody v teorii uprugosti i plastichnosti (in Russ., Numerical methods in the theory of elasticity and plasticity), 2nd ed. Moscow University Publishing House
[64] Pobedria BE, Georgievskii DV (2006) Fundamentals of Continuum Mechanics. Moscow: FIZMATLIT, (in Russian)
[65] Pobedria BE, Georgievskii DV (2020) Lektsii po teorii uprugosti (in Russ., Lectures on Elasticity Theory). Moscow: LENAND
[66] Polizzotto C (2003) Gradient elasticity and nonstandard boundary conditions. International Journal of Solids and Structures 40(26):7399–7423, DOI 10.1016/j.ijsolstr.2003.06.001
[67] Polizzotto C (2012) A gradient elasticity theory for second-grade materials and higher order inertia. International Journal of Solids and Structures 49(15):2121–2137, DOI 10.1016/j.ijsolstr.2012.04.019
[68] Polizzotto C (2013) A second strain gradient elasticity theory with second velocity gradient inertia – Part II: Dynamic behavior. International Journal of Solids and Structures 50(24):3766–3777, DOI 10.1016/j.ijsolstr.2013.07.026
[69] Polizzotto C (2013) A second strain gradient elasticity theory with second velocity gradient inertia – part i: Constitutive equations and quasi-static behavior. International Journal of Solids and Structures 50(24):3749–3765, DOI 10.1016/j.ijsolstr.2013.06.024
[70] Polizzotto C (2016) A note on the higher order strain and stress tensors within deformation gradient elasticity theories: Physical interpretations and comparisons. International Journal of Solids and Structures 90:116–121, DOI 10.1016/j.ijsolstr.2016.04.001

[71] Ru CQ, Aifantis EC (1993) A simple approach to solve boundary-value problems in gradient elasticity. Acta Mechanica 101:59–68, DOI 10.1007/BF01175597

[72] Toupin RA, Gazis DC (1965) Surface effects and initial stress in continuum and lattice models of elastic crystals. In: In R. F. Wallis (Ed.), Lattice Dynamics: Proceedings of the International Conference Held at Copenhagen, Denmark, August 5–9, 1963, Oxford: Pergamon Press., pp 597–605

[73] Vekua IN (1967) New Methods for Solving Elliptic Equations, North-Holland Series in Applied Mathematics and Mechanics, North-Holland Publishing Company

[74] Vekua IN (1978) Osnovy tenzornogo analiza i teorii kovariantov (in Russ., Fundamentals of Tensor Analysis and Theory Covariants). Moscow: Nauka

[75] Vekua IN (1982) Nekotoryye obshchiye metody postroyeniya razlichnykh variantov teorii obolochek (in Russ., Some general methods for constructing various variants of the theory of shells). Moscow: Nauka

[76] Yang Y, Misra A (2010) Higher-order stress-strain theory for damage modeling implemented in an element-free galerkin formulation. Computer Modeling in Engineering & Sciences 64(1):1–36, DOI 10.3970/cmes.2010.064.001

[77] Yang Y, Misra A (2012) Micromechanics based second gradient continuum theory for shear band modeling in cohesive granular materials following damage elasticity. International Journal of Solids and Structures 49(18):2500–2514, DOI 10.1016/j.ijsolstr.2012.05.024

[78] Zhang X, Jiao K, Sharma P, Yakobson B (2006) An atomistic and non-classical continuum field theoretic perspective of elastic interactions between defects (force dipoles) of various symmetries and application to graphene. Journal of the Mechanics and Physics of Solids 54(11):2304–2329, DOI 10.1016/j.jmps.2006.06.007

Chapter 25
On the Equilibrium Shapes and Stability of Rotating Fluid Mass

Alexander G. Petrov

Abstract Equilibrium figures of a rotating fluid mass subjected to gravity or electric forces or the influence of an electric field, as well as surface tension, are considered. Using the Rouse method, the cyclic angular coordinate is eliminated, and the problem of equilibrium and stability is reduced to finding the minimum of a function depending on two dimensionless axes of a triaxial ellipsoid. The suggested approach may be verified by comparing the results it yields to the known exact solutions for a gravitating liquid mass: the well-known series of Maclaurin ellipsoids, the bifurcation point from which the series of triaxial Jacobi equilibrium figures branches off. For a droplet with surface tension, an approximate solution to the problem of the equilibrium and stability of a rotating drop is presented in elementary functions. Comparing the partial results obtained with the numerical calculations of Chandrasekhar (Nobel laureate) we see an agreement to within 3 significant digits. Solutions to many similar problems of equilibrium and stability of rotating charged droplets in electric fields are given, generalizing the classical solution of Taylor and solutions of other authors.

Key words: Drop, Surface tension, Rotation, Equilibrium, Stability, Electrostatic field

25.1 General Method for Studying Equilibrium Figures

1. Compose a function of the three semi-axes of a triaxial ellipsoid

---

Alexander G. Petrov
Ishlinsky Institute for Problems in Mechanics RAS, Prospekt Vernadskogo 101-1, 119526 Moscow, Russian Federation, e-mail: petrovipmech@gmail.com

H. Altenbach et al. (eds.), *Stability Problems in Mechanics of Continua*,
Advanced Structured Materials 260, https://doi.org/10.1007/978-3-032-31032-3_25

$$U(a_1,a_2,a_3) = \Pi(a_1,a_2,a_3) + \frac{K^2}{2J}, \quad J = \frac{4\pi\rho}{15}a_1a_2a_3(a_1^2+a_2^2),$$
$$K = J_0\Omega, \quad J_0 = \frac{4\pi\rho}{15}a_{10}a_{20}a_{30}(a_{10}^2+a_{20}^2) \tag{25.1}$$

where $\Pi(a_1,a_2,a_3)$ - potential energy, $J(a_1,a_2,a_3)$ - moment of inertia, K - kinetic moment, which is conserved due to the conservation law, ρ - density, Ω - angular velocity of drop rotation, a_1,a_2,a_3 - semi-axes of the triaxial ellipsoid, a_{10},a_{20},a_{30} - semi-axes at the equilibrium point, to be determined.

2. To determine the equilibrium shapes, we write the function $U(a_1,a_2,a_3)$ in dimensionless form

$$\tilde{U}(x,y) = \tilde{\Pi}(x,y) + \frac{\tilde{K}^2}{2\tilde{J}(x,y)}, \quad \tilde{K} = \tilde{J}_0\tilde{\Omega},$$
$$\tilde{\Pi}(x,y) = \frac{\Pi(a_1,a_2,a_3)}{\Pi_*}, \quad \tilde{\Omega} = \frac{\Omega}{\Omega_*}, \quad \tilde{J}(x,y) = \frac{J(a_1,a_2,a_3)}{J_*} \tag{25.2}$$

with $x = a_3/a_1, y = a_3/a_2, J_* = \Pi_*/\Omega_*^2$. Here $\Pi_* = \Pi(a,a,a)$ is the potential energy of a sphere of radius

$$a = (a_1a_2a_3)^{1/3},$$

whose volume coincides with the volume of the ellipsoid. The characteristic quantity Ω_* is chosen for convenience, and then J_* is found by the formula $J_* = \Pi_*/\Omega_*^2$.

3. The axisymmetric series of equilibrium shapes is found from the extremum condition of a function of one variable $U(x) = \tilde{U}(x,x)$

$$\tilde{\Omega}^2(x) = 2\Pi'(x)/J'(x), \tag{25.3}$$

4. Stability with respect to perturbations of the triaxial ellipsoid can be obtained from the sufficient condition for a minimum of a function of two variables. To do so, we expand the function $\tilde{U}(x,y)$ in small perturbations u, v up to second order

$$\tilde{U}(x+u,x+v) = \tilde{U}(x,x) + U_x(u+v) + A_{11}(u^2+v^2) + 2A_{12}uv \tag{25.4}$$

From the condition of positive definiteness of the quadratic form we obtain the stability condition

$$A_{11} - A_{12} > 0 \tag{25.5}$$

5. Stability with respect to axisymmetric perturbations reduces to the problem of finding the minimum of a function of one variable

$$U(x) = \tilde{U}(x,x) = \tilde{\Pi}(x,x) + \frac{\tilde{K}^2}{2\tilde{J}(x,x)}.$$

Comparing the expansion of this function

$$U(x+u) = U(x) + U'(x)u + \frac{1}{2}U''(x)u^2$$

to the expansion (25.4) we obtain

$$U''(x) = 4(A_{11} + A_{12}),$$

and from the sufficient condition for the minimum of a function of one variable we obtain the following stability condition

$$A_{11} + A_{12} > 0 \tag{25.6}$$

Conditions (25.5) and (25.6) imply that for $A_{12} > 0$ it is sufficient to use the stability condition with respect to triaxial perturbations (25.5), and condition (25.6) will be satisfied automatically. Conversely, for $A_{12} < 0$ it is sufficient to use condition (25.6). As we see below, this condition is rather simple

$$U''(x) = \tilde{\Omega}^2 \left(-\frac{J''(x)}{2} + \frac{J'(x)^2}{J(x)} \right) + \Pi''(x) > 0 \tag{25.7}$$

We formulate this result as a theorem.

Theorem 25.1. *If for the mixed derivative U_{xy} holds*

$$U_{xy} = \frac{\partial^2 U}{\partial x \partial y} < 0$$

then a sufficient condition for the minimum of the function $U(x, y)$ at the point $x = y$ is (25.7).

6. The non-axisymmetric series is found using equations

$$-\frac{1}{2}\tilde{\Omega}^2 \tilde{J}_x(x,y) + \tilde{\Pi}_x(x,y) = 0, \quad -\frac{1}{2}\tilde{\Omega}^2 \tilde{J}_y(x,y) + \tilde{\Pi}_y(x,y) = 0 \tag{25.8}$$

by eliminating $\tilde{\Omega}^2$

$$\tilde{J}_y \tilde{\Pi}_x - \tilde{J}_x \tilde{\Pi}_y = 0. \tag{25.9}$$

Setting the ratio $y/x < 1$, we may find x from this equation, and then $\tilde{\Omega}^2$ from the previous one. The condition of positive definiteness of the quadratic form of the second variation is a condition for the stability of such an equilibrium shape with respect to perturbations of the triaxial ellipsoid.

25.2 Testing the Accuracy of the Method

The accuracy of the method was tested in [18] on two known solutions:

1. For a rotating gravitating liquid mass, the formula for the potential energy has the form

$$\Pi = -\frac{8\pi^2}{15} G\rho^2 a_1 a_2 a_3 \int_0^\infty \frac{a_1 a_2 a_3 du}{\sqrt{(a_1^2+u)(a_2^2+u)(a_3^2+u)}} \qquad (25.10)$$

The problem is reduced to solving the problem of minimizing a function of two variables $\tilde{U}(x,y)$

$$\tilde{U} = \tilde{\Pi} + \frac{\tilde{K}^2}{2\tilde{J}}, \quad \tilde{J} = \frac{1}{4}\left(\frac{y^{2/3}}{x^{4/3}} + \frac{x^{2/3}}{y^{4/3}}\right), \quad x = \frac{a_3}{a_1}, \quad y = \frac{a_3}{a_2},$$

$$\tilde{\Pi} = -\frac{1}{2}(xy)^{-1/3} \int_0^\infty \frac{dt}{\sqrt{(1+t)(x^{-2}+t)(y^{-2}+t)}}, \quad \tilde{K} = \tilde{J}_0\tilde{\Omega}, \quad \tilde{\Omega} = \frac{\Omega}{\sqrt{\pi G\rho}}.$$

The solution of this problem implies that the axisymmetric series of Maclaurin ellipsoids is exactly determined. It is shown in Fig. 25.1 by a solid line. The thick line is the stable series. The stability boundary is the bifurcation point, from which the series of triaxial Jacobi ellipsoids branches off, shown by dots. This series is also an exact solution.
2. The problem of equilibrium figures of a rotating drop with surface tension was also solved in accordance with the general methodology outlined in Sect.25.1. The potential energy of a drop is determined by the Laplace formula $\Pi = \sigma S$, where σ is the surface tension coefficient, S is the surface area of the drop. The surface area of a triaxial ellipsoid S is expressed in a rather complicated way through elliptic integrals. It can be replaced by a simpler approximation proposed by the Danish mathematician Knud Thomsen

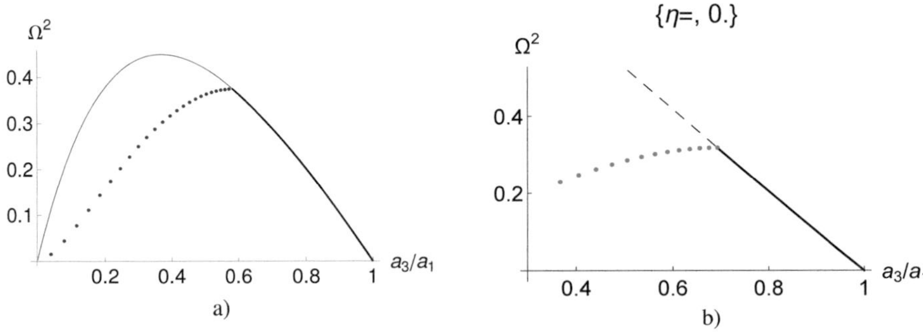

Fig. 25.1: Dependence of the square of the angular velocity of rotation of the ellipsoid $\tilde{\Omega}^2$ on the axis ratio a_3/a_1: a) for the series of axisymmetric Maclaurin ellipsoids (thick line) and triaxial Jacobi ellipsoids (dots); b) symmetric drop shapes a_3/a_1 in the stable region (thick line) and triaxial ellipsoids (dots).

$$S = 4\pi \left(\frac{a_1^p a_2^p + a_1^p a_3^p + a_3^p a_2^p}{3} \right)^{1/p}$$

As shown in [18] for $p = 1.6075$ the relative error does not exceed 1.061%. In [18] the problem is reduced to determining the minimum of a function of two variables $\tilde{U}(x, y)$ just as in the previous section

$$\begin{aligned} &\tilde{U} = \tilde{S} + \frac{\tilde{K}^2}{2\tilde{J}}, \quad \tilde{J} = \frac{J}{J_*} = \frac{8}{15}\left(\frac{x^{2/3}}{y^{4/3}} + \frac{y^{2/3}}{x^{4/3}}\right), \quad x = \frac{a_3}{a_1}, \quad y = \frac{a_3}{a_2} \\ &\tilde{S} = \frac{S}{S_*} = \left(\frac{1}{3x^{p/3}y^{p/3}} + \frac{x^{2p/3}}{3y^{p/3}} + \frac{y^{2p/3}}{3x^{p/3}}\right)^{1/p}, \quad \tilde{K} = \tilde{J}_0\tilde{\Omega}, \quad \tilde{\Omega}^2 = \frac{\Omega^2}{\Omega_*^2}, \\ &S_* = 4\pi a^2, \quad \Omega_*^2 = 8\sigma/(\rho a^3), \quad a^3 = a_1 a_2 a_3. \end{aligned} \tag{25.11}$$

where a and S_* are the radius and surface area of a sphere whose volume equals the volume of the ellipsoid.

The difference between the drop problem and the problem for a gravitating mass is that the ellipsoidal drop is not an exact solution. At first we are talking about an approximate solution. The chosen approximation of the surface area allows solving the problem of minimizing function $\tilde{U}$ in elementary functions. Consider variations of function $\tilde{U}$ in the form

$$\tilde{U} = U_0 + \delta U + \delta^2 U$$

with respect to small deviations $X+\delta X, Y+\delta Y$. For an axisymmetric ellipsoid $X = Y$. As the first variation equals zero, we find the dependence of the dimensionless angular velocity on the axis ratio of the ellipsoid

$$\tilde{\Omega}^2 = C(1 - X^p)(1 + 2X^p)^{1/p-1}, \quad C = \frac{15}{8 \times 3^{1/p}} = 0.946655. \tag{25.12}$$

The condition for positive definiteness of the quadratic form of the second variation is

$$D > 0, \quad D = \frac{(1 + 2X^p)^{1/p-1}}{2 \times 3^{1/p} X^{8/3}}((2 + p)X^p - 2)$$

This condition implies that the axisymmetric series is stable for $X = a_1/a_3 < X_*$, where

$$X_* = \left(\frac{2}{2+p}\right)^{1/p} = 0.692846.$$

Thus, using (25.12), we derive the range of variation of the angular velocity of the stable branch $0 < \tilde{\Omega}^2 < \Omega_*^2$,

$$\Omega_*^2 = \frac{Cp}{(6 + p)^{1-1/p}(2 + p)^{1/p}} = 0.318184.$$

In [5] the problem of equilibrium and stability of a rotating drop was solved numerically by the virial method developed therein. For the stable branch, Chandrasekhar obtained the following inequality

$$\Sigma = \frac{\Omega^2 \rho a_1^3}{8\sigma} < 0.4587$$

Taking into account the definition of the parameter $\tilde{\Omega}^2$ in (25.11), the Chandrasekhar parameter Σ can be expressed in terms of $\tilde{\Omega}^2$

$$\Sigma = \tilde{\Omega}^2/X \rightarrow \Sigma_* = \Omega_*^2/X_* = 0.4592.$$

The value $\Sigma_* = 0.4587$ calculated by Chandrasekhar [5] differs from the one we obtain in the fourth digit. This point in the theory of equilibrium figures of rotating liquids is called neutral. From it branches off a series of equilibrium triaxial ellipsoids. Comparison with the numerical solution of the Nobel laureate Chandrasekhar shows that the approximate analytical results determined by this method differ from the numerical ones in the third digit.

Thus, from the problem of minimizing the function $\tilde{U}(x, y)$ in (25.11), the axisymmetric series of equilibrium figures, its stability with respect to perturbations of the triaxial ellipsoid, the bifurcation point, and the series of non-axisymmetric equilibrium figures branching from it are determined. The axisymmetric series of equilibrium figures is shown in Fig. 25.1 b) by a solid line, in the stable region (thick line) and triaxial ellipsoids (dots). The axis ratios at these points are $a_2/a_1 = 1 - 0.05i$, $i = 0, 1, 2, \ldots 12$. The value $i = 0$ corresponds to the bifurcation point $a_3/a_1 = X_*$, $\Omega = \Omega_*$.

25.3 Energy of a Drop in an Electrostatic Field

The energy of a drop in the form of a triaxial ellipsoid consists of three terms

$$\Pi = \sigma S + \Pi_E + \Pi_Q \tag{25.13}$$

Here S is the surface area of the drop, defined in (25.11). The energy of a triaxial ellipsoid in an electric field of strength E_∞, directed along the semi-axis a_3 has the form

$$\Pi_E = -\frac{a_1 a_2 a_3 E^2 (\varepsilon - 1)}{6(1 + (\varepsilon - 1) n_3)}$$

where $\varepsilon-$ dielectric permittivity. For a water drop $\varepsilon \approx 81$. Therefore, one can use the limit $\varepsilon \rightarrow \infty$

$$\Pi_E = -\frac{a_1 a_2 a_3 E^2}{6 n_3}, \quad n_3 = \frac{a_1 a_2 a_3}{2} \int_0^\infty \frac{dt}{(t + a_1^2)\sqrt{(t + a_1^2)(t + a_2^2)(t + a_3^2)}} \tag{25.14}$$

The depolarization coefficient n_3 can be represented as a function of two dimensionless variables $x = a_3/a_1$ and $y = a_3/a_2$ and expressed in terms of the elliptic integral

$$
\begin{aligned}
n_3(x, y) &= \frac{1}{2xy} \int_0^\infty \frac{dt}{\sqrt{(1+t)^3(x^{-2}+t)(y^{-2}+t)}} \\
&= \frac{1}{1-x^2} - \frac{x}{(1-x^2)\sqrt{1-y^2}} E\left(\arccos y, \frac{x^2-y^2}{x^2(1-y^2)}\right),
\end{aligned}
\tag{25.15}
$$

where

$$
F(\varphi, m) = \int_0^\varphi (1 - m\sin^2\varphi')^{-1/2} d\varphi', \quad E(\varphi, m) = \int_0^\varphi (1 - m\sin^2\varphi')^{1/2} d\varphi'
$$

In the case of axial symmetry $x = y$ the depolarization coefficient is an elementary function of x

$$
\begin{aligned}
n_3(x, x) &= \frac{2\sqrt{1-x^2} - \pi x + 2x\sin^{-1}(x)}{2\left(1-x^2\right)^{3/2}}, \quad x < 1, \\
n_3(x, x) &= \frac{1}{1-x^2} + \frac{x\log\left(\sqrt{x^2-1}+x\right)}{\left(x^2-1\right)^{3/2}}, \quad x > 1.
\end{aligned}
\tag{25.16}
$$

The potential energy of a charged dielectric drop is represented by an integral similar to the integral for a self-gravitating ellipsoid (25.10)

$$
\Pi_Q = \frac{Q^2}{2} \int_0^\infty \frac{dt}{\sqrt{(t+a_1^2)(t+a_2^2)(t+a_3^2)}}
$$

It can also be expressed through an elliptic integral

$$
\begin{aligned}
\Pi_Q &= \frac{Q^2 a_3^2}{2a_1a_2a_3} \frac{1}{xy} \int_0^\infty \frac{dt}{\sqrt{(t+1)(t+x^{-2})(t+y^{-2})}} \\
&= \frac{Q^2 a_3}{a_1 a_2} \frac{1}{x\sqrt{(1-y^2)}} F\left(\arccos y, \frac{x^2-y^2}{x^2(1-y^2)}\right).
\end{aligned}
\tag{25.17}
$$

Dividing U by the potential energy of a sphere of the same volume σS_0

$$
S_0 = 4\pi(a_1a_2a_3)^{2/3},
$$

we obtain the dimensionless function $\tilde{U}$

$$
\begin{aligned}
&\tilde{U}(x,y) = \tilde{S} - k_E f(x,y) + k_Q f_Q(x,y) + \frac{\tilde{K}^2}{2\tilde{J}}, \quad \tilde{J} = \frac{8}{15}\left(\frac{x^{2/3}}{y^{4/3}} + \frac{y^{2/3}}{x^{4/3}}\right),\\
&\tilde{S} = \left(\frac{1}{3x^{p/3}Y^{p/3}} + \frac{x^{2p/3}}{3y^{p/3}} + \frac{y^{2p/3}}{3y^{p/3}}\right)^{1/p}, \quad \tilde{K} = \tilde{J}_0\tilde{\Omega},\\
&f_Q(x,y) = \frac{1}{x^{5/3}y^{2/3}\sqrt{1-y^2}} E\left(\arccos y, \frac{x^2-y^2}{x^2(1-y^2)}\right),\\
&x = \frac{a_3}{a_1}, \quad y = \frac{a_3}{a_2}, \quad f(x,y) = 1/n(x,y)
\end{aligned}
\tag{25.18}
$$

Consider the dimensionless parameters

$$
\tilde{\Omega}^2 = \frac{\Omega^2 \rho a^3}{8\sigma}, \quad k_E = \frac{aE_\infty^2}{24\pi\sigma}, \quad k_Q = \frac{Q^2}{8\pi a^3\sigma}, \quad a = (a_1 a_2 a_3)^{1/3}.
$$

The extremum points of function $U(X,Y)$ (25.13) for given parameters k_E and k_Q determine the series of equilibrium figures, and the minimum points determine the series of stable equilibrium figures of the drop. For axisymmetric ellipsoids $x = y$ the dimensionless surface area is expressed through elementary functions

$$
\tilde{S} = \frac{1}{2x^{2/3}}\left(1 + \frac{x \arcsin\sqrt{1-\frac{1}{x^2}}}{\sqrt{1-\frac{1}{x^2}}}\right), \quad x > 1,
$$

$$
\tilde{S} = \frac{1}{2x^{2/3}}\left(1 + \frac{x^2 \log\left(\frac{\sqrt{1-x^2}+1}{x}\right)}{\sqrt{1-x^2}}\right), \quad 0 < x < 1
$$

25.4 Axisymmetric Equilibrium Shapes of a Drop in an Electrostatic Field Without Rotation

Axisymmetric equilibrium shapes of a drop in an electrostatic field were determined by Taylor [24] and then by Saranin [19] without addressing the question of their stability. We will show not only how to obtain the solution to the problem of equilibrium shapes, but also how to find the stability boundary. Without rotation $\Omega = 0$ and drop charge $Q = 0$ the potential energy (25.17) of the drop takes the form $\tilde{U}(x,y) = s(x,y) - k_E f(x,y)$, and for the axisymmetric shape $x = y$ may be simplified

$\frac{a_3}{a_1}$	η_T	η_S
1.02	0.46	0.471
1.154	1.14	1.155
1.512	1.575	1.549
1.666	1.622	1.591
1.898	1.635	1.602
2.105	1.623	1.585
2.294	1.588	1.558
2.55	1.542	1.515
3.202	1.424	1.394
5.025	1.086	0.919

a)

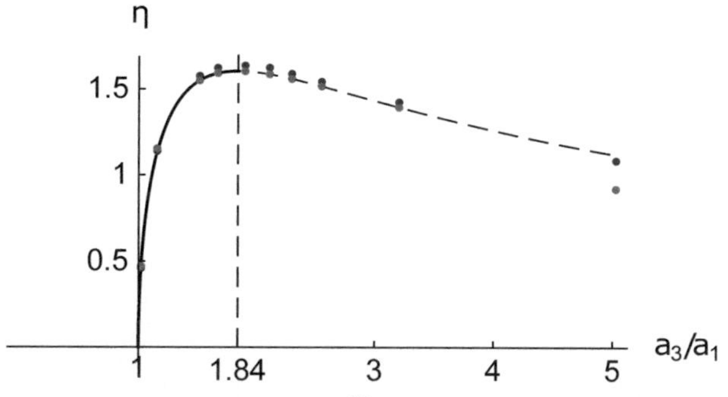

b)

Fig. 25.2: a) Table of numerical values η_T [24] and η_S [19]; b) Series of equilibrium axisymmetric shapes: in the stable region (thick line), and triaxial ellipsoids (dots).

$$w(x) = s(x) - k_e f(x), \quad f(x) = 1/n_3(x), \quad s(x) = \frac{1}{2x^{2/3}}\left(1 + \frac{x \arcsin\sqrt{1-\frac{1}{x^2}}}{\sqrt{1-\frac{1}{x^2}}}\right),$$

$$n_3 = \frac{1}{1-x^2} + \frac{x\log\left(\sqrt{x^2-1}+x\right)}{\left(x^2-1\right)^{3/2}}, \qquad k_E = \frac{\eta^2}{24\pi}, \qquad \eta = \frac{E_\infty\sqrt{a}}{\sqrt{\sigma}} \tag{25.19}$$

The parameter η was introduced by Taylor [24] and he provided a table of values of x, η. This table is reproduced in Fig. 25.2 a). The first column shows the axis ratio a_3/a_1, the second and third columns show the corresponding η values found in [24] and [19].

Condition $w'(x) = 0$ for the extremum of function $w(x)$ implies the dependence

$$k_e(x) = s'(x)/f'(x) \tag{25.20}$$

Thus, we find the dependence of the Taylor number

$$\eta(x) = \sqrt{24\pi k_e(x)}$$

on the drop deformation $x = a_3/a_1$. This dependence is shown in Fig. 25.2 b) by solid and dashed lines, the solid line shows the stable part, the dashed line shows the unstable part, the blue dots and red dots are the points η_T and η_S from the table 25.2 a). From Fig. 25.2 b) it can be seen that the function $\eta(x)$ increases monotonically on the interval $x \in (0, x_0)$ and decreases on the rest of the axis $x \in (x_0, \infty)$. At the point $x_0 = 1.839$ the function reaches its maximum value $\eta_{max} = 1.603$.

We show that on the interval $x \in (0, x_0)$ the extremum point of function $w(x') = s(x') - k_e(x) f(x')$ is a minimum point, and for $x \in (x_0, \infty)$ the extremum point of function $w(x')$ is a maximum point. Accordingly, the interval $x \in (0, x_0 \approx 1.84)$ corresponds to the stable series of drop equilibria. This series is shown in Fig. 25.2

b) by a solid line. The remaining part of the curve $\eta(x)$ corresponds to the unstable series of drop equilibria.

Indeed, the function $k_e(x)$ increases if

$$\frac{d}{dx}\ln k_e(x) = \frac{s''(x)}{s'(x)} - \frac{f''(x)}{f'(x)} > 0 \tag{25.21}$$

Identity

$$w''(x) = s''(x) - k_e(x)f''(x) = s'(x)\frac{d}{dx}\ln k_e(x),$$

the inequality $s'(x) > 0$ and (25.21) imply the condition $w''(x) > 0$, which is a sufficient condition for the minimum of the function $w(x)$ and hence the stability condition of the equilibrium shape. Thus, the stability boundary is the root $x_0 = 1.839$ of the equation $k_e'(x) = 0$. It is marked in Fig. 25.2 b) by a dashed line.

In Fig. 25.3 the dependence on x of the mixed derivatives $f_{xy}(x)$ (dashed line) and $s_{xy}(x)$ (solid line) is shown. The values of both functions for prolate ellipsoids $x > 1$ are negative. It follows that the mixed derivative

$$\tilde{U}_{xy}(x) = s_{xy}(x) - k_E f_{xy} < 0.$$

Thus, by the theorem in Section 25.1, if $x \in (1, x_0 = 1.839)$ then the equilibrium drop shapes are stable with respect to both axisymmetric and non-axisymmetric perturbations.

25.5 Equilibrium Shapes of a Rotating Drop and its Stability in an Electrostatic Field

In the absence of drop charge $Q = 0$ the minimization problem for the function $\tilde{U}(x, y)$ (25.17) takes the form

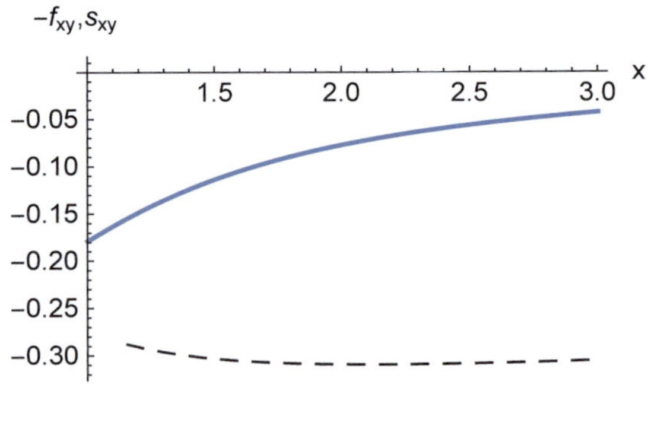

Fig. 25.3 Dependencies on x of the mixed derivatives $f_{xy}(x)$ (dashed line) and $s_{xy}(x)$ (solid line).

$$
\begin{aligned}
&\tilde{U}(x,y) = \tilde{S} - k_E f(x,y) + \frac{\tilde{K}^2}{2\tilde{J}}, \quad \tilde{J} = \frac{8}{15}\left(\frac{x^{2/3}}{y^{4/3}} + \frac{y^{2/3}}{x^{4/3}}\right),\\
&\tilde{S} = \left(\frac{1}{3x^{p/3}Y^{p/3}} + \frac{x^{2p/3}}{3y^{p/3}} + \frac{y^{2p/3}}{3y^{p/3}}\right)^{1/p}, \quad \tilde{K} = \tilde{J}_0\tilde{\Omega},\\
&f(x,y) = \left(\frac{1}{1-x^2} - \frac{x}{(1-x^2)\sqrt{1-y^2}} E\left(\arccos y, \frac{x^2-y^2}{x^2(1-y^2)}\right)\right)^{-1},\\
&x = \frac{a_3}{a_1}, \quad y = \frac{a_3}{a_2}, \quad k_E = \frac{\eta^2}{24\pi}, \quad \eta = \frac{E_\infty\sqrt{a}}{\sqrt{\sigma}}.
\end{aligned}
\tag{25.22}
$$

It differs from the Chandrasekhar problem (25.11) by the presence in the minimized function $\tilde{U}(x,y)$ of the term $-k_E f(x,y)$, which accounts for the action of the electric field.

To determine the axisymmetric equilibrium shapes, we find the potential energy (25.22) for $y = x$

$$
w(x) = s(x) - k_E f(x) + \frac{K^2}{2j(x)}, \quad j(x) = \frac{16}{15x^{2/3}}.
$$

where $f(x)$ is defined in (25.19). It differs from the function of the Taylor problem (25.19) by the presence of the term $\tilde{K}^2/(2\tilde{J})$, which accounts for rotation.

The axisymmetric series of equilibrium shapes are determined by the formula

$$
\tilde{\Omega}^2 = \tilde{\Omega}_0^2 - \eta^2\omega_1, \quad \omega_1 = \frac{2f'(x)}{24\pi j'(x)}
$$

where $\tilde{\Omega}_0^2$– is the solution of the Chandrasekhar problem (25.12).

The axisymmetric series of equilibrium figures are shown in Fig. 25.4 by dashed and thick lines for various values of the Taylor parameter: $\eta =$: 0.5; 1; 1.5; 1.6; 2 – Fig. 25.4 a), b), c), d) and e) respectively. Thick solid lines are stable regions, and dashed lines are unstable ones. The dots correspond to shapes of triaxial ellipsoids. The axis ratios at these points are $a_2/a_1 = 1 - 0.05i$, $i = 0, 1, 2, \ldots 12$. The value $i = 0$ corresponds to the bifurcation point $a_3/a_1 = X_*$, $\Omega = \Omega_*$. The bifurcation points were determined by the method of Sect. 25.1 from the equation $A_{11} - A_{12} = 0$, where A_{11}, A_{12} are the coefficients of the quadratic forms of the second variation of the function $U(x,y)$. The obtained bifurcation points are in Table 25.1.

The first row of the table shows the values of the Taylor parameter. The second row shows the axis ratio at the bifurcation point $x_1 = a_3/a_1$. The third row shows the maximum value of the axis ratio $x_2 = a_3/a_1$ of the stable axisymmetric series. For values $\eta < 1.6$ the upper stability boundary is determined from the solution of the Taylor problem

$$
\eta(x) = \sqrt{24\pi k_e(x)}.
$$

For large values $\eta > 1.6$ the upper stability boundary is determined from the stability condition with respect to symmetric perturbations $U''(x) > 0$.

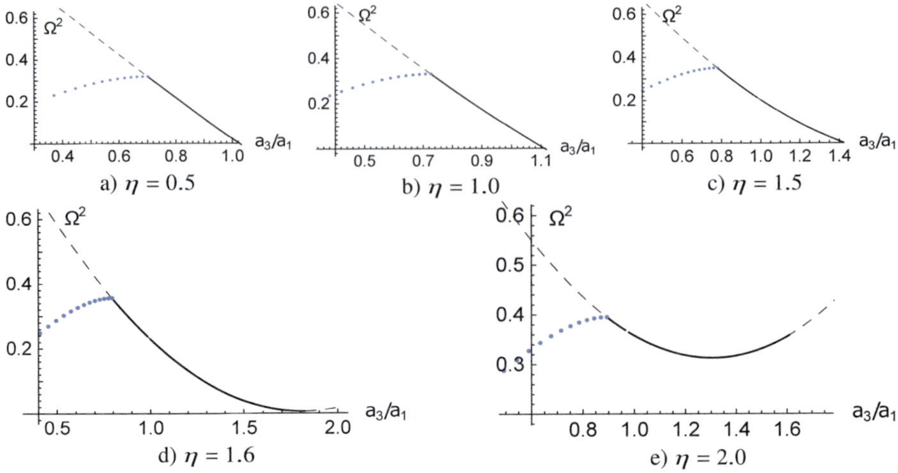

Fig. 25.4: Dependence of the square of the angular velocity of rotation of the ellipsoid $\tilde{\Omega}^2$ on the axis ratio a_3/a_1. Series of equilibrium axisymmetric shapes: in the stable region – thick line, in the unstable region – dashed line. Series of triaxial ellipsoids – dots for various values of the Taylor parameter η.

Table 25.1: Bifurcation points.

η	x_1	x_2
0.5	0.701	1.06
1	0.726	1.11
1.5	0.779	1.4
1.6	0.796	1.82
2	0.892	1.6

Acknowledgements The author thanks Mariana Lopushanski for helping prepare the article. The study was supported by the Government program (contract 124012500443-0).

References

[1] Behera N, Mandal S, Chakraborty S (2019) Electrohydrodynamics settling of drop in uniform electric field: beyond Stokes flow regime. Journal of Fluid Mechanics 881:498-523, DOI 10.1017/jfm.2019.744

[2] Beroz J, Hart AJ, Bush JWM (2019) Stability limit of electrified droplets. Physical Review Letters 122(24):244501, DOI 10.1103/PhysRevLett.122.244501

[3] Borisov AV, Mamaev IS (2010) Dinamika zhidkikh i gazovykh ellipsoidov (in Russ., Dynamics of Liquid and Gas Ellipsoids. SIC "Regular and Chaotic Dynamics", Moscow-Izhevsk
[4] Brown RA, Scriven LE (1980) The shape and stability of rotating liquid drops. Proceedings of the Royal Society of London. A. Mathematical and Physical Sciences 371(1746):331-357, DOI 10.1098/rspa.1980.0084
[5] Chandrasekhar S (1965) The stability of a rotating liquid drop. Proceedings of the Royal Society of London. A. Mathematical and Physical Sciences 286(1404):1-26, DOI 10.1098/rspa.1965.0127
[6] Chandrasekhar S (1969) Ellipsoidal Figures of Equilibrium. Yale University Press, New Haven and London
[7] Fikhtengolts GM (1963) Kurs differencial'nogo i integral'nogo ischesleniya (in Russ., Course of Differential and Integral Calculus), Vol 3. Fizmatgiz, Moscow
[8] Grigoriev AI (1990) Neustoychivosti zaryazhennykh kapel' v elektricheskikh polyakh (obzor) (in Russ., Instabilities of charged drops in electric fields (review)). Elektronnaya Obrabotka Materialov (6):23-32
[9] Han Y (2022) Droplets in an electric field: surface rheology, coalescence and rebound. PhD thesis, City College of New York
[10] Holgate J, Coppins M (2018) Shapes, stability, and hysteresis of rotating and charged axisymmetric drops in a vacuum. Physics of Fluids 30(6):064107, DOI 10.1063/1.5023437
[11] `http://www.numericana.com/answer/ellipsoid.htm`. Accessed 2026
[12] Hyer AP, McMillin RE, Ferri JK (2024) The shape of things to come: amisymmetric drop shape analysis using deep learning. Journal of Colloid and Interface Science 653B:1188-1195, DOI 10.1016/j.jcis.2023.09.120
[13] Kochin NE, Kibel IA, Roze NV (1963) Teoreticheskaya gidrodinamika (in Russ., Theoretical Hydrodynamics), Vol 1. Fizmatgiz, Moscow
[14] Kondratyev BP (2003) Teoriya potentsiala i figury ravnovesiya (in Russ., Potential Theory and Figures of Equilibrium). Moscow-Izhevsk (in Russian)
[15] Lamb H (1945) Hydrodynamics. Dover, New York
[16] Landau LD, Lifshitz EM (1975) Course of Theoretical Physics. Vol 2: The Classical Theory of Fields. Pergamon, Oxford)
[17] Merkin DR (1974) Giroskopicheskiye sistemy (in Russ., Gyroscopic Systems). Nauka, Moscow
[18] Petrov AG (2018) Problems of hydrodynamics for a triaxial ellipsoid. Journal of Applied Mechanics and Technical Physics 59(4):618-630, DOI 10.1134/S0021894418040077
[19] Saranin VA (2000) Bifurcation of the equilibrium droplet geometry in electric field. Technical Physics 45(4):505-506, DOI 10.1134/1.1259664
[20] Shchukin SI, Grigoriev AI (1998) Stability of a charged drop having the form of a triaxial ellipsoid. Technical Physics 43(11):1314-1317, DOI 10.1134/1.1259190
[21] Shchukin SI, Grigoriev AI (1999) Critical equilibrium spheroidal deformation of a drop of dielectric liquid in a uniform electrostatic field. Technical Physics 44(7):758-762, DOI 10.1134/1.1259342

[22] Shiryaeva SO, Grigoriev AI, Moksheev PV (2006) Ravnovesnaya forma zaryazhennoy kapli, vrashchayushcheysya vokrug svoyey osi simmetrii (in Russ., Equilibrium shape of a charged drop rotating about its axis of symmetry). Elektronnaya Obrabotka Materialov (4):46-52

[23] Sretensky LN (1946) Teoriya N'yutonovskogo potentsiala (in Russ., Theory of the Newtonian Potential). OGIZ, Moscow

[24] Taylor J (1964) Disintegration of water drops in an electric field. Proceedings of the Royal Society of London. A. Mathematical and Physical Sciences 280(1382):383-397, DOI 10.1098/rspa.1964.0151

Chapter 26
Modeling of Dynamics of Piecewise Homogeneous Partially Saturated Poroelastic Bodies by the Boundary Element Method

Andrey Petrov, Leonid Igumnov, Aleksandr Belov, and Svetlana Litvinchuk

Abstract This work develops a boundary element formulation for three-dimensional dynamic problems in piecewise homogeneous, partially saturated poroelastic media. The mathematical framework employs a five-function model to describe the partially saturated porous continuum, with the governing equations of dynamic poroelasticity solved in the Laplace domain. A direct boundary integral equation method, utilizing full-space fundamental solutions, is applied to transform the boundary value problem into integrals over the subdomain boundaries and interfaces. Interface conditions between dissimilar materials ensure continuity of displacements, pore pressures, tractions, and fluid fluxes. The numerical implementation is validated through the analysis of a prismatic composite column subjected to a surface load. The computed displacements and pore pressures show excellent correlation with a reference numerical-analytical solution. A detailed parametric investigation into the effect of partial saturation reveals critical insights.

Key words: Partially saturated porous media, Piecewise homogeneous bodies, Boundary element method, Laplace transform

26.1 Introduction

Fluid-saturated porous media, composed of a deformable solid skeleton and interstitial pore fluid, are ubiquitous in both natural and engineered environments. As established by Biot's theory [3], the dynamic interaction between the solid and fluid phases necessitates a biphasic continuum approach, which provides a more physically

Andrey Petrov (corresponding author) · Leonid Igumnov · Aleksandr Belov · Svetlana Litvinchuk
Research Institute for Mechanics, National Research Lobachevsky State University of Nizhny Novgorod, 23, Gagarin av., bld. 6, 603950 Nizhny Novgorod, Russian Federation, e-mail: andrey.petrov@mech.unn.ru, igumnov@mech.unn.ru, belov_a2@mech.unn.ru, litivnchuk@mech.unn.ru

H. Altenbach et al. (eds.), *Stability Problems in Mechanics of Continua*,
Advanced Structured Materials 260, https://doi.org/10.1007/978-3-032-31032-3_26

representative description of material behavior than classical single-phase elasticity. In recent decades, Biot's original formulation for full saturation has been extended to accommodate partially saturated conditions, introducing a third, gaseous phase. Foundational multiphase models applicable across a wide range of gas volume fractions were proposed by Garg and Nayfeh [4] and Berryman et al. [2]. Subsequently, numerous authors have developed frameworks for analyzing the isodynamic behavior of multiphase porous media; comprehensive reviews of these developments are provided by Zienkiewicz et al. [18] and Schanz [16].

Recent contributions have advanced the field in several directions. Nenning and Schanz [14] introduced an infinite element formulation for wave propagation problems, while Heider et al. [6] presented numerical solutions for dynamic wave propagation in infinite half-spaces with incompressible constituents. Detmann [1] conducted a detailed analysis of wave propagation in both saturated and partially saturated porous media. Parallel efforts have focused on coupled thermo-hydro-mechanical processes, with or without material nonlinearities, dating back to the foundational work of Lewis and Schrefler [11].

Research into wave propagation in poroelastic media has proceeded along two primary avenues: analytical and numerical methods. Analytical approaches offer profound insight into the underlying wave mechanisms. For instance, Li and Schanz [12] derived transient response governing equations for unsaturated media based on mixture theory and, employing the Laplace transform, provided a solution for a single-layer medium, thereby confirmng the existence of three distinct compressional waves. Shan et al. [17] contributed an exact time-domain solution for unsaturated media using variable separation and eigenfunction expansion. Ma et al. [13] extended this work to layered configurations, establishing a fluid-unsaturated porous medium double-layer model and presenting analytical and semi-analytical solutions under various dynamic permeability conditions.

Despite their value, analytical methods are inherently limited to problems with relatively simple geometries and boundary conditions, rendering them impractical for the complex partial differential operators and irregular domains encountered in many real-world applications. Numerical methods, by contrast, offer far greater flexibility in handling material heterogeneity and geometric complexity. As reviewed in Igumnov [8], numerical approaches for modeling wave processes in partially saturated media can be broadly categorized into domain-type methods, boundary-type methods, and hybrid schemes.

Among these, boundary-type methods – particularly the boundary element method (BEM) – offer distinct advantages. These include a reduction in problem dimensionality by one, automatic satisfaction of the Sommerfeld radiation condition in unbounded domains, absence of high-frequency dispersion errors, and comparable precision in the computation of both primary fields (displacements, pore pressures) and their derivatives (stresses, fluxes). Consequently, BEM has been extensively applied to wave motion simulation in infinite and semi-infinite poroelastic domains. However, its application to the dynamic analysis of partially saturated, piecewise homogeneous composite structures remains remarkably limited.

This paper addresses this gap by developing a three-dimensional boundary element formulation for dynamic problems in piecewise homogeneous, partially saturated poroelastic media. The proposed approach is validated through a comprehensive numerical study of a prismatic composite column subjected to a transient surface load, with particular attention devoted to the influence of saturation degree and mesh resolution on the accuracy of the computed response.

26.2 A Poroelasticity Model for Partially Saturated Porous Materials

Porous media is modeled by the three-phase mixture consisting solid skeleton, wetting and nonwetting fluids. The stress-strain-pore pressure in the framework of Biot's theory can be formulated as

$$\sigma_{ij} = 2G\varepsilon_{ij} + \left(K - \frac{2}{3}G\right)\varepsilon_{kk}\delta_{ij} - \alpha\left(S_w p^w + S_a p^a\right)\delta_{ij}, \tag{26.1}$$

where σ_{ij} is the stress tensor, ε_{ij} the strain tensor, p^w and p^a the pore pressures of the wetting and nonwetting fluids, respectively, δ_{ij} the Kronecker delta, G the shear modulus, K the drained bulk modulus, α the Biot-Willis coefficient. In Eq. (26.1), S_w and S_a are the degrees of saturation for the wetting and nonwetting fluids defined by

$$\begin{aligned} S_w &= \frac{V_w}{V_p}, \\ S_a &= \frac{V_a}{V_p}, \\ S_w &+ S_a = 1, \end{aligned} \tag{26.2}$$

where V_p is the volume of interconnected pores in the specimen, where V_w and V_a are the volumes of the wetting and nonwetting fluids in the pore, respectively.

The strains ε_{ij} have the following relationships with the displacements u_i in solid skeleton

$$\varepsilon_{ij} = \frac{1}{2}\left(u_{i,j} + u_{j,i}\right), \tag{26.3}$$

where the comma denotes a partial derivative with respect to a spatial variable.

The governing equations for the porous media can be written in five major formulations, namely, the conservation of momentum for the whole porous media, the conservation of momentum for the wetting phase, the conservation of momentum for the nonwetting phase, the conservation of mass for the wetting, and the conservation of mass for the nonwetting phase.

The mass balance equation of the mixture is written as

$$\frac{\partial n}{\partial t} - \zeta\left(S_{ww}\frac{\partial p^w}{\partial t} + S_{aa}\frac{\partial p^a}{\partial t}\right) - \zeta K_s \frac{\partial u_{i,i}}{\partial t} = 0, \tag{26.4}$$

where

$$\begin{aligned} \zeta &= \frac{\alpha - n}{K_s}, \\ S_{ww} &= S_w + p^c \frac{\partial S_w}{\partial p^c}, \\ S_{aa} &= S_a - p^c \frac{\partial S_w}{\partial p^c} \end{aligned}$$

and n, K_s,

$$p^c = p^a - p^w$$

are porosity, bulk modulus of the solid skeleton, capillary pressure, respectively.

The mass balance equations of the of the wetting and nonwetting fluids in the absence of fluid sources/sinks are written as

$$\begin{aligned} &\alpha S_w \frac{\partial u_{i,i}}{\partial t} + \left(\zeta S_{ww} S_w + \frac{n}{K_w} S_w - n\frac{\partial S_w}{\partial p^c}\right)\frac{\partial p^w}{\partial t} + \\ &\qquad + \left(\zeta S_{aa} S_w + n\frac{\partial S_w}{\partial p^c}\right)\frac{\partial p^a}{\partial t} + n S_w \frac{\partial u^w_{i,i}}{\partial t} = 0 \end{aligned} \tag{26.5}$$

$$\begin{aligned} &\alpha S_a \frac{\partial u_{i,i}}{\partial t} + \left(\zeta S_{ww} S_a + n\frac{\partial S_w}{\partial p^c}\right)\frac{\partial p^w}{\partial t} + \\ &\qquad + \left(\zeta S_{aa} S_a + \frac{n}{K_a} S_a - n\frac{\partial S_w}{\partial p^c}\right)\frac{\partial p^a}{\partial t} + n S_a \frac{\partial u^a_{i,i}}{\partial t} = 0 \end{aligned} \tag{26.6}$$

where K_w, K_a, u^w, u^a are bulk moduli of wetting and nonwetting fluids, relative displacements of wetting and nonwetting fluids with respect to solid, respectively.

The momentum balance equation for the mixture is the sum of the equations for each individual constituent. Taking into account the constitutive assumption for the total stress Eq. (26.1), the momentum balance equation for the mixture in the absence of body forces is given by

$$\begin{aligned} &G u_{i,jj} + \left(K + \frac{G}{3}\right) u_{j,ij} - \alpha\left(S_w p^w_{,i} + S_a p^a_{,i}\right) = \\ &\qquad = \rho \frac{\partial^2 u_i}{\partial t^2} + n S_w \rho_w \frac{\partial^2 u^w_i}{\partial t^2} + n S_a \rho_a \frac{\partial^2 u^a_i}{\partial t^2} \end{aligned} \tag{26.7}$$

where $\rho = (1-n)\rho_s + nS_w\rho_w + nS_a\rho_a$, ρ_s, ρ_w, ρ_a stand densities of the mixture, solid skeleton, wetting and nonwetting fluids, respectively.

The balance equation of momentum for each fluid phase can be written as

$$nS_w \frac{\partial u_i^w}{\partial t} = -\frac{k_{rw}k}{\mu_w}\left(p_{,i}^w + \rho_w \frac{\partial^2 u_i}{\partial t^2} + \rho_w \frac{\partial^2 u_i^w}{\partial t^2}\right), \tag{26.8}$$

$$nS_a \frac{\partial u_i^a}{\partial t} = -\frac{k_{ra}k}{\mu_a}\left(p_{,i}^a + \rho_a \frac{\partial^2 u_i}{\partial t^2} + \rho_a \frac{\partial^2 u_i^a}{\partial t^2}\right), \tag{26.9}$$

where k is the intrinsic permeability of the porous medium, k_{rw} and k_{ra} are the relative permeability values for the wetting and non-wetting fluids, respectively, and μ_w and μ_a are the fluid viscosities. The relative permeability can be determined by the following expressions [5]:

$$k_{rw} = S_e^{(2+3\theta)/\theta}, k_{ra} = (1 - S_e)^2 \left(1 - S_e^{(2+\theta)/\theta}\right), \tag{26.10}$$

where θ is the pore size distribution index, S_e represents the effective wetting fluid saturation as

$$S_e = \frac{S_w - S_{rw}}{S_{ra} - S_{rw}} \tag{26.11}$$

with S_{rw} the residual wetting fluid saturation and S_{ra} the non-wetting fluid entry saturation.

In a partially saturated porous material containing two non-mixing fillers the interface surface between then is curved as a result of intermolecular interaction forces. The accompanying pressure difference in both of the interfacing phases is called capillary pressure p^c. Capillary pressure can be represented as a function of the saturation degree as follows:

$$p^c = p^a - p^w = p^d S_e^{-1/\theta}, \tag{26.12}$$

where p^d is the entry pressure required for driving non-wetting fluid out of the pores.

Equations (26.5), (26.6), (26.7), (26.8) and (26.9) provide a complete basis for solving the partially saturated poroelasticity problem. From a physical standpoint, it is convenient to choose the solid displacement and the two pore pressures as primary unknowns. Doing so, however, requires eliminating the relative displacements. This elimination is not feasible directly in the time domain, as these quantities appear in the equations with different orders of time derivatives. To overcome this, a Laplace transform is employed to remove the time derivatives and facilitate the solution. The final differential equations in the Laplace domain are arranged into matrix product form as follows:

$$\begin{bmatrix} B_1\delta_{ij} + B_2\partial_i\partial_j & B_3\partial_i & B_4\partial_i \\ B_5\partial_j & B_6 & B_7 \\ B_8\partial_j & B_9 & B_{10} \end{bmatrix} \begin{bmatrix} \hat{u}_i \\ \hat{p}^w \\ \hat{p}^a \end{bmatrix} = \begin{bmatrix} 0 \\ 0 \\ 0 \end{bmatrix}, \qquad \mathbf{x} \in \Omega, \Omega \subset \mathbf{R}^3, \tag{26.13}$$

where

$$
\begin{aligned}
B_1 &= G\nabla^2 - (\rho - \beta S_w \rho_w - \gamma S_a \rho_a)\, s^2, \\
B_2 &= K + \frac{G}{3}, B_3 = -\,(\alpha - \beta)\, S_w, \\
B_4 &= -\,(\alpha - \gamma)\, S_w, B_5 = -\,(\alpha - \beta)\, S_w s, \\
B_6 &= -\left(\zeta S_{ww} S_w + \frac{\phi}{K_w} S_w - n \frac{\partial S_w}{\partial p^c}\right) + \frac{\beta S_w}{\rho_w s} \nabla^2, \\
B_7 &= -\left(\zeta S_{aa} S_w + n \frac{\partial S_w}{\partial p^c}\right) s, \\
B_8 &= -\,(\alpha - \gamma)\, S_a\, s, \\
B_9 &= -\left(\zeta S_{ww} S_a + n \frac{\partial S_w}{\partial p^c}\right) s, \\
B_{10} &= -\left(\zeta S_{aa} S_a + \frac{\phi}{K_a} S_a - n \frac{\partial S_w}{\partial p^c}\right) s + \frac{\gamma S_a}{\rho_a s} \nabla^2, \\
\frac{\partial S_w}{\partial p^c} &= -\frac{\theta\,(S_{ra} - S_{rw})}{p^d} S_e^{(\theta+1)/\theta}, \\
p^c \frac{\partial S_w}{\partial p^c} &= -\theta\,(S_w - S_{rw}), \\
\beta &= \frac{\kappa_w \phi \rho_w s}{\phi S_w + \kappa_w \rho_w s}, \\
\gamma &= \frac{\kappa_a \phi \rho_a s}{\phi S_a + \kappa_a \rho_a s}, \\
\kappa_w &= \frac{k_{rw} k}{\mu_w}, \kappa_a = \frac{k_{ra} k}{\mu_a}
\end{aligned}
$$

and symbol ^ denotes Laplace transform with complex variable s.

A generalized displacement vector and a generalized force vector are additionally introduced as

$$\mathbf{u}(\mathbf{x}, s) = (\hat{u}_1, \hat{u}_2, \hat{u}_3, \hat{p}^w, \hat{p}^a)\ , \tag{26.14}$$

$$\mathbf{t}(\mathbf{x}, s) = (\hat{t}_1, \hat{t}_2, \hat{t}_3, \hat{q}^w, \hat{q}^a)\ , \tag{26.15}$$

where

$$
\begin{aligned}
\hat{t}_i &= \hat{\sigma}_{ij} n_j, \\
\hat{q}^w &= \frac{s}{n S_w} \hat{u}_i^w n_i, \\
\hat{q}^a &= \frac{s}{n S_a} \hat{u}_i^a n_i
\end{aligned}
$$

and n_i is the outward unit normal vector to the boundary $\partial\Omega$.

Equation (26.13), supplemented with boundary conditions:

$$\mathbf{u}(\mathbf{x}, s) = \widetilde{\mathbf{u}},\ \ \mathbf{x} \in \Gamma^{\mathrm{u}}\ , \tag{26.16}$$

$$\mathbf{t}(\mathbf{x}, s) = \widetilde{\mathbf{t}}, \ \mathbf{x} \in \Gamma^{\sigma}, \tag{26.17}$$

where Γ^{u} is the Dirichlet's boundary and Γ^{σ} is the Neumann's boundary, fully describe the boundary problem in the representations of the 3D isotropic dynamic theory of poroelasticity.

26.3 Boundary Element Method Framework

Boundary-value problem (26.13), (26.16), (26.17) is solved using the direct boundary element method, based on a combined use of integral Laplace transform and boundary integral equation (BIE) of the 3D isotropic theory of poroelasticity:

$$\mathbf{C}(\mathbf{y})\mathbf{u}(\mathbf{y}, s) + \int_{\Gamma} \mathbf{T}(\mathbf{x}, \mathbf{y}, s)\mathbf{u}(\mathbf{x}, s)d\Gamma = \int_{\Gamma} \mathbf{U}(\mathbf{x}, \mathbf{y}, s)\mathbf{t}(\mathbf{x}, s)d\Gamma, \mathbf{x}, \mathbf{y} \in \Gamma, \tag{26.18}$$

where $\mathbf{U}(\mathbf{x}, \mathbf{y}, s)$ and $\mathbf{T}(\mathbf{x}, \mathbf{y}, s)$ are matrices of fundamental and singular solutions, respectively, $\mathbf{x}$ is an integration point, $\mathbf{y}$ is an observation point. The values of the coefficients of matrix $\mathbf{C}$ are defined by the geometry of boundary Γ. A procedure for obtaining BIE's, based on the weighted residual method can be found in [15].

Equations (26.18) comprise singular integrals in the sense of Cauchy, which are quite difficult to compute. Use of the boundary properties of retarded potentials makes it possible, based on [9], to write down a regular representation of equation (26.18):

$$\int_{\Gamma} \Big(\mathbf{T}(\mathbf{x}, \mathbf{y}, s)\mathbf{u}(\mathbf{x}, s) - \mathbf{T}^0(\mathbf{x}, \mathbf{y}, s)\mathbf{u}(\mathbf{y}, s) - \mathbf{U}(\mathbf{x}, \mathbf{y}, s)\mathbf{t}(\mathbf{x}, s)\Big) d\Gamma = 0, \mathbf{x}, \mathbf{y} \in \Gamma, \tag{26.19}$$

where $\mathbf{T}^0(\mathbf{x}, \mathbf{y}, s)$ is singularity matrix. Using Eq. (26.19), it is possible to construct a boundary-element solution of the BIE.

In the result of spatial discretization, boundary Γ is represented with a set of K_E quadrangular eight-node boundary elements. The geometry of each element E_K is defined by biquadratic functions of form N_m and the global coordinates of nodes $\mathbf{x}_m^k$:

$$\mathbf{x}(\xi) = \sum_{m=1}^{8} N_m(\xi)\mathbf{x}_m^k, \ k = 1..K \tag{26.20}$$

where $\xi = (\xi_1, \xi_2) \in (-1, 1)^2$ are local coordinates. The displacements are described using bilinear elements with the related bilinear functions of form $R_l(\xi)$, and surface generalized forces are described with constant boundary elements:

$$\mathbf{u}(\xi) = \sum_{l=1}^{4} R_l(\xi)\mathbf{u}_l^k, \tag{26.21}$$

$$\mathbf{t}(\xi) = \mathbf{t}^k , \tag{26.22}$$

where $\mathbf{u}_m^k$ and $\mathbf{t}^k$ are nodal values of displacements and tractions, respectively, over element E_k .

A discrete representation of BIE's written at the nodes of the approximation of boundary functions $\mathbf{y}^i$, using the collocation method and accounting for (26.20)–(26.22), is of the following form:

$$\sum_{k=1}^{K_E} \sum_{m=1}^{4} \Delta\mathbf{T}_{mi}^k \mathbf{u}_m^k = \sum_{k=1}^{K} \Delta\mathbf{U}_{mi}^k \mathbf{t}^k , \tag{26.23}$$

$$\Delta\mathbf{U}_{mi}^k = \int_{-1}^{1} \mathbf{U}(\mathbf{x}^k(\xi), \mathbf{y}^i, s) J^k(\xi) \mathrm{d}\xi , \tag{26.24}$$

$$\Delta\mathbf{T}_{mi}^k = \int_{-1}^{1} \left(R_m(\xi)\mathbf{T}(\mathbf{x}^k(\xi), \mathbf{y}^i, s) - \mathbf{I} \cdot \mathbf{T}^0(\mathbf{x}^k(\xi), \mathbf{y}^i) \right) J^k(\xi) \mathrm{d}\xi , \tag{26.25}$$

where $\mathbf{I}$ is unit matrix, J^k is Jacobian of the local coordinates into global ones.

The elements of matrices $\Delta\mathbf{U}_{mi}^k$, $\Delta\mathbf{T}_{mi}^k$ are computed using numerical integration schemes depending on the kind of integral (nonsingular or singular). Nonsingular integrals arise, when the collocation point does not belong to the element. Here, standard Gaussian-type formula is used in combination with a hierarchical subdivision of the elements [10]. Singular integrals arise, when the collocation point is situated on the element being integrated over. In this case, new local coordinates are introduced, making it possible to avoid a singularity in the integrand and to use Gaussian integration.

26.4 Laplace Transform Inversion

Consider a method based on the fundamental integration theorem – the stepped method of numerical inversion of Laplace transform. Consider the following integral:

$$y(t) = \int_{0}^{1} f(\tau) \mathrm{d}\tau \tag{26.26}$$

Integral (26.26) gives rise to Cauchy problem for an ordinary differential equation:

$$\frac{\mathrm{d}}{\mathrm{d}t} x(t) = s x(t) + C , x(0) = 0 . \tag{26.27}$$

Integral (26.26) is substituted for by a quadrature sum, weighting factors of which are determined using Laplace representation and the linear multistep method [7].

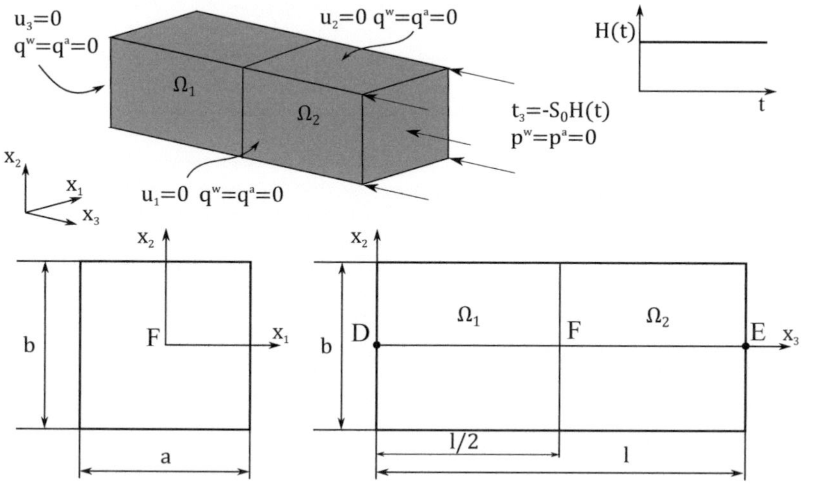

Fig. 26.1: Problem description.

Further derivation is based on the results of this work. The traditional stepped method of integrating the original is that integral (26.26) is calculated using the following relation:

$$y(0) = 0 \,, y(n\Delta t) = \sum_{k=1}^{n} \omega_k(\Delta t) \,, n = 1..N \,, \tag{26.28}$$

$$\omega_n(\Delta t) = \frac{R^{-n}}{L} \sum_{l=0}^{L-1} \hat{f}\left(\frac{\gamma(R e^{il\frac{2\pi}{L}})}{\Delta t}\right) e^{-inl\frac{2\pi}{L}} \,, \tag{26.29}$$

where Δt is time step, $\gamma(z) = 3/2 - 2z + z^2/2$, N is number of time steps, R is the parameter of the method.

26.5 Numerical Results and Analysis

The 3D poroelastic column loaded by a Heaviside function $H(t)$ is investigated as example. The width of the column is $a = b = 0.5m$, the height $l = 6m$. A vertical uniform impact force is exerted on the upper boundary and the rest boundaries are sliding. the column consists of two columns Ω_1 and Ω_2 of the same length $3m$ and poroelastic material. The sketch of problem is shown in Fig. (26.1) The material parameters used in calculation are summarized in Table 26.1. The total time duration T for the analysis is 80ms. Firstly, in this section a comparison with the known numerical-analytical solution was performed. Also a grid convergence study was performed to ensure the precision of the results and validation of the employed numerical model, considering the evolution of the longitudinal displacement and pore

Table 26.1: General material parameters.

Parameter type	Symbol	Value	Unit
Porosity	n	0.45	-
Density of the solid skeleton	ρ_s	2650	kg/m^3
Density of the wetting fluid	ρ_w	997	kg/m^3
Density of the non-wetting fluid	ρ_a	1.1	kg/m^3
Drained bulk modulus of the mixture	K	1.02×10^6	N/m^2
Shear modulus of the mixture	G	1.44×10^6	N/m^2
Bulk modulus of the solid grains	K_s	3.5×10^{10}	N/m^2
Bulk modulus of the wetting fluid	K_w	2.25×10^9	N/m^2
Bulk modulus of the non-wetting fluid	K_a	1.1×10^5	N/m^2
Intrinsic permeability	k	2.5×10^{-12}	m^2
Viscosity of the wetting fluid	η_w	1.0×10^{-3}	$N \times s/m^2$
Viscosity of the non-wetting fluid	η_a	1.8×10^{-7}	$N \times s/m^2$
Pore entry pressure	p^d	5×10^{-4}	m^2/N
Wetting fluid saturation	S_w	0.9	-
Residual wetting fluid saturation	S_{rw}	0	-
Residual non-wetting fluid saturation	S_{ra}	1	-
Pore size distribution index	θ	1.5	-

pressures in time. To assess convergence, four discretization levels were employed. The computational model leverages symmetry, representing only one-quarter of the full geometry. Consequently, the mesh for each subdomain, Ω_1 and Ω_2, consists of 104, 234, 416, and 650 boundary elements for the coarse, medium, fine, and extra-fine meshes, respectively.

The parameters N, R, and Δt for the CQM are selected as 1000, 0.997, and $4.0 \cdot 10^{-5} s$. The graphs of displacement and pore pressures at point F are shown in Figs. (26.2)-(26.4). Also shown in the figures for comparison are graphs of displacements and pore pressures calculated numerically and analytically for the corresponding one-dimensional problem.

The convergence and accuracy of the proposed formulation are further examined through the time histories of key field variables. Figure 26.2 shows the evolution of the displacement u_3 at point F. A close correlation between the BEM results and the numerical-analytical reference solution is observed. While the displacement solutions obtained with the four meshes are nearly coincident for early times (up to $t = 0.025$ s), a slight mesh-dependent divergence becomes apparent at later stages.

The corresponding pore pressure histories at point F are depicted in Figs. (26.3) and (26.4). The results accurately capture the physical phenomenon of stepwise pressure changes triggered by the arrival of the longitudinal wavefront. Between

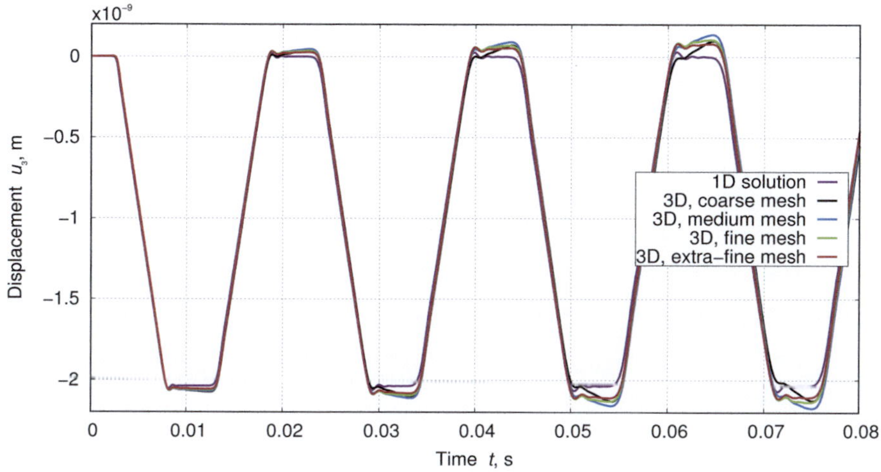

Fig. 26.2: Time history of u_3 at point F.

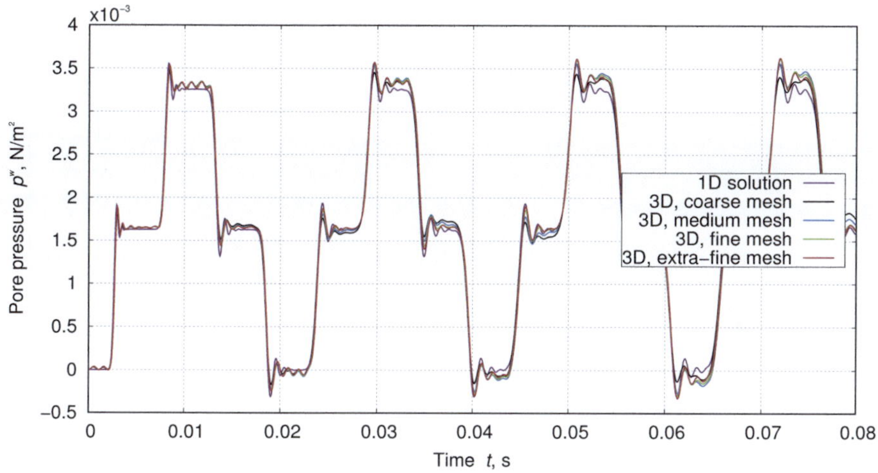

Fig. 26.3: Time history of pore pressure p^w at point F.

these arrivals, the pressure maintains a constant value. The computed pressures are in excellent agreement with the reference solution.

The influence of the saturation degree in the poroelastic material of subdomain Ω_1 on the dynamic response of the composite column is examined next. The corresponding results are presented in Figs. (26.5)-(26.7).

A preliminary analysis revealed that an increase in the water saturation S_w from 0.9 to 0.901 produces an insignificant change in the displacement u_3; consequently, the corresponding graphs are not presented here. However, as illustrated in Fig. 5, the same incremental change in saturation has a substantial impact on the pore

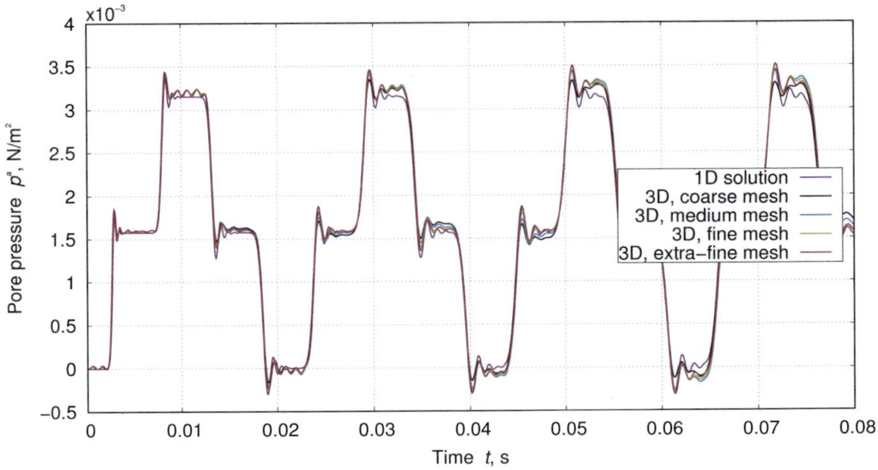

Fig. 26.4: Time history of pore pressure p^a at point F.

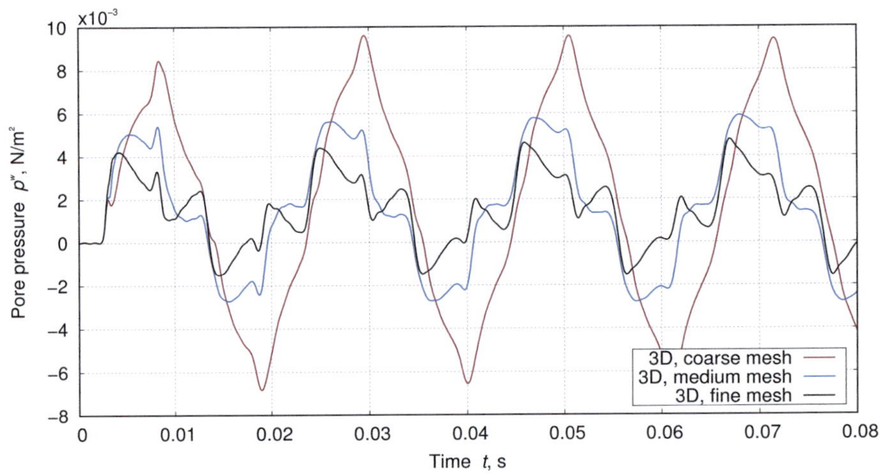

Fig. 26.5: Time history of pore pressure p^w at point F for $Sw = 0.901$: comparison of different mesh densities.

pressure p^w at point F. The pore pressure history computed with the coarse mesh for $S_w = 0.901$ deviates significantly from that obtained for $S_w = 0.9$, exhibiting notable differences in both quantitative values and the overall character of its time-dependent behavior.

The results obtained with the medium and fine meshes are capable of resolving the expected stepwise pattern in the pore pressure response, which is characteristic of the column under the specified loading. While the amplitude of the pore pressure variation decreases as the mesh is refined, it remains surprisingly large considering

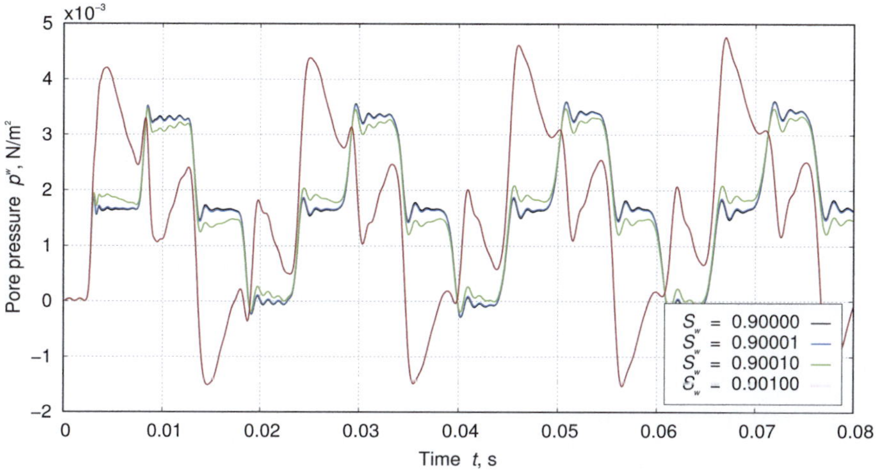

Fig. 26.6: Effect of saturation degree on the time history of pore pressure p^w at point F: fine mesh results for $0.9 \leq S_w \leq 0.901$.

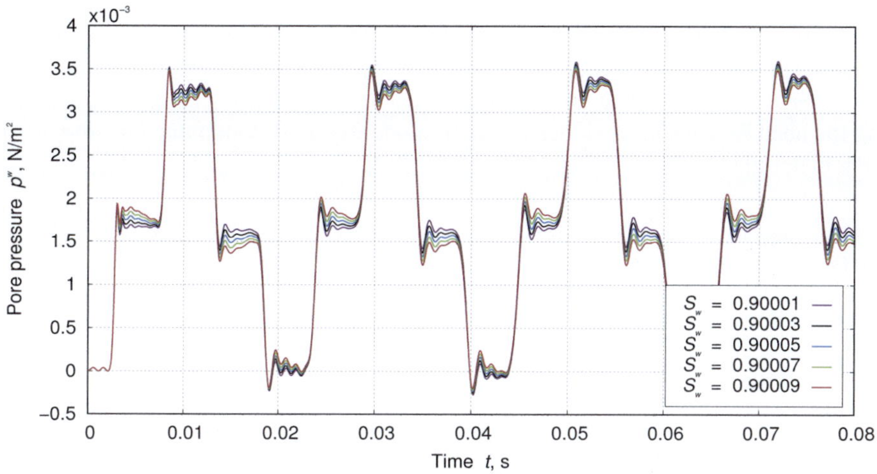

Fig. 26.7: Effect of saturation degree on the time history of pore pressure p^w at point F: fine mesh results for $0.90001 \leq S_w \leq 0.90009$.

the minute adjustment in saturation. This underscores the heightened sensitivity of the pore pressure field to changes in saturation compared to the displacement field.

Further numerical investigations employing the fine mesh were conducted to identify the threshold at which the spatial discretization no longer significantly influences the computed pore pressure p^w at point F (Fig. (26.6)). The results indicate that the solutions obtained for saturation values of $S_w = 0.9$ and $S_w = 0.90001$ are virtually indistinguishable. A minor quantitative effect due to the increased saturation, accompanied by a slight alteration in the shape of the pore pressure curve, becomes observable for $S_w = 0.9001$.

To examine the transition more systematically, Fig. (26.7) presents the pore pressure pwpw histories computed for saturation values incrementally increasing from $S_w = 0.90001$ to $S_w = 0.90009$ in steps of 0.00002. The figure illustrates a smooth and progressive evolution in the pore pressure curve as saturation increases, reflecting a gradual change in the dynamic response of the composite poroelastic column. It is evident that a more detailed characterization of this transition would benefit from extending the observation time interval. The trends observed for p^w are equally applicable to the behavior of the companion pore pressure p^a.

26.6 Concluding Remarks

In conclusion, this study successfully demonstrates the application of a direct boundary element method to three-dimensional problems of dynamic poroelasticity in piecewise homogeneous, partially saturated media. The numerical implementation, validated against a reference solution, accurately reproduces the dynamic response of a composite column subjected to a surface load. The parametric analysis reveals that while displacements are relatively insensitive, pore pressures exhibit a pronounced dependence on both the degree of saturation and the spatial discretization, underscoring the need for careful mesh design when modeling fluid-dominated phenomena in such materials.

Acknowledgements The work was carried out with the financial support of the Ministry of Science and Higher Education of the Russian Federation (project No. FSWR-2026-0019).

References

[1] Albers B (2008) Modeling and numerical analysis of wave propagation in partially saturated porous media. In: Manganaro N, Monaco R, Rionero S (eds) Waves and Stability in Continuous Media, Proceedings of the 14th Conference on WASCOM 2007. Held 30 June-6 July 2007 in Baia Samuele, Sicily, Italy. World Scientific Publishing, pp 7–12, DOI 10.1142/9789812772350_0002
[2] Berryman JG, Thigpen L, Chin RCY (1988) Bulk elastic wave propagation in partially saturated porous solids. The Journal of the Acoustical Society of America 84(1):360–373, DOI 10.1121/1.396938
[3] Biot MA (1962) Mechanics of deformation and acoustic propagation in porous media. Journal of Applied Physics 33(4):1482–1498, DOI 10.1063/1.1728759
[4] Garg SK, Nayfeh AH (1986) Compressional wave propagation in liquid and/or gas saturated elastic porous media. Journal of Applied Physics 60(9):3045–3055, DOI 10.1063/1.337760

[5] van Genuchten MT (1980) A closed-form equation for predicting the hydraulic conductivity of unsaturated soils. Soil Science Society of America Journal 44(5):892–898, DOI 10.2136/sssaj1980.03615995004400050002x

[6] Heider Y, Markert B, Ehlers W (2012) Dynamic wave propagation in infinite saturated porous media half spaces. Computational Mechanics 49(3):319–336, DOI 10.1007/s00466-011-0647-9

[7] Igumnov L, Petrov A (2016) Modelirovaniye dinamiki chastichno nasyshchennykh porouprugikh tel na osnove metoda granichno-vremennykh elementov (in Russ., Dynamics of partially saturated poroelastic solids by boundary-element method). PNRPU Mechanics Bulletin (3):47–61, DOI 10.15593/perm.mech/2016.3.03

[8] Igumnov L, Petrov A (2022) Chislennoe modelirovanie volnovykh processov v chastichno nasyshchennykh porouprugikh sredakh (obzor) (in Russ., Numerical simulation of wave processes in partially saturated poroelastic media (review)). Problems of Strength and Plasticity 84(4):441–467, DOI 10.32326/1814-9146-2022-84-4-441-467

[9] Igumnov LA, Litvinchuk SY, Petrov AN, Ipatov AA (2016) Numerically analytical modeling the dynamics of a prismatic body of two- and three-component materials. In: Parinov IA, Chang SH, Topolov VY (eds) Advanced Materials. Springer Proceedings in Physics, vol 175. Springer International Publishing, Cham, pp 505–516, DOI 10.1007/978-3-319-26324-3_35

[10] Lachat JC, Watson JO (1976) Effective numerical treatment of boundary integral equations: A formulation for three-dimensional elastostatics. International Journal for Numerical Methods in Engineering 10(5):991–1005, DOI 10.1002/nme.1620100503

[11] Lewis R, Schrefler B (1998) The Finite Element Method in the Static and Dynamic Deformation and Consolidation in Porous Media (2nd ed), Numerical Methods in Engineering, vol 34. Chichester et al, Wiley

[12] Li P, Schanz M (2011) Wave propagation in a 1-D partially saturated poroelastic column. Geophysical Journal International 184(3):1341–1353, DOI 10.1111/j.1365-246X.2010.04913.x

[13] Ma R, Shan Z, Jing L (2024) Transient response of an unsaturated single-layer seabed subjected to a vertical compressional wave. International Journal of Geomechanics 24(7):04024133, DOI 10.1061/IJGNAI.GMENG-8185

[14] Nenning M, Schanz M (2011) Infinite elements in a poroelastodynamic FEM. International Journal for Numerical and Analytical Methods in Geomechanics 35(16):1774–1800, DOI https://doi.org/10.1002/nag.980

[15] Schanz M (2001) Wave Propagation in Viscoelastic and Poroelastic Continua: A Boundary Element Approach. Berlin:Springer-Verlag, DOI 10.1007/978-3-540-44575-3

[16] Schanz M (2009) Poroelastodynamics: Linear models, analytical solutions, and numerical methods. Applied Mechanics Reviews 62:030,803, DOI 10.1115/1.3090831

[17] Shan Z, Ling D, Ding H (2013) Exact solutions for the one-dimensional transient response of unsaturated single-layer porous media. Computers and Geotechnics 48:127–133, DOI 10.1016/j.compgeo.2012.08.001
[18] Zienkiewicz O, Chan A, Pastor M, Schrefler B, Shiomi T (1999) Computational Geomechanics with Special Reference to Earthquake Engineering. Chichester: Wiley

Chapter 27
Periodic Contact Problems for a Three-Dimensional Elastic Wedge-Shaped Domain

Dmitry Pozharskii and Elizaveta Pozharskaya

Abstract Within the framework of classical linear elasticity theory, we study the spatial contact problem of normal contact for a wedge, one face of which is subject to sliding support. The other face of the wedge interacts with an infinite periodic rectilinear system of rigid punches located along the edge of the wedge. The problem is reduced to an integral equation for the contact pressure. The infinite system of punches causes an infinite normal displacement of the wedge face. Therefore, the kernel of the integral equation diverges. To regularize it, an additional periodic system of normal forces acting outside the contact area is used. Two regularization options are considered: the chain of forces is applied outside the edge of the wedge (option one) or on the edge (option two). The contact area is unknown in advance. To solve the regularized integral equation, B.A. Galanov's numerical method of nonlinear boundary integral equations is used. A comparison is made with a previously studied case in which one face of the wedge is stress-free.

Key words: Elasticity, Periodic contact, Wedge, Integral equation, Regularization

27.1 Introduction

The relevance of contact problems, including problems of multiple and periodic contact, is confirmed by numerous studies, in particular, monographs [2, 6, 7, 10, 14]. Pioneering studies in the field of periodic contact are [20, 23]. Periodic contact problems taking into account friction, adhesion, wear, and anisotropy were studied in [13, 18, 21]. Doubly periodic contact [1] and contact of three-dimensional wave-like surfaces [22, 25] were considered. The relationship between periodic contact problems and problems of fracture mechanics was shown [24]. Clusters of adhesive

Dmitry Pozharskii (corresponding author) · Elizaveta Pozharskaya
Don State Technical University, Gagarin Square 1, 344000, Rostov-on-Don, Russian Federation,
e-mail: pozharda@rambler.ru, pozharskaya.elizaveta@rambler.ru

H. Altenbach et al. (eds.), *Stability Problems in Mechanics of Continua*,
Advanced Structured Materials 260, https://doi.org/10.1007/978-3-032-31032-3_27

microcontacts [4] and periodic contact of a system of spherical indenters with a viscoelastic half-space [11] were studied. To solve multiple contact problems, the method of integral transforms, asymptotic methods, the localization method, the fast Fourier transform method, the method of nonlinear boundary integral equations, the finite element method, etc. were used [1, 2, 7–11, 13, 18, 21, 22, 25]. Three-dimensional contact problems for wedge-shaped elastic bodies were studied on the basis of the obtained fundamental solutions [2, 3, 5]. For a spatial elastic wedge (in a particular case, a half-space), one face of which is fixed, periodic contact was considered in works [15, 16, 26] (regularization is not required). If one face of a three-dimensional wedge is stress-free, then with periodic contact on the other face, regularization of the kernel of the integral equation is required, since the type of symbol of the kernel of the integral equation changes (from tangent to cotangent hyperbolic) [17].

Here, we consider periodic contact on the face of a three-dimensional elastic wedge, where the other face is subject to sliding support. Regularization of the integral equation kernel is also required. Two kernel regularization options are proposed (on the edge and outside the edge of the wedge).

As known [2, 12], the general Papkovich–Neuber solution of Navier's equations of elastic equilibrium in cylindrical coordinates r, φ, z can be expressed by means of three arbitrary harmonic functions $\Phi_n = \Phi_n(r, \varphi, z)$, $n = 0, 1, 2$, in the form

$$u_r = \frac{\partial \Phi_0}{\partial r} + \frac{1}{4(1-\nu)} \frac{\partial}{\partial r}(r\chi_1) - \chi_1, \quad \chi_1 = \sin\varphi\, \Phi_1 - \cos\varphi\, \Phi_2\,, \tag{27.1}$$

$$u_\varphi = \frac{1}{r}\frac{\partial \Phi_0}{\partial \varphi} + \frac{1}{4(1-\nu)} \frac{\partial \chi_1}{\partial \varphi} - \chi_2, \quad \chi_2 = \cos\varphi\, \Phi_1 + \sin\varphi\, \Phi_2\,, \tag{27.2}$$

$$u_z = \frac{\partial \Phi_0}{\partial z} + \frac{r}{4(1-\nu)} \frac{\partial \chi_1}{\partial z}\,, \tag{27.3}$$

where ν is Poisson's ratio.

Bu using Hooke's law and the second elasticity parameter G (shear modulus), one can derive relations for the elastic stresses from (27.1)–(27.3) [2]:

$$\begin{aligned}\frac{\sigma_\varphi}{G} &= \frac{2}{r}\frac{\partial \Phi_0}{\partial r} + \frac{2}{r^2}\frac{\partial^2 \Phi_0}{\partial \varphi^2} + \frac{1-2\nu}{2(1-\nu)r}\left(\frac{\partial \chi_2}{\partial \varphi} + \frac{\partial}{\partial r}(r\chi_1)\right)\\ &\quad - \frac{2}{r}\left(\chi_1 + \frac{\partial \chi_2}{\partial \varphi}\right) - \frac{1}{2(1-\nu)r}\frac{\partial}{\partial \varphi}\left(\chi_2 - \frac{\partial \chi_1}{\partial \varphi}\right)\,,\end{aligned} \tag{27.4}$$

$$\frac{\tau_{r\varphi}}{G} = \frac{\partial}{\partial r}\left(\frac{2}{r}\frac{\partial \Phi_0}{\partial \varphi} + \frac{1}{2(1-\nu)}\frac{\partial \chi_1}{\partial \varphi} - \chi_2\right) + \frac{1}{r}\left(\chi_2 - \frac{\partial \chi_1}{\partial \varphi}\right)\,, \tag{27.5}$$

$$\frac{\tau_{\varphi z}}{G} = \frac{\partial}{\partial z}\left(\frac{2}{r}\frac{\partial \Phi_0}{\partial \varphi} + \frac{1}{2(1-\nu)}\frac{\partial \chi_1}{\partial \varphi} - \chi_2\right)\,. \tag{27.6}$$

The harmonic functions Φ_n can be presented as Kontorovich–Lebedev integrals with respect to r and Fourier integrals with respect to z:

$$\Phi_n(r,\varphi,z) = \frac{1}{\sqrt{2\pi}} \int_{-\infty}^{\infty} \exp(\mathrm{i}\beta z)\mathrm{d}\beta$$

$$\times \int_{0}^{\infty} [B_1(\beta,\tau)\sinh(\varphi\tau) + B_2(\beta,\tau)\cosh(\varphi\tau)] K_{\mathrm{i}\tau}(|\beta|r)\mathrm{d}\tau \,. \quad (27.7)$$

Here, $K_{\mathrm{i}\tau}(r)$ is the cylindrical Bessel function [19].

The method of Fourier and Kontorovich–Lebedev integral transforms for constructing fundamental solutions for a three-dimensional elastic wedge is described in [2]. For special cases, these solutions reduce to well-known classical Boussinesq, Cerruti, Mindlin and Lord Kelvin solutions in the theory of elasticity. Of course, for periodic problems one takes superposition of those solutions.

27.2 Periodic Force System on a Wedge Face

Let us consider a three-dimensional wedge-shaped domain

$$\{r \in [0,\infty)\,, \quad \varphi \in [0,\alpha]\,, \quad z \in (-\infty,\infty)\} \quad (27.8)$$

filled with a linear-elastic material with elasticity parameters G and ν. The z-axis of the cylindrical coordinate system is directed along the edge of the dihedral region. The face $\varphi = 0$ of the wedge-shaped region is subject to sliding support (it is in frictionless contact with a rigid base). An infinite periodic system of normal forces F with a period of $2l$ is applied to the face $\varphi = \alpha$ of the wedge at the points $r = h,\ z = 2kl,\ (k = 0, \pm 1, \pm 2, \ldots)$, the line of action of which is parallel to the edge of the wedge (Fig. 27.1).

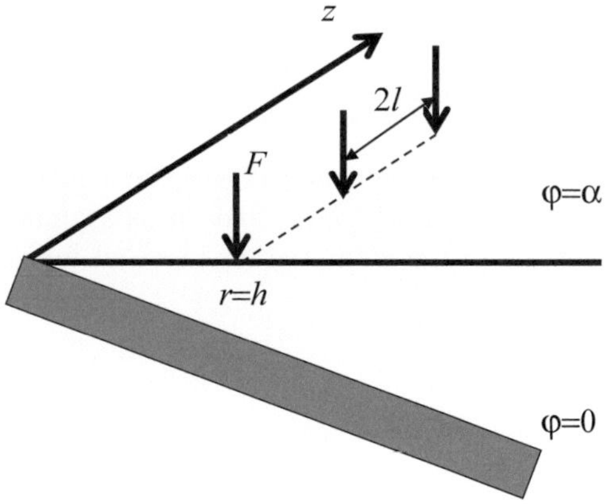

Fig. 27.1 Periodic system of forces F at $\varphi = \alpha$. Sliding support at $\varphi = 0$.

The superposition of fundamental solutions [2, 3] gives the expression for the normal elastic displacement of the face $\varphi = \alpha$

$$u^*_\varphi(r, \alpha, z) = -\frac{F}{2\pi\theta} U_h(r, z) , \quad \theta = \frac{G}{1-\nu} , \tag{27.9}$$

where

$$U_h(r, z) = \frac{4}{\pi^2} \int_0^\infty \int_0^\infty \sinh(\pi u) K_{iu}(\beta r) W(u, \beta h) \times \sum_{k=-\infty}^{\infty} \cos(\beta(z + 2lk)) \mathrm{d}u \mathrm{d}\beta , \tag{27.10}$$

$$W(u, \beta h) = W(u) \left[K_{iu}(\beta h) + \frac{\Psi(u, \beta h)}{\cosh(\pi u/2)} \right] , \tag{27.11}$$

$$W(u) = \frac{\cosh(2\alpha u) - \cos(2\alpha)}{\sinh(2\alpha u) + u \sin(2\alpha)} . \tag{27.12}$$

For a fixed value of βh, the function $\Psi(u, \beta h)$ satisfies the Fredholm integral equation of the second kind ($0 \leq u < \infty$)

$$\Psi(u, \beta h) = (1 - 2\nu) \int_0^\infty L(u, \tau) \left[\Psi(\tau, \beta h) + \cosh \frac{\pi\tau}{2} K_{iu}(\beta h) \right] \mathrm{d}\tau , \tag{27.13}$$

where

$$L(u, \tau) = 2 \cosh \frac{\pi u}{2} \sinh \frac{\pi\tau}{2} W(\tau) \times \int_0^\infty \frac{\sinh(\pi t) g(t) \mathrm{d}t}{[\cosh(\pi t) + \cosh(\pi u)][\cosh(\pi t) + \cosh(\pi\tau)]} , \tag{27.14}$$

$$g(t) = \frac{\coth(\alpha t) \sin^2(2\alpha)}{\cosh(2\alpha t) - \cos(4\alpha)} . \tag{27.15}$$

The symbol function (27.12) coincides with that for the corresponding plane problem for a wedge [2] and, in its asymptotic behavior at zero and infinity, is a function of the $\coth(Au)$ type (except for the case $\alpha = \pi$, which was considered separately [26]),

$$\frac{1}{A} = \lim_{u \to 0} uW(u) = \frac{1 - \cos(2\alpha)}{2\alpha + \sin(2\alpha)} . \tag{27.16}$$

Note that for the case of one stress-free face of the wedge instead of (27.5), we have [17]

$$\frac{1}{A} = \lim_{u \to 0} uW(u) = \frac{2\alpha + \sin(2\alpha)}{2(\alpha^2 - \sin^2 \alpha)} . \tag{27.17}$$

The divergence of the series-integral (27.10) becomes obvious for an elastic quarter-space, when $\alpha = \pi/2$, $W(u) = \coth(\pi u/2)$, $\Psi(u, \beta h) = 0$, and on the basis of the integral [19]

$$\frac{4}{\pi^2}\int_0^\infty\int_0^\infty [\cosh(\pi u) + 1]K_{iu}(\beta r)K_{iu}(\beta h)\cos(\beta z)\mathrm{d}u\mathrm{d}\beta = \frac{1}{\sqrt{(r-h)^2+z^2}} + \frac{1}{\sqrt{(r+h)^2+z^2}}, \tag{27.18}$$

we arrive at the divergent series

$$U_h(r,z) = \sum_{k=-\infty}^{\infty}\left[\frac{1}{\sqrt{(r-h)^2+(z+2kl)^2}} + \frac{1}{\sqrt{(r+h)^2+(z+2kl)^2}}\right]. \tag{27.19}$$

An infinite constant C_∞ can be extracted from the series-integral (27.10) if we use the theory of generalized functions and formulae [1, 17]

$$\sum_{k=-\infty}^{\infty}\cos(\beta(z+2lk)) = 2\pi\cos(\beta z)\sum_{k=-\infty}^{\infty}\delta(2\beta l - 2\pi k), \tag{27.20}$$

$$\sinh(\pi u)K_{iu}(\beta r) = \frac{\pi}{2l}[I_{-iu}(\beta r) - I_{iu}(\beta r)], \tag{27.21}$$

$$\lim_{\beta\to 0} K_{iu}(\beta h)I_{iu}(\beta r) = \frac{1}{2iu}\left(\frac{r}{h}\right)^{iu}, \tag{27.22}$$

$$\int_0^\infty \frac{\coth(Au)}{u}\cos(\gamma u)\mathrm{d}u = \ln\left(2\sinh\frac{\gamma\pi}{2A}\right) + C_\infty \quad (A>0,\ \gamma>0). \tag{27.23}$$

Here, $\delta(\beta)$ is the Dirac δ-function, $I_\mu(r)$ is the cylindrical Bessel function [19]. The 'plane' part of the displacement (27.9) is related to the term at $k = 0$ on the right-hand side of (27.20).

By passing in (27.9)–(27.13) to the limit $h \to 0$, for the system of forces on the wedge edge we get the displacement (27.10), where $h = 0$ and

$$\begin{aligned} U_0(r,z) &= \frac{\pi}{A}\sum_{k=-\infty}^{\infty}\frac{1}{\sqrt{r^2+(z+2kl)^2}} \\ &+ \frac{8}{\pi^2}\int_0^\infty\int_0^\infty \sinh\frac{\pi u}{2}K_{iu}(\beta r)W(u)\Psi^*(u)\sum_{k=-\infty}^{\infty}\cos(\beta(z+2lk))\mathrm{d}u\mathrm{d}\beta. \end{aligned} \tag{27.24}$$

The function $\Psi^*(u)$ satisfies the Fredholm integral equation of the second kind ($0 \le u < \infty$)

$$\Psi^*(u) - (1-2\nu)\int\limits_0^\infty L(u,\tau)\Psi^*(\tau)d\tau = (1-2\nu)L^*(u)\,, \tag{27.25}$$

where

$$L^*(u) = \frac{\pi^2}{2A}\cosh\frac{\pi u}{2}\int\limits_0^\infty \frac{\tanh(\pi t/2)g(t)dt}{\cosh(\pi t)+\cosh(\pi u)}\,. \tag{27.26}$$

The smaller the wedge angle, the smaller the first term on the right-hand side of (27.24). In contrast to the case of the stress-free face, where $A = O(\alpha^3)$ as $\alpha \to 0$ (see (27.17)), here on the basis of (27.16) we have $A = O(\alpha^{-1})$. For $\alpha = \pi/2$, we have $A = \pi/2$, $\Psi^*(u) = 0$ and formula (27.24) coincides with (27.19) if we put $h = 0$ in it.

The values of $1/A$ for sliding support (formula (27.16)) and the stress-free wedge face (formula (27.17)) are given in Table 27.1. The values in Table 27.1 demonstrate the difference between two types of boundary conditions which increases dramatically for small α.

Table 27.1: Values of $1/A$ from formulae (27.16) and (27.17).

α	$\pi/8$	$\pi/4$	$\pi/2$	$3\pi/4$	π	$5\pi/4$	$3\pi/2$	$7\pi/4$
(27.16)	0.20	0.39	0.64	0.27	0	0.11	0.21	0.10
(27.17)	96	11	1.1	0.37	0.32	0.30	0.22	0.17

27.3 Integral Equation of the Contact Problem and its Regularization

Let us consider the periodic frictionless contact problem on a linear chain of identical rigid punches impressed into the face $\varphi = \alpha$ of the wedge (period $2l$, the axis of the chain is parallel to the wedge edge and spaced from it by a distance c, Fig. 27.2). The face $\varphi = 0$ is subject to sliding support. The embedding of the punches is δ, there is no skew, symmetry with respect to z is assumed. Normal forces F are applied to the punches at a distance H from the wedge edge. Due to the periodicity, it is enough to study a unit contact area Ω symmetrical about the semi-axis r ($z = 0$). Suppose that the shape of the base of the punch in Ω is determined by the formula (elliptical paraboloid, curvature radii A_1 and A_2)

$$f(r,z) = \frac{(r-c)^2}{2A_1} + \frac{z^2}{2A_2}\,. \tag{27.27}$$

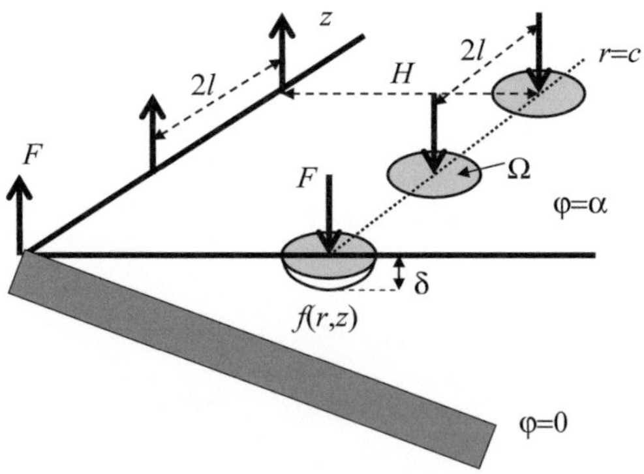

Fig. 27.2 Periodic systems of punches and forces at $\varphi = \alpha$. Sliding support at $\varphi = 0$

.

Boundary conditions of the contact problem have the form (Ω^* is combined contact area)

$$\varphi = \alpha : \quad u_\varphi = -(\delta - f(r, z)) \ , \quad (r, z) \in \Omega^* \ ; \tag{27.28}$$

$$\sigma_\varphi = 0 \ , \quad (r, z) \notin \Omega^* \ ; \quad \tau_{r\varphi} = \tau_{rz} = 0 \ ; \tag{27.29}$$

$$\varphi = 0 : \quad u_\varphi = \tau_{r\varphi} = \tau_{rz} = 0 \ . \tag{27.30}$$

The integral equation with respect to the contact pressure $q(r, z)$ is derived by using the boundary condition (27.28) (contact condition). To derive the integral equation, replace F by $q(x, y)$, h by x, z by $z - y$ in (27.9) and integrate over the domain Ω over the variables x, y. In this case, the series (27.10) in the kernel of the integral equation will diverge. To regularize the kernel, we turn to the idea of relative displacement. Replace the boundary condition (27.28) with the following:

$$\varphi = \alpha : \quad u_\varphi - u_\varphi^* = -(\delta - f(r, z)) \ , \quad (r, z) \in \Omega^* \ , \tag{27.31}$$

where u_φ^* is the elastic displacement of the wedge face $\varphi = \alpha$ due to the action of the periodic system of normal forces F applied on the face $\varphi = \alpha$ outside the contact domain on the axis $r = h$ parallel to the edge wedge. The period of the system of forces is the same, equal to $2l$. These forces are directed opposite to the forces that act on the punches. For $h = 0$, the forces are applied to the wedge edge (Fig. 27.2). The displacement is determined by formulae (27.9), (27.10) or (27.24).

For given values of $G, \nu, \alpha, c, l, \delta, h$ and the function $f(r, z)$ (27.27), it is required to determine the contact region Ω and contact pressure $\sigma_\varphi(r, \alpha, z) = -q(r, z)$, $(r, z) \in \Omega$. Then, taking into account the integral equilibrium conditions of the punches, the values of F and H can be found. In particular,

$$\int_\Omega q(x, y) dx dy = F \ . \tag{27.32}$$

For $h \neq 0$, the regularized contact integral equation based on the boundary conditions (27.29)–(27.31) will be (the first regularization option)

$$\int_{\Omega} q(x,y)Q(x,y,r,z)\mathrm{d}x\mathrm{d}y = 2\pi\theta[\delta - f(r,z)]\,, \qquad (r,z) \in \Omega\,, \tag{27.33}$$

where

$$Q(x,y,r,z) = \frac{1}{R} - \frac{1}{R_h} + \sum_{k=1}^{\infty}\left[\frac{1}{R_k^+(x,y)} + \frac{1}{R_k^-(x,y)} - \frac{1}{R_k^+(h,0)} - \frac{1}{R_k^-(h,0)}\right]$$
$$+ \frac{A_*}{2l}\ln\frac{x}{h} + \frac{1}{l}\int_0^{\infty} W_*(u)\left[\cos\left(u\ln\frac{x}{r}\right) - \cos\left(u\ln\frac{h}{r}\right)\right]\frac{du}{u}$$
$$+ \frac{4}{\pi l}\int_0^{\infty}\sinh(\pi u)\left\{W_*(u)\Sigma(u) + \frac{W(u)\Sigma_1(u)}{\cosh(\pi u/2)}\right\}\mathrm{d}u\,, \tag{27.34}$$

and

$$R = \sqrt{(r-x)^2 + (z-y)^2}\,, \qquad R_h = \sqrt{(r-h)^2 + z^2}\,, \tag{27.35}$$

$$R_k^{\pm}(x,y) = \sqrt{(r-x)^2 + (z-y \pm 2kl)^2}, \tag{27.36}$$

$$A_* = 1 - \pi A^{-1}\,, \qquad W_*(u) = W(u) - \coth(\pi u)\,, \tag{27.37}$$

$$\Sigma(u) = \sum_{k=1}^{\infty} K_{\mathrm{i}u}(\pi_k r)[K_{\mathrm{i}u}(\pi_k x)\cos(\pi_k(z-y)) - K_{\mathrm{i}u}(\pi_k h)\cos(\pi_k z)]\,, \quad \pi_k = \frac{\pi k}{l}\,, \tag{27.38}$$

$$\Sigma_1(u) = \sum_{k=1}^{\infty} K_{\mathrm{i}u}(\pi_k r)[\Psi(u,\pi_k x)\cos(\pi_k(z-y)) - \Psi(u,\pi_k h)\cos(\pi_k z)]\,. \tag{27.39}$$

The function $\Psi(u,\pi_k x)$ in (27.39) satisfies the Fredholm integral equation of the second kind (27.13).

In the second option of regularization, when $h = 0$, the kernel of the integral equation (27.33) takes the form

$$Q(x,y,r,z) = \frac{1}{R} - \frac{\pi}{AR_0} + \sum_{k=1}^{\infty}\left[\frac{1}{R_k^+(x,y)} + \frac{1}{R_k^-(x,y)} - \frac{\pi}{AR_k^+(0,0)}\right.$$
$$\left. - \frac{\pi}{AR_k^-(0,0)} - \frac{A_*}{kl}\right] + \frac{A_*}{l}\ln\frac{|r-x|}{4l|\ln(r/x)|} + \frac{A_* C}{l}$$
$$+ \frac{1}{l}\int_0^{\infty}\left[(W^*(u) - A_*)\cos\left(u\ln\frac{x}{r}\right) + A_*\exp(-u)\right]\frac{du}{u}$$

$$+\frac{4}{\pi l}\int\limits_0^\infty \sinh(\pi u)\left\{W_*(u)\Sigma_0(u)+\frac{W(u)\Sigma_1^*(u)}{\cosh(\pi u/2)}\right\}du\,, \qquad (27.40)$$

where

$$W^*(u) = W(u) - \pi A^{-1}\coth(\pi u), \qquad (27.41)$$

$$\Sigma_0(u) = \sum_{k=1}^\infty K_{iu}(\pi_k r)K_{iu}(\pi_k x)\cos(\pi_k(z-y))\,, \qquad (27.42)$$

$$\Sigma_1^*(u) = \sum_{k=1}^\infty K_{iu}(\pi_k r)[\Psi(u,\pi_k x)\cos(\pi_k(z-y)) - \Psi^*(u)\cos(\pi_k z)] \qquad (27.43)$$

The function $\Psi^*(u)$ in (27.43) satisfies the Fredholm integral equation of the second kind (27.25), C is the Euler constant. To derive (27.40), the following formulae were used [2, 19]

$$\int\limits_0^\infty \frac{\cos(ut)-\exp(-u)}{u}du = -\ln|t|\,, \qquad (27.44)$$

$$\int\limits_0^\infty \cosh(\pi u)K_{iu}(tx)K_{iu}(tr)du = \frac{\pi}{2}K_0(t|r-x|)\,, \qquad (27.45)$$

$$\frac{2}{l}\sum_{k=1}^\infty K_0(\pi_k r)\cos(\pi_k z) = \frac{1}{R}+\frac{1}{l}\left(C+\ln\frac{r}{4l}\right)$$
$$+\sum_{k=1}^\infty\left[\frac{1}{R_k^+(0,0)}+\frac{1}{R_k^-(0,0)}-\frac{1}{kl}\right]. \qquad (27.46)$$

Integrals and series in (27.34) and (27.40) converge taking into account the asymptotic behavior of the Bessel function and the symbol $W(u)$. The logarithmic term in the kernel (27.40) has a removable singularity as $r \to x$. The kernels (27.34) and (27.40) contain both the kernel of the integral equation of the Hertz contact problem R^{-1} for a half-space, and the regularized kernel of the integral equation of the plane contact problem for the wedge, one face of which is subject to sliding support.

For the case of a quarter-space ($\alpha = \pi/2$), the kernels in both options of regularization can be represented in equivalent forms without quadratures using the integral (27.18). The kernel (27.34) for $\alpha = \pi/2$ is converted to the form

$$Q(x,y,r,z) = \frac{1}{R}+\frac{1}{R_+}-\frac{1}{R_h}-\frac{1}{R_h^+}+\sum_{k=1}^\infty\left[\frac{1}{R_k^+(x,y)}+\frac{1}{R_k^-(x,y)}+\frac{1}{R_k^+(-x,y)}\right.$$
$$\left.+\frac{1}{R_k^-(-x,y)}-\frac{1}{R_k^+(h,0)}-\frac{1}{R_k^-(h,0)}-\frac{1}{R_k^+(-h,0)}\right.$$

$$
-\left.\frac{1}{R_k^-(-h,0)}\right] , \tag{27.47}
$$

where

$$
R_+ = \sqrt{(r+x)^2+(z-y)^2}\,, \quad R_h^+ = \sqrt{(r+h)^2+z^2}\,. \tag{27.48}
$$

The kernel (27.40) for $\alpha = \pi/2$ takes the form

$$
Q(x,y,r,z) = \frac{1}{R} + \frac{1}{R_+} - \frac{2}{R_0} + \sum_{k=1}^{\infty}\left[\frac{1}{R_k^+(x,y)} + \frac{1}{R_k^-(x,y)} + \frac{1}{R_k^+(-x,y)}\right.
$$

$$
\left. + \frac{1}{R_k^-(-x,y)} - \frac{2}{R_k^+(0,0)} - \frac{2}{R_k^-(0,0)}\right] . \tag{27.49}
$$

Numerical verification of the coincidence of the kernels (27.47) and (27.49), respectively, with the kernels (27.34) and (27.40) was carried out for $\alpha = \pi/2$, $\nu = 0.5$, $l = 1$, $z = y = 0$, $r = x = 0$. In the case of $h \neq 0$ it is necessary to check the equality

$$
\frac{1}{2x} - \frac{1}{x+h} + \sum_{k=1}^{\infty}\left[\frac{1}{\sqrt{x^2+k^2}} - \frac{2}{\sqrt{(x+h)^2+4k^2}}\right] = -\frac{1}{2}\ln\frac{x}{h}
$$

$$
+ \int_0^{\infty}\frac{1-\cos(u\ln(x/h))}{u\sinh(\pi u)}\mathrm{d}u + 2\sum_{k=1}^{\infty}[K_0(2\pi kx) - K_0(2\pi k(x+h))] \,. \tag{27.50}
$$

For $h = 0$, the test is reduced to establishing equality

$$
\frac{1}{2x} + \sum_{k=1}^{\infty}\left[\frac{1}{\sqrt{x^2+k^2}} - \frac{1}{k}\right] = -\ln\frac{x}{4} - C
$$

$$
+ \int_0^{\infty}\left[1 - \tanh\frac{\pi u}{2} - \exp(-u)\right]\frac{\mathrm{d}u}{u} + 2\sum_{k=1}^{\infty}K_0(2\pi kx) \,. \tag{27.51}
$$

Here, it is used the value of the integral [19]

$$
\int_0^{\infty} K_{\mathrm{i}u}(a)K_{\mathrm{i}u}(b)\mathrm{d}u = \frac{\pi}{2}K_0(a+b) \,. \tag{27.52}
$$

27.4 Method of Solving

To solve the integral equation (27.33), we use the numerical method of Galanov, which allows us to simultaneously determine the contact pressures and the contact area [2, 9]. Enclose the domain Ω containing the initial touch point $r = c$, $z = 0$ in a rectangle $S = \{|r - c| \leq d, |z| \leq l\}$. We extend in (27.33) the integration to S, taking into account that $q(r, z) = 0$ in the additional domain $S\backslash\Omega$. To the integral equation we add the integral non-contact inequality in $S\backslash\Omega$ and the non-negativity condition for $q(r, z)$ in Ω. By introducing special non-linear operators, the system of the integral equation and inequality can be reduced to one non-linear integral equation of the Hammerstein type, for the solution of which the modified Newton method is used. In this case, the integral inequality is automatically satisfied. The boundary of the domain Ω is determined by the points at which the search function vanishes.

We introduce the dimensionless notation (and omit the primes in what follows)

$$r' = \frac{r - c}{l}, \; x' = \frac{x - c}{l}, \; z' = \frac{z}{l}, \; y' - \frac{y - c}{l}, \; \lambda = \frac{c}{l}, \tag{27.53}$$

$$A_0 = \frac{l}{2A_1}, \; B_0 = \frac{l}{2A_2}, \; h' = \frac{h}{l}, \; \delta' = \frac{\delta}{l}, \; \varepsilon = \frac{d}{l}, \tag{27.54}$$

$$q'(r', z') = \frac{q(r, z)}{2\pi\theta}, \; F' = \frac{F}{2\pi\theta l^2}, \; S' \leftrightarrow S, \; \Omega' \leftrightarrow \Omega. \tag{27.55}$$

The parameter λ characterizes the relative distance of the punch chain from the wedge edge.

Calculations are carried out for circular paraboloids $A_0 = B_0 = \varepsilon = 1$, $\nu = 0.5$ for sliding support (case $\mathcal{A}$) or stress-free [17] (case $\mathcal{B}$) conditions on one wedge face. Table 27.2 shows the force $F(h)$ (the integral characteristic (27.32)) for $\delta = 1$, $\lambda = 1.5$. It can be seen from Table 27.2 that the integral characteristic of the contact pressures F decreases as the regularizing system of forces moves away from the chain of punches. The value of F increases together with the wedge angle.

Table 27.2: Values of $F(h)$ for $\delta = A_0 = B_0 = \varepsilon = 1$, $\lambda = 1.5$.

Case	h	4	6	8	10	12
$\mathcal{A}$	$\alpha = \pi/2$	0.229	0.179	0.157	0.143	0.134
$\mathcal{A}$	$\alpha = 3\pi/2$	0.308	0.249	0.226	0.212	0.203
$\mathcal{B}$	$\alpha = \pi/2$	0.181	0.136	0.117	0.106	0.0979
$\mathcal{B}$	$\alpha = 3\pi/2$	0.305	0.247	0.224	0.210	0.200

Table 27.3 gives the force $F(\lambda)$ for $\delta = 1$, $h = 10$. One can conclude from Table 27.3 that the pressing force increases when the chain of punches moves away from the wedge edge.

Table 27.4 shows the force $F(\delta)$ for $\lambda = 2$, $h = 4$. With an increase in the punch settlement, the pressing force and the contact area increase too, which can lead to the merger of adjacent contact spots (percolation [25]). For large wedge angles, the difference in F values for cases $\mathcal{A}$ and $\mathcal{B}$ is small (see Tables 27.2–27.4 for $\alpha = 3\pi/2$).

Table 27.3: Values of $F(\lambda)$ for $\delta = A_0 = B_0 = \varepsilon = 1$, $h = 10$.

Case	λ	2	3	4	5	6
$\mathcal{A}$	$\alpha = \pi/2$	0.152	0.167	0.181	0.197	0.217
$\mathcal{A}$	$\alpha = 3\pi/2$	0.212	0.214	0.220	0.230	0.243
$\mathcal{B}$	$\alpha = \pi/2$	0.116	0.134	0.152	0.171	0.194
$\mathcal{B}$	$\alpha = 3\pi/2$	0.210	0.213	0.219	0.228	0.243

Table 27.4: Values of $F(\delta)$ for $A_0 = B_0 = \varepsilon = 1$, $\lambda = 2$, $h = 4$.

Case	δ	0.5	1	1.5	2	2.5
$\mathcal{A}$	$\alpha = \pi/2$	0.0958	0.269	0.494	0.765	1.052
$\mathcal{A}$	$\alpha = 3\pi/2$	0.114	0.337	0.657	1.032	1.445
$\mathcal{B}$	$\alpha = \pi/2$	0.0827	0.220	0.394	0.602	0.826
$\mathcal{B}$	$\alpha = 3\pi/2$	0.113	0.335	0.652	1.024	1.432

27.5 Conclusion

Possibility of regularization of the divergent integral equation kernel is demonstrated in the three-dimensional periodic contact problem on an infinite rectilinear chain of punches on one face of an elastic wedge, the other face of which is subject to sliding support. For this, an additional system of normal concentrated forces with the same period and the same magnitude modulo as the forces applied to the punches is used. The regularizing system can be applied either outside the wedge edge or on the edge. Due to the periodicity, the contact problem investigated for the infinite wedge is equivalent to a problem for a wedge with a finite edge (length is equal to period) in contact with only one punch.

Acknowledgements This work was supported by the Russian Science Foundation in frames of project 24-21-00014, https://rscf.ru/project/24-21-00014/.

References

[1] Aleksandrov VM (2002) Doubly periodic contact problems for an elastic layer. Journal of Applied Mathematics and Mechanics 66(2): 297-305, DOI 10.1016/S0021-8928(02)00036-9
[2] Alexandrov VM, Pozharskii DA (2001) Three-Dimensional Contact Problems, Solid Mechanics and Its Applications, Vol 93. Kluwer, Dordrecht, DOI 10.1007/978-94-010-9893-9

[3] Alexandrov VM, Bach M, Pozharskii DA (2001) To three-dimensional contact problems with friction for an elastic wedge. Mechanics of Solids 36(5):18-24
[4] Argatov I, Li Q, Popov VL (2019) Cluster of the Kendall-type adhesive microcontacts as a simple model for load sharing in bioinspired fibrillar adhesives. Archive of Applied Mechanics 89(8):1447-1472, DOI 10.1007/s00419-019-01516-1
[5] Bach M, Pozharskii DA (2004) 3-D Contact problems for elastic wedges with Coulomb friction. Mathematical Methods in the Applied Sciences 27(2):193-220, DOI 10.1002/mma.451
[6] Barber JR (2018) Contact Mechanics, Solid Mechanics and Its Applications, Vol 250. Springer, Berlin, DOI 10.1007/978-3-319-70-939-0
[7] Cai H, Lu J (2000) Mathematical Theory in Periodic Plane Elasticity, Modern Analysis Series. Gordon and Breach Scientific Publisher, Amsterdam, DOI 10.1201/9781482287530
[8] Gaidzhurov PP, Savel'eva NA (2023) Application of the double approximation method for constructing stiffness matrices of volumetric finite elements. Advanced Engineering Research 23(4):365-375, DOI 10.23947/2687-1653-2023-23-4-365-375
[9] Galanov BA (1985) The method of boundary equations of the Hammerstein-type for contact problems of the theory of elasticity when the regions of contact are not known. Journal of Applied Mathematics and Mechanics 49(5):634-640, DOI 10.1016/0021-8928(85)90084-X
[10] Makhovskaya Y, Goryacheva I (2022) Discrete Contact Mechanics with Applications in Tribology. Elsevier, Amsterdam
[11] Yakovenko A, Goryacheva I (2021) The periodic contact problem for spherical indenters and viscoelastic half-space. Tribology International 161: Article No 107078, DOI 10.1016/j.triboint.2021.107078
[12] Hahn HG (1983) Elastizitätstheorie. Teubner, Stuttgart
[13] Kuznetsov EA (1976) Periodic contact problem for half-plane allowing for forces of friction. Soviet Applied Mechanics 12(10):1014-1019, DOI 10.1007/BF00885047
[14] Popov VL, Heß M (2015) Method of Dimensionality Reduction in Contact Mechanics and Friction. Springer, Berlin, DOI 10.1007/978-3-642-53876-6
[15] Pozharskaya ED, Pozharskii DA, Sobol BV (2023) Periodic contact problems for a wedge with friction forces taken into account. Mechanics of Solids 58(5):1578-1586, DOI 10.3103/S0025654423700218
[16] Pozharskii DA (2015) Periodic contact problem for an elastic wedge. Journal of Applied Mathematics and Mechanics 79(6):604-610, DOI 10.1016/j.jappmathmech.2016.04.007
[17] Pozharskii DA, Pozharskaya ED (2025) Periodic contact problems for a wedge with one traction-free face. Journal of Applied Mechanics and Technical Physics 66(3):575-582, DOI 10.1134/S0021894425700142
[18] Pozharskii DA, Zolotov NB (2022) Periodic contact problems for a transversely isotropic layer. Journal of Applied Mechanics and Technical Physics 3(6):1065-1072, DOI 10.1134/S0021894422060207

[19] Prudnikov AP, Brychkov YA, Marichev OI (1986) Integral and Series, Vol 2 Special Functions. Gordon and Breach Scientific Publisher, New York
[20] Sadowsky M (1928) Zweidimensionale Probleme der Elastizitätstheorie. ZAMM - Journal of Applied Mathematics and Mechanics / Zeitschrift für Angewandte Mathematik und Mechanik 8(2):107-121, DOI 10.1002/zamm.19280080203
[21] Soldatenkov IA (2013) The periodic contact problem of the plane theory of elasticity. Taking friction, wear and adhesion into account. Journal of Applied Mathematics and Mechanics 77(2):245-255, DOI 10.1016/j.jappmathmech.2013.07.017
[22] Tsukanov IY (2019) An extended asymptotic analysis for elastic contact of three-dimensional wavy surfaces. Tribology Letters 67(4):Article No 107, DOI 10.1007/s11249-019-1220-5
[23] Westergaard HM (1939) Bearing pressure and cracks. ASME Journal of Applied Mechanics E 6(1):43-53, DOI 10.1115/1.4008919
[24] Xu Y, Jackson RL (2018) Periodic contact problems in plane elasticity: The fracture mechanics approach. ASME Journal of Tribology 140(1):Article No 011404, DOI 10.1115/1.4036920
[25] Yastrebov VA, Anciaux G, Molinari J-F (2014) The contact of elastic regular wavy surfaces revisited. Tribology Letters 56(1):171-183, DOI 10.1007/s11249-014-0395-z
[26] Zolotov NB, Pozharskii DA (2022) Periodic contact problems for a half-space with a partially fixed boundary. Mechanics of Solids 57(7):152-159, DOI 10.3103/S0025654422070202

Chapter 28
Dynamic Problem for a Layered Cylinder under Non-Axisymmetric Loading

Sergey G. Pshenichnov

Abstract The paper is devoted to constructing a solution to the problem of the propagation of transient waves in an infinitely long circular cylinder, on the lateral surface of which non-axisymmetric normal and tangential dynamic loads act, constant along the generatrix. A cylinder consists of an arbitrary number of coaxial homogeneous isotropic viscoelastic cylindrical layers with continuity conditions at the contact between them. The problem is considered within the framework of the state of plane deformation, while the hereditary properties of the layer materials are described by integral linear viscoelasticity relations in the Boltzmann-Volterra form. To obtain the solution, the Fourier trigonometric series expansion in angular coordinate and the Laplace transform in time are used. After constructing the Fourier component images, their originals are presented in a special form, obtained and justified by the author in previous studies. The special case when all layers are linearly elastic is considered separately.

Key words: Dynamic problem, Non-stationary waves, Viscoelasticity, Layered cylinder, Non-axisymmetric loading, Laplace transform, Analytical solution

28.1 Introduction

With the development of modern technologies, when using new materials, it becomes necessary to study the dynamics of inhomogeneous deformable media containing components with hereditary properties. One of the important directions in this field is the study of unsteady wave processes in piecewise homogeneous viscoelastic structures using analytical and numerical- analytical methods. The analytical results obtained in this area began to be published several decades ago [9, 11, 12, 20]. At

Sergey G. Pshenichnov
Institute of Mechanics, Lomonosov Moscow State University, Michurinsky prospect 1, 119192 Moscow, Russian Federation, e-mail: serp56@yandex.ru

H. Altenbach et al. (eds.), *Stability Problems in Mechanics of Continua*,
Advanced Structured Materials 260, https://doi.org/10.1007/978-3-032-31032-3_28

a later time, analytical methods were used in combination with numerical methods for such studies. One of the examples of this combination is the application of the method of boundary integral equations [2, 3, 13]. The results obtained by other methods are contained, for example, in [7, 8, 10, 19, 21]. Note that most works on the dynamics of piecewise homogeneous viscoelastic media are devoted not to transient wave processes, but to harmonic vibrations and waves (for example, [1, 4, 6]). Only a part of the publications from a much broader list is mentioned here, but an analysis of the achievements in the field under consideration as a whole shows that many issues still remain relevant. Among them is the development of analytical and numerical-analytical methods for solving non–stationary viscoelasticity problems for piecewise homogeneous bodies with an arbitrary number of homogeneous components with unequal hereditary properties.

This work continues the cycle of studies of unsteady processes in piecewise homogeneous viscoelastic media [15–18]. Its purpose is to construct a solution to a non-one-dimensional initial and boundary value problem on transient wave processes in a layered homogeneous viscoelastic cylinder under non–axisymmetric loading of its lateral surface. The hereditary properties of the layers are not assumed to be the same. The integral Laplace transform in time is applied and then the previously developed elements of mathematical theory [15–18] are used, which make it possible to present the solution in the originals in a special form.

28.2 Formulation of the Problem

Consider an infinitely long circular cylinder consisting of N coaxial homogeneous isotropic viscoelastic cylindrical layers. In the cylindrical coordinate system $R,\ \theta,\ Z\ (-\pi < \theta \leq +\pi)$, the geometric parameters of such a layered structure will be considered to be a set of radii R_n $(n = 1, 2, \ldots, N)$: $0 < R_1 < R_2 < \ldots < R_N$. Herewith, each layer with number n is determined by the ratio

$$R_{n-1} \leq R \leq R_n, \quad n = 1, 2, .., N, \quad R_0 = 0,$$

and its material is characterized by instantaneous values of the shear modulus $G_0^{(n)}$ and Poisson's ratio $\nu_0^{(n)}$, the kernels of volumetric and shear relaxation $T_{\mathrm{v}}^{(n)}(t)$, $T_{\mathrm{s}}^{(n)}(t)$, as well as the propagation velocities of longitudinal and transverse elastic waves $c_1^{(n)}$, $c_2^{(n)}$. Assuming the deformations are small, we construct a solution to a non-stationary dynamic problem for such a cylinder in the case when normal $P_{RR}(\theta, t)$ and tangential $P_{R\theta}(\theta, t)$ loads, constant along the generatrix, are applied to its lateral surface $R = R_N$ at time $t = 0$. In this case, the corresponding external load vector is orthogonal to the axis of the cylinder. Continuity conditions are assumed to be met at the contact of adjacent layers. Up to the moment $t = 0$, the cylinder was at rest. Thus, at each moment of time, a state of plane deformation is realized in the cylinder (Fig. 28.1).

We assume that the relations $P_{RR}(\theta,t) = P(t)Q_1(\theta), P_{R\theta}(\theta,t) = P(t)Q_2(\theta)$ are valid, where Q_1, Q_2 are dimensionless functions. Let us introduce dimensionless values $(n = 1, 2, \ldots, N)$:

$$u_1^{(n)}(r,\theta,\tau) = u_R^{(n)}/R_N,\;\; u_2^{(n)}(r,\theta,\tau) = u_\theta^{(n)}/R_N,$$
$$\sigma_{11}^{(n)}(r,\theta,\tau) = \sigma_{RR}^{(n)}/\left(2G_0^{(n)}\right),\quad \sigma_{22}^{(n)}(r,\theta,\tau) = \sigma_{\theta\theta}^{(n)}/\left(2G_0^{(n)}\right),$$
$$\sigma_{12}^{(n)}(r,\theta,\tau) = \sigma_{R\theta}^{(n)}/\left(2G_0^{(n)}\right),\quad P_0 f(\tau) = P(t)/\left(2G_0^{(N)}\right),$$
$$\tau = t/t_0,\quad r = R/R_N,\quad e_j^{(n)} = c_1^{(N)}/c_j^{(n)}\;\;(j = 1,2),$$
$$r_n = R_n/R_N\;\;(r_N = 1),\;\; r_0 = 0, \gamma_{\mathrm{v}}^{(n)}(\tau) = t_0 T_{\mathrm{v}}^{(n)},\quad \gamma_{\mathrm{s}}^{(n)}(\tau) = t_0 T_{\mathrm{s}}^{(n)},$$
$$f_1^{(n)}(\tau) = [(1+\nu_0^{(n)})\gamma_{\mathrm{v}}^{(n)}(\tau) + 2(1-2\nu_0^{(n)})\gamma_{\mathrm{s}}^{(n)}(\tau)]/[3(1-\nu_0^{(n)})],$$
$$f_2^{(n)}(\tau) = \gamma_{\mathrm{s}}^{(n)}(\tau),\; f_3^{(n)}(\tau) = [(1+\nu_0^{(n)})\gamma_{\mathrm{v}}^{(n)}(\tau) - (1-2\nu_0^{(n)})\gamma_{\mathrm{s}}^{(n)}(\tau)]/(3\nu_0^{(n)}),$$

where $t_0 = R_N/c_1^{(N)}$, and P_0 is a dimensionless constant; $u_R^{(n)}, u_\theta^{(n)}$ are the radial and circumferential displacements; $\sigma_{RR}^{(n)}, \sigma_{\theta\theta}^{(n)}, \sigma_{R\theta}^{(n)}$ are the corresponding physical components of the stress tensor. The superscript in parentheses for the quantities denotes the number of the layer to which they belong, while the dimensionless coordinates of the layers boundaries (as well as the dimensional ones) are provided with subscripts: $r_1, r_2, \ldots, r_N$.

We introduce dimensionless potentials $\varphi_1^{(n)}(r,\theta,\tau), \varphi_2^{(n)}(r,\theta,\tau)$ $(n = 1, 2, \ldots, N)$

$$u_1^{(n)}(r,\theta,\tau) = \frac{\partial\varphi_1^{(n)}}{\partial r} + \frac{1}{r}\frac{\partial\varphi_2^{(n)}}{\partial\theta},\quad u_2^{(n)}(r,\theta,\tau) = \frac{1}{r}\frac{\partial\varphi_1^{(n)}}{\partial\theta} - \frac{\partial\varphi_2^{(n)}}{\partial r}. \tag{28.1}$$

As a result, the equations of cylinder dynamics will take the form

$$(1 - \hat{f}_j^{(n)})\Delta\varphi_j^{(n)}(r,\theta,\tau) - (e_j^n)^2\frac{\partial^2\varphi_j^{(n)}(r,\theta,\tau)}{\partial\tau^2} = 0, \tag{28.2}$$

where

$$\Delta = \frac{1}{r}\frac{\partial}{\partial r}(r\frac{\partial}{\partial r}) + \frac{1}{r^2}\frac{\partial^2}{\partial\theta^2},\;\; j = 1,2,\;\; n = 1, 2, \ldots, N.$$

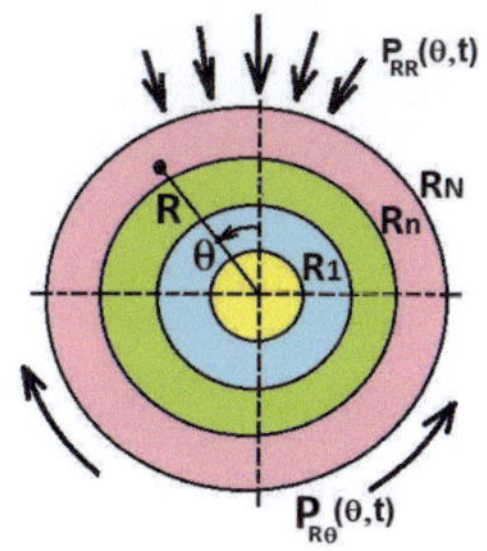

Fig. 28.1 Cylinder cross section.

A variable with a lid denotes an integral operator with a corresponding difference-type kernel corresponding to the constitutive relations of linear viscoelasticity in the Boltzmann-Volterra form, for example

$$\hat{f}_j^{(n)}\varphi(r,\theta,\tau) = \int_0^\tau f_j^{(n)}(\tau-\xi)\varphi(r,\theta,\xi)d\xi.$$

Taking into account the constitutive relations of linear viscoelasticity and the Cauchy kinematic relations, we present expressions of dimensionless stresses in terms of dimensionless displacements ($n = 1, 2, \ldots, N$):

$$\begin{aligned}
\sigma_{11}^{(n)}(r,\theta,\tau) &= w^{(n)}(1-\hat{f}_1^{(n)})\frac{\partial u_1^{(n)}}{\partial r} + \frac{w^{(n)}-1}{r}(1-\hat{f}_3^{(n)})\left(u_1^{(n)} + \frac{\partial u_2^{(n)}}{\partial\theta}\right),\\
\sigma_{22}^{(n)}(r,\theta,\tau) &= \frac{w^{(n)}}{r}(1-\hat{f}_1^{(n)})\left(u_1^{(n)} + \frac{\partial u_2^{(n)}}{\partial\theta}\right) + (w^{(n)}-1)(1-\hat{f}_3^{(n)})\frac{\partial u_1^{(n)}}{\partial r},\\
\sigma_{12}^{(n)}(r,\theta,\tau) &= \frac{1}{2}(1-\hat{f}_2^{(n)})\left(\frac{1}{r}\frac{\partial u_1^{(n)}}{\partial\theta} + \frac{\partial u_2^{(n)}}{\partial r} - \frac{u_2^{(n)}}{r}\right),
\end{aligned} \tag{28.3}$$

where $w^{(n)} = (1-\nu_0^{(n)})/(1-2\nu_0^{(n)}),\quad n = 1, 2, \ldots, N.$

Next, we will consider the case when the external load function has the form of a Heaviside function $f(\tau) = h(\tau)$. The boundary conditions at $r = 1$, have the form

$$\sigma_{1j}^{(N)}(1,\theta,\tau) = -P_0 h(\tau) Q_j(\theta),\quad j = 1, 2,\quad \tau > 0. \tag{28.4}$$

We assume the existence of finite limits of the displacements as r tends to zero:

$$\lim_{r\to 0} u_j^{(1)},\quad j = 1, 2. \tag{28.5}$$

for each value of $\tau \in (0; \infty)$.

The continuity conditions are fulfilled at the contact $r = r_\nu$ of the adjacent layers:

$$\begin{aligned}
&u_j^{(\nu)}(r_\nu,\theta,\tau) = u_j^{(\nu+1)}(r_\nu,\theta,\tau), \\
&G_0^{(\nu)}\sigma_{1j}^{(\nu)}(r_\nu,\theta,\tau) = G_0^{(\nu+1)}\sigma_{1j}^{(\nu+1)}(r_\nu,\theta,\tau),\quad \nu = 1, 2, \ldots, N-1,\quad j = 1, 2.
\end{aligned} \tag{28.6}$$

The initial conditions correspond to a state of rest:

$$\varphi_j^{(n)}(r,\theta,0) = \frac{\partial\varphi_j^{(n)}}{\partial\tau}(r,\theta,0) = 0,\quad j = 1, 2. \tag{28.7}$$

Decompose the functions $Q_j(\theta)$, $\varphi_j^{(n)}(r,\theta,\tau)$, $u_j^{(n)}(r,\theta,\tau)$, $\sigma_{kj}^{(n)}(r,\theta,\tau)$ into Fourier series by the angular coordinate θ $(i^2 = -1)$, $k, j = 1, 2$, $n = 1, 2, \ldots, N$:

$$Q_j = (-i)^{j-1} \sum_{m=0}^{\infty} Q_{jm} e^{im\theta}, \quad \varphi_j^{(n)}(r,\theta,\tau) = (-i)^{j-1} P_0 \sum_{m=0}^{\infty} \varphi_{jm}^{(n)}(r,\tau) e^{im\theta},$$

$$u_j^{(n)}(r,\theta,\tau) = (-i)^{j-1} P_0 \sum_{m=0}^{\infty} u_{jm}^{(n)}(r,\tau) e^{im\theta}, \tag{28.8}$$

$$\sigma_{kj}^{(n)}(r,\theta,\tau) = (-i)^{|k-j|} P_0 \sum_{m=0}^{\infty} \sigma_{kjm}^{(n)}(r,\tau) e^{im\theta}.$$

Herewith, the following relations are fulfilled ($n = 1, 2, .., N,\ m = 0, 1, 2, 3, ...$):

$$u_{1m}^{(n)}(r,\tau) = \frac{\partial \varphi_{1m}^{(n)}}{\partial r} + \frac{m}{r}\varphi_{2m}^{(n)}, \qquad u_{2m}^{(n)}(r,\tau) = \frac{-m}{r}\varphi_{1m}^{(n)} - \frac{\partial \varphi_{2m}^{(n)}}{\partial r},$$

$$\sigma_{11m}^{(n)}(r,\tau) = w^{(n)}(1 - \hat{f}_1^{(n)})\frac{\partial u_{1m}^{(n)}}{\partial r} + \frac{w^{(n)} - 1}{r}(1 - \hat{f}_3^{(n)})(u_{1m}^{(n)} + m u_{2m}^{(n)}),$$

$$\sigma_{22m}^{(n)}(r,\tau) = \frac{w^{(n)}}{r}(1 - \hat{f}_1^{(n)})(u_{1m}^{(n)} + m u_{2m}^{(n)}) + (w^{(n)} - 1)(1 - \hat{f}_3^{(n)})\frac{\partial u_{1m}^{(n)}}{\partial r},$$

$$\sigma_{12m}^{(n)}(r,\tau) = \frac{1}{2}(1 - \hat{f}_2^{(n)})\Big[\frac{\partial u_{2m}^{(n)}}{\partial r} - \frac{1}{r}(u_{2m}^{(n)} + m u_{1m}^{(n)})\Big]. \tag{28.9}$$

As a result, the problem (28.1) - (28.7) reduces to finding functions $\varphi_{1m}^{(n)}(r,\tau)$, $\varphi_{2m}^{(n)}(r,\tau)$, ($n = 1, 2, .., N$) for each $m = 0, 1, 2, 3, ...$ The formulation of the problem for finding these functions includes the equations:

$$(1 - \hat{f}_j^{(n)})\Big[\frac{\partial^2 \varphi_{jm}^{(n)}}{\partial r^2} + \frac{1}{r}\frac{\partial \varphi_{jm}^{(n)}}{\partial r} - \frac{m^2}{r^2}\varphi_{jm}^{(n)}\Big] - (e_j^{(n)})^2 \frac{\partial^2 \varphi_{jm}^{(n)}}{\partial \tau^2} = 0, \quad j = 1, 2, \tag{28.10}$$

and the boundary conditions at $r = 1$:

$$\sigma_{1jm}^{(N)}(1,\tau) = -h(\tau) Q_{jm}, \quad j = 1, 2, \tag{28.11}$$

the assumption of the existence of finite limits for the Fourier components of the displacements as r tends to zero:

$$\lim_{r \to 0} u_{jm}^{(1)}(r,\tau), \quad j = 1, 2 \tag{28.12}$$

for each value $\tau \in (0; \infty)$; conditions on the contact of adjacent layers:

$$u_{jm}^{(\nu)}(r_\nu,\tau) = u_{jm}^{(\nu+1)}(r_\nu,\tau), \tag{28.13}$$
$$G_0^{(\nu)}\sigma_{1jm}^{(\nu)}(r_\nu,\tau) = G_0^{(\nu+1)}\sigma_{1jm}^{(\nu+1)}(r_\nu,\tau), \quad \nu = 1, 2, \dots, N-1, \quad j = 1, 2$$

and initial conditions

$$\varphi_{jm}^{(n)}(r,0) = \frac{\partial \varphi_{jm}^{(n)}}{\partial \tau}(r,0) = 0, \quad j = 1,2. \tag{28.14}$$

Herewith, the relations (28.9) should be taken into account.

28.3 Research Method

Let us apply to the problem (28.9)–(28.13) the integral Laplace transform in time τ, denoting by $\Phi_{jm}^{(n)}(r,s)$, $U_{jm}^{(n)}(r,s)$, $S_{kjm}^{(n)}(r,s)$, $F_p^{(n)}(s)$ the images of the functions $\varphi_{jm}^{(n)}$, $u_{jm}^{(n)}$, $\sigma_{kjm}^{(n)}$, $f_p^{(n)}$ $(k,j = 1,2;\ p = 1,2,3;\ n = 1,2,\dots,N;\ m = 0,1,2,...)$ Taking into account conditions (28.14), we obtain the formulation of the problem in the image space for each $m = 0,1,2,\dots$ includes the equations with complex parameter s:

$$\frac{d^2\Phi_{jm}^{(n)}}{dr^2} + \frac{1}{r}\frac{d\Phi_{jm}^{(n)}}{dr} - \left[\frac{m^2}{r^2} + (\tilde{e}_j^{(n)})^2 s^2\right]\Phi_{jm}^{(n)} = 0, \quad j = 1,2, \tag{28.15}$$

$$\tilde{e}_j^{(n)} = e_j^{(n)}/\sqrt{1 - F_j^{(n)}(s)}$$

and boundary conditions at $r = 1$

$$S_{1jm}^{(N)}(1,s) = -Q_{jm}/s, \quad j = 1,2, \tag{28.16}$$

the assumption of the existence of finite limits for the Fourier components of the displacements in the images as r tends to zero:

$$\lim_{r\to 0} U_{jm}^{(1)}(r,s), \quad j = 1,2 \tag{28.17}$$

for each value s as well as the contact conditions of the layers $(\nu = 1,2,\dots,N-1)$:

$$U_{jm}^{(\nu)}(r_\nu,s) = U_{jm}^{(\nu+1)}(r_\nu,s); \tag{28.18}$$

$$G_0^{(\nu)}S_{1jm}^{(\nu)}(r_\nu,s) = G_0^{(\nu+1)}S_{1jm}^{(\nu+1)}(r_\nu,s), \quad j = 1,2.$$

Herewith, the following relations are fulfilled $(n = 1,2,\dots,N,\quad m = 0,1,2,3,...)$:

$$U_{1m}^{(n)} = \frac{d\Phi_{1m}^{(n)}}{dr} + \frac{m}{r}\Phi_{2m}^{(n)},$$

$$U_{2m}^{(n)} = \frac{-m}{r}\Phi_{1m}^{(n)} - \frac{d\Phi_{2m}^{(n)}}{dr},$$

$$S_{11m}^{(n)} = w_n(1 - F_1^{(n)})\frac{dU_{1m}^{(n)}}{dr} + \frac{w_n - 1}{r}(1 - F_3^{(n)})(U_{1m}^{(n)} + mU_{2m}^{(n)}), \tag{28.19}$$

$$S_{22m}^{(n)} = \frac{w_n}{r}(1 - F_1^{(n)})(U_{1m}^{(n)} + mU_{2m}^{(n)}) + (w_n - 1)(1 - F_3^{(n)})\frac{dU_{1m}^{(n)}}{dr},$$

$$S_{12m}^{(n)} = \frac{1}{2}(1 - F_2^{(n)})\left[\frac{dU_{2m}^{(n)}}{dr} - \frac{1}{r}(U_{2m}^{(n)} + mU_{1m}^{(n)})\right].$$

We find the functions $U_{1m}^{(n)}$ $U_{2m}^{(n)}$ by solving Eqs. (28.15) and using conditions (28.16)–(28.18), as well as relations (28.19). The general solution of Eqs. (28.15) has the form ($n = 1, 2, \ldots, N$)

$$\Phi_{jm}^{(n)}(r, s) = A_{jm}^{(n)}(s)I_m(y_j^{(n)}) + B_{jm}^{(n)}(s)K_m(y_j^{(n)}), y_j^{(n)} = rs\tilde{e}_j^{(n)}, j = 1, 2, \quad (28.20)$$

where I_m, K_m are the Bessel functions of the imaginary argument and the MacDonald functions of the index m, and the functions $A_{jm}^{(n)}$, $B_{jm}^{(n)}$ are determined from conditions (28.16)–(28.18).

According to relations (28.19), (28.20) as well as the properties of the functions of the Bessel functions of the imaginary argument and the MacDonald functions, to satisfy condition (28.17), we should take

$$B_{1m}^{(1)}(s) \equiv B_{2m}^{(1)}(s) \equiv 0, \quad m = 0, 1, 2, \ldots \quad (28.21)$$

The contact conditions (28.18), taking into account equalities (28.19), make it possible to find the relations between the sets

$$\{A_{1m}^{(\nu+1)}, B_{1m}^{(\nu+1)}, A_{2m}^{(\nu+1)}, B_{2m}^{(\nu+1)}\} \text{ and } \{A_{1m}^{(\nu)}, B_{1m}^{(\nu)}, A_{2m}^{(\nu)}, B_{2m}^{(\nu)}\} \ (\nu = 1, 2, .., N - 1)$$

for each $m = 0, 1, 2, \ldots$ After cumbersome mathematical transformations, these relations can be represented in the matrix form:

$$\mathbf{A}^{(\nu+1)(m)} = \frac{1}{\delta^{(\nu)}}\boldsymbol{\eta}^{(\nu)(m)} \cdot \mathbf{A}^{(\nu)(m)}, \quad \nu = 1, 2, .., N - 1. \quad (28.22)$$

The following designations are introduced here:

$$\delta^{(\nu)} = -r_\nu(\tilde{e}_2^{(\nu+1)})^2 s/2,$$

$$\mathbf{A}^{(n)(m)} = \begin{pmatrix} A_{1m}^{(n)} \\ B_{1m}^{(n)} \\ A_{2m}^{(n)} \\ B_{2m}^{(n)} \end{pmatrix}, \quad n = 1, 2, \ldots N; \quad \boldsymbol{\eta}^{(\nu)(m)} = \left(\eta_{ql}^{(\nu)(m)}\right), \quad q, l = 1, 2, 3, 4.$$

Hereafter, the index m, corresponding to the number of the Fourier component, for some quantities is located not at the bottom, but at the top in parentheses after the index corresponding to the number of the cylinder layer. The elements of the matrices $\boldsymbol{\eta}^{(\nu)(m)}$ have the following form ($\nu = 1, 2, .., N - 1$, $m = 0, 1, 2, \ldots$):

$$\eta_{11}^{(\nu)(m)} = \tilde{e}_1^{(\nu+1)}\{K_{m+1}(b_1'^{(\nu)})[g_1^{(\nu)} b_1^{(\nu)} I_{m+1}(b_1^{(\nu)}) + g_2^{(\nu)(m)} I_m(b_1^{(\nu)})] +$$
$$+\frac{K_m(b_1'^{(\nu)})}{b_1'^{(\nu)}}[g_3^{(\nu)(m)} b_1^{(\nu)} I_{m+1}(b_1^{(\nu)}) + g_4^{(\nu)(m)} I_m(b_1^{(\nu)})]\},$$
$$\eta_{12}^{(\nu)(m)} = \tilde{e}_1^{(\nu+1)}\{K_{m+1}(b_1'^{(\nu)})[g_1^{(\nu)} b_1^{(\nu)} K_{m+1}(b_1^{(\nu)}) + g_2^{(\nu)(m)} K_m(b_1^{(\nu)})] +$$
$$+\frac{K_m(b_1'^{(\nu)})}{b_1'^{(\nu)}}[g_3^{(\nu)(m)} b_1^{(\nu)} K_{m+1}(b_1^{(\nu)}) + g_4^{(\nu)(m)} K_m(b_1^{(\nu)})]\}, \tag{28.23}$$
$$\eta_{21}^{(\nu)(m)} = \tilde{e}_1^{(\nu+1)}\{I_{m+1}(b_1'^{(\nu)})[g_1^{(\nu)} b_1^{(\nu)} I_{m+1}(b_1^{(\nu)}) + g_2^{(\nu)(m)} I_m(b_1^{(\nu)})] -$$
$$-\frac{I_m(b_1'^{(\nu)})}{b_1'^{(\nu)}}[g_3^{(\nu)(m)} b_1^{(\nu)} I_{m+1}(b_1^{(\nu)}) + g_4^{(\nu)(m)} I_m(b_1^{(\nu)})]\},$$
$$\eta_{22}^{(\nu)(m)} = \tilde{e}_1^{(\nu+1)}\{I_{m+1}(b_1'^{(\nu)})[g_1^{(\nu)} b_1^{(\nu)} K_{m+1}(b_1^{(\nu)}) + g_2^{(\nu)(m)} K_m(b_1^{(\nu)})] -$$
$$-\frac{I_m(b_1'^{(\nu)})}{b_1'^{(\nu)}}[g_3^{(\nu)(m)} b_1^{(\nu)} K_{m+1}(b_1^{(\nu)}) + g_4^{(\nu)(m)} K_m(b_1^{(\nu)})]\},$$

where denoted

$$\begin{aligned} g_1^{(\nu)} &= g^{(\nu)} - 1, \quad g_2^{(\nu)(m)} = m(m-1)g_1^{(\nu)} - 0,5g^{(\nu)}(b_2^{(\nu)})^2, \\ g_3^{(\nu)(m)} &= m(m-1)g_1^{(\nu)} - 0,5(b_2'^{(\nu)})^2, \; g_4^{(\nu)(m)} = (g_3^{(\nu)(m)} - g_2^{(\nu)(m)})m, \\ g^{(\nu)} &= G_0^{(\nu)}(1-F_2^{(\nu)})/[G_0^{(\nu+1)}(1-F_2^{(\nu+1)})], \\ b_j^{(\nu)} &= r_\nu \tilde{e}_j^{(\nu)} s, \quad b_j'^{(\nu)} = r_\nu \tilde{e}_j^{(\nu+1)} s, \quad (j=1,2). \end{aligned} \tag{28.24}$$

We also introduce the notations:

$$\begin{aligned} g_5^{(\nu)} &= -g_1^{(\nu)}, \qquad g_6^{(\nu)(m)} = (1-m)g_1^{(\nu)}, \\ g_7^{(\nu)(m)} &= -g_6^{(\nu)(m)}, \; g_8^{(\nu)(m)} = g_3^{(\nu)(m)} - g_2^{(\nu)(m)}. \end{aligned} \tag{28.25}$$

The formula for the element $\eta_{k\; j+2}^{(\nu)(m)}$ $(k, j = 1, 2)$ is obtained from the corresponding formula (28.23) for the element $\eta_{kj}^{(\nu)(m)}$ by simultaneously formally replacing $\tilde{e}_1^{(\nu+1)}$ with $m\tilde{e}_1^{(\nu+1)}$, as well as $g_1^{(\nu)}$ with $g_5^{(\nu)}$, $g_l^{(\nu)(m)}$ with $g_{l+4}^{(\nu)(m)}$ $(l = 2, 3, 4)$, and $b_1^{(\nu)}$ with $b_2^{(\nu)}$. The formula for the element $\eta_{k+2\; j}^{(\nu)(m)}$ $(k, j = 1, 2)$ is obtained from the corresponding formula (28.23) for the element $\eta_{kj}^{(\nu)(m)}$ by simultaneously formally replacing $\tilde{e}_1^{(\nu+1)}$ with $m\tilde{e}_2^{(\nu+1)}$, as well as $g_1^{(\nu)}$ with $g_5^{(\nu)}$, $g_l^{(\nu)(m)}$ with $g_{l+4}^{(\nu)(m)}$ $(l = 2, 3, 4)$, and $b_1'^{(\nu)}$ with $b_2'^{(\nu)}$. The formula for the element $\eta_{k+2\; j+2}^{(\nu)(m)}(k, j = 1, 2)$ is obtained from the corresponding formula (28.23) for the element $\eta_{kj}^{(\nu)(m)}$ by simultaneously formally replacing $\tilde{e}_1^{(\nu+1)}$ with $\tilde{e}_2^{(\nu+1)}$, as well as $b_1^{(\nu)}$ with $b_2^{(\nu)}$, and $b_1'^{(\nu)}$ with $b_2'^{(\nu)}$.

Based on the relation (28.22), the column vector $\mathbf{A}^{(n)(m)}$ is expressed in terms of the column vector $\mathbf{A}^{(1)(m)}$ using the formula

$$\mathbf{A}^{(n)(m)} = \lambda^{(n)}\mathbf{H}^{(n)(m)} \cdot \mathbf{A}^{(1)(m)}, \quad n = 1, 2, \ldots, N, \quad m = 0, 1, 2, \ldots,$$

where

$$\lambda^{(1)} = 1; \ \lambda^{(n)} = (\delta^{(1)}\delta^{(2)} \ldots \delta^{(n-1)})^{-1} \ (n = 2, 3, \ldots, N),$$

herewith, the matrix $\mathbf{H}^{(n)(m)}$ for $n = 1$ is a unit of size 4x4, and for $n \geq 2$ it is determined by the product of the matrices

$$\mathbf{H}^{(n)(m)} = \boldsymbol{\eta}^{(n-1)(m)} \cdot \boldsymbol{\eta}^{(n-2)(m)} \cdot \ldots \cdot \boldsymbol{\eta}^{(1)(m)}.$$

According to conditions (28.21), the second and fourth elements of the column vector $\mathbf{A}^{(1)(m)}$ are zero, so in addition to the matrix $\mathbf{H}^{(n)(m)}$ of size 4×4, it is convenient to consider the matrix

$$\boldsymbol{\xi}^{(n)(m)} = \left(\xi_{lj}^{(n)(m)}\right), \quad l = 1, 2, 3, 4, \ j = 1, 2$$

of size 4×2. Herewith, we can write the formula

$$\mathbf{A}^{(n)(m)} = \lambda^{(n)}\boldsymbol{\xi}^{(n)(m)} \cdot \begin{pmatrix} A_{1m}^{(1)} \\ A_{2m}^{(1)} \end{pmatrix}, \quad n = 1, 2, \ldots, N, \ m = 0, 1, 2, \ldots,$$

where the following recurrence relation holds for the matrix $\boldsymbol{\xi}^{(n)(m)}$:

$$\boldsymbol{\xi}^{(1)(m)} = \begin{pmatrix} 1 \, 0 \\ 0 \, 0 \\ 0 \, 1 \\ 0 \, 0 \end{pmatrix}, \ \boldsymbol{\xi}^{(\nu+1)(m)} = \boldsymbol{\eta}^{(\nu)(m)} \cdot \boldsymbol{\xi}^{(\nu)(m)}, \ \nu = 1, 2, \ldots, N-1. \quad (28.26)$$

Using the two boundary conditions (28.16), taking into account the relations (28.19), (28.20), we find for each $m = 0, 1, 2, ..$ both functions of $A_{1m}^{(1)}(s)$, $A_{2m}^{(1)}(s)$. As a result, omitting intermediate mathematical transformations, we write expressions for images $U_{jm}^{(n)}(r, s)$ $(j = 1, 2, \ n = 1, 2, \ldots, N, \ m = 0, 1, 2, \ldots)$ in the following form:

$$\begin{pmatrix} U_{1m}^{(n)}(r, s) \\ U_{2m}^{(n)}(r, s) \end{pmatrix} = \frac{\Pi^{(n)}}{s^2 Z_m (1 - F_2^{(N)})} \begin{pmatrix} z_{11}^{(n)(m)} & z_{12}^{(n)(m)} \\ z_{21}^{(n)(m)} & z_{22}^{(n)(m)} \end{pmatrix} \cdot \begin{pmatrix} Q_{1m} \\ Q_{2m} \end{pmatrix}, \quad (28.27)$$

where denoted

$$\Pi^{(n)} = \delta^{(n)}\delta^{(n+1)} \cdot \ldots \cdot \delta^{(N-1)} \ (n \leq N-1), \ \Pi^{(N)} \equiv 1,$$

$$Z_m(s) = \det \mathbf{D}^{(m)}, \ \mathbf{D}^{(m)} = \left(D_{kj}^{(m)}(s)\right) = \boldsymbol{\Lambda}^{(m)} \cdot \boldsymbol{\xi}^{(N)(m)},$$

$$\boldsymbol{\Lambda}^{(m)} = \left(\Lambda_{kl}^{(m)}(s)\right), \ k, j = 1, 2, \ l = 1, 2, 3, 4,$$

$$z_{j1}^{(n)(m)}(r,s) = \det\begin{pmatrix} C_{j1}^{(n)(m)} & C_{j2}^{(n)(m)} \\ D_{21}^{(m)} & D_{22}^{(m)} \end{pmatrix},$$

$$z_{j2}^{(n)(m)}(r,s) = \det\begin{pmatrix} D_{11}^{(m)} & D_{12}^{(m)} \\ C_{j1}^{(n)(m)} & C_{j2}^{(n)(m)} \end{pmatrix},$$

herewith

$$\mathbf{C}^{(n)(m)} = \left(C_{kj}^{(n)(m)}(r,s)\right) = \mathbf{\Gamma}^{(n)(m)} \cdot \boldsymbol{\xi}^{(n)(m)},$$

$$\mathbf{\Gamma}^{(n)(m)} = \left(\Gamma_{kl}^{(n)(m)}(r,s)\right), \quad k,j = 1,2, \quad l = 1,2,3,4,$$

the elements of the matrices $\mathbf{\Lambda}^{(m)}$ and $\mathbf{\Gamma}^{(n)(m)}$ have the form:

$$\begin{aligned}
\Lambda_{11}^{(m)}(s) &= I_m(\tilde{e}_1^{(N)}s)[\frac{(\tilde{e}_2^{(N)})^2}{2} + \frac{m(m-1)}{s^2}] - \frac{\tilde{e}_1^{(N)}}{s} I_{m+1}(\tilde{e}_1^{(N)}s), \\
\Lambda_{13}^{(m)}(s) &= \frac{m}{s^2}[\tilde{e}_2^{(N)} s I_{m+1}(\tilde{e}_2^{(N)}s) + (m-1) I_m(\tilde{e}_2^{(N)}s)], \\
\Lambda_{21}^{(m)}(s) &= -\frac{m}{s^2}[\tilde{e}_1^{(N)} s I_{m+1}(\tilde{e}_1^{(N)}s) + (m-1) I_m(\tilde{e}_1^{(N)}s)], \\
\Lambda_{23}^{(m)}(s) &= -I_m(\tilde{e}_2^{(N)}s)[\frac{(\tilde{e}_2^{(N)})^2}{2} + \frac{m(m-1)}{s^2}] + \frac{\tilde{e}_2^{(N)}}{s} I_{m+1}(\tilde{e}_2^{(N)}s), \qquad (28.28) \\
\Gamma_{11}^{(n)(m)}(r,s) &= \tilde{e}_1^{(n)}[I_{m+1}(y_1^{(n)}) + \frac{m}{y_1^{(n)}} I_m(y_1^{(n)})], \\
\Gamma_{13}^{(n)(m)}(r,s) &= \frac{m}{rs} I_m(y_2^{(n)}), \\
\Gamma_{21}^{(n)(m)}(r,s) &= -\frac{m}{rs} I_m(y_1^{(n)}), \\
\Gamma_{23}^{(n)(m)}(r,s) &= -\tilde{e}_2^{(n)}[I_{m+1}(y_2^{(n)}) + \frac{m}{y_2^{(n)}} I_m(y_2^{(n)})].
\end{aligned}$$

Formulas for the elements $\Lambda_{j2}^{(m)}$, $\Lambda_{j4}^{(m)}$, $\Gamma_{j2}^{(n)(m)}$, $\Gamma_{j4}^{(n)(m)}$ are obtained from formulas (28.28) for the corresponding elements $\Lambda_{j1}^{(m)}$, $\Lambda_{j3}^{(m)}$, $\Gamma_{j1}^{(n)(m)}$, $\Gamma_{j3}^{(n)(m)}$ $(j = 1,2)$ by formally replacing the Bessel functions of the imaginary argument I_m and I_{m+1} with the MacDonald functions K_m and K_{m+1} with the same arguments.

Thus, the images $U_{jm}^{(n)}(r,s)$ are determined by formulas (28.23) – (28.28). Let us formally consider the functions $U_{jm}^{(n)}$ on the entire complex plane C. Analysis of the formulas for these functions shows that each of the functions $U_{jm}^{(n)}(r,s)$ for each r, $0 \le r \le 1$ has the following singular points on the complex plane C: point $s = 0$ and the roots $s = s_{km}$ $(k = 1,2,..,\ m = 0,1,2,..)$ of the characteristic equation

$$Z_m(s) = 0. \qquad (28.29)$$

Besides, the the roots of the equations

$$1 - F_1^{(n)}(s) = 0, \quad 1 - F_2^{(n)}(s) = 0 \tag{28.30}$$

can be the limiting points of the set $E_m = \{s_{km}\}$ of roots of the characteristic equation (28.29). The roots of the equation (28.29) for each $m = 0, 1, 2, ...$ are the eigenvalues of the spectral problem of free oscillations of the considered layered viscoelastic cylinder with a free lateral surface. Their set E_m is countable.

Let us consider the following options for describing the hereditary properties of the materials of the cylinder layers.

Case 1

Hereditary kernels $\gamma_{\mathrm{v}}^{(n)}(\tau)$ $\gamma_{\mathrm{s}}^{(n)}(\tau)$ (if they are nonzero) belong to the class of functions that are finite sums of exponentials

$$\sum_{k=1}^{K} a_k \mathrm{e}^{-b_k \tau}, \quad \sum_{k=1}^{K} a_k / b_k < 1, \quad a_k > 0, \quad b_k > 0 \quad (k = 1, 2, ..., K). \tag{28.31}$$

Case 2

$\gamma_{\mathrm{v}}^{(n)} \equiv 0$ for every n, and all non-identically zero $\gamma_{\mathrm{s}}^{(n)}$ are Rzhanitsyn-Koltunov kernels

$$A\mathrm{e}^{-B\tau}\tau^{\delta-1} \quad \text{where} \quad A\Gamma(\delta)/B^{\delta} < 1, \quad A > 0, \quad B > 0, \quad 0 < \delta < 1, \tag{28.32}$$

$\Gamma(\delta)$ is a gamma-function with an appropriate argument. The constants a_k, b_k and K, as well as the constants A, B, δ for each layer of a cylinder, are, generally speaking, unique, so hereditary properties in different layers can differ significantly. Moreover, in Case 1, within the same layer, the kernels $\gamma_{\mathrm{v}}^{(n)}(\tau)$ and $\gamma_{\mathrm{s}}^{(n)}(\tau)$ may differ from each other, which means that the Poisson's ratio of the layer material is inconstant over time (In the Case 2 this is obvious). The first of inequalities (28.31) and the first of inequalities (28.32) correspond to the condition of limited creep of the material of all viscoelastic layers.

The paper [15] contains the results of general theoretical studies of non-stationary viscoelasticity problem for piecewise homogeneous bodies based on the application of the integral Laplace transform in time and the subsequent construction of a solution in the originals. It was established the relationship between the properties of the eigenvalues set of the free oscillations problem for the piecewise homogeneous body and the branching points of the solution of the non–stationary problem in the space of Laplace images, as well as the relationship between the elements of the eigenvalues set and the poles of the solution in the images. For relaxation kernels corresponding to Cases 1 and 2 (formulas (28.31) and (28.32)), new forms of the solution to the linear viscoelasticity problem for a piecewise homogeneous bodies are presented in [18].

In the first case, the images of the relaxation kernels have no branching points in the complex plane. In the second case, the image of each of the shear relaxation kernels has exactly one branching point on the negative part of the real axis. The image of the external load function $f(\tau) = h(\tau)$ has no branching points. Then, according to the statement proved in the work [15], the images $U_{jm}^{(n)}$ for any r, $0 \le r \le 1$ in the Case 1 have no branching points on the complex plain, and in the Case 2, such points are only the branching points of the images of the shear relaxation kernels (despite the presence in formulas (28.27), (28.28) of functions with square root signs, as well as MacDonald functions with a logarithmic singularity). Note that this fact can also be directly detected using the properties of the Bessel functions of the imaginary argument and the MacDonald functions after a series of cumbersome mathematical transformations.

In Case 1, both Eqs. (28.30) for each n are reduced to polynomial equations, all of whose roots are simple real negative ones. In Case 2, for each n, both Eqs. (28.30) have one real negative root.

Thus, the set of poles of the images of the mth Fourier components $U_{jm}^{(n)}(r, s)$ is a union $E_m \cup \{0\}$. Moreover, E_m may have finite limit points on the negative part of the real axis, corresponding to the roots of Eqs. (28.30). The method for finding elements of the spectral set E_m was discussed in [16, 17]. In the vicinity of the finite limit points E_m, these values are sought on the real axis. The complex elements of E_m can be calculated by iterative methods, taking as an initial approximation, for example, the purely imaginary elements of the corresponding spectral set for an elastic piecewise homogeneous cylinder.

An examination of the formulas (28.27) shows that the functions $U_{jm}^{(n)}$ satisfy sufficient conditions [14] for "favorable" asymptotic behavior as s tends to infinity. The "favorable" asymptotic behavior of $U_{jm}^{(n)}$ as $s \to \infty$ means the existence on the complex plane of an infinite sequence of continuous curves $\ell_1, \ell_2, \ell_3, \ldots$ closing to the left the Bromwich contour so that the distance from the point $s = 0$ to ℓ_k increases unboundedly as $k \to \infty$, and, furthermore,

$$\lim_{k \to \infty} \int_{\ell_k} U_{jm}^{(n)}(r, s) e^{s\tau} ds = 0.$$

Note that the conditions of the problem considered here allow the possibility of moving the layered cylinder as a rigid whole. This occurs because the Fourier components Q_{20}, Q_{11}, Q_{21} of the functions $Q_1(\theta)$, $Q_2(\theta)$, which determine the dependence of the normal and tangential external loads on the angle (28.4), do not correspond to the self-equilibrated loading of the cylinder. Displacement as a rigid whole can occur in the form of axisymmetric rotation, corresponding to the Fourier component of the circumferential displacement $u_{20}^{(n)}$ (for $m = 0$ and $Q_{20} \ne 0$), as well as in the form of translational displacement determined by the projections of the Fourier components $u_{11}^{(n)}$, $u_{21}^{(n)}$ (for $m = 1$ and $Q_{11} \ne 0$ or $Q_{21} \ne 0$) onto directions parallel or perpendicular to the polar axis. The general forms of the solution in the originals, presented in [18], are preserved with a slight remark. Now, the residues of the functions $U_{20}^{(n)} e^{s\tau}, U_{j1}^{(n)} e^{s\tau}$ $(j = 1, 2)$ at $s = 0$ will be an expressions that include the time-dependent quantities that determine the displacement of the cylinder as a rigid whole. Under the condition $Q_{20} = Q_{11} = Q_{21} = 0$ the external load is self-balanced and there will be no displacement as a rigid whole. In this paper, we will use the form of the solution in the originals [18], which is suitable for both choices of relaxation kernels (28.31), (28.32). Thus, in accordance with the results of works [15, 18], the originals $u_{jm}^{(n)}(r, \tau)$ for both Case 1 and Case 2 can be represented in the following form ($n = 1, 2, ..., N;\ j = 1, 2;\ m = 0, 1, 2, 3, ...$):

$$u_{jm}^{(n)}(r,\tau) = \mathrm{Res}_{s=0}[U_{jm}^{(n)}(r,s)\mathrm{e}^{s\tau}] + 2\sum_{k=1}^{\infty}\mathrm{Re}\{\mathrm{Res}_{s=s_{km}}[U_{jm}^{(n)}(r,s)\mathrm{e}^{s\tau}]\} +$$

$$+\frac{1}{\pi}\mathrm{e}^{\alpha_{0m}\tau}\int_{0}^{\omega_{0m}}\mathrm{Re}\{U_{jm}^{(n)}(r,\alpha_{0m}+i\omega)\mathrm{e}^{i\omega\tau}\}d\omega - \tag{28.33}$$

$$-\frac{1}{\pi}\int_{-\infty}^{\alpha_{0m}}\mathrm{Im}\{U_{jm}^{(n)}(r,\alpha+i\omega_{0m})\mathrm{e}^{i\omega_{0m}\tau}\}\mathrm{e}^{\alpha\tau}d\alpha.$$

Here $s_{km} = \alpha_{km} + i\omega_{km}$ $(\alpha_{km},\ \omega_{km} \in R)$ are the complex poles of the images $U_{jm}^{(n)}$, i.e. the roots of Eqs. (28.29) for $\omega_{km} > 0$; note that all α_{km} are negative. The values α_{0m}, ω_{0m} are chosen arbitrarily on the intervals $s^* < \alpha_{0m} < 0$ and $0 < \omega_{0m} < \omega_{1m} = \mathrm{Im}(s_{1m})$. Herewith s_{1m} is a complex root of Eqs. (28.29) with the least positive imaginary part; s^* is the largest of the roots of Eqs. (28.30) (Fig. 28.2 explains the integration process). In expression (28.33) the integral over the segment is multiplied by a function tending to zero as τ increases ($\alpha_{0m} < 0$). The multiplier $e^{\alpha\tau}$ under the sign of improper integral is not a rapidly oscillating function of α for any τ; with time, the convergence of this integral improves and the integral itself tends to zero. This is an essential advantage of formula (28.33) compared to the one presented in [15] and including the improper integral over the imaginary axis. The formula (28.33) is especially important when the number of limit points in the spectrum E_m is large and finding the real elements of the spectrum from their neighborhoods requires a high accuracy of computations. The series in terms of residues in (28.33) represents the spectral decomposition over a subset of the eigenforms of the cylinder's free oscillations corresponding only to complex

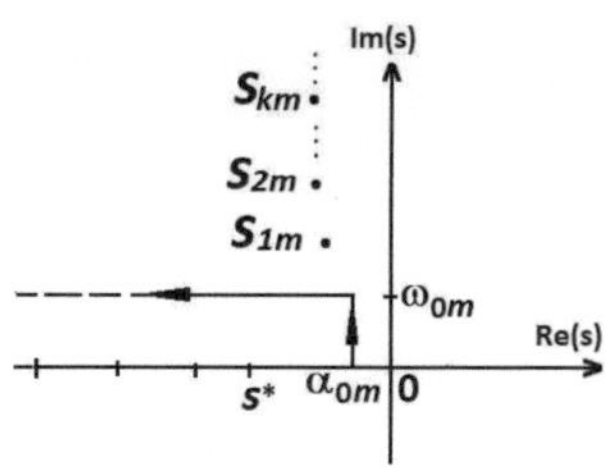

Fig. 28.2 Illustration to the formula (28.33).

spectral elements with a nonzero imaginary part. The formula (28.33) is essentially a modification of the spectral decomposition in nonstationary viscoelasticity problems.

So, if the load function $f(\tau) = h(\tau)$ is a Heaviside function, and the hereditary properties of the cylinder layers correspond to Cases 1 or 2, then the solution of the non-stationary problem (28.1) – (28.7) is determined by relations (28.8), (28.23) – (28.28) and formula (28.33). Fourier components of stresses $\sigma_{qjm}^{(n)}$ are expressed in terms of its images $S_{qjm}^{(n)}$ (28.19) similarly to (28.33). For other functions $f(\tau)$, the solution is obtained on the basis of the one presented here using the Duhamel integral.

Based on the constructed solution, it is possible to obtain a solution to the corresponding non-stationary problem for a layered cylinder with a free coaxial cavity. To do this, we must make instantaneous shear modulus $G_0^{(1)}$, velocities of longitudinal and transverse elastic waves $c_1^{(n)}$, $c_2^{(n)}$ and the density in the first (inner) layer of the cylinder tend to zero and herewith consider the solution of the non-stationary problem for all layers except the first. Then r_1 will be the radius of the cavity, and the number of viscoelastic layers will be $N - 1$.

28.4 Special Case. Layered Elastic Cylinder

Let us consider the case when the hereditary properties of the layers are not taken into account and all the layers of the cylinder are linearly elastic. The external load function is still a Heaviside function: $f(\tau) = h(\tau)$. For $\gamma_{\text{v}}^{(n)} \equiv \gamma_{\text{s}}^{(n)} \equiv 0$, the roots of the characteristic equations (28.29) are purely imaginary: $s_{km} = i\omega_{km}$, $\omega_{km} \in R$, and there are no integrals in formula (28.33). This problem for an elastic layered cylinder was considered in [5], where its solution was presented in a very concise form, and some components of the solution were not written out explicitly. Here, the solution is presented in sufficient detail. The originals $u_{jm}^{(n)}(r, \tau)$ are decompositions into series by residues and, after appropriate calculations, have the following form ($n = 1, 2, .., N,\ \ m = 0, 1, 2, 3, ...$):

$$
\begin{aligned}
u_{1m}^{(n)}(r,\tau) &= Y_{1m}^{(n)(0)}(r) + \delta_m^1 \Omega_{11}\tau^2 + p^{(n)} \sum_{k=1}^{\infty} Y_{1m}^{(n)}(r, \omega_{km}) \cos(\omega_{km}\tau), \\
u_{2m}^{(n)}(r,\tau) &= Y_{2m}^{(n)(0)}(r) + [\delta_m^1 \Omega_{21} + \delta_{m+1}^1 \Omega_{20} r]\tau^2 + \\
&\quad + p^{(n)} \sum_{k=1}^{\infty} Y_{2m}^{(n)}(r, \omega_{km}) \cos(\omega_{km}\tau),
\end{aligned}
\tag{28.34}
$$

where denoted:

$\delta^1_m, \delta^1_{m+1}$ – Kronecker symbols,

$$p^{(n)} = 2\pi^{N-n} \prod_{\lambda=n}^{N-1} r_\lambda (e_2^{(\lambda+1)})^2 \ \ (n \le N-1), \ \ p^{(N)} \equiv 2,$$

$$Y^{(n)}_{jm}(r,\omega) = \omega^{N-n-2}[Q_{1m}X^{(n)(m)}_{j1}(r,\omega) + Q_{2m}X^{(n)(m)}_{j2}(r,\omega)] \Big/ \left(\frac{dX_m(\omega)}{d\omega}\right),$$

$$X^{(n)(m)}_{j1}(r,\omega) = \det\begin{pmatrix} d^{(n)(m)}_{j1} & d^{(n)(m)}_{j2} \\ V^{(m)}_{21} & V^{(m)}_{22} \end{pmatrix},$$

$$X^{(n)(m)}_{j2}(r,\omega) = \det\begin{pmatrix} V^{(m)}_{11} & V^{(m)}_{12} \\ d^{(n)(m)}_{j1} & d^{(n)(m)}_{j2} \end{pmatrix},$$

$$X_m(\omega) = \det \mathbf{V}^{(m)}, \ \ \mathbf{V}^{(m)} = \left(V^{(m)}_{qj}\right) = \mathbf{L}^{(m)}\chi^{(N)(m)},$$

$$\mathbf{L}^{(m)} = \left(L^{(m)}_{ql}\right), \ \ \chi^{(n)(m)} = \left(\chi^{(n)(m)}_{lj}\right),$$

$$j,q = 1,2, \ \ l = 1,2,3,4,$$

herewith

$$\mathbf{d}^{(n)(m)} = \left(d^{(n)(m)}_{qj}(r,\omega)\right) = \mathbf{K}^{(n)(m)}\chi^{(n)(m)},$$

$$\mathbf{K}^{(n)(m)} = \left(K^{(n)(m)}_{ql}(r,\omega)\right),$$

$$q,j = 1,2, \ \ l = 1,2,3,4.$$

The elements of the matrices $\mathbf{L}^{(m)}(\omega)$, $\mathbf{K}^{(n)(m)}(r,\omega)$ have the form:

$$\begin{aligned}
&L^{(m)}_{11}(\omega) = J_m(\omega)\Big[\frac{(e_2^{(N)})^2}{2} + \frac{m(1-m)}{\omega^2}\Big] - \frac{J_{m+1}(\omega)}{\omega};\\
&L^{(m)}_{13}(\omega) = \frac{m}{\omega^2}[(1-m)J_m(e_2^{(N)}\omega) + e_2^{(N)}\omega J_{m+1}(e_2^{(N)}\omega)];\\
&L^{(m)}_{21}(\omega) = -\frac{m}{\omega^2}[(1-m)J_m(\omega) + \omega J_{m+1}(\omega)];\\
&L^{(m)}_{23}(\omega) = -J_m(e_2^{(N)}\omega)\Big[\frac{(e_2^{(N)})^2}{2} + \frac{m(1-m)}{\omega^2}\Big] + \frac{e_2^{(N)}}{\omega}J_{m+1}(e_2^{(N)}\omega);\\
&K^{(n)(m)}_{11}(r,\omega) = e_1^{(n)}\Big[\frac{m}{x_1^{(n)}}J_m(x_1^{(n)}) - J_{m+1}(x_1^{(n)})\Big];\\
&K^{(n)(m)}_{13}(r,\omega) = \frac{me_2^{(n)}}{x_2^{(n)}}J_m(x_2^{(n)});\\
&K^{(n)(m)}_{21}(r,\omega) = -\frac{me_1^{(n)}}{x_1^{(n)}}J_m(x_1^{(n)});\\
&K^{(n)(m)}_{23}(r,\omega) = e_2^{(n)}\Big[J_{m+1}(x_2^{(n)}) - \tfrac{m}{x_2^{(n)}}J_m(x_2^{(n)})\Big],\\
&x_j^{(n)} = r\omega e_j^{(n)} \ \ (j=1,2).
\end{aligned} \tag{28.35}$$

Formulas for the elements $L_{j2}^{(m)}$, $L_{j4}^{(m)}$, $K_{j2}^{(n)(m)}$, $K_{j4}^{(n)(m)}$ are obtained from formulas (28.35) for $L_{j1}^{(m)}$, $L_{j3}^{(m)}$, $K_{j1}^{(n)(m)}$, $K_{j3}^{(n)(m)}$ $(j = 1, 2)$ by formally replacing Bessel functions of the 1st kind J_m J_{m+1}, respectively, with Bessel functions of the 2nd kind (Neumann functions) N_m, N_{m+1} with the same arguments.

The recurrence relation is performed for the matrix $\chi^{(n)(m)}$:

$$\chi^{(1)(m)} = \begin{pmatrix} 1\,0 \\ 0\,0 \\ 0\,1 \\ 0\,0 \end{pmatrix}, \qquad \chi^{(\nu+1)(m)} = \mathbf{h}^{(\nu)(m)} \chi^{(\nu)(m)}, \quad \nu = 1, 2, \ldots, N-1,$$

where the elements of the matrix $\mathbf{h}^{(\nu)(m)} = \left(h_{pl}^{(\nu)(m)} \right)$, $p, l = 1, 2, 3, 4$ have the following form ($\nu = 1, 2, .., N-1$, $m = 0, 1, 2, \ldots$):

$$\begin{aligned}
h_{11}^{(\nu)(m)} &= e_1^{(\nu+1)} \{ N_{m+1}(\beta_1'^{(\nu)}) [a_1^{(\nu)} \beta_1^{(\nu)} J_{m+1}(\beta_1^{(\nu)}) - a_2^{(\nu)(m)} J_m(\beta_1^{(\nu)})] + \\
&+ \frac{N_m(\beta_1'^{(\nu)})}{\beta_1'^{(\nu)}} [a_3^{(\nu)(m)} \beta_1^{(\nu)} J_{m+1}(\beta_1^{(\nu)}) - a_4^{(\nu)(m)} J_m(\beta_1^{(\nu)})] \}, \\
h_{12}^{(\nu)(m)} &= e_1^{(\nu+1)} \{ N_{m+1}(\beta_1'^{(\nu)}) [a_1^{(\nu)} \beta_1^{(\nu)} N_{m+1}(\beta_1^{(\nu)}) - a_2^{(\nu)(m)} N_m(\beta_1^{(\nu)})] + \\
&+ \frac{N_m(\beta_1'^{(\nu)})}{\beta_1'^{(\nu)}} [a_3^{(\nu)(m)} \beta_1^{(\nu)} N_{m+1}(\beta_1^{(\nu)}) - a_4^{(\nu)(m)} N_m(\beta_1^{(\nu)})] \}, \\
h_{21}^{(\nu)(m)} &= -e_1^{(\nu+1)} \{ J_{m+1}(\beta_1'^{(\nu)}) [a_1^{(\nu)} \beta_1^{(\nu)} J_{m+1}(\beta_1^{(\nu)}) - a_2^{(\nu)(m)} J_m(\beta_1^{(\nu)})] + \\
&+ \frac{J_m(\beta_1'^{(\nu)})}{\beta_1'^{(\nu)}} [a_3^{(\nu)(m)} \beta_1^{(\nu)} J_{m+1}(\beta_1^{(\nu)}) - a_4^{(\nu)(m)} J_m(\beta_1^{(\nu)})] \}, \\
h_{22}^{(\nu)(m)} &= -e_1^{(\nu+1)} \{ J_{m+1}(\beta_1'^{(\nu)}) [a_1^{(\nu)} \beta_1^{(\nu)} N_{m+1}(\beta_1^{(\nu)}) - a_2^{(\nu)(m)} N_m(\beta_1^{(\nu)})] + \\
&+ \frac{J_m(\beta_1'^{(\nu)})}{\beta_1'^{(\nu)}} [a_3^{(\nu)(m)} \beta_1^{(\nu)} N_{m+1}(\beta_1^{(\nu)}) - a_4^{(\nu)(m)} N_m(\beta_1^{(\nu)})] \} \\
a_1^{(\nu)} &= g^{(\nu)} - 1; \;\; a_2^{(\nu)(m)} = m(m-1) a_1^{(\nu)} + \frac{1}{2} g^{(\nu)} (\beta_2^{(n)})^2, \\
a_3^{(\nu)(m)} &= m(m-1) a_1^{(\nu)} + \frac{1}{2} (\beta_2'^{(n)})^2; \;\; a_4^{(\nu)(m)} = m(a_3^{(\nu)(m)} - a_2^{(\nu)(m)}), \\
\beta_j^{(\nu)} &= r_\nu e_j^{(\nu)} \omega; \;\; \beta_j'^{(\nu)} = r_\nu e_j^{(\nu+1)} \omega, \;\; (j = 1, 2),
\end{aligned} \tag{28.36}$$

here $g^{(\nu)} = G_0^{(\nu)} / G_0^{(\nu+1)}$ in accordance with (28.24), since all layers are linearly elastic. We also introduce the notations:

$$\begin{aligned}
a_5^{(\nu)} &= -a_1^{(\nu)}, \quad a_6^{(\nu)(m)} = (1-m) a_1^{(\nu)}, \\
a_7^{(\nu)(m)} &= -a_6^{(\nu)(m)}, \quad a_8^{(\nu)(m)} = a_3^{(\nu)(m)} - a_2^{(\nu)(m)}.
\end{aligned}$$

The formula for the element $h_{q\,j+2}^{(\nu)(m)}$ with $q, j = 1, 2$ is obtained from the corresponding formula (28.36) for the element $h_{qj}^{(\nu)(m)}$ by simultaneously formally replacing

$e_1^{(\nu+1)}$ with $me_1^{(\nu+1)}$, as well as $a_1^{(\nu)}$ with $a_5^{(\nu)}$, $a_l^{(\nu)(m)}$ with $a_{l+4}^{(\nu)(m)}$ $(l = 2, 3, 4)$ and $\beta_1^{(\nu)}$ with $\beta_2^{(\nu)}$. The formula for $h_{q+2\ j}^{(\nu)(m)}$ is obtained from $h_{qj}^{(\nu)(m)}$ by simultaneously replacing $e_1^{(\nu+1)}$ with $me_2^{(\nu+1)}$, as well as $a_1^{(\nu)}$ with $a_5^{(\nu)}$, $a_l^{(\nu)(m)}$ with $a_{l+4}^{(\nu)(m)}$ $(l = 2, 3, 4)$ and $\beta_1'^{(\nu)}$ with $\beta_2'^{(\nu)}$. The formula for $h_{q+2\ j+2}^{(\nu)(m)}$ is obtained from $h_{qj}^{(\nu)(m)}$ by simultaneously replacing $e_1^{(\nu+1)}$ with $e_2^{(\nu+1)}$, as well as $\beta_1^{(\nu)}$ with $\beta_2^{(\nu)}$ and $\beta_1'^{(\nu)}$ with $\beta_2'^{(\nu)}$.

The values of ω_{km} in the formula (28.34) are the positive roots of the characteristic equation

$$X_m(\omega) = 0, \tag{28.37}$$

which determines the eigenfrequencies of free vibrations of a layered elastic cylinder with an unloaded lateral surface. The values of ω_{km} can be taken as an initial approximation for searching for the complex roots of the corresponding characteristic equation (28.29) for the case when at least one of the cylinder layers is viscoelastic.

The expressions

$$Y_{1m}^{(n)(0)} + \delta_m^1 \Omega_{11} \tau^2, \quad Y_{2m}^{(n)(0)} + [\delta_m^1 \Omega_{21} + \delta_{m+1}^1 \Omega_{20} r] \tau^2$$

represent the residues of the functions $\mathrm{e}^{s\tau} U_{1m}^{(n)}$ and $\mathrm{e}^{s\tau} U_{2m}^{(n)}$ at the point $s = 0$. Herewith values Ω_{11}, Ω_{21}, Ω_{20} are constant and they determine the displacement of a cylinder as a rigid whole in the case of a non-self-balanced external load. The indicated residues are determined by the following formulas.

Case $m \geq 2$

$$\begin{aligned} Y_{jm}^{(n)(0)} &= -p^{(n)(0)} \{ Q_{1m} \det \begin{pmatrix} \epsilon_{j1}^{(n)(m)}(r) & \epsilon_{j2}^{(n)(m)}(r) \\ V_{21}^{(m)(0)} & V_{22}^{(m)(0)} \end{pmatrix} + \\ &+ 2Q_{2m} \det \begin{pmatrix} V_{11}^{(m)(0)} & V_{12}^{(m)(0)} \\ \epsilon_{j1}^{(n)(m)}(r) & \epsilon_{j2}^{(n)(m)}(r) \end{pmatrix} \} / D^{(m)(0)}, \end{aligned} \tag{28.38}$$

with $n = 1, 2, \ldots, N$, $j = 1, 2$, and where denoted:

$$\begin{aligned} &p^{(n)(0)} = (-4)^{N-n} (1 - \nu_0^{(n+1)})(1 - \nu_0^{(n+2)}) \cdot \ldots \cdot (1 - \nu_0^{(N)}) \ (n \leq N - 1), \\ &p^{(N)(0)} = 1, \\ &D^{(m)(0)} = \det \mathbf{V}^{(m)(0)}, \ \mathbf{V}^{(m)(0)} = \left(V_{qj}^{(m)(0)} \right) = \mathbf{L}^{(m)(0)} \cdot \chi^{(N)(m)(0)}, \\ &\mathbf{L}^{(m)(0)} = \left(L_{ql}^{(m)(0)} \right), \\ &\chi^{(n)(m)(0)} = \left(\chi_{lj}^{(n)(m)(0)} \right), \ q, j = 1, 2, \ l = 1, 2, 3, 4, \end{aligned}$$

herewith

$$\boldsymbol{\epsilon}^{(n)(m)} = \left(\epsilon_{qj}^{(n)(m)}(r)\right) = \mathbf{K}^{(n)(m)(0)}\chi^{(n)(m)(0)},$$
$$\mathbf{K}^{(n)(m)(0)} = \left(K_{ql}^{(n)(m)(0)}(r)\right), \quad q, j = 1, 2, \quad l = 1, 2, 3, 4.$$

The elements of the matrices $\mathbf{L}^{(m)(0)}$ and $\mathbf{K}^{(n)(m)(0)}$ have the form:

$$\mathbf{L}^{(m)(0)} = \begin{pmatrix} m-1 & (m+1)(m-2) & -(m+1) & -(m-1)(m+2) \\ -(m-1) & -m(m+1) & -(m+1) & -m(m-1) \end{pmatrix},$$
$$\mathbf{K}^{(n)(m)(0)} = \begin{pmatrix} r^{m-1} & (m-2+4\nu_0^{(n)})r^{m+1} & r^{-(m+1)} & (m+2-4\nu_0^{(n)})r^{1-m} \\ -r^{m-1} & (4\nu_0^{(n)}-m-4)r^{m+1} & r^{-(m+1)} & (m+4\nu_0^{(n)}-4)r^{1-m} \end{pmatrix}.$$

The recurrence relations hold for the matrix $\chi^{(n)(m)(0)}$:

$$\chi^{(1)(m)(0)} = \begin{pmatrix} 1 & 0 \\ 0 & 0 \\ 0 & 1 \\ 0 & 0 \end{pmatrix},$$
$$\chi^{(\mu+1)(m)(0)} = \mathbf{h}^{(\mu)(m)(0)}\chi^{(\mu)(m)(0)}, \quad \mu = 1, 2, \ldots, N-1,$$

where the elements of the matrix $\mathbf{h}^{(\mu)(m)(0)} = \left(h_{pl}^{(\mu)(m)(0)}\right)$, $p, l = 1, 2, 3, 4$ have the following form ($\mu = 1, 2, \ldots, N-1$, $m = 2, 3, \ldots$):

$$h_{11}^{(\mu)(m)(0)} = (4\nu_0^{(\mu+1)} - 3)g^{(\mu)} - 1;$$
$$h_{12}^{(\mu)(m)(0)} = 2(m+1)[(2\nu_0^{(\mu+1)} - 1)g^{(\mu)} - 2\nu_0^{(\mu)} + 1]r_\mu^2,$$
$$h_{13}^{(\mu)(m)(0)} = (m+1)(g^{(\mu)} - 1)r_\mu^{-2m},$$
$$h_{14}^{(\mu)(m)(0)} = [(g^{(\mu)} - 1)(m^2+2) + 4(\nu_0^{(\mu)} - g^{(\mu)}\nu_0^{(\mu+1)})]r_\mu^{2-2m},$$
$$h_{21}^{(\mu)(m)(0)} = 0; \quad h_{22}^{(\mu)(m)(0)} = 4\nu_0^{(\mu)} - 3 - g^{(\mu)},$$
$$h_{23}^{(\mu)(m)(0)} = (1 - g^{(\mu)})r_\mu^{-2-2m}; \quad h_{24}^{(\mu)(m)(0)} = (1-m)(g^{(\mu)} - 1)r_\mu^{-2m},$$
$$h_{31}^{(\mu)(m)(0)} = -h_{24}^{(\mu)(m)(0)}r_\mu^{4m}, \quad h_{32}^{(\mu)(m)(0)} = -h_{14}^{(\mu)(m)(0)}r_\mu^{4m},$$
$$h_{33}^{(\mu)(m)(0)} = h_{11}^{(\mu)(m)(0)}, \quad h_{34}^{(\mu)(m)(0)} = h_{12}^{(\mu)(m)(0)}(m-1)/(m+1),$$
$$h_{41}^{(\mu)(m)(0)} = (g^{(\mu)} - 1)r_\mu^{2m-2}, \quad h_{42}^{(\mu)(m)(0)} = (g^{(\mu)} - 1)(m+1)r_\mu^{2m},$$
$$h_{43}^{(\mu)(m)(0)} = 0, \quad h_{44}^{(\mu)(m)(0)} = h_{22}^{(\mu)(m)(0)}.$$

Case $m = 0$

$$Y_{10}^{(n)(0)}(r) = -Q_{10}\mu^{(n)}\frac{r\chi_{11}^{(n)(0)(0)} + \chi_{21}^{(n)(0)(0)}/r}{(2w^{(N)} - 1)\chi_{11}^{(N)(0)(0)} - \chi_{21}^{(N)(0)(0)}}, \qquad (28.39)$$

where $\mu^{(N)} = 1, \quad \mu^{(n)} = 2^{N-n} w^{(n+1)} w^{(n+2)} \cdot \ldots \cdot w^{(N)} \quad (n \le N-1, n = 1, 2, .., N)$. The recurrence relations are fulfilled for the column vector

$$\boldsymbol{\mathcal{X}}_1^{(n)(0)(0)} = \begin{pmatrix} \mathcal{X}_{11}^{(n)(0)(0)} \\ \mathcal{X}_{21}^{(n)(0)(0)} \end{pmatrix},$$

$$\boldsymbol{\mathcal{X}}_1^{(1)(0)(0)} = \begin{pmatrix} 1 \\ 0 \end{pmatrix}, \quad \boldsymbol{\mathcal{X}}_1^{(\nu+1)(0)(0)} = \mathbf{h}^{(\nu)(0)(0)} \boldsymbol{\mathcal{X}}_1^{(\nu)(0)(0)}, \quad \nu = 1, 2, \ldots, N-1,$$

the elements of the matrix $\mathbf{h}^{(\nu)(0)(0)} = \left(h_{qj}^{(\nu)(0)(0)} \right)$, $q, j = 1, 2$ have the form:

$$\mathbf{h}^{(\nu)(0)(0)} = \begin{pmatrix} 1 + g^{(\nu)}(2w^{(\nu)} - 1) & (1 - g^{(\nu)})/r_\nu^2 \\ r_\nu^2[2w^{(\nu+1)} - 1 - g^{(\nu)}(2w^{(\nu)} - 1)] & 2w^{(\nu+1)} + g^{(\nu)} - 1 \end{pmatrix},$$
$$\nu = 1, 2, \ldots, N-1.$$

Besides that $(n = 1, 2, \ldots, N)$

$$Y_{20}^{(n)(0)}(r) = -Q_{20}[r\mathcal{X}_{12}^{(n)(0)(0)} + \mathcal{X}_{22}^{(n)(0)(0)}/r + (e_2^{(n)})^2 r^3/4]/a, \quad (28.40)$$

$$\Omega_{20} = -4Q_{20}/a, \quad a = \sum_{n=1}^{N} (e_2^{(n)})^2 \frac{G_0^{(n)}}{G_0^{(N)}} (r_n^4 - r_{n-1}^4), \quad r_0 = 0.$$

The recurrence relations are fulfilled for the column vector

$$\boldsymbol{\mathcal{X}}_2^{(n)(0)(0)} = \begin{pmatrix} \mathcal{X}_{12}^{(n)(0)(0)} \\ \mathcal{X}_{22}^{(n)(0)(0)} \\ 1 \end{pmatrix}, \quad \boldsymbol{\mathcal{X}}_2^{(1)(0)(0)} = \begin{pmatrix} 0 \\ 0 \\ 1 \end{pmatrix},$$

$$\boldsymbol{\mathcal{X}}_2^{(\nu+1)(0)(0)} = \tilde{\mathbf{h}}^{(\nu)(0)(0)} \boldsymbol{\mathcal{X}}_2^{(\nu)(0)(0)}, \quad \nu = 1, 2, \ldots, N-1,$$

where the elements of the matrix $\tilde{\mathbf{h}}^{(\nu)(0)(0)} = \left(\tilde{h}_{pl}^{(\nu)(0)(0)} \right)$, $p, l = 1, 2, 3$ have the form $(\nu = 1, 2, \ldots, N-1)$:

$$\tilde{\mathbf{h}}^{(\nu)(0)(0)} = \begin{pmatrix} 1 & (1 - g^{(\nu)})/r_\nu^2 & [(e_2^{(\nu)})^2(1 + g^{(\nu)}) - 2(e_2^{(\nu+1)})^2]r_\nu^2/4 \\ 0 & g^{(\nu)} & [(e_2^{(\nu+1)})^2 - g^{(\nu)}(e_2^{(\nu)})^2]r_\nu^4/4 \\ 0 & 0 & 1 \end{pmatrix}.$$

The expression $\Omega_{20} r\tau^2$ represents the corresponding component of the rotational displacement of the cylinder as a rigid whole.

Case $m = 1$

$$Y_{j1}^{(n)(0)} = \frac{1}{cV_{11}^{(1)(0)}}\left\{Q_{11}\det\begin{pmatrix}\epsilon_{j1}^{(n)(1)}(r) & \epsilon_{j2}^{(n)(1)}(r)\\ V_{21}^{(1)(0)} & V_{22}^{(1)(0)}\end{pmatrix}+\right.$$

$$\left.+\ Q_{21}\det\begin{pmatrix}V_{11}^{(1)(0)} & V_{12}^{(1)(0)}\\ \epsilon_{j1}^{(n)(1)}(r) & \epsilon_{j2}^{(n)(1)}(r)\end{pmatrix}\right\}, \quad n=1,2,\ldots,N,\ j=1,2, \quad (28.41)$$

$$\Omega_{11} = \frac{Q_{21}-Q_{11}}{c}, \quad \Omega_{21} = -\Omega_{11}, \quad c = \sum_{n=1}^{N}\frac{G_0^{(n)}}{G_0^{(N)}}(e_2^{(n)})^2(r_n^2 - r_{n-1}^2), \quad r_0 = 0.$$

The expressions $\Omega_{11}\tau^2$; $-\Omega_{11}\tau^2$ represent the contribution of the corresponding Fourier components external load Q_{11}, Q_{21} (at $m = 1$) to the translational displacement of the cylinder as a rigid whole. Here denoted:

$$V_{11}^{(1)(0)} = V_{21}^{(1)(0)} = (w^{(N)} - \frac{1}{2})\chi_{31}^{(N)(1)(0)} - 2\chi_{21}^{(N)(1)(0)},$$

$$V_{12}^{(1)(0)} = w^{(N)} - 2\chi_{22}^{(N)(1)(0)} + (w^{(N)} - \frac{1}{2})\chi_{32}^{(N)(1)(0)} + (4w^{(N)} - 1)\chi_{42}^{(N)(1)(0)},$$

$$V_{22}^{(1)(0)} = V_{12}^{(1)(0)} - 2w^{(N)}(1 + 2\chi_{42}^{(N)(1)(0)}),$$

$$\boldsymbol{\epsilon}^{(n)(1)} = \left(\epsilon_{qj}^{(n)(1)}(r)\right) = \mathbf{K}^{(n)(1)(0)}\boldsymbol{\chi}^{(n)(1)(0)},$$

$$\mathbf{K}^{(n)(1)(0)} = \left(K_{ql}^{(n)(1)(0)}(r)\right), \quad q = 1,2,\ l = 1,2,3,4,5.$$

The elements of the matrix $\boldsymbol{K}^{(n)(1)(0)}$ have the form ($n = 1,2,\ldots,N$):

$$K_{11}^{(n)(1)(0)} = 1, \quad K_{12}^{(n)(1)(0)} = \frac{1}{r^2}, \quad K_{13}^{(n)(1)(0)} = \frac{r^2}{4}(e_1^{(n)})^2(3 - 2w^{(n)}),$$

$$K_{14}^{(n)(1)(0)} = (e_1^{(n)})^2[1 + (2w^{(n)} + 1)\ln r], \quad K_{15}^{(n)(1)(0)} = \frac{r^2}{2}(e_1^{(n)})^2 w^{(n)},$$

$$K_{21}^{(n)(1)(0)} = -1, \quad K_{22}^{(n)(1)(0)} = \frac{1}{r^2}, \quad K_{23}^{(n)(1)(0)} = -\frac{r^2}{4}(e_1^{(n)})^2(1 - 6w^{(n)}),$$

$$K_{24}^{(n)(1)(0)} = -(e_1^{(n)})^2[2w^{(n)} + (2w^{(n)} + 1)\ln r], \quad K_{25}^{(n)(1)(0)} = -\frac{3}{2}r^2(e_1^{(n)})^2 w^{(n)},$$

The recurrence relations are fulfilled for the matrix

$$\boldsymbol{\chi}^{(n)(1)(0)} = \left(\chi_{lj}^{(n)(1)(0)}\right), \quad j = 1,2,\ l = 1,2,3,4,5;$$

$$\boldsymbol{\chi}^{(1)(1)(0)} = \begin{pmatrix}0 & 0\\ 0 & 0\\ 1 & 0\\ 0 & 0\\ 0 & 1\end{pmatrix},$$

$$\boldsymbol{\chi}^{(\nu+1)(1)(0)} = \mathbf{h}^{(\nu)(1)(0)}\boldsymbol{\chi}^{(\nu)(1)(0)}, \quad \nu = 1,2,\ldots,N-1,$$

where the elements of the matrix

$$\mathbf{h}^{(\nu)(1)(0)} = \left(h_{pl}^{(\nu)(1)(0)}\right), \; p, l = 1, 2, 3, 4, 5$$

have the following form ($\nu = 1, 2, \ldots, N-1$):

$$
\begin{aligned}
h_{21}^{(\nu)(1)(0)} &= h_{31}^{(\nu)(1)(0)} = h_{41}^{(\nu)(1)(0)} = h_{42}^{(\nu)(1)(0)} = h_{43}^{(\nu)(1)(0)} = 0,\\
h_{51}^{(\nu)(1)(0)} &= h_{52}^{(\nu)(1)(0)} = h_{53}^{(\nu)(1)(0)} = h_{54}^{(\nu)(1)(0)} = 0,\\
h_{11}^{(\nu)(1)(0)} &= 1, \quad h_{12}^{(\nu)(1)(0)} = (1-2g^{(\nu)})(1-2w^{(\nu+1)})/(2w^{(\nu+1)}r_\nu^2),\\
h_{13}^{(\nu)(1)(0)} &= -\frac{(e_1^{(\nu)})^2 r_\nu^2}{8w^{(\nu+1)}}\{4w^{(\nu+1)}(2w^{(\nu)}-1)+\\
&+(1-2w^{(\nu+1)})[2w^{(\nu)}(1+g^{(\nu)})+1-g^{(\nu)}]\},\\
h_{14}^{(\nu)(1)(0)} &= -\frac{(e_1^{(\nu)})^2}{4w^{(\nu+1)}}\{4[(2w^{(\nu+1)}+1)w^{(\nu)}g^{(\nu)} - w^{(\nu+1)}(2w^{(\nu)}+1)]\ln r_\nu +\\
&+4(w^{(\nu)}g^{(\nu)} - w^{(\nu+1)}) + 2(w^{(\nu+1)}g^{(\nu)} - w^{(\nu)}) + 1 - g^{(\nu)}\},\\
h_{15}^{(\nu)(1)(0)} &= -\frac{r_\nu^2}{8w^{(\nu+1)}}\{((e_1^{(\nu+1)})^2 w^{(\nu+1)} - (e_1^{(\nu)})^2 w^{(\nu)})(1+6w^{(\nu+1)})-\\
&-(1-2w^{(\nu+1)})[(e_1^{(\nu)})^2 w^{(\nu)}(1-2g^{(\nu)}) + (e_1^{(\nu+1)})^2 w^{(\nu+1)}]+\\
&+4(g^{(\nu)}(e_1^{(\nu)})^2 w^{(\nu)} - (e_1^{(\nu+1)})^2 w^{(\nu+1)})[2w^{(\nu+1)} + (2w^{(\nu+1)}+1)\ln r_\nu]\}\\
h_{22}^{(\nu)(1)(0)} &= \frac{1}{4w^{(\nu+1)}}[4w^{(\nu+1)} - (1-g^{(\nu)})(1+2w^{(\nu+1)})],\\
h_{23}^{(\nu)(1)(0)} &= \frac{(e_1^{(\nu)})^2 r_\nu^4}{16w^{(\nu+1)}}\{4w^{(\nu+1)}(2w^{(\nu)}+1)-\\
&-(1+2w^{(\nu+1)})[2w^{(\nu)}(1+g^{(\nu)})+1-g^{(\nu)}]\}\\
h_{24}^{(\nu)(1)(0)} &= \frac{(e_1^{(\nu)})^2 r_\nu^2}{8w^{(\nu+1)}}\{(1-2w^{(\nu)})[4w^{(\nu+1)} - (1+2w^{(\nu+1)})(1-g^{(\nu)})]-\\
&-4(1-2w^{(\nu+1)})g^{(\nu)}w^{(\nu)}]\},\\
h_{25}^{(\nu)(1)(0)} &= \frac{r_\nu^4}{8w^{(\nu+1)}}(1-2w^{(\nu+1)})[(e_1^{(\nu)})^2 w^{(\nu)}(1-2g^{(\nu)}) + (e_1^{(\nu+1)})^2 w^{(\nu+1)}],\\
h_{32}^{(\nu)(1)(0)} &= (1-g^{(\nu)})/(w^{(\nu+1)}(e_1^{(\nu+1)})^2 r_\nu^4),\\
h_{33}^{(\nu)(1)(0)} &= [1-g^{(\nu)}+2w^{(\nu)}(1+g^{(\nu)})](e_1^{(\nu)})^2/(4w^{(\nu+1)}(e_1^{(\nu+1)})^2),\\
h_{34}^{(\nu)(1)(0)} &= (e_1^{(\nu)})^2(1-g^{(\nu)})(1-2w^{(\nu)})/(2w^{(\nu+1)}(e_1^{(\nu+1)})^2 r_\nu^2),\\
h_{35}^{(\nu)(1)(0)} &= ((e_1^{(\nu+1)})^2 w^{(\nu+1)} - (e_1^{(\nu)})^2 w^{(\nu)})/(2w^{(\nu+1)}(e_1^{(\nu+1)})^2),\\
h_{44}^{(\nu)(1)(0)} &= (e_1^{(\nu)})^2 g^{(\nu)} w^{(\nu)}/(w^{(\nu+1)}(e_1^{(\nu+1)})^2),\\
h_{45}^{(\nu)(1)(0)} &= r_\nu^2[g^{(\nu)}(e_1^{(\nu)})^2 w^{(\nu)} - (e_1^{(\nu+1)})^2 w^{(\nu+1)}]/(2w^{(\nu+1)}(e_1^{(\nu+1)})^2),\\
h_{55}^{(\nu)(1)(0)} &= 1.
\end{aligned}
$$

Thus, in the particular case considered, when all layers of the cylinder are linearly elastic, the solution to the non-stationary problem (28.1) – (28.7) is determined by formulas (28.8), (28.34) – (28.41) for the external load function, which is the Heaviside function $f(\tau) = h(\tau)$. For other functions $f(\tau)$, the solution is obtained on the basis of the one presented here using the Duhamel integral.

28.5 Conclusion

A plane dynamic problem of the transient waves propagation in the cross section of an infinitely long circular layered viscoelastic cylinder under non-axisymmetric loading of its lateral surface is considered. Continuity conditions are assumed at the contact between adjacent layers. The number of homogeneous layers is arbitrary, and their hereditary properties, generally speaking, are not identical. These hereditary properties are described by the relations of the Boltzmann-Volterra model, both with regular relaxation kernels in the form of finite sums of exponentials and with singular kernels in the Rzhanitsyn-Koltunov form.

The solution is constructed for the case where the time dependence of the external load has the character of a Heaviside function (the solution for a different type of load-time dependence can easily be expressed in terms of the constructed one). After expansion into a trigonometric Fourier series in the angular coordinate, the integral Laplace transform in time is applied. After constructing the Fourier components in the images, their originals are represented using a special formula based on previously developed elements of the corresponding mathematical theory. This formula for each Fourier component contains an expansion in series in terms of residues. The series represents the spectral decomposition over a subset of the eigenforms of the cylinder's free oscillations corresponding only to complex spectral elements with a nonzero imaginary part. Furthermore, it includes a term in the form of a residue at the zero point of the complex plane. If the external load is not self-equilibrated, this term, among other things, takes into account the corresponding component of the cylinder's displacement as a rigid body. Finally, the formula for each Fourier component in the originals contains a term in the form of a special contour integral. This integral has number of advantages and is very convenient for calculating the parameters of nonstationary processes over the entire time range. The formula is essentially a modification of the spectral decomposition in nonstationary viscoelasticity problems with relaxation kernels of a wide class.

A special case where viscosity is absent and all cylinder layers are linearly elastic is considered separately. For this case, all details of the solution in the originals are presented explicitly. We emphasize that the constructed solution is valid over the entire time range, herewith the hereditary properties of the layers can be described by the constitutive relations with relaxation kernels of various types.

Acknowledgements The study was conducted under the state assignment of Lomonosov Moscow State University.

References

[1] Alizadeh V, Amirkhizi AV (2021) Overall dynamic properties of locally resonant viscoelastic layered media based on consistent field integration for

oblique anti-plane shear waves. Mechanics of Materials 160:103981, DOI 10.1016/j.mechmat.2021.103981
[2] Bazhenov VG, Belov AA, Igumnov LA (2009) Granichno-elementnoye modelirovaniye dinamiki kusochno-odnorodnykh sred i konstruktsiy (in Russ., Boundary-element modeling of the dynamics of piecewise homogeneous media and structures). NNGU, Nizhnii Novgorod
[3] Belov AA (2008) Granichno-elementnyy raschet dinamiki sostavnykh vyazkouprugikh tel. (in Russ., Boundary-element analysis of dynamics of compound viscoelastic bodies). Problemy prochnosti i plastichnosti (70):162–168
[4] Borcherdt RD (2020) Viscoelastic Waves and Rays in Layered Media. Cambridge Univ Press, Cambridge
[5] Bulychev GG, Pshenichnov SG (1988) Propagation of elastic waves in a layered cylinder. Soviet physics. Doklady 33(12):967-969
[6] Djalilov ML(2023) Equation of transverse vibration of a piecewise homogeneous viscoelastic plate. International Journal of Advance Scientific Research 32(10):181–189
[7] Hosseini SH, Khaniki HB (2017) Dynamic behavior of multi-layered viscoelastic nanobeam system embedded in a viscoelastic medium with a moving nanoparticle. Journal of Mechanics 33(5):559–575, DOI 10.1017/jmech.2016.91
[8] Khudoynazarov K (2024) Longitudinal-radial vibrations of a viscoelastic cylindrical three-layer structure. Facta Universitatis, Series: Mechanical Engineering 22(3):473–484, DOI 10.22190/FUME231219010K
[9] Kozlov VI, Kucher NK (1980) Dynamic behavior of multilayer cylindrical structures with transient loads. Strength of Materials 12(5):639–648, DOI 10.1007/BF01118046
[10] Lee HS (2014) Viscowave — a new solution for viscoelastic wave propagation of layered structures subjected to an impact load. International Journal of Pavement Engineering 15(6):542-557, DOI 10.1080/10298436.2013.782401
[11] Lokshin AA (1994) The head wave at the boundary of two hereditary-elastic half-spaces. The case of a linear source. Journal of Applied Mathematics and Mechanics 58(1):171–176, DOI 10.1016/0021-8928(94)90043-4
[12] Nuriev BR (1985) Udar po vyazkouprugomu sloistomu kompozitu (in Russ., Impact on a viscoelastic layered composite). Izvestiya Akademii Nauk Azerbaidzhanskoi SSR Seriya Fiziko-Tekhnicheskikh i Matematicheskikh Nauk 4:35–41
[13] Petrov AN, Ermolaev MD (2011) Raschet metodom granichnykh elementov dinamiki sostavnykh vyazkouprugikh tel (in Russ., Boundary element analysis of the dynamics of composite viscoelastic bodies). Mechanics of deformed solid. Vestnik Nizhegorodskogo universiteta N.I. Lobachevskii 4(4):1694–1696. (in Russian)
[14] Pshenichnov SG (1997) Postroyeniye originala dlya transformanty Laplasa s pomoshch'yu teorii vychetov v zadachakh dinamiki lineyno-vyazkouprugikh tel (in Russ., Construction of the original for the Laplace transform using residues theory in problems of dynamics of linearly viscoelastic bodies). In:

Nonlinear phenomena in open systems. A collection of scientific papers, State Institute of Physics and Technology, Moscow 8:79–87

[15] Pshenichnov SG (2016) Dynamic linear viscoelasticity problems for piecewise homogeneous bodies. Mechanics of Solids 51(1):65–74, DOI 10.3103/s0025654416010076

[16] Pshenichnov SG (2025) Spectral decompositions in dynamic linear viscoelasticity problems for piecewise homogeneous bodies. Moscow University Mechanics Bulletin 80(2): 63-72, DOI 10.3103/S0027133025700128

[17] Pshenichnov SG, Korovaytseva EA, Bazhlekova E, and Datcheva MD (2023) Dynamic problems for piecewise homogeneous viscoelastic bodies. In: Altenbach H, Eremeyev VA, Igumnov LA, and Bragov A (ed) Deformation and Destruction of Materials and Structures Under Quasi-static and Impulse Loading. Advanced Structured Materials, Vol 186. Springer, Cham 186:227–243, DOI 10.1007/978-3-031-22093-7_16

[18] Pshenichnov SG, Korovaytseva EA, Razorenova AM (2025) Nestatsionarnyye zadachi dlya neodnorodnykh vyazkouprugikh tel (in Russ., Nonstationary problems for inhomogeneous viscoelastic bodies). Vestnik Moskovskogo Universiteta. Seriya 1: Matematika, Mekhanika (6):59-67, DOI 10.55959/MSU0579-9368-1-66-6-9

[19] Shamaev AS, Shumilova VV (2023) Homogenization of equations of dynamics of a medium consisting of viscoelastic material with memory and incompressible Kelvin–Voigt fluid. Journal of Mathematical Sciences 270(6):827–841, DOI 10.1007/s10958-023-06393-2

[20] Sabodash PF (1971) Propagation of longitudinal viscoelastic waves in a three-layer medium. Polymer Mech 7(1):124–128, DOI 10.1007/BF00856626

[21] Zhu JB, Zhao XB, Wu W, Zhao J (2012) Wave propagation across rock joints filled with viscoelastic medium using modified recursive method. Journal of Applied Geophysics 86:82–87, DOI 10.1016/j.jappgeo.2012.07.012

Chapter 29
On the Properties of Semigroups Generated by Volterra Integro-Differential Equations Arising in Viscoelasticity Theory

Nadezhda Rautian

Abstract We study abstract integro-differential equations that are operator models of problems in viscoelasticity. We present results based on an approach related to the study of one-parameter semigroups for linear evolution equations. The presented approach can also be used to study other integro-differential equations containing integral terms of the Volterra convolution form. A method for reducing the original initial value problem for a model integro-differential equation with operator coefficients in a Hilbert space to the Cauchy problem for a first-order differential equation in an extended function space is presented. The existence of a contraction C_0-semigroup is proved under certain assumptions about the kernels of integral operators. As a corollary, we establish the well-posed solvability of the resulting Cauchy problem for the first order differential equation in an extended function space and the initial value problem for the original abstract integro-differential equation and indicate a relation between their solutions. Examples of exponential and fractional-exponential (Rabotnov functions) kernels of integral operators satisfying the above assumptions are given.

Key words: Volterra integro-differential equation, Linear differential equation, Hilbert space, Operator semigroup, Viscoelasticity

29.1 Introduction

A major role in the theory of constitutive relations of linear viscoelasticity is played by the stress-strain relations in the form of integral operators with kernels that are material functions that can be found from homing experiments (see [5, 7]). If the material functions depend only on one time variable, then viscoelastic media are said

Nadezhda Rautian
Faculty of Mechanics and Mathematics, Moscow State University, Leninskie gory, 119992 Moscow, Russian Federation, e-mail: nrautian@mail.ru

H. Altenbach et al. (eds.), *Stability Problems in Mechanics of Continua*,
Advanced Structured Materials 260, https://doi.org/10.1007/978-3-032-31032-3_29

to be nonaging. The use of these constitutive relations in the postulates of continuum mechanics leads, in particular, to equations of motion that are Volterra integro-differential equations with unbounded operator coefficients in Hilbert spaces. We will call them *abstract integro-differential equations*.

Below, using an example of an abstract integro-differential equation arising in the linear theory of viscoelasticity, we present a general approach to the study that can be applied to many other linear models containing Volterra integral operators. The indicated abstract integro-differential equation (equation of motion of a viscoelastic medium) can be realized as the integro–partial differential equation

$$\begin{aligned} u_{tt}(x,t) &= \rho^{-1}\left[\mu\Delta u(x,t) + (\mu+\lambda)\mathrm{grad}(\mathrm{div} u(x,t))\right] \\ &- \int_0^t K(t-\tau)\rho^{-1}\mu\left[\Delta u(x,\tau) + \mathrm{grad}(\mathrm{div} u(x,\tau))\right] d\tau \\ &- \int_0^t Q(t-\tau)\rho^{-1}\lambda\left[\mathrm{grad}(\mathrm{div} u(x,\tau))\right] d\tau + f(x,t), \end{aligned} \tag{29.1}$$

where $u = \vec{u}(x,t) \in \mathbb{R}^3$ is the vector of small displacements of a viscoelastic isotropic medium filling a bounded domain $\Omega \subset \mathbb{R}^3$ with smooth boundary, $\rho > 0$ is constant density, and λ and μ which are analogs of the Lamé coefficients in elasticity theory (see [5, 7]).

It is assumed that the Dirichlet conditions

$$u|_{\partial\Omega} = 0$$

are satisfied on the boundary of the domain Ω. The kernel functions $K(t)$ and $Q(t)$ of the integral operators are positive nonincreasing integrable functions characterizing the hereditary properties of the medium.

The analysis of the integro-differential equation (29.1) becomes much simpler if we assume that the kernels $K(t)$ and $Q(t)$ of the integral operators are sums of decaying exponentials with positive coefficients; in particular,

$$K(t) = \sum_{k=1}^{m} c_k e^{-\beta_k t},\ t \geqslant 0,\ c_k > 0,\ \beta_k > 0, \tag{29.2}$$

because in this case, the original equation (29.1) can be reduced by a standard change of variables to a system of first-order differential equations with constant coefficients (see [10]). However, when processing experimental data, it became clear that kernels of the form (29.2) fail to sufficiently adequately describe the behavior of real materials. Illustrative examples of the kind can be found in the monograph (see [13]). The above shortcoming can be eliminated if one uses the so-called singular kernels (tending to infinity as $t \to +0$).

The simplest example of singular kernels integrable on any finite interval of the positive real halfline and suitable for use in the linear viscoelasticity theory is the so-called Abel kernels

$$K(t) = ct^{-\alpha}, \quad t \geqslant 0, \quad c > 0, \quad 0 < \alpha < 1.$$

However, unlike exponential kernels of the form (29.2), the Abel kernels do not decay rapidly enough as $t \to +0$; this limits the possibilities of their usage in modeling the behavior of real materials. Therefore, from the viewpoint of applications, one needs singular kernels that decrease sufficiently rapidly as $t \to +0$; collections of such kernels were constructed in a number of publications (see the bibliography in [13]). Kernels of the following form have become widespread in theoretical and applied mechanics:

$$K(t) = \sum_{k=1}^{m} c_k \exists_{\alpha-1}(-\beta_k, t),\ t > 0,\ c_k > 0,\ \beta_k > 0,\ 0 < \alpha < 1, \qquad (29.3)$$

where $\exists_\alpha(\beta, t)$ is a fractional-exponential Rabotnov function (see [13]). Note that for $\alpha = 1$ the representation (29.3) transforms into the representation (29.2), while for $\beta_k = 0$ it becomes an Abel kernel. At present, there is an extensive list of papers dealing with Rabotnov functions; tables of these functions and their integrals have been compiled (see the bibliography in [13]).

The Gurtin–Pipkin integro-differential equations (see [6, 8, 11]) are Volterra integro-differential equations as well. They describe heat propagation in media with memory at a finite speed. In addition, these equations arise in homogenization problems in multiphase media (the Darcy law), see [23].

The above-listed problems can be integrated into a sufficiently broad class of integro-partial differential equations; therefore, it is more natural to consider integro-differential equations with unbounded operator coefficients in Hilbert spaces (abstract integro-differential equations), which can be realized as integro–partial differential equations. Currently, there is a vast amount of references dealing with Volterra integro-differential equations and related problems arising in numerous applications (see [1–8, 10–12, 15–23] and the bibliography therein).

The results in the present chapter continue the research published in [17–22], where the spectral analysis of operator functions that are symbols of integro-differential equations was carried out. The semigroup theory based approach to studying Volterra integro-differential equations was developed in (see [1, 3, 12, 14–16]).

29.2 Definitions, Notations, Statement of the Problem.

Let H be a separable Hilbert space, and let A be a positive self-adjoint operator on H, $A^* = A \geqslant \kappa_0 I$ ($\kappa_0 > 0$), with bounded inverse. Let B be a self-adjoint nonnegative operator on H with domain $D(B) \supseteq D(A)$ such that, for some $\kappa > 0$, the inequality $\|Bx\| \leqslant \kappa \|Ax\|$ holds for any $x \in D(A)$. Here I is the identity operator on H.

Consider the following problem for a second-order integro-differential equation on the positive half-line $\mathbb{R}_+ = [0, \infty)$:

$$\frac{d^2u(t)}{dt^2}+(A+B)\,u(t)-\int_0^t K_1(t-s)Au(s)ds-\int_0^t K_2(t-s)Bu(s)ds = f(t),\ \ t\in\mathbb{R}_+, \tag{29.4}$$

$$u(+0) = \varphi_0, \quad u^{(1)}(+0) = \varphi_1, \tag{29.5}$$

Suppose that the function $K_i(t)$, $i = 1, 2$ has the representation

$$K_i(t) = \int_0^{+\infty} e^{-t\tau} d\mu_i(\tau), \quad i = 1, 2. \tag{29.6}$$

where $d\mu_i$, $(i = 1, 2)$ is the positive measure generated by a nondecreasing right continuous function μ_i on $\mathbb{R}_+$. The integral is understood in the sense of Stieltjes. In addition, we will assume that the conditions

$$\int_0^{+\infty} \frac{d\mu_i(\tau)}{\tau} < 1, \quad i = 1, 2 \tag{29.7}$$

are satisfied. Let us introduce the notation

$$M_i(t) := \int_0^{+\infty} \frac{e^{-t\tau}d\mu_i(\tau)}{\tau}, \quad t \in \mathbb{R}_+, \quad i = 1, 2. \tag{29.8}$$

Set

$$A_0 := \left(1 - \int_0^{+\infty} \frac{d\mu_1(\tau)}{\tau}\right) A + \left(1 - \int_0^{+\infty} \frac{d\mu_2(\tau)}{\tau}\right) B. \tag{29.9}$$

It follows from the self-adjointness of the positive operator A and the nonnegative operator B as well as from conditions (29.7) that the operator A_0 is self-adjoint and positive.

Note that problems of the form (29.4) and (29.5) are operator models of problems arising in the theory of viscoelasticity (see [2, 7]) and thermal physics (see [1, 6, 11]). The results on the spectral analysis of (29.4) for the case in which the kernels $K_i(t)$ are decaying exponentials can be found in the monograph [17]. Let us transform the domain $D(A_0^\beta)$ of the operator A_0^β, $\beta > 0$ into a Hilbert space H_β by introducing on $D(A_0^\beta)$ a norm equivalent to the graph norm of the operator A_0^β.

Remark 29.1. It follows from the properties of the operators A and B and Heinz's inequality (see [9, pp. 177–178]) that the operator A_0 is invertible, the operators A_0^{-1}, $Q_1 := A^{1/2}A_0^{-1/2}$ and $Q_2 := B^{1/2}A_0^{-1/2}$ are bounded in H.

Definition 29.1. A vector function $u(t)$ is called a *classical solution* of problem (29.4), (29.5), if $u(t) \in C^2(\mathbb{R}_+, H)$, $Au(t),\ Bu(t) \in C(\mathbb{R}_+, H)$ and $u(t)$ satisfies Eq. (29.4) for each $t \in \mathbb{R}_+$ and the initial conditions (29.5).

By Ω_k we denote the spaces $L^2_{\Omega_k}(\mathbb{R}_+, H)$ of vector functions on $\mathbb{R}_+$ ranging in H, equipped with the norms

$$||u||_{\Omega_k} = \left(\int_0^{+\infty} ||u(s)||_H^2 d\mu_k(s) \right)^{1/2}.$$

29.3 Reduction of the Original Problem to a First-Order Differential Equation

Applying the integration by parts formula to the integrals on the left-hand side in Eq. (29.4), we obtain the equation

$$\begin{aligned} \frac{d^2u(t)}{dt^2} + A_0 u(t) &+ \int_0^t \left(\int_0^{+\infty} \frac{e^{-(t-s)\tau}}{\tau} d\mu_1(\tau) \right) A \frac{du(s)}{ds} ds \\ &+ \int_0^t \left(\int_0^{+\infty} \frac{e^{-(t-s)\tau}}{\tau} d\mu_2(\tau) \right) B \frac{du(s)}{ds} ds \\ = f(t) &- \left(\int_0^{+\infty} \frac{e^{-t\tau}}{\tau} d\mu_1(\tau) A + \int_0^{+\infty} \frac{e^{-t\tau}}{\tau} d\mu_2(\tau) B \right) \varphi_0 \end{aligned} \tag{29.10}$$

Note that $A = A_0^{1/2} Q_1^* Q_1 A_0^{1/2}$ and $B = A_0^{1/2} Q_1^* Q_1 A_0^{1/2}$ and so Eq. (29.10) can be formally written as

$$\begin{aligned} &\frac{d^2u(t)}{dt^2} + A_0^{1/2} \left[A_0^{1/2} u(t) + \sum_{k=1}^{2} \int_0^{+\infty} \frac{1}{\sqrt{\tau}} Q_k^* \left(\int_0^t \frac{e^{-(t-s)\tau}}{\sqrt{\tau}} Q_k A_0^{1/2} \frac{du(s)}{ds} ds \right) d\mu_k(\tau) \right] \\ &= f_1(t), \end{aligned}$$

where

$$f_1(t) = f(t) - \left(\int_0^{+\infty} \frac{e^{-t\tau}}{\tau} d\mu_1(\tau) A + \int_0^{+\infty} \frac{e^{-t\tau}}{\tau} d\mu_2(\tau) B \right) \varphi_0. \tag{29.11}$$

Let us introduce the new variables

$$v(t) := u'(t), \quad \xi_0(t) := A_0^{1/2} u(t),$$

$$\xi_k(t,\tau) = \int_0^t \frac{e^{-(t-s)\tau}}{\sqrt{\tau}} Q_k A_0^{1/2} \frac{du(s)}{ds} ds, \quad t \in \mathbb{R}_+, \quad \tau \in \bigcup_{k=1}^{2} \operatorname{supp} \mu_k, \quad k = 1, 2. \tag{29.12}$$

In these variables, problem (29.4), (29.5) is formally reduced to the Cauchy problem for the system of first-order differential equations

$$\begin{cases} \dfrac{dv(t)}{dt} + A_0^{1/2}\left[\xi_0(t) + \sum\limits_{k=1}^{2} \int\limits_0^\infty \dfrac{1}{\sqrt{\tau}} Q_k^* \xi_k(t,\tau) d\mu_k(\tau)\right] = f_1(t), \\ \dfrac{d\xi_0(t)}{dt} = A_0^{1/2} v(t), \\ \dfrac{d\xi_1(t,\tau)}{dt} = \dfrac{1}{\sqrt{\tau}} Q_1 A_0^{1/2} v(t) - \tau \xi_1(t,\tau), \\ \dfrac{d\xi_2(t,\tau)}{dt} = \dfrac{1}{\sqrt{\tau}} Q_2 A_0^{1/2} v(t) - \tau \xi_2(t,\tau), \end{cases} \tag{29.13}$$

where $t \in \mathbb{R}_+$, $\tau \in \bigcup\limits_{k=1}^{2} \operatorname{supp} \mu_k$ and the function $f_1(t)$ is defined by Eq. (29.11) with the initial conditions

$$v(t)|_{t=0} = \varphi_1, \ \xi_0(t)|_{t=0} = A_0^{1/2} \varphi_0, \ \xi_k(t,\tau)|_{t=0} = 0, \tau \in \bigcup_{k=1}^{2} \operatorname{supp} \mu_k, k = 1, 2. \tag{29.14}$$

Now, first, we must turn problem (29.13), (29.14) into a Cauchy problem in some extended function space in which this problem will be well posed, and second, we must establish a correspondence (not only formal) between the solutions of problem (29.13), (29.14) and the solutions of the original problem (29.4), (29.5).

29.4 Cauchy Problem in an Extended Function Space. Statement of the Results

First, let us define the operator $\tau\xi(\tau)$, which occurs in the third and fourth equations of system (29.13).

Definition 29.2 (see [4] p. 31.). The multiplication operator on the independent variable $\mathbb{T}_k : \Omega_k \to \Omega_k$ $(k = 1, 2)$ is defined as follows:

$$\mathbb{T}_k \xi(\tau) = \tau \xi(\tau), \quad \xi(\tau) \in D(\mathbb{T}_k), \quad \tau \in \operatorname{supp} d\mu_k, \tag{29.15}$$

with domain

$$D(\mathbb{T}_k) = \{\xi \in \Omega_k : \tau\xi(\tau) \in \Omega_k\}. \tag{29.16}$$

Remark 29.2. 1. For any $\xi(\tau) \in \Omega_k$ the inequality

$$\int_0^{+\infty} \left\| \frac{e^{-t\tau}}{\sqrt{\tau}} \xi(\tau) \right\|_H d\mu_k(\tau) \leqslant \left(\int_0^{\infty} \frac{e^{-2t\tau} d\mu_k(\tau)}{\tau} \right)^{1/2} ||\xi(\tau)||_{\Omega_k}. \tag{29.17}$$

holds for $t \geqslant 0$.

2. The estimate

$$\left| \langle \tau\xi(\tau), \xi(\tau) \rangle_{\Omega_k} \right| \leqslant ||\tau\xi(\tau)||_{\Omega_k} ||\xi(\tau)||_{\Omega_k} \tag{29.18}$$

holds for any $\xi \in D(\mathbb{T}_k)$.

Indeed, it suffices to apply the Hölder inequality to the integrals on the left-hand side in inequalities (29.17) and (29.18).

Let us introduce operators $\mathbb{B}_k : H \to \Omega_k$ $(k = 1, 2)$ by the formulas

$$\mathbb{B}_k v = \frac{1}{\sqrt{\tau}} Q_k v \quad k = 1, 2, \quad \tau \in \operatorname{supp} \mu_k.$$

Then the adjoint operators $\mathbb{B}_k^* : \Omega_k \to H$ $(k = 1, 2)$ have the form

$$\mathbb{B}_k^* \xi(\tau) = Q_k^* \int_0^{\infty} \frac{1}{\sqrt{\tau}} \xi(\tau) d\mu_k(\tau) \quad k = 1, 2.$$

Indeed, for any $v \in D(\mathbb{B}_k)$ and $\xi(\tau) \in \Omega_k$ one has the chain of equalities

$$\begin{aligned} \langle \mathbb{B}_k v, \xi(\tau) \rangle_{\Omega_k} &= \left\langle \frac{1}{\sqrt{\tau}} Q_k v, \xi(\tau) \right\rangle_{\Omega_k} = \int_0^{+\infty} \left\langle \frac{1}{\sqrt{\tau}} Q_k v, \xi(\tau) \right\rangle_H d\mu_k(\tau) \\ &= \left\langle v, Q_k^* \int_0^{+\infty} \frac{1}{\sqrt{\tau}} \xi(\tau) d\mu_k(\tau) \right\rangle_H = \left\langle v, \mathbb{B}_k^* \xi(\tau) \right\rangle_H. \end{aligned}$$

Let us introduce the Hilbert space $\mathbb{H} = H \oplus H \oplus \left(\oplus_{k=1}^2 \Omega_k \right)$, with the natural inner product; in particular, the norm on $\mathbb{H}$ is given by

$$||(v, \xi_0, \xi_1(\tau), \xi_2(\tau))||_{\mathbb{H}}^2 = ||v||_H^2 + ||\xi_0||_H^2 + \sum_{k=1}^{2} ||\xi_k(\tau)||_{\Omega_k}^2, \ \tau \in \bigcup_{k=1}^{2} \operatorname{supp} \mu_k,$$

This space will be called the *extended function space*.

Consider the linear operator $\mathbb{A}$ on $\mathbb{H}$ with domain

$$D(\mathbb{A}) = \Big\{ (v, \xi_0, \xi_1(\tau), \xi_2(\tau)) \in \mathbb{H} :$$

$$v \in H_{1/2},\ \xi_0 + Q_k^* \int_0^{+\infty} \frac{1}{\sqrt{\tau}} \xi_k(\tau) d\mu_k(\tau) \in H_{1/2}, \xi_k(\tau) \in D(\mathbb{T}_k), k = 1, 2\Big\}$$

$$= \{(v, \xi_0, \xi_1(\tau), \xi_2(\tau)) \in \mathbb{H} :$$
$$v \in H_{1/2},\ \xi_0 + \mathbb{B}_k^* \xi_k(\tau) \in H_{1/2},\ \xi_k(\tau) \in D(\mathbb{T}_k),\ k = 1, 2\},$$

acting by the formula

$$\mathbb{A}(v, \xi_0, \xi_1(\tau), \xi_2(\tau))^T = \Bigg(-A_0^{1/2}\Bigg[\xi_0 + \sum_{k=1}^{2} Q_k^* \int_0^{+\infty} \frac{1}{\sqrt{\tau}} \xi_k(\tau) d\mu_k(\tau)\Bigg],$$
$$A_0^{1/2} v,\ Q_k A_0^{1/2} \frac{1}{\sqrt{\tau}} v(t) - \tau \xi_k(t, \tau),\ k = 1, 2\Bigg)^T$$
$$= \Bigg(-A_0^{1/2}\Bigg[\xi_0 + \sum_{k=1}^{2} \mathbb{B}_k^* \xi_k(\tau)\Bigg],\ A_0^{1/2} v,\ \mathbb{B}_k A_0^{1/2} v(t) - \mathbb{T}_k \xi_k(t, \tau),\ k = 1, 2\Bigg)^T.$$

Thus, $\mathbb{A}$ can be represented by the operator matrix

$$\mathbb{A} = \begin{pmatrix} A_0^{1/2} & 0 & 0 & 0 \\ 0 & I & 0 & 0 \\ 0 & 0 & I & 0 \\ 0 & 0 & 0 & I \end{pmatrix} \begin{pmatrix} 0 & -I & -\mathbb{B}_1^* & -\mathbb{B}_2^* \\ I & 0 & 0 & 0 \\ \mathbb{B}_1 & 0 & -\mathbb{T}_1 & 0 \\ \mathbb{B}_2 & 0 & 0 & -\mathbb{T}_2 \end{pmatrix} \begin{pmatrix} A_0^{1/2} & 0 & 0 & 0 \\ 0 & I & 0 & 0 \\ 0 & 0 & I & 0 \\ 0 & 0 & 0 & I \end{pmatrix}.$$

where I is the identity operator on the corresponding space.

Let us introduce the four-component vectors

$$Z(t) = (v(t), \xi_0(t), \xi_1(t, \tau), \xi_2(t, \tau)) \in \mathbb{H}, \quad z = (v_0, \xi_{00}, \xi_{10}(\tau), \xi_{20}(\tau)) \in \mathbb{H}.$$

Now we can write system (29.13), (29.14) as a first-order differential equation in an extended function space. Consider the following Cauchy problem in $\mathbb{H}$:

$$\frac{d}{dt} Z(t) = \mathbb{A} Z(t), \tag{29.19}$$

$$Z(0) = z. \tag{29.20}$$

Definition 29.3. A vector function

$$Z(t) = (v(t), \xi_0(t), \xi_1(t, \tau), \xi_2(t, \tau)),$$

$t \in \mathbb{R}_+$, ranging in $\mathbb{H}$, is a *classical solution* of problem (29.19), (29.20) if it belongs to the class $C^1(\mathbb{R}_+, \mathbb{H}) \cap C(\mathbb{R}_+, D(\mathbb{A}))$ for any

$$\tau \in \bigcup_{k=1}^{2} \operatorname{supp} d\mu_k$$

and satisfies Eq. (29.19) for any $t \in \mathbb{R}_+$ and the initial condition (29.20).

Definition 29.4 (see [9]). A linear operator A with dense domain on a Hilbert space is said to be dissipative if $\mathrm{Re}\,(Ax, x) \leqslant 0$ for any $x \in D(A)$ and *maximally dissipative* if it is dissipative and has no nontrivial dissipative extensions.

Theorem 29.1. *Let conditions* (29.7) *be satisfied. Then the linear operator* $\mathbb{A}$ *on* $\mathbb{H}$ *with dense domain* $D(\mathbb{A})$ *is maximally dissipative.*

Theorem 29.2. *Let conditions* (29.7) *be satisfied. Then the linear operator* $\mathbb{A}$ *is the generator of a contraction* C_0*-semigroup* $S(t) = e^{t\mathbb{A}}$ *on* $\mathbb{H}$*, the solution of problem* (29.19), (29.20) *can be represented in the form* $Z(t) = S(t)z$, $t \geqslant 0$, *and the energy equality*

$$\frac{d}{dt}||S(t)z||_{\mathbb{H}}^2 = -2\left(\int_0^{+\infty} \tau||\xi_1(t,\tau)||_H^2 d\mu_1(\tau) + \int_0^{+\infty} \tau||\xi_2(t,\tau)||_H^2 d\mu_2(\tau)\right). \quad (29.21)$$

holds for any $z \in D(\mathbb{A})$.

Consider the Cauchy problem for the inhomogeneous equation

$$\frac{d}{dt}Z(t) = \mathbb{A}Z(t) + F(t), \quad (29.22)$$

$$Z(0) = z. \quad (29.23)$$

We will assume that the vector function $F(t)$ has the form $F(t) := (f_1(t), 0, 0, 0)$, where $f_1(t) = f(t) - (M_1(t)A + M_2(t)B)\,\varphi_0$, and the functions $M_k(t)$, $k = 1, 2$ are given by formulas (29.8), and that the vector z has the form $z = \left(\varphi_1, A_0^{1/2}\varphi_0, 0, 0\right)$.

Theorem 29.3. *Let conditions* (29.7) *and any of the following conditions hold:*

1. *The vector function* $f(t)$ *belongs to the space* $H_{1/2}$ *and* $A_0^{1/2} f(t)$ *belongs to the space* $C(\mathbb{R}_+, H)$, *and the vectors* φ_0 *and* φ_1 *belong to the spaces* $H_{3/2}$ *and* $H_{1/2}$, *respectively.*
2. *The vector function* $f(t)$ *belongs to the space* $C^1(\mathbb{R}_+, H)$, *the functions* $M_k(t)$, $k = 1, 2$ *belong to the space* $C^1(\mathbb{R}_+)$, *and the vectors* φ_0 *and* φ_1 *belong to the spaces* H_1 *and* $H_{1/2}$, *respectively.*

Then problem (29.22), (29.23) *has a unique classical solution*

$$Z(t) = (v(t), \xi_0(t), \xi_1(t,\tau), \xi_2(t,\tau)),$$

where $v(t) := u'(t)$, $\xi_0(t) := A_0^{1/2}u(t)$, $u(t)$ *is a classical solution of problem* (29.4), (29.5), *and the estimate*

$$E(t) := \frac{1}{2}\left(\|u'(t)\|_H^2 + \left\|A_0^{1/2}u(t)\right\|_H^2\right) \leqslant \frac{1}{2}\|Z(t)\|_{\mathbb{H}}^2$$
$$\leqslant d\left[\left(\|\varphi_1\|_H^2 + \left\|A_0^{1/2}\varphi_0\right\|_H^2\right) + \left(\int_0^t \|f(s) - (M_1(s)A + M_2(s)B)\varphi_0\|_H ds\right)^2\right] \tag{29.24}$$

holds with constant d independent of the vector function f and the vectors φ_0 and φ_1.

29.5 Proofs of Theorems 29.1 and 29.2

To prove Theorems 29.1 and 29.2 we use the following two well-known theorems in the monograph [9, pp. 109–110].

Theorem 29.4. *([9, Theorem 4.3]) Any dissipative operator can be extended to a maximally dissipative operator. A dissipative operator $\mathcal{A}$ is maximally dissipative if and only if, for any $\lambda \in \mathbb{C}$ with positive real part, the range $R(\mathcal{A}-\lambda I)$ of the operator $\mathcal{A} - \lambda I$ coincides with the entire space.*

Theorem 29.5. *([9, Theorem 4.5]) There exists a contraction C_0-semigroup corresponding to the Cauchy problem for the equation $\dot{x} = \mathcal{A}x$ with a closed operator $\mathcal{A}$ in a Hilbert space if and only if $\mathcal{A}$ is a maximally dissipative operator.*

By $\mathcal{B}(H_1, H_2)$, where the H_i, $i = 1, 2$, are Banach spaces, we denote the Banach algebra of all bounded linear operators from H_1 to H_2. In particular, if $H_1 = H_2$, then the algebra $\mathcal{B}(H_1, H_2)$ is denoted by $\mathcal{B}(H_1)$.

Proof of Theorem 29.1. Let us show that the following statements hold for the operator $\mathbb{A}$:

1. The inequality $\operatorname{Re}\langle \mathbb{A}Z, Z\rangle_{\mathbb{H}} \leqslant 0$ holds for any $Z \in D(\mathbb{A})$;
2. The range of the mapping $(\mathbb{A} - \lambda\mathbb{I}) : D(\mathbb{A}) \to \mathbb{H}$ coincides with the space $\mathbb{H}$ for any λ with positive real part. Here $\mathbb{I}$ is the identity operator on $\mathbb{H}$.

Then the operator $\mathbb{A}$ will be maximally dissipative by Theorem 29.4.

Let us prove 1. Indeed, by the definition of $\mathbb{A}$,

$$\langle \mathbb{A}Z, Z\rangle_{\mathbb{H}} = \left\langle A_0^{1/2}\left[\xi_0 + \sum_{k=1}^{2} \mathbb{B}_k^*\xi_k(\tau)\right], v\right\rangle_H + \left\langle A_0^{1/2}v, \xi_0\right\rangle_H$$
$$+ \sum_{k=1}^{2}\left\langle \mathbb{B}_k A_0^{1/2}v, \xi_k(\tau)\right\rangle_{\Omega_k} - \sum_{k=1}^{2}\langle \tau\xi(\tau), \xi(\tau)\rangle_{\Omega_k}$$

$$= -\left\langle A_0^{1/2}\left[\xi_0 + \sum_{k=1}^{2} \mathbb{B}_k^* \xi_k(\tau)\right], v\right\rangle_H + \overline{\left\langle \xi_0, A_0^{1/2} v\right\rangle_H}$$

$$+ \sum_{k=1}^{2} \overline{\left\langle \mathbb{B}_k^* \xi_k(\tau), A_0^{1/2} v\right\rangle_H} - \sum_{k=1}^{2} \langle \tau\xi(\tau), \xi(\tau)\rangle_{\Omega_k}$$

$$= -\left\langle A_0^{1/2}\left[\xi_0 + \sum_{k=1}^{2} \mathbb{B}_k^* \xi_k(\tau)\right], v\right\rangle_H + \overline{\left\langle A_0^{1/2}\left[\xi_0 + \sum_{k=1}^{2} \mathbb{B}_k^* \xi_k(\tau)\right], v\right\rangle_H}$$

$$- \sum_{k=1}^{2} \langle \tau\xi(\tau), \xi(\tau)\rangle_{\Omega_k}$$

$$= -2i \operatorname{Im} \left\langle A_0^{1/2}\left[\xi_0 + A_0^{1/2}\mathbb{B}_k^* \xi_k(\tau)\right], v\right\rangle_H - \sum_{k=1}^{2} \langle \tau\xi(\tau), \xi(\tau)\rangle_{\Omega_k}. \tag{29.25}$$

Hence it follows that

$$\operatorname{Re}\langle \mathbb{A}Z, Z\rangle_{\mathbb{H}} = -\sum_{k=1}^{2} \langle \tau\xi_k(\tau), \xi_k(\tau)\rangle_{\Omega_k} = -\sum_{k=1}^{2} \int_0^{+\infty} \tau\|\xi_k(\tau)\|_H^2 d\mu_k(\tau) \leqslant 0. \tag{29.26}$$

To prove 2, we will show that the operator $(\mathbb{A} - \lambda\mathbb{I})$ is continuously invertible on $\mathbb{H}$ and $(\mathbb{A} - \lambda\mathbb{I})^{-1} \in \mathcal{B}(\mathbb{H})$ for any λ with positive real part. To this end, we need the following assertion, which can be proved by a straightforward verification.

Proposition 29.1. *Let $\tilde{H}_k$ $(k = 1, 2)$ be Hilbert spaces. Assume that $A_{11}^{-1} \in \mathcal{B}(\tilde{H}_1)$, $A_{22}^{-1} \in \mathcal{B}(\tilde{H}_2)$, $A_{12} \in \mathcal{B}(\tilde{H}_2, \tilde{H}_1)$, $A_{21} \in \mathcal{B}(\tilde{H}_1, \tilde{H}_2)$, and $D_1 := A_{11} - A_{12}A_{22}^{-1}A_{21}$, $D_1^{-1} \in \mathcal{B}(\tilde{H}_1)$, and consider the linear operator defined on the space $\tilde{H}_1 \oplus \tilde{H}_2$ by the formula*

$$\mathcal{A} = \begin{pmatrix} A_{11} & A_{12} \\ A_{21} & A_{22} \end{pmatrix}.$$

Then the operator $\mathcal{A}^{-1}$ belongs to the space $\mathcal{B}\left(\tilde{H}_1 \oplus \tilde{H}_2\right)$.

Proof. If the assumptions of Proposition 29.1 are satisfied, then one can readily verify the representation (a Schur–Frobenius type factorization, see [24, Sect. 1.6])

$$\mathcal{A}^{-1} = \begin{pmatrix} A_{11} & A_{12} \\ A_{21} & A_{22} \end{pmatrix}^{-1} = \left[\begin{pmatrix} I & A_{12}A_{22}^{-1} \\ 0 & I \end{pmatrix}\begin{pmatrix} D_1 & 0 \\ 0 & A_{22} \end{pmatrix}\begin{pmatrix} I & 0 \\ A_{22}^{-1}A_{21} & I \end{pmatrix}\right]^{-1}$$

$$= \begin{pmatrix} D_1^{-1} & -D_1^{-1}A_{12}A_{22}^{-1} \\ -A_{22}^{-1}A_{21}D_1^{-1} & A_{22}^{-1}\left[I + A_{21}D_1^{-1}A_{12}A_{22}^{-1}\right] \end{pmatrix} \in \mathcal{B}\left(\tilde{H}_1 \oplus \tilde{H}_2\right). \tag{29.27}$$

This proves Proposition 29.1.

We represent the operator $\mathbb{A} - \lambda\mathbb{I}$ as the product

$$\mathbb{A}-\lambda\mathbb{I}=\begin{pmatrix} A_0^{1/2} & 0 & 0 & 0\\ 0 & I & 0 & 0\\ 0 & 0 & I & 0\\ 0 & 0 & 0 & I\end{pmatrix}\begin{pmatrix} -\lambda A_0^{-1} & -I & -\mathbb{B}_1^* & -\mathbb{B}_2^*\\ I & -\lambda I & 0 & 0\\ \mathbb{B}_1 & 0 & -\mathbb{T}_1-\lambda I & 0\\ \mathbb{B}_2 & 0 & 0 & -\mathbb{T}_2-\lambda I\end{pmatrix}\begin{pmatrix} A_0^{1/2} & 0 & 0 & 0\\ 0 & I & 0 & 0\\ 0 & 0 & I & 0\\ 0 & 0 & 0 & I\end{pmatrix}$$
$$=: \mathbb{A}_0\mathbb{A}_1(\lambda)\mathbb{A}_0. \tag{29.28}$$

Applying the notation of Proposition 29.1 to the operator function $\mathbb{A}_1(\lambda)$, we obtain $\tilde{H}_1 = H$ and $\mathbb{H} = \tilde{H}_1 \oplus \tilde{H}_2 = H \oplus \mathbb{H}_0$, where $\mathbb{H}_0 := H \oplus \left(\oplus_{k=1}^2 \Omega_k\right)$ $A_{11} := -\lambda A_0^{-1}$, $A_{12} := \left(-I, -\mathbb{B}_1^*, -\mathbb{B}_2^*\right)$, $A_{21} := (I, \mathbb{B}_1, \mathbb{B}_2)^T$, and

$$A_{22}=\begin{pmatrix} -\lambda I & 0 & 0\\ 0 & -\mathbb{T}_1-\lambda I & 0\\ 0 & 0 & -\mathbb{T}_2-\lambda I\end{pmatrix}.$$

Then the relations

$$A_{22}^{-1}=\begin{pmatrix} -\lambda^{-1}I & 0 & 0\\ 0 & -(\mathbb{T}_1+\lambda I)^{-1} & 0\\ 0 & 0 & -(\mathbb{T}_2+\lambda I)^{-1}\end{pmatrix}=\begin{pmatrix} -\lambda^{-1}I & 0 & 0\\ 0 & -(\tau+\lambda)^{-1}I & 0\\ 0 & 0 & -(\tau+\lambda)^{-1}I\end{pmatrix},$$

$$A_{12}A_{22}^{-1} = \left(\lambda^{-1}I, \mathbb{B}_1^*(\mathbb{T}_1+\lambda I)^{-1}, \mathbb{B}_2^*(\mathbb{T}_2+\lambda I)^{-1}\right),$$

$$A_{22}^{-1}A_{21} = \left(-\lambda^{-1}I, -(\mathbb{T}_1+\lambda I)^{-1}\mathbb{B}_1, -(\mathbb{T}_2+\lambda I)^{-1}\mathbb{B}_2\right)^T,$$

$$D_1 = A_{11} - A_{12}A_{22}^{-1}A_{21} = -\lambda A_0^{-1} - \lambda^{-1}I - \sum_{k=1}^{2}\mathbb{B}_k^*(\mathbb{T}_k+\lambda I)^{-1}\mathbb{B}_k$$

$$= -\lambda A_0^{-1} - \lambda^{-1}I - \sum_{k=1}^{2}\int_0^{+\infty}\frac{d\mu_k(\tau)}{\tau(\lambda+\tau)}Q_k^*Q_k$$

$$= -\lambda^{-1}A_0^{-1/2}\left[\lambda^2 + A_0 + \left(\int_0^{+\infty}\frac{d\mu_1(\tau)}{\tau} - \int_0^{+\infty}\frac{d\mu_1(\tau)}{\lambda+\tau}\right)A\right.$$
$$\left. + \left(\int_0^{+\infty}\frac{d\mu_2(\tau)}{\tau} - \int_0^{+\infty}\frac{d\mu_2(\tau)}{\lambda+\tau}\right)B\right]A_0^{-1/2}$$

$$= -\lambda^{-1}A_0^{-1/2}\left[\lambda^2 + \left(1 - \int_0^{\infty}\frac{d\mu_1(\tau)}{\tau}\right)A + \left(1 - \int_0^{\infty}\frac{d\mu_2(\tau)}{\tau}\right)B\right.$$

$$+\left(\int_0^{+\infty}\frac{d\mu_1(\tau)}{\tau}-\int_0^{+\infty}\frac{d\mu_1(\tau)}{\lambda+\tau}\right)A+\left(\int_0^{+\infty}\frac{d\mu_2(\tau)}{\tau}-\int_0^{+\infty}\frac{d\mu_2(\tau)}{\lambda+\tau}\right)B\Bigg]A_0^{-1/2}$$

$$=-\lambda^{-1}A_0^{-1/2}\left[\lambda^2+A+B-\int_0^{+\infty}\frac{d\mu_1(\tau)}{\lambda+\tau}A-\int_0^{+\infty}\frac{d\mu_2(\tau)}{\lambda+\tau}B\right]A_0^{-1/2}=:M(\lambda),\tag{29.29}$$

where the operator function $M(\lambda)$ holds for all $\lambda\neq 0$, $\lambda\neq-\tau$, $\tau\in\bigcup_{k=1}^{2}\operatorname{supp}\mu_k$.

Let us show that the assumptions of Proposition 29.1 are satisfied for the operator function $\mathbb{A}_1(\lambda)$ for all λ such thatRe $\lambda>0$.

Lemma 29.1. *In the above notation, the following statements hold:*

a) For all λ such that $\operatorname{Re}\lambda>0$, *the operator function $M^{-1}(\lambda)$ belongs to space $\mathcal{B}(H)$;*

b) For all $\lambda\neq 0$, $\lambda\neq-\tau$, where $\tau\in\bigcup_{k=1}^{2}$ supp μ_k one has the inclusion

$$-(\mathbb{T}+\lambda)^{-1}:=\begin{pmatrix}-\lambda^{-1}I & 0 & 0\\ 0 & -(\mathbb{T}_1+\lambda I)^{-1} & 0\\ 0 & 0 & -(\mathbb{T}_2+\lambda I)^{-1}\end{pmatrix}\in\mathcal{B}(\mathbb{H}_0),$$

c) One has the inclusions

$$\mathbb{B}^*=\left(I,\mathbb{B}_1^*,\mathbb{B}_2^*\right)\in\mathcal{B}(\mathbb{H}_0,H),\mathbb{B}=(I,\mathbb{B}_1,\mathbb{B}_2)^T\in\mathcal{B}(H,\mathbb{H}_0).$$

Proof. Note that the estimate

$$\operatorname{Re}\int_0^{+\infty}\frac{d\mu_k(\tau)}{\tau(\lambda+\tau)}=\int_0^{+\infty}\frac{(\operatorname{Re}\lambda+\tau)d\mu_k(\tau)}{\tau|\lambda+\tau|^2}\leqslant\int_0^{+\infty}\frac{d\mu_k(\tau)}{\tau|\lambda+\tau|}\leqslant\frac{1}{|\lambda|}\int_0^{+\infty}\frac{d\mu_k(\tau)}{\tau},\quad k=1,2\tag{29.30}$$

holds for all λ with positive real part by virtue of condition (29.7).

Let us prove a). By the definition of the operator function in (29.29),

$$M(\lambda):=-\lambda A_0^{-1}-\lambda^{-1}I-\sum_{k=1}^{2}\mathbb{B}_k^*(\mathbb{T}_k+\lambda I)^{-1}\mathbb{B}_k.$$

Therefore, in view of the estimate (29.30), the chain of inequalities

$$\|M(\lambda)v\|_H\geqslant\frac{|(M(\lambda)v,v)|_H}{\|v\|_H}$$

$$=\frac{1}{\|v\|_H}\left|-\lambda\left(A_0^{-1}v,v\right)_H-\frac{1}{\lambda}(v,v)_H-\sum_{k=1}^{2}(\mathbb{B}_k^*(\tau+\lambda)^{-1}\mathbb{B}_kv,v)_H\right|$$

$$\geqslant \frac{1}{||v||_H}\left(\operatorname{Re}\lambda\left(A_0^{-1/2}v, A_0^{-1/2}v\right)_H+\frac{\operatorname{Re}\lambda}{|\lambda|^2}(v,v)_H+\sum_{k=1}^{2}\operatorname{Re}\int_0^{+\infty}\frac{d\mu_k(\tau)}{\tau(\lambda+\tau)}(Q_kv,Q_kv)_H\right)$$

$$=||v||_H\left(\operatorname{Re}\lambda\frac{||A_0^{-1/2}v||_H^2}{||v||_H^2}+\frac{\operatorname{Re}\lambda}{|\lambda|^2}+\sum_{k=1}^{2}\operatorname{Re}\int_0^{+\infty}\frac{d\mu_k(\tau)}{\tau(\lambda+\tau)}\frac{||Q_kv||_H^2}{||v||_H^2}\right)\geqslant ||v||_H\frac{\operatorname{Re}\lambda}{|\lambda|^2}.$$

holds for all λ such that $\operatorname{Re}\lambda > 0$.

b) Let us show that the operator $\mathbb{T}(\lambda)$ belongs to the space $\mathcal{B}(\mathbb{H}_0)$ for all $\lambda \neq 0$ such that $\lambda \neq -\tau$, where $\tau \in \bigcup_{k=1}^{2} \operatorname{supp}\mu_k$.

Indeed, the estimate

$$\frac{||\mathbb{T}(\lambda)Z_0||_{\mathbb{H}_0}}{||Z_0||_{\mathbb{H}_0}}=\frac{||\left(\lambda^{-1}\xi_0,(\tau+\lambda)^{-1}\xi_1(\tau),(\tau+\lambda)^{-1}\xi_2(\tau)\right)||_{\mathbb{H}_0}}{||(\xi_0,\xi_1(\tau),\xi_2(\tau))||_{\mathbb{H}_0}}$$

$$=\frac{|\lambda|^{-2}||\xi_0||_H^2+\sum_{k=1}^{2}\int_0^{+\infty}\frac{||\xi_k(\tau)||_H^2d\mu_k(\tau)}{|\lambda+\tau|^2}}{||\xi_0||_H^2+\sum_{k=1}^{2}\int_0^{+\infty}||\xi_k(\tau)||_H^2d\mu_k(\tau)}\leqslant\frac{1}{|\lambda|^2}.$$

holds for any vector $Z_0=(\xi_0,\xi_1(\tau),\xi_2(\tau))^T\in\mathbb{H}_0$.

c) For any vector $Z_0=(\xi_0,\xi_1(\tau),\xi_2(\tau))^T\in\mathbb{H}_0$, we obtain

$$\frac{||\left(I,\mathbb{B}_1^*,\mathbb{B}_2^*\right)Z_0||_H^2}{||Z_0||_{\mathbb{H}_0}^2}=\frac{\left\|\xi_0+\sum_{k=1}^{2}Q_k^*\int_0^{+\infty}\frac{\xi_k(\tau)}{\sqrt{\tau}}d\mu_k(\tau)\right\|_H^2}{||\xi_0||_H^2+\sum_{k=1}^{2}\int_0^{+\infty}||\xi_k(\tau)||_H^2d\mu_k(\tau)}$$

$$\leqslant 2\frac{||\xi_0||_H^2+\sum_{k=1}^{2}\left(\int_0^{+\infty}\frac{||Q_k^*\xi_k(\tau)||_H}{\sqrt{\tau}}d\mu_k(\tau)\right)^2}{||\xi_0||_H^2+\sum_{k=1}^{2}\int_0^{+\infty}||\xi_k(\tau)||_H^2d\mu_k(\tau)}$$

$$\leqslant 2\frac{||\xi_0||_H^2+\sum_{k=1}^{2}\left(\int\limits_0^{+\infty}\frac{d\mu_k(\tau)}{\tau}\right)\left(\int\limits_0^{+\infty}||Q_k^*\xi_k(\tau)||_H^2 d\mu_k(\tau)\right)}{||\xi_0||_H^2+\sum_{k=1}^{2}\int\limits_0^{+\infty}||\xi_k(\tau)||_H^2 d\mu_k(\tau)}$$

$$\leqslant 2\frac{||\xi_0||_H^2+\sum_{k=1}^{2}||Q_k^*||_H^2\int\limits_0^{+\infty}||\xi_k(\tau)||_H^2 d\mu_k(\tau)}{||\xi_0||_H^2+\sum_{k=1}^{2}\int\limits_0^{+\infty}||\xi_k(\tau)||_H^2 d\mu_k(\tau)}\leqslant 2\max\left\{1,||Q_1||_H^2,||Q_2||_H^2\right\}.$$

Since the operators Q_k, $k=1,2$, are bounded, we have

$$\frac{||(I,\mathbb{B}_1,\mathbb{B}_2)^T v||_{\mathbb{H}_0}^2}{||v||_H^2}=\frac{||v||_H^2+\sum_{k=1}^{2}\int\limits_0^{+\infty}\frac{d\mu_k(\tau)}{\tau}||Q_k||_H^2||v||_H^2}{||v||_H^2}\leqslant 1+\sum_{k=1}^{2}||Q_k||_H^2.$$

for any vector $v\in H$. The proof of Lemma 29.1 is complete.

Lemma 29.1 implies that the assumptions of Proposition 29.1 are satisfied for the operator $\mathbb{A}_1(\lambda)$ for $\operatorname{Re}\lambda>0$, and hence the operator $(\mathbb{A}-\lambda\mathbb{I})^{-1}$ $\mathcal{B}(\mathbb{H})$. Thus, the range $R(\mathbb{A}-\lambda\mathbb{I})=\mathbb{H}$ coincides with the entire space $\mathbb{H}$ for $\operatorname{Re}\lambda>0$ and hence the operator $\mathbb{A}$ is maximally dissipative in this space. The proof of Theorem 29.1 is complete.

Proof of Theorem 29.2. The assertion of Theorem 29.2 follows from Theorem 29.5 and Theorem 29.1. Let $z\in D(\mathbb{A})$. Then $S(t)z\in D(\mathbb{A})$ for any $t>0$, and Eq. (29.19) implies the equality

$$\frac{d}{dt}||S(t)z||_{\mathbb{H}}^2=2\operatorname{Re}\langle\mathbb{A}S(t)z,S(t)z\rangle_{\mathbb{H}}.$$

On the other hand, we obtain

$$\operatorname{Re}\langle\mathbb{A}Z,Z\rangle_{\mathbb{H}}=-\sum_{k=1}^{2}\langle\tau\xi_k(\tau),\xi_k(\tau)\rangle_{\Omega_k}$$

by (29.26), whence the energy equality (29.21) follows. The proof of Theorem 29.2 is complete.

29.6 Proof of Theorem 29.3

In this section, we use the definitions and statements from the monograph [9, Sect. 1.2].

Definition 29.5. (see [9]) Let $\mathcal{A}$ be a linear operator with dense domain $D(\mathcal{A})$ on a Banach space $\mathcal{H}$. The Cauchy problem

$$\frac{d}{dt}Z(t) = \mathcal{A}Z(t), \tag{29.31}$$

$$Z(0) = z. \tag{29.32}$$

is said to be *well posed (uniformly well posed)* if

1. For each $z \in D(\mathcal{A})$ there exists a unique solution of problem (29.31), (29.32);
2. This solution continuously depends on the initial data in the following sense: if a sequence $Z_n(0) \to 0$ ($Z_n(0) \in D(\mathcal{A})$) converges to zero, then the sequence $Z_n(t) \to 0$ of the corresponding solutions converges to zero for each $t \in [0, T]$ (uniformly with respect to $t \in [0, T]$) on any finite interval $[0, T]$.

Remark 29.3. If the Cauchy problem (29.31), (29.32) generates a contraction semigroup on the space $\mathcal{H}$, then this problem is uniformly well posed.

We will use the following results in [9, Secs. 1.1, 1.6].

Theorem 29.6. *([9, Theorem 1.1]) If the Cauchy problem* (29.31), (29.32) *is well posed, then its solution is given by the formula $Z(t) = S(t)z$, ($z \in D(\mathcal{A})$) where $S(t)$ is an operator semigroup strongly continuous for $t > 0$.*

Theorem 29.7. *([9, Theorem 6.5]) If the Cauchy problem* (29.31), (29.32) *is uniformly well posed, then the formula*

$$Z(t) = S(t)z + \int_0^t S(t-p)F(p)dp \tag{29.33}$$

gives the solution of the Cauchy problem for the inhomogeneous equation

$$\frac{d}{dt}Z(t) = \mathcal{A}Z(t) + F(t), \tag{29.34}$$

$$Z(0) = z. \tag{29.35}$$

where $z \in D(\mathcal{A})$ and the vector function $F(t)$ satisfies one of the following two conditions:

1. *The set of values of the function $F(t)$ is contained in $D(\mathcal{A})$ and the function $\mathcal{A}F(t)$ belongs to the class $C([0, +\infty), \mathbb{H})$.*
2. *The function $F(t)$ belongs to the class $C^1([0, +\infty), \mathbb{H})$.*

Proof of Theorem 29.3.

The Cauchy problem (29.22), (29.23) written coordinatewise has the form (29.13), (29.14). Consider the last two equations in system (29.13). Let us apply the variation of arbitrary constants to these equations. By the definition of the operators $\mathbb{T}_k\xi(\tau) = \tau\xi(\tau)$, $k = 1, 2$, where $D(\mathbb{T}_k) = \{\xi \in \Omega_k : \tau\xi \in \Omega_k\}$, the corresponding homogeneous equations have the form

$$\frac{d\xi_k(t,\tau)}{dt} = -\mathbb{T}_k\xi_k(t,\tau).$$

Therefore, the general solutions of the homogeneous equations can be written in the form $\xi_k^O(t,\tau) = e^{-t\mathbb{T}_k}C_k(\tau)$, where the $C_k(\tau) \in \Omega_k$. Applying formula(29.33) to solve the inhomogeneous equations under the condition

$$\xi_k(t,\tau)|_{t=0} = 0, \quad \tau \in \bigcup_{k=1}^{2} \operatorname{supp} \mu_k, \quad k = 1, 2,$$

we obtain

$$\xi_k(t,\tau) = \int_0^t e^{-(t-s)\mathbb{T}_k} \frac{1}{\sqrt{\tau}} Q_k A_0^{1/2} v(s)ds = \int_0^t e^{-(t-s)\tau} \frac{1}{\sqrt{\tau}} Q_k A_0^{1/2} v(s)ds.$$

From the first equation in the system, in accordance with the domain of $\mathbb{A}$, we have

$$\xi_0(t) + \sum_{k=1}^{2} \int_0^\infty \frac{1}{\sqrt{\tau}} Q_k^* \xi_k(t,\tau) d\mu_k(\tau) \in D(A_0^{1/2}). \tag{29.36}$$

Substituting the resulting expressions for $\xi_k(t,\tau)$ into (29.36), we obtain

$$\int_0^t A_0^{1/2} v(s)ds + A_0^{1/2}\varphi_0 + \sum_{k=1}^{2} \int_0^\infty \frac{1}{\sqrt{\tau}} Q_k^* \int_0^t e^{-(t-s)\tau} \frac{1}{\sqrt{\tau}} Q_k A_0^{1/2} v(s)ds d\mu_k(\tau)$$

$$= \int_0^t \left[A_0^{1/2} + \sum_{k=1}^{2} \left(\int_0^{+\infty} \frac{e^{-(t-s)\tau}}{\tau} d\mu_k(\tau) \right) Q_k^* Q_k A_0^{1/2} \right] v(s)ds + A_0^{1/2}\varphi_0 \in D(A_0^{1/2}).$$

By the assumptions of Theorem 29.3, either $\varphi_0 \in H_{3/2}$ and $\varphi_1 \in H_{1/2}$ or $\varphi_0 \in H_1$ and $\varphi_1 \in H_{1/2}$. Thus, $z = \left(\varphi_1, A_0^{1/2}\varphi_0, 0, 0\right) \in D(\mathbb{A})$; i.e. the assumptions of Theorem 29.7 are satisfied. Hence

$$\int_0^t \left[A_0^{1/2} + \sum_{k=1}^{2} \left(\int_0^{+\infty} \frac{e^{-(t-s)\tau}}{\tau} d\mu_k(\tau) \right) Q_k^* Q_k A_0^{1/2} \right] v(s)ds \in D(A_0^{1/2}).$$

Thus, problem (29.22), (29.23) is uniformly well posed by Theorem 29.7 and one has the estimate

$$\|Z(t)\|_{\mathbb{H}} \leqslant d\left(\|z\|_{\mathbb{H}} + \int_0^t \|F(s)\|_{\mathbb{H}} ds\right) \tag{29.37}$$

with a constant d independent of the vector function F and the vectors φ_0 and φ_1. The estimate (29.37) follows from formula (29.33), applied to problem (29.22), (29.23), in the notation of Theorem 29.3 because $Z(t) = S(t)z$, where $S(t)$ is a contraction semigroup by Theorem 29.1.

Let us show that $v(t) := u'(t)$, $\xi_0(t) := A_0^{1/2}u(t)$, where $u(t)$ is the classical solution of problem (29.4), (29.5). Consider the vector function

$$-A_0^{1/2}\int_0^t \left[A_0^{1/2} + \sum_{k=1}^{2}\left(\int_0^{+\infty} \frac{e^{-(t-s)\tau}}{\tau} d\mu_k(\tau)\right) Q_k^* Q_k A_0^{1/2}\right] v(s)ds$$

$$= -A_0 \int_0^t A_0^{-1/2}\left[I + \sum_{k=1}^{2}\left(\int_0^{+\infty} \frac{e^{-(t-s)\tau}}{\tau} d\mu_k(\tau)\right) Q_k^* Q_k\right] A_0^{1/2} v(s)ds, \quad t > 0. \tag{29.38}$$

Relation (29.38) implies the inclusion

$$\int_0^t A_0^{-1/2}\left[I + \sum_{k=1}^{2}\left(\int_0^{+\infty} \frac{e^{-(t-s)\tau}}{\tau} d\mu_k(\tau)\right) Q_k^* Q_k\right] A_0^{1/2} v(s)ds \in D(A_0).$$

Let us introduce the notation

$$R(t) := A_0^{-1/2}\left[\sum_{k=1}^{2}\left(\int_0^{+\infty} \frac{e^{-t\tau}}{\tau} d\mu_k(\tau)\right) Q_k^* Q_k\right] A_0^{1/2}, \quad t > 0.$$

Then the vector function in (29.38) can be written as

$$\int_0^t v(s)ds + \int_0^t R(t-s)v(s)ds =: y(t) \in D(A_0). \tag{29.39}$$

After integrating by parts in (29.39) and introducing the new vector function $w(t) = \int_0^t v(s)ds$, we obtain the Volterra integral equation of the second kind

$$(I + R(0))w(t) + \int_0^t R'(t-s)w(s)ds = y(t), \quad y(t) \in C(\mathbb{R}_+; H_1). \tag{29.40}$$

where

$$R'(t) = -A_0^{-1/2} \sum_{k=1}^{2} \left(\int_0^{+\infty} e^{-t\tau} d\mu_k(\tau) \right) Q_k^* Q_k A_0^{1/2}$$

$$= -\left(\int_0^{+\infty} e^{-t\tau} d\mu_1(\tau) \right) A_0^{-1} A - \left(\int_0^{+\infty} e^{-t\tau} d\mu_2(\tau) \right) A_0^{-1} B,$$

$$R(0) = A_0^{-1/2} \left[\sum_{k=1}^{2} \left(\int_0^{+\infty} \frac{d\mu_k(\tau)}{\tau} \right) Q_k^* Q_k \right] A_0^{1/2}$$

$$= \left(\int_0^{+\infty} \frac{d\mu_1(\tau)}{\tau} \right) A_0^{-1} A + \left(\int_0^{+\infty} \frac{d\mu_2(\tau)}{\tau} \right) A_0^{-1} B.$$

Let us show that the operator function $R'(t)$ belongs to the space $C(\mathbb{R}_+; \mathcal{B}(H_1))$. Indeed,

$$||R'(t)z||_{H_1} = \left\| \left[\left(\int_0^{+\infty} e^{-t\tau} d\mu_1(\tau) \right) A + \left(\int_0^{+\infty} e^{-t\tau} d\mu_2(\tau) \right) B \right] A_0^{-1} (A_0 z) \right\|_H$$

$$\leqslant \left\| \left(\int_0^{+\infty} e^{-t\tau} d\mu_1(\tau) \right) A A_0^{-1} + \left(\int_0^{+\infty} e^{-t\tau} d\mu_2(\tau) \right) B A_0^{-1} \right\|_H ||z||_{H_1}$$

$$\leqslant \left[\left(\int_0^{+\infty} e^{-t\tau} d\mu_1(\tau) \right) ||A A_0^{-1}||_H + \left(\int_0^{+\infty} e^{-t\tau} d\mu_2(\tau) \right) ||B A_0^{-1}||_H \right] ||z||_{H_1}.$$

for each $z \in H_1$. Thus, $R'(t) \in \mathcal{B}(H_1)$. In addition, the inequality

$$||R'(t_1) - R'(t_2)||_{H_1} \leqslant \left(\int_0^{+\infty} (e^{-t_1\tau} - e^{-t_2\tau}) d\mu_1(\tau) \right) ||A_0^{-1} A||_H$$

$$+ \left(\int_0^{+\infty} (e^{-t_1\tau} - e^{-t_2\tau}) d\mu_2(\tau) \right) ||A_0^{-1} B||_H.$$

holds for any $t_1, t_2 > 0$. Consequently, the operator function $R'(t)$ belongs to the space $C(\mathbb{R}_+; \mathcal{B}(H_1))$. It follows from Eq. (29.40) that

$$w(t) = (I + R(0))^{-1} \left(y(t) - \int_0^t R'(t-s) w(s) ds \right) =: S w(t),$$

where the operator S acts in the spaces $S : H_1 \to H_1$. Let us show that $||S||_{C(\mathbb{R}_+;L(H_1))} < +\infty$.

Remark 29.4. For any functions $w_1(t), w_2(t) \in H_1$ and any $T > 0$ for $t \in [0,T]$ one has the estimate

$$\sup_{t\in[0,T]} ||Sw_1(t) - Sw_2(t)||_{H_1} \leqslant \kappa \sup_{t\in[0,T]} ||w_1(t) - w_2(t)||_{H_1},$$

where $\kappa = \left\|(M_1(0)A + M_2(0)B)\,(A+B)^{-1}\right\|_H < +\infty$ and the functions $M_k(t)$ $(k = 1, 2)$, are defined by formulas (29.8).

Indeed, for any $w_1(t), w_2(t) \in H_1$ for $t > 0$ we have

$$||Sw_1(t) - Sw_2(t)||_{H_1} = \left\|(I + R(0))^{-1} \int_0^t R'(t-s)\,(w_1(s) - w_2(s))\,ds\right\|_{H_1}. \tag{29.41}$$

Further, for any $z \in H_1$ we obtain

$$\left\|(I + R(0))^{-1}z\right\|_{H_1} = \left\|A_0(I + R(0))^{-1}A_0^{-1}(A_0z)\right\|_H = \left\|\left(A_0(I+R(0))A_0^{-1}\right)^{-1}(A_0z)\right\|_H$$

It is easy to verify using the definition of the operator A_0 that

$$A_0(I + R(0))A_0^{-1} = (A + B)A_0^{-1}.$$

Thus,

$$\left\|\left(A_0(I + R(0))A_0^{-1}\right)^{-1}(A_0z)\right\|_H = \left\|A_0(A + B)^{-1}(A_0z)\right\|_H. \tag{29.42}$$

Substituting $z = \int_0^t R'(t-s)\,(w_1(s) - w_2(s))\,ds$ into formula(29.42) and taking into account the representation (29.41), for any $T > 0$ we obtain

$$\sup_{t\in[0,T]} ||Sw_1(t) - Sw_2(t)||_{H_1}$$

$$= \sup_{t\in[0,T]} \left\|A_0(A + B)^{-1}A_0 \int_0^t R'(t-s)\,(w_1(s) - w_2(s))\,ds\right\|_H$$

$$\leqslant \left\|A_0(A + B)^{-1}A_0 \int_0^t R'(t-s)A_0^{-1}ds\right\|_H \sup_{t\in[0,T]} ||w_1(t) - w_2(t)||_{H_1}. \tag{29.43}$$

One can establish the estimate

$$\sup_{t\in[0,T]}\left\|A_0(A+B)^{-1}A_0\int_0^t R'(t-s)A_0^{-1}ds\right\|_H$$
$$= \sup_{t\in[0,T]}\left\|A_0(A+B)^{-1}A_0\left(R(t)-R(0)\right)A_0^{-1}\right\|_H \leqslant \left\|A_0(A+B)^{-1}A_0R(0)A_0^{-1}\right\|_H. \tag{29.44}$$

It can readily be verified using the definition of the operator A_0 that

$$\left\|A_0(A+B)^{-1}A_0R(0)A_0^{-1}\right\|_H = \left\|(M_1(0)A+M_2(0)B)\,(A+B)^{-1}\right\|_H < +\infty. \tag{29.45}$$

Indeed, we have the chain of equalities

$$A_0(A+B)^{-1}A_0R(0)A_0^{-1} = A_0(A+B)^{-1}A_0(M_1(0)A_0^{-1}A+M_2(0)A_0^{-1}B)A_0^{-1}$$

$$= A_0(A+B)^{-1}(M_1(0)A+M_2(0)B)A_0^{-1} = A_0(A+B)^{-1}(A+B-A_0)A_0^{-1}$$

$$= A_0(A+B)^{-1}((A+B)A_0^{-1}-I) = I - A_0(A+B)^{-1}$$

$$= I-(A+B-M_1(0)A-M_2(0)B)(A+B)^{-1} = (M_1(0)A+M_2(0)B)\,(A+B)^{-1}.$$

It follows from the earlier established estimates (29.43)–(29.45) that

$$\sup_{t\in[0,T]}||Sw_1(t)-Sw_2(t)||_{H_1}$$

$$\leqslant \left\|(M_1(0)A+M_2(0)B)\,(A+B)^{-1}\right\|_H \sup_{t\in[0,T]}||w_1(t)-w_2(t)||_{H_1}$$

$$= \kappa \sup_{t\in[0,T]}||w_1(t)-w_2(t)||_{H_1},$$

where $\kappa = \left\|(M_1(0)A+M_2(0)B)\,(A+B)^{-1}\right\|_H < +\infty$. Hence for each $n\in\mathbb{N}$ we obtain the inequality

$$\sup_{t\in[0,T]}||S^nw_1(t)-S^nw_2(t)||_{H_1} \leqslant \frac{\kappa^nT^n}{n!}\cdot \sup_{t\in[0,T]}||w_1(t)-w_2(t)||_{H_1}.$$

Thus, the value of n can be chosen so large that $\dfrac{\kappa^nT^n}{n!} < 1$. Consequently, $||S^n||_{C([0,T];\mathcal{B}(H_1))} < 1$ i.e., the mapping $S^n : C([0,T],H_1) \to C([0,T],H_1)$ is contractive, and Eq. (29.40) has a unique solution $w(t)\in C\left([0,T],H_1\right)$. By virtue of the arbitrariness $T>0$, based on this, we conclude that $w(t)\in C\left([0,+\infty),H_1\right)$ and $v(t)\in D(A_0)$.

Let us return to the first equation in system (29.13) and use the fact that $v(t)\in D(A_0)$. We have chain of equalities

$$-A_0^{1/2}\left[\xi_0(t)+\sum_{k=1}^{2}Q_k^*\int_0^{+\infty}\frac{1}{\sqrt{\tau}}\xi_k(t,\tau)d\tau\right]$$

$$+f(t)-\left(\int_0^{\infty}\frac{e^{-t\tau}}{\tau}d\mu_1(\tau)A+\int_0^{\infty}\frac{e^{-t\tau}}{\tau}d\mu_2(\tau)B\right)\varphi_0$$

$$=-A_0^{1/2}\left[\int_0^t A_0^{1/2}v(s)ds+A_0^{1/2}\varphi_0+\sum_{k=1}^{2}Q_k^*\int_0^{+\infty}\int_0^t\frac{e^{-(t-s)\tau}}{\tau}Q_kA_0^{1/2}v(s)dsd\mu_k(\tau)\right]$$

$$+f(t)-\left(\int_0^{\infty}\frac{e^{-t\tau}}{\tau}d\mu_1(\tau)A+\int_0^{\infty}\frac{e^{-t\tau}}{\tau}d\mu_2(\tau)B\right)\varphi_0$$

$$=-\left[\int_0^t A_0v(s)ds+A_0\varphi_0+\int_0^{+\infty}\int_0^t\frac{e^{-(t-s)\tau}}{\tau}Av(s)dsd\mu_1(\tau)\right.$$

$$\left.+\int_0^{+\infty}\int_0^t\frac{e^{-(t-s)\tau}}{\tau}Bv(s)dsd\mu_2(\tau)\right]+f(t)-\left(\int_0^{\infty}\frac{e^{-t\tau}}{\tau}d\mu_1(\tau)A+\int_0^{\infty}\frac{e^{-t\tau}}{\tau}d\mu_2(\tau)B\right)\varphi_0. \tag{29.46}$$

Making the change of variables $v(t):=u'(t)$ in the expression (29.46) and applying the integration by parts formula, we obtain

$$=-\left[\int_0^t A_0u'(s)ds+A_0\varphi_0+\int_0^{+\infty}\int_0^t\frac{e^{-(t-s)\tau}}{\tau}Au'(s)dsd\mu_1(\tau)\right.$$

$$\left.+\int_0^{+\infty}\int_0^t\frac{e^{-(t-s)\tau}}{\tau}Bu'(s)dsd\mu_2(\tau)\right]+f(t)-\left(\int_0^{\infty}\frac{e^{-t\tau}}{\tau}d\mu_1(\tau)A+\int_0^{\infty}\frac{e^{-t\tau}}{\tau}d\mu_2(\tau)B\right)\varphi_0$$

$$=-\left[A_0u(t)+\int_0^t\int_0^{+\infty}\frac{e^{-(t-s)\tau}}{\tau}d\mu_1(\tau)Au'(s)ds+\int_0^t\int_0^{+\infty}\frac{e^{-(t-s)\tau}}{\tau}d\mu_2(\tau)Bu'(s)ds\right]$$

$$+f(t)-\left(\int_0^{\infty}\frac{e^{-t\tau}}{\tau}d\mu_1(\tau)A+\int_0^{\infty}\frac{e^{-t\tau}}{\tau}d\mu_2(\tau)B\right)\varphi_0$$

$$=-A_0u(t)-\left(\int_0^{+\infty}\frac{e^{-(t-s)\tau}}{\tau}d\mu_1(\tau)\right)Au(s)\Bigg|_0^t+\int_0^t\int_0^{+\infty}e^{-(t-s)\tau}d\mu_1(\tau)Au(s)ds$$

$$-\left(\int_0^{+\infty}\frac{e^{-(t-s)\tau}}{\tau}d\mu_2(\tau)\right)Bu(s)\Bigg|_0^t+\int_0^t\int_0^{+\infty}e^{-(t-s)\tau}d\mu_2(\tau)Bu(s)ds$$

$$+f(t)-\left(\int_0^{\infty}\frac{e^{-t\tau}}{\tau}d\mu_1(\tau)A+\int_0^{\infty}\frac{e^{-t\tau}}{\tau}d\mu_2(\tau)B\right)\varphi_0$$

$$=-A_0u(t)-\left(\int_0^{+\infty}\frac{d\mu_1(\tau)}{\tau}\right)Au(t)+\left(\int_0^{+\infty}\frac{e^{-t\tau}}{\tau}d\mu_1(\tau)\right)Au(0)$$

$$+\int_0^t\int_0^{+\infty}e^{-(t-s)\tau}d\mu_1(\tau)Au(s)ds-\left(\int_0^{+\infty}\frac{d\mu_2(\tau)}{\tau}\right)Bu(t)+\left(\int_0^{+\infty}\frac{e^{\ t\tau}}{\tau}d\mu_2(\tau)\right)Bu(0)$$

$$+\int_0^t\int_0^{+\infty}e^{-(t-s)\tau}d\mu_2(\tau)Bu(s)ds-\left(\int_0^{+\infty}\frac{d\mu_1(\tau)}{\tau}\right)Au(t)$$

$$+\left(\int_0^{+\infty}\frac{e^{-t\tau}}{\tau}d\mu_1(\tau)\right)A\varphi_0+\int_0^t\int_0^{+\infty}e^{-(t-s)\tau}d\mu_1(\tau)Au(s)ds$$

$$-\left(\int_0^{+\infty}\frac{d\mu_2(\tau)}{\tau}\right)Bu(t)+\left(\int_0^{+\infty}\frac{e^{-t\tau}}{\tau}d\mu_2(\tau)\right)B\varphi_0+\int_0^t\int_0^{+\infty}e^{-(t-s)\tau}d\mu_2(\tau)Bu(s)ds$$

$$+f(t)-\left(\int_0^{\infty}\frac{e^{-t\tau}}{\tau}d\mu_1(\tau)A+\int_0^{\infty}\frac{e^{-t\tau}}{\tau}d\mu_2(\tau)B\right)\varphi_0$$

$$=-\left(A+B\right)u(t)+\int_0^t K_1(t-s)Au(s)ds+\int_0^t K_2(t-s)Bu(s)ds+f(t).$$

Thus, the first equation in system (29.13) coincides with the integro-differential equation (29.4):

$$\frac{d^2u(t)}{dt^2}=-\left(A+B\right)u(t)+\int_0^t K_1(t-s)Au(s)ds+\int_0^t K_2(t-s)Bu(s)ds+f(t),\ t\in\mathbb{R}_+,$$

with the initial conditions $u(+0)=\varphi_0$ and $u^{(1)}(+0)=\varphi_1$. Therefore, $u(t)$ is a classical solution of problem (29.4) and (29.5). Moreover, the validity of the assumptions of Theorem 29.3 ensures that of the assumptions of Theorem 29.7, and then the estimate (29.24) follows from the estimate (29.37). The proof of Theorem 29.3 is complete.

29.7 Examples

29.7.1 Example 1

Consider the step functions

$$\mu_1(\tau) = \sum_{j=1}^{N} a_j \chi[\beta_{j-1}, \beta_j)(\tau) \quad \text{and} \quad \mu_2(\tau) = \sum_{j=1}^{N} b_j \chi[\beta_{j-1}, \beta_j)(\tau),$$

where $a_j > 0, b_j \geqslant 0, j = 1, 2, ..., N, \chi[\beta_{j-1}, \beta_j)(\tau)$ are the characteristic functions of half-intervals $[\beta_{j-1}, \beta_j)$, $0 \leqslant \beta_{j-1} < \beta_j$, $j = 1, 2, ..., N$ and $\beta_0 = 0$. Then the kernels of the integral operators have the representations:

$$K_1(t) = \sum_{j=1}^{N} a_j e^{-\beta_j t}, \quad K_2(t) = \sum_{j=1}^{N} b_j e^{-\beta_j t}$$

and conditions (29.7) become

$$\sum_{j=1}^{N} \frac{a_j}{\beta_j} < 1, \quad \sum_{j=1}^{N} \frac{b_j}{\beta_j} < 1.$$

We introduce the new variables $v(t) := u'(t)$, $\xi_0(t) := A_0^{1/2} u(t)$ and

$$\xi_{1j}(t) = \int_0^t \frac{\sqrt{a_j} e^{-(t-s)\beta_j}}{\sqrt{\beta_j}} Q_1 A_0^{1/2} \frac{du(s)}{ds} ds,$$

$$\xi_{2j}(t) = \int_0^t \frac{\sqrt{b_j} e^{-(t-s)\beta_j}}{\sqrt{\beta_j}} Q_2 A_0^{1/2} \frac{du(s)}{ds} ds, \qquad t > 0,\ j = 1, 2, ..., N.$$

In this notation, problem (29.4), (29.5) is reduced to the following Cauchy problem for a system of first-order differential equations:

$$\begin{cases} \dfrac{dv(t)}{dt} + A_0^{1/2} \left[\xi_0(t) + \displaystyle\sum_{j=1}^{N} \frac{\sqrt{a_j}}{\sqrt{\beta_j}} Q_1^* \xi_{1j}(t) + \sum_{j=1}^{N} \frac{\sqrt{b_j}}{\sqrt{\beta_j}} Q_2^* \xi_{2j}(t) \right] = f_1(t), \\ \dfrac{d\xi_0(t)}{dt} = A_0^{1/2} v(t), \\ \dfrac{d\xi_{1j}(t)}{dt} = \dfrac{\sqrt{a_j}}{\sqrt{\beta_j}} Q_1 A_0^{1/2} v(t) - \beta_j \xi_{1j}(t, \tau),\ j = 1, ..., N, \\ \dfrac{d\xi_{2j}(t)}{dt} = \dfrac{\sqrt{b_j}}{\sqrt{\beta_j}} Q_2 A_0^{1/2} v(t) - \beta_j \xi_{2j}(t, \tau),\ j = 1, ..., N, \end{cases}$$

$$v(t)|_{t=0} = \varphi_1, \quad \xi_0(t)|_{t=0} = A_0^{1/2}\varphi_0, \quad \xi_j(t)\big|_{t=0} = 0, \quad j = 1, ..., N,$$

where

$$f_1(t) = f(t) - \left(\sum_{j=1}^{N} \frac{a_j}{\beta_j} e^{-\beta_j t} A + \sum_{j=1}^{N} \frac{b_j}{\beta_j} e^{-\beta_j t} B\right)\varphi_0.$$

One can readily see that the estimate (29.24) acquires the form

$$E(t) = \frac{1}{2}\left(\|u'(t)\|_H^2 + \left\|A_0^{1/2}u(t)\right\|_H^2\right) \leqslant \frac{1}{2}\|Z(t)\|_{\mathbb{H}}^2 \leqslant d\left[\left(\|\varphi_1\|_H^2 + \left\|A_0^{1/2}\varphi_0\right\|_H^2\right)\right.$$
$$\left. + \left(\int_0^t \left\| f(s) - \left(\sum_{j=1}^{N} \frac{a_j}{\beta_j} e^{-\beta_j s} A + \sum_{j=1}^{N} \frac{b_j}{\beta_j} e^{-\beta_j s} B\right)\varphi_0 \right\| ds\right)^2\right].$$

29.7.2 Example 2

Consider the kernel functions

$$K_1(t) = \sum_{j=1}^{N} a_j \exists_{\alpha-1}(-\beta_j, t), \quad K_2(t) = \sum_{j=1}^{N} b_j \exists_{\alpha-1}(-\beta_j, t),$$

where $0 < \alpha < 1$, $a_j > 0$, $b_j \geqslant 0$, $0 \leqslant \beta_j < \beta_{j+1}$, $j = 1, ..., N$,

$$\exists_{\alpha-1}(-\beta_j, t) := t^{\alpha-1} \sum_{n=0}^{\infty} \frac{(-\beta_j)^n t^{n\alpha}}{\Gamma[(n+1)\alpha]}, \quad j = 1, ..., N$$

is a fractional-exponential Rabotnov function (see [10, 13]) $\Gamma(\cdot)$ is the Euler gamma function. Rabotnov function has the following integral representation (see [13]):

$$\exists_{\alpha-1}(-\beta_j, t) = \frac{\sin \pi\alpha}{\pi} \int_0^{+\infty} \frac{e^{-t\tau} d\tau}{\tau^\alpha + 2\beta_k \cos \pi\alpha + \beta_k^2 \tau^{-\alpha}}, \quad 0 < \alpha < 1.$$

Then

$$d\mu_1(\tau) = \frac{\sin \pi\alpha}{\pi}\left(\sum_{j=1}^{N} \frac{a_j}{\tau^\alpha + 2\beta_j \cos \pi\alpha + \beta_j^2 \tau^{-\alpha}}\right) d\tau,$$

$$d\mu_2(\tau) = \frac{\sin \pi\alpha}{\pi}\left(\sum_{j=1}^{N} \frac{b_j}{\tau^\alpha + 2\beta_j \cos \pi\alpha + \beta_j^2 \tau^{-\alpha}}\right) d\tau,$$

$$M_1(t) = \int_0^{+\infty} \frac{e^{-t\tau} d\mu_1(\tau)}{\tau} = \sum_{j=1}^{N} a_j \int_t^{+\infty} \exists_{\alpha-1}(-\beta_j, s) ds,$$

$$M_2(t) = \int_0^{+\infty} \frac{e^{-t\tau} d\mu_2(\tau)}{\tau} = \sum_{j=1}^{N} b_j \int_t^{+\infty} \exists_{\alpha-1}(-\beta_j, s) ds.$$

Conditions (29.7) become

$$\sum_{j=1}^{N} \frac{a_j}{\beta_j} < 1, \quad \sum_{j=1}^{N} \frac{b_j}{\beta_j} < 1.$$

In this notation, problem (29.4), (29.5) is reduced to the following Cauchy problem for a system of first-order differential equations:

$$\begin{cases} \dfrac{dv(t)}{dt} + A_0^{1/2}\left[\xi_0(t) + \dfrac{\sin\pi\alpha}{\pi}\displaystyle\sum_{j=1}^{N}\int_0^{+\infty}\dfrac{\left(a_j Q_1^*\xi_1(t,\tau) + b_j Q_2^*\xi_2(t,\tau)\right)d\tau}{\sqrt{\tau}(\tau^\alpha + 2\beta_j\cos\pi\alpha + \beta_j^2\tau^{-\alpha})}\right] \\ = f_1(t),\ t > 0, \\ \dfrac{d\xi_0(t)}{dt} = A_0^{1/2} v(t), \\ \dfrac{d\xi_1(t,\tau)}{dt} = \dfrac{1}{\sqrt{\tau}} Q_1 A_0^{1/2} v(t) - \tau\xi_1(t,\tau), \\ \dfrac{d\xi_2(t,\tau)}{dt} = \dfrac{1}{\sqrt{\tau}} Q_2 A_0^{1/2} v(t) - \tau\xi_2(t,\tau), \quad \tau > 0, \end{cases}$$

$$v(t)|_{t=0} = \varphi_1, \quad \xi_0(t)|_{t=0} = A_0^{1/2}\varphi_0, \quad \xi_k(t,\tau)|_{t=0} = 0, \quad k = 1, 2,$$

$$f_1(t) = f(t) - \sum_{j=1}^{N}\left(\left(\int_t^{+\infty} \exists_{\alpha-1}(-\beta_j, s) ds\right)(a_j A + b_j B)\varphi(0)\right).$$

The estimate (29.24) acquires the form

$$E(t) := \frac{1}{2}\left(\|u'(t)\|_H^2 + \left\|A_0^{1/2}u(t)\right\|_H^2\right) \leqslant \frac{1}{2}\|Z(t)\|_{\mathbb{H}}^2 \leqslant d\left[\|\varphi_1\|_H^2 + \left\|A_0^{1/2}\varphi_0\right\|_H^2\right.$$

$$\left. + \left(\int_0^t \left\| f(s) - \sum_{j=1}^{N}\left(\int_s^{+\infty} \exists_{\alpha-1}(-\beta_j, \tau) d\tau\right)(a_j A + b_j B)\varphi(0)\right\|_H ds\right)^2\right].$$

References

[1] Amendola G, Fabrizio M Golden JM (2021) Thermodynamics of Materials with Memory. Theory and Applications. Springer Nature, Cham, DOI 10.1007/978-3-030-80534-0

[2] Christensen RM (1971) Theory of Viscoelasticity. An Introduction. Academic Press, New York–London

[3] Dafermos CM (1970) Asymptotic stability in viscoelasticity. Archive for Rational Mechanics and Analysis 37(4):297-308, DOI 10.1007/BF00251609

[4] Engel KJ, Nagel R (2000) One-Parameter Semigroups for Linear Evolution Equations, Graduate Texts in Mathematics, Vol 194. Springer, New York, DOI 10.1007/b97696

[5] Georgievskii DV (2023) Modeli teorii vyazkouprugosti (in Russ., Models of Viscoelasticity Theory). Lenand, Moscow

[6] Gurtin ME, Pipkin AC (1968) General theory of heat conduction with finite wave speed. Archive of Rational Mechanics and Analysis 31(2):113-126, DOI 10.1007/BF00281373

[7] Il'yushin AA, Pobedrya BE (1970) Osnovy matematicheskoi teorii termovyazkouprugosti (in Russ., Mathematical Theory of Thermoviscoelasticity). Nauka, Moscow

[8] Ivanov S, Pandolfi L (2009) Heat equations with memory: Lack of controllability to the rest. Journal of Mathematical Analysis and Applications 355(1):1-11, DOI 10.1016/j.jmaa.2009.01.008

[9] Krein SG (1967) Lineinye differentsial'nye uravneniya v banakhovykh prostranstvakh (in Russ., Linear Differential Equations in Banach Spaces). Nauka, Moscow

[10] Kopachevsky ND, Krein SG (2003) Operator Approach to Linear Problems of Hydrodynamics. Vol. 2: Nonself-Adjoint Problems for Viscous Fluids, Operator Theory: Advances and Applications, Vol 146. Birkhäuser, Basel

[11] Miller RK (1978) An integrodifferential equation for rigid heat conductors with memory. Journal of Mathematical Analysis and Applications 66(2):313-332, DOI 10.1016/0022-247X(78)90234-2

[12] Pata V (2009) Stability and exponential stability in linear viscoelasticity. Milan Journal of Mathematics 77(1):333–360, DOI 10.1007/s00032-009-0098-3

[13] Rabotnov YN (1980) Elements of Hereditary Solid Mechanics. Mir, Moscow

[14] Rautian NA (2020) Semigroups generated by Volterra integro-differential equations. Differential Equations 56(9):1193-1211, DOI 10.1134/S0012266120090098

[15] Rautian NA (2021) On the properties of semigroups generated by Volterra integro-differential equations with kernels representable by Stieltjes integrals. Differential Equations 57():1231–1248, DOI 10.1134/S0012266121090111

[16] Rautian NA (2021) Eksponentsial'naya ustoychivost' polugrupp, porozhdayemykh vol'terrovymi integro-differentsial'nymi uravneniyami (in Russ., Exponential stability of semigroups generated by Volterra integro-differential equations). Ufimskii Matematicheskii Zhurnal 13(4):65-81

[17] Vlasov VV, Rautian NA (2016) Spektral'nyi analiz funktsional'no-differentsial'nykh uravnenii (Spectral Analysis of Functional Differential Equations). Maks Press, Moscow
[18] Vlasov VV, Rautian NA (2018) Well-posedness and spectral analysis of integrodifferential equations arising in viscoelasticity theory. Journal of Mathematical Sciences 233(4):555-577, DOI 10.1007/s10958-018-3943-5
[19] Vlasov VV, Rautian NA (2019) Well-posed solvability and the representation of solutions of integrodifferential equations arising in viscoelasticity. Differential Equations 55(4):561-574, DOI 10.1007/s10958-018-3943-5
[20] Vlasov VV, Rautian NA (2019) Spectral analysis of integrodifferential equations in Hilbert spaces. Journal of Mathematical Sciences 239(5):771–787, DOI 10.1007/s10958-019-04325-7
[21] Vlasov VV, Rautian NA (2020) A study of operator models arising in problems of hereditary mechanics. Journal of Mathematical Sciences 244(2):170-182, DOI 10.1007/s10958-019-04612-3
[22] Vlasov VV, Rautian NA (2021) On Volterra integro-differential equations with kernels representable by Stieltjes integrals. Differantial Equations 57(4):517-532, DOI 10.1134/S001226612104008X
[23] Sanchez-Palencia E (1980) Non-Homogeneous Media and Vibration Theory, Lecture Notes in Physics, Vol 127. Springer-Verlag, Berlin–Heidelberg, DOI 10.1007/3-540-10000-8
[24] Tretter C (2008) Spectral Theory of Block Operator Matrices and Applications. World Scientific, DOI 10.1142/p493

Chapter 30
Wave Motions in Rods

Vasily Saurin

Abstract This article investigates the dynamic behaviour of elastic rods with a constant cross-section. The study aims to construct rigorous mathematical models that detail the kinematics of individual points within a rod, with a primary focus on understanding the propagation and interaction patterns of elastic waves. A comprehensive consideration of both conservative and non-conservative loads enables a complete and accurate description of the system's dynamics. These models are fundamentally governed by the conservation laws of energy, momentum, and angular momentum. Compliance with these laws ensures the physical adequacy of the obtained results and their consistency with the fundamental principles of mechanics. The paper examines two key problems in the nonstationary dynamics of rods, related to their longitudinal and transverse motion. For longitudinal motion, it is shown that the wavefront propagates along the rod at a constant velocity, with accelerations of all points remaining zero throughout the process. Furthermore, analysis of the nonstationary dynamics reveals that an infinite number of linearly independent vibration modes correspond to the first natural frequency. This finding fundamentally diverges from classical theory, where each natural frequency corresponds to a unique vibration mode. Investigation of the transverse motion demonstrates that, in contrast to the constant-velocity propagation of longitudinal waves, bending waves travel at a variable speed. This speed is a decreasing function of time, asymptotically tending to zero as the wavefront recedes to infinity. This result underscores the complex, nonlinear nature of beam dynamics and highlights the necessity of analyzing different vibration modes separately. It was also established that the nature of the load significantly influences the wave speed: the wavefront from a concentrated moment propagates almost twice as fast as the one caused by a concentrated transverse force. These findings are fundamental for understanding the wave dynamics of rods and open new avenues for more accurate design and analysis of related engineering structures.

Vasily Saurin
Ishlinsky Institute for Problems in Mechanics RAS, Prospekt Vernadskogo 101-1, 119526 Moscow, Russian Federation, e-mail: saurin@ipmnet.ru

H. Altenbach et al. (eds.), *Stability Problems in Mechanics of Continua*,
Advanced Structured Materials 260, https://doi.org/10.1007/978-3-032-31032-3_30

Key words: Elastic rod, Initial-boundary value problem, Conservation laws, Wave front, Natural frequency

30.1 Introduction

In modern science and engineering, understanding the dynamic response of structures and their components is a critical field of study. The primary motivation is to guarantee the reliability and safety of various engineered systems exposed to dynamic loads. Despite extensive research on structural dynamics, the focus remains uneven. The majority of studies investigate natural vibrations, aiming to identify the characteristic frequencies and modes at which a structure oscillates freely [4, 5, 13].

This is undoubtedly an important problem, as understanding natural frequencies is crucial for avoiding resonant phenomena that may cause structural failure. However, significantly less attention has been devoted to studying wave processes that arise in structures subjected to pulsed or impact loads. These wave processes are key to forming the stress-strain state within a structure, particularly under short-term dynamic effects. In such scenarios, stresses and strains propagate as waves through the material, reflecting at boundaries and interacting with one another. Neglecting these wave effects can result in an inadequate assessment of structural strength and reliability, as well as in the flawed design of elements intended to damp or dissipate wave energy. A classic problem within dynamics is the study of longitudinal wave motion in rods. This well-developed area for both homogeneous and inhomogeneous rods under various boundary conditions provides fundamental insights into wave propagation in elastic media [1, 2, 6–12, 15].

Numerous studies on this topic have shown that under dynamic loading, a wave front arises in a rod and propagates along its length at the speed of sound in the material. Simultaneously, the point of load application – for instance, the rod's free end – also begins to move at a certain velocity. These results, derived from classical elasticity theory, are well-established among researchers and widely applied in engineering practice.

However, significantly less attention has been paid to investigating the dynamic behavior of rods subjected to impulsive loading, where the load acts for a short duration with high intensity. In this regard, the work of [3] can be highlighted as having made a significant contribution to the understanding of rod dynamics under such conditions.

A notable and counterintuitive result is found in the analytical solution for an infinitely long rod subjected to a concentrated transverse force. This solution exhibits a distinctive feature: the complete absence of a propagating wave front. Instead, it suggests an instantaneous – or infinitely fast – propagation of changes in the stress-strain state throughout the rod. Such a conclusion stands in direct contradiction to the fundamental principles of wave dynamics in elastic solids, thereby necessitating a careful re-examination of its physical interpretation.

This deviation from the classical wave propagation model may be attributed to several factors [8, 10]. First, the mathematical model used to describe rod dynamics might possess inherent limitations, failing to accurately capture physical processes under transverse loading. For instance, it may neglect transverse deformation or cross-sectional rotation, which can be significant under dynamic conditions. Alternatively, the solution itself [8–10] might accurately represent the phenomenon's physical essence, suggesting that under specific conditions, stress-strain changes propagate at a very high – though not infinite – velocity. This could stem from the complex stress-strain distribution near the load application point and the inherently more intricate deformation pattern induced by a transverse force compared to a longitudinal one.

To resolve this contradiction, a more detailed investigation into the dynamics of rods under impulsive transverse loading is essential. This requires analyzing various mathematical models that incorporate the different factors influencing wave propagation. Additionally, experimental studies are necessary to obtain data on actual rod behavior under such conditions for comparison with theoretical predictions. A comprehensive approach combining these efforts will lead to a deeper understanding of the underlying dynamic processes and support the development of more accurate and reliable design methodologies for structures exposed to impulsive loads.

30.2 Preliminary Comments

At the beginning of the article, we will examine the simplest problem in oscillation theory, a topic typically introduced to 9th-grade students. Although well-known, it presents aspects worthy of further discussion.

Consider a simple spring pendulum: a rigid body (mass m) resting on a horizontal plane and attached by a spring to a wall (Fig. 30.1). This mechanical oscillatory system – consisting of a spring with elasticity coefficient k, fixed at one end and carrying the mass at the other – is assumed to be free from significant frictional effects. Initially, the spring is undeformed (neither stretched nor compressed), so no net horizontal force acts on the mass. Point 0 marks the equilibrium position of the mass.

Let us move the mass to the right. In doing so, the spring will stretch, generating an elastic force F_{E} directed to the left, toward the equilibrium position, with a magnitude equal to:

$$F_{\mathrm{E}} = \kappa x \,. \tag{30.1}$$

Fig. 30.1 Vibrations of a spring pendulum.

Here x is the load shift.

According to d'Alembert's principle (the principle of kinetostatics), the elastic force F_{E} must be balanced by an inertial force F_{I}

$$F_{\mathrm{I}} = m\ddot{x} \ . \tag{30.2}$$

The equation of dynamic equilibrium is

$$F_{\mathrm{E}} + F_{\mathrm{I}} = m\ddot{x} + \kappa x = 0 \ . \tag{30.3}$$

The solution of equation (30.3) can be represented as

$$x = c_1 \sin \omega t \ , \quad \dot{x} = c_1 \omega \cos \omega t \ , \quad \ddot{x} = -c_1 \omega^2 \sin \omega t \ . \tag{30.4}$$

Here c_1 is an indefinite constant, ω is the natural frequency of the spring pendulum, which is equal to

$$\omega = \sqrt{\frac{\kappa}{m}} \ . \tag{30.5}$$

This formula remains valid even when additional external forces, such as gravity, act on the mass, as in the case of a vertical spring pendulum. In such a system, the oscillating body's acceleration is independent of the gravitational force. Gravity serves only to shift the equilibrium position of the system.

This analysis outlines critical shortcomings in this conventional model that undermine its predictive reliability. The model's foundation rests on a pronounced idealization: a strict division of the system into two independent components. It attributes all inertia to a point mass m, while confining all elasticity to a massless spring. While often justified when the spring's mass is negligible compared to the attached mass, this simplification significantly compromises the model's dynamic fidelity. Consequently, attributing zero mass to the spring (and thus zero momentum density) prevents the model from describing internal wave dynamics. As a result, it entirely misses the key wave phenomenon of elastic energy propagation along the spring during oscillation.

Third, the question of how to properly account for inertial forces in the dynamic equations, while less central to the present discussion, remains open. In many continuum mechanics models, inertial forces are currently assumed to depend solely on the second time derivative of the displacement. This approach is typically justified by an appeal to Newton's second law. However, it is critical to note that this law was originally formulated for the dynamics of perfectly rigid bodies, not for continuous media. Analytical results for beam bending in [8–10] were derived from the compliance with fundamental conservation laws: energy, linear momentum, and angular momentum. The resulting displacement fields violate the wave equation for beam bending. Indeed, it is unreasonable to expect a system of three coupled, nonlinear differential equations derived from conservation laws to coincide with a single, linear wave equation.

Naturally, addressing the full scope of these questions within a single article is challenging. This work, therefore, focuses on several of the most foundational aspects.

30.3 Dynamic Behavior of the Rod

Consider a uniform rod of constant cross-section, extending from a point $x = 0$ to infinity (Fig. 30.2). Under the assumption that longitudinal stresses dominate, all other stress tensor components can be neglected. The rod's material is defined by a modulus of elasticity E. Consequently, the problem reduces to a single non-zero stress component and a corresponding strain component

$$\sigma(x,t) = E\varepsilon(x,t)\ , \tag{30.6}$$

where $\sigma(x,t)$ is a function of internal stresses, and

$$\varepsilon(x,t) = \frac{\partial u(x,t)}{\partial x}$$

is a function of the strains, which depend on the displacements $u(x,t)$ within the rod.

Consider the momentum density function

$$p(x,t) = \rho\frac{\partial u(x,t)}{\partial t}\ .$$

Thus, the equation of dynamic equilibrium for the rod – expressing the balance between the inertial force and the longitudinal variation in stress – can be written as:

$$\frac{\partial\sigma(x,t)}{\partial x} - \frac{\partial p(x,t)}{\partial t} = 0\ . \tag{30.7}$$

Relations (30.6), (30.7) leads to the wave equation:

$$E\frac{\partial^2 u}{\partial x^2} - \rho\frac{\partial^2 u}{\partial t^2} = 0\ , \tag{30.8}$$

that determines the longitudinal elastic wave speed in a rod

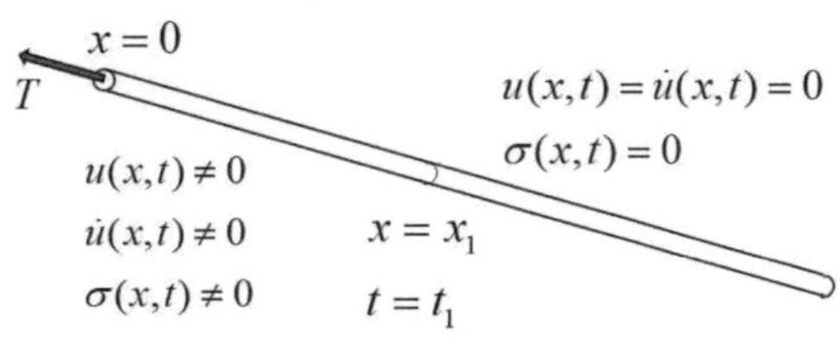

Fig. 30.2 Semi-infinite homogeneous rod.

$$v_s = \sqrt{E/\rho}\ . \tag{30.9}$$

Here ρ is the bulk density.

First, let us consider the following initial boundary value problem: find a solution to Eq. (30.8) for the space-time region

$$\Omega = \{0 < x < l, \quad t > 0).$$

The boundary conditions are given in the form

$$u|_{x=0} = 0, \quad u|_{x=L} = 0 \tag{30.10}$$

and the initial conditions

$$u|_{t=0} = u_0(x), \quad \left.\frac{\partial u}{\partial t}\right|_{t=0} = u_1(x)\ . \tag{30.11}$$

We solve problem (30.8) , (30.10) , (30.11) using the method of separation of variables, assuming a solution of the form:

$$u(x,t) = X(x)Y(t)\ . \tag{30.12}$$

This approach is perhaps the most developed method for solving the initial-boundary value problem. Accordingly, we transform equation (30.8) into the following form:

$$\frac{X''}{X} = \frac{Y''}{{v_s}^2 Y} = -\lambda^2\ . \tag{30.13}$$

Taking into account the boundary conditions (30.10), two differential equations are obtained

$$X'' + \lambda^2 X = 0, \qquad Y'' + \lambda^2 {v_s}^2 Y = 0 \tag{30.14}$$

with the boundary conditions

$$X(0) = X(L) = 0.$$

The boundary value problem for the first equation has a nontrivial solution at

$$\lambda_n = \frac{\pi n}{L}$$

of the form:

$$X_n = \sin\frac{\pi n x}{L}\ . \tag{30.15}$$

Then the second equation has a solution

$$Y_n(t) = A_n \cos\frac{\pi n v_s t}{L} + B_n \sin\frac{\pi n v_s t}{L}\ . \tag{30.16}$$

Assuming the convergence of series

$$u(x,t) = \sum_{n=1}^{\infty} \left(A_n \cos \frac{\pi n v_s t}{L} + B_n \sin \frac{\pi n v_s t}{L} \right) \sin \frac{\pi n x}{L}, \tag{30.17}$$

it is straightforward to verify that it satisfies equation (30.8) and conditions (30.10). The unknown constants A_n and B_n are then determined by substituting the initial conditions (30.11) into (30.16), yielding:

$$\begin{aligned} u|_{t=0} &= u_0(x) = \sum_{n=1}^{\infty} A_n \sin \frac{\pi n x}{L}, \\ \left.\frac{\partial u}{\partial t}\right|_{t=0} &= u_1(x) = \sum_{n=1}^{\infty} \frac{\pi n v_s}{L} B_n \sin \frac{\pi n x}{L} \end{aligned} . \tag{30.18}$$

It follows from this

$$A_n = \frac{2}{l} \int_0^l u_0(x) \sin \frac{\pi n x}{l} dx, \quad B_n = \frac{2}{\pi n v_s} \int_0^l u_1(x) \sin \frac{\pi n x}{l} dx . \tag{30.19}$$

While the solution in (30.16) and (30.18) applies to a broad range of rod dynamics governed by (30.10) and (30.11), many other initial-boundary value problems are not addressed by it. A key example is the case of a semi-infinite rod that is initially at rest $t = 0$, as shown in Fig. 30.2

$$u|_{t=0} = u_0(x) = 0, \quad \left.\frac{\partial u}{\partial t}\right|_{t=0} = u_1(x) = 0 . \tag{30.20}$$

At that instant, an axial force T is applied to the rod's left end ($x = 0$)

$$E \left.\frac{\partial u}{\partial x}\right|_{t=0} = T . \tag{30.21}$$

We assume the force is constant during the rod's motion. The boundary condition at a right end will be addressed later.

Homogeneous initial conditions (30.20) impose significant restrictions on the applicability of formula (30.17). Indeed, if the functions $u_0(x) = u_1(x) = 0$, then the solution is identically equal to zero, regardless of the type of homogeneous boundary conditions (30.10). The use of eigenfunctions in the solution assumes that the resulting function has nonzero values along the entire length of the rod.

Another fundamental property of eigenfunctions is their orthogonality. This property greatly facilitates the integration of the solution. Furthermore, the resulting eigenfunctions explicitly describe standing waves, which, by definition, imply a fixed spatial pattern without net energy transport along the rod. In real-world scenarios, however, energy is often transferred longitudinally, as in the case of a longitudinal impact at the rod's end (e.g., hammering a nail). To model such energy transfer, the simple standing wave solution becomes inadequate. In more complex, non-conservative systems, this leads to a description involving non-self-adjoint operators, where the eigenmodes are not orthogonal and their shapes can evolve over time.

Of course, during the process of motion, the energy stored in a particular form can change. However, this happens in a similar manner. For example, if the maximum amplitude in this form changes by a factor of α, then only the amplitude changes, and the form itself remains unchanged. However, the energy changes by a factor of α^2.

We now examine the evolution of the rod's state following the application of a force. The initial response involves motion and elastic stretching. Concurrently, the mechanical disturbance from the force propagates as a wave along the rod. This results in a continuous expansion of the region affected by the disturbance.

The motion of the rod is such that at any given moment $t = t_1$, one can identify a distinct point $x = x_1$ – a propagating wavefront – that separates two regions (see Fig. 30.2). To the left of this point, the rod is in a perturbed state, characterized by

$$u(x,t) \neq 0, \quad \dot{u}(x,t) \neq 0, \quad \sigma(x,t) \neq 0, \quad x < x_1, \quad t < t_1 . \tag{30.22}$$

To the right, the rod remains in its original, unperturbed state

$$u(x,t) = 0, \quad \dot{u}(x,t) = 0, \quad \sigma(x,t) = 0, \quad x > x_1, \quad t > t_1 . \tag{30.23}$$

Over time, this boundary point x_1 propagates with a characteristic wavefront velocity v_{f}. Consequently, the solution must be sought within the class of non-analytic functions, as the moving discontinuity precludes a representation by smooth, analytic forms.

The dynamic behavior of a rod must obey fundamental conservation laws for energy and momentum. While previous studies [8, 10] have formulated solutions based on these principles, the present work also employs them as a foundation. Consequently, we seek a solution that satisfies the equation of motion (30.8), which inherently embodies these conservation laws.

The conditions in (30.22) and (30.23) define two distinct zones. In the first zone, we seek a solution in the form of a linear function of the spatial coordinate and time:

$$u_1(x,t) = a + bx + ct , \tag{30.24}$$

where a, b and c are unknown constants. It is obvious that function (30.24) identically satisfies equation (30.8) at $x < x_1$. At $x > x_1$ (the second zone), the displacement function is equal to zero: $u_2(x,t) = 0$. At the point $x = x_1$, it is necessary to ensure the joining of two solutions $u_1(x_1,t) = u_2(x_1,t) = 0$. Let us also assume that the wave front velocity is constant:

$$v_{\mathrm{f}} = \frac{\partial x_1}{\partial t} = \text{const} . \tag{30.25}$$

It follows from (30.25) that

$$x_1 = v_{\mathrm{f}} t_1 . \tag{30.26}$$

The matching condition for the second zone, derived from equation (30.26), is:

$$u_1(x_1, t_1) = a + (bv_f + c)t_1 = 0\,. \tag{30.27}$$

To satisfy Eq: (30.27), the coefficients of t_1^0 and t_1^1 must be set to zero.

$$\begin{aligned} a &= 0, \\ c &= -bv_f. \end{aligned} \tag{30.28}$$

The undefined coefficient b if found from the boundary condition (30.21). Thus the solution of equation (30.8) is

$$u_0(x,t) = \frac{T}{E}\,(x - v_f t)\,, \quad x < x_1\,. \tag{30.29}$$

The wave front velocity v_f remains uncertain. To do this, equation (30.8) must be satisfied at the point $x = x_1$

$$E\frac{\partial^2 u}{\partial x^2} - \rho\frac{\partial^2 u}{\partial t^2}\bigg|_{\substack{x = x_1 \\ t = t_1}} = 0 \quad \Rightarrow \quad \frac{\sigma_0}{E}\left(-\rho v_f \delta\left(\frac{x}{v_f}\right) + E\delta(x)\right) = 0\,. \tag{30.30}$$

It follows from this

$$v_f = \sqrt{E/\rho}\,. \tag{30.31}$$

This solution involves the delta function, denoted by $\delta(x)$. Consequently, the propagation velocity of longitudinal elastic waves in the rod v_s, given by formula (30.9), is shown to coincide with the wavefront velocity derived from equality (30.31).

The complete nonsmooth solution of the problem has the form

$$u_T^{(1)}(x,t) = \begin{cases} \dfrac{T}{E}\left(x - \sqrt{\dfrac{E}{\rho}}t\right), & x \le x_1, \\ \qquad 0, & x > x_1. \end{cases} \tag{30.32}$$

A fundamental parameter for describing the rod's behavior is its rate of deformation v_d. We define this quantity as:

$$v_d = \frac{T}{\sqrt{E\rho}} = \text{const}\,. \tag{30.33}$$

The dynamic response is characterized by three principal features:

1. the wavefront propagates from the point of impact at a constant velocity v_f;
2. material points in the disturbed zone translate toward the force at a uniform deformation rate v_d; and
3. the deformation within this zone remains homogeneous and constant

$$\varepsilon_d = \frac{T}{E}\,.$$

Remarkably, this implies that every aspect of the rod's state evolves at a constant speed, even under the action of an external force.

30.4 Natural Vibrations of Finite Rods

At the beginning of this section, we examine the classical methods for determining the natural frequencies and mode shapes of straight rods with a constant cross-section. As an example, consider a rod of length L that is clamped at one end $u(0)$ and free of loads at the other

$$\left.\frac{\partial u}{\partial x}\right|_{x=L} = 0\,,$$

as shown in Fig. 30.3.

It is consider that the stress-strain state of the rod is described by equation (30.8). A solution is assumed in the form:

$$u(x,t) = u_0(x)e^{i\omega t}\,. \tag{30.34}$$

Applying the method of finding eigenvalues and eigenforms described in the previous section, we obtain the following expressions:

$$\omega_n = \frac{\pi}{2L}\sqrt{\frac{E}{\rho}}(1+2n), \quad u_n = \sin\left(\frac{\pi(2n+1)}{2L}x\right), \quad n = 0, 1, 2, \ldots. \tag{30.35}$$

For the subsequent analysis, we designate the zero natural frequency as ω_0 and its corresponding mode shape:

$$\omega_0 = \frac{\pi}{2L}\sqrt{\frac{E}{\rho}}, \quad u_0 = \sin\left(\frac{\pi x}{2L}\right)\,. \tag{30.36}$$

Let us consider, as in the previous case, a straight rod of length L, shown in Fig. 30.2. The rod is assumed to be at rest, i.e., the displacements $u(x) = 0$ and velocities

$$\frac{\partial u}{\partial x} = 0$$

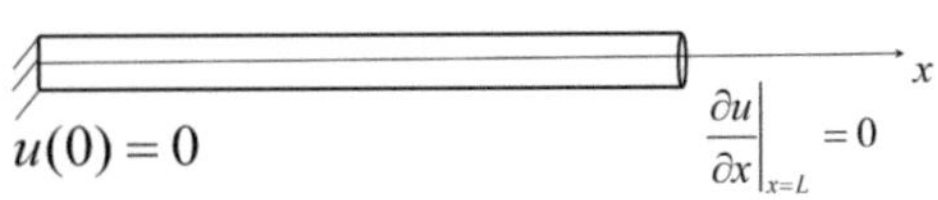

Fig. 30.3 A rod clamped at one end.

of all points on the rod are zero. Initially $t = 0$, a concentrated longitudinal force $T < 0$ is applied to the end of the rod with coordinate $x = 0$, which acts on the rod for the entire time ($t > 0$). The rod is fixed at the point $x = L$, i.e., $u(L) = 0$. Following the application of the load, the rod's points are set in motion, oscillating with a frequency ω_1 about the new equilibrium position–its statically deformed state under the force T.

We now examine how these rod oscillations can be formally described. The motion is initiated by a propagating wavefront, suggesting that the overall oscillation can be decomposed into a series of independent elementary motions. The superposition of these components constitutes the complete wave pattern in the rod. The first stage of the rod's motion corresponds to the previously obtained solution (30.32). At the instant $t = L/v_{\mathrm{f}}$ when the wavefront reaches the clamped end, all points on the rod are moving at velocity and the rod itself is uniformly stretched. This state is the initial condition for analyzing the second stage of the motion.

30.4.1 Evolution of the Vibration Process of the Rod (Stage II)

During motion, a region of zero displacement emerges near the anchor point, as displacement at that point $x = L$ is fixed at zero. This region expands until it reaches the load application point, marking the second stage of the rod's motion.

We now proceed to quantitatively characterize the rod's motion during this stage. To this end, we introduce a set of assumptions analogous to those employed in the analysis of the first stage. Specifically, a wave front is assumed to propagate at constant velocity $v_{f2}(t) = \text{const}$ from the fixed end toward the point of load application. The current position of the wave front is defined by a coordinate $x_2(t)$, $0 \le x_2(t) \le L$ that varies with time (see Fig. 30.4). Behind the front (on the side of the anchor), all points $x_2(t) \le x$ are at rest, and the stress in this region is uniform with a constant magnitude $T_2(t) = \text{const}$. Ahead of the front (toward the point of load application), every point moves with a constant velocity v_{d} and experiences a uniform tensile strain $T(t) = \text{const}$ throughout the segment $0 \le x \le x_2(t)$.

Based on the last assumption, the displacement field in the region $x_2(t) \le x \le L$ can be approximated by a linear function of the coordinate, expressed as:

$$u_1(x,t) = a_1 + b_1 x\,. \tag{30.37}$$

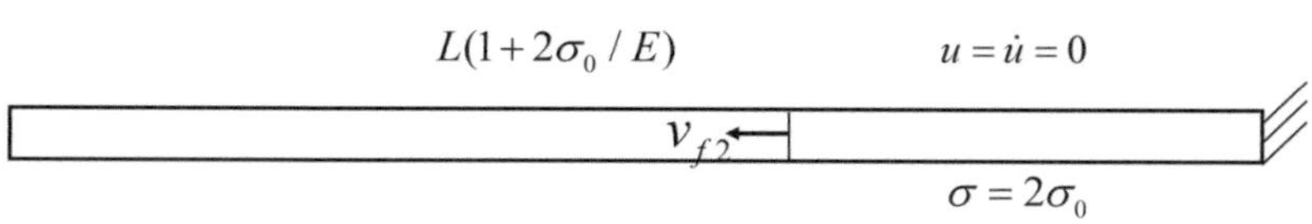

Fig. 30.4: Cyclic movements of a rod of finite length under the action of a constant tensile force (stage 2).

Applying the boundary condition $u_1(L, t) = 0$ together with the condition of uniform strain $T_2(t) = const$ along the rod, Eq. (30.37) simplifies to:

$$u_1(x, t) = -\frac{T_2}{E}(L - x)\,. \tag{30.38}$$

In the region $0 \le x \le x_2(t)$, we represent the displacement function as:

$$u_2(x, t) = a_2 + b_2 x + c_2 t\,. \tag{30.39}$$

Here a_2, b_2, and c_2 – unknown coefficients. Using the previous solution (30.32) formula (30.39) changes

$$u_2(x, t) = a_2 + \frac{T}{E}x - v_{\rm d} t\,. \tag{30.40}$$

It is necessary to satisfy the conditions for joining the displacement functions (30.38) and (30.40) at point $x = x_2$, $t = t_2$

$$u_1(x_2, t_2) - u_2(x_2, t_2) = 0\,. \tag{30.41}$$

Taking into account $x_2 = v_{f2} t_2$, equality (30.41) is reduced to two linear equations

$$\begin{gathered} \frac{T}{E} - \frac{v_{\rm d}}{v_{f2}} - \frac{T_2}{E} = 0\,, \\ L\frac{T}{E} - a_2 = 0\,. \end{gathered} \tag{30.42}$$

The unknown parameter v_{f2} for the second stage is found from the momentum balance. Let the wave front be $x_2(t_2)$. Taking a slightly different time $t_2 + \Delta t_2$, the resulting variation in momentum takes the form:

$$\Delta p = p(t_2 + \Delta t_2) - p(t_2) = \rho v_{\rm d} \Delta x_2, \quad \Delta x_2 = \frac{\Delta t_2}{v_{f2}}\,. \tag{30.43}$$

And to change the impulse of force

$$\Delta f = T \Delta t_2\,. \tag{30.44}$$

Applying the law of momentum change

$$Deltap = \Delta f$$

yields an expression for the wave front velocity v_{f2}

$$v_{f2} = -\sqrt{\frac{E}{\rho}}\,. \tag{30.45}$$

Thus, the wave front speeds in the first and second stages are identical in magnitude but opposite in direction.

Substituting (30.45) and (30.33) into (30.42), we will obtain a solution to this system

$$\begin{aligned} T_2 &= 2T\,, \\ a_2 &= \frac{2T}{E}L\,. \end{aligned} \tag{30.46}$$

The resulting displacements are written as follows:

$$u_T^{(2)}(x,t) = \begin{cases} -\dfrac{2T}{E}(L-x), & x_2 \le x \le L, \quad t > 0\,, \\ \\ \dfrac{2T}{E}L + \dfrac{T}{E}x - v_\mathrm{d}t, & 0 \le x \le x_2, \quad t > 0\,. \end{cases} \tag{30.47}$$

Thus, all parameters describing the rod's motion during the second stage have been determined. To verify the principle of energy conservation, we now write the respective increments of elastic energy ΔW, kinetic energy ΔK, and the work done by external forces ΔA.

$$\begin{aligned} \Delta W &= \frac{3T^2 \Delta x_2}{2E}\,, \\ \Delta K &= -\frac{\rho {v_\mathrm{d}}^2 \Delta x_2}{2}\,, \\ \Delta A &= T v_\mathrm{d} \Delta t_2\,. \end{aligned} \tag{30.48}$$

Taking into account $\Delta x_2 = v_\mathrm{f} \Delta t_2$ and (30.31), the expression $\Delta K - \Delta A$ have a form

$$\Delta K - \Delta A = -\frac{T^2}{2E}\Delta x_2 - \frac{T^2}{E}\Delta x_2 = -\frac{3T^2}{2E}\Delta x_2\,. \tag{30.49}$$

It can be seen that expression (30.49) is equivalent to the first equality from (30.48) taken with a minus sign. Consequently, the relation of the law of conservation of energy

$$\Delta W + \Delta K - \Delta A = 0\,. \tag{30.50}$$

satisfied as an identity.

At the moment

$$t = \frac{L}{v_\mathrm{f}}$$

when the wave front reaches the point of load application, all points on the rod are at rest, and the rod itself is uniformly stretched with a deformation of

$$\varepsilon = 2\frac{v_\mathrm{d}}{v_\mathrm{f}}.$$

This state is the initial condition for analyzing the third stage of motion.

30.4.2 Evolution of the Vibration Process of the Rod (Stage III)

Upon reaching the load application point, the wave front causes that end of the rod to begin moving toward the fixed support. We now proceed to derive the quantitative parameters characterizing the rod's motion during this third stage. Let us first formulate several hypotheses similar to those employed in the analysis of the preceding stages. In particular, it is supposed that a wave front emerges and advances with constant velocity $v_{f3}(t) = \text{const}$ from the load application point toward the fixed support (Fig. 30.5).

Using the last condition and using (30.47), we write the linear approximation of displacements for $x_3(t) \le x \le L$ in the form:

$$u_2(x,t) = -\frac{2T}{E}(L-x)\,. \tag{30.51}$$

In the region $0 \le x \le x_3(t)$, we represent the displacement function as:

$$u_1(x,t) = a_1 + b_1 x + c_1 t\,. \tag{30.52}$$

Here a_1, b_1, and c_1 - unknown coefficients. Using the assumptions made about the velocities $v_{d3} = \text{const}$ and $T_3(t) = const$ relation (30.52) can be represented as:

$$u_1(x,t) = a_1 + \frac{T_3}{E}x + v_{d3}t\,. \tag{30.53}$$

It is also necessary to satisfy the conditions for joining the displacement functions (30.51) and (30.53) in the point $x = x_3$, $t = t_3$:

$$u_1(x_3,t_3) - u_2(x_3,t_3) = 0\,. \tag{30.54}$$

Taking into account $x_3 = v_{f3}t_3$ equality (30.54) is reduced to two linear equations:

$$\begin{aligned} &\frac{v_{d3}}{v_{f3}} - \frac{T}{E} = 0\,,\\ &\frac{2LT}{E} + a_1 = 0\,. \end{aligned} \tag{30.55}$$

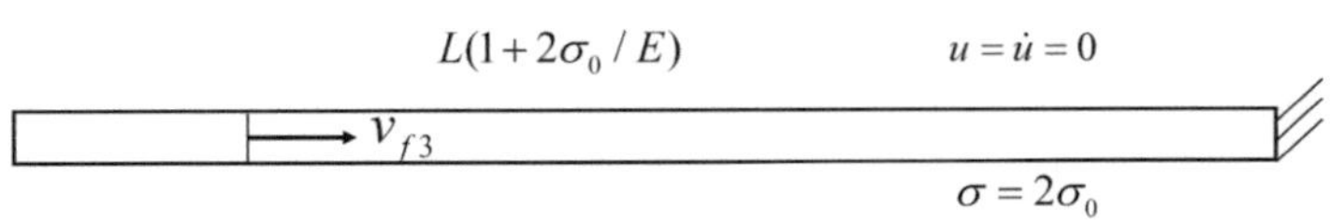

Fig. 30.5: Cyclic movements of a rod of finite length under the action of a constant tensile force (stage 3).

To determine the unknown parameter v_{f3} characterizing the third stage of the rod's motion, we again employ the law of momentum change. Consider a time instant t_3 at which the wave front is located at $x_3(t_3)$. Choosing a neighboring instant $t_3 + \Delta t_3$, the corresponding variation in the rod's momentum can be expressed as follows:

$$\Delta p = p(t_3 + \Delta t_3) - p(t_3) = \rho v_{\mathrm{d}} \Delta x_3, \quad \Delta x_3 = \frac{\Delta t_3}{v_{f3}}. \tag{30.56}$$

And to change the impulse of force:

$$\Delta f = T \Delta t_3 . \tag{30.57}$$

Enforcing the law of momentum change $\Delta p = \Delta f$ yields an expression for the wave front velocity v_{f3}

$$v_{f3} = \frac{T}{\rho v_{d3}} . \tag{30.58}$$

To satisfy the law of conservation of energy, we write out the increments of elastic ΔW and kinetic ΔK energies, as well as the increment of the work of external forces ΔA, respectively.

$$\begin{aligned} \Delta W &= -\frac{3T^3 \Delta t_3}{2E\rho v_{d3}}, \\ \Delta K &= \frac{v_{d3} T \Delta t_3}{2}, \\ \Delta A &= -T v_{d3} \Delta t_3 . \end{aligned} \tag{30.59}$$

In order to comply with the principle of energy conservation, the ensuing relation must hold:

$$dW + dK - dA = -\frac{3T dt_3}{2E\rho v_{d3}} \left(-v_{d3}{}^2 E\rho + T^2 \right) = 0 . \tag{30.60}$$

It is follows from (30.60):

$$v_{d3} = \frac{T}{\sqrt{E\rho}} . \tag{30.61}$$

Thus, the wave front speeds in the second and third stages are equal in magnitude but opposite in direction.

To find a_1 we have second equation of (30.55)

$$a_1 + \frac{2T}{E} L = 0 , \tag{30.62}$$

the solution of which is trivial.

The resulting displacements are written as follows:

$$u_T^{(3)}(x,t) = \begin{cases} \dfrac{T}{E}x + \dfrac{T}{\sqrt{E\rho}}t - \dfrac{2T}{E}L, & 0 \le x \le x_3, \quad t > 0 \\ -\dfrac{2T}{E}(L-x), & x_3 \le x \le L, \quad t > 0. \end{cases} \tag{30.63}$$

Thus, all the characteristics of the rod motion in the third stage are obtained.

At the instant

$$t = \frac{L(v_{\mathrm{f}} + v_{\mathrm{d}})}{v_{\mathrm{f}}^2},$$

when the wave front reaches the rod's attachment, every part of the rod is moving at a constant speed v_{d}, and the rod is uniformly stretched with deformation $\epsilon = v_{\mathrm{d}}/v_{\mathrm{f}}$.

If the load is removed at the point of application $x = 0$ and the rod is released at the fastening point $x = L$, the rod would move as a perfectly rigid body. However, because the displacement at the fastening point $x = L$ remains zero, a stagnation zone – where no displacement occurs – develops near that point during motion. This zone grows until it reaches the load application point. The resulting motion closely resembles that of the rod in the second stage.

30.4.3 Evolution of the Vibration Process of the Rod (Stage IV)

We quantify the fourth stage of motion under assumptions similar to those used for the second stage. A wave front is taken to propagate at constant velocity $v_{f4}(t) = \mathrm{const}$ from the anchor point toward the load application point (see Fig. 30.6). The position of the front on the rod is specified by the coordinate $0 \le x_4(t) \le L$. All points after the wave front $x_4(t) \le x$ are at rest, and in this zone, all stresses are equal to zero. All points on the rod before the front move at a constant velocity v_{d} and are uniformly stretched with a stress $T(t) = \mathrm{const}$ of $0 \le x \le x_4(t)$.

Based on the final assumption, we express the approximation of displacements for $x_4(t) \le x \le L$ as:

$$u_1(x,t) = 0\,. \tag{30.64}$$

In the region $0 \le x \le x_4(t)$, we represent the displacement function as

$$u_2(x,t) = a_2 + b_2 x + c_2 t\,. \tag{30.65}$$

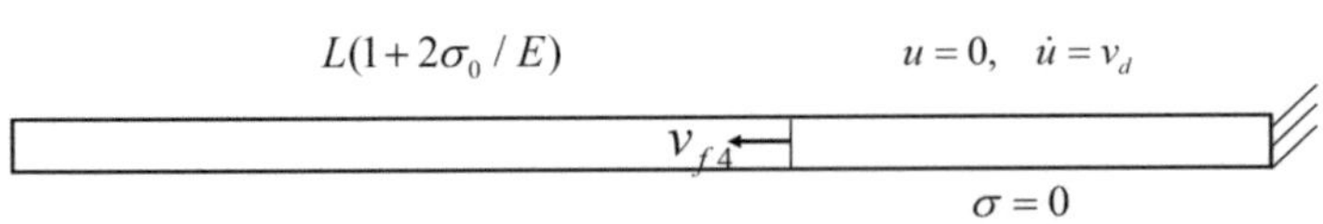

Fig. 30.6: Cyclic movements of a rod of finite length under the action of a constant tensile force (stage 4).

Here a_2, b_2, and c_2 – unknown coefficients. Using the previous solution, relation (30.65) will change

$$u_2(x,t) = a_2 + \frac{T}{E}x + v_{d4}t\,. \tag{30.66}$$

It is also necessary to satisfy the displacement function condition (30.66) at the point $x = x_4, t = t_4$

$$u_2(x_4, t_4) = 0\,. \tag{30.67}$$

Taking into account $x_4 = v_{f4}t_4$, equality (30.67) is reduced to two linear equations

$$\begin{aligned} \frac{v_\mathrm{d}}{v_{f4}} + \frac{T}{E} &= 0\,,\\ a_2 &= 0\,. \end{aligned} \tag{30.68}$$

To determine the unknown parameter v_{f4} for the fourth stage of the rod's motion, we apply the impulse-momentum principle. The resulting change in the rod's momentum can then be expressed as:

$$\Delta p = p(t_4 + \Delta t_4) - p(t_4) = \rho v_\mathrm{d}\Delta x_4, \quad \Delta x_4 = \frac{\Delta t_4}{v_{f4}}\,. \tag{30.69}$$

And to change the impulse of force

$$\Delta f = T\Delta t_4\,. \tag{30.70}$$

Compliance with the law of change in momentum $\Delta p = \Delta f$ will allow us to find an expression for the velocity of the wave front v_{f4}

$$v_{f4} = -\sqrt{\frac{E}{\rho}}\,. \tag{30.71}$$

Thus, the wave front velocities for all stages of the rod's motion are equal to each other in absolute value, but different in direction.

The resulting displacements are written as follows:

$$u_T^{(4)}(x,t) = \begin{cases} 0, \quad x_2 \le x \le L, \quad t > 0\,, \\ \dfrac{T}{E}\left(x + \sqrt{\dfrac{E}{\rho}}t\right), \quad 0 \le x \le x_2, \quad t > 0\,. \end{cases} \tag{30.72}$$

Thus, all characteristics of the rod's motion in the fourth stage are now determined. To verify the law of conservation of energy, we calculate the increments of elastic ΔW and kinetic energy ΔK, as well as the work done ΔA by external forces.

$$\Delta W = -\frac{T^2 \Delta x_4}{2E}\,,$$
$$\Delta K = -\frac{\rho {v_\text{d}}^2 \Delta x_4}{2}\,, \qquad (30.73)$$
$$\Delta A = -T v_\text{d} \Delta t_4\,.$$

Following the same procedure as for the second stage of motion, one can show that the displacement function (30.72) exactly satisfies the law of conservation of energy.

30.5 Natural Wave Vibrations of the Rod

In the previous paragraphs, the possible movements of the rod were discussed in detail. Firstly, these are natural oscillatory motions described by the wave equation of the rod (30.8) with natural frequencies

$$\omega_n = \frac{\pi}{2L}\sqrt{\frac{E}{\rho}}(1+2n), \quad n = 0, 1, 2, \ldots$$

and shapes (30.35). Let us write out in explicit form the natural frequency and shape for $n = 0$

$$\omega_0 = \frac{\pi}{2L}\sqrt{\frac{E}{\rho}}, \quad u_0 = \sin\left(\frac{\pi}{2L}x\right). \qquad (30.74)$$

These movements occur in the absence of a wave front.

The vibratory motion of a rod with a wavefront is described in detail in Section 6.4. The analysis reveals that the motion within a single period consists of four stages. During the first two stages, the rod undergoes tension; in the third and fourth stages, it undergoes compression. At the end of the fourth stage, the rod returns to its original, unperturbed state. Over the course of a full period, the wavefront traverses the rod four times.

A schematic representation of the wave process is given in the space-time diagram (see Fig. 30.7). In this diagram, the wave front position is depicted by straight lines, while critical points on the time axis – where the wave front velocity reverses – are clearly marked

$$\left\{L\sqrt{\frac{\rho}{E}},\ 2L\sqrt{\frac{\rho}{E}},\ 3L\sqrt{\frac{\rho}{E}},\ 4L\sqrt{\frac{\rho}{E}}\right\}.$$

The analysis shows that all points of the rod move with velocities that are constant in absolute value. Only the signs of the wavefront velocity v_f and strain rate v_d change, as illustrated in Fig. 30.7.

Note that the lengths of the deformed rod are different for different stages of motion. In the first and fourth stages of motion, the length traveled by the wave front is L. In the second and third stages of motion, the front travels

$$L\left(1+2\frac{v_{\mathrm{d}}}{v_{\mathrm{f}}}\right).$$

Thus, taking deformation into account, the distance traveled by the wave front over one oscillation period is equal to

$$L_{\pi}=4L\left(1+\frac{v_{\mathrm{d}}}{v_{\mathrm{f}}}\right). \tag{30.75}$$

This solution falls within the framework of geometrically nonlinear theory, since it incorporates an explicit dependence on the rod's deformation.

If we assume that the deformations during rod vibrations are small, then the change in the rod's length can be neglected. Consequently, the distance traveled by the wave front over one oscillation period can be taken as

$$L_{\pi}=4L. \tag{30.76}$$

Since the wave front travels this distance at a constant speed v_{f}, the period of oscillations is given by

$$T_{\pi}=\frac{4L}{v_{\mathrm{f}}}=4L\sqrt{\frac{\rho}{E}}. \tag{30.77}$$

Considering that the frequency of vibrations is related to the period by

$$\omega=\frac{2\pi}{T},$$

the natural frequency is given by

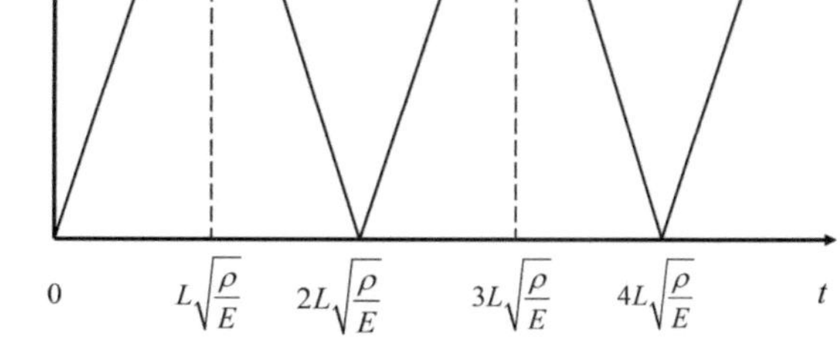

Fig. 30.7 Space-Time Diagram of the Wave Vibration Motion of the Rod

$$\omega_\pi = \frac{\pi}{2L}\sqrt{\frac{E}{\rho}}. \tag{30.78}$$

Comparing (30.74) and (30.78) , we see that

$$\omega_0 = \omega_\pi . \tag{30.79}$$

Thus, one frequency corresponds to two different forms of oscillation.

Consider a situation where, at some instant $\tau > 0$, the external load applied to the end of the rod is removed. How will the rod continue its motion? It can be shown that the rod will either oscillate or come to rest. The latter occurs if the load is removed at the moment corresponding to the end of the fourth stage of motion, described in Sect. 30.4. At this instant, the rod is at rest, and if the external load is removed, it will remain at rest indefinitely.

We now examine the following problem. A rod of length L, clamped at one end ($u(0) = 0$) and free at the other

$$\left(\left.\frac{\partial u}{\partial x}\right|_{x=L} = 0\right)$$

as illustrated in Fig. 30.3, is subjected to a force applied at its free end at time $t = 0$. The solution to this problem was constructed in Sect. 30.4. The displacement functions for the four stages of motion are given by the formulas (30.32), (30.47), (30.63) and (30.72). Let us assume that at the moment of time $t = \tau$,

$$0 < \tau < \frac{L}{v_\mathrm{f}},$$

a force T is applied to the end of the rod. The solution to the new problem is described by the expressions (30.32), (30.47), (30.63) and (30.72) taking into account the substitutions and, accordingly, have the form

$$u^{(1)}_{-T}(x,t) = \begin{cases} \dfrac{-T}{E}\left(x - \sqrt{\dfrac{E}{\rho}}(t-\tau)\right), & x \le v_\mathrm{f}(t-\tau) \\ 0 & x > v_\mathrm{f}(t-\tau) \end{cases}, \tag{30.80}$$

$$u^{(2)}_{-T}(x,t) = \begin{cases} \dfrac{2T}{E}(L-x), & x_2 \le x \le L, \quad t-\tau > 0 \\ -\dfrac{2T}{E}L - \dfrac{T}{E}x - v_\mathrm{d}(t-\tau), & 0 \le x \le x_2, \quad t-\tau > 0 \end{cases}, \tag{30.81}$$

$$u_{-T}^{(3)}(x,t) = \begin{cases} -\dfrac{T}{E}x - \dfrac{T}{\sqrt{E\rho}}(t-\tau) + \dfrac{2T}{E}L, & 0 \le x \le x_2, \quad t-\tau > 0 \\ \dfrac{2T}{E}(L-x), & x_2 \le x \le L, \quad t-\tau > 0 \end{cases}, \tag{30.82}$$

$$u_{-T}^{(4)}(x,t) = \begin{cases} 0, & x_2 \le x \le L, \quad t-\tau > 0 \\ \dfrac{-}{E}\left(x + \sqrt{\dfrac{E}{\rho}}(t-\tau)\right), & 0 \le x \le x_2, \quad t-\tau > 0 \end{cases}. \tag{30.83}$$

Then, the resulting displacements for this problem can be expressed as the sum:

$$u(x,t) = u_T(x,t) + u_{-T}(x,t)\,, \tag{30.84}$$

where

$$\begin{aligned} u_T(x,t) &= \left\{u_T^{(1)}(x,t), u_T^{(2)}(x,t), u_T^{(3)}(x,t), u_T^{(4)}(x,t)\right\}, \\ u_{-T}(x,t) &= \left\{u_{-T}^{(1)}(x,t), u_{-T}^{(2)}(x,t), u_{-T}^{(3)}(x,t), u_{-T}^{(4)}(x,t)\right\}. \end{aligned}, \tag{30.85}$$

The wave oscillation process is shown schematically in the space-time diagram of Fig. 30.8, where the wave fronts are represented by straight lines. The specific front that formed after the load was removed is highlighted in red. A series of critical points

$$\left\{L\sqrt{\frac{\rho}{E}},\ 2L\sqrt{\frac{\rho}{E}},\ 3L\sqrt{\frac{\rho}{E}},\ 4L\sqrt{\frac{\rho}{E}}\right\},$$

at which the wave front velocity changes direction, are marked accordingly on the time axis.

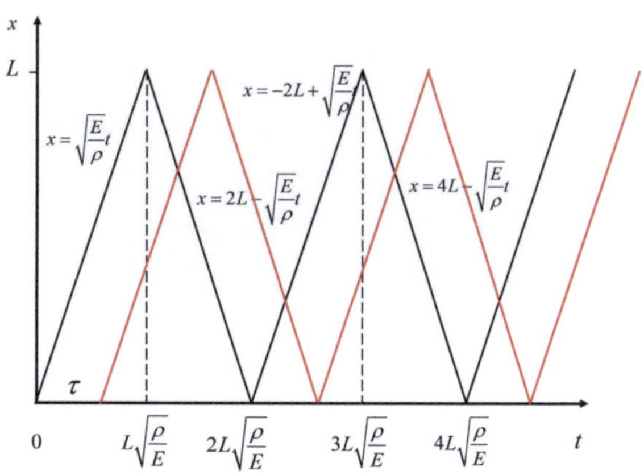

Fig. 30.8 Space-Time Diagram of the Wave Oscillatory Motion of the Rod Under Impulse Loading

From the given dependence, it is evident that the oscillation period (30.78), originally derived for a rod under force loading, remains unchanged in the case of impulsive loading. Furthermore, the period is independent of the duration τ of the force application. It is also worth noting that the vibration modes corresponding to different time intervals $\tau_1 \neq \tau_2$ are linearly independent. In other words, a single natural frequency (30.79) corresponds to an infinite number of linearly independent oscillation modes.

30.6 Dynamic Bending of a Beam Under Concentrated Moment

Consider a semi-infinite rectilinear rod, as shown in Fig. 30.9. The rod is initially assumed to be at rest, meaning that the displacements $u(x) = 0$ and velocities $\dot{u}(x) = 0$ of all its points are zero. At the initial moment $t = 0$, a concentrated bending moment $M_0 > 0$ is applied to the end of the rod at coordinate $x = 0$, and this moment continues to act on the rod indefinitely. The basic ideas of the approach solving this problem where discussed in [14].

Inertial (linear density of the material ρ) and rigidity (flexural rigidity EJ) characteristics of the rod are also introduced. Here E is Young's modulus and J is the moment of inertia of the section. It is considered that all characteristics of the rod are constant and do not change, both in length and in time.

The problem is to determine the stress-strain state of the rod at any arbitrary time. To solve it, we introduce two key assumptions. First, we assume that at any given time, there exists a point $x = x_1$ (see Fig. 30.9) that divides the rod into two distinct regions. For points $x > x_1$, the rod remains at rest; for points $x < x_1$, all points of the rod exhibit non-zero displacements and velocities. Second, we assume that for all points in the region $x < x_1$ the stress-strain state is uniform and governed by the following equation of state:

$$EJu_{xx} - M_0 = 0\,. \tag{30.86}$$

It is further assumed that the position of the point x_1 varies continuously with time t. These assumptions are analogous to those introduced in the previous sections.

Integrating equation (30.86) twice with respect to the coordinate x yields the following expression for the displacement function:

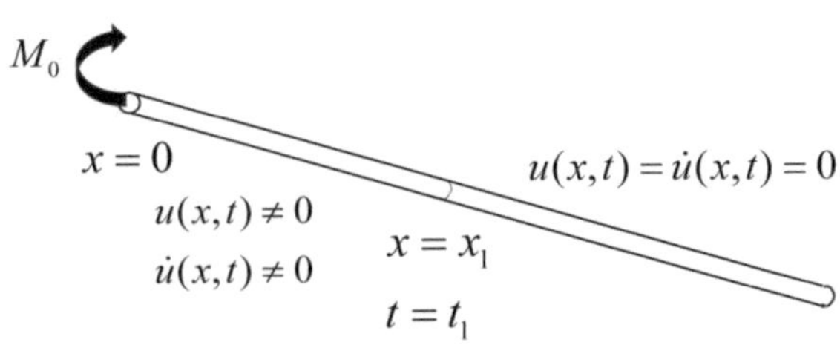

Fig. 30.9 Bending Rod with Concentrated Moment

$$u(x,t) = -\frac{M_0 x^2}{2EJ} - a_1(t)x - b_1(t)\,. \tag{30.87}$$

Here $a_1(t)$ and $b_1(t)$ are the unknown functions of time. The arguments of the functions are omitted in what follows.

Satisfying the continuity condition of the displacement function at the point $x = x_1$ and excluding the function $b_1(t)$, the following displacement function is obtained

$$u = \frac{M_0\left(x_1{}^2 - x^2\right)}{2EJ} + a_1(x_1 - x)\,. \tag{30.88}$$

The velocity function u_{t} is found by simply differentiating the displacement function with respect to time

$$u_{\mathrm{t}} = a_{1t}\,(x_1 - x) + \left(\frac{M_0 x_1}{EJ} + a_1\right) x_{1t}\,. \tag{30.89}$$

With these assumptions in place, the problem reduces to finding two unknown functions of time $x_1(t)$ and $a_1(t)$. To determine these functions, we employ the conservation laws – specifically, the conservation of energy, linear momentum, and angular momentum.

The law of conservation of energy is written in the form

$$T_0(t) + \Pi_0(t) - A_0(t) = 0\,. \tag{30.90}$$

Where $T_0(t)$ is the kinetic energy of the beam, $\Pi_0(t)$ is its potential energy, $A_0(t)$ is the work of external forces performed on the beam. All these quantities are calculated at the moment of time t.

For kinetic energy we have the following expression

$$\begin{aligned} T_0 = \int\limits_0^{x_1} \frac{\rho u_{\mathrm{t}}{}^2}{2} dx = \frac{\rho a_t^2 x_1^3}{6} + \frac{\rho a x_1^2 a_t x_{1t}}{2} + \\ \frac{\rho a_1{}^2 x_1 x_{1t}^2}{2} + \frac{\rho M_0 x_1^3 a_{1t} x_{1t}}{2EJ} + \frac{\rho M_0 a_1 x_1^2 x_{1t}^2}{EJ} + \frac{\rho M_0^2 x_1^3 x_{1t}^2}{2EJ^2}\,. \end{aligned} \tag{30.91}$$

The potential energy has the following representation

$$W_0 = \int\limits_0^{x_1} \frac{M_0^2}{2EJ} dx = \frac{M_0^2 x_1}{2EJ} \tag{30.92}$$

and the work of external forces

$$A_0 = \int\limits_0^{x_1} u_x|_{x=0} dx = -M_0 a_1\,. \tag{30.93}$$

Substituting expressions (30.91) – (30.93) into equality (30.90) allows us to compose one ordinary nonlinear differential equation depending on two unknown functions x_1 and a_1.

The expression for the momentum density of the rod is

$$p(x,t) = \rho u_t (x,t)\,. \tag{30.94}$$

Then, the condition of dynamic equilibrium, given by the law of momentum change, can be written as:

$$\frac{dP(t)}{dt} = F = 0\,.$$

Consequently, $P(t) = C = \text{const}$. According to the law of momentum change, the integral of the beam velocities over its length must be equal to zero

$$\int_0^{x_1} u_t dx = \frac{\rho {x_1}^2 a_{1t}}{2} + \rho x_1 x_{1t} a_1 + \frac{\rho M_0 {x_1}^2 x_{1t}}{EJ} = 0\,. \tag{30.95}$$

The expression for the angular momentum density of the rod is

$$l_0 = \rho u_t x\,. \tag{30.96}$$

The change in moments relative to the point $x = 0$ is calculated and takes the form:

$$\begin{aligned} &\frac{d}{dt}\left[\int_0^{x_1} l_0 dx\right] - M_0 = \frac{\rho {x_1}^3 a_{1tt}}{6} + \frac{\rho {x_1}^2 x_{1tt} a_1}{2} + \\ &\rho {x_1}^2 x_{1t} a_{1t} + \rho x_1 a_1 x_{1t}^2 + \frac{\rho M_0 {x_1}^3 x_{1tt}}{2EJ} + \frac{\rho M_0 {x_1}^2 x_{1t}^2}{2EJ} - M_0 = 0 \end{aligned}\,. \tag{30.97}$$

The problem of finding the two unknown functions $\{x_1, a_1\}$ is therefore governed by three nonlinear ordinary differential equations: (30.90), (30.94), and (30.96). Exact solution of this system is problematic for two reasons: its nonlinearity and its overdetermined character (three equations for two unknowns). For a solution to exist, the equations must be dependent. While a complete analysis of the system is unlikely to be feasible, it is nevertheless possible to construct a particular solution.

First, we attempt to transform this system into a more convenient form through a special substitution that makes it homogeneous with respect to the powers of the unknown functions. To this end, we introduce two new functions x_2 and a_2

$$x_2 = {x_1}^2, \quad a_2 = a_1 x_1\,. \tag{30.98}$$

Using substitution (30.98), equations (30.90), (30.94), and (30.96) are respectively transformed to the form

$$\begin{array}{l} 4\rho EJ^2 a_{2t}^2 x_2^2 + 2\rho EJ^2 a_{2t} x_{2t} a_2 x_2 + \rho EJ^2 x_{2t}^2 a_2^2 + 6\rho M_0 EJ a_{2t} x_{2t} x_2^2 + \\ 3\rho M_0 EJ x_{2t}^2 a_2 x_2 + 3\rho {M_0}^2 x_{2t}^2 x_2^2 + 24 M_0 EJ^2 a_2 x_2 + 12 {M_0}^2 EJ x_2^2 = 0 \end{array}, \tag{30.99}$$

$$2M_0 x_{2t} x_2 + 2EJ a_{2t} x_2 + EJ x_{2t} a_2 = 0\,, \tag{30.100}$$

$$\begin{array}{l} \dfrac{\rho a_{2tt} {x_2}^3}{6} + \dfrac{\rho x_{2tt} {x_2}^2 a_2}{2} + \rho a_{2t} x_{2t} {x_2}^2 + \\ \rho {x_{2t}}^2 x_2 a_2 + \dfrac{\rho M_0 x_{2tt} {x_2}^3}{2EJ} + \dfrac{3\rho M_0 x_{2t}^2 {x_2}^2}{2EJ} - M_0 = 0 \end{array}. \tag{30.101}$$

From the representation of equations (30.99) – (30.100) it is clear that the variables $x_2(t)$ and $c_2(t)$ enter uniformly (with the same powers) into the equations. Thus, the substitution $\{x_2 = b_1 t, \quad c_2 = b_2 t\}$ reduces the system of differential equations to a system of nonlinear algebraic equations with respect to the parameters $\{b_1, \quad b_2\}$

$$\begin{array}{l} 7\rho EJ^2 b_1 {b_2}^2 + 9\rho EJ M_0 {b_1}^2 b_2 + \\ 3\rho {M_0}^2 {b_1}^3 + 24 EJ^2 M_0 b_2 + 12 EJ {M_0}^2 b_1 = 0 \end{array}, \tag{30.102}$$

$$2M_0 b_1 + 3EJ b_2 = 0\,, \tag{30.103}$$

$$\frac{\rho M_0}{4EJ} {b_1}^2 + \frac{\rho b_1 b_2}{3} - M_0 = 0\,. \tag{30.104}$$

It is clear that the easiest way to solve equations (30.103) and (30.104). As a result, we have

$$\left\{ b_1 = 6\sqrt{\frac{EJ}{\rho}}, \quad b_2 = -\frac{4M_0}{\sqrt{EJ\rho}} \right\}. \tag{30.105}$$

It can be verified that substitution (30.105) identically satisfies equation (30.102).

It follows from solution (30.105) that the desired functions $x_1(t)$ and $a_1(t)$ have the form

$$x_1 = \sqrt{6t\sqrt{\frac{EJ}{\rho}}}, \quad a_1 = -\frac{2M_0}{3}\sqrt{\frac{6t}{EJ\sqrt{\rho EJ}}}\,. \tag{30.106}$$

Thus, the beam displacement function (30.89) can be represented as

$$u(x,t) = \begin{cases} -M_0\left(\dfrac{x^2}{2} + \dfrac{x\sqrt{8t}}{\sqrt{3}} + t\right), & x < \sqrt{6t\sqrt{\dfrac{EJ}{\rho}}} \\ 0, \quad x \geq \sqrt{6t\sqrt{\dfrac{EJ}{\rho}}} \end{cases}. \tag{30.107}$$

The characteristic distribution of displacements along the beam length is shown in Fig. 30.10. Notably, a wave front is present at point x_1). The appearance of negative displacements is explained by the fact that the total momentum of the beam is always zero.

For the velocities of the beam points we have the following representation, respectively

$$u_{\mathrm{t}}(x,t) = \begin{cases} M_0\left(\dfrac{\sqrt{2}x}{\sqrt{3t\sqrt[4]{\rho EJ^3}}} - \dfrac{1}{\sqrt{\rho EJ}}\right), & x < \sqrt{6t\sqrt{\dfrac{EJ}{\rho}}} \\ 0, \quad x \geq \sqrt{6t\sqrt{\dfrac{EJ}{\rho}}} \end{cases}. \tag{30.108}$$

In this case, the wave front velocity $v_{\mathrm{f}}(t)$ is a variable quantity and is equal to

$$v_{\mathrm{f}}(t) = \sqrt{\frac{3}{2t}\sqrt{\frac{EJ}{\rho}}}. \tag{30.109}$$

It is worth emphasizing two key points regarding velocities in this problem. First, the wave front velocity is governed exclusively by the beam's rigidity and inertial characteristics, and is unaffected by the magnitude of the bending moment M_0; it decreases over time, vanishing as the disturbance propagates infinitely. Second, the velocity of the point of load application remains constant throughout the motion and is equal to:

$$u_{\mathrm{t}}(0,t) = -\frac{M_0}{\sqrt{\rho EJ}}. \tag{30.110}$$

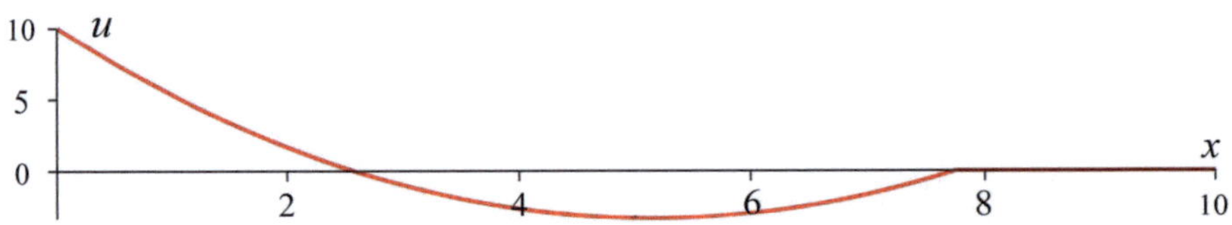

Fig. 30.10: Beam Displacements at a Gven Moment in Time $t = 10$ with the Following Parameter Values $EJ = \rho = M_0 = 1$.

30.7 Dynamic Bending of a Beam Under Concentrated Shear Force.

Consider again the semi-infinite rectilinear rod shown in Fig. 30.11. Initially, the rod is at rest, with zero displacements $u(x) = 0$ and velocities $\dot{u}(x) = 0$ at all points. At time $t = 0$ a concentrated transverse force $P_0 > 0$ is applied to the rod end at $x = 0$ and it acts on the rod continuously. The problem is to determine the stress-strain state of the rod at any given time.

To solve this problem while satisfying the conservation laws – as in the previous section – we introduce two important assumptions. First, at any given instant $t = t_1$ there exists a point $x = x_1$ (see Fig. 30.11) that divides the rod into two distinct regions. For points $x > x_1$ the rod remains at rest; for points $x < x_1$ all points of the rod exhibit non-zero displacements and velocities.

Secondly, for all points $x < x_1$ the stress-strain state of the rod is uniform and satisfies the following governing equation

$$u_{xxx} - \frac{P_0}{EJ} = 0\,. \tag{30.111}$$

Assuming the position of the point x_1 is a continuous function of time t, we integrate equation (30.111) three times with respect to the coordinate x to obtain the following expression for the displacement function:

$$u\,(x,t) = \frac{P_0 x^3}{6EJ} + \frac{c_1(t)x^2}{2} + c_2(t)x + c_3(t)\,. \tag{30.112}$$

Here $c_1(t)$, $c_2(t)$, and $c_3(t)$ are unknown functions of time. In what follows, the arguments of the functions will be omitted. After satisfying the continuity condition for the displacement function at the point $x = x_1(t)$ and eliminating the function c_3, we obtain the following expression for the displacement function:

$$u = P_0\frac{x^3 - {x_1}^3}{6EJ} + c_1\frac{x^2 - {x_1}^2}{2} + c_2\,(x - x_1)\,. \tag{30.113}$$

The velocity function u_t is found by simply differentiating the displacement function with respect to time

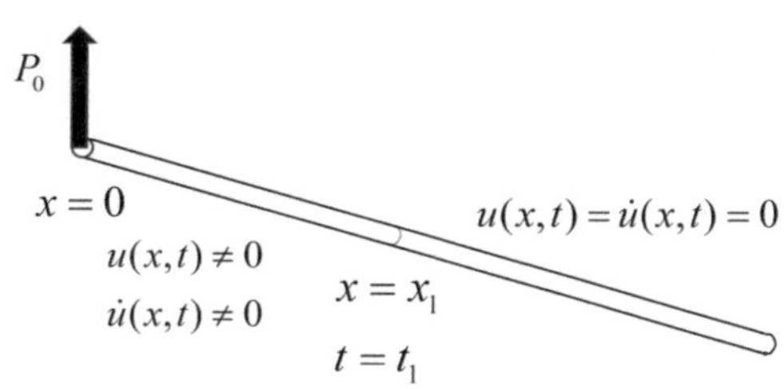

Fig. 30.11 Bending Rod with Concentrated Force

$$u_t = c_{1t}\frac{x^2 - x_1{}^2}{2} - c_{2t}\,(x - x_1) - \frac{P_0 x_1{}^2 x_{1t}}{2EJ} - x_{1t}c_2 - x_1 c_{2t}\,. \tag{30.114}$$

With these assumptions in place, the problem reduces to finding three unknown functions of time: x_1, c_1, c_2. To determine these functions, we employ the conservation laws – specifically, the conservation of energy, linear momentum, and angular momentum.

The law of conservation of energy has the form

$$T_0(t) + \Pi_0(t) - A_0(t) = 0\,. \tag{30.115}$$

For kinetic energy we have the following expression

$$T_0(t) = \frac{\rho x_1{}^5 c_{1t}}{15} + \frac{\rho x_1{}^4 x_{1t} c_{1t}}{3} + \frac{\rho x_1{}^3 x_{1t}^2 c_1{}^2}{2} + \frac{\rho P_0 x_1{}^5 c_{1t} x_{1t}}{6EJ} + \frac{\rho P_0 x_1{}^4 x_{1t}{}^2 c_1}{2EJ} +$$

$$\frac{\rho P_0{}^2 x_1{}^5 x_{1t}^2}{8EJ^2} + \frac{5\rho x_1{}^4 c_1 c_2}{24} + \frac{\rho x_1{}^3 x_{1t} c_{1t} c_2}{3} + \frac{\rho x_1{}^3 x_{1t} c_{2t} c_1}{2} + \rho x_1{}^2 x_{1t}{}^2 c_1 c_2 +$$

$$\frac{\rho P_0 x_1{}^4 c_{2t} x_{1t}}{4EJ} + \frac{\rho P_0 x_1{}^3 x_{1t}^2 c_{2t}}{2EJ} + \frac{\rho x_1{}^3 c_{2t}^2}{6} + \frac{\rho x_1{}^2 c_{2t} x_{1t} c_2}{2} + \frac{\rho x_1 x_{1t}^2 c_2{}^2}{2}\,. \tag{30.116}$$

It is clear that any qualified researcher can write this relationship explicitly. Expression (30.116) is presented here merely to illustrate the complexity with which the unknown functions enter into it.

Potential energy has the following representation

$$W_0 = \int_0^{x_1(t)} \frac{(P_0 x)^2}{6EJ}\,dx = \frac{P_0{}^2 x_1{}^3}{6EJ} \tag{30.117}$$

and the work of external forces

$$A_0 = P_0 u(t,0) = -P_0 x_1 \left(\frac{x_1 c_1}{2} + \frac{x_1{}^2}{6EJ} + c_2\right). \tag{30.118}$$

Substituting expressions (30.116) – (30.118) into equality (30.115) allows us to formulate a single nonlinear ordinary differential equation involving the three unknown functions x_1, c_1, c_2.

The expression for the density of the rod's momentum has the form

$$p(x,t) = \rho u_t\,. \tag{30.119}$$

The law of change of momentum (the condition of dynamic equilibrium) gives

$$\left(\int_0^{x_1} \rho u_t dx\right)_t - P_0 = -2\rho x_1{}^2 c_{1t} x_{1t} - \rho x_1 c_1 x_{1t}^2 - \frac{\rho x_1{}^3 c_{1tt}}{3} - \rho x_1{}^2 c_1 x_{1tt} -$$

$$-\frac{\rho P_0 x_1{}^2 x_{1t}{}^2}{EJ} - \frac{\rho P_0 x_1{}^3 x_{1t}}{2EJ} - 2\rho x_1 c_{2t} x_{1t} - \frac{\rho x_1{}^2 c_{2tt}}{2} - \rho x_1 c_2 x_{1tt} - P_0 = 0\,. \tag{30.120}$$

The expression for the density of the rod angular momentum is

$$l_0 = \rho u_t x\,. \tag{30.121}$$

We calculate the change in moments relative to the point $x = 0$. The moment from the force P_0 is equal to zero. Then the expression leads to

$$\frac{dL}{dt} = 0\,.$$

At the initial moment of time $L = 0$. It follows that $L = 0$. Consequently, the moment of momentum does not depend on time and is equal to zero

$$L = \int_0^{x_1} l_0 dx = 3x_1{}^2 c_{1t} + 12x_1 c_1 x_{1t} + \frac{6P_0}{EJ} x_1{}^2 x_{1t} + 4x_1 c_{2t} + 12c_2 x_{1t} = 0\,. \tag{30.122}$$

Thus, we have 3 nonlinear differential equations, namely, (30.115) with substitutions (30.116) – (30.118) , and equations (30.120) and (30.122) with respect to three variable functions x_1, c_1, and c_2.

Introduce two new functions into consideration x_2 and c_3

$$x_2 = x_1{}^2, \quad c_3 = c_1 x_1 \tag{30.123}$$

substituting this into the system of equations renders the components homogeneous with respect to the powers of the functions x_2, c_2, and c_3. Then, using this transformation

$$\{x_2 = b_1 t, \quad c_3 = b_2 t, \quad c_2 = b_3 t\}\,, \tag{30.124}$$

we obtain the following system of nonlinear algebraic equations with respect to the parameters $\{b_1, b_2, b_3\}$.

$$108\rho b_1 b_2{}^2 + 330\rho b_1 b_2 b_3 + 260\rho b_1 b_3{}^2 + \frac{80\rho P_0 b_1{}^2 b_2}{EJ} +$$

$$\frac{120\rho P_0 b_1{}^2 b_3}{EJ} + \frac{15\rho P_0{}^2 b_1{}^3}{EJ^2} + 240P_0 b_2 + 480P_0 b_3 + \frac{160P_0{}^2 b_1}{EJ} = 0\,, \tag{30.125}$$

$$10\rho EJ b_1 b_2 + 18\rho EJ b_1 b_3 + 3\rho P_0 b_1{}^2 + 24P_0 EJ = 0\,, \tag{30.126}$$

$$15EJ b_2 + 20EJ b_3 + 6P_0 b_1 = 0\,. \tag{30.127}$$

This system of equations has a unique analytical solution

$$
\begin{aligned}
b_1 &= \sqrt{\frac{10EJ\left(385\sqrt{13}-1373\right)}{27\rho}},\\
b_2 &= \frac{8\sqrt{30}P_0\left(40-11\sqrt{13}\right)}{3\sqrt{\rho EJ\left(385\sqrt{13}-1373\right)}},\\
b_3 &= -\frac{P_0\left(1027-275\sqrt{13}\right)}{\sqrt{30\rho EJ\left(85\sqrt{13}-1373\right)}}
\end{aligned}
\tag{30.128}
$$

or an approximate value

$$
\left\{b_1 = 2.368\sqrt{\frac{EJ}{\rho}}, \quad b_2 = 1.272\frac{P_0}{\sqrt{\rho EJ}}, \quad b_2 = -1.665\frac{P_0}{\sqrt{\rho EJ}}\right\}. \tag{30.129}
$$

From solution (30.129) it follows that the desired functions $x_1(t)$, $c_1(t)$, and $c_2(t)$ have the form

$$
x_1 = 1.539\left(\frac{EJ}{\rho}\right)^{1/4}\sqrt{t}, \quad c_1 = \frac{0.827P_0}{\sqrt{EJ\sqrt{\rho EJ}}}\sqrt{t}, \quad c_2 = -\frac{1.665P_0}{\sqrt{\rho EJ}}. \tag{30.130}
$$

Thus, the displacement function (30.112) of the beam can be represented for $x < x_1$

$$
u(x,t) = P_0\left(\frac{0.167}{EJ}x^3 - \frac{1.665}{\sqrt{\rho EJ}}xt + \frac{0.975}{\sqrt{\rho\sqrt{\rho EJ}}}t^{3/2} + \frac{0.413}{\sqrt{EJ\sqrt{\rho EJ}}}x^2\sqrt{t}\right) \tag{30.131}
$$

and $u(x,t) = 0$ for $x > x_1$.

The characteristic distribution of displacements along the length of the beam is shown in Fig. 30.12, where a distinct wave front is visible at point (x_1). The occurrence of negative displacements arises from the fact that the total moment of momentum of the beam is always zero.

Hence, the velocities of the beam points are given for $x < x_1$ by

$$
u_t(x,t) = P_0\left(-\frac{1.664}{\sqrt{\rho EJ}}x + \frac{0.207}{\sqrt{EJ\sqrt{\rho EJ}}}\frac{x^2}{\sqrt{t}} + \frac{1.462\sqrt{t}}{\sqrt{\rho\sqrt{\rho EJ}}}\right) \tag{30.132}
$$

and $u_t(x,t) = 0$ at $x > x_1$.

In this case, the wave front velocity is variable and given by

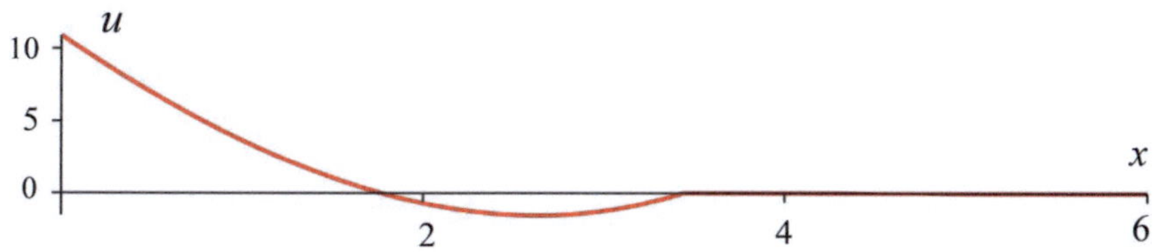

Fig. 30.12: Beam displacements at a given moment in time $t = 5$ with the following parameter values.

$$u_t(0,t) = \frac{1.462 P_0}{\sqrt{\rho\sqrt{\rho EJ}}}\sqrt{t} \tag{30.133}$$

It is important to note that the wave front velocity depends on the beam's rigidity and inertial characteristics but is independent of the shear force P_0 magnitude. Over time, this velocity decreases and asymptotically approaches zero as the disturbance propagates infinitely. Meanwhile, the velocity of the load application point follows a root function of time during motion and is given by

$$u_t(0,t) = \frac{1.462 P_0}{\sqrt{\rho\sqrt{\rho EJ}}}\sqrt{t}\,. \tag{30.134}$$

Acknowledgements The work was carried out within the framework of State Assignment No. 124013000674-0.

References

[1] Abramson HN, Piass HJ, Ripperger EA (1958) Stress wave propagation in rods and beams. Advances in Applied Mechanics 5:111-194, DOI 10.1016/S0065-2156(08)70019-X

[2] Bayanov EV, Gulidov AI. (2011), Propagation of elastic waves in circular rods homogeneous over the cross section. Journal of Applied Mechanics and Technical Physics 52(5): 806–814, DOI 10.1134/S0021894411050166

[3] Belyaev AK, Tovstik PE, Tovstik TP (2017), Thin rod under longitudinal dynamic compression. Mechanics of Solids 52(4):364–377, DOI 10.3103/S0025654417040021

[4] Clough RW, Penzien J (1993) Dynamics of Structures. McGraw-Hill, 1993, New York

[5] Cortes F, Elejabarrieta MJ (2006) Longitudinal vibration of a damped rod. Part I: Complex natural frequencies and mode shapes. International Journal of Mechanical Sciences 48: 969–975, DOI 10.1016/j.ijmecsci.2006.03.010

[6] Gan C B, Wei YM, and Yang SX. (2014), Longitudinal wave propagation in a rod with variable cross-section. Journal of Sound and Vibration 333: 434–445,

DOI 10.1016/j.jsv.2013.09.010
[7] Hsueh WJ. (1999), Free and forced vibrations of stepped rods and coupled system. Journal of Sound and Vibration 226(5): 891–904, DOI 10.1006/jsvi.1999.2249
[8] Kostin GV, Saurin VV (2018) Dynamics of Solid Structures: Methods using Integrodifferential Relations. De Gruyter, Berlin
[9] Kostin GV, Saurin VV (2012) Integrodifferential relations in linear elasticity. De Gruyter, Berlin
[10] Kostin GV, Saurin VV (2011) Modelling and analysis of the natural oscillations of a prismatic elastic beam based on a projection approach. Journal of Applied Mathematics and Mechanics 75(6):700–710, DOI 10.1016/j.jappmathmech.2012.01.010
[11] Krawczuk M, Grabowska J, Palacz M. (2006), Longitudinal wave propagation, part I: comparison of rod theories. Journal of Sound and Vibration 295: 461–478, DOI 10.1016/j.jsv.2005.12.048
[12] Morse RV (1950) The velocity of compressional waves in rods rectangular cross sections. Journal of the Acoustical Society of America 22: 219–223, DOI 10.1121/1.1906592
[13] Nayfeh AH, Abdelrahman WG (2000) An approximate model for wave propagation in rectangular rods and their geometric limits. Journal of Vibration and Control 6(1): 3–17, DOI 10.1177/107754630000600101
[14] Saurin VV (2024) Dynamic Bending of a Beam. Solid Body Mechanics. Mechanics of Solids. 59(5): 2706–2718, DOI 10.1134/S0025654424603422
[15] Yang, K. (2008), A unified solution for longitudinal wave propagation in an elastic rod. Journal of Sound and Vibration 314: 307–329, DOI 10.1016/j.jsv.2008.01.007

Chapter 31
Detection of an Inclusion in 2D Elastic Body by Means of Cauchy Data Given on a Part of the Body Boundary

Efim I. Shifrin and Alexander V. Kaptsov

Abstract A plane isotropic linear elastic body is considered. It is assumed that the elastic body contains an inclusion or a cavity that occupies a quadrature domain. It is also assumed that, as a result of a single static experiment, on the part of the body boundary the applied loads and displacements are determined. A method of determination of the nodal points of the inclusion using the known Cauchy data is developed. Numerical examples are considered.

Key words: Plane elasticity, Inverse problem, Cauchy data, Quadrature inclusion, Nodal point

31.1 Introduction

The problem of reconstruction of non-homogeneous Lame coefficients by means of overdetermined boundary data has been discussed in a number of publications. The case when the Lame coefficients are smooth functions was considered in [2, 23, 24]. Many methods have been developed for detecting and identifying cavities and inclusions in elastic bodies. In these problems, the Lame coefficients are piecewise constant functions. Let us recall some methods for solving geometric inverse problems. To solve these problems, the method of fundamental solutions (MFS) was applied in [30, 31]. In publications [4, 28], the method for identifying small inclusions in an elastic body is based on far-field analysis. The enclosure method was developed in [21, 22]. The application of reciprocity gap functional method to elastostatic inverse problems can be found in [5, 38, 41, 42]. There are some publications where the linear sampling method [15] and the factorization method [33, 37], developed initially for inverse scattering problems, were applied to the solution of

Efim I. Shifrin (corresponding author) · Alexander V. Kaptsov
Ishlinsky Institute for Problems in Mechanics RAS, Prospekt Vernadskogo 101-1, 119526 Moscow, Russian Federation, e-mail: shifrin@ipmnet.ru, kaptsov@ipmnet.ru

H. Altenbach et al. (eds.), *Stability Problems in Mechanics of Continua*,
Advanced Structured Materials 260, https://doi.org/10.1007/978-3-032-31032-3_31

inverse static problems. The listed methods require measuring the applied loads and displacements on the entire outer boundary of the body in one or multiple static experiments.

Optimization methods are also widely used in solving geometric inverse problems. The level set method was applied to solving the problems in [10, 11]. In [16] the genetic algorithm was used to identify an ellipsoidal defect. In [20] the authors applied the topological gradient method for the cavity identification. The immersed finite elements based parameterized optimization algorithm was developed in [18]. The quadratic misfit functional with a regularization term that penalizes the perimeter of the inclusion combined with a phase-field approach was used in [7]. A phase-field approach for a Kohn-Vogelius type functional was proposed in [6]. Some of the optimization methods can be used for solving geometric inverse problems by means of incompletely overdetermined boundary data. However, the implementation of these approaches is quite complicated, and the objective function can have many extrema.

The possibilities of detecting and identifying cavities and inclusions in a linearly elastic body using incompletely overdetermined data at the outer boundary have been studied to a much lesser extent. Let us note some of the publications devoted to this problem. In [9, 26, 27] methods for identifying cavities by displacements and only the normal component of applied loads are developed. In the paper [17] several approaches are proposed for identification cracks using incomplete boundary data. All approaches proposed in this paper use an idea based on the application of the reciprocity gap functional. One method first solves the Cauchy problem, which is used to determine the missing data at the boundary. After that the reciprocity gap functional is applied. In other methods the crack and the missing boundary data are determined simultaneously.

In [32, 40], a method for detecting inclusions using Cauchy data on a part of the outer boundary of an elastic body is proposed. The proposed method is based on the following idea. An expanding system of subsets is constructed that includes the part of the boundary where the Cauchy data are given. A functional of the Kohn-Vogelius type is constructed for every subset of the system. The value of the functional can be used to determine whether there is an inclusion. Paper [40] shows that if a subset does not intersect a defect, then the value of the functional is zero. If the subset contains the inclusion inside, then the value of the functional is greater than zero. The disadvantage of this approach is that it does not allow one to accurately determine the position of the barrier, upon crossing which the value of the Kohn-Vogelius functional changes.

Papers [29, 39] consider the problem of an inclusion in an infinite elastic plane. Constant stresses are specified at infinity. It is shown that when the inclusion occupies a quadrature domain, the solution to the problem can be extended from the matrix into the inclusion, excluding the inclusion's nodal points. The nodal points themselves are determined using a reciprocity gap functional method.

The aim of this paper is to develop a fairly simple method for determining the nodal points of a quadrature inclusion in a 2D elastic body by means of the Cauchy data given on a part of the outer boundary. The paper is organized as follows. In Sect.

31.2 we will recall some mathematical concepts and results that will be used later in determining nodal points. We first illustrate the developed method using the example of the 2D Laplace equation. Section 31.3 examines the Laplace equation in a doubly connected domain. It is assumed that Cauchy data are specified on part of the outer boundary. Neumann conditions are specified on the inner boundary. A method for determining the nodal point of a cavity occupying a quadrature domain is proposed. Numerical examples for determining the nodal point in the case of Laplace equation are considered in Sect. 31.4. Section 31.5 formulates the problem of a linear elastic inclusion in a linearly elastic body. It is assumed that applied loads and displacements are known on a part of the body's outer boundary. A method for determining the nodal point of the inclusion using the given Cauchy data is proposed. Numerical examples of determining the nodal point of the elastic inclusion are considered in Sect. 31.6.

31.2 Basic Notions and Notations Used in the Paper

31.2.1 Quadrature Domains

Below we will denote by Ω a plane bounded simply connected domain with the boundary $\partial\Omega$. We assume that $D \subset \Omega$ is a simply connected quadrature domain bounded by smooth, closed, Jourdan curve ∂D. Let us briefly recall the definition and basic properties of quadrature domains. The foundations of the theory of quadrature domains are given in [1, 14]. Let us denote by $AL^1(D)$ the set of functions that are holomorphic in D and belong to $L_1(D)$.

Definition 31.1. D is a quadrature domain if there exist a finite number of points $z_1, \cdots, z_m$, where $z_i \in D$, such that for any $f(z) \in AL^1(D)$ the following equality is valid

$$\int_D f(z)\, dS = \sum_{k=1}^{m} \sum_{j=0}^{n_k-1} a_{kj} f^{(j)}(z_k) \tag{31.1}$$

Here Ox_1x_2 - Cartesian coordinates, $z = x_1 + ix_2$, a_{kj} are complex numbers. The points $z_k = x_{k1} + ix_{k2}$ are called nodal points. The quantity $n = \sum_{k=1}^{m} n_k$ is called the order of the quadrature identity (31.1).

This definition of the quadrature domain is given in [1]. There are also other equivalent definitions of the quadrature domains. First we recall the definition of the Schwarz function.

Definition 31.2. Let ∂D be an analytic curve. The holomorphic function $S(z)$ defined in some neighborhood of the curve ∂D is a Schwarz function if the following equality is valid

$$S(z) = \bar{z}, \quad \bar{z} = x_1 - ix_2, \quad z \in \partial D \tag{31.2}$$

Now we can give another definition of the quadrature domain [14].

Definition 31.3. A domain D bounded by an analytic curve ∂D is a quadrature domain if the Schwarz function $S(z)$ defined in the neighborhood of the curve ∂D can be continued into the domain D as a meromorphic function.

Let us note that the poles of the function $S(z)$ coincide with nodal points z_k and the orders of the poles are equal to the corresponding values n_k in the formula (31.1). The most famous examples of quadrature domains are the disc, cardioid, and Neumann oval. It is important to note that there are plenty quadrature domains. Moreover, as proved in [1, 14, 19], the following theorem is valid, which for simply connected domains considered in the paper, has the following form.

Theorem 31.1. *Suppose D is a bounded domain bounded by a regular analytic curve ∂D. Then there are quadrature domains arbitrarily close to D and conformally equivalent to D.*

The problem of uniqueness of a quadrature domain, that is, the problem of determining the geometry of a quadrature domain by its nodal points, has not been sufficiently studied to date. Only a few particular results are known in this area. For example, it is proved that a quadrature domain of order 1 ($m = 1$, $n_1 = 1$) is a disc. Other results concerning the uniqueness problem for quadrature domains can be found in [1, 3].

31.2.2 Padé Approximates

The idea behind Padé approximates is to re-expand a function defined by a Taylor series over a basis of rational functions. This allows for a good approximation of the function in the region where its series no longer converges, making it possible to determine poles on the function's extension.

Definition 31.4. Let the function $f(z)$ be represented by a Taylor series

$$f(z) = \sum_{i=0}^{\infty} c_i z^i \tag{31.3}$$

A Padé approximation of order $[L/M]$ is a rational function of the form

$$[L/M] = \frac{a_0 + a_1 z + \cdots + a_L z^L}{b_0 + b_1 z + \cdots + b_M z^M}, \quad b_0 = 1 \tag{31.4}$$

Here coefficients a_i and b_i are selected so that the expansion (31.3) of the function $f(z)$ coincides with the expansion in a Taylor series of the function $[L/M]$ with an accuracy of terms of order $L + M + 1$.

$$\sum_{i=0}^{\infty} c_i z^i = [L/M] + O\left(z^{L+M+1}\right) \tag{31.5}$$

Equality (31.5) allows us to determine the unknown coefficients a_i and b_i, which are included in the definition of function $[L/M]$, defined by formula (31.4). Function $[L/M]$ makes it possible to continue the function $f(z)$ beyond the convergence boundaries of series (31.3) and determine the poles on the continuation of the function. Details of the theory of Padé approximates can be found in [8]. There are a number of theorems clarifying the conditions for the convergence of a sequence of Padé approximates $[L_k/M_k]$ to the original function [8]. In the absence of additional information about the function to be approximated using Padé approximates, it is most appropriate to use diagonal $[M_k/M_k]$ or paradiagonal $[M_k \pm I/M_k]$, $M_k \to \infty$ sequences. Such sequences ensure convergence, for example, to meromorphic functions. Let us now consider a generalization of Padé approximates to the case where the function $f(z)$ is given by the Laurent series

$$f(z) = \sum_{i=-\infty}^{\infty} c_i z^i \tag{31.6}$$

In this case, two power series can be considered.

$$f^+(z) = \frac{c_0}{2} + \sum_{i=1}^{\infty} c_i z^i, \quad f^-(z) = \frac{c_0}{2} + \sum_{i=1}^{\infty} c_{-i} z^{-i}, \quad f(z) = f^+(z) + f^-(z) \tag{31.7}$$

When searching for poles on the continuation of the function $f(z)$, one can consider Padé approximates separately for the functions $f^+(z)$ and $f^-(z)$, see, for example, [13]. The theory of Padé approximates for Laurent series is presented in [12]. Let us now consider a complex-valued function $g(x)$ defined on the interval $-l \le x \le l$. We assume that this function is a restriction to the interval $[-l, l]$ of some meromorphic function. The function $g(x)$ can be expanded in a Fourier series.

$$g(x) = \sum_{k=-\infty}^{\infty} c_k e^{\frac{ik\pi x}{l}} \tag{31.8}$$

Let us make the substitution $z = e^{\frac{i\pi x}{l}}$ in (31.8). As a result, we obtain the Laurent series. Applying the Padé approximation to this Laurent series, we get continuation of the function $g(x)$ into complex plane. Thus, it becomes possible to determine the poles of a function that is a continuation of the function $g(x)$.

31.3 Detection of a Nodal Point of a Cavity by Means of the Cauchy Data of a Harmonic Function on the Part of Outer Boundary

Consider a harmonic function in the domain $\Omega\backslash\bar{D}$, where $\bar{D}$ is a closure of the domain D, see Fig. 31.1

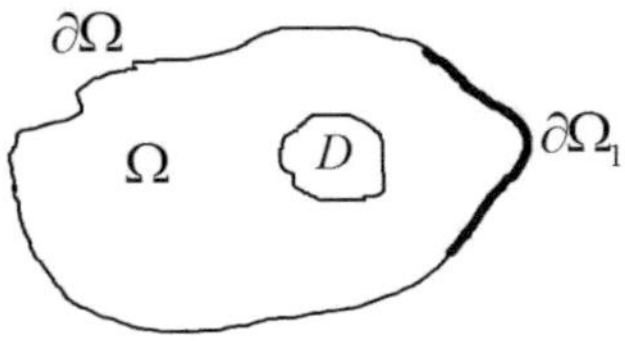

Fig. 31.1 Plane domain with a cavity.

$$\Delta u(x) = 0, \quad x = (x_1, x_2) \in \Omega \backslash \bar{D} \tag{31.9}$$

Equation (31.9), for example, corresponds to the problem of steady-state heat conduction. We assume that the boundary ∂D is thermally insulated

$$\frac{\partial u(x)}{\partial n} = 0, \quad x = (x_1, x_2) \in \partial D \tag{31.10}$$

Let us also assume that the function $u(x)$ and its normal derivative are known on a part of the outer boundary $\partial \Omega_1 \subset \partial \Omega$

$$\frac{\partial u(x)}{\partial n} = u_1(x), \quad u(x) = u_0(x), \quad x = (x_1, x_2) \in \partial \Omega_1 \tag{31.11}$$

We will study the problem of continuing the solution of problem (31.9) - (31.11) outside the domain $\Omega \backslash \bar{D}$. Let $u(x)$ be the real part of a function $f(z)$, that is holomorphic in the domain $\Omega \backslash \bar{D}$

$$u(x) = \frac{1}{2}\left(f(z) + \overline{f(z)}\right) \tag{31.12}$$

According to [14], the normal derivative of the function $u(x)$ at the boundary ∂D can be written in the form

$$\frac{\partial u(x)}{\partial n} = \frac{i}{2}\left(f'(z)\frac{1}{\sqrt{S'(z)}} - \bar{f}'(S(z))\sqrt{S'(z)}\right) \tag{31.13}$$

Here $S(z)$ is the Schwarz function of the curve ∂D, the function $\bar{f}(z)$ is defined as follows

$$\bar{f}(z) = \overline{f(\bar{z})} \tag{31.14}$$

It follows from the Eqs. (31.10) and (31.13) that at the curve ∂D the following equality is valid

$$f'(z) = \bar{f}'(S(z))\, S'(z) = \frac{d\left(\bar{f}(S(z))\right)}{dz}, \quad z \in \partial D \tag{31.15}$$

From the Eq. (31.15) it follows

$$f(z) = \bar{f}(S(z)) + \text{const}, \quad z \in \partial D \tag{31.16}$$

Let us assume that the function $f(z)$ can be continued as a holomorphic function into the domain $C\backslash\Omega$, where C is the complex plane. In this case, from Eq. (31.16) and the fact that the domain D is quadrature, it follows that the function $f(z)$ can be continued into the domain D as a meromorphic function. Boundary conditions (31.11) determine the analytical function $f(z)$ on $\partial\Omega_1$. Thus, we consider a problem of determination poles of the meromorphic function using the data on the open curve $\partial\Omega_1$. If the continuation of the function $f(z)$ into the region $C\backslash\Omega$ has singular points, then the continuation of this function from the curve $\partial\Omega_1$ can also determine these singularities. In particular, the continuation of the function $f(z)$ can determine the singular points of the boundary $\partial\Omega$. There are some publications where the problem of detecting poles of a meromorphic function from its values on an open curve were considered [25, 36]. In all examples below, we will use Padé, Laurent-Padé and Fourier-Padé approximations for identification of nodal points of quadrature inclusions.

31.4 Numerical Examples for the Laplace Equation

First, let us consider an approach to determining the poles of a function that is harmonic in a half-strip, may be except discrete set of points, and satisfies the following mixed boundary conditions. We designate the half-strip $0 < x < \infty$, $0 < y < H$ as Ω. In this part, it is more convenient for us to use variables x, y instead of x_1, x_2. The function $u(x, y)$ satisfies Eq. (31.9) in Ω may be except some singular points. We consider the following boundary conditions:

$$u(x,0)=0, \quad \frac{\partial u(x,H)}{\partial y}=0, \quad u(0,y)=u_0(y), \quad \frac{\partial u(0,y)}{\partial x}=0 \tag{31.17}$$

Here $u_0(y)$ is a given function. Solution of this problem near the boundary $x = 0$ can be presented by a series

$$u(x,y)=\sum_{k=0}^{\infty} c_k \cosh(\alpha_k x)\sin(\alpha_k y), \quad \alpha_k=\frac{1}{H}\left(\frac{\pi}{2}+\pi k\right) \tag{31.18}$$

where the coefficients c_k represent the coefficients of the Fourier series for the function $u_0(y)$.

$$c_k=\frac{2}{H}\int_0^H u_0(y)\sin(\alpha_k y)\,dy \tag{31.19}$$

The function $u(x, y)$ presented in Eq. (31.18) can be considered as a real part of a holomorphic function

$$u(x,y)=\mathrm{Re}(f(z)), \quad f(z)=-\frac{i}{2}\sum_{k=0}^{\infty} c_k\left(\mathrm{e}^{\alpha_k z}-\mathrm{e}^{-\alpha_k z}\right) \tag{31.20}$$

Let us now make a change of variables in the function $f(z)$, taking into account the values of α_k presented in Eq. (31.18)

$$f(z) = \tilde{f}(w) = -\frac{i}{2}\sum_{k=0}^{\infty} c_k \left(w^{2k+1} - w^{-(2k+1)}\right), \quad w = \mathrm{e}^{\frac{\pi z}{2H}} \tag{31.21}$$

Equation (31.21) is a Laurent expansion of the function $\tilde{f}(w)$. The function $\tilde{f}(w)$ we represent as a sum of power series in positive and negative degrees.

$$\tilde{f}(w) = \tilde{f}_+(w) + \tilde{f}_-(w), \quad \tilde{f}_+(w) = -\frac{i}{2}\sum_{k=0}^{\infty} c_k w^{2k+1}, \quad \tilde{f}_-(w) = \frac{i}{2}\sum_{k=0}^{\infty} c_k w^{-(2k+1)} \tag{31.22}$$

Using Padé approximates, we can study the poles of the functions $\tilde{f}_+(w)$ and $\tilde{f}_-(w)$. However, the problem of the poles determination can be simplified even further. Let us consider the functions $f_+^0(w) = \tilde{f}_+(w)/w$, $f_-^0(w) = \tilde{f}_-(w)\,w$. The poles that are interesting for us of the function $\tilde{f}_+(w)$ and $f_+^0(w)$ are the same. Similarly the poles of the functions $\tilde{f}_-(w)$ and $f_-^0(w)$ are coincide also. Finally, let us make a change of variable $w^2 = \zeta$.

$$f_+^0(w) = \hat{f}_+(\zeta) = -\frac{i}{2}\sum_{k=0}^{\infty} c_k \zeta^k, \quad f_-^0(w) = \hat{f}_-(\zeta) = \frac{i}{2}\sum_{k=0}^{\infty} c_k \zeta^{-k} \tag{31.23}$$

Using Padé approximates of the functions $\hat{f}_+(\zeta)$ and $\hat{f}_-(\zeta)$ we find the poles. After that we return to the variable z.

Example 1. Here and below in all examples we consider a square $|x| \le 10$, $|y| \le 10$ as a domain Ω. The cavity D is a disc of the radius 1 with the center located at the point $(3, 4)$. $u(x, y)$ is a harmonic function in the domain $\Omega \backslash D$ satisfying the following boundary conditions

$$u(x, -10) = 20, \quad \frac{\partial u(x, 10)}{\partial y} = -1, \quad \frac{\partial u(-10, y)}{\partial x} = \frac{\partial u(10, y)}{\partial x} = 0 \tag{31.24}$$

Besides the conditions (31.24) it is assumed that the function $u(10, y)$ is measured on the right boundary. The problem is to determine the nodal point of the cavity $(3, 4)$ using the function $u(10, y)$. We determined the function $u(10, y)$ by solving the formulated problem by finite element method (FEM). To solve the posed inverse problem, we first determine the harmonic function in region Ω without a cavity that satisfies the boundary conditions (31.24). We denote this harmonic function by $u^0(x, y)$. Let us consider the function $u_d(x, y) = u(x, y) - u_0(x, y)$. The function $u_d(x, y)$ is harmonic in $\Omega \backslash D$ and satisfies the boundary conditions

$$u_d(x, -10) = 0, \quad \frac{\partial u_d(x, 10)}{\partial y} = 0, \quad \frac{\partial u_d(-10, y)}{\partial x} = \frac{\partial u_d(10, y)}{\partial x} = 0 \tag{31.25}$$

Besides, it is known the function on the right boundary $u_d(10, y)$. Let us change the variables $\xi = -x + 10, \eta = y + 10$. The function $\tilde{u}_d(\xi, \eta) = u_d(10 - \xi, \eta - 10)$ satisfies the boundary conditions (31.17) with $H = 20$. Thus, we can apply the above method of finding poles in a half-strip. Let is present the results of diagonal Padé approximations of order $[M/M]$ of the function $\widehat{f}_+(\xi + i\eta)$. Of course, with a Padé approximation of order $[M/M]$, we should obtain M poles. Therefore, we formulate a criterion for eliminating spurious poles. After returning to the variable z in the Padé approximation of order $[M/M]$, the Padé approximation of order $[M/M]$ can be represented as follows:

$$[M/M] = a_0 + \sum_{k=1}^{M} \frac{a_k}{z - z_k} \tag{31.26}$$

The spurious poles can be excluded using the following criteria:

1. The pole z_k is located outside the domain Ω.
2. The value $|a_k|$ is much less than the maximal value of the $|a_i|$ among the poles located in the domain Ω.
3. The position of the pole should be stable with respect to the increase of the order of Padé approximation.

As an example, let us consider the solution obtained for the Padé approximation of order $[9/9]$. The poles z_k and coefficients a_k are presented in Table 31.1. We see from the Table 31.1, that the poles z_1, z_3, z_6 and z_8 are located outside the domain Ω. The poles z_2, z_5, z_7 and z_9 are located inside the domain Ω but the corresponding values of $|a_k|$ are much less than the value $|a_4|$. Thus, the only pole remaining is z_4. The coordinates of this pole coincide with the center of the disc cavity with a very high accuracy. In all Padé approximates, starting with $[2/2]$ and up to $[9/9]$, similar to Table 31.1, a single pole remains. The coordinates of the single remaining pole coincide with the coordinates of the center of the circle with very high accuracy. The coordinates of the pole of the $[2/2]$ approximation are (2.9986, 4.0109). Starting with the $[3/3]$ approximation and up to $[9/9]$, the coordinates of the pole are (3.0004, 4.0004).

Example 2. In this example, all conditions are the same as in example 1, except

Table 31.1: Poles of the approximation $[9/9]$ for the disc cavity.

k	1	2	3	4	5
z_k	$-24.13 + 10i$	$-7.09 + 10i$	$3.00 - 24i$	$3.00 + 4.00i$	$8.61 + 10i$
$\|a_k\|$	126.11	0.010	5.51	5.51	$6.31 \cdot 10^{-10}$
k	6	7	8	9	10
z_k	$5.16 - 23i$	$5.16 + 3.0i$	$8.46 - 11.6i$	$8.46 - 8.4i$	–
$\|a_k\|$	$2.02 \cdot 10^{-7}$	$2.02 \cdot 10^{-7}$	$1.79 \cdot 10^{-9}$	$1.79 \cdot 10^{-9}$	–

that the cavity has the shape of a cardioid. Let us recall that a cardioid is a quadrature domain that can be obtained by shifting and rotating a domain whose boundary is described by the following equations

$$x(t) = a\left(\cos t + \frac{1}{2}\cos 2t\right), \quad y(t) = a\left(\sin t + \frac{1}{2}\sin 2t\right), \quad 0 \leq t < 2\pi \qquad (31.27)$$

Here a is a parameter. The example considers the case when the cardioid is defined by the parameter $a = 0.5$ and is shifted by the vector $(2, -5)$. According to [34], for any function $f(z) \in AL^1(D)$, in the case when the region is determined by Eqs. (31.27) and shifted by the vector z_0, the following equality is valid

$$\int_D f(z)\, dS = \frac{3\pi a^2}{2} f(z_0) + \frac{\pi a^3}{2} f'(z_0) \qquad (31.28)$$

where $z_0 = x_0 + i y_0$. From (31.28) it follows that the cardioid has a single nodal point z_0 of order 2. Consequently, the Schwarz function has one pole at a point z_0 of order 2. We consider the solution obtained for the Padé approximation of order [9/9] of the function $\widehat{f}_+(\xi + i\eta)$. The poles z_k and coefficients a_k are presented in Table 31.2. It follows from the Table 31.2 that the poles z_1, z_2, z_4, z_6 and z_8 are located outside the domain Ω. The coefficients $|a_3|$ and $|a_5|$ are very small. Thus, only two poles z_7 and z_9 remain. So, instead of one pole of multiplicity 2, we got two simple poles, which are, however, very close to a multiple pole. The dependence of the position of the poles, that are remained after eliminating spurious solutions, on the order of the Padé approximations is presented in Table 31.3.

For Padé approximates of order [2/2], [3/3], and [4/4], there is a simple pole close to the existing multiple pole. Starting with the approximation of order [5/5], two simple poles appear close to the multiple pole.

Example 3. In this example, all conditions are the same as in examples 1 and 2, except that the cavity has the shape of a Neumann oval. The Neumann oval is a domain that can be obtained by shifting and rotating the domain, whose boundary is described by the following curve

Table 31.2: Poles of the approximation [9/9] for the cavity having the shape of a cardioid.

k	1	2	3	4	5
z_k	$-24.77 + 10i$	$9.48 - 29.07i$	$9.48 + 9.07i$	$10.37 - 10.0i$	$9.04 - 10.0i$
$\lvert a_k \rvert$	24.809	$4.89 \cdot 10^{-11}$	$4.89 \cdot 10^{-11}$	$7 \cdot 10^{-11}$	$5.057 \cdot 10^{-9}$
k	6	7	8	9	10
z_k	$2.03 - 15.01i$	$2.027 - 4.990i$	$1.96 - 14.98i$	$1.961 - 5.014i$	–
$\lvert a_k \rvert$	6.502	6.502	4.585	4.585	–

Table 31.3: Dependence of the poles positions on the order of Padé approximates. The case of cardioid.

$[M/M]$	[2/2]	[3/3]	[4/4]	[5/5]
z_1	$2.163 - 4.999i$	$2.163 - 4.999i$	$2.154 - 4.998i$	$2.081 - 5.004i$
z_2	–	–	–	$1.824 - 4.975i$
$[M/M]$	[6/6]	[7/7]	[8/8]	[9/9]
z_1	$2.147 - 4.520i$	$2.006 - 4.986i$	$2.085 - 4.973i$	$2.028 - 4.990i$
z_2	$2.155 - 5.043i$	$1.990 - 5.015i$	$1.829 - 5.124i$	$1.962 - 5.014i$

$$x_1(t) = \sqrt{a^2 + 4\varepsilon^2 \cos^2(t)} \cos t,$$
$$x_2(t) = \sqrt{a^2 + 4\varepsilon^2 \cos^2(t)} \sin t, \quad 0 \le t < 2\pi \tag{31.29}$$

Here a and ε are constants. The Schwarz function for the region bounded by the curve (31.29) has the form [14]

$$S(z) = \frac{z\left(a^2 + 2\varepsilon^2\right) + z\sqrt{a^4 + 4\left(a^2 + z^2\right)\varepsilon^2}}{2\left(z^2 - \varepsilon^2\right)} \tag{31.30}$$

Thus, the Schwarz function of the curve (31.29) has two simple poles $z = \pm\varepsilon$. In this Example we took the following parameters of the Neumann oval: $a = 0.5$, $\varepsilon = 0.75$ and the domain described by the Eq. (31.18) is shifted by the vector $(-2, -3)$. Thus, the poles corresponding to considered domain are located at the points $(-1.25, -3)$ and $(-2.75, -3)$.

The poles of the Padé approximation of order [9/9] are presented in Table 31.4. The Table 31.4 shows that the poles z_2, z_4, z_6, z_8 are located outside the domain

Table 31.4: Poles of the approximation [9/9] for the cavity having the shape of a Neumann oval.

k	1	2	3	4	5
z_k	$-8.993 + 10i$	$8.955 - 29.14i$	$8.955 + 9.14i$	$9.915 - 10.29i$	$9.915 - 9.71i$
$\lvert a_k \rvert$	0.295	$1.672 \cdot 10^{-10}$	$1.672 \cdot 10^{-10}$	$1.701 \cdot 10^{-10}$	$1.701 \cdot 10^{-10}$
k	6	7	8	9	10
z_k	$-1.322 - 16.98i$	$-1.322 - 3.024i$	$-2.544 - 16.93i$	$-2.544 - 3.074i$	–
$\lvert a_k \rvert$	7.183	7.183	9.786	9.786	–

Ω and therefore must be excluded. The pole z_1 is on the boundary $\partial\Omega$. So it also should be excluded. The coefficients $|a_3|$ and $|a_5|$ are very small. Thus, the poles z_3 and z_5 should be excluded also. As a result, two poles remain, z_7 and z_9, which approximately correspond to the nodal points $(-1.25, -3)$ and $(-2.75, -3)$. The y

Table 31.5: Dependence of the poles positions on the order of Padé approximation. The case of Neumann oval.

$[M/M]$	[2/2]	[3/3]	[4/4]	[5/5]
z_1	$-1.817 - 3.001i$	$-1.753 - 3.030i$	$-1.428 - 3.064i$	$-1.337 - 3.034i$
z_2	–	–	$-2.862 - 3.240i$	$-2.579 - 3.100i$
$[M/M]$	[6/6]	[7/7]	[8/8]	[9/9]
z_1	$-1.427 - 3.042i$	$-1.320 - 3.022i$	$-1.346 - 3.025i$	$-1.322 - 3.024i$
z_2	$-2.869 - 3.170i$	$-2.540 - 3.071i$	$-2.613 - 3.077i$	$-2.544 - 3.074i$

coordinates of the poles are determined quite accurately. The x coordinates are determined less accurately, likely due to increase of their distance, in comparison of previous examples, from the right boundary of region Ω. Table 31.5 shows how the poles approach to the nodal points with the increase of the order of Padé approximates. The results presented in Table 31.5 are quite stable with respect to the order of the Padé approximations, but contain some small error.

31.5 Detection of Nodal Points of an Elastic Inclusion

Let us start with mathematical formulation of 2D problem for a linear elastic body containing linear elastic inclusion. We assume that the domain $\Omega\backslash D$ occupied with a linear elastic material with the shear modulus μ_M and Poisson ratio ν_M. Let us return again to the designation of the Cartesian coordinate system Ox_1x_2. We mark the stress-strain state in the matrix $\Omega\backslash D$ with the superscript M: $\mathbf{u}^M(x)$ is a displacement vector, $e^M_{\alpha\beta}(x)$ is the strain tensor, and $\sigma^M_{\alpha\beta}(x)$ is a stress tensor. Equations of a 2D linear elasticity in the domain $\Omega\backslash D$ have the form

$$\begin{aligned}
&e^M_{\alpha\beta}(x) = \frac{1}{2}\left(u^M_{\alpha,\beta}(x) + u^M_{\beta,\alpha}(x)\right), \quad \alpha = 1,2, \ \beta = 1,2, \quad x = (x_1, x_2) \in \Omega\backslash D \\
&\sigma^M_{\alpha\beta}(x) = 2\mu_M\left(\frac{\nu_{M*}}{1-2\nu_{M*}}\theta^M(x)\,\delta_{\alpha\beta} + e^M_{\alpha\beta}(x)\right), \quad \theta^M(x) = \sum_{k=1}^{2} e^M_{kk}(x) \\
&\sigma^M_{\alpha\beta,\beta}(x) = 0
\end{aligned} \tag{31.31}$$

Here $\delta_{\alpha\beta}$ is the Kronecker delta, $\nu_* = \nu$ in the case of plane strain state and $\nu_* = \nu/(1+\nu)$ in the case of plane stress state.

The elastic field in the inclusion D is marked with the superscript I: $\mathbf{u}^I(x)$ is the displacement vector, $e^I_{\alpha\beta}(x)$ is the strain tensor and $\sigma^I_{\alpha\beta}(x)$ is the stress tensor. The elastic field satisfies the following equations inside the inclusion

$$e^I_{\alpha\beta}(x) = \frac{1}{2}\left(u^I_{\alpha,\beta}(x) + u^I_{\beta,\alpha}(x)\right), \quad \alpha = 1,2, \quad \beta = 1,2, \quad x = (x_1, x_2) \in D$$
$$\sigma^I_{\alpha\beta}(x) = 2\mu_I\left(\frac{\nu_{I*}}{1-2\nu_{I*}}\theta^I(x)\,\delta_{\alpha\beta} + e^I_{\alpha\beta}(x)\right), \quad \theta^I(x) = \sum_{k=1}^{2} e^I_{kk}(x)$$
$$\sigma^I_{\alpha\beta,\beta}(x) = 0 \tag{31.32}$$

We assume complete bonding between the matrix and inclusion.

$$\mathbf{u}^M(x^*) = \mathbf{u}^I(x^*), \quad \sigma^M_{\alpha\beta}(x^*)\,N_\beta(x^*) = \sigma^I_{\alpha\beta}(x^*)\,N_\beta(x^*), \quad x^* \in \partial D \tag{31.33}$$

Here $\mathbf{N}(x^*) = (N_1(x^*), N_2(x^*))$ is a unit vector normal to the inclusion boundary ∂D at point x^*. Let us assume that the outer boundary of an elastic body $\partial\Omega$ is divided into two parts $\partial\Omega_1$ and $\partial\Omega_2$. $\partial\Omega_1 \cap \partial\Omega_2 = \emptyset$, $\overline{\partial\Omega_1} \cup \overline{\partial\Omega_2} = \partial\Omega$. We suppose that Cauchy data are given on $\partial\Omega_1$. For simplicity, we assume that part of the boundary $\partial\Omega_1$ is free from the loads.

$$\sigma^M_{\alpha\beta}(x')\,n_\beta(x') = 0, \quad \mathbf{u}^M(x') = \mathbf{u}^0(x'), \quad x' \in \partial\Omega_1 \tag{31.34}$$

Here $n(x') = (n_1(x'), n_2(x'))$ is the unit outward normal to the boundary $\partial\Omega_1$ at the point x'. In the general case, we will assume that part of the boundary $\partial\Omega_2$ is also divided into two parts $\partial\Omega_{2N}$ and $\partial\Omega_{2D}$. $\partial\Omega_{2N} \cap \partial\Omega_{2D} = \emptyset$, $\overline{\partial\Omega_{2N}} \cup \overline{\partial\Omega_{2D}} = \overline{\partial\Omega_2}$. The following boundary conditions are given on $\partial\Omega_{2N}$ and $\partial\Omega_{2D}$:

$$\sigma^M_{\alpha\beta}(x')\,n_\beta(x') = t^N_\alpha(x'), \quad x' \in \partial\Omega_{2N}, \quad \mathbf{u}^M(x') = \mathbf{u}^D(x'), \quad x' \in \partial\Omega_{2D} \tag{31.35}$$

The problem is to identify the nodal points of the quadrature inclusion D.

Proposition 31.1. *Let the components of the vector-function* $\mathbf{u}^M(x)$ *be*

$$\mathbf{u}^M(x) = \left(u^M_1(x), u^M_2(x)\right).$$

For $x = x' \in \partial\Omega_1$, *the function* $u^M_1(x') + iu^M_2(x')$ *is the restriction to* $\partial\Omega_1$ *of some holomorphic function in* $\Omega\backslash D$.

Proof. Due to the fact, that we have assumed complete bonding between the matrix $\Omega\backslash D$ and the inclusion D, the stress-strain state in the region $\Omega\backslash D$ can be represented using two holomorphic functions [35].

$$2\mu_M\left(u^M_1(x) + iu^M_2(x)\right) = \kappa_M\varphi_M(z) - z\overline{\varphi'_M(z)} - \overline{\psi_M(z)} \tag{31.36}$$

Here $z = x_1 + ix_2$, $z \in \Omega\backslash D$, $\varphi_M(z)$ and $\psi_M(z)$ are holomorphic functions,

$$\kappa_M = 3 - 4\nu_M$$

in case of plane strain state and

$$\kappa_M = \frac{3 - \nu_M}{1 + \nu_M}$$

in case of plane stress state. The vector of loads acting on the boundary $\partial\Omega_1$ we denote by $\mathbf{t}(x') = (t_1(x'), t_2(x'))$, where $t_j(x') = \sigma_{jk}(x') n_k(x')$. The following function is introduced in [35]

$$f_M(x_*) = \varphi_M(z_*) + z_* \overline{\varphi'_M(z_*)} + \overline{\psi_M(z_*)} = i \int_{z^0}^{z_*} (t_1(x') + it_2(x'))\, ds + \text{const} \tag{31.37}$$

Here

$$z^0 = x_1^0 + ix_2^0, \quad z_* = x_{*1} + ix_{*2}.$$

The points (x_1^0, x_2^0), (x_{*1}, x_{*2}) and (x'_1, x'_2) belong to the curve $\partial\Omega_1$. It follows from the Eqs. (31.34) and (31.37)

$$f_M(x_*) = \text{const}, \quad x_* \in \partial\Omega_1 \tag{31.38}$$

From the Eqs. (31.36) - (31.38) it follows

$$2\mu_M \left(u_1^M(x) + iu_2^M(x)\right) = (\kappa_M + 1)\varphi_M(z) + \text{const}, \quad z \in \partial\Omega_1 \tag{31.39}$$

Thus, the proposition is proved.

The stress-strain state in the inclusion D can also be represented using two holomorphic functions:

$$2\mu_I \left(u_1^I + iu_2^I\right) = \kappa_I \varphi_I(z) - z\overline{\varphi'_I(z)} - \overline{\psi_I(z)} \tag{31.40}$$

Here μ_I is the shear modulus of the inclusion, ν_I is the Poisson ratio,

$$\kappa_I = 3 - 4\nu_I$$

in case of plane strain state and

$$\kappa_I = \frac{3 - \nu_I}{1 + \nu_I}$$

in case of plane stress state. The functions $\varphi_I(z)$ and $\psi_I(z)$ are holomorphic in D. The condition of complete bonding between the matrix and inclusion (31.33) in terms of introduced holomorphic functions has the form [35]:

$$\begin{aligned} f_I(x_1, x_2) = \varphi_I(z) + z\overline{\varphi'_I(z)} + \overline{\psi_I(z)} = f_M(x_1, x_2) = \\ = \varphi_M(z) + z\overline{\varphi'_M(z)} + \overline{\psi_M(z)} \end{aligned} \tag{31.41}$$

$$\frac{1}{2\mu_I}\left[\kappa_I\varphi_I(z) - \overline{z\varphi'_I(z)} - \overline{\psi_I(z)}\right] =$$
$$= \frac{1}{2\mu_M}\left[\kappa_M\varphi_M(z) - \overline{z\varphi'_M(z)} - \overline{\psi_M(z)}\right], \quad (x_1,x_2) \in \partial D \tag{31.42}$$

From Eqs. (31.41) and (31.42) it follows that the following equality is valid on the curve ∂D, see [29, 39].

$$\varphi_M(z) = \frac{(\kappa_I+1)\varphi_I(z) + (\gamma-1)\left[\overline{z\varphi'_M(z)} + \overline{\psi_M(z)}\right]}{\gamma\kappa_M + 1}, \quad z \in \partial D, \tag{31.43}$$

where

$$\gamma = \frac{\mu_I}{\mu_M}.$$

Let us consider the possibility of continuing the function $\varphi_M(z)$ into the region D. The right side of Eq. (31.43) can be transformed at the boundary ∂D

$$\overline{\varphi'_M(z)} = \overline{\varphi'_M}(\bar{z}) = \overline{\varphi'_M}(S(z)), \quad \overline{\psi_M(z)} = \overline{\psi_M}(\bar{z}) = \overline{\psi_M}(S(z)) \tag{31.44}$$

Let us assume that the functions $\varphi_M(z)$ and $\psi_M(z)$ can be continued as holomorphic functions into the domain $C\backslash\Omega$. It follows from the Eqs. (31.43) and (31.44) that the function $\varphi_M(z)$ can be continued as a meromorphic function in the domain D. The poles of the function $\varphi_M(z)$ are nodal points of the domain D. If the functions $\varphi_M(z)$ and $\psi_M(z)$ cannot be continued in $C\backslash\Omega$, then in addition to the nodal points of the region D, the function $\varphi_M(z)$ can have singular points on the boundary of the region Ω and in the region $C\backslash\Omega$.

31.6 Numerical Examples for the Elasticity Problem

In all considered examples the domain Ω is the same square as above $|x_1| \le 10$ cm $|x_2| \le 10$ cm. The elastic constants of the material are as follows: the Young modulus $E_M = 68.5$ GPa, the Poisson ratio $\nu_M = 0.36$. The domain D is a disc cavity of radius 0.75 cm. It is assumed that the uniaxial tension is applied in the direction of the axis x_2

$$\sigma_{22}(x_1, \pm 10) = \sigma, \quad \sigma_{12}(x_1, \pm 10) = 0, \quad \sigma = 100\text{MPa}.$$

The lateral sides of domain Ω are free of the loads. We assume that the displacements are measured on the right edge of Ω. The values of the displacements we get by solving the direct problem. In the numerical procedure we used not the displacements $\mathbf{u}^M(10, x_2)$, but the difference in displacements

$$\mathbf{u}^d(10, x_2) = \mathbf{u}^M(10, x_2) - \mathbf{u}^h(10, x_2),$$

where $\mathbf{u}^h(x_1, x_2)$ are displacements corresponding to the solution of the direct problem for the domain Ω without cavity. After this we make a change of coordinates $\xi = x_2$, $\eta = 10 - x_1$. Let us denote

$$\mathbf{u}^d(x_1, x_2) = \tilde{\mathbf{u}}^d(\eta, \xi).$$

The function

$$g(\xi) = \tilde{u}_2^d(0, \xi) - i\tilde{u}_1^d(0, \xi)$$

is the restriction of the holomorphic function on the line $\eta = 0$. Finally we apply the Fourier – Padé approximation to the function $g(\xi)$, see Eqs. (31.6) - (31.8). The obtained Laurent series is presented as a sum, see Eq. (31.7),

$$g(\zeta) = g^+(\zeta) + g^-(\zeta), \qquad \zeta = \xi + i\eta.$$

The purpose of the examples considered below is to find out the accuracy of determining the location of defects depending on their distance from the part of the boundary on which the Cauchy data are given. The results presented below were obtained by Padé approximates of the function $g^-(\zeta)$.

Example 4. The cavity D is a disc with the center at the point $(2, -2)$, see Fig. 31.2. The poles obtained by the Padé approximates of order [9/9] of the function $g^-(\zeta)$ are presented in Table 31.6 Table 31.6 shows that there exists only one pole z_9. All other poles are located near the boundary and corresponding coefficients $|a_k|$ are small. The position of the pole z_9 is very close to the center of the disc cavity. When considering Padé approximates of other orders, one can also verify that there is only one pole. Dependence of the position of the pole on the order of the Padé approximates presented in the Table 31.7.

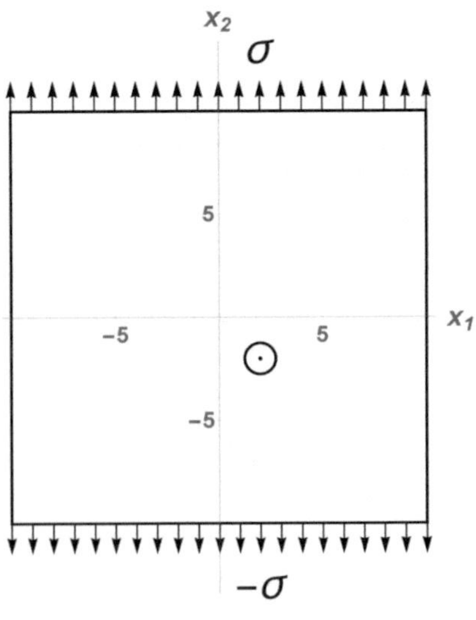

Fig. 31.2 Elastic plate with a disc cavity with the center at the point $(2, -2)$.

Table 31.6: Poles of the approximation for the disc cavity with the center at the point (2,-2).

k	1	2	3	4	5
z_k	$1.548-9.366i$	$5.033-9.596i$	$7.020-9.717i$	$8.308-9.792i$	$9.160-9.870i$
$\lvert a_k\rvert$	0.00299	0.00038	$7.510\cdot10^{-5}$	$1.993\cdot10^{-5}$	$7.079\cdot10^{-6}$
k	6	7	8	9	10
z_k	$9.679-9.945i$	$9.941-9.990i$	$8.925-1.151i$	$2.009-2.003i$	–
$\lvert a_k\rvert$	$3.251\cdot10^{-6}$	$1.257\cdot10^{-6}$	$1.033\cdot10^{-13}$	0.00644	–

Table 31.7: Dependence of the poles positions on the order of Padé approximates. The case of the disc with a center at the point (2,-2).

$[M/M]$	[4/4]	[5/5]	[6/6]
z	$2.106-2.319i$	$2.060-2.090i$	$2.027-2.022i$
$[M/M]$	[7/7]	[8/8]	[9/9]
z	$2.012-2.006i$	$2.011-2.006i$	$2.009-2.003i$

Example 5. The cavity D is a disc with the center at the point $(0,-2)$. The poles obtained by the Padé approximates of order $[9/9]$ of the function $g^-(\zeta)$ are presented in Table 31.8.

It follows from the Table 31.8 that the main pole z_9 is located close enough to the center of a disc cavity. The coefficients corresponding to the poles z_7 and z_8 are very small. Thus, these poles should be excluded. All other poles have small coefficients and located close to the body boundary. So there is only one pole left. The proximity of this pole to the center of the circle cavity is slightly lower than in the case of a cavity centered at point (2,-2). In case of Padé approximates of other orders after eliminating the spurious poles a single pole remains also. The dependence of the position of this pole on the order of the Padé approximate is presented in Table 31.9. The results presented in Table 31.9 show the stability in the determination of the nodal point of the disc cavity.

Table 31.8: Poles of the approximation $[9/9]$ for the disc cavity with the center at the point (0,-2).

k	1	2	3	4	5
z_k	$2.132-9.373i$	$5.564-9.673i$	$7.437-9.632i$	$9.091-9.873i$	$9.643-9.965i$
$\lvert a_k\rvert$	0.00219	0.00016	$1.56\cdot10^{-5}$	$3.82\cdot10^{-6}$	$3.29\cdot10^{-6}$
k	6	7	8	9	10
z_k	$9.934-9.995i$	$8.877+7.003i$	$8.407+0.090i$	$0.043-2.137i$	–
$\lvert a_k\rvert$	$1.50\cdot10^{-6}$	$7.12\cdot10^{-11}$	$3.61\cdot10^{-12}$	0.01207	–

Table 31.9: Dependence of the poles positions on the order of Padé approximates. The case of the disc with a center at the point (0,-2).

$[M/M]$	[4/4]	[5/5]	[6/6]
z	$0.164 - 3.276i$	$0.129 - 2.383i$	$0.119 - 2.200i$
$[M/M]$	[7/7]	[8/8]	[9/9]
z	$0.073 - 2.028i$	$0.063 - 2.111i$	$0.043 - 2.137i$

Example 6. In this example the cavity is a disc with the center at the point $(-2,-2)$. The poles obtained by the Padé approximates of order [9/9] of the function $g^{-}(\zeta)$ are presented in Table 31.10. Analysis of Table 31.10 shows that after eliminating spurious poles, only one pole remains, z_9. This pole approximately corresponds to the center of the circular cavity, however, because it is located further from the right boundary than in the previous examples, it is determined less accurately. In case of Padé approximates of other orders after eliminating the spurious poles a single pole remains also.

The dependence of the position of this pole on the order of the Padé approximate is presented in Table 31.11. From Table 31.11 it can be seen that the position of the pole slowly approaches the center of the circular cavity, but the center is determined with an error.

31.7 Conclusion

The paper proposes a method for determining the nodal points of a quadrature inclusion using Cauchy data specified only on a part of the outer boundary of an elastic body. The proposed approach can be applied to various elliptic equations and systems of equations. In particular, the paper shows that the proposed method works successfully for both the Laplace equation and the equations of plane elasticity. The examples considered show that the proposed method can be used to find nodal points of inclusion even in the case where the inclusion is located quite far from the part

Table 31.10: Poles of the approximation [9/9] for the disc cavity with the center at the point (-2,-2).

k	1	2	3	4	5
z_k	$1.781 + 9.782i$	$5.343 + 9.427i$	$8.216 + 9.430i$	$9.033 + 9.785i$	$9.621 + 9.929i$
$\lvert a_k \rvert$	0.00188	$8.810 \cdot 10^{-5}$	$5.245 \cdot 10^{-6}$	$7.062 \cdot 10^{-6}$	$4.653 \cdot 10^{-6}$
k	6	7	8	9	10
z_k	$9.930 + 9.988i$	$9.637 + 5.417i$	$9.948 - 0.949i$	$-2.228 - 2.411i$	-
$\lvert a_k \rvert$	$1.965 \cdot 10^{-6}$	$8.172 \cdot 10^{-13}$	$7.388 \cdot 10^{-15}$	0.03061	-

Table 31.11: Dependence of the poles positions on the order of Padé approximates. The case of the disc with a center at the point (-2,-2).

$[M/M]$	[4/4]	[5/5]	[6/6]
z	$-3.328-4.652i$	$-2.711-3.559i$	$-2.387-2.665i$
$[M/M]$	[7/7]	[8/8]	[9/9]
z	$-2.326-2.344i$	$-2.172-2.596i$	$-2.228-2.411i$

of the boundary on which the Cauchy data are given. Since any domain can be approximated by a quadrature domain, there is reason to believe that the developed approach can be applied in the case where the inclusion is not quadrature.

Acknowledgements The support of Ministry of Science and Higher Education within the framework of the Russian State Assignment under contract No 124012500437-9 is gratefully acknowledged.

References

[1] Aharonov D, Shapiro HS (1976) Domains on which analytic functions satisfy quadrature identities. Journal d'Analyse Mathematique 30(1):39-73, DOI 10.1007/BF02786704

[2] Akamatsu M, Nakamura G, Steinberg S (1991) Identification of Lame coefficients from boundary observations. Inverse Problems 7(3):335-354, DOI 10.1088/0266-5611/7/3/003

[3] Ameur Y, Helmer M, Tellander F (2021) On the uniqueness problem for a quadrature domains. Computational Methods and Function Theory 21(3):473-504, DOI 10.1007/s40315-021-00373-w

[4] Ammari H, Kang H, Nakamura G, Tanuma K (2002) Complete asymptotic expansions of solutions of the system of elastostatics in the presence of an inclusion of small diameter and detection of an inclusion. Journal of Elasticity 67(2):97-129, DOI 10.1023/A:1023940025757

[5] Andrieux S, Ben Abda A, Bui H (1999) Reciprocity principle and crack identification. Inverse Problems 15(1):59-65, DOI 10.1088/0266-5611/15/1/010

[6] Aspri A (2022) A phase-field approach for detecting cavities via a Kohn-Vogelius type functional. Inverse Problems 38(9):094001, DOI 10.1088/1361-6420/ac82e4

[7] Aspri A, Beretta E, Cavaterra C, Rocca E, Verani M (2022) Identification of cavities and inclusions in linear elasticity with a phase-field approach. Applied Mathematics and Optimization 86(3):32, DOI 10.1007/s00245-022-09897-6

[8] Baker Jr GA, Graves-Morris P (1996) Pade approximants. Encyclopedia of Mathematics and its Applications (2nd ed), Vol 59, Cambridge University Press

[9] Ben Abda A, Jalem E, Khalfallah S, Mejri B (2018) Voids identification from partially overspecified boundary data. New Trends in Mathematical Sciences 6(4):96-110, DOI 10.20852/ntmsci.2018.320
[10] Ben Ameur H, Burger M, Hackl B (2004) Level set methods for geometric inverse problems in linear elasticity. Inverse Problems 20(3):673-696, DOI 10.1088/0266-5611/20/3/003
[11] Ben Ameur H, Burger M, Hackl B (2007) Cavity identification in linear elasticity and thermoelasticity. Mathematical Methods in Applied Sciences 30(6):625-647, DOI 10.1002/mma.772
[12] Bultheel A (1987) Laurent series and their Pade approximations. Vol 27, Operator Theory: Advances and Applications. Birkhauser
[13] Capcelea M, Capcelea T (2020) Laurent-Padé approximation for locating singularities of meromorphic functions with values given on simple closed contours. Buletinul Academiei di Ştiinţe a Republicii Moldova, Matematica 93(2):76-87
[14] Davis PJ (1974) The Schwarz Function and its Applications. The Carus Mathematical Monographs, Vol 17. Mathematical Association of America, DOI 10.5948/9781614440178
[15] Dehghan Manshadi SH, Khaji N, Rahimian M (2014) Cavity/inclusion detection in plane linear elastic bodies using linear sampling method. Journal of Nondestructive Evaluation 33(1):93-103, DOI 10.1007/s10921-013-0206-8
[16] Engelhardt M, Schanz M, Stavroulakis GE, Antes H (2006) Defect identification in 3-D elastostatics using a genetic algorithm. Optimization and Engineering 7(1):63-79, DOI 10.1007/s11081-006-6591-4
[17] Ferrier R, Kadri M, Gosselet P (2021) Crack identification with incomplete boundary data in linear elasticity by the reciprocity gap method. Computational Mechanics 67(6):1559-1579, DOI 10.1007/s00466-021-02006-4
[18] Guo R, Lin T, Lin Y (2020) Recovering elastic inclusions by shape optimization methods with immersed finite elements. Journal of Computational Physics 404:109123, DOI 10.1016/j.jcp.2019.109123
[19] Gustafsson B (1983) Quadrature identities and the Schottky double. Acta Applicandae Mathematicae 1(3):209-240, DOI 10.1007/BF00046600
[20] Hrizi M, Hassine M, Abdelwahed M, Chorfi N (2019) Fast and accurate algorithm for cavities reconstruction in an elasticity problem. Mathematical Methods in the Applied Sciences 42(18):6083-6100, DOI 10.1002/mma.5706
[21] Ikehata M (2002) A regularized extraction formula in the enclosure method. Inverse Problems 18(2):435-440, DOI 10.1088/0266-5611/18/2/309
[22] Ikehata M, Itou H (2007) Reconstruction of a linear crack in an isotropic elastic body from a single set of measured data. Inverse Problems 23(2):589-607, DOI 10.1088/0266-5611/23/2/008
[23] Imanuvilov OY, Uhlmann G, Yamamoto M (2012) On uniqueness of Lamé coefficients from partial Cauchy data in three dimensions. Inverse Problems 28:125002, DOI 10.1088/0266-5611/28/12/125002
[24] Imanuvilov OY, Yamamoto M (2011) On reconstruction of Lamé coefficients from partial Cauchy data. Journal of Inverse and Ill-posed Problems 19(6):881-891, DOI 10.1515/JIIP.2011.060

[25] Ito S, Aishima K, Nara T, Sugihara M (2015) Orthogonal polynomial approach to estimation of poles of rational functions from data on open curves. Journal of Computational and Applied Mathematics 273:326-345, DOI 10.1016/j.cam.2014.05.028
[26] Jaïem E (2016) Shape derivative of an energy error functional for voids detection from sub-Cauchy data. Electronic Journal of Differential Equations (265):1-13,
[27] Jalem E, Abda AB, Rjaibi B (2016) Cavities identification from partially overdetermined boundary data in linear elasticity. Journal of Applied and Computational Mathematics 5(1):1000295, DOI 10.4172/2168-9679.1000295
[28] Kang H, Kim E, Lee J-Y (2003) Identification of elastic inclusions and elastic moment tensors by boundary measurements. Inverse Problems 19:703-724, DOI 4172/2168-9679.1000295
[29] Kaptsov AV, Shifrin EI (2023) Plane problem of the theory of elasticity on the identification of nodal points of a quadrature inclusion. Mechanics of Solids 58(6):1967–1985, DOI 10.3103/S0025654423600435
[30] Karageorghis A, Lesnic D, Marin L (2012) The method of fundamental solutions for the detection of rigid inclusions and cavities in plane linear elastic bodies. Computers & Structures 106:176-188, DOI 10.1016/j.compstruc.2012.05.001
[31] Karageorghis A, Lesnic D, Marin L (2016) The method of fundamental solutions for three-dimensional inverse geometric elasticity problems. Computers & Structures 166:51-59, DOI 10.1016/j.compstruc.2016.01.010
[32] Kasparova EA, Shifrin EI (2020) Solution for the geometric elastostatic inverse problem by means of not completely overdetermined boundary data. Mechanics of Solids 55(8):1298–1307, DOI 10.3103/S0025654420080117
[33] Kirsch A (2005) The factorization method for a class of inverse elliptic problems. Mathematische Nachrichten 278(3):258-277, DOI 10.1002/mana.200310239
[34] Lee S-Y, Makarov NG 2016 Topology of quadrature domains. Journal of the American Mathematical Society 29(2):333-369, DOI 10.1090/jams828
[35] Muskhelishvili NI (1977) Some Basic Problems of the Mathematical Theory of Elasticity. Springer Netherlands, Dordrecht
[36] Nara T, Ando S (2010) Direct localization of poles of a meromorphic function from measurements on an incomplete boundary. Inverse Problems 26(1):015011, DOI 10.1088/0266-5611/26/1/015011
[37] Shifrin EI (2016) Factorization method in the geometric inverse problem of static elasticity. Mechanics of Solids 51(5):562-570, DOI 10.3103/S0025654416050083
[38] Shifrin EI, Kaptsov AV (2018) Identification of multiple cracks in 2D elasticity by means of the reciprocity principle and cluster analysis. Inverse Problems 34(1):015009, DOI 10.1088/1361-6420/aa9d74
[39] Shifrin EI, Kaptsov AV (2023) Identification of nodal points of an elastic inclusion in elastic plane. Doklady Mathematics 107(1):64-68, DOI 10.1134/S1064562423700527

[40] Shifrin EI, Kasparova EA (2022) Defect detection using Cauchy data on a part of outer boundary of an elastic body. Zeitschrift für Angewandte Mathematik und Physik (ZAMP) 73(4):134, DOI 10.1007/s00033-022-01765-1
[41] Shifrin EI, Shushpannikov PS (2010) Identification of a spheroidal defect in an elastic solid using a reciprocity gap functional. Inverse Problems 26(5), 055001, DOI 10.1088/0266-5611/26/5/055001
[42] Shifrin EI, Shushpannikov PS (2011) Identification of an ellipsoidal defect in an elastic solid using boundary measurements. International Journal of Solids and Structures 48(7):1154-1163, DOI 10.1016/j.ijsolstr.2010.12.016

Chapter 32
The Influence of the Fractal Dimension of the Rough Surface Model on Tribological Tests

Evgeny Sosenushkin, Elena Yanovskaya, and Andrey Zhelnov

Abstract When computer modeling fractal surfaces to predict wear, an interesting question is the transition from fractal dimension to real physical quantities. Based on the constructed models and fractal geometry for validation of the results obtained, confirmation by means of a full-scale experiment is required. This paper considers a methodological approach to a coherent transition from a fractal description of surface micro-relief to real physical and mechanical parameters of friction and wear, as well as to tribological experiments. The analytical transformation is carried out in several stages, starting with the specification of the fractal dimension range. Next, the real contact area is modeled from the given dimension, considering the influence of the contact area on the amount of surface wear. The contact area is considered as a function for constructing a friction model. Wear is calculated using Archard's formula, which links fractal dimension, hardness, and friction coefficient. As a result, numerical modeling was performed to transition to real parameters, wear modeling, and calculations for various materials. The influence of the main parameters on the amount of wear was determined using the SHAP and Partial Dependence methods. The main areas of fractal dimension were identified and the influence of dimension on wear in these areas was considered. Calculations were performed for specific material to find the area of minimum wear, which can be used in the future when conducting a full-scale tribological experiment on wear.

Key words: Fractal surfaces, Artificial intelligence, Generative models, Mathematical and computer modeling, Numerical modeling, Wear, Tribology

Evgeny Sosenushkin · Elena Yanovskaya (corresponding author) · Andrey Zhelnov
Moscow State Technological University "STANKIN", Vadkovskiy Pereulok 1, 127055 Moscow, Russian Federation, e-mail: sen@stankin.ru, elena_yanovskaya@bk.ru, a.zhelnov@stankin.ru

H. Altenbach et al. (eds.), *Stability Problems in Mechanics of Continua*,
Advanced Structured Materials 260, https://doi.org/10.1007/978-3-032-31032-3_32

32.1 Transition to Dimensional Physical Parameters and Numerical Modeling of Wear

In [12, 13], an analytical model of the normalized wear rate Q^* was obtained based on a fractal representation of contact surfaces. However, to apply the model in engineering practice, it is necessary to transition from dimensionless expressions to real physical parameters—the volume of wear at given loads, material hardness, and friction conditions. Let us consider the complete numerical algorithm and the results of parametric analysis demonstrating the influence of the fractal dimension D and the normalized true contact area A_r^* on the actual wear volume. Let us rewrite the analytical model obtained earlier in explicit form:

$$Q^* = \left[k_e - \left(k_e - k_p\right) \frac{D\,G^{*2}}{A_r^*(2-D)} \varphi^{\frac{2}{D-1}}\right] A_r^* \sqrt{1+3\mu^2} \tag{32.1}$$

where A_r^* — normalized real contact area, G^* — normalized amplitude of micro-irregularities, φ — dimensionless material characteristic, μ — friction coefficient, D — fractal dimension of the profile, k_e and k_p — elastic and plastic wear coefficients. Archard's law is used to convert to dimensional wear volume [3]:

$$V = Q\frac{WL}{H}, \quad Q \approx C_{\text{scale}} Q^* \tag{32.2}$$

where W — normal load, L — friction path length, H — material hardness, C_{scale} — calibration factor selected according to an experimental reference.

As an example, let us take the parameters of structural steel 45: $\mu = 0.2$, $N = 3$, $\sigma_y = 1 \cdot 10^9$ Pa, $E' = 2.1 \cdot 10^{11}$ Pa, $\varphi \approx 0.0143$, $G^* = 1$ (which is typical for an average level of irregularities), $A_r^* = 0.05$ (true contact ~5% of the nominal area), $k_e = 10^{-5}$, $k_p = 5 \cdot 10^{-5}$ — we will assume these values conditionally. Let us set the range of D: 1.15–1.9 [12]. Using the function $Q^*(D, A_r^*)$, we determine the normalized wear rate $Q^*(D)$.

Next, we use Archard's law described earlier. At this stage, Q^* can be considered either as Q, or as a quantity proportional to Q with some calibration factor C. Let us set $C = 1$ (for real models, the parameter is determined based on experiment). Let the normal load be $W = 10$ kN, hardness $H = 7 \cdot 10^9$ Pa, friction path $L = 1000$ m. Then the characteristic wear [2] volume reaches the order of $(7.6 - 8.5) \cdot 10^{-10}$ m^3 per 1000 m of run at the given values. These values correspond to the real scale of wear of steel samples under high loads and relatively short friction paths. Figure 32.1 shows how the volume changes at different values of D.

The dependence is monotonic because the values of A_r^* and G^* are fixed; if the value of the real contact area A_r^* is varied, the picture will be closer to that obtained in [1]. To get a more complete picture of the influence of two factors simultaneously — fractal dimension and actual contact area — a two-dimensional simulation was performed on a rectangular grid with given values $D = 1.15 \ldots 1.90$, $A_r^* = 0.01 \ldots 0.20$. Based on the module, a map of the wear volume $V(D, A_r^*)$ was constructed, shown in Fig. 32.2.

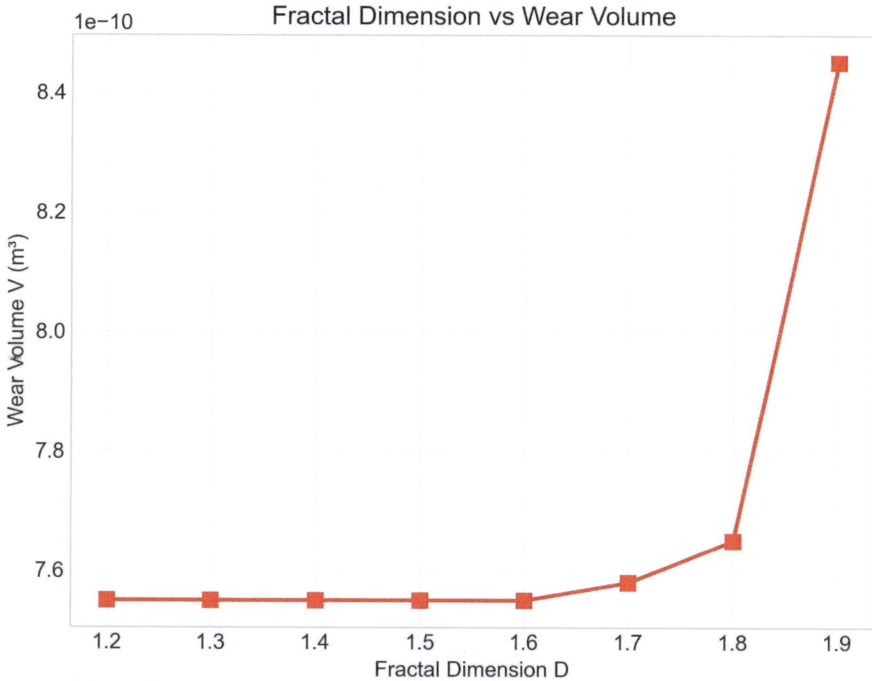

Fig. 32.1: Dependence of Q^* on fractal dimension D.

From the obtained map, it can be seen that the main influence on the wear volume is exerted by the value of the true contact area. When A_r^* increases from 0.01 to 0.2, the calculated wear volume increases by more than an order of magnitude. At the same time, the influence of D becomes noticeable precisely at large A_r^*; in the region of low loads, when the actual contact is small, changes in D practically do not affect wear.

To demonstrate the practical application of the proposed fractal model [10], calculations of the predicted wear volume V were performed for a number of materials: structural steel, titanium, aluminum alloy, and high-temperature polymer. In all examples, the same external conditions were used: normal load $W = 10$ kN, friction path $L = 1000$ m, wear coefficients: $k_e = 10^{-5}$, $k_p = 5 \cdot 10^{-5}$, $G^* = 1$, $N = 3$, scale factor $C_{\text{scale}} = 1$. For each material, the characteristic mechanical and tribological properties are approximately specified, presented in Table 32.1.

Calculations were performed for three fractal dimensions $D = 1.2; 1.5; 1.8$ and three normalized true contact area $A_r^* = 0.02;\ 0.05;\ 0.10$. Table 32.2 shows the calculated wear volume V in mm^3 per 1000 m of path for structural steel, titanium, aluminum alloy, and high-temperature polymer, respectively.

From the obtained results, it can be seen that under the same external conditions, the wear volume increases in the series steel, titanium, aluminum, polymer. This is

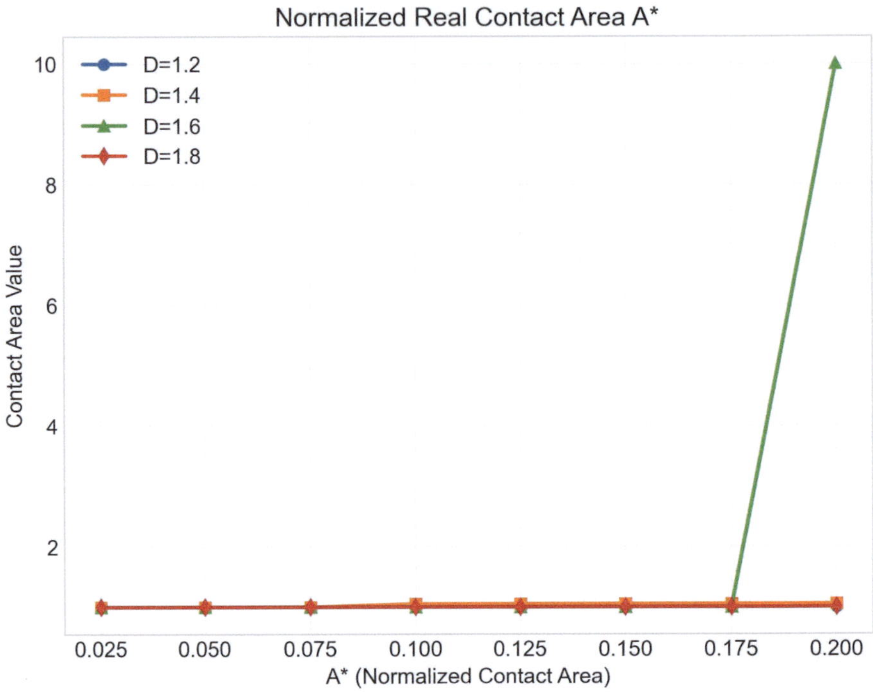

Fig. 32.2: Wear volume $V(D, A_r^*)$.

Table 32.1: Materials and their main characteristics.

Material	σ_y, GPa	E', GPa	H, GPa	μ
Steel 45	1.0	210	7.0	0.20
Ti–6Al–4V	0.88	110	3.5	0.25
Al 6061-T6	0.28	70	1.0	0.30
PEEK	0.10	4	0.25	0.35

due to the decrease in hardness and the increase in the friction coefficient for the respective materials. The influence of fractal dimension D for hard metals (steel, titanium) is weaker than the influence of A_r^*, however, for soft materials (PEEK) at large D, a sharp increase in wear is observed. The presented numerical model allows quantitatively comparing different combinations and selecting the least wear-prone options.

Table 32.2: Calculated wear volume V (mm^3) for various materials.

D	A_r^*	Steel 45	Ti–6Al–4V	Al 6061-T6	PEEK
$A_r^* = 0.02$					
1.2	0.02	0.302	0.623	2.253	9.355
1.5	0.02	0.302	0.623	2.254	9.533
1.8	0.02	0.316	0.672	2.318	35.296
$A_r^* = 0.05$					
1.2	0.05	0.756	1.557	5.634	23.388
1.5	0.05	0.756	1.557	5.635	23.566
1.8	0.05	0.769	1.606	5.699	49.329
$A_r^* = 0.10$					
1.2	0.10	1.512	3.113	11.269	46.776
1.5	0.10	1.512	3.114	11.269	46.954
1.8	0.10	1.525	3.163	11.333	72.717

32.2 Method of Characteristics

In practice, for interpreting and measuring the relationship between model parameters, third-party tools such as SHapley Additive exPlanations and/or Partial dependence are used. SHAP (SHapley Additive exPlanations) is a method for interpreting predictions of machine learning models. It helps to understand how the model makes decisions and to assess how much each input feature influences the final prediction [3, 11]. SHAP is based on the concept of game theory developed by Lloyd Shapley in the 1950s. In turn, Partial dependence is a method for measuring the relationship between an individual predictor and the outcome in a prediction model. It allows estimating what part of the changes in the model is associated with a specific input feature. The method is based on recalculating the average value of the machine learning model prediction after changing one or more variables, while fixing all other input data as constant [14]. As a result, it becomes clear how much each variable affects the overall model prediction, as well as how sensitive it is to any given variable.

In the presented work, Partial dependence was used to estimate the average wear volume as a function of one given parameter. Let us plot the dependencies of the average wear volume V on the fractal dimension D, the normalized real contact area A_r^*, and the normalized amplitude of micro-irregularities G^* over a grid of values, averaging 2000 random combinations of the remaining parameters. The considered graphs are illustrated in Figs. 32.3, 32.4, and 32.5, respectively.

From the graphs, it can be seen that when D changes in the range 1.15–1.7, the average wear hardly changes. A sharp increase begins only at $D \gtrsim 1.8$. In the

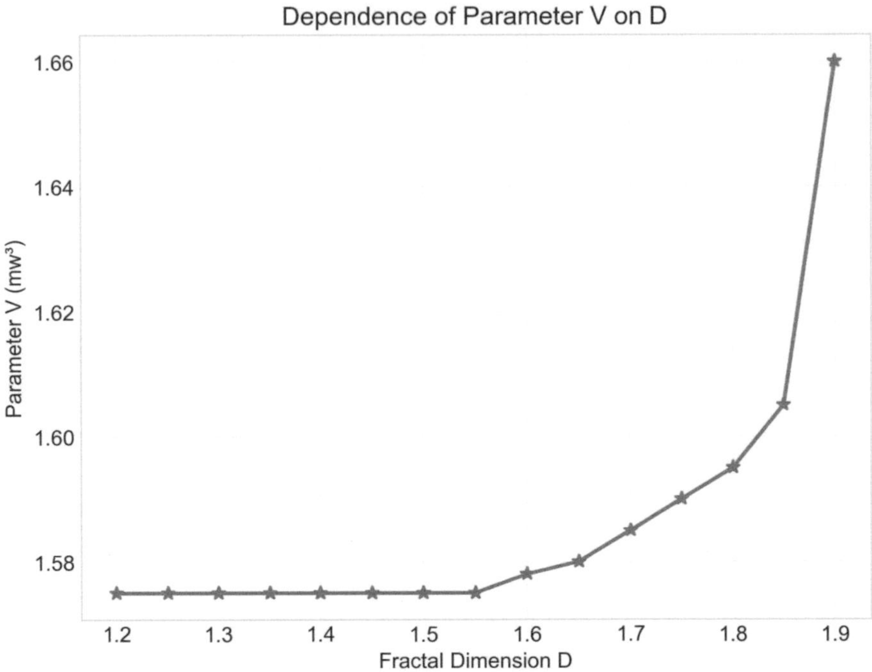

Fig. 32.3: PDP dependence of the average wear volume V on D.

"normal" dimension zone, the influence of D is secondary compared to A_r^*. At the same time, Fig. 32.5 shows an almost perfect linear dependence for A_r^*. When A_r^* increases from 0.01 to 0.2, the average wear grows approximately from 0.15 to 3.0 mm^3. It can be concluded that the real contact area A_r^* is the main controlling parameter of the model, setting the wear scale, considering that the amplitude of micro-irregularities weakly affects the wear volume. For G^*, when the value changes from 0.8 to 1.2, the average wear changes by less than 0.5%. With fixed values of D and A_r^*, changing the amplitude of irregularities within reasonable limits has almost no effect on the final result.

The assessment of each parameter's contribution to the wear model was carried out for 300 randomly selected points from the given dataset relative to the base state $x_0 = (D, A_r^*, G^*)$. Analytically, the formula for 3 features can be given in the following form:

$$\phi_j = \frac{1}{3}(f(x_j) - f(x_0)) + \frac{1}{6}\sum_{k \neq j}(f(x_{j,k}) - f(x_k)) \tag{32.3}$$

where x_j is a vector in which the j-th parameter is taken from the considered instance [5, 7, 17], the others are from the base one; $x_{j,k}$ means that j and k are taken from the instance, the others are from the base one; $f(x_j)$ is the wear model V.

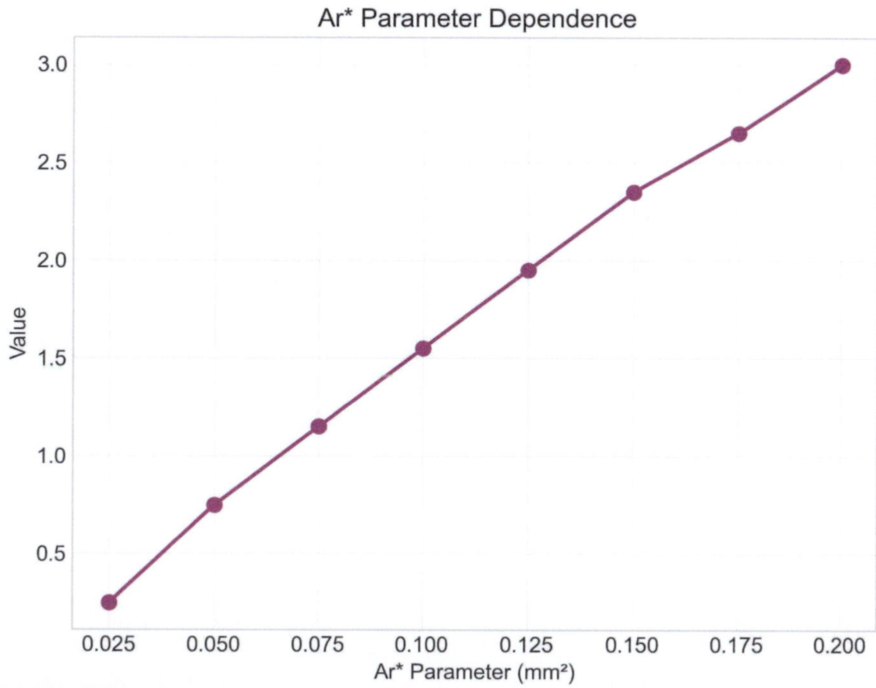

Fig. 32.4: PDP dependence of the average wear volume V on A_r^*.

Next, we determine the importance of features as the average absolute value over the sample using the formula:

$$I_j = \mathbb{E}_i \left[\left| \phi_j^{(i)} \right| \right] \tag{32.4}$$

As a result one gets $I_D = 0.0042$, $I_{A_r^*} = 0.4591$, $I_{G^*} = 0.00023$.

From the presented estimates, it can be seen that the real contact area A_r^* contributes an order of magnitude more to the variation in wear than the parameters D and G^*. This result confirms the conclusions obtained using Partial dependence. The fractal dimension D influences the wear volume significantly less (about 1% of the influence of A_r^*), but its effect is nonlinear and manifests itself mainly in the region of large D values. The influence of G^* against the background of other parameters can be considered secondary in the considered ranges.

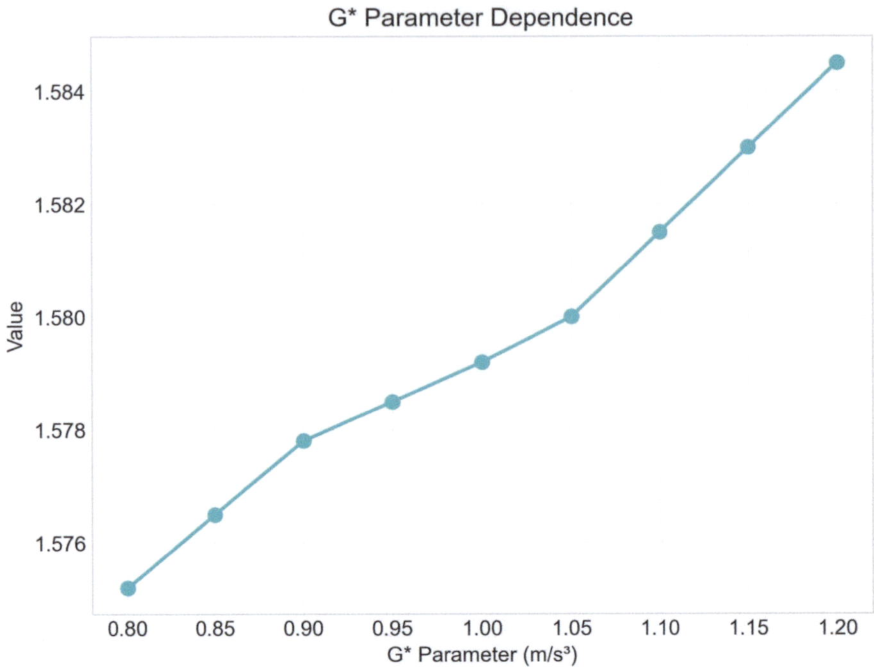

Fig. 32.5: PDP dependence of the average wear volume V on G^*.

32.3 Analysis of Fractal Dimension Ranges

The analytical model of the normalized wear rate 32.1 contains three terms nonlinear in D, which makes the behavior of wear in different ranges of fractal dimension significantly different. To systematically explain the results of the article [12], as well as the numerical analysis data obtained using Partial dependence and SHAP, we divide the range of D into two regions: $D = 1.15 - 1.5$ (low–medium fractal dimension) and $D = 1.6 - 1.9$ (high fractal dimension).

Consider the range of low–medium fractal dimension. The exponent in formula 32.1 in the given range is:

$$\frac{2}{D-1} \gg 1 \tag{32.5}$$

i.e., for $D = 1.2$, the exponent is approximately 10, for $D = 1.5$, approximately 4. Since the material-dependent quantity, for example, for steels, titanium:

$$\varphi = \frac{N\sigma_y}{E'} \approx 10^{-2} \tag{32.6}$$

then

$$\varphi^{\frac{2}{D-1}} \ll 1 \tag{32.7}$$

Consequently, the "plastic" correction becomes much less than 1:

$$\left(k_e - k_p\right) \frac{DG^{*2}}{A_r^*(2-D)} \varphi^{\frac{2}{D-1}} \ll 1 \tag{32.8}$$

and thus the dependence on D disappears:

$$Q^*(D) \approx k_e A_r^* \sqrt{1 + 3\mu^2} \tag{32.9}$$

In fact, at low fractal dimension [1, 15], the surface is almost smooth, the profile is low-irregular. The real contact area is determined mainly by macroscale geometry, and the influence of the "fractal structure" has not yet appeared. Microcontacts have relatively wide platforms, therefore we are talking about a quasi-stationary elastic mode, in which the role of sharp peaks is small. This is also confirmed by numerical analysis; SHAP showed near-zero average significance of D. This is precisely because for a significant part of the range (1.15–1.5), the contribution of D is mathematically suppressed by the small factor $\varphi^{2/(D-1)}$. The Partial dependence $V(D)$ is almost horizontal in this zone, which also confirms the obtained results.

At the same time, for the range $D \geq 1.6$:

$$2 < \frac{2}{D-1} < 3.5 \tag{32.10}$$

that is, now the exponent is a moderate value, so $\varphi^{2/(D-1)}$ is no longer so small, for example, for $D = 1.6$, the value $\varphi^{3.3} \approx 10^{-5}$, for $D = 1.8$, the value $\varphi^{2.5} \approx 10^{-4.5}$. That is, the quantity becomes comparable to or greater than the small quantities used in the coefficients k_e, k_p. It turns out that as the value of D increases, $Q^*(D)$ increases. This means that the surface sharply becomes self-similar at different scales, the number of peaks per unit area increases, they become sharper, as local gradients increase. The load over the area is redistributed onto many small microcontacts, which leads to an increase in local pressures [16]. This leads to an increase in the proportion of plastic microcontacts, an increase in local stresses, and an intensification of micro-chipping and adhesive "tear-offs" [9].

It can be concluded that in the range $D = 1.6-1.9$, an increase in fractal dimension causes a noticeable increase in wear, since the fractal model enters a zone where high local pressures and plastic contacts dominate.

32.4 Boundary Value Problem of Plastic Layer Flow in a new Formulation with Constraints in One or Several Directions

Let us determine how, in practice, one can find the optimal region of fractal dimension, guided by the results obtained earlier. For example, consider structural steel "Steel 45". Let us numerically calculate the derivative of the wear volume with respect to the fractal dimension using the formula [6]:

$$\frac{dV}{dD} \approx \frac{V(D+h) - V(D-h)}{2h} \tag{32.11}$$

with a step $h = 10^{-3}$, using the same model as before ($W = 10$ kN, $L = 1000$ m, $H = 7$ GPa).

In Fig. 32.6, in the region of low–medium fractal dimension, the graph is practically close to zero. The growth of the derivative begins closer to 1.45–1.5, but with derivative values on the order of 10^{-4} mm^3 per unit D, a change in D hardly affects wear. The model confirms the previously made theoretical conclusion.

In the region of high fractal dimension, as seen in Fig. 32.7, the derivative increases sharply [4].

That is, a small increase in D in the upper part of the range leads to a noticeable increase in wear. If we consider the entire dimension range in more detail with a step of 0.0015 at fixed $A_r^* = 0.05$, $G^* = 1$, we find that the minimum wear volume is achieved at the edge of the range, i.e., at $D_{\text{opt}} \approx 1.15$, the wear volume $V(D_{\text{opt}}) \approx 0.756$ mm^3. Thus, for structural steel, the optimal range of fractal dimension from the point of view of wear resistance is $D \lesssim 1.4$, while at $D \to 1.9$, the predicted wear increases by more than 10%.

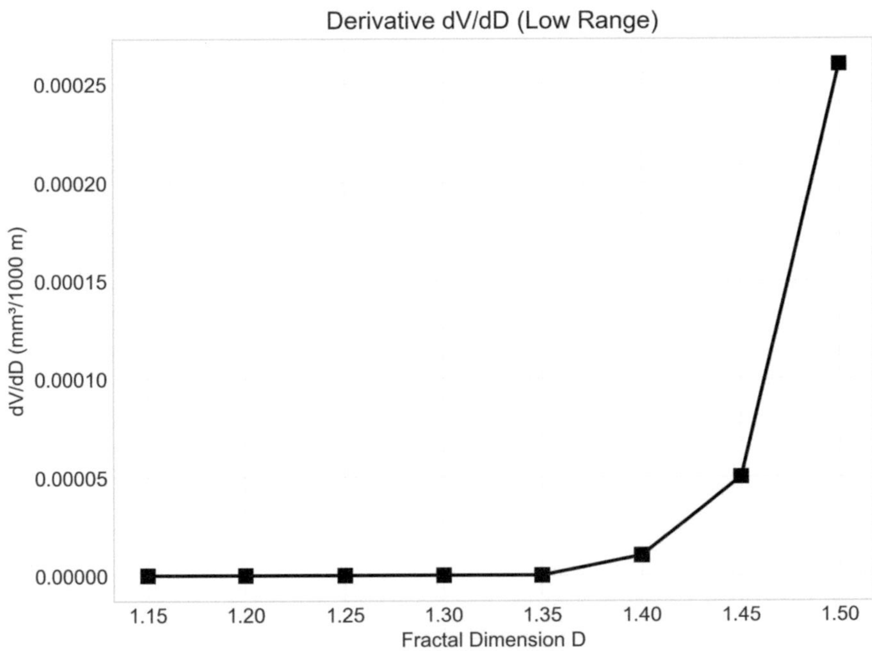

Fig. 32.6: Derivative of wear volume with respect to D in the zone 1.15–1.5 for Steel 45.

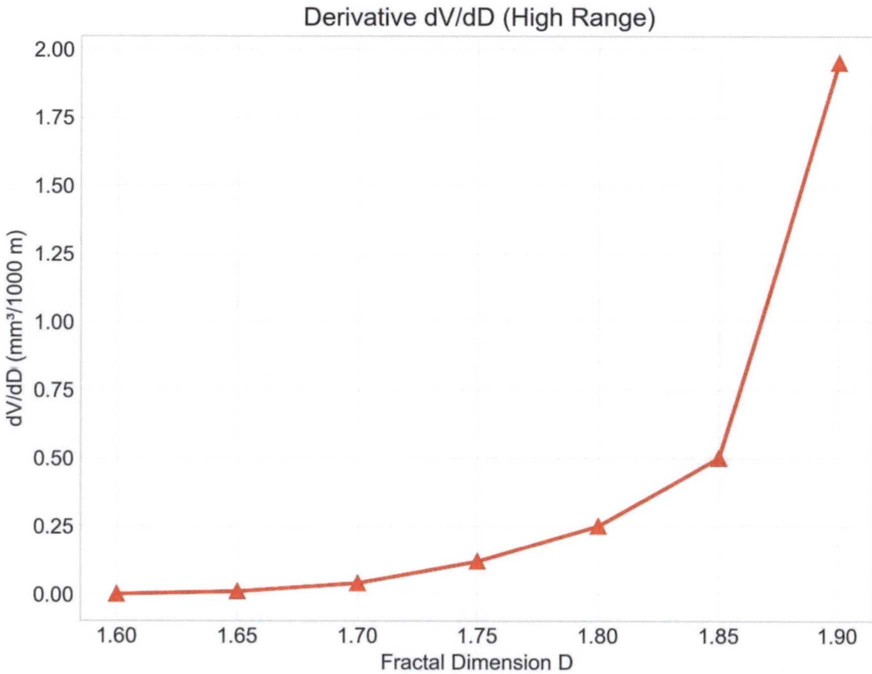

Fig. 32.7: Derivative of wear volume with respect to D in the zone 1.6–1.9 for Steel 45.

References

[1] Aizerman MA, Braverman EM, Rozooner LI (1970) Metod potentsial'nykh funktsiy v teorii mashinnogo obucheniya (in Russ., The Method of Potential Functions in Machine Learning Theory). Moscow: Nauka

[2] Artes AE (1991) Gruppovoye proizvodstvo detaley metodom kholodnoy ob"yemnoy shtampovki (in Russ., Group Production of Parts by Cold Volume Stamping). Moscow: Mashinostroenie

[3] Atkinson R, Bower GH, Crothers EJ (1965) Introduction to Mathematical Learning Theory. New York: John Wiley & Sons

[4] Dorofeyuk AA (1971) Algoritmy avtomaticheskoy klassifikatsii (in Russ., Algorithms for Automatic Classification). In: Problemy rasshireniya vozmozhnostey avtomatov (in Russ., Problems of Expanding the Capabilities of Automata). Moscow: Institute of Control Problems of the USSR Academy of Sciences, Issue 1:5-41

[5] Evstifeev BB, Artes AE (1987) Klassifikatsiya tekhnologicheskikh protsessov kholodnoy obyemnoy shtampovki (in Russ., Classification of Technological Processes for Cold Volume Stamping). Issues of Group Technology. Moscow: Mashinostroenie

[6] Hunt EB (1975) Artificial Intelligence. New York: Acedemic Press
[7] Isachenko NE (1981) Klassifikatsiya detaley pri reshenii zadachi otsenki urovnya tekhnologicheskikh protsessov obrabotki metalla (in Russ., Classification of parts in solving the problem of assessing the level of technological processes for metal forming). Forging and Stamping Production (4):32-35
[8] Ishlinsky AY, Ivlev DD (2001) Matematicheskaya teoriya plastichnosti (in Russ., The Mathematical Theory of Plasticity). Moscow, FIZMATLIT
[9] Ilyushin AA Mekhanika nepreryvnykh sred (in Russ., Mechanics of Continuous Media). Textbook. Moscow: Moscow State University Press
[10] Kraev LF (1975) Konstruktsiya klassifikatora dlya kovanykh i shtampovannykh zagotovok (in Russ., Construction of a classifier for forged and stamped forgings). Blacksmithing and stamping production (2):20-22
[11] Lansky EG, Sosenushkin EN (1989) Avtomatizatsiya proyektirovaniya gruppovykh protsessov kholodnoy i polugoryachey obyemnoy shtampovki v mnogokomponentnom proizvodstve detaley (in Russ., Automation of the design of group processes for cold and semi-hot volume stamping in multi-product production of parts). Machine Building Production. Series. Technology and Equipment for Forging and Stamping Production: Review Information/VNIITEMR, Issue 6
[12] Sosenushkin EN, Yanovskaya EA, Zhelnov AS (2024) Computer modeling of fractal surfaces constructed using artificial intelligence methods for wear prediction. News of the Samara Scientific Center of the Russian Academy of Sciences 26(4):143–149, DOI 10.37313/1990-5378-2024-26-4-143-149
[13] Sosenushkin EN, Yanovskaya EA, Zhelnov AS (2025) Classification of parts using artificial intelligence methods when selecting a mathematical model for solving plastic flow problems. News of the Samara Scientific Center of the Russian Academy of Sciences 27(2):170–179, DOI 10.37313/1990-5378-2025-27-2-170-179
[14] Sosenushkin EN (2011) Progressivnyye protsessy obyemnoy shtampovki (in Russ., Progressive Processes of Volume Stamping). Moscow: Mashinostroenie
[15] Sosenushkin EN, Khromenkov AV, Melnik YA (2014) Mathematical model of adhesive wear of dies for volume stamping. Journal of Friction and Wear 35(6):752-758, DOI 10.3103/S1068366614060130
[16] Sosenushkin EN (1990) Automated material selection system for cold and semi-hot volume stamping dies. In: Actual problems of plastic processing of metals. Varna, pp 379–385
[17] Zagoruiko HG (1972) Metody raspoznavaniya i ikh primeneniye (in Russ., Recognition Methods and Their Application). Moscow: Soviet Radio

Chapter 33
Worst-Case Impact Problems for Mechanical Systems with Distributed Parameters

Polina P. Tkachenko, Dmitry V. Balandin, Tatiana V. Ryabikova, and Alexander K. Lyubimov

Abstract The problem of determining worst-case disturbances acting on a mechanical structure is fundamental to structural design. A worst-case disturbance is an external perturbation that leads to maximum deflections or stresses at specified points of the structure. This paper considers worst-case disturbances for an elastic beam fixed on a moving base. The Kelvin-Voigt viscoelastic model is used within the Euler–Bernoulli beam hypotheses [1, 6]. The Fourier method of separation of variables reduces the distributed-parameter system to ordinary differential equations. A unified performance index is introduced as the ratio of the target output to the L_2-norm of the disturbance [5]. Numerical simulations analyze the dependence of the normalized response on the temporal profile of the disturbance, investigate the influence of the viscosity parameter, and determine optimal points of application and temporal characteristics of perturbations. The results have practical significance for designing vibration protection systems and assessing seismic resistance of structures.

Key words: Worst-case disturbance, Euler–Bernoulli beam, Distributed parameter systems, Fourier method, Modal analysis, Kelvin–Voigt model

33.1 Introduction

The problem of determining extreme external loads capable of inducing the greatest response in a mechanical system is fundamental to the mechanics of deformable solids and optimal control theory. Interest in this problem stems from both its

Polina P. Tkachenko · Dmitry V. Balandin (corresponding author) · Tatiana V. Ryabikova
Sirius University of Science and Technology, Sochi, Russian Federation, e-mail: PTkachen@gmail.com, dbalandin@yandex.ru, tanya.dovid@gmail.com

Alexander K. Lyubimov
Lobachevsky State University of Nizhni Novgorod, Nizhni Novgorod, Russian Federation, e-mail: ljubimov@mm.unn.ru

H. Altenbach et al. (eds.), *Stability Problems in Mechanics of Continua*,
Advanced Structured Materials 260, https://doi.org/10.1007/978-3-032-31032-3_33

theoretical significance and its wide range of applications, from assessing the seismic resistance of building structures to designing vibration protection systems for high-precision equipment.

A worst-case disturbance is understood as an admissible perturbation (kinematic or force) that leads to the maximum value of a given quality functional – deflection, stress, or internal force at a control point of the structure. Mathematically, this is an optimal control problem with distributed parameters, where the goal is to find a disturbance from a given class that maximizes the objective function under a constraint on the norm of the perturbation.

In the present work, an elastic beam fixed on a moving base undergoing displacements according to the law $v(t)$ is considered. This formulation models a wide class of practical situations: seismic action on building structures, vibration loading of machine elements and mechanisms, and kinematic excitation of elastic systems.

The Kelvin–Voigt viscoelastic model is used as the material model within the framework of the Euler–Bernoulli beam hypotheses. This model accounts for internal friction proportional to the rate of change of curvature, which introduces a dissipative term with a fifth mixed derivative into the equation of motion. This allows for a more realistic description of the dynamic behavior of structures made of polymer composites, metals at elevated temperatures, and other materials with pronounced viscoelastic properties.

The aim of the work is to develop an analytical method for constructing worst-case disturbances for a viscoelastic beam and to investigate the influence of system parameters on the characteristics of the extreme response.

33.2 Mathematical Model

Consider a straight beam of length L fixed on a moving base. The beam performs transverse vibrations in the Oxz plane under the action of a kinematic excitation $v(t)$, which represents the displacement of the base.

Within the Kelvin-Voigt model, the stress-strain relation takes the form [1, 6]:

$$\sigma_x = E\varepsilon_x^z + \vartheta \frac{\partial \varepsilon_x^z}{\partial t}, \tag{33.1}$$

where E is Young's modulus and ϑ is the material viscosity coefficient. According to the plane section hypothesis, the strain in a layer at a distance z from the neutral axis is expressed through the deflection $w(x, t)$ [3]:

$$\varepsilon_x^z = -z \frac{\partial^2 w}{\partial x^2}. \tag{33.2}$$

Substituting (33.2) into (33.1) and calculating the bending moment

$$M = \int_A z\sigma_x dA,$$

we obtain:

$$M(x,t) = -EI\frac{\partial^2 w}{\partial x^2} - \vartheta I\frac{\partial^3 w}{\partial x^2 \partial t}, \tag{33.3}$$

where

$$I = \int_A z^2 dA$$

is the second moment of area of the cross-section.

Using d'Alembert's principle and considering that the shear force $Q = \partial M/\partial x$, we arrive at the equation of motion:

$$\rho A\frac{\partial^2 w}{\partial t^2} + \vartheta I\frac{\partial^5 w}{\partial x^4 \partial t} + EI\frac{\partial^4 w}{\partial x^4} = v(t), \tag{33.4}$$

where ρ is the material density and A is the cross-sectional area.

For analysis, it is convenient to switch to dimensionless variables. We introduce the following scales:

$$x = L\hat{x}, \quad t = \frac{L^2}{a}\hat{t}, \quad w = L\hat{w}, \quad v = \rho A\frac{a^2}{L^3}\hat{v}, \tag{33.5}$$

where $a^2 = EI/(\rho A)$. Substituting (33.5) into (33.4) and omitting the ^ sign, we obtain the dimensionless equation [2]:

$$\frac{\partial^2 w}{\partial t^2} + \beta\frac{\partial^5 w}{\partial x^4 \partial t} + \frac{\partial^4 w}{\partial x^4} = v(t). \tag{33.6}$$

Here $\beta = \vartheta a/(EL^2)$ is the dimensionless viscosity parameter.

The dimensionless internal force factors are given by:

$$M = -\frac{\partial^2 w}{\partial x^2} - \beta\frac{\partial^3 w}{\partial x^2 \partial t}, \tag{33.7}$$

$$Q = -\frac{\partial^3 w}{\partial x^3} - \beta\frac{\partial^4 w}{\partial x^3 \partial t}. \tag{33.8}$$

In this work, three types of boundary conditions are considered:

- cantilever beam: $w(0) = w_x(0) = 0$, $w_{xx}(1) = w_{xxx}(1) = 0$;
- simply supported beam: $w(0) = w_{xx}(0) = 0$, $w(1) = w_{xx}(1) = 0$;
- clamped-clamped beam: $w(0) = w_x(0) = 0$, $w(1) = w_x(1) = 0$.

33.3 Expansion in Eigenfunctions

The solution of equation (33.6) is sought in the form of an expansion in the eigenfunctions of vibration [7]:

$$w(x,t) = \sum_{i=1}^{\infty} X_i(x) q_i(t). \tag{33.9}$$

The eigenfunctions $X_i(x)$ satisfy the Sturm–Liouville problem:

$$X_i''''(x) = \lambda_i^4 X_i(x), \tag{33.10}$$

with the corresponding boundary conditions.

33.3.1 Cantilever Beam

For a cantilever beam, the boundary conditions $X(0) = X'(0) = 0$, $X''(1) = X'''(1) = 0$ lead to the characteristic equation:

$$1 + \cos\lambda \cosh\lambda = 0. \tag{33.11}$$

The eigenfunctions have the form:

$$X_i(x) = \cos(\lambda_i x) - \cosh(\lambda_i x) - \sigma_i[\sin(\lambda_i x) - \sinh(\lambda_i x)], \tag{33.12}$$

where $\sigma_i = (\cos\lambda_i + \cosh\lambda_i)/(\sin\lambda_i + \sinh\lambda_i)$. For $i \geq 2$, the asymptotics $\lambda_i \approx \pi(i - 1/2)$ are valid.

33.3.2 Simply Supported Beam

In the case of simple supports, $X(0) = X''(0) = 0$, $X(1) = X''(1) = 0$, the characteristic equation takes the simple form $\sin\lambda = 0$, whence $\lambda_i = \pi i$. The eigenfunctions are:

$$X_i(x) = \sin(\pi i x), \quad i = 1, 2, \ldots \tag{33.13}$$

33.3.3 Clamped-Clamped Beam

For clamped ends, $X(0) = X'(0) = 0$, $X(1) = X'(1) = 0$, we obtain the characteristic equation:

$$\cos\lambda \cosh\lambda = 1, \tag{33.14}$$

and the eigenfunctions are described by expression (12) with $\sigma_i = (\cos\lambda_i - \cosh\lambda_i)/(\sin\lambda_i - \sinh\lambda_i)$. For large λ_i, the asymptotics $\lambda_i \approx \pi(i + 1/2)$ hold.

The eigenfunctions satisfy the orthogonality condition:

$$\int_0^1 X_i(x)X_j(x)dx = N_i^2\delta_{ij}, \tag{33.15}$$

where N_i^2 is the squared norm and δ_{ij} is the Kronecker delta.

Substituting expansion (33.9) into equation (33.6) and using orthogonality, we obtain a system of ordinary differential equations for the generalized coordinates:

$$\ddot{q}_i + \beta\lambda_i^4\dot{q}_i + \lambda_i^4 q_i = b_i v(t), \quad i = 1, 2, \ldots \tag{33.16}$$

Here, b_i are the modal participation factors:

$$b_i = \frac{1}{N_i^2}\int_0^1 X_i(x)dx. \tag{33.17}$$

For a simply supported beam, $b_i = 2(1 - (-1)^i)/(\pi i)$, i.e., only odd modes are non-zero. For cantilever and clamped-clamped beams, the participation factors are:

$$b_i = \frac{1}{N_i^2}\frac{1}{\lambda_i}\left[\sin\lambda_i - \sinh\lambda_i + \sigma_i(\cos\lambda_i + \cosh\lambda_i - 2)\right]. \tag{33.18}$$

33.4 The Worst-Case Disturbance Problem

Consider the problem of finding a disturbance $v(t) \in L_2$ that maximizes the functional [4, 5]

$$J_T = \frac{\max_{t\in[0,T]}|z(t)|}{\|v\|_2}, \tag{33.19}$$

where $\|v\|_2^2 = \int_0^T v^2(t)dt$, and $z(t)$ is the target output.

The target output can be:

- deflection at point x_0: $z = w(x_0, t) = l^T q$, $l_i = X_i(x_0)$;
- bending moment: $z = M(x_0, t) = -l^T(q + \beta\dot{q})$, $l_i = X_i''(x_0)$;
- shear force: $z = Q(x_0, t) = -l^T(q + \beta\dot{q})$, $l_i = X_i'''(x_0)$.

We introduce the impulse response matrix $\Psi(t) = \mathrm{diag}(\psi_i(t))$, where $\psi_i(t)$ is the solution to the equation

$$\ddot{\psi}_i + \beta\lambda_i^4\dot{\psi}_i + \lambda_i^4\psi_i = 0, \quad \psi_i(0) = 0,\ \dot{\psi}_i(0) = 1. \tag{33.20}$$

Depending on the relationship between β and λ_i, three scenarios are possible:

- underdamped ($\beta^2\lambda_i^4/4 < 1$): $\psi_i(t) = e^{-\alpha_i t}\sin(\omega_i t)/\omega_i$;

- overdamped ($\beta^2\lambda_i^4/4 > 1$): $\psi_i(t) = (e^{-\alpha_i^{(1)}t} - e^{-\alpha_i^{(2)}t})/(\alpha_i^{(2)} - \alpha_i^{(1)})$;
- critically damped ($\beta^2\lambda_i^4/4 = 1$): $\psi_i(t) = te^{-\alpha_i t}$.

The solution to system (33.16) is represented as a convolution:

$$q(t) = \int_0^t \Psi(t-\tau)Bv(\tau)d\tau, \quad B = (b_1, \ldots, b_n)^T. \tag{33.21}$$

Then the target output can be written as

$$z(t) = \int_0^t h(t-\tau)v(\tau)d\tau, \tag{33.22}$$

where the weight function $h(t)$ is determined by the type of output. For deflection, $h(t) = l^T\Psi(t)B$; for moment and force, $h(t) = l^T\Psi(t)B + \beta l^T\dot{\Psi}(t)B$.

Applying the Cauchy–Schwarz inequality to the integral (33.22), we obtain the estimate:

$$|z(t)| \le \|v\|_2 \left(\int_0^t h^2(\theta)d\theta\right)^{1/2}. \tag{33.23}$$

Since the right-hand side increases monotonically with t, the maximum on the interval $[0, T]$ is achieved at $t = T$:

$$\frac{\max_t |z(t)|}{\|v\|_2} \le \left(\int_0^T h^2(\theta)d\theta\right)^{1/2}. \tag{33.24}$$

The estimate (33.24) is attained by choosing

$$v_*(t) = h(T-t), \tag{33.25}$$

which is the desired worst-case disturbance. The corresponding value of the functional is

$$J_T^* = \left(\int_0^T h^2(\theta)d\theta\right)^{1/2}. \tag{33.26}$$

33.5 Asymptotic Convergence Analysis

We investigate the rate of decay of the contribution of higher vibration modes to the functional (33.26) [4, 5]. For deflection, we have $h(\theta) = \sum_i h_i(\theta)$, where $h_i(\theta) = l_i\psi_i(\theta)b_i$.

The asymptotic behavior of the factors as $\lambda_i \to \infty$ is:

- $|l_i| \le C$ (boundedness of eigenfunctions);
- $|b_i| \sim C_1/\lambda_i$ (decay of participation factors);
- $|\psi_i(t)| \le 1/(\beta\lambda_i^4)$ (uniform estimate).

From this, we obtain $|h_i(\theta)| \le C_2/\lambda_i^5$, and consequently,

$$\|h_i\|_{L_2} = O(1/\lambda_i^5). \tag{33.27}$$

Given the linear growth $\lambda_i \sim \pi i$ (for simple supports) and $\lambda_i \sim \pi(i \pm 1/2)$ (for clamped and cantilever ends), we conclude that the series converges. The contribution of the third mode is less than 10^{-4} of the total value, which allows for limiting the analysis to the first one or two modes in engineering calculations.

33.6 Numerical Results and Discussion

To verify the developed analytical approach and investigate the effectiveness of the synthesized worst-case disturbances, a series of numerical experiments was conducted. All calculations were performed in the MATLAB environment. The simulation parameters were chosen to cover different types of boundary conditions and to demonstrate the influence of viscosity on the dynamic response of the system.

33.6.1 Simulation Parameters

The dimensionless viscosity parameter was set to $\beta = 0.01$, corresponding to moderate damping and allowing for the observation of the influence of both elastic and dissipative properties. The time interval $[0, T]$ was chosen individually for each type of boundary condition, considering their fundamental periods: $T = 50$ for the cantilever beam, $T = 30$ for the simply supported beam, and $T = 10$ for the clamped-clamped beam. Since the decay rate of transients and the time to reach the peak response depend significantly on the support stiffness and material viscosity, a specific time interval T is required for each type of boundary condition.

The control points for evaluating the target outputs were selected based on physical considerations: deflection was analyzed at sections where it reaches a maximum (the free end for the cantilever, the mid-span for simply supported and clamped-clamped beams); the bending moment and shear force were analyzed at the support sections ($x_0 = 0$), where these factors are maximal. According to the results of the asymptotic analysis (Section 33.5), only the first two modes of vibration were considered in the calculations.

33.6.2 Profiles of Worst-Case Disturbances

Numerical calculation showed that the shape of the worst-case disturbance does not depend significantly on the chosen target output. Fig. 33.1 presents the shapes of the worst-case disturbances $v_*(t)$, obtained using formula (33.25) for the three types of boundary conditions. For comparison, harmonic disturbances at the first resonant

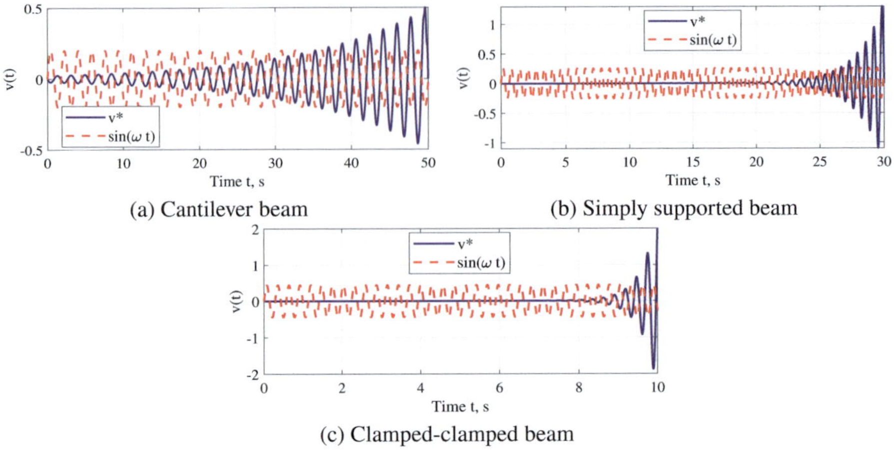

Fig. 33.1: Shape of the worst-case kinematic disturbance $v_*(t)$ maximizing the beam deflection for three types of boundary conditions (solid lines) compared to a harmonic disturbance at the first resonant frequency (dashed lines) with the same energy norm: (a) cantilever beam; (b) simply supported beam; (c) clamped-clamped beam.

frequencies with the same energy norm are shown:

$$v_{\text{harm}}(t) = \frac{\sin(\omega_1 t)}{\|\sin(\omega_1 t)\|_2}, \tag{33.28}$$

where $\omega_1 = \lambda_1^2$ is the first natural frequency.

Analysis of the obtained dependencies shows that in all three cases, the disturbance is a diverging oscillatory process determined by the weight function $h(t)$. Furthermore, the nature of the divergence corresponds to the damping properties of the system: the higher the dissipation, the slower the amplitude growth.

It can be seen that the worst-case disturbance is a diverging oscillatory process, the shape of which is determined by the system's weight function.

33.6.3 Value of the Criterion J_T

For a quantitative assessment of the "destructive" potential of various disturbances, a comparison was made of the values of the functional $J_T^{(2)}$ achieved by the found worst-case disturbance with the value yielded by the harmonic disturbance (33.28). Figures 33.2 present the functional values obtained from formula (33.26) for the cantilever and for three types of target output: deflection at the tip, bending moment and shear force at the fixed end. The values of the functionals $J_T^{(2)}$, showing the maximum normalized response for each output type for the simply supported beam, are given in Figures 33.3. The values of the functionals $J_T^{(2)}$, showing the maximum normalized response for the clamped-clamped beam, are given in Figures 33.4.

Analysis of the obtained graphs shows that the value of the functional J_T for different outputs and different boundary conditions varies significantly. The shear force at the support point exhibits the highest sensitivity to external disturbances, which is in good agreement with the physical understanding of the stress-strain state of the beam.

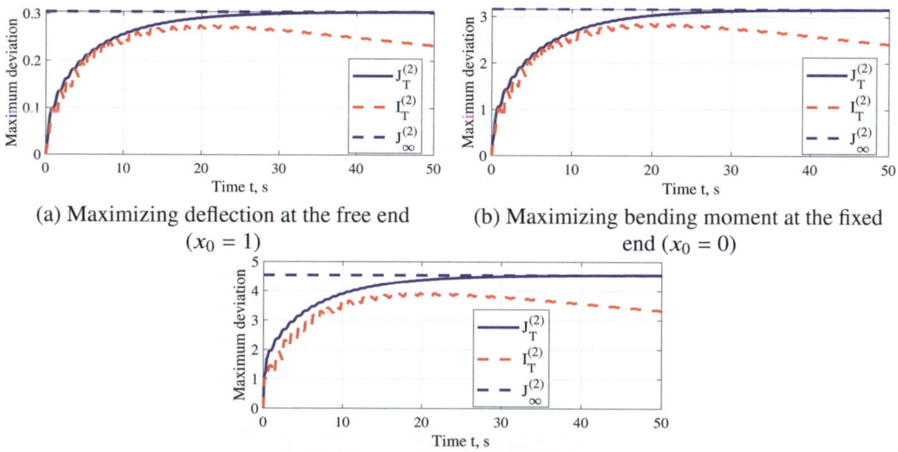

(a) Maximizing deflection at the free end ($x_0 = 1$)

(b) Maximizing bending moment at the fixed end ($x_0 = 0$)

(c) Maximizing shear force at the fixed end ($x_0 = 0$)

Fig. 33.2: Comparison of the quality criterion values J_T for a cantilever beam under the action of the worst-case disturbance (solid lines) and a harmonic disturbance at the first resonant frequency (dashed lines) as a function of the observation time T. Results are shown for three types of target output: (a) deflection at the free end ($x_0 = 1$); (b) bending moment at the fixed end ($x_0 = 0$); (c) shear force at the fixed end ($x_0 = 0$).

The worst-case disturbance, constructed according to the proposed theory, leads to a greater structural response compared to the harmonic disturbance at the resonant frequency. This underscores the importance of the problem of finding the worst-case disturbance: an energetically equivalent but specially tailored perturbation can cause more damage than the traditionally considered resonant one.

33.6.4 Spatial Distribution of the Response

To visualize the spatial distribution of deformations and internal forces caused by different types of disturbances, Figures 33.5-33.7 present the diagrams of displacements $w(x, T)$, bending moments $M(x, T)$, and shear forces $Q(x, T)$. The diagrams are constructed for the time instant $t = T$. Figure 33.5 shows the diagrams of dis-

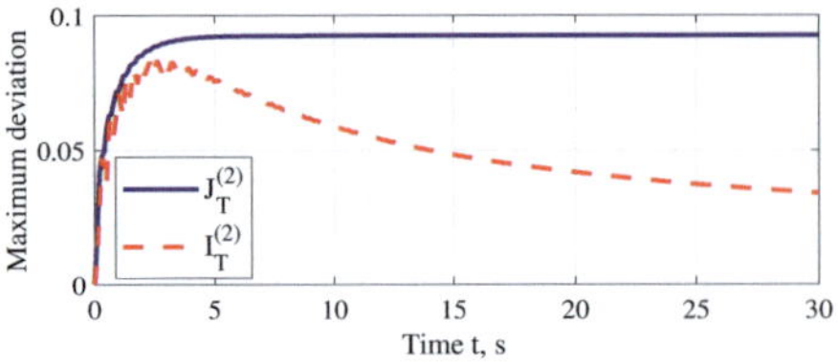

(a) Maximizing deflection at mid-span ($x_0 = 0.5$)

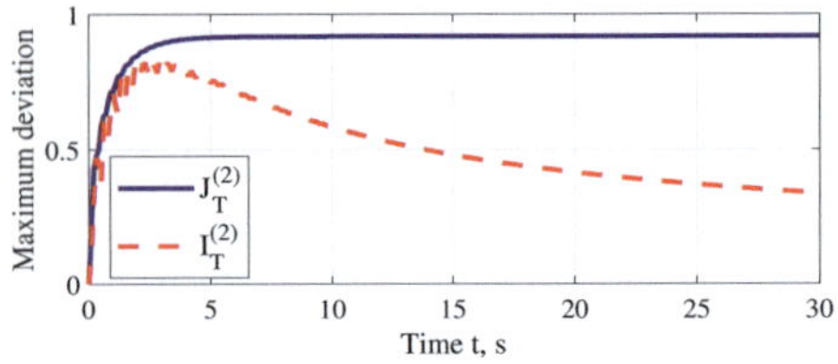

(b) Maximizing bending moment at the support ($x_0 = 0$)

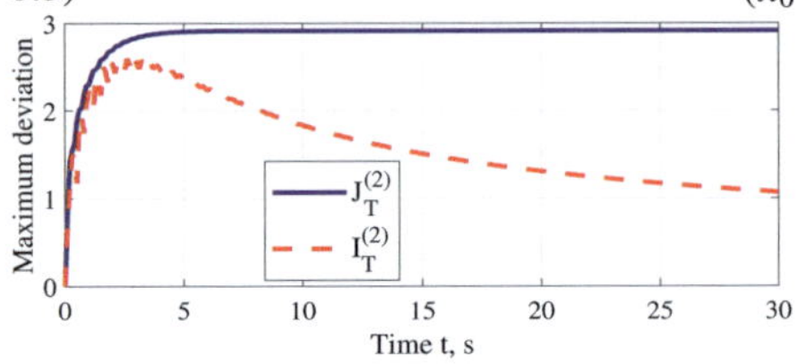

(c) Maximizing shear force at the support ($x_0 = 0$)

Fig. 33.3: Comparison of the quality criterion values J_T for a simply supported beam under the action of the worst-case disturbance (solid lines) and a harmonic disturbance (dashed lines): (a) deflection at mid-span; (b) bending moment at the support; (c) shear force at the support.

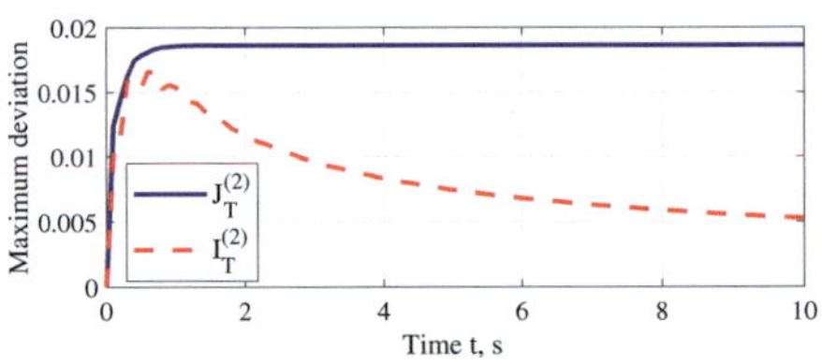

(a) Maximizing deflection at mid-span ($x_0 = 0.5$)

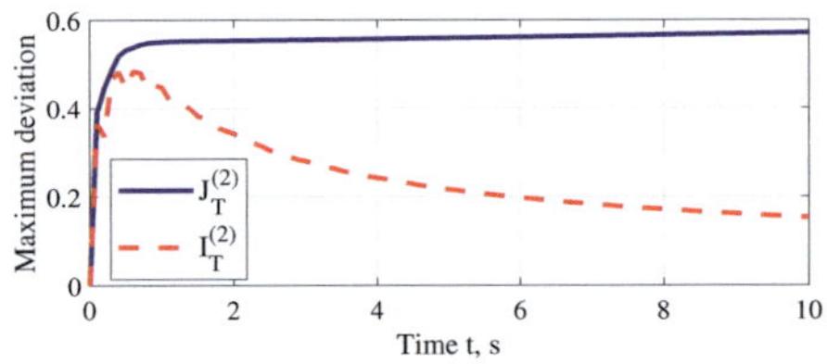

(b) Maximizing bending moment at the support ($x_0 = 0$)

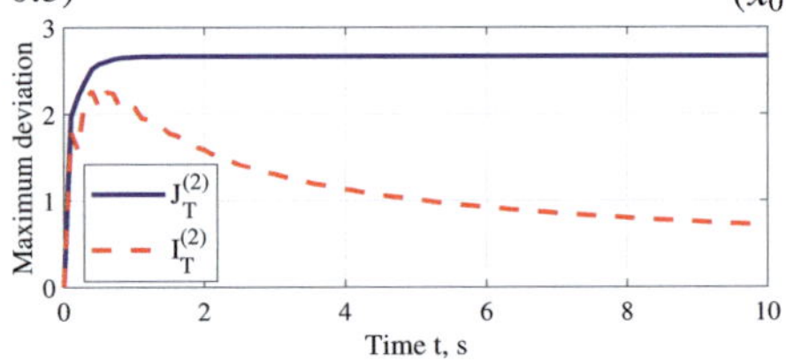

(c) Maximizing shear force at the support ($x_0 = 0$)

Fig. 33.4: Comparison of the quality criterion values J_T for a clamped-clamped beam under the action of the worst-case disturbance (solid lines) and a harmonic disturbance at the first resonant frequency (dashed lines): (a) deflection at mid-span; (b) bending moment at the support; (c) shear force at the support. .

placements, bending moments, and shear forces at the time $t = T$, when the target output reaches its maximum value under the action of the worst-case disturbance.

As expected, the shapes of the diagrams for both types of disturbances (worst-case and harmonic) are qualitatively similar. The difference is that the diagrams constructed for the worst-case disturbance show larger values in all sections of the beam. This clearly confirms that the worst-case disturbance leads to a more intense stress-strain state.

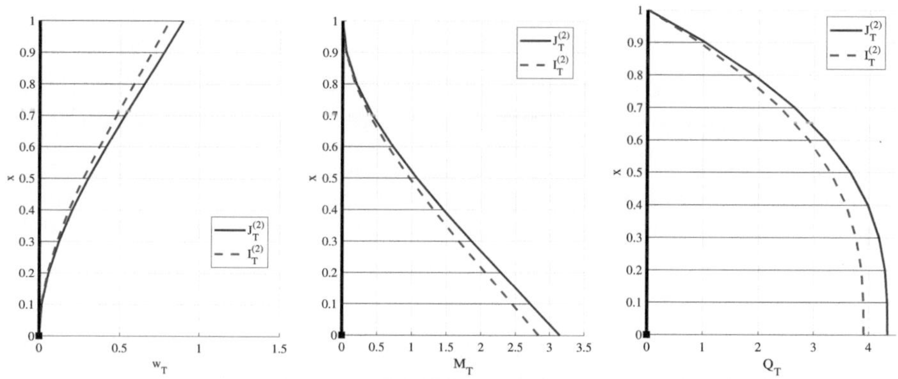

(a) Deflection (b) Bending moment (c) Shear force

Fig. 33.5: Diagrams of the distribution of displacements $w(x, T)$, bending moments $M(x, T)$, and shear forces $Q(x, T)$ along the length of the cantilever beam at the time instant $t = T$, corresponding to the peak response. Solid lines represent the result of the worst-case disturbance; dashed lines represent the harmonic disturbance.

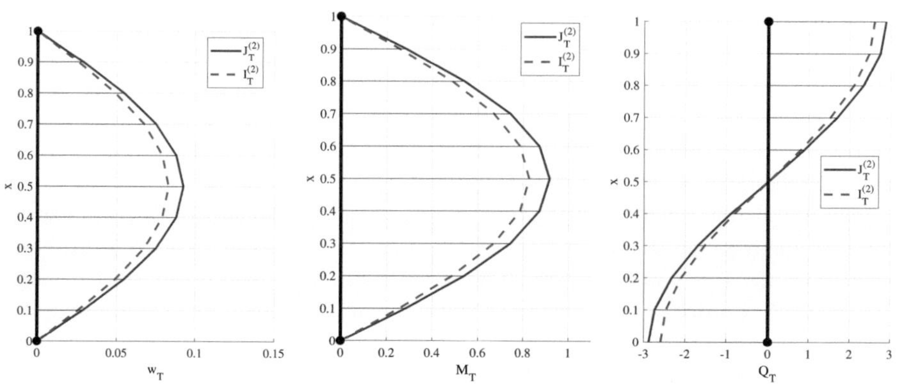

(a) Deflection (b) Bending moment (c) Shear force

Fig. 33.6: Diagrams of the distribution of displacements $w(x, T)$, bending moments $M(x, T)$, and shear forces $Q(x, T)$ along the length of the simply supported beam at the time instant $t = T$. Solid lines represent the worst-case disturbance; dashed lines represent the harmonic disturbance.

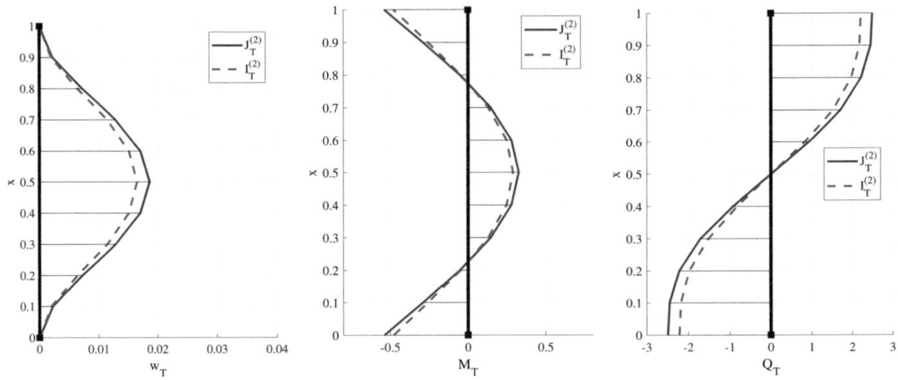

(a) Deflection (b) Bending moment (c) Shear force

Fig. 33.7: Diagrams of the distribution of displacements $w(x,T)$, bending moments $M(x,T)$, and shear forces $Q(x,T)$ along the length of the clamped-clamped beam at the time instant $t = T$. Solid lines represent the worst-case disturbance; dashed lines represent the harmonic disturbance. .

The worst-case disturbance induces a more intense stress-strain state in all sections. The qualitative agreement of the diagram shapes for both types of disturbances confirms that the difference lies in the scale (intensity) of the response.

33.6.5 Influence of the Viscosity Parameter

To investigate the influence of the material's dissipative properties on the system's susceptibility to worst-case disturbances, a parametric analysis was performed. Figures 33.8 show the dependence of the maximum normalized deflection J_T^* on the dimensionless viscosity parameter β for all three types of boundary conditions. The calculations were performed for four values of the parameter:

$$\beta = 0.001; 0.01; 0.1; 1.0,$$

covering the range from weak to strong damping.

The following dependence is observed: as the viscosity parameter β increases, the maximum achievable normalized deflection $J_T^{(2)}$ decreases. At $\beta = 0.001$ (low viscosity), the functional value is maximal. This is explained by the fact that the parameter β is proportional to the material viscosity coefficient ϑ. According to the Kelvin-Voigt model, the higher the material viscosity, the more vibration energy is converted into heat and, consequently, the smaller the amplitude of the system's response to an external disturbance.

This result quantitatively confirms that materials with higher internal friction (e.g., polymers, composites with a viscous matrix, rubber-like materials) possess a greater capacity for vibration absorption and are less sensitive to dynamic loads.

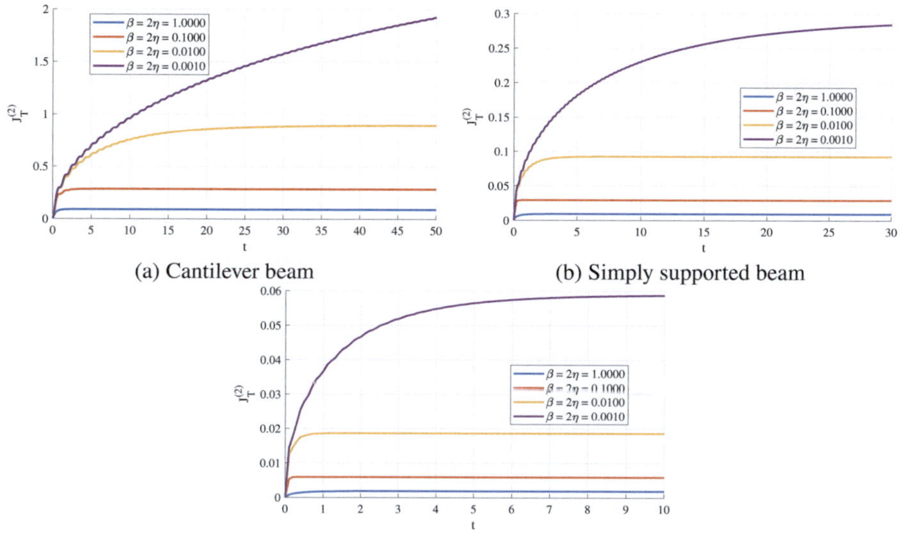

(a) Cantilever beam

(b) Simply supported beam

(c) Clamped-clamped beam

Fig. 33.8: Dependence of the maximum normalized deflection J_T on the dimensionless viscosity parameter β for various boundary conditions: (a) cantilever beam; (b) simply supported beam; (c) clamped-clamped beam.

For materials with low β, the problem of finding the worst-case disturbance is more relevant.

Thus, the conducted analysis shows that the viscosity parameter β is a factor determining the vulnerability of a viscoelastic system to worst-case dynamic disturbances. A monotonic decrease in response with increasing material viscosity is observed, confirming the role of internal friction in dissipating vibration energy and reducing the system's vulnerability to dynamic loads.

33.7 Conclusion

The conducted numerical simulation fully confirms the theoretical results obtained in Sects. 33.4 and 33.5. The following key points can be highlighted:

- convergence of the modal expansion: The fast convergence of the series is confirmed – considering two modes yields an error of less than 10^{-4}, which justifies truncation in engineering calculations.
- efficiency of synthesized disturbances: The worst-case disturbances constructed according to formula (33.25) indeed deliver the maximum to the quality functional, surpassing resonant harmonic disturbances [2, 4].

- role of damping: The viscosity parameter β is a key factor determining the system's vulnerability. This dependence is monotonic and manifests itself for all types of boundary conditions.
- practical significance: The obtained results can serve as a basis for developing methodologies for testing mechanical systems for resistance to extreme loads, as well as for verifying numerical methods for calculating viscoelastic systems.

Acknowledgements The work was supported by the Russian Science Foundation (Project No. 24-11-20023).

References

[1] Georgievskii DV (2023) Modeli teorii vyazkouprugosti (in Russ., Models of Viscoelasticity Theory). Moscow: LENAND

[2] Petrakov EV (2019) Optimal'noe gashenie poperechnykh kolebanii konsol'noi balki (in Russ., Optimal transverse vibrations damping of a console beam). Problemy Prochnosti i Plastichnosti 81(1):81–89, DOI 10.32326/1814-9146-2019-81-1-94-102

[3] Slivker VI (2005) Stroitel'naya mekhanika. Variatsionnye osnovy (in Russ., Structural Mechanics. Variational Basics). Moscow: Izdatel'stvo Assotsiatsii stroitel'nykh vuzov (ASV)

[4] Tkachenko P, Balandin D, Ryabikova T (2026) The worst-case disturbances acting on multi-mass mechanical systems. In: Balandin D, Meyerov I (eds) Mathematical Modeling and Supercomputer Technologies (MMST 2026). Communications in Computer and Information Science, Vol 2815, Cham: Springer, pp 184-197, DOI 10.1007/978-3-032-15761-4_13

[5] Tkachenko PP, Balandin DV, Ryabikova TV (2025) Zadachi o naikhudshikh vozdeystviyakh na mnogomassovuyuuprugyuyu sistemu (in Russ., The problems of the worst-case disturbances acting on multi-mass elastic system). Zhurnal Srednevolzhskogo Matematicheskogo Obshchestva 27(4):451–470

[6] Vinogradov GV, Malkin AY (1977) Rheology of Polymers: Viscoelasticity and Flow of Polymers. Moscow: Mir Publisher

[7] Vladimirov VS, Zharinov VV (2003) Uravneniya matematicheskoi fiziki (in Russ., Equations of Mathematical Physics), 2nd edn. Moscow: Fizmatlit

Chapter 34
Dynamics of a Flexible Kapitsa Cantilever Pendulum with Three Degrees of Freedom

Tatyana P. Tovstik and Alexander K. Belyaev

Abstract The problem of the dynamics of an inverted flexible cantilever pendulum in a gravity field under the influence of vertical vibration of the base is solved. The resonances of the higher-order bending modes arising during this motion are investigated. The pendulum is assumed to be a homogeneous deformable rod, whose bending is described using the Bernoulli-Euler beam model. The rod's bending modes are found by solving an auxiliary boundary-value problem for the rod's equilibrium modes under the action of gravity. The problem is reduced to solving a system of ODEs with rapidly varying coefficients. Using an asymptotic two-scale expansion, the problem is reduced to a system of ODEs with constant coefficients. The system coefficients are functions of the pendulum's base vibration frequency. The behavior of the unaveraged system is studied over a wide frequency range, including three resonances of the second and third bending modes.

Key words: Kapitsa pendulum, Rod under longitudinal load, Bending vibrations of rods, Stability, Resonance

34.1 Introduction

Kapitsa's classic pendulum stabilization problem [11], involving the vertical vibration of the pendulum's lower hinge point under the force of gravity, is a classic problem in mechanics with applications in many fields of science and engineering [4, 6]. Equations describing the motion of such a pendulum are encountered in modern physics problems [8], in phase transition problems, in calculating the stability of anthropomorphic robots, and in high-energy physics problems, as P.L. Kapitsa

Tatyana P. Tovstik (corresponding author) · Alexander K. Belyaev
Institute for Problems in Mechanical Engineering of the Russian Academy of Sciences, St. Petersburg, Russian Federation, e-mail: tovstik_t@mail.ru, vice.ipme@gmail.com

H. Altenbach et al. (eds.), *Stability Problems in Mechanics of Continua*,
Advanced Structured Materials 260, https://doi.org/10.1007/978-3-032-31032-3_34

once wrote about [11]. Many theoretical and experimental studies of the Kapitsa pendulum problem and its generalizations have been published [2, 3, 7, 9, 12–16].

The effect of fulcrum vibration on the stability of an inverted pendulum was first studied in 1906-1908 by Stephenson [18], who examined small oscillations of a rigid pendulum. In this case, the problem is reduced to the Mathieu equation, and the stability conditions for the zero solution of this equation are known [1].

This work presents a linear approach to describing experiments with Kapitsa's flexible pendulum [10], made from rope. The video recording clearly shows that the pendulum behaves differently as the vibration frequency of the base changes. At some frequencies, the pendulum remains vertical, while at others, it performs large-amplitude bending oscillations at the same frequency as the base. Two major bending modes can be distinguished, which the pendulum sequentially passes through at different frequencies. Between these modes, there is a range of frequencies at which the pendulum's transverse oscillations are minimal.

The aim of this work is to determine the resonances of the highest transverse vibration modes of a cantilever-mounted flexible Kapitsa pendulum and their influence on the dynamics of its slow swings. A previously solved problem [20] on the influence of bending vibrations on the stability and attractive field of a hinged flexible Kapitsa pendulum over a wide range of rod stiffness parameters revealed certain patterns concerning the influence of rod bending on the frequency of the pendulum's slow swings before and after resonances. This work continues the previously initiated research [20], but here the highest bending vibration modes are superimposed not on the rectilinear mode of a deflected hinged pendulum, but on the curved first mode of a cantilever pendulum.

In the classical problem of the Kapitsa pendulum, it is assumed that the pendulum is unstable in the absence of vibration at the base, and that vibration at the base brings the pendulum to a stable position. Here, we depart from the classical formulation and, to study the influence of resonances in higher-order flexural modes, consider a pendulum with a stiffness slightly exceeding the Euler critical value, allowing it to remain in a vertical equilibrium position even without support vibration. This will allow us to solve the problem in a linear approximation for a wide range of base vibration frequencies. This problem is conceived as a first step toward studying the geometrically nonlinear motion of a pendulum with resonances in higher-order flexural modes.

The pendulum's flexural modes are determined by solving an auxiliary boundary value problem for the bifurcation of the vertical equilibrium of a rod under its own weight. A small parameter of the problem - the ratio of the base oscillation amplitude to the pendulum length - allows us to analyze the solution outside of resonances using asymptotic methods.

34.2 Model of a Flexible Cantilevered Pendulum

We assume that the pendulum is a homogeneous flexible inextensible rod, obeying the Bernoulli-Euler beam model. Point A of the pendulum's pivot point moves along the vertical coordinate axis $O\zeta$ according to the harmonic law

$$\hat{\zeta}_A = a\sin(\omega\hat{t} + \beta)$$

with a given frequency ω, initial phase β, and amplitude a. We introduce a small parameter ε,

$$\frac{a}{L} \overset{\text{def}}{=} \varepsilon \ll 1, \tag{34.1}$$

where L is the pendulum length, and then we move on to dimensionless parameters for length and time, $\zeta = \hat{\zeta}/L, \quad t = \omega\hat{t}$ (Fig. 34.1).

We denote the angle of inclination of the pendulum axis by $\varphi(s,t)$, where $s \in [0,\ 1]$ is the dimensionless coordinate of the pendulum axis points, t is the dimensionless time. The function $\varphi(s,t)$ completely defines the shape of the pendulum axis bending; we represent it as the sum of three terms

$$\varphi(s,t) = \varphi_1(s,t) + \varphi_2(s,t) + \varphi_3(s,t), \qquad ()' = \frac{d()}{ds}, \tag{34.2}$$

$$\varphi_1(s,t) = \Psi_1{}'(s)\,u(t), \qquad \varphi_2(s,t) = \Psi_2{}'(s)\,v(t), \qquad \varphi_3(s,t) = \Psi_3{}'(s)\,w(t), \tag{34.3}$$

where $\Psi_n(s)$ are functions defining the deviation of the rod along the first three bending modes; are eigenfunctions of the boundary value problem of the equilibrium of a deformable rod under the action of gravity, with cantilever fixation at the lower end and a free upper end [19]

$$\frac{d^4\Psi_n}{ds^4} + \lambda_n \frac{d}{ds}\left((1-s)\frac{d\Psi_n}{ds}\right) = 0, \quad \Psi''_n(1) = \Psi'''_n(1) = \Psi'_n(0) = \Psi''_n(0) = 0. \tag{34.4}$$

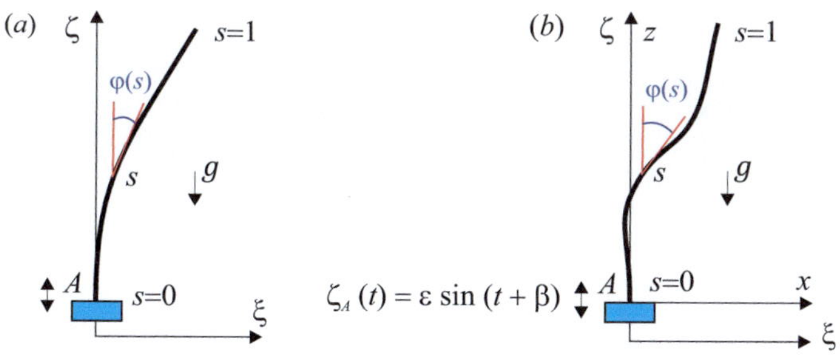

Fig. 34.1: Kapitsa pendulum. (*a*) pendulum deflection according to the first mode, (*b*) general case of pendulum deflection.

The eigenvalues of the boundary value problem (34.4) correspond to the first static bending modes and are equal to

$$\lambda_1 = 7.8373, \ \ \lambda_2 = 55.98, \ \ \lambda_3 = 148.5. \qquad (34.5)$$

Also, the orthogonality conditions are satisfied for the eigenfunctions

$$\int_0^1 (1-s)\,\frac{d\Psi_n}{ds}\,\frac{d\Psi_m}{ds}\,ds = 0, \qquad \int_0^1 \frac{d^2\Psi_n}{ds^2}\,\frac{d^2\Psi_m}{ds^2}\,ds = 0, \qquad k \neq n. \qquad (34.6)$$

We find the functions $\Psi_n(s)$ numerically by integrating Eq. (34.4) taking into account the boundary conditions. Denoting

$$V(s, \lambda_n) = \Psi_n{}'(s),$$

we arrive at the Airy equation

$$V''(s,\lambda) + \lambda(1-s)\,V(s,\lambda) = 0, \qquad V'(1,\lambda) = 0, \qquad (34.7)$$

the solution of which is written as a tabular series [1]

$$V(s,\lambda) = 1 + \sum_{k=0}^{\infty} \frac{\left(-\lambda(1-s)^3\right)^k}{a_k}, \qquad a_0 = 1, \quad a_k = 3k\,(3k-1)\,a_{k-1}. \qquad (34.8)$$

From this we find

$$\Psi_n(s) = \int_0^s V(s,\lambda_n)\,ds, \qquad \varphi_n(s) = V(s,\lambda_n). \qquad (34.9)$$

The results of the numerical calculation of the functions Ψ_n are shown in Fig. 34.2. Calculating the function Ψ_1 required 10 terms of the series (34.8), and calculating Ψ_3 required 30 terms of the series.

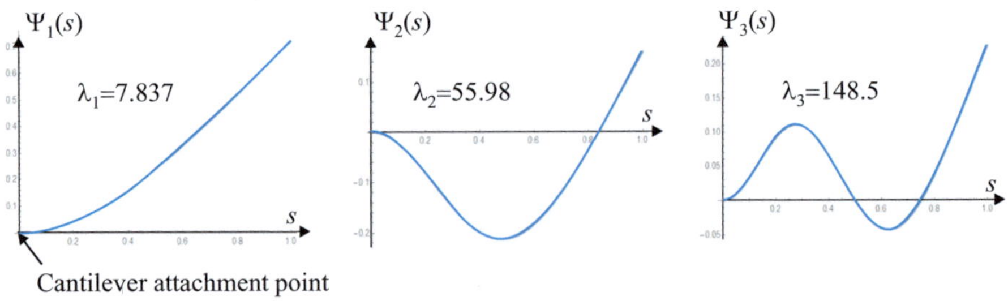

Fig. 34.2: Graphs of the eigenfunctions $\Psi_n(s)$ of the boundary value problem (34.4).

The coordinates of the rod axis points in the coordinate system associated with the lower end of the rod are

$$x(s,t) = \int_0^s \sin\varphi(\sigma,t)\, d\sigma, \qquad z(s,t) = \int_0^s \cos\varphi(\sigma,t)\, d\sigma. \tag{34.10}$$

To obtain the dynamic equations in the linear approximation, we retain the approximate values of the trigonometric functions

$$\sin\varphi \approx \varphi, \quad \cos\varphi \approx 1 - 0.5\,\varphi^2.$$

In a fixed coordinate system (Fig. 34.1), we have

$$\xi(s,t) = x(s,t), \qquad \zeta(s,t) = z(s,t) + \varepsilon \sin(t+\beta). \tag{34.11}$$

34.3 System of Dynamic Equations

The system has three degrees of freedom. Three unknown functions of time determine the amplitudes of the rod's vibration modes. We apply Lagrange's equations of the second kind, taking $u,\ v,\ w$ as generalized functions. In dimensionless form, the system of equations of motion of a pendulum model with three degrees of freedom has the form

$$\frac{d}{dt}\left(\frac{\partial L}{\partial \dot{u}}\right) - \frac{\partial L}{\partial u} = 0, \quad \frac{d}{dt}\left(\frac{\partial L}{\partial \dot{v}}\right) - \frac{\partial L}{\partial v} = 0, \quad \frac{d}{dt}\left(\frac{\partial L}{\partial \dot{w}}\right) - \frac{\partial L}{\partial w} = 0, \quad L = T - \Pi, \tag{34.12}$$

where T is the kinetic energy and Π is the potential energy of the pendulum

$$T = \frac{1}{2}\int_0^1 \left(\dot{x}^2 + (\dot{z} + \varepsilon\cos(t+\beta))^2\right) ds, \qquad \Pi = q_1 \int_0^1 \left(\frac{\varphi'^2}{2P_D} + z\right) ds, \tag{34.13}$$

$$q_1 = \frac{g}{L\omega^2} \stackrel{\text{def}}{=} q\varepsilon^2, \qquad P_D = \frac{mgL^2}{D},$$

where the order of smallness of the dimensionless constant q_1 is assumed to be the same as in the problem of the motion of the classical Kapitsa pendulum [11, 13]. We find expressions for the coordinates $x,\ z$ from (34.2), (34.3) and (34.10)

$$x(s,t) = u(t)\,\Psi_1(s) + v(t)\,\Psi_2(s) + w(t)\,\Psi_3(s), \tag{34.14}$$

$$z(s,t) = s - \frac{1}{2}\int_0^s \left(u(t)\,\Psi_1{}'(\sigma) + v(t)\,\Psi_1{}'(\sigma) + w(t)\,\Psi_1{}'(\sigma)\right)^2 d\sigma \tag{34.15}$$

and substitute into the dynamic equations (34.12). Keeping only the linear terms in the functions u, v, w, we obtain a system that can be written in matrix form with respect to the vector function $\mathbf{Y}(t)$, $\mathbf{Y} = \{u, v, w\}^{\mathrm{T}}$

$$\mathbf{A} \cdot \frac{d^2\mathbf{Y}}{dt^2} + \frac{q_1}{P_D}\mathbf{C} \cdot \mathbf{Y} + (\varepsilon \sin(t+\beta) - q_1)\,\mathbf{B} \cdot \mathbf{Y} = \mathbf{0}, \tag{34.16}$$

where the elements of the matrices $\mathbf{A}$, $\mathbf{C}$, $\mathbf{B}$ depend only on the fixing conditions and, for a cantilever pendulum, are numerical parameters, the constant P_D is inversely proportional to the cylindrical stiffness of the rod, and the constant q_1 is inversely proportional to the square of the vibration frequency of the base. We will rewrite the expressions for the matrix components in view of the boundary value problem (34.4)

$$a_{mn} = \int_0^1 \Psi_m(s)\,\Psi_n(s)\,ds,$$

$$b_{mn} = \int_0^1 \int_0^s \Psi_m{}'(\sigma)\,\Psi_n{}'(\sigma)\,d\sigma ds = \int_0^1 (1-s)\,\Psi_m{}'(s)\,\Psi_n{}'(s)ds$$

$$= \delta_{mn} \int_0^1 (1-s)\,\Psi_m{}'^2(s)ds,$$

$$c_{mn} = \int_0^1 \Psi_m{}''(s)\,\Psi_n{}''(s)\,ds = \delta_{mn} \int_0^1 \lambda_m (1-s)\,\Psi_m{}'^2(s)ds = \delta_{mn}\lambda_m b_{mm},$$

where δ_{mn} is the Kronecker delta. Numerical calculations give

$$\begin{aligned}
&a_{11} = 0.1276, && a_{12} = -0.01429, && a_{22} = 0.01786, && a_{13} = 0.01703,\\
&a_{23} = -0.00053, && a_{33} = 0.00673, && &&\\
&b_{11} = 0.20217, && b_{22} = 0.14384, && b_{33} = 0.12193, &&\\
&c_{11} = 1.5845, && c_{22} = 8.0526, && c_{33} = 18.126. &&
\end{aligned}$$

Let the pendulum be initially deflected by u_0 according to the first form and released without initial velocity.

$$u(0) = u_0, \quad \dot{u}(0) = v(0) = \dot{v}(0) = w(0) = \dot{w}(0) = 0. \tag{34.17}$$

Equation (34.16) with initial conditions (34.17) can be solved numerically and also investigated using the two-scale asymptotic expansion method.

34.4 Asymptotic Expansion

We introduce the slow time θ, $\theta = \varepsilon t$, and represent the desired functions $u,\ v,\ w$ as a sum of slow-time functions θ and rapidly varying small functions, retaining only the following first approximations [5]

$$u(t) = U(\theta) + \varepsilon u_1(t,\theta), \quad v(t) = V(\theta) + \varepsilon v_1(t,\theta), \quad w(t) = W(\theta) + \varepsilon w_1(t,\theta), \tag{34.18}$$

$$\langle u_1 \rangle = \langle v_1 \rangle = \langle w_1 \rangle = 0, \quad \langle f \rangle \overset{\text{def}}{=} \int_0^{2\pi} f(t,\theta)\,dt. \tag{34.19}$$

Total time derivatives have the form

$$\frac{du}{dt} = \varepsilon \dot{u}_1 + \varepsilon U' + \varepsilon^2 {u_1}' + \ldots, \quad \frac{d^2u}{dt^2} = \varepsilon \ddot{u}_1 + \varepsilon^2 U'' + 2\varepsilon^2 {\dot{u}_1}' + \ldots, \quad \ldots,$$

where the prime denotes the derivative with respect to θ, and the dot denotes the partial derivative with respect to t.

Let us rewrite equation (34.16), replacing q_1 with $q\varepsilon^2$

$$\mathbf{A} \cdot \frac{d^2\mathbf{Y}}{dt^2} + \frac{q\varepsilon^2}{P_D}\mathbf{C} \cdot \mathbf{Y} + \left(\varepsilon \sin(t+\beta) - q\varepsilon^2\right)\mathbf{B} \cdot \mathbf{Y} = \mathbf{0}. \tag{34.20}$$

Leaving the terms at ε^1, we obtain the system

$$\mathbf{A} \cdot \mathbf{Y}_1(t,\theta) + \sin(t+\beta)\,\mathbf{B} \cdot \mathbf{Y}_0(\theta) = \mathbf{0}, \quad \mathbf{Y}_1 = \{u_1, v_1, w_1\}^T, \quad \mathbf{Y}_0 = \{U, V, W\}^{\mathrm{T}},$$

having a rapidly changing solution

$$\mathbf{Y}_1(t,\theta) = \sin(t+\beta)\,\mathbf{A}^{-1} \cdot \mathbf{B} \cdot \mathbf{Y}_0(\theta). \tag{34.21}$$

Next, we substitute expression (34.21) into system (34.20). Taking into account terms only at ε^2, we integrate over the period 2π, taking into account that the integral of rapidly changing functions (34.19) is equal to zero, and obtain

$${\mathbf{Y}_0}'' + \mathbf{H} \cdot \mathbf{Y}_0 = \mathbf{0}, \qquad ()' = \frac{d\,()}{d\theta}, \tag{34.22}$$

$$\mathbf{H} = \frac{q}{P_D}\mathbf{A}^{-1} \cdot \mathbf{C} - q\mathbf{A}^{-1} \cdot \mathbf{B} + \frac{1}{2}\mathbf{A}^{-1} \cdot \mathbf{B} \cdot \mathbf{A}^{-1} \cdot \mathbf{B}. \tag{34.23}$$

The initial conditions of the averaged equation (34.22) are obtained from the initial conditions (34.17)

$$\mathbf{Y}_0(0) = \{u_0, 0, 0\}^{\mathrm{T}}, \tag{34.24}$$

$$\left.\frac{d\mathbf{Y}}{dt}\right|_{t=0} = \varepsilon {\mathbf{Y}_0}'|_{\theta=0} + \varepsilon \dot{\mathbf{Y}}_1|_{t,\theta=0} = \mathbf{0} \quad \Rightarrow \quad {\mathbf{Y}_0}'(0) = -\cos\beta\,\mathbf{A}^{-1} \cdot \mathbf{B} \cdot \mathbf{Y}_0(0). \tag{34.25}$$

The matrix **H** in equation (34.22) can also be rewritten as

$$\mathbf{H} = q\mathbf{A}^{-1} \cdot (\mathbf{P} - \mathbf{E}) \cdot \mathbf{B} + \frac{1}{2}\mathbf{A}^{-1} \cdot \mathbf{B} \cdot \mathbf{A}^{-1} \cdot \mathbf{B}, \tag{34.26}$$

where $\mathbf{P} - \mathbf{E}$ is a diagonal matrix with elements

$$\lambda_1/P_D - 1, \quad \lambda_2/P_D - 1, \quad \lambda_3/P_D - 1$$

along the main diagonal, and the stability of the Kapitsa pendulum in the asymptotic approximation follows from the positive definiteness of the matrix **H**. The solution (34.22) will be a superposition of oscillations with three vector amplitudes and the corresponding natural frequencies p_i, $i = \overline{1, 3}$ of the system (34.22), calculated from the characteristic equation

$$\|\mathbf{H} - p^2\mathbf{E}\| = 0.$$

The obtained solutions do not provide a complete picture of all processes in the problem. In particular, the solution to the conservative system (34.22) does not reflect the amplitude growth caused by energy pumping into the system near resonances. Another inaccuracy of the averaged solution obtained results from a difference of several orders of magnitude between the numerical values of the corresponding diagonal elements of the matrices **A** and **C**. As a result, the averaged function may have a higher frequency than the foundation vibration itself. However, at high foundation vibration frequencies, the solutions of the averaged and unaveraged equations outside resonances are close (Fig. 34.3).

34.5 Discussion of Numerical Results

The resonance pattern of the system under consideration is beautiful and varied. If we take a pendulum slightly stiffer than required to lose stability in the absence of vibrations ($P_D \leq \lambda_1$), then such a pendulum, in terms of the behavior of the second and third modes, will differ little from the Kapitsa pendulum, which is unstable in the upper position in the absence of vibrations. However, we can consider slightly slower vibrations of the base ($q > 1$) , producing resonances in the second and third modes.

To conveniently find the resonant frequencies, we will use Krylov's formula [17] for the natural frequencies of the first modes of vibration of a cantilever beam in the absence of gravity. In our notation, the n-th dimensionless natural frequency is equal to

$$\omega_n = \alpha_n^2 \frac{1}{\omega}\sqrt{\frac{g}{P_D L}}, \quad \alpha_1 = 1.875, \quad \alpha_2 = 4.694, \quad \alpha_3 = 7.855, \quad \alpha_4 = 10.996,$$

where α_n are solutions of the transcendental equation $\cos\alpha\ \cosh\alpha = -1$ and for large n tend to the values $(2n-1)\,\pi/2$.

We present numerical calculations for the values $P_D = 6$, $\varepsilon = 0.01$ and for various values of $q = gL/a^2\omega^2$ in the range $q \in [1,\ 120]$, where the resonances for the second and third modes of oscillations for the selected P_D and ε. Within the framework of the linear model, fairly wide regions of resonant growth were obtained near the values $\omega_m = 0.5$, $m = 2.3$ (the widest zone) and $\omega_m = 1$, and near the values $\omega_m = 2$ relatively narrow resonance zones were found. For $\omega_3 = 2$ the resonant increase in the amplitude w of oscillations along the third bending mode does not extend to infinity, but rather to the initial values of the deviation in the variable u. The amplitude u decreases accordingly, and the period of evolution of the amplitudes of the functions $u,\ v,\ w$ lasts 10,000 dimensionless time units, where one oscillation of the pendulum base occurs in a time of 2π.

Table 34.1 shows the boundaries of the considered resonance zones. The boundaries of the resonance zones for resonances along the second oscillation mode are given for the three-mode model; for the two-mode model, the resonance boundaries are slightly wider.

In many cases, taking into account the intense vibration of the rod along the higher bending modes reduces the period of the pendulum oscillations. The ratios of the periods $T_{(2)}/T_{(1)}$ and $T_{(3)}/T_{(1)}$ are given in Table 34.2. Here $T_{(3)}$, $T_{(2)}$, $T_{(1)}$ denote the periods of the pendulum oscillations along the first mode, taking into account three modes of oscillation, two and one, respectively. We see that strong changes in the periods of oscillations are located in the regions of the main resonances $\omega_3 = 1$ and $\omega_2 = 1$. Proximity to other resonances does not change the result. Note that the approximate values of the natural frequencies ω_2 given in the last two columns of the table, marked with an asterisk, are actually on opposite sides of the resonance $\omega_2 = 1$. This is because the approximate values of the natural frequencies were found without taking into account the influence of gravity.

For high-frequency vibration of the foundation (for small q), far from resonances, the oscillations in three modes found from Eqs. (34.16) and (34.22) coincide quite

Table 34.1: Values of the parameter q in resonant zones.

Resonance specification	$\omega_3 = 0.5$	$\omega_3 = 1$	$\omega_2 = 0.5$	$\omega_3 = 2$	$\omega_2 = 1$
Resonance boundaries, q	1.39-5.74	13.44-16.58	27.80-39.98	63.19-63.20	115-118

Table 34.2: Change in the pendulum's oscillation period near resonances.

q	1	6	12	13	16.6	17	27.65	39.99	63.20	113	114	119
ω_2	.09	.22	.31	.32	.37	.37	.473	.562	.72	.956	.96*	.98*
ω_3	.25	.62	.87	.91	1.03	1.04	1.32	1.57	2.002	2.67	2.68	2.75
$\frac{T_{(2)}}{T_{(1)}}$ %	98	99	99	99	99.5	99.5	99.5	99	99.5	97	96	102
$\frac{T_{(3)}}{T_{(1)}}$ %	90	89	82	76	900	200	103	101	100	98	97	102

accurately (Fig. 34.3). As the frequency of foundation oscillations decreases and q increases, a resonance zone begins near the value $\omega_3 = 0.5$. For $q \in [1.39,\ 5.74]$, the linear equations give infinite growth of solutions. The value of the frequency parameter q at which resonance occurs is independent of the initial phase of the foundation vibrations, but near resonance the oscillation pattern is highly dependent on β . At $\beta \neq \pm\pi/2$, an increase in the amplitude of the rod oscillations according to the third mode $w\,(t)$ and an increase in the evolution period of this amplitude are observed. At $\beta = \pm\pi/2$, an increase in $w\,(t)$ does not occur outside the resonant zone (Fig. 34.4). The evolution period of the amplitude $w\,(t)$ at $\beta \neq \pm\pi/2$ depends only on q and is independent of β .

Charts $u_{(1)}\,(t)$, $u_{(2)}\,(t)$, $u_{(3)}\,(t)$ pendulum oscillations in the first mode, found using the one-, two-, and three-mode models, respectively, are shown in Fig. 34.5. Figure 34.6 shows the graphs of the oscillation amplitudes $u,\ v,\ w$ near the main resonance for the third mode. Figure 34.6 (*a*) shows the consistency of the amplitudes of small vibrations caused by the support vibration. The third mode (green) oscillates out of phase with the first mode, while the second (red) oscillates in phase with the

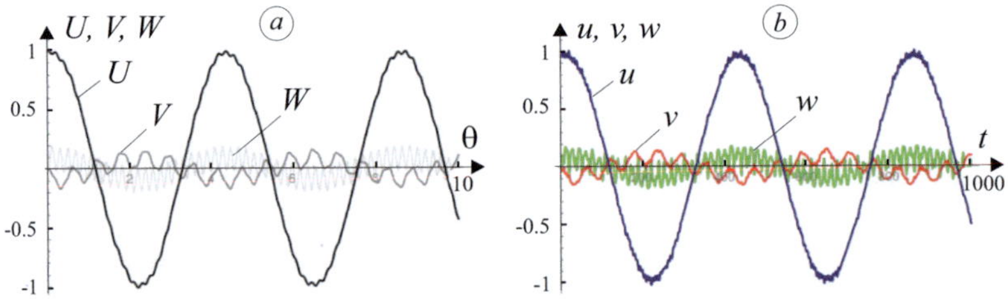

Fig. 34.3: Comparison of the averaged and unaveraged solutions for $q = 1,\ \ \beta = \pi/2\ +0.15$. (*a*) solution $U,\ V,\ W$ of the averaged system (34.22), (*b*) solution $u,\ v,\ w$ of the system (34.16).

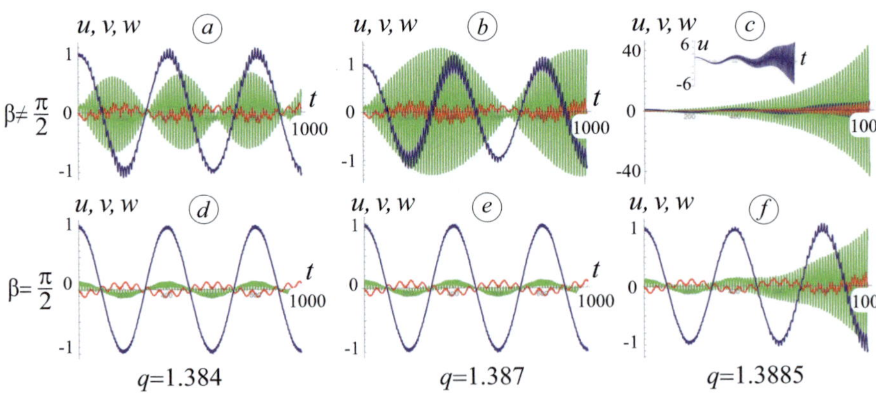

Fig. 34.4: Behavior of functions near resonance. (*a*) $q = 1.384,\ \ \beta = \pi/2 + 0.1$, (*b*) $q = 1.387,\ \ \beta = \pi/2 + 0.1$, (*c*) $q = 1.3885,\ \ \beta = \pi/2 + 0.1$, (*d*) $q = 1.384,\ \ \beta = \pi/2$, (*e*) $q = 1.387,\ \ \beta = \pi/2$, (*f*) $q = 1.3885,\ \ \beta = \pi/2$.

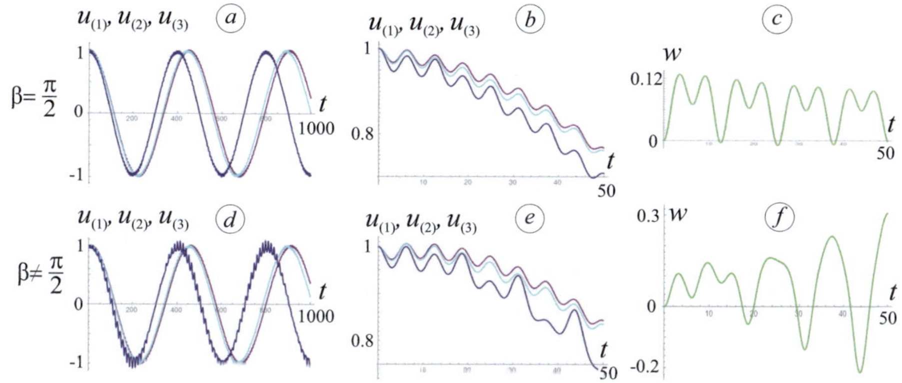

Fig. 34.5: Graphs of the functions $u_{(n)}\ (t)$, blue — for the three-mode model, cyan — for the two-mode model, purple — for the single-mode model, green — graphs of the functions $w\ (t)$ at $q = 1.384$. (*a*) several periods of pendulum swings, $\beta = \pi/2$, (*b*), (*c*) the beginning of the motion, $\beta = \pi/2$; (*d*), (*e*), (*f*) — $\beta = \pi/2 + 0.1$.

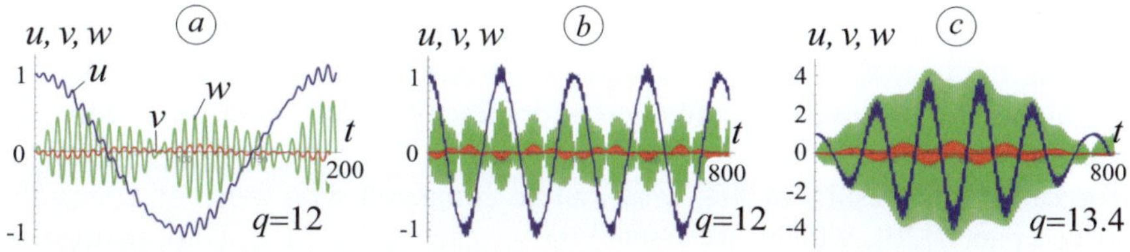

Fig. 34.6: Graphs of the functions $u,\ v,\ w$ as they approach the main resonance in the third mode. $\beta = \pi/2$. (*a*)-(*b*) $q = 12$, (*c*) $q = 13.4$.

first. Figures 34.6 (*b*)-(*c*) show the gradual increase in oscillations in the third mode as resonance is approached. The evolution period of the oscillation amplitudes $w\ (t)$ gradually increases. When the amplitude w (green line) exceeds the pendulum swing amplitude (blue line), the maximum values of u and v begin to increase following w.

Near resonance, the frequency of the pendulum's slow swings changes. For a rod less rigid than required for exact coincidence with the resonant value, the pendulum's swing frequency increases (Fig. 34.7 (*a*)). On the other side of resonance, the pendulum's swing frequency in the linear model decreases to zero. Such a decrease in the pendulum's swing frequency is only obtained in the linear approximation, that is, for very small deviations from the vertical. If we add to the system (34.16) the largest of the nonlinear terms of the potential energy, included in the expression

$$q_1 \int_0^1 z(s)ds,$$

then for small initial deflections of the pendulum, for example, for $u_0 = 0.02$, we obtain a graph almost exactly similar to Fig. 34.7 (*b*), and for nonlinear initial

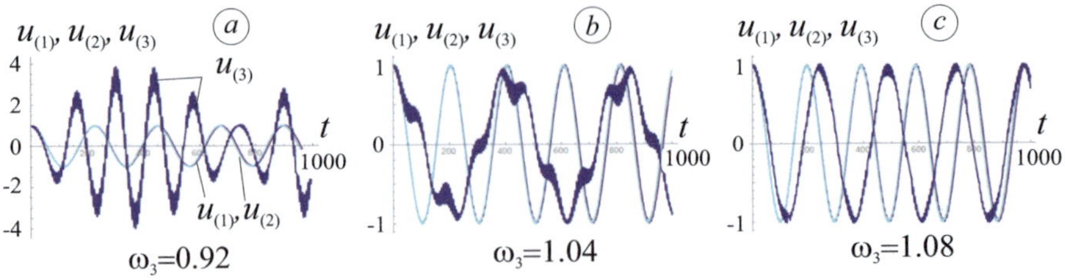

Fig. 34.7: Graphs of $u_{(n)}\ (t)$ before and after resonance in the third form at $\beta = \pi/2$. (*a*) $q = 13.4,\ \ \omega_3 = 0.922,\ T_{(3)}/T_{(1)} = 64\,\%$, (*b*) $q = 17,\ \ \omega_3 = 1.039,\ T_{(3)}/T_{(1)} = 200\,\%$, (*c*) $q = 18.5,\ \ \omega_3 = 1.083,\ T_{(3)}/T_{(1)} = 120\,\%$.

deflections, for example, for $u_0 = 0.2$, the pendulum swing frequency will be the same or slightly higher than in the single-mode model. The nonlinear model will not give infinite resonant growth here, and for $q \in [15,\ 15.5]$ the pendulum swings according to the first form are so non-uniform in amplitude that if we consider only large deflections as swings, the pendulum will swing more slowly, and even if the smallest ones, then faster than the swings of the single-mode model. Figure 34.7 shows that the pendulum's oscillation frequency for the dual-mode model is only 0.55% higher than the oscillation frequency for the single-mode model (two adjacent regular sine waves in cyan and purple).

The calculation results of this work can be compared with those of a recently published paper [20], which investigated the geometrically linear deformations of a hinged Kapitsa pendulum with finite deflection angles. It turned out that for a cantilever pendulum, there is a similar curve of changes in the period of slow pendulum swings depending on the vibration frequency of the base. Specifically, at almost all points on the graph, introducing higher-order oscillation modes into the pendulum model slightly reduces the oscillation period, and only near the main resonance, for $\omega_3 < 1$, the pendulum swings accelerate significantly, while for $\omega_3 > 1$, they slow down significantly. In [20], an explicit asymptotically approximate formula for the pendulum's oscillation period was found as a function of the base vibration frequency. So far, only numerical calculations have been obtained, which are generally consistent with the case of a Kapitsa pendulum on a hinged support.

34.6 Conclusions

The problem of the oscillation of a flexible cantilever rod subjected to vertical vibration of the base in a gravity field is solved in three formulations - taking into account three, two, and one transverse vibration mode. The pendulum axis bending mode is represented as a sum over three degrees of freedom - the bending modes of the cantilever rod under its own weight in a gravity field. In the linear formulation, the problem is reduced to a system of three linear equations. In the dimensionless notation, the system contains three parameters, one of which is responsible for the base vibration frequency, the second for the rod's stiffness, and the third for the initial phase of the base vibration. The coefficients of the averaged system of equations contain two parameters, and the initial phase is included in its initial conditions. It was found that within the linear model, the second and third frequencies of the rod's natural oscillations each have two broad resonant zones and one narrow zone of resonant increase in oscillation amplitude for the corresponding mode. Taking into account the second and third natural oscillation modes changes the pendulum's oscillation period for the first mode. As the main resonance is approached from the side of the high-frequency oscillations of the base, ($\omega_m < 1$), the pendulum's oscillation period decreases, while on the opposite side, it increases.

Acknowledgements The work was carried out within the framework of the state assignment of the Institute of Mechanical Engineering of the Russian Academy of Sciences No 124041100005-4.

References

[1] Abramovits M, Stigan I (eds.) (1964) Handbook of Mathematical Functions. National Bureau of Standards. Applied Mathematics series, Vol 55

[2] Belyaev AK, Morozov NF, Tovstik PE, Tovstik TM, Tovstik TP (2022) On an Attraction Basin of the Generalized Kapitsa's Problem. In: Polyanskiy VA, Belyaev AK (eds.) Mechanics and Control of Solids and Structures. Advanced Structured Materials, Vol 164 Springer, Cham, pp 23-48, DOI 10.1007/978-3-030-93076-9_2

[3] Belyaev AK, Morozov NF, Tovstik PE, Tovstik TP (2018) Stability of a flexible vertical rod on a vibrating support, Vestnik St Petersburg University. Mathematics. Mechanics. Astronomy 51(3):296-304, DOI 10.3103/S1063454118030020

[4] Blekhman II (2000) Vibrational Mechanics. World Scientific, Singapore

[5] Bogoliubov NN, Mitropolski YA (1961) Asymptotic Methods in the Theory of Non-Linear Oscillations. Gordon and Breach, New York

[6] Chelomey SV (1999) On two problems of dynamical stability of oscillatory systems formulated by academicians P.L. Kapitsa and V.N. Chelomei. Mechanics of Solids 34(6):134-140

[7] Fraser WB, Champneys AR (2002) The "Indian rope trick" for a parametrically excited flexible rod: nonlinear and subharmonic analysis. Proceedings of the Royal Society A: Mathematical, Physical and Engineering Sciences 458(2022):1353-1373, DOI 10.1098/rspa.2001.0907

[8] Gerving CS, Hoang TM, Land BJ, Anquez, M, Hamley CD, Chapman MS (2012) Non-equilibrium dynamics of an unstable quantum pendulum explored in a spin-1 Bose-Einstein condensate. Nature Communications 3:1169, DOI 110.1038/ncomms2179

[9] Gribkov VA, Gordin YaD (2024) Ob avtorskoy matematicheskoy modeli mayatnika Chelomeya. (in Russ., On the authors' mathematical model of the Che-

lomey's pendulum). Herald of the Bauman Moscow State Technical University, Series Mechanical Engineering 2(149):42-62

[10] IPME (2026) Institute for Problems in Mechanical Engineering of the Russian Academy of Sciences. Articles Laboratorys. Indian rope trick. https://ipme.ru/en/183-indiiskaja-magicheskaja-veryovka.html, accessed 21 March 2026.

[11] Kapitsa PL (1965) Collected papers of P.L. Kapitsa edited by D. TerHaar. Pergamon, New York 2:714-726

[12] Markeev AP (2018) On the accuracy problem for pendulum clock on a vibrating base. Mechanics of Solids 53(5):573-583, DOI 10.3103/S0025654418080113

[13] Morozov NF, Belyaev AK, Tovstik PE, Tovstik TP (2018) Stability of a Vertical Rod on a Vibrating Support. Doklady Physics 63(9):380-384, DOI 10.1134/S1028335818090069

[14] Morozov NF, Belyaev AK, Tovstik PE, Tovstik TP, Ma CC (2022) Dynamics and Stability of Axially Loaded Elastic Rods. In: Irschik H, Krommer M, Matveenko VP, Belyaev AK (eds) Dynamics and Control of Advanced Structures and Machines. Advanced Structured Materials, Vol 156. Springer, Cham, pp 147-160, DOI 10.1007/978-3-030-79325-8_13

[15] Neishtadt AI, Sheng, K (2017) Bifurcations of phase portraits of pendulum with vibrating suspension point. Communications in Nonlinear Science and Numerical Simulation 47:71-80, DOI 10.1016/j.cnsns.2016.11.003

[16] Petrov AG (2021) Areas of attraction of a pendulum under the influence of oblique vibration of a suspension point. Vestnik St Petersburg University, Mathematics 54(3):273-281, DOI 10.1134/S1063454121030067

[17] Birger IA, Panovko YG (eds) (1968) Prochnost'. Ustoychivost'. Kolebaniya. Spravochnik v trekh tomakh. Tom 3 (in Russ., Strength. Stability. Oscillations. Handbook in three volumes. Vol 3) Mashinostroenie, Moscow (1968)

[18] Stephenson A (1908) XX. On an induced stability. The London, Edinburgh, and Dublin Philosophical Magazine and Journal of Science, Series 6 15(86):233-236, DOI 10.1080/14786440809463763

[19] Svetlitsky VA (2005) Dynamics of Rods. Series Foundations of Engineering Mechanics, Springer, Berli, Heidelberg, DOI 10.1007/b137699

[20] Tovstik TP (2026) Dynamics of the final deflections of a flexible Kapitsa pendulum on a hinged support. Vestnik St Petersburg University, Mathematics 59(2):224-233 (in print)

Chapter 35
On Simplified Models for Determining Inhomogeneous Elastic Properties

Alexander Vatulyan and Victor Yurov

Abstract Dynamic coefficient inverse problems in elasticity are studied using iterative Newton-type methods requiring sensitivity analysis. In the static case, coefficient inverse problems of identifying spatially varying material characteristics in an isotropic elastic rectangle are analyzed. Using the Kantorovich method, the two-dimensional problem is reduced to one-dimensional applied models. An asymptotic WKB solution accounting for the boundary layer is constructed for normal loading, while an analytical solution is derived for shear loading. Reconstruction of the shear modulus is performed by approximating the unknown function and determining its coefficients via the least squares method using displacement data from finite element simulations. The proposed approach demonstrates good accuracy for monotonic and non-monotonic inhomogeneity laws.

Key words: Coefficient inverse problems, Elasticity, WKB method, Kantorovich method, Identification of inhomogeneity parameters

35.1 Introduction

Modern technological advances, particularly the manufacturing of functionally graded materials, necessitate the improvement of methods for determining their physical properties, which are typically spatially varying (functions of coordinates). Unlike direct experimental approaches, which are applicable only to homogeneous media, determining these variable characteristics requires the application of spe-

Alexander Vatulyan
Southern Federal University, 344049 Rostov-on-Don, Russian Federation, e-mail: aovatulyan@sfedu.ru

Victor Yurov (corresponding author)
Southern Mathematical Institute, Vladikavkaz Scientific Center of the Russian Academy of Sciences, 362027 Vladikavkaz, Russian Federation, e-mail: vyurov@sfedu.ru

H. Altenbach et al. (eds.), *Stability Problems in Mechanics of Continua*,
Advanced Structured Materials 260, https://doi.org/10.1007/978-3-032-31032-3_35

cialized loading schemes and data processing techniques based on computational approaches for solving inverse problems.

Coefficient inverse problems (CIPs) in mathematical physics are a rapidly developing area of mathematics devoted to the study of problems related to determining the coefficients of a boundary-value problem for a mathematical model from additional information. For example, additional information for models in elasticity theory and coupled-field mechanics is typically provided in the form of displacement fields, temperature, and electric potential/field distributions, measured either on the boundary or within the domain. Model characteristics in mechanics (elastic moduli, density, coupling coefficients of the models) can be constant or coordinate-dependent coefficients of differential operators. As a rule, such problems are classified as ill-posed problems, which are characterized by possible non-uniqueness and strong instability of the constructed solution with respect to small perturbations of the input information [1].

The formulation and development of methods for solving coefficient inverse problems [2–9, 28] in various branches of mathematical physics have been extensively covered in the modern literature. Of particular interest to researchers are inverse problems in the mechanics of deformable solids [10–17], where the goal is to determine the spatially varying elastic properties of a material.

An analysis of approaches to solving such problems indicates that the key to constructing solutions is an iterative process, at each stage of which it is necessary to solve the direct problem and then find a correction to the initial approximation. Among the iterative algorithms, the most common are Newton's operator method and its modifications. In general, when the unknown characteristics depend on coordinates, an analytical solution to the direct problem is unavailable, and it is solved only numerically – for example, by the finite element method, the finite difference method, or projection methods. This is because it is impossible to construct explicit operator relationships linking the input and output data of the problem (the prescribed and sought-for functions). Note that, for example, in [11–15], some features of the implementation of iterative processes are discussed.

An important factor determining the success of iterative algorithms such as Newton's method is the choice of an appropriate initial approximation. In practice, two main approaches are used to construct it. The first approach is based on solving a series of auxiliary problems for a homogeneous body model, which allows one to determine the effective or averaged characteristics of the heterogeneous structure under study. The second approach involves seeking the desired coefficient-functions not in cones of positive summable functions, but in a narrower class of parameterized functions (e.g., linear, exponential, power-law, or linear combinations thereof). At this stage, the model parameters are selected by minimizing the misfit functional between the experimental and calculated data. This requires solving the direct problem multiple times, but allows one to obtain a physically meaningful starting point for the subsequent iterative refinement procedure. A review of the literature shows that many authors confine themselves to minimizing the misfit functional as the primary tool for solving the inverse problem. Minimization is typically implemented using gradient-based schemes or heuristic (evolutionary) algorithms.

When solving dynamic inverse problems of inhomogeneous elasticity, in the simplest case of a continuously inhomogeneous (or piecewise homogeneous) isotropic body, three functions are required: the Lamé moduli and the density of the medium. In the static case, only the elastic moduli are required. Most often, the functions describing the inhomogeneity are assumed to be one-dimensional. A common method for determining them is based on analyzing the response of the object to applied loads. Here, the response refers to the displacement fields or, in the case of dynamic loading, their amplitudes, measured on the surface of the body. When studying the response, the concept of sensitivity is usually introduced. Sensitivity analysis allows one to identify uninformative loading configurations for the chosen geometry of the object for which there is little or no change in the response when the desired functions are varied. For example, the Lamé parameter λ does not affect the torsional deformation of rod structures and, accordingly, cannot be recovered from a torsion test.

This work is devoted to some aspects of the formulation and study of a number of inverse problems for models of inhomogeneous elasticity, the application of asymptotic methods for simplifying the original models when solving direct problems, and the use of these methods for the analysis of inverse problems.

35.2 Statement of the Direct and Inverse Problem in General Form

We consider steady-state oscillations of a bounded domain V with a piecewise-smooth boundary $S = S_u \cup S_\sigma$, where displacements are prescribed on S_u and external loads are prescribed on S_σ.

The boundary value problems for the oscillation equations at frequency ω are formulated using the equations of motion:

$$\sigma_{ij,j} + \rho\omega^2 u_i = 0 \quad i = 1, 2, 3 \tag{35.1}$$

and the constitutive relations for an anisotropic body:

$$\sigma_{ij} = c_{ijkl} u_{k,l} \tag{35.2}$$

Mixed boundary conditions are imposed:

$$u_i | S_u = 0, \quad \sigma_{ij} n_j \big| S_\sigma = p_i. \tag{35.3}$$

Here c_{ijkl} are the components of the elastic moduli tensor, which are piecewise-smooth functions of coordinates satisfying the usual symmetry and positive definiteness conditions [18], u_j are the components of the displacement vector, σ_{ij} are the components of the stress tensor, ρ is the density of the medium (also a function of coordinates), and n_j are the components of the unit outward normal vector to S.

We formulate the statement of the coefficient inverse problem of determining the spatially varying material characteristics (coefficients of the differential operator of elasticity theory) in the case where additional information is specified on the boundary of the object:

$$u_i|S_\sigma = f_i(x,\omega), \quad \omega \in [\omega_-, \omega_+]. \tag{35.4}$$

This formulation corresponds to measuring displacement fields on the part of the boundary S_σ, where the loading is applied, over a certain frequency range. Note that in this general formulation, the problem is nonlinear, and the construction of its solution is based on an iterative process, as described in the introduction.

In practice, a different formulation of the problem is more realistic. Let $S_\sigma = S_p \cup S_0$, where the external load p_j vanishes on the boundary part S_0, and is non-zero on S_p. Then, the probing experiment consists of measuring, for a given loading, the additional information on the unloaded part of the boundary $S_f \subset S_0$:

$$u_i|_{S_f} = f_i(x,\omega), \quad \omega \in [\omega_-, \omega_+] \tag{35.5}$$

In accordance with the classification introduced in [8], we will call formulation (35.1)–(35.4) an inverse problem of the second type, and (35.1)-(35.3) with additional information of the form (35.5) a problem of the third type. Note that in the problem of the first type from [8], the additional information is specified in the entire domain V, which seems feasible only in certain situations, such as thin plates or prismatic rods, for which displacements within the domain can be reconstructed with sufficient accuracy from surface displacements.

We present the formulation for the case of a problem of the second type and an isotropic body, for which Hooke's law (35.2) is represented as

$$\sigma_{ij} = \lambda u_{k,k}\delta_{ij} + \mu(u_{i,j} + u_{j,i}), \tag{35.6}$$

where λ, μ are the Lamé parameters.

To solve the inverse problem of reconstructing three spatially varying characteristics λ, μ, ρ, several approaches are used, which are based on the Newton-type iterative method [8, 14] and its modifications. We present the corresponding operator equations. To determine the three characteristics, we will use three probing modes, characterized by different regions of load application and different methods of its application, referring to these loads as $p_i^{(m)}$, $m = 1, 2, 3$; the corresponding displacement fields will be denoted by $u_i^{(m)}$.

For elastic bodies, it is possible to use the generalized reciprocity relation and the operator relation derived from it [8]. In the study of coefficient inverse problems, which involve determining certain functions that characterize the inhomogeneity, one can also rely on a weak formulation for general linear models of continuum mechanics. This formulation is also very often used for studying direct problems.

We now turn to the weak formulation, whereby we project the equation of motion (35.1) onto the element $\mathbf{v}$, which is a vector function with continuously differentiable components and satisfies the boundary conditions on S_u. Using the Gauss–

Ostrogradsky theorem and taking into account the boundary conditions in (35.3), we reduce the resulting equality to the form $(a, \mathbf{u}, \mathbf{v}) = b(\mathbf{v})$, where $(a, \mathbf{u}, \mathbf{v})$ is a trilinear form (linear in each argument) of the variables $a, \mathbf{u}, \mathbf{v}$, and the right-hand side $b(\mathbf{v})$ is a linear form.

Thus, a weak formulation for the theory of anisotropic elasticity is presented in [8] and consists in determining the components of the displacement vector, and for the trilinear form and the linear form we have, respectively, the following representations:

$$(a, \mathbf{u}, \mathbf{v}) = \int\limits_V 2L(u_i, v_i, c_{ijkl}, \rho)dV, \quad b(\mathbf{v}) = \int\limits_{S_\sigma} p_i v_i dS, \quad \omega \in [\omega_-, \omega_+],$$
$$2L(u_i, v_i, c_{ijkl}, \rho) = c_{ijkl} u_{k,l} v_{i,j} - \rho\omega^2 u_i v_i. \tag{35.7}$$

In the isotropic case, for the constitutive relations of the form (35.6), from the weak formulation (35.7), setting $v_i = u_i^{(m)}$, we obtain the following system of nonlinear Urysohn-type integral equations [15]:

$$\int\limits_V \left[\lambda(u_{k,k}^{(m)})^2 + \mu(u_{i,j}^{(m)} + u_{j,i}^{(m)})(u_{i,j}^{(m)} + u_{j,i}^{(m)}) - \rho\omega^2 u_i^{(m)} u_i^{(m)}\right] dV$$
$$= \int\limits_{S_\sigma} p_i^{(m)} f_i^{(m)} dS. \tag{35.8}$$

A characteristic feature of this system of three operator equations is the sign-definiteness of the kernels multiplying the unknown functions λ, μ, ρ. A possible loss of sign-definiteness, depending on the components of the strain tensor, occurs in the case of an incompressible material, where the kernel multiplying λ vanishes identically; the same situation arises in the torsion problem.

The system of nonlinear operator equations is studied on the basis of the operator Newton method. This requires finding the Fréchet derivatives of the main operator. Using a linearization procedure, it was previously possible to construct a linear system of Fredholm integral equations of the first kind with smooth kernels, which allows one to determine corrections to some initial approximation. The initial approximation is usually sought in the class of simple functions – linear or piecewise linear – by minimizing the misfit functional.

$$J_n = \sum_j \int\limits_{S\sigma} \left(f_i\left(x, \omega_j\right) - u_i^{(n)}\left(x, \omega_j\right)\right)^2 dS, \; \omega_j = \omega_- + \left(\omega_+ - \omega_-\right) jN^{-1}, \; j = 0..N. \tag{35.9}$$

Research within the framework of the proposed approach can be used in the analysis of vibrations of a circular cylinder and in solving coefficient inverse problems (CIPs) for general two-dimensional inhomogeneity laws. To date, studies of coefficient inverse problems have been carried out for the following cylindrical structures in the case of one-dimensional inhomogeneity laws of two types:

1. For *radial* inhomogeneity laws in the study of the deformation of a finite cylinder [11, 12], where the corresponding iterative processes are formulated and a num-

ber of problems are solved for an inhomogeneous cylindrical waveguide, where the situation is more complex and involves finding residues for meromorphic functions specified at a set of points [13].

2. For *axial* inhomogeneity laws in the study of vibrations of cylindrical rods of circular cross-section – longitudinal, flexural, and torsional. Note that when the radius-to-length ratio satisfies $\varepsilon = aL^{-1}$, one-dimensional theories can be used for modeling. Note that in this case, the inverse problem separates into three subproblems. From the combined analysis of the problems for longitudinal and flexural vibrations, Young's modulus and density are determined [14, 15], while from the problem for torsional vibrations, the shear modulus $G = \mu$ is found [19].

For a wide class of thin-walled structures, plate and shell models are effective for studying CIPs, as they make it possible to significantly reduce computational costs compared to the full three-dimensional formulation. Inhomogeneity laws in such structures are often averaged over the thickness, which, in general CIPs, leads to the need to identify two-dimensional functions.

In the literature, the Kantorovich method in its extended version [20] is widely employed to reduce three-dimensional problems for thin-walled structures to two dimensions. This approach is based on formulating hypotheses and applying the Lagrange variational principle [21], which leads to a separation of variables. The resulting problems are usually solved iteratively. For example, in [22], a simplified microplate model accounting for size effects was constructed within the framework of gradient elasticity theory. The problem is reduced to a system of ordinary differential equations. Various types of boundary conditions are considered. The resulting ODEs are solved iteratively. Fast convergence is reported. A comparison with literature data is performed. In [23], a model of an FGM plate shaped as a parallelogram, resting on a Winkler foundation and subjected to a uniform load, was constructed. Various boundary condition options are considered. In [24], a dynamic problem for a quartz plate was examined; the extended Kantorovich method was used for its solution, and an accuracy assessment was carried out.

A three-dimensional problem of elasticity theory can be reduced to a two-dimensional one by considering the generalized plane stress state that arises in thin flat plates under the action of static planar loads. Problems described by such simple models sometimes admit analytical solutions in particular cases. This opens up new possibilities for constructing property identification schemes using static tests, without resorting to the procedure of solving integral equations of the first kind. Thus, in the present work, the Kantorovich method is applied to reduce the dimensionality of a two-dimensional problem, which makes it possible to obtain boundary value problems described by systems of ordinary differential equations. Previously, in [16], the problem of tension-compression of a homogeneous rectangle was considered; the Lagrange variational principle was used to construct simplified problems. First- and second-order models were constructed. Constant values of Poisson's ratio and Young's modulus were reconstructed.

Based on the literature review, the idea of applying the Kantorovich method combined with subsequent reconstruction of elastic properties using applied models appears to be new and promising.

35.3 Formulation for the Case of Static Loading

Note that static problems are also quite often used in solving CIPs in elasticity theory. One of the approaches being developed recently is the asymptotic approach, which makes it possible to simplify the formulation and construct semi-analytical solutions based on boundary value problems for ordinary differential operators with constant or variable coefficients [16], depending on the Lamé parameters. We describe the asymptotic approach as applied to solving the direct and inverse problems for an elastic isotropic rectangular plate, in the case of a generalized plane stress state.

Equations (35.1) for $\omega = 0$, the constitutive relations (35.6), and the boundary conditions (35.3) describe the equilibrium problem for an inhomogeneous elastic isotropic body. When passing to the two-dimensional problem within the framework of the generalized plane stress state, the modulus λ is replaced by λ^*, and the dimensionality of the problem is reduced by one. The force and deformation characteristics introduced below will be quantities averaged over the thickness of the plate (integral quantities) in relation to the three-dimensional body.

In formulating the inverse problem in the static formulation, additional information on displacement fields of the type (35.5) is transformed into the following form:

$$u_i(x) = f_i(x), \quad x \in S_f. \tag{35.10}$$

In this case, it is important that the additional information is specified precisely in the region free from loading. In the dynamic case, this leads to the fact that it is not possible to apply the solvability condition and write integral equations for finding the corrections, and in that case the problem is solved by narrowing the search set for the corrections to linear combinations of known functions and constructing a simplified iterative process. In the static case, the main issue will be to determine the entire displacement field based on the measured components on a part of the boundary, for which purpose applied models are constructed.

35.3.1 Formulation of the Direct and Inverse Problems for a Rectangular Domain

Consider the plane problem of elasticity theory for a rectangular domain V with variable properties, where $V = [-a/2 \leq x_1 \leq a/2] \times [0 \leq x_2 \leq L]$. The lower part of the boundary ($x_2 = 0$) is clamped, the lateral parts ($x_1 = \pm a/2$) are stress-free, and a normal load p_2 and a tangential load p_1 act on the upper part of the boundary ($x_2 = L$).

We assume that the parameter $\varepsilon = aL^{-1}$ is small. The governing equations of the problem consist of the equilibrium equations

$$\begin{cases} \sigma_{11,1} + \sigma_{12,2} = 0 \\ \sigma_{21,1} + \sigma_{22,2} = 0 \end{cases} \tag{35.11}$$

and the constitutive relations

$$\begin{aligned}\sigma_{11} &= \lambda^* \left(u_{1,1} + u_{2,2}\right) + 2\mu u_{1,1},\\ \sigma_{12} &= \mu \left(u_{1,2} + u_{2,1}\right),\\ \sigma_{22} &= \lambda^* \left(u_{1,1} + u_{2,2}\right) + 2\mu u_{2,2},\end{aligned} \tag{35.12}$$

where $\lambda^* = \kappa\mu$, and μ are smooth positive functions of the coordinate x_2.

The boundary conditions of the problem are

$$\begin{aligned}u_1 (x_1, 0) = u_2 (x_1, 0) = 0,\\ \sigma_{11} (\pm a/2, x_2) = \sigma_{12} (\pm a/2, x_2) = 0,\\ \sigma_{12} (x_1, L) = p_1 (x_1), \quad \sigma_{22} (x_1, L) = p_2 (x_1).\end{aligned} \tag{35.13}$$

Problem (35.11)–(35.13) can be split into two subproblems: tension/compression (DP1 – Direct problem 1) and bending (DP2 – Direct problem 2). For simplicity, we assume that the loads $p_1(x_1)$ and $p_2(x_1)$ are constant. Moreover, due to the linearity of the direct problem, one may choose a loading amplitude that is convenient for further calculations (even if unrealistically high). Let tension be generated by $p_1(x_1) = 0,\ p_2(x_1) = \mu_*$, and bending by $p_1(x_1) = \mu_*,\ p_2(x_1) = 0$, where μ_* is a characteristic value of the shear modulus.

In accordance with the main idea of simplifying boundary-value problems, we seek the displacement components in the following form:

$$u_1 = x_1 s_2(x_2), \quad u_2 = s_1(x_2) \tag{35.14}$$

or

$$u_1 = w_1(x_2), \quad u_2 = x_1 w_2(x_2). \tag{35.15}$$

Moreover, in DP1, hypotheses of the form (35.14) will be used, and in DP2, hypotheses of the form (35.15). Using the Lagrange variational principle and the Kantorovich method [20], we obtain the equations of the first-order models. When writing them, we will use the dimensionless coordinate $x = x_2 L^{-1} \in [0, 1]$ and the following dimensionless functions: $\tilde{s}_2 (x) = s_2 (x_2)$, $\tilde{s}_1 (x) = s_1 (x_2) L^{-1}$, $\tilde{w}_1 (x) = w_1 (x_2) L^{-1}$, $\tilde{w}_2 (x) = w_2 (x_2)$, $\tilde{\mu} (x) = \mu (x_2) (\mu_*)^{-1}$.

Thus, for example, in the case of DP1 we obtain the boundary value problem

$$\begin{aligned}&\tilde{\mu} \left((\kappa + 2)\, \tilde{s}_1' + \kappa \tilde{s}_2\right) = 1,\\ &\frac{\varepsilon^2}{12} \left(\tilde{\mu}\tilde{s}_2'\right)' - \tilde{\mu} \left(\kappa \tilde{s}_1' + (\kappa + 2)\, \tilde{s}_2\right) = 0,\\ &\left(\tilde{\mu}\tilde{s}_2'\right)\big|_{x=1} = 0, \tilde{s}_1 (0) = \tilde{s}_2 (0) = 0.\end{aligned} \tag{35.16}$$

Furthermore, the boundary value problem (35.16) can be reduced to solving a single second-order equation for $\tilde{s}_2$, after which $\tilde{s}_1$ can be expressed in terms of $\tilde{s}_2$:

$$\begin{aligned}
&\frac{\varepsilon^2}{12}\left(\tilde{\mu}\tilde{s}_2'\right)' - 4\tilde{\mu}\left(\frac{\kappa+1}{\kappa+2}\right)\tilde{s}_2 = \frac{\kappa}{(\kappa+2)},\\
&\left(\tilde{\mu}\tilde{s}_2'\right)\Big|_{x=1} = 0, \quad \tilde{s}_2(0) = 0,\\
&\tilde{s}_1 = \int_0^x \frac{1 - \tilde{\mu}(t)\kappa(t)\tilde{s}_2(t)}{\tilde{\mu}(t)(\kappa(t)+2)}dt.
\end{aligned} \tag{35.17}$$

The boundary value problem for DP2 is formulated in a similar manner:

$$\begin{aligned}
&\tilde{\mu}\left(\tilde{w}_1' + \tilde{w}_2\right) = 1,\\
&\tilde{\mu}\left(\kappa+2\right)\tilde{w}_2' = \frac{12}{\varepsilon^2}\left(x-1\right)\\
&\tilde{w}_1\left(0\right) = \tilde{w}_2\left(0\right) = 0
\end{aligned} \tag{35.18}$$

For an arbitrary inhomogeneity law $\tilde{\mu}(x)$, it is not possible to construct an exact analytical solution to problem (35.16); however, a solution to problem (35.18) can be obtained in the form

$$\begin{aligned}
&\tilde{w}_2 = \frac{12}{\varepsilon^2}\int_0^x \frac{(t-1)}{\tilde{\mu}\left(t\right)\left(\kappa\left(t\right)+2\right)}dt\\
&\tilde{w}_1 = \int_0^x \left(\frac{1}{\tilde{\mu}\left(t\right)} - \tilde{w}_2\left(t\right)\right)dt
\end{aligned} \tag{35.19}$$

Henceforth, we will consider the case $\kappa = \text{const}$, which corresponds to a constant Poisson's ratio ν. We restrict ourselves to constructing an approximate solution to (35.17). Consider the boundary value problem for the function $\tilde{s}_2(x)$ on the interval $x \in [0,1]$ under the assumption $\kappa = \text{const}$:

$$\frac{\varepsilon^2}{12}\frac{d}{dx}\left(\tilde{\mu}(x)\frac{d\tilde{s}_2(x)}{dx}\right) - 4v_1\tilde{\mu}(x)\tilde{s}_2(x) = v_2, \tag{35.20}$$

with the boundary conditions (taking into account that $\tilde{\mu}(x) > 0$)

$$\tilde{s}_2(0) = 0, \quad \tilde{s}_2'(1) = 0, \tag{35.21}$$

where

$$v_1 = \frac{\kappa+1}{\kappa+2}, \qquad v_2 = \frac{\kappa}{\kappa+2}, \qquad \kappa = \frac{2\nu}{1-\nu}.$$

Analyzing problem (35.20) – (35.21), we note that the ordinary differential operator under consideration, with variable coefficients, contains a term with a small parameter multiplying the highest derivative. Consequently, the solution contains a boundary layer term, and its construction can be carried out using the WKB approximation with a boundary layer [25, 26].

We assume that the material inhomogeneity law is given by the function $\tilde{\mu}(x) = P^{-1}(x)$. For $\varepsilon \ll 1$, the solution $\tilde{s}_2^a(x)$ (asymptotic) is sought as a sum of the leading term of the asymptotics $\tilde{s}_2^r(x)$ (regular) and boundary layer solutions in the vicinity of the points $x = 0$ and $x = 1$:

$$\tilde{s}_2^a(x) = \tilde{s}_2^r(x) - \tilde{s}_2^r(0)\, e^{-qx/\varepsilon} - \frac{\varepsilon}{q}\left(\tilde{s}_2^r\right)'\Big|_{x=1} e^{q(x-1)/\varepsilon}, \tag{35.22}$$

where the parameter q is determined by the coefficients of the equation:

$$q = \sqrt{48 \upsilon_1}.$$

Letting the parameter ε tend to zero in (35.20), we find the leading term of the asymptotics:

$$\tilde{s}_2^r(x) = -\frac{\upsilon_2}{4\upsilon_1} P(x). \tag{35.23}$$

Note that in (35.22) the third term is an order of magnitude smaller than the first two and can be neglected. Computational experiments confirm this. Thus, henceforth we will not take into account the boundary layer present in the region close to the loading zone. Expressing $\tilde{\mu}(x)$ in terms of $P(x)$ does not narrow down the class of functions that can be considered when solving the direct problem. However, the WKB solution will be most accurate if $\tilde{\mu}(x)$ varies slowly in the vicinity of the points $x = 0$ and $x = 1$.

Now let us consider solving inverse problems and conduct several computational experiments. We assume that, instead of full-scale tests on the object under study, problem (35.11) – (35.13) is solved using the finite element method (FEM). Let Poisson's ratio be given as $\nu = 0.3$ and the dimensionless shear modulus as $\tilde{\mu}(x)$. Two types of probing loads are applied. For $p_1(x_1) = 0,\ p_2(x_1) = \mu_*$, the displacements u_1 are measured on the boundary $x_1 = a/2$ and denoted by $2u_1(a/2, xL)a^{-1} = f_1(x)$. For $p_1(x_1) = \mu_*,\ p_2(x_1) = 0$, the displacements u_2 are similarly measured on the same boundary and denoted by $2u_2(a/2, xL)a^{-1} = f_2(x)$. Based on (35.14) and (35.15), we obtain a connection with the first-order applied models:

$$\tilde{s}_2(x) \approx f_1(x), \quad \tilde{w}_2(x) \approx f_2(x). \tag{35.24}$$

Information of the form (35.24) allows the reconstruction of $\tilde{\mu}(x)$.

In the case of DP1, we have the approximate equality $\tilde{s}_2^a(x) \approx \tilde{s}_2(x) \approx f_1(x)$. Let $P(x)$ be a polynomial of degree N with coefficients b_j, then $\tilde{s}_2^a(x)$ will be a linear combination of the b_j. From the expression $\tilde{s}_2^a(x) = f_1(x)$, we find the unknown coefficients using the least squares method by introducing a functional analogous to (35.9):

$$\sum_{j=1}^{M-1} \left(f_1(x_j) - \tilde{s}_2^a(x_j)\right)^2 \to \min, \quad x_j = jM^{-1}, \quad N < M. \tag{35.25}$$

In the case of DP2, the approximation $\tilde{\mu}(x) = P^{-1}(x)$ is also convenient. From (35.19), taking into account $\kappa = \text{const}$, we have:

$$\tilde{w}_2(x) = \frac{12}{\varepsilon^2(\kappa+2)} \int_0^x P(t)(t-1)\,dt. \tag{35.26}$$

When $P(x)$ is a polynomial, we obtain an explicit form of the function $\tilde{w}_2(x)$, whose unknown coefficients are easily found from (35.24) using the collocation method or the least squares method.

For example, with $P(x) = b_0 x + b_1$, we obtain

$$\tilde{w}_2 = \frac{(4x-6)\,x^2 b_0 + (6x-12)\,x b_1}{\varepsilon^2\,(\kappa+2)},$$

$$\sum_{j=1}^{M-1} \left(f_2\left(x_j\right) - \tilde{w}_2\left(x_j\right)\right)^2 \to \min, \quad x_j = jM^{-1}, \quad N < M. \tag{35.27}$$

From the minimization condition of functional (35.27), we obtain a system of linear algebraic equations (SLAE) for the coefficients b_j.

Thus, both problems reduce to solving a SLAE for the coefficients of the polynomial $P(x)$. Of particular interest are the cases where $P(x)$ is a first- or second-order polynomial, as this yields fairly simple relations that may find engineering applications. Note that any linear combination of a priori known functions can be taken as the function $P(x)$. In that case, the inverse problem also reduces to determining the coefficients of the linear combination. The main requirement is its positivity.

35.3.2 Computational Experiments

Let us consider DP1. For $\nu = 0.3$ we have $v_1 = 0.65$, $v_2 = 0.3$. Let $\varepsilon = 0.2$ and $P(x) = b_0 x + b_1$, with $b_0 = 0.2$, $b_1 = 0.9$. We compare the WKB solution and the numerical solution $\tilde{s}_2 = \tilde{s}_2^n$ of problem (35.17), which is constructed using the shooting method. The conducted series of computational experiments showed that $\tilde{s}_2^n$ and $\tilde{s}_2^a$ are sufficiently close (one example of the implementation is shown in Fig. 35.1a). Verification of solution (35.26) of problem (35.18) was also carried out using the shooting method.

After verifying the solutions for the applied models, we proceed to a comparison with more accurate (in terms of describing the deformation of real objects) FE solutions. Figure 35.1b compares the analytical solution (35.26) and the FE solution of DP2. The largest discrepancy is observed at the boundary $x = 1$ and reaches 10%. The calculations were performed for $\tilde{\mu}(x) = 2 - x^2$. In the case of DP1, comparison with FEM revealed discrepancies that, for various inhomogeneity laws, do not exceed 5%.

Since the difference between $\tilde{s}_2^n$ and the *conditionally exact* FEM solution is larger than the difference between $\tilde{s}_2^n$ and $\tilde{s}_2^a$, one can use the WKB approximation with a boundary layer on an equal footing with the analytical solution (35.26). Moreover,

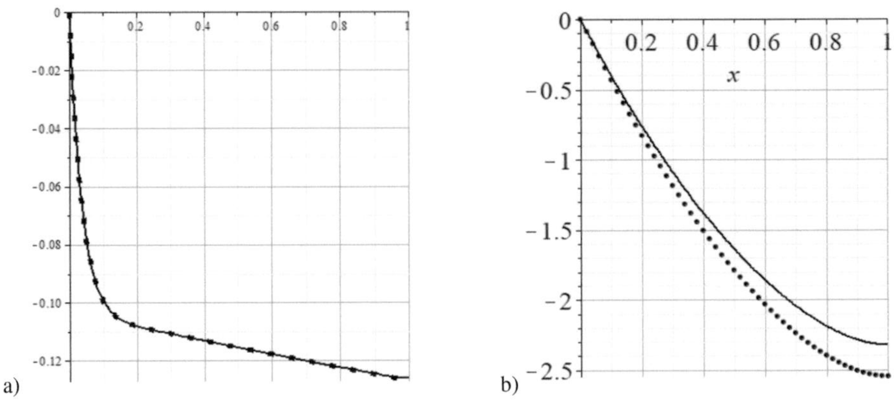

Fig. 35.1: Left – Comparison of the numerical solution $\tilde{s}_2^n$ (solid line) and the asymptotic WKB approximation $\tilde{s}_2^a$ (dots); right – Comparison of the analytical solution (35.26) (solid line) and the FE solution in the case of DP2 (dots).

there is reason to believe that the errors that will arise in the reconstruction from data obtained using FEM are precisely related to the insufficient accuracy of the first-order models.

35.3.3 Inverse Problem 1 (Data from the Direct Problem (35.17))

We also set $b_0 = 0.2$, $b_1 = 0.9$. Solving problem (35.17) yields the numerical solution $\tilde{s}_2 = \tilde{s}_2^n$. The parameters b_0, b_1 are then found from the condition that $\tilde{s}_2^a$ coincides with $\tilde{s}_2^n$ at the points $x = 0.2$ and $x = 0.4$ (collocation method). From the system $\tilde{s}_2^a(0.2) = \tilde{s}_2^n(0.2)$, $\tilde{s}_2^a(0.4) = \tilde{s}_2^n(0.4)$ we obtain $b_0 = 0.20038$, $b_1 = 0.89979$. Note that the reconstruction error in the two parameters is less than 1% in determining the two parameters.

Let us consider a stronger inhomogeneity. Setting $b_0 = 0.9$, $b_1 = 0.6$, $\varepsilon = 0.3$ gives $b_0 = 0.912$, $b_1 = 0.592$, again very close to the exact values; the error is less than 2%.

35.3.4 Inverse Problem 2 (Data from the FEM)

Inverse Problem 1 is a model problem and allows the reconstruction of functions of the form $\tilde{\mu}(x) = 1/(b_0 x^N + \ldots + b_{N-1} x + b_N)$ with an error of less than 2%. This result motivates us to proceed to reconstruction using additional information taken from an FEM experiment. The results of computational experiments also showed good agreement when reconstructing via (35.22) and (35.25). We consider the process of reconstructing functions of general form (monotonic, non-monotonic).

Figure 35.2 shows the results of reconstructing the monotonically decreasing function $\tilde{\mu}(x) = 2 - x^2$ in the class of functions $\tilde{\mu}(x) = P^{-1}(x)$, where

$$P(x) = b_0 x^2 + b_1 x + b_2.$$

Using the displacement data from DP1, we obtained $b_0 = 0.557, b_1 = -0.297$, $b_2 = 0.647$, while using the data $\tilde{w}_2(x) \approx f_2(x)$ from (35.27), we obtained $b_0 = 0.560, b_1 = -0.103, b_2 = 0.494$. Figure 35.3 shows the results of reconstructing the monotonically increasing function $\tilde{\mu}(x) = 0.6 + x^2$. The maximum error is observed at the right end. Figure 35.4 presents the results of reconstructing the non-monotonic function

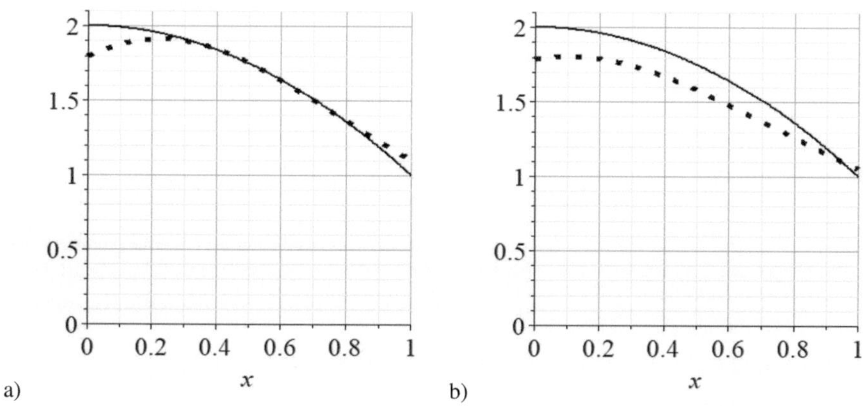

Fig. 35.2: The exact function $\tilde{\mu}(x)$ (solid line) and the result of its reconstruction (dots). Left – DP1, right – DP2.

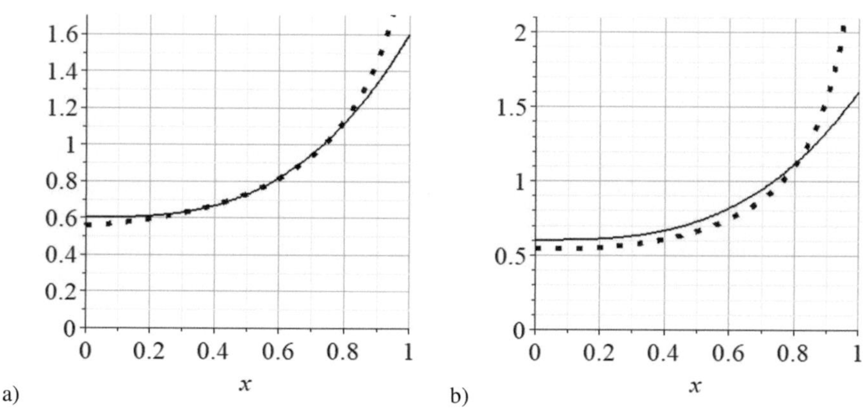

Fig. 35.3: The exact function $\tilde{\mu}(x)$ (solid line) and the result of its reconstruction (dots). Left – DP1, right – DP2.

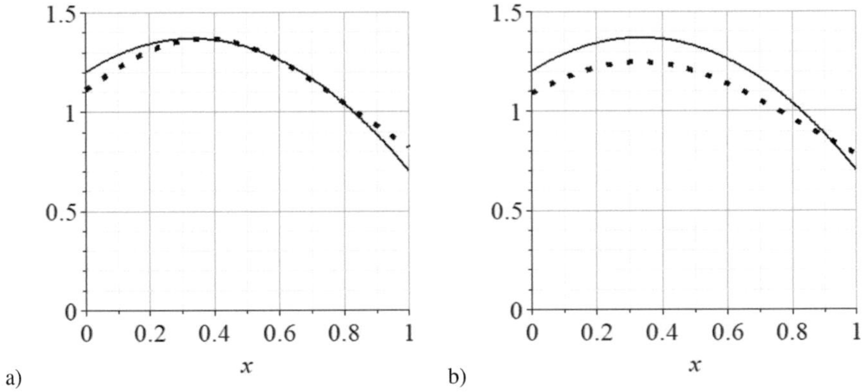

Fig. 35.4: The exact function $\tilde{\mu}(x)$ (solid line) and the result of its reconstruction (dots). Left – DP1, right – DP2.

$$\tilde{\mu}(x) = 1.2 + x - 1.5x^2.$$

The largest differences between the exact and the reconstructed function occur at the ends of the interval.

When using the bending model DP2 (35.18) under the action of a tangential load (Figs. 35.2b–35.4b), the reconstruction is always worse than when using model DP1 (35.16) (Figs. 35.2a–35.4a). The results turn out to be underestimated by about 10% on average. The reason is that the solution of the direct problem within the framework of model DP2 differs from the FE solution by an amount of the order of 10%. This is because model (35.18) is analogous to the Timoshenko beam bending model, in which the shear correction factor is absent [27].

35.4 Conclusion

Various formulations of coefficient inverse problems are analyzed. A model two-dimensional problem for a rectangular domain is considered. Using the WKB method, an asymptotic solution to the problem under a normal load is constructed, allowing for the boundary layer in the vicinity of the clamped part of the boundary. For the case of shear loading, an analytical solution is derived. For both types of loading, a scheme for solving the inverse problem of reconstructing the shear modulus is implemented. An approximation of the desired function is performed, and the unknown coefficients are found by solving a system of linear algebraic equations using the least-squares method.

Acknowledgements The study was supported by the grant of the Russian Science Foundation No. 22-11-00265-P, https://rscf.ru/project/22-11-00265/ at the Southern Federal University.

References

[1] Tikhonov AN, Arsenin VYa (1986) Metody resheniya nekorrektnykh zadach (in Russ., Methods for Solving Ill-Posed Problems). Nauka, Moscow

[2] Gladwell GML (2008) Obratnye zadachi teorii kolebaniy (in Russ., Inverse Problems in Vibration Theory). Regulyarnaya i khaoticheskaya dinamika, Institut komp'yuternykh issledovaniy, Moscow-Izhevsk

[3] Bal G (2012) Introduction to Inverse Problems. Columbia University, New York

[4] Neto FDM, Neto AJS (2013) An Introduction to Inverse Problems with Applications. Springer, Berlin, DOI 10.1007/978-3-642-32557-1

[5] Kabanikhin SI (2018) Obratnye i nekorrektnye zadachi (in Russ., Inverse and Ill-Posed Problems). Izdatel'stvo SO RAN, Novosibirsk

[6] Tikhonov AN, Goncharsky AV, Stepanov VV, Yagola AG (1990) Chislennye metody resheniya nekorrektnykh zadach (in Russ., Numerical Methods for Solving Ill-Posed Problems). Nauka, Moscow

[7] Samarsky AA, Vabishchevich PN (2004) Chislennye metody resheniya obratnykh zadach matematicheskoy fiziki (in Russ., Numerical Methods for Solving Inverse Problems of Mathematical Physics). Editorial URSS, Moscow

[8] Vatulyan AO (2019) Koeffitsientnye obratnye zadachi mekhaniki (in Russ., Coefficient Inverse Problems of Mechanics). Fizmatlit, Moscow

[9] Hutson V, Pym JS (1980) Applications of functional analysis and operator theory. Mathematics in Science and Engineering, Vol 146. Academic Press, London-New York

[10] Perkowski Z, Czabak M (2019) Description of behaviour of timber-concrete composite beams including interlayer slip, uplift, and long-term effects: Formulation of the model and coefficient inverse problem. Engineering Structures 194:230–250, DOI 10.1016/j.engstruct.2019.05.058

[11] Dudarev VV, Vatulyan AO, Mnukhin RM, Nedin RD (2020) Concerning an approach to identifying the Lamé parameters of an elastic functionally graded cylinder. Mathematical Methods in the Applied Sciences 43(11):6861-6870, DOI 10.1002/mma.6428

[12] Vatulyan AO, Dudarev VV, Mnukhin RM (2021) Identification of characteristics of a functionally graded isotropic cylinder. International Journal of Mechanics and Materials in Design 17(2):321-332, DOI 10.1007/s10999-020-09527-5

[13] Vatulyan AO, Yurov VO (2021) On the reconstruction of material properties of a radially inhomogeneous cylindrical waveguide. Mathematical Methods in the Applied Sciences 44(6):4756-4769, DOI 10.1002/mma.7067

[14] Vatulyan AO, Yurov VO (2020) On the Determination of the Mechanical Characteristics of Rod Elements Made of Functionally Graded Materials. Mechanics of Solids 55(6):907-917, DOI 10.3103/S0025654420660036

[15] Vatulyan AO, Yurov VO (2023) On the estimation of the sensitivity of the coefficients of models for inhomogeneous solids. Mechanics of Solids 58(3):793-801, DOI 10.3103/S0025654422601768

[16] Vatulyan AO, Yurov VO (2025) Ob odnoy postanovke obratnykh koefficientnykh zadach dlya uprugikh tel i metodakh ee issledovaniya (in Russ., On one formulation of inverse coefficient problems for elastic bodies and methods of their study). Problems of Strength and Plasticity 87(4):483-493, DOI 10.32326/1814-9146-2025-87-4-483-493
[17] Vatulyan AO, Yurov VO, Gusakov IV (2025) On the Applied Theory of Rectangle Stretching. Mechanics of Solids 60(6):4424-4433, DOI 10.1134/S0025654425603052
[18] Lekhnitsky SG (1977) Teoriya uprugosti anizotropnogo tela (in Russ., Theory of Elasticity of an Anisotropic Body). Nauka, Moscow
[19] Vatulyan AO, Plotnikov DK (2019) Obratnye koefficientnye zadachi v mekhanike (in Russ., Inverse coefficient problems in mechanics). PNRPU Mechanics Bulletin (3):37–47, DOI 10.15593/perm.mech/2019.3.04
[20] Kantorovich LV, Krylov VI (1962) Priblizhonnyye metody vysshego analiza (in Russ., Approximate Methods of Higher Analysis). Fizmatgiz, Leningrad
[21] Mikhlin SG (1962) Variatsionnyye metody v matematicheskoy fizike (in Russ., Variational Methods in Mathematical Physics). Fizmatgiz, Moscow
[22] Movassagh AA, Mahmoodi MJ (2013) A micro-scale modeling of Kirchhoff plate based on modified strain-gradient elasticity theory. European Journal of Mechanics - A/Solids 40:50-59, doi10.1016/j.euromechsol.2012.12.008
[23] Joodaky A, Joodaky I, Hedayati M, Masoomi R, Farahani EB (2013) Deflection and stress analysis of thin FGM skew plates on Winkler foundation with various boundary conditions using extended Kantorovich method. Composites Part B: Engineering 51:191-196, DOI 10.1016/j.compositesb.2013.03.010
[24] Liu B, Xing YF, Eisenberger M, Ferreira AJM (2014) Thickness-shear vibration analysis of rectangular quartz plates by a numerical extended Kantorovich method. Composite Structures 107:429-435, DOI 10.1016/j.compstruct.2013.08.021
[25] Stoer J, Bulirsch R (2002) Introduction to Numerical Analysis. Springer, New York, DOI 10.1007/978-0-387-21738-3
[26] Maslov VP (1977) Kompleksnyy metod VKB v nelineynykh uravneniyakh (in Russ., Complex WKB Method in Nonlinear Equations). Nauka, Moscow
[27] Filippov AP (1970) Kolebaniya deformiruemykh sistem (in Russ., Vibrations of Deformable Systems). Mashinostroenie, Moscow
[28] Yakhno VG (1990) Obratnye koeffitsientnye zadachi dlya differentsial'nykh uravneniy uprugosti (in Russ., Inverse Coefficient Problems for Differential Equations of Elasticity). Nauka, Novosibirsk

Chapter 36
Evanescent Lamb Waves and Their Complex Wavenumbers' Spectrum in Layered Plates

Dmitry Zakharov and Anatolij Nikonov

Abstract A new method that combines asymptotic and iterative approaches is proposed to study the total wavenumber spectrum of a plate composed of isotropic layers. In this work, propagator matrices are employed to derive the dispersion equation and its static limit in closed form using symbolic calculations. The static roots, analyzed in the first step, play a crucial role in the development of the method. Based on these roots, low-frequency approximations of the dispersion curves are obtained. The proposed approach is designed to evaluate dispersion curves of arbitrary order n and to represent them in closed form through asymptotic expressions. In particular, the method is well suited for the case of large n, which typically poses significant challenges for purely numerical approaches. Building on the presented approach, the wavenumber spectrum is evaluated. First, for the nth complex dispersion curve, the asymptotic behavior of the static wavenumber is examined as $n \rightarrow +\infty$. Two different approaches for deriving this asymptotic expression are compared. Then, for arbitrary n, the exact static value is computed using an iterative procedure, with the asymptotic result serving as the initial approximation. A parametric analysis reveals the existence of a specific (critical) value of a geometrical parameter at which a redistribution of the wavenumbers occurs. Next, low-frequency asymptotic expressions for the dispersion curves are obtained. These exhibit a flat initial segment for each complex dispersion curve, whose length increases with the curve index n. As a result, this part of the spectrum can be described using simple analytical formulas. The exact dispersion curves are subsequently evaluated using a separate iterative algorithm to refine the approximations. In both stages, the numerical results show good agreement with the asymptotic predictions.

Dmitry Zakharov (corresponding author)
Institute of Computer Aided Design, Russian Academy of Sciences, 2nd Brestskaya Str, Moscow, 123056, Russian Federation, e-mail: dmitrii.zakharov@gmail.com

Anatolij Nikonov
Faculty of Mechanical Engineering, University of Ljubljana, Aškerčeva 6, 1000 Ljubljana, Slovenia, e-mail: anatolij.nikonov@gmail.com

H. Altenbach et al. (eds.), *Stability Problems in Mechanics of Continua*,
Advanced Structured Materials 260, https://doi.org/10.1007/978-3-032-31032-3_36

Key words: Lamb waves, Structural acoustics, Complex dispersion curves, Asymptotics

36.1 Introduction

The use of eigenwaves for solving a wide range of problems in structural members has become increasingly common and now extends to progressively more complex composite structures. This is particularly relevant to structural health monitoring and non-destructive evaluation of solids, which can be modeled as elastic or viscoelastic waveguides. It is well known that, at a given frequency, an elastic waveguide with a finite cross-sectional size may possess a finite number of real wavenumbers and an infinite (countable) set of complex wavenumbers. In the viscoelastic case, all wavenumbers are complex. However, the problem of determining the complete complex spectrum has been fully resolved only for a limited number of cases, primarily involving simple isotropic and homogeneous waveguides such as plates or circular cylindrical solids. The present chapter focuses mainly on Lamb waves in layered plates.

In [31, 32], the appropriate analytical tools were first introduced, and the resulting solutions demonstrated consistency with earlier static analyses [8, 66]. With regard to complex dispersion curves, it has been shown that the higher is the curve index n, the wider is the frequency interval over which the curve remains nearly constant and the wavenumber k is close to its static value [28, 29]. These static values can be efficiently approximated by asymptotic expressions as $n \to +\infty$ or $|k| \to +\infty$ and subsequently computed exactly for any finite index n. A similar behavior of complex wavenumbers has also been observed for circular rods [9, 64, 66].

For other types of waveguides, which may exhibit layered configurations and/or material anisotropy, comparable results have not yet been obtained. At present, numerical methods are commonly employed to compute real, imaginary, and selected complex branches of dispersion curves for isotropic, anisotropic and/or layered plates within a limited frequency range. A major limitation of these approaches is their fundamental inability to describe the complete infinite spectrum, characterized by n. While, for many engineering applications, information about only the few wavenumbers may be sufficient, this is not the case for all problems, nor does it address the lack of fundamental understanding in this area.

For example, the expansion of stress fields into modal series using the complete complex spectrum provides a natural and rigorous framework for analyzing wave interaction with edges, wave diffraction by inhomogeneities (see, e.g., [7, 11, 14, 36, 37, 41, 42]), and the implementation of non-reflecting boundary conditions to reduce the computational domain size [55]. Moreover, access to the complete spectrum enables the efficient use of the mode orthogonality relations for both two-dimensional waves [3, 5, 12, 34, 43] and three-dimensional waves, originally introduced in [48, 54, 58] and partially revisited in [24]. It should be noted that three-dimensional wave propagation in an infinite plate can be described using the spectra of two-dimensional waves [48, 54, 58], thereby extending the applicability of these results.

With regard to existing approaches, attention can be drawn to methods that combine direct numerical techniques with variational formulations (see the comprehensive reviews in [16, 17]), [30], as well as to approaches based on modal expansion and the collocation method [6].

For multilayered media, the most widely used analytical approach, as proposed by Thompson and Haskell [15, 45], is based on direct and recursive formulations of propagator matrices. These matrices relate the amplitudes of waves in adjacent layers and allow the calculation of reflection and transmission coefficients, as well as dispersion relations.

An alternative, numerically more stable approach based on the determinant decomposition method was introduced and later adapted for inelastic media by Knopoff and Schwab [21, 44]. Exact and approximate analyses of propagator matrices, developed by Lowe [25], are now frequently combined with direct discretization of the waveguide across its thickness, such as in the SAFE method (see reviews in [27, 35]) or the scaled boundary finite element method [13, 22]. A generalization of the spectral collocation method implemented in DISPERSE [25] for viscoelastic and anisotropic layered waveguides is described in [18–20].

Furthermore, a stable recursive algorithm for multilayered media, known as the recursive asymptotic stiffness matrix method, has been extended to account for piezoelectric and functionally graded materials [38–40, 47].

A method that avoids the need to precompute wavenumbers within a limited frequency range is presented in [33]. This approach projects the system of differential equations governing Lamb waves onto a specially chosen basis of orthogonal functions. Unlike the standard method, it does not require solving the transcendental dispersion equation; instead, it computes the solutions as a classical eigenvalue problem for a projection matrix derived from the previous step. The method enables the approximation of complex branches of the dispersion curves, with its accuracy and the required matrix dimension depending on the number of complex branches: the greater the number of branches, the more basis functions are retained. Results for the symmetrical modes are demonstrated for a single isotropic layer. A generalization of this classical analytical and asymptotic framework to linearly viscoelastic isotropic layers is presented in [57].

Our investigation has several objectives: to propose a complementary analytical tool to existing methods for obtaining the complete spectrum of wavenumbers in isotropic layered solids and to clarify its regular behavior; to demonstrate the operation of this tool using a relatively simple model problem, i.e., to evaluate dispersion curves of any order n and describe them using a relatively simple asymptotic formula - particularly for large n - along with estimates of the error and the interval of validity. Finally, we would like to highlight the key features of this method for potential application to other structural members.

First, we introduce a new method for a layered plate in the low-frequency range and describe our approach. It is well known that, over any finite frequency interval, the spectrum of wavenumbers $\{k\}$ in an isotropically layered plate is infinite. For SH waves, determining the spectrum is relatively straightforward, as it can be reduced to an problem with a real spectrum.

However, for waves with P- and SV- polarization, the situation is more complicated: there exists a finite number of real dispersion curves, a finite number of purely imaginary ones, and an infinite (countable) set of complex-valued curves, with $\Re(k)$ and $\Im(k)$ increasing to infinity together with the curve index n. The Lamb waves provide a classical example of this case.

As a model problem, we consider the propagation of antisymmmetrical and symmetrical modes in a three-layered plate. The choice of a symmetrical layup is illustrative and does not represent a general limitation.

The limiting behavior of the wavenumbers in the static case plays a central role. To this end, we first derive the explicit form of the dispersion equation and its static limit using propagator matrices. We then examine the asymptotic behavior of the wavenumbers, first in statics and subsequently in the low-frequency dynamic regime.

Next, we illustrate how the complete complex spectrum is organized and proposing improvements to the iterative calculation method. It should be noted that we do not consider interactions between dispersion curves, which occur at higher frequencies; rather, we focus on the initial segments of the curves. Our analysis shows that the length of these "initial" segments increases with the order of the curve. The key advantage of this result is that the larger the modulus $|k|$, the smaller correction is required in the asymptotic solution of the dispersion equation.

This infinite subset of the spectrum poses a natural challenge for existing numerical methods. In the case of a thin coating, approximate corrections to the dispersion curves of the core can be applied [39, 40, 52, 53, 56], but the overall effect of the coating is generally more significant. Direct numerical treatment of this infinite subset requires increasingly finer meshes, more finite elements, additional terms in the approximation basis [28, 33], or other strategies that result in a rapid growth of computational dimension.

Our alternative approach is to come from infinity using asymptotic methods. These methods have proven effective for describing Lamb waves in single layers [28, 29, 57]. For three-layered plates, an analytical and iterative method along these lines was proposed in [61, 62]. In the following, we present its analytical and numerical results and discuss potential extensions to other composite structural members.

36.2 Mathematical Foundations

In what follows, we consider a plate made of N linearly elastic layers where each jth layer occupies a region $-\infty < x, y < +\infty$, $z_j \le z \le z_{j+1}$ $(j = 1, 2, \ldots, N)$ in Cartesian coordinates x, y, z. The plate configuration is shown in Fig. 36.1. Hence, the layer thickness is $h_j = z_{j+1} - z_j$. The physical properties are characterized by the Lamé constants λ_j, μ_j and mass density ρ_j.

We consider waves propagating according to the exponential form $\exp(ikx - i\omega t)$, where k denotes the wavenumber, ω the angular frequency, and t time. In cases where

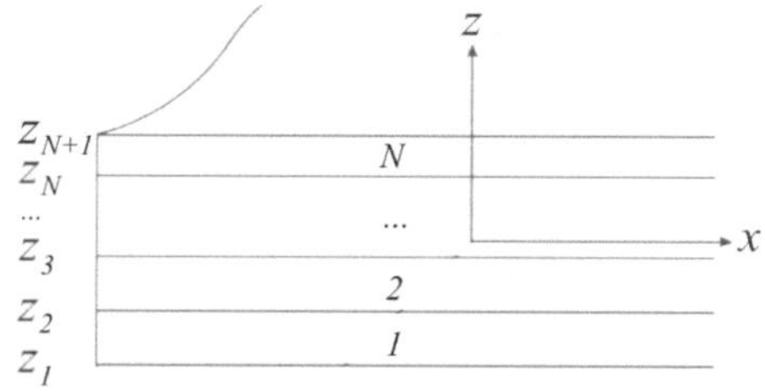

Fig. 36.1 Configuration of laminate.

no ambiguity arises, this factor, together with the layer index, may be omitted for simplicity. The wave is assumed to exhibit P- and SV-type polarization, such that the displacement vector is given by $\mathbf{u} = (u_x, 0, u_z)$, depending only on the spatial coordinates x and z.

The respective stresses satisfy the Hooke law

$$\sigma_{xx} = \lambda \operatorname{div} \mathbf{u} + 2\mu\partial_x u_x \qquad \sigma_{xz} = \mu(\partial_z u_x + \partial_x u_z) \quad (x \leftrightarrow z), \tag{36.1}$$

the tensor equation of motion

$$\operatorname{div} \boldsymbol{\sigma} + \rho\omega^2 \mathbf{u} = 0\,, \tag{36.2}$$

with the stress tensor $\boldsymbol{\sigma}$ under the continuity conditions of the displacements and stresses on each interface

$$[\sigma_{xz}] = 0,\ [\sigma_{zz}] = 0,\ [u_x] = 0,\ [u_z] = 0\,. \tag{36.3}$$

The upper and lower faces of the plate are supposed to be stress free

$$\sigma_{xz} = \sigma_{zz} = 0. \tag{36.4}$$

Remark 36.1. The conditions (36.4) can be also reformulated in the form for rigidly clamped faces

$$u_x = u_z = 0. \tag{36.5}$$

or for the combination where one face is clamped and another is stress free. It is also possible that on the face the displacement in one direction equals zero and in the complimentary direction the stress is zero

$$\sigma_{xz} = 0, \quad u_z = 0 \quad \text{or} \quad \sigma_{zz} = 0, \quad u_x = 0. \tag{36.6}$$

All possible combinations of homogeneous boundary conditions ensure zero power flow through the faces and complete energy reflection at each face within the plate. The approach described below can be applied in an analogous manner.

Our subject is the spectrum of wavenumbers $\{k\}$, i.e., the generalized Lamb waves in a multilayer plate [1, 4, 46]. The main attention is focused on the infinite set of complex dispersion curves $k(\omega)$ with $\operatorname{Re} k(\omega), \operatorname{Im} k(\omega) \neq 0$.

36.3 Matrix Relations for a Single Layer and Their Asymptotics

The relations in a single layer can be conveniently rewritten in the explicit form using propagator matrix. Let us denote the coordinates of the layer faces as $z^+ = z_2$, $z^- = z_1$, and $h = z_2 - z_1$.

Introduce the 4×1 motion-stress vector [2]

$$\boldsymbol{\xi} = \left(-ik^{-1}u_x,\ k^{-1}u_z,\ k^{-2}\sigma_{zz},\ -ik^{-2}\sigma_{zx}\right)^{\mathrm{T}},$$

containing the scaled displacements and stress components, and the following diagonal matrix

$$\boldsymbol{\Lambda} = \operatorname{diag}\left[E_P^-,\ E_P^+,\ E_S^-,\ E_S^+\right].$$

Here

$$E_P^{\mp} = \mathrm{e}^{\mp\gamma_P kh}, \qquad E_S^{\mp} = \mathrm{e}^{\mp\gamma_S kh},$$

$$\gamma_{P,S} = \left(1 - k_{P,S}^2 k^{-2}\right)^{1/2}, \qquad \gamma = 2k^2 k_S^{-2} = 2c^{-2}c_S^2, \qquad \gamma_1 = \gamma - 1$$

are dimensionless quantities,

$$c = \frac{\omega}{k}$$

is a phase speed and

$$k_{P,S} = \frac{\omega}{c_{P,S}}$$

are wavenumbers, corresponding to bulk waves the speed of which are

$$c_P = \left(\frac{\lambda + 2\mu}{\rho}\right)^{\frac{1}{2}}$$

and

$$c_S = \left(\frac{\mu}{\rho}\right)^{\frac{1}{2}},$$

respectively. Let us also mention the additional simplifications, such as

$$\begin{aligned}
&\rho\gamma c^2 = \rho 2k^2 k_S^{-2}\omega^2 k^{-2} = 2\mu,\\
&\rho\gamma_1 c^2 = \rho(\gamma - 1)c^2 = 2\mu - \mu k_S^2 k^{-2} = 2\mu\kappa,\\
&\kappa = 2^{-1}(2 - k_S^2 k^{-2}) = \gamma_1\gamma^{-1},\\
&-(2\rho c^2)^{-1} = -k^2(2\rho\omega^2)^{-1} = -k^2(2\mu k_S^2)^{-1} = -\gamma(4\mu)^{-1}.
\end{aligned}$$

The values of vector $\boldsymbol{\xi}$ on the layer faces are related using the propagator matrix $\mathbf{P}_\xi(h)$ as follows. The components of the vector $\boldsymbol{\xi}$ at the layer faces are related through the propagator matrix $\mathbf{P}_\xi(h)$ as follows:

$$\boldsymbol{\xi}(z_2) = \mathbf{P}_\xi(h) \times \boldsymbol{\xi}(z_1), \tag{36.7}$$

$$\mathbf{P}_\xi(h) = \mathbf{M}_\xi \times \boldsymbol{\Lambda} \times \mathbf{M}_\xi^{-1}. \tag{36.8}$$

with the explicit expressions for $\mathbf{M}_\xi$ and $\mathbf{M}_\xi^{-1}$ presented below

$$\mathbf{M}_\xi = \begin{bmatrix} 1 & 1 & -\gamma_S & \gamma_S \\ -\gamma_P & \gamma_P & 1 & 1 \\ 2\mu\kappa & 2\mu\kappa & -2\mu\gamma_S & 2\mu\gamma_S \\ -2\mu\gamma_P & 2\mu\gamma_P & 2\mu\kappa & 2\mu\kappa \end{bmatrix},$$

$$\mathbf{M}_\xi^{-1} = \begin{bmatrix} \gamma 2^{-1} & \gamma_1(2\gamma_P)^{-1} & -\gamma(4\mu)^{-1} & -\gamma(4\mu\gamma_P)^{-1} \\ \gamma 2^{-1} & -\gamma_1(2\gamma_P)^{-1} & -\gamma(4\mu)^{-1} & \gamma(4\mu\gamma_P)^{-1} \\ \gamma_1(2\gamma_S)^{-1} & \gamma 2^{-1} & -\gamma(4\mu\gamma_S)^{-1} & -\gamma(4\mu)^{-1} \\ -\gamma_1(2\gamma_S)^{-1} & \gamma 2^{-1} & \gamma(4\mu\gamma_S)^{-1} & -\gamma(4\mu)^{-1} \end{bmatrix}.$$

The obtained matrices correspond to those presented in [2, 65] but differ due to the scaling introduced for the components of vector $\boldsymbol{\xi}$. The order of these components corresponds to the waves with P- and SV-polarization.

If we single out the exponential term $\exp(kh)$ the matrix $\mathbf{\Lambda}(h)$ yields in the explicit formula

$$\mathbf{\Lambda}(h) = e\mathbf{E}_+ + e^{-1}\mathbf{E}_-, \tag{36.9}$$

where

$$\mathbf{E}_+ = \text{diag}\left[0 \;\; F_P^+ \;\; 0 \;\; F_S^+\right], \qquad \mathbf{E}_- = \text{diag}\left[F_P^- \;\; 0 \;\; F_S^- \;\; 0\right],$$

$$e = \mathrm{e}^{kh}, \qquad F_{P,S}^{\pm} = \mathrm{e}^{\pm(\gamma_{P,S}-1)kh}.$$

Thus, the propagator matrix $\mathbf{P}_\xi(h)$ in the formula (36.8) can be represented in the following form:

$$\mathbf{P}_\xi = e\mathbf{P}_+ + e^{-1}\mathbf{P}_-, \qquad \mathbf{P}_\pm = \mathbf{M}_\xi \times \mathbf{E}_\pm \times \mathbf{M}_\xi^{-1}, \tag{36.10}$$

where

$$\mathbf{P}_+ =$$

$$\begin{bmatrix} \frac{1}{2}(\gamma F_P^+ - \gamma_1 F_S^+) & \frac{1}{2}\left(\gamma\gamma_S F_S^+ - \frac{\gamma_1 F_P^+}{\gamma_P}\right) & \frac{\gamma}{4\mu}(F_S^+ - F_P^+) & \frac{\gamma}{4\mu}\left(\frac{F_P^+}{\gamma_P} - \gamma_S F_S^+\right) \\ \frac{1}{2}\left(\gamma\gamma_P F_P^+ - \frac{\gamma_1 F_S^+}{\gamma_S}\right) & \frac{1}{2}(\gamma F_S^+ - \gamma_1 F_P^+) & \frac{\gamma}{4\mu}\left(\frac{F_S^+}{\gamma_S} - \gamma_P F_P^+\right) & \frac{\gamma}{4\mu}(F_P^+ - F_S^+) \\ \mu(\kappa\gamma F_P^+ - \gamma_1 F_S^+) & \mu\left(\gamma\gamma_S F_S^+ - \frac{\kappa\gamma_1 F_P^+}{\gamma_P}\right) & \frac{\gamma}{2}(F_S^+ - \kappa F_P^+) & \frac{\gamma}{2}\left(\frac{\kappa F_P^+}{\gamma_P} - \gamma_S F_S^+\right) \\ \mu\left(\gamma\gamma_P F_P^+ - \frac{\kappa\gamma_1 F_S^+}{\gamma_S}\right) & \mu(\kappa\gamma F_S^+ - \gamma_1 F_P^+) & \frac{\gamma}{2}\left(\frac{\kappa F_S^+}{\gamma_S} - \gamma_P F_P^+\right) & \frac{\gamma}{2}(F_P^+ - \kappa F_S^+) \end{bmatrix},$$

and

$$\mathbf{P}_- =$$
$$\begin{bmatrix} \frac{1}{2}(\gamma F_P^- - \gamma_1 F_S^-) & \frac{1}{2}\left(\frac{\gamma_1 F_P^-}{\gamma_P} - \gamma\gamma_S F_S^-\right) & \frac{\gamma}{4\mu}(F_S^- - F_P^-) & \frac{\gamma}{4\mu}\left(\gamma_S F_S^- - \frac{F_P^-}{\gamma_P}\right) \\ \frac{1}{2}\left(\frac{\gamma_1 F_S^-}{\gamma_S} - \gamma\gamma_P F_P^-\right) & \frac{1}{2}(\gamma F_S^- - \gamma_1 F_P^-) & \frac{\gamma}{4\mu}\left(\gamma_P F_P^- - \frac{F_S^-}{\gamma_S}\right) & \frac{\gamma}{4\mu}(F_P^- - F_S^-) \\ \mu(\kappa\gamma F_P^- - \gamma_1 F_S^-) & \mu\left(\frac{\kappa\gamma_1 F_P^-}{\gamma_P} - \gamma\gamma_S F_S^-\right) & \frac{\gamma}{2}(F_S^- - \kappa F_P^-) & \frac{\gamma}{2}\left(\gamma_S F_S^- - \frac{\kappa F_P^-}{\gamma_P}\right) \\ \mu\left(\frac{\kappa\gamma_1 F_S^-}{\gamma_S} - \gamma\gamma_P F_P^-\right) & \mu(\kappa\gamma F_S^- - \gamma_1 F_P^-) & \frac{\gamma}{2}\left(\gamma_P F_P^- - \frac{\kappa F_S^-}{\gamma_S}\right) & \frac{\gamma}{2}(F_P^- - \kappa F_S^-) \end{bmatrix}.$$

Consider the asymptotic behavior of matrices $\mathbf{P}_\pm$ as $|k| \to +\infty$. We obtain

$$\mathbf{P}_\pm = k^1\mathbf{P}_\pm^{(1)} + \mathbf{P}_\pm^{(0)} + k^{-1}\mathbf{P}_\pm^{(-1)} + k^{-2}\mathbf{P}_\pm^{(-2)} + \dots. \tag{36.11}$$

and two leading terms of the matrix $\mathbf{P}_-$ look as follows

$$\mathbf{P}_-^{(1)} = \frac{h}{2}\left(1 - \frac{c_S^2}{c_P^2}\right)\begin{bmatrix} \mathbf{J} & -\frac{1}{2\mu}\mathbf{J} \\ 2\mu\mathbf{J} & -\mathbf{J} \end{bmatrix},$$

$$\mathbf{P}_-^{(0)} = -\frac{k_s^2h^2}{8}\left(1 - \frac{c_S^4}{c_P^4}\right)\begin{bmatrix} \mathbf{J} & -\frac{1}{2\mu}\mathbf{J} \\ 2\mu\mathbf{J} & -\mathbf{J} \end{bmatrix} + \frac{c_S^2}{2c_P^2}\begin{bmatrix} \hat{\mathbf{I}} & -\frac{1}{2\mu}\hat{\mathbf{I}} \\ 2\mu\hat{\mathbf{I}} & -\hat{\mathbf{I}} \end{bmatrix} + \frac{1}{2}\begin{bmatrix} \mathbf{I} & -\frac{1}{2\mu}\hat{\mathbf{I}} \\ -2\mu\hat{\mathbf{I}} & \mathbf{I} \end{bmatrix},$$

where

$$\mathbf{J} = \begin{bmatrix} 1 & 1 \\ -1 & -1 \end{bmatrix}, \qquad \mathbf{I} = \begin{bmatrix} 1 & 0 \\ 0 & 1 \end{bmatrix}, \qquad \hat{\mathbf{I}} = \begin{bmatrix} 0 & 1 \\ 1 & 0 \end{bmatrix}.$$

It can be shown that $\mathbf{P}_-^{(-s)} = O(\omega^2), \quad s \geq 1$ as $\omega \to 0$. Hence

$$\lim_{\omega\to 0} \mathbf{P}_-(\omega, k) = \lim_{\omega\to 0}\left(k\mathbf{P}_-^{(0)} + \mathbf{P}_-^{(1)}\right) =$$

$$= \frac{kh}{2}\left(1 - \frac{c_S^2}{c_P^2}\right)\begin{bmatrix} \mathbf{J} & -\frac{1}{2\mu}\mathbf{J} \\ 2\mu\mathbf{J} & -\mathbf{J} \end{bmatrix} + \frac{c_S^2}{2c_P^2}\begin{bmatrix} \hat{\mathbf{I}} & -\frac{1}{2\mu}\hat{\mathbf{I}} \\ 2\mu\hat{\mathbf{I}} & -\hat{\mathbf{I}} \end{bmatrix} + \frac{1}{2}\begin{bmatrix} \mathbf{I} & -\frac{1}{2\mu}\hat{\mathbf{I}} \\ -2\mu\hat{\mathbf{I}} & \mathbf{I} \end{bmatrix}.$$

The leading components of matrix $\mathbf{P}_+$ result in a similar form

$$\mathbf{P}_+^{(1)} = \frac{h}{2}\left(1 - \frac{c_S^2}{c_P^2}\right)\begin{bmatrix} \mathbf{J}^T & -\frac{1}{2\mu}\mathbf{J}^T \\ 2\mu\mathbf{J}^T & -\mathbf{J}^T \end{bmatrix}, \qquad \mathbf{J}^T = \begin{bmatrix} 1 & -1 \\ 1 & -1 \end{bmatrix},$$

$$\mathbf{P}_+^{(0)} = -\frac{k_s^2h^2}{8}\left(1 - \frac{c_S^4}{c_P^4}\right)\begin{bmatrix} \mathbf{J}^T & -\frac{1}{2\mu}\mathbf{J}^T \\ 2\mu\mathbf{J}^T & -\mathbf{J}^T \end{bmatrix} - \frac{c_S^2}{2c_P^2}\begin{bmatrix} \hat{\mathbf{I}} & -\frac{1}{2\mu}\hat{\mathbf{I}} \\ 2\mu\hat{\mathbf{I}} & -\hat{\mathbf{I}} \end{bmatrix} + \frac{1}{2}\begin{bmatrix} \mathbf{I} & \frac{1}{2\mu}\hat{\mathbf{I}} \\ 2\mu\hat{\mathbf{I}} & \mathbf{I} \end{bmatrix},$$

with $\mathbf{P}_+^{(-s)} = O(\omega^2), \quad s \geq 1$, and we obtain

$$\lim_{\omega\to 0} \mathbf{P}_+(\omega, k) = \lim_{\omega\to 0}\left(k\mathbf{P}_+^{(0)} + \mathbf{P}_+^{(1)}\right)$$

$$= \frac{kh}{2}\left(1 - \frac{c_s^2}{c_p^2}\right)\begin{bmatrix} \mathbf{J}^T & -\frac{1}{2\mu}\mathbf{J}^T \\ 2\mu\mathbf{J}^T & -\mathbf{J}^T \end{bmatrix} - \frac{c_s^2}{2c_p^2}\begin{bmatrix} \hat{\mathbf{I}} & -\frac{1}{2\mu}\hat{\mathbf{I}} \\ 2\mu\hat{\mathbf{I}} & -\hat{\mathbf{I}} \end{bmatrix} + \frac{1}{2}\begin{bmatrix} \mathbf{I} & \frac{1}{2\mu}\hat{\mathbf{I}} \\ 2\mu\hat{\mathbf{I}} & \mathbf{I} \end{bmatrix}.$$

These limit matrices coincide with respective matrices for statics.

36.4 Asymptotic Approach to Spectral Problems for Laminates

For a layered plate we arrive at the matrix relation for the boundary values of displacements and stresses in the following form

$$\boldsymbol{\xi}(z_{n+1}) = \mathbf{P}_* \times \boldsymbol{\xi}(z_1), \qquad \mathbf{P}_* = \mathbf{P}_\xi(h_N) \times \mathbf{P}_\xi(h_{N-1}) \times \cdots \times \mathbf{P}_\xi(h_1), \quad (36.12)$$

$$\mathbf{P}_\xi(h_j) = e_j \mathbf{P}_+^j + e_j^{-1}\mathbf{P}_-^j, \qquad e_j = \mathrm{e}^{kh_j}. \quad (36.13)$$

Thus [15, 45], the boundary value problems (BVP) described in Sect. 36.2, is finally reduced to the analysis of blocks in matrix $\mathbf{P}_*$. For example, the BVP for a laminates with conditions (36.4) for free faces and nonzero displacements on the faces

$$\boldsymbol{\xi}^+ \equiv \boldsymbol{\xi}(z_{n+1}) = \left(-ik^{-1}u_x^+,\ k^{-1}u_z^+,\ 0,\ 0\right)^{\mathrm{T}},$$

$$\boldsymbol{\xi}^- \equiv \boldsymbol{\xi}(z_1) = \left(-ik^{-1}u_x^-,\ k^{-1}u_z^-,\ 0,\ 0\right)^{\mathrm{T}}$$

immediately leads to the dispersion equation

$$\Delta(k, \omega) \equiv \det \mathbf{P}\begin{pmatrix} 34 \\ 12 \end{pmatrix} = 0, \quad (36.14)$$

where the upper indices 3,4 denote rows and the lower indices 1,2 denote columns of the respective minor in matrix $\mathbf{P}_*$. Another choice of boundary conditions in the form of (36.5) or (36.6), or any combination of them, just leads to the changing of rows and columns in the respective minor of matrix $\mathbf{P}_*$.

Note, that the propagator matrix $\mathbf{P}_*$ can be rewritten as a linear combination of exponential terms as follows

$$\mathbf{P}_* = \sum_{s_1,\ldots,s_N} \mathrm{e}^{k(s_N h_N + \cdots + s_1 h_1)} \mathbf{P}_{s_N}^N(h_N) \times \cdots \times \mathbf{P}_{s_1}^1(h_1). \quad (36.15)$$

Due to equality (36.14), the symbol s_j means "+" or "-", respectively.

Upon the relation (36.11), we may conclude that as $\omega \to 0$, or $\omega k^{-1} \to 0$ (or $\omega = \text{const}$, $|k| \to +\infty$), just two terms remain in the left hand side. Consequently, the right hand side of equality (36.15) can be expressed as a linear combination of a finite number of exponential-and-power terms like $k^m \mathrm{e}^{kl_m(h_1,h_2,\ldots,h_n)}$ with matrix factors of the order $O(1)$, where m runs some integer values, and $l_m(h_1, h_2, \ldots, h_n)$ are just linear combinations of the thicknesses. Evidently, a similar exponential-and-power expansion appears as a limit representation of the dispersion equation (36.14) or in the case of another BVP. Such limit form coincides with the equation with respect to k in statics.

Let us rewrite this limit (static) equation as

$$\Delta_0(k) = 0. \tag{36.16}$$

Note, that despite the appearance of radicals, the function $\Delta(k,\omega)$ (in fact, depending on k^2 and ω^2) is always an entire function. After collecting the exponents we may arrive at the expression with cosine and sine of radicals; in addition, the sine is multiplied or divided by its radical.

In addition, the function $\Delta(k,\omega)$ can be represented in the form of convergent power series with respect to the even powers of radicals and with real coefficients, i.e., the wavenumbers possess a symmetry with respect to changing sign and complex conjugating: $\Delta(k,\omega) = 0 \ \left(k \leftrightarrow -k, \bar{k}, -\bar{k}\right)$. So, the root searching can be restricted by the first quadrant $\mathrm{Re}(k) \geq 0, \mathrm{Im}(k) \geq 0$.

Of course, $\Delta_0(k)$ is also an entire function and possesses the same symmetry. This function is single valued with respect to k. The spectral problems is reduced to the problem to search zeroes of the entire function of complex variable k, i.e., the roots of the static form of the dispersion equation. When such function cannot be represented as the product of a polynomial and another entire function with no zeroes, we are facing a general case of an infinite countable set of complex zeroes with an accumulation point at infinity $|k| \to +\infty$. This means that for any finite distance $R_0 > 0$ the number of roots satisfying the inequality $|k| > R_0$ is infinite and the number of roots inside the circle $|k| \leq R_0$ is finite. For waves with P- and SV-polarization the roots are complex (except finite numers of pure real and pure imaginary roots at given frequency), and both $\mathrm{Re}(k) \to +\infty$, $\mathrm{Im}(k) \to +\infty$.

More details about these mathematical aspects can be found in [23]. We just mention that $|\Delta_0(k)|$ function cannot increase faster than respective exponent as $|k| \to +\infty$. This consideration permits us to suggest the following algorithm 1 to obtain complex roots k as $|k| \to +\infty$ with its main steps.

36.5 Three Layered Plate as an Example of Laminate

As an example, let us consider a three layered plate with a core and two symmetrical coatings as shown in Fig. 36.2 where we can see an upper half of a plate. The plate faces are assumed to be stress free. The basic results for such plates have been obtained in [59–63].

When the coating is made of equal plies and the laminates possesses a middle plane of symmetry, we may divide the consideration of the symmetrical waves (S) and antisymmetrical waves (A). On considering the half of a plate, the conditions on the interfaces (36.3) remain in force as well as the stress free conditions (36.4)

Algorithm 1 Algorithm to obtain complex roots

1. As the first step we deduce functions $\Delta(k,\omega)$ and $\Delta_0(k)$;
2. Secondly, introduce dimensionless quantities, namely, by normalizing elastic moduli, mass densities, wavenumber $\xi = kh$ and ply thicknesses $\beta_j = h_j h^{-1}$, $h = \sum h_l$.
3. Third, rewrite $\Delta_0(k)$ in the dimensionless exponential-and-power form $\sum_\alpha T_\alpha = 0$ with the terms $T_\alpha = C_\alpha \zeta^{m_\alpha} e^{\zeta l_\alpha(\beta_1,\beta_2,...,\beta_m)}$, $C_\alpha = \text{const}$ [63]. Here m_α runs some integer values and $l_\alpha(\beta_1,\beta_2,...,\beta_m)$ are linear functions of β_j.
4. Due to this form of function [61, 63] we may expect the asymptotic behavior of large ζ (zeroes of entire function) in the form

$$\zeta_n^{as} = i(ap+b) + c\ln p + \ln d, \qquad p = 2\pi n \gg 1. \tag{36.17}$$

The families of asymptotics (36.17) can be obtained from the analysis of terms T_α which yields the different leading order power $p^{f_\alpha(c)}$ after substitution: $T_\alpha \sim C_\alpha(d) p^{f_\alpha(c)} e^{i\Psi_\alpha(a,b)}$. The order-functions $f_\alpha(c)$ are linear and $F(c) = \max f_\alpha(c)$ is a piece-wise continuous linear function. Each corner point c_* of $F(c)$ gives the necessary value c for ζ_n^{as} . Each terms T_α, which have the same orders $f_\alpha(c_*)$ in the corner point, should cancel the leading-order components of each other to provide the asymptotically accurate equation in the form

$$\left(\sum_\alpha T_\alpha\right) \times p^{-F(c_*)} = O(p^{-\varepsilon}), \qquad \varepsilon > 0. \tag{36.18}$$

Then, we obtain the additional equation to deduce a, b and d . For example, for two such terms T_1 and T_2: $T_1(\zeta_n^{as}(c_*)) \sim -T_2(\zeta_n^{as}(c_*))$. Such particular values β_j for which two corner points merge and there are more than two terms T_α of equal order give us the critical values of geometrical parameters and they can be studied in a separate way. This set is of less dimension than the total set of $\{\beta_j\}$. This sensitivity to geometrical parameters is a flip side of the analytical description.
5. The asymptotic values (36.17) can be used to obtain the accurate values ζ_n for each nth branch by iterative method due to the asymptotically accurate form of Eq. (36.18).
6. Finally, we proceed to solving equation

$$\Delta(\zeta,\Omega) = 0, \tag{36.19}$$

with dimensionless frequency Ω for each complex branch $\zeta_n(\Omega)$ with $\zeta_n(0)$ obtained from previous step 5 for statics. The necessary derivatives of implicit function $\zeta_n(\Omega)$ can be obtained directly from Eq. (36.19).

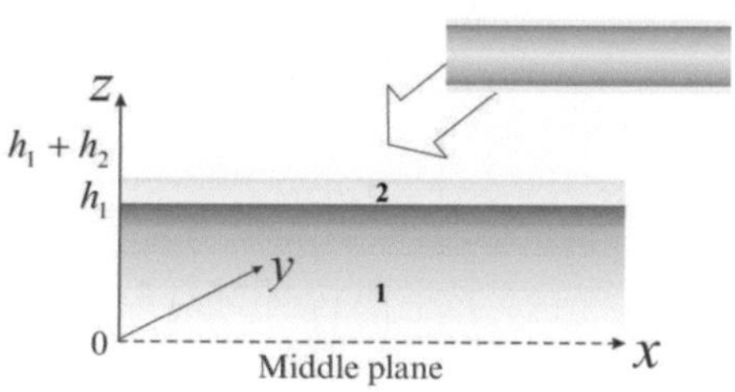

Fig. 36.2 Configuration of a three-layered plate.

for $z = z^+$ but other conditions are set at $z = 0$ in the form

$$u_z = 0,\ \ \sigma_{zx} = 0 \quad \text{(S-waves);} \qquad u_x = 0,\ \ \sigma_{zz} = 0 \quad \text{(A-waves),} \tag{36.20}$$

due to the symmetry or antisymmetry of longitudinal displacements.

The respective dispersion equations are

$$\Delta(k,\omega) \equiv \det \mathbf{P}\binom{34}{13} = 0 \ \text{(S-waves);} \quad \Delta(k,\omega) \equiv \det \mathbf{P}\binom{34}{24} = 0 \ \text{(A-waves).} \tag{36.21}$$

Suppose the thickness of a half of core equals h_1 and the thickness of coating is h_2 with the dimensionless notations

$$\zeta = 2k(h_1+h_2), \quad \beta = \frac{h_2}{h_1+h_2}, \quad 1-\beta = \frac{h_1}{h_1+h_2}, \quad 1-2\beta = \frac{h_1-h_2}{h_1+h_2}, \tag{36.22}$$

$$\tilde{B}_0 = \frac{2B_0}{A_0}, \quad \tilde{h}_2^1 = \frac{H_2^1}{2A_0(h_1+h_2)^2}, \quad \tilde{h}_0^1 = \frac{2H_0^1}{A_0}, \quad \tilde{h}_1^2 = \frac{H_1^2}{A_0(h_1+h_2)}, \tag{36.23}$$

$$\tilde{h}_3^0 = \frac{H_3^0}{8A_0(h_1+h_2)^3}, \quad \tilde{h}_1^0 = \frac{\hat{H}_1^0}{2A^0(h_1+h_2)}. \tag{36.24}$$

and the quantities below appear in static limits of dispersion equation (36.16)

$$\begin{aligned}
A_0 &= \frac{1-(b_2/a_2)^2}{8\mu_1}\left[2\mu_1\mu_2\left(1+\frac{b_1^2b_2^2}{a_1^2a_2^2}\right)+\mu_1^2\left(1-\frac{b_1^2}{a_1^2}\right)\left(1+\frac{b_2^2}{a_2^2}\right)\right.\\
&\quad\left.+\mu_2^2\left(1+\frac{b_1^2}{a_1^2}\right)\left(1-\frac{b_2^2}{a_2^2}\right)\right],\\
B_0 &= \frac{1-(b_2/a_2)^2}{8\mu_1}\left[2\mu_1\mu_2\left(-1+\frac{b_1^2b_2^2}{a_1^2a_2^2}\right)+\mu_1^2\left(1-\frac{b_1^2}{a_1^2}\right)\left(1+\frac{b_2^2}{a_2^2}\right)\right.\\
&\quad\left.+\mu_2^2\left(1+\frac{b_1^2}{a_1^2}\right)\left(1-\frac{b_2^2}{a_2^2}\right)\right],\\
H_2^1 &= -\frac{h_2^2}{2\mu_1}(\mu_2-\mu_1)\left(1-\frac{b_2^2}{a_2^2}\right)^2\left[\frac{b_1^2}{a_1^2}(\mu_2-\mu_1)+\mu_2+\mu_1\right],\\
H_0^1 &= \frac{1}{4\mu_1}\left[-2\mu_1\mu_2\frac{b_1^2b_2^2}{a_1^2a_2^2}\left(1-\frac{b_2^2}{a_2^2}\right)+\mu_1^2\left(1-\frac{b_1^2}{a_1^2}\right)\left(1+\frac{b_2^4}{a_2^4}\right)\right.\\
&\quad\left.-\mu_2^2\left(1+\frac{b_1^2}{a_1^2}\right)\left(1-\frac{b_2^2}{a_2^2}\right)^2\right],\\
H_1^2 &= \frac{h_1}{2\mu_1}(\mu_2-\mu_1)\left(1-\frac{b_1^2}{a_1^2}\right)\left(1-\frac{b_2^2}{a_2^2}\right)\left[-\frac{b_2^2}{a_2^2}(\mu_2-\mu_1)+\mu_2+\mu_1\right],
\end{aligned}$$

$$H_3^0 = -\frac{2h_1 h_2^2}{\mu_1}(\mu_2-\mu_1)^2\left(1-\frac{b_1^2}{a_1^2}\right)\left(1-\frac{b_2^2}{a_2^2}\right)^2,$$

$$H_1^0 = -\frac{1}{\mu_1}\left[\mu_2^2 h_1\left(1-\frac{b_1^2}{a_1^2}\right)\left(1-\frac{b_2^2}{a_2^2}\right)^2 + 2\mu_1\mu_2 h_1\left(1-\frac{b_1^2}{a_1^2}\right)\frac{b_2^2}{a_2^2}\left(1-\frac{b_2^2}{a_2^2}\right)\right.$$
$$\left. + \mu_1^2 h_1\left(1-\frac{b_1^2}{a_1^2}\right)\left(1+\frac{b_2^4}{a_2^4}\right) + 2\mu_1\mu_2 h_2\left(1-\frac{b_2^2}{a_2^2}\right)\right].$$

Here and in what follows, the indices 1, 2 are used for the quantities of a core and a coating, respectively, denote μ_1, μ_2 the shear moduli and a_1, a_2 and b_1, b_2 denote the speeds of P- and S- bulk waves in the layers.

The limiting dispersion equation itself acquires the dimensionless power-and-exponential form

$$\begin{aligned}\Delta_0(\zeta) \equiv \sinh\zeta + \frac{\tilde{B}_0}{2}\sinh\zeta(1-2\beta) + \frac{\zeta^2\tilde{h}_2^1 + \tilde{h}_0^1}{2}\sinh(\beta_1\zeta) \\ \pm\frac{\zeta\tilde{h}_1^2}{2}\cosh(\beta\zeta) \pm \frac{\zeta^3\tilde{h}_3^0 + \zeta\tilde{h}_1^0}{2} = 0,\end{aligned} \tag{36.25}$$

where the upper sign corresponds to A-waves and the lower sign - to S-waves. The equivalent exponential or logarithmic form yields

$$\begin{aligned}\mathrm{e}^{\zeta} = f(\zeta) \quad \Leftrightarrow \quad \zeta = \ln f(\zeta) = ip + \ln f(\zeta) \quad \Leftrightarrow \\ \Leftrightarrow \quad -T_0 + T_1 + T_2 + T_3 + T_4 = 0 \qquad (p = 2\pi n),\end{aligned} \tag{36.26}$$

$$f(\zeta) = T_1 + T_2 + T_3 + T_4, \tag{36.27}$$

$$\begin{aligned}T_0 = \mathrm{e}^{\zeta}, \qquad T_1 = \mp\zeta^3\left|\tilde{h}_3^0\right|, \\ T_2 = -\zeta^2\tilde{h}_2^1\sinh(\beta_1\zeta) = T_2^+ + T_2^- \quad \left(T_2^{\pm} = \mp\frac{1}{2}\zeta^2\tilde{h}_2^1\mathrm{e}^{\pm\beta_1\zeta}\right),\end{aligned} \tag{36.28}$$

$$\begin{aligned}T_3 = \mp\zeta\tilde{h}_1^2\cosh(\beta\zeta) = T_3^+ + T_3^- \quad \left(T_3^+ = \mp\frac{1}{2}\zeta^2\tilde{h}_1^2\mathrm{e}^{\beta\zeta}\right), \\ T_4 = \mp\zeta\tilde{h}_1^0 + \mathrm{e}^{-\zeta} + \tilde{B}_0\sinh(2\beta-1)\zeta - \tilde{h}_0^1\sinh(\beta_1\zeta).\end{aligned} \tag{36.29}$$

Let us also introduce specific values describing the difference of shear moduli

$$s = \mathrm{sign}(\tilde{h}_1^2) = \mathrm{sign}(\mu_2-\mu_1) = -\,\mathrm{sign}(\tilde{h}_2^1), \quad s_1 = \frac{(s+1)}{2}. \tag{36.30}$$

When the difference between parameters of plies disappear we arrive at the known relations for a single layer [57]. The relations (36.26)-(36.28) also give us the natural numeration of static roots ζ_n and of the forthcoming dispersion curves with this starting point as $\omega = 0$. Further we assume that $\mu_1 \neq \mu_2$ because this particular case is not so interesting its analysis is more similar to the analysis of roots for a single layer [57, 59].

36.5.1 Asymptotical Behavior of Large Wavenumbers

Let us substitute the approximation (36.17) into our tiers T_α (36.26)-(36.29) as $p \to +\infty$ and estimate the orders of quantities. We obtain

$$\begin{aligned} T_0 &= \mathrm{e}^{\zeta_n^{as}} = d\, \mathrm{e}^{i(ap+b)}\, p^{f_0} \quad (f_0 = c), \\ T_1 &= a^3 \mathrm{e}^{\mp i\pi/2}\, p^{f_1} \times (1 + t_1) \quad (f_1 = 3), \\ T_2^+ &= 2^{-1} \left|\tilde{h}_2^1\right| a^2 d^{\beta_1} \mathrm{e}^{i\beta_1(ap+b) - i\pi s_1} p^{f_2} \times (1 + t_2) \quad (f_2 = 2 + \beta_1 c), \\ T_3^+ &= 2^{-1} \left|\tilde{h}_1^2\right| a d^{\beta} \mathrm{e}^{i\beta(ap+b) \mp i\pi s/2} p^{f_3} \times (1 + t_3) \quad (f_3 = 1 + \beta c), \end{aligned}$$

where the order of t_α ($\alpha = 1 \div 3$) is not less than $O(p^{-\varepsilon})$ at certain $\epsilon > 0$.

The plots of the order functions $f_\alpha(c)$ ($\alpha = 0 \div 3$) are depicted in Fig. 36.3 (a) for $0 < \beta < 2/3$ and Fig. 36.3 (b) for $2/3 < \beta < 1$. The value $\beta_* = 2/3$ is a critical value of geometrical parameters separating the different hierarchy of the order functions. Its meaning is an equivalence of the total core thickness to the thickness of coating $(2h_1 = h_2)$. The respective intervals can be called as subcritical (a) and supercritical (b).

Consider the piece wise linear function $F(c) = \max f_\alpha(c)$ presenting the maximal value of the orders $f_\alpha(c)$ and corner points $c = c_*$. In the subcritical interval

$$c_* = 1/\beta_1 : \quad f_1 = f_2 = 3 \quad \left(-T_1^{-1} \sim T_2^+\right), \tag{36.31}$$

$$c_* = 2/\beta : \quad f_0 = f_2 = 2/\beta \quad \left(T_0^{-1} \sim T_2^+\right) \tag{36.32}$$

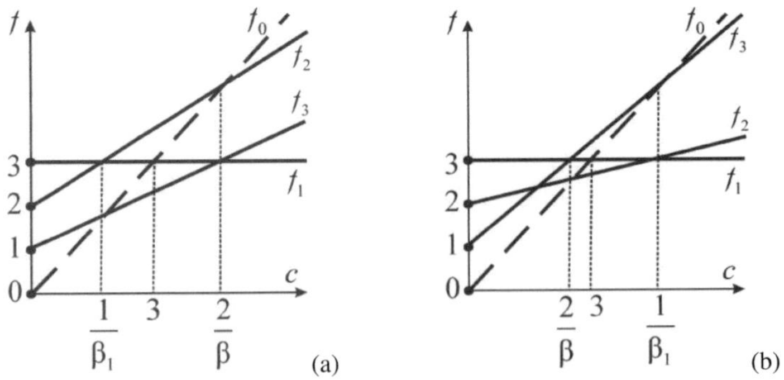

Fig. 36.3: Order functions $f_\alpha(c)$ of members T_α.

and we call them as β_1-case and β-case.

Comparing the orders, absolute values and phases of terms for which $-T_1^{-1} \sim T_2^+$ we arrive at the expression for β_1-case

$$\zeta_n^{as} = \frac{i}{\beta_1}\left[p \mp \frac{\pi s}{2}\right] + \frac{1}{\beta_1}\ln p + \frac{1}{\beta_1}\ln\left(\frac{2\left|\tilde{h}_3^0\right|}{\beta_1\left|\tilde{h}_2^1\right|}\right), \tag{36.33}$$

and for β-case with requirement $T_0^{-1} \sim T_2^+$ we arrive at the expression

$$\zeta_n^{as} = \frac{i}{\beta}\left[p \mp \pi s_1\right] + \frac{2}{\beta_1}\ln p + \frac{1}{\beta}\ln\left(\frac{2\left|\tilde{h}_2^1\right|}{2\beta^2}\right). \tag{36.34}$$

As previously, the upper sign corresponds to A-waves and the lower sign - to S-waves.

In the supercritical interval we obtain similarly

$$c_* = 2/\beta: \quad f_1 = f_3 = 3 \quad \left(-T_1^{-1} \sim T_3^+\right), \tag{36.35}$$

$$c_* = 1/\beta_1: \quad f_0 = f_3 = 1/\beta_1 \quad \left(T_0^{-1} \sim T_3^+\right) \tag{36.36}$$

with the asymptotics for $\beta-$case

$$\zeta_n^{as} = \frac{i}{\beta}\left[p \mp \pi s_1\right] + \frac{2}{\beta}\ln p + \frac{1}{\beta}\ln\left(\frac{2\left|\tilde{h}_3^0\right|}{\beta^2\left|\tilde{h}_1^2\right|}\right), \tag{36.37}$$

and for β_1 -case

$$\zeta_n^{as} = \frac{i}{\beta_1}\left[p \mp \frac{\pi s}{2}\right] + \frac{1}{\beta_1}\ln p + \frac{1}{\beta_1}\ln\left(\frac{\left|\tilde{h}_1^2\right|}{2\beta_1}\right) \tag{36.38}$$

with the same rule for the signs. It should be also noted, that

$$\frac{2\left|\tilde{h}_3^0\right|}{\beta^2\left|\tilde{h}_1^2\right|} = \frac{|\mu_1 - \mu_2|}{(\mu_1 + \mu_2)}\left(1 + \frac{2\mu_1\mu_2}{(\mu_1 + \mu_2)(\lambda_2 + \mu_2)}\right)^{-1} = \frac{\left|\tilde{h}_2^1\right|}{2\beta^2},$$

$$\frac{\left|\tilde{h}_1^2\right|}{2\beta_1} = \frac{2|\mu_1 - \mu_2|}{(\mu_1 + \mu_2)}\left(1 + \frac{2\mu_1\mu_2}{(\mu_1 + \mu_2)(\lambda_1 + \mu_1)}\right)^{-1} = \frac{2\left|\tilde{h}_3^0\right|}{\beta_1\left|\tilde{h}_2^1\right|}.$$

36.5.2 Recurrent Procedure to Calculate Exact Values

Let us justify the obtained expressions (36.33), (36.34) and (36.37), (36.38) by estimating the asymptotic error of difference

$$\theta_n = \zeta_n - \zeta_n^{as} \qquad (n \to +\infty)\,. \tag{36.39}$$

In the subcritical interval $0 < \beta < 2/3$ we consider first the $\beta-$case. To this end we rewrite the Eq. (36.26) as follows

$$\zeta_n = ip + \left(\zeta_n^{as} - ip\right) + \ln\left[\mathrm{e}^{-\zeta_n^{as}+ip}\, f(\zeta_n)\right] = \zeta_n^{as} + g(\theta_n), \tag{36.40}$$

$$\theta_n = g(\theta_n), \qquad g(\theta_n) = \ln\left(\mathrm{e}^{-\zeta_n^{as}}\, f(\zeta_n^{as} + \theta_n)\right). \tag{36.41}$$

If we suppose that $\theta_n \to 0$ $(n \to +\infty)$, the following estimations yield

$$\mathrm{e}^{-\zeta_n^{as}} f(\zeta_n^{as} + \theta_n) = \begin{cases} \mathrm{e}^{\beta_1\theta_n}\left[1 + O\left(\frac{\ln p}{p}\right) + O\left(\frac{1}{p}\right)\right] + o\left(\frac{1}{p}\right), & 0 < \beta \le \frac{1}{2}, \\ \mathrm{e}^{\beta_1\theta_n}\left[1 + O\left(\frac{\ln p}{p}\right) + O\left(p^{3-\frac{2}{\beta}}\right)\right] + o\left(p^{3-\frac{2}{\beta}}\right), & \frac{1}{2} < \beta < \frac{2}{3}, \end{cases} \tag{36.42}$$

$$g(\theta_n) = \begin{cases} \beta_1\theta_n + O\left(\frac{\ln p}{p}\right), & 0 < \beta \le \frac{1}{2}, \\ \beta_1\theta_n + O\left(p^{-3-\frac{2}{\beta}}\right) + O\left(\frac{\ln p}{p}\right), & \frac{1}{2} < \beta < \frac{2}{3}. \end{cases} \tag{36.43}$$

The relations (36.42) and (36.43) immediately lead to the consequences below.

Statement 1 For $p \gg 1$ there exist constants $\epsilon, C > 0$ for which the completely continuous function $g\,(\theta_n)$ maps the circle

$$|\theta_n| \le C p^{-\epsilon}, \tag{36.44}$$

on the complex plane into itself. Namely, any value of ϵ can be chosen in the interval $0 < \epsilon < 1$ for $0 < \beta \le 1/2$ or $0 < \epsilon < 2\beta^{-1} - 3$ for $1/2 < \beta \le 2/3$.

Statement 2 For $p \gg 1$ there exists a value for which the operator in the circle (36.44) determines a contraction mapping

$$|g\,(\theta_n)| \le \beta'\,|\theta_n|\,. \tag{36.45}$$

Namely, any value of β' can be chosen in the interval $\beta_1 < \beta' < 1$.

Statement 3 The exact value ζ_n is a limit of sequence

$$\zeta_n^{(\ell+1)} = \zeta_n^{as} + \ln\left[\mathrm{e}^{-\zeta_n^{as}}\, f\left(\zeta_n^{(\ell)}\right)\right], \quad \zeta_n^{(0)} = \zeta_n^{as}, \tag{36.46}$$

as $\ell \to +\infty$.

This statement follows from the general theorems of functional analysis: the Schauder theorem about the existence of fixed point of operator and the Banach theorem about the contractive mapping [10]. The convergence of sequence (36.46) to a limit $\lim\limits_{\ell\to\infty} \zeta_n^{(\ell)} = \zeta_n$ is exponential. Thus, the β-asymptotics (36.34) approximate the roots ζ_n with the error not larger than $O(p^{-\epsilon})$ as $(n \to +\infty)$.

Secondly, for β_1-case in the subcritical interval $0 < \beta < 2/3$ we transform Eq. (36.26) to the form

$$e^{\eta} = \psi(\eta) \;\Leftrightarrow\; \eta = \ln\psi(\eta) = ip + \ln\psi(\eta), \tag{36.47}$$

$$\psi(\eta) = 2\zeta\frac{|\tilde{h}_3^0|}{\tilde{h}_2^1} + \frac{2}{\zeta^2\tilde{h}_2^1}\left(-e^{\zeta} + T_2^- + T_3 + T_4\right), \qquad \eta = \zeta\beta_1,\ \zeta = \frac{\eta}{\beta_1}, \tag{36.48}$$

with respect to the more convenient variable η. Similarly, introduce the asymptotic error and proceed to the equation with respect to it

$$\kappa_n = \eta_n - \eta_n^{as} \quad (n \to +\infty), \tag{36.49}$$

$$\kappa_n = q(\kappa_n), \qquad q(\kappa_n) = \ln\!\left(e^{-\eta_n^{as}}\psi(\eta_n^{as} + \kappa_n)\right). \tag{36.50}$$

Under assumption $\kappa_n \to 0$ $(n \to +\infty)$ we obtain the estimate

$$e^{-\eta_n^{as}}\psi(\eta_n^{as} + \kappa_n) = 1 + O(\kappa_n p^{-1}) + O(p^{-1}\ln p) + O(p^{-3+\frac{1}{\beta_1}}). \tag{36.51}$$

and similar statements.

Statement 4 For $p \gg 1$ there exist constants $\epsilon, C > 0$ for which the completely continuous function $q(\kappa_n)$ maps the circle

$$|\kappa_n| \le Cp^{-\epsilon}, \tag{36.52}$$

on the complex plane into itself. Due to the estimation (36.52) we can choose any value of ϵ in the interval $0 < \epsilon < min\left(1, 3 - \beta_1^{-1}\right)$ when $0 < \beta < 2/3$.

Statement 5 For $p \gg 1$ there exists a value $0 < \beta' < 1$ for which the operator $q(\kappa_n)$ in the circle (36.52) determines a contraction mapping

$$|q(\kappa_n)| \le \beta' |\kappa_n|. \tag{36.53}$$

Namely, any value of β' can be chosen in the interval $0 < \beta' < 1$.

Statement 6 The exact value η_n is a limit of sequence

$$\eta_n^{(\ell+1)} = \eta_n^{as} + \ln\left[e^{-\eta_n^{as}}\Psi(\eta_n^{(\ell)})\right], \quad \eta_n^{(0)} = \eta_n^{as}, \tag{36.54}$$

as $\ell \to +\infty$. This is quite similar to the Statement 4 with the same reasoning and the same speed of convergence of the limit: $\lim\limits_{\ell\to\infty} = \eta_n$. Let us consider the supercritical interval $2/3 < \beta < 1$.

For the β_1-case (36.33), the error (36.39) and the operators (36.40) and (36.41), we derive the estimations

$$e^{-\zeta_n^{as}} f(\zeta_n^{as} + \theta_n) = \begin{cases} e^{\beta\theta_n}\left[1 + O\left(\dfrac{\ln p}{p}\right) + O\left(\dfrac{1}{p}\right)\right] + o\left(\dfrac{1}{p}\right), & \dfrac{3}{4} \le \beta < 1, \\ e^{\beta\theta_n}\left[1 + O\left(\dfrac{\ln p}{p}\right) + O\left(p^{3-\frac{1}{\beta_1}}\right)\right] + o\left(p^{3-\frac{1}{\beta_1}}\right), & \dfrac{2}{3} < \beta < \dfrac{3}{4}, \end{cases} \tag{36.55}$$

$$g(\theta_n) = \begin{cases} \beta\theta_n + O\left(\dfrac{\ln p}{p}\right), & \dfrac{3}{4} \le \beta < 1, \\ \beta\theta_n + O\left(p^{3-\frac{1}{\beta_1}}\right) + O\left(\dfrac{\ln p}{p}\right), & \dfrac{2}{3} < \beta < \dfrac{3}{4}. \end{cases} \tag{36.56}$$

Consequently, for the β_1-case the Statement1 remains in force for any ϵ in the interval $0 < \epsilon < \beta_1^{-1} - 3$ $(2/3 < \beta < 3/4)$ or $0 < \epsilon < 1$ $(3/4 \le \beta < 1)$ and both Statement 2 and Statement 3 also hold for the supercritical case $(2/3 < \beta < 1)$ for any β' in the interval $\beta < \beta' < 1$.

For β–asymptotics (36.37) the transformation of the Eq. (36.26) and another respective function $\Psi(\eta)$ looks as follows

$$e^{\eta} = \psi(\eta) \iff \eta = \ln\psi(\eta) = ip + \ln\psi(\eta). \tag{36.57}$$

$$\psi(\eta) \equiv \frac{2\zeta^2\left|\tilde{h}_3^0\right|}{\tilde{h}_1^2} + \frac{2}{\zeta\tilde{h}_1^2}\left(-e^{\zeta} + T_2 + T_3^- + T_4\right), \qquad \eta = \zeta\beta, \ \ \zeta = \frac{\eta}{\beta}, \tag{36.58}$$

and for the asymptotic error (36.49) we obtain the estimation

$$e^{-\eta_n^{as}}\psi\left(\eta_n^{as} + \kappa_n\right) = 1 + O\left(\kappa_n p^{-1}\right) + O\left(p^{-1}\ln p\right) + O\left(p^{-3+2/\beta}\right). \tag{36.59}$$

So, the Statements 4-6 remain in force for the β-asymptotics (36.37) where ϵ is from the interval $0 < \epsilon < \min\left(1, 3 - 2\beta^{-1}\right)$ and β' is from the interval $\beta < \beta' < 1$.

As a result of this subsection, we prove the asymptotic representations (36.33), (36.34), (36.37) and (36.38), estimate the error of asymptotic approximations ζ_n^{as} and justify the iteration formulas (36.46) and (36.54) to calculate the exact values ζ_n. As we can see below, this approach also holds for the case when the assumption $n \gg 1$ is no longer valid.

36.6 Basic Numerical Results

36.6.1 Complex Wavenumbers

Let us set the parameters of core (layer 1 in Fig. 36.2) as $\rho = 2700\,\mathrm{kg/m^3}$, $c_P = 6400\,\mathrm{m/s}$, $c_S = 3150\,\mathrm{m/s}$ and the parameters of coating (layer 2 in Fig. 36.2) as $\rho = 1170\,\mathrm{kg/m^3}$, $c_P = 2610\,\mathrm{m/s}$ and $c_S = 1100\,\mathrm{m/s}$. Consider the subcritical value $\beta = 1/2 < 2/3$ $(h_1 = h_2)$ and the supercitical value $\beta = 3/4 > 2/3$ $(3h_1 = h_2)$.

The roots ζ_n are calculated using iteration method until the inequalities

$$\left|\zeta_n^{(\ell+1)} - \zeta_n^{(\ell)}\right| \le 10^{-10}, \quad \left|\Delta_0\left(\zeta_n^{(\ell)}\right)\right| \le 10^{-12}$$

are satisfied.

The results for the *subcritical* case and *antisymmetrical* waves are presented in Fig. 36.4 (a, b) for $\Re(\zeta_n)$ and $\Im(\zeta_n)$, respectively. The circles and squares correspond to ζ_n calculated and lines show the asymptotics for n =1, 2,. . . ,15. The β−case and β_1−case are indicated by letters.

The results for the *subcritical* case and *symmetrical* waves are depicted in Fig. 36.5 (a, b) in a similar way. The number of iterations for all cases does no exceed some dozens and rapidly decreases with increased n.

The behavior of the relative error of asymptotics is shown in Fig. 36.6 (a). The errors have maximal values for $n = 1$ (20%–10%) and rapidly decrease for $n > 1$.

The results for the *supercritical* case and *antisymmetrical* waves are presented in Fig. 36.7 (a, b) for $\Re(\zeta_n)$ and $\Im(\zeta_n)$, respectively. The circles and squares correspond to ζ_n calculated and lines show the asymptotics for n =1,2,...,15. The β−case and β_1−case are indicated by letters. The error is shown in Fig. 36.6 (b).

Similar results for *symmetrical* waves (*supercritical* case) are depicted in Figs. 36.8 (a, b). The relative errors of asymptotics for *supercritical* interval are shown in Fig. 36.9 (a, b) and again we see a fast decreasing of error for $n > 1$.

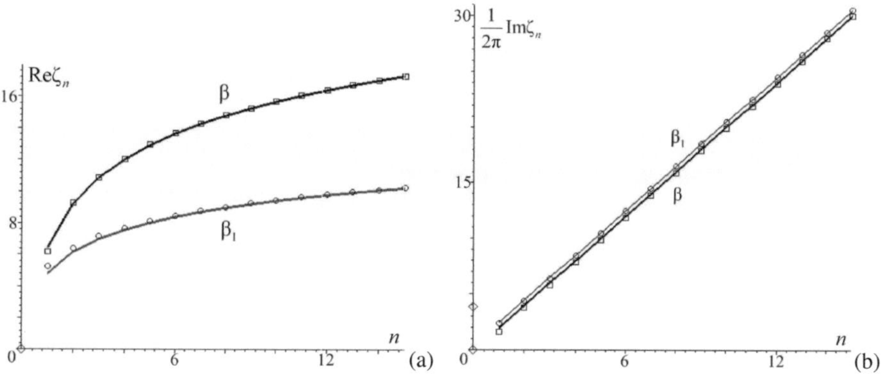

Fig. 36.4: The subcritical values ζ_n and their asymptotics for antisymmetrical waves.

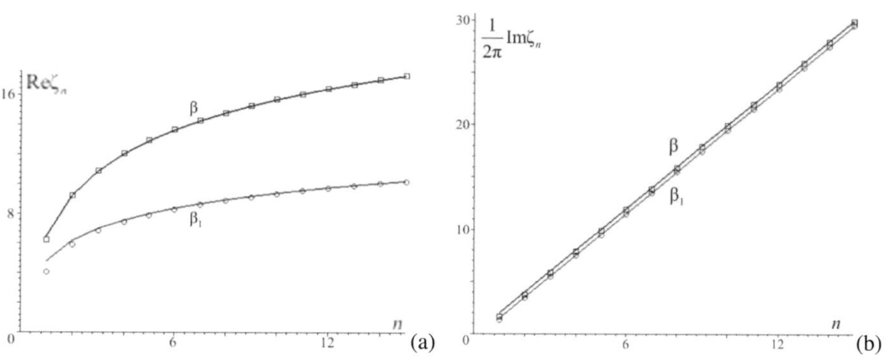

Fig. 36.5: The subcritical values ζ_n and their asymptotics for symmetrical waves.

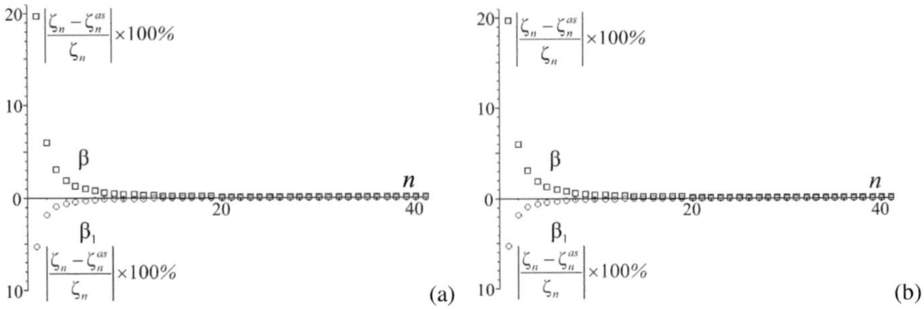

Fig. 36.6: The subcritical relative error of asymptotics for antisymmetrical (a) and symmetrical (b) waves.

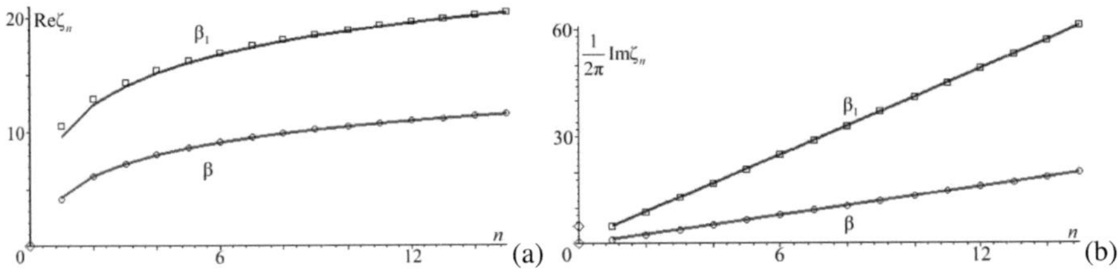

Fig. 36.7: The supercritical values ζ_n and their asymptotics for antisymmetrical waves.

36.6.2 Pure Imaginary Wavenumbers

An additional interesting fact consists in the appearance of pure imaginary root caused by the appearance of interface. Introducing the function

$$g(y) = -i\left[e^{iy} - f(iy)\right] \quad (\zeta = iy)\,, \tag{36.60}$$

we can prove that for antisymmetrical waves

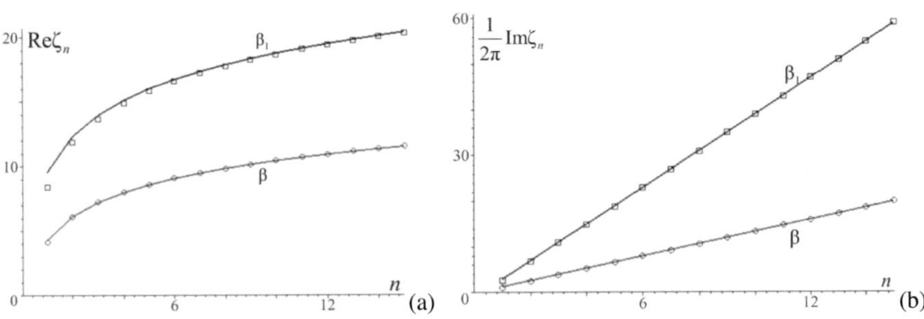

Fig. 36.8: The supercritical values ζ_n and their asymptotics for symmetrical waves.

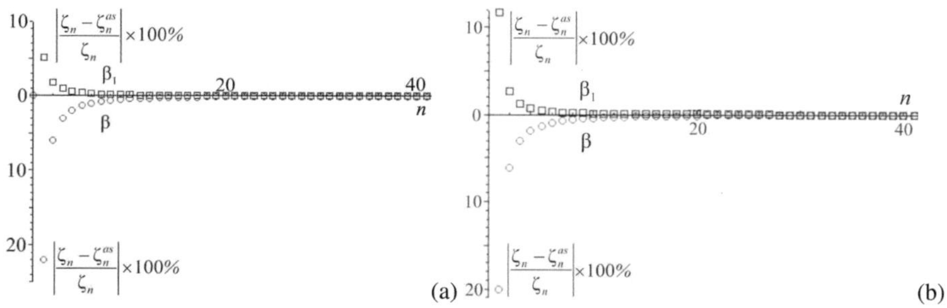

Fig. 36.9: The supercritical relative error of asymptotics for antisymmetrical (a) and symmetrcial (b) waves.

$$g(y) = -Gy^3\left[\frac{1}{6} + O(y)\right] < 0 \quad (y \to +0),$$

$$g(y) = \left|\tilde{h}_3^0\right| y^3 \left[1 + O\left(y^{-1}\right)\right] > 0 \quad (y \to +\infty),$$

$$G = 2 + \tilde{h}_0^1\beta^3 + 6\tilde{h}_2^1\beta + 3\tilde{h}_1^2\beta^2 + 6\tilde{h}_3^0 + 6(\beta - \beta_1)^3 > 0,$$

and for symmetrical waves

$$g(y) = Gy\left[1 + O(y^2)\right] > 0 \quad (y \to +0),$$

$$g(y) = -\left|\tilde{h}_3^0\right| y^3 \left[1 + O\left(y^{-1}\right)\right] < 0 \quad (y \to +\infty),$$

$$G = 2 + \beta(1 - 2\beta)\left|\tilde{h}_1^0\right| + \tilde{h}_0^1\beta_1 - \tilde{h}_1^2 > 0.$$

By this reason, the pure imaginary root always exists. It acquires the value $\zeta = i4.82534157$ ($\beta = 3/4$) and $\zeta = i3.88908628$ ($\beta = 1/2$) for antisymmetrical wave. For symmetrical waves $\zeta = i2.4128948589$ ($\beta = 3/4$) and $\zeta = i3.457398198$ ($\beta = 1/2$). These roots are shown in Figs. 36.4 and 36.5 (b) and Figs. 36.7 and 36.8 (b) by diamonds on the vertical axis.

36.6.3 Additional Control of Roots

The additional control of our appoach can be performed using the principle of argument for entire functions [26]. Let us choose a loop Γ in the form of square in which the length of side equals $p = 2\pi n$ as shown in Fig. 36.10 (a). The loop has punctured points at the origin $\zeta_0 = 0$ and in the pure imaginary root and these roots are surrounded by arcs of small radius r_0 (see Fig. 36.10 (b)). The starting point is denoted as ζ_* . If there are exactly m roots inside the loop Γ , the principle of argument for the entire function $e^{\zeta} - f(\zeta)$ results in the equality

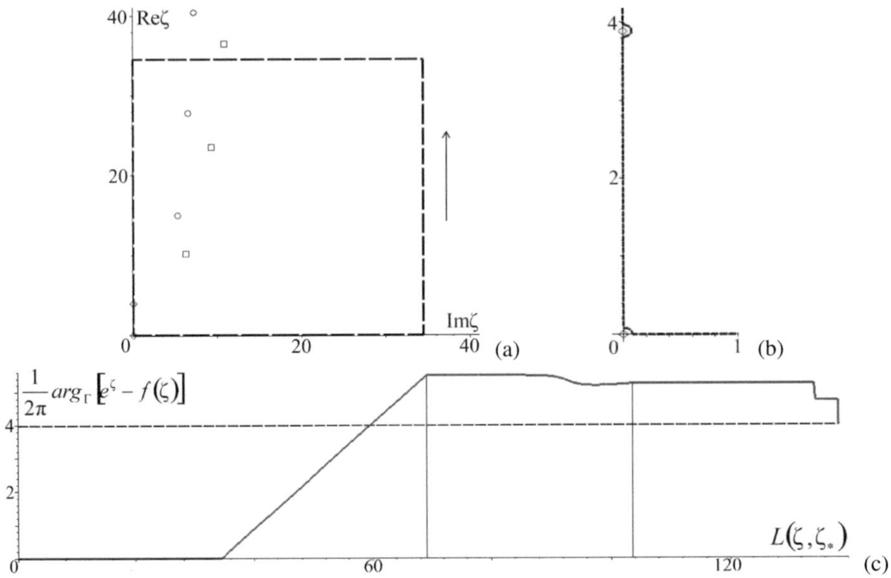

Fig. 36.10: The loop Γ (a) at $\beta = \frac{1}{2}$, $r_0 = \frac{\pi}{40}$, $m = 4$, $n = 5$; the increased left fragment of the loop Γ (b) nearly the origin and pure imaginary root; the increment of argument along the loop (c).

$$\frac{1}{2\pi}\Delta_\Gamma \arg\left[e^{\zeta} - f(\zeta)\right] = m, \tag{36.61}$$

where Δ_Γ is the increment of argument while going counterclockwise.

The example of loop Γ at $\beta = 1/2$ and $n = 5$, $r_0 = \pi/40$ has $m = 4$ wavenumbers of the antisymmetrical waves inside the loop. These wavenumbers are shown in Fig. 36.10 (a) by circles and squares.

The increment of argument

$$\frac{1}{2\pi}\arg\left[e^{\zeta} - f(\zeta)\right]$$

with respective function (36.27) for axisymmetrical waves is depicted in Fig. 36.10 (c) to demonstrate the equation of coincidence (36.61). The horizontal axis represent the natural coordinate, i.e., the arc length $L(\zeta, \zeta_*)$ along the loop between the starting point ζ_* and the current point ζ. All the calculations presented in the Subsection 36.6.1 are checked in this manner.

36.7 Dispersion Curves

Due to our consideration of initial static values value ζ_n we subdivide all the complex dispersion curves into $\beta-$ and β_1-sets where

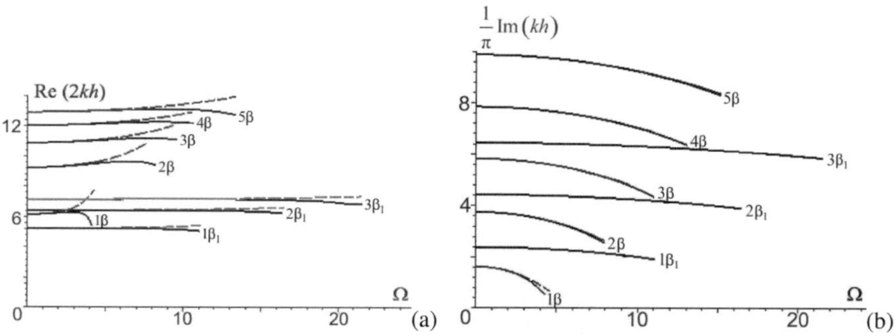

Fig. 36.11: Real (a) and imaginary (b) parts of kh against Ω at the subrcritical $\beta = \frac{1}{2}$ for β- and β_1 case (antisymmetrical waves).

$$k_n(0) = \frac{\zeta_n}{2\,(h_1 + h_2)}. \tag{36.62}$$

Actually, the functions $\Delta(k, \omega)$ for antisymmetrical or symmetrical waves depend on k^2 and ω^2 . Hence, the low frequency asymptotics of dispersion curve $k\,(\omega)$ can be derived as the approximation of implicit function determined by dispersion equation

$$k_n^{\mathrm{as}}(\omega) = k_n(0)\left\{1 - \frac{\omega^2}{k_n^2(0)}\frac{\partial\Delta/\partial\left(k_n^2\right)}{\partial\Delta/\partial\left(\omega^2\right)}\bigg|_{\omega=0}\right\}^{1/2} \sim k_n(\omega), \tag{36.63}$$

where $\frac{\partial\Delta/\partial(k_n^2)}{\partial\Delta/\partial(\omega^2)} = O\left(k_n^{-1}(0)\right)$.

This also means that the approximation (36.63) of dispersion curve has a flat segment which is the longer the larger is n and $k_n(0)$. It is physically consistent because we deal with the slower and slower wave as $\omega/k \to 0$. The values $k_n(\omega)$ are evaluated using the standard Newton iteration method.

We start with the low frequency asymptotics (36.63) and for a larger frequency with a step $\delta\omega$, the initial value of k is extrapolated by the Lagrange polynomial using three previous exact points of the dispersion curve calculated at previous steps. The wavenumber step is $\delta k = -\Delta(k, \omega)\left[\frac{\partial\Delta(k, \omega)}{\partial k}\right]^{-1}$, and the process is repeated with the correction value of $k + \delta k$. This iteration procedure is performed until $|\delta k| \leq 10^{-10}$ and $|\Delta\,(k, \omega)| \leq 10^{-12}$.

The results for antisymmetrical waves and their subcritical and supercritical dispersion curves are depicted in Fig. 36.11 and Fig. 36.12. The solid lines correspond to the exact values and dashed lines correspond to the asymptotics (36.63). Due to the different scales we choose n =1, 2, 3, 4, 5 for $\beta-$ case and n =1, 2, 3 for β_1-case in one figure. On the horizontal axis the dimensionless frequency $\Omega = \frac{2\omega h}{c_S}$ is shown. Here c_S is the speed of shear waves for the core (the layer 1 in Fig. 36.2). The similar results for the symmetrical waves are presented in Fig. 36.13 and Fig. 36.14 for $n = 1, 2, 3, 4, 5$ for $\beta-$ case and $n = 1, 2, 3$ for β_1-case.

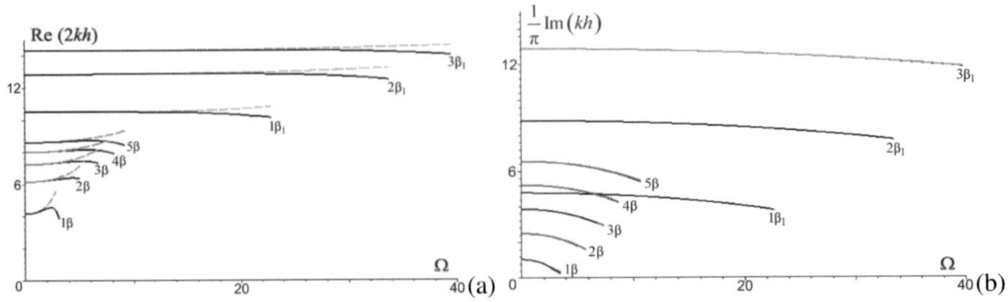

Fig. 36.12: Real (a) and imaginary (b) parts of kh against Ω at the supercritical $\beta = \frac{3}{4}$ for β- and β_1-case (antisymmetrical waves).

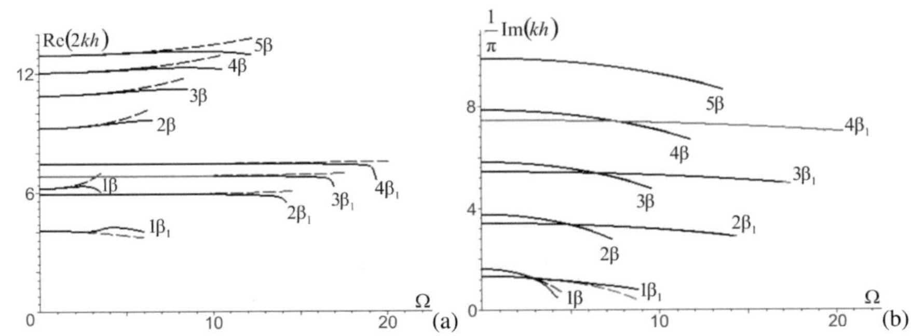

Fig. 36.13: Real (a) and imaginary (b) parts of kh against Ω at the subrcritical $\beta = \frac{1}{2}$ for β- and β_1-case (symmetrical waves).

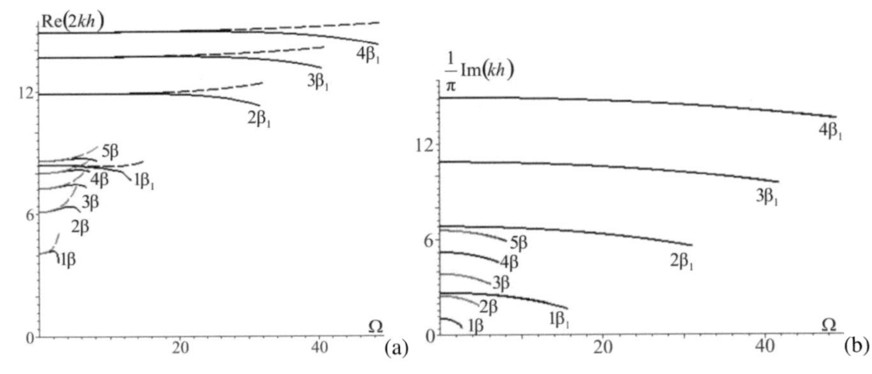

Fig. 36.14: Real (a) and imaginary (b) parts of kh against Ω at the supercritical $\beta = \frac{3}{4}$ for β- and β_1-case (symmetrical waves).

As we can see in Figs 36.11-36.14, the asymptotics (36.63) approximate better the imaginary part with the interval of validity twice larger then for real part. In each $\beta-$ and β_1-family of dispersion curves there are initial flat segment for low frequency and the larger the index n the longer is this flat segment. The curve families themselves may have different scales and the superciritical curves are approximated by asymptotics ((36.63) slightly better.

36.8 Pure Real and Pure Imaginary Dispersion Curves

To complete our previous consideration, let us describe the initial segments of the fundamental pure real and pure imaginary dispersion curves. At low frequency, the laminate is well described by the averaged plate equation of bending and stretching [49–51], which yield the familiar dispersion relations

$$\omega^2\left[\sum_j \rho_j\left(z_{j+1}-z_j\right)\right]=\frac{2}{3}k^4\left[\sum_j \frac{\mu_j\left(z_{j+1}^3-z_j^3\right)}{1-\nu_j}\right], \tag{36.64}$$

$$\omega^2\left[\sum_j \rho_j\left(z_{j+1}-z_j\right)\right]=2k^2\left[\sum_j \frac{\mu_j\left(z_{j+1}-z_j\right)}{1-\nu_j}\right]. \tag{36.65}$$

where (36.64) corresponds to antisymmetrical (bending) waves and (36.65) is for symmetrical waves. Here ν_j are Poisson's ratios, $z_{j+1}-z_j=h_j$ and $j=1,2,3$ for all layers of the total three layered plate.

The results for antisymmetrical waves are shown in Fig. 36.15. The notation A_0 indicate the pure real (fundamental mode) and A_{0i} is its pure imaginary twin due to the Eq. (36.64). The pure imaginary curve I_0 appears due to the additional imaginary root of the limit form of the dispersion equation in statics (see the Subsect. 36.6.2). This curve can be approximated by the same asymptotic formula (36.63) when substituting the respective imaginary value $k(0)$. Everywhere, the dashed curves correspond to the asymptotics (36.64), (36.65) or (36.63). The solid curves are exact and they are obtained by the standard Newton iteration method described in Sect. 36.7.

Similar results for symmetrical waves are presented in Fig. 36.16 for the fundamental (pure real) curve S_0 , and the pure imaginary curve I_0.

36.9 Conclusion

The described approach and detailed examples explain how one can obtain the infinite (countable) set of the complex-valued wavenumbers for isotropically-layered plates. This approach is focused on the so-called generalized Lamb waves with P- and SV-polarization.

Actually, this is the most difficult subset of the wavenumber spectrum even for a finite frequency interval because both the real and imaginary parts of wavenumbers

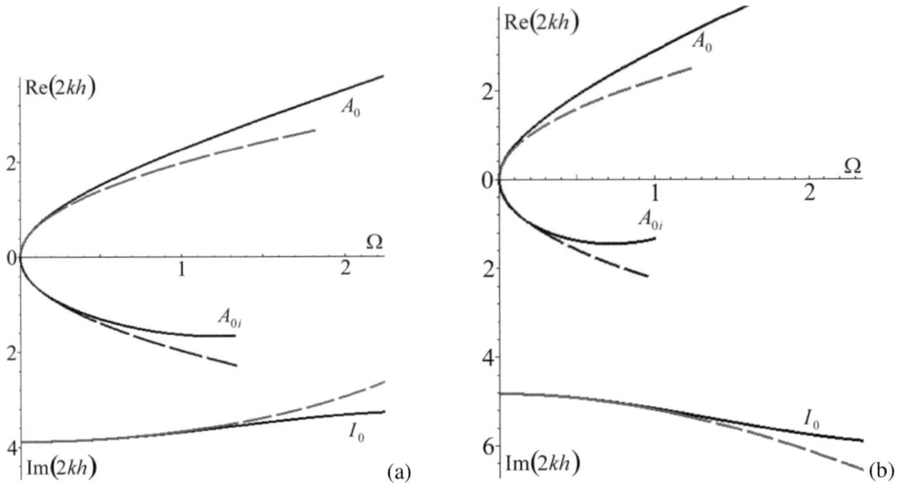

Fig. 36.15: The fundamental curves A_0 and pure imaginary curves A_0 and I_0 for subcritical $\beta = \frac{1}{2}$ (a) and supercritical $\beta = \frac{3}{4}$ (b).

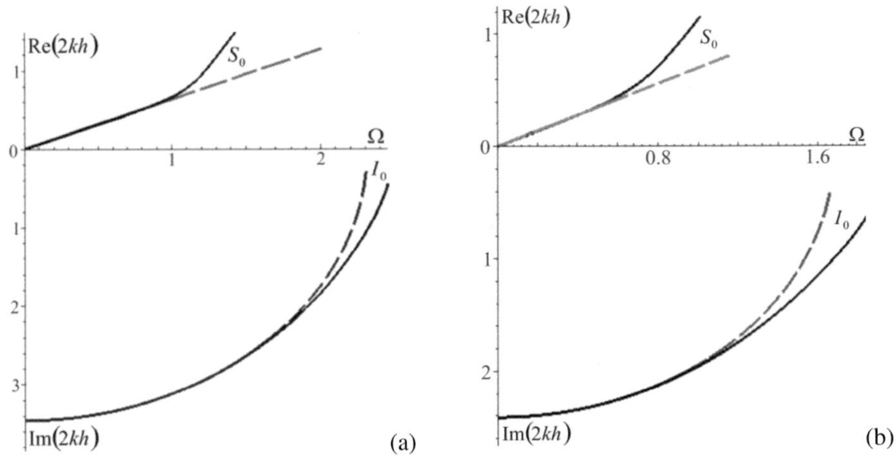

Fig. 36.16: The fundamental curves S_0 and the imaginary curves I_0 for subcritical $\beta = \frac{1}{2}$ (a) and supercritical $\beta = \frac{3}{4}$ (b).

tend to infinity. By this reason one cannot evaluate this subset in total using pure numerical method based on the discretization of plate.

Our approach suggests to come from infinity and to obtain preliminary the asymptotics of large wavenumbers. This asymptotics is caused by the power-and-exponential type of the limit form of the dispersion equation for large wavenumber or/and small frequency. The asymptotic value becomes more accurate as the static (limit) value of the complex wavenumber increases. Thus, for the dispersion branch with large index we obtain its analytical formula.

The layered nature of a plate leads to the appearance of several types of dispersion curves corresponding to the difference of the proper scale in each layer. There also exist critical values of geometrical parameters separating the curve description. Additionally, each interface causes a complementary pure imaginary dispersion curve at low frequency.

At low frequency ranges, the suggested approach permits one to evaluate the total wavenumber spectrum. For larger frequency, this approach may serve as an efficient complementary analytical tool to existing numerical methods and codes.

Acknowledgements The work of the fist author (DZ) was carried out as part of the State Assignment for the Institute of Computer Aided Design, Russian Academy of Sciences.

References

[1] Achenbach JD (1984) Wave Propagation in Elastic Solids. Amsterdam: Elsevier
[2] Aki K, Richards PG (1980) Quantitative Seismology, Theory and Methods, Vol 1. San Francisco: Freeman
[3] Auld A, Kino GS (1971) Normal mode theory for acoustic waves and its application tointerdigital transducer. IEEE Transaction. Electronic Devision 18(10):898-908, DOI 10.1109/T-ED.1971.17303
[4] Auld BA (1990) Acoustic Fields and Waves in Solids, Vol 2. Krieger Inc
[5] Bobrovnitskii YI (1973) Sootnosheniya ortogonal'nosti dlya voln Lemba (in Russ., Orthogonality relations for Lamb waves). Soviet Acoustic Physics 18(4):432-433
[6] Bondarenko AA (2007) Normal'nyye mody v pryamougol'nom uprugom volnovode (in Russ., Normal modes of the elastic waveguide with rectangular crosssection). Akustichnyi visnik 10(4):12-27
[7] Crane LJ, Gilchrist MD, Miller JJH (1997) Analysis of Rayleigh-Lamb wave scattering by a crack in an elastic plate. Computational Mechanics 19:533-537, DOI 10.1007/s004660050162
[8] Dougall J (1904) VIII. An analytical theory of the equilibrium of an isotropic elastic plate. Transactions of the Royal Society of Edinburgh 41(1):129-228, DOI 10.1017/S0080456800080091
[9] Dougall J (1914) XVII. An analytical theory of the equilibrium of an isotropic elastic rod of circular section. Transactions of the Royal Society of Edinburgh 49(4):895-978, DOI 10.1017/S008045680001718X
[10] Dunford N, Schwartz JT (1958) Linear Operators: General Theory. New York: Wiley - Interscience Publishers
[11] Flores-Lopez MA, Gregory RD (2006) Scattering of Rayleigh-Lamb waves by a surface breaking crack in an elastic plate. Journal der Acoustical Society of America 119(4):2041-2049, DOI 10.1121/1.2178720

[12] Fraser WB (1976) Orthogonality relations for Rayleigh-Lamb modes of vibration of a plate. Journal of the Acoustic Society of America 59:215-216, DOI 10.1121/1.380851

[13] Gravenkamp H, Birk C, Song C (2015) Simulation of elastic guided waves interacting with defects in arbitrarily long structures using Scaled Boundary Finite Element Method. Journal of Computational Physics 295:438–455, DOI 10.1016/j.jcp.2015.05.053

[14] Gregory RD, Gladwell I (1983) Reflection of a symmetric Rayleigh-Lamb wave at a fixed or free edge of a plate. Journal of Elasticity 13(2):185–206, DOI 10.1007/BF00044820

[15] Haskel NA (1990) The dispersion of surface waves on multilayered media. In: Ben-Menahem A (ed) Vincit Veritas: A Portrait of the Life and Work of Norman Abraham Haskell, 1905–1970. American Geophysical Union, pp 86-103, DOI 10.1029/SP030p0086

[16] Hayashi T, Song WJ, Rose JL (2003) Guided wave dispersion curves for a bar with an arbitrary cross section, a rod and rail example. Ultrasonics 41(3):175–183, DOI 10.1016/S0041-624X(03)00017-0

[17] Hayashi T, Tamayama C, Murase M (2006) Wave structure analysis of guided waves in a bar with an arbitrary cross-section. Ultrasonics 44(1):17–24, DOI 10.1016/j.ultras.2005.02.007

[18] Hernando Quintanilla F, Fan Z, Lowe MJS, Craster RV (2015) Guided waves' dispersion curves in anisotropic viscoelastic single- and multi-layered media. Proceedings of the Royal Society A: Mathematical, Physical and Engineering Sciences 471(2183):1–23, DOI 10.1098/rspa.2014.0850

[19] Hernando Quintanilla F, Fan Z, Lowe MJS, Craster RV (2015) Modeling guided elastic waves in generally anisotropic media using a spectral collocation method. The Journal of the Acoustical Society of America 137:1180–1194, DOI 10.1121/1.4913911

[20] Hernando Quintanilla F, Fan Z, Lowe MJS, Craster RV (2016) Full 3D dispersion curve solutions for guided waves in generally anisotropic media. Journal of Sound and Vibrations 363:545–559, DOI 10.1016/j.jsv.2015.11.009

[21] Knopoff AL (1964) A matrix method for elastic wave problem. Bulletin of the Seismological Society of America 54(1):431–438, DOI 10.1785/BSSA0540010431

[22] Krome F, Gravenkamp H (2016) Analyzing modal behavior of guided waves using high order eigenvalue derivatives. Ultrasonics 71:75–85, DOI 10.1016/j.ultras.2016.05.015

[23] Levin BY, in collaborations with Lyubyanskii Y, Sodin M, Tkachenko V (2000) Lectures on Entire Functions Translations of Mathematical Monographs, vol 150. Providence: American Mathematical Society

[24] Li L, Haider MF, Mei H, Giurgiutiu V, Xia Y (2020) Theoretical calculation of circularcrested Lamb wave field in single- and multi-layer isotropic plates using the normal mode expansion method. Structural Health Monitoring 19(2):357–372, DOI 10.1177/1475921719848149

[25] Lowe MJS (1995) Matrix techniques for modeling ultrasonic waves in multilayered media. IEEE Transactions on Ultrasonics Ferroelectrics and Frequency Control 42:525–541, DOI 10.1109/58.413830
[26] Markushevich AI (1965) Theory of Functions of a Complex Variable, Vol 1. Prentice-Hall
[27] Mazzotti M, Marzani A, Bartoli I, Viola E (2012) Guided waves dispersion analysis for prestressed viscoelastic waveguides by means of the SAFE method. International Journal of Solids and Structures 49:2359–2372, DOI 10.1016/j.ijsolstr.2012.03.004
[28] Merkulov LG, Rokhlin SI, Zobnin OP (1970) Raschet spektra volnovykh chisel dlya voln Lemba v plastine (in Russ., Calculation of the spectrum of wave numbers for Lamb waves in a plate). Soviet Journal of Nondestructive Testing (6):369–373
[29] Merkulov LG, Firsov IP (1972) Excitation spectrum of Lamb waves in a plate (in Russ., Spectrum of Lamb's wave generated in a plate). Soviet Journal of Nondestructive Testing (5):115–118
[30] Miamoto T, Yasuura K (1977) Numerical analysis on isotropic elastic waveguides by mode-matching method - II. Particle velocities and dispersion characteristics in rods of rectangular cross section. IEEE Transactions on Sonics and Ultrasonics 24:359–375, DOI 10.1109/T-SU.1977.30967
[31] Mindlin RD, Onoe M, Medick MA (1957) Mathematical theory of vibrations of elastic plates. In: Proceedings of the 11th Annual Symposium on Frequency Control, Fort Monmouth, Army Signal Engineering Laboratories (USA)
[32] Mindlin RD (2006) An Introduction to the Mathematical Theory of Vibrations of Elastic Plates. World Scientific, Singapore, DOI 10.1142/6178
[33] Pagneux V, Maurel A (2001) Determination of Lamb mode eigenvalues. The Journal of the Acoustical Society of America 110(3):1307–1314, DOI 10.1121/1.1391248
[34] Prakash BG (1978) Generalized orthogonality relations for rectangular strips in elastodynamics. Mechanics Research Communications 5(4):251–255, DOI 10.1016/0093-6413(78)90019-8
[35] Predoi MV, Castaings M, Moreau L (2008) Influence of material viscoelasticity on the scattering of guided waves by defects. The Journal of the Acoustical Society of America 124(5):2883–2894, DOI 10.1121/1.2977604
[36] Rokhlin S (1980) Diffraction of Lamb waves by a finite crack in an elastic layer. The Journal of the Acoustical Society of America 67:1157–1165, DOI 10.1121/1.383941
[37] Rokhlin S (1981) Resonance phenomena of Lamb waves scattering by a finite crack in a solid layer. The Journal of the Acoustical Society of America 69:922–928, DOI 10.1121/1.385592
[38] Rokhlin SI, Wang L (2002) Stable recursive algorithm for elastic wave propagation in layered anisotropic media: Stiffness matrix method. The Journal of the Acoustical Society of America 112:822–834, DOI 10.1121/1.1497365
[39] Rokhlin SI, Wang L (2004) Recursive geometric integrators for wave propagation in a functionally-graded multilayered elastic medium. Journal of the Me-

chanics and Physics of Solids 52(11):2473–2506, DOI 10.1016/j.jmps.2004.05.001

[40] Rokhlin SI, Wang L (2004) Modeling of wave propagation in layered piezoelectric media by a recursive asymptotic method. IEEE Transactions on Ultrasonics Ferroelectrics and Frequency Control 51(9):1060–1071, DOI 10.1109/TUFFC.2004.1334839

[41] Shkerdin G, Glorieux C (2004) Lamb mode conversion in a plate with a delamination. The Journal of the Acoustical Society of America 116:2089–2100, DOI 10.1121/1.1784446

[42] Shkerdin G, Glorieux C (2005) Lamb mode conversion in an absorptive bilayer with a delamination. The Journal of the Acoustical Society of America 118:2253–2264, DOI 10.1121/1.2011397

[43] Slepyan LI (1979) Teorema Betti i sootnosheniya ortogonal'nosti dlya sobstvennykh funktsiy (in Russ., Betti theorem and orthogonality relations for eigenfunctions). Izvestiya AN SSSR. Mekhanika Tverdogo Tela (1):83–87

[44] Schwab F, Knopoff AL (1971) Surface waves in multilayered anelastic media. Bulletin of the Seismological Society of America 61(4):893–912, DOI 10.1785/BSSA0610040893

[45] Thomson WT (1950) Transmission of elastic waves through a stratified solid medium. Journal of Applied Physics 21:89–93, DOI 10.1063/1.1699629

[46] Viktorov IA (1967) Rayleigh and Lamb Waves. New York, Plenum Press

[47] Wang L, Rokhlin SI (2002) Recursive asymptotic stiffness matrix method for analysis of surface acoustic wave devices on layered piezoelectric media. Applied Physical Letters 81(21):4049–4051, DOI 10.1063/1.1522831

[48] Zakharov DD (1988) Sootnosheniya obobshchennoy ortogonal'nosti dlya sobstvennykh funktsiy v prostranstvennykh zadachakh dinamiki uprugogo sloya (in Russ., Generalised orthogonality relations for eigenfunctions in three-dimensional dynamic problems for an elastic layer). Izvestiya AN SSSR. Mekhanika Tverdogo Tela 23(6):62–68

[49] Zakharov DD (1992) Asymptotic analysis of three dimensional dynamic elastic equations for a thin multilayered anisotropic plate of arbitrary structure. Journal of Applied Mathematics and Mechanics 56(5):637–644, DOI 10.1016/0021-8928(92)90049-E

[50] Zakharov DD (1992) Asymptotic integration of 3D dynamic equations for thin multilayered anisotropic plates. Comptes rendus de l'Académie des sciences II 315(8):915–920

[51] Zakharov DD, Becker W (2000) 2D problems of thin asymmetric laminates. Zeitschrift für angewandte Mathematik und Physik 51(4):49–66, DOI 10.1007/s000330050017

[52] Zakharov DD, Nikonov AV (2005) Approximate description of the dynamics of thin isotropic elastic coatings and interlayers by using asymptotics of high order of accuracy. Mechanics of Composite Materials 41(6):527–534, DOI 10.1007/s11029-006-0006-7

[53] Zakharov DD (2006) High order approximate dynamic theory of elastic anisotropic lining and coating. The Journal of the Acoustical Society of America 119(4):1961–1970, DOI 10.1121/1.2169922

[54] Zakharov DD (2008) Orthogonality of 3D guided waves in viscoelastic laminates and far field evaluation to a local acoustic source. International Journal of Solids and Structures 45(6):1788–1803, DOI 10.1016/j.ijsolstr.2007.10.019

[55] Zakharov DD (2010) Dirichlet-Neumann conditions and the orthogonality of 3D guided waves in layered solids. Computational Mathematics and Mathematical Physics 50(9):1522–1535, DOI 10.1134/S0965542510090094

[56] Zakharov DD (2010) Effective high order approximation of layered lining and coating of elastic, viscoelastic and nematic materials. Journal of Applied Mathematics and Mechanics 74(3):286–296, DOI 10.1016/j.jappmathmech.2010.07.004

[57] Zakharov DD, Castaings M, Singh D (2011) Numerical and asymptotic approach for evaluating complex wave-numbers of guided modes in viscoelastic plates. The Journal of the Acoustical Society of America 130(2):764–771, DOI 10.1121/1.3607426

[58] Zakharov DD (2013) The property of orthogonality and energy transfer by 3D eigenwaves in transversely isotropic laminated plates with and without contact with a fluid. Journal of Applied Mathematics and Mechanics 77(1):39–50, DOI 10.1016/j.jappmathmech.2013.04.006

[59] Zakharov DD (2017) Low-frequency asymptotics of complex dispersion curves for Lamb waves in layered elastic plate. Acoustical Physics 63(4):497–507, DOI 10.1134/S1063771017040150

[60] Zakharov DD (2018) Low frequency spectra of layered plates and their parametrical study. In: Proceedings of International Conference Days on Diffraction, pp 305–310, DOI 10.1109/DD.2018.8556763

[61] Zakharov DD (2018) Parametric analysis of complex dispersion curves for flexural Lamb waves in layered plates in the low-frequency range. Acoustical Physics 64(4):387–401, DOI 10.1134/S1063771018040125

[62] Zakharov DD, Nikonov AV (2019) Evaluation of complex spectrum of the symmetrical Lamb waves for coated plates at low frequency. Part I: Foundations. Composite Structures 230:111429, DOI 10.1016/j.compstruct.2019.111429

[63] Zakharov DD, Nikonov AV (2019) Evaluation of complex spectrum of the symmetrical Lamb waves for coated plates at low frequency. Part II: Asymptotics/numerical results. Composite Structures 230:111428, DOI 10.1016/j.compstruct.2019.111428

[64] Zemanek J (1972) An experimental and theoretical investigation of elastic wave propagation in a cylinder. The Journal of the Acoustical Society of America 51:265–283, DOI 10.1121/1.1912838

[65] Zhang B (2007) Elastic waves excited by a plane source on the surface of a multilayered medium. The Journal of the Acoustical Society of America 121(3):1440–1448, DOI 10.1121/1.2431331

[66] Zlatin AN (1980) Roots of several transcendental equations encountered in elasticity theory. Soviet Applied Mechanics 16(12):1068–1073, DOI 10.1007/BF00884687

Chapter 37
Metal Fatigue Resistance Under Multiaxial Loading of Materials on Various Scale-Structural Levels

Eleonora Zavoychinskaya and Georgii Lavrikov

Abstract The analysis of well-known phenomenological approaches and fatigue criteria of structural materials leads to formulate of fatigue probabilistic multilevel models. According to the proposed approach, the probability of the ultimate states at the micro-, meso-, and macroscale levels is determined by the loading process in the form of a hierarchical system of recurrent constitutive relations on three linear operators applied to this process. A method of identifying of the eigenvalues based on fatigue macroexperiments, taking into account the results of physical studies of brittle evolution, is described. A computational algorithm has been developed and its numerical implementation for the presented model has been carried out. High cycle fatigue curves by defect levels at proportional multiaxial loading are obtained. The model was verified under symmetric tension-compression, shear, and axial loading with torsion at different amplitude ratios for 42CrMo4, S355, SAE1045 steels and 075-T651 aluminum alloy. Ranges of amplitude ratios for which a proportional process can be calculated on axial loading or torsion have been obtained.

Key words: Multiaxial stress-strain state, High-cycle fatigue, Proportional loading, Scale-structural defect levels, Micro-level, Meso-level, Macro-level, Failure probability, Brittle failure, Fatigue curves by defect levels, Symmetrical uniaxial loading with torsion

37.1 Introduction

The current state of industrial development is associated with the use of artificial intelligence technologies. One of the important areas of their application is the forecasting of the durability of engineering systems [2, 6, 9, 21, 34, 36, 61]. At the

Eleonora Zavoychinskaya (corresponding author) · Georgii Lavrikov
Faculty of Mechanics and Mathematics, Moscow State University, Leninskie gory, 119991 Moscow, Russian Federation, e-mail: elen@velesgroup.com, george.lavrikov@yandex.ru

H. Altenbach et al. (eds.), *Stability Problems in Mechanics of Continua*,
Advanced Structured Materials 260, https://doi.org/10.1007/978-3-032-31032-3_37

stages of training and testing of neural networks, it is necessary to create a database of fatigue of materials and typical structural elements at various stress-strain states. Experimental studies are very expensive and require considerable efforts, and it is almost impossible to cover the entire variety of loading modes of different design schemes at combinations of multiple alternating axial loading, internal pressure and torsion. As a result, the creation of a database should be based on the accumulated theoretical material.

According to modern concepts, fatigue is a random step-by-step process of joint brittle and viscous failure evolution [54]. The process of brittle failure consists in the sequential passage of the following stages:

- the origin of band-shaped defect substructures and submicroscopic cracks as a result of the evolution of metallurgical and technological defects and the emergence of new defects due to the structural heterogeneity of the material;
- the evolution of the band structure, the origin of microcracks as a result of mergers and evolution of nanoscale defects at the microlevel;
- mergers of microdefects with the origin of short cracks in the average of grain size at the mesolevel;
- the origin and grow of brittle cracks at the macrolevel of grain ensembles;
- the formation of one or more macrocracks-leaders, and their evolution until the brittle fracture of the material by the mechanisms of intergranular or transcrystalline chipping.

The process of viscous fracture is described by the following sequence of stages:

- the occurrence of plastic distortion at a critical curvature of the crystal lattice with the generation and evolution of dislocations by twinning and sliding mechanisms at the nanoscale;
- further evolution of dislocations with the creation of sliding planes and the nucleation of pores and dimpled microrelief at the microlevel;
- pore growth with the origin of a cellular substructure at the mesolevel of the grains;
- intensive sliding at the macrolevel at a plastic deformation, the porosity making;
- nucleation and growth of single viscous macrocracks to viscous macrofracture.

An analysis of numerous explorations shows that in the areas of high-cycle ($N_f \in (5\,10^4, 10^6)$ cycles to macrofracture) and very high-cycle ($N_f \in (10^6, 10^{11})$ cycles) fatigue, small maximum values of plastic deformations not exceeding the values of elastic deformations (for most materials by several orders of magnitude less) are observed, the process of plastic deformation slows down the spread of brittle cracks, making dominant the mechanisms of brittle failure. In these areas the time before the origin of a macrocrack-leader is the crucial for the overall durability of the material. Despite the high accuracy of modern flaw detection tools, to detect of operational defects is a difficult problem, including due to the inaccessibility of the hazardous structure areas. In this regard, the creation of mathematical models of fatigue by defect levels and the conduct of numerical experiments is an urgent and developing area in the design of engineering systems. At the same time, with the intensive development of nondestructive testing methods (the ultrasound analysis, the eddy current testing, the full-scale strain gauge, the acoustic emission analysis, the magnetic flaw detection, etc.), the importance of residual resource estimation is increasing.

The phenomenological approaches of solid mechanics to the description of fatigue are represented by classical strength criteria, continuum damage theories, approaches of continuum mechanics of the damaged medium and fracture mechanics [54, 55]. Classical strength criteria for fatigue strength under proportional loads (among the most well-known are the criteria of Gough, Polard, Nishihara, Kawamoto, Lebedev and Shkanov, McDiarmid, Matake, Troost, El-Magd, Mielke, Birger, Kakuno, Kawada, Sijander, Kurath, FLawrence, Carpinteri, Spagnoli, Krempl, Mankovsky, Crossland, Sines, Ohgi, Grubisic, Simburger, Deperrois and Papadopoulos) assume the existence of a universal function of extreme values of stress tensor invariants [1, 8, 10, 12, 18, 19, 24–26, 28, 29, 32, 33, 42, 43, 51], they also include criteria considering states on critical planes selected by the authors in accordance with the mechanisms of brittle or viscous fracture [7, 31, 45, 58]. Classical energy approaches consider the potential energy or specific work of deformation (Findley [15], Marin [27], Garud [16], Troshchenko [48], Glinka with colleagues [17], Pallarés-Santasmartas [35]), which in the areas of high- and very high cycle fatigue at elastic macrodeformation are expressed via the second invariant of the stress tensor. The extensive database of calculation results on these models has been collected for specific groups of materials and stress-strain states (see, for example, reviews [3, 5, 13, 22, 30, 37, 40, 41, 46, 50, 59]). These models are widely used in engineering design due to their simplicity and experimental confirmation for the corresponding groups of materials and loading processes and are included in the majority of normative documents and standards. The most probabilistic models are formulated for uniaxial loading (see, for example, reviews [46, 47]). Some variants for multiaxial loading are presented in the works of Troshchenko and colleagues. To determine durability according to the well-known approach of Rzhanitsyn uses the condition of equality of the average values and variances of the distribution of random loads and random strength properties at fracture (all functions are distributed according to the normal law). There are a number of models based on the Monte-Carlo method and Markov chains. Probabilistic approaches are developing, but difficulties in applying of these models at variable loading, including the finding of material constants, do not allow them to be actively used in engineering calculations.

Continuum damage theories [14, 20, 29, 39, 44] in the area of high-cycle fatigue mainly use the linear or nonlinear scalar damage summation models and energetic approaches of continuum mechanics of the damaged medium [55]. However, fatigue is a multistage, multilevel process in which the material structure undergoes various levels on different physical mechanisms, and cannot be described as a continuous accumulation of damage throughout the whole process. This leads to the need to consider different damage functions on the various levels and to formulate the criteria for the transition between levels. The problems associated with this, as well as the difficulties to determine of material functions for these theories, significantly reduce their applicability in real engineering problems.

Modern modeling of fatigue at multiaxial stress-strain state requires to apply of an interdisciplinary approach of physics, materials science and solid mechanics based on a probabilistic description. First of all, it is necessary to note the physical approaches [54, 55], including the well-known models of dislocation motion by Walgraef and Aifantis and works on the evolution of dislocation structures by Klesnil and Lukas. One of the main approaches to describing of the origin and growing of brittle fatigue cracks was developed by Yokobori [52], Tanaka, Mura, Nakasone and Venkataraman and colleagues, who develop the ideas of Lin and Ito. The significant number of structure parameters and fatigue physical mechanisms are not determined in macroexperiments, the lack of consideration of the loading process and the difficulty in determining of the fatigue macrocharacteristics do not allow these models to reach the engineering level. Models of Papadopoulos [38] and Dang Van [49], according to which the formation of mesolevel cracks is determined by achieving of the maximum plasticity of the most stressed crystal in the grain, can be to assigned to the approaches for determining of durability at multiaxial fatigue. It is important to emphasize the results of many experimental studies indicate that at high cycle fatigue the most sliding systems are not activated. The process of viscous failure proceeds, but is not decisive for macrofrcture. According the stochastic models of Afanasyev the probability distribution function of microcracks in a certain macrovolume depending on a stress tensor is determined.

According to the authors, the most promising direction of the fatigue modeling is the construction of multilevel probabilistic failure models with the possibility of their numerical implementation in order to determine the durability in engineering applications. In this paper, one of these approaches is proposed to describe the process of brittle failure at proportional loading.

37.2 Basic Concepts and Hypotheses of the Scale-Structural Fatigue Model

When analyzing the process of brittle failure, six scale-structural levels are identified. For each level, the concept of a defect at the j-th scale level within a certain representative volume V_c ($V_c = L^3$ is the volume of a solid body in which the initiation of an initial macrocrack-leader of length L is possible) is determined [55, 57]. A defect at the j-th level is characterized by a linear size $l_j = l_j(\tau)$ over the time interval $\tau \in [0, t]$ and a defect density $q_j = q_j(\tau)$ in the volume V_c, given by the formula: $q_j(\tau) = \lim_{\Delta V \to V_c} \Delta q_j(\tau)/\Delta V$, where $\Delta q_j = \Delta q_j(\tau)$ is the average number of defects in the volume ΔV at time $\tau \in [0, t]$, $j = 1, \ldots, 6$. A certain continuous increasing function of the j-th level $l_j^* = l_j^*(\tau)$ of the linear size $l_j = l_j(\tau)$ and density $q_j = q_j(\tau)$ is introduced. An example of such a function is: $l_j^*(\tau) = l_j(\tau)\left(q_j(\tau)V_c^{\gamma}\right)$, $\gamma = \text{const}$, $j = 1, \ldots, 6$, $\tau \in [0, t]$, it has the dimension of length. The j-th scale level defect origin begins when the function $l_{j-1}^* = l_{j-1}^*(\tau)$ reaches the limiting value $l_{f,j-1}^*$ at the time τ_{j-1}, $j = 1, \ldots, 6$, $\tau_0 = 0$ (at the time τ_j the j-th level defects reach their ultimate state). The formation of defects of each scale level occurs as a result of the successive generation, growth and fusion of defects at previous levels.

The evolution of the material structure is studied by various physical methods, such as the analysis of density changes, the measurement of magnetic and electrical characteristics, study of acoustic, thermal, and X-ray fields. One of them is the widely used method of studying Vickers microhardness during the failure process, which a correspondence between a certain ultimate state of each level defects and the Vickers microhardness value is made [56]. The representative volume V_c is characterized by a microhardness $HV = HV(\tau)$ on the time interval $\tau \in [0, t]$, and when the function $l^*_{j-1} = l^*_{j-1}(\tau)$ reaches its limiting value $l^*_{f,j-1}$ the microhardness $HV = HV(\tau)$ obtains of a certain characteristic value $HV(\tau_{j-1})$ at time τ_{j-1}, $j = 1, \ldots, 6$, $\tau_0 = 0$ (at time τ_j defects of the j-th level get their ultimate state). In the engineering practice, non-destructive testing methods based on various physical principles are used to monitor of structural failure, such as electron defectoscopy, ultrasonic testing, eddy current testing, magnetic flux leakage testing, X-ray diffraction analysis, acoustic emission analysis, radiation defectoscopy, digital speckle interferometry and others, which are described in many scientific publications.

Due to the random distribution of defects in the volume V_c, the fatigue process is stochastic, and the values of $l^*_j(\tau)$, $j = 1, \ldots, 6$, and $HV(\tau)$ for each moment of time τ, $\tau \in [0, t]$ are random variables. For each level a reliability distribution function $F_j = F_j(\tau)$ is defined. This function describes the probability that the value of the function l^*_j at time τ will be less than the critical value $l^*_{f,j}$, i.e., $F_j(\tau) = P(l^*_j(\tau) < l^*_{f,j})$. The j-th level failure probability distribution function, denoted as $Q_j = Q_j(\tau)$, is defined in the view: $Q_j(\tau) = 1 - F_j(\tau)$, $\tau \in [0, t]$, $j = 1, \ldots, 6$, namely

$$Q_j(\tau) = P(l^*_j(\tau) \geq l^*_{f,j}), \quad j = 1, \ldots, 6. \tag{37.1}$$

The fact that the origin of the each level defect occurs due to the sequential origin, growth and fusion of defects of all previous levels, should be reflected at the construction of the constitutive relations for functions $Q_j = Q_j(\tau)$, $j = 1, \ldots, 6$. The macrofailure probability distribution function $Q = Q(\tau)$, $\tau \in [0, t]$, $0 \leq Q(\tau) \leq 1$, is written as follows:

$$Q(\tau) = Q_5(\tau) + Q^*_4(\tau) + Q^*_5(\tau) + Q_6(\tau), \tag{37.2}$$

when the function $Q^*_j = Q^*_j(\tau)$ is the failure probability by the j-th level defect failure probability distribution function during the formation and growing of $(j + 1)$ level defects, $j = 4, 5$. The fatigue curves by defect levels are given by the following equations:

$$Q_j(\tau_j) = Q_{j,\text{th}}, \quad j = 1, \ldots, 6, \tag{37.3}$$

$$Q^*_j(\tau_{j+1}) = 1 - Q_{j+1,\text{th}}, \quad Q^*_j(\tau_j) = Q_{j,\text{th}}, \quad j = 4, 5, \tag{37.4}$$

when the value τ_j is the durability of the j-th level, $Q_{j,\text{th}} \in [0, 1]$ is some given value of the probability (here is $Q_{j,\text{th}} = 1$). The function $Q = Q(\tau)$,

$\tau \in [0,t]$, $0 \le Q(\tau) \le 1$, determines the fatigue curve by the origin of one or more macrocracks-leaders of the certain length in the form:

$$Q(t_{\mathrm{f}}) = Q_{\mathrm{th}}, \tag{37.5}$$

when t_{f} is the durability by the origin of a macrocrack-leader with the possibility Q_{th}. Its further evolution is described by the methods and approaches of fracture mechanics.

Constitutive relations for the function $Q_j = Q_j(\tau)$ in Eq. (37.1), $j = 1, \ldots, 6$, $\tau \in [0,t]$, is written as follows:

$$Q_j(\tau) = \mu_j(\tau) \max_{z\in[0,t]} \left|\varphi_j(z)\right|, \quad j = 1, \ldots, 6, \tag{37.6}$$

and for the function $Q_j^* = Q_j^*(\tau)$, $j = 4, 5$, $\tau \in [0,t]$ in the (37.2) they have the view:

$$Q_j^*(\tau) = \mu_j^*(\tau) \max_{z\in[0,t]} \left|\varphi_j^*(z)\right|, \quad j = 4, 5, \tag{37.7}$$

when $\varphi_j = \varphi_j(z)$, $\varphi_j^* = \varphi_j^*(z)$ are the functions to be determined later. The functions $\mu_j = \mu_j(\tau)$, $\mu_j^* = \mu_j^*(\tau)$ in Eqs. (37.6) and (37.7) are presented in the forms:

$$\mu_j(\tau) = \left(\lg \frac{\tau}{\tau_{j-1}}\right)^{\chi}, \quad \tau > \tau_{j-1}, \quad j = 1, \ldots, 6, \tag{37.8}$$

$$\mu_j^*(\tau) = \left(\lg \frac{\tau}{\tau_{j-1}}\right)^{\chi^*}, \quad \tau > \tau_j, \quad j = 4, 5, \tag{37.9}$$

the values τ_{j-1} are the durability on the $(j-1)$-th level, χ, χ^* are material constants.

37.3 Constitutive Relations of the Scale-Structural Fatigue Model in the Case of Proportional Loading

The simple (proportional) process by Ilyushin [23] of variable loading (by Moskvitin) in the point of a body is set via the principal values (eigenvalues) of the stress tensor $\sigma_{kk} = \sigma_{kk}(\tau)$ in the principal axes (eigenvectors) $\{e_k\}_{k=1}^3$ on the time interval $\tau \in [0,t]$ as follows:

$$\sigma_{kk}(\tau) = S_k(\tau) + \sigma_0(\tau) = S_1\left(\alpha_k f(\tau) + \alpha f_1(\tau)\right), \; k = 1, 2, 3, \tag{37.10}$$

$$S_k(\tau) = S_1 \alpha_k f(\tau), \; \sigma_0(\tau) = \frac{1}{3}\sum_{k=1}^{3} \sigma_{kk}(\tau) = S_1 \alpha f_1(\tau), \; \tau \in [0,t],$$

$$\max_{\tau\in[0,t]} |f(\tau)| = 1, \quad \max_{\tau\in[0,t]} |f_1(\tau)| = 1,$$

when $S_k = S_k(\tau)$, $k = 1,2,3$ are the principal values of the stress deviator, $S_k = \max_{\tau\in[0,t]} S_k(\tau)$, $k = 1,2,3$, $\sigma_0 = \sigma_0(\tau)$ is the mean stress (the first invariant of the stress tensor), the functions $f = f(\tau)$, $f_1 = f_1(\tau)$ are continuous functions on a time interval $\tau \in [0,t]$, a value t is a loading time. The principal axes of stress tensor are numerated so that $|S_1| \geq |S_2| \geq |S_3|$. It follows from equality that it is possible to fulfill the conditions From the equality $\sum_{k=1}^{3} S_k(\tau) = 0$ it follows that $\sum_{k=1}^{3} \alpha_k = 0$ and it is possible to fulfill the following conditions:

$$\alpha_1 = 1, \quad -1 \leq \alpha_2 \leq -\frac{1}{2}, \quad -\frac{1}{2} \leq \alpha_3 \leq 0. \tag{37.11}$$

From the inequalities (37.11) we have the relation: $\alpha_3 = -1 - \alpha_2$. For the process (37.10) the stress vector $\mathbf{P}^{(\nu)} = \mathbf{P}^{(\nu)}(\tau)$ on an infinitesimal plane of the coordinate tetrahedron at a point of the body with the unit normal ν has the following known form:

$$\begin{aligned} \mathbf{P}^{(\nu)}(\tau) = \sum_{k=1}^{3} \sigma_{kk}(\tau)\nu_k \mathbf{e}_k &= \sum_{k=1}^{3} \left(S_k(\tau) + \sigma_0(\tau)\right) \nu_k \mathbf{e}_k \\ &= S_1 \sum_{k=1}^{3} \left(\alpha_k f(\tau) + \alpha f_1(\tau)\right) \nu \mathbf{e}_k, \end{aligned} \tag{37.12}$$

when $\nu = \sum_{k=1}^{3} \nu_k \mathbf{e}_k$, $\nu_1^2 = 1 - \nu_2^2 - \nu_3^2$. After the notations $S_n^{(\nu)} = S_n^{(\nu)}(\tau)$ and $S_\tau^{(\nu)}$ of normal and shear components of the deviator stress vector

$$\mathbf{S}^{(\nu)}(\tau) = \sum_{k=1}^{3} S_k(\tau)\nu_k \mathbf{e}_k$$

correspondingly it can be written:

$$S_n^{(\nu)}(\tau) = S_1 f(\tau) \sum_{k=1}^{3} \alpha_k \nu_k^2, \tag{37.13}$$

$$S_\tau^{(\nu)}(\tau) = S_1 f(\tau) \sqrt{\sum_{k=1}^{3} \alpha_k^2 \nu_k^2 - \left(\sum_{k=1}^{3} \alpha_k \nu_k^2\right)^2}. \tag{37.14}$$

At the loading (37.10) the durability τ_j on the j-th level in the expressions (37.3) is the function of the following variables: $\tau_j = \tau_j(S_1, \alpha_2, \alpha, f(\tau), f_1(\tau))$.

The functions $\varphi_j = \varphi_j(z)$, $j = 1, \ldots, 6$, in (37.6) are proposed according the approach via three linear operators [53] in the form:

$$\varphi_j(z) = \max_{\nu_2,\nu_3:\ \nu_2^2+\nu_3^2 \leq 1} \left\{ \sum_{m=0}^{2} \left| \Pi_j^{(m)} \left(z, (g_m(\tau))_{\tau=0}^{t}\right) \right| \right\}, \quad z \in [0,t], \tag{37.15}$$

when

$$\Pi_j^{(m)} = \Pi_j^{(m)} \left(z, (g_m(\tau))_{\tau=0}^t\right), \qquad m = 0, 1, 2, \quad j = 1, \ldots, 6,$$

are the linear operators of the functions $g_m = g_m(\tau)$, $\tau \in [0, t]$ and there are introduced the following notations: $g_0 = \sigma_0(\tau)$, $g_1 = S_n^{(\nu)}(\tau)$, $g_2 = S_\tau^{(\nu)}(\tau)$. The operators

$$\Pi_j^{(m)} = \Pi_j^{(m)} \left(z, (g_m(\tau))_{\tau=0}^t\right), \qquad m = 0, 1, 2, \quad j = 1, \ldots, 6,$$

are given by the Hilbert-Schmidt integral operators:

$$\Pi_j^{(m)} \left(z, (g_m(\tau))_{\tau=0}^t\right) = \int_0^t K_j^{(m)}(z, \tau) g_m(\tau) d\tau, \quad m = 0, 1, 2, \quad j = 1, \ldots, 6, \tag{37.16}$$

when $K_j = K_j(z, \tau)$ are the positive definite symmetric kernels of the corresponding integral operators, $g_m(\tau) \in L_2[0, t]$, $K_j(z, \tau) \in L_2[0, t]^2$. According the Hilbert-Schmidt theorem [57] the operators

$$\Pi_j^{(m)} = \Pi_j^{(m)} \left(z, (g_m(\tau))_{\tau=0}^t\right), \qquad m = 0, 1, 2, \quad j = 1, ..., 6,$$

in (37.16) can be presented as the following rows:

$$\Pi_j^{(m)} \left(z, (g_m(\tau))_{\tau=0}^t\right) = \sum_{n=0}^{\infty} \frac{D_n^{(m)} \Omega_n(z)}{\lambda_{j,n}^{(m)}}, \quad D_n^{(m)} = \frac{\int_0^t g_m(\tau) \Omega_n(\tau) d\tau}{\int_0^t \Omega_n^2(\tau) d\tau}, \tag{37.17}$$

when $\{\lambda_{j,n}^{(m)}\}_{n=0}^{\infty}$ are the eigenvalues of the operators, corresponding to the eigenfunctions $\{\Omega_n(z)\}_{n=0}^{\infty}$. In the relation (37.17) the series converges absolutely and uniformly on the segment $[0, t]$. A complete closed orthogonal system of functions in $L_2[0, t]$ is chosen as the eigenfunctions $\{\Omega_n(z)\}_{n=0}^{\infty}$. The functions $g_m = g_m(\tau)$, $m = 0, 1, 2$, can be decomposed into a system of functions $\{\Omega_n(z)\}_{n=0}^{\infty}$: $g_m = g_m(\tau)$, $m = 0, 1, 2$, in the view:

$$g_m(\tau) = \sum_{n=0}^{\infty} D_n^{(m)} \Omega_n(z), \quad D_n^{(m)} = \frac{\int_0^t g_m(\tau) \Omega_n(\tau) d\tau}{\int_0^t \Omega_n^2(\tau) d\tau}. \tag{37.18}$$

Equation (37.18) is the Fourier series of the function $g_m = g_m(\tau)$, $m = 0, 1, 2$. Here is considered the kernels of the integral operators which the system of eigenfunctions $\{\Omega_n(z)\}_{n=0}^{\infty}$ is the system of trigonometric functions of the following type:

$$\Omega_0 = 1, \Omega_1 = \sin\frac{2\pi\tau}{t}, \Omega_2 = \cos\frac{2\pi\tau}{t}, \ldots, \Omega_{2n-1} = \sin\frac{2\pi n\tau}{t}, \Omega_{2n} = \cos\frac{2\pi n\tau}{t}, \ldots \tag{37.19}$$

A decomposition of the functions $g_m = g_m(\tau)$, $m = 0, 1, 2$, in the Fourier series on the segment $[0, t]$ on the eigenfunctions (37.19) has the following view:

$$g_m(\tau) = \sum_{n=0}^{\infty} D_n^{(m)}\Omega_n(\tau),\ D_0^{(m)} = \frac{1}{t}\int_0^t g_m(\tau)d\tau,\ D_n^{(m)} = \frac{2}{t}\int_0^t \Omega_n^{(m)}(\tau)g_m(\tau)d\tau. \tag{37.20}$$

In this paper, let us confine ourselves to considering of symmetric processes of proportional loading, so $D_0^{(m)} = 0$, $m = 0, 1, 2$.

A maximum value on the coordinates of normal ν_2, ν_3 in the expression (37.15) can be found analytically. There are introduced the following notations:

$$y_1 = \sum_{k=1}^{3} \alpha_k \nu_k^2,\ y_2 = \sum_{k=1}^{3} \alpha_k^2 \nu_k^2. \tag{37.21}$$

So the expressions (37.13) and (37.14) take the following view:

$$S_n^{(\nu)}(\tau) = S_1 f(\tau) y_1,\ S_\tau^{(\nu)}(\tau) = S_1 f(\tau)\sqrt{y_2 - y_1^2}. \tag{37.22}$$

In the notations (37.21) the equality (37.15) is written so:

$$\begin{aligned}\varphi_j(z) = |S_1| \max_{\nu_2,\nu_3:\ \nu_2^2+\nu_3^2\le 1} &\Big\{ |\alpha| \left|\Pi_j^{(0)}\left(z, (f_1(\tau))_{\tau=0}^t\right)\right| \\ &+ |y_1| \left|\Pi_j^{(1)}\left(z, (f(\tau))_{\tau=0}^t\right)\right| + \sqrt{y_2 - y_1^2}\left|\Pi_j^{(2)}\left(z, (f(\tau))_{\tau=0}^t\right)\right|\Big\},\ z \in [0, t].\end{aligned} \tag{37.23}$$

For the convenience of the maximum value finding in Eq. (37.23), it is proposed to represent it in this form:

$$\begin{aligned}\varphi_j(z) &= |S_1|\, p_j^{(2)}(z)\left(|\alpha|\, p_j^{(0)}(z) + \max_{\nu_2,\nu_3:\ \nu_2^2+\nu_3^2\le 1}\left\{|y_1|\, p_j^{(1)}(z) + \sqrt{y_2 - y_1^2}\right\}\right),\\ p_j^{(2)}(z) &= \left|\Pi_j^{(2)}\left(z, (f(\tau))_{\tau=0}^t\right)\right|,\ p_j^{(1)}(z) = \frac{\left|\Pi_j^{(1)}\left(z, (f(\tau))_{\tau=0}^t\right)\right|}{\left|\Pi_j^{(2)}\left(z, (f(\tau))_{\tau=0}^t\right)\right|},\\ p_j^{(0)}(z) &= \frac{\left|\Pi_j^{(0)}\left(z, (f_1(\tau))_{\tau=0}^t\right)\right|}{\left|\Pi_j^{(2)}\left(z, (f(\tau))_{\tau=0}^t\right)\right|},\ j = 1, ..., 6.\end{aligned} \tag{37.24}$$

In the expression (37.24) it is supposed, that $p_j^{(2)}(z) \neq 0$ for any $z \in [0, t]$. The case $p_j^{(2)}(z) = 0$ will be considered below. So it is necessary to solve the following problem:

$$\begin{aligned}&\max_{v_2,v_3:\ v_2^2+v_3^2\le 1}\left\{\left|y_1 p_j^{(1)}(z)+\sqrt{y_2-y_1^2}\right|\right\},\quad z=\text{const},\\&y_1=1-(1-\alpha_2)v_2^2-(1-\alpha_3)v_3^2,\ \ y_2=1-(1-\alpha_2^2)v_2^2-(1-\alpha_3^2)v_3^2,\\&\alpha\in[-1,-0.5],\ \alpha_3=-1-\alpha_2,\ p_2(z)\neq 0.\end{aligned}\tag{37.25}$$

The points of the global maximum of the problem (37.25) have the following form:

$$v_2^{*2}=\frac{1}{2}\left(1-\frac{p_1}{\sqrt{1+p_1^2}}\right),\quad v_3^*=0.\tag{37.26}$$

After substituting (37.26) into (37.24), is it obtained the following:

$$\varphi_j(z)=|S_1|\,p_j^{(2)}(z)\left[|\alpha|\,p_j^0(z)+\frac{(1+\alpha_2)p_j^{(1)}(z)+(1-\alpha_2)\sqrt{1+p_1^{(2)}(z)}}{2}\right],z\in[0,t].\tag{37.27}$$

At $\alpha_2=-0.5$ the symmetry at $\alpha_2=\alpha_3$ causes the presence of following additional global maximum points, in addition to those already mentioned in (37.26):

$$v_2^*=0,\ v_3^{*2}=\frac{1}{2}\left(1-\frac{p_1}{\sqrt{1+p_1^2}}\right).\tag{37.28}$$

After substituting of the function $p_j^{(m)}=p_j^{(m)}(z)$ from the relation (37.24) the expression (37.27) takes the form:

$$\begin{aligned}\varphi_j(z)=|S_1|\Bigg[&|\alpha|\left|\Pi_j^{(0)}\left(z,(f_1(\tau))_{\tau=0}^t\right)\right|+\frac{1+\alpha_2}{2}\left|\Pi_j^{(1)}\left(z,(f(\tau))_{\tau=0}^t\right)\right|\\&+\frac{1-\alpha_2}{2}\sqrt{\left(\Pi_j^{(1)}\left(z,(f(\tau))_{\tau=0}^t\right)\right)^2+\left(\Pi_j^{(2)}\left(z,(f(\tau))_{\tau=0}^t\right)\right)^2}\Bigg],\ j=1,...,6.\end{aligned}\tag{37.29}$$

Using the relation (37.17) here can obtained that

$$\begin{aligned}\varphi_j(z)=|S_1|\Bigg(&|\alpha|\left|\sum_{n=1}^{\infty}\frac{\xi_n^{(1)}\Omega_n(z)}{\lambda_{j,n}^{(0)}}\right|+\frac{1+\alpha_2}{2}\left|\sum_{n=1}^{\infty}\frac{\xi_n^{(0)}\Omega_n(z)}{\lambda_{j,n}^{(1)}}\right|\\&+\frac{1-\alpha_2}{2}\sqrt{\left(\sum_{n=1}^{\infty}\frac{\xi_n^{(0)}\Omega_n(z)}{\lambda_{j,n}^{(1)}}\right)^2+\left(\sum_{n=1}^{\infty}\frac{\xi_n^{(0)}\Omega_n(z)}{\lambda_{j,n}^{(2)}}\right)^2}\Bigg),\quad j=1,\dots,6,\\\Omega_1&=\sin\frac{2\pi z}{t},\ \Omega_2=\cos\frac{2\pi z}{t},\dots,\ \Omega_{2n-1}=\sin\frac{2\pi nz}{t},\ \Omega_{2n}=\cos\frac{2\pi nz}{t},\dots\end{aligned}\tag{37.30}$$

$$\xi^{(0)} = f(\tau), \xi^{(1)} = f_1(\tau), \xi_n^{(l)} = \frac{2}{t}\int\limits_0^t \Omega_n(\tau)\xi^{(l)}d\tau, l = 0, 1, \; n = 1, 2, 3, \ldots \tag{37.31}$$

And finally substituting (37.28) in (37.6), the constitutive relations for the function

$$Q_j = Q_j(\tau), \quad j = 1, \ldots, 6,$$

have the final form ($j = 1, \ldots, 6$):

$$Q_j(\tau) = |S_1|\left(\lg\frac{\tau}{\tau_{j-1}\left(S_1, \alpha_2, \alpha, f(\tau), f_1(\tau)\right)}\right)^{\chi}\max_{z\in[0,t]}\left[|\alpha|\left|\sum_{n=1}^{\infty}\frac{\xi_n^{(1)}\Omega_n(z)}{\lambda_{j,n}^{(0)}}\right|\right.$$
$$\left.+\frac{1+\alpha_2}{2}\left|\sum_{n=1}^{\infty}\frac{\xi_n^{(0)}\Omega_n(z)}{\lambda_{j,n}^{(1)}}\right| + \frac{1-\alpha_2}{2}\sqrt{\left(\sum_{n=1}^{\infty}\frac{\xi_n^{(0)}\Omega_n(z)}{\lambda_{j,n}^{(1)}}\right)^2 + \left(\sum_{n=1}^{\infty}\frac{\xi_n^{(0)}\Omega_n(z)}{\lambda_{j,n}^{(2)}}\right)^2}\right]. \tag{37.32}$$

37.4 Determination of the Model Material Function

The problem of the criteria formulation involves the complete identification of the model, i.e. the method of experimental determination of all material functions and constants. To find the eigenfunctions

$$\left\{\lambda_{j,n}^{(m)}\right\}_{n=1}^{\infty}, \qquad m = 0, 1, 2, \quad j = 1, \ldots, 6,$$

which are the material functions of the model, here can be considered the following three stress-strain states of the form (37.10):

- pure shear:

$$\sigma_{11}(\tau) = -\sigma_{22}(\tau) = \tau_a\Omega_n(\tau), \; \sigma_{33}(\tau) = 0, \tag{37.33}$$
$$S_1 = \tau_a, \; \alpha_1 = 1, \; \alpha_2 = -1, \; \alpha_3 = \alpha = 0, \; f_1(\tau) = f(\tau) = \Omega_n(\tau),$$

- one-dimensional loading process:

$$\sigma_{11}(\tau) = \sigma_a\Omega_n(\tau), \; \sigma_{22}(\tau) = \sigma_{33}(\tau) = 0, \tag{37.34}$$
$$S_1 = \frac{2}{3}\sigma_a, \; \alpha_1 = 1, \; \alpha_2 = \alpha_3 = -\frac{1}{2}, \; \alpha = \frac{1}{2}, \; f_1(\tau) = f(\tau) = \Omega_n(\tau),$$

- uniform two-dimensional loading:

$$\sigma_{11}(\tau) = 0, \; \sigma_{22}(\tau) = \sigma_{33}(\tau) = \sigma_a^*\Omega_n(\tau), \tag{37.35}$$

$$S_1 = -\frac{2}{3}\sigma_a^*,\ \alpha_1 = 1,\ \alpha_2 = \alpha_3 = -\frac{1}{2},\ \alpha = -1,\ f_1(\tau) = f(\tau) = \Omega_n(\tau),$$

when τ_a, σ_a, σ_a^* are the amplitudes of the corresponding loadings, $\omega = 2\pi n/t$ is the circular frequency, n is the number of cycles.

At loadings (37.33) – (37.35) the equation (37.32) leads to the following:

$$Q_j(n) = \tau_a \left(\lg \frac{n}{n_{j-1}(\tau_a, -1, 0)}\right)^{\chi} \sqrt{\frac{1}{\lambda_{j,n}^{(1)\,2}} + \frac{1}{\lambda_{j,n}^{(2)\,2}}}, \tag{37.36}$$

$$Q_j(n) = \frac{1}{3}\sigma_a \left(\lg \frac{n}{n_{j-1}\left(\frac{2}{3}\sigma_a, -\frac{1}{2}, \frac{1}{2}\right)}\right)^{\chi} \left(\frac{1}{\lambda_{j,n}^{(0)}} + \frac{1}{2}\frac{1}{\lambda_{j,n}^{(1)}} + \frac{3}{2}\sqrt{\frac{1}{\lambda_{j,n}^{(1)\,2}} + \frac{1}{\lambda_{j,n}^{(2)\,2}}}\right), \tag{37.37}$$

$$Q_j(n) = \frac{1}{3}\sigma_a^* \left(\lg \frac{n}{n_{j-1}\left(-\frac{2}{3}\sigma_a^*, -\frac{1}{2}, -1\right)}\right)^{\chi} \left(\frac{2}{\lambda_{j,n}^{(0)}} + \frac{1}{2}\frac{1}{\lambda_{j,n}^{(1)}} + \frac{3}{2}\sqrt{\frac{1}{\lambda_{j,n}^{(1)\,2}} + \frac{1}{\lambda_{j,n}^{(2)\,2}}}\right). \tag{37.38}$$

In expressions (37.36) – (37.38) and further, the corresponding values of parameters S_1, α_2 and α of the process (37.10) are indicated in the arguments for the function of the number of cycles $n_j = n_j(S_1, \alpha_2, \alpha)$, at which defects of the j-th level reach their ultimate state (a durability at the j-th level). Using equations (37.3), it can be obtained a system of eigenvalues $\{\lambda_{j,n}^{(m)}\}_{n=0}^{\infty}$:

$$\tau_j \left(\lg \frac{N_j}{n_{j-1}(\tau_a, -1, 0)}\right)^{\chi} \sqrt{\frac{1}{\lambda_{j,n}^{(1)\,2}} + \frac{1}{\lambda_{j,n}^{(2)\,2}}} = 1, \tag{37.39}$$

$$\frac{1}{3}\sigma_j \left(\lg \frac{N_j}{n_{j-1}\left(\frac{2}{3}\sigma_j, -\frac{1}{2}, \frac{1}{2}\right)}\right)^{\chi} \left(\frac{1}{\lambda_{j,n}^{(0)}} + \frac{1}{2}\frac{1}{\lambda_{j,n}^{(1)}} + \frac{3}{2}\sqrt{\frac{1}{\lambda_{j,n}^{(1)\,2}} + \frac{1}{\lambda_{j,n}^{(2)\,2}}}\right) = 1, \tag{37.40}$$

$$\frac{1}{3}\sigma_j^* \left(\lg \frac{N_j}{n_{j-1}\left(-\frac{2}{3}\sigma_j^*, -\frac{1}{2}, -1\right)}\right)^{\chi} \left(\frac{2}{\lambda_{j,n}^{(0)}} + \frac{1}{2}\frac{1}{\lambda_{j,n}^{(1)}} + \frac{3}{2}\sqrt{\frac{1}{\lambda_{j,n}^{(1)\,2}} + \frac{1}{\lambda_{j,n}^{(2)\,2}}}\right) = 1, \tag{37.41}$$

when

$$\tau_j = \tau_j\left(n, N_j\right), \quad \sigma_j = \sigma_j\left(n, N_j\right), \quad \sigma_j^* = \sigma_j^*\left(n, N_j\right), \quad j = 1, \ldots, 6,$$

are the material functions, at which defects of the$j$-th level reach their ultimate state at symmetrical shear, one-dimensional and two-dimensional uniform loadings respectively, and the function $l_j^* = l_j^*(\tau)$ is the limiting value $l_{f,j}^*$ at the corresponding loading.

It is necessary to conduct the series of corresponding fatigue macroexperiments on the each level at the given frequency $\omega = 2\pi n/t$ with the surface cutting processing using microstructure analysis methods to find the material functions $\eta_j = \eta_j(n, N_j)$, $j = 1, \ldots, 6$, $n = 1, 2, 3, \ldots$, when η_j are the functions $\tau_j, \sigma_j, \sigma_j^*$, at which the defect of the j-th level reach their ultimate state at symmetrical shear, one-dimentional and two-dimentional uniform loadings respectively. Well-known works, reference books, and other sources contain a limited amount of information. In some special cases, the dependences of the density of defects and their sizes on the stress amplitude may be known for materials. An analysis of the known results of Vickers microhardness studies makes it possible to compare the lines of the defect ultimate states to the lines of equal microhardness (isolevels of microhardness) [56]. In general, currently, due to the complexity to conduct such experiments and the insufficient amount of known data, the problem of determining of fatigue characteristics at the microlevel based on test data at the macrolevel arises. A certain method of finding of the function $\eta_j = \eta_j(n, N_j)$ was proposed [55, 57]. According to experimental data, a pair (η_4, N_4) is determined, the value η_4 is assumed to be equal to the endurance η_{-1} for the number of cycles N_4, at which macrofracture by short cracks is observed. Significant change of the micro- and macrostructure observed at changes of the number of loading cycles for decimal orders, so you can put: $\lg N_j = \lg N_4 + 4 - j$, $j = 1, 2, 3$, and $\lg N_j = \lg N_4 + 3 - j$, $j = 5, 6$. In the work the amplitudes η_j, $j = 1, 2, 3$, of microlevel ultimate state reaching are determined via the value η_4 as follows: $\eta_1 = 0.65\eta_4$, $\eta_2 = 0.76\eta_4$, $\eta_3 = 0.86\eta_4$. Mark if it is existed the relevant experimental data, other values can be considered as material constants. For example, if the material sensitivity η_{th} this known then it is assumed: $\eta_1 = \eta_{th}$. The stress amplitude σ_5 can be determined from the experimental fatigue curve by the formation of single macrocracks at N_5. In the absence of any data, it is assumed: $\eta_6 = \sigma_s$, where η_s is the material yield strength.

After the introducing of the following notations:

$$\begin{aligned}
\mu_j^{\tau}(N_j) &= \left(\lg \frac{N_j}{n_{j-1}\left(\tau_a, -1, 0\right)}\right)^{\chi}, \\
\mu_j^{\sigma}(N_j) &= \left(\lg \frac{N_j}{n_{j-1}\left(\frac{2}{3}\sigma_j, -\frac{1}{2}, \frac{1}{2}\right)}\right)^{\chi}, \\
\mu_j^{\sigma^*}(N_j) &= \left(\lg \frac{N_j}{n_{j-1}\left(-\frac{2}{3}\sigma_j^*, -\frac{1}{2}, -1\right)}\right)^{\chi}.
\end{aligned} \tag{37.42}$$

The solution of the system (37.39) – (37.41) takes the form:

$$\frac{1}{\lambda_{j,n}^{(0)}} = 3\left(\frac{1}{\sigma_j^*\left(n, N_j\right)\mu_j^{\sigma^*}(N_j)} - \frac{1}{\sigma_j\left(n, N_j\right)\mu_j^{\sigma}(N_j)}\right), \tag{37.43}$$

$$\frac{1}{\lambda^{(1)}_{j,n}} = \left(-\frac{3}{\tau_j\left(n,N_j\right)\mu^{\tau}_j(N_j)} + \frac{12}{\sigma_j\left(n,N_j\right)\mu^{\sigma}_j(N_j)} - \frac{6}{\sigma^*_j\left(n,N_j\right)\mu^{\sigma^*}_j(N_j)}\right), \tag{37.44}$$

$$\frac{1}{\lambda^{(2)}_{j,n}} = \sqrt{\frac{1}{\left(\tau_j\left(n,N_j\right)\mu^{\tau}_j\left(N_j\right)\right)^2} - \frac{1}{\lambda^{(1)\,2}_{j,n}}}. \tag{37.45}$$

In the expressions (37.43) – (37.45) it is assumed that $\lambda^{(m)}_{j,n} > 0$, $j = 1,\ldots,6$, $m = 0, 1, 2$.

So the constitutive relations for the functions $Q_j = Q_j(\tau)$, $j = 1,\ldots,6$, at loading (37.10) according the expressions (37.31) taking into account (37.29), (37.30), (37.42) – (37.45) are represented in the following final form ($j = 1,\ldots,6$):

$$Q_j(\tau) = |S_1|\left(\lg\frac{\tau}{\tau_{j-1}\left(S_1,\alpha_2,\alpha,f(\tau),f_1(\tau)\right)}\right)^{\chi}\max_{z\in[0,t]}\left\{|\alpha|\left|\sum_{n=1}^{\infty}\frac{\xi^{(1)}_n\Omega_n(z)}{\lambda^{(0)}_{j,n}}\right|\right.$$
$$\left.+\frac{1+\alpha_2}{2}\left|\sum_{n=1}^{\infty}\frac{\xi^{(0)}_n\Omega_n(z)}{\lambda^{(1)}_{j,n}}\right| + \frac{1-\alpha_2}{2}\sqrt{\left(\sum_{n=1}^{\infty}\frac{\xi^{(0)}_n\Omega_n(z)}{\lambda^{(1)}_{j,n}}\right)^2 + \left(\sum_{n=1}^{\infty}\frac{\xi^{(0)}_n\Omega_n(z)}{\lambda^{(2)}_{j,n}}\right)^2}\right\}, \tag{37.46}$$

$$\Omega_1 = \sin\frac{2\pi z}{t},\ \Omega_2 = \cos\frac{2\pi z}{t},\ldots,\ \Omega_{2n-1} = \sin\frac{2\pi nz}{t},\ \Omega_{2n} = \cos\frac{2\pi nz}{t},\ldots$$

$$\xi^{(0)} = f(\tau),\ \ \xi^{(1)} = f_1(\tau),\ \ \xi^{(l)}_n = \frac{2}{t}\int_0^t \Omega_n(\tau)\xi^{(l)}d\tau,\ \ l = 0, 1,\ n = 1, 2, 3, \ldots$$

$$\frac{1}{\lambda^{(0)}_{j,n}} = 3\left(\frac{1}{\sigma^*_j\left(n,N_j\right)\mu^{\sigma^*}_j(N_j)} - \frac{1}{\sigma_j\left(n,N_j\right)\mu^{\sigma}_j(N_j)}\right),$$
$$\frac{1}{\lambda^{(1)}_{j,n}} = \left(-\frac{3}{\tau_j\left(n,N_j\right)\mu^{\tau}_j(N_j)} + \frac{12}{\sigma_j\left(n,N_j\right)\mu^{\sigma}_j(N_j)} - \frac{6}{\sigma^*_j\left(n,N_j\right)\mu^{\sigma^*}_j(N_j)}\right), \tag{37.47}$$

$$\frac{1}{\lambda^{(2)}_{j,n}} = \sqrt{\frac{1}{\left(\tau_j\left(n,N_j\right)\mu^{\tau}_j\left(N_j\right)\right)^2} - \frac{1}{\lambda^{(1)\,2}_{j,n}}},$$

$$\mu^{\tau}_j(N_j) = \left(\lg\frac{N_j}{n_{j-1}\left(\tau_a,-1,0\right)}\right)^{\chi},\ \ \mu^{\sigma}_j(N_j) = \left(\lg\frac{N_j}{n_{j-1}\left(\frac{2}{3}\sigma_j,-\frac{1}{2},\frac{1}{2}\right)}\right)^{\chi}, \tag{37.48}$$

$$\mu_j^{\sigma^*} = \left(\lg \frac{N_j}{n_{j-1}\left(-\frac{2}{3}\sigma_j^*, -\frac{1}{2}, -1\right)} \right)^{\chi}.$$

37.5 The Case of a Material with the same Fatigue Characteristics at Uniaxial Loading, Shear, and Biaxial Uniform Loading

Let uss consider the model case of a material under the following condition:

$$\tau_{-1} = \sigma_{-1} = \sigma_{-1}^*.$$

In this case, the functions $\varphi_j = \varphi_j(z)$, $j = 1, \ldots, 6$, have the following form:

$$\varphi_j(z) = |S_1| \max_{\nu_2,\nu_3:\ \nu_2^2+\nu_3^2 \le 1} \left\{ |\alpha| \left| \Pi_j^{(0)} \left(z, (f_1(\tau))_{\tau=0}^t\right) \right| + |y_1| \left| \Pi_j^{(1)} \left(z, (f(\tau))_{\tau=0}^t\right) \right| \right\}. \tag{37.49}$$

After finding the maximum value the expression (37.49) takes the view:

$$\varphi_j(z) = |S_1| \left(|\alpha| \left| \Pi_j^{(0)} \left(z, (f_1(\tau))_{\tau=0}^t\right) \right| + \left| \Pi_j^{(1)} \left(z, (f(\tau))_{\tau=0}^t\right) \right| \right). \tag{37.50}$$

Using the relation (37.17) it is obtained that

$$\varphi_j(z) = |S_1| \left(|\alpha| \left| \sum_{n=1}^{\infty} \frac{\xi_n^{(1)} \Omega_n(z)}{\lambda_{j,n}^{(0)}} \right| + \left| \sum_{n=1}^{\infty} \frac{\xi_n^{(0)} \Omega_n(z)}{\lambda_{j,n}^{(1)}} \right| \right) \tag{37.51}$$

$$\Omega_1 = \sin\frac{2\pi z}{t},\ \Omega_2 = \cos\frac{2\pi z}{t}, \ldots,\ \Omega_{2n-1} = \sin\frac{2\pi n z}{t},\ \Omega_{2n} = \cos\frac{2\pi n z}{t}, \ldots$$

$$\xi^{(0)} = f(\tau),\ \xi^{(1)} = f_1(\tau),\ \xi_n^{(l)} = \frac{2}{t}\int_0^t \Omega_n(\tau)\xi^{(l)} d\tau,\ l = 0, 1,\ n = 1, 2, 3, \ldots$$

and substituting (37.51) into (37.6) the following expressions for the functions

$$Q_j = Q_j(\tau), \quad j = 1, \ldots, 6,$$

can be written:

$$Q_j(\tau) = |S_1| \left(\lg \frac{\tau}{\tau_{j-1}\left(S_1, \alpha_2, \alpha, f(\tau), f_1(\tau)\right)} \right)^{\chi} \max_{z\in[0,t]} \left\{ |\alpha| \left| \sum_{n=1}^{\infty} \frac{\xi_n^{(1)} \Omega_n(z)}{\lambda_{j,n}^{(0)}} \right| + \left| \sum_{n=1}^{\infty} \frac{\xi_n^{(0)} \Omega_n(z)}{\lambda_{j,n}^{(1)}} \right| \right\}. \tag{37.52}$$

For the finding the material functions

$$\left\{ \lambda_{j,n}^{(m)} \right\}_{n=1}^{\infty}, \quad m = 0, 1, \quad j = 1, \dots, 6,$$

it is enough to consider one of regimes of loading (37.33) – (37.35). Using the equation (37.3) and the formula (37.52), it can be written the following:

$$\lambda_{j,n}^{(0)} = \lambda_{j,n}^{(1)} = \sigma_j\left(n, N_j\right) \left(\lg \frac{N_j}{n_{j-1}\left(\tau_a, -1, 0\right)} \right). \tag{37.53}$$

So the constitutive relations for the functions $Q_j = Q_j(\tau)$, $j = 1, \dots, 6$, in the case at loading (37.10) are as follows:

$$Q_j(\tau) = |S_1| \left(\lg \frac{\tau}{\tau_{j-1}\left(S_1, \alpha_2, \alpha, f(\tau), f_1(\tau)\right)} \right)^{\chi} \max_{z\in[0,t]} \left\{ |\alpha| \left| \sum_{n=1}^{\infty} \frac{\xi_n^{(1)} \Omega_n(z)}{\lambda_{j,n}^{(0)}} \right| + \left| \sum_{n=1}^{\infty} \frac{\xi_n^{(0)} \Omega_n(z)}{\lambda_{j,n}^{(1)}} \right| \right\}, \tag{37.54}$$

$$\Omega_1 = \sin \frac{2\pi z}{t}, \ \Omega_2 = \cos \frac{2\pi z}{t}, \dots, \ \Omega_{2n-1} = \sin \frac{2\pi n z}{t}, \ \Omega_{2n} = \cos \frac{2\pi n z}{t}, \dots,$$

$$\xi^{(0)} = f(\tau), \ \xi^{(1)} = f_1(\tau), \xi_n^{(l)} = \frac{2}{t} \int_0^t \Omega_n(\tau) \xi^{(l)} d\tau, \ l = 0, 1, \ n = 1, 2, 3, \dots$$

$$\lambda_{j,n}^{(0)} = \lambda_{j,n}^{(1)} = \sigma_j\left(n, N_j\right) \left(\lg \frac{n}{n_{j-1}\left(\sigma_a, -1, 0\right)} \right).$$

37.6 Calculation Results for Symmetrical Axial Loading with Torsion

The process of symmetrical axial loading with an amplitude σ_{a} with torsion of the amplitude τ_{a} in the principal axes (37.10) is written as follows:

$$\sigma_{11}(\tau) = \frac{\sigma_a + \sqrt{\sigma_a^2 + 4\tau_a^2}}{2}\sin\omega\tau, \qquad \sigma_{22}(\tau) = \frac{\sigma_a - \sqrt{\sigma_a^2 + 4\tau_a^2}}{2}\sin\omega\tau,$$
$$\alpha = -\alpha_3 = \frac{2\sigma_a}{\sigma_a + 3\sqrt{\sigma_a^2 + 4\tau_a^2}}, \qquad \alpha_2 = \frac{\sigma_a - 3\sqrt{\sigma_a^2 + 4\tau_a^2}}{\sigma_a + 3\sqrt{\sigma_a^2 + 4\tau_a^2}}. \qquad (37.55)$$

Experimental data on macrofailure for 42CrMo4 (AISI 4140) steel (thermally treated by austenization, quenching and holding) under axial tension-compression, shear and axial torsional loading were obtained in [4]. The addition of chromium and molybdenum to the steel provides the significant strength (static strength $\sigma_b = 1100$ MPa, yield strength $\sigma_s = 980$ MPa) and ductility (residual elongation at break is $\delta = 16\%$). It is widely used in automotive components (crankshafts, front wheels, steering elements). For 42CrMo4 steel, Fig. 37.1 shows the results of calculations of 1-6 defect levels at axial loading (steak stipple lines) and torsion (solid lines). The corresponding calculated fatigue curves by macrofracture well describe the experimental data [4] (crosses at axial loading and points for torsion). In Fig. 37.2a and 37.2b represent the calculated fatigue curves for 1-6 defect levels (the solid lines) at $\sigma_a/\tau_a = 1$ and $\sigma_a/\tau_a = 3$, respectively. Calculated fatigue curves and experimental data for axial loading (streak stipple lines and crosses) are also shown here. The results of calculations based on the model well describe the experimental data [4] (dots).

Figure 37.3 for 42CrMo4 steel shows the calculated $N_{f,t}$ and experimental $N_{f,e}$ durability for the five proportional loading processes considered. It can be seen from the figure that the calculation results satisfy the corresponding experimental data, the variation in durability does not exceed one order of magnitude. Calculations based

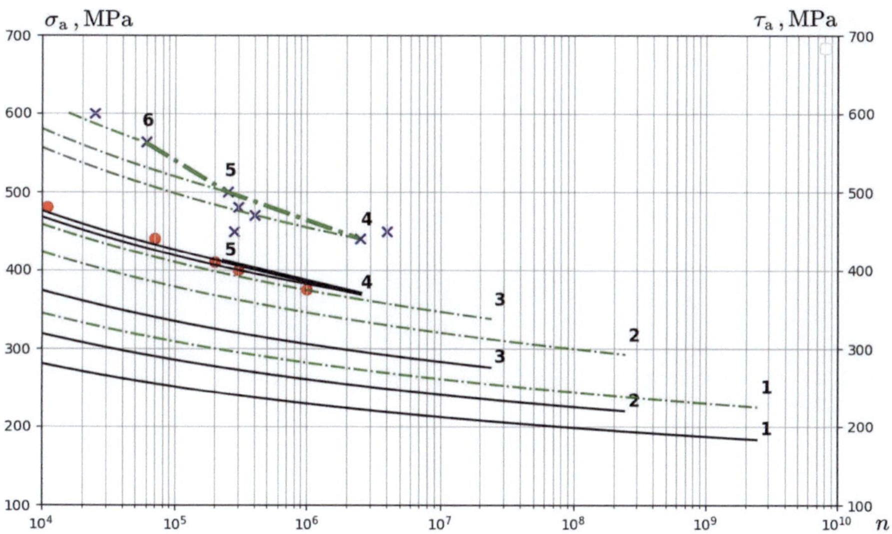

Fig. 37.1: The fatigue curves by the 1-6 defect levels and test data [4] for 42CrMo4 steel at axial loading and shear.

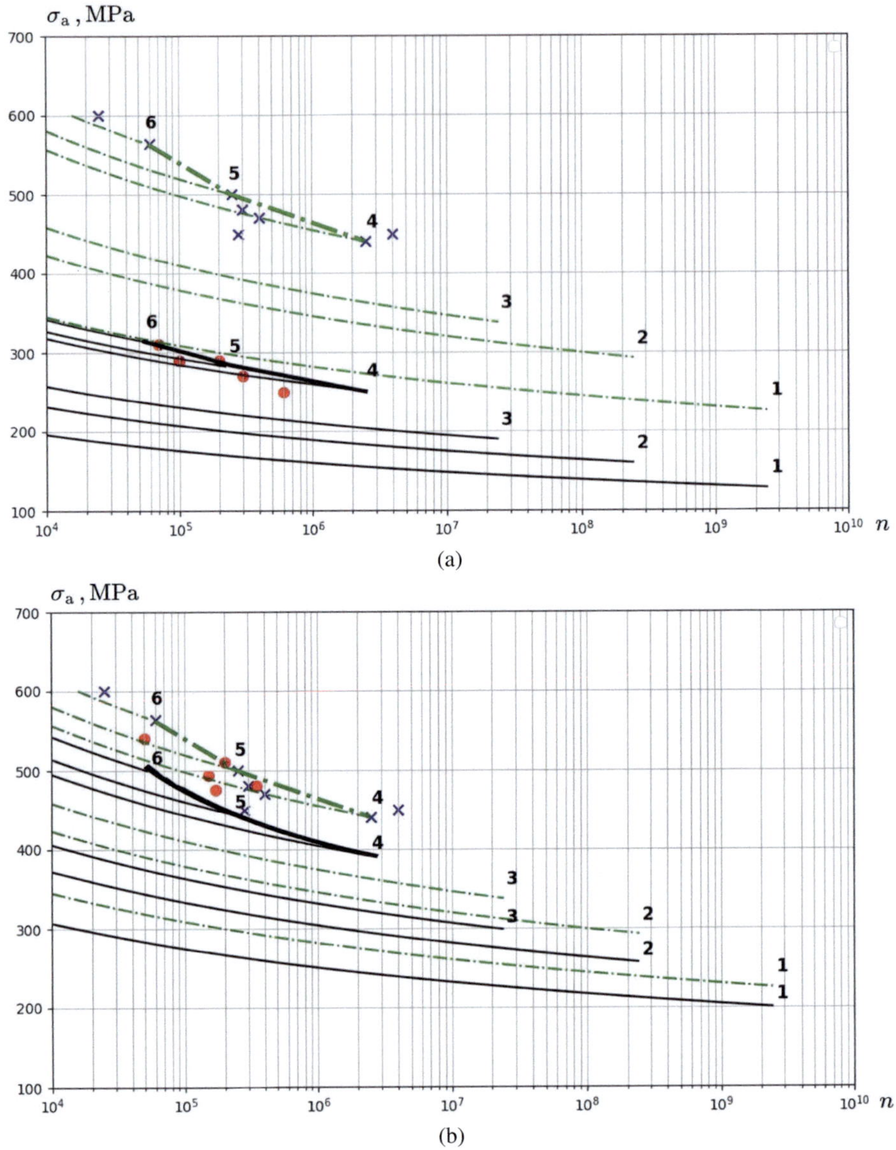

Fig. 37.2: The fatigue curves by the 1-6 defect levels and test data [4] for 42CrMo4 steel at axial loading and axial loading with torsion: (a) $\sigma_a/\tau_a = 1$; (b) $\sigma_a/\tau_a = 3$.

on the model have been performed for the following aluminium alloys 42CrMo4, S355, SAE1045 steels and 075-T651 at axial loading with torsion and different amplitude ratios, their results satisfactorily correspond to known experimental data [4, 11, 50, 60]. The calculated values of the amplitudes were determined for each

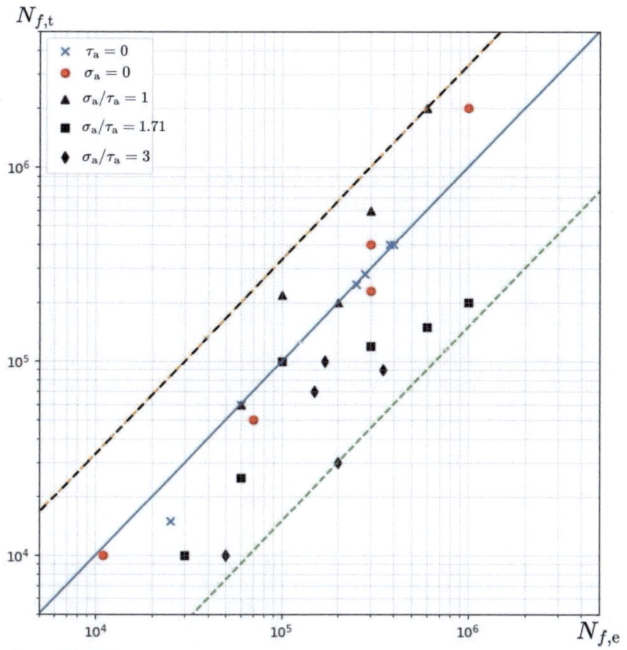

Fig. 37.3: Correlation between the calculated $N_{\mathrm{f,t}}$ and experimental $N_{\mathrm{f,e}}$ durability.

of the studied materials at $N_{\mathrm{f}} = 4\ 10^6$ cycles to macrofracture for various levels of macrodeficiency (4-6), which are represented by the corresponding curves depending on $\sigma_{\mathrm{a}}/\tau_{\mathrm{a}} \geq 1$ in Fig. 37.4a and $\tau_{\mathrm{a}}/\sigma_{\mathrm{a}} \geq 1$ in Fig. 37.4b (the values $\sigma_{\mathrm{a}}^{\mathrm{sin}}$ and $\tau_{\mathrm{a}}^{\mathrm{sin}}$ are the fatigue limits by defect levels at axial loading and torsion, respectively). A characteristic feature is that for each material, all these curves, regardless of the number of cycles and the level of macrodeficiency, practically coincide. Figure 37.4a shows that, for example, the fatigue strength of 42CrMo4 steel (the calculation results are represented by a streak stipple line) at proportional loading (37.53) can be calculated on the axial loading at $\sigma_{\mathrm{a}}/\tau_{\mathrm{a}} \geq 3$ (with an accuracy of 10%), aluminum alloy 075-T651 (dotted line) and SAE1045 steel (stipple line) at $\tau_{\mathrm{a}}/\sigma_{\mathrm{a}} \geq 3.5$, and for S255 steel (solid line) at $\sigma_a/\tau_a \geq 4$. An analysis of the calculations in Fig. 37.4b for steel 42CrMo4 shows: at $\tau_{\mathrm{a}}/\sigma_{\mathrm{a}} \geq 3.5$ the estimating of the strength at the proportional (37.53) by symmetrical shear gives an error of less than 5%. Fatigue of aluminum alloy 075-T651 and steel SAE1045 at $\tau_{\mathrm{a}}/\sigma_{\mathrm{a}} \geq 2$, and for S355 steel at $\tau_a/\sigma_a \geq 1.3$ can be determined by shear (with an accuracy of 10%).

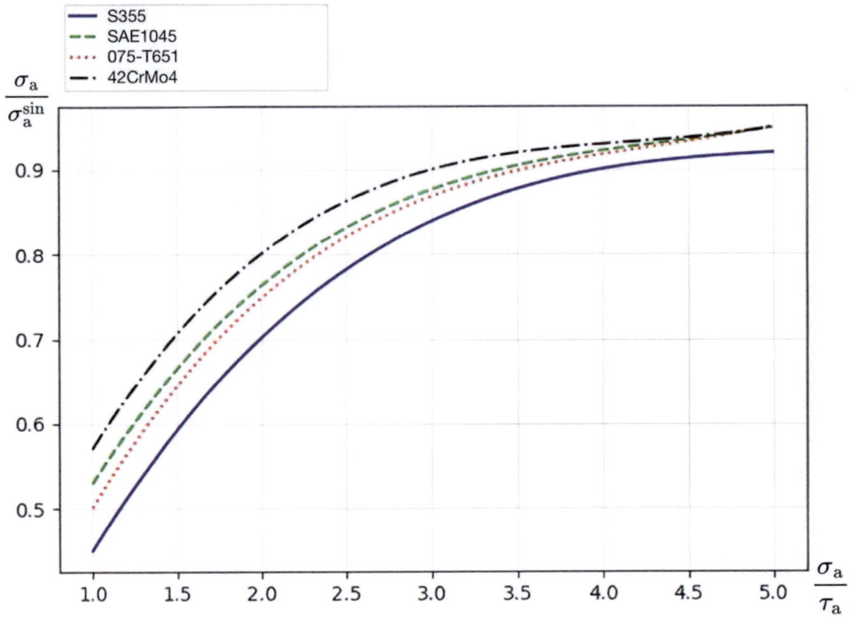

(a)

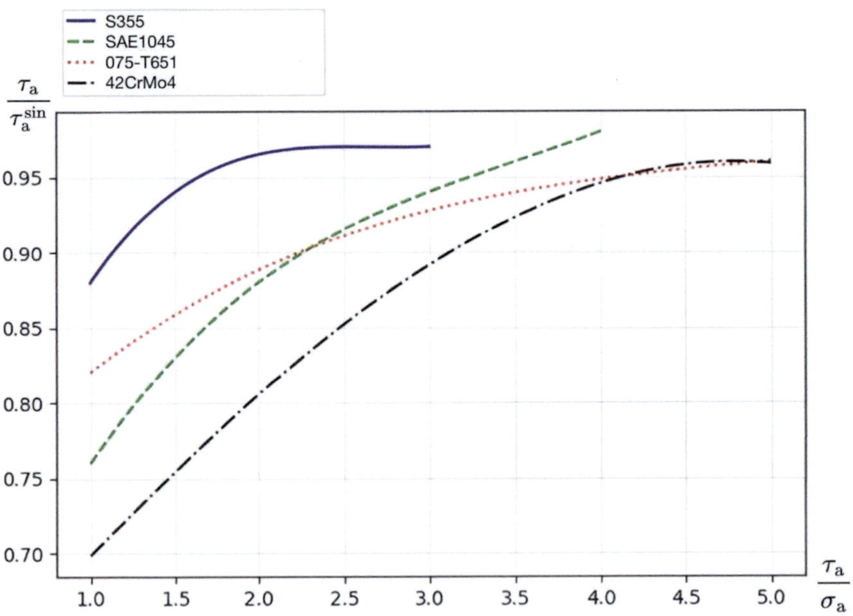

(b)

Fig. 37.4: Graphs (a) σ_a/σ_a^{sin} and (b) τ_a/τ_a^{sin} from the relation of amplitudes of axial loading with torsion.

37.7 Conclusion

High- (on average, $N_f \in (10^4, 10^6)$ cycles to fracture) and very high cycle ($N_f \in (10^6, 10^{11})$ cycles) fatigue at multiaxial stress-strain state is the most important fundamental and engineering problem for various fields of applications in the automotive, aviation, and aerospace industries, civil engineering. According to literature, more than 80% of all destructions with significant economic and social consequences are related to fatigue problems. In a variety of design schemes, a multiaxial stress-strain state is realized at combinations of alternating tension, bending and torsion, and so at uniaxial loading due to geometric concentrators and an inhomogeneous stress state. Sustained interest to the very high cycle fatigue is associated with the creation of new (nano- and submicrostructural) materials and new technologies due to this range can be reached. It is provided the economic relevance of the very high cycle resource estimation.

The relevance of fundamental research, modern methods of modeling of the failure process taking into account the physical patterns at different scale levels and the construction of fatigue curves by defect levels at multiaxial stress-strain state is determined by the solution of the problems of engineering system safe operation at long service life and the need to extend it, including using artificial intelligence methods.

In this work, within the framework of an interdisciplinary approach, a fatigue criterion is proposed based on an analysis of the brittle failure at different scale-structural levels. The probability of the ultimate states at the micro-, meso- , and macroscale levels (before the macrocrack-leader) is determined as a hierarchical system of recurrence relations based on three linear operators on the proportional loading. Based on them, a multilevel series of fatigue curves is constructed at proportional loading. A method for identifying of the material functions is proposed.

Verification of the proposed model was carried out for 42CrMo4, S355, SAE1045 steels and 075-T651 aluminum alloy at symmetrical axial loading with torsion and different amplitude ratios. The calculated fatigue curves by macrofracture correspond satisfactorily to the known experimental data.

One of the results is the justification of the possibility to replace the proportional loading the uniaxial loading and torsion. Thus, for 42CrMo4 steel in the range of axial load-to–torsion amplitudes $\sigma_a/\tau_a \geq 3$, for alloy 075–T651 and SAE1045 steel at $\sigma_a/\tau_a \geq 3.5$, and for steel S255 at $\sigma_a/\tau_a \geq 4$ it becomes possible to calculate the durability on the axial loading with the amplitude σ_a. Analysis of the calculation results shows that for steel 42CrMo4 at $\tau_a/\sigma_a \geq 3.5$, for alloy 075-T651 and steel SAE1045 at $\tau_a/\sigma_a \geq 2$ and for steel S355 at $\tau_a/\sigma_a \geq 1.3$ the strength can be estimated by shear. In all cases considered the difference between the values of the corresponding stresses does not exceed 10%. Finally, it should be noted that the calculation formulas are based on a general approach allowed to consider of various cases of proportional loadings.

References

[1] Abasolo M, Santasmartas LP, Eizmendi M (2024) A new critical plane multiaxial fatigue criterion with an exponent to account for high mean stress effect. Metals 14(9):964, DOI 10.3390/met14090964
[2] Adachi Y, Chen T-T, Sun F, Maruyama D, Sawai K, Fukatsu Y, Wang Z (2024) A review on inverse analysis models in steel material design. Materials Genome Engineering Advances 2(4):e71 DOI 10.1002/mgea.71
[3] Almamoori MK, Alizadeh Y, Abolghasemzadeh M (2021) A review of multi-axial fatigue tests. IOP Conference Series: Materials Science and Engi 1094:012058, DOI 10.1088/1757-899X/1094/1/012058
[4] Anes V, Reis L, Li B, Fonte M, de Freitas M (2014) New approach for analysis of complex multiaxial loading paths. International Journal of Fatigue 62:21–33, DOI 10.1016/j.ijfatigue.2013.05.004
[5] Apodaka Manzarbeitia P, Abasolo Bilbao M, Tabares Lorenzo E, Orue Llona X, Pallarés Santasmartas L (2025) Comparación de métodos de fatiga multiaxial de ciclos altos. Anales De Ingeniería Mecánica 1(24), DOI 10.63450/aim.1.83. 2025
[6] Bartsch H, Voelkel J, Feldmann M (2025) Developing artificial neural networks to estimate the fatigue strength of structural steel details using the new European database. Steel Construction 18(1):1-13, DOI 10.1002/stco.202400029
[7] Brown M, Miller KJ (1973) A theory for fatigue under multiaxial stress-strain conditions. Proceedings of the Institution of Mechanical Engineers 187(65):745–755
[8] Carpinteri A, Spagnoli A (2001) Multiaxial high-cycle fatigue criterion for hard metals. International Journal of Fatigue 23(2):135–145, DOI 10.1016/S0142-1123(00)00075-X
[9] Chen J, Liu Y (2022) Fatigue modeling using neural networks: A comprehensive review. Fatigue & Fracture of Engineering Materials & Structures 45(4):945–979, DOI 10.1111/ffe.13640
[10] Crossland B (1956) Effect of large hydrostatic pressures on torsional fatigue strength of an alloy steel. In: The third International Conference on Fatigue of Metals, London, pp 138–149
[11] Dantas R, Correia J, Lesiuk G, Rozumek D, Zhu S-P, de Jesus A, Susmel L, Berto F (2021) Evaluation of multiaxial high-cycle fatigue criteria under proportional loading for S355 steel. Engineering Failure Analysis 120:105037, DOI 10.1016/j.engfailanal.2020.105037
[12] Deperrois ARC (1991) Sur le calcul des limites d'endurance des aciers. Thèse de l'Ecole Polytechnique Palaiseau
[13] Fajri A, Rio Prabowo A, Muhayat N, Smaradhana DF, Bahatmaka A (2021) Fatigue analysis of engineering structures: State of development and achievement. Procedia Structural Integrity 33:19–26, DOI 10.1016/j.prostr.2021.10.004
[14] Fatemi A, Yang L (1998) Cumulative fatigue damage and life prediction theories: A survey of the state of the art for homogeneous materials. International Journal of Fatigue 20(1):9–34, DOI 10.1016/S0142-1123(97)00081-9

[15] Findley WN (1953) Combined-Stress Fatigue Strength of 76S-T61 Aluminum Alloy with Superimposed Mean Stresses and Corrections for Yielding. Natural Advisory Committee for Aeronautics Technical Note #2924
[16] Garud YS (1981) Multiaxial fatigue: a survey of the state of the art. Journal of Testing and Evaluation 9(3):165–178, DOI 10.1520/JTE11553J
[17] Glinka G, Shen G, Plumtree A (1995) A multiaxial fatigue strain energy density parameter related to the critical fracture plane. Fatigue & Engineering Materials & Structures 18(1):37–46, DOI 10.1111/j.1460-2695.1995.tb00140.x
[18] Gonçalves CA, Araujo JA, Mamiya EN (2005) Multiaxial fatigue: a stress based criterion for hard metals. International Journal of Fatigue 27(2):177–187, DOI 10.1016/j.ijfatigue.2004.05.006
[19] Gough HJ, Pollard HV (1935) The strength of metals under combined alternating stresses. Proceedings of the Institution of Mechanical Engineers 131(1):3-103, DOI 10.1243/PIME_PROC_1935_131_008_02
[20] Hectors K, De Waele W (2021) Cumulative damage and life prediction models for high-cycle fatigue of metals: A review. Metals 11(2):204, DOI 10.3390/met11020204
[21] Huang Z, Yan J, Zhang J, Han C, Peng J, Cheng Ju, Wang Z, Luo M, Yin P (2024) Deep learning-based fatigue strength prediction for ferrous alloy. Processes 12(10):2214, DOI 10.3390/pr12102214
[22] Ibrahim AU, Woldemichael DE (2024) Review of the techniques for life prediction and cumulative damage of multiaxial fatigue loading. International Journal of Engineering Trends and Technology 72(2):203–221, DOI 10.14445/22315381/IJETT-V72I2P122
[23] Il'yushin AA (1961) Ob osnovaniyakh obshchey matematicheskoy teorii plastichnosti (in Russ., On the foundations of the general mathematical theory of plasticity). In: Problems in the Plasticity Theory, pp. 3–29. Moscow: Izdatel'stvo Akademii Nauk SSSR
[24] Kakuno H, Kawada Y (1979) A new criterion of fatigue strength of a round bar subjected to combined static and repeated bending and torsion. Fatigue &Fracture of Engineering Materials & Structures 2(2):229–236, DOI 10.1111/j.1460-2695.1979.tb01358.x
[25] Liu Y, Mahadevan S (2005) Multiaxial high-cycle fatigue criterion and life prediction for metals. International Journal of Fatigue 27(7):790–800, DOI 10.1016/j.ijfatigue.2005.01.003
[26] Mamiya EN, Araujo JA (2002) Fatigue limit under multiaxial loadings: on the definition of the equivalent shear stress. Mechanics Research Communications 29(2-3):141–151, DOI 10.1016/S0093-6413(02)00239-2
[27] Marin J (1962) Behavior of Engineering Materials. New York: Prentice-Hall
[28] Matake T (1977) An explanation on fatigue limit under combined stress. Bulletin of JSME 20(141):257–263
[29] McDiarmid DL (1972) Failure criteria and cumulative damage in fatigue under multiaxial stress conditions. PhD thesis, The City University, London
[30] Mendes P, Correia J, Mourão A, Dantas R, de Jesus A, Horas C., Fantuzzi N, Manuel L (2024) High-cycle fatigue design curves of mild- and high-strength

steels for offshore applications. Structures 67:106827, DOI 10.1016/j.istruc.2024.106827
[31] Miller KJ, de Los Rios ER (2005) The Behaviour of Short Fatigue Cracks. John Wiley & Sons Inc
[32] Nishihara T, Kawamoto M (1940) The fatigue test of steel under combined bending and torsion (in Japanese). Transactions of the Japan Society of Mechanical Engineers 6(24-1):8-11, DOI 10.1299/kikai1938.6.24-1_8
[33] Nishihara T, Kawamoto M (1944) A new criterion for the strength of metals under combined alternating stresses. Memoirs of the College of Engineering, Kyoto Imperial University 11(4):65–83
[34] Opasanon N, Stahr LJ, Suchy L, Hasse A (2025) Augmenting fatigue datasets for improved multiaxial fatigue strength prediction with neural networks. Metals 15(5), 528, DOI 10.3390/met15050528
[35] Pallarés-Santasmartas L (2021) Development of Two Multiaxial Fatigue Methods for Ductile Metals Based on Energy Considerations. PhD Thesis, University of the Basque Country, Leioa, Spain
[36] Pan R, Gao J, Yuan Y, Zhou J, Meng L, Ding H, Kong W (2025) A novel physical neural network based on transformer framework for multiaxial fatigue life prediction. Fatigue & Fracture of Engineering Materials & Structures 48(5):2381–2405, DOI 10.1111/ffe.14618
[37] Pan Q, Lu L (2025) Fatigue in metals and alloys. Nature Materials 25(3):357–365, DOI 10.1038/s41563-025-02308-5
[38] Papadopoulos IV (1987) Fatigue polycyclique des métaux une nouvelle approche. Thèse de l'Ecole Nationale des Ponts et Chaussées
[39] Papuga J (2005) Mapping of fatigue damages – Program shell of FE calculation. PhD tesis, Czech Technical University, Prague, `http://www.pragtic.com/docu/Papuga_Thesis.zip`
[40] Robert JL, Fogue M, Bahuaud J (1994) Fatigue life prediction under periodical or random multiaxial stress states. In: Amzallag C (ed) Automation in Fatigue and Fracture: Testing and Analysis, ASTM STP 1231, pp. 369–387, DOI 10.1520/STP13959S
[41] Sakai T (2023) Historical review and future prospect for researches on very high cycle fatigue of metallic materials. Fatigue & Fracture of Engineering Materials & Structures 46(4):1217–1255, DOI 10.1111/ffe.13885
[42] Serensen SV (1975) Soprotivleniye materialov ustalostnomu i khrupkomu razrusheniyu (in Russ., Strength of Materials Fatigue and Brittle Fracture). Moscow: Atomizdat
[43] Sines G (1955) Failure of materials under combined repeated stresses with superimposed static stress. National Advisory Committee for Aeronautics (NACA) Washington, Technical note 3495
[44] Socie D (1987) Multiaxial fatigue damage models. Journal of Engineering Materials and Technology 109(4):293–298, DOI 10.1115/1.3225980
[45] Susmel L, Taylor D (2012) A critical distance/plane method to estimate finite life of notched components under variable amplitude uniaxial/multiaxial fa-

tigue loading. International Journal of Fatigue 38:7-24, DOI 10.1016/j.ijfatigue.2011.11.015

[46] Tridello A, Niutta CB, Rossetto M, Berto F, Paolino DS (2025) Fatigue design curves for industrial applications: A review. Fatigue & Fracture of Engineering Materials & Structures 48(3):1001–1021, DOI 10.1111/ffe.14545

[47] Tridello A, Niutta CB, Rossetto M, Berto F, Paolino DS (2022) Statistical models for estimating the fatigue life, the stress life relation, and the P-S–N curves of metallic materials in Very High Cycle Fatigue: A review. Fatigue & Fracture of Engineering Materials & Structures 45(2):332–370, DOI 10.1111/ffe.13610

[48] Troshchenko VT (2010) Fatigue and inelasticity of metals under nonuniform stressed state. Strength of Materials 42(5):494–505, DOI 10.1007/s11223-010-9241-1

[49] Van KD, Papadopoulos IV (1999) High Cycle Metal Fatigue. CISM International Centre for Mechanical Sciences, Vol 392. Vienna: Springer, DOI 10.1007/978-3-7091-2474-1

[50] Yang J, Gong Y, Jiang L, Lin W, Liu H (2022) A multiaxial and high-cycle fatigue life prediction model based on critical plane criterion. Journal of Materials Research and Technology 18:4549–4563, DOI 10.1016/j.jmrt.2022.04.069

[51] Yankin A, Mugatarov AI, Wildemann VE (2020) Influence of different loading paths on multiaxial fatigue behavior of 2024 aluminum alloy under the same amplitude values of the second invariant of the stress deviator tensor. Frattura ed Integrità Strutturale 15(55):327–335, DOI 10.3221/IGF-ESIS.55.25

[52] Yokobori T (1963) Unified engineering theory of metal fatigue. Technological Reports of Tohoku University 27(2):53–73

[53] Zavoychinskii BI (1987) On the variational principle of the description of limiting processes of complex loading. Mechanics of Solids 22(2):113–122

[54] Zavoichinskaya EB (2025) Durability of structural materials under variable loading. Moscow University Mechanics Bulletin 80(1):11–22, DOI 10.3103/S0027133025700074

[55] Zavoychinskaya EB (2022) Obshchiye zakonomernosti i kriterii razrusheniya tverdykh tel na raznykh masshtabno-strukturnykh urovnyakh pri dlitel'nom nagruzhenii (in Russ., General principles and criteria of failure of solids on different scale-structural levels under static and alternating loading). Industrial laboratory. Diagnostics of materials 88(7):48–62, DOI 10.26896/1028-6861-2022-88-7-48-62

[56] Zavoichinskaya EB, Lavrikov GE (2025) Brittle failure modelling at cyclic loading of variable amplitude and asymmetry on different structural levels. Mechanics of Solids 60(8):6627–6634, DOI 10.1134/S0025654425605658

[57] Zavoychinskaya EB, Rautian NA, Lavrikov GE (2025) Mathematical modeling of scale-structural failure at program cyclic loading of metals and alloys. Differential Equation 61(5):650–667, DOI 10.1134/S0012266125050088

[58] Zhang CC, Yao WX (2011) An improved multiaxial high cycle fatigue criterion based on critical plane approach. Fatigue & Fracture of Engineering Materials & Structures 34(5):337-344, DOI 10.1111/j.1460-2695.2010.01523.x

[59] Zhang L, Jiang B, Zhang P, Yan H, Xu X, Liu R, Tang J, Ren C (2023) Methods for fatigue-life estimation: A review of the current status and future trends. Nanotechnology and Precision Engineering 6(2):025001, DOI 10.1063/10.0017255

[60] Zhao T, Jiang Y (2008) Fatigue of 7075-T651 aluminum alloy. International Journal of Fatigue 30(5):834–849, DOI 10.1016/j.ijfatigue.2007.07.005

[61] Zhou S, Wei Z, Feng F, Yang B, Henrich M, Münstermann S (2025) A general physics-informed neural network framework for fatigue life prediction of metallic materials. Engineering Fracture Mechanics 322:111136, DOI 10.1016/j.engfracmech.2025.111136

MIX
Papier aus verantwortungsvollen Quellen
Paper from responsible sources
FSC® C105338

If you have any concerns about our products,
you can contact us on
ProductSafety@springernature.com

In case Publisher is established outside the EU,
the EU authorized representative is:
Springer Nature Customer Service Center GmbH
Europaplatz 3, 69115 Heidelberg, Germany

Printed by Libri Plureos GmbH
in Hamburg, Germany